# Plant Intelligent Automation and Digital Transformation

# Plant Intelligent Automation and Digital Transformation

## Volume III: Control Intelligence and Communication (Field and Network)

**Swapan Basu (Retired)**
Founder Member, Systems and Controls, Kolkata, India

ELSEVIER

ACADEMIC PRESS
An imprint of Elsevier

Academic Press is an imprint of Elsevier
125 London Wall, London EC2Y 5AS, United Kingdom
50 Hampshire Street, 5th Floor, Cambridge, MA 02139, United States



ISBN: 978-0-323-89892-8

For information on all Academic Press publications visit our website at https://www.elsevier.com/books-and-journals

*Publisher:* Priscilla Braglia
*Acquisitions Editor:* Dan Egan
*Editorial Project Manager:* Asmi Bharati
*Production Project Manager:* Maria Bernard
*Cover Designer:* Mark Rogers

Typeset by TNQ Tech

# Dedication

In the name of Gurudeb I trust,
I dedicate this volume of the book to everyone who dedicated their lives to world peace and maintaining harmony in life

# Contents

# Foreword

In this modern world, there is hardly any device that is devoid of digital electronics. In our everyday life, even mobile phones or any domestic appliances that we use make use of embedded digital electronics. However, each one of them makes use of different aspects of control system applications. It is becoming extremely difficult to use and maintain each one of them properly. As a result, there arose a need to integrate them on a common platform. One can now use a mobile phone not only as a phone but also as a social networking and communicating device and also as a remote-controlling device for various home appliances.

In this book, an effort has been made to cover extensively the entire range of theory and implementation technology required to design, develop, manufacture, and maintain such control and automation devices or systems. This book not only covers the topics on system integration and migration but also covers IoT/IIoT-based remote control requirements as per Industry 4.0 and digital implementation standards. This evidently makes up an extensive volume of information to be put between two covers, and so the book entitled *Plant Intelligent Automation and Digital Transformation* has been divided into the following four volumes:

Volume I—Process and Factory Automation
Volume II—MIS and HW-SW Implementation
Volume III—Control Intelligence
and Communication (Field and Network)
Volume IV—Security, IoT/IIoT, and Applications

Beginning with the overall issues, a slow build-up of in-depth discussions with appropriate detailing, as and when needed, offers a smooth reading and understanding of different technical concepts. The same philosophy has been followed in every book volume and in every chapter of each volume. Each volume starts with the basic working principles describing the mathematical background, and detailed discussions on the functional aspects, software, hardware specifications, applications, and implementation options have also been covered—step by step.

The four volumes, as a whole, maintain a fine balance between fundamental concepts, theoretical studies with mathematical details, and in-depth discussions on all practical issues related to design, implementation, testing, and fulfillment of application requirements. As far as my knowledge goes, no single book can be found today where such detailed studies on intelligent control of systems, implemented with digital devices and techniques, have been covered so thoroughly. The author of this monumental work, Mr. Swapan Basu, a highly reputed engineer in the field of instrumentation and control engineering, having experience both in India and abroad, is an expert in the field of process design, commissioning, and testing. He has been able to maintain a continuing interest in the latest developments found in the field of digital instrumentation and control. I truly feel that the book written by an author who has such a long industrial experience will be extremely helpful, not only to practicing engineers but also to the students and fresh professionals entering the field.

I wish all success to the author, and his efforts will be highly appreciated by all the readers.

**A.M. Ghosh**

# Preface

In this competitive market, industrial production plants globally are struggling to sustain their market and have to opt for integrated intelligent controls with big data analysis for cooperative production processes linking supply chains, with the help of cloud computing in a digital transformation approach. This book has been developed to address all these issues in four volumes. In this volume of the book with 18 sections, theoretical and practical approaches for several subsystems pertinent to intelligent plant controls, automation, and the digital transformation approach have been dealt with, keeping a proper balance between two extremes. Instead of a far-reaching theoretical approach with complicated, advanced mathematical deductions, the book provides basic theoretical details followed by physical explanations, application areas with implementation options, relevant figures, tables, and detailed specifications that create a simple, clear understanding of these complex topics both from a hardware and software point of view.

This volume has been dedicated to discussing control intelligence and communication both in the field and in network. The discussions start with the basics of control and process intelligence (PI) to cover various features and benefits of control intelligence, as well as various design requirements for various levels of intelligence, also covering the basics of process intelligence and robotics to open the gateway for further detailing. The discussions here have been presented in both conceptual and detailed design forms and have been covered to include PI and Robotic Process Automation (RPA) in separate dedicated sections. In order to have a strong base, the discussions started from adaptive gain, autotuning, and model-based approaches for controls and automation, covering controller tuning methods and auto- and self-tuning details. This book discusses expert systems, including explanations, technologies, characteristics, etc., for knowledge bases, inference engines, and inference drawings. This book also discusses fuzzy logic and automation starting from the basics. Artificial intelligence and machine learning details for an intelligent automation approach are also presented. Also, comprehensive discussions have been presented on artificial neural networks (ANNs) to cover types, human issues, resemblance, etc. All these issues have been discussed at a basic level at the beginning, followed by comprehensive discussions in a dedicated section spread over five big chapters. Right from the beginning, detailed discussions have been presented on industrial data communication fundamentals as recapitulations for a dedicated chapter before presenting details on several fieldbus systems and protocols. Dedicated discussions have been presented most popular field communications in industrial automation, i.e., Foundation Fieldbus, Profibus, HART, and BRAIN protocol. Detailed coverage includes , the technology, characteristic features, and details on the design and implementation along with challenges. This book also covers MODBUS types and technology characteristics. This book also discusses AS Interface, WORLD FIP, FIPIO, SERCOS, Interbus, and CC-Link. In each case, technology, application design, and implementation in industrial controls have been covered. Also, a chapter with several sections has been dedicated to FDDI, various protocol families, remote controls, and security in field communication. This book discusses wireless communication with a special emphasis on the ISA100 standard. A dedicated chapter spanning several sections has been used to discuss Dynamic Data Exchange (DDE), OPC, UCA, DNP, and discussions on smart grid and associated IEC standards. Prior to beginning discussions on networks, the basics of communication and networking issues to cover fundamentals of network transmission and communication, network and interface hardware, wireless devices, transmission media, etc., have been discussed. These books also include sections on the network control system, the ISO–OSI model, the TCP/IP model, etc., as well as common important network types like Time-Sensitive Network (TSN) and Common Industrial Protocol (CIP). A dedicated chapter has been allotted for discussions on software-defined networks. This book also covers Ethernet, Ethernet types based on speed, Power over Ethernet, APL, etc., in detail. Networking with OPC UA, network integration, and migration issues are also covered comprehensively. There has been detailed discussions of industrial Ethernet to cover EtherCAT, EtherNet/IP, Profinet, etc. This book also covers topics on CAN, CANopen, ControlNet, CompoNet, and DeviceNet along with their characteristics, applications, safety, and security issues.

A comprehensive idea about intelligent control technologies, with detailing on hardware, software, and MIS has been presented in Volume II. Volume III deals with extensive details on artificial intelligence and communication both at field and network levels. Volume IV deals with security, IoT, IIoT, and application details in various industrial plants. All four volumes complement each other for comprehensive details.

In order to correlate various figures, tables, and text references, the following principles are followed:

**Fig:** Fig. A N.Mb: A = Volume no. (I, II, III, IV); N= Chapter no. of the volume; M= Section no. of chapter "N" of the volume "A"; — a, b, c are unique numbers for the drawing (if two or more sketches are in the figure, these are designated by 1,2. after the unique number, viz., Fig. III 4.2a or II 4.2b1, Fig. III 4.2b2.

**Table:** Table A N.Mb: A = Volume no. (I, II, III, IV); N = Chapter no. of the volume; M = Section no. of Chapter "N" of the volume "A"; a, b, c are unique numbers for the table, viz., Table I 4.2a.

**Equation:** Eq. (A N.Mb): A = Volume no. (I, II, III, IV); N= Chapter no. of the volume; M = Section number of Section "N" of the volume "A"; a, b, c are unique numbers for the eq., viz., Eq. (I 1.2b)

**Text reference:** A N.M.P.n: A = Volume no. (I, II, III, IV); N = Chapter no.; M = Section of chapter "N" of the volume "A"; P = Subsection of the main section, and n is the point under that subsection, viz., I 6.0.3.2 stands to represent point number 2 in subsection 3 of section 0 in Chapter 6 in Volume I.

The book is extremely helpful for professionals working or practicing in modern intelligent control and automation systems. Engineers starting their careers in automation and control engineering, postgraduate students, andresearch scholars in this area as well as in computer science/electrical engineering will be greatly benefited from the comprehensiveness and practical approaches in this book.

In an attempt to incorporate this extensive subject area in the form of a book, a great deal of research has been carried out over the years so as to include the knowledge gained during the decades-long experience in industrial automation and control engineering. I would feel rewarded only when my research efforts are able to benefit practicing engineers by providing proper guidance in industrial automation in the era of fast-changing technologies and customer demands.

Swapan Basu
Sr. Member IEEE and Sr. Member ISA

# Acknowledgment

At the outset, I wish to convey my sincerest thanks and gratitude to my lovely wife Bani, who sacrificed her quality time and companionship with me for these time-consuming efforts for the development of a book volume of this standard. I am indeed thankful to my son and daughter and her friend, who not only encouraged me in this effort but also helped me in different areas even from abroad. The book cover has been designed and developed by my lovely son Piku. Various chapters were overseen and edited by my lovely daughter Idai and lovely Dhruba, who are very close to me.

During my old B.E. college days, he was an idol to many of us through his profound knowledge, limitless energy, and encouragement to his students—he is none other than Dr. Ananda Mohan Ghosh. In spite of many odds, he always came out with full vigor to help me whenever I approached him. As usual, even over the phone, he explained many difficult issues and queries and never felt disturbed. Not only that, he also oversaw some difficult portions of the book. Also, Dr. Ghosh spared his valuable time to go through the book and agreed to write the Foreword.

I would also like to thank many of my friends and ex-colleagues, especially Debnathda, for their encouragement and technical support from time to time during the development of the book.

Last but not least, the author would like to thank and acknowledge the support he has received from the entire team of Elsevier, without which it would have been impossible to publish the book.

Chapter 1

# Discussions on process and control intelligence

When a conventional control system (control systems—smart but without control intelligence) is developed for any process, first the goal or setpoint is fixed. Then, variables to be manipulated are chosen so that desired set point can be achieved. In the designed control loop, measured variables and set point are compared and manipulated to generate error signal. Based on the error signal, controller has to take corrective actions. After that, we would evaluate the dynamic "personality" of the process, its gains, time constants, etc., which is information needed to tune the controller [1]. In conventional control system, based on designer's knowledge and intelligence, the system models are developed and decide on controller type (say, cascade, ratio …). Naturally, the intelligence lies on the designer not really on controller. In intelligent control systems, software does the models for the system based on analyzed data and synthesized knowledge. So, the intelligence (with knowledge to modify) lies with the software for intelligent controller—making a shift of intelligence from the designer toward the software. Here, basics of process and control intelligence shall be covered.

## 1.0 Basics of control and process intelligence

On account of its immense capability intelligent automation is finding its application in almost every industry. Intelligent automation makes it possible to analyze, validate vast amount of data, synthesize them and convert huge data/information into *knowledge* so that it can take part in decision making (subject to final human approval). So, it could face challenges like: high labor cost and labor scarcity, meet competition through greater efficiency, enhanced production and services with customer satisfaction yet making profit. Intelligent control makes use of various artificial intelligence computing approaches. So, *what is Artificial intelligence? Artificial intelligence (AI) is a branch of computer science. It is concerned with the development of smart device/machine or system which can mimic and perform tasks that typically require human intelligence (through deep learning process). This is an interdisciplinary science with multiple approaches.* Two forms of artificial intelligence are Artificial narrow intelligence (ANI) and Artificial general intelligence (AGI). ANIs are programmed for single specific task which could be wide varied in nature. There has been immense myth and misconception about reality of AGI. *Machine learning* and *Deep learning* are two major classifications of ANI. Globally around over 25% industrial enterprises depend on AI for their mission to Industrie 4.0 and Digital transformation (DT).

### 1.0.0 Control intelligence features and benefits

According to Report of the Task Force on Intelligent Control IEEE Control Systems Society in intelligent control problems there may not be a clear separation of the plant and the controller—having the control laws embedded and a part of the system to be controlled. Intelligent systems can act appropriately in an uncertain environment, with appropriate set of actions to increase the probability of success, and success is the achievement of behavioral subgoals that support the system's ultimate goal [2]. Intelligent control/automation convert data/information into knowledge for application to remove uncertainty of the specific task to achieve probability of success. Intelligent systems can analyze vast data and can synthesize them to form organized knowledge base. Therefore, system intelligence decides and use the goal and control moves the system toward the goal and does the refinement to adjust the goal in new changed circumstances. So, one can argue that intelligence adds high degree of autonomous behavior in control system. Major Characteristic features and dimension of Intelligent control shall include but not limited to [2]:

- Adapting and learning
- Autonomy and intelligence
- Structures and Hierarchies.

The degree or level of intelligences depends on:

- Analysis and computation power of the system, that is, brain of the system
- Memorizing, that is, memory storage capability

Plant Intelligent Automation and Digital Transformation. https://doi.org/10.1016/B978-0-323-89892-8.00007-7

- Sophistication of software algorithm required for vast data/information processing, generalized modeling, synthesizing, development of behavior keeping parity with the process, value judgment and communication.

Intelligent control is a generalized-features in many systems but our concentration will be focused on plant control systems. In plant control systems there are many ways to achieve these goals viz. fuzzy/neural controllers, expert controllers, machine learning controllers, etc. All of these are heuristic and based on artificial intelligence. These controllers obviously show better performance and high degree of autonomy when compared with conventional controllers. For a given process, in conventional control system requirements of specification controller needs to meet are restricted to stability, rise time, and overshoot specifications. On the contrary, in cases of complex plants, there are vast data and wide varieties of uncertainty. Intelligent controls make own set of models to make the system perform well and act with high degrees of autonomy. Therefore, from control point of view, mathematical models, systematic design procedure, nonlinear analysis to study stability, boundedness, convergence issues, limit cycles, controllability, observability, robustness, etc., performance analysis—simulation techniques for intelligent systems and implementation issues [2] are extremely important.

### 1.0.1 Control intelligence levels and requirements

In an integrated system, enterprise network is combined with plant automation network. Also, the number of levels in hierarchical network structure are:

1. **Process level:** It consists of the main process with process equipment (e.g., Pump valve, heat exchangers as other associated equipment) which is the heart of any process to operate of any operation.
2. **Control level:** Control systems include DCS/PLC and associated supervisory HMI SCADA required to control processes automatically. Depending on applicability this shall include both BPCS and SIS as well. For large complex systems such as refineries consider this as the baseline control systems-to be the baseline. Control intelligence may be a part of this level but it depends highly on functioning of two other levels.
3. **Information level (local supervision level):** As discussed, information helps to control system to extract intelligence and knowledge from the process. Historian and associated application process in any control system helps to collect plant parameter over time so that process can be tracked. It can compare different conditions and their influences on the system Pre and post upset data trends helps in tracing process upsets, to gather knowledge for future retrieval. So, this subsystem enables to develop centralized database across different areas of plant operation. Also, it is very important to monitor various plant parameters as a part of conditioning monitoring. With the help of trending tool historian provides current and historical trend. Here, an important issue is the cost of historian. Cost highly depend on how much data can be stored and for how long and it depends on the process. To limit data storage capacity use of data compression algorithm in Historian is common. Also, there will be suitable setting to distinguish the collection frequency for *slow changing* process (say) level and *fast changing* process (say) pressure spikes. Memory space talked about earlier for intelligent control systems depends on space in historian greatly. In some systems such collections are based on set percentage change in the parameter, that is, if (say) there is a change of over 2% then it will be collected and stored [3]. Unless plant data are co-related with outside control data it may not be possible to generate proper *information* from these data. In such case, trend alone will be refined raw data. So, the raw plant data, must be co related with outside data through statistical process or other means so that real information can be generated.
4. **Data Analysis—knowledge base:** Trends alone all the time provide contexts of a number of the things happen in the plant. Therefore, suitable reports need to be used to support trends and historian data. Also, time stamped Sequence of event (SOE) helps to understand sequential events in the plant including, how many times certain equipment or plant itself started/stopped/tripped and associated reason thereby. This helps engineers to quickly identify the weakness in the process equipment and process operations and they may come with solutions. Exactly in the similar way it could help control intelligence to assimilate data analysis to gain knowledge hence intelligence. So, reports, SOE can add the contexts missing in data trends. Analyzing data can maximize the value of information systems. In fact, from information knowledge is developed through data analysis in conjunction with various of other records and events. In data analysis suitable strategies need to be implemented so that process conditions and health can be well understood equipment data and effectiveness and other key performance indicators (KPIs) and reporting tools to share information with all levels of an organization [3]. As stated, before through data analysis, information also turns in to knowledge required for control intelligence. While collecting data there will be many *data noise* are created on account of many reasons (say) data corruption. Effective an useful data analysis systems can remove most of the noise

from data and help answer whatever questions might be posed [3]. In modern days there are many web-based data analysis methods are available and these are quite flexible and powerful tool. These not only help in generating knowledge but also can help create dashboards at different level based on requirements. When statistical process control is implemented in the system, then correlating various data are easy as it provides statistical analysis. All these make raw data into information and information into knowledge base for control intelligence which in turn helps in decision making process (for human as well as inference engines).

5. **Enterprise level (ERP section):** This part is mainly concerned with business strategy, planning, sales, inventory, marketing and other administrative and commercial issues beyond the purview of this discussions. Let the discussions be concluded with the view that this kind of structure not only help in implementing control intelligence but also is helpful for operator in dealing with raw data otherwise operator needs to deal with.

### 1.0.2 Robotics and process intelligence—conceptual discussions

Here, we discuss the conceptual idea about robot, robotics and process intelligence, so that subsequent discussions can be meaningful. Let us start the discussions with robot and robotics.

1. **Robot:** According to dictionary definition of Robot is a machine resembling living creatures, with capability to move independently (as by walking or rolling on wheels) meant to perform complex actions. When we have a closer look one can also say that robot—an autonomous machine with capability of sensing its environment, learning and carrying out computation to make decisions and carrying out complex tasks in real world. Therefore, robots are diverse bunch with diversity in movement, for example, some can walk with four/six legs or may have wheel for movement. Also, the size of robots may vary from as small as coin to a big refrigerator or more. Robots can swim in underwater application. Similarly, as drone can fly in the sky.
2. **Robotics:** The branch of engineering involved in the conception, design, manufacture and operation of robots are referred to as robotics. The main goal of the robotics is to create intelligent machines that can assist humans in a variety of ways. Robotics is an interdisciplinary sector of science and engineering dedicated to the design, construction and use of mechanical robots. Based on design basis and applications, several types of robotics are as follows: *Industrial*, *Service*, *Medical*, *Military*, *Space research*, *Education*, *Underwater/Marine*. *Search and Resue*, *Autonomous vehicle,* etc. From here one could understand that application areas shall include but not limited to: Manufacturing, *Healthcare*, *Service industry (including customer service)*, *Entertainment*, *Agriculture*, *Transportation*, *Construction*, *Food Preparation*, *Space exploration*, *Military and Defense*, Security, *Under water exploration*. Robotics and Artificial intelligence (AI) are closely related to each other with some areas of overlap and integration. Application of AI technology in robotics enable robots to perceive and sense the environment, make decisions, and carry out tasks autonomously. Some of use of AI in robotics include but not limited to: *Computerized vision*, *Natural* and *Machine language*, *Motion planning* and *Robotic process automation* (RPA). Robots will continue evolving as a collaborative partners with cognitive capabilities on account of sustained developments in sensor technology and advancements in machine learning and artificial intelligence. Such advancements, would impact different application area differently. A few are included here to get some conceptual idea. Logistic robots would not only be used for stacking in warehouse shelves, retrieve products, and even make local deliveries. Self-driving cars become more intelligent to undertake delivery service as well. However, popularity of this heavily depends on development of actual road traffic. In defense/marine use, there may be Unmanned ground vehicle (UGV), or Unmanned aerial vehicle (UAV). Also, the robot can perform a greater variety of duties on the battlefield. In health applications, in addition carrying out operations, they will converse with patients, monitor their health, and decide whether or not they require follow-up sessions. Therefore, it is needless to mention that robotics has huge prospects in all its application area including the same in industries. However, it does have serious impacts on social and economy.
3. **Process Intelligence (PI):** Process intelligence helps for optimum RPA and IA, to keep business at top position. Therefore, realize the best business value, developing and implementation of automation strategy will be of immense importance especially for effective use of digital workers in the form of RPA. Conceptually, PI is the use of business intelligence methods and tools in business process management to remove business bottlenecks, improve operational efficiency, and design guide of business process for improved performance. Technology driven analysis of the of process data, in PI helps in focusing on actionable insights for process improvement, that is, represents the overlapping area of business intelligence and process management for the business. Process discovery, process improvement, root cause analysis and process optimization are major reasons to opt for process intelligence with the help of:

*Process* and *task discovery*, *Process monitoring*, *Process mapping*, *Process reengineering* and *process implementation*. Therefore, PIs consists *Raw data extraction* (i.e., systematic data collection), *Data transformation into information* and *Data loading for analysis*. Process intelligence software—useful for analyzing and managing nonlinear processes, can help by monitoring and analyzing processes on a historic or real-time basis.

PI is a combination of process mining, task mining, and business intelligence-like metrics to provide deep insights into business processes. Two important issues here are:

- **Task mining:** This allows an organization to capture process and user interactions, identify repetitive/time-consuming tasks and recommend any potential opportunity for automation to resolve these sticking points.
- **Process mining:** Process mining software uses system data for discovering, monitoring and improvement of real digital processes. It uses event log data to create accurate process models indicating how the processes are executed. Itis an ideal tool for analyzing processes retrospectively and making data-driven process decisions about your automation strategy.

These are further elaborated with overlap in III 5.0.1.1 also.

## 1.1 Adaptive gain, auto tuning and model based approaches

Adaptive control is one of the widely used control strategies used in advanced control systems for better accuracy and performance. Let the discussions start with recapitulations of the discussions made in I 1.2.8. When the parameters of the dynamic model of the plant in the question are well known, but constant are unpredictable with time, adaptive control systems—kind of nonlinear stochastic control problems associated with the control of processes with unknown and time-varying parameters, are used to achieve and maintain an acceptable level of control system performance [3]. An important parameter is Performance index (PI) defined in Fig. I 1.2h. PI is often referred to as Key performance index (KPI). KPI/PI takes in to account changes in process dynamic and/or changes in disturbance parameters. Two sets of loops in an adaptive controller are as follows: feedback control or adjustable system and adaptation loop with adaptation mechanism to adjust controller parameters for adapting under changed circumstances. Model Reference Adaptive Control (MRAC) for a plant in the question, is based on the concept to follow or "track" a reference system (model) characterized by a state, Input/Output (I/O) model employing an adaptive optimization algorithm to adjust the controller parameters in real-time. For control systems to act, controller is of utmost importance. There exists a number of factors like, dynamic behavior of the process to be controlled, performance specified and understanding of tuning process. So, it is not only scientific but also artistic feature covering feeling of the process. So, in auto and self-tuning methods controllers can automatically select own tuning parameters sound good, but face challenges. These will also be covered here.

### 1.1.0 Adaptive control: nature and function

Two actions to perform by adaptive controller are as follows: Learning of the process and regulate control behavior. Adaptive controls are more effective mainly suitable for nonlinear and stochastic, hence difficult to control. Also, there are no general design procedures for designing adaptive controllers as each application calls for case-study. Hence success of adaptive controllers are always point of debate. When the parameters of the dynamic model of the plant under question are unknown but constant (over regimes of operation). In such situations even though structure of controller does not depend generally on the values of the plant model parameters yet for correct tuning of the controller plant parameters will be necessary. With deployment of adaptive control techniques can provide an automatic tuning procedure in closed loop for the controller parameters as shown in Fig I 1.2j. With time as the controller adapts itself more requirement for adaption diminishes, but changing operating condition again may require a restart of the adaptation procedure. In another situation if parameters of the dynamic model of the plants unpredictable with time. Such change may occur either due to the environmental conditions change or choosing linear model for a nonlinear model which changes due to changed condition. To maintain an acceptable level of control system performance when large and unknown changes in model parameters and/or disturbances occur, an *adaptive control* approach could be considered. Therefore, one can infer that adaptive control are applied for mainly two cases:

- Plant with unknown but constant parameter
- Unpredictable change in plant dynamic (disturbances) model with time.

  For designing adaptive controller issues like: control loop *performance specification*, plant *dynamic model* and *suitable design method* should be known/established. Adaptive controls may be *direct* or *indirect* type. Apart from such clarification, different approaches for adaptive controls include the following: MRAC, Gain scheduling adaptive control (GSAC) and self-tuning controller (III 1.1.1). Many of these issues once have been discussed briefly in I 1.2.8, so, there may be some overlap. Let the discussions start with adaptive gain controller features.

1. **Characteristics:** Characteristics of adaptive control systems shall mainly include the following:
   - No prior knowledge of the process dynamic parameters to for adaptive control function
   - The stability and robustness of the algorithm
   - Algorithm can cope up with parameter variations/external disturbances and nonlinearity
   - Quick response to disturbances.
   - The adaptive gains can start from an initial value to speed up convergence.
   - It has the capability to lock the gain (act as fixed gain) and work in parallel with linear controller in retrofit applications.
2. **Adaptive, Conventional and Robust loops:** Adaptive loop is compared with conventional loops and robust loops here.
   - **Adaptive** versus **conventional loop:** Table III 1.1a lists out major difference between them
   - **Comparison with robust controller**: Robust controls deal with uncertainty. Robust control systems function properly when uncertain parameters or disturbances are within set range. Uncertainty presents a challenge to issues like *observability*, *controllability* and *stability*. In robust controller design, the focuses are on reliability/robustness, of the control algorithm. Robust feedback controllers designed to maintain the desired performance without adapting changing conditions like adaptive controller. So, prior to opting for adaptive control, often robust (with "fixed parameter") control design is developed first to improve in general the adaptation transients. The tuning of a robust control system using an adaptive control, will improve in extension of the operational range and the performance of the robust control system design.
3. **Adaptive control and regulation:** Based on plant model for each input there will be some output. As noted in volume I there also exists disturbance model. Depending on adaption, there are as follows: adaptive control, adaptive regulation and combination of both. These are shown in Fig III 1.1a.
   - **Adaptive control:** Here, Plant model is unknown but time varying disturbance model as shown in Fig III 1.1a1 is known and constant. In adaptive control objective is to track disturbance attenuation performance by focusing on adaptation related to plant model parameters variations without any effort to estimate the model of the disturbance at the same time, assuming model of disturbance is constant.
   - **Adaptive Regulation:** In contrast to adaptive control, as shown in Fig III 1.1a2 here, plant model is known and constant but the *disturbance* model is *unknown* and *time varying* with objective to suppress the effect of the unknown disturbance with rational power spectrum by focusing on adaptation disturbance model parameters variation with no effort to estimate the plant model at the same time.

   *Combination of both is very difficult problem.* It is extremely hard to distinguish in the performance error, parts coming from plant model error and parts coming from disturbance model error.
4. **MRAC:** MRAC strategies the design the adaptive controller working on the principle of adjusting the controller parameters so that the output of the actual plant tracks the output of a reference model having the *same* reference input. This means that the controller parameters are adapted in a way such that the plant response *matches* the response of the reference model. Here, the possibility of correcting *not only* the *manipulated variables* but also *the underlying model parameters* as new process data become available. As shown in I 1.2i2, major components are:
   - **Reference Model:** Meant to provide an ideal peaceful response of the adaptive control system to the reference input in all conditions. Since its response is to be precisely matched, it must be stable with minimum phase (poles/zeros in left-half plane) and should be as close as possible to the true *representative* of the process.
   - **Controller:** It must satisfy: the "perfect model matching" in closed-loop response that matches that of the reference model; and *direct* adaption implies a linear function of the control parameters, that is, linear control law.
   - **Adjustment mechanism:** As the name suggests, it is used to adjust the controller parameters to enable actual plant to track the reference model. The adaptive rules: sensitivity models, Lyapunov stability; and Popov hyperstability—to guarantee a stable, closed-loop system. We will only consider the first two here [4].
5. **MRAC discussions:** Following are a few important issue pertinent to MRAC
   - **MIT rule Recapitulation:** MIT rule are often used for designing MRAC control system. Let us first recapitulate discussions from I 1.2.8.4. From eq. III 1.2d we get, $J(\theta) = 0.5\ e^2$ with reference to Fig I 1.2i2, where $e$ is the error, between the outputs of plant and reference model ($y$-$y_m$).

     [Used eq, ref from volume I of the book which should be read in conjunction with this].

     Adjustable parameter θ is adjusted in such a fashion so that the cost function can be minimized to zero. Parameter θ, is in negative direction to gradient of function J, so that we get from Eq. III 1.2f (after

**TABLE III 1.1A Adaptive control versus Normal Feedback control.**

| Issues | Adaptive controller | Normal feedback controller |
|---|---|---|
| Objective | Monitoring of PI of the control system for unknown and varying parameters | Monitoring of control variables of the control system against specified performance for known parameters |
| Gain | Variable | Fixed |
| Monitoring | Performance index | Control variable |
| Measurements | IP measurement | Parameter measurement |
| Reference | Performance index | Reference input |
| Robust and stable | Much better | Comparatively low |
| Error | Comparison decision block | Comparison block |
| Loop | Control and adaption loops | Control loop |
| Elimination | Effect of disturbances | Effect of parameter disturbances upon performance |

computation): $\frac{d\theta}{dt} = -\Upsilon \frac{\delta J}{\delta\theta}$. Here, partial derivative is referred to as Sensitivity derivative and ϒ as adaptive gain.

- **The Lyapunov Rule:** To improve efficiency, robustness and stability Lyapunov rule can be applied in MRAC. Lyapunov rule can be used to prove stability and improve convergence of tracing error. The Lyapunov stability theory can be used to describe the algorithms for adjusting parameters. It uses state space equations to describe the rule.
- **MRAC Mathematically:** In Vol I we discussed state space model which is well applicable for MRAC. Redrawing MARC with all component in Fig III 1.1b we can write:
  - Reference model in state space equation is as follows:

$$x\dot{m} = A_m x_m(t) + B_m r(t) \text{ and } y_m(t) = C_m x_m(t) \quad \text{Eq (III 1.1a)}$$

  - Controller model in state space equation is as follows:

$$U(\dot{t};\theta) = \theta_x\, x_p(t) - \theta_r\, r(t) \quad \text{Eq (III 1.1b)}$$

  - Process model in state space equation is as follows:

$$x\dot{p} = A_p x_p(t) + B_p U(t;\ \theta) \text{ and } y_p(t) = C_p x_p(t) \quad \text{Eq (III 1.1c)}$$

  - Adaption in state space equation: MIT rule (see vol I): $\frac{d\theta}{dt} = -\frac{\delta J}{\delta\theta}$ (I 1.2f) and in Lyapunov rule:

$$\text{Lyapunov rule:} \frac{d\theta}{dt} = \Upsilon\, \Theta(t) * e(t;\ \theta) \quad \text{Eq (III 1.1d)}$$

  - Error in state space equation is as follows:

$$e\ (t;\ \theta) = x_p(t;\ \theta) - x_m(t) \quad \text{Eq (III 1.1e)}$$

6. **Gain scheduling:** From I 1.2.8 we understand that conceptually, gain scheduling involves an intuitive simplification of the problem into parallel decompositions of the total system including controlling nonlinear systems. Gain scheduling facilitates control when the gains and the time constants *vary* with the current value of the process variable. So, it is appropriate for processes that speed up or slow down with rises and falls of the process variable. Also, it can be well applied for the processes sensitive to the controller's efforts with process variable changes. There are several ways gain scheduling can be divided viz. decomposing nonlinear design problems into *linear* subproblems and decomposing nonlinear design problems into simpler *nonlinear* subproblems. Also, gain scheduling methods can be decomposition of continuous gain-scheduling and *discrete*, that is, switched gain-scheduling. As is clear from I 1.2.8, various approach based gain scheduling are as follows: classical gain scheduling, Linear parameter varying method (LPV)/Linear fractional transformation method (LFT) and Fuzzy gain scheduling.

With this let us now conclude preliminary discussions on adaptive control and move on to auto-tuning and self-tuning controls. Also, see III 2.0.0.

### 1.1.1 Discussions on auto-tunning and self-tuning

We know that Proportional (P), Integral (I) and Derivative (D) controller (referred to as PID controller) are most common/de facto standard for in feedback control of industrial control systems. Out of various tuning approaches,

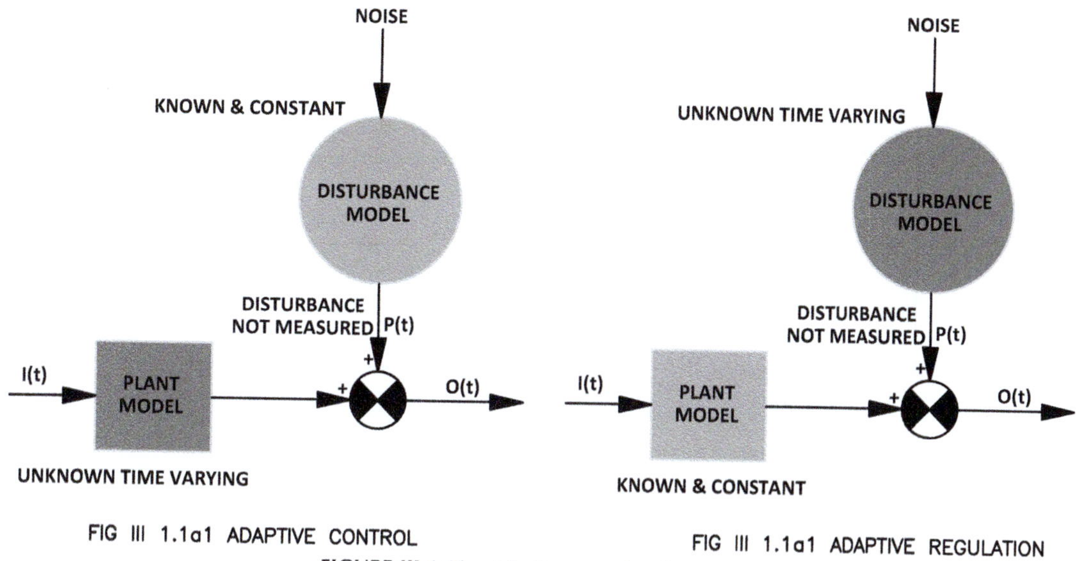


**FIGURE III 1.1A** Adaptive control and regulation.

autotuning is the most preferred one in real-time applications. This is due to the fact that autotuning can perform a reliable and quick tuning of PIDs in simple procedure, with "on-the-spot test-and-tune" in a couple of seconds. The entire tuning procedure can be implemented through programming on an onboard chip/centralized processor in a multiply-connected process. As robotic systems are prone to uncertainty with varying operating conditions, each requiring ideal a unique configuration of PID, so in robotic applications, are of great advantage. As discussed before, self-tuning controller is a kind of adaptive controller where control system adjusts controller parameters to ensure required performance. These kind of PID controllers were Though they sound very similar, yet they are different with main difference is the number of them applied in adjustment. *Auto-tuning* tunes the controller's PID parameters *once*, for optimal performance achievement. By using process model, *self-tuning systems, estimate the optimal PID* values and *continuously* adjusts them to achieve the desired setpoint and minimize. It is important to note that self-tuning in its generalized term is a method, not only applicable for standalone controller but is also applicable to DCS/software-based applications also. Since self-tuning is a special type of adaptive control, so, after adaptive controls let the discussions start with concept of self-tuning.

1. **Self-tuning controller concept:** Self-tuning as the name suggests, self-tuning controllers can readjust automatically readjusting the controller tuning parameters as necessary due to changes in process, setpoint and disturbances. It does so continuously. With relay-based test in 1984 also facilitated the system. Self-tuning controllers are often referred to as Auto-tuning controllers but two are not same as already described. Self-tuning" refers to such procedures continuously executed while the controller is online regulating the process. "Autotuning" refers to on demand procedures executed while the controller is online. So, basically, self-tuning or auto-tuning are functions in the control system, so, these can be found in standalone controller as well as in DCS/PLCs.
2. **Self-tuning controllers:** Self-tuning control function consist of a typical PID control block as well as a self-tuning functional block which always tries to maintain optimum closed-loop performance by continuously updating the controller's P, I, and D tuning parameters based on changes in process, setpoint and disturbances as shown in Fig III 1.1c. Most of modern commercial controllers and DCS include both auto-tuning and self-tuning functions. As both tuning and control operations are done simultaneously, self-tuning functions are much more challenging than auto-tuning.
3. **Self-tuning control types:** Self-tuning controllers can be implemented in various ways including deployment of control intelligence also like: fuzzy logic, Artificial neural network (ANN), etc. Model based self-tuning controllers are very popular especially for handling robust control (with uncertainty). Model based self-tuning controllers are two types viz. optimal and response type, that is, *optimal* type and *input/output matching* types. Out of these, as the name indicates, optimal type makes use of self-tuning algorithm based on optimization theory in control system. As shown in Fig. III 1.1c the *controller* continues to maintain PV at specified SP level and learns PV interacts with controller output. Here, there are two issues. More efforts to maintain PV at SP, tuning function is not

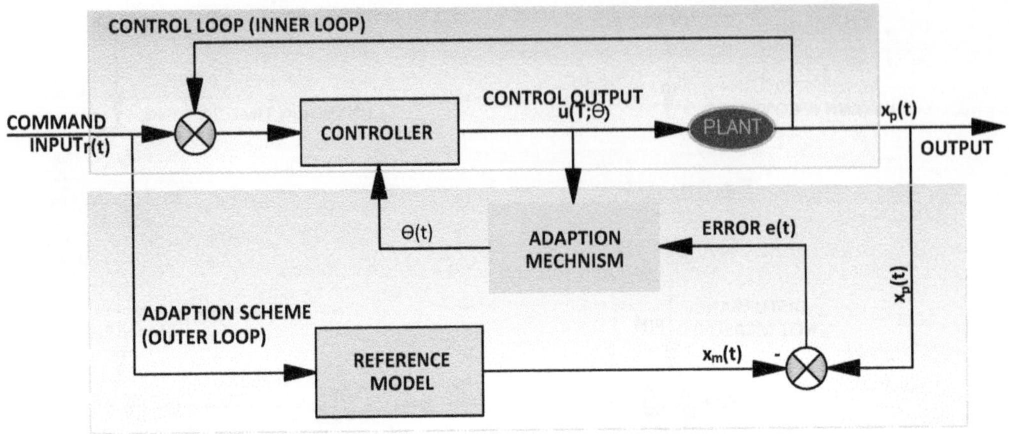


**FIGURE III 1.1B** Model reference adaptive control scheme.

properly attended any useful insight of the process may be missing. If it goes to look after watching insight of the process and its interaction with controller output, then there may be more error in the system. There are times during the normal course of closed-loop operation when the control effort and the process variable fluctuate anyway, and most self-tuners are designed to take advantage of those situations [5]. There are various methods used for self-tuning as listed below:

- **Set point tuning:** Self-tuning control functional block and/or adaption block can observe interaction of PV to control output, for each setpoint using present setting of tuning parameter. Depending on the fastness and sluggishness of the reaction, the adaption block changes the tuning parameters to make the control to respond optimally. The entire closed-loop system is a single process and the setpoint change as a step test, naturally this would yield the overall gain and time constant(s) of open loop system (discussed in controller tuning). Setpoint tuning is more relevant in changes in batches in *batch controls.*
- **Disturbance tuning:** This is re-tuning of control loop based on a process's reaction to a disturbance. When the control function tries to compensate for the disturbance, both control output as well as PV would fluctuate. Here measurement noise, any other disturbances could influence tuning parameters, so that parameters may not achieve the optimal value. In such situation operator's *observation* is important otherwise it is necessary to enable the tuning function only when the control effort and process variable have begun a "limit cycle" or "hunting." A *limit cycle* occurs when overly aggressive tuning causes the control function to over-react to an error between the process variable and setpoint [5]. This could result oscillation on account of over reaction of control function.
- **Limit cycle tuning and mathematical model-based tuning:** Limit cycle tuning discussed is applied in YEL's UT 100 is another way of achieving self-tuning. In this method it is difficult to distinguish between actual limit cycles and periodic oscillations due to interloop coupling, dead band in a control valve, or cyclical disturbances [5]. Mathematical model-based system is more sophisticated alternative study process behavior with different the input/output data. Here, a mathematical model is created from analysis of input/output data. Based on that appropriate tuning parameters can be derived in Internal model-based control (IMC). Delta V of Emerson and/or many other vendors use this process.

4. **Auto-tuning concept:** An auto-tuning is a process by which controller parameters can be computed and set automatically, with main objective to achieve the best control result under given process conditions, for the performance specified. Often default parameter are taken as staring point, then following tuning rules discussed later, controllers are tried to tune manually for the process conditions. This is a time-consuming process. An autotuning is run on the user's initiative to complete the job in a limited time, including testing. With autotuning performed the control loop resumes normal operation with the new parameters immediately. Like adaptive controllers, an auto-tuner follows the same approach that a control engineer would when tuning a controller. Associated major steps are:
   - Process model by choosing a suitable model for the process
   - Process identification—by collecting process data and fit it to the above model
   - Design method—computation of the new parameter.

The tuning objectives are process specific and must be determined before implementation and its usability must be checked prior to use.

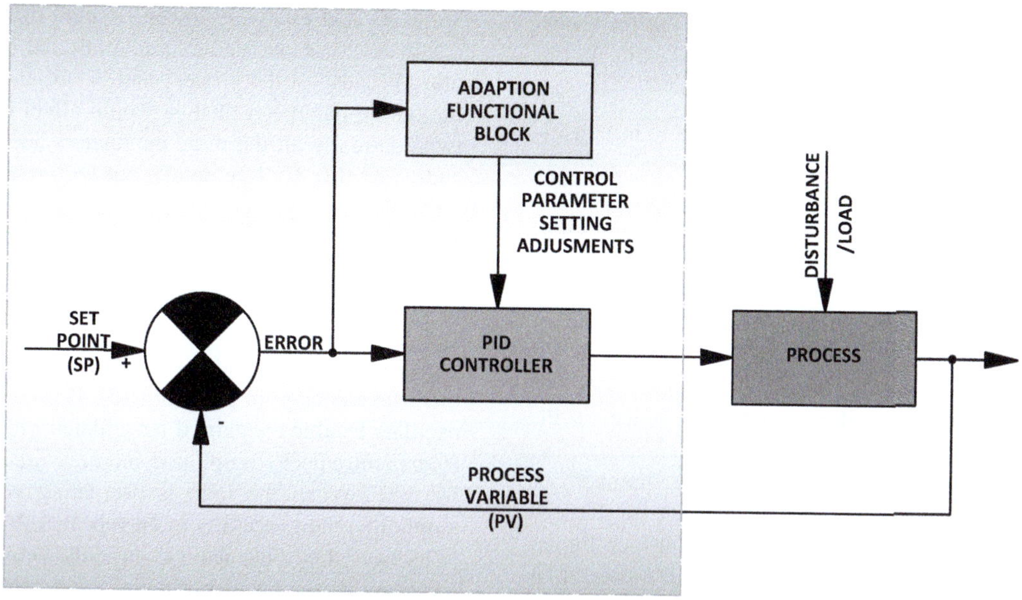


**FIGURE III 1.1C** Self-tuning concept.

5. **Auto tuning controller:** In Auto-tuning controllers can generate suitable tuning parameters, when all control functions are disabled and tuning command is initiated. Controller performs the functions and continue till it could understands the process's natural behavior through collection of enough I/Os.

### 1.1.2 Controller tuning

Controller behavior is determined by the controller parameter. So, in controller tuning we adjust these controller parameters, so that it can produce the matching the controller's response to the process. Thus, controller tuning could be conceived as the process of adjusting a controller's response to a process, by determining the controller parameters which produce the desired output. Therefor we need to find the goodness of control to ensure optimum response. This goodness of control comes from the observation of Controlled variable (CV) trend over time to eliminate overshoot and undershoot material as well as the integral of the error (IE). Goodness of control also consider the impact on the manipulated variable (MV), the controller output. Here, a few important issues are:

Timing, understanding and skill of personnel responsible for tuning is very important.

Tuning procedure should be simple, fast, and robust.

Tuning procedure should not be invasive, so that the process is upset.

Tuning must accommodate what the controller might encounter. Good control is subjective issue, someone's concept of good control can be of poor control for another. We always view good control as an individual control loop's ability to achieve and maintain the desired control objective. Dynamic view of process could be learned through testing. Even manual open loop tests provides lots of data. However, close loop(auto-mode) tests can also be performed with the objective to move the controller output both far enough and fast enough so that the dynamic character of the process is revealed through the response of the measured process variable (PV) [6]. Various tests performed are as follows: Ste[/pulse/Droplet/PRBS, etc. Loop disturbance representing many details must also be considered, while performing the tests. A good test establishes a clear correlation between the planned change in controller output with the observed change in measured process variable. For various controller parameter assessment details Refer to I 1.1.3. Here, it is worth noting that selection of proper PID structure (Process parameter dependent) and to follow tuning rule to tune all the loops in the right order with suitable software as applicable. Process data collection rate is very important and some guideline for different parameters have been provided by Ref. [6]. In this connection, Table III 13.5a [Flow/Pressure < 2sec; Level: 1–5sec Temp: 5–15/15–30 sec] may be referenced.

In this section brief discussions on controller tuning methods shall be covered. In this connection I 13.5.1

may also be referenced for advance loop tuning issues already discussed. Let the discussions start with preparatory stage.

1. **Preparatory stage:** Following stage/steps to be considered prior to start of main tuning process:
   - **Loop devices:** Apart from controller for devices like sensors and actuators issues like: range, calibration, dynamics and ensure all are working.
   - **Process model:** Development of model by trial -error method/autotune method.
   - **Setting of objectives:** Accuracy, speed of response and robustness.
   - **Selection of controller structure:** Choice of PI, PID, etc.
   - **Method selection:** There exists a number of methods, one needs to select the best one for the process. Here, issues like take into account loop limits.
   - **Additional issues:** a few additional issues, like simulation process by varying setpoint, etc., should be planned. While simulating issue like: observations like unexpectedness, change in equipment. Also, proper documentations are important.
2. **Controller tuning methods:** From I 1.1.3 and I 1.2.8 we have gained some knowledge about controller tuning, that is, setting controller parameter. Proportional integral differential (PID) controllers in different control loop configurations must perform up to their potential to yield best results. The controllers are first be tuned to fit each particular application. As the plant conditions and disturbances changes, they periodically re-tuned to maintain the desired closed-loop behavior. Before, we proceed further, it is better to understand controller tuning. Best method of controller tuning is Zeigler- Nichols controller tuning of 1942. For this refer Fig III 1.1d.
   As already discussed in Eq. I 1.1i, standard formula for PID controller is given by

$$m(u) = k\_(p)\left[e + \frac{1}{Ti}\int_0^t \text{edt} + T_d.\frac{de}{dt}\right]$$

   Tuning means selecting appropriate values for parameters $K_p$, $T_i$ and $T_d$ in the above equation, that is, based on particular application through tuning, one fixes the *relative weightage* of tuning parameters referenced above to eliminate errors quickly keeping the system stable. For a sluggish process the controller can be tuned to take immediate action whenever a random disturbance changes the process variable or for set point change. Thus, in such cases, larger $K_p$ and relatively lower $T_i$ could offer better result. As these parameters are connected to each other because of the process and disturbances, each of the tuning parameter change would affect the other two. Naturally all the three parameters need to be addressed. Fig III 1.1e details out Different tuning methods.
3. **Ziegler–Nichols tuning:** This method is applied when the process stabilizes. If the loop oscillates it is to be taken to manual to make sure Process variable (PV) does not oscillate. Then, the loop is taken in Auto mode with Integral and derivative actions are switched off. Make change in setpoint (SP). If process does not oscillate Double the gain, if it oscillates with decreasing (*increasing*)/decay amplitude increase (*decrease*) gain by 50%. When the loop is oscillating with variable amplitude, the process is to be repeated. If the oscillations have died out, again change the setpoint to note the result. If the PV is oscillating with *constant* amplitude, and neither the PV nor the Controller Output hits its limits then, ultimate gain with ultimate period defined as illustrated in Fig III 1.1e1. The smallest controller gain that can cause such an oscillation is called the *ultimate gain* $\mathbf{K_u}$. The period of those sinusoidal oscillations is called the *ultimate period* $\mathbf{T_U}$. Parameters can be calculated using ZN rule as depicted Table III 1.1b. ZN Parameters of Controller types are given in Table III 1.1c
4. **Relay based tuning:** In 1984, Karl Åström and Tore Hägglund of the Lund Institute of Technology Sweden proposed an alternative to the ZN method of test using On-Off relay In *relay method* generates sustained oscillation of the process variable with the amplitude restricted within a safe range, that is, forcing the process variable excursion into *limit cycle* as shown in Fig III 1.1e2. In this method test is carried out by disabling PID terms, and use an On-Off relay to apply a step-like control output. It is held constant at this position, till the PV exceeds the Setpoint (SP). When PV exceeds SP, it applies a negative step and again waits for PV to drop below SP. Repeating this procedure, PV to oscillate (out of sync with the control output) at the same frequency. Although the process variable's oscillations aren't strictly sinusoidal, their period turns out to be a close approximation of the ultimate period that Ziegler and Nichols' tuning rules [7]. Then measures TU directly from the square wave control effort and computes the value of $K_u$ with measurements of a and b as shown in Fig III 1.1e2. $K_u$ can be approximated from

$$K_u = (4/\pi) * (a/b) \qquad \text{eq. III 1.1f}$$

By limiting a, PV swing can be limited hence limit cycle which ZN method may have limitation can be achieved.

5. **Tuning mistakes** [8]: A few common mistakes encountered during controller tuning are:
   - Choice/use of wrong control action
   - Use of Default setting for PID
   - ISA standard form: using parallel form and series tuning settings
   - Use of wrong units: Tuning setting/output limits and antireset limits
   - Tuning level controller
   - Violating the window of allowable controller gains
   - Recognition failure: Sensor lag, transmitter damping, or filter-setting effect
   - Failing to do tuning tests at different times, set points, and production rates
   - Failing to increase PID gain to decrease backlash limit cycle amplitude

Let us now move our focus on Concept of Expert systems.

## 1.2 Concept of expert system

The concept of expert systems was developed by E. Feigenbaum, professor computer science at Stanford University in 1970. Expert system (ES) represents a reliable interactive computer-based system deployed to solve complex decision-making problems, and for decision-making based mainly on both facts and heuristics. An expert system program normally runs on a set of rules in the form: "IF" conditions, "THEN" actions. One of the major objectives of the expert system is to replicate the expertise of the best practitioners (not always available) in the field. Once the logic is encapsulated in the rules. ES can be conceived in multiple dimensions ways like: a kind of *AI methodology*, *high performance*, *greater flexibility*, *understandability for reasoning* are a few to name. ES technology allows the heuristic knowledge for solving complex problems. Also, it helps in mechanization of reasoning process with knowledge from of the problem domain separated from inference tool. Expert programming is different from conventional computer programming. To represent varying behavior of process plants in an expert system requires a knowledge representation with different rules as well as differential equation models such as material balances, energy balances and reaction kinetics. The object-oriented paradigm is used, with a hierarchical frame structure. The most efficient representation, behavior may be in generic way with as much abstraction as possible. Major issues are as follows: *generic representation of behavior*, *plant structure*, representation of *temporal and integration knowledge* of the plant. Also, for real time reasoning associated knowledge is essential. In complex systems like controls of reactors, Fuzzy logics are often applied with expert systems. With developments of AI over the years, rule based expert systems plays a big role. For implementation of expert system, along with standard automation at certain level is necessary. Then, upgradation of instrumentation in terms of reliable process sensors and actuators as well as upgraded control and automation systems with advanced software will be necessary. At the same time upgradation of mechanical/electrical equipment for which expert systems will be applied, is necessary. In this section, fundamentals on expert systems will be discussed.

*Zeigler Nichols Controller Tuning:* **Out of ma y rules for tuning of PID controller Zeigler-Nichols' (B of then *Taylor now ABB*) controller tuning rule is the first (19[illegible] d probably best ad most popula. ZN method does not n ed a mathematical model but it is set empiricly through experimentation. In this method better combination of Gain (P), Reset (I) and Rte (D) timings are found for specific application. Tuning done in this method would result less oscillation, better reset time with lower overshot. Manully this is done by (thumb rule) setting reset time to maimum (long reset time), with ra e set to zero ad gain is increaed from zero to a value where oscillation just stats. Then the ga n is hav ed and reset time is dj usted to get offset correction within a ceptable time value. Finl y,ra e is increased till overshoot is minimum.**

**FIGURE III 1.1D** Zeigler-nichols controller tuning.

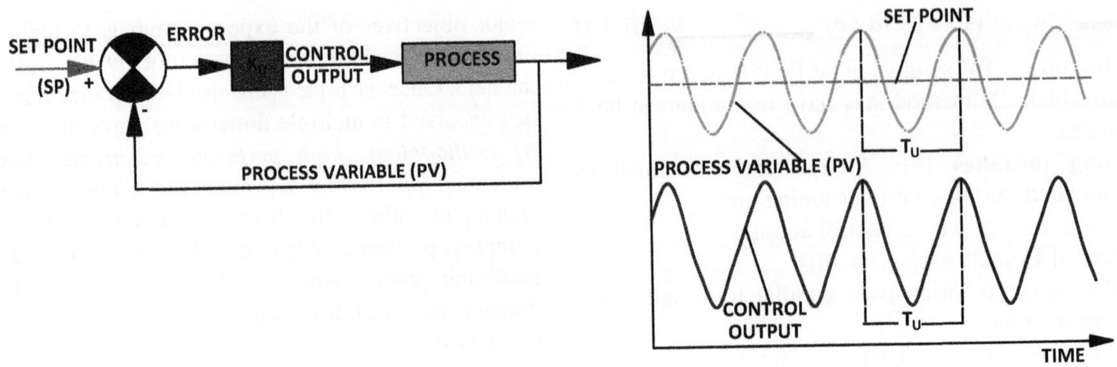


FIG I 1.1e1 ZEIGLER–NICHOLS CLOSED LOOP CONTROL LOOP TUNING METHOD

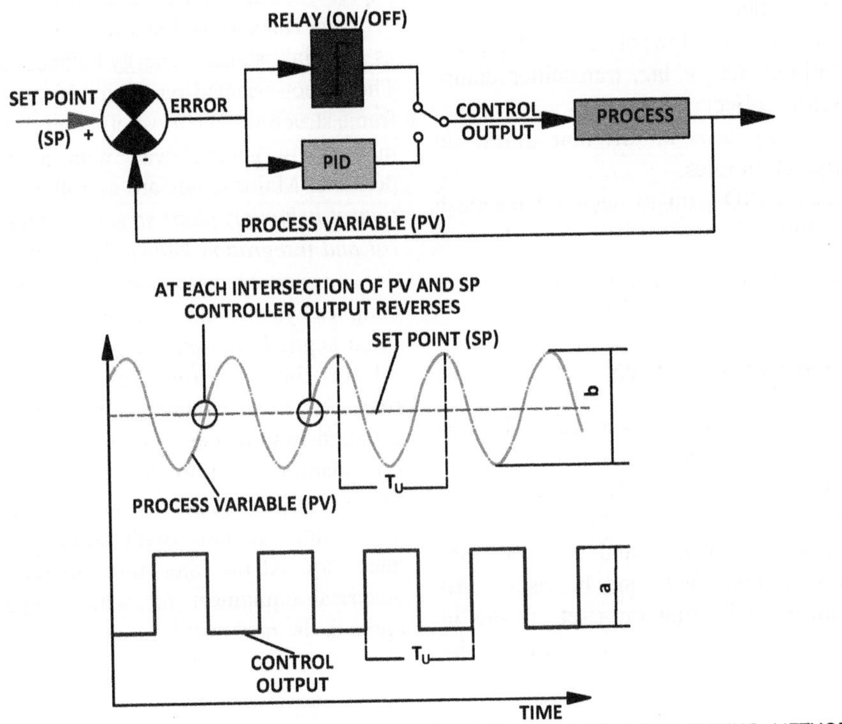


FIG I 1.1e2 INTEGRATED RELAY TEST IN CLOSED LOOP CONTROL LOOP TUNING METHOD

**FIGURE III 1.1E** Controller tuning.

**TABLE III 1.1B** ZN Tuning Parameters for PID controller.

| Parameter | ZN rule | Robust controller | Level control |
|---|---|---|---|
| **Controller gain ($K_p$)** | $K_p = 0.6K_u$ | $K_p = 0.3K_u$ | |
| **Proportional band ($1/K_p$)** | 1.66 $PB_u$ | 3.33 $PB_u$ | |
| **Integral time ($T_i$) time/repeat** | $T_i = 0.5\ T_u$ | | $T_i = 1.0\ T_u$ |
| **Integral gain ($1/T_i$) Repeat/time** | $K_i = 2.0/T_u$ | | $K_i = 1.0/T_u$ |
| **Derivative rate ($T_d$)** | $T_d = 0.125T_u$ | | |

Suffix u for ultimate gain/time period.

**TABLE III 1.1C** ZN Tuning parameters for different controller types.

| Controller types | $K_p$ | $T_i$ | $T_d$ |
|---|---|---|---|
| P | 0.5 $K_u$ | – | – |
| PI | 0.45 $K_u$ | $T_u$/1.2 | – |
| PID | 0.6 $K_u$ | 0.5$T_u$ | 0.125$T_u$ |

## 1.2.0 Expert system (ES): explanation, structure technology and development

ES is a Software (SW) program, and as a branch of AI, it uses AI technologies to simulate the judgment and behavior of a human expertise having experience and knowledge in a particular field. So, expert systems are deployed to replicate expertized personnel for:

- Development of an automation system which could produce consistent results under wide variations of operating condition of the plant ensuring safety.
- Preserve knowledge about the plant operations under different conditions, in this plant as well as similar plants and other expert operator knowledge;
- This is software based knowledge of expert system is permanent not perishable like human expertise;
- Totally new procedure development for the exercise during the knowledge base building, including analysis process, fitting operational limits, etc.
- Development of Inference engine based on analysis of knowledge base with ÏF" and "THEN"logic
- Explanation the results

Apart from these other issues which calls for expert systems are as follows: High efficiency. Coverage of all situations, No Dependence and limitation from memory and Security. In this section fundamentals pf expert systems will be covered. Let the discussions start with explanation of expert system with examples.

1. **ES explained:** Expert systems are interactive and reliable computer-based applications which can mimic human experts to solve complex problems in a particular domain. For very complex decision-making and other complex problems, expert systems provide solutions at the level of extra-ordinary human intelligence and expertise. It is based on knowledge acquired from an expert. Expert system solutions are heuristic (i.e., they provide reasons with judgmental knowledge), transparent (explanation available) and flexible, that is, can integrate new knowledge. In computer non-numeric and heuristic problems are encoded in the program as logical and mathematical algorithms. *MYCIN*, *CaDet* DENDRAL, etc., are very common expert systems available in the market. However, Expert systems can be used for process/plant controls also. Fig III 1.2a illustrates Expert system structure and technology.
2. **Basic structure:** In order for the computer to be able retrieve and effectively use heuristic knowledge, the knowledge must be organized in an easily accessible format that distinguishes among data, knowledge, and control structures. As shown in Fig III 1.2a1, expert systems are organized in following distinct levels (detailed in III 1.2.2).
   - **User interface:** This refers to the interface through which nonexpert user interfaces with the expert system.
   - **Knowledge base:** This refers to the problem-solving rules procedures, and intrinsic data relevant to the problem domain. So, the knowledge base constitutes the problem-solving rules, facts, or intuition that a human expert might use in solving problems in a given problem domain.
   - **Working memory:** This is a part of knowledge base, and refers to task-specific data for the problem under consideration.
   - **Inference engine:** IE is a software based task specific control mechanism to impart the axiomatic knowledge in the knowledge base to arrive at some conclusion which can be directly used in the control/may be used as operator's guide (advice).
3. **Expert technology:** Expert technology comprises three parts as shown in Fig III 1.2a2. The software that helps to build expert systems is often referenced as *Expert system Shell* which can manage the user interface as well. Inner shell consists of knowledge acquisition, inference engine design and user interface, that is, it is the expert system without knowledge base. To develop this inner shell normally technology takes the help of various tools such editor, debugging unit, etc., ad detailed out in Fig III 1.2a2. Both shell and tools can work in an environment of computer/work station with High level Symbolic program language (SPL), for example, LISP, Prolog, etc. They work on data base of the problem domain.
4. **Comparison with conventional control:** Table III 1.2a provides comparison of conventional control systems versus control system with expert system.
5. **Expert system development:** Expert systems being an interactive computerized decision-making tool use both facts and heuristics to solve difficult decision problems based on acquired knowledge. Starting from basic *back ground, concept,* problem *definition and selection, knowledge acquisition*, suitable *tool selection, developmental process, validation, testing, verification, implementation, integration, training* and *maintenance.* Main technical parts of the expert system development

consisting of several stages has been detailed out in Fig III 1.2b. First suitable expert system in the problem domain is identified. It must be suitable for solving the problem in hand and should be cost effective. After that suitable technology is identified along with necessary data integration then, inference engine is developed with "IF THEN ELSE" logic. Based on the same initially prototype system is developed and tested. If test show any deficiency then necessary refinements are introduced in to the system. After Prototype is tested successfully, complete expert system as detailed out in Fig III 1.2b. Training of people for expert system and maintenance of it is also important and these should be taken care suitably as a part of development.

### 1.2.1 Expert system: control operation, characteristics, Miscellaneous issues

Computerized software-based, expert systems serve as a permanent storehouse of knowledge with easy accessibility. Fault diagnosis, process troubleshooting, quality control, control system synthesis, statistical process control and alarm management are major application area of expert systems in process controls. There are certain distinct features of expert system from human expert Additional element "time" needs to be included in expert system for its functioning in real-time applications. This time element should be there in the reasoning process also. Software vendors have come out with such solution in the market to meet the requirements of process controls. This calls for integration and exchange of data between Distributed control system (DCS) and expert system. This is elaborated here. Expert systems are always very domain specific requiring special tools/programming languages to accomplish the specific objectives of the system. As against that, expert systems never forget crucial details of a problem like human and always offer consistent decisions always. Applications that are computational or deterministic in nature are not good candidates for expert systems. In this subsection control operation, characteristics along with pros and cons of expert systems have been discussed.

1. **Plant control:** As already discussed above in III 1.2.0.2 Experts systems mainly comprise of three major components viz. *Knowledge base*, *Inference engine* and *user interface*. Experts put their experience and knowledge which are inputted to the system to develop knowledge base by Knowledge engineer (*say* software engineer through proper encoding). Based on data base of knowledge bank, utilizing IF THEN ELSE logic inference engine makes the decision to pass it on to user interface. Through user interface, user (nonexpert) can exchange data for plant operations. ES in conjunction with control system (DCS) and plant have elaborated in Fig III 1.2c. Here, detailing on knowledge engineer, as well as operators has been depicted. This also elaborates how expert system can interface with control system/DCS. As is seen that there will be continuous data exchange between control system/DCS and expert system. Expert system can be an integral part of the system for direct action or it may have separate interface. As shown in Fig III 1.2c, there are two options for expert system viz. direct control through communication link (integrated operation) and advices to operator through messaging (operator's guide mode).
2. **Tools and programming:** Expert systems tools can filter knowledgebase; they can organize storage and proper retrieve procedure. Also, tools can update and expand knowledgebase on regular basis. Expert systems normally use special programming languages, for example, LISP/PROLOG for simplified the coding process. LISP has many advantages like more procedural than prolog, and is more generalize in nature. However, LISP is less portable. Some characteristics of programming languages are:
   - High degree of memory-management, tagged memory architecture and extensive data manipulation.
   - Mixing of integer and real variables efficiently
   - Organization of system environment with efficient search procedure
   - Incremental compilation.
3. **Reasoning and inference:** Expert systems generally, work in interactive way like: *seeking clarification, responds to query, issuance of recommendations/decision making.* Here, expert systems do these with *remembrance* of logical chain of reasonings (heuristic or old case basis) and then set out recommendations with detailed explanation and factors to be considered. These would enhance confidence of user. Inferences are arrived at with simple reasoning mechanism. Also, knowledge base must be able to exploit the knowledge stored, to make it useful. Thus, for inference, expert systems exploit stored knowledge to help in making decision or provide advice.
4. **Characteristics of expert systems:** Following are a few characteristic features:
   - **Performance:** Expert systems can offer solutions at the highest level of performance with enhanced efficiency, accuracy in the specified field for complex problem.
   - **Challenges and understanding:** Expert systems can take challenges for provide optimum solutions for very complex decision-making problems. They put forward suitable explanation for better understanding of user. Some unique features are:
     - *Knowledge base:* They utilize mainly heuristic (rather than algorithmic) knowledgebase

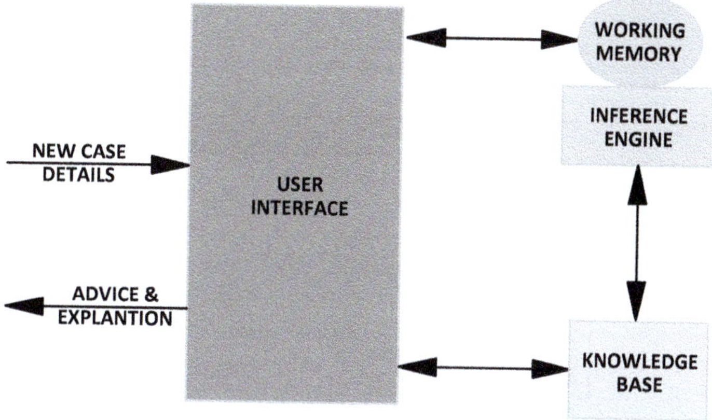


FIG III 1.2a1 EXPERT SYSTEM BASIC STRUCTURE

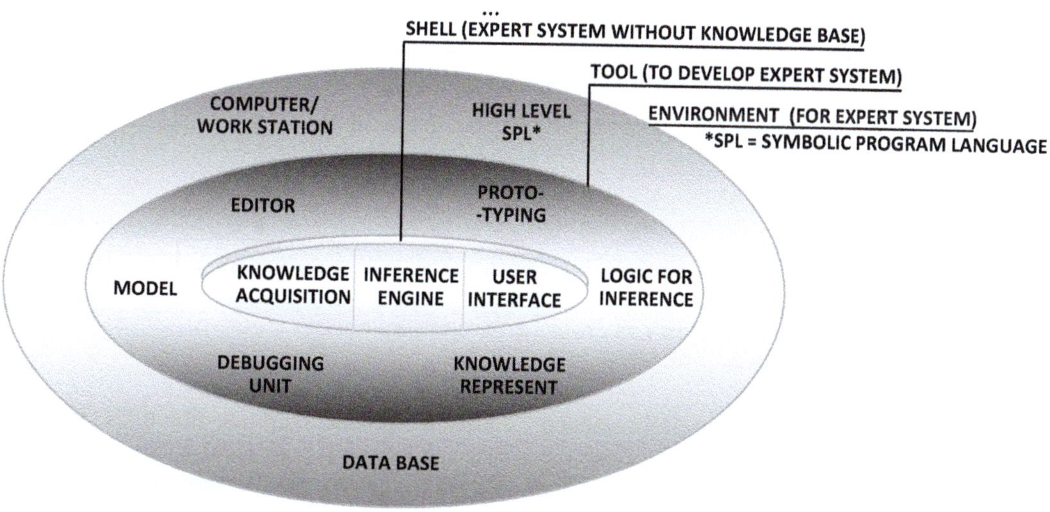


FIG III 1.2a2 EXPERT SYSTEM TECHNOLOGY

**FIGURE III 1.2A** Expert system structure and technology.

rather than data which can be from old cases as well.

- *Separation:* Encoded knowledge can keep it separate from control programs.
- *Knowledge Representation:* Rules, network and frame are major symbolic representation of their knowledge base which they learned and gathered from past in the related area
- *Optimum configuration:* They would find optimal configuration to solve complex issues/problems for variety of tasks and can reason incomplete uncertain facts.

**TABLE III 1.2A** Comparison of Conventional control with expert system.

| Issues | Conventional control system | Expert control system |
|---|---|---|
| Objective | Data, program and problem expertise are preset | In data structure problem related expertise present |
| Data base | Fact based knowledge | Fact and procedural knowledge |
| Query | Human to machine for reply | Machine to human query |
| Problem | Wide/broad and complex | Narrow domain complex |
| Manipulation | Numeric normal program | Symbolic special program |
| Decision | Direct decision making | Expert transfer and advices |

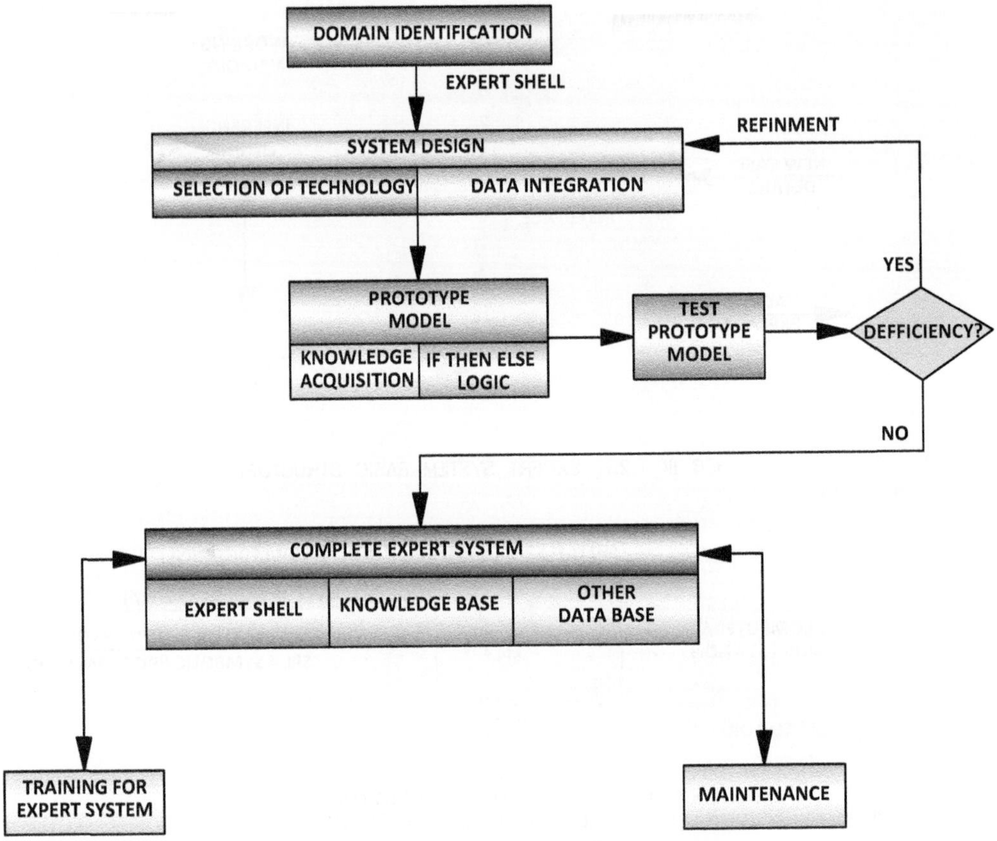


**FIGURE III 1.2B** Expert system development process.

- **Responsiveness:** Expert Systems generally respond with accuracy within reasonable time to make the systems very responsive. Normally time taken by them is less than that by human expert to give accurate solution.
- **Reliability:** Highly reliable and hardly commit any mistake when there is proper knowledge base.
- **Effectiveness:** Domain based Expert System must offer efficient way to solve the issue, that is, they must be very effective.
- **Flexible:** High flexibility is one of the important characteristics of Expert System.

5. **Pros and cons:** Table III 1.2b shows various pros and cons of expert system.

## 1.2.2 Expert system: integration and component details with functional issues

In 1970s the notion of a production system as a model for psychological behavior of human, was introduced by A. Newell and H.A. Simon. Here, part of the human knowledge is being represented in *separate units* containing information concerning actions of a person to be taken upon the perception of certain stimuli from the environment. These are production/production rules. Such actions could affect a person's perception about the environmental reality. This can happen because of two issues viz. *previous assumptions may have to be revised*, or a possible new phenomenon to be explained. The model closely resembles the two-process theory of memory in cognitive psychology. Many present-day expert systems use the production-rule formalism as a knowledge representation scheme. Practical experience with production rules has proven this formalism to be particularly suitable in solving various kinds of problems. For arriving at a good solution quickly and efficiently, ES base their reasoning process on symbolic manipulation and heuristics inference procedures. This very similar to the thinking process of human expert. However, there are many techniques which are deployed to make these intensive searches more efficient. *Branch and bound*, *pruning*, *depth first search* and *breath first search techniques* are often used. In this subsection, some detailing about various components of ES has been covered.

1. **Structural integration:** A good expert system is expected to grow as it learns from user feedback by incorporating feedback into the knowledge base

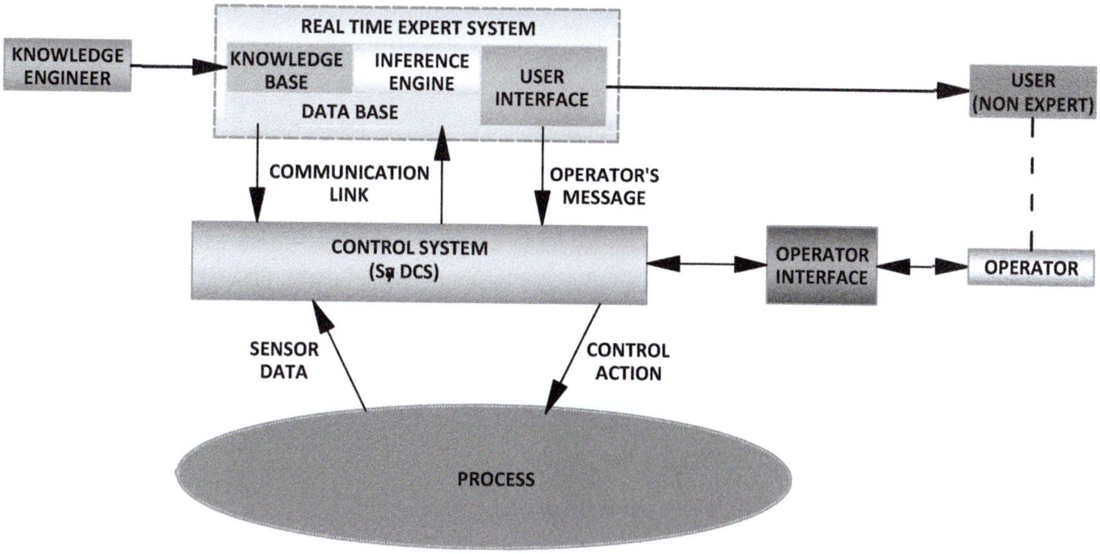


**FIGURE III 1.2C** Expert system in plant controls.

appropriately to make the expert system smarter as detailed in Fig III 1.2d.

The dynamism of the application environment for expert systems are:

- **Most dynamic:** *working memory:* Working memory contents, (referred to as data structure), changes with each other problem situations.
- **Moderately dynamics***:* *knowledge base:* No change in the knowledge is needed unless a new piece of information arises to indicate a change in the problem solution procedure which must be carefully evaluated before implementing changes in procedure in knowledge base.
- **Least dynamic:** *inference engine:* Because of the strict control and coding structure of an inference engine, changes are made only if absolutely necessary/to correct a bug or enhance the inferential process.

2. **Knowledge representation:** Out of many available formalisms, the production-rule formalism is the most important as it expresses the actual problem-solving knowledge. This set of production rules in a production system is called its rule base discussed later. A production system also provides a means for defining the objects. The rule base and the domain declaration together constitute the knowledge base for the production system. Let us now look into the following related to knowledge representation.
   - **Variables and facts:** During a consultation of the knowledge base in a production system, information is constantly is added/removed/modified because of the production rules, data entry by the user, or as a result of querying some database. The system learns during a consultation and stores in a so-called fact set (global database) or working memory.
     - *Factual information*: This can be represented in a number of ways such as by means of "variables" which can take either a single constant or a set of constants as a value. These variables in a production system with their possible values, presents a picture of the information to be for field modeled in the system.
     - *Variables:* Two types of variables are as follows: *single-valued variables* (one constant value at a time) and *multivalued variables* (a set of constants).
   - **Condition and conclusion:** Mostly, this type of knowledge is in the form of heuristic rules or rules of thumb, and are the informal, real-life analogies of production rules and follow "IF" and "THEN" logic.
   - **Object-attributes:** In many problem domains for model, one can often discern separate subdomains, or objects, which are interrelated in some way. Each subdomain then is described by a number of properties or attributes which are specific for that subdomain.
3. **Discussions on knowledgebase:** The knowledge base is the library or a repository of facts and rules to store all necessary available or acquired (from different experts) knowledge about the problem domain in an organized manner. The success of the ES is highly dependent on the accurate and precise knowledge acquired and stored in knowledge base. There are types of knowledge such as factual and heuristic.
   - **Factual knowledge:** Well accepted piece of information in the problem domain accepted by knowledge engineer/experts.

TABLE III 1.2B Pros and cons of expert system.

| Issue | Explanation | Pros | Cons |
|---|---|---|---|
| **Enhanced output and productivity** | Higher than human being | X | |
| **Better product quality** | Lower error in size and rate | X | |
| **Lower decision-making time** | Efficient decisions rapidly | X | |
| **Flexibility** | Much than human experts | X | |
| **Easy operation round the clock** | Makes easy to operate | X | |
| **Difficult environment** | Helps to avoid by human | X | |
| **Lower downtime** | Ability to detect and repair fault | X | |
| **No need for expensive device** | Prompt and in detail analyses | X | |
| **Knowledge accessibility predictable modeling** | Accept query with expert level replies. Problem solving model | X | |
| **Immediate feedback** | Improved quality and productivity | X | |
| **Training** | On job to become expert | X | |
| **No memory limit** | No applicable like human | X | |
| **Situational assessment** | Hard—not easy development | | X |
| **Expensive knowledge engineer** | Costly proposition | | X |
| **Different ways, costly -No independent verification** | Different approaches may be costly but no ways to verify | | X |
| **Task specific different uses** | No generalized approach costly | | X |
| **Limited vocabulary to** | Express of relations and facts | | X |

courtesy: [9].

- **Heuristic knowledge:** These are judgmental or evaluation, interpretations, proven practices and some guess work, and experience
- **Knowledge representation:** It is mainly based on IF -THEN LOGIC, etc., to organize the and formalize the knowledge in the knowledge base.
- **Knowledge acquisition:** The knowledge base is developed by acquiring knowledge from different experts, scholars, and the Knowledge engineers and past incidents. The knowledge base quality highly depends on the quality, completeness, and accuracy of the information.
- **Knowledge engineer:** The knowledge engineer needs to have ability to quickly learning, and analysis skills with sound knowledge on logic formation. Knowledge engineer can acquire information from subject expert. Knowledge engineer must be able to categorize and organizes the information in a meaningful way so that it is possible to develop IF-THEN-ELSE rules (the backbone). Major functions of knowledge engineer shall include but not limited to: defining the *domain* of expert system, *Elicit* desired information from human experts, *incorporation of the knowledge* in acceptable form and test the system to check its *robustness and accuracy.*

4. **Issues related IE:** IE makes it possible to take decision in solving and/or diagnosing a problem or guide operator (advice). Inference engine is purely "Ruled-Based" system to predict what would happen or why it happened. With reference to knowledge base, it selects facts and rules to apply to answer the user's query. In case multiple rules are applicable, it can resolve rules conflict—if necessary. Also, it provides reasoning about the information in the knowledge base. This component is responsible for formulating conclusions by applying efficient procedure, utilizing rules to come out with correct, flawless solution. It can add new knowledge into the knowledge base if required. As discussed, recommended solutions can be two types viz. *Forward chaining* (to find what would happen next, that is, solution to a problem) and *Backward chaining* (to find why it happened i.e., diagnostic issues) Both systems have been depicted in Fig III 1.2e.
   - **Forward chaining IE:** Here, the IE follows the chain of conditions, facts and derivations to deduce find final conclusion. It considers and rules, and

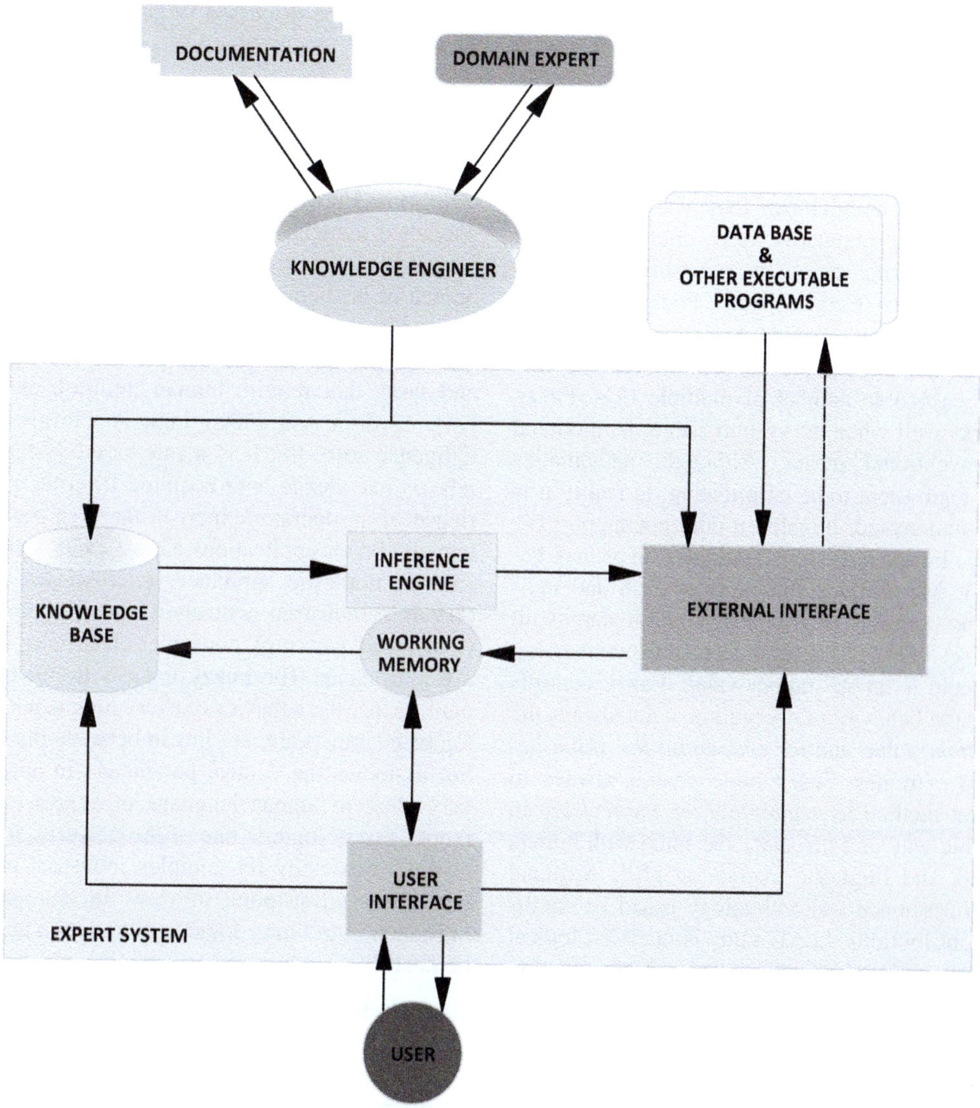


**FIGURE III 1.2D** Structural integration in expert system.

sorts in arriving at a solution. Basic philosophy is as follows: "If the *condition is true*, then *the action part* of the rule is also true." This is illustrated in Fig III 1.2e1. This follows decision tree principles; share *market prediction* could be a good example to understand the issue. This is applied in plant controls also.

- **Backward chaining IE:** With this strategy is utilized for fact finding, that is, why it happened? Here, IE tries to which conditions could have caused in the past for the current result. IE does so, based on what has already happened to find the probable result. This finds its application in plant controls systems for diagnosis or analysis of the issue after it has happened. This is also called "goal driven." Here, a goal is specified and the expert system tries to determine possible conditions are needed to arrive at the specified goal. This is illustrated in Fig III 1.2e2.

5. **User interface**: User interface is a means through which user can access to the expert system. User engine explains in details derivation of Expert recommendations and verbal narrations in natural language. It also lists out rule numbers displayed on the screen so that it is easy to trace the credibility of the deductions. So, user interface should be capable to accomplish their goals in shortest possible way utilizing inputs efficiently. Also, technology should be able to adapt user interface should be designed to work with the existing system.

## 1.3 Fuzzy logic, artificial intelligence and machine language

Fuzzy logic is a part of AI—a subset of AI. Rule-based Fuzzy logic system can rely on the practical experience of an operator, particularly useful to capture experienced operator knowledge. Though Fuzzy logic was known since mid-1960s, it gained popularity only after mid-1970s. Apart from engineering areas like: washing machine, camera, process plants (Cement plant), Fuzzy logics find its usage in stock market, medical uses also. Fuzzy logic systems can nicely handle problems in nonlinear systems/ systems with reasonable number of multiple I/Os. Fuzzy logic also works well when the system cannot be modeled easily by conventional means. Although mathematics behind Fuzzy logic seem to be intimidating, in reality it is not so hard to understand. It reality it does not support 0/1 binary directly, Fuzzy logic is a continuum of values between 0 and 1. Membership functions of each and individual input and output is very important and it starts with this. Then, a set of rules are applied to the membership functions to yield a "crisp" output value. Fuzzy controls knows/understand behavior of everything is not always the *top* or the *bottom* value and for *evaluating the value between* the two extremes. Fuzzy logic goal is always to create an easier method to *manipulate the imprecision* in the best possible way and *associate* the same with human vague thoughts and linguistic expression [10]. Artificial intelligence is a science and technology based on multidisciplinary contributions. In AI, study is made on logical thinking process of human brain and how humans learn, gather experience and make decision to try to solve any problem in different areas. By using the outcomes of such studies as the basis of developing intelligent software and systems. So, it could be conceived as branch of science and engineering for developing intelligent computer programs to make intelligent machines and controls of various other devices. Development of AI started with the intention of creating similar intelligence in machines/devices that we find in humans. So, basically this calls for creation of expert system and implanting human intelligence in machines/devices. Major AI techniques are as follows: Rules and logic based, machine/deep learning and hybrid system. Pattern based AI is powerful, used in automated language translation, driving, etc., to name a few. Machine learning (ML) is a kind of software algorithms used to find patterns in data for inferring rules of their own, "Learn" from data to initiate improvement over time. Also, ML is used extensively in automating various applications as well as for prediction issues. ML is used the most applications of AI. Although ML and AI are used interchangeably, yet they do not mean the same thing. While all ML are AI yet all AIs are *not* ML. As is clear from the above discussions that in this section three different topic viz. Fuzzy logic, AI and ML will be discussed to gain preliminary knowledge about them.

### 1.3.0 Basics of fuzzy Logic

In 1965, Professor Dr. Lotfi. A. Zadeh of the University of California, Berkeley first came out with Fuzzy theory and published associated paper on the same. Soon it gained popularity around the globe and a new mathematical branch of mathematics developed based on Fuzzy system. According to B.G. Liptak "Fuzzy logic is a very human concept, potentially applicable to a wide range of processes and tasks that require human intuition and experience" [11]. As discussed, Fuzzy logic is a form of artificial intelligence software. It is a rule-based system that mainly rely on knowledge base acquired from the practical experience of an operator/expert in the area. Fuzzy logic systems find their applications almost every sphere of our life such as domestic appliance (microwave), environmental (HVAC), Industrial controls (Cement plants), automotive (four-wheel steering), consumer electronics (camera), etc., a few to name. The Fuzzy logic is useful because it recognizes thar the behavior of everything is not always on the top or bottom value, but lies in between the two extremes. So, it studies the control parameters in approximate way very close to human language of expression (e.g., Little more). Fuzzy logic is one of the powerful tool for control process, especially for complex industrial processes from cost optimization point of view. In this subsection brief introduction to Fuzzy logic in automation application shall be dealt.

1. **Introduction to fuzzy concepts:** A control system working on fuzzy logic is Fuzzy control system. In classical/conventional system or digital logics only discrete values of either 1 or 0 (as true or false), is considered. Fuzzy logic—a mathematical system that analyzes analog input values in terms of logical variables that take on continuous values between 0 and 1. This means that fuzzy logic is an approach to computing based on "degrees of truth" rather than discrete value. In place of discrete values, it considers degree of truth or membership function. In reality whenever human takes a decision it does not function like computer instead in human decision there exists a range of possibilities between YES and NO, such as: *Certainly* "YES," *Possibly* "YES" *Cannot tell*; *Possibly* "NO" *Certainly* "NO."

   In Fuzzy logic YES and NO has been arranged between value 0 and 1 so that other possibilities gets chance to be represented. The Fuzzy logic works on the levels of possibilities of input to achieve the definite output. Therefore, Fuzzy logic is well accepted because it can provide acceptable reasoning which may not be very

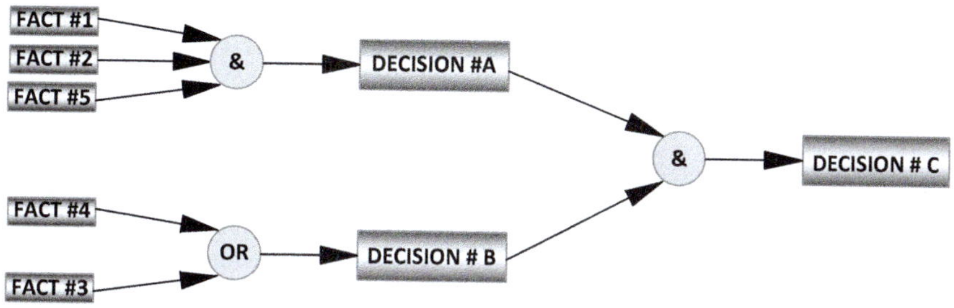


III 1.2e1 INFERENCE ENGINE FOR SOLUTION (FORWARD CHAIN)

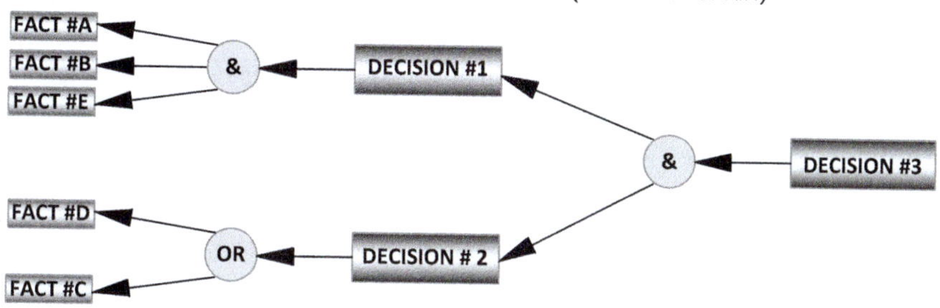


III 1.2e2 INFERENCE ENGINE FOR DIAGNOSTICS (BACKWARD CHAIN)

**FIGURE III 1.2E** Inference engine solutions and diagnostics.

precise and accurate, but it always helps in handling uncertainty in engineering. Let us take an example from industries. Chemical reaction inside cement kiln in dry cement process is not very clear till today. Therefore, many plants may not totally depend on auto-control especially when there are chances of raw material quality variations. In such cases experience kiln operators acts as expert to operate the kiln. Many advanced control suppliers, for example, F. L. Smidth take the advantage in implementing expert system in controls using Fuzzy logic. To understand the system better let us refer to Fig III 1.3a Fuzzy concept. This will also help in understanding Fuzzy system with associated terms.

2. **Crisp and fuzzy set: crisp variable and fuzzy set:** As per conventional set theory, an element either belongs to or does not belong to a given set based on premise set for it. *When Crisp logic is binary* in nature *Fuzzy logic takes into account nearby areas also. Hence in* Fig III 1.3a1 *crisp is represented by black and white circle. Fuzzy logic is represented by another set of circles which include gray areas also. This has been shown conceptualize the idea.* If we look at Fig III 1.3a1. it could be noticed that a ring representing a set. In conventional system any element within this circle belong to the set. So, it is quite clear that variables a, b is within the set and d is out of the set. Fuzzy set theory is not that much rigid like conventional set and or binary systems (recognizing either 0 or 1 Fuzzy allows "*some-what*" or "*very near*" to be in the set. Crisp number/input measurements are those which have some uncertainty. In this case the variable "c" which is in the boundary. So, variable "c" is falling under category of "somewhat" uncertain. Since *Crisp* logic acts binary *values (0/1),* that is, either "true" or "false," so in crisp representation it is 0 as shown. On the contrary, as Fuzzy allows "some-what" so, it has been given a membership value and included in fuzzy set as shown. Let us take an example from Heating ventilation and air conditioning (HVAC); assuming that, above 24°C is the uncomfortable. Any temperature (below) **nearby** temperature of 24°C is comfortable. In crisp *logic* there will be a single line as shown in Fig III 1.3a2. It is artificial that any temperature that falls slightly outside this range would not be a member of the set, even though human being may not be able to differentiate. In fuzzy set theory, that is not the case, there is not such strict limit, instead there is "*membership" function* to cover nearby region. Membership function in Fuzzy system is used to describe degree of participation of the variable, that is, the relationship between a variable and the degree of membership for the fuzzy logic. Such degree of membership (or participation in the logic) is usually defined in terms of a number between 0 and 1 representing 0%–100% of participation. Thus, 0 (0%) implies total absence of membership and 1 (100%) implies complete membership. So, any value in between signifies partial membership. This may be written as follows: If $\boldsymbol{\mu_A}$ is the membership function $\mu_A = [0; 1]$ for $X \in U$, and $U$ is

the universe of discourse that defines the total range of interest. In our case, $\mu_A$ is defined as

$$\mu_A = 0: X < 19°C;\ \mu_A = \frac{X - 19}{(24 - 19)}: 19°C < X < 24°C;\ \mu_A = 1: X > 24°C \quad \text{(Eq. III 1.3a)}$$

This has been graphically represented in Fig III 1.3a2. So, variables in 19°C–24°C are crisp variable not crisp logic. Let us take a simple example from mathematics. Suppose there is a set of real numbers $X = [1,2,3,4,5,6]$;
Let us find the membership function A of "near 2" in this set then it will be given by $\mu_A$ (1) = 0.5; $\mu_A$ (2) = 1, $\mu_A$ (3) = 0.5 and $\mu_A$ (4) = 0 ....

3. **Basic fuzzy system structure:** Basically, there are four main parts that constitute Fuzzy system. Typical structure has been depicted in Fig III 1.3a3. Main control and other systems mainly handle crisp input/output sets. Therefore, when these crisp I/Os are handled by fuzzy systems there will be need for fuzzification of crisp input so that fuzzy system can work and finally fuzzy output needs to be transformed in to crisp output so that other systems can work. Typical fuzzy control scheme has been detailed out in Fig III 1.3a3. Let us now look into details of main parts as follows:
   - **Fuzzification module:** It transforms the system inputs, which are crisp numbers, into fuzzy sets. Depending on the need input crisp numbers/inputs are divided into some odd numbers. Let us consider five splitting of crisp inputs X, here, let middle be at (say) 0 denoted by small, then two in the positive side two in negative side (as large and medium in each side) and as shown in Table III 1.3a.
   - **Intelligent system (fuzzy):** it consists of two parts as knowledgebase and inference engine. Intelligent system work on "IF THEN ELSE" rules base
     - *Knowledge base:* Like expert system, knowledge base is acquired expert knowledge stored
     - *Inference engine:* It simulates the human reasoning process by making fuzzy inference on the inputs. Based on the rules inference engine calculates the degree to which each rule fires for a given fuzzified input as detailed later.
   - **Rule base:** This is the logic which is based on IF THEN ELSE as in expert system discussed. Such collection of rules forms the rule base.
   - **Defuzzification module**: As shown, function of defuzzification module is to transform the fuzzy set obtained by the inference engine of intelligent system into a crisp output value. Mostly commonly used method is centroid which takes a weighted sum of the designated consequences of the rules according to the firing strengths of the rules [11].
4. **Membership function:** It is membership functions that allows to quantify linguistic terms to be represented as a fuzzy set graphically. A membership function for a fuzzy *set A* on the universe of discourse X is defined as $\mu$:X → [0,1]. This means within this discourse each X has a some value, referenced as membership value, which is lying between 0 (0%) and 1 (100%). It quantifies the degree of membership of the element in *X* to the fuzzy set *A*. As shown in Fig III 1.3a4, X axis represents the universe of discourse while $\mu_A$ (Y) axis represents the degrees of membership in the [0, 1] interval. In our example, let the crisp numbers be divided into five divisions as discussed above. Let Small be at 0 and others are placed with an *interval* of "a" in X axis, on either side of small, as shown. Here most popular *triangular* membership function shapes are chosen for simplicity. Other common forms are *trapezoidal*, *singleton*, and *Gaussian.* So, for any value in X axis there will be a value corresponding to associated membership function, for example, (say) for any value lying within -a and -2a there will be two membership values corresponding to MN and LN. These two values need to be projected to similar membership graph for output. Then, one needs to calculate the centroid for the combined area. This is well explained at length in subsequent discussions.
5. **Pros and cons of fuzzy logic:** Pros and cons of Fuzzy logic systems are as follows:
   - **Advantage of fuzzy logic:** Following are advantages of Fuzzy logic
     - Fuzzy system be in linguistic form very easy to understand and develop
     - Fuzzy reasonings in mathematical representations are very simple.
     - Fuzzy logic systems are very flexible and can be modified easily.
     - Fuzzy logic Systems can work with imprecise, distorted, noisy and crisp inputs
     - Fuzzy logic can work on systems with uncertainty and provide solutions to complex problems in almost all areas of life.
   - **Disadvantages:** Following are a few disadvantages of fuzzy logic
     - Fuzzy logic does not have any systematic approach.
     - Too complex systems are difficult to work with accuracy/precisions may not be great
     - Fuzzy logic does not support machine learning.

### 1.3.1 Artificial intelligence basics

In history as early as 1923, the word "Robot" was found, meaning that there was at least some thinking about AI

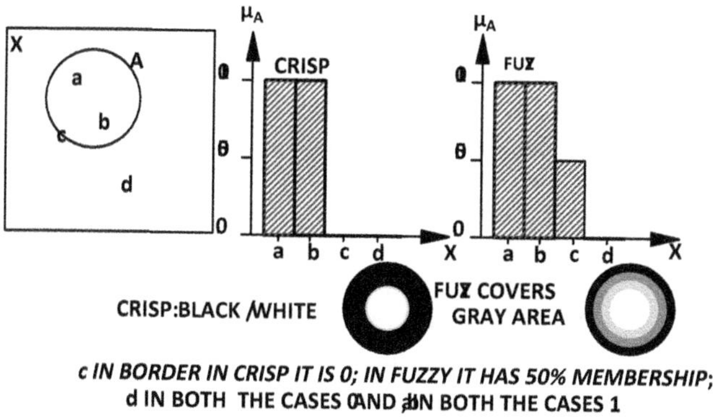


FIG III 1.3a1 DEFINITION OF FUZZY SET

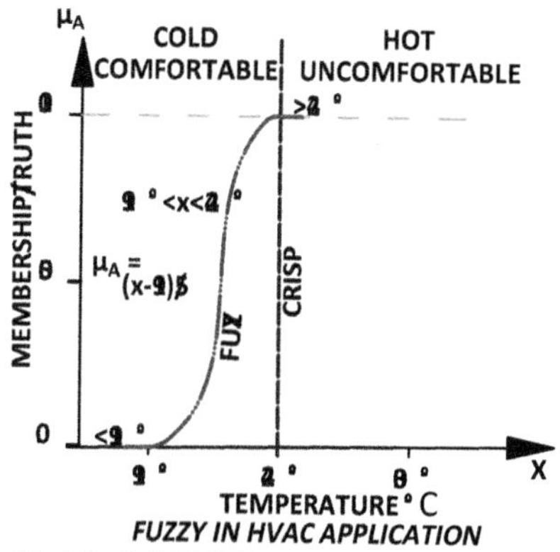


FIG III 1.3a2 REPRESENTATION OF FUZZY SET

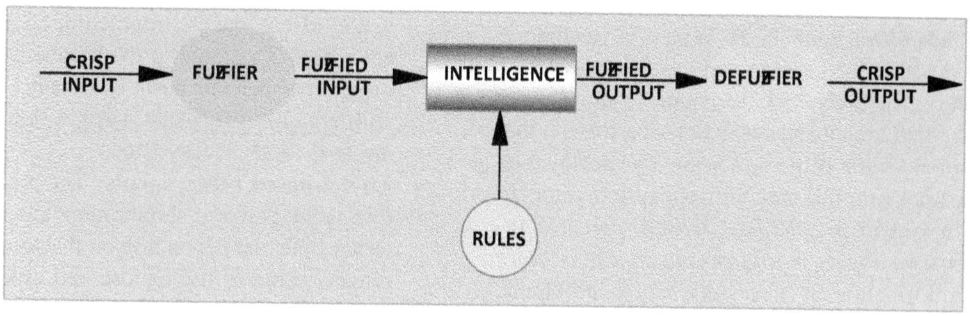


FIG III 1.3a3 FUZZY CONTROL SCHEME CONCEPT

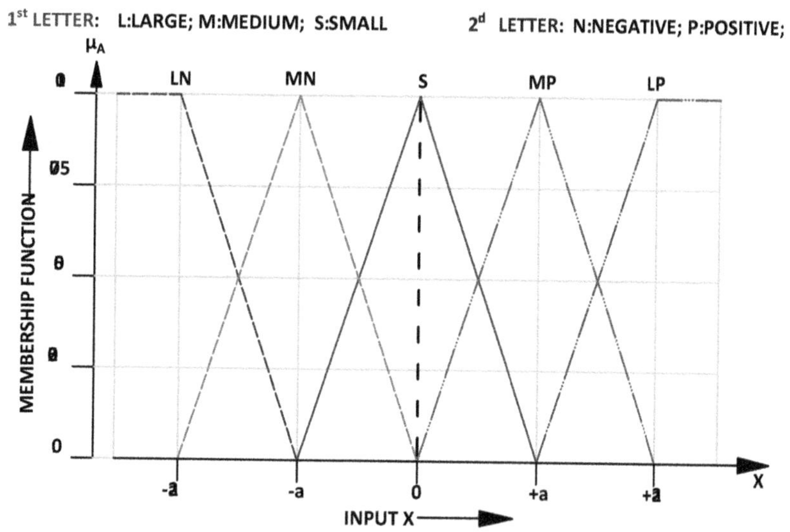


FIG III 1.3a4 FUZZY MEMBERSHIP FUNCTION CONCEPT

**FIGURE III 1.3A** Fuzzy concept.

which is an umbrella term for a various methods of control intelligences including fuzzy logic, ANN, expert systems, etc., with their highly interconnected processing elements using the *human-like technique of learning*. A smart device executes tasks automatically following instructions without user inputs. An intelligent device extracts knowledge from data and experience (may be through learning) to improve its behavior and can suggest new things also. So, control systems can be programmed with or without. When such programming is done without AI, then these are mainly meant to address specific issues and any modification is not easy as it mostly modifies the structure. On the contrary

**TABLE III 1.3A** Example of crisp numbers splitting.

| Symbol | Description |
|---|---|
| LN | X is large negative |
| MN | X is medium negative |
| S | X is small (0) |
| MP | X is medium positive |
| LP | X is large positive |

programs with AI can be meant for generic purpose as it is easier to modify in AI programming. In real life, there is huge volume of the knowledge which is ever changing, and not well-organized/well-formatted. With AI technique adaption it is possible to organize and use the knowledge effectively and efficiently. AI techniques allow easy modification for betterment and correction of errors, so that complete solutions with better precision are achievable. Therefore, one may infer that an intelligent system refers to the ability of a system to *calculate*, *reason*, *perceive relationships*, *draw analogies*, *learn from past/analysis*, *store* and *retrieve* information as necessary, *solve* problems, *comprehend complex ideas*, use *natural* language, *classify*, *generalize* with possibility to *modify* for specific use, and *adapt*. So, intelligence composed of components like: Perception, reasoning, linguistics, ability to learn problem solving. Here, reasoning can be "inductive" (generalization from specific information) and "deductive"(from general information deduced to specific information). Similarly, there could be different learning process like: *auditory*, *perceptual*, *relational*, etc., to name a few. Types of intelligence include but not limited to: Linguistic, musical, logical-mathematical, spatial, intra and inter personal, etc. Another Important issues are speech and voice recognition. Apparently, they seem to be same, yet in reality they have differences. While first one refers to "what is said" latter one stands for recognition of who said it. ANN, ES, Fuzzy logic, natural language processing and robotics are major AI types. Fig III 1.3b shows AI types, agents and Environment. Since our focus on control system natural language processing will not be covered here. Fig III 1.3b1 shows various AI types covered and used for control system. In task domain AI fall under three categories viz. *Mundane* (e.g., computer vision, speech voice), Formal (mathematical/logic) and Expert task (Engineering). Here, basics of AI will be covered with details discussed later.

1. **Agent and environment:** In an AI system an *agent* acts in its *environment* which refers to the *surrounding of agent(s)*. So, the job of AI is to design an *agent program* for implementing the agent function— the mapping from design philosophy to actions. It is obvious that the program (an implementation of an agent function) will run on some computing device with sensors and actuators. This is architecture. Therefore, an agent is composed of program plus architecture. Also, an agent senses/perceives its environment through *sensors* and acts through *effectors/actuator***.** These are detailed out in Fig III 1.3b2. There could be human, robotic and software agents with different sensors and actuators as depicted in Fig III 1.3b3.
2. **AI Agent types:** Various of AI agents have been explained with associated schematics. Four basic kinds of agent programs in AI systems are:
   - **Simple reflex agents:** This is the simplest kind of agent. These agents select actions on the basis of the *current* percept, ignoring the rest of the percept history. Here, the environment is completely observable. This is rational, provided the decision is made on *current* percept basis as detailed in Fig III 1.3c with condition rule that *maps a state (condition) to an action*—IF THEN Rule.
   - **Model-based reflex agents:** The philosophy behind this agent, is that it should maintain some sort of internal state that depending on the percept history and reflects some of the unobserved aspects of the current state. Updating this internal state information with time requires some information about how the world evolves independently of the agent and some information about how the agent's own actions affect the world. Here, the model represents "how the things happen in the world" as shown in Fig III 1.3d.
   - **Goal-based agents:** Fig III 1.3e details out goal-based agent where actions are chosen with intention to achieve goals.
     Goal-based actions are mostly straightforward but could be tricky also. may be less efficient but is more flexible than reflex agent as the knowledge supporting a decision is explicitly modeled, hence modifiable.
   - **Utility-based agents:** Fig III 1.3f shows utility agent details. Goals alone cannot generate high-quality behavior in most environments, especially when there are conflicting goals.

   An agent's utility function is an internalization of the performance measure. If the internal utility function and the external performance measure are in agreement, then an agent that chooses actions to maximize its utility. They choose actions based on a preference (utility) for each state.
3. **Environment Issues:** Many computer programs operate in the artificial environment of keyboard input, database, computer file systems and output on a screen. There exists different kinds of environments as listed below.

- **Full/partial observable:** If agent sensor can determine the complete state of the environment at each time point from the percepts it is *observable.* Otherwise, *partial observable.* If the agent does not have sensor in any environment, it is *unobservable.*
- **Deterministic or stochastic:** In deterministic, the next state of the environment is completely determined by the current state and the agent's actions. Otherwise, it is nondeterministic/stochastic. The stochastic environment is random in nature very difficult to completely determine by the agent.
- **Static or dynamic:** When the environment does not change during agent in action it is static, if changes it is dynamic.
- **Accessible or Inaccessible:** In accessible environment, agent's sensory apparatus have access to the complete state of the environment. Otherwise, it is inaccessible.
- **Discrete or continuous:** When there are a (limited) number of distinct, clearly defined, states of the environment, the environment is discrete. The environment, where the actions performed cannot be numbered it is continuous.
- **Single or multi agent:** An environment comprising only one agent, it is single-agent environment. Naturally, environment involving more than one agent is a multiagent environment.
- **Episodic or sequential (nonepisodic):** In simpler episodic environment, agent perceives then acts and independent of the actions in the previous episodes. In a *Sequential environment,* the previous decisions can affect all future decisions as the next action of the agent depends on the action taken previously.
- **Competitive or collaborative:** An agent is in a competitive environment when it competes against another agent for optimum output. On the contrary, when multiple agents cooperate to produce the desired output.
- **Known or unknown:** In known environment, the output for all probable actions is predictable. In unknown environment, for an agent to make a decision for actions.

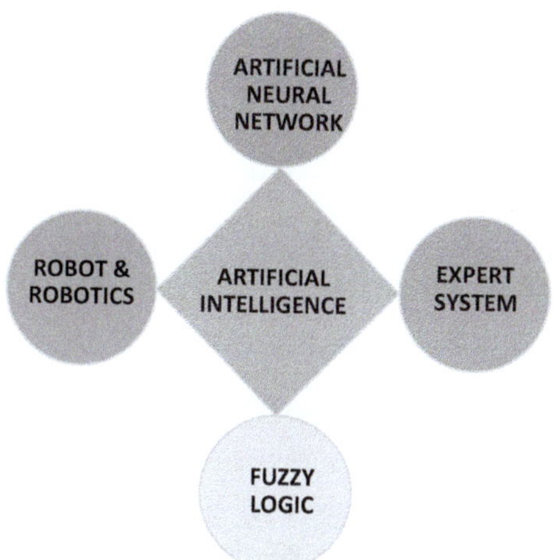


FIG III 1.3b1 AI TYPES IN CONTROL SYSTEMS

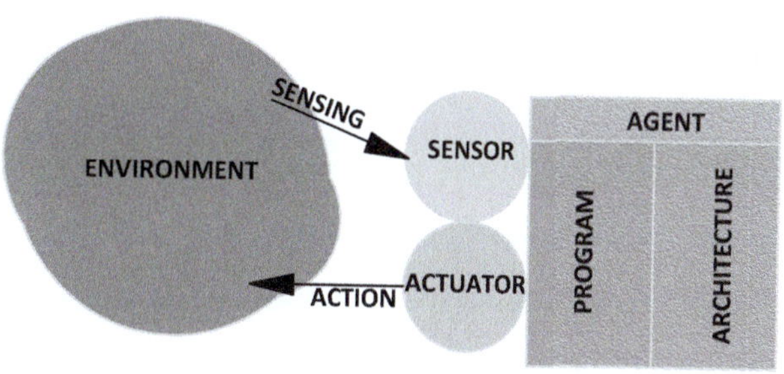


FIG III 1.3b2 AI AGENT AND ENVIRONMENT

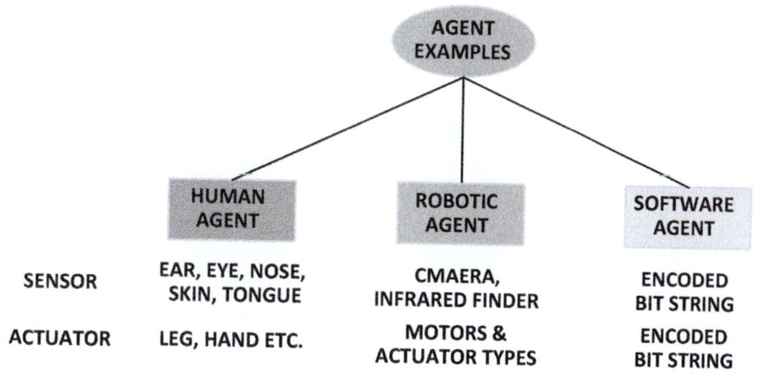


FIG III 1.3b3 AI AGENT EXAMPLES WITH SENSORS & ACTUATORS

**FIGURE III 1.3B** Ai types agent and environment concept.

4. **AI techniques:** There are a range of AI techniques: Fuzzy logic, ANN, multiagent systems, Case-based reasoning (CBR), Rule-based systems, Genetic algorithms, Cellular automata, hybrid systems, etc.
   - **Fuzzy logic:** Fuzzy logic use fuzzy sets to deal with nonprecise and incomplete data.
   - **ANN:** ANNs employ mimics the way the human brain processes information. An ANN can be highly interconnected by links (synapses) with suitable weightages.
   - **Multiagent system (MAS):** MAS comprises a network of agents interacting to arrive at desired destination. MAS can solve the problem assigned to it of its own.
   - **Case-based reasoning (CBR):** CBR as detailed in Fig III 1.3g, solves a problem with experience from past similar problems.
     CBR recognizes that problems are *easier* to solve by repeated attempts, acquiring knowledge through following steps:
     - *retrieve* the most relevant past cases from the database;
     - *utilization* of the retrieved case for a solution of the new problem;
     - *revise* the proposed solution by simulation or test;
     - *retain* the solution for future use after successful adaptation.
   - **Rule based system (RBS):** RBS solves problems by rules derived from expert knowledge [74]. The rules have conditions and actions with "If and Then" they conclude/infer in an inference engine. Inference engine with working memory has information about the problem, a pattern matching (to decide relevant rules) and a rule application (what rule to apply) software.
   - **Generic algorithm (GA):** GA is a search technique mimicking natural selection [24]. The algorithm evolves until it satisfactorily solves the problem, through the fitter solutions in a population surviving and passing their traits to offspring which replace the poorer solutions.

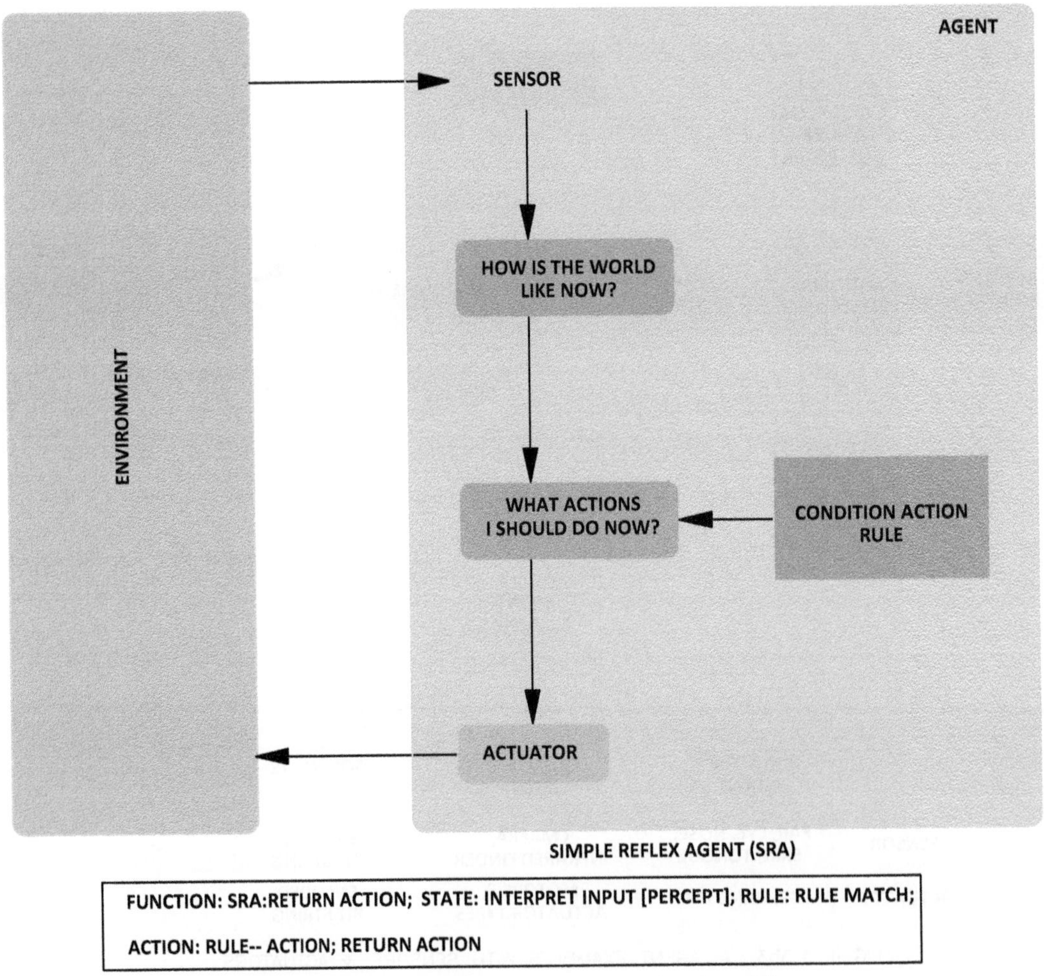


**FIGURE III 1.3C** Simple reflex agent- schematic and conditions.

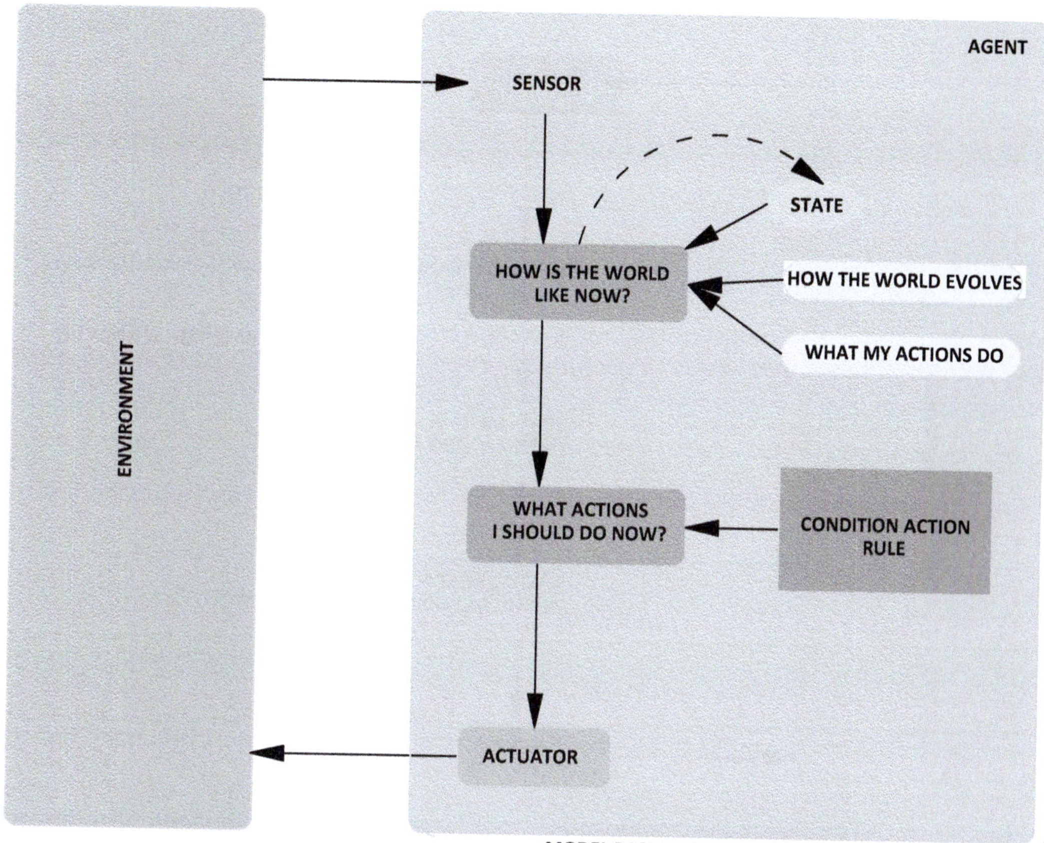


MODEL BASED REFLEX AGENT (MBRA)

| FUNCTION : MBRA: RETURN ACTION; |
|---|
| STATE: AGENT'S CURRENT CONCEPTION OF WORLD STATE WITH UPDATES |
| BASED ON HOW WORLD EVOLVES AND THE ACTIONS DO. |
| MODEL: DESCRIPTION OF HOW NEXT STATE DEPENDS ON CURRENT STATE & ACTION |
| RULE: RULE MATCH; |
| ACTION: RULE-- ACTION; RETURN ACTION |

**FIGURE III 1.3D** Model based reflex agent- schematic and conditions.

- **Cellular automata:** Cellular automata are dynamic models, discrete in space, time and state with a regular lattice of cells to interact with their neighbors.
- **Hybrid intelligent systems:** Each AI technique has its own strengths and weaknesses. The advantages of these technologies can be combined together to produce hybrid intelligent systems to work in a complementary manner. Their Synergy allows a hybrid system to accommodate *common sense*, extract knowledge from raw data, use human-like reasoning mechanisms, deal with uncertainty and imprecision, and learn to adapt to a rapidly changing and unknown environment.

5. **AI benefits: discussions:** AI offers numerous benefits across applications. Therefore, commonly cited benefits of AI include the following:
   - **Repetitive tasks:** AI automates routine, repetitive and often tedious tasks—including data collection, entering and preprocessing, and physical tasks.
   - **Insight data:** Artificial intelligence can improve data analysis processes by automating tasks, analyzing large data-sets efficiently, identifying patterns and trends, and providing insights for decision-making.
   - **Fewer errors:** AI reduces error but AI reduces human errors in various ways: *guiding* people through the proper steps of a process, *predicting* and *flagging*

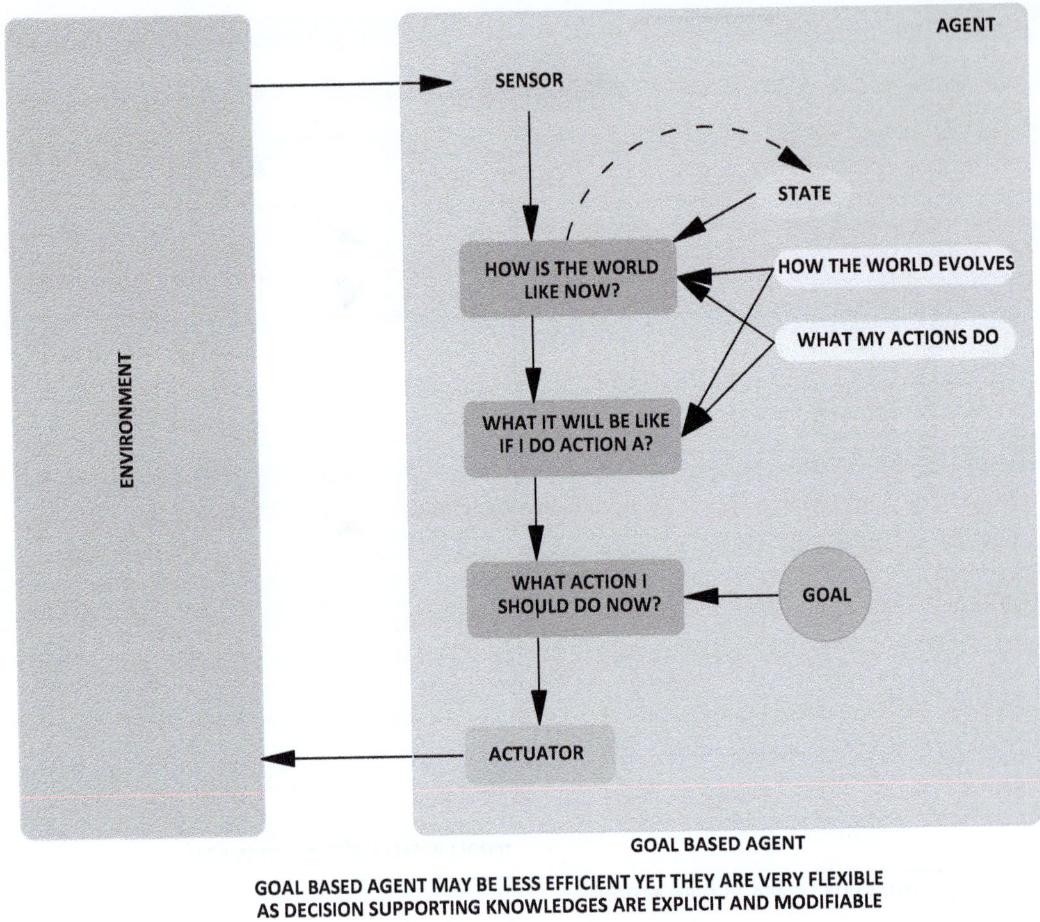


**FIGURE III 1.3E** Goal based agent- schematic.

potential errors before occurrence, through complete automation.

- **Reduced risks:** As discussed in safety system (Vol II), by automating dangerous work like handling explosives, working in hazardous environment and exposure, tasks in deep ocean/high altitudes, etc. AI can eliminate the need to put human workers at risk of injury or worse and they often can make better decision in meeting greater challenges from environment.
- **Round the clock:** AI being automated system, is always on, ensuring round the clock services. Also, AI does not face any fatigue like human.
- **Decision making:** AI can help in decision making support or for fully automated decision-making, AI provides faster, accurate predictions and reliable, data-driven decisions.

## 1.3.2 Machine learning (ML) in industries

All MLs are AI but all Ais are not ML. Machine Learning (ML) as a subset of AI focuses on the development of algorithms to improve the associated application automatically through experience and by the use of data. This implies that in its core, ML is all about creating and implementing algorithms that facilitate these decisions and predictions. Many times, AI and ML are used interchangeably, creating some confusion. The two are not quite the same thing. As indicated, AI is the broader concept of machines/systems being able to carry out tasks in a way that we would consider "intelligent." ML is a current application of AI based around the idea that we should really just be able to give machines/systems access to data and let them learn for themselves. AI and ML play a number of major roles in manufacturing and process industries. In the present scenario of Industrie 4.0 and DT, all these are connected to internet of things (IoT) and Industrial IoT (IIoT) cloud computing to drive value in the industry. AI and ML, industrial plants can provide:

- better efficiencies with curtailment of waste to save money and better product quality
- improved safety with reduced environmental impact meeting industry standards more accurate root cause analysis to improve processes

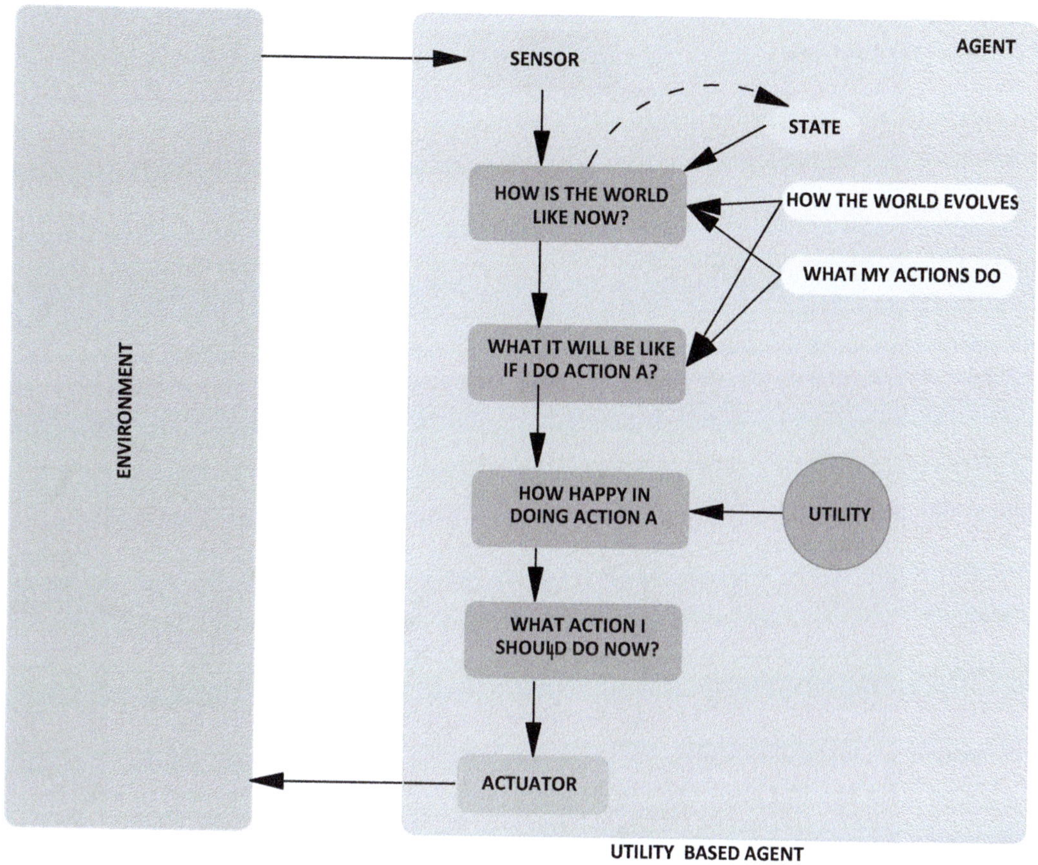


**FIGURE III 1.3F** Utility based agent- schematic.

- means to remove production bottlenecks
- prediction of failure/anomalies to cut downtime with quicker repairing
- improved visibility into supply chain and distribution networks
- understanding of the market trends and changes
- optimization equipment lifecycle

By leveraging advanced algorithms and predictive analytics, machine learning can help organizations make better decisions, improve operations, and enhance the bottom line.

*Supervised*, *unsupervised*, and *reinforcement* learning are 3 ML types with which machines/systems can provide valuable insights, through analysis of complex patterns and identify hidden structures. In this subsection brief discussions will be presented on impact of ML in industrial plants including manufacturing plants and process plants.

1. **Improved manufacturing:** Four major impacts of ML on manufacturing to improve the process are:
   - **Planning optimization:** ML can help a manufacturing process through greater productivity. Multiple ML software algorithm can be deployed for data collection, analysis for production schedules, inventory levels, machine performance, supply chain logistics, and market demand. With the help of these data, ML models can assist user with demand forecasting, inventory levels optimization, resource allocation, and streamlining production processes. Advanced ML algorithms in conjunction with IoT/IIoT, manufacturers can go for predictive maintenance from data analysis and different performance indicators.
   - **Energy: efficiency and sustainability:** ML powered data analysis, provides details on energy consumption during production processes. AI automates processes and can dynamically adjust equipment settings for optimal efficiency, even automate equipment shutdowns during idle periods and optimization of power usage by eliminating unnecessary energy consumption. The AI's impact also reaches the implementation of renewable energy sources for production.
   - **Optimized supply chain:** ML-powered supply chain management opens multiple ways to effectively meet production demands and adapt to dynamic market conditions [12]. As stated above, ML helps user with highly accurate demand

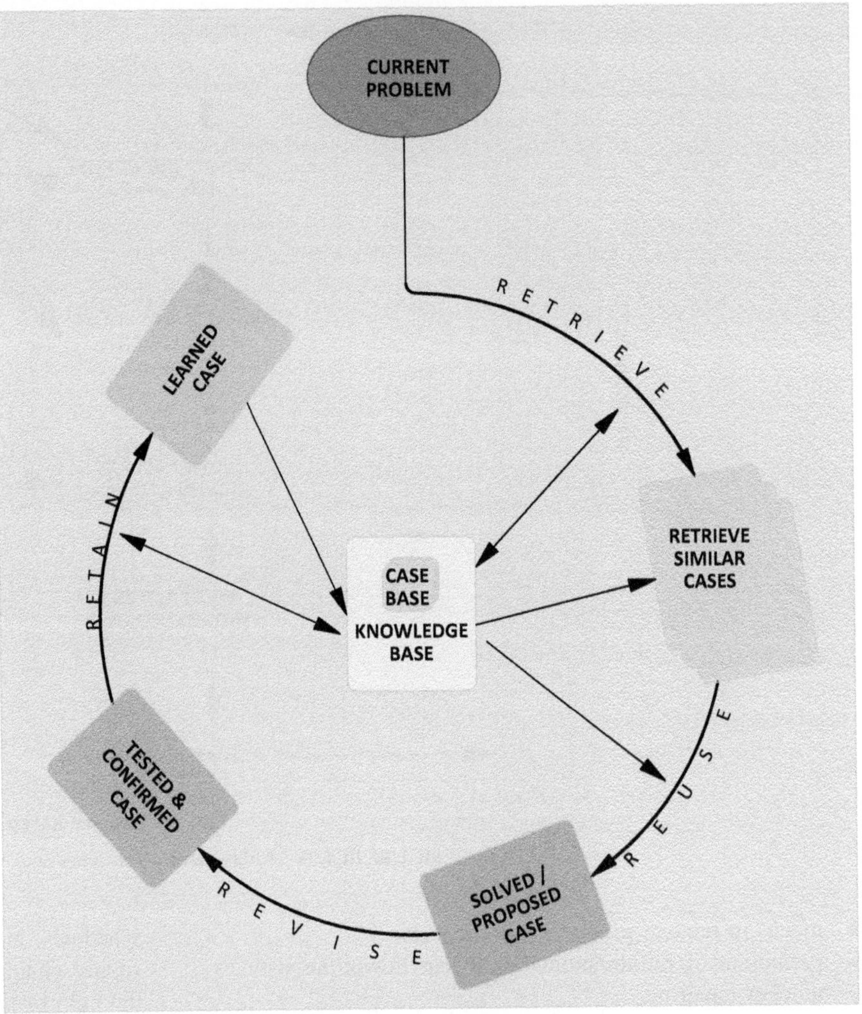


**FIGURE III 1.3G** AI case based reasoning.

forecasts to optimize inventory levels, preventing stockouts or overstocking. Also, AI streamlines supplier selection and management, identifying reliable partners by analyzing quality, reliability, and cost.

2. **Manufacturing benefits:** As discussed, ML algorithms can help to optimize multiple processes in the manufacturing industry. Some of the major benefits are:
   - **Cost reduction:** In manufacturing, ML along with AI, contribute to automate entire production processes including supply chain management, and repetitive tasks. ML eliminates slower production process, error by human and reduced waste. All these contribute in cost reduction. Also, by early detection of defects, and helping in taking predictive maintenance, ML plays critical role in cost reduction. ML-powered visual inspection systems surpass human inspectors both in precision, speed and very detection of defects in the production line. Therefore, early action could be initiated to avoid downtime—hence cost reduction.
   - **Quality production:** ML technologies bring about huge benefit by bringing precision and eliminating defects/reducing waste. Digital twins in ML play a major role in maintaining high precision quality. Powered by ML, digital twins create *virtual replicas* of products, in terms of designs and characteristics for evaluation and testing. ML can analyze data precisely to simulate performance of product under different changing conditions.
   - **Time-to-market:** It is obvious that with ML powered automated tasks and predictive maintenance, the manufacturing processes become faster, would ensure the timely product launching. As seen above, ML can significantly improve supply chains and optimize logistics with machine learning allowing engineers to utilize ML for analysis of transportation routes, supplier locations, exploring logistics

routes and real-time traffic conditions. All these drastically reduce time for product delivery in the market.

- **Customer preference:** ML in manufacturing sector is very successful in identifying customer sentiment and preferences. ML powered models can analyze huge datasets to easily identify the needs and the emotional tone of customer from social media posts, online reviews, and customer surveys. Also, ML technology can forecast customer response to potential product features or designs. So, it is possible to understand how customers feel and interest about the products.

3. **Usage: process industries:** There are a number of cases in process industries where ML are used. Here some major usages have been listed and these are:
   - **Quality assurance:** ML technology can improve and maintain product quality by reducing defect as well as wastage. This is possible by analyzing data from many sources. With the analysis, ML algorithms can identify patterns and anomalies to address the quality issues. So, with real time adjustments, the quality can be maintained and improved.
   - **Process optimization:** With the help of huge data analysis from sensors and other sources in analogous cases, ML technology can identify inefficiencies and bottlenecks in production processes. Therefore, applying recommendations from ML, entire process can be optimized. Thus, ML can offer opportunity for the industrial process industry, enabling organizations to optimize production, improve efficiency, and reduce costs.
   - **Predictive maintenance:** This is common use found in process industries. As discussed before, ML analyzes data from sensors and many other sources to early predict impending equipment failure. This helps user to organize perform maintenance prior to actual failure reducing downtime and minimizing the risk of sudden equipment failures.
4. **Process industries benefits:** Major benefits of ML in process industries are:
   - **Automation repeated work:** Repetitive work in process industries are not uncommon. These repetitive jobs create fatigue in human. ML can automate this repetitive work like, data collection, analysis in changing conditions, etc., for less error, lesser time and increased efficiency.
   - **Identification of patter:** As discussed, through analysis of data from sensors and other sources, it is possible to identify patterns and anomalies that may indicate problems or opportunities for improvement. This can help in two ways. By this it can optimize the production process and opt for predictive maintenance to avoid downtime.
5. **Pros and cons:** From the above discussions we find that there are a number of obvious advantages of ML. However, there are a number of limitations of the systems as well. Table III 1.3b shows pros and cons of ML.
6. **Challenges for ML:** There are a few challenges for ML. These are:
   - **Intelligent assistant:** Main issue here is that ML like AI, should be a good assistant and should not rule over human. From this it transpires that at AI, and ML, are only as intelligent as we teach them to be.
   - **Quality data:** One of the major issues associated with ML is the need for high-quality time-series or process data since ML directly impact the accuracy and effectiveness of these algorithms.
   - **Specialized data:** Another important issue in ML is the need for specialized skills and expertise. Machine learning is a complex and technical field that requires specialized skills and knowledge to implement effectively.
   - **Unlabeled data:** Capability of ML to handle unlabeled data as ML relies on data to learn, hence calls for labeled data. However, in many cases data aren't available or is unlabeled.

## 1.4 Artificial neural network (ANN) basics

The combining of computer with theoretical bases like nonlinear dynamics/chaos theory allows the creation of "intelligent" agents, like ANNs with capability to adapt dynamically to problems of high complexity. Here, ANN is an artificial adaptive system which is inspired by the human brain functioning. Fundamentals of ANN has been elaborated in Fig III 1.4a. ANN can modify their internal structure in relation to any functional objective. Conventional approaches are successful in certain well-constrained environments; however, these may not be flexible enough to perform well outside its domain. Here, ANNs enjoy an edge by providing exciting alternatives, and many applications could benefit from using them. *Adaptability to different environment*, *contextual information processing capability*, *learning and generalization ability*, *distributed computation*, *massive parallelism*, *fault tolerance* and *lower energy consumption* are major desirable human brain characteristics which are absent in normal computer/controls. So, for nonlinear and complex problems ANNs are preferred choice. An artificial neuron is a mathematical function conceived as *a simple model of a real (biological) neuron*. Simplified model of real neuron is Threshold logic unit. A set of input connections, associated with *weight*, brings in activations from other neurons. Fig III 1.4a1 shows processor acting as *summator of inputs to apply nonlinear activation function*. Output line sends the result. So, one can conclude, that there are three basic components –weights, thresholds and a single activation function. As shown in Fig III 1.4b there exists hidden layers which are the intermediary stages between input and output in ANN. Hidden layers responsible for learning the intricate structures in data and making neural networks a powerful tool.

This biologically inspired method brings about revolution in computing, in solving very complex problems. As discussed, there are some characteristics of ANN described in Fig III 1.4a2. ANN can be conceived as a robust approach to approximating *real-valued, discrete-values*, and *vector-valued* target functions for solutions pertinent to complex and hard to interpret input data. Therefore, ANN can be conceived as a machine learning algorithm with a specific architecture for solving complex issues in real life. In this subsection basics about ANN shall be covered so that detailed discussions in subsequent sections could be understood easily.

## 1.4.0 Conceptual discussions on ANN

According to Dr. R. Hecht-Nielsen " ... *a computing system made up of a number of simple, highly interconnected processing elements, which process information by their dynamic state response to external inputs*" [13]. From IEEE one would get "*Artificial neural network (ANN) is a statistical data modeling tool inspired by the functionality and the structure of the biological nervous system.*" As per Technopodia, "*An artificial neuron network (ANN) is a computational model based on the structure and functions of biological neural networks*". In nutshell, ANN are composed of multiple nodes that imitate neurons of human brain. The neurons interact with each other through connected links which are associated with weight. The nodes can accept number of input data as weighted sum and perform simple operations on the data to pass on the result to other neurons. The output at each node is called its *activation or node value* already explained in Fig III 1.4a. ANNs are trainable through proper learning process which could help them to accurate predictions. Thus, on account of its interconnection like neurons of brain, these are referenced as neural networks. From the discussions it is clear that connections with individual strength are storehouses of knowledge in ANN architectures. It is obvious that learning process plays a great role in ANN for adjusting connection strengths to yield correct result. Initially, all the connection strength are set randomly, and the ANN is in a blank slate. So, when it is initialized, the responses could be random. When output is compared with correct picture, it reveals that it needs training. Here, the important issue "*back-propagation steps*" come into action. The entire process of input, processing, getting output for comparing with correct answer, and adjusting connection strengths is called one "back-propagation cycle." In ANN apart from It is an iteration process hence referred to as "iteration." Fig III 1.4b shows the detailing of ANN concept, in this subsection brief discussions have been presented on conceptual details on ANN. Based on numbers of hidden layers, discussed before, ANN can be shallow or deep. Let the discussions start with *shallow and deep neural network*.

1. **Shallow and deep neural network:** Fig III 1.4b1 details out *shallow* and *deep* neural network. Basically, ANN is a supervised learning system (discussed later) built from a large number of simple elements referred to as "Perceptrons" (A simple model and consists of

**TABLE III 1.3B** Pros and cons of machine learning.

| Advantages of ML | Disadvantages of ML |
|---|---|
| Automating repetitive work | Data centric dependability |
| Improved precision and better accuracy | Expert dependency |
| Better decision making | Complexity and interpretation |
| Predictive analytics: Future from past data | Over and under fitting |
| Cost reduction | Higher computational cost |
| Improved planning and high scalability | Lack in generalization |
| Threat detection and improved security | Adverse attack vulnerability |
| Quick marketability and customer preference | Ethical concern |
| Innovation and competitive edge | Legal and regulatory challenge |
| Human capability improvement | Maintenance and updates |

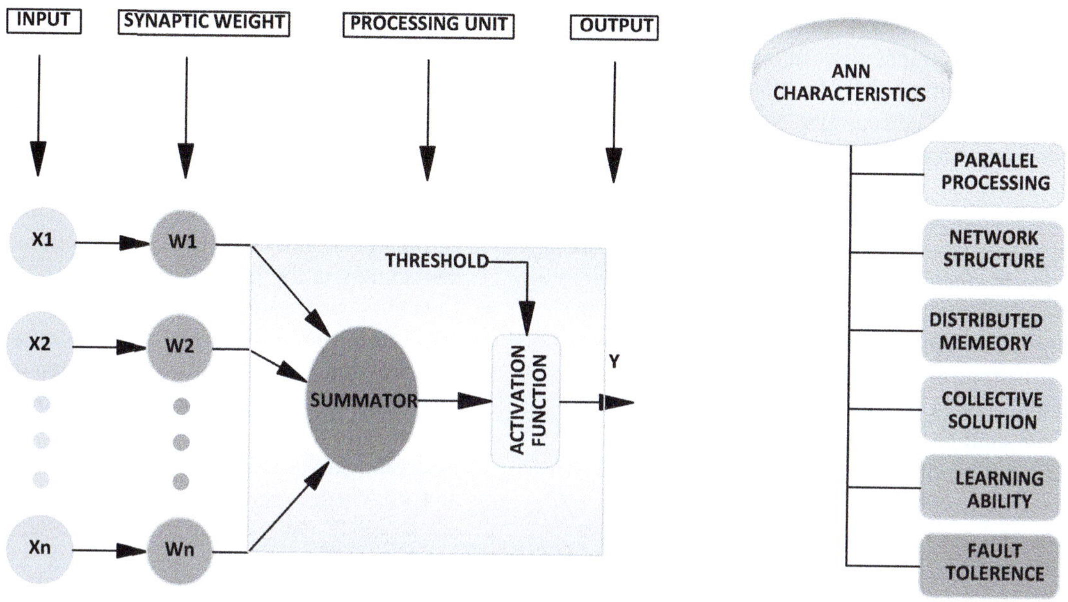


FIG III 1.4a1 ANN BASIC SCHME WITH ELEMENTS FIG III 1.4a2 ANN BASIC CHARACTERISTICS

**FIGURE III 1.4A** Artificial neural network conceptual idea.

a single "trainable" neuron *elaborated below*). Each perceptron or neutron is capable of taking decisions and feed decisions to subsequent neurons in organized interconnected layers. Together they form Neural network (NN) to provide solutions to many problems when they go through training samples and computing power. Based on number of hidden layers, two types of ANN are:

- **Shallow ANN**: A "shallow" neural network comprises of three layers of neurons:
  - *Input layer:* One input layer that accepts the independent variables or inputs.
  - *One hidden layer:* It is located between input and output layer to perform nonlinear transformations of the inputs entered into the network.
  - *Output layer:* An output layer that generates predictions or output
- **Deep neural network**: Deep neural network has similar structure that of shallow neural network only, it has >2 hidden layers. More hidden layers offer better solutions. However, it has been observed that after 9–10 hidden layers prediction power not only stop increasing but may start declining. So, presently deep neural networks use between 3–10 neuron layers.

2. **Perceptron explained:** As shown in Fig III 1.4b2, the perceptron which possesses the ability to learn and solve very complex problems, has a simple structure. Units of NN are called nodes. The link connecting two nodes/neutrons (or input to neuron) is referred to as edge. It accepts inputs $x_1, x_2, \ldots, x_n$ which could be real numbers or Boolean values to produce y is Boolean output. Each input has associated edge weightage (real valued) associated with them. These are $w_1, w_2, \ldots, w_n$. Perceptron also has a *threshold* T which is real value. It produces output $y = 1$ when the weighted sum of inputs, that is, $w_1 {*} x_1 + w_2\, x_2 + \ldots.\, w_n\, x_n$ is greater than the threshold T, otherwise $y = 0$. One multiplied by *Bias weight* is added (Ref: Fig III 1.4a1) to the sum so that it is possible to move the output function of each perceptron (the activation function) up, down, left and right on the number graph. When perceptron *wasn't trained* for particular input, then it should try to find the *best possible* output depending based on its training. When the output is wrong, then we *need to change its weights and/or threshold* properly to avoid similar mistake later. Another important issue may be noted that inputs are not directly fed to trainable perceptron but through processing units. So, from the above generalized equation could be:

$$F(x) = \begin{cases} \mathbf{1}\ \boldsymbol{if}\ \boldsymbol{w} * \boldsymbol{x} + \boldsymbol{b} > \mathbf{0} \\ \mathbf{0}\ \boldsymbol{otherwise} \end{cases} \qquad \text{Eq. III 1.4a}$$

Here, *w*, *x* defined above and b is the bias (discussed above).

These perceptrons can be used for logical operation of AND, OR, NOT. Here, with NOT there is little trick. If weightage is set (−)1, then for binary input it should produce (−)1/0. When threshold is set 0, output will be 1 for zero input. It is more difficult in resolving XOR logic. When edge weight is set at 0.5.

3. **Multilayer perceptron (MLP):** Multilayer perceptron (MLP) is a group of perceptrons, organized in multiple layers, that can accurately answer complex questions. From earlier discussions it has been noted that MLP contains an input layer, at least one hidden layer, and an output. This is shown in Fig III 1.4b3. In this case since there is single hidden layer it is *shallow* ANN. Like single perceptron, here also various edges connecting neurons have weightage associated to them. Again, at Output layers weighted sum of various data from hidden layers produces output. In case of deep ANN there shall be 2 or more hidden layers connected by edges with weights. So, here there are numbers of outputs.
4. **Learning and training issues:** These are two important issues for ANN. There are a number of learning methods as detailed in subsequent chapter. Here preliminary discussions on the same has been presented.
   - **Training:** It is the process of teaching ANN to perform a task—in this way, the network is taught to *change its weight and bias.*
   - **Learning:** It is the internal process of training in ANN to update/adapt the weights and biases, that is, modification of weights of connections between the neurons in the network.
   - **Learning procedures:** ANN learning algorithms are divided in to three main classes viz. supervised, unsupervised and reinforced; as shown in Fig III 1.4c. These procedures have been elaborated later in this volume.
   - **Learning law:** Basic understanding of the system and knowledge of mathematics and programming are required. Other issues pertinent to learning law requirements are:

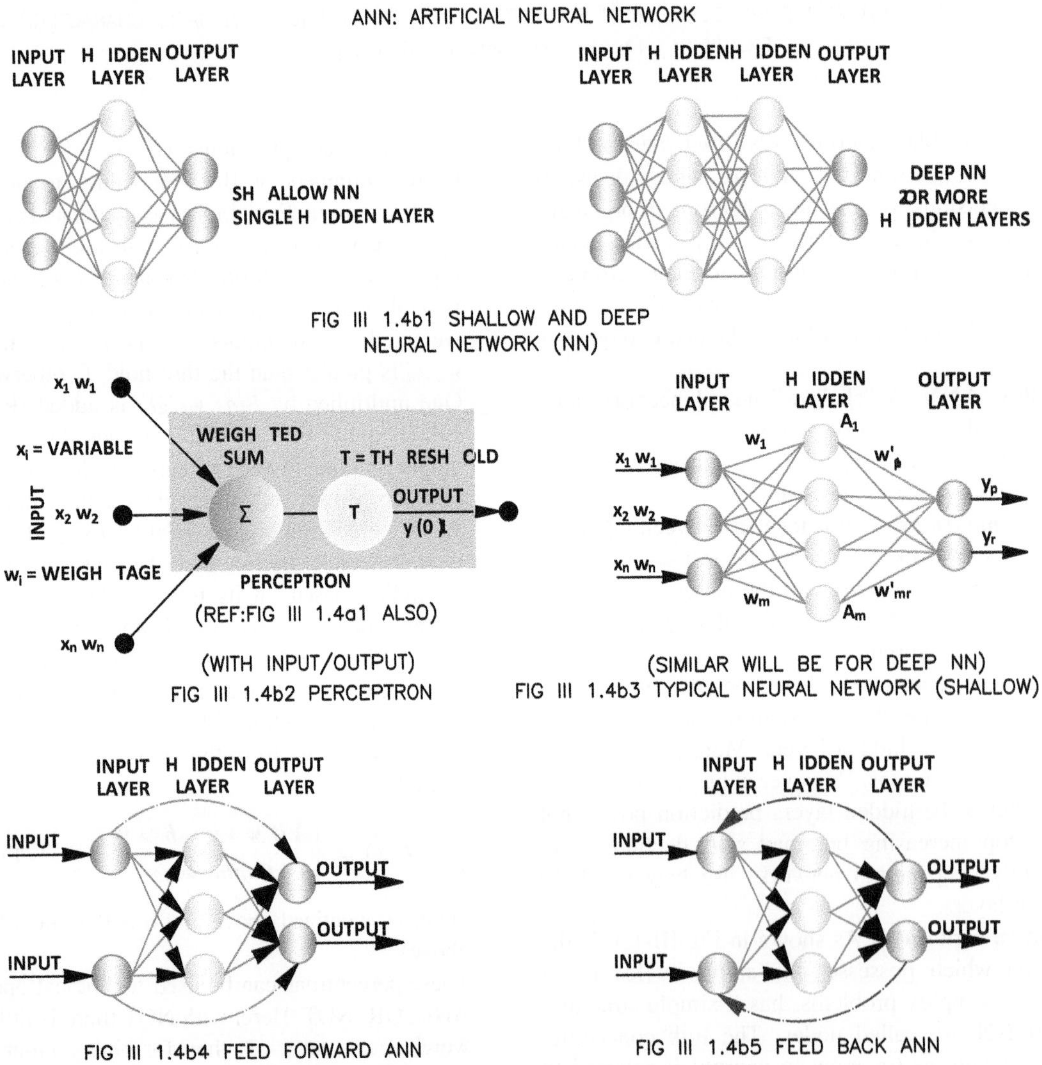


**FIGURE III 1.4B** Artificial neural network concept detailing.

- Training time in acquiring information from training tool should be low
- Learning process should be capable of capturing as many as patterns as possible and should use local information also.
- Learning process should be able to capture the complex nonlinear mapping in I/O pair
- Storage of pattern information's gathered, during learning should be high.

5. **Distinguishing characteristics:** There are a few specific characteristics for ANN as detailed in Fig III 1.4a2 which shows six specific characteristics of ANN. These are as follows: *parallel processing, network structure, distributed memory, collective solution, learning ability* and *fault tolerance.* Brief comparison of normal computer with ANN has been presented in Table III 1.4a.

### 1.4.1 ANN types and common issues

The increasing complexity of clinical data requires the use of mathematical models that are able to capture the key properties of the entire group, including the linkages and hubs. The complexity and nonlinearity processes and problem have shown that are ubiquitous in nervous system of living organisms. ANN is an information manager model that is similar to biological nervous systems function of the man brain. In recent times ANNs has become popular and helpful model for classification, clustering, pattern recognition and prediction in many disciplines. Various artificial intelligence like ANN, ML, DL and robotic, information security, big data, cloud computing, internet, and forensic science are all hotspots and exciting topics of information and communication technology (ICT) [14]. The ANN is found to be very novel and useful models that can be used easily and more accurately for complex systems with huge data. McCullochand Pitts4 proposed a binary threshold unit as a computational model for an artificial neuron already discussed. It is also seen different connectivity types yield different network behaviors viz. static feed-forward networks and dynamic recurrent, or feedback, networks. In this subsection both these basic types will be addressed along with other common issues like perceptron training, types of activation functions and various signal propagation types. Let the discussions start with basic ANN types.

1. **Basic ANN types:** Starting with single *perceptron*, currently there are as many as 27 types of ANN topologies are in use for deep learning detailed out later. All are not required for plant controls. Out of them seven types are most popular. Here, let us discuss two simple fundamental types viz. Feed Forward and Feedback types. Feedforward ANN process signals in a one-way direction with no inherent temporal dynamics, hence referred to as *static.* On the contrary, recurrent/feedback networks have feedback and loops and referred to as dynamic system with a state space *stable* and *unstable equilibria.*
   - **Feed forward:** ANN The information flow is unidirectional, that is, one unit sends information to (next) wherefrom it does not receive any information. No feedback loop exists here. This implies that in graphical representation, it is acyclic having no path along the connections can end where it started. Multilayer networks (*hidden neurons* can be arranged in a *variable* number of layers) operate sequentially meaning that neurons in a layer will receive the signals only when the units in the previous layer have produced them. The output vector of a layer is used as input for the successive layer. This is shown in Fig III 1.4b4 where it is shown that output can get information from input directly also. Pattern recognition, classifications are common application areas.
   - **Feedback (feed backward) ANN:** This is shown in Fig III 1.4b5. Here, feedback loops are allowed to adjust output. Compared with feedforward, feedback ANNs are a bit complex because of signals traveling as signals are constantly traveling from both sides, that is, signals in both directions. Feedback ANNs aim to attend a state of equilibrium, by constantly changing themselves and by comparing the signals and units. As long as no change in input, the state of equilibrium is maintained. When the input changes, the network tries to achieve a new point of equilibrium. They are commonly used in content addressable memories and in deep learning system with huge data.
2. **Training and learning process of perceptron:** The fundamental issue is a Perceptron—an essential building block of ANNs, for accepting both real and Boolean inputs. By learning the appropriate weights and bias it can derive the desired function. During training, by using a learning algorithm, the perceptron adjusts its weights based on observed errors. It may be noted that during the training period, inputs one at a time are presented to perceptron to see what output it gives. If the output is wrong, then it has committed a mistake, so, then its weights and/or threshold are adjusted properly to avoid making the same mistake later. Thus, the training of the perceptron means feeding it multiple training samples. In perceptron learning algorithm, the objective is to determine the optimal weight vector, w, that can effectively separate positive and negative inputs within given dataset. The algorithm mainly covers: *initialization of weight vector with random values, application of them iteratively for each training*

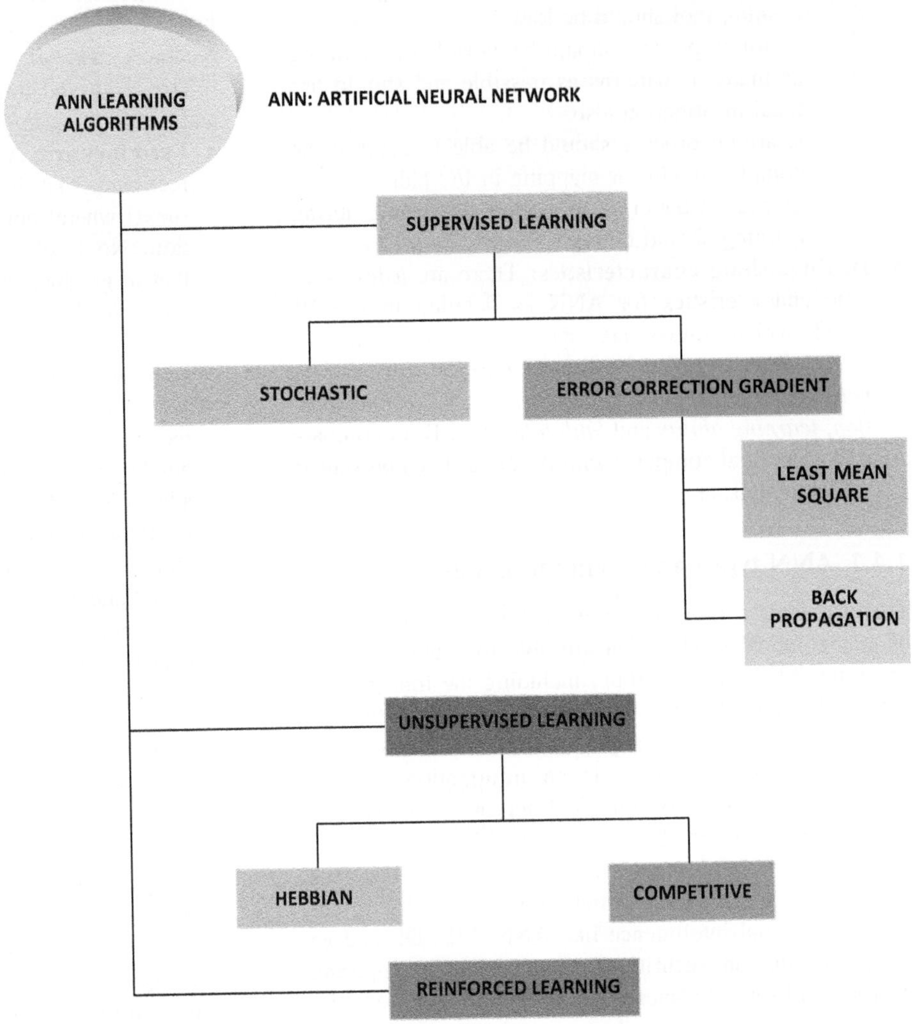


**FIGURE III 1.4C** Artificial neural network learning procedures.

*case, when it is misclassified, weight needs update* and iteration of the process till getting desired output/classification. The key principle for weight adjustments is given by:

$$W_i = W_i + (n * (t - o)) \qquad \text{Eq III 1.4f}$$

Where; "$W_i$" to be updated, "$n$" is the learning rate, "$t$" is the target output and "$o$" is the actual output. Therefore, the training of the perceptron means

**TABLE III 1.4A** Comparison of ANN with conventional computing.

| Issue | ANN computing | Conventional computing |
|---|---|---|
| **Computing** | Distributed, parallel and Self-learning | Central, sequential and programmed |
| **Memory** | Integrated, distributed content addressable | Separate from processor noncontent addressable |
| **Reliability** | Robust | Not robust |
| **Processing** | Low speed, complex large nos. | High speed; less in number |
| **Environment** | Poorly defined | Well defined |
| **Expertise** | Perpetual problem | Numeric, symbolic |

feeding it multiple training samples and calculating the output for each of them. Another important issue here is Activation function.

3. **Activation function types:** Activation function in ANNs is a mathematical function applied to the output of a neuron with the purpose of introducing nonlinearity into the model, allowing the network to learn and represent complex patterns in the data.
Fig III 1.4d shows some of the common activation function as discussed below:
   - **Linear:** This is the simplest form of function and an identity—referred to as identity function with as illustrated in Fig III 1.4d1 and expressed in

$$f(x) = x \text{ for all values of } x \qquad \text{Eq. III 1.4b}$$

   - **Step function:** Binary step function depends on a threshold value that decides whether a neuron should be activated or not. As shown in Fig III 1.4d2, here it is a binary step with threshold θ). Function is represented as

$$f(x) = 1 \text{ for } x > \theta;$$
$$= 0 \text{ for } x < \theta \qquad \text{Eq III 1.4c}$$

   - **Binary sigmoid:** This is also known as logistic function and is graphically represented in Fig III 1.4d3. The functional representation is

$$f(x) = \left[\frac{1}{(1 + e^{-ax})}\right] \qquad \text{Eq III 1.4d}$$

   - **Bipolar sigmoid:** This is also referred to as tanh function, bounded function values lying between the range of (−)1 to +1 as shown in Fig III 1.4d4.

$$f(x) = \left[\frac{(1 - e^{-ax})}{(1 + e^{-ax})}\right] \qquad \text{Eq III 1.4e}$$

4. **Back propagation network (BPN):** Backpropagation is one of the important concepts of ANN. The main features of BPN are the iterative, recursive and efficient method to calculate the updated weight to improve the networks, it can perform the task for which it is trained. Using supervised training process BPN has a systematic procedure (extended version of perceptron training law) for training the network and error detection and correction.
As shown in Fig III 1.4e the weights between input and the hidden portion is considered as Wij and the weight between first hidden to the next layer is considered as Vjk. Three stages involved for backpropagation training process are as follows: *feedforward of input training pair, calculation and backpropagation error*, and *weight adjustments*. Other issues are:
   - **BPN algorithm:** BPN algorithm can be divided into four steps: *Initialization of bias, weights, feedforward process, back propagation of errors*, and finally *updating of weights and biases.*
   - **Advantages of BPN:** Following are advantages of BPN:
     - *Simplicity, flexibility, scalability, efficiency* and *ease of implementation*
     - *Smooth effect on weight correction with little time for small weights*
     - *Systematic weight updating procedure*
     - *Much faster than perceptron model*
   - **Disadvantage of BPN:** A few disadvantage points are:
     - *Intensive calculations during learning phase*
     - *Number of Hidden layer neurons is an issue*
     - *Temporal Instability with network paralysis*
     - *Training time is more for complex problems*
5. **Counter propagation network (CPN):** Counter propagation network (CPN) are multilayer network based on the combinations of the input, output, and clustering layers. CPNs are used for data compression, function approximation and pattern association. Main types of CPN are Full CPN and Forward only CPN. CPN functions are like a Look up table generalization. The training pairs may be Binary or Continuous. CPN produces a correct output even when input is partially incomplete or incorrect. It has advantages like: Easy look up table, fast coarse approximation. However, leaning phase calls for intensive calculations and dependent on numbers of hidden layers.

### 1.4.2 ANN discussion: human and other Miscellaneous issues

1. **Human brain and ANN:** We are yet to learn how the mind works! While many scientists agree that ANN are a rough imitation of the brain's structure but many believe that ANNs are statistical inference engines, hence they do not mirror many functions of the brain which is extremely complex. Once we learn to swim, we never forget swimming, it is hard-coded into our brain. However, it is common at times we cannot remember full name of a very close friend of ours after a long gap. So, we understand and remember certain things completely, partially and may forget.
   - **Forget:** However, in case of neural network, it can never forget. So, big difference is humans can forget but neural networks cannot.
   - **Memory retention:** As discussed, our retention could be partial, that is, human's knowledge is volatile and may not become permanent. Since brain cells may die hence information stored there is lost

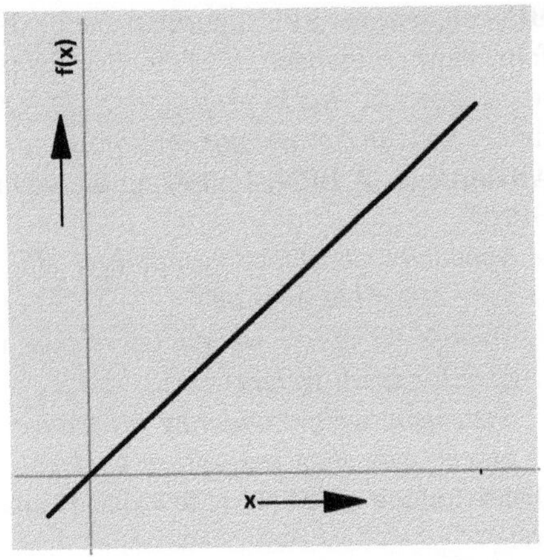


FIG III 1.4d1 LINEAR (IDENTITY) FUNCTION

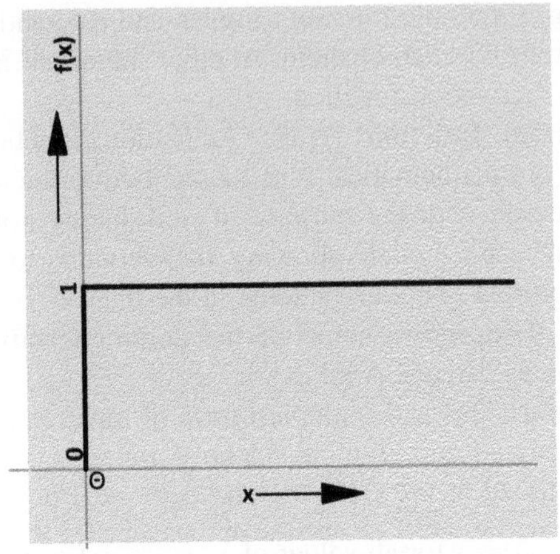


FIG III 1.4d2 STEP FUNCTION

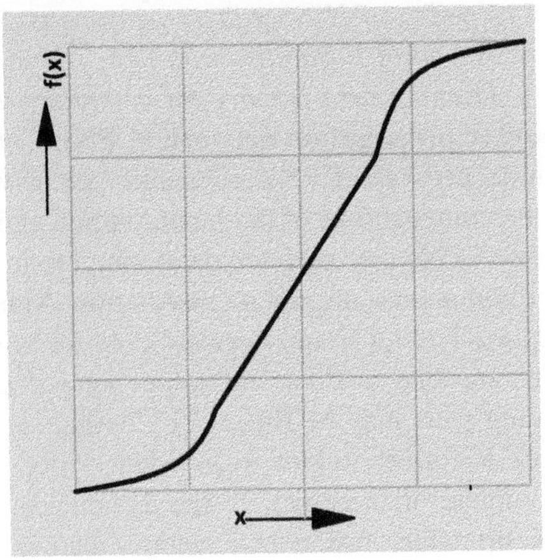


FIG III 1.4d BINARY SIGMOID

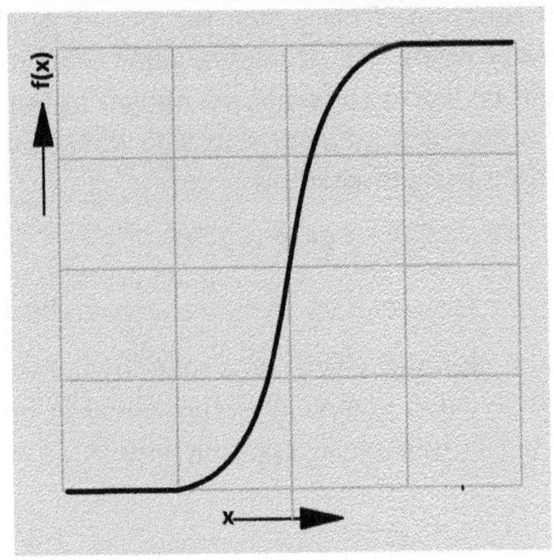


FIG III 1.4d2 BIPOLAR SIGMOID

**FIGURE III 1.4D** Common ANN activation functions.

(memory loss) ANN learnings are always hard-coded and permanent.

- **Accuracy and repeatability:** When a particular application or process is automated through ANN result will be same accuracy even in millions of repeats. This is not the case with human brain. After a few numbers of repetition, same accuracy will not be available (May be simply for fatigue)
- **Speed:** Since ANN work with HW and SW, naturally it can perform the action with speed much faster than humans in processing data and information.
- **Interconnection issue:** Starting with initial interconnections—mainly dictated by generic factor interconnections of neurons in brain constantly changes (even some may die). This makes it possible human brain to tackle many types of decision making. For ANN it is difficult really to determine the neural network structure to cope up with changing interconnections of brain. This could be one of the reasons why it is difficult to conceive AGI discussed initially.
- **Edge Weight and Threshold change:** In neural network during learning process if a mistake is made, weight value of edge and threshold are changed. This is possible for known case. In real life the mistake usually becomes obvious only after a long time. In ANN, in a generalized way it is difficult to know which input is responsible for the mistake. Thus, weight of edge and threshold are very difficult to adjust, that is, structural changes in the network to handle the mistake.

2. **ANN working principles:** The human brain is the inspiration behind ANN, like brain cells called neurons, form the nodes in here. The association between the neurons outputs and neuron inputs can be viewed as the directed connections with weights as shown in Fig III 1.4f. From the external sources, ANN receives assigns the inputs as (say) $x_1$—— $x_n$. Each of the input is multiplied by its corresponding weights (depending utilization). Therefore, these weights represent the strength of the interconnection between neurons inside ANN. All the weighted inputs are summarized inside the computing unit. If the weighted sum is equal to

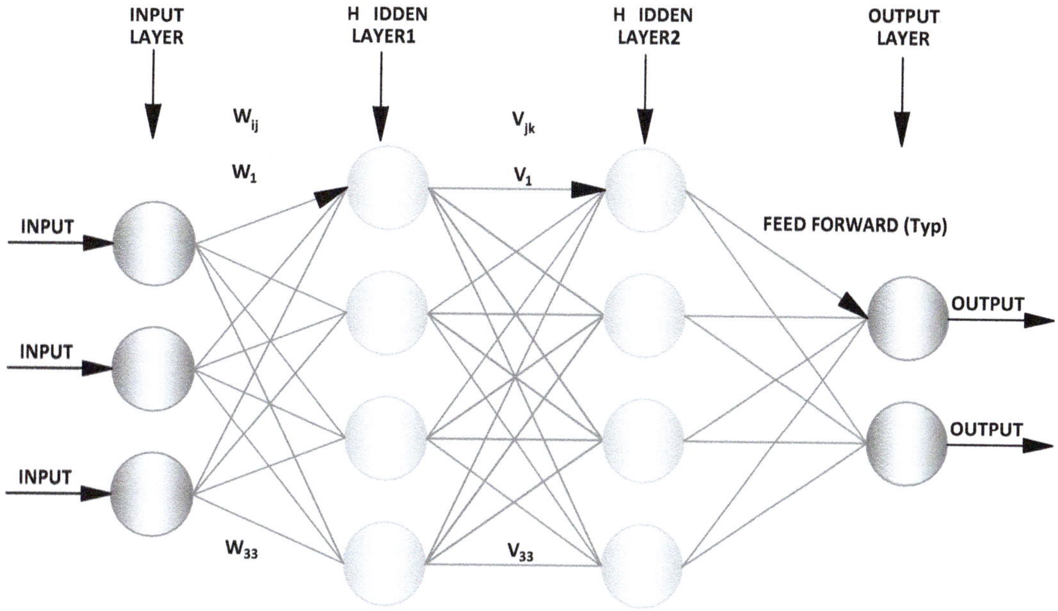


FIGURE III 1.4E Back propagation network architecture.

zero, then bias is added to make the output nonzero or something else to scale up to the system's response. Bias has same input with unity. Here, weighted sum could be any positive values between 0 and + ∞-. To limit the desired value, a certain maximum value is benchmarked, and the total of weighted inputs is passed through the activation function discussed. Some commonly used sets of activation functions have been discussed before.

3. **ANN property** [15]**:** ANNs are very much capable of pattern recognition, decision-making and robust classifiers as well as possess ability to generalize issue from large fuzzy inputs data. Mathematically, a function is linear, when these two equations are satisfied. $f(cx) = cf(x)$ and $f(x_1+x_2) = f(x_1) + f(x_2)$. ANN is used to solve complex, nonlinear systems which violate one or both of these conditions. Some of the distinct problems of ANN are:
   - **Hidden rule**: ANNs reveal the hidden rules in a problem. ANNs do not follow specific rules for data processing but use the received data to discover the governing rule from data. However, do not know these data are elated.
   - **Generalization, and recognition**: Once trained with suitable data to discover the hidden rules for a phenomenon, ANNs can then correctly generalize to data even when it has never seen before data.
   - **Fuzzy rules for data sets**: When ANNs receive certain data they focus on certain rules; If they later receive new and different data, ANNs will adjust their rules in accordance, integrating the old data with the new, a without any external instruction. With the continuous updating of data, a dynamic bank is created with rules which are automatically refined by the ANNs.
4. **ANN solution** [15]: Theoretically, ANNs can solve limitless problems. In reality, it is not so. Following criteria when are satisfied, then ANN can be utilized for problem solutions.
   - The problem is a nonlinear complex one
   - Acceptable solution (s) offers great savings in human and/or economic terms
   - Other physical methods have not produced satisfactory results
   - It is an approximate solution and is valuable, but may not only or best solution
   - There exists a large case history demonstrating the "strange" behavior of the problem pertains.
5. **Commonly used terms in ANN**: There are a few terms frequently encountered during the discussions on ANN. These have been elaborated in Table III 1.4b.

## 1.5 Process intelligence and robotic automation: basic discussions

Regardless of industry, a process automation strategy is mostly complex in some cases the Return on Investment (ROI) can be challenging. So, it is essential to understand the process and select automations as necessary. Other important aspects are as follows: the exploration and optimization of the process with focus on cost saving based on the volume of the process as well as unit cost and suitable implementation of the solution to mitigate regulatory and financial risks. Therefore, it is transformation from a people-driven system to a process-driven system which should have a focus on optimizing human intervention for maximum impact. This also needs to ensure increased consistency, redeployment of workforce. A high-performing business needs to be able to transform, optimize and control its processes across the entire enterprise. For building an agile and future-ready enterprise, effective process management plays an important role. However, there some organizational challenges like: *information blind spots*, *constant vulnerability* and *inconsistent customer experience* to overcome. Simply bolting intelligence onto processes one-by-one is a piecemeal approach. So, it is important for process intelligence to identify bottlenecks and improve operational efficiency.

Industrial robotics as technology has changed dramatically in the past few decades, as has the volume and variety of deployments. Because of many bottom-line benefits mainly like: *efficiency*, *very high uptime* and (associated) lower operating cost, industrial robots have transformed the manufacturing industry. There are many types of automation applications that industrial robots are equipped to handle. Some of the most popular ones are as follows: *pick and place*, *dispensing*, *material handling*, *welding*, *cutting*, etc.

Industrial robots normally ensure greater ROI may be with high initial costs. The operating cost curtailments come from consistency; the higher productivity coupled with higher efficiency. Naturally, combined effect of all of them made industrial robotics so popular among manufacturers and other industrial plant owners. As long as a robot is programmed properly and suited to fit the unique needs of a particular application, it will almost certainly outperform manual labor.

This section has been dedicated for basic discussions on both process and control intelligence. Apparently, they seem to be same, yet in reality is not so. In this section both process intelligence and control intelligence in the form of robot and robotics have been elaborated.

## 1.5.0 Fundamentals of process intelligence (PI)

Process intelligence (PI) allows improvement of the processes, betterment of operational quality. To achieve this goal, business leaders are thus willing to adopt a variety of tools which are popular as various names like process intelligence/mining software. These tools have distinct capabilities, they also share many features, creating debate among business analysts about the best ways to implement

**FIGURE III 1.4F** Artificial neural network working principle.

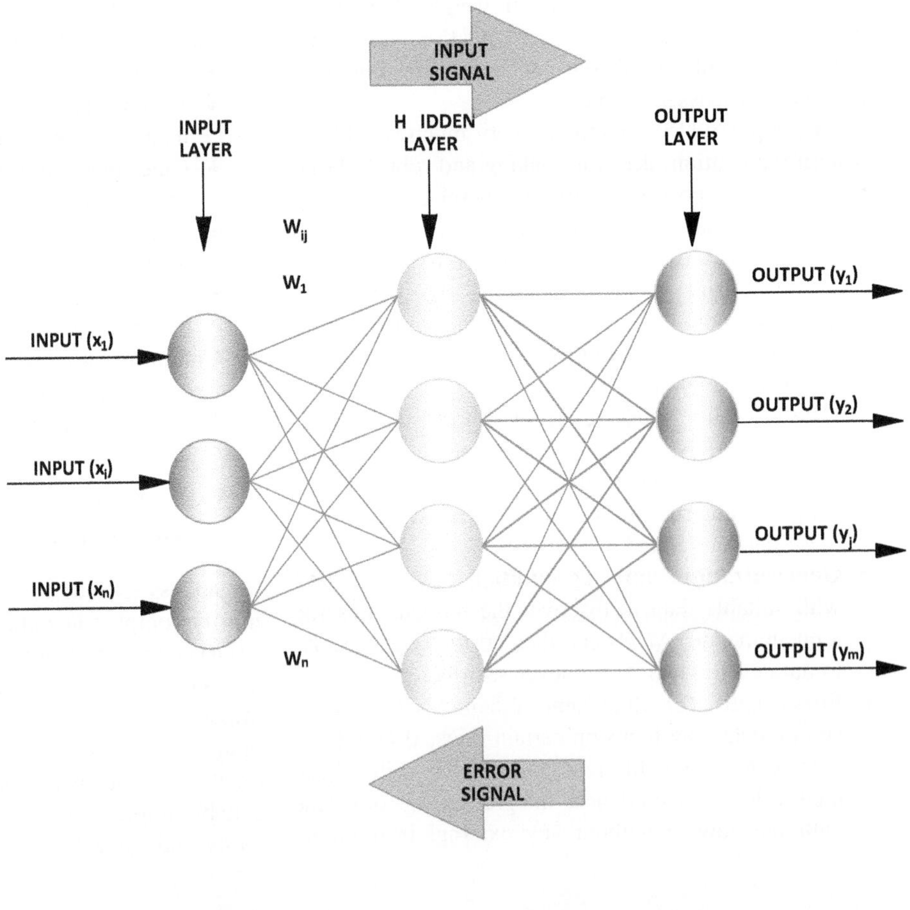

process improvement initiatives. Gaining complete visibility of business operations can be challenging, taking intense manual effort by analysts and teamwork. So, it is extremely important to put all efforts to move from a limited process discovery tool and manual documentation to understanding all the right processes to automate and availing digital transformation. Process intelligence allows analyze multiple processes within business to deliver a digital process model for operations. It creates a visual model, enabling identification of deviations and find the root cause of problems that may be costing time and money. Process intelligence can improve processes end-to-end, regardless of complexity, with its ability to extract data from almost any system, including legacy systems. Here, discussions have been presented on the basics of PI including its power, benefits and other basic details.

1. **Basic issues:** PI is the use of methods and normally software tools to provide greater control over process management with primary goal is to improve operational efficiency, scalability by removing bottlenecks. Various related terms are:
   - **Process mapping:** This relies on tracking a process from beginning to end by closely monitoring the actions workers perform to complete the process.
   - **Process mining:** This relies on backend software to record workflows step-by-step, based on user interaction with systems.
   - **Process discovery:** This is a more recent alternative to process mining that tracks workflows via user interfaces with less manual evaluation of the results.
   - **Task mining:** Task mining utilizes user interaction data, to assess the efficiency of a task within a larger process.
   - **Process discovery** versus **Task mining:** Process discovery is combination of different technologies including ML, it needs the data to find automation opportunities more effectively. While task mining is a mapping tool, used to identify process inefficiencies process discovery.

   PI *automatically and continuously* acquires data at scale, resulting in clear visibility into an organization's journey toward improving its processes. PI connects the best aspects of process mining, task mining and business intelligence with monitoring to realize organization's value faster. PI tools investigate every level of a workflow to find bottlenecks and dysfunctional processes. They extract this data and summarize it in a detailed process definition document. PI can quickly identify inefficiencies and peak process performance of applications hence can scale automation fast. Also, PI helps to identify and edit process functions through discovery, analysis and process optimization. The approach to process discovery calls for modeling to analyze process performance and identify scope for optimization potential.
2. **PI benefits:** There are a number of benefits offered by PI including creation of digital twins for the operation, Instant AI derivation and improved customer experience. Major benefits are:
   - **Continuous improvement:** PI ensures process improvement by suggesting the most suitable for automation, complexity and business impact through data analysis with task mining and process discovery. When PI tools are implemented properly, improvements come *before* workflows stop running smoothly.
   - **Improved efficiency and effectiveness**: PI tools can help analysts to automate processes—improving operational efficiency through effective departmental collaborations for data gathering, mapping and analysis. Therefore, results are easy obtainable without human error and faster implementation of changes.
   - **Enhanced decision-making**: Today's industrial plants generate huge data including those from IIoT/IoT. With PI it is possible to identify data patterns to get detailed insight view about the process for making decisions. On account of centralized location of data storage, user can view from a single dashboard.
   - **Cost reduction:** PI tools break down the process into parts for automating, and stream line them. Also, deviations, bottlenecks, and silos can also be identified and eliminated by PI tools. As a result there will be huge cost reduction for the running processes.
   - **Compliance and risk management**: PI helps in meeting government regulations and industry standards and do the risk management for plant safety.
   - **End-to-end visualization**: PI tools are effective from beginning to end. First, waste is identified and eliminated, then automation is applied to appropriate steps. After a few rounds of testing and adjustments, the improvement can be visualized.
3. **PI cycle and components:** As shown in Fig III 1.5a, PI is a cyclic process with five major phases like: *planning and direction*, data *collection*, *processing*, *analysis and production* and *dissemination with feedback.* The main components are:
   - **Data collection:** As shown, PI collects data from various sources viz. enterprise databases, sensors employee input, etc., to understand the process and integrate necessary data together.
   - **Process mining:** As already discussed, process mining monitors the current condition of processes rather than their ideal state.

**TABLE III 1.4B** Explanations of commonly used terms in ANN.

| Term | Explanation |
|---|---|
| **Backpropagation** | To obtain optimal weights for the neuron(s) backward pass is performed that is, backward movement from network's prediction to the neurons which generated that prediction. This tracks the derivatives of the activation functions in each successive neuron, to find weights to bring the loss function to a minimum. Also, it is referred to as *gradient descent.* |
| **Bias and variance** | Bias is a measure of how well the model fits the training set, so that it is able to correctly predict the known outputs during training. Variance measures how well the model copes up with unknown inputs (not available during training). *Bias* also refers to a "bias neuron" (holding the number 1), used in every layer of the NN to makes it possible to move the activation function. |
| **Error function** | This defines the gap between the actual output of the current model and the correct output. |
| **Forward pass** | Forward pass basically refers to calculation process for values of the output from the inputs data traversing through all perceptrons/neurons from input to last layer. perceptrons/neurons generate their outputs passing through layers until eventually the network generates an output. |
| **Hyperparameter** | Hyperparameter settings has the ability to modify structure or operation of the neural network. It is used in deep learning projects. |
| **Inputs** | These refers to source data fed into the neural network, for making a decision/prediction about the data. Inputs are typically a set of real values. |
| **Outputs** | Neural networks generatedecisions/predictions in the form of a set of real values or Boolean decisions. These are referenced as output which are generated at the output layer. |
| **Perceptron/Neuron** | The basic unit of the neural network is referred to as Perceptron/Neuron which on acceptance of input can generates a decision/prediction. These are trainable that is, their threshold and input weights are modifiable. Inputs are presented to the perceptron/neuron to produce a desired (within acceptable range) output passing through the activation function. Nonlinear form of activation functions are critical during training for the network. |
| **Training** | A set of inputs with the known correct outputs are used during learning process to train the neural network. This process is referred to as training. |
| **Weightage** | Each neuron/input is given a numeric weight. The weights, together with the activation function, define each neuron's output. ANNs are trained by fine-tuning weights, to find the optimal set of weights to produce accurate decision/ prediction. |

- **Analysis and visualization:** This stands for the graphical representation of process data/model to help in understanding of the process and improve process efficiency. Visualizations mainly refers to dashboards and data tables for scrutiny and analysis.
- **Process model:** Process model is a kind of practices for visualization a business's workflows and processes. This also involves detailing of each step in a process, for a thorough understanding of the activities and tasks comprising that process.
- **Digital twins:** Digital twin is the data-based replica of the process, that is, a virtual model of a process, used to study various aspects of the process, for example, objective, performance/KPI

4. **PI automation:** PI refers to systematically collect data—concerning accurate information about each work item, to analyze the individual steps within operational workflow to identify bottlenecks and improve efficiency. PI automation refers to the integration of PI with automation technologies. It combines the analysis and insights provided by PI with automation tools like RPA, workflow automation, etc. This synergy allows organizations to not only understand their processes but also to automate and optimize them based on data-driven insights. It is a business process automation. Thus, PI uses, AI, ML, computer vision, etc., to continuously observe and interpret digital activities across all applications and departments. This results a

real-time detailed map of every digital process across the enterprise.

5. **PI: Business process management (BPM)** A business process refers to activities deployed to accomplish business goals. BPM refers to a structured approach to improving the business processes So, BPM refers to the ability to gather, analyze, and interpret data from various processes. It involves use of advanced analytics and tools to gain insights of the processes for identifying bottlenecks, inefficiencies, and opportunities for improvement and optimization of business strategy and processes. Three main types of BPM are as follows: *integration-centric*, *human-centric*, and *document-centric*.

   Fig. III 1.5b details out BPL lifecycle and controls. As shown in Fig III 1.5b1, major five phases of BPM lifecycle are:
   - **Design.** Analyze the existing process to strategies improvements through developed business process using standardization and automation. Clear definition is important for
   - process optimization and tracking improvements.
   - **Model.** Study the redesigned process operation in different scenarios utilizing BPM software.
   - **Execute/implement:** Execution of improvements, inclusive of standardization and process automation.
   - **Monitor:** Tracking of improvements by measuring improvements in efficiency and identifying any additional bottlenecks if any.
   - **Optimize.** Final adjustments, to improve the business process on an ongoing basis.

Fig. III 1.5b2 shows various *BPM controls* to achieve the desired business goal.

### 1.5.1 Fundamentals of robotic process automation

Robots are developed with the objective of manipulating the objects by picking, moving, modifying the physical properties of object, even destroying it, or any other effect thereby freeing manpower from doing these or other repetitive functions without getting bored, distracted, or exhausted. The branch of AI, comprising electronics electrical, mechanical and computer engineering, engaged in designing, construction, and application of robots is referred to as robotics. Robots are normally developed for specific services and contain some program that determines the working and behavior of the robot. There are many components and joints in robots. Major components are:

- **Power supply: energy to carry out and action,** various power sources which could be batteries, solar power, hydraulic and others, are required.
- **Actuators and motors:** These are responsible for any action or movement for which AC/DC motors are deployed.
- **Sensors:** These are responsible for gathering and providing knowledge of real time information on the task environment viz. vision sensors.

  As the robots can move in its environment, it has some locomotion. There are several kinds of such locomotion like:
- **Leg type:** These are meant for walk, jump, trot, hop, climb up or down, etc. This type calls for more power and motors to accomplish a movement.
- **Wheeled motion:** Various wheeled motion viz. standard, spherical wheel, etc., requires fewer number of motors (and lesser power) to accomplish a movement. Though this is easier to implement, but with increase in number of wheels instability sets in.
- **Skid type:** In this type mainly used in vehicles which use tracks, for example, a tank. The robot is steered by moving the tracks with different speeds in the same or opposite direction and is more stable.

In robot vision is referred to as computer vision which can automatically extracts, analyzes, and comprehends useful information. The computer vision plays vital role in the domains of safety, security, health, access, and entertainment. Apart from industrial use, robots cater to wide application areas like, agriculture, medical. Remote sensing, etc. Here, short discussions have been presented basics of robots in industries and approach to Robotic process automation (RPA).

1. **Basics of robot and robotics:** From the above discussions, we gather some knowledge on them. Let us now try to detail out them with reference to common industrial usage.
   - **Robot:** IEEE article what is robot? provides very precise definition of Robot as "*A Robot is an autonomous machine capable of sensing environment, carrying our computations to make decision and perform actions in real world*".
   - **Robotics:** The branch of engineering and technology that deals with robots is referenced as Robotics. Obviously, it follows from the above that robotics is a multidisciplinary subject involving mechanical, electrical, electronics, computer science involving artificial-intelligence, mechatronics, nanotechnology and bioengineering, etc. Robotics is combination of engineering and technology that produces machines, called robots, which is capable of substituting or replicating major human actions. This branch of engineering involves the conception, design, manufacture, and operation of robots defined above.
   - **Artificial Intelligence in robot:** Modern robots incorporate AI which provides computerized vision

FIGURE III 1.5A Process intelligence cycle.

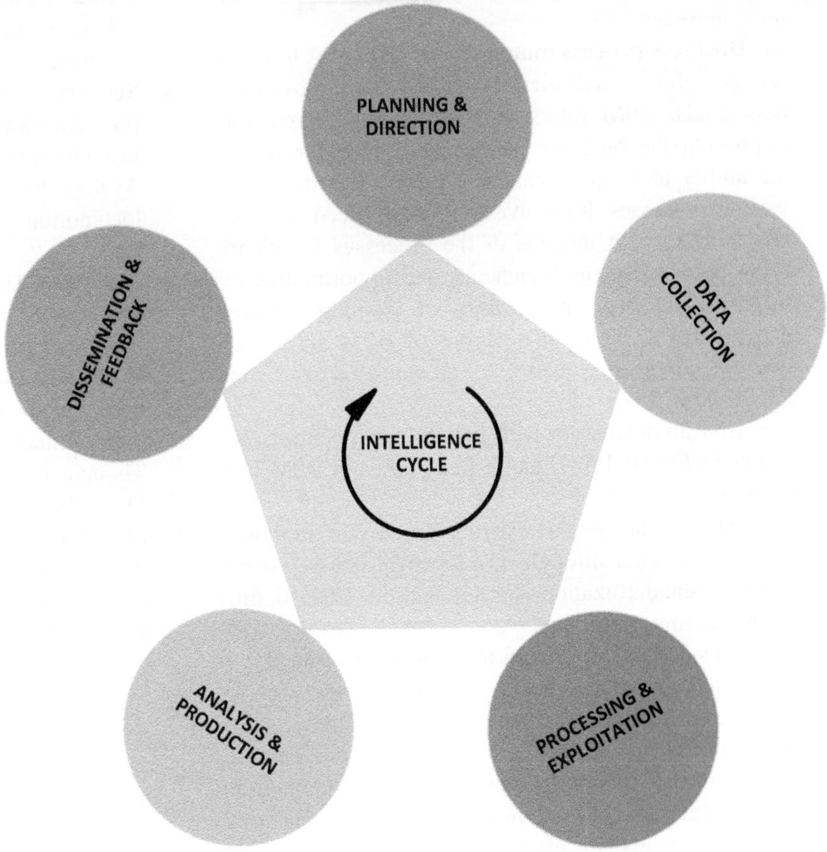


to navigate, sense and calculate their reaction accordingly. With the help of AI robots learns to perform their tasks from humans through ML/DL as the case may be.

- **AI programming and LISP:** List processing (LISP), which was designed by John McCarthy in 1959 and developed by Steve Russell, Timothy P. Hart, and Mike Levin is used for AI programming. LISP supports software computation with the help of symbolic languages.
- **Common use of robots in industries:** Common use of robots in many industries are for repetitive tasks and various tasks in very hazardous environment to humans. In many cases robots may be equipped with the equivalent of human senses such as vision, touch, and the ability to sense temperature [https://www.britannica.com/technology/robotics].
- **Common characteristics of robots:** Generally, robots have following characteristics.
  - *Mechanical construction:* Robot involves some special mechanical construction, that helps it to complete tasks in the environment for which it is designed.
  - *Electrical connections and components:* These involve power and control components for the machinery.
  - *Computing and artificial intelligence:* All Robots have some level of computing and decision-making capability. So, robots contain some level of computer programming which generates codes to tell what to do, etc. By inserting computer program and artificial intelligence into a robot, a simple machine could replicate human actions.

2. **RPA concept:** Robotic process automation (RPA) helps to move toward digital transformation. RPA even though is not really part of IT infrastructure of the company but sits on the top of it. In many enterprises there are several IT systems to carry out different tasks in practically uncoordinated way. All such tasks are not only labor intensive but are repetitive and time-consuming affair. So, sitting at the top of IT infrastructure RPA mimics human activities. So, it is possible to automate those types of repetitive, time-consuming and labor oriented tasks through its rule-based activities. This would not only save time, but will be performed with more accuracy in tireless

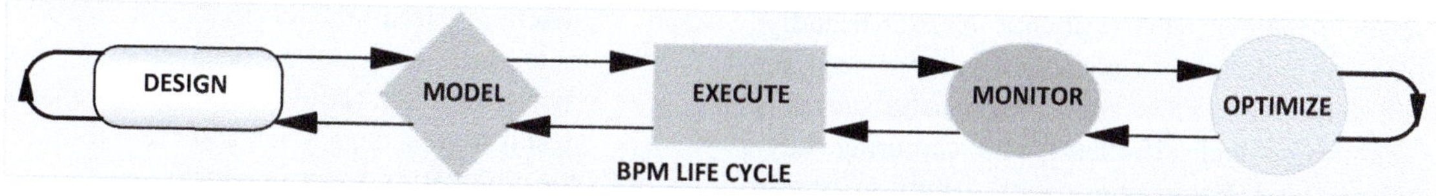


Fig III 1.5b1 BUSINESS PROCESS MANAGEMENT LIFECYCLE

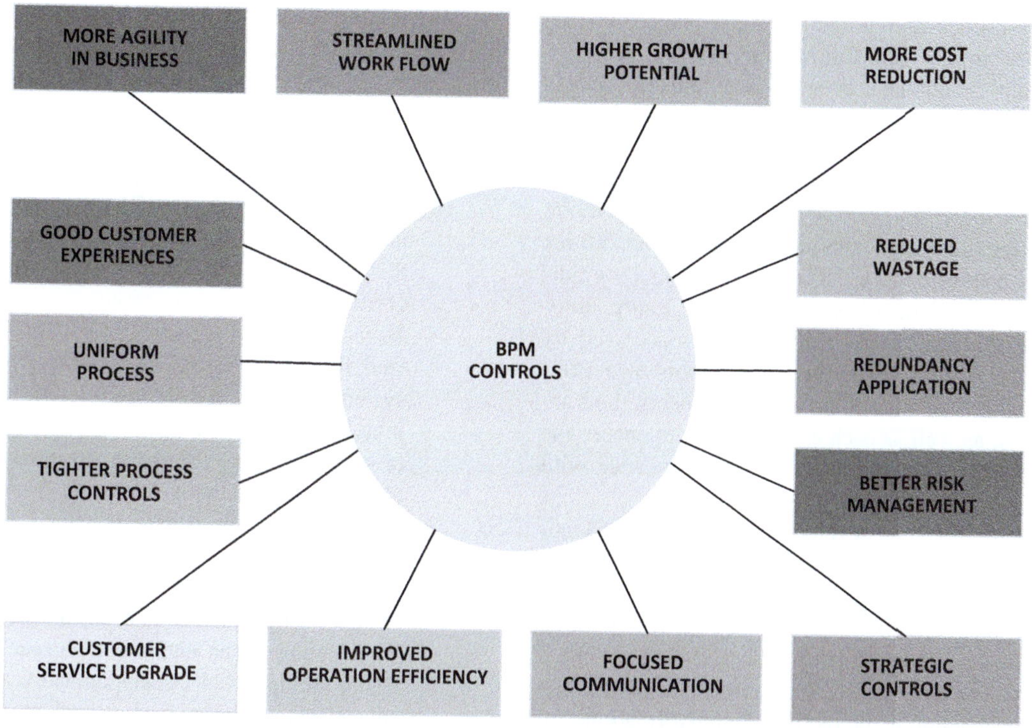


Fig III 1.5b2 BUSINESS PROCESS MANAGEMENT CONTROLS

**FIGURE III 1.5B** Business process management lifecycle and controls.

mode. So, RPA helps human beings to focus on the tasks requiring emotional intelligence, reasoning, judgment, and interactions with the customers [16].

3. **Functional definition:** It is interesting to note that each of three words, for example, *Robotic*, *Process*, *Automation* actually constitutes RPA with functional details and relevance of each detailed out in Fig III 1.5c.
RPA involves utilization of specialized computer programs to automate and standardize repeatable business processes across multiple systems. Through RPA execution process is transferred from humans to bots (*interactive program with user/system interface*). Robots perform tasks and associated actions directly across the user interfaces of applications. They mainly mimic human actions for the tedious and repetitive activities. However, when compared with human, properly designed robot performs all these tasks at a much higher speed with no mistake and fatigue. Robot never sleeps and can interact with in-house applications, websites, user portals, etc. [16]. RPA can interact existing IT architecture with no complex system integration required. So, robots act like virtual assistant and this would result free time for employee for productive work. As a code-free, user-friendly RPA is fast becoming an important automation tool driving digital transformation and the future of work [17]. As shown in Fig III 1.5c, there are three basic types of RPA technologies viz. Probots, Knowbots and Chatbots [16].
4. **RPA Pros and cons:** Here, both pros and cons of RPA have been presented.
   - **Advantages:** RPA is a growing technology with several distinct benefits and is an excellent enabler for Industrie 4.0 and digital transformation, changing human machine interface in a better working environment. A few advantages are:
     - *Code free*: RPA doesn't require any coding/programming knowledge. The workforce requires training in RPA to develop robots using Graphical user interface (GUI) and other intuitive wizards.

- *User friendly*: RPA software is user-friendly, easy to understand, easy to use without need for IT skill. RPA tools allow users to create bots seamlessly and effortlessly by capturing keystrokes and mouse clicks with a built-in screen recorder component.
- *Nondisruptive*: RPA is very simple, seamless streamlined and straightforward. The RPA software robots can follow the existing security, quality, and data integrity standards to access the end-user system in the same manner as human beings preventing disruption of any kind and maintain functionality and protections. RPA.
- *Security*: It is important that RPA tools provide options to assign role-based security capabilities to ensure action specific permissions. To eradicate any malevolent changes, the entire automated instructions, data and audits accessed by the robots are encrypted. RPA tools also offer detailed statistics of the logging of users, their actions, as well as each executed task to ensure the internal security and maintains compliance with industry regulations.
- *Debugging*: For development, one major advantages of RPA is debugging which lets, dynamic interaction, that is, validating different situations by augmenting the variable values without pausing the running process allowing seamless developments and resolutions in the production phase without prerequisite transformations to the procedure.
- *Exception handling*: RPA system allows users to deploy bots using rules-based exception handling: The *same process* is re-assigned to a *different bot*, the current bot *retries and removes* the bot from production, for *successful retry* the server to maintain the reassignment and issue an *alert* to report, for *unsuccessful retry* the server to stop the current bot and raises *an alert.*
- *Analytical suite*: RPA software contains an integrated analytical suite to assess the bot's workflow performance. It helps monitoring or managing the automated functions from a central console. The analysis executed by the suite helps users to track the operations and determine operational issues.
- *Intelligent decisions*: This is the ability to acquire and apply knowledge as skills, by converting data into information, and subsequently changing the same into actionable details for users. Thus, commonly, RPA solutions use AI and Cognitive Intelligence in enhancing decision-making.
- *Hosting and implementation*: The RPA system provides deployment options across virtual machines, terminal services, and cloud which is one of the best among because of its flexibility and scalability. RPA tools can be installed on desktops and deploy it on servers to access data for completing repetitive tasks.

- **Disadvantages:** In spite of so many advantages, there is a negative side to the method, especially when implemented in nonstrategic fashion. A few disadvantages are:
  - *Initial investment:* Since RPA is in innovation stage and not fully developed, so it is not easy to take decision on to invest in automation or wait for it to fully develop.
  - *Long term stability*: RPA can be a serious inveigle from the long-run task required. So, there will be risks that the things are not done correctly.
  - *Maintenance*: RPA solutions have to be customized for the business. Even little change in the system may call for reconfiguration in RPA and may be costly. So, it may not be worth investing in such a system *if* the way business runs could transform significantly in future.
  - *Skilled staff*: Many believe that for RPA, the staff must have significant technical knowledge of automation force organizations to either hire a skilled staff or train existing employees to expand their skills. It is worth noting that an automation firm can be of a little a help during initial installation and set-up but can only adapt and control the bots in the long run.
  - *Resistance for Job loss*: With completion of numerous tasks at a swift turnaround time by RPA with no need for human operations. This creates apprehension for loss of potential job loss, creating stress to the employees. As a result there will be resistance from employees.
  - *Process selection*: It is always best to select rule-based tasks without any need for human judgment. The nonstandard processes are difficult to automate, and human interaction is necessary to complete the processes.

5. **Implementation process:** For best RPA implementation practices, it is important to make sure that due importance and considerations have been given on: *impact on business and people, effect of compliance on standards on RPA, defined focus on desired ROI, scope of RPA and automation with due effect including attended and unattended type* and *alternative arrangement for poor design or change in management.* All these will have impact on different of phases of RPA implementation. Keeping these in mind, implementation should be initiated. RPA implementation generally covers phases like: *planning*, development, testing, and support—maintenance.

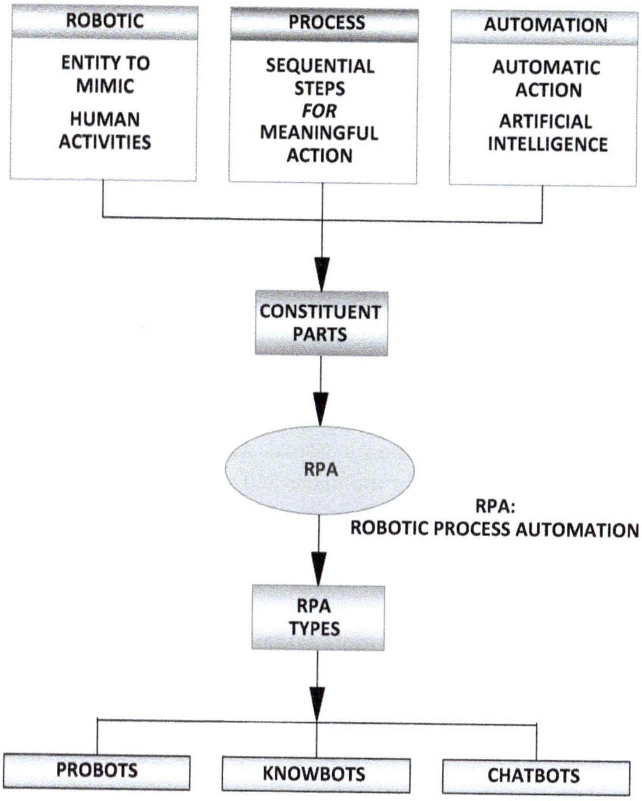


**FIGURE III 1.5C** RPA constituent parts and RPA technology types.

- **Planning:** Issues like: *rule-based process, input readable data,* requirements *for any change in the system,* etc., help to ascertain the correct process. Major activities are:
  - *Team setup and fixing timeline*
  - *Agreement on design for RPA processes.*
  - *Identification of most time and effort consuming manual tasks for the entire process*
  - *identify bottlenecks of key processes mainly manual tasks*
  - *If necessary, plant to Implement a pilot project to tackle a high-value discrete task.*
  - *Planning for the issues for clear roadmap for running bots and subsequent scale up.*
- **Development:** This is time for development of the automation workflows agreed upon during planning. Wizard driven the implementation is quick.
- **Testing:** This is test cycle to detect and correct defects
- **Support maintenance:** Continuous support for the live project. General maintenance guidelines with business and IT support teams to be initiated.

*With this let us conclude the fundamental discussions on RPA, with to begin the discussions on impact and applications of RPA in industries in next subsection.*

### 1.5.2 RPA: impact and applications in industries

The roots of RPA can be found in the use of mechanical devices to perform repetitive tasks in the beginning of the industrial era. Today computers and robots have a prominent place in virtually all manufacturing operations. Material transfer, machine loading/unloading, arc welding, spray coating, assembly and inspection activity are some of the examples of use of robot, and RPA in industries to carry out activities of human being especially in the area of repetitive work cycle, work in hazardous area/difficult handling for humans, multishift operation, infrequent changeovers, etc. Also, tasks which are intensively manual, high volume, otherwise prone to human error, having standardized and well-documented workflows are well benefited with RPA. In many cases the process often lacks the flexibility or scalability to adjust to the increasing number of tasks/functions. This is where RPA comes into the picture. Also, RPA can automate the task of document creation and modifications. Thus, employees are freed up the from manually updating and keeping track of tiny updates. Because of the simple implementation, RPA can be implemented as a standalone product in an existing system—without costly adjustments to the software or changes to an enterprise's IT infrastructure. RPA bots can be used in different systems, interact with other software systems, and are not limited to specific applications. Based on use, these applications can be divided into different sectors:

- Common business processes and activities viz. *Pay roll/invoice preparation*,
- Activities in commercial functions, for example, *sales*
- Activities in support functions like: *Technology, tech support*
- Industry-specific activities viz. *Manufacturing industries*
- RPA in personal use, for example, *digital assistants.*

Since, the book is meant for plant controls and automation, naturally, discussions will be mainly on industrial application. In this subsection brief discussions have been presented on positive impact of RPA, application of RPA in different industries.

1. **Positive industrial impact** Major positive impact of robots/RPA in industrial plants are:
   - **Increased productivity.** Bots can function tirelessly without fatigue, around the clock. Employees need only to intervene only for certain decisions, and resolving errors if any.
   - **Cost reduction:** As jobs are completed at higher speed, hence there will be efficient process cycle time to reduce the cost of production.
   - **Better customer Service:** With deployment of RPA, there will be more free time for employee to offer better customer services and would be able to satisfy customers in a better way.
   - **Employee Morale:** Since employees are relieved of tedious jobs and gets time to spend the same for interesting jobs, employee's morale boosts up.
   - **Process analysis and audit:** Actions of Robots are software driven naturally these can be very easily be recorded for retrievals and any bottleneck in the system could be more easily detected. All these help business operations smoother
   - **Compliance to standard:** As actions of robots are software driven, hence its actions are controlled in nature. Therefore, robots can be instructed strictly as per standard norms.
   - **Reliability:** Bots can work tirelessly all the time $24 \times 7$ and they are very reliable in nature.
   - **Accuracy and consistency:** As actions of robots are software driven, hence Bots offer very high accuracy in consistent manner without error normally committed by human.
   - **Low technical barrier:** There is no need for advanced programming knowledge for configuring robots now a days. Even nontechnical employees can configure software robots specified automation challenges, through graphical modes.
   - **Flexibility and scalability:** Robots can be employed or shutdown as per owners' choice (deployment at high work load otherwise may be shut down). This provides owner flexibility and there will be no need to hire temporary staff and then lay them off in lean period. Also, in general bots/RPA offer more flexibility and scalability.
   - **Noninvasive in nature:** RPA work on higher layer of ISO OSI model, just like another application so RPA does not involve in any disruption to underlying systems.
2. **Industrial applications:** RPA has wide application area such as banks, financial institutions, and insurance companies process large numbers of operations daily. RPA can be used as a virtual worker, as replacement human workers in mundane and repetitive tasks. RPA also finds application in health care systems also. The retail industry and logistics find good use of RPA. In utility and telecommunication companies RPA is used to monitor CRM subscriber feeds, fraud management data, and customer data updates, etc. However, RPA can also be utilized in manufacturing industries too to reach the goal of Industrie 4.0. RPA offers new way. Two major robot application in plants are:
   - **Manufacturing:** In manufacturing robots can perform a number of tasks very well like:
     - *Material transfer/handling:* Robots are utilized to load and unload machine tools, products, etc. Robots may be utilized to service *a single machine* or *several machines.*
     - *Robotic packaging:* Robots are used for packaging, palletizing dispensing pick and place.
     - *Welding:* Robots can perform *spot welding, arc welding* and *plasma welding* very well.
     - *Spray painting:* The unhealthy and unpleasant environment of industrial painting, robot is an ideal candidate for handling toxic solvent materials.
     - *Machining and cutting:* Out of various machining operations, drilling can be successfully done with robots. Similarly, for deburring metal parts, robots can be deployed. Robotic cutting, mechanical cutting, waterjet cutting are common.
     - *Assembly:* Assembling with industrial robots are mainly used for small products. Robotized assembly systems are programmable hence a cost-effective solution.
   - **Industrial applications:** Although manufacturing is the major area of robotic industrial application, in other industries also, robots find their applications for automation. Also, there are a myriad of other types of collaborative and service robots, the applications. In most of the cases robots are designed or integrated and tailored for specific task with

unique needs of that task. Some of the common industrial areas are:
- *Soldering robots, robot for grinding, deburring* and *polishing*
- *Robotic nondestructive inspection*
- *Gluing, adhesive sealing* and *material spray*
- *Collaborative robots*—built to work together with other robots.
- *Miscellaneous applications*: *3D Laser vision robots, appliance automation robots, refueling robot, cleanroom robots, fiberglass cutting robots, foundry robots,* etc.

3. **Functional areas in manufacturing:** In manufacturing, RPA finds most of its applications in the following areas:
   - **Bill of Materials (BOM):** In manufacturing industry central document BOM has a wide list of raw materials, components, subcomponents and other products for the new product creation. The main aim of the document is to provide employees with all the data they require regarding new products and its journey. People can get an idea of where to purchase, what to purchase, when to purchase and how to purchase.
     Any mistakes committed at any of these stages can create an adverse chain impact resulting a massive loss in the industry. RPA, equipped with its ability to conduct error free task function can play a crucial role. RPA will give the companies create the product much faster, with better data accuracy, and the product creation in time.
   - **Administration and reporting:** Implementation RPA in the administration department would not relieve high work load and assist the department to get all the reports related to production and allied work. Bots would generate required reports precisely and timely. RPA helps in regular effective upgrades and help in taking decision to manage the show in most effective manner—essential in manufacturing. RPA could accelerate the performance and reduce the possibility of error. All these would result more productivity and efficiency.
   - **Customer Support Services:** RPA is very helpful for an enhanced interaction and communication with the customer and alerting reminder for employee in following up customers relations, maintain the contact schedule and/or customer complaints, etc. Through RPA, multiple data can be stored at one place to access to serve customer in a better way. Virtual chat could be arranged.
   - **Data Migration:** Data migration means moving data from an old system to newly replaced systems and this could be related to introduction of new improved product/service/business line. RPA is very much apt for such application with s proper planning procedure. Any mistake can cause huge loss. Since RPA works on ruled based system it can do proper planning and execution definitely in reduced time and cost.
   - **Regulatory compliance:** There are various kinds of data such as company sensitive information and data of consumers and suppliers, etc. The actions of the bots are entirely transparent, and could be utilized for analysis. So, central log and collection can be done almost in no time. Naturally audit and regulatory compliances are much easier when RPA is deployed.
   - **RPA integration with ERP and WEB:** Integration of ERP with RPA is easier to generate and distribute reports, like inventory, accounts pricing and other reports. When there are branches and offices and factories are located apart, normally data manipulations are web based so as to get better system integration. Therefore, when RPA is deployed in this area in this area such manipulation of data are not only effective and easier but also these data can be easily be monitored and updated accurately in time. Therefore, integration of RPA with Web is a next step solution to better customer servicing too.
   - **Invoice processing:** Invoice processing calls for interaction between various people belonging to various departments for inspection authorization, etc. Even when an enterprise uses IT system but invoicing is done and checked manually, then there will be every possibility of making some wrong entry which could be very difficult to trace. As RPA is ruled based can it can fill the form very promptly without mistake. So, RPA can generate error free invoice quickly for essential department for authorization.
   - **Inventory control:** This is labor-intensive time-consuming work. Variations in supplier's and r customer's demand could make the issue further complicated. Obviously, when bots are employed auto emails, acquisition and other processing activities relating to inventory controls, mistakes and time consumption will be highly reduced. Therefore, deployment of RPA would much better result in inventory controls.
   - **Logistics automation and control:** All manufacturing industries need to manage and control logistics department for transportation of finished products to their customers. In some cases where semifinished products need to travel different workshops which may not be located nearby. The issue becomes more complicated then. When integrating RPA transport management system, better and efficient transportation is easily available and

this can be monitored for its effectiveness. The RPA will give the report of which choice is giving the best cost, insurance and transit time [18].

4. **Manufacturing: RPA benefits:** From above, it is clear that industrial robots offer huge benefit especially in terms of efficiency accuracy and tireless 24 × 7 working at high speed without mistake/error even for repetitive job without fatigue. On account of ruled based artificial intelligence, they could maintain consistent quality production with reducing waste, fetching higher ROI. RPA also offers facilities like: lesser time to market, maintenance of audit trails. Major benefits are:
   - **Cost saving:** According to the latest report, implementing RPA typically results in up to 20% cost saving. Implementation of RPA in manufacturing involves initial capital cost but offers long-term cost savings in terms of:
     - *Labor Costs:* automated processes lower manual labor, and associated labor cost including costs for healthcare, insurance, administration maintenance and upkeep.
     - *Higher productivity:* Also, for higher 24 × 7 operational ability of RPA bots companies can enhance productivity hence associated cost saving. Also, there will be lowering of cost for cutting cycle times and simplifying operations.
   - **Labor shortage:** Finding skilled manpower to perform certain processes is not easy and often calls for more time, effort, and money. However, RPA *can* lower such dependence on manpower.
   - **Productivity, quality and efficiency:** 24 × 7 operation of RPA bots, leads to continuous production and process execution. RPA can do tasks like data entry, reporting, and transaction processing much faster. Also, tasks are completed consistently to maintain a consistent level of quality. So, efficiency is enhanced by reducing the number of human errors. In place of these, workforce can focus on higher value operations boosting overall productivity without delays or downtime.
   - **Reduced wastage and errors:** The RPA bots are extremely efficient in fixing manufacturing hurdles like manual lapses, inventory wastage, and process wastage. Also, the rule-based executions of RPA bring down occurrences of errors to increases accuracy.
   - **Scalability and flexibility:** With RPA it is easier to readily scale up/down the process, that is, to handle different job types. Also, ruled RPA can easily adapted to the modification for new criteria/process.
   - **Customer experience:** By automating customer support, faster order processing and inquiry handling are possible to satisfy customer. Similarly, ensuring timely updates and correct order fulfillment increases client trust and happiness.
   - **Back-office operation:** These includes customer support and functional area services discussed. RPA can be utilized to attain up to 40% cost saving in back-office.
5. **RPA trends:** According to International Federation of Robotics, By 2024, the global robot installations are expected to level off at 541,000 units. Some factors for the growth could be: quality maintenance exported cars, promotion of remote/programmed vehicles, governmental incentive plans fostering a homegrown robotics industry for rising wages changing the metrics of human robot deployment, shortage of capable factory workers, etc. A few issues here to look into are:
   - **Investment plan**: In a survey, it has been revealed that in future, enterprises will spend more on automating tasks. Hence, there will be more investment on RPA. Even there will investments for robotic workers.
   - **Intelligent automation:** RPA technology allows computer robots to learn and perform repetitive tasks, such as filling out forms and processing paperwork. Enterprises automate tasks in collaboration with human and RPA. Some of such automated tasks are as follows: email, answering a customer query-reply, etc.
   - **Collaborative robotics**: In present time in many industries like car industry, conventional industrial robots are replaced with a combination of humans. These co-bots offer benefits like flexibility, safety, rapid deployment capability, and ease of training. Co-bots could be the stepping stone for robot deployment in the process.
   - **Robotics as a service (RAAS)**: As the name suggests RAAS offers services. RAAS is a recent trend used in many ways like introduction of untested products into the marketplace, and finding economies of scale benefit the service provider. These can be used for interpretation of unmanned aerial vehicles sensor/camera data and analysis and develop actionable plan. This at times better and lower cost vision systems, particularly low-cost 3D vision, navigation and mobility, etc.

## List of abbreviations

**AGI** Artificial general intelligence
**AI** Artificial intelligence
**ANI** Artificial narrow intelligence
**ANN** Artificial neural network
**BOM** Bill of material
**BPM** Business process management
**BPN** Back propagation network

**CBR** Case-based reasoning
**CPN** Counter Propagation network
**CV** Control variable
**DT** Digital transformation
**ES** Expert system
**GSAC** Gain scheduling adaptive control
**GUI** Graphical user Interface
**HVAC** Heating ventilation and air conditioning
**I/O** Input/Output
**IIoT** Industrial IoT
**IoT** Internet of things
**KPI** Key performance index
**LFT** Linear fractional transformation method
**LISP** List processing (AI)
**LPV** Linear parameter varying method
**MAS** Multiagent system
**ML** Machine learning
**MLP** Multilayer perceptron
**MRAC** Model reference adaptive control
**NN** Neural network
**PI** Process intelligence/Performance index
**PID (Controller)** Proportional Integral and Derivative (controller)
**PV** Process variable
**RAAS** Robotics as a service
**RBS** Rule based system
**ROI** Return on investment
**RPA** Robotic process automation
**SOE(R)** Sequence of event (recording)
**SP** Set point
**SPL** Symbolic program language
**SW** Software

## References

[1] B. Lipták, Can we control artificial intelligence?–*As in process automation, the first step is to define the goals*, Control Global, 2020. https://www.controlglobal.com/articles/2020/can-we-control-artificial-intelligence/.

[2] Defining Intelligent Control; Report of the Task Force on Intelligent Control IEEE Control Systems Society (Panos Antsaklis, Chair), December 1993. https://www3.nd.edu/~pantsakl/Publications/150-ISIS94.pdf.

[3] L. Rogelio, Adaptive Control; Chapter 1 Introduction, Springer, 1998. https://link.springer.com/book/10.1007/978-0-85729-664-1.

[4] J. V. Candy; Model Reference Adaptive Control (MRAC) for Additive Manufacturing; LLNL-TR-823797; Lawrence Livermore National Laboratory; https://www.osti.gov/servlets/purl/1798434.

[5] V. Vandoren, Fundamentals of Self-Tuning Control, Control Engineering, 2007. https://www.controleng.com/articles/fundamentals-of-self-tuning-control/.

[6] PID Tuning Guide; R.C. Rice; Control Station; Yokogawa Electric (CS Tuner); https://web-material3.yokogawa.com/Yokogawa_PID_Tuning_Guide_-_csTuner.pdf.

[7] V. Vandoren, Relay Method Automates PID Loop Tuning, Control Engineering, 2009, in: www.controleng.com/articles/relay-method-automates-pid-loop-tuning/#:~:text=Automating%20the%20tuning%20processandtext=Their%20relay%20method%20generates%20a,the%20%22relay%20test%22%20graphic.

[8] G. McMillan; How to Avoid Common Tuning Mistakes with PID Controller; ISA Interchange; https://blog.isa.org/avoid-common-tuning-mistakes-pid.

[9] Advantages and Disadvantages of Expert System; Free Learning Platform; Java point Blog; https://www.javatpoint.com/advantages-and-disadvantages-of-expert-system.

[10] F.A. Chung, T. Lee, Fuzzy competitive learning, Neural Network Magazine 7 (3) (1994) 539–551.

[11] D.H.F. Liu, J. Abonyi, Software for Fuzzy Logic Control (Section 2.31), in: B.G. Liptak (Ed.), Instrument Engineers' Handbook: Process Control and Optimization, 3rd ed. 2, CRC Press, Boca Raton, 1995.

[12] S. Leleko, Y. Holoborodko, AI and Machine Learning in the Manufacturing Industry: Shaping the next Industrial Revolution, SPD Technology, 2024. https://spd.tech/machine-learning/ai-and-ml-in-manufacturing-industry/#:~:text=What%20Is%20Machine%20Learning%20in,and%20ensure%20consistent%20quality%20control.

[13] Artificial Intelligence - Neural Networks; Tutorials Point; https://www.tutorialspoint.com/artificial_intelligence/artificial_intelligence_neural_networks.htm.

[14] O.I. Abiodun, A. Jantan, A.E. Omolara, K.V. Dada, N.A. Mohamed, H. Arshad, State-Of-The-Art in Artificial Neural Network Applications: A Survey, National Library of Medicine; Pub med Center (PMC), 2018. https://pmc.ncbi.nlm.nih.gov/articles/PMC6260436/.

[15] E. Grossi, M. Buscema, Introduction to artificial neural networks, Eur. J. Gastroenterol. Hepatol. 19 (2008) 1046–1054.

[16] Robotic tutorials; robotic process automation; Java point; Internet document. https://www.javatpoint.com/rpa.

[17] The guide to robotic process automation; e-book; laserfiche; internet document. https://www.laserfiche.com/ecmblog/what-is-robotic-process-automation-rpa/.

[18] C. Gogineni, Top 7 RPA Use Cases in Manufacturing Industry, Comptecinfo, 2019. https://www.comtecinfo.com/rpa/rpa-use-cases-in-manufacturing-industry/.

## Further reading

[1] Principles of Expert Systems, P.J.F. Lucas, L.C. van der Gaag, Centre for Mathematics and Computer Science, Addison-Wesley, Amsterdam, 1991. Internet document, https://www.cs.ru.nl/P.Lucas/proe.pdf.

Chapter 2

# Adaptive gain, autotuning, and model-based control

An adaptive control (AC) system automatically adjusts its parameter to compensate for variations in system dynamics so that the overall system performance index (PI) remains unaltered/an optimum level is assured/maintained. So, for its operation, it takes into various issues like *Identification of system dynamics*, *Decision*, and *Modification*. Self-tuning controller (STC) was originally conceived in 1958 by Kalman and later introduced in Europe in1970s. STC is one of the techniques of adaptive control. This adaptive controller is developed to overcome the shortcomings of the nonadaptive controllers. As per control theory, STC comprising *expectations*, *measurement*, *analysis*, and *actions* can modify its own internal tuning parameters to optimize the control and achieve the objectives. In this chapter, discussions have been presented separately for AC & STC. The discussions also include various modeling for Artificial intelligence (AI) with details for better understanding.

## 2.0 Adaptive gain: functional and design issues

On account of plant variations and nonlinear actuators or sensors, there will be nonlinearities in plants/processes which act as the environment for a control system. In contrast to dealing of conventional controls of dealing with priori information, adaptive controls deal with posteriori information. Therefore, adaptive controls deal with finding parameter adjustment algorithms that offer global stability and convergence guarantees especially in complex systems exhibiting even nonlinear time-varying dynamics. Modern controllers mostly include autotuning/self-tuning function, enabling operator to automatically tune a controller to a process even with no knowledge of that process. A few key advantages for this include thw following: *No need for process knowledge*, *automatic processes detection for use*, *execution of tuning under feedback control*, and *shorter tuning time than loop response time*.

Linear parameter-varying (LPV) systems are *time-varying* plants whose state space matrices are *fixed* functions of some vector of varying parameters $\theta$ $(t)$. These can be used for LPV systems and can be interpreted as they can be viewed as *linear time invariant* (LTI) with *time-varying known parameters* $\theta(t) \in (\theta_1\ \theta_2)$, also can be model for LTI and model resulting from linearization of the nonlinear plants along trajectories of the parameter $\theta$.

Model free adaptive (MFA) controllers are also general-purpose controllers developed in numbers to control a variety of problematic industrial loops viz. antidelay MFA for processes with large time delays; multiple input multiple output (MIMO). MFAs are also developed for multivariable processes; feedforward MFA to deal with large measurable disturbances; and robust MFA to force the process variable to stay within defined bounds.

Here, discussions have been presented on functional details and detailed discussions on adaptive control (AC) and gain scheduling (GS), including discussions on MFA.

### 2.0.0 Functional details: adaptive gain control (AC) and gain scheduling (GS)

We know that three major steps for AC are as follows: identification, decision, and modification. So, once the difficult process of system identification is over, decision function operates to activate the modification function for altering parameters to maintain optimum performance which is evaluated mainly by model comparison and performance criteria. In the model comparison, a model having resemblance to the desired system characteristics is selected [1]. In it, all effects of the system characters and effects of disturbances are known. The characteristics of the control system variable parameters are "slaved" to the response characteristics of the reference model for system adaptation. Gain scheduling is a practical and powerful method for the control of nonlinear systems. A gain-scheduled controller is formed by interpolating between a set of linear controllers derived for a corresponding set of plant linearization associated with several operating points. The resulting controller is a linear system whose parameters (gains) are scheduled & adjusted) as a function of the exogenous scheduling variables [2]. As discussed, many of the processes have nonlinearity in the system causing

Plant Intelligent Automation and Digital Transformation. https://doi.org/10.1016/B978-0-323-89892-8.00013-2

problems for traditional Proportional integral derivative (PID) controllers. Majority of process variables and control valves, to some degree, are nonlinear. Such nonlinearities are mostly inherent to the process; because of changes in product specifications, production rates, parameter nature/measurement principle as well as for operating conditions. As a result, loops may get detuned to provide stable control under all process conditions. In all such cases, AC continuously adjusts PID tuning for optimal control to accommodate changes in process conditions. In cases of advanced intelligent ACs, best tuning parameters under same operating region are memorized for next-time applications. Hence achieving desired objective is done very quickly. Here, functional parts of AC & GS shall be discussed.

1. **AC: functions & application:** For understanding of AC, functional details, classification, and applications are important. Let the discussions on AC start with functional details on AC.
   - **Functional details:** From the above discussions, we understand that AC mainly comprises functions like
     - *Identification:* This calls for determination of current performance whose current value is measured with the help of feedback data.
     - *Decision:* Based on identification, the decision procedure is carried out by preprogrammed logic for the selection of suitable control mechanism along with its adjustment to improve process performance.
     - *Modification:* This is the implementation stage for decision. This is concerned with physical changes in the control system and modifying system parameters and variables to drive the process closer to the optimal desired state.
   - **Applications:** In certain special cases ACs can be used based on characteristic curves for the steady-states or to extremum value control, optimizing the steady state. So, there are several ways to apply adaptive control algorithms. A few application issues are as follows:
     - Adaptive *control* of linear controllers for *nonlinear or time-varying* processes

   *Adaptive control* (/self-tuning control) of nonlinear controllers for nonlinear processes or for control for Multivariable processes (MIMO).
2. **Discussions on AC & GS:** In Section I 1.2.8.6 gain scheduling has been discussed initially. Here, application of the same will be studied in *open-loop adaptive gain* and multiple-model AC. In linear process, characteristics do not change significantly with time or load conditions. However, in nonlinear processes, it can change with time/load. So, setting fixed parameters can become problematic. This means that the controller has to change parameter to ensure stability in the system at different process gains. Generally, in adaptive controller mathematical model is created for the process to estimate the optimum controller settings for the process. This model is continually updated utilizing the data achieved form the comparison of the desired performance of the model and actual operating conditions, as depicted in Fig III 2.0a. Adaptive control can also be combined with gain scheduling to deliver benefits for processes which are nonlinear/which vary with time or other conditions. Fig III 2.0a 1 &2 shows adaptive gain control and gain scheduling, respectively. Let us now look into the advantages of AC as well as pros & cons of GS.
   - **Advantages of AC:** Some advantages of AC are as follows:
     - AC helps control of processes with changing dynamics with fast responses. They provide better control for nonlinear process without process disruption.
     - Fully integrated with autotuner to automatically provide initial process knowledge.
     - Creates and updates an internal model of the process it is trying to control
     - Filters the control output and the process variable signals so that unwanted information from noise and local disturbances is removed.
     - Signals are analyzed in a least-squares estimator, with the resulting information used to update the process model
     - Works out the PID settings that would work best with the internal model
   - **Advantages of GS:** Some advantages of GS are as follows:
     - Improves control of nonlinear processes
     - Eliminates the need to have prior estimation as the values of controller parameters are programmed in advance. With gain scheduling, the controller parameters are changed quickly in response to the changes in process dynamics
     - Is based on a measurable variable that correlates well with the process dynamics
     - Three independent sets of PID parameters can be defined
     - The reference signal can be any analog signal
     - Two dedicated limits can be programmed in engineering units for ease of use
     - Capable of responding to rapid changes in operating conditions
   - **Limitations of GS**: A few limitations of GS are as follows:
     - It is time consuming, as the controller parameters are determined for all possible operating conditions and need validation through simulations.

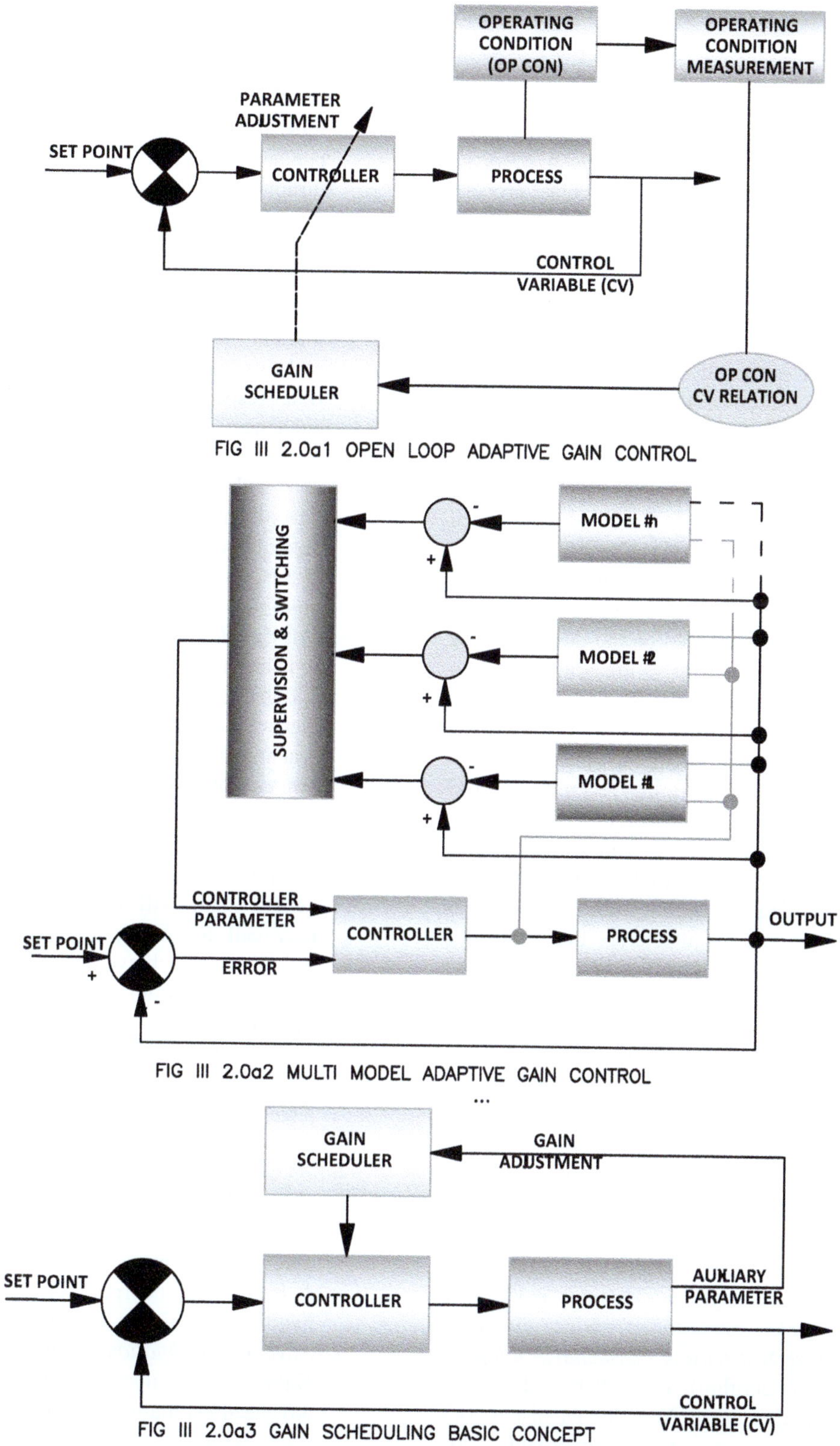


**FIGURE III 2.0A** Adaptive gain controls and gain scheduling.

- No feedback compensation for an incorrect adaptation, making it comparable to feed-forward or open-loop compensation.

3. **Process adaptation:** We understand that Adaptive control is a new and enhanced model-based approach enabling embedded process learning, which identifies

process models from normal day-to-day operation. These process models are used to calculate optimal tuning, and remember the best tuning for different regions of operation, that is, fast adaptation during changing conditions, providing users with familiar tuning rules for defined desired performance. Advanced adaptive controllers use learning algorithms to calculate process models and diagnostics for every control loop in the system. Process models are assessed based on process changes both manually by operator or automatically viz. batch control sequences. For a change in the setpoint, it would capture the process input and output data to identify the process dynamics and best process model suitable. These process models are stored in a database for evaluation & analyze for process nonlinearities without configuration change/downloads for the process learning.

4. **Adaptive tuning discussions:** In predictable advanced ACs, adaptive tuning can be made available for all PID blocks with embedded process learning enabled. For each new model is identified; adaptive tuning provides optimal tuning parameters based on the desired tuning criteria. Before use, they can be validated using identification quality metrics that take into account factors such as signal/noise ratio, model error, convergence, and variability. These adaptive tuning calculations can be used for both advisory tuning recommendations and for closed-loop adaptive control. For improvement in reliability, robustness in the controller, features like diagnostics from smart devices to detect fault, *calculation of process models and tuning* from *known trigger events* only, etc. Some manufacturers like Emeson's Delta V follow this type and offer different versions of adaptive control types like *full adaptive mode*, partial adaptive mode, and learning mode only.
5. **Multivariable AC:** Multivariable AC normally refers to a control system with MIMO, where process parameters or dynamics may vary over time or be uncertain. Adaptive control systems in one sense are "self-tuning," that is., it can adjust own control tuning parameters to meet the desired performance target even if there are changes in system behavior/environment. Like other ACs MIMO type ACs also have components like *process model*, *reference/desired model* (to compare), *parameter estimator/scheduler*, *operating condition measurement/sensing*, *associated adaptive law* (set of rules/algorithm), *controller*, and *control actuator* for output actions. Now let us discussions challenges and solutions for multivariable AC implementations.
   - **Challenges:** Number of challenges for multivariable adaptive controls are as follows:
     - *Cross coupling*: In multivariable system, one input may impact a number of outputs, or interact other input (viz. interaction among feed flow, steam flow and air flow in boiler). This gives rise to interdependence and cross-coupling.
     - *Modeling & estimation*: On account of nonlinearities, time variations of parameters, and interdependence, modeling of multivariable systems is difficult. As a result, real-time estimation of multiple parameters becomes complex.
     - *Complex computation:* These control systems generally demand significant computational resources, for real-time operations, parameter estimation, and feedback calculations. Therefore, real-time implementation with required response time (especially for fast dynamics) is a real challenge.
     - *Performance convergence*: Because of multivariable with associated interactions cause delay in adjustments the system parameters there will be delay in convergence to the desired performance, that is, quick meeting desired performance is not possible.
     - *Noise scalability:* Adaptive controls generally are sensitive to noise, and this is especially in multivariable systems because of interdependencies. Also, scalability of multivariable adaptive control systems for large systems is real challenge because of cross-coupling, computational load, and sensitivity to modeling inaccuracies.
     - *Design issue:* For interactions, there will be complexities in tuning of controller parameters; hence, there will challenges for design. This may call for multiple iterations and advanced techniques to achieve desired performance under various operating conditions and maintain stability.
     - *Stability:* It is quite difficult to ensure stability of the controls, as stability with respect to one variable may cause instability in the other part.
   - **Remedy for challenges:** A few suggestions for solution are as follows:
     - *Sub problems*: Main complex multivariable AC problems may be divided into smaller manageable subproblems.
     - *Robustness:* Incorporation of robustness to handle uncertainties and complexities.
     - *Advanced estimation*: Recursive least square or Kalman filter algorithm may be deployed for advanced estimation purposes.

### 2.0.1 Discussions on adaptive controls and gain scheduling

Adaptive control is an advanced control systems for better performance and accuracy. Model reference adaptive control (MRAC) is a *direct adaptive* strategy with some adjustable controller parameters and an adjusting mechanism to adjust them. Adaptive controllers are very effective to handle nonlinearity and time-varying system parameter, that is, the unknown parameter variations and environmental changes. As discussed, an adaptive controller consists of two loops, an outer loop or normal feedback loop and an inner loop or parameter adjustment loop. So, adaptive controls consist of *identification* of system dynamics, *decision,* and *modification.* There are two types of direct and indirect adaptive controls which mainly refers to self-tuning control. Control-system design is greatly influenced by the number of nonlinearities present within the process. Pattern recognition is used for data analysis in AI and Machine learning (ML) as well as in control system. On account of seasonal differences, building load changes or even variations in the performance of some Heating ventilation and air conditioning (HVAC) devices would call for a re-tuning process of the Proportional & integral (PI) parameters to achieve optimal control. In gain-scheduled controller gains are automatically adjusted as a function of time, operating condition, or plant parameters. Gain scheduling is a common strategy for controlling systems whose dynamics change with time or operating condition. In this part short discussions is presented on adaptive control and gain scheduling.

1. **Classification of adaptive control:** There could be several ways adaptive controls could be classified viz. *Feedback*: There are several broad categories of feedback adaptive control (classification can vary):
    - **Dual adaptive controllers**—based on dual control theory: *optimal dual controllers* –and *suboptimal dual controllers.*
    - **Nondual adaptive controllers**: *adaptive pole placement, extremum-seeking controllers, iterative learning control, gain scheduling*, and Model reference adaptive controllers (MRACs) *with a reference model defining desired closed-loop performance.*
    - **Gradient optimization** MRACs—using adjustment local rule e,g, "MIT rule." *Stability-optimized MRACs*: Model identification adaptive controllers (MIACs)—for identification during running condition
    - **Cautious adaptive controllers**
    - **Certainty equivalent adaptive controllers**
    - **Nonparametric adaptive controllers**
    - **Parametric adaptive controllers**
    - **Explicit parameter adaptive controllers**
    - **Implicit parameter adaptive controllers**
    - **Feed forward:** Similarly, ACs can be classified as
    - **Direct:** In direct methods, the *estimated parameters* are *directly* used in AC.
    - **Indirect:** In indirect methods the *estimated parameters* are used *to calculate* required controller parameters.
    - **Hybrid:** Hybrid method is a blending of direct MRAC and indirect AC (recursive least square).
2. **Pattern recognition adaptive control (PRAC) discussion:** PRAC approach (termed "EXACT) is an Artificial intelligence (AI) controller which offers superior adaptation through direct performance feedback. PRAC can re-adjust the proportional gain (/band) and the integral time of the digital PI controller while under closed-loop control.
    Fig. III 2.0b shows PRAC in HVAC application where the Variable air volume (VAV) box damper to air-condition the room while the prefilter measures the room temperature and balances the signal. The whole system consisting of the actuator (the damper position), and the environment viz. the room, which is subject to change from time to time due to various reasons. As shown in Fig III 2.0b1, the PRAC monitoring the controlled variable, (room temperature), the setpoint, and the controller output, for example, the damper position angle. To combat noise and obtaining a smoother rate of signal change numerical method may be deployed. In PRAC, determination of quantum of disturbance for the setpoint or the process is necessary to determine the tuning band accordingly. The flow diagram for the control has been detailed in Fig III 2.0b2 in terms of various blocks. The PRAC is also smart enough to avoid any unnecessary tuning under nontypical situations, especially when the controller is saturated, when the process responds erratically or when the controller is manually overridden, because further tuning in these situations would result in no better or even worse performance. The decision to halt any unreasonable updating of the controller settings is handled by block 4. Finally, the PRAC determines new gains which are represented by the proportional band and the integral time based on the characteristics learned in block 3 [3].
3. **MARC explained:** As explained in III 1.1.0.4 &5, MRAC strategizes to design and develop the adaptive controller to work on the principle of adjusting the controller parameters so that it controls the plants in such a way, that plant process output tracks the output of a reference model with the same reference input. It is in fact a class of direct type with no explicit estimate or identification of the plant. So, there is no room for checking the model quality. Therefore, MRAC can be developed starting with a time-varying matrix of gains $k(t)$ as $J(\theta) = 0.5e2\ K(t)$ [ref III 1.1.0.5 & Eq. I 1.2d]. This methodology applies to several approaches in which the MIT rule discussed in I 1.2.8.4 is classic one. For discussions on MIT rule, I 1.2.8.4 may be

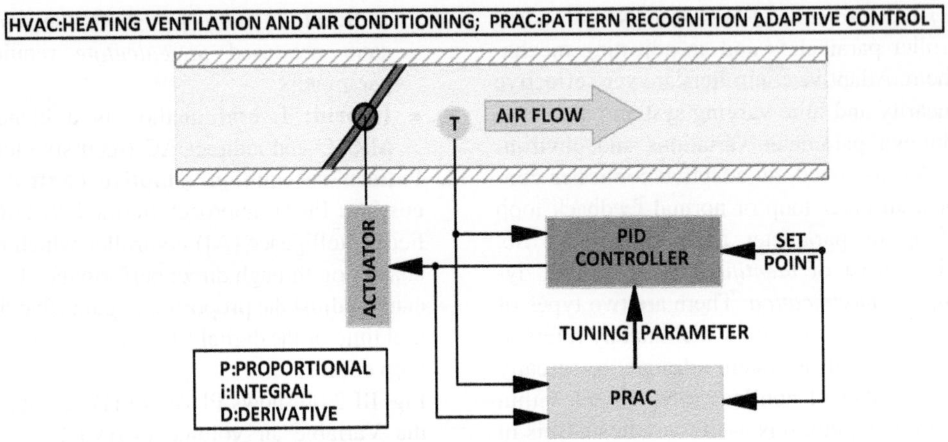


FIG III 2.0b1 PRAC BLOCK DIAGRAM IN HVAC [3] [COURTESY: JOHNSON CONTROLS].

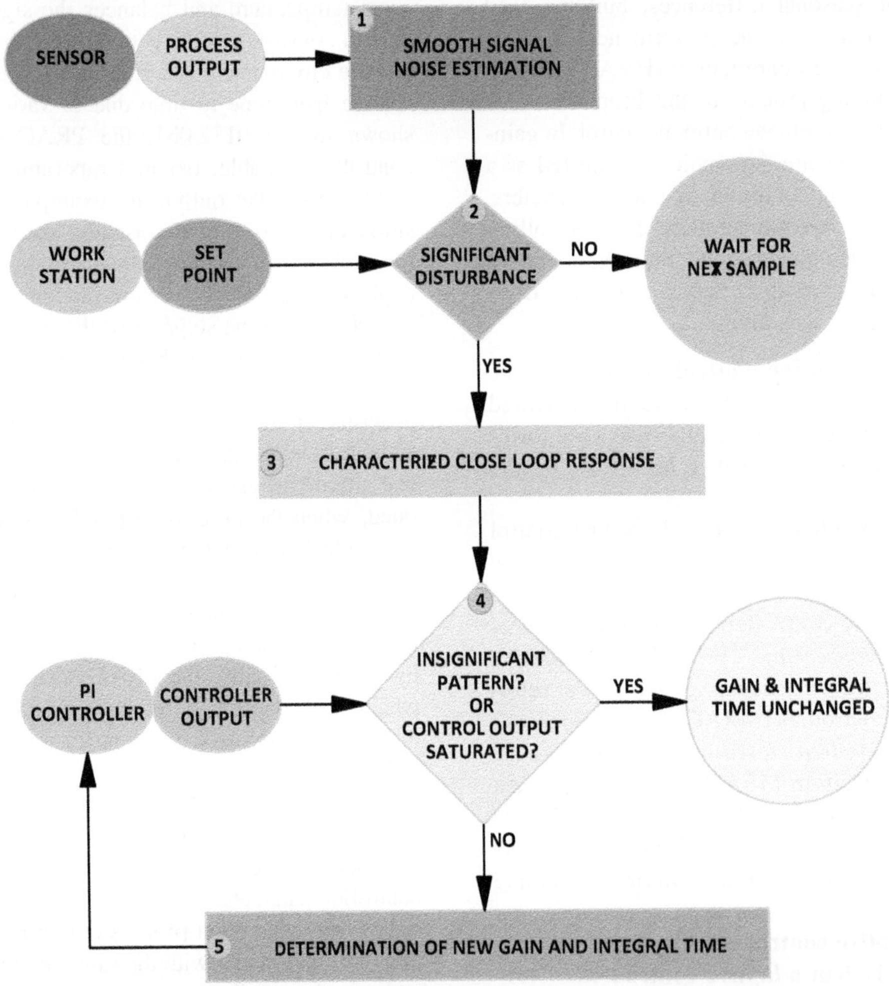


FIG III 2.0b2 PRAC FLOW CHART [3] [COURTESY: JOHNSON CONTROLS].

**FIGURE III 2.0B** PRAC application in HVAC.

referenced. In the industrial world, there were reports of instability based on the basic MIT rule. So, there will be modification of MIT rule [ref: Eq. I 1.2j] with necessary adjustments to modify the following equation.

$$\frac{d\theta}{dt} = \gamma \frac{\delta e(t)}{\delta \theta} e(t) \qquad \text{Eq. III 2.0a}$$

4. **Gain scheduling types:** Gain scheduling (GS) can be

classified in different ways. In this connection, I 1.0.0.6 may be referenced. Various methods are as follows:

- **Main approaches:** There are several ways of classifications
  - *Classical GS:* A family of LTI approximates a nonlinear plant at constant operating points parameterized by constant values of convenient plant variables/exogenous signals is computed using linearization-based scheduling. Then, LTI controllers of the previously derived local LTI models are designed for the specified performance and stability at each operating point. The family of LTI controllers are implemented in such a way that the controller coefficients are scheduled according to the current value of the scheduling variable. As last step; local performance assessment may be performed analytically, but overall/global performance and robustness are established through simulation. Here, other syntheses to derive a parameter-dependent model are as follows: *off-equilibrium* or *velocity-based linearization*, *quasi-LPV approach,* and *direct LPV modeling.* This enjoys main advantage that it inherits the benefits of linear controller design including classical design tools and time as well as frequency domain performance specifications. However, global robustness, performance and especially stability are not guaranteed.
  - *LPV & Linear fractional transformation (LFT) Synthesis:* A family of LTI approximations as discussed above. LPV models, in specific cases an LFT description (which can be the basis for subsequent LFT controller synthesis).is possible. LPV and LFT control synthesis directly develop a gain-scheduled controller with stability and performance specifications guaranteed a priori as the time-varying parameter instead of its corresponding frozen-value.
  - *Fuzzy gain scheduling:* This overcomes disadvantages of classical type and involves techniques from classical gain scheduling and LPV techniques. For fuzzy model, weighting/scaling functions are designed, corresponding to the above-mentioned operating regions. Utilizing weighting functions, the local models are blended similar to set of local models with utilization of same weighting functions. This blending yields scheduling of the controller outputs rather than scheduling of the controllers. Stability and performance are established by extensive simulations.

There are other classification methods including modern GS techniques. Some of other types are as follows:

- **Decomposition:** GS methods *decomposing nonlinear* design problems into *linear subproblems* and GS methods decomposing *nonlinear design problems* into nonlinear (affine) subproblems.
- **Signal processing:** continuous *GS Discrete GS* and *hybrid or switched GS* methods.

5. **Gain scheduling uses:** As explained in I 1.2.8.6, GS is a practical method for control of nonlinear systems. Gain scheduling is attractive because it exploits linear control design methods that are theoretically mature, are well understood, and involve computationally efficient synthesis procedures [4]. Since, it is based on linear approximation, it easier to ensure stability as well as performance properties. Again, as gains are adjusted as per operating condition its response would be rapid for changing operating condition. However, gain scheduling values are accepted after suitable simulations as discussed above.
   - **Open-loop adaptive gain control:** In open-loop adaptive control technique main presumption is that there exists a definite relationship between some of the measurable variables which characterizes the environment or operating condition and the plant parameters. Let us take an example of for a boiler air fuel ratio control, where steam temperature could be used as a parameter which holds direct relation combustion control of the boiler. Also, there exists a relationship of plant operating condition load (Air/steam flow) and superheater temperature in utility boilers. Using this relationship, it is then possible to reduce (or to eliminate) the effect of parameter variations upon the performance of the system by changing the parameters of the controller accordingly. Open-loop adaptive control system is a type of "gain-scheduling" scheme as shown in Fig III 2.0a2. Here, the adaptation mechanism is based on a look-up table stored in DCS. For a given set of environment measurements, this look-up table provides the suitable controller parameters. Since there is no measurement and feedback in decision making this is an open-loop adaptive control system as gain changes with operating condition. This system can fail if for some reason or another the rigid relationship between the environment measurements and plant model parameters changes (in our case reheater temperature can also influence SH temperature control relationship).
   - **Multiple-model adaptive control:** For large and rapid variations of the plant model parameters, typical indirect adaptive control schemes discussed

in I 1.2.8 often produce unsatisfactory results. For better results and to improve the adaptation transients, often multiple-model adaptive controls are used. The main philosophy behind this approach is to select the best model of the plant in real time, from previously known and tested models and apply the output of the associated controller to the plant. To improve performance on-line adaptation of an adjustable model is used through switching as shown in Fig III 2.0a3.

### 2.0.2 Model-free adaptive control discussions

Classic model-based adaptive control methods can be applied only when a convenient model of the plant or environment is provided. Model-free adaptive controllers (MFAC) is that MFAC does not require a precise model of the system. Mostly, the advanced control methods are developed based on good understanding of the process and its environment. When due to complexity, the physical process is not well understood and the quantitative knowledge of the process is not available. This is referred to as a "black box" problem. When the quantitative knowledge of the process is available, we call it a "white box" problem. However, when one is not sure if the process knowledge is accurate or not, it is called a "gray box" problem. Although MFA control can actually deal with black, gray, and white box problems, it is more suitable to deal with the *gray box* problem. In traditional adaptive controls, if the quantitative knowledge of the process is not available, an on-line/off-line identifier is required to obtain the process dynamics. However, these may not well suit process controls, hence.

MFA controls are applied. As the name suggests, it does not require process models. MFA controllers are available in various platform and can be embedded easily in equipment. Also, they find applications in Programmable logic controller, Distributed control systems (DCS) as well as Process automation control (PAC). MFA controls can address various specific control problems like

- Single input single output (SISO) MFA can replace PID without manual tuning. Also, MIMO MFA handles multivariable processes
- Nonlinear, antidelay, and feedforward MFA can deal with nonlinear, large time delay processes and disturbances, respectively.
- With robust MFA process variable can be restricted within boundary.
  1. **MFA Types:** Block diagram for both SISO & MIMO controls have been depicted in Fig III 2.0c and explained below. After the configuration is over, MFA can start control the process immediately without need for process identification and any dynamic model. Also, it is easy to switch control between manual and automatic mode.

- **SISO MFA:** As shown in Fig III 2.0c1, a single-loop MFA control includes a SISO, process, MFA controller in a feedback loop. The control objective is for the controller to produce an output *u* (*t*) to force *y*(*t*) the Process variable (PV) to track the given setpoint *r*(*t*) to minimize the error *e*(*t*) through regulation of MFA controller, under variations of setpoint, disturbance, and process dynamics. In SISO MFA controller only, a few parameters need to be configure. These are as follows: *sample interval* (based on min sampling law—it could as low as be @ 1 mS), *controller gain* (default gain may be used to start with and acting type (direct/reverse).
- **MIMO MFA:** This system comprises MIMO process, a set of controllers, and a set of algebraic adders. MIMO system has controller setpoints $\mathbf{r_n(t)}$, error signals $\mathbf{e_n(t)}$, controller outputs $\mathbf{u_n(t)}$, process variables $\mathbf{y_m(t)}$, and disturbance signals $\mathbf{d_n(t)}$. Here. Suffix n,& m stands for number of input output. For simplicity and balanced in representation two input and outputs have been considered in Fig III 2.0c2 with parameters detailed in Table III 2.0a.

  Here, *e*1 or *e*2, is combined with the output of the controller (+compensator), with the other input to produce control signals *u*1 and *u*2. In real applications, the outputs from the subprocesses may not measurable directly, and only their combined signals *y*1 and *y*2 can be measured. A MIMO MFA control system should follow these pairing rules [5]:
  - Each process of the main loops has to be controllable, open-loop stable, and either reverse or direct acting.
  - A faster process, or a process with a large static gain should be included in the main loop. While a slower process, processes with time delays and a process with a small static gain should be treated as a subprocess. Any conflict in this, a tradeoff is the option.
- **Feedforward MFA:** Control philosophy behind feedforward control, is to take advantage of *measurable* disturbance/influencing signals. In feedforward control, this measurable disturbing signal can be used to reduce its effect before the feedback loop can take corrective action. Air flow/steam flow as feedforward signal in combustion control loop or in steam temperature control loop in boiler control is glaring example. This this measurable disturbing signal can be

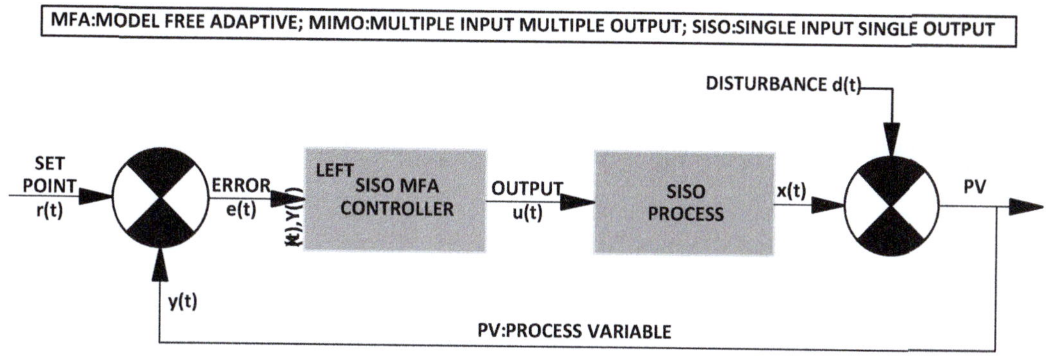


FIG III 2.0c1 SISO TYPE MFA CONTROLLER

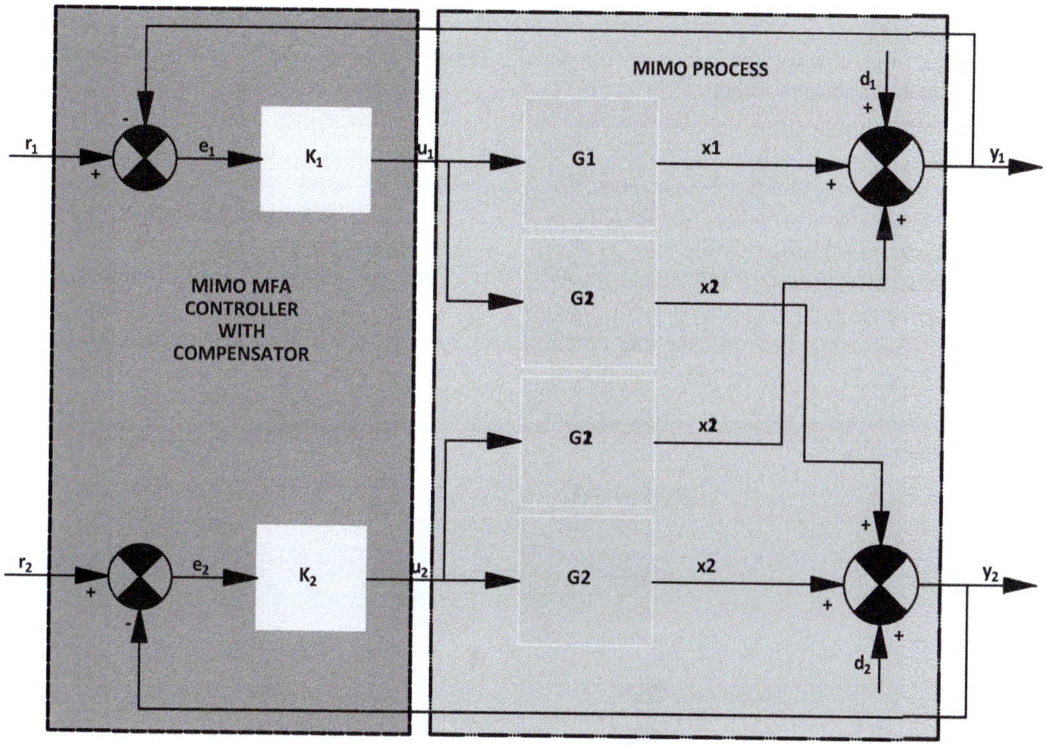


FIG III 2.0c2 MIMO TYPE MFA CONTROLLER

**FIGURE III 2.0C** SISO and MIMO type MFA controller.

used directly (with feedback controller through a summator) or it can be used as ratio or derivative. The feedforward MFA is a general-purpose feedforward controller without making any attempt for total cancellation of the disturbances (very difficult to implement). Disturbance feedforward signal can be added with the output of MFA controller to get better response.

**TABLE III 2.0A** Parameters in MIMO MFA.

| Symbol | Description | Remarks |
|---|---|---|
| $y_1$, $y_2$ | Process variables (PV) | Process output |
| $r_1$, $r_2$ | Set points (real one may be different) | For process variables |
| $e_1$, $e_2$ | Error (for each controller) | |
| $u_1$, $u_2$ | Controller with decoupling compensator | Process input |

2. **MFA discussions:** MFA are general purpose controller requiring no specific design procedure. MFA control technology enables a major impact on the automatic control industry by successfully solving many industry-wide control problems in various applications. This brought significant economic benefits. As discussed, the control objective is to produce an output for forcing the PV to track the setpoint under variations of setpoint, disturbances, and process dynamics. Major issues are as follows:
   - **MFA properties:** MFA control system has the following properties:
     - No precise quantitative knowledge of the process is available.
     - No process identification mechanism or identifier is included in the system.
     - No controller design for a specific process is needed.
     - No need for manual tuning of controller.
     - To guarantee system stability, stability analysis and criteria for are available
   - **MFA requirements:** Feedback MFA control system demands for the following behavior: The process is *open-loop stable*, *controllable*, and *either direct or reverse*. For nonopen-loop stability, it must be stabilized first. When the process is not controllable, improvement of the process structure or its variable pairing is required [5].
   - **MFA architecture:** SISO MFA controller can be conceived as a Multilayer perceptron (MLP) of artificial neural network (ANN) used in the design of the controller, for this III 1.4.0.3 may be referred to. As discussed, in ANN a group of weighting factors are updated as needed to vary the behavior of the dynamic block. The algorithm for updating the weighting factors is based on the goal of minimizing the error.
3. **Nonlinear MFA:** In control theory, nonlinear control is very challenging because a nonlinear process with so many variations, is difficult to control and prescribe any single controller configuration to deal with so many variations. In traditional approach, piecewise linearization is used first to linearize the process, then to apply controller. As described, nonlinear MFA controller is a general-purpose controller ready to offer relatively uniform solution for nonlinear process operations comprising nonlinear sensors, actuators. Some additional important parameters for nonlinear MFA controllers are as follows: *sample interval*, *time constant*, *controller gain*, *direct/reverse*,

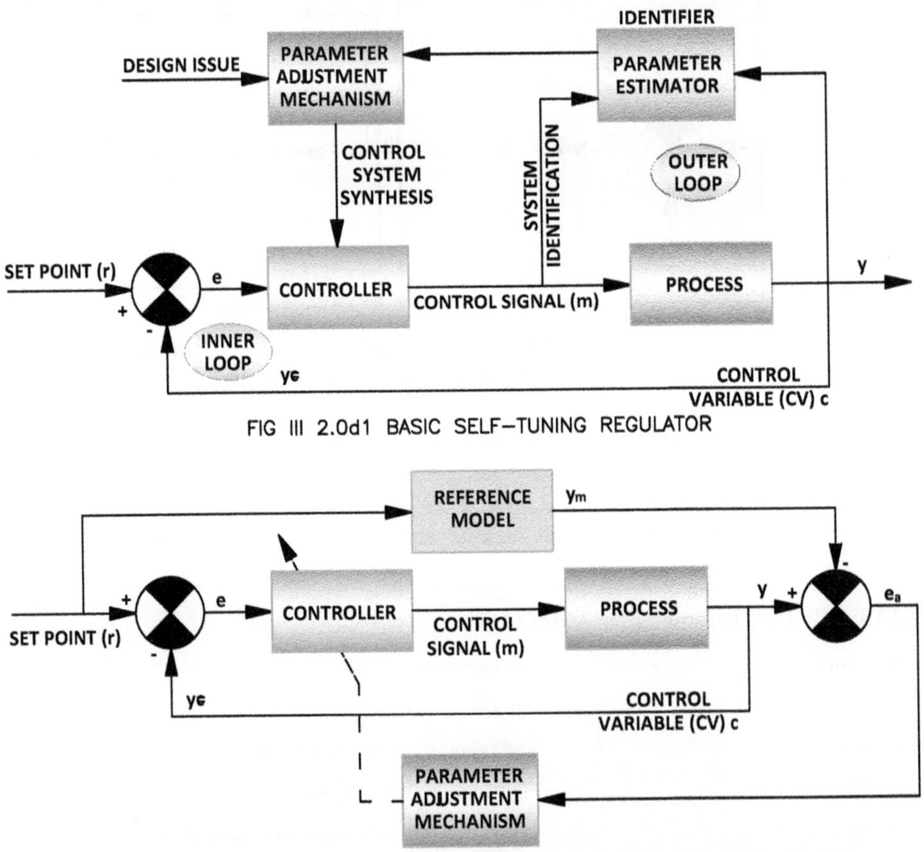


FIG III 2.0d1 BASIC SELF–TUNING REGULATOR

FIG III 2.0d2 REFERENCE MODEL BASED SELF–TUNING CONTROL SYSTEM

**FIGURE III 2.0D** Self Tuning control system details.

and process *nonlinearity factor*. This nonlinearity factor value lies between 0 and 10, for linear to extremely nonlinear process, respectively. Also, these nonlinearities could have upper or lower curvature about linear line. These both side curvatures indicate that MFA does not care about the nonlinear characteristics are for this process. A Linear matrix inequality (LMI) which can be expressed as linear mathematical expression, is quite popular in control systems especially for MFA controllers for design development, robustness, stability analysis, and improved performance.

4. **MFA: Robust design challenges: C**omplex robust MFA design challenges are as follows:
   - For large system dynamics change prompt control action is required to meet the desired control performance.
   - Dominant system disturbance (s) may be feasible to measure, and offer feedforward control for design advantage.
   - For large disturbance/change in process dynamics, the control may be lost.
   - The dynamic behavior or load change cannot provide trigger to switch operating modes.
5. **Comparison: MFA & other controls:** Table III 2.0b provides the comparison of other technologies with MFA to show why it has gain so much interests for control designs. Various control systems used for comparisons are as follows: PID; Model adaptive controller (MAC), Robust controller (RC), Model-based predictive controller (MBPC) & MFA.

## 2.1 Autotuning and self tuning controller (STC)

There were existence of Lyapunov stability theory, MRAC and self-tuning concepts were developed in 1960s. Early 1970s the self-tuning regulators and robust adaptive controllers came into existence in Europe through K. Åström and B. Wittenmark. However, these were in conceptual stage. Controls of continuous-time adaptive systems and nonlinear systems were seen in 1980s when practical MRAC have been developed. Then, microprocessor systems programmable systems were developed much in 1990–2000. Modern Self-tuning controller (STC)s use AI and machine learning. Autotuning controller (ATC) or STCs are designed and developed for choosing own tuning parameters based on analysis of the behavior of the process controlled and process dynamics. These automatic procedures include mathematical model of the process with input/output relationship derived from process data as well as operational augmented by experience by operator. Both ATC and STC use same principles for calculating and setting own tuning parameters. There exists difference between ATC & STC. STC automatically and continuously updates the its tuning parameters (by observing the normal control response waveform to calculate and sets tuning parameter), when STC is online. On the contrary, ATC in offline carries out its tuning operation once with manual intervention. In this section, brief discussions will be covered for self and autotuning controllers in modern industrial plants.

### 2.1.0 Discussions on self-tuning controller

From previous discussions, we understand that except single operation & manual intervention, ATC & STC are similar. Many modern controllers include both. STC includes a traditional PID function and a self-tuning function required to maintain optimal closed-loop performance by continuously updating the controller's tuning parameters. Self-tuning controllers' operation without need for any form of perturbations is preferred in process plants. Normally, these controllers do not introduce any upsets for evaluation of the controller's response to set-point changes

**TABLE III 2.0B** Comparison of MFA with other control technologies (X: Available).

| Control issues | MFA | MAC | MBPC | RC | PID |
|---|---|---|---|---|---|
| **General control** | X | | | | X |
| **Adaptivity capability** | X | X | | | |
| **No process model** | X | | | | X |
| **No identification** | X | | X | X | X |
| **No controller design** | X | | | | X |
| **No controller manual tuning** | X | X | X | X | |
| **Available stability criteria** | X | *May be* | *May be* | X | X |
| **Easy use and maintain** | X | | | | |
| **Controller of future** | X | | | | |

or to other, naturally occurring load variations and other upsets as they take place. In these cases, self-tuning algorithm is inactive, till an error reaches a given predetermined value (usually> 1%). After the error reaches/just crosses the predetermined value, self-tuning subroutine is activated. After error is evolved, the self-tuning subroutine checks the response of the controller in terms of the controlled variable's period of oscillation, damping, and overshoot [6]. Maintaining a constant process variable deprives the tuning function of any useful insight into the behavior of the process, whereas exercising the process to see how it will react to a control effort defeats the purpose of the control function [7].

There is a large gap between simulation/academics and practical algorithms implementation of STC/ATC. In the ideal environment, it is easy to perform well for varieties adaptive algorithms. In practice, it is quite difficult handle nonlinearities, unmodelled dynamics, and unmodelled disturbances over a wide range of operating conditions by the adaptive or self-tuning controllers. Therefore, actual implementation is not easy. Starting with working principle of STC, In this subsection various aspects of STC shall be discussed. In this connection, Fig III 2.0d may be referenced.

1. **Self-tuning: working principle: self-tuning controller working principle:** As shown in Fig III 2.0d1, STC actually consists of *two* loops. *Inne*r loop consisting of *process* and a *conventional feedback controller* whose control parameters are varied by outer control loop. *Outer* control loop consists of a *plant model parameters* or system identifier/recursive parameter estimator and an *adjustment mechanism* for the controller parameters. So essential parts are as follows:
   - **System identifier/recursive parameter estimator:** This element consists of a process parameter estimation algorithm to estimate the parameters of the process, that is, identification of the system concerned.
   - **Parameter adjustment mechanism:** This unit/element is responsible for calculation of new controller parameters based on the estimated process parameters estimated by the estimator mentioned above.
   - **Controller block or feedback controller:** This is the feedback controller whose parameters are updated at periodic intervals through recursive estimator via adjustment mechanism to take care of plant dynamics.
2. **Model-based tuning:** Reference model-based STC has been depicted in Fig III 2.0d2. Here, it is seen that there is a conventional controller to control main control variable at desired set point. Naturally, on account of disturbance, there will be variations in performance of the loop. After simulations, the plant reference model for is selected to obtain the desired performance from the same set point (fed to the reference model). This performance $y_m$ is compared with the plant output y to observe amount of variation in the process due to disturbance. Based on the difference between the two performances, controller parameters are modified with the help of parameter adjustment mechanism, so that controller can readjust itself to face the challenges of disturbances. In self-tuning controller design, linear, quadratic, minimum variance, gain phase margin design, pole assignment (pole placement), and pole shifting, etc., are used. For plant/system identification/estimation recursive least squares and Kalman filtering, etc., are applied [8]. Self-tuning controllers are identified by their estimators and adjustment mechanism. Popular varieties include the minimum variance, generalized minimum-variance, detuned minimum variance, dead-beat, and generalized pole-placement controllers [6]. Two types of model-based STC are as follows:
   - **Optimal self-tuning:** They use optimal regulation theory for design rule in the controller parameter calculations. This involves minimum variance method and general predictive control.
   - **Transient response self-tuning:** They use the transient response characteristics of the loop as the basis for operating the self-tuning algorithm with the use of pole-placement/pole-assignment technique.

   In mathematical modeling approach for ATC/STC complex, numerical curve fitting is used for computing the parameters of a process model fitting best. A curve-fitting helps in continuously updates its process model according to the latest input/output data should be able to identify the process gain Also, often Smith predictors are implemented for addressing deadtime/transport lag, etc.
3. **Starting and performance:** The start STC in any control loop may two different things viz. STC introduction for a particular process or it could mean how to start implementing STC in a particular control loop initially. The performance of STC needs to be evaluated on different types of processes and compared with conventional PIDs. Start up and performance issues have been discussed in sequence. Basic steps in both the issues required steps have been detailed out in Fig III 2.1a.
   - **STC introduction:** This involves starting STC for any control system. Naturally, this involves integrating adaptive/self-tuning mechanisms in the system for safe operation of the system with effective improved performance. As shown in Fig III 2.1a1 major steps involved are as follows: *Loop assessment, STC mechanism, Testing & introduce control, monitoring & process tuning* and *steady-state operation.* In this also, adaptation should be validated.

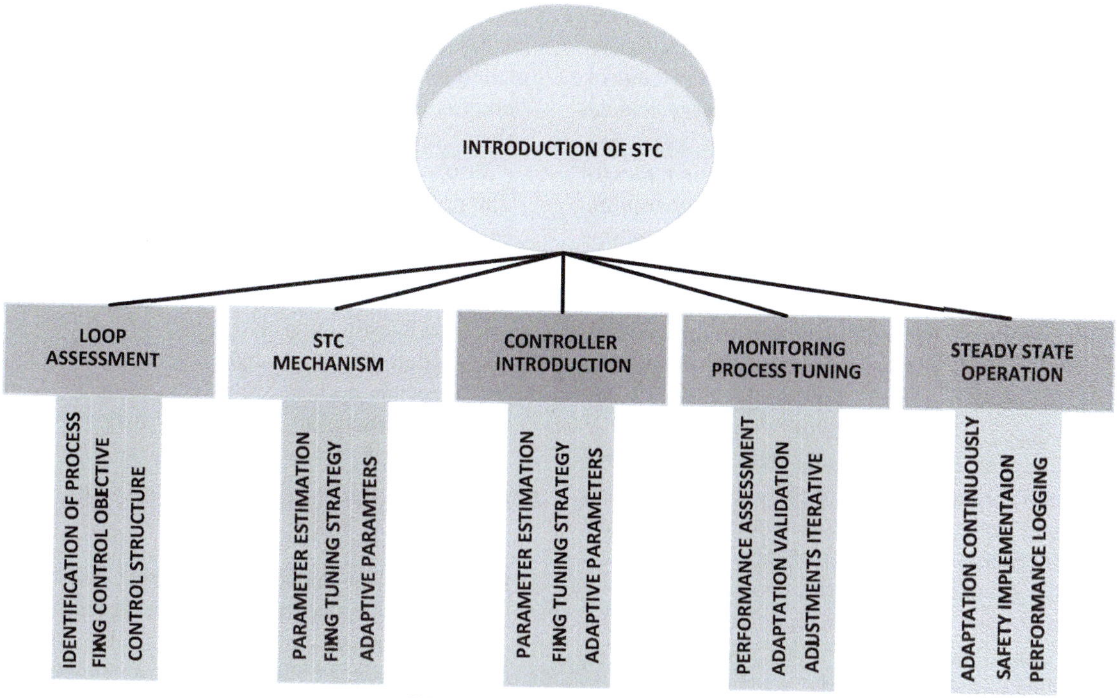


FIG III 2.1a1 STC INTRODUCTION

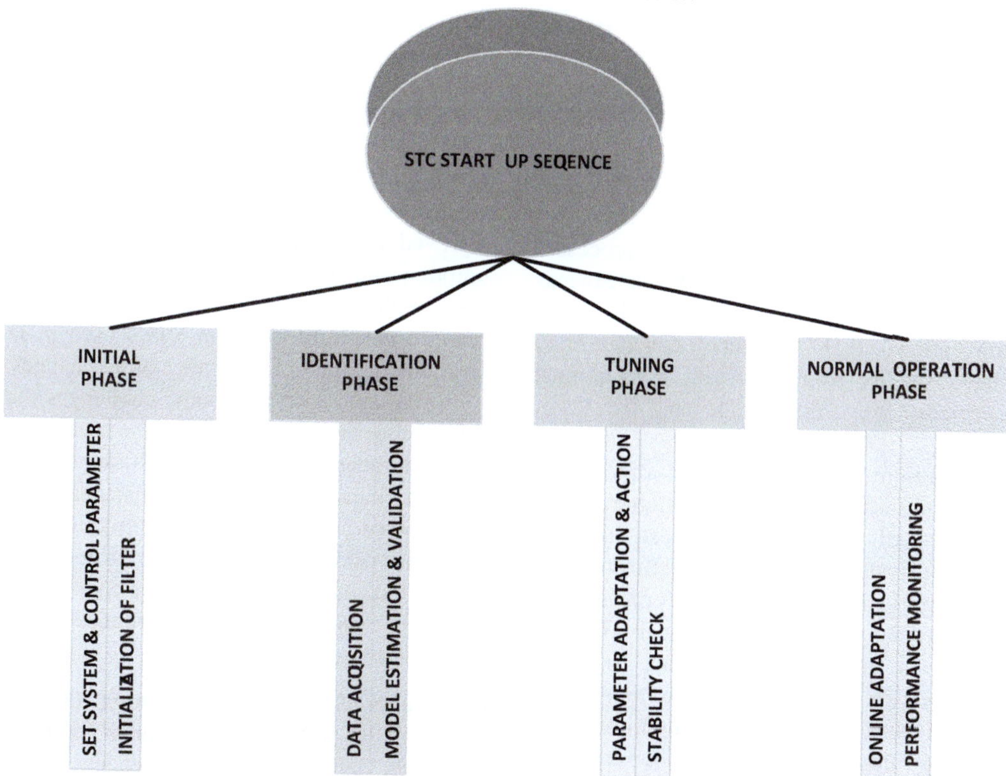


FIG III 2.1a2 STC START UP SEQUENCE

**FIGURE III 2.1A** STC introduction and STC start up sequence.

- **Staring sequence:** The starting sequence in an STC deals with implementing STC from initialization to assurance for accurate performance of the control under different environments. So, it is very critical for STC. As shown in III 2.1a2 major steps and phases involved are as follows: *initialization, identification, tuning,* and *transient to steady operation.* Evaluation of period of oscillation (T), the damping ratio, and the amount of overshoot are important to determine proper controller tuning and transient state response. The controller under normal load, is first manually brought and an adjustable-size step change in its output signal to the control value is introduced. This will cause the controlled variable (*c*) to draw an S-shaped steady-state reaction curve normally seen and discussed in Fig I 1.01. Then, again and from there the controller output is brought back to original value by step change back to collect data like the *steady-state gain, dead time*, and *process sensitivity* in response to load decrease [6]. Also, noise band is identified. After start up controller is over STC needs reevaluate its performance after each upset.
- **Performance:** The performance of STC is compared with conventional PID for different processes. Tuning at worst condition, conventional PID's performance is evaluated on the basis of absolute error. When compared with STC, one will find that STCs provide improved performance for processes with low gain little dead time. However, such improvement starts decreasing for processes with higher gain or dead time. If dead time exceeds system time constant this performance may be worse than conventional PID. One may conclude that STCs have smaller overshoot than conventional PID, especially for process with demand for slower responses.

4. **Pattern recognition:** For dealing with disturbance rejection and robustness STC based on pattern recognition, is widely used in industrial applications. This helps in data analysis and parameter estimation. In this method, perturbation of a specified form is introduced into the system viz. step/pulse signal. The process response is analyzed to extract transient performance data for the process (rise time, dead time, peak-overshoot, etc.). In the algorithm, the parameter tuning rules of the PID controller are innovated utilizing pattern recognition method of analysis. Since in pattern recognition method data analysis is used hence no model of the process is necessary. In this type, tuning constants are adjusted as per some tuning rules, and these are restricted to PID type controllers.
5. **Noise and disturbances:** Disturbance could be very annoying, and can make a simple, step response a challenging one, if a disturbance happens to intrude on the process variable while applying a step signal. An experience operator can make out changes, similarly, STC with observational capacity can meet the similar goal. The problem is acute when the process variable is subject to measurement noise and STC may not be in a position to distinguish between phantom noise and real disturbances as a result there may be corruption in calculation/estimation associated with STCs by introducing deformation in reaction curve. A precision of STC depends on how well it can calculate & estimate the process model to fit the noisy data.

### 2.1.1 STC: benefits challenges and operational issues

Self-tuning control can adapt to different operating modes, nonlinearities, and uncertainties in the process. In spite of the fact that majority of the industrial automation has accepted and adopted PID algorithm as the de facto standard for closed-loop process control, there is no standard means for tuning a PID loop to achieve the goal with optimal performance. Dynamic behavior of the controlled process, the performance objectives, and the operator's understanding of how tuning works are a few issues on which the art of tuning of PID depends. Controller tuning allows for optimization of a process and minimizes the error between the variable of the process and its set point. Improved consist product quality, reduced energy consumption and wastage, enhanced reliability safety, along with simplified operation and maintenance are major advantages generally found in STCs used in industrial applications. Despite of a number of advantages, self-tuning control systems also face some challenges that limit their applicability and effectiveness. The trade-off between the speed and accuracy of the adaptation process is the most common challenge of STC. As discussed in Chapter 1 in Vol I, for very fast adaptation there will be instability and may cause oscillations. Similarly, for too slow adaptation, the loop may not catch up with the changes in the process. So, there could be some compromise in STC between performance and adaptability.

1. **Benefits of STC and autotuning:** One of the obvious benefits of STC is that it reduces the need for manual tuning, calibration, or maintenance. With STC it is easier bring about improvement in system performance and robustness. This is because of the fact that STC can adapt the variations in the system dynamics, disturbances, and uncertainties. Also, STCs increase the efficiency and reliability of the system. As discussed, nonlinearities, time delays, and interactions to the

complex systems difficult to model can well be tackled much better by STC than conventional PIDs. All these enable the system to operate in a wider range of conditions and achieve more complex tasks. Some of the common benefits are as follows:
- No requirement of detailed knowledge about the process
- Automatic detection of the processes without need for use of derivative
- Tuning process done with strict feedback control
- Short tuning time compared with loop response time.

2. **STC challenges:** Major challenges faced by STC include but not limited to:
   - **Computational complexity**: It becomes a real challenge to Implement/update the models and algorithms for large and computationally complex processes.
   - **Speed-accuracy trade off**: There is a trade-off between speed & accuracy of action
   - **Adaptability & performance**: Modification of tuning parameters/strategies frequently or very quickly/too drastically may cause system instability or poor response.
   - **Measurement noise**: At times STCs cannot distinguish between real disturbances and phantom noise.
   - **Stability**: Under all noise disturbances and feedback conditions, STC stability cannot be ensured
3. **Design discussions:** STC design involves a number of steps to ascertain that the design system properly identifies the process with associated characteristics & dynamics for accurately adjusting the control parameters in real-time to achieve the performance objective. Based on the identified model parameters, the tuning process of PID parameters, needs to address both set-point tracking and disturbance regulation characteristics. As shown in Fig III 2.1b, there are five basic steps each with associated design issues to be addressed. Implementation of STC mainly refers to adaptation of algorithm, testing, keeping safety in protected. Common algorithms used are as follows: *Recursive least squares (RLS), Kalman filter type, MRAC* and *GS.*

   Models are meant to behave like real plants for which the controllers are designed. In this connection, essential issues to be addressed are.
   - Identifiable controlled object so that it could be modeled by analysis of measurable input and output variables.
   - Determination of parameters through identification for developing parametric models (a function of independent variables).
   - A criterion for comparing the differences between various types of models in a class. For the criterion major issues are as follows: it is *easily computable, sensitive to random effects & errors* and measurable (for detecting the behavioral differences in model and object).

   For a model, a function "*f*" describing the plant output $y(t)$ as a function of input variables is chosen as shown in Eq. III 2.1a.

$$y(t) = f[u(t), v(t), t] + n(t) \quad \text{(III 2.1a)}$$

   Here, controller output = $u(t)$, disturbance variable = $v(t)$ and $n(t)$ is the term describing stochastic influences.
4. **Nonlinearity and tuning tool:** Linear control system responds predictably with scalable relation between output and input. In contrast to this nonlinear system is not only unpredictable but there may not be any scalable relation between output and input. Major reasons for various nonlinearities in controllers and in control loops include the following:
   - Control system nonlinearity
   - The dynamic behavior of the control loop affected by disturbances.
   - Nonlinear characteristics of final control element
   - Nonlinearities in the actuator (dead zone, hysteresis, and saturation).
   - Measurement noise of various sensors.

   To bring about optimized operation of control loop, the right tool is extremely helpful to ensure control system and process are running at peak performance. "EnTech Toolkit" of Emerson is an example. Based on understanding of process dynamics, the analysis and recommendations provided by the toolkit, and effective tuning strategies can be framed. Normally, the tool comprises collection and tuner parts to carry out flexible, data collection, necessary for tuner operation (analysis part). Analysis/tuner part, With the help of database of control loop dynamics and controller algorithms, assesses open-loop step tests recommends tuning parameters to help meet process objectives. Additional benefits listed below are also available [9]:
   - Improved performance for difficult control loops.
   - Easy data collection and model identifications.
   - Optimum tuning and associated performance improvement.
5. **Estimation and target:** Fig III 2.1c, shows functional block diagram and control response for estimating process characteristics completing.

   Intelligent STCs carefully observe the variations and fluctuation in PV and Manipulated variable (MV) for estimating process characteristics from resultant waveform in response to set point variations. When STC finds that the PV is exceeding preset level for various

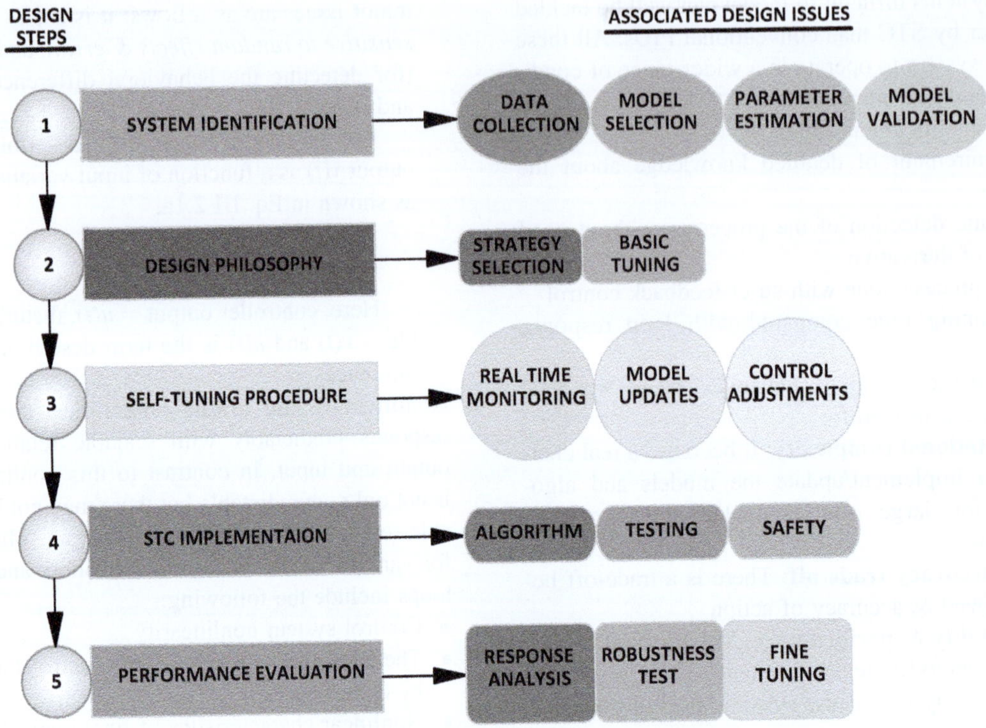


**FIGURE III 2.1B** STC design step details.

process abnormalities, it starts re-estimating process characteristics such as dead time, time constant, etc., from the response of both MV and PV by relating them. It is evident that PV will be affected by MV, noise, and disturbances. Every time for estimating and computing process characteristics, STC keeps observing the responses. Appropriate re-parametrization of the control loop is often useful in recursively to estimate the controller parameters in implicit manner, also, a synthesis from estimates of the process model can be used as explicit manner for same purpose. Estimation process is detailed out in functional diagram shown in Fig III 2.1c1. A target is set for STC, and actual control response waveform different from the target is shown in Fig III 2.1c2. Various important terms detailed out in Table III 2.1a, are used for evaluating performance criteria in controller tuning especially for STC tuning. These important terms are as follows: Integral of absolute error (*IAE*), Integral square of error (*ISE*), and Integral of time-weighted absolute error (*ITAE*). All these terms deal with error e(t). The target for STC refers to the desired performance goals. Major issues in these targets include the following: *accuracy*, *response speed*, *minimum error*, *stability,* and *robustness*. Target are set mainly to address issues like set point tracking, disturbance rejection, overall control, overshoot control, fast response, stability, energy efficiency, and robustness to ensure permanent adaptive adjustment, model optimization, and robust design of control.

### 2.1.2 STC types, predictive approaches and applications

Different variants under several names, predictive control is one of successful methodologies drawing attention and interest over recent decades. In a systematic way, predictive controls can handle on-line constraints. Also, the predictive control strategy eliminates drawbacks of the other optimal methods. There are various types of control systems in use. Each of them with its advantages and disadvantages has been studied and implemented in industrial applications. There have been hybrid control system consisting of more than one type of control technique also. In STC applications often the gain scheduling principle is replaced by using a neural network-based model or Fuzzy controls with capability to capture the predictable dynamics variations of the process. So, a few types of STC with special reference to hybrid type and predictive type self-tuning control shall be discussed. Also, application area in them shall be covered.

In 1975 Ebrahim Mamdani introduced Mamdani model where the output of each rule is a membership function in a Fuzzy set, is a Fuzzy interference method to create a control system by synthesizing a set of linguistic control rules obtained from experienced operators. After the

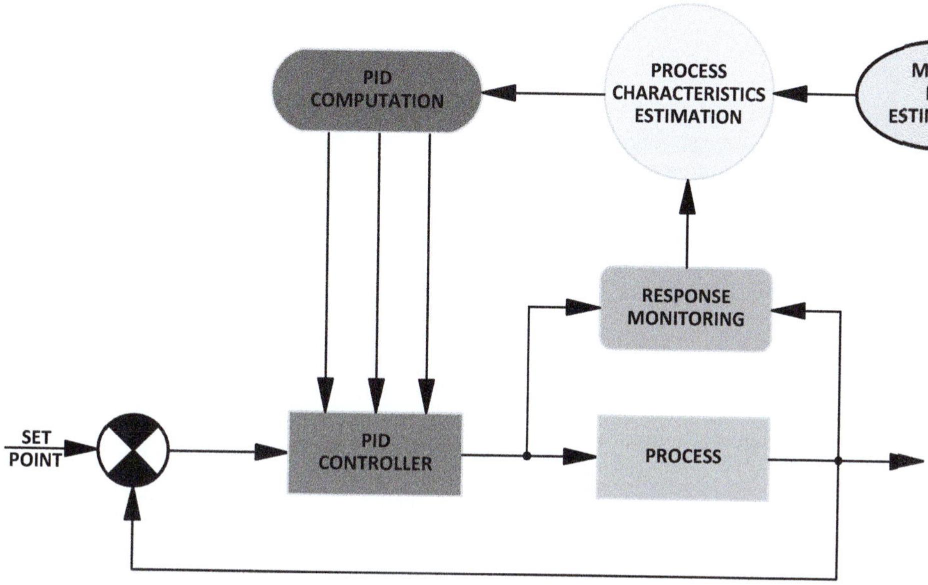


FIG III 2.1c1 STC TUNING FUNCTIONAL BLOCK DIAGRAM

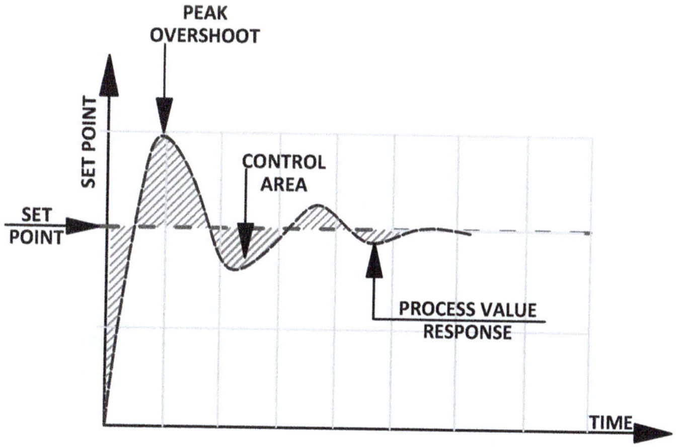


FIG III 2.1c2 STC TUNING CONTROL REPONSE DIAGRAM

**FIGURE III 2.1C** STC tuning block and control response diagram.

aggregation process, the fuzzy set for each output variable that needs defuzzification. These systems have more intuitive and easier to understand rule bases and are used for STC and expert systems.

1. **STC: fuzzy design:** STC—Fuzzy design refers to STC with Fuzzy tuner, to tune the three parameters $K_P$, $K_I$, and $K_D$ of PID controller in self-tuning mode. The tuning parameters of the conventional PID controller are often not property tuned for the nonlinear plant/plants with unpredictable parameter variations. It is a good approach for them. Mamdani model as detailed in Fig III 2.1d, is applied. As shown in Fig III 2.1e, error (e) and derivation of error Δe are two inputs to Fuzzy tuner (interface) and there are three outputs for PID controller tuning parameters viz. $K'_P$, $K'_I$, and $K'_D$ for *proportional*, *integral*, and *derivative*, respectively.

Mamdani model applied, is with some modification to obtain the optimum value for $K_P$, $K_I$, and $K_D$. PID controller tuning parameter ranges between min and max, that is, $[K_{Pmin}, K_{Pmax}]$, $[K_{Imin}, K_{Imax}]$, and $[K_{Dmin}, K_{Dmax}]$ for *proportional*, *integral* and *derivative,* respectively. Hence $K'_P$, $K'_I$, and $K'_D$ can be calculated from

$$K'P = \frac{K_P - K_{P\,\min}}{K_{P\,\max} - K_{P\,\min}},\ K'I = \frac{K_I - K_{I\,\min}}{K_{I\,\max} - K_{I\,\min}},\ K'D = \frac{K_D - K_{D\,\min}}{K_{D\,\max} - K_{D\,\min}}$$

Eq III 2.1b

With reference to III 1.3.0 and Fig III 1.3a, the membership functions of these inputs fuzzy sets could be obtained. The linguistic variable levels have been

**TABLE III 2.1A** Explanation of important terms for Tuning (STC).

| Issue | IAE | ISE | ITAE |
|---|---|---|---|
| **Expression e(t) is the error** | $IAE = \int_0^\infty \lvert e(t) \rvert dt$ | $ISE = \int_0^\infty e^2(t) dt$ | $ITAE = \int_0^\infty t.\lvert e(t) \rvert dt$ |
| **Technical issue addressed** | Total error measurement for smooth performance. Slower response | Impact more for large error, used to minimize error with little oscillation with faster response; | Weighing later error more than start, priority for long term error reduction and faster making 0 error |
| **Purpose** | For smooth stable operation | Reducing peak error | Quick stability & long-term performance |
| **STC tuning** | Balanced performance without excess oscillation | Minimizing large deviation under noise/disturbance | Reduction long term error & quick recovery |
| **Influenced parameter** | Medium overshoot (10%) and rise time | Large overshoot (15%) & fast rise time | Small overshoot (5%) & short settling time |

assigned as: Large negative (LN), Medium negative (MN), small/zero (Z), Medium positive (MP), and large positive big (LP). Based on these Fuzzy rules for $K_P$, $K_I$, and $K_D$ have been presented in Table III 2.1b.

2. **Hybrid system:** In this discussion, hybrid of Fuzzy with PID & hybrid of Fuzzy with Fuzzy STC have been presented. Hybrid controllers have been detailed in Fig III 2.1f.
   - **Fuzzy with PID:** Response of PID controllers to variations in the system parameters is sensitive when compared with fuzzy controllers which do not need call for precise information about the system variables for its operation. Fig III 2.1f1 shows a hybrid system of PID controller with Fuzzy controller. Switching between the PID controller and the fuzzy controller is carried out based on the error between the actual value and set point value. When modulus of velocity error is *higher* than a preset threshold $|e0|$ measured in a comparator, then the hybrid system would select the fuzzy controller, for its faster rise time and a smaller overshoot, to the system to correct the velocity with respect to the set point. Similarly, when the error is *lower* than a preset threshold $|e0|$, then PID controller will be switched in for better accuracy near setpoint. The modulus of error has been considered for error in both directions.
   - **Fuzzy with STC-Fuzzy:** A hybrid of fuzzy and fuzzy self-tuning PID controller, as shown in Fig III 2.1f2, was developed to combine the advantages of both fuzzy and STC PID controller together. Operational details are same as described above.
3. **Model predictive control (MPC) & STC:** From detailed discussions in I 1.3.3 we understand that MPCs use the power of dynamic matrix control to easily address process interaction and difficult dynamics. It is possible to implement multivariable model-based control strategies very easily. In MPC, because of use of process dynamic model, it is easier to predict outputs over a defined domain hence for addressing optimization problems in cost, energy, and wastage saving. Here, short discussions have been provided for benefits of MPC and its comparison with STC. Both these can be combined for advanced applications.
   - **Benefits:** Apart from advantages given in I.3.3.2, a few more advantages of MPC are as follows:
     - *Adaptability:* MPCs can predict and adapt to the dynamic changes, uncertainties. Also, reduce tuning needs with variations in operating conditions
     - *Improved consistent product quality:* MPCs can automatically take care process interactions and difficult process dynamics. So, with advanced control techniques, the variability in process can be reduced drastically.
     - *Operation at limits*: As MPCs can predict and reduce process variability, the process can be operated at limits to achieve increased throughput with optimal control performance without increased risk of violating operating constraints.

- *Quicker model development:* MPC with STC, it is easier to develop model easily through proper identification.
- **Feature comparison:** Like STCs, MPCs also address the applications include *linear and nonlinear systems, MIMO, constrained, hybrid* systems. The performance of data-driven MPC depends on the accuracy of the model requiring precise careful tuning. Therefore, MPC & STC are interrelated. Also, STC can be integrated to develop powerful data-driven MPC. Table III 2.1c enlists comparison of features STC & MPC.

4. **Integration of STC and MPC:** Integration of both STC and MPC open new avenue of controls which would result *adaptive parameter estimation* combined with *predictive optimization.* MIMO control is a good example for the application of this. MPC is one of the most effective control approaches for MIMO system for its capability to handle MIMO in a straightforward manner using future values of the reference signals. This is an example of adaptive predictive controller handling constraints for MIMO. For adaptation part, the controller would realize self–tuning control with recursive identification of the model of the process. The recursive least squares method with the directional forgetting is used in the identification part. Similarly, the performance of data-driven MPC depends on the accuracy of the model requiring precise careful tuning. As discussed before, MPCs can *handle constraints better*, so, when it is integrated with STC, them same shall be done on a continuously *updated model.* In the following ways, integration MPC would enhance STC operation:
   - **Model updating:** STC continuously updates its model with real-time parameter estimation techniques for example, recursive least squares. The updated model can then feed into an MPC for *optimization* of control actions.
   - **Precise prediction:** Beauty of MPC relies on an accurate forecast future output based on process model. When self-tuning/adaptation mechanism is incorporated, the model used for prediction would be constantly updated as per system dynamics to account for changes in the system. Obviously, this would enhance control performance under varying operating conditions due to changes in load as well as for aging equipment
   - **Cost-effectiveness:** MPC the cost function, highly depends on model parameters viz. gains or time constants, etc. Self-tuning mechanism with adaptive capability can effectively adjust these parameters, ensuring the true behavior of the system—hence make the system more cost-effective.
   - **Robustness:** STC continuously adjusts model parameters eliminating mismatch, and MPC tries to optimize actions to minimize disturbances: Therefore, combination of STC and MPC offer robust system to disturbances/errors.
5. **Application issues:** Here, discussions on STC are concluded with a few examples of applications of STC in various control applications for better understanding. Some of these example loops have been detailed in Fig III 2.1g, with all legends explained in Fig III 2.1g1 common for all the sketches.
   - **Interfering loop:** As shown in Fig III 2.1g 2, there are two separate control loops, one is maintaining flow in the line by throttling FCV as [er flow measured by FE & FT in conjunction with FIC. Line pressure due to flow is maintained by another controller PIC and control valve PCV operating based on PT. If STC is used then there will be interference and may give rise to oscillations. So, here STC in PIC is permitted to be stopped because flow is the primary reason for causing pressure change.
   - **Cascade control:** With reference to I 1.2.2 we understand that there are two loops in a cascade control where it is recommended to use STC in secondary loop for finding optimum control parameters keeping cascade loop open. General recommendation calls for PID on secondary loop with STC in primary loop mainly unless there is frequent changes in

The Mand ani model is a Fuzzy inferece system (FIS) developed in 1975 for used in control appl ications to traslate hu man experiece and expert kowledg e into liguis tic rules. This is a very useful tool for approximatig fun ctions based on IF-THEN rules.Man dani-type fuzzy controllers provide a natural fram ework to iclud e expert kowled ge in the form of linguistic rules with locl adp tability to define rules tha are fired only a some specific regios of the input space. Man dani systems ha e more itui tive and easie r to understd rule b ses hece more suited for human iput iterp retale rule b se.

FIGURE III 2.1D Mamdani Model in Fuzzy Control.

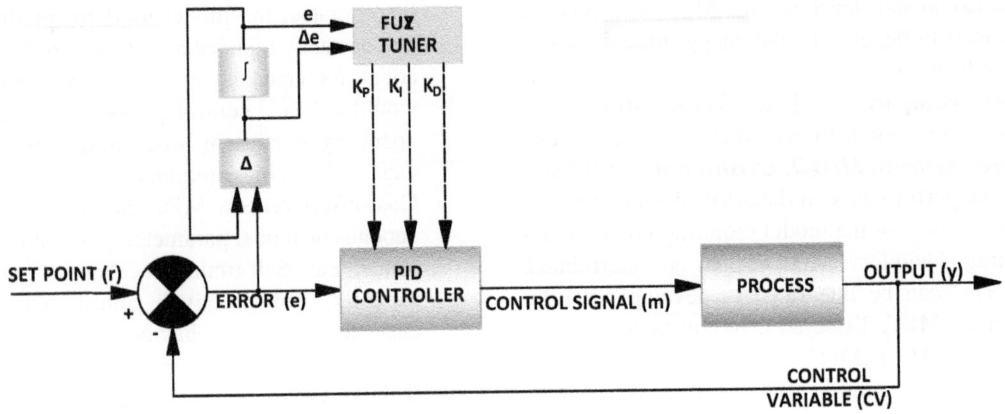


**FIGURE III 2.1E** STC with fuzzy tuner for PID.

process characteristics in the other loop. Common cascade loop in boiler superheater steam control has been shown in Fig III 2.1g 3.

- **Fluctuating gain control:** Fig. 2.1g4 shows the control loop with changing gain. STC is applied for to take care of large fluctuation in input water flow to the tank. Fluctuation of input flow for long time, there will be changes in level which is regulated by LIC through LCV. To maintain outflow, part of the flow is by passed though FCV, controlled by FIC. STC can follow the fluctuations. If change in flow–PV response changes due to fluctuation, STC can regulate the same.
- **Control loop with hysteresis:** Control loop with hysteresis as has been detailed in Fig. 2.1g5, is common in inverter/temperature control. It is a controller with nonlinear controller loop with hysteresis comparators; an adaptive band of the controller, has to be designed. Naturally, PID parameters fluctuates and STC is applied for this purpose.
- **Control with dead time**: Even though both the deadtime and time constant both describe aspects of the process' dynamic response affecting control loop, yet there are several fundamental differences. On account of the following processes with significant dead times are difficult to control: the *effect*

**TABLE III 2.1B** Fuzzy Rules for PID parameters.

| Fuzzy rule | Status | LN | MN | S (Z) | MP | LP |
|---|---|---|---|---|---|---|
| Applicable for $K_P$ | LN | LN | LN | LN | MN | S (Z) |
| Applicable for $K_P$ | MN | LN | MN | MN | MN | S (Z) |
| Applicable for $K_P$ | S (Z) | LN | MN | S (Z) | MP | LP |
| Applicable for $K_P$ | MP | S (Z) | MP | MP | MP | LP |
| Applicable for $K_P$ | LP | S (Z) | MP | LP | LP | LP |
| Applicable for $K_I$ | LN | LP | LP | LP | MN | LN |
| Applicable for $K_I$ | MN | LP | MP | MP | S (Z) | LN |
| Applicable for $K_I$ | S (Z) | MP | MP | S (Z) | MN | LN |
| Applicable for $K_I$ | MP | S (Z) | MP | MN | MN | LN |
| Applicable for $K_I$ | LP | S (Z) | MN | LN | LN | LN |
| Applicable for $K_D$ | LN | LN | LN | LN | MP | LP |
| Applicable for $K_D$ | MN | LN | MN | MN | S (Z)_ | LP |
| Applicable for $K_D$ | S (Z) | MN | MN | S (Z) | MP | LP |
| Applicable for $K_D$ | MP | S (Z) | MN | MP | MP | LP |
| Applicable for $K_D$ | LP | S (Z) | MP | LP | LP | LP |

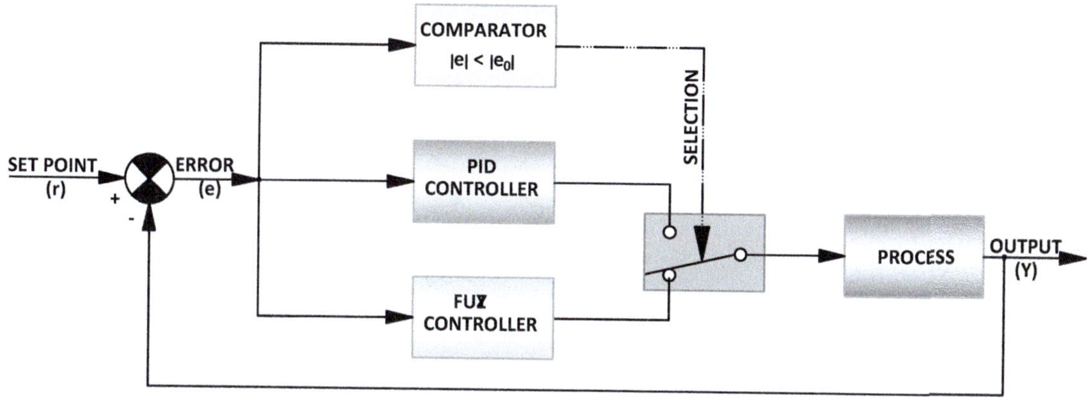


FIG III 2.1f1 HYBRID: PID WITH FUZZY CONTROLLER

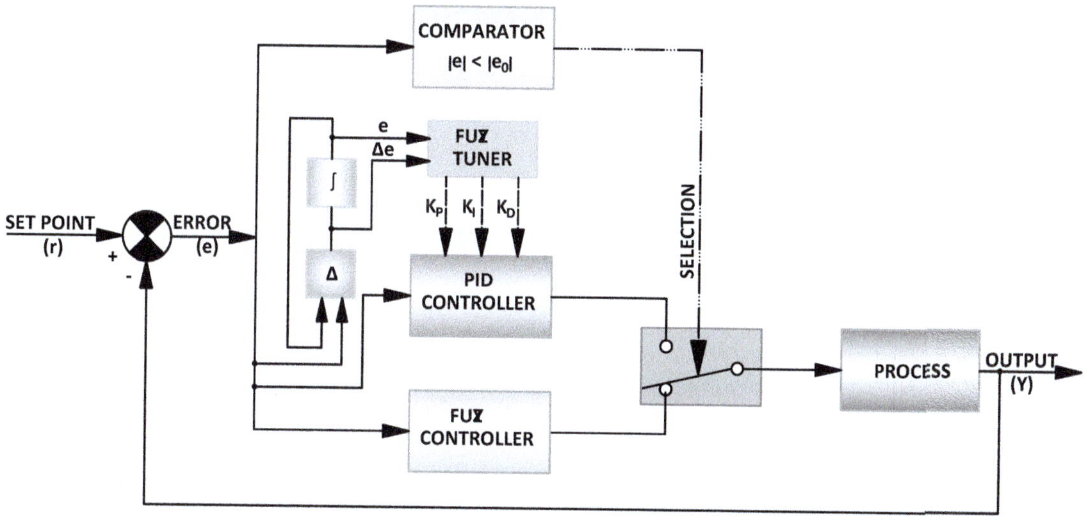


FIG III 2.1f2 HYBRID: FUZZY WITH STC –FUZZY CONTROLLER

**FIGURE III 2.1F** Hybrid STC systems.

*of the disturbance* are felt after a *considerable time lapsed*, *control action* is felt only *after a delay* and the control action applied is based on the actual error tries to correct a situation that originated some time before. When the dead time to time constant ratio exceeds certain value, it is difficult to control with conventional PID. Here, MPC offers good result and STC with MPC could be a better choice.

## 2.2 Mathematical model and control intelligence

Mathematical model generation from physical descriptions of systems is an issue of great importance. Mathematical modeling is the activity devoted to the study of the simulation of physical phenomena by computational processes with the aim to predict the behavior of some the system

**TABLE III 2.1C** Comparison of features of STC & MPC.

| Feature | STC | MPC |
|---|---|---|
| **Adaptability** | Continuous updating of model for new parameter | Fixed model, dynamic recalculation of inputs |
| **Computation** | Moderate complex computation | For optimization complex computation |
| **Optimization** | No future prediction | By future prediction |
| **Complexity** | Not complexity handling | Can handle complexity |

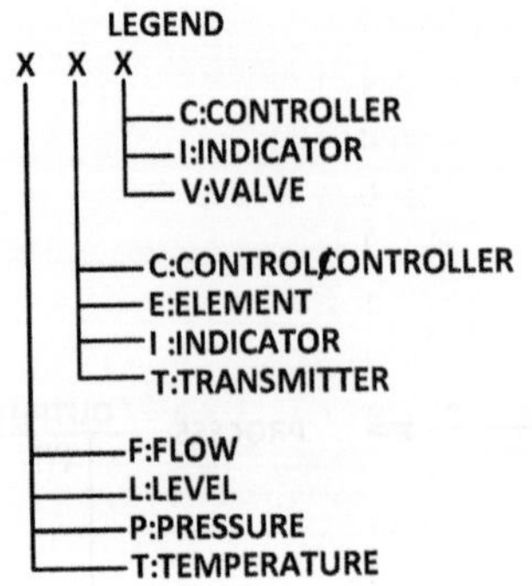


FIG III 2.1g1 LEGEND DETAILS

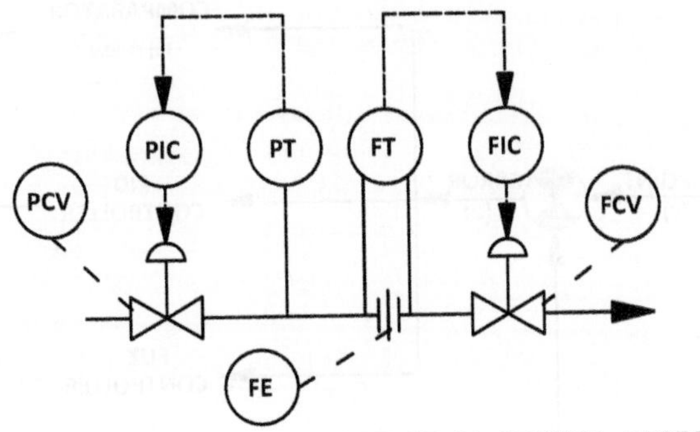


FIG III 2.1g2 INTERFERENCE IN CONTROL LOOPS

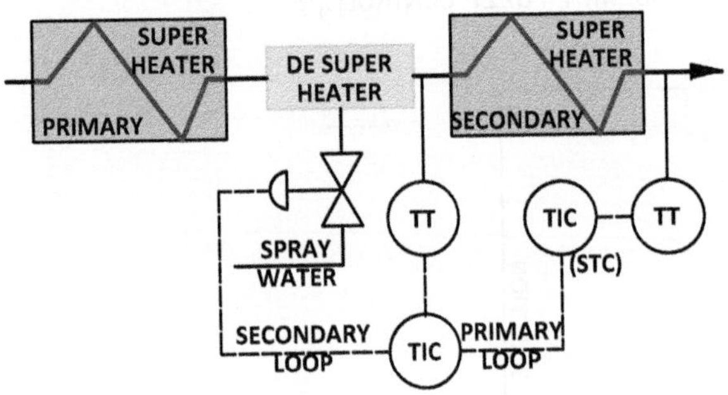


FIG III 2.1g3 CASCADE CONTROL LOOPS

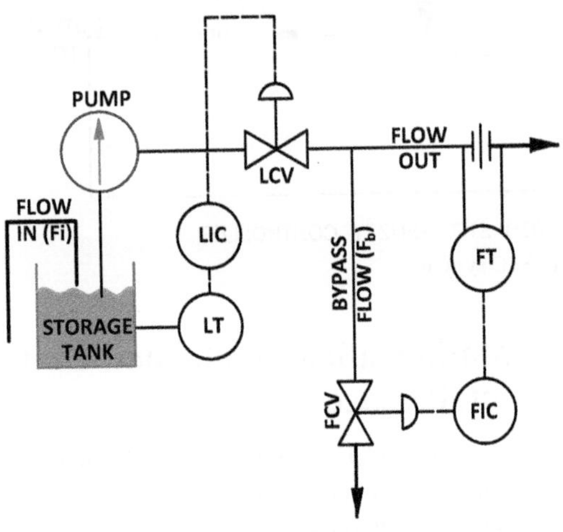


FIG III 2.1g4 CONTROL LOOP WITH GAIN CHANGE

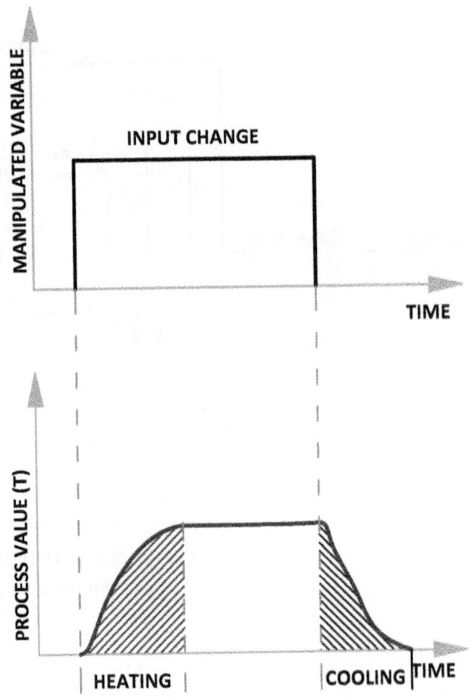


FIG III 2.1g5 CONTROL LOOPS WITH HYSTERESIS

**FIGURE III 2.1G** Some examples of STC systems.

within its environment. Modeling as a technique uses four instruments: *language*, *structure*, *argumentation*, and *presentation*. A modeling language (can be formal, with strict syntax and rules) is a common terminology for clearly communicating an abstract idea that the model captures viz. Systems Modeling Language (SysML) and Unified Modeling Language (UML), and specialized languages Architecture Analysis Design Language (AADL) [10]. A model must be a well-structured one especially for complex systems so that the model would be understandable, usable, and maintainable, etc. The aim of structured model is to ensure that the presented design satisfies the system's requirements. Easy comprehensibility visualization are important issues for model development. Model verification and validation cannot be overestimated. Model-based system engineering (MBSE) is not meant for any specific process, but must cover four systems-engineering domains: requirements/capabilities, behavior, architecture/structure, verification and validation. Four major components of a model are as follows: *Conceptual problem description* through operational logic. This gives rise to two aspects viz. *conceptual solution description* in terms of system logic & *actual problem description* through operational physical issues. Both these are aimed to provide *allocated solution description,* that is, system physical. AI can impact mathematical modeling with a host of ideas and techniques. Knowledge-based systems and environments can offer representations with associated problem-solving methods to be used to encode domain-specific knowledge and strategies. Advanced AI can provide high-level mechanisms for numerical models and solutions. For all these model interpretation and validation are very important issues.

### 2.2.0 Model and control intelligence: basic issues

Automatic control/automation has developed in a close connection to the industry and the evolution technology. The advanced networks and communications have significant impact on future control systems and advanced methodologies associated with control. The increase in complexity of processes, together with high performance and safety demands in functioning imposes new criteria for an integrated process and control design. So, model-based MBSE approach is now common in this area. MBSE is an engineering method to support the requirements, design, analysis, certification, and validation related to the development of complex systems keeping models at the center of system design. Cybersecurity is one of the main issues in modern complex system, so in MBSE, systems are secure by design in place of adding security feature later after the design. Capturing system attributes in models, it is possible to analyze threat-model at an early stage so that mitigation strategies can be incorporated in design to inherently reducing the system's overall security-related risks. Digital-modeling environment creates a standards-based approach with documentation which can be validated to remove inconsistencies. *Model, systems thinking,* and systems engineering are the pillars of MBSE. *Model* is a simplified form of representation of abstracts reality in graphical, mathematical, or physical representation form eliminating complexities of *language, structure,* etc. *Systems thinking* is the way of looking at a system as part of a larger system. *Systems engineering* is a multidisciplinary integrative approach to enable the successful realization, use, etc., using systems principles and concepts, and technological, and management methods. Based on these, in this section, discussions have been put forward on application of models in control intelligence as well as use of Artificial intelligence (AI) in modeling.

1. **Control intelligence model:** Mathematical models are very important tool in designing control system especially for designing control systems like Fuzzy control, ANN MPC based self-tuning controls, etc. Brief processes and components are as follows:
   - **System representation:** Mathematical models are used to represent the dynamics of the system in the question. For this, *state-space equations* for modeling & analysis, *differential equations* (continuous systems), and *difference equations* for discrete systems are deployed.
   - **Identification & analysis:** For mathematical modeling identification of the process/system is of prime importance. Mathematical approaches like *Kalman filtering, least-square* estimation, etc., are some of the common approaches. Similarly for control stability, Bode and Nyquist plots (frequency domain), Eigenvalue analysis (linear system) & Lyapunov analysis (nonlinear system) discussed before are deployed.
   - **Control strategy:** Mathematical models are deployed for strategizing control systems as well as for optimization of control system. Some common strategies are as follows: *PID* parameter tuning, predictive control through *MPC, adaptive control* to meet the challenges of dynamic changes and optimal controls for cost factor.
   - **Postdesign test:** After design, model/control algorithms should be tested based on simulation for validation of design.
   - **Implementation:** Any real-time control system calls for computation and numerical methods for mathematical model with control algorithms.
2. **AI in Model development:** A host of ideas and techniques of AI can impact the practice of mathematical modeling especially with knowledge-based systems and environments for problem solving viz. AI programming and methodologies for implementing numerical models and solutions. In mathematical modeling of complex systems, AI/ML [even Deep learning (DL)] can learn system patterns and laws in data-driven approaches. Some AI uses in modeling are as follows:
   - **Simplification process:** AI can help in simplifying large complex process, retaining basic dynamical requirements. ANN based autoencoder, component analysis process drastically can simplify large complex process.
   - **Hybrid model:** Combining strengths of AI with traditional physics-based modeling hybrid models can be developed.
   - **Identification process:** As discussed, data-driven AI approach can analyze large datasets to identify patterns and relationships, leading to the development of mathematical models for process viz. linear, polynomial regression analysis, ANN with DL. Also, AI can identify system parameters with associated dynamics from observed data.

- **Estimation & optimization:** AI is extremely efficient in estimating and optimizing parameters using Bayesian optimizing technique.
- **Time series:** AI is excellent in time-series data analysis and same can be utilized in modeling and forecasting very important for dynamic systems.
- **Noise suppression:** AI methods can manage and model uncertainty, disturbances and noise in data. Therefore, AI can take care the development of robust mathematical models by using probabilistic and other approaches.

3. **Automated model:** AI including ML & DL can be used for various tasks like prediction & classification, anomaly detection, optimization, and decision-making in mathematical modeling. For complex systems, AI can learn from large amounts of data, and can be applied to time series analysis and other areas. Integration of AI in the mathematical model development enhances their accuracy, efficiency, and adaptability by reducing model complexity, and handle uncertainty. A few areas for automated model development by AI are as follows:
   - **ML automation**: Total automation for ML applications to actual complex problems. This also includes *symbolic regression* with the aim to produce interpretable results. Symbolic regression can model and map data for easy understanding by scientists.
   - **Reinforcement learning (RL):** RL is a type ML which learns optimal behavior through trial and error and is focused on decision-making to maximize cumulative rewards. So, it learns optimal policies for control problems, adapting the model based on interactions with the environment.
   - **Online algorithms**: Continuous updating of model parameters with new data streams.
   - **Adaptive modeling:** AI can be used for the development of adaptive models capable of updating in real-time with new data.
4. **Predictive model:** MPC has been introduced to process industry to enhance its operational efficiency. Now AI-based models replace linear model for reduced the computational burden along with increased the speed of optimization and accuracy of the model. For developing a predictive intelligent model, a few key steps are as follows:
   - **Definition & objective:** Define exact prediction requirements with specific objectives and goals for the desired *accuracy*, *speed*, and *scalability* requirements of the model.
   - **Data handling:** Data handling refers to data *collections* from various sources, data *processing* from raw data to create *temporal*, *relevant*, *knowledge-driven*, *statistical,* and *domain-specific* features. This also includes detection of missing values, errors.
   - **Data analysis:** This includes correlation analysis, visualization with descriptions.
   - **Development of model:** Another issue to address here is as follows: system integration, Application program interface (API), etc., to name. This also should include suitable training for data splitting, cross validation, etc.
   - **Model selection:** There are various models for MPC to choose from. These are as follows: *linear*, *neural network*, *tree based,* etc., to name a few.
   - **Model evaluation:** Models are evaluated based on validation, metrics selection, and performance assessment.
   - **Model management.** This covers management of version control & model and maintenance.

   **Predictive intelligence** This refers to specialized technique used in predictive models. Some of the common techniques used are as follows: Time series analysis, *reinforcement learning* (models to learn from environmental interactions for sequential decisions), Introduction of AI to meet compliance and transparency.

### 2.2.1 Discussions on model and artificial intelligence

In this subsection discussions have been presented on AI techniques and mathematical models in complex systems. Complex systems modeling involves studying the structure & behavior of complex systems, understand interactions with environment and identifying nonlinear effects in the system. System dynamics, network theory, algebraic methods, etc., are commonly used modeling methods. Application of AI is rather new addition. AI technologies include ML and DL, which can be used for tasks such as prediction and classification, anomaly detection, optimization, and decision-making. In complex system math modeling, AI can study to learn system patterns and laws from large amounts of data and these can be used in different applications including image recognition, time series analysis, and other fields. DL & ML realize the modeling and analysis of complex systems by building neural network models. AI-based systems are data-driven modeling, where actual data are combined in traditional theoretical modeling, to better describe and predict the behavior of complex systems. Self-control systems can realize its own optimization and adjustment through adaptive control algorithms and feedback mechanisms. AI has broad application prospects in complex system mathematical modeling and act as new tools and methods for understanding and solving problems in complex systems.

Mathematical modeling of complex systems use mathematical equations and techniques to understand, analyze, and predict the behavior of systems comprising many interconnected components [11]:

1. **Data modeling:** Mathematical modeling refers to mathematical methods & techniques that can describe and analyze the systems to identify the system components and the interrelations among them including nonlinear relations and dynamics. Various mathematical modeling procedures include mainly:
   - **Algebraic equations**: These mostly describe static relationships in a system, that is, the state of the system at a certain point in time—hence useful for stability check for the system.
   - **Control theory:** This is used to design controllers to regulate the state of a system making system controllable, stable to achieve the performance goals.
   - **AI technology**: AI, ML & DL are widely used in generating models from complex systems, predicting system behavior, optimizing system performance, etc.
   - **Optimization methods**: This is used for determining system parameters and associate adjustments for the system structure to achieve specific goals.
   - **Chaos theory:** This is used in complex systems with chaotic phenomena, to describe and analyze system behavior. Nonlinear dynamics and chaos models can be used to capture the complex and unpredictable nature of systems.
   - **Graphical**: Used to establish interrelation among components
   - **Differential and difference equation:** These are used to describe the system dynamic behavior including interactions and changes between components
   - **System dynamics:** System dynamics method is helpful in quantitatively analyzing cause-and-effect relationships in systems.
   - **Statistical models:** Useful for complex systems to check uncertainty factors.
   - **Multiagent modeling:** For systems containing multiple individuals or agents.
2. **Model and AI discussions:** AI includes a variety of methods and algorithms designed, enabling intelligent control systems to perceive, understand, learn, reason, and perform tasks. In mathematical modeling of control systems, AI technology can learn system patterns and laws from large amounts of data, and then can be utilized for various applications like image recognition, time series analysis, etc. ML & DL can realize the model and analyze complex systems also. There exists a large number of ideas and techniques in AI which have the potential to impact the practice of mathematical modeling especially, knowledge-based systems and environments for problem-solving methods. These methods can be used to encode domain-specific knowledge and strategies for a variety of ill-structured problems. Advanced AI programming, methodologies, etc., provide necessary intelligent tools to implement numerical models and solutions, with comparatively cleaner, easier to write, and adaptable computational mechanics codes. These AI algorithms can provide effective and rigorous mechanisms for addressing problems. Major modeling activities include model generation, interpretation (of numerical results), validation, etc.
   - **Generation:** Model generation refers to all activities related to the generation of models of physical systems suitable as input for a computational program. Major Model generation issues in three levels are as follows: *Intelligent systems, customized preprocessing* and *model generation tool.*
   - **Interpretation:** This is the AI potential to assist in postanalysis operations.
   - **Validation:** Model validation determines if the mathematical model's numerical results accurately reflect the modeled system's real behavior.
3. **ML-DL issues:** ML & DL also contribute tools for development of mathematical models especially for complex system with data-rich environments. Various issues are as follows:
   - **Data requirements**: *Quality, quantity data, and labeled data*
   - **Hardware resources:** *High-performance computing, memory,* and *storage*
   - **Software tools:** *ML & DL framework* and *Data processing libraries*
   - **Algorithm selection:** *Supervised learning, unsupervised learning,* and *neural network*
   - **Model deployment:** *APIs, services, monitoring and maintenance*
   - **Advanced issues:** *Reinforced learning, transfer learning, etc.*
   - **Training & evaluation:** *training techniques, cross validation,* and *evaluation metrics.*
   - **Expandability & interpretation:** *Model interpretation tool, ethics, etc.*
4. **ML -DL in model:** AI technology, especially ML & DL finds its usage and applications in generating models from complex systems, predicting system behavior, optimizing performance, etc. Various tasks such as prediction and classification, anomaly detection, optimization and decision-making.
   - **Prediction:** ML and DL can predict future states or values based on historical data, applicable through learning in many applications.
   - **Classification:** Mathematical models can be created to describe system behavior based on data, useful in engineering and control systems. In this way, this

can be used for various identification and classification issues.

- **Detection:** ML and DL can identify deviations from normal value/behavior in time-series data, network security, fraud detection, and industrial monitoring.
- **Optimization:** By learning from data, applicable in logistics, resource allocation, etc.ML and DL can solve complex optimization problems
- **Decision:** Complex process of decision-making refers to the strategy or method of decision. For decision-making problems, AI & ML can help with selecting the best choices. Artificial intelligence and learning machine-driven decision-making are widely applied in computer science, engineering, physics, mathematics, and control systems.

5. **AI in complex modeling:** AI has extensive applications in mathematical modeling of complex systems. The discussions are mainly based on implementation and application area of AI-based mathematical model as shown in Fig III 2.2a. AI Implementation in mathematical model has been detailed in Fig III 2.2a1. Application area of AI in mathematical model has been depicted in Fig III 2.2a2.

AI can best be used for model optimization and parameter adjustment. From previous discussions we have understood that AI for adjusting the parameters of a model to minimize a cost function (a measure of actual outcome), can be used for image/speech recognition, time series analysis, etc. *Bayesian optimization*, *Conjugate gradient*, *Genetic algorithms*, *Gradient descent*, *Particle swarm algorithms*, and *Simulated annealing* are some of the common algorithms used.

### 2.2.2 Artificial intelligence in model-based system

There exists a number of concepts and techniques of AI having the potential to impact and influence mathematical modeling. Knowledge-based systems and environments can provide many problem-solving methods with domain knowledge to encode domain-specific strategies for varieties poorly structured problems in model generation as well as for result interpretation. Advanced AI programming languages and methodologies offer improved method for implementing AI in modeling with numerical models and solutions which would help to write, and have more adaptable computational mechanics codes. From previous chapter, we understand that knowledge-based expert systems are interactive computerized programs incorporating judgment, experience, rules of thumb, intuition, and other expertise. Knowledge-based systems provide knowledgeable advice about a variety of tasks. problems can be State space approach can be used for many problem-solving and searching methods. Various approaches in this connection are as follows: *Search reduction*, *Problem reduction*, *Adaptive search techniques*, and *Use of domain knowledge*. Mathematical modeling refers to the activity devoted to the study of the simulation of physical phenomena with computational processes. AI is rapidly gaining prominence; by leveraging AI technologies, it is possible to improve the efficiency and effectiveness of model-based system and associated engineering. We understand that AI can be implemented for *Intelligent requirement analysis*, *Automated model generation*, *Intelligent model validation & verification*, *Optimization*, and *Decision making*. Also, from the discussions we understand that use of AI offers benefits like Better accuracy, consistency & decision making, increased efficiency, flexibility & adaptability. However, it faces challenges like *data quality*, *human experts*, *complexity*, *lack of definite standard*, *integration issue*, *ethics*, and *security*. In this subsection short discussions have been provided on potential impacts of AI technology on various mathematical modeling activities.

1. **Model generation:** Model generation refers to all activities related to the generation of models of physical systems suitable as input for a computational process. The generation of mathematical models from physical descriptions of systems is an importance issue. In all disciplines & areas, the needs to model are in a rising trend at every stage of system from conceptual design to detailed engineering, etc., calling for detailed performance evaluation requirements. On account of the availability of so many computational mechanisms and codes, model generation is a primary activity of the analysts who interact with programs at various levels to help making decisions. So, there exists need for higher-level program interfaces that will free analysts from details to generate models in terms of behavioral descriptions. AI-based techniques and knowledge-based systems promises to provide an opportunity to respond to the needs discussed. Some common model generation tools are as follows:
   - **Help systems:** This is an intelligent system to address the consulting advice to nonexpert engineers in an interactive mode. With enquiring the user with some key issues of the problem, based on the key problem features, provide the user some advices in the form of appropriate sequence of commands, program options to select, analysis strategies to invoke, numerical parameters to assign, etc.
   - **Customized preprocessors:** These are knowledge-based programs integrated into the environment. Customized preprocessors extract relevant features from a database to describe the physical object for modeling so that the user intervention in the modeling process is minimum.

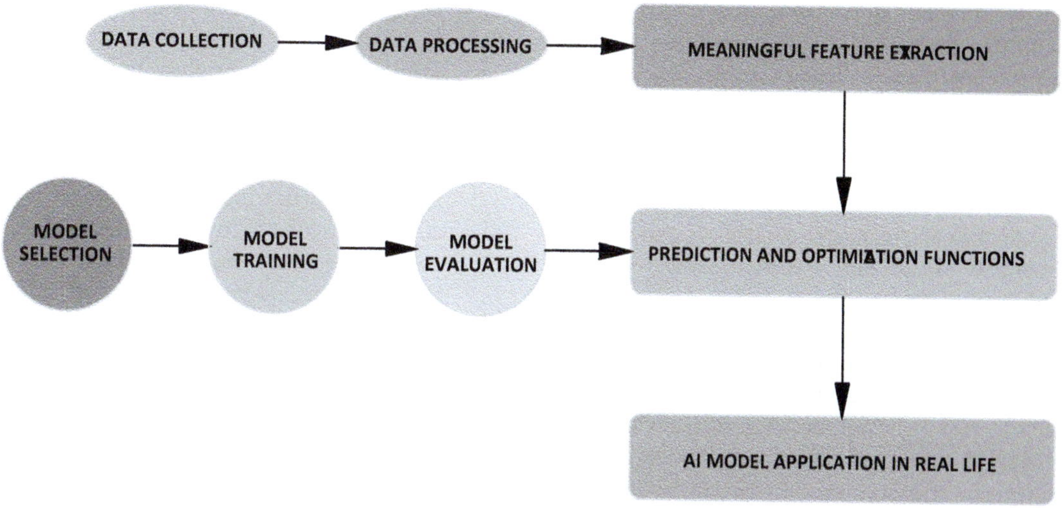


FIG III 2.2a1 AI IMPLEMENTATION IN MODEL

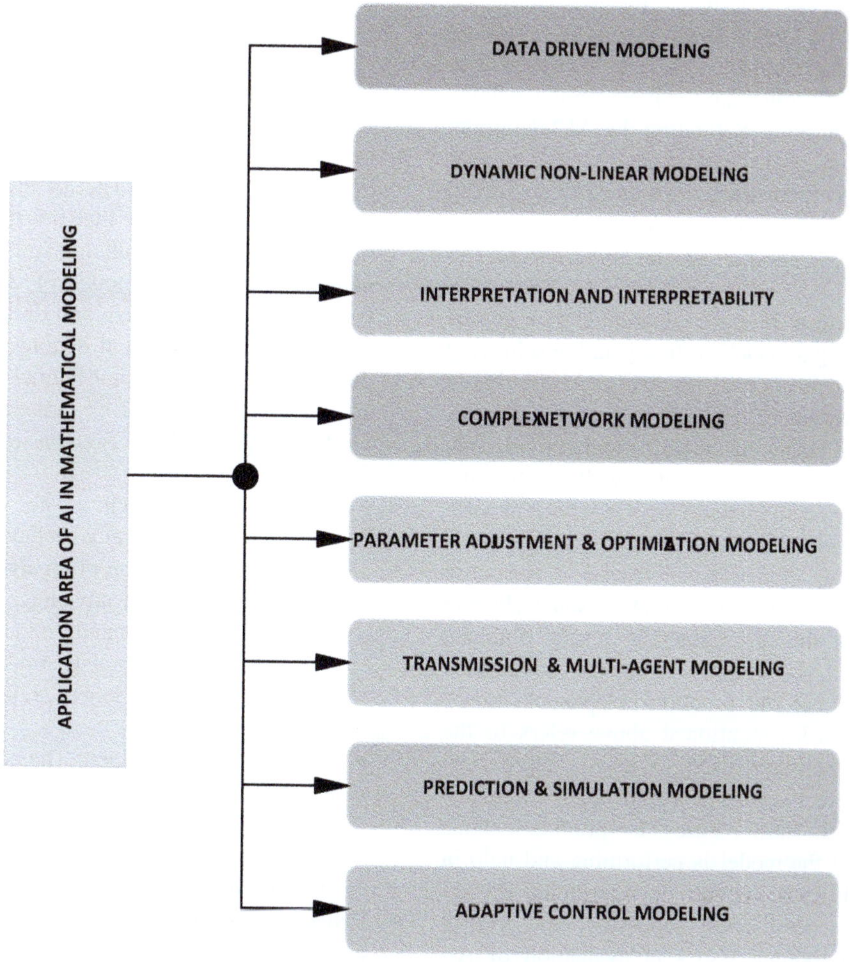


FIG III 2.2a2 AI APPLICATIONS IN MODEL

**FIGURE III 2.2A** Implementation and applications of AI based model.

- **High-level tools:** High-level tools incorporate techniques that are highly flexible compared with the heuristic approaches. These are very helpful for model generation.

2. **Post analysis:** AI has the potential to assist in post analysis operations which generally refer to as interpretation. It also includes other processes like model validation, response abstraction & evaluation, and redesign recommendations. Major issues involved are as follows:
   - **Model validation:** With model validation one can determine whether numerical results of the mathematical model can accurately describe the modeled system's real behavior under given environment. Knowledge-based techniques can provide practical mechanisms to represent and characterize the idealization errors viz. structural/modeling/rounding errors. This process that helps for evaluating the performance of a trained model including validating the ML model performance on new or unseen data. This helps in affirming that the model could achieve the intended purpose. Validating the AI/ML model during the training and development stages is crucial for ensuring accurate predictions. In addition, model validation provides benefits like *Enhancement of model quality*, *Prevention of overfitting and underfitting* and *discovery of error types*.
   - **Model evaluation:** Model evaluation plays a pivotal role in assessing a model's efficacy and in achieving the defined business goal. Statistical/AI/ML based model's performance and efficacy are measured quantitatively using evaluation metrics so that various models can be compared well-functioning, etc. Therefore, with evaluation, it possible to verifying that the computed results satisfy design specifications and functional criteria. Response abstraction is the task of generating some abstract description of the raw numerical results obtained from the analysis program. Response abstraction can be functional and function independent types. Evaluation metrics mentioned above refers to the quantitative measures used to assess the performance and effectiveness of a statistical or machine learning model. These metrics also provide insights into how well the model is performing and help in comparing different models or algorithms. Predictive ability generalization capability, and overall quality are important issues in evaluation of AI-based models. The choice of evaluation metrics dependent on the specific problem domain. Some common metrics for evaluation are as follows: *True positive/negative*, *False positive/negative*, *Predictive value-positive/negative*, *Accuracy*, *Specificity*, *Sensitivity* and *Rate*.
   - **Design integration:** Integration in design for postanalysis in AI for mathematical modeling focuses on extracting deeper insights, improving model reliability, optimizing decision-making and recommendation for redesign based on the model's results. It faces many challenges like Computational demand, Data dependency and Interpretability. Major issues addressed are as follows: *Model validation & verification*, *Sensitivity analysis*, *Uncertainty quantification*, *Data augmentation and enhancement*, Interpretability, *Error analysis with correction*, *Pattern recognition* insight *extraction*, *postoptimization analysis*, *Scalability* and *Feedback loop integration*. Analysis is rarely, if ever, an end in itself. Another major use of analysis is to guide and confirm design decisions. Often AI techniques recommend redesign when the analyzed system's response is not satisfactory. Important issue here is whether the nature of the knowledge base that can generate redesign recommendations or should it be separate and independent. The second approach formulates redesign as a goal-oriented problem.

3. **Model formulation:** Formulating good models is more of an art than science. The art lies with the ability to develop simple models that are good approximations of reality. Model formulation refers to the process of generating a simple model of a real-world system from the choices of behavioral design pattern, structure and assumptions. Thus, it denotes the process of producing a computational mechanics—a set of numerical software routines offering practical benefit for real world problem. Here, two basic approaches are as follows:
   - **Template approach**: The Template Method approach is based on a behavioral design pattern to defines the skeleton of an algorithm or operations in a superclass. The advantage of this approach is that it does not demand for technical background of the user when template model closely resembles the real situation. Hence preferred by users with less analytical skills.
   - **Constructive approach**: The constructive approach is more fundamental & generalized with defined set of guidelines for constructing a model from the ground. This approach demands for some analytical skill but may be applied to any practical situation.

4. **Processing & reusability:** Let us discuss symbolic processing and reusability subroutine.
   - **Symbolic processing:** A branch of AI is symbolic processing used for processing of symbols, which represent objects or concepts in the world, and their relationships. This approach is for logic-based programming, with rules and axioms used for inferences & deductions. Out of many usage, this is

used for symbolic computation in mathematical modeling. In modeling, this is used to overcome time consuming program development and error proneness in programming viz. MACSYMA and Mathematica. Studies on symbolic processing suggest that this can significantly assist the generation of computational model components to be incorporated in the source code. *Interpretability*, *Flexibility*, and *Knowledge representation* are major benefits of symbolic processing which has limitations like *Difficulty in handling ambiguous information*, *Knowledge limitation*, *Scalability*, etc.
- **Model reusability:** This is numerical subroutines with monolithic structure to perform function evaluations, domain and boundary integrations, linear and nonlinear equation solving, etc., abound in computational mechanics codes.

5. **Next generation programming:** New generation of programming types are object oriented and use data abstraction technique. In object-oriented programming, abstraction is one of three central principles (others encapsulation and inheritance). Through the process of abstraction, programmer hides all but the relevant data about an object to reduce complexity and increase efficiency. So, *data abstraction* is the programming process of creating a data type, usually a class, to hide the details of the data representation to make the data type easier to work with. In data abstraction, a representation is created for data to separate the interface from the implementation so user only has to understand the interface, the commands to use (hiding the internal structure of the data and its implementation). Therefore, for the next generation of software, it is important that the new tools, concepts, and languages that have emerged in the interim be properly evaluated and that the software be built using the best appropriate tools. One potential application of AI methodology here is an expert system to assist in synthesizing computational programs tailored to particular problems.

## List of abbreviations used

**AC** Adaptive control
**AI** Artificial intelligence
**ANN** Artificial neural network
**API** Application programming interface
**ATC** Auto-tuning controller
**DCS** Distributed control system
**DL** Deep learning
**GS** Gain scheduling
**HVAC** Heating ventilation and air conditioning
**IAE** Integral of absolute error
**ISE** Integral square of error
**ITAE** Integral of time-weighted absolute error
**LFT** Linear fractional transformation
**LMI** Linear matrix inequality
**LPV** Linear parameter varying
**LTI** Linear time invariant
**MAC** Model adaptive controller
**MBPC** Model-based predictive controller
**MBSE** Model-based system engineering
**MFA (C)** Model-free adaptive (controller)
**MIAC** Model identification adaptive controllers
**MIMO** Multiple input multiple output
**ML** Machine learning
**MLP** Multilayer perceptron
**MPC** Model predictive control
**MRAC** Model reference adaptive control
**MV** Manipulated variable
**PAC** Process automation control
**PI** Performance index
**PID (Controller)** Proportional Integral and Derivative (controller)
**PLC** Programmable logic controller
**PRAC** Pattern recognition-based adaptive control
**PV** Process variable
**RC** Robust controller
**RL** Reinforcement learning
**RLS** Recursive Least Squares
**SISO** Single input single output
**STC** Self-tuning controller
**VAV** Variable air volume

| Cl. | Table no. | TABLE Title |
|---|---|---|
| 2.0 | III 2.0a | Parameters in MIMO MFA |
| 2.0 | III 2.0b | Comparison of MFA with Other Control Technologies |
| 2.1 | III 2.1a | Explanation of important terms for Tuning (STC) |
| 2.1 | III 2.1b | Fuzzy Rules for PID parameters |

| Cl. | Dwg. no. | TITLE | Dwg. type |
|---|---|---|---|
| 2.0 | III 2.0a | Adaptive Gain Controls and Gain Scheduling | AUTOCAD |
| 2.0 | III 2 0b | PRAC application in HVAC | AUTOCAD |
| 2.0 | III 2 0c | SISO and MIMO Type MFA Controller | AUTOCAD |
| 2.0 | III 2.0d | Self-tuning Control System Details | AUTOCAD |
| 2.1 | III 2 1a | STC Introduction and STC Start up Sequence | AUTOCAD |
| 2.1 | III 2 1b | STC Design Step Details | AUTOCAD |
| 2.1 | III 2 1c | STC Tuning Block and Control Response Diagram | AUTOCAD |

| | | | |
|---|---|---|---|
| 2.1 | III 2.1d | Mamdani Model in Fuzzy Control | MS WORD |
| 2.1 | III 2 1e | STC with Fuzzy Tuner for PID | AUTOCAD |
| 2.1 | III 2 1f | Hybrid STC Systems | AUTOCAD |
| 2.1 | 2.1g | Some Examples of STC Systems | AUTOCAD |
| 2.2 | III 2 2a | Implementation and Applications of AI-Based Model | AUTOCAD |

## References

[1] A. Gupta, D. Yan, Mineral Processing Design and Operations: An Introduction, 2nd ed., Elsevier, 2016. https://www.sciencedirect.com/topics/chemical-engineering/adaptive-control-systems.

[2] M.R. Mojallizadeh, C. Prieur, Modeling and control of overhead cranes: a tutorial overview and perspectives, Ann. Rev. Control 56 (2023) 100877. https://www.sciencedirect.com/topics/engineering/gain-scheduling.

[3] Regional Controls Team, Asia. . Johnson Controls, Inc.; Internet document; Accessed May 5, 2026. https://www.johnsoncontrols.com.au/-/media/jci/be/australia/air-systems/variable-air-volume-terminals/files/be_pracpatternrecognitionadaptivecontrol_white-paper.pdf?la=en?la=en.

[4] C.J. Bett, Gain schedule controller, in: W.K. Chen (Ed.), The Electrical Engineering Handbook, Elsevier, 2004. https://shop.elsevier.com/books/the-electrical-engineering-handbook/chen/978-0-12-170960-0.

[5] G.S. Cheng, Model-Free Adaptive (MFA) Control, in: B.G. Lipták (Ed.), Instrument Engineers' Handbook, Vol. 2: Process Control and Optimization, 4th ed., CRC Press, 2006, pp. 301–312.

[6] P.E. Wellsted, B.G. Lipták, S. Renganathan, Self-Tuning Controllers, in: B.G. Lipták (Ed.), Instrument Engineers' Handbook, Vol. 2: Process Control and Optimization, 4th ed., CRC Press, 2006, pp. 468–477.

[7] V. Vandoren, Fundamentals of Self-Tuning Control; Control Engineering, 2007. https://www.controleng.com/articles/fundamentals-of-self-tuning-control/.

[8] R. Gianto, Review of self-tuning controller and its application in electrical power system, Int. J. Sci. Tech. Res. 4 (11) (2015). http://www.ijstr.org/final-print/nov2015/-Review-Of-Self-tuning-Controller-And-Its-Application-In-Electrical-Power-System-.pdf.

[9] Emerson Electric, EnTech Toolkit: DeltaV Distributed Control System [Product Data Sheet], June 2017. https://www.emerson.com/documents/automation/entech-toolkit-en-57642.pdf.

[10] N. Shevchenko, An Introduction to Model-Based Systems Engineering (MBSE), SEI Blog, Software Engineering Institute, December 21, 2020. https://insights.sei.cmu.edu/blog/introduction-model-based-systems-engineering-mbse/.

[11] N.V. Martyushev, B.V. Malozyomov, S.N. Sorokova, E. A. Efremenkov, D.V. Valuev, M. Qi, Review models and methods for determining and predicting the reliability of technical systems and transport, Mathematics 11 (15) (2023) 3317.

Chapter 3

# Expert systems and fuzzy logic in control

Computerized program expert system (ES) is a part of artificial intelligence (AI), built and maintained utilizing knowledge engineering. This is used to simulate the behavior and judgment of human meant to complement, not replace, human experts. Some important levels of ES shall include the following: *Specific ES, ES shells, Domain-specific shells, ES development environment, High-level programming,* etc. Out of several available levels and types of ES technologies, selection should be made keeping in mind that the selected tool must match the capability and sophistication of the ES and the tool should be commensurate with the qualifications of the project team. Generically ES can be applied for *Design, Planning and Scheduling, Classification, Process Control, Monitoring, Diagnosis,* and *Option generation*. Despite various challenges discussed, expert systems remain valuable for automating decision-making in specialized domains, helping improve efficiency and accuracy in various fields. So, ES efficiently can solve complex problems using a structured knowledge base and inference engine. In nature, most systems and concepts are naturally unpredictable and fuzzy. Based on *Fuzzy set theory* and *linguistic variables*, Fuzzy logic is an intelligent control method to imitate human's reasoning and decision-making in such systems. Four major components: *fuzzification, defuzzification, rules,* and *inference engine* work in conjunction with membership functions make Fuzzy control to happen. In Fuzzy logic, it is intended to capture the essential concept of vagueness; hence, it is probabilistic. However, these two are not same and have a number of differences. Apart from natural language processing, decision-making support, Fuzzy logics are used in modern control systems, for example, Kiln control. There are (at least) four sources of uncertainties in type-1Fuzzy logic set (T1FLS). These are as follows: the *antecedents and consequents of rules* can be uncertain, *Consequents* with histogram of values, *Measurements noisy* and *uncertain*, the *data noisy*. With all of these uncertainties, T1FL may not be able to directly model. Hence, T2FLS has been created to model. Type 2 Fuzzy logic (T2FL) which is a generalized version of Type 1 Fuzzy set has been introduced to handle more uncertainty. The implementation of this type-2 FL involves the operations of fuzzification, inference, and output processing. Membership functions of type-1 fuzzy sets are two-dimensional, whereas membership functions of type-2fuzzy sets are three-dimensional. In this chapter, detailed discussions shall be presented on both expert system as well as Fuzzy control systems.

## 3.0 Discussions on expert and fuzzy systems

ES is important branch of AI with capability to simulate the decision-making ability of a human expert. As has been seen that Ess makes use of *Knowledge base, Inference engine, User interface, Explanation module, Knowledge acquisition* and *Learning module* for any decision-making, that is, these systems use a knowledge base filled with domain-specific information and rules to interpret and solve complex problems. Even though the system highly depends on *knowledge acquisition* but this knowledge acquisition could be a bottleneck in the implementation of expert systems. However, it has been found that it is easier for the expert to directly enter the knowledge into the expert system shell discussed at the beginning. ES is an open new dimension for the computers which can be manufacturing batch control and process control. Also, the technology could be integrated into programmable logic controller (PLC), distributed control systems (DCS), and networked systems.

Fuzzy sets are important in understanding fuzzy logic. Basically, a Fuzzy set can be conceived as a collection of elements in a universe of information where the boundary of the set contained in the universe is *ambiguous, vague,* and *otherwise fuzzy*. Fuzzy sets are introduced by first reviewing the elements of classical (crisp) set theory. Fuzzy set theory was developed by comparison the precepts and operations of fuzzy sets with respect to classical set theory. Fuzzy sets contain the vast majority of the definitions, precepts, and axioms—define classical sets. Fuzzy set theory can be conceived as broader theory compared with the classical set theory as it takes into account an infinite number of degrees of membership in a set other than the canonical values of 0 and 1. Having gained certain knowledge about fuzzy logic in III 1.3, here the discussions will be more toward Fuzzy controller issues. In this section, we cover discussions on ES and fuzzy from the point where we left the same discussions in Chapter I.

Plant Intelligent Automation and Digital Transformation. https://doi.org/10.1016/B978-0-323-89892-8.00014-4

### 3.0.0 Discussions on expert system types

We know that an ES is a kind of AI software that uses human expert's knowledge stored in a knowledge base to solve complex problems. ES can not only emulate the decision-making abilities of human experts but also provide detailed explanations about how they reached a particular conclusion or advice. Knowledge engineering is the base of an ES, so knowledge engineers make sure that the computer possesses all the knowledge required in desired form so that ES can solve the problem. Their application areas with underlying technologies can be used to categorize ES in various types, and these are as follows:

- **Rule-Based ES:** ES uses a set of <u>*if-then rules*</u> to process data and make decisions viz. heath care systems.
- **Frame-based ES:** Knowledge is represented by data structures known as "frames" each of which contains attributes and associated values allowing the system to make decisions based on the current state of the attributes to represent viz. natural language processing.
- **Neural network-based ES:** Here, ESs use Artificial neural networks (ANN) to learn from data and improve and make predictions or decisions based on learned patterns viz. Image recognition.
- **Fuzzy logic ES:** These are used to handle uncertain or imprecise information, by allowing degrees of truth rather than binary true/false decisions viz. washing machine.
- **Neuro-Fuzzy ES:** This is integrated ANN and fuzzy logic to enhance the learning capabilities of neural networks and handling of uncertainty and imprecision from Fuzzy logic viz. use in engineering.

In this subsection, detailed discussions on various types of ES shall be addressed.

1. **Representation requirement:** <u>*Knowledge representation*</u> is a key aspect of expert systems and involves formalizing knowledge in a way that allows the system to reason and make decisions. <u>*Frame representation*</u> involves organizing knowledge into frames that represent objects, concepts, or situations with attributes and slots. <u>*Logical representation*</u> involves formal languages to represent facts and relationships. *Consistency, completeness, robustness, transparency,* etc., are major design considerations for all expert systems. Apart from above, other knowledge representation requirements are as follows: *precision, Causal Models, Temporal relations, strategies.* For some specific problems, it may be necessary to represent and reason with temporal relations, spatial models, compound objects. There a few requirements for knowledge Basic requirements for representation scheme in an expert system are as follows:
    - **Extendibility:** The data structures and access programs must be flexible enough for allowing extensions to the knowledge base containing heuristics built on experts' experience. Also, there should be a necessary arrangement to cater to issue new distinctions for new cases. Also, for treating the knowledge base of an expert system as an open-ended set of facts and relations, and keeping the items of knowledge as modular as possible.
    - **Simplicity:** Simplicity is very much required for representation. The simplicity should use the same terminology as the experts use. Once the syntax of the knowledge base is fixed, the access routines can be fixed to a large extent. Ingenious alternative way for representing and coding may offer more efficiency, yet any further modification could be extremely difficult. So, this should be kept in mind.
    - **Explicitness:** The representation an expert's knowledge Should provide the system a rich enough knowledge base for high-performance problem solving for this it is necessary to provide means for inspecting and debugging it easily.

    We understand that expert systems and knowledge representation are essential components of the rapidly evolving field of AI.
2. **Rule-based ES:** Rule-based systems could have many antecedents joined by logical gates. Major issues related to rule-based ES are as follows:
    - **Structural details:** As shown in Fig. III 3.0a various components of rule-based ES are as follows: the *knowledge base*, the *database*, the *inference engine*, the *explanation facility,* and the *user interface.* The knowledge base and data base contain the knowledge about the domain and the set of facts, respectively. Database is subjected to match the against the "*IF-THEN*" rules. External and developer interface are two other additional components. Text editor, bookkeeping facilities, debugging aids, and run-time knowledge acquisition are part of them. Run time knowledge acquisition facility enables to include additional knowledge/facts. The rule interpreter decides the rules to apply. The strategy used by the rule interpreter is called *the control strategy.*
    - **Inference issue:** Rule-based ES uses "IF-THEN" rules to represent the domain knowledge and for set of facts representing current situation. Inference engine compares each rule in knowledge base with the facts in the database. When the rule matches with a fact, the rule is fired, action as stated "THEN" is initiated. The fired rule may be *changed* by adding new facts in database. Inference types and techniques have been discussed in III 1.2.2.4. The

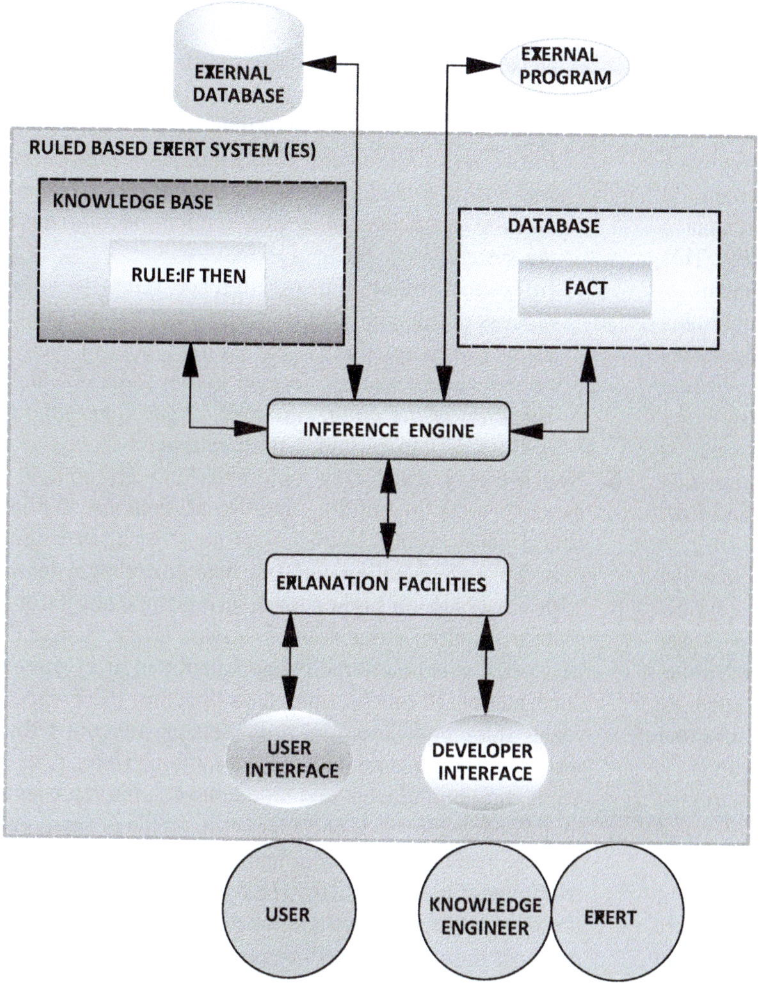


**FIGURE III 3.0A** Structural details of rule based ES.

performance of expert systems is mostly determined by the amount and organization of the knowledge possessed and partly by the strategies employed by the inference. Three commonly used strategy types are as follows:

- *Data-driven control:* Data-driven control is very popular, is known by many different names, viz. bottom-up, forward chaining, etc. Different data-driven strategies differ greatly in the amount of problem-solving effort. These rules are applied when left-hand-side conditions are satisfied. Select-Rule has the responsibility for making this decision. The information about the current problem as facts in the database is necessary for this strategy implementation. If there are no applicable rules, no action could be initiated, but to return to the user for some additional information.
- *Goal-driven control*: This strategy focuses its efforts by only considering rules that are applicable to achieve some particular goal. So, it works backward from a hypothesized goal, using backward chaining from a goal. Rules are selected with conclusions matching the goal, then to replace the goal by the rule's premises to create subgoals. Stating from the goal, to work backward till all subgoals are known to be true.
- *Mixed strategies* [1]: Data-driven and goal-driven strategies represent two extreme approaches to control. Various mixtures of these approaches have been investigated in an attempt to secure their various advantages while minimizing their disadvantages.

3. **Frame-based ES:** Fig III 3.0b details out frame-based and neural ES systems. Let us look into details on this from following issues:

- **Frame details:** Fig III 3.0b1 depicts the schematic details showing frame, facet slot value, etc., with interconnections for frame-based ES. Frame represents a data structure with particular object or concept knowledge. Here, knowledge is in hierarchical manner. So, frames are used to capture knowledge also in a frame-based ES. This provides a structured and concise representation of knowledge to treat a particular object or concept in a single entity. This cannot learn to adjusting in the new environment.
- **Slot and frame:** Conceptually may be thought as a collection of slots which are used to store values needed to describe a particular *attribute* or *operation* of the frame. A slot may contain a default value or a pointer to another frame. Generally, slot information is as follows: *Frame name, Frame interrelationship, Slot/default slot value, Range of slot value,* and *Procedural information.*
- **Facet, method and demon:** Here, facet refers to the extended knowledge about an attribute of a frame, used to establish the attribute value, end-user queries, and the processing details of the attribute. Method is a procedure pertinent to a frame attribute executed on request. Demons are IF-THEN structure executed when IF statement changes its value.
- **Inference engine:** Here, the inference engine searches for the goal/specific attribute. The inference engine looks out for the rules having consequents toward the goal of interest. So, they examine them one by one in order of rule base. When all the rules are valid, then only. Inference engine concludes that goal is reached.

4. **Neural ES:** Major issues for hybrid system *neural ES*, which is a combination of a neural network and a rule-based expert system, are as follows:
   - **Structural details:** Basic structure of neural ES has been elaborated in Fig III 3.0b2. The combination of rule extraction and neural network provides justification and explanation facility for its conclusion. Because of the capabilities of generalization, neural network can deal with noisy and incomplete data.
   - **Rule extraction:** Each neuron has a numerical weight, and they are connected by links in the network. So, this allows approximate reasoning. Neurons in the input layer transmit external signals to the next layer. The middle layer is the conjunction layer (hidden layer). Here activation function is used to calculate weights for neurons. "*IF-THEN*" rules are mapped into multilayer neural network. The last represents the action parts of the rules [2].
5. **Fuzzy ES:** We understand that Fuzzy logic is based on the idea that all things are described on a sliding scale, so as to differentiate among membership functions. Major issues related to multivalued Fuzzy ES are as follows:
   - **Development process:** Basic development processing steps are as follows: *Problem specification and definition of linguistic variables, Determination of Fuzzy sets along with construction of fuzzy sets and rules, Encoding Fuzzy sets andrules for performing fuzzy inference,* and *Evaluation and tune the system.*
   - **Inference engine:** Fuzzy inference refers to as a process of mapping from a given input to an output, using Fuzzy set theory. Like Fuzzy control, here, inference process is performed in four steps viz. *fuzzification of the input, rule evaluation, aggregation of the rule output,* and *defuzzification.*
6. **Neuro-Fuzzy ES:** Fuzzy logic and ANN are complementary tools in building intelligent systems. While neural networks are low-level computational structure to perform well dealing with raw data, fuzzy logic deals with reasoning on higher level, using linguistic information acquired from domain experts. Here, a neural network that is functionality equivalent to a fuzzy inference model. It can be trained to develop "*IF-THEN*" fuzzy rules and determine membership functions for input and output variables of the system. Here, it has input and output layers and hidden layers that represent membership functions and fuzzy rules. This is based on a fuzzy system trained by a learning algorithm derived from neural network theory. Here, the learning procedure of a neuro-fuzzy system takes the semantical properties of the underlying fuzzy system into account.

## 3.0.1 Basic fuzzy control discussions

As already discussed in III 1.3.0, the characteristic function of a fuzzy set is called the *membership function*, for it gives the degree of membership for each element of the universe of the given problem. Based on definition, there can be many different shapes. Like crisp set, various logical operations are applicable for fuzzy sets also. Fuzzy relations, are developed with the relationship between elements of two or more sets to take an infinite number of degrees of relationships.

$$\text{UXV} = ((u, v)|u \in U, v \in V) \qquad \text{(eq. III 3.0a)}$$

Where $\forall u \in U$ and $\forall v \in V$ for an ordered pair to form unconstrained matches between $u$ and $v$. Thus, Fuzzy relations can map elements of one universe to another universe through product of the two universes. In Fuzzy logic, the true Fuzzy proposition P in the close interval (0,1) is given by

$$\text{T(P)} = \mu_{\text{A}}(\text{x}) \text{ with } 0 < \mu_{\text{A}} < 1 \qquad \text{(eq. III 3.0b)}$$

Among the many applications of Fuzzy sets, logic and fuzzy control is quite common. Most industrial fuzzy logic

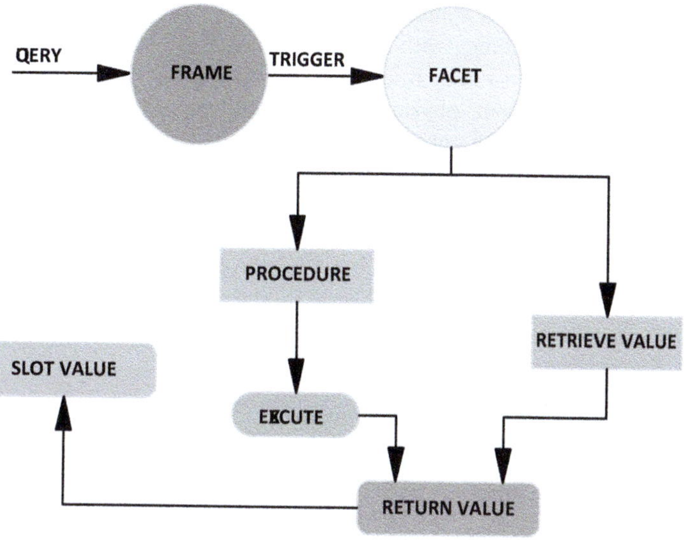


FIG III 3.0b1 FRAME BASED EXPERT SYSTEM SCHEME

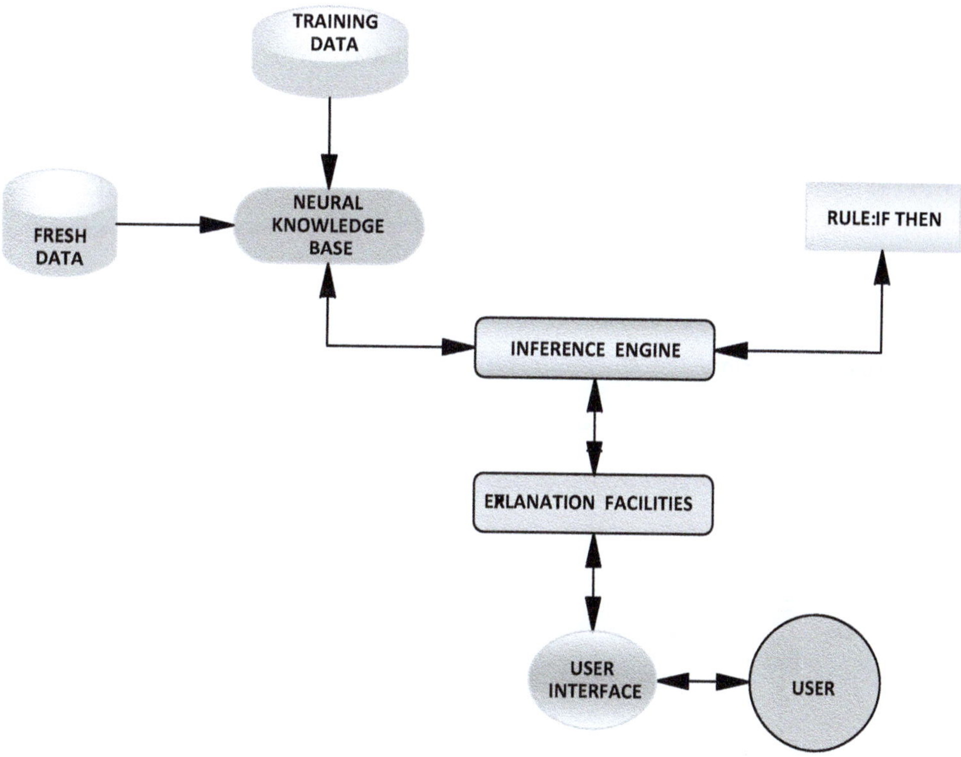


FIG III 3.0b2 NEUTRAL EXPERT SYSTEM SCHEME

**FIGURE III 3.0B** Frame based and neural expert systems.

applications in many places like come from Fuzzy controls both theoretical and practical. In this subsection further details on Basics of Fuzzy systems shall be discussed.

1. **Process and inference:** Fuzzy control system can emulate a human expert, when the human operator's knowledge would be considered as a Fuzzy set linguistic rules to produce an approximate decision, similar to a human would have done. Fig III 3.0c elaborates the same. Two distinct parts here are as follows:
   - **Processing details:** The human operator observes quantities from the inputs, that is, meter reading/ measuring a chart, then performs a definite action (a crisp action) to produce the output variable $y(t)$.

The human operator can be replaced by a combination of a Fuzzy rule-based system along with a *defuzzifier*. The input data are fed into Fuzzy rule-based system to convert and represent physical quantities into linguistic variables with suitable membership functions. Linguistic variables mentioned above are used in the *antecedents* (IF) part for fuzzy rules in an inference engine to generate a new set fuzzy linguistic variables used as *consequent* (THEN). This is elaborated Fig III 3.0c1. As shown, this "z" is crisp approximate output $y^*(t)$ to actual output y(t). Typical structure of a fuzzy control in a digital system with operators shall be as shown in Fig III 3.0c2 to include digital-to-analog and analog-to-digital converter. Fig III 3.0c3 details out interconnection of a fuzzy inference to a knowledge base and other structural details. Here, keeping existing crisp controller, through a fuzzy inference engine/fuzzy adaptation, the crisp controller is altered to meet the challenges

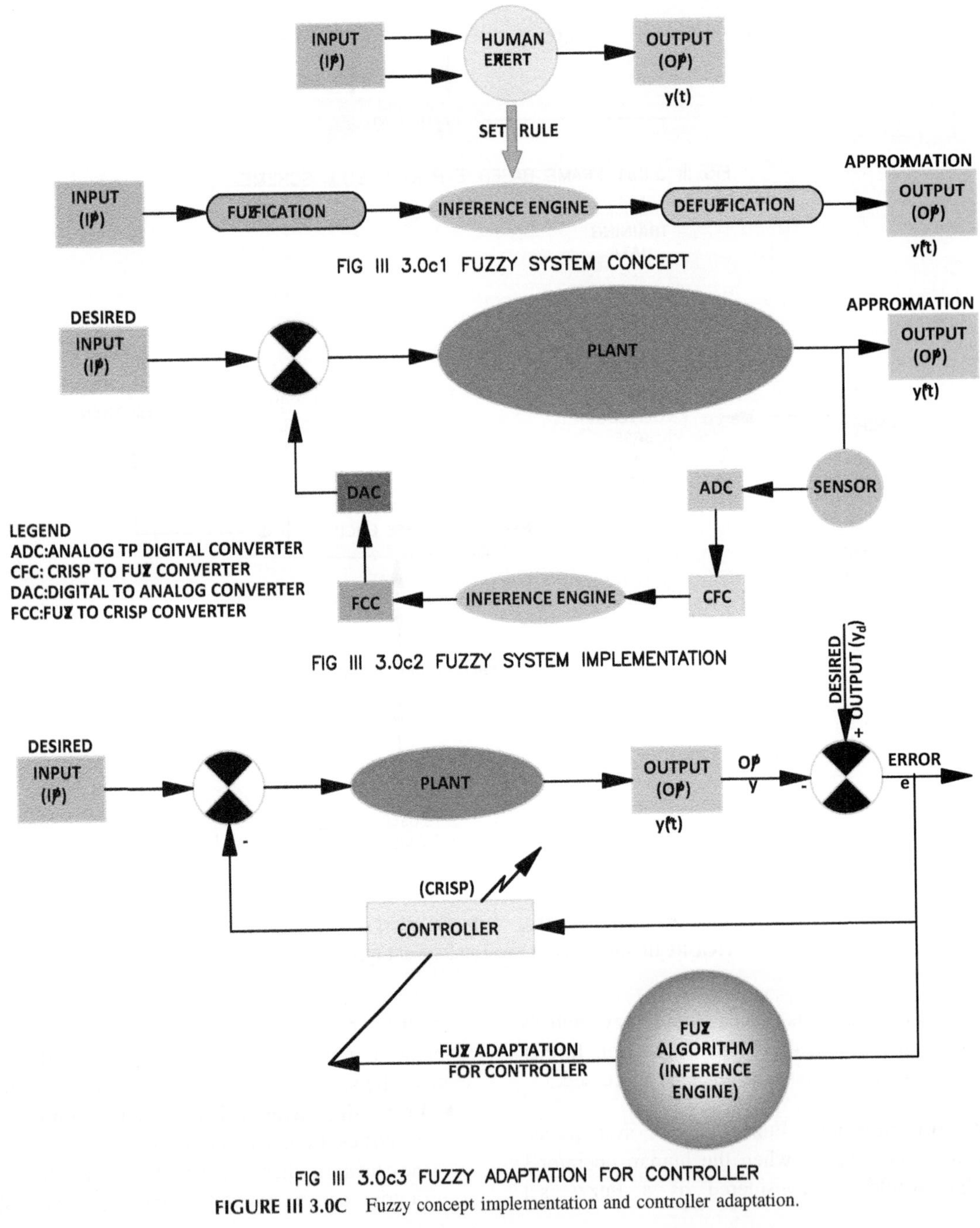


FIG III 3.0c3 FUZZY ADAPTATION FOR CONTROLLER

**FIGURE III 3.0C** Fuzzy concept implementation and controller adaptation.

of the system's unmodeled dynamics, disturbances, etc. The fuzzification operation represents a mapping from a crisp point $x = (x_1\ x_2\ \ldots\ x_n)^T \in X$ into a fuzzy set $A \in X$, where $X$ is the universe area and $T$ denotes matrix transposition. Singleton represented by one-point (value) $x_p$ and nonsingleton are two types of fuzzifiers.

- **Inference engine** [3]: Inference engine is the heart of a rule-based nonlinear fuzzy controller. IF $A$ is $A1$ AND $B$ is $B1$ OR $C$ is $C1$ THEN $U$ is $U1$; where $A$, $B$, $C$, and $U$ are fuzzy variables, $A1$, $B1$, $C1$, and $U1$ are fuzzy linguistic values with membership functions. Operators "AND," "OR," and "NOT" are connectives of the rule. In the above example, there are three antecedents and one consequent. In process controls these may represent process variables like "temperature," "pressure," etc., with fuzzy linguistic values (labels) may be "hot," "very high," "low," or "much," "slightly," "more," etc. As discussed before, in Mamdani rules the antecedents and the consequent parts of the rule are expressed using linguistic labels. However, there can be many forms and variations of fuzzy rules. *Takagi-Sugeno* rules are another popular form of Fuzzy rule.

2. **Fuzzy controller types:** There are variations in Fuzzy logic controllers (FLC)s. Some are PID controllers whereas some are non-PID controllers. Fig III 3.0d elaborates classification of Fuzzy controllers.
   - **PID FLC:** As the PID controller is linear is not suitable for nonlinear systems. The parameterization using rules and fuzzy membership functions makes it easy to add nonlinearities, logic, and additional input signals to control law. So, majority of industrial Fuzzy controllers used have the structure of incremental PI or PID controllers. In modern time, Fuzzy PID controllers have been widely used for industrial processes for their heuristic nature associated with simplicity and effectiveness for both linear and nonlinear systems.
     - *Direct action*: In this, FLC is constructed with set of heuristic control rules and control signal is directly deduced from the knowledge base and the fuzzy *inference as it is done* in Mc Vicar-Whelan/diagonal rule-base/generation approaches, that is, the FLC directly drives the process. This can be with single/double/three input types.
     - *Gain scheduling*: In gain scheduling type, the gains of the conventional PID controller (generating control signal) are tuned on-line in terms of the knowledge base and fuzzy inference. Fig III 3.0c3 may be an example.
     - *Hybrid*: This is combination of PID with DA FLC.
   - **Non-PID FLC:** The Mamdani controller, De-fuzzification and Takagi-Sugeno are some of the non-PID Fuzzy controller. De-fuzzification has been explained here as technical processes require crisp control actions.
     - *Mamdani controller*: The idea of the Mamdani controller as elaborated in Fig III 2.1d, is to describe process states with the help of linguistic variables and to use these variables as inputs to control rules. The base variable is an input variable that can be measured or derived from a measured signal or an output variable of the controller [4].
     - *Defuzzification*: Necessity of and various types of defuzzification has been explained in III 3.0.1.3 below. Defuzzification is an important element of any fuzzy controller. The weighted average of the membership function or the center of gravity of the area bounded by the membership function curve is computed as the most typical crisp value of the union of all output fuzzy sets:

$$Y_c = \frac{\int y.\mu A(y)dy}{\int \mu A(y)dy} \qquad \text{(eq. III 3.0c)}$$

       Defuzzification methods are very often based on heuristic ideas, and can be characterized by their formal (mathematical) properties. The crisp value selection should generally be an element of the Fuzzy sets and the criteria, can depend on very different bases viz. it can be the area below the membership functions or it can be other indicators.
     - *Sugeno type*: This is modified version of Mamdani rule. In this, Fuzzy antecedents, are equivalent to the Mamdani controller, and crisp consequences are functions of the input variables. The results are obtained from the weighted sums of the control actions corresponding to each rule. Here, the weight of each rule is the degree of membership of the input value in the rule antecedent as computed in the Mamdani controller. The variables are defined as in the Mamdani case, and the rule is as follows:
       If XI is $A_1^{j1}$ and $X^2$ is $A_2^{j2}$ and … and $X_n$ is $A_n^{jn}$ then $u$ is $!_r(X_1, X_2, \ldots, X_n)$.
3. **Control design**: Identification (input, output, and the state variables) and configuring (determine the fuzzy subsets for the variables) the variables for Fuzzification, is the primary step for designing FLC. Different optional shapes are available for membership functions to constitute the *fuzzifier* to convert the crisp inputs. Fuzzy subsets and associated membership functions also need to be configured for the output of the

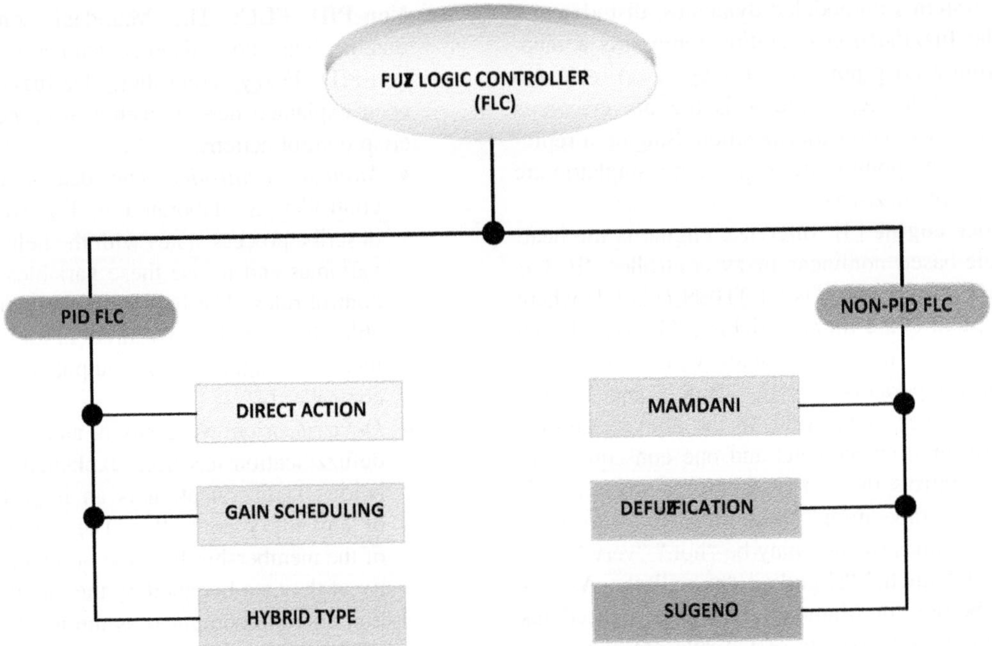


**FIGURE III 3.0D** Fuzzy logical control classification.

*inference engine* which can combine the fuzzy output from the rule base for defuzzifier. So, FLC design process has steps like: *identification* of Input/output (I/O). *tasks* and *linguistic variables* for Fuzzy definitions and memberships. Fuzzy logic refers to a numerical strategy permitting I/Os. Normally, three linguistic variables chosen viz. +ve (P), -ve (N) and 0 (Z) in this connection Table III 1.3a may be referenced. Using fuzzy sets and linguistic variables procedural rules are decided followed by defuzzification, validation, and modification. The membership function can be chosen from different shaped for different functionalities, for example, error is trapezoidal shaped, error change is Gaussian shaped output as the triangular shape, etc. With reference to Fig III 3.0e issues listed below are important for the control design. Also, Fig III 3.0f illustrates defuzzification and fuzzy set parameters.

- **Assumption:** Normally, there are a few assumptions need recognition for the FLC design. These are as follows: *A solution exists*, *I/Os, and state variables can be used*, *Knowledge exists in linguistic rules/ set of input-output data*, *"good enough" solution has acceptability* and *An acceptable range of precision.*
- **Preprocessing:** Main preprocessing issues are as follows: *Quantization*, *Normalization/Scaling*, *Filtering*, *Averaging* (for long/short term trend), *Measurements* (combined), *Differentiation*, and *Integration.*
- **Design rule:** The rules may use several variables both in the condition and the conclusion of the rules. The controller equally applicable for Single-input-single-output (SISO) and Multiinput–multioutput (MIMO) problems. As already discussed, the linguistic controller contains rules in the *"If Then"* format mainly, *if and only if or* modification with *not* are also applied.
- **Fuzzification:** Shown as first issue (in the controller) is *fuzzification*, to convert each piece of crisp input data to degrees of membership by a lookup in one or several membership functions. So, here, matching of data with the conditions of the rules and determine how closely the input instance matches the condition of each rule.
- **Membership function:** The grade of membership for all its members describes a fuzzy set. In fuzzy sets, elements are assigned *a grade of membership* such that the transition from membership to nonmembership is gradual rather than abrupt.
- **Inference engine:** For each rule, the inference engine looks up the membership values.
  - *Aggregation*: This is used for calculating the *degree of fulfillment or firing strength.*
  - *Activation:* The activation refers to the deduction of the conclusion.
  - *Accumulation:* Here, all activated conclusions are accumulated using MAX operation.
- **Defuzzification design:** in reality, the controller does not understand fuzzified values like slightly low, but can understand any crisp value say 0.5°C, so it is necessary to have defuzzification to obtain crisp value. There are many types of defuzzification

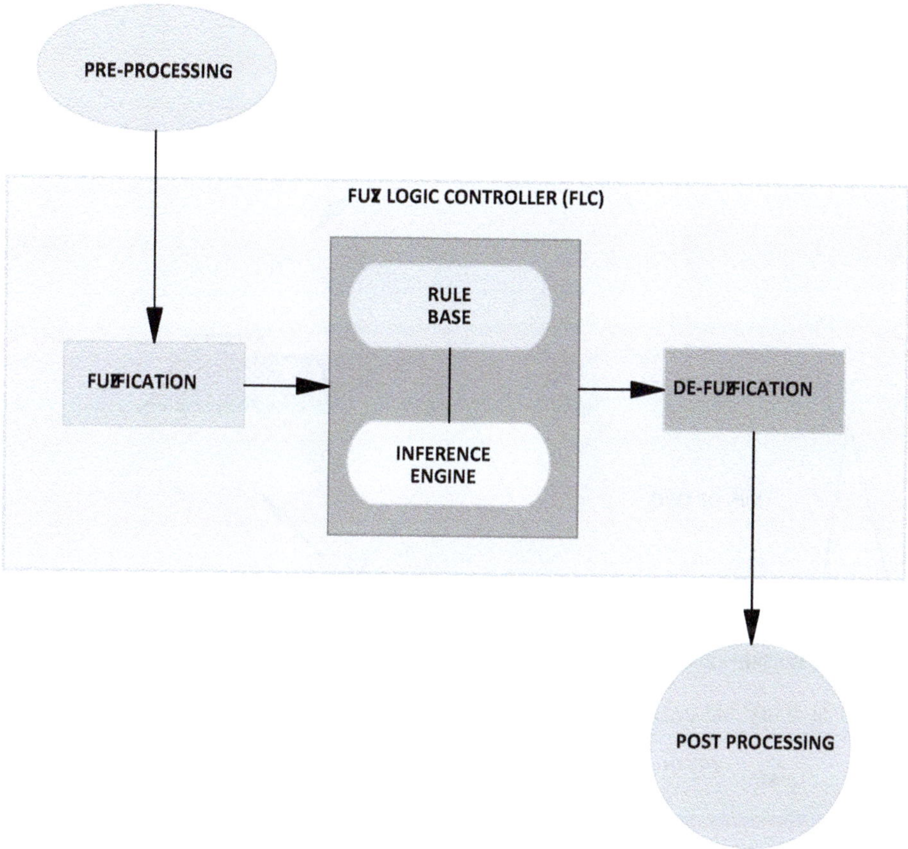


**FIGURE III 3.0E** Fuzzy logical controller block diagram.

methods some of them are as follows: *Max membership* principle, *Centroid* (Center of gravity), *Weighted average*, *mean-max membership*, *center of sum*, *Center of largest area*, *first and last maxima*, etc., out of these, centroid type is very popular for which III 3.0.1.2 may be referenced. Defuzzification with example has been illustrated in Fig III 3.0f1.

- **Post processing:** This refers to scaling of output for defining on a standard universe.
- **Performance:** Performance parameters are Rise time (sec), Overshoot %, settling time (s), and final steady state value. For these, *error*, *error change*, *output variable* control, and *membership* function chosen with some different shapes. Performance depends heavily on Fuzzy controller structure and parameter choice, choice of membership function, choice of the rules, decision of the defuzzification techniques.

4. **Design parameters:** It is extremely important to consider and decide on a number of design parameters and several critical parameters that can influence the performance and effectiveness of FLC. Major issues here are as follows:

- **Fuzzy Sets**: Several other parameters need adjustment for determining the shape of the fuzzy sets describing terms of linguistic variables. Suitable shape for the membership functions are to be selected and should match with the control applications. The peak value of a membership function is the value of the base variable where the membership function is equal to one. The *left* and *right width* of the membership function is the first value of the base variable on the left or right side of the peak value, respectively, that has a zero membership. The *cross point* between two membership functions is the value of the base variable where both membership functions assume the same membership value greater than zero. The *cross point level* is the membership at the cross point. Fig III 3.0f2 depicts a linguistic variable with three fuzzy sets and the corresponding parameters [4]. *Fuzzification strategy* including *input variables, membership function, and universe of discourse* are important issues need consideration.
- **Rules:** The development of rules is time-consuming, and designers often have to translate process

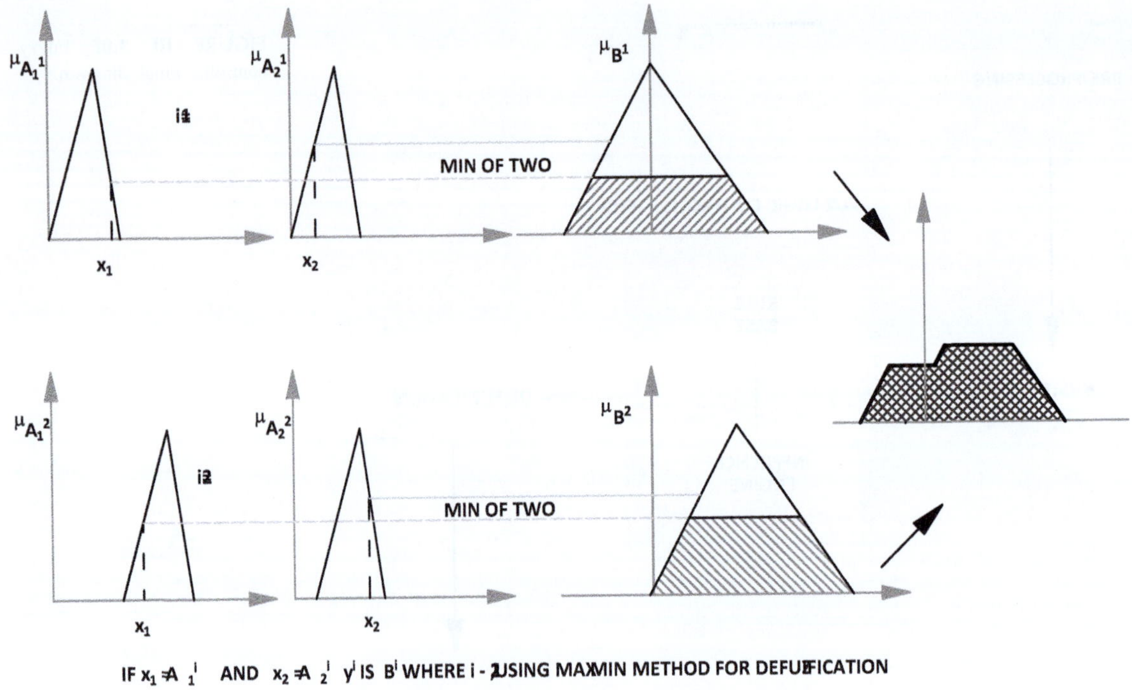


FIG III 3.0f1 DEFUZZIFICATION PROCESS EXAMPLE

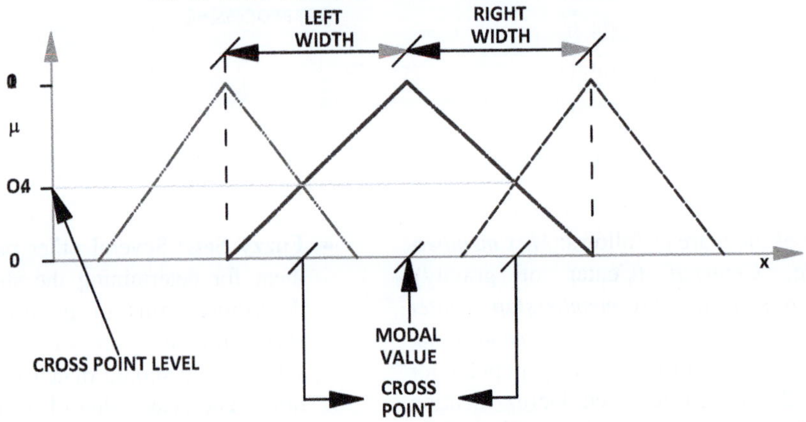


FIG III 3.0f2 FUZZY SET PARAMETER

**FIGURE III 3.0F** Defuzzification and fuzzy set parameter.

knowledge into appropriate rules. This should focus on "*If Then*" rule definition and rule structure. Major influencing factor are as follows: *Engineer's knowledge, Operator's experience, Crisp and Fuzzy modeling of the process, Fuzzy modeling of operator actions, Heuristic design and on-line adaptation rules.*

- **Scale and other factors:** With scale factors, the *base* variables in the linguistic variables are scaled. Scaling factors allow the definition of normalized base variables of the corresponding linguistic variables. Also, rule weighting (based on importance), adaptive mechanism (dynamic update) are also important issues.
- **Influencing factors:** Defuzzification strategy, inference mechanism including type of inference, rule evaluation are important influencing function.

5. **Control stability:** Stability in a control system is defined as its ability to produce a bounded output in response to a bounded input, that is, stability ensures that the system can reach and maintain a steady state for a given input, even when there are variations in the system parameters. Therefore, stability refers to the ability of the control system to reach and maintain equilibrium and remain there despite variations in parameters or external disturbances. The stability is more important when the control is connected to safety of lives, machinery, and environment. The stability test

for FLC systems, or lack of it, has been a subject of criticism. FLC can act either a lower-level conventional controller or as a supervisory control. Depending on the existence and nature there are four following classes of fuzzy control stability problems. So, there will be different treatment for stability.

- *Class* 1: Crisp linear process model with FLC at low level.
- *Class* 2: Crisp nonlinear process model with FLC at low level.
- *Class* 3: Linear/nonlinear crisp process model with fuzzy tuner/adaptive controller at high level.
- *Class* 4: Process model is fuzzy, and fuzzy controller is low level.

Thus, we find that stability analysis in fuzzy control systems is challenging. Lyapunov's direct method with Lyapunov function, energy-like function is the most commonly used one for stability analysis. A few other types are as follows: *Nonlinear stability theory*, *Root locus and frequency response* (Equivalent linear model, Gain and Phase margins), *Extended circle criterion, Phase plane analysis* (Phase plot, limit cycles), *Linear matrix inequality*-based designs, *Bounded input, bounded output* stability, and *Zadeh's stability criterion.*

### 3.0.2 Realtime expert system (RTES) and fuzzy

Real-time expert system (RTES) refers to on-line knowledge-based system which can combine analytical process models with conventional process control with reasoning about the past, present, and future to assess on-going developments and plan appropriate actions. RTES can reduce the cognitive load on users and enable them to increase their productivity without increase in the cognitive load. Architecture and life cycle of RTES are important issues, and these are discussed here along with application of RTES. Many processes have time-variant parameters on account of continuous changes in the process itself. This is one of the motivating factor for the development of adaptive control that can adapt itself to cater to the requirements of the process changes. Similarly, most of the adaptive FLCs change the shape of the membership functions, the scaling factors, etc. Adaptive FLCs are designed with embedded mechanism for adjusting parameters. For improving the controller performance adaptive FLCs are often used. The combination of Fuzzy logic and ES is a fundamental technique flowing directly from the nature of fuzzy logic. Fuzzy ES are currently the most popular with many applications. In this subsection, both expert systems adaptive fuzzy controls, and Fuzzy logic in expert systems shall be discussed.

1. **RTES architecture:** RTES can be both "Hard" as well as "Soft" types. "Hard" RTES have features guarantee a response within a fixed amount of real-time. However, "Soft" RTES can offer faster response. The components of RTES architecture are real-time software elements. In this architecture shown in Fig III 3.0g, the components of intelligent subsystem—the expert system are shown separately in the right side.
   The components on the left side representing real-time subsystem and are responsible for the real-time control behavior. As shown, the active (tasks, processes/the data- or knowledge base manager) and passive (data files/knowledge base), are usually separated by an interface for data exchange and synchronization. Data exchange between active and passive elements is shown by solid line arrows whereas the communication between active elements is shown by dashed arrows. G2RTES detailed in Fig III 3.0h, often use event-driven processing to enhance fast response efficiently.
2. **RTES lifecycle:** Actually, expert system development is a long process comprising: *Problem identification, Feasibility study, Project planning, Knowledge acquisition,* Knowledge representation then comes the implementation stage.
   With reference to Fig III 3.0i, main stages in lifecycle of RTES are as follows:
   - **Task analysis and specification:** In identification stage, goal of the systems and associated system is established. After engineering the ideal system, it is necessary to study the software requirement for the system in feasibility study phase. In this phase, it is necessary to visit customer and study the system to collect and collate various requirements. They investigate the need for possible software automation for the given system. At this stage apart from technical issue, cost/economic and operational feasibilities are also evaluated. After such study, a report with specific recommendations including elaborate specification for the said systems need to be developed and released the candidate system.
   - **Software design:** This is the developmental stage when system's overall structure and its nuances are defined. Design stage is extremely critical as in this stage, design of all components in expert systems is frozen, that is, crucial issues like *System architecture design, Knowledge base design, Inference engine design,* and *User interface design* are developed in this phase. Any problem in the design phase could be very expensive to solve in the later stage of the software development.
   - **Development and coding phase:** The software design issues are translated into a machine-readable form/coding. This is knowledge acquisition, when gathering and codifying expert knowledge into the knowledge base is done. Implementation of knowledge base, inference engine, **and**

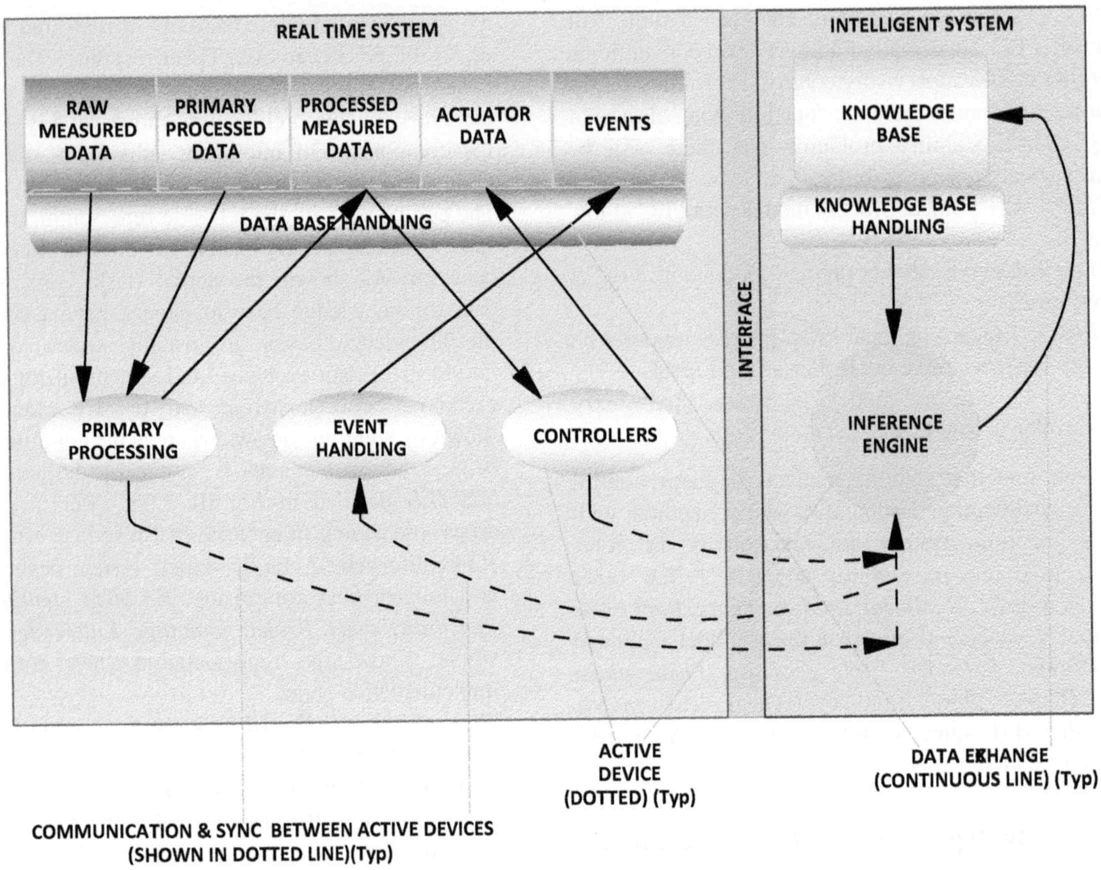


**FIGURE III 3.0G** Architecture for real time expert system.

the user interface for interaction with the system are developed at this stage.

- **Testing of system:** Once the code is generated, the system testing begins to carry out tests like unit testing, performance testing, integration testing, etc., to name a few.
- **Documentation and operation:** This is a staged rollout of the new application, involving *System installation*, *Documentation*, *User training* and *support*, *Knowledgebase updates.* Also, this may involve hardware and network upgrades.
- **Support and maintenance:** This phase comprises mainly, *Monitoring and support*, *Knowledge base update*, *System upgrade,* and *Performance tuning.*

*Design prototype*, *Special test tool,* and *Test plan* are a few special steps which help in the development of RTES.

3. **Adaptive fuzzy controller (AFC):** As the name suggests, the adaptive fuzzy control is a fuzzy control with adaptive functions, that is, Fuzzy controller can self-regulate the fuzzy control rules based on plant status and conditions.

**FIGURE III 3.0H** G2 real time expert system (G2RTES).

**GRTE S: GRT ES is a dvaed softwae pla form, with excellent graphica , object-oriented environmet, desiga d for developing & deploying rea time intelligen systems. G2 RTES can offer functioa lities like: *Knowledge representation, Reliability, Scalability, interoperability* etc. The event driven use of GRT ES eh ances capability to monitor, process & response efficiently to reatim e events as they occur in reatim e.**

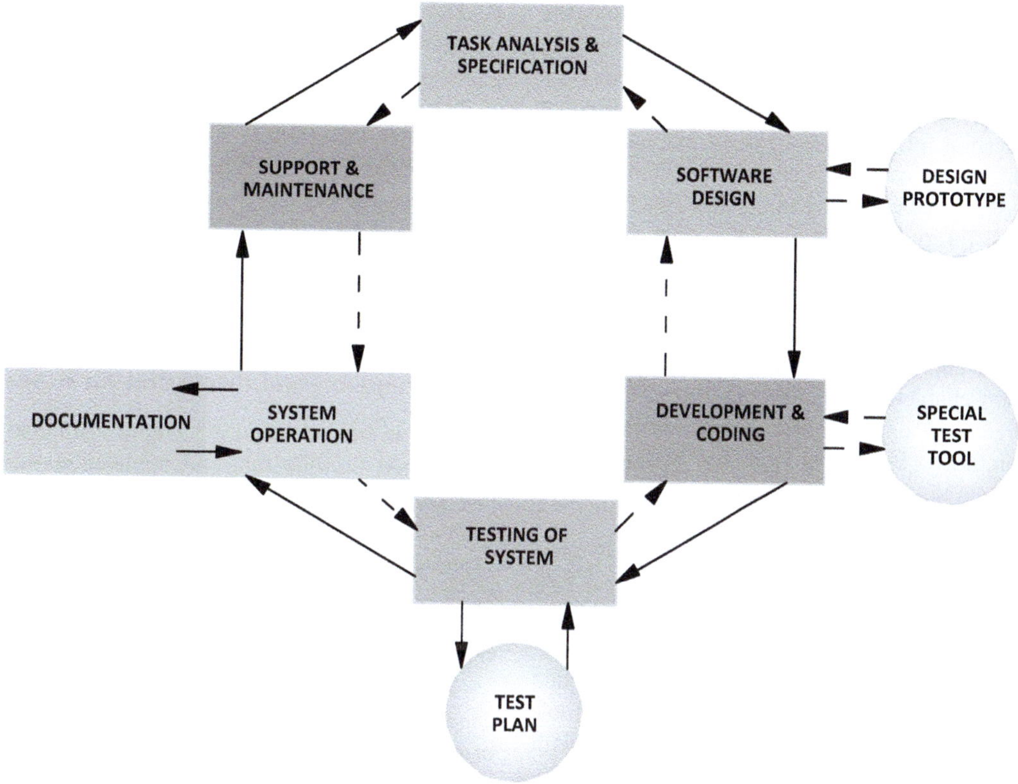


FIGURE III 3.01 Real time expert system life cycle.

A fuzzy controller consists of *Fuzzification*, *Rule base*, *Inference engine*, and *Defuzzification* for a rule base/ fuzzy relation between the fuzzy sets for the plant states and control variables.

If the plant is not known/predictable, then it is difficult to relate the desired performance to the possible states of the plant.

So, adaptive controllers with capability to monitor have to be found that detect changes in the process and adapt to the changes. Two common approaches for adaptations are *performance measure* and *parameter estimator*. Therefore, by combining the two above controller types above, in adaptive Fuzzy controller, the parameters of the FLC adjusted in real-time, that is, involves adjustments of *Membership functions*, *Rule base parameters*, and *Inference mechanisms*. With reference to Fig III 3.0c2 & 3.0c3, *Input*, *Fuzzification*, *Rule base*, *Inference engine*, *Defuzzification*, *Control output*, *Plant/System*, and *Adaptive mechanism blocks* with *feedback loop*. Therefore, one may conclude that Adaptive Fuzzy controller can offer the flexibility of fuzzy logic with the adaptability of real-time parameter adjustment, to suit complex systems with varying dynamics.

4. **AFC: selection and implementation:** Two issues for AFC here are as follows:

- **AFC: Selection:** Parameters listed below are important, and needs need to be considered for selecting an adaptive fuzzy controller:
  - Can a fuzzy model approximate entirely the system? If Fuzzy logic can approximate the system, then is the Fuzzy available readily on can found online?
  - If a fuzzy model, can be approximated piecewise by a set of fuzzy models?
  - When a system can be approximated by a set of fuzzy models, then it is important to consider whether same format with different parameters will be used or there will be different format altogether. Now, in case of approximation by a set of fuzzy models with same format, but with a different set of parameters, then it is important to look for these parameter sets readily available or must they be determined online?
- **AFC: Implementation:** Pertinent issues for AFC implementations are as follows:
  - The observable data are collected and collated to calculate the performance of controller.
  - Based on controller performance, controller parameters are calculated and adjusted.
  - The controller parameter adjustments are done to improvize controller performance.

5. **Fuzzy logic and expert system:** ES applications use AI techniques to find solutions which normally require human expertise and bring about a number of other advantages. Again, Fuzzy systems can store expert knowledge in a form of rules or mathematical expressions which are easy to visualize, implement, and modify. Also, because of membership functions and parameters, Fuzzy logic allows the system writing in a more abstract level for achieving better performance. So, Fuzzy logic with its capability to handle vague, incomplete, or conflicting information can boost the performance and flexibility of the systems. On account of common characteristics and complementary nature Fuzzy logic and expert system technologies can be easily combined. Both techniques are very good for decision-making. Important issues on Fuzzy ES are as follows:
   - **Fuzzy ES**: **Design** A Fuzzy expert system comprises Fuzzy knowledge base with facts and rule as discussed, inference engine/decision-making logic, and defuzzification module.
   - **Fuzzy ES: Implement:** Fuzzy ES is implemented using various programming languages and tools that support fuzzy logic, such as MATLAB with its Fuzzy Logic Toolbox, Python, etc. Each of these offers a range of features to create and test fuzzy systems.

## 3.1 Detailed discussions on expert systems

From previous section, we have seen various ES types to cater to wide range of application areas. Thus. ESs offer various capabilities to find its applications in diagnostics, advice, and assistance to decision-making, demonstration, Understanding, interpreting input to predict results, offers alternatives, deriving solutions with explanation and justification for conclusion are major capabilities and/or functions of expert systems. ESs do these with high accuracy. However, in spite of all these, they cannot be a substitute for human decision-making as expert systems work in narrow domain and do not possess all human quality (e.g., farsightedness) and refinement of own knowledge they depend on knowledge base. ESs have some limitations as well. Major system limitations are as follows:

- **Creativity:** Unable to make a creative response in an extraordinary situation.
- **Common sense:** Errors in the knowledge base can lead to wrong decision as it lacks in common sense which human can use to overcome a situation.
- **Maintenance cost:** The maintenance cost of an expert system is high.
- **Learning:** Human automatically can adapt new environment for the problem, but it can only reproduce what it has learned so even in case-based reasoning revision part due change in environment is limited.
- **Sensory experience:** Human have wide range of sensory experience but expert system depends on symbolic inputs only [5].

With these in mind in this section, we would cover detailed discussions on various issues pertinent to expert systems.

### 3.1.0 Fundamental discussion on expert system

An ES can be conceived as computer simulation of a human expert in wide range of applications in automation like: *Interpreting* and *identification*, *Prediction*, *Diagnosis*, *Planning*, *Designing*, *Monitoring*, *Debugging and testing*, *Training*, *Controlling*, etc. Expert systems are necessitated by the limitations of human expert in decision-making processes arising out of physical or mental workload, short memory, death/lie, etc.

ES very much domain specific and is not a good candidate for applications which are deterministic in nature or calling for high computation. From earlier discussions, one can conclude that inference engine is the mechanism by which search for conclusion or decision and reasoning is conducted utilizing search strategy on the knowledge built in with knowledge base [5]. Also, we have seen that ESs with fuzzy-logic capabilities allow for more flexible and creative handling of problems. These systems are used, for example, to control manufacturing processes. Fundamental details of ES have been detailed out in Fig III 3.1a.

In Expert system inference engine is the brain of expert system which again functions on the basis of knowledge base. Inference engine actually integrates all components together as shown in Fig III 3.1a1. Knowledge base can be two types rule based and case based. ES revolutionizes, processing in the system by bringing about transition from data processing to knowledge processing as shown in Fig III 3.1a2. Here, these fundamental details on ES have been elaborated. Let our discussions here start with these two types of knowledge bases.

1. **Rule and case-based reasoning and decision:** IF there are cases of "COVID-19" infection in locality THEN Locality may be could be declared as contentment zone. This kind of decision is based on certain rule. So, this is Rule based. On the contrary, there are cases when the decision is based on where there is limited evidence but decision is made on the basis of past experience. When someone having dry cough, some respiratory problem in pandemic situation, he could be affected by "COVID-19" hence should be in isolation and expert (doctor) should be consulted. Here,

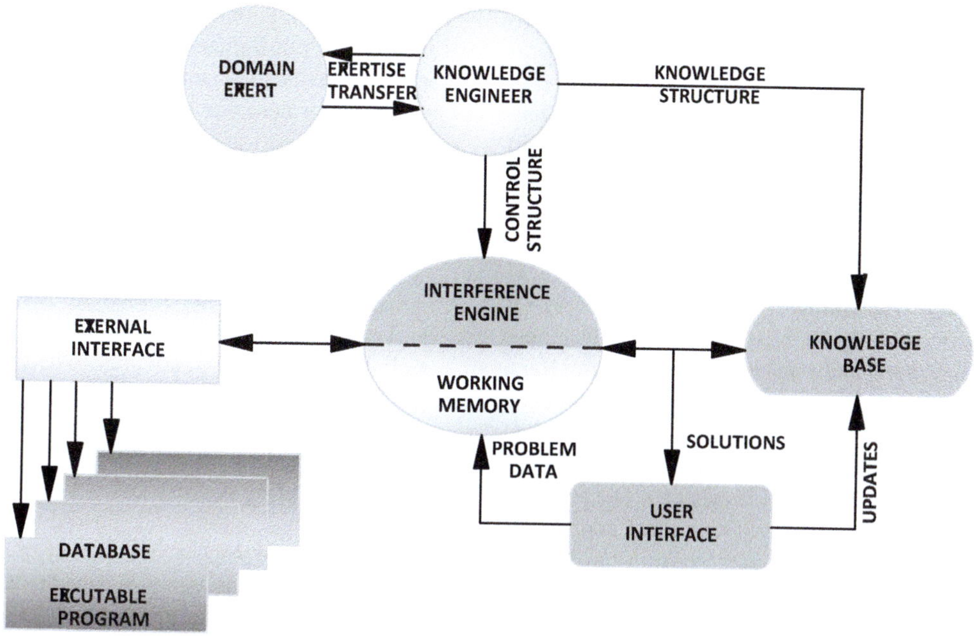


FIG III 3.1a1 EXPERT SYSTEM: INTEGRATION OF COMPONENTS

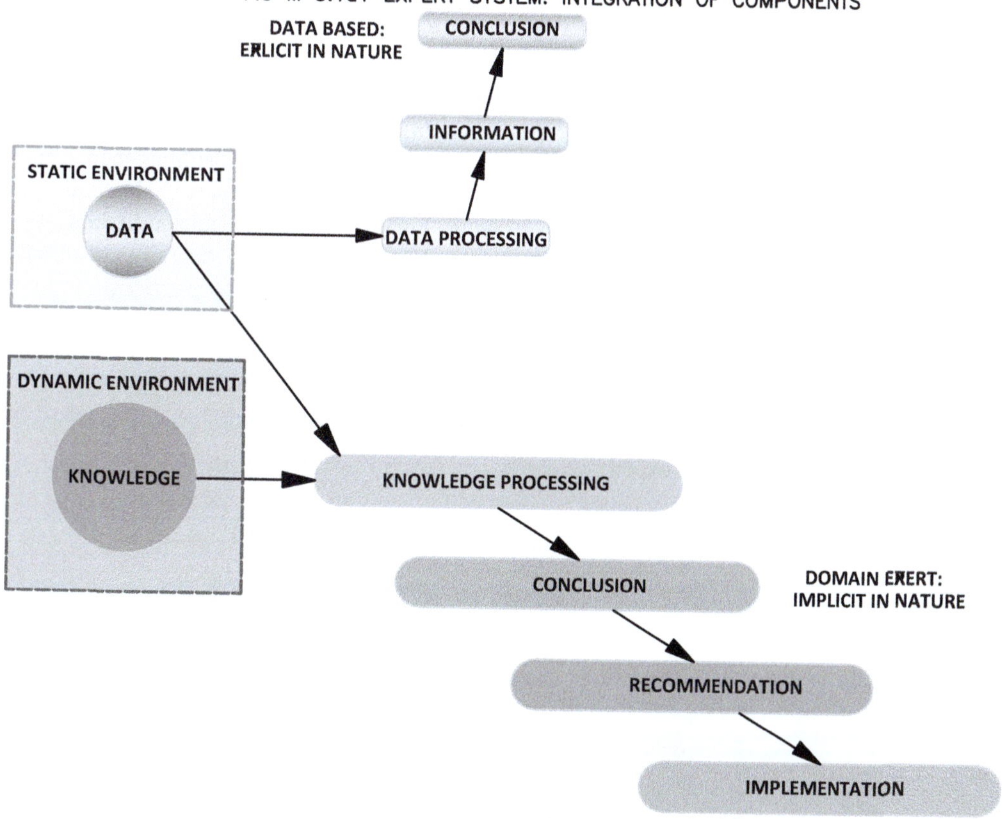


FIG III 3.1a2 DATA AND KNOWLEDGE PROCESSING

**FIGURE III 3.1A** Expert system fundamental details.

there is no concrete solution but recommendation for expert consult. Fig III 3.1b illustrates both the systems.

- **Rule-based reasoning:** In expert system when the knowledge based is represented by set of rules is often referred to as rule-based reasoning. Fig III 3.1b 1 and 2 depicts the types and rule-based reasoning flow chart, respectively. As shown in rule-based reasoning decision/solution is arrived at through set (s) of rules which follows IF THEN-ELSE logic. There can be five types of rules as depicted. These can be thumb/heuristic rule, or may be priori type rules. Many of these follow natural law, for example, IF it is *rainy season* AND if *sky* is *cloudy* THEN there is *chance of rain.* In flow chart it is seen it tries to find solutions based of rules. If not possible then, further expert opinion will have to be sought to update the knowledge base.

- **Case-based reasoning:** When Fig III 3.1b3 is compared with Fig III 3.1b2, it is clear that these two types vary only in their approaches. Here, instead of rule, human analogical/experimental reasoning is used to learn for solving a new problem based on prior knowledge acquired through experiments/analogy. Many legal issues could be resolved by this than plant controls. Here, most appropriate case is *retrieved* for *reuse* with *revision* for the proposed solution and *retain* part of this experience for future use [5]. From the above, it is clear that there are four elements viz. Case base (set of cases), *Case index* (to facilitate searching), *Similarity metrics* (to draw similarity), and *Adoption module* (adoption to find solution). Google Search is based on case-based reasoning. In fact, they use both the above reasoning in arriving at solutions and they have necessary patent for case base reasoning as well as specific inference engine.

2. **ES application areas:** Expert systems are part of artificial intelligence and are extensively used in different areas. The situations under which they are used shall include but not limited to:
   - Cost justification analysis for many developmental works for complex systems
   - Human expert are either not available/very difficult to find all possible solutions/unable to memorize previous solution
   - Problems are well structured, and/or can be solved with reasoning with many steps in time-consuming manner
   - Traditional computing solution may not be possible
   - Following are the major areas where Expert systems of AI are used:
   - Planning and project scheduling, Maintenance management
   - Manufacturing process and industry applications—-Production planning and scheduling
   - Plant control supervision management and operation—comparing with prescribed data and condition. If necessary, alerting operator and recommend action.
   - Process plant control—control a physical process based on monitoring
   - Generation of Options—generate alternative solutions to a problem
   - Information, help desk management
   - Hospitals, medical facilities management
   - Employee performance evaluation
   - Financial analysis decision-making
   - Virus detection and diagnosis of diseases

**TABLE III 3.1A Comparison between human expert and expert system.**

| Issue | Human expert | Expert system |
|---|---|---|
| Knowledge transfer | Difficult to transfer | Easily reproducible |
| Training cost | Time consuming & costly | Comparatively easier |
| Documentation | Difficult to document | Self-documenting |
| Consistency | Not always consistent | Always consistent |
| Overall cost | Expensive | In expensive (good ROI) |
| Adaptability | Adaptable to change | Rigid—not adaptable |
| Environment | Sensitive to change | Insensitive to change |
| Volatility | Volatile | Permanent |
| Common sense | Possesses common sense | Not possible |
| Creativity | Creative | Un inspired |
| Capacity with time | Decays with age | With more knowledge base, capacity increases |
| Efficiency & speed | Comparatively lower | Much higher |
| Area of focus | Broad | Narrow |

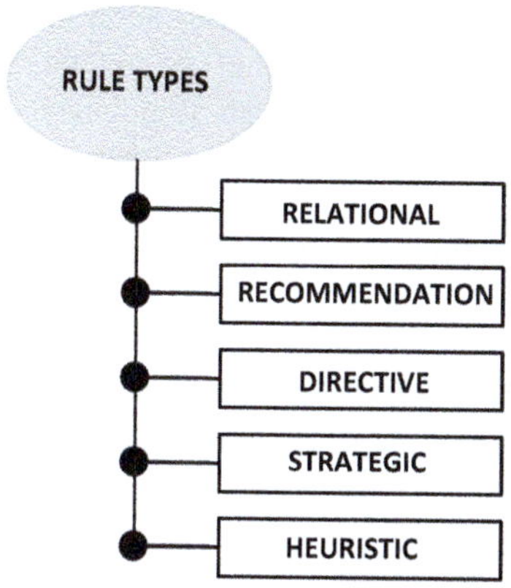


FIG III 3.1b1 RULE TYPES

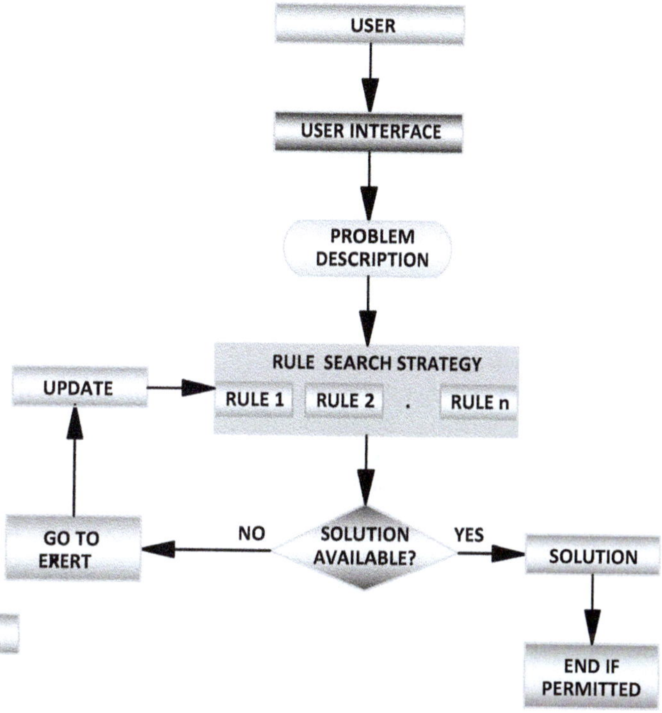


FIG III 3.1b2 FLOW CHART FOR RULE BASED REASONING

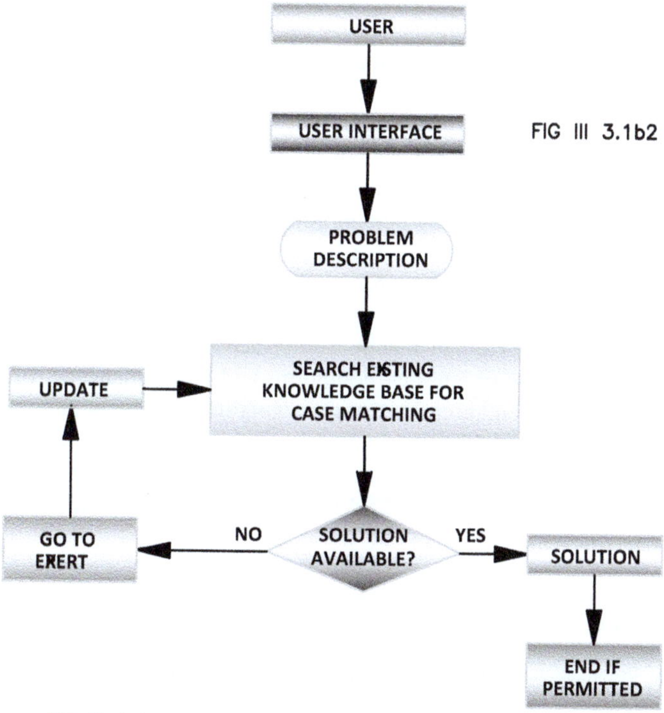


FIG III 3.1b3 FLOW CHART FOR CASE BASED REASONING

**FIGURE III 3.1B** Expert systems reasoning.

3. **Humann expert and expert system:** It is interesting to compare human expert with expert systems as in Table III 3.1a to get to know pros and cons of the systems.
4. **ES discussions:** The issue of connecting expert systems to on-line processes actually started in 1983. Time element in reasoning process is an important issue for RTES. Software vendors have recognized this. So, expert system shells are currently available on the market tailored to meet the requirements of process control applications ensuring close integration with the underlying DCS/PLC for exchange of sensor data and communication with the operator [6]. Based on detailing Fig III 1.2a2 and Fig III 1.2b, ES can be developed for process control systems. In the case of a process plant, it is required is an integration of heuristic and analytic forms of knowledge. The object-oriented

paradigm is used, with a hierarchical frame structure [7]. Several kinds of knowledges in expert systems shall not only include rules but also models for material balances, energy balances, and reaction kinetics, etc., in process controls. Detailed discussions on structure and technology have already presented in I 1.2.0. Generally following are major headings that should be included in knowledge representation in an RTES especially for plant control and monitoring:

- Generic Representation of Process Behavior
- Representation of the Plant Structure
- Temporal Knowledge Representation
- Integration of Knowledge Representations.
- Integration of Knowledge Representations.

ES possesses the explanation facility by which ES can explain how it arrives at its conclusions. When asked by the user on "what"/"how"/"why" aspects of a problem, ES will then provide the user with a trace of the consultation process, pointing out the key reasoning paths followed. Truth maintenance system (TRS) as depicted in Fig III 3.1c is used RTES discussed below.

5. **RTES: Reasoning:** Another important issue in RTES is "Reasoning in real time." Initially snap shot of plant data and using a static expert system paradigm to perform inference. Static expert systems use pattern matching of a set of "facts" and a set of rules, using some combination of forward and backward chaining to combine the inferences [7]. However, with time constraints for real-time operations, this approach has to be modified. Different inference approaches applicable for real-time systems are tried. Here, human expert and a peripheral awareness across the domain, watching for performance exceptions are important. Major steps considered for RTES are as follows:
   - **Performance issues and temporal validity:** In static expert systems, truth maintenance detailed in Fig III 3.1c involves changing inferences when data changes but in real-time system, there would be changes in inference even if no new data are available, since measurements cannot be assumed to remain valid *indefinitely*. For this temporal validity of information plays a major role.
   - **Scheduler:** A real-time scheduler maintains the overall operation.
   - **Coherence:** The asynchronous data arrival time is available through its time stamping for events.
   - **Time limit:** Tasks have priorities, and within each basic interval the tasks are executed by priority, so that systems act within time limit.
   - **Knowledge validation:** To check consistency and completeness of knowledge.

### 3.1.1 Expert systems in automation systems

Since ES simulates human expertise, one can conceived ES as computer applications with embodied non-algorithmic expertise for solving certain types of problem in a specific domain. The characteristics of ESs make them different from conventional programming and traditional decision-making. ES technologies include ES development environment with work station, mini/microcomputer, LISP, PROLOG Tools (in built definitions of model, editor, debugging tool, etc. Shell (ES minus knowledge base, e.g., Java ES shell). Expert systems are an emerging technology with many areas for potential applications including in automation area such as Human machine interface (HMI) (for user interfaces for ubiquitous computing). As the quantities of recommendations are huge, available guidelines allow eliciting guidelines for evaluating already implemented interfaces. Like Robotic process automation (RPA), Expert process automation (EPA) is a technology that enables the

Truth Maintenance System (TRS): TMS are used for problem solving in expert systems in conjunction with Inference Engine (IE) to mange a dependency network the Inference Engine's belief in a set of rule/fact statements. To choose their actions, reasoning programs must be able to make certain assumptions and subsequently revising their belief when at subsequent discoveries contradicts their assumption. TMS as a problem solver subsystem performs these function by recording and maintaining reasons for beliefs

FIGURE III 3.1C Truth maintenance system.

digitization and automation of expert knowledge and processes. Using EPA, a company can build software without programming skills that emulate the work of a subject-matter expert by using a *no-code* graphical user interface (GUI) to implement content, if-then logic, and actions. EPA is like Robotic process automation. Most applications of expert systems will fall into various categories mentioned at the beginning of the section. Here, design and automation application issues shall be addressed.

1. **Process control:** ES in process control can be applied for control system design as well as for real-time process control. ES for design tries to capture the expertise of the design engineers to reduce the design time necessary to develop optional design strategy. Expert control engineers precisely know the questions and exact control system configuration to apply under the given situation. For design problems, the expert gathers all the necessary information on the particular problem at hand. Expert then uses expertise and experience to organizing this knowledge to encode it into a computer program for the suggested control system design. Interactive program in this context is helpful. In such cases, by supplying relevant data for the specific problem, the ES can suggest an appropriate control configuration. Experts are well aware of different types of process behavior. Therefore, ES requires knowledge representation to integrate several kinds of knowledge, including not only rules, differential equation models for material balances, energy balances, and reaction kinetics, etc. For plant objects, the generic knowledge at the various class levels is combined with the parameters, states, inputs, and specific behaviors of the specific instance to determine the overall object behavior. For inference, ES in process control takes "snap-shot" of plant data, to use with paradigm of static expert system which uses pattern matching of a set of "facts" and "rules," in combination with forward and backward chaining. For a time constraint issue, more efficient inference paradigms are called for. There are different inference approaches for real-time problems. Consistency and completeness of knowledge are extremely important in the process control domain, especially for safety issues. Failure scenarios are a powerful way of validating a knowledge base [8]. However, there exist several dynamic models for this.
2. **Other process controls:** Applications of ES in other common process control types viz. Batch process control and Statistical process control (SPC) types have been discussed. There are a number of areas where ES supports these process controls, which have been discussed below:
   - **Batch process control:** *Process monitoring*, *Diagnosis*, *Control optimization*, *Decision support* (at different levels), Quality assurance (QA), *Energy and resource management* and other system integration, etc., are application areas of ES in this control type. The objective of ES in this industry is to capture the expert knowledge and encode that the same in a software package that could be used by the operators to optimize the system and diagnose routine problems. Some pertinent rules are often used to detect, control, and optimize the control process to increase production/improve product quality. Batch process control is improved by defining a step endpoint condition in an expert system knowledge base helpful in monitoring the occurrence of the endpoint in the batch process to trigger a change in a batch process condition at the endpoint.

     It is convenient and preferable to implement ES, and the batch process condition change as module. Multiple instances of modular expert systems allow parallel process units to be easily controlled. ES can help in integrating of procedural and expert system capabilities into one software application environment. *Chemical*, *Food and beverage* and *Pharmaceutical* Industry are common examples for use of Batch process controls.
   - **Statistical process control (SPC):** As the name suggests SPC is defined as the use of statistical techniques to control a process or production method. Here, quality data are collected in the form of product/process measurements to evaluate, monitor, and control a process. *Automated data collection and analysis*, *Root cause analysis* (RCA), *Monitoring and Control*, *Decision support*, *QA*, *Integration and reporting, etc.*, are major application areas in SPC for ES. SPC is a very good tool for monitoring the process both during normal as well as abnormal operation. Many DCS can automatically generate quality control charts for the operator to decide the action to be taken during abnormality. This can be done through proper analysis assisted by ES. The DCS can feed the current state of the process to the expert system at prespecified time intervals. Using the knowledge contained in the expert system, further analysis of the data can be performed to provide advice and guidance to the operator so that the operator can take appropriate action [6]. Automotive industry, Electronics manufacturing, Textile industry, etc., where SPC is used commonly and ES can be applied.
3. **Fault detection:** Software based expert system can offer assistance in monitoring, detecting abnormal conditions, and fault diagnosis to provide safety guards against unexpected abnormal conditions. The

diagnostic process consists of fault detection and show suggestions to the operation and maintenance team. At times for prognosis or fault diagnosis requires services from experts familiar with the operation are required. Therefore, expert systems can serve as an additional tool for the inexperienced operator in diagnosis of faults. Objectively, expert system can not only diagnose the cause of the fault, but also can suggest remedial actions to bring the process back to normal. The Diagnostic expert system for turbo (DIGEST) machinery developed by Siemens in Germany illustrated the potential of integration of the rule-based Fuzzy logic. There are three-levels of hierarchy in system design. The lowest level represents measured data acquisition and conditioning, the middle level represents module-specific analysis and mathematical diagnostics [9].

4. **ES: plant application:** There are huge scope of applications of ES in various plants. To limit the discussions, we discuss ES in power plants. For power plants general application areas are as follows: *Plant data management*, *Training*, *Safety/condition monitoring*, *Alarm analysis*, *Sequence of event* (SOE), *Fault diagnostics*, *Pre and post trip analysis*, especially for nuclear plant: *Accident management Emergency planning*. Many other areas of applications of ES include but not limited to: *design*, *schedule optimization*, *overload clearing* and *load forecasting*. Apart from these, operator support, start-up scheduling are important areas for application of application of ES in power (and process) plants. Functionally, operator support system needs to perform the following:
   - Prediction of functional disorders
   - Detecting the primary causes of functional disorders
   - Compression of information display, suppression of the secondary fault displays for functional disorder;
   - Support operator in decision-making, predicting the consequences of decisions made by the operator.

   Accordingly, a real-time system must perform—response activities within the specified (limited) time, activities of different priorities—event-controlled and scheduled activities. A few operator support areas where ES can be very helpful are as follows:
   - **Design:** Knowledge of standard manuals and expert operators are stored in the knowledge base for production rules, in ES technologies in the design process.
   - **Start-up schedule:** In power plant start-up schedule is an important issue as it is highly nonlinear and involves interaction among complex process variables making difficult to model. ES with proper knowledge base can be applied for the issue to automatically optimize the speed-up and load-up pattern of the plant. Quantitative knowledge of dynamics models and qualitative Fuzzy knowledge can be applied in ES for optimization.
   - **Overload and load forecasting:** It is required to reduce the magnitude and time of overloads on the systems. The overload clearing calls for the skill and experience of operators. So, ES can be applied for this to enhance the decision-making capacity for appropriate control actions. Also, based on the situation, trained fuzzy logic based ES can be used for making load forecasting in short pipe.
   - **Data analysis:** ES can be deployed Intelligent system Expert system for performance evaluation and diagnosis (ESPD) in nuclear power plant to support operators in improving plant performance and to enhance operational flexibility.
   - **Alarm analysis:** ES based alarm processing can detect the location of equipment faults, types of faults and possible malfunctions of the protection system and will give explanation as applicable.
5. **HMI Design:** Some of the area where ESs find their applications are as follows: *sensitive decision-support*, *specific local customization*, *updating changes in circumstances and experience*, *uncertain inferences and facts* as well as for *ubiquitous computing* which relates to human-computer interfaces whereby systems must adapt themselves to users (not the opposite). Three areas covered in ES are *Knowledge*, *Environment* and *Interaction*.
   - **Knowledge**: Users register anywhere its knowledge, experiences, or memories by means of various ways.
   - **Environment:** This refers to the computer and physical environment information and dealing with it. Applications can gather data from the place where they are and dynamically build computational models to adapt themselves to users' needs.
   - **Interaction:** This refers to an interaction closer to humans, providing multiple ways of interactions.

### 3.1.2 Detailed discussions on expert system and programming languages

Expert system (ES) features are achieved by utilizing human expertise/knowledge in control systems. Expert control pertains to a general class of intelligent control meant to increase the autonomy of technological systems viz. process control, manufacturing and robotic systems, autonomous vehicles, etc. Three intelligent control levels are as follows: *Organization*, *Coordination* and *Execution*. So, when looked at expert system very closely one would find that it comprises:

- **Development environment:** This includes hardware and tools like *Workstations*, *minicomputers*, High level symbolic programming languages like LISP and large databases.

- **Tools:** Tools including powerful editors and debugging tools are used to reduce the effort and cost for the developing an expert system. Also, they are helpful for knowledge representation, and inference design.
- **Shells**—This is an expert system without knowledge base, provide the developers with knowledge acquisition, inference engine, user interface, and explanation facility. Some of the popular expert systems are as follows:
- **DENDRAL:** As an artificial intelligence project can be used for chemical analysis expert system especially for organic chemistry for detection of unknown organic molecules.
- **MYCIN:** It was backward chaining expert systems to find the bacteria causing infections.

  There are several programming languages LISP, Prolog used ES. Staring with basic features and expert control architecture, discussions will be presented on various technical issues and programming languages shall be covered in this subsection.

1. **Expert control architecture:** Expert control systems can control large varieties of systems with uncertainty, nonlinearity, time varying, etc., using control intelligence with available a priori knowledge. Often the control system is meant for meeting tough qualitative specifications and to enhance system performance by applying knowledge database. These expert systems are often used for fault detection/diagnosis. The controls can communicate and interact with the user/system operator nicely. Knowledge-based ES comprises parts like: *knowledge of facts, knowledge of relations between the facts,* and a suitable technique for *acquiring* and *storing* this information. Let us discuss in details the ES control system. Fig III 3.1d generic ES functionality and control architecture. Fig III 3.1e shows ES control components and control process.
   - **Components:** With reference to Fig III 1.2d, various components of expert systems are as follows: *Knowledge base* (KB), *User interface* (UI), *Inference engine* (IE), *Work space* (WS), *knowledge acquisition* and data base.
   - **ES control schematic:** Functional details of ES has been explained in Fig III 3.1d1 ES control scheme has been detailed in Fig III 3.1d2 [10]. For better understanding of the system, expert operator with operative knowledge has been shown, to point out how digital expert system is updated. This shows that expert system influences both control algorithm as well as supervisory level meant for of parameter estimation and for fault detection. ES and control systems need to communicate and exchange data.
   - **ES architecture:** The KB contains the symbolic knowledge about the problem. Knowledge acquisition helps to extract the knowledge. WS refers to the memory for storing. Data base The DB contains numeric information like input data. IE does the job of matching the left-hand side rules with the available general knowledge of the problem for succeed the goals. Explanation component in conjunction with IE informs the user about the conclusions with detailed explanation. These components are interconnected and related as shown Fig III 3.1e1. The knowledge engineer extracts the knowledge from the expert and embeds it into the expert system tool used to construct the expert system. A schematic view of The ES building process has been schematically shown in shown Fig III 3 1e2.
2. **Categorized applications:** With various examples and description applications of expert system in different categories have been detailed in Table III 3.1b.
3. **Developmental steps:** Initial discussions on ES development has been covered in III 1.2.0.5 with elaboration in Fig III 1.2b. Detailed iterative process of ES development comprises following steps:
   - **Identification and domain:** *Definition of the problem, determination of the problem characteristic, so that it is suitable for ES, establish ES task domain for cost-effectiveness of the system.*
   - **System design:** *Identification of the ES technology* and *identification and establishment of the degree of integration with the other systems and databases.*
   - **Knowledge engineering:** Knowledge engineer *translates the knowledge into a computer-understandable language.* Knowledge engineering also includes finding the *ways to integrate and use uncertain knowledge in the reasoning process with suitable explanation* form. Also, *inference engine with reasoning structure, is developed.*
   - **Prototype development:** From knowledge base, *domain knowledge is acquired* to represent it *in the form of If-THEN-ELSE rules.*
   - **Test and refine the prototype:** *The sample cases are tested as prototype for any deficiencies in performance* and *end users to test the prototypes of the ES.*
   - **Complete ES:** *Testing and establishing the interaction of the ES with environment, including end users, databases, other information systems, Documentation the ES project* and *User training.*
   - **Maintenance of ES:** *Regular periodic updating of the knowledge base* and *Cater to the new interfaces requirements as it comes.*
4. **LISP basics and comparison with Prolog: LISP** (*John McCarthy in 1958*) is an abbreviation for *LISt processing* which is strong in symbol manipulation

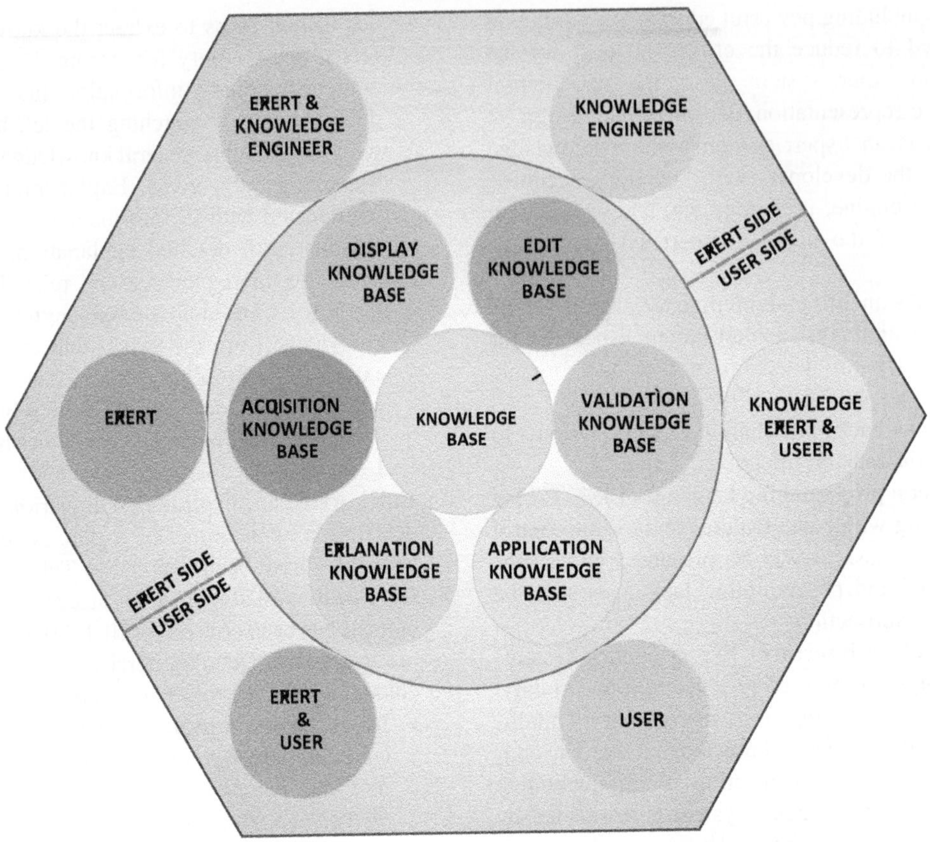


FIG III 3.1d1 EXPERT SYSTEM FUNCTIONAL DETAILS

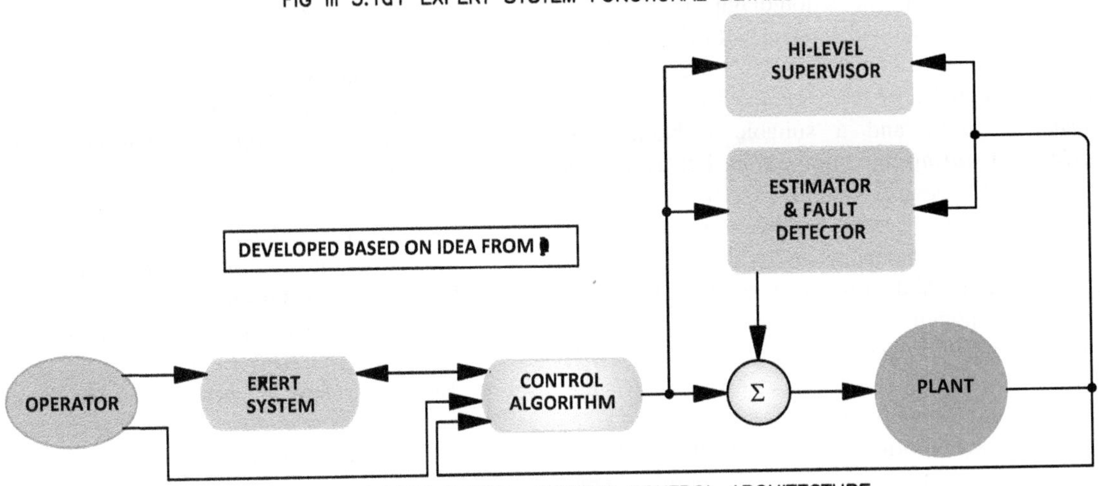


FIG III 3.1d2 EXPERT SYSTEM: GENERIC CONTROL ARCHITECTURE

**FIGURE III 3.1D** ES: Functionality and generic control architecture.

computer programming language family. From the very beginning, LISP has been applied in AI in general, especially for the development of ES. A few essential details for LISP are as follows:

- **LISP process and components:** Every LISP procedure is a function, when called, it returns a data object as its value. LISP program can be conceived as a collection of expressions to be evaluated by a LISP interpreter. To distinguish these expressions from expressions representing data, the former ones are often called forms. LISP generates an interactive environment to alter the form, the user to insert the main program as a sequence of expressions. It can use calls to communicate with the other functions running simultaneously, but avoids block structure/other type of complex syntactic organization. Two main components are as follows:

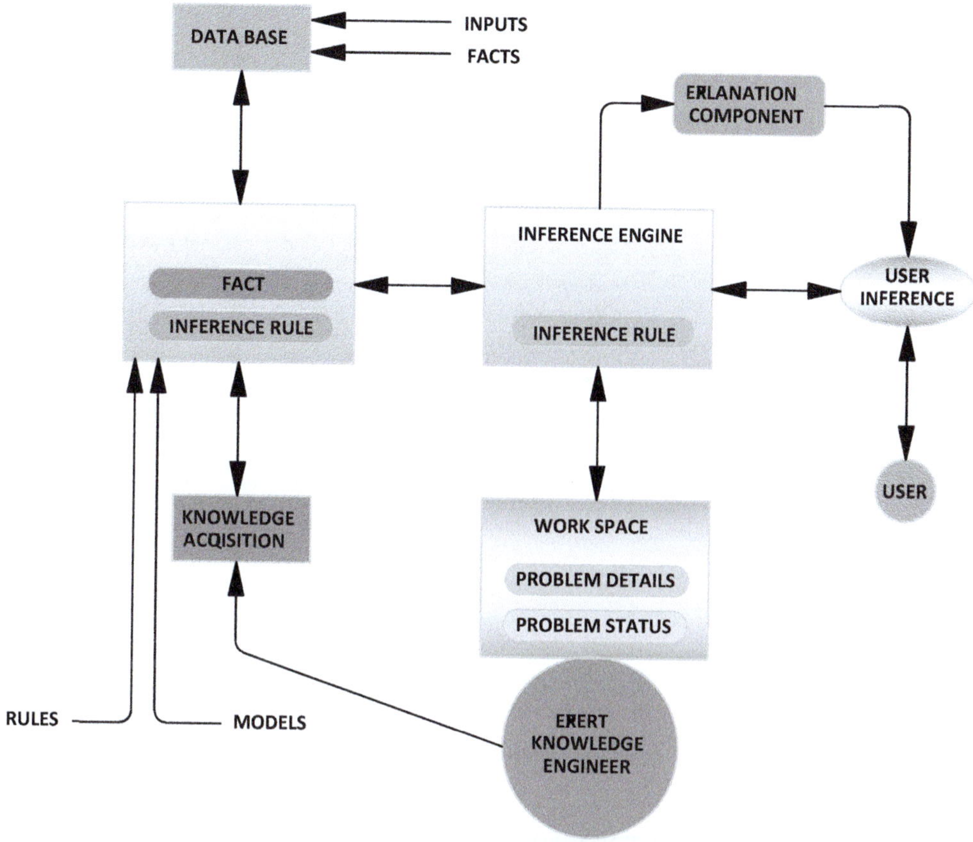


FIG III 3.1e1 ES: CONTROL COMPONENT DETAILS

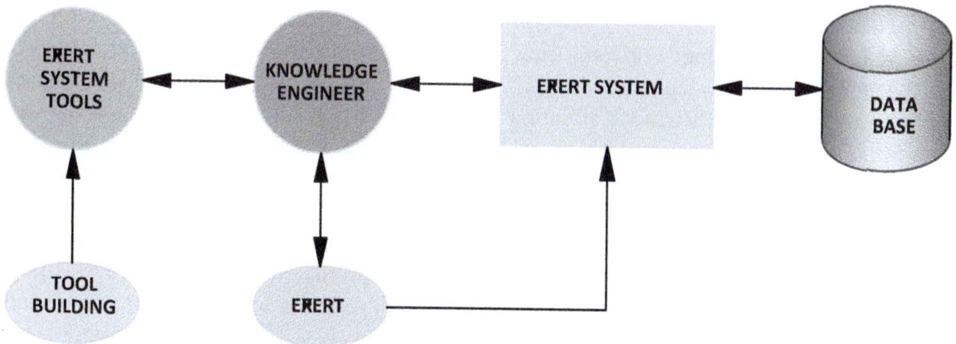


FIG III 3.1e2 ES: CONTROL PROCESS DETAILS

DEVELOPED BASED ON IDEA FROM

**FIGURE III 3.1E** ES: Control components and control process.

- *Atoms:* Atoms including symbols and numeric values are the types identifiers, utilized in imperative languages.
- *Lists*: As the name signifies these are essential components of LISP programming. Here, lists are defined by delimiting their components within parentheses rather than utilizing the insert and delete operations.

- **Common LISP features:** A few common features are as follows:
  - *It is Machine-independent, expression-based supporting different decision-making statements.*
  - *It uses iterative design methodology with easy extensibility.*
  - *It allows create and updating the programs dynamically.*
  - *It provides advanced object-oriented programming and condition with high level debugging.*
  - *It provides wide-range data types (symbols, objects, structures, etc.) and convenient macros.*

**TABLE III 3.1B** Some expert system categories and applications.

| Category | Description with examples |
|---|---|
| Classification | Based on features and attributes viz. Dendroid |
| Control | Interpretation, prediction, monitoring & repairs: Knowledge based ES, RTES |
| Debugging | Incremental solutions for complex problems: Simulation |
| Design of product | Design & product development: Camera lens, automobile |
| Diagnosis | Medical: Cause of disease from observed data with remedial measure. Engineering/knowledge domain: Cause of failure with remedy viz. Faults in vehicles, computers or in finance: Detection of fraud, suspicious transactions. |
| Forecasting | Forecasting results, e.g., Gas price, share trend |
| Intelligent monitoring | Continuously comparing data observation with prescribed item viz. Leakage monitoring in midstream pipeline. |
| Process control | Based on physical measurement controlling a physical process |
| Repair | Schedule repair, maintenance schedule |
| Selection | Selection of tools & action (kind of decision-making): Tools for computer numerical control (CNC) machine. |

 - *It provides input and output functions, complete I/O library, extensive control structures.*
- ***Pros and cons*:** Various pros and cons of LISP has been detailed in Table III 3.1c.
- **Applications area:** Apart from ES, large successful applications of LISP are as follows: *Emacs, G2, AutoCAD, Igor Engraver,* etc.
- **Comparison of LISP and prolog:** For long time, LISP was the lone language deployed for AI, recently with introduction of PROLOG situation has changed. So, it is better to compare the two languages as detailed in Table III 3.1d.

5. **Introduction to prolog:** Prolog (*Alain Colmerauer* and *Robert Kowalski in 1972)* is an acronym for "PROgrammation en LOGique." Based on the principles of first order predicate logic, Prolog is a simple but powerful programming language with following essential details:
   - **Description:** PROLOG is highly strong in solving problems characterized by requiring complex symbolic computations. This is makes Prolog very much suitable as the programming language in AI where *symbol* and *inference manipulations* are the fundamental tasks. In comparison with conventional program, PROLOG programs are often shorter and easier to grasp. As seen from Table III 3.1d, Prolog is a logical programming language. PROLOG supports stepwise refinement in developing programs because of its modular nature. Three elements of logical programming are as follows:
     - *Facts:* The fact is predicate that is true.
     - *Rules:* Rules are extinctions of facts containing conditional clauses.
     - *Questions:* Questions are needed for answering by the given facts and rules.
   - **Common prolog features:** Some major common features are as follows:
     - A declarative language, with programming in logic, that is, program consists of data based on the facts and rules (logical relationship).
     - The solution is based on a question by the user. The answer is based on the database of facts and rules. Also, relies heavily on recursive definitions.
     - Supports a backtracking strategy to search for proofs, a pattern-matching facility, mathematical variable, and input and out are interchangeable.
     - A basic logic programming environment with type less basic Prolog elements.
   - **Functional vs logical programming:** In Functional programming, we have to define the procedures, and the rule how the procedures work. In logic programming, the problem description and the method for solution are separated from each other. So, algorithm is the sum of logic and control. Here, "logic" indicates the descriptive component of the algorithm (problem) the term "control" refers to the component for a solution. So, a logic component may be developed using a method of stepwise refinement and the changes to the control component affect the efficiency of the algorithm; they do not influence the solutions produced. Table III 3.1e depicts the difference between two programming types.

**TABLE III 3.1C Pros and cons of LISP.**

| Advantages | Disadvantages |
|---|---|
| Clear, concise syntax and semantics with range of primitives for creating, and editing lists | Time consuming for getting accustomed to notation and parentheses. |
| Supports to create faster apps and assists for app development faster | Less job prospects for lisp developers |
| Straightforward control structures, | No common lisp environment. |
| Complete & well-defined functions in the form of expressions | Comparatively small ecosystem with respect to Java, C++ etc. |

**TABLE III 3.1D** Comparison of LISP and prolog.

| Feature | LISP | Prolog |
|---|---|---|
| Language type | Functional programming | Logical programming |
| Programming details | Functions, conditional evaluation, loops, recursion, and iteration. | Directed control search, patterns, loops, and backtracking. |
| Based on | Functions in a global Environment. | Rules & facts |
| Data types | Symbolic & numeric atoms, lists & list structure, property and association lists | Symbolic & numeric atoms, predicates, lists, & list structures |
| Variables | Dynamic scoping of free variables, bound variables, and local variables using let blocks. | Variable scope scoping is confined to a Single rule or fact |
| Data transmission | Program function returns, local & global variables binding, parameter passing by value | Binds the data/numbers with the variable via unification. |
| Supports paradigm | The functional, reflective, procedural, and meta | Logical computing |

- **Pros and cons:** Various pros and cons of Prolog has been detailed in Table III 3.1f.
- **Applications area:** The applications of prolog are as follows: *Specification language, Natural language understanding, Robotics, Machine learning* (ML), *Intelligent database retrieval, Automated reasoning* and *ES.*

## 3.2 Detailed discussions on fuzzy control systems

As already discussed, the fundamental concept of Fuzzy logic is the membership function defining the degree of membership for an input to a particular set/category. The membership function refers to a membership degree between 0 and 1 for any input. So, one can conclude that Fuzzy Logic is a mathematical method for representing vagueness and uncertainty in decision-making, allowing for partial truth for use in a wide range of applications. Therefore, it is implemented based on Fuzzy rules. While applying rule, FLCs use some knowledge base/expertise of artificial in a decision-making system, in direct contrast to conventional controllers. So, the merits of computation, expert knowledge, could be converted into functions to be applied in automatic control. Thus, FLC divides the measured parameters into subsets in the range between 0 and 1 to establish the interrelationship between subsets, for estimating the manipulated variable based on rules. Therefore, FLC is a rule-based robust algorithm to address complex nonlinear problems. FLC offers advantages like: *Flexible operation, Convenient user interface, Ease of implementation. Qualified validation,* etc. On account of these FLC is preferred in many controls like Kiln control or Maximum power point *tracing* (MPPT). Different Fuzzy control schemes are in use. These include but not limited to cascade, feedforward, variable structure, self-tuning, adaptive and quality control leading to hybrid classical/fuzzy control systems. Fuzzy logic is a philosophy with degrees of imperfect truth to arrive at a decision to solve complex problems. Probability that way is crisp to assertions which can be either true or false with certain chances. So, probability is the degree of confidence in its veracity. Fuzzy logic essentially attempts to capture the essence of ambiguity. It largely refers to the degree of truth. Probability theory cannot be used to reason with propositions that are partially true. Fuzzy modeling is a mathematical technique for creation of models with imprecise information. So, it can be conceived as an extension of classical logic utilizing Fuzzy set theory for creation of models having flexible decision boundaries, better adaptation to specific applications with precise reflection of their characteristics. Network congestion control is a complex problem and is critical for Internet Protocol (IP) networks. People are looking for alternative nonanalytical control system design and modeling schemes. Some of these schemes employ fuzzy logic control for this purpose. Based on the properties, FLC finds potentials applications in process automation. In this section detailed discussions have been presented on all these various aspects including Fuzzy logic computational details, Fuzzy modeling as well as Fuzzy in automation.

### 3.2.0 Fuzzy set theory, logic in plant automation

For typical conventional controls with crisp logic, an accurate mathematical model is necessary. However, on account of high non linearity, time variance of the process and/or complicated process dynamics it is not possible to achieve the same. FLC which incorporates uncertainty and abstract nature of human can provide a good solution. Better solutions are achievable when fuzzy comes with expert systems. Fuzzy logic can handle uncertainty very well as probability can. *Does it mean it is a kind of probability*? No both probability and fuzzy logic handle those problems where we have limited data about the problem is at our disposal. Randomness actually represents *uncertainty*. Fuzziness describes ambiguity for an event. Fuzziness *quantifies* (through membership function) the *vagueness, ambiguity* and/or *imprecision* of data. As per to Prof Bhanot. "*Fuzzy in Fuzzy logic applies to the imprecision in the data and not in the logic*" [5]. As discussed at the beginning of the section, there are differences which will be clear from the following:

**TABLE III 3.1E** Functional programming vs logical programming.

| Functional programming | Logical programming |
|---|---|
| Logics and controls are mixed | Logics and controls can be separated. |
| Von-neumann architecture/uses the sequential steps. | Abstract model/deals with objects and their relationships. |
| The computations are done by executing the statements sequentially | The computations are done by clause deduction |
| The syntax are the sequence of statements | The syntax are the logic formulae |

**TABLE III 3.1F** Pros and cons of Prolog.

| Advantages | Disadvantages |
|---|---|
| Execution is as per the definition of predicates and supports pattern matching and backtracking | Negative statements/"*NOT*"/"*OR*" logical conditions not supported support. |
| Simple coding to store & operate data lists with data base data structures in language and human thinking. | The prologue sequence impacts the programming efficiency. First-order logic programming language not supportive to second-order logic |
| Its rules are created using recursive thinking | Without graphic capabilities; for visuals use turbo prolog. |
| Declarative, compact, rational, interpretive, and modular by definition | Not easy to understand the various algorithms and code for input and output processes |

- **Probability:** It handles uncertainty, lac of data overtime in mechanized way. When mutual added up it must be unity and it could be described in probability distribution function through prolonged observation.
- **Fuzzy logic:** Fuzzy logic also handles lac of data but this is based on vagueness in defining an event and fuzzy logic is handled through human cognitive process. Unlike probability, mutual exclusive events are not added to get unity, but its membership functions varies between 0 and 1. Membership function is decided based on expert opinion and not as per any distribution function.

With these ideas in mind, let us look into fuzzy set theory which is the basis of Fuzzy logic in artificial intelligence for solving various plant automation issues.

1. **Fuzzy set theory and mathematical representation:** By this time, it is clear that unlike crisp set Fuzzy sets classify an element into continuous set in terms of its membership, with varying degree between 0 and 1.
   Fuzzy set is represented by $A = \{x, \mu_A(x)\};\ x \in X$ (discourse), where x is the element and $\mu_A$ is the membership already discussed in I 1.3.0. Some Fuzzy operators have been elaborated in Fig III 3.2a. Various Fuzzy operators are as follows:
   - **Fuzzy union:** Fuzzy union has been elaborated in Fig III 3.2a1.

$$\mu_{A \cup B\,(x)} = \mu_A(x) \vee \mu_B(x) = \text{Max of } \{\mu_A(x), \mu_B(x)\} \quad \text{(Eq. III 3.2a)}$$

   - **Fuzzy Intersection:** Fig III 3.2a2 depicts Fuzzy intersection

$$\mu_{A \cap B\,(x)} = \mu_A(x) \wedge \mu_B(x) = \text{Min of } \{\mu_A(x), \mu_B(x)\} \quad \text{(Eq. III 3.2b)}$$

   - **Complement:** Fuzzy complement has been elaborated in Fig III 3.2a3.

$$\overline{\mu_A(x)} = 1 - \mu_A(x) \quad \text{(Eq. III 3.2c)}$$

   - **Product of Fuzzy sets:**

$$\mu_{AB}(x) = \mu_A(x).\mu_B(x) \quad \text{(Eq. III 3.2d)}$$

   - **Fuzzy set Power:**

$$\mu_A^{\alpha}(x) = [\mu A(x)]^{\alpha} \quad \text{(Eq. III 3.2e)}$$

   - **Equality of Fuzzy set:**
     Fuzzy set A = Fuzzy set B if

$$\mu_A(x) = \mu_B(x) \quad \text{(Eq. III 3.2f)}$$

2. **Centroid and solution:** Here, discussions have been presented on centroid important in arriving at solutions in Fuzzy systems.
   - **Centroid of an area:** Centroid of an area is important to find control solutions through Fuzzy logic controller. Similar to center of mass, the *centroid* of an area can be thought of as the geometric center of that area. The centroid is often denoted by "C" having coordinates $\bar{x}$ and $\bar{y}$. They are represented by $\bar{x}$ (distance from y axis) and $\bar{y}$ (distance from x axis) to denote the average x and y coordinate for the area. Following are two methods to find centroid of an area:
     - *First moment integral* and *Method of composite part.*
     - Formula to find the coordinate is given below:

$$\bar{x} = \frac{\int x dA}{A} \text{ and } \bar{y} = \frac{\int y dA}{A} \quad \text{where, } A = \int f(x)dx \quad \text{(Eq. III 3.2g)}$$

   In case it is complex shape, the shape is divided into known shapes then to use

$$\bar{x} = \frac{\sum_n Ai.\overline{xi}}{\sum Ai}, \bar{y} = \frac{\sum_n Ai.\overline{yi}}{\sum Ai} \quad \text{(Eq. III 3.2h)}$$

   With reference to Fig III 3.2b, let us first see what is centroid of an area and method for defuzzification. Now let us see how to calculate centroid of an area.
   - **Calculation of centroid of area:** In previous sub section methods to calculate centroid has been described. Now let us consider a simple example shown in Fig III 3.2b1. Objective is to find the coordinate of centroid of this triangular area. Let the coordinate of centroid be $(\bar{x}, \bar{y})$. Let us choose a small stripe (shown by red [dark gray in print version] hatch in Fig III 3.2b1) with height y and width dx as shown. Naturally the centroid of this will be a $x_c$, $y_c$ for the small stripe as shown. Now with the help of Eq. I 5.1f we can get centroid for the entire space. Since we have chosen origin at one of the vertices of triangle, so we can write the equation of the hypotenuse line as

$$y = (a\,/\,b).x \quad \text{(Eq. III 3.2i)}$$

   - Also, area of the stripe be $dA = y.dx$ ———————— (Eq. III 3.2j)
   - If we integrate small stripe from 0 to b we get the total area—hence the limit of integration (also, as this is a right angled triangle hence area $A = \int_0^b dA = 0.5$ a.b).

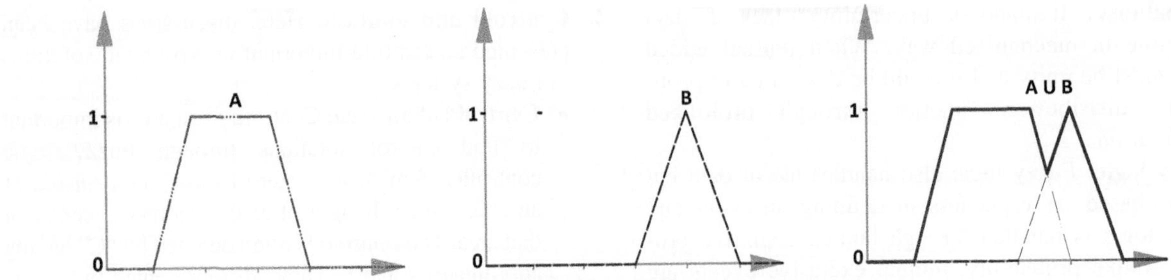


FIG III 3.2a1 FUZZY SET UNION

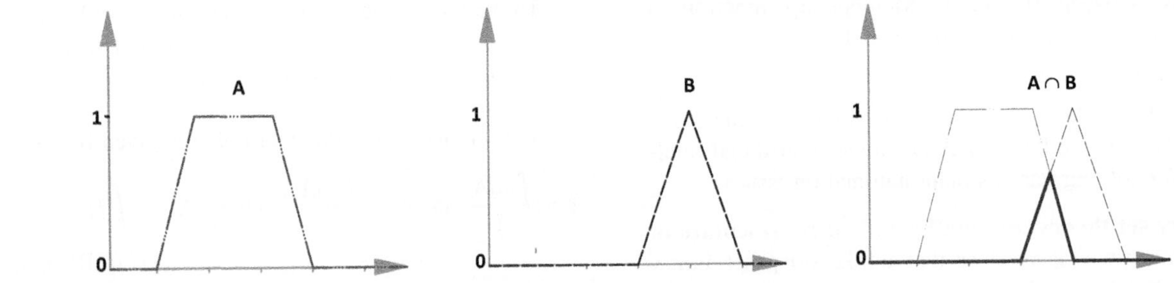


FIG III 3.2a2 FUZZY SETS NTERSECTION

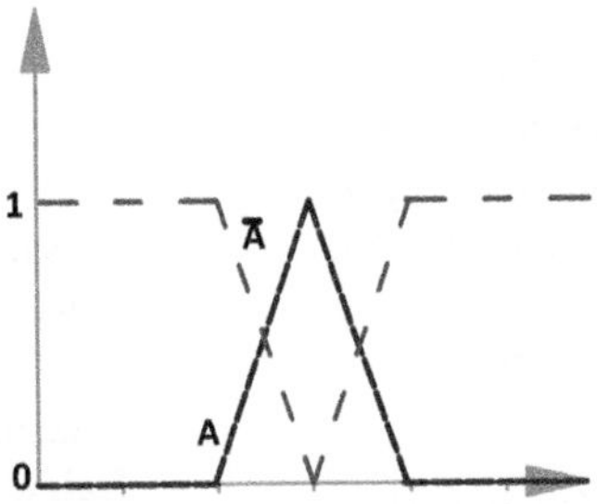


FIG III 3.2a3 FUZZY COMPLEMENT

**FIGURE III 3.2A** Some fuzzy logic operations.

Therefore, $\bar{x} = \frac{\int_0^b XcdA}{\int_0^b dA} = \frac{\int_0^b X.ydx}{\int_0^b y.dx} = \frac{\int_0^b X.\frac{a}{b}.xdx}{\int_0^b \frac{a}{b}.xdx} \frac{\int_0^b x^2dx}{\int_0^b xdx} = \frac{2b}{3}$
——————————— (Eq. III 3.2k)

By applying Eq. I 5.1f, substituting values by Eq. I 5.1 h and I and evaluating the integral.

Similarly,

$\bar{y} = \frac{\int_0^b ycdA}{\int_0^b dA} = \frac{\int_0^b 0.5y.ydx}{\int_0^b y.dx} = \frac{1a}{3}$ ———————————

(Eq. III 3.2L)

Here, y has been substituted from Eq. I 5.1i and intermediate steps are skipped.

In the similar way centroid for different areas could be calculated. In case of complex area, it is divided into several areas of know figures to calculate as shown in Fig III 3.2b2.

Typical values of a few standard areas are available in any standard university book on mathematics/physics.

3. **Logic handling in fuzzy system:** If we recall the discussions in subsection III 1.3.0.3 and Fig III 1.3a3, it is clear that there will be a few steps followed in Fuzzy logic which will be seen little closely below:
Matching the facts with the rule premises to get membership of the fact to the premise of the rule. In case there are several premises then the logical connection between then has to be established through fuzzy union/fuzzy intersection discussed above. The evaluation is performed by t-norms or t-conorms to get the fulfillment of fact with (complete premise) [11].
[Here T norm and T co norm (also called S norm are similar to binary operations in fuzzy mathematics. They normally for Commutativity, Monotonicity and Associativity. For further details any standard book on mathematics for Fuzzy system may be consulted].

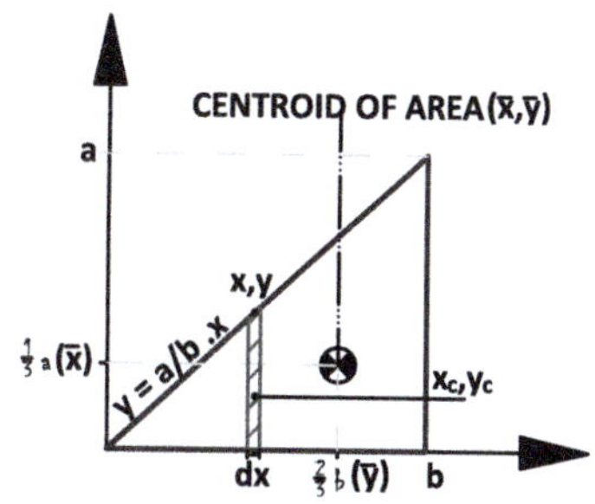


FIG III 3.2b1 CENTROID OF AN AREA

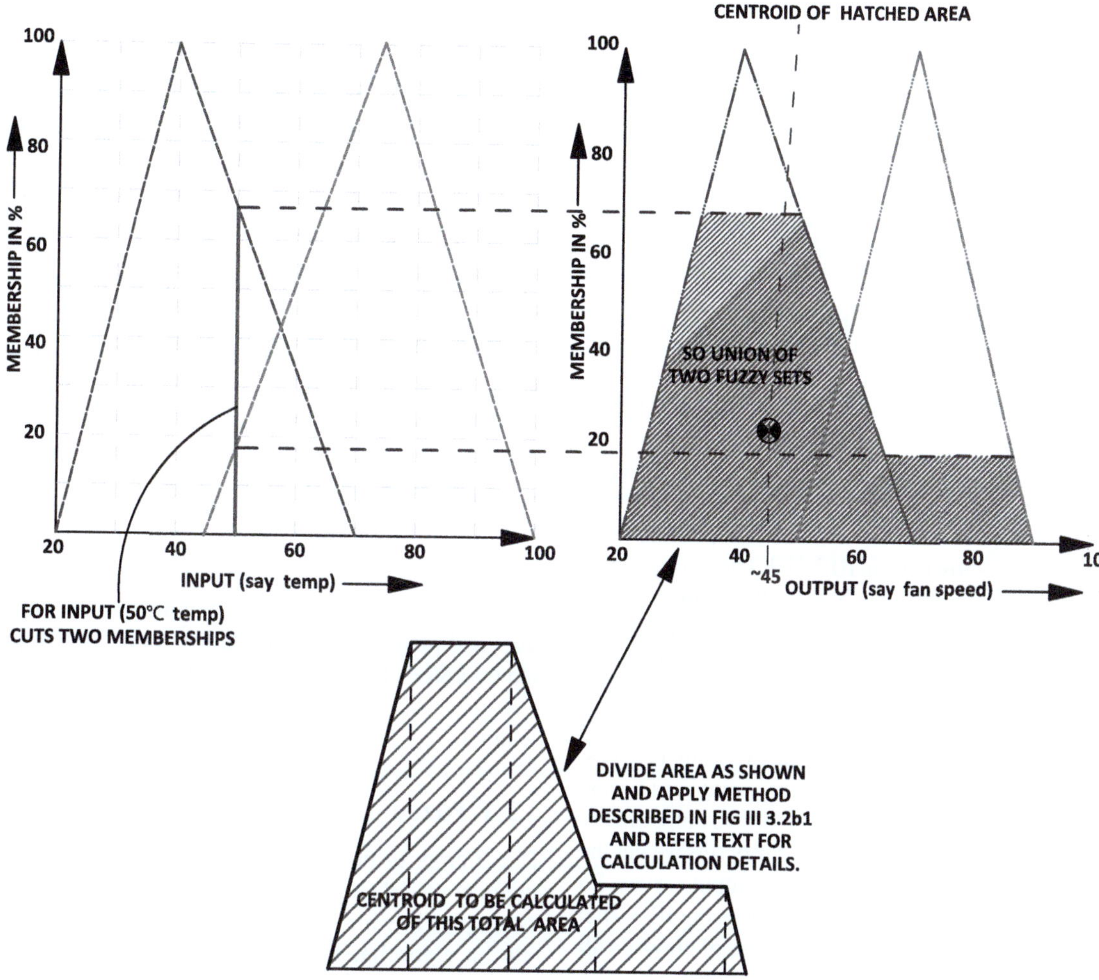


FIG III 3.2b2 DEFUZZIFICATION BY CENTROID METHOD

**FIGURE III 3.2B** Centroid of an area and fuzzy logic.

Then all conclusion of all fuzzy sets is accumulated if there are several rules to get final fuzzy conclusion. Obviously next step is defuzzification discussed below. If the number of fuzzy sets increase, then the computational effort of a fuzzy-rule-based system increases heavily. So, to avoid complications at times special assumptions are considered to simplify the issue.

4. **Discussions on defuzzification:** Defuzzification method with example has been elaborated below:
   - **Defuzzification methods:** From the discussions in sub section III 1.3.0 it is clear that Fuzzy logics evaluate in linguistic manner. In this method results are in fuzzy form. Now these results are required to be converted into crisp output. This is

defuzzification described earlier. Following are a few methods used for defuzzification:

- Centroid of Area (COA)/Center of gravity (COG) Method
- Center of Sums Method (COS)
- Center of Area/Bisector of Area Method (BOA)
- Weighted Average Method
- Maxima Methods: *First of Maxima Method* (FOM), *Last of Maxima Method* (LOM) and *Mean of Maxima Method* (MOM).

Out of these first two are indicated in equation Eqs. III 3.2g & III 3.2h. Also, typical calculation method for centroid of area has been elaborated in the above. Now we shall see how it is applied for defuzzification.

- **Defuzzification example:** If we recall the discussion of sub section III 1.3.0.3 it could be seen that there are several parts or sections in Fuzzy logic. This has been well depicted in Fig III 1.3a3. Now let us take the example of a cooling system with rule that if temperature is hot speed will be fast, when warm speed will be medium and when cold, speed will be low. Based on the same let us take the example of two inputs one between 20 and 70°C and another between 40 and 100°C as shown in Fig III 3.2b2. On the contrary, let two output fuzzy sets be such that one is between one is between 20% and 70% speed and another be between 50% and 90% speed for two input functions as per rule as shown in Fig III 3.2b2. These are just hypothetical for explanation. Now let the temperature be 50°C. Then by plotting that in input curve we see that it intersects two inputs, with membership of approximately, 70% and 20% for first and second inputs respectively as shown. From the intersection points we draw horizontal lines to cut the two output curves. So It is clear that our desired control point lies at some point in the area formed by union of two output fuzzy sets. On account of union of two fuzzy sets we get the hatched portion of area of both the output curves. So total area has been shown separately in Fig III 3.2b2. Now the centroid of this area or center of gravity will give the crisp output. Referring to the hatched portion separately we find that it is a complex area. The centroid in such cases can be found by dividing the area into regular shapes as shown and then finding the centroid of each area then summing them up as per Eq. Eq. III 3.2g & Eq. III 3.2h. In this case desired point will be at ~ 44% speed.

5. **Fuzzy system in plant automation:** Fuzzy controller can be used both as direct feedback control as well as supervisory control viz. regulating parameter of controller used as main feedback controller to get better performance of control system, that is, better PI. In cement kiln it is often used as supervisory control through operator actions. As already discussed, Fuzzy logic is developed based on "*IF THEN*" rules which can be obtained from many sources such as:
   - Working experience of Expert
   - Control engineering knowledge on the process
   - Operator's control actions.
   - Model developed for the process following fuzzy mathematics and logics
   - Development of knowledge base through learning.

   As result these can be used in expert systems, self-tuning/adaptive control systems and artificial neural network in artificial intelligence of control systems.

   The neuro-fuzzy model can initialize and learn linguistic rules, the modeling framework can be considered as a direct transfer of knowledge [12]. Fuzzy control can work with conventional feedback controllers, that is, typical PID controllers. As stated above it can be operated as supervisory controller to adjust tuning parameters of feedback controller for better PI. Also, operator can use this as a parallel controller with switching. It can used as feedforward controller also. When used as feed forward controller it acts as to linear PID controller which take care of the operation is steady state condition near to normal conditions. When sudden changes occur or if the process enters abnormal situations, then this (*supplementary* controller) configuration may be useful to bring the process back to normal operation as fast as possible [12]. Fuzzy systems can be used for *monitoring and supervision* system also. Fuzzy system finds its application in all levels of plant automation viz. control level, monitoring and supervision level and management information level. System principle and interface of FLC with plant has been detailed in Fig III 3.2c [13] where human thinking of Fuzzy model has been detailed.

### 3.2.1 Fuzzy control: models and tools

The attempt to utilize the imprecise information in mathematical models led to the development of fuzzy modeling techniques in which a linguistic description is transformed into an algorithm for an action. The Fuzzy logic model (FLM) is an approach of multiple-valued logic to imitate the uncertainty of the human brain and the form for unknown or undetermined models. As the name suggests, it is based Fuzzy set and membership function for handle *fuzzy relations* to solve uncertain problems. Model-based fuzzy controllers using rules, are meant to model the operators response to particular process situations. These controllers call for considerable "knowledge engineering." So, some means should be there to gather a collection of rules from knowledgeable operators and then condense them. FLC rules to handle ambiguities, are experience oriented and might be obtained from time-consuming trial-and-error

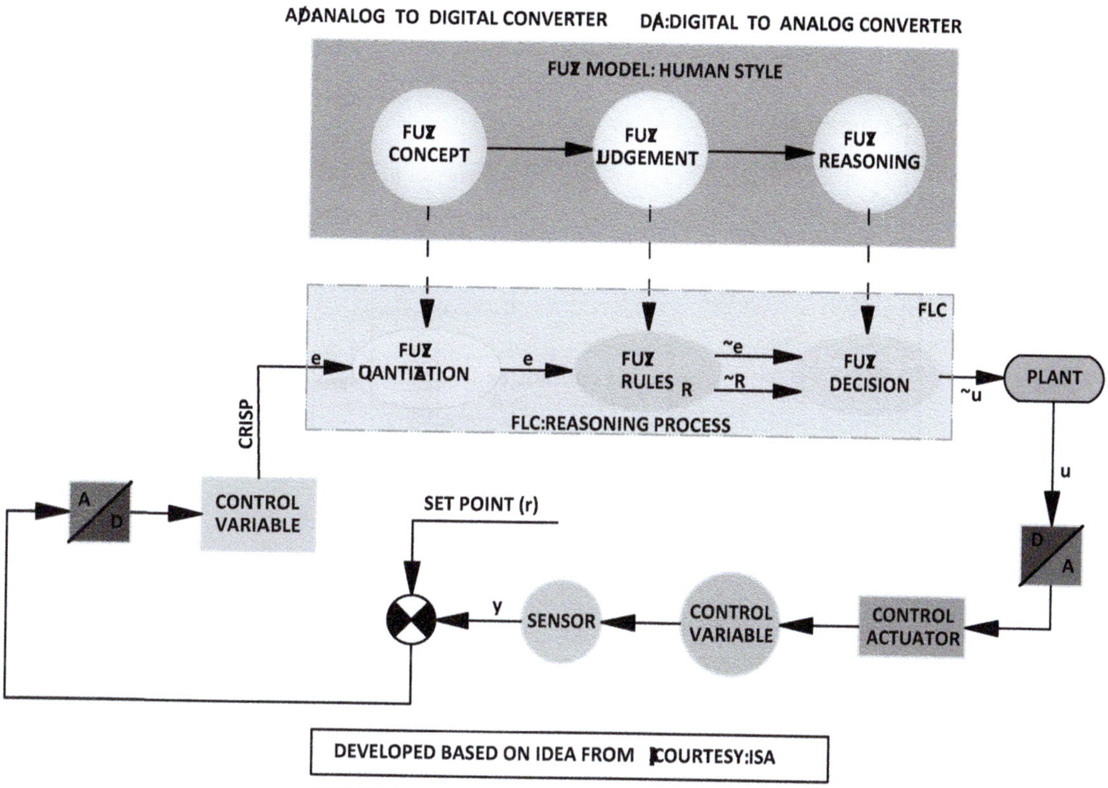


**FIGURE III 3.2C** FLC: System principle and interface.

experiments. Thus, for any large deviation/significant change in the system, beyond operator's experience, re-tuning of FLC will be necessary. FLCs can adapt the changes in process and environment dynamics. The membership functions associated with the process controller output are adjusted based on the response/past performance of the feedback control system. A linearized process model may be identified and this can be used to predict the future performance to the controller. For these reasons there is recent surge of interest in model-based designs of fuzzy controllers [12]. Out of many applications of fuzzy modeling, the most distinguished ones are in Fuzzy control system. A fuzzy combined model and a fuzzy combined controller to handle nonlinear systems. In this sub section discussions shall be presented on Fuzzy models and their applications in control systems.

1. **FLC: Functional discussions:** Conventional PID controller being linear is not suited for strongly nonlinear systems. On account of *adaptive behavior, improved performance*, reduced tuning effort, etc. Fuzzy PID is often found as an alternative to PID control. Also, the algorithms used to tune the PID parameters use behavior similar to skilled human operator to tackle imprecise information during the application of the controller to a process. A Fuzzy PID controller is a combination of conventional PID with Fuzzy logic principles of membership to enhance the performance of control systems, especially in complex, nonlinear, or uncertain environments with imprecise information. Typical FLC with PID and gain scheduling has been detailed in Fig III 3.2d.
   - **Fuzzy PID functioning**: Here, *error processing* is very important. The system calculates the error (e)—as the difference between the desired setpoint and the actual output. It also calculates the change rate of the change quantity $e_c$, Both these e and $e_c$ are sent to Fuzzy controller which using membership function and rule base produces the three outputs for adjustments as shown in Fig III 3.2d1.
   - **FLC performance**: FLC provides no overshoot and responds faster than an aggressively tuned PID loop. Loop performance of FLC could be about 30%–40% better than normal PID on account of much lower overshoot with decreased settling or response time. FLC can take more aggressive control action on large control errors compared with small ones. It is possible for FLC to take suitable action to eliminate oscillating behavior. This is achieved by balancing high gain response on large errors as well as with strong damping when there exists oscillatory nature in the loop. As a result—tighter, more precise control is possible in uncertain scenario also. Fig III

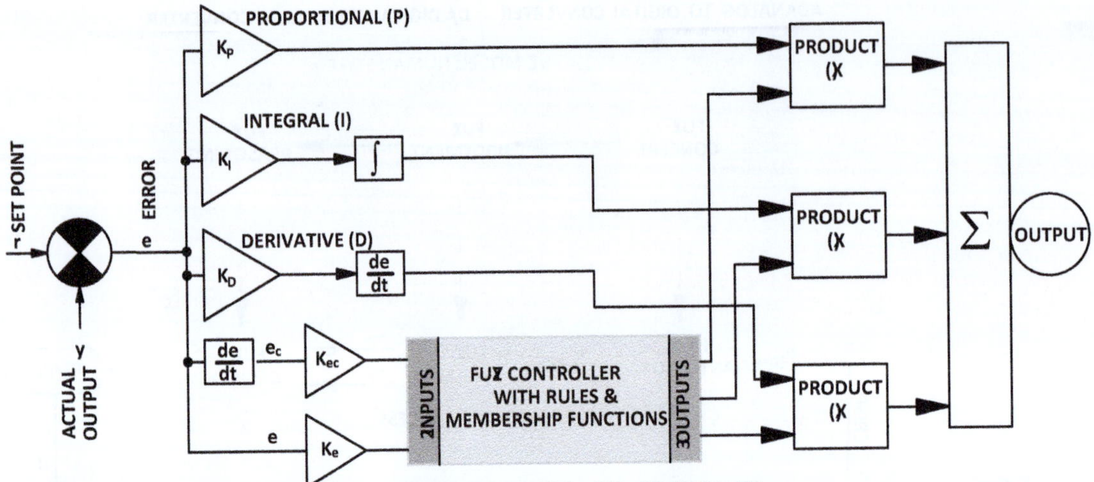


FIG III 3,2d1 FLC: PID CONTROLLER

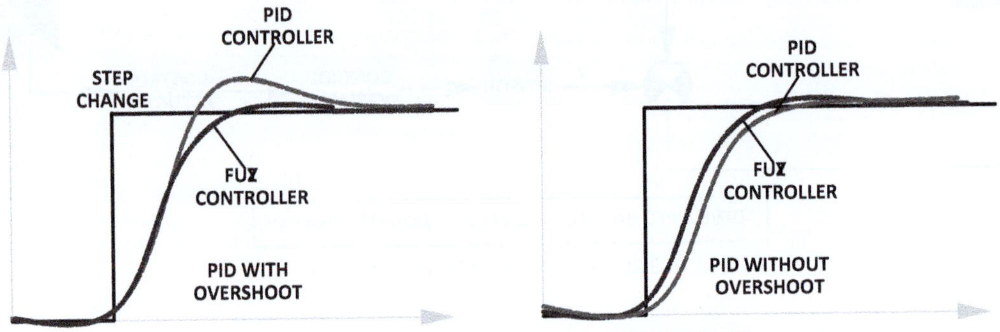


FIG III 3,2d2 FLC: PID PERFORMANCE

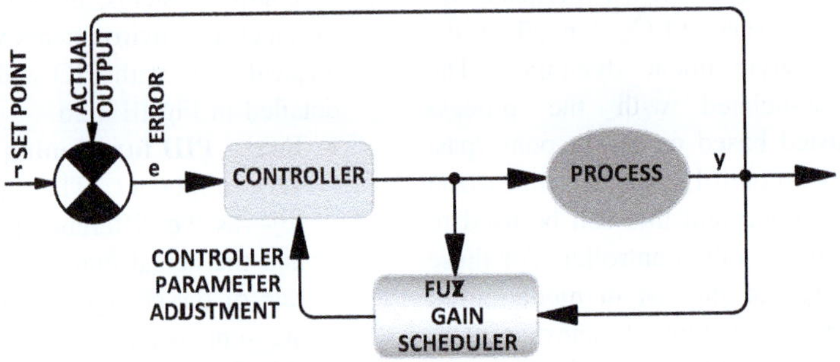


FIG III 3,2d3 FLC: GAIN SCHEDULING

**FIGURE III 3.2D** FLC: PID performance and gain scheduling.

3.2d2 compares the FLC response curves with PID control settings.

- **Gain scheduling (GS)**: GS in FLC/Fuzzy PID, dynamically adjusts the different gains based on operating condition of the system, as shown in Fig III 3.2d3. It uses interface to determine the appropriate gains based on error e and change in error $e_C$. Then based on a set of *if-then rules* to describe how the PID gains should be adjusted.

2. **Model based Fuzzy:** Popular Fuzzy modeling can be effectively used for process dynamics identification for nonlinear and complex systems. Fuzzy combined model combining different local fuzzy models to represent nonlinear systems. With the fuzzy combined model, a fuzzy combined controller can be used control the system. Several Fuzzy control and modeling approaches are as follows:
   - **Model free:** This is suited for the plant is a conventional system without a mathematical description, signals like setpoint, output and error are crisp in nature. The aim is to track the setpoint as already discussed in I 1.3.0.
   - **Model based**: FLCs are to be designed without enough/no information about the structure of the

system for setpoint tracking problems, only the system input-outputs are measurable. This approach is also adopted when mathematical model of the system, or a fairly good approximation of it, is available, that is, It is possible to design an FLC for better results in performance specifications as discussed earlier. When suitable Fuzzy model is not available meaning that no complete knowledge about the local linear system matrices is available. In this case, fuzzy system, identification/*fuzzy system modeling*, becomes necessary.

- **Adaptive approach:** In direct adaptive FLC (ref: Fig III 3.2d3), parameters are directly adjusted according to some adaptive law, to reduce the difference between the output of the plant and that of the reference model.
- **Fuzzy model description** [14]**:** Fuzzy models can be nonlinear dynamic models. Fuzzy model can utilize both qualitative and quantitative information. Qualitative information utilizes human expertise approach for knowledge engineering. Here, knowledge engineering is used in information capturing and utilized to form fuzzy sets/logic and rules. The basic objective of modeling is to establish an input output representative mapping to describe the system behaviors satisfactorily in all operational conditions. Two major identification issues of system modeling are as follows:
  - *Structure*: Structure identification of a fuzzy model, selects some appropriate input variables from the collection of system inputs to determine the number of membership functions for each input variable. This closely related to input partitioning in several approaches like: *Grid partitioning* (to generate fuzzy rules based on system *input-output training* data), *Tree partitioning (guillotine cuts)* and *Scatter partitioning* (extraction of Fuzzy rules directly from numerical data).
  - *Parameter:* This refers to the identification of optimal parameters of Fuzzy sets in the "IF and the THEN" rule using optimization techniques.

3. **Fuzzy modeling: explained:** Fuzzy modeling has been recognized as one of the nonlinear black-box modeling techniques which call for identification of its qualitative properties on the basis of experimentally measured data, while neither its structure nor its parameters are known. When processes grow in complexity, then implementation of straight mathematical modeling built on physical laws may be impractical on account of: *Nonlinear*, *stochastic*, *time varying* and *distributed*, *Exposure to environmental disturbances*, *difficulties in measurements of parameters*, having *unidentified/causal relations*, or *costly to estimate in real-time*. The practical relevance of fuzzy modeling, various types of rule-based fuzzy models and different fuzzy inference mechanisms are huge and important.
   In a wider sense the fuzzy modeling can be interpreted as a qualitative modeling structure for the system whose behavior can be described qualitatively using a natural language. Two methods for developing fuzzy model are as follows:
   - The expert knowledge in a verbal form/natural language is used to develop series of "IF-THEN" rules to definite Fuzzy model structure. Using this structure with input-output data parameters can be fine-tuned with issues like membership functions, etc.

   For building the model instead of *prior* technical data/information about the system, expertise knowledge is deployed. Posteriori explanation for the behavior of the systems can be estimated from the derived rules and membership functions. So, Fuzzy techniques are dependent on the specific application. Based on the above one can infer that Fuzzy modeling is *the task of identifying the parameters* of a fuzzy *inference system* so that a desired behavior is attained. The parameters of a fuzzy inference system can be classified as *Logical parameters*, *Structural parameters*, *Connective parameters* and *Operational parameter*. There exist a good numbers of fuzzy modeling methods differing mainly in the search strategy. Some common approaches are as follows: *Direct approach*, *Classic identification algorithm*, *Constructive learning approach*, *Bio-inspired approach*, etc. When the available knowledge is inadequate/problem space is too complex/large this modeling task becomes very difficult. One of the major problems in fuzzy modeling is the *curse of dimensionality*, meaning that the computation requirements grow exponentially with the number of variables [14].
4. **Miscellaneous models:** As shown in Fig III 3.2e, some common forms of models based Fuzzy controls are discussed below. Apart from these, there are a few other common types viz. Takagi–Sugeno Fuzzy model, operating regime-based model, etc.
   - **Inverse fuzzy:** Inverse Fuzzy model as a controller at times are efficient ways to control nonlinear systems. As Fuzzy can represent a nonlinear system, so, inverse fuzzy model can be a good choice for controlling nonlinear plants. Out of several methods, to obtain the inverse fuzzy model two common types are as follows: *Identification of the inverse model from input/output data* and *Inversion of the original model* when an ideal model is available for, perfect control. With reference to Fig III 3.2e1, the Internal model control (IMC) structure consists of a controller, a process model, a filter, and the process itself. Here, the process model is placed in parallel with a real process to identify the difference between

the actual system and its model, that is, the modeling error and process disturbances. By using the feedback filter, the *filtered difference* is used by the to compensate for the disturbance and the effects of the modeling error. This is a direct design method for the feedback controllers. For concept of IMC chapter I of volume I may be referenced.

- **Predictive fuzzy:** This uses Model predictive control (MPC) concept detailed in Chapter 1 of Volume I. A highly nonlinear system controlled by a linear MPC may not provide satisfactory performance. For nonlinear MPC (NMPC) has been developed. However, NMPC may require tremendous computational effort for optimization. Here, Fuzzy MPC approach with optimizer could provide better result by incorporating two layered optimizer to reduce prediction errors and control energy. However, in nonlinear fuzzy prediction model discussed above, may call for time-consuming optimization with no guarantee for an optimal solution. An approach to

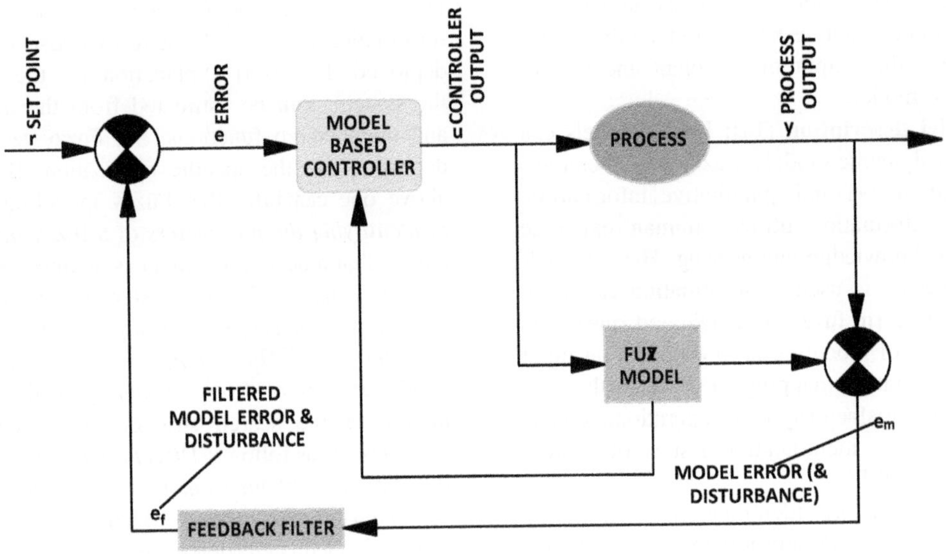


III 3.2e1 FUZZY MODEL: INTERNAL MODEL CONTROL (IMC)

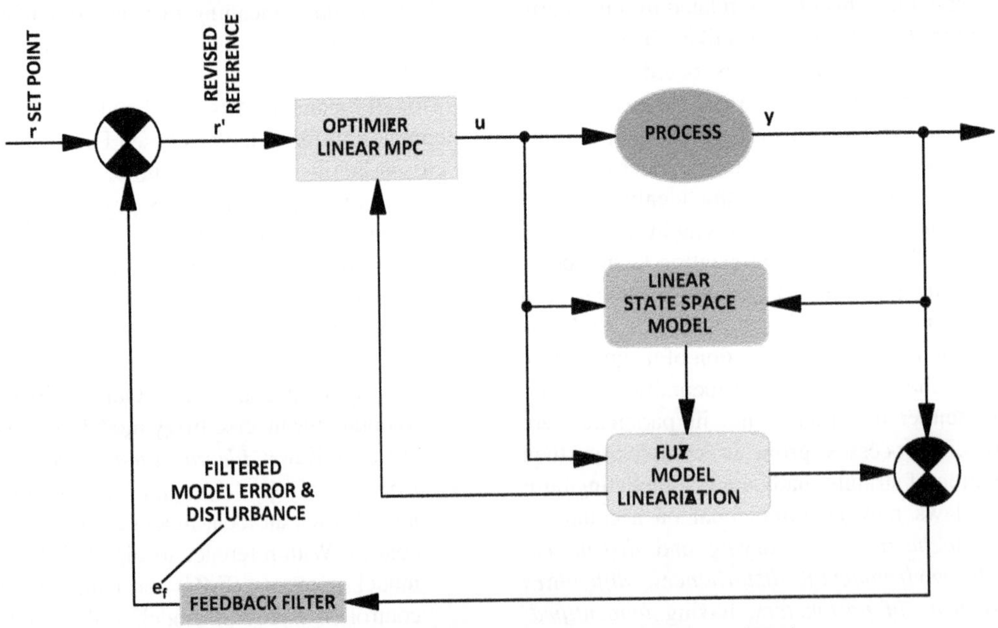


III 3.2e2 FUZZY MODEL: MODEL PREDICTIVE CONTROL (MPC)

**FIGURE III 3.2E** Model based fuzzy control.

circumvent the problem is to linearize the fuzzy model at the current operating point and use linear MPC. Linearizing the fuzzy model is done along the predicted input and output trajectories. Model prediction is further improved by iteratively applying the optimized control sequence to the fuzzy model and linearizing along the so obtained simulated trajectories. This is detailed in Fig III 3.2e2.

5. **Hardware and software Tool:** For the development of FLCs, intensive interaction and interface with the user. Naturally, various suppliers and developers have introduced various hardware and software tools. Some of the common such tools include but not limited to the following:
   - **Project editor:** The project editors are mainly Graphical user interface (GUI) with primary aim is to design and develop GUI for the specific control system. This assists the user to build FLC from basic blocks. Input and output variables can be identified/defined and connected to the Fuzzy inference unit. These connections can be direct or via processing units viz. filter integrators/differentiators as detailed in Fig III 3.2d &e. Defining these functional blocks are done by user in various languages including Fuzzy programming language (FPL). FPL is a type of *programming* language specifically designed to implement fuzzy logic, dealing with vagueness or uncertain information in Fuzzy programing method which involves iterative algorithms to search for solutions with a prescribed accuracy, used in computer science.
   - **Rule base and membership Functions:** A rule-based controller is easy to understand and maintain by the end-user. As discussed, Fuzzy logic involves the degree of membership which is represented mathematically by a membership function viz. Triangular/Trapezoidal membership functions enabling developers to model complex systems where inputs are not strictly defined, enhancing the flexibility of the applications. So, *membership functions editor* supporting different shapes and positions is used for the graphical environment.
   - **Analysis and simulation:** The function of the fuzzy controller can be tested using tools for *analysis* and *dynamic simulation* the best way to dynamic behavior of the closed loop system and control loop performance.
   - **Code and communication:** After testing FLC can be deployed for controlling the plant either directly from the environment/through generating a run-time code.
   - **FLC development tools types:** These are mostly used for FLC development and also, can take over the implementation problem. Most of the tools can simulate and debug the designed fuzzy systems and are able to generate executable code as discussed. Some common ones are as follows: Fuzzy Logic Toolbox for *MATLAB*, PID and Fuzzy Logic Toolkit for *NI* (LabVIEW), fuzzyTECH, Fuzzy-COPE, SCILAB, etc.

### 3.2.2 Advanced discussions on fuzzy systems

Starting with *complex robotic manipulation*, *adaptive autonomous vehicle navigation*, *production process optimization*, *multiobjective optimization complex fault diagnosis*, *intelligent power grid management*, etc., are some of the examples of advanced applications areas for Fuzzy controls. *Handling uncertainty*, *Highly dynamic environments*, *Multiobjective optimization* and *Hybrid approaches* are some of the major issues mainly addresses. In Fuzzy controls, the degree of vagueness of variables, process details including automation functions are considered. Fuzzy-rule-based systems at different levels of rules in hierarchy can form the basis for different automation functions in multilevel automation system. Fuzzy controls are well suited for normal and for special operating conditions/environments. A wide varieties of control types and systems can be designed and developed based on Fuzzy rule-based systems viz. *cascade*, *feedforward*, *variable structure*, *self-tuning*, *adaptive* and *quality control* leading to hybrid classical/fuzzy control systems. Based on Fuzzy logic approaches there exists huge potential in application of Fuzzy logic in complete process automation. In a single fuzzy system, to obtain better accuracy, as to maintain a specified accuracy, the number of rules grows exponentially with the number of input variables/Fuzzy sets. To reduce numbers of rules Hierarchical fuzzy system (HFS) consisting of a number of low-dimensional fuzzy systems in a hierarchical form, was introduced. In this subsection, detailed discussions have been presented on advanced issues pertinent to Fuzzy based controls. Let the discussions start with discussions on software tools with which previous subsection was concluded.

1. **Software tools** On account of use of natural language, simplicity and effectiveness, made FLC very popular and there is clear noticeable trend in the development of FLC algorithms in control applications. As mentioned before that there are a number of software products available that would allow the user to design FLC some cases including interactive GUI. Some free and commercial software tools have been tabled in Table III 3.2a.
2. **HFS discussions:** In traditional Fuzzy Logic Systems (FLS) the number of fuzzy rules required grows exponentially with the numbers of inputs. To form an FLS with *n input* variables and *m membership functions*

for each input variable, $m^n$ rules are needed. As discussed before, HFS approach addresses the complexity issue in a better way and enhances interpretability. Unlike FLS, in this approach. The rule increase will be a linear instead of exponential growth. In HFSs, assuming *two* input *variables* per low dimensional fuzzy system and *m fuzzy sets* per variable, each low-dimensional fuzzy system comprises $m^2$ rules. Therefore, for n number input variables, the total number of rules is $(n - 1)m^2$. So, rule increase is in a linear function. Thus, there will be significant advantages of HFS which also offers enhanced interpretability through its hierarchical structure and reduces system complexity by decomposing FLS into multiple levels.

- **HFS functioning:** The benefit of HFS is its ability to reduce the number of fuzzy rules while maintaining a high level of system accuracy. For managing a large-scale complex system, utilizing an HFS structure is highly effective. This hierarchical arrangement of rules in the hierarchical structure are reduced by dividing any level with *three or more* system variables into two levels, one containing only two system variables. This process can be repeated for all levels, resulting in the minimum number of rules when each level consists of only two system variables. This in turn also reduces computational complexity for fuzzy systems, especially for those with large number of input variables.
- **HFS advantages:** Mostly HFS are used to cope up with the complexity of problems. *Rule reduction, Enhancement of the interpretability, universal approximation* and *Trade-off* (may implicate selection of FLS for specific system) *between accuracy and interpretability* are some other advantages of the implementation of the HFS over traditional FLS.

3. **Automation vagueness:** Based on Purdue level discussed in Fig II 16.0d2, we understand that that there are several levels in multilevel process/plant automation. In a multilevel process automation, the lowest level refers to measurement and manipulation work. Next above level pertains to feedback/feed forward controls system where some variables will be adjusted in line with reference variables/certain condition. Supervision level sits above control level does the monitoring and supervision to indicate undesired or unpermitted process states and directs to take appropriate preventive/fail safe actions viz. protective failsafe action, shut-down, or triggering of redundance or reconfiguration schemes. Upper management level is meant to perform tasks like optimization, coordination or general management. These are mainly deployed to meet economic demands or scheduling. vague variables and knowledge. "Vagueness" in an automatic control system can come from *Imprecise information*, *Uncertain objective*, Complex system dynamics, etc., that is, it refers to the automatic control system having uncertain desired state/improperly defined input parameters. Naturally, there would be uncertainty in the system response resulting in potential instability or inconsistent outputs.
4. **Multilevel automation:** Based on the above discussions let us explore the following:
   - **Functional components:** Process class, input and output variables and automation functions are major functional components of automation by FLC.
     - *Process class*: This refers to the mathematical models, by which the static and dynamic behavior of the process can be described. In this case again there can be several situations relating to the knowledge about the associated parameters (Known/Partially known/Unknown). This is very vital in designing a technical process including designing of high-performance control systems. However, when the internal behavior cannot be expressed (by physical laws) in the form of equations, it is "not well defined" process. To tackle these situations, Fuzzy logics come in. The *condition* part then contains as *inputs the memberships* to the premise and possibly their logical connections with the logical consequence on the output as conclusion viz. IF (condition) THEN (conclusion).
     - *Input Variables*: These refer to input signals, distinguished precise and imprecise values. Normal measuring instruments are designed to deliver precise signals. Nondirectly measurable variables like slip, material stress, etc., can only be calculated/estimated based on other measurement variables and models. Also, linguistic values like the color of fire in a cement kiln/the type of noise of an engine are examples of input variables.
     - *Output variables*: In control systems, actuators need crisp inputs. Outputs to the human operator through the man-machine interface can be either crisp, if numbers are indicated or fuzzy, if they are linguistic. So, defuzzification is done.
     - *Automation functions*: The functions of process automation are usually organized according to different levels in a multilevel automation. Various control types at different levels have been discussed in III 3.2.2.3 above. In automation both feedforward anticipatory action and feedback controls can be deployed.
   - **Control intelligence:** As discussed, model-based, adaptive control, ANN, etc., are referred to as "*intelligent*" *control.* The intelligent controllers adapt the control to the mostly nonlinear/complex process behavior (adaptation) and stores its controller parameters. Operation is carried out depending on situations (learning), supervises all relevant elements.

**TABLE III 3.2A** List of some available Software Developmental Tools for FLC.

| Software tool | Software features | Development platform |
|---|---|---|
| **Free software** | | |
| **Fuzzy COPE** | For teaching/research & intelligent system development | MS Window 95/98 |
| SCILAB | Logic simulation and code<br>Generation | Linux, MS Window XP/Vista/7/8, MAC OS |
| Fuzzy CLIP | An enhanced version of CLIPS<br>Supporting fuzzy expert systems development | Mackintosh, Unix, Windows PC |
| Fuzzy logic inference engine | Performs fuzzy inferences from given fuzzy rules—well suited for class exercises. | Mackintosh |
| **Commercial software** | | |
| MATLAB fuzzy logic tool box | Fuzzy logic simulation—MATLAB installed | MS Windows/UNIX/Macintosh |
| PID and fuzzy<br>Logic toolkit<br>NI LabVIEW | Fuzzy logic simulation and<br>Code generation for embedded controls | Windows/Linux |
| FIDE/FLDE | Fuzzy logic simulation and<br>Code generation | MS Windows |
| Fuzzy TECH | Fuzzy logic simulation and<br>Code generation | MS Windows |

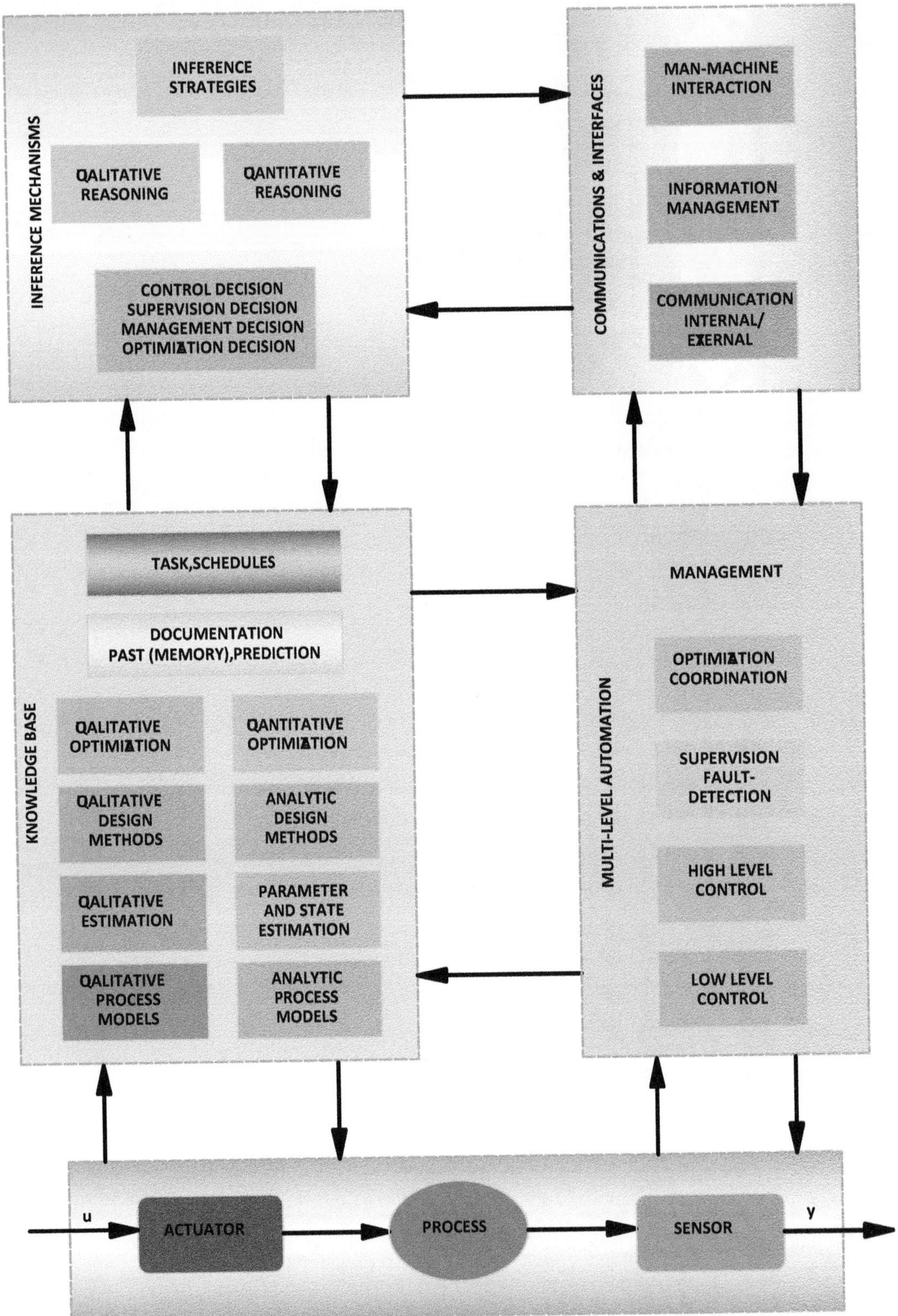


DEVELOPED BASED ON IDEA FROM 3

**FIGURE III 3.2F** Control intelligence in multilevel automation.

They are capable of performing a fault diagnosis (supervision) and issue advices maintenance, to shut off faulty components, or to fail safe (decision on actions). Typical "advanced intelligent" control scheme has been detailed in Fig III 3.2f developed based on ideas from Ref. [11].

5. **Configuration tool:** The configuration tool is used to configure and generate the fuzzy systems. The runtime software present in data block executes the system during operation. There are a number of efficient configuration tools available in the market for development and planning of process automation with Fuzzy logic. FuzzyControl++ of Siemens, MATLAB's Fuzzy Logic Toolbox, FuzzyTECH ST6 Explorer Edition, etc., are a few examples of the same. *User friendliness, Membership function editing, Creation of rule base, Inference engine selection, Defuzzification methods, Visualization and simulation, Platform compatibility,* are some of the features they possess. There are a few advance features in some of them also. Each of them provides comprehensive function sets with necessary graphical interfaces. As indicated in Table III 3.2a, these are meant for specific development platform and compatible software environment (Fuzzy Logic Toolbox is suitable in MATLAB environment). Empirical process knowledge/experience described verbally is implementable as fuzzy rules directly in open loop and closed-loop control, parameter adaptation of controllers, fault compensation. These are specified in terms of system input/output, number of variables, and number of rules supportive in the given platform. Also, these data are important to assess Graphic designer editors. Supports both Online (Real-time visualization, Modification Data recording) and Offline optimization (to include but not limited to: Plot transfer, simulation, debugging, etc). Memory size, number of rules, input/output, etc., are responsible for performance assessment of developmental tools.

## List of abbreviations used

**AFC** Adaptive fuzzy controller
**AI** Artificial intelligence
**ANN** Artificial neural network
**CNC** Computer numerical control
**DIGEST** Diagnostic expert system for turbo
**ES** Expert system
**ESPD** Expert system for performance evaluation and diagnosis
**FLC** Fuzzy logic controller
**FLM** Fuzzy logic model
**FLS** Fuzzy Logic Systems
**FOM** First of maxima method
**FPL** Fuzzy program languages
**GS** Gain scheduling
**GUI** Graphical user Interface
**HFS** Hierarchical fuzzy system
**HMI** Human machine interface
**I/O** Input/Output
**IE** Inference engine
**IMC** Internal model control
**KB** Knowledge base
**LISP** List processing (AI programming)
**MIMO** Multiple input multiple output
**MPC** Model predictive control
**MPPT** Maximum power point tracking
**QA** Quality assurance
**RCA** Root cause analysis
**RTES** Real-time expert system
**SISO** Single input single output
**SOE** Sequence of event
**SPC** Statistical process control
**UI** User interface
**WS** Work space

## References

[1] B.G. Buchanan, R.O. Duda, Principles of Rule-Based Expert Systems (Report No. HPP-82-14, STAN-CS-82-926), Stanford University, Heuristic Programming Project, 1982. http://i.stanford.edu/pub/cstr/reports/cs/tr/82/926/CS-TR-82-926.pdf.

[2] V. Nagor, B. Trivedi, Types of Expert System: Comparative Study, Asian Journal of Computer and Information Systems 2 (02) (2014).

[3] A. Zilouchian, M. Jamshidi, Intelligent Control Systems Using Soft Computing Methodologies, CRC Press, 2001. www.crcpress.com.

[4] H.-J. Zimmermann, Fuzzy Set Theory and Its Applications, 4th ed., Springer Science+Business Media, 2001.

[5] S. Bhanot, Process Control—Principles and Applications, 3rd impression, Oxfort University press, 2010.

[6] B.A. Jensen, B. Joseph, B.G. Lipták, Expert Systems ( Section 2.8), in: B.G. Liptak (Ed.), Instrumentation Engineers Handbook — Process Control and Optimization (Vol II), 3rd ed., CRC Press, 2003.

[7] R. Moore, H. Rosenof, G. Stanley, Process control using a real time expert system, in: Proc. IFAC 1990, USSR, 1990, pp. 234–239. https://www.gregstanleyandassociates.com/IFAC_Estonia_1990_Paper_Reformatted.pdf.

[8] R. Moore, H. Rosenof, G. Stanley, Process control using a real time expert system, in: Proc. IFAC, 1990. Internet document, https://www.gregstanleyandassociates.com/IFAC_Estonia_1990_Paper_Reformatted.pdf.

[9] N. Mayadevi, S.S. Vinodchandra, S. Ushakumari, A review on expert system applications in power plants, International Journal of Electrical and Computer Engineering 4 (1) (2014).

[10] S.G. Tzafestas, Expert control systems: an introduction with case studies, Control systems, Robotics and Automation 17 (2011) 343–356.

[11] R. Isermann, On fuzzy logic applications for automatic control, supervision, and fault diagnosis, IEEE Transaction on Systems, man and Cybernetics—PartA: Systems and Humans 28 (2) (1998).

[12] D.H.F. Liu, J. Abonyi, Software for Fuzzy Logic control;(section 2.31), in: B.G. Liptak (Ed.), Instrument Engineers' Handbook—Process Control and optimization, 3rd ed. 2, CRC Press, 2003.

[13] K. Chang, K. Chu, Y.C. Lin, J.S. Pan, Overview of Some Intelligent Control Structures and Dedicated Algorithms, Intech Open, 2020. https://www.intechopen.com/chapters/71711.

[14] D. Sharma, Designing and modeling fuzzy control systems, International Journal of Computer Applications 16 (1) (2011).

Chapter 4

# Neural network artificial intelligence and future trends

This chapter has been divided into three different section each dealing with a system different from the other. Historically, Artificial neural network (ANN) is as old as 1938 for conceptual idea. However, major work on this started from late eighties of last century. ANN is a type of Artificial intelligence (AI) technique that was inspired by the learning algorithm of the human brain i.e., after the neurons in the human brain. An ANN is composed of a large number of processing elements with their connections, and it has three distinctive layers, namely input, hidden and output layers ANN are modeled and are trained using a training set. There are a few common types of ANNs viz. *Feedforward Neural network (NN)*, *Convolutional NN*, *Radial basis function NN*, and *Recurrent NN*. In recent time, AI, Machine learning (ML), and Deep learning (DL) have revolutionized almost all field of our life especially in the field of advanced robotics. Some applications of AI, ML, and DL include autonomous navigation, object recognition and manipulation, natural language processing, and predictive maintenance. AI is a general umbrella term for anything that has the capacity to perform a human-like skill on its own. ML is a subset of AI. According to MIT, ML algorithms use statistics to find patterns in massive amounts of data in various applications such as speech recognition, natural language etc. In automation area they include process automation and manufacturing to support Industrie 4.0 and Digital transformation (DT). DL is a subset of ML concerned with algorithms inspired by the structure and function of the brain — ANN. Blockchain is a distributed database or ledger shared across a computer network's nodes. They are best known in cryptocurrency systems, maintaining a secure and decentralized record of transactions. However, these are useful in automation as well especially for Distributed control system (DCS)/Integrated control system (ICS). The last section of the chapter deals with Blockchain. However, the last section also have recapitulative discussions on Industrie 4.0 and DT under section heading of *futuristic digital control system.* Since Industrie 4.0 and DT have been covered in Vol II of the book these are discussed briefly for complete discussions. Let the discussions start with ANN.

## 4.0 Neural network (NN) in industrial controls

An ANN consists of a network of connected neural units referred to as nodes, divided into the input, output, and hidden layers. The input layer are often independent variables having size based on input data set dimension. The output layer generating the ANN outputs mostly the predicted value of a dependent variable in a regression/classification groups. AI automation, and ANN are indispensable mainly for intelligent software enabling machines to perform tasks such as decision-making and various human-like activities. Major issues involved include but not limited: *Software intelligent automation* (for system enhancement and optimization), *Predictability* (foretelling plant/machine conditions), *Predictability* (foretelling plant/machine conditions), *Pattern recognition and DL* (for large datasets), *Image identification*, *Natural language interpretation*, *Time series* etc. Main emphases have been given on the discussions on ANN in plant automation in both manufacturing as well as in process industries. There are two kinds of approaches in NN control topologies viz. *Direct* type (controller itself is an NN) and *Indirect* type (controller developed based on NN model of the process). *System identification* (development NN model) and *Control design*: use the NN plant model to design (and train) are two major steps in using NN in control systems. Apart from various control loop types, ANN are applied for Proportional integral differentiating (PID) controller tuning also viz. NN-based PID auto-tuner where Multi layered perceptrons (MLP) are used. ANN needs suitable training for successful result so that it can understand and mimic plant properties. *Generalized learning architecture*, *Specialized learning architecture*, and *Inverse dynamics* are common ways for training. Now, ANNs have been reckoned as an established versatile algorithmic and information processing tool for use in automation of manufacturing process where the role of the ANN is to perform various functions such as: *signal processing*, *pattern recognition*, *system identification* and *control*, *optimization* and *multisensors data fusion*. Product quality of most processes cannot be measurable in an on-

Plant Intelligent Automation and Digital Transformation. https://doi.org/10.1016/B978-0-323-89892-8.00021-1

line manner. There have been tremendous research efforts in monitoring of manufacturing processes ANN can be used here for quality check viz. Perceptron type ANN can be deployed for quality variables Surface finish in turning or for acceptability of weld/weld pool geometry in Arc welding. When ANN is applied for maintaining quality in manufacturing, the major steps are: Getting raw signal data through measurement, Signal processing, Feature extraction and classification pattern recognition, evaluation to provide feedback to manufacturing process for control action. Fundamental details of ANN have already been discussed in I 1.4, so, discussions in this section mainly cover detailed discussions on ANN in plant automation.

## 4.0.0 Neural network fundamentals

From the discussions in I 1.4, it is clear that Artificial neural networks (ANNs) are becoming very popular and helpful model for a number of issues, which include but not limited to classification, clustering, pattern recognition and prediction in many disciplines. From the earlier discussions we had seen that utility of ANNs can be assessed from its enormous capabilities such as accuracy, processing speed, latency, performance, fault tolerance, volume, scalability and convergence. A good advantage of ANNs application is that it can make models easy to use and more accurate from complex natural systems with large inputs [1]. Application areas of ANN, in plant automation shall include but not limited to *Process control* and *monitoring*, *Energy* (generation, transmission and distribution), *Manufacturing industries*, *Automotive*, *Factory automation*, *Electronics*, *Aerospace*, *Mechanics*, *robotics*, *Telecommunication* (e.g., ATM control, auto information) *Engineering* and *technology* etc. Broadly their applications can be categorized as *Classification*, *Clustering* (recognition of unique feature of data to classify without knowledge of prior data) *Prediction* (model-based) and *Pattern recognition*. Since working of ANN is little different from conventional computing, there are some difference between traditional computing machines and neural machines. While former ones process data sequentially, latter ones can do many things at once. Major differences between traditional and neural machines are [2]:

- **Following Instructions versus Learning Capability** (neural networks continuously adapt their programming and essentially program themselves to find solutions).
- **Rules versus Concepts and Imagery** (ANN use logic functions and use abstract concepts, graphics, and photographs. They perform tasks and then learn from them).
- **Complementary, Not Equal** (they complement each other).

It is clear that while some tasks are well done by ANN, where conventional systems fail, but for some tasks are not suitable for ANN. Let us now explore ANN in details.

1. **Application suitability:** ANNs can learn and adapt complex functional relations in a generalized form through its training to be applied for the following:
   - When the solutions are sought for the cases where there exist highly interdependent variables. From process control point of view highly interacting loops where in many cases precise quantifications are not available. Energy sector could be a good example.
   - **Complex** unknow interaction of variables or reactions where precise solution may not be possible but fuzzy modeling, and/or prior expertise knowledge would be helpful. Kiln control is an example
   - The problem requires qualitative or complex quantitative reasoning [3].
   - **Issues** involving good reasoning but some imprecision is allowed.
   - **Development** of new project/product line in short time, even though sufficient data and training/ML time is available.
   - **Readily** available data but are multivariate and intrinsically noisy or error-prone [3].
   - **Complex** problems for classifying and clustering.

   However, in cases where mathematically accurate results are required and/or for the control where step by step logics/sequential operations are required, ANN is not suitable as ANN does not follow sequential logic but go by training.
2. **Model concept and classes: ANN Model Concept and Classes**: Let us get the concept of ANN modeling by referring to Fig III 1.4b3 with single layer perceptron for model. In this there are "n" numbers of elements in input vector with "m" numbers of neurons in a layer. So, output will be given by

$$y = f(Wx + A) \quad \text{(Eq. III 4.0a)}$$

Where,

$$y = \begin{bmatrix} y1 \\ y2 \\ yr \end{bmatrix};\ A = \begin{bmatrix} A1 \\ A2 \\ Am \end{bmatrix};\ W = \begin{bmatrix} W1,1 & W2,1 & Wm,1 \\ W1,2 & W2,2 & Wm,2 \\ W1,n & W2,n & Wm,n \end{bmatrix};\ X = \begin{bmatrix} x1 \\ x2 \\ xn \end{bmatrix}$$

In this manner, models for higher layers can be formed. Basically, there are *three* kinds of model could be conceived in ANN. In *White box* well defined knowledge based are available. *Fuzzy type* is that some vagueness described in linguistic form and last one is where internals are not known it is defined in term of I/O.

- **White Box Type model:** White-box models are developed from prior experience and knowledge-base. In white box models machine language learning is transparent and features are well understandable. Behavior of the models, influencing factors, their prediction and prediction pattern can be explained easily.
- **Fuzzy type model:** Fuzzy-neural network model are often used for pattern recognition. A linguistically interpretable rule-based model is formed, which is based on the available expert knowledge and measured data [3]. In these models there are three layers viz. input layer, fuzzy membership layer (which maps input features to corresponding membership value) and third layer to implement inference engine.
- **Black box model:** As the name suggests here, physical insight is neither available nor used for modeling. Here the systems are viewed in terms of inputs and outputs. Such models have structures with flexibility used successfully in similar applications in previous occasions. In these models' parameters are fixed on the basis of measurements. For accurate modeling this approach. Therefore, nonlinear black-box modeling of process systems is a challenging and promising research field [3].

3. **ANN Classification: Classification of ANN:** ANN broadly can be classified as feed forward and Feed backward as shown in III 4.0a [1] and as discussed in III 1.4.1.1. The concepts of feed forward and Feed backward have been well depicted in Figs. III 1.4b4 and III 1.5. Basically, each of feed forward and feed backward has several divisions based on the network types as shown. Commonly used ANN types have been detailed in III 4.0.1.4.

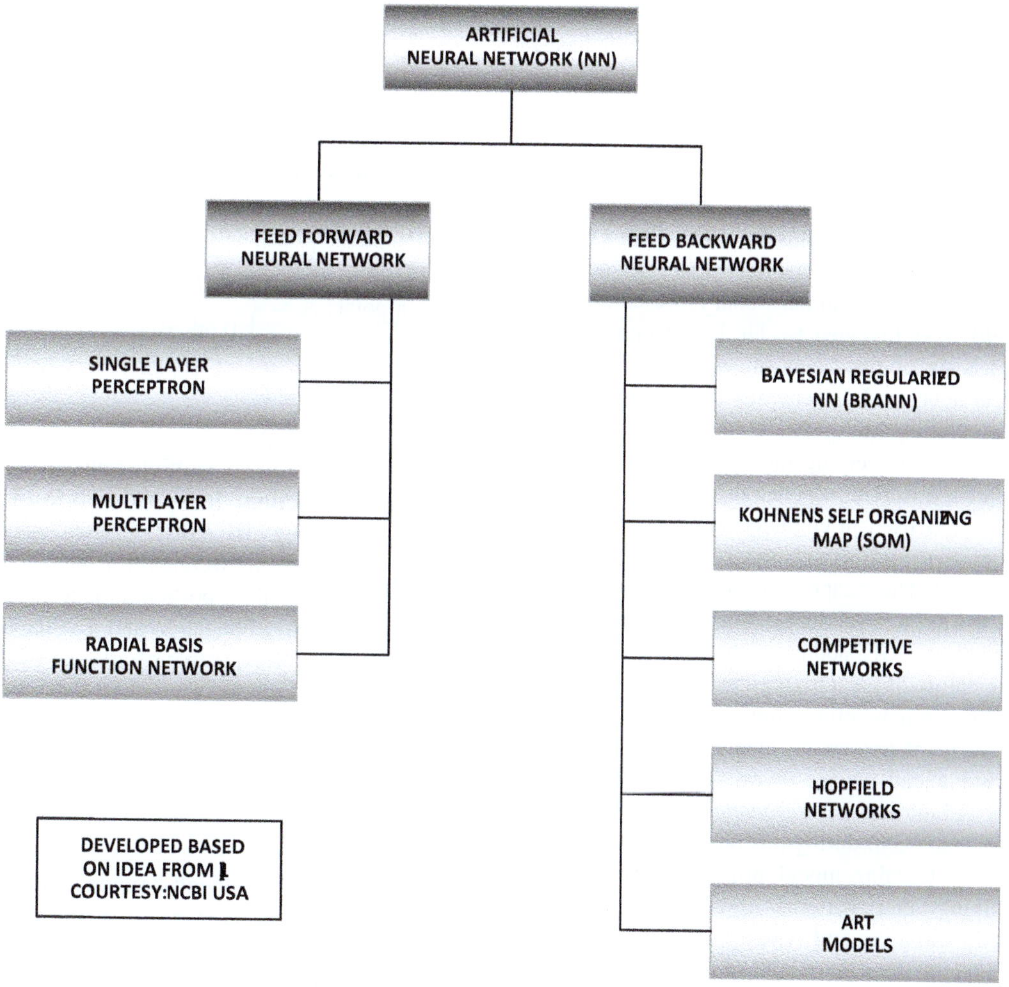


FIGURE III 4.0A Basic classification of ANN [1].

From discussion is I 1.4.1.1, it is clear that in feed forward NN units are in layer (s) relates to all the other units in the layers with unidirectional flow of information. Feed-backward NN can use internal state "memory" (store information) to process sequence of data inputs. Let us now look into various NN classes shown in Fig III 4.0a. Out of all these, *Radial basis function* and *Kohonen NN* shown in Fig. III 4.0a have been detailed in III 4.0.1.4. Other NN types are as follows:

- **Bayesian regularized NN (BRNN):** As the name suggests, this is a mathematical process that converts a nonlinear regression into a "well-posed" statistical problem in the manner of a ridge regression. BRNNs are very robust with capability to reduce/r eliminate the need for lengthy cross-validation. BRNN calculates and trains on a number of effective network parameters or weights, effectively turning off those that are not relevant. It is difficult to overfit/over train.
- **Competitive NN:** This is classical example of NN where natural rule survival for the fittest apply is followed. In a competitive neural network, the neurons "compete" to be activated, which is done based on a function of distance from the selected data point. The neuron closest to the data point—that is, with the highest activation—"wins," and is moved toward the data point, attracting some of its neighborhood. In this sense, competitiveness allows learning topology, thus, in effect when a reconstruction is done, higher-dimensional shapes are formed from a lower-dimensional representation.
- **Hopfield NN:** Invented by Dr J. J. Hopfield, this NN, consists of one layer of "*n*" fully connected recurrent neurons. It is generally used in performing auto-association and optimization tasks. It is calculated using a converging interactive process and it generates a different response than our normal neural nets. It is used in engineering and life science, and cognitive studies.
- **ART model:** The Adaptive resonance theory (ART) refers to as a hypothesis for human cognitive data handling. ART addresses the stability-plasticity i.e., the nature of memorizing the learning with flexibility to gain new information. In this NN, the data in the form of processing elements output *reflects back* and ahead among layers. If an *appropriate pattern* is build-up, *the resonance is reached,* then adaption can occur during this period. This is an unsupervised learning model primarily designed for recognizing binary patterns.

4. **Activation Functions and types:** The activation function of a node defines the output of that node given an input or set of inputs. This output will be used as an input in the next layer of the neural network and so on. Without it the ANN would be a linear regression model. Usually, the activation function does the nonlinear transformation to the input making it capable to learn and perform more complex tasks [4]. Activation functions are a crucial component for DL. For DL models, activation function decides the quality, which mainly includes: the output, accuracy, efficiency, convergence ability with its speed etc.

- **Activation function Explained:** Activation function is attached to each neuron/node to determine whether it should be activated/"fired" or not, on the basis of relevance of the input for prediction. Activation functions help to normalize the output of each neuron to a range between 1 and 0 (unipolar) or between −1 and 1 Bipolar. Activation functions should be computationally efficient because they are calculated across thousands or even millions of neurons for each data sample [5].
Various activations functions has been depicted in III 4.0b.
There are a number of activation functions. With reference to above figure a few of them are discussed here. For input "x" activation function is represented by A.
- **Step function:** This is Threshold oriented discussed earlier. If the input value is above or below a certain threshold, the neuron is activated and sends exactly the same signal to the next layer. As shown in Fig III 4.0b1, this function is defined as

$$\text{Unipolar: } A = \begin{cases} 1 \; if \; x > 0 \\ 0 \text{ otherwise} \end{cases}$$
$$\text{Bipolar: } A = \begin{cases} 1 \; if \; x > 0 \\ -1 \text{ (oherwise)} \end{cases} \qquad \text{(Eq. III 4.0b)}$$

It does not support multiple values.
- **Linear function:** It is in the form of $A = C{\cdot}x$ as shown in Fig. III 4.0b2. It uses weighted inputs (multiplied by the weights) for each neuron to create an output signal proportional to the input. It does not support back propagation and it collapses all layers in to one as irrespective of number of layers O/P is proportional to I/P.
- **Sigmoid (or Logistic) function:** This is shown in Fig. III 4.0b3 where,

$$A = \frac{1}{(1 + e^{-x})} \qquad \text{(Eq. III 4.0c)}$$

It has Smooth gradient; bounded output and offers clear prediction. However, it is not zero centered output and computationally expensive [5].
- **Tanh Function:** Since the of Tanh function is from (−1 to 1), its output is zero centered. Tanh is also

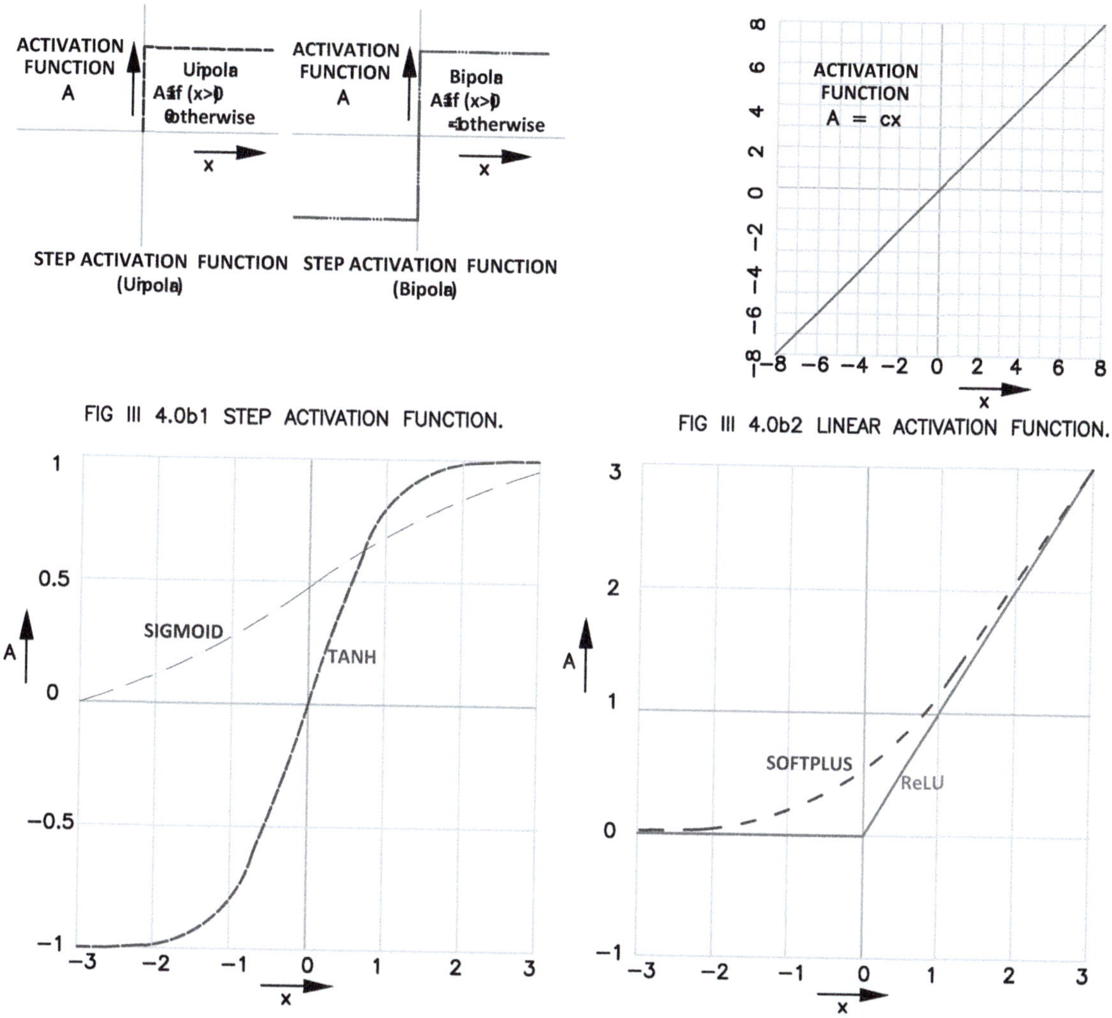


FIG III 4.0b1 STEP ACTIVATION FUNCTION. FIG III 4.0b2 LINEAR ACTIVATION FUNCTION.

FIG III 4.0b3 SIGMOID & TANH ACTIVATION FUNCTION. FIG III 4.0b4 SOFTPLUS & ReLU ACTIVATION FUNCTION.

**FIGURE III 4.0B** ANN activation functions.

sigmoidal s – shaped. This is shown in Fig III 4.0b3 and is defined as

$$A = \frac{(e^x - e^{-x})}{(e^x + e^{-x})} \quad \text{(Eq. III 4.0d)}$$

So, by function definition it is also called the hyperbolic tangent function. It is also expensive in computation.

- **Rectified Linear Unit (ReLU):** As shown in III 4.0b4, this function is defined as

$$A = x^+ = \text{Max}(0, x) \quad \text{(Eq. III 4.0e)}$$

This has been indirect way discussed in connection with NOT function in III 1.4. This is quite popular now. Different types of ReLu functions are like Leaky ReLu, Noisy ReLu, Exponential Linear Units. In some cases results are inconsistent.

- **Softplus function:** As shown in III 4.0b4, Softplus function is defined as

$$A = \ln(1 + e^x) \quad \text{(Eq. III 4.0f)}$$

$$\dot{A} = \frac{e^x}{(1 + e^x)} = \frac{1}{1 + e^{-x}} \quad \text{(Eq. III 4.0j)}$$

So, $\dot{A}$ called the logistic function, refer Eq. III 4.0c.

5. **ANN: Various aspects: Discussions on ANN:** In this section we shall briefly cover various aspects of ANN.
   - **Supervised and unsupervised network:** These are types of learning.

- *Supervised Network:* Here network learns by comparing network output with correct answer. Here network receives feedback about the error to adjust weights.
- *Unsupervised Network:* As the name suggests, here, there is no teacher— meaning that the network does not get any feedback on error. Network has to discover features in inputs and adjust weights. Unsupervised networks need to recognize input patterns.

- **Perceptron Learning rules:** Neutron is an adaptive element. Weight of inputs are modifiable, hence in supervised system, based on feedback weightages are changed. In case of unsupervised system neutrons need to modify weight based on input and output. There is learning rule in ANN. Learning rule represents modification of the weights and biases of a network. Perceptron learning rule (PLR)/algorithms can be categorized as single-layer and multilayer perceptrons. Former type organizes neurons in a single layer while in latter type arranges neurons in multiple layers. The initial choice of weights is very important and it should be small random numbers and unequal for better convergence. A number of iterations may be necessary for better results.
- **Determination of layers:** More the number of layers better will be the power of ANN. However, this calls for highly complicated training with good numbers of examples. According to B. G. Liptak "In general, the more complex the relationship, the more processing elements are required in the hidden layer. Depending upon the number of total nodes in a network, a sufficient number of training sets must be provided in order to train the system adequately" [3].
- **ANN lifecycle:** Like any other system the phases of ANN lifecycle are concept, design, implementation, and maintenance with stress on experimentation.
  - *Concept:* The concept phase in fact is a planning stage for developing application. Starting with selection of paradigm, associated steps are: (a) Identification of project objectives, and project constraints (b) Determine the length of the study period (c) data collection and preparation (d) Cost breakdown structure (e) Variable selection, Training and Validation (f) time constrains.
  - *Design:* At this stage initial parameters and conditions for the selected neural paradigm are developed. Also, processing elements, number of layers with connectivity are determined. Also transfer functions suitable learning algorithms are fixed.
  - *Implementation:* The neural network involves applying the design criteria using customized or commercial environments [24]. Here, major thrust is on experimentation.
  - *Maintenance:* The network must undergo retraining and reevaluation in an iteration of the implementation phase. If reevaluation shows that the network performance is not to specification with the new data set, the design phase must be repeated [3].

### 4.0.1 Treatise of neural network in industries

ANN are computational models that try to adopt and function like human brain. ANN along with some other systems such as Fuzzy, Bayesian reasoning form *Soft computing area.* Intersection of all these provides Neuro-Fuzzy systems and technique. Neural approach provides a practical way to create virtual sensors for measurements as an alternative to lab/online analysis. Based on AI several Data-driven models (DDMs) have been developed and all these pave the way for the development of powerful ML algorithms. These give rise to regression models including ANN. All thesehelp to achieve superior efficiency by unearthing complex relationships amongst inputs, hidden layers and outputs. As a result of these, early prediction systems for the plants could be used. A rapid early warning system can contribute to effective risk mitigation strategies. ANNs function like brain in production system and have become a cornerstone in the realm of plant/production automation in guiding industries' operation and associated process management. In Production process ANNs are increasingly playing a pivotal/vital role in enhancing efficiency, optimizing processes, and ensuring quality. ANNs can learn from data, recognize patterns/trends, to actively participate in decision making process. Traditional automation systems cannot adapt and learn from new data. However, ANNs being intelligent systems can adapt and learn from data then adapt to the changes in environment continuously for system optimization. So, ANNs can be deployed for *Training ANNs with manufacturing data*, *Tasks performed* of trained ANN–*Process optimization*, *Quality control* and *Predictive maintenance*, Also, ANN can be deployed for *Robot control*, *Scheduling* etc. These make them indispensable in plant automation. ANNs are poised to play more significant role in the future of manufacturing automation with advances in deep learning techniques, the potential for self-learning and autonomous manufacturing systems. Also, *ANNs can be used for* fault detection and diagnostics as *fault detectors* and as *fault classifiers*. In fault detection, ANNs can compare the actual outputs of a process with the expected outputs (based on a reference model/historical data) to flag a fault when the value goes beyond set limits. ANNs can identify the type and location of a fault, based on the features or patterns of the input data and classify the faults. In this section let us cover a few characterized issues related to ANN including soft sensing. Let us look into pros and cons of ANN.

1. **Pros and cons**: Let us examine some of the advantages and limitation of ANN.
   - **Advantages:** A few major system advantages are:
     - Process availability: On account of Provisions for backup with crosscheck on a measurement, ensures better process availability.
     - *Improves product quality with reduced variations:* On account of continuous "virtual" measurement of qualitative and analytical parameters there will much tighter control for process parameters with automatic compensation for unmeasured disturbances.
     - *Adaptive learning:* An ability to learn based on the data given for training or initial experience.
     - *Self-Organization:* An ANN can organize the information receipt during learning time.
     - *Real time (RT)_operation:* ANN computations can be carried out in parallel by using special hardware devices.
     - *In the Controller:* Some vendors offer NN running in the controller for fast action.
     - *Pattern recognition:* This powerful technique is for harnessing the information in the data and generalizing about it by learning to recognize the pattern.
     - *Adaptability:* ANNs are excellent at constantly changing information, and adapting to changing environment.
     - *Model building:* ANNs build models with reflective of the structure of the data in significantly in less time. ANN informative models are better than traditional systems.
   - **Limitation**: Some of the limitations of the system are:
     - *Methodology:* ANN does not have structured methodology
     - *Standardization:* There exists any standardized paradigm for ANN
     - *Problem solving:* ANN is not a general-purpose problem solver.
     - *Unpredictability:* The Output Quality of an ANN may be unpredictable.
     - *Black box:* ANNs do not describe how they solve problems, and behave like black box
     - *Nature:* Empirical nature of model development.
     - *Computation:* Greater computational burden and possesses proneness to over fitting.
2. **Neuro Fuzzy:** As the name suggests "Neuro-fuzzy" is a type of artificial intelligence that combines elements of both NNs and Fuzzy logic. These two technologies are used for creating systems, which are more flexible and efficient than systems using any one technology alone. Neuro-fuzzy system may be conceived NN as "hardware" representing structure of human brain, for emulating its function and Fuzzy logic systems concentrating on "software", emulating fuzzy and symbolic reasoning. The functioning of Neuro fuzzy can be conceived as detailed in Fig. III 4.0c, which shows that the system consists of four sections: *fuzzification, multiplication, summation*, and *division.* Fuzzification block translates the input into Fuzzy variables through membership functions. Then, product operations are performed on Fuzzy variables to generate smoother control surface. Here, the summation and division layers are used as defuzzification translation. Neuro-fuzzy systems are characterized by their ability to adaptively learn and draw inferences from imprecise or incomplete information, imitating human cognitive process as discussed. Neuro fuzzy functional elements are:
   - **Hybrid learning**: Neuro-fuzzy systems integrate both supervised and unsupervised learning for assimilating and data processing, for learning patterns with relationships.
   - **Rule-based systems:** Neuro-fuzzy systems employ flexible rule-based frameworks to evolve and adapt to changing environments, enhancing capability to handle dynamic and unpredictable scenarios.
   - **Fuzzy inference:** With fuzzy logic, neuro-fuzzy systems can interpret and reason over uncertain or vague inputs, enabling them to generate meaningful outputs.

   Neuro-fuzzy systems find a variety of applications, including control systems, data mining etc. Working with both *structured* and *unstructured data*, capability to handle *imprecise/incomplete data, pattern recognition*, use in a *variety of app* etc. are major benefits of Neuro-fuzzy system.
3. **Intelligent soft sensor:** Hard sensors refer to physical sensor, whereas soft sensors are developed based on other information using mathematical equations/calculations. Soft sensors may exist based on measurements from other sensors and leveraging benefit from Digital signal processing (DSP). Soft sensors often are used for estimating 'hard-to-measure' quality-related parameters in industrial processes. There are mainly two soft sensor categories: model-driven and data-driven soft sensors. Model-driven soft sensors (first-principles) are developed using equations derived from physical or chemical principles pertinent to the specific process. Data-driven soft sensors, are developed based on empirical models -built using real process data for which these are capable of accurately modeling the

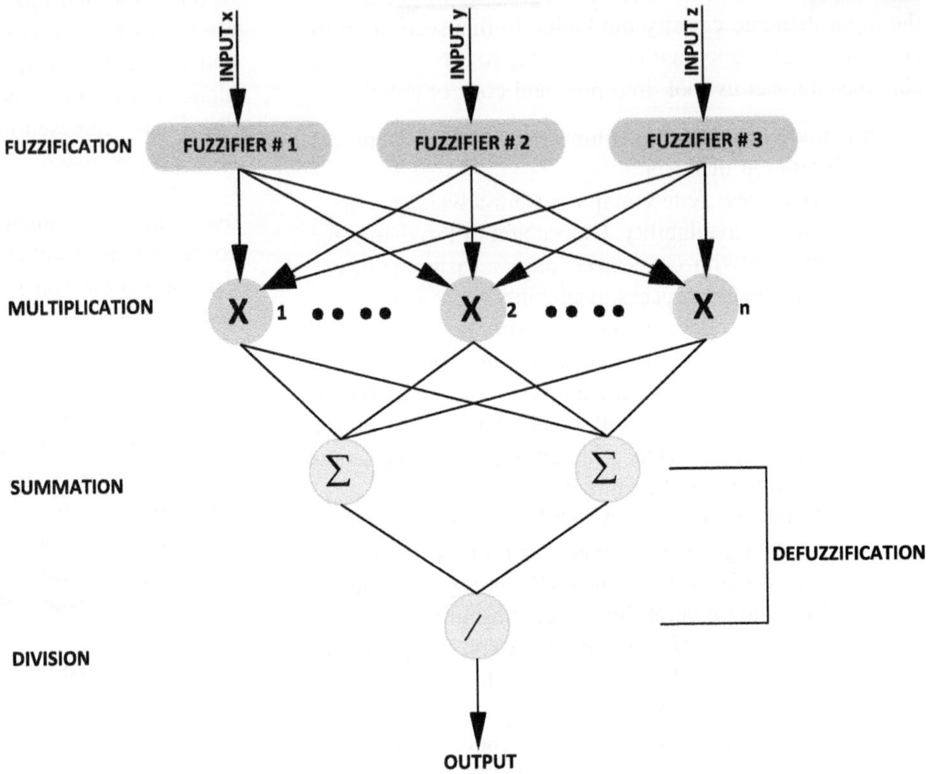


**FIGURE III 4.0C** Neuro fuzzy system: conceptual working.

process dynamics offering better predictive performance compared with the other one. model-driven soft sensors. Initially, data driven models were based on statistical models. With the increasing integration of technologies such as AI, ML, Internet of things (IoT), Industrial IoT (IIoT) the potential of soft sensors is growing exponentially. An *Intelligent Soft Sensor* (ISS) is a soft sensor to estimates/predict process variables using mathematical models based on AI/ML techniques. Therefore, Data collection, *Data preprocessing* (DSP), *Modeling* (e.g., Fuzzy, ML), *Prediction* and *estimation*, *Adaptation* and *Feedback* are commonly used process involved in developing ISS. The AI-based ISSs ensure consistent manufacturing/production quality and efficiency with minimum waste. This not only meets consumer expectations, but also optimizes resource use and contributes to sustainable practices. Also, AI-based soft sensor is designed to automatically predict key process parameters and adjust process setpoints to optimize overall process quality. So, *Cost-effective, better Reliability, Real time estimate, Complex problem handling, Adaptability* etc. are some of the advantages of the system. As, ANNs can learn from historical and real-time data, ANNs find their wide applications in ISS. *Nonlinear mapping, Data-driven approach, Robustness to noise* etc. made ANNs as natural choice for ISS.

4. **Popular ANN detail:** Classification of ANN has been discussed in III 4.0.0.3. Here we discuss common popular ANN types used for ML. Most commonly used ANN types used for learning are:
   - **Feed Forward NN:** From III 1.4.1.1 and III 4.0.0.3, we understand that feed forward NN is the simplest and straight forward type. Here, the information flow only in one direction. Thus, the data passes through the input nodes and exit on the output nodes with or without in between hidden layers. These are found commonly in computer vision and speech recognition with classification is complicated. Even though this is less complex and easy to design, yet not possible to use in deep learning tasks without dense layers and backpropagation. Also, the size and shape of the data must be known as it cannot handle surprises in the data.
   - **Convolutional NN (CNN)**: CNN are similar to feed forward NN, where the neurons have learnable weights and biases. CNNs, are designed for processing grids of data for images. CNNs use matrix filters for extracting information in their view and relating them together making it well suited for images. CNN can learn to hierarchically identify small, patterned features contained within a larger dataset and aggregating and reidentifying features until the entire dataset has been examined. So, CNN can

identify patterns in datasets requiring fewer parameters than feedforward NN hence specialized for image processing with fewer parameters. However, it is complex to design, cannot generalize and it requires a lot of computational power.

- **Radial basic functions NN:** These functions have *two* layers, first the inner layer where the features are combined with the radial basis function and the output of these features are taken into consideration while computing the same output in the next time-step, which is basically a memory. So, here, the input layer performs no computations. Instead, it passes the data directly to the hidden layer. Applications of these NNs are found in time series prediction and function approximation.
- **Kohonen Self Organizing NN:** It is often refer to as Self-organizing map (SOM) The objective of SOM is to input vectors of arbitrary dimension to discrete map comprised of neurons. The map needs to be trained to create its own organization of the training data in one/two dimensions. SOM process has different phases. When training the map the location of the neuron remains constant but the weights differ depending on the value. SOMs use competitive learning algorithms in which neurons must compete to be represented in the output. This is decided by which neurons best represent the input. SOM applications include displaying voting trends for analysis and organizing complex data collected for interpretation.
- **Recurrent Neural Network (RNN) – Long Short Term Memory (LSTM):** RNN works on the principle of saving the output of a layer and feeding this back to the input to help in predicting the outcome of the layer. In this way each neuron act like a memory cell in performing computations. RNNs take a single, complicated function and apply it repeatedly to a sequence of data i.e., RNNs learn one function very well and repeatedly apply it. The network keeps track of its understanding of the data it has read, and updates its understanding as new data is ingested with each application of the function it has learned. It is extremely effective at ingesting and generating sequential data. Memory loss is a big challenge for RNN. Also, RNNs are notoriously sensitive to how the model is 'initialized'
- **Modular Neural Networks** [6]: These have a collection of different networks working independently and contributing toward the output. Each neural network has a set of inputs that are unique compared with other networks constructing and performing subtasks. These networks do not interact or signal each other in accomplishing the tasks.
- **Generative adversarial networks (GAN):** GAN consists of two networks viz. a generator – meant for creation synthetic data and a discriminator for evaluation. This arrangement allows GANs to generate high-quality synthetic data (images). However, training is difficult and unstable.

5. **ML and AI difference:** AI refers to something that is made by human or nonnatural thing and Intelligence means ability to understand or think. It is implanted in system but it is not a system. It is a way or study to train any system so that it can do any thing/ perform any action better than at human can do better incessantly without fatigue. Ai can be vertical type (services focus on the single job) and Horizontal type (focuses on multiple tasks). ML is the learning where machine can learn by its own through software algorithm without being explicitly programmed. So, it is a subset of AI that provide system the ability to automatically learn and improve from experience. So, we find that there exists some difference between them, which is tabled in Table III 4.0a.

**TABLE III 4.0A Difference between artificial intelligence and machine learning.**

| Artificial intelligence | Machine learning |
|---|---|
| Intelligence in AI refers to ability to acquire knowledge, i.e., acquisition of knowledge and application knowledge. | ML refers to the acquisition of knowledge or skill to provide required services |
| Aimed for success not accuracy in achieving optimum solution | Aimed for accuracy not success in finding solution, which may not be optimum. |
| Simulation of natural intelligence to Complex problem solving | Learn from data on a task To maximize the performance of machine on the task. |
| It leads to wisdom to make decision | Learns from data leading to knowledge |
| It develops system to mimic human to Respond behave in circumstances. | Creation self-learning algorithms to find solution |

### 4.0.2 Neural network discussions: plant automation and application

Neural networks have been applied in a number of areas including plant automation with great success. Proper training to lift the veil from the technology, it can be widely applied to solve a number of complex problem in process and manufacturing industries. Like human nervous system, NN are composed of neurons/nodes (core computational units to drive the learning process) programmed to produce a response output from external stimuli in the form of input. The learning ability of NN provides, it great flexibility to grasp/understand the underlying the operatioanal process, even with attributes that are not readily obvious. ANNs the brains behind intelligent manufacturing/production processes, learn from vast amounts of data to carry out various tasks, including *decision making*, *predictive maintenance*, *process optimization*, and quality regulations. Also, ANNs are used for controlling robots in production environments and optimize production schedules (based on real-time data for assuring efficient resource allocation with reduced idle time). Thus, major benefits assured are: *Increased productivity and efficiency*, *Improved product quality*, *Reduced cost*, *Enhanced decision making*. Major purposeful features behind the use of ANN in industrial plants include but not limited to: The purpose of using neural network models stem from their ability to: *Model a linear or nonlinear process even based on indirect measurements*, Ability to *model the processes*, which are difficult to *understand/model based on first-principles* equations and shorter *model development time for a complex process*. Also, it is possible to do some historical datamining to acquire a data set for training and select the architecture depending on the application. Nowadays, there are many types of neural networks in deep learning, which are used for different special purposes. In this subsection discussions on ANN shall be concluded with throwing some light on ANN in plant automation issues, including controlling and monitoring as well as applications of ANN in industries.

1. **Control baseline:** As discussed, it is possible to do some historical datamining for training. Process conditions and the system states aren't always known from when the data was generated. Before acquiring a data record set, it is recommended to create a measurement and control baseline to ensure that *all transmitters* to be used as inputs to the model are *well calibrated* and to check and ensure that the valves/drives/heaters are in good working condition. Before parametric testing, prerequisites to be considered are:
   - Have a safe condition to place the process in case of an unusual event.
   - Know safety limits and constraints when executing the parametric test
   - Defining necessary limits of valves, speed controller etc. to be used in the test.
   - To record any noise filter settings.
   - Defining settling time after each parametric test.
   - All measurement/transport lag/delays should be noted.
   - All operations are as per agreed test procedure. and all data points are recorded.
   - As applicable, to perform a hazard analysis before manipulating the process.
2. **Industrial automation issues:** Major process control issues are:
   - **Architectural issue:** Based on application, corresponding system architecture is chosen. In industrial control applications, generally speaking, Models are developed in such a way to follow the function of important/influencing process variables; based on the process conditions, to set the target; and the inputs surrounding it. There is no set rule for determination of the hidden layers and number of neurons in the hidden layer. The entire exercise is done on the basis of the complexity of the applications as well as the target function.
   - **Measurement and Control space:** In application of NN, in intelligent industrial control systems, the terms such as measurement/control space as well as input vector are important issues to describe different system aspects.
     - *Measurement space*: This refers to the set of possible sensor measurements/observed states. In NN it is related to input feature, which describes the multidimensional limits defined by the measurement range for the input.
     - *Control space*: This refers to the set of all possible control actions/outputs, in NN this is related to output layer. Within measurement space, It is used for the limits and shape (dependent on the point vector distribution of the input used for training).
     - *Input vector:* This multidimensional vector stands for the numerical representation of the current state of the system as derived from the measurement space.
   - **Input element selection:** Selection of the process variables for the model input vector data elements is very crucial in creating an NN model. Its inputs should be robust, independent variables with effective and high influence on the target. For comparatively simpler systems, it is easier to select inputs from experience. However, in some complex cases, with more variables, and choosing the right ones are challenging.

- **Data type:** As the name suggests, it stands for *binary*, *integer*, and *floating point* are the primary data types recommended.
- **Data set selection:** It is important for selecting records for training to ensure the record set covers the range and the target response throughout the input range.

3. **Automation requirements:** Brief discussions on automation requirements involving ANN in industrial plants been presented as follows:
   - **Technological issues:** For applying NN, it is necessary to go in for NN model, which must have *Real-time processing capability*, *High-accuracy and reliability*, *Capability to meet computational complexities*, *adaptability* etc.
   - **Operational conditions:** This calls for *interface with existing control*, *robustness with fault tolerance capability*, *training and easy interpretability*.
   - **Security and safety:** Suitable guard against *cyber-attack and provision for security*, *failsafe mechanism and following international standards* viz. IEC 61508, IEC 62443 for safety and security respectively.
   - **NN model:** Use of various models such as CNN, RNN. discussed. Sensor data, Historical process data, set points and operator data as input to the model. Control action, discrepancy detection, prediction for quality and maintenance requirements are associated model outputs.
4. **Control and monitoring:** Common issues related to control and monitoring in industrial plants shall include but not limited to:
   - **Control issue:** Major implementation issues to be considered are: *NN configuration*, *selection of suitable NN control algorithm with required inputs and instruction sets*, *Input quality*, *valid control space*, *measurement and control strategy* and *suitable interfaces*. In control systems, an NN is trained to estimate the unknown nonlinear process and based on NN, controller is formulated. The control signal formulation, is determined from the NN model. It is, necessary to develop an approach to simplify the control signal formulation.
   - **Operator interface:** Well-designed Human machine interface (HMI) should cover: *Overall control space with alarm*, *Auto/manual mode*, *model output value* and *alternate strategy for output*.
   - **Monitoring and Surveillance:** The inferential control uses measurements of secondary process parameters/outputs, to infer the effect of unmeasurable disturbances on primary process outputs, in other words, variables being controlled viz. product quality. Inferential control addresses the effect of measurable disturbances on the product quality for regulating the control effort to counteract the effect of the measured disturbances. Fault detection and diagnosis is another use of NN for operation of industrial plants. Here, the inputs to the network include the symptoms present or absent in the system being diagnosed, and the outputs of the network represent the present or absence of particular fault causes [7].
5. **ANN Types and Applications:** From the discussions on ANN types, It could be noted that there are various types of ANN in industrial plants and other purposes. Table III 4.0b has been presented to list put these ANN types with common usages.

## 4.1 Discussions on artificial intelligence

As discussed, AI is a branch of engineering built on computers and machines that can *reason*, *learn*, and *act* very similar to normal human intelligence. These are done with the help of data whose volume is too big for humans to analyze. AI is a large and complicated field with an extended ecosystem. Also, previous discussions we know Machine learning (ML) is a subset of artificial intelligence. Similarly, Deep learning (DL) is a subset of ML as depicted in Fig. III 4.1a. Therefore, ML practices with a set of algorithms to analyze and interpret data, learn from it, and based on those learnings, makes the best possible decisions. However, DL being related to ANN depends upon layers of ANN. Recent advancements in AI-based approaches have shifted the manufacturing practices toward Industrie 4.0 practices with consistent improved decision making, productivity and system performance. A positive impact of AI-based techniques on sustainability can be seen in manufacturing organizations' at the system, product and process levels.

We know that various ML types are: *Supervised learning (*a task drive ML model using labeled training data Viz.. *Linear regression, Decision tree), Unsupervised learning* (a data driven ML model using unlabeled data viz. *Principal component analysis*, *Fuzzy means*) and *Reinforcement learning* (ML model described as "learn by doing"/learn from mistake viz. learning playing game). *Automated surveillance*, *Quality management*, *Computer vision*, *Quantifying Customer Experience*, *Document classification*, *IoT* (smart home, smart city) etc. are major application area of ML. Broadly application areas of ML are: Detection: *Image interpretation*, *Human behavior and identity*, *Fraud/fault detection* etc. Prediction: *Classification*, *Analysis*, *Recommendation* etc. and Generation: *Design*. *Text*, *Musk* etc.

DL is used in ANN. Common NN using DL are: *Feedforward NN*, *Recurrent neural networks* (RNN), *Long/short term memory* (LSTM), *Convolutional neural networks* (CNN) and *Generative adversarial networks* (GAN).

**TABLE III 4.0B Artificial neural network types with applications.**

| Name | Description |
|---|---|
| **Feed forward [III 4.0.1.4]** | Data compression, pattern recognition, sonar target recognition. |
| **CNN [III 4.0.1.4]** | Autonomous vehicles, industrial automation, natural language processing, video/image recognition |
| **RNB [III 4.0.1.4]** | Function approximation problems, classification, time series prediction |
| **RNN [III 4.0.1.4]** | Robot control, time series prediction, music speech synthesis |
| **Kohonen [III 4.0.1.4]** | Dimensionality reduction, assessment and prediction of water quality |
| **ART [III 4.0.0.3]** | Target recognition, face recognition, medical diagnosis, signature verification, mobile control robot. |
| **GAN [III 4.0.1.4]** | Generation of image and training |
| **GRU** | Gated recurrent unit (GRU): these are variation of LSTMs. Speech and music modeling, natural language processing |
| **VAE** | Variational autoencoder (VAE): Probabilistic approach<br>*Application:* Auto image generation |
| **AE** | Auto encoder (AE): Unsupervised ML<br>Clustering feature compression |
| **HN [III 4.0.0.3]** | Image detection and recognition, medical imaging [III 4.0.0.3] |
| **Competitive NN [III 4.0.0.3]** | Image clustering processes and specific image segmentation. |
| **DAE** | Denoising autoencoder (DAE): forces the hidden layer to learn more robust features for refined output against noisy input<br>*Application:* Feature extraction, dimensionality reduction. |
| **DN** | CNN in reverse process [ III 4.0.1.4]<br>Surface depth estimation from an image. Optical flow estimation. |
| **SAE** | Sparse autoencoder (SAE): SAE helps to construct loss function by<br>*Application:* Feature extraction, hand written digit recognition |
| **MC** | Markov chain (MC): Mathematical system with probabilistic rules.<br>*Application:* Statistics and speech recognition |
| **LSM** | Liquid state machine (LSM): **A**n extensive collection of neurons each receives inputs from an external source/and other nodes<br>*Application:* Computer vision, speech recognitions |

DL algorithms-based on the workings of the human brain functioning, are used for analysis of data with a logical structure. DL can make nonlinear, complex correlations within datasets though more training data and computational resources than machine learning. Nowadays, DL is applied in almost all fields. Some are: *Universal learning technique* (useful technique to almost every application field), *Robust deep learning technique* (needing no precise designing feature its process of automatically learning and representing the best features for any task make it robust), *The generalized DL approach* (the same DL method can be applied with different types of datasets or in different applications), The extremely scalable DL Technique (highly scalable in terms of data and computation). However, there are plenty of challenges in DL such as Big data analytics with DL. The key motives in the acceptance of deep learning nowadays are because of enlarged chip processing capabilities such as Graphics processing units (GPUs), cost effective computer hardware, and create current advancements in machine learning projects.

In this section, all the three issues have been discussed at length with required relationship amongst them.

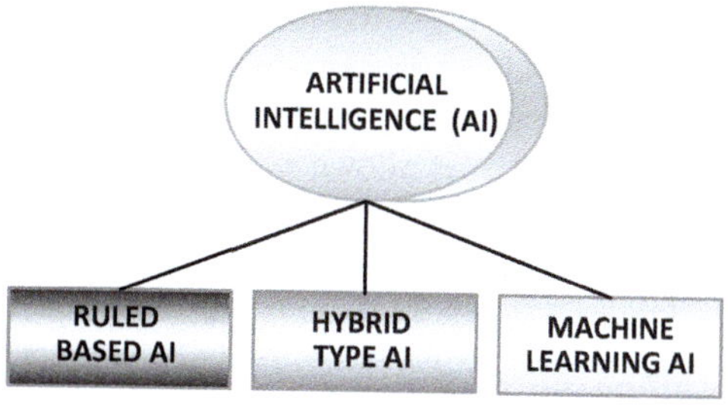


FIG III 4.1a1 ARTIFICIAL INTELLIGENCE (AI) TYPES

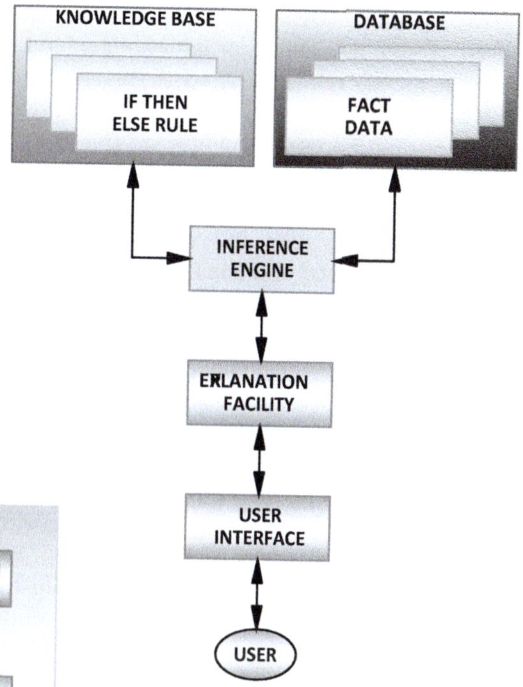


FIG III 4.1a2 RULES BASED AI STRUCTURE

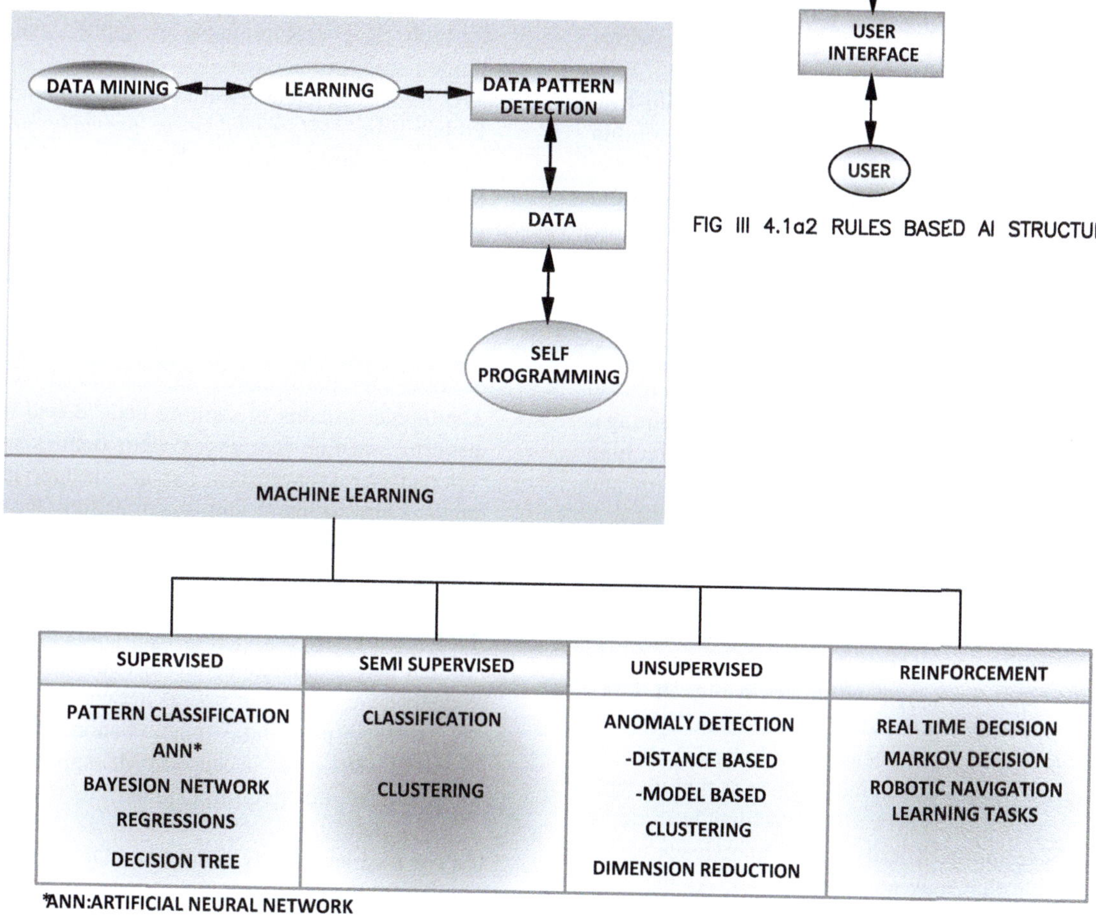


FIG III 4.1a3 MACHINE LEARNING AI STRUCTURE AND TYPES

**FIGURE III 4.1A** Artificial intelligence (AI) approaches.

## 4.1.0 Artificial intelligence outlines

Like human, through intelligence, systems can gain the capability to adapting to the environment. Like human, systems through intelligence and computational power can also learn, and after conceiving it, can establish relation between information. Alike human, systems can store the learning in classified manner in memory so that depending on applicability, same can be adapted and reused to adjust in new condition. However, AI using Fuzzy logic and ANN may use linguistic means/cognitive way similar to human being. Through learning and training—Machine learning

(ML) and Deep learning (DL). Learning, pattern detection, data and self-programing are major issues related to ML. Whereas DL is mainly related to ANN to solve the problems. AI usually provide highly reliable and more precise results in lesser time because machines are free from all emotions, they can work tirelessly and possess more computing power. *No thinking*, *No abstract reasoning*, *Limitation in capturing data* and *some challenges to interpret* systems are some of the major limitation of AI. AI machines are all programmed machines, naturally they need to follow the program so they cannot exceed in knowledge from their programmer [8]. Therefore, it is clear that the programmer must have thorough process knowledge as well as programming. Since this is not humanly possible to possess knowledge in different discipline, so, artificial intelligence requires multi discipline approach. Although AIs take part in many decisions making but for many important decisions making, they only recommend for human to finally approve. Industrial units such Boiler, chemical reactors, distillation column etc. are well-understood and established process. These can be mathematically described or model by their heat and material balance. On the contrary in some area e. g., Chemical process in cement kiln, it is really difficult to predict reactions, in such cases Fuzzy logic is well suited. In cases such as pattern recognition, prediction as in manufacturing industries, ANN can be applied. Robotics is another area that also helps in bring about sea changes in industries especially in manufacturing industries helping it to move toward digital transformation. Robotics offer high ROI, when it is seen from accuracy, consistency time and efficiency point of view.

1. **AI approach discussions:** Details on various types of these have been depicted in Fig. III 4.1a. In AI approach discussions two major issues are:
   - **AI types:** Artificial intelligence can be broadly be divided into three types as shown in Fig. III 4.1a1. AI types are: *Rule-based*, *Hybrid* and *Machine leaning type*. Fundamentally rule-based AIs (e. g., fuzzy logic) leads to intelligence to find the optimum solutions. Whereas for machine learning (e. g., ANN) approach leads toward knowledge to find a solution that may not be optimum. As the name suggests, Hybrid systems utilize both approaches. Both of ruled-based and machine learning approaches have some advantages and disadvantages. Therefore, it is very important to decide which approach would be correct for the particular application.
   - **Rule-based AI:** Rules-based AI systems are deterministic. Following definite sets of rules and instructions, rule-based AI systems produce preset outputs. This means that this approach offers somewhat "fixed intelligence". As the name signifies, rule-based AI produces preset outputs based on a set of certain rules coded by humans. This AI consists of four main parts as shown in Fig. III 4.1a2. These parts are:
     - *Facts:* This is the knowledge base
     - *Rules: If* then *else* logic (rule)
     - *Inference Engine:* This used for inferencing information and acts according to the interactions between input and rule base.
     - *Working Memory:* For storage of data.

     Rule-based systems also have explanation, and as discussed earlier user interact to the system through user interface as shown. As stated above these are coded by human so, for large sets of conditional rules a minor mistake demands huge manual efforts for recheck. Therefore, these are applied mainly for limited scale projects/job.
   - **ML in AI:** Machine Learning (ML) methods are Probabilistic (*ML certification provides practical training of large datasets*). Without explicitly programmed system it performs tasks by trying to simulate human intelligence through machine deep learning (learning model). This human intelligence is introduced into machine learning models as datasets to train machines. Instead of definite outputs, machine learning models constantly "adapt" and "evolve" its performance in accordance with the continuous streams of training data. Based on data pattern, machine learning system defines and develops its own set of rules. Typical machine learning method has structure and classifications have been depicted in Fig I 5.1a3. As shown in the figure there are four classes of Machine learning methods viz. Supervised, Semi supervised, unsupervised and reinforcement for decision makings. Unlike rule-based system here explanations are difficult at times. The hidden layers of Deep neural network (DNN)s are where the "black box" phenomenon of machine learning takes place because in ML approach, algorithms make it difficult to explain how really output is generated [9].
2. **AI: Pros and cons:** Advantages and Disadvantages of AI are as follows:
   - **AI Advantages:** According to published report, upto ~18% cost savings McKinsey, companies using AI have witnessed and revenue growth.16and average around a 6%–10% increase in overall revenue. Also, as per report, with AI in manufacturing, can offer facilities such as self-optimizing machines to automate production, AI-enabled design for better efficiency and reduction in waste forecast of efficiency losses and quick detection of quality defects for predictive maintenance. Major benefits of AI in

control and automation shall include but not limited to: *Gain process insight through learning, Quick identification of underperforming control with quick loop tuning for better control, Test tuning and simulation, Easy adaptable for use, Tracking control performance with easy reporting* etc. Now let us jot down major AI advantages in generalized way.
  - Precise unbiased decision making with reduced human error
  - High risk avoidance with all-time availability
  - Digital assistance everywhere with new inventions and automating the process
  - Enhanced productivity, efficiency and safeguard
  - Reduction in waste, time and cost
  - Advanced data analysis and complex problem solutions
  - Fault diagnosis, detection and prediction including predictive maintenance
  - Improvement of Human workflow, avoidance of human fatigue
  - Personalization Better customer experience
  - Smarter surveillance, fair and unbiased activity without emotion.
- **AI Disadvantages:** In spite of so many advantages listed above, AI is not free from some disadvantages, which are listed as follows:
  - Lackin intrinsic creativity
  - Lack of common sense and emotional intelligence (not suitable for all human interfaces).
  - Promoting laziness in human and over technology dependence
  - Privacy may not be maintained properly to meet huge data requirements.
  - Environmental issue and algorithm concern (pace and direction)
  - Sticking to norms and experience –lack in flexibility
  - Interpretability and Transparency—especially decision making

3. **AI application challenges:** In order to implement AI in any industrial plant/enterprise, there are a number of challenges to be faced for implementation. Major challenges are:
   - **Insufficient data:** There is requirement of huge data for AI for learning and bringing AI to life. Therefore, to source huge data, as well as the structuring and processing of the siloed data pose big challenge.
   - **Right data set:** Full benefit of AI can only be developed by leveraging quality data hence importance of determination of right data cannot be overestimated.
   - **Data security and storage:** AI calls for utilizing large volumes of data for which the best and right data management environment is essential to implement AI. This data management environment refers to create a storage issue as well as issues related to **data security.**
   - **AI infrastructure:** Majority of AI-based systems call for high computational speed, so, in addition right data management environment, suitable substantial infrastructure and high-end processors along with robust environment and flexible infrastructure fully compatible with AI solutions should be made available.
   - **AI algorithm:** The function and performance of business intelligence operations heavily rely on AI algorithms. Therefore, user should have a clear idea about AI-solutions/technologies along with idea about AI algorithm.
   - **AI integration:** This is the most common challenges most businesses face for AI.
   - **Company culture.** This is a challenge to integrate AI solutions into their day-to-day activities, unless there is established company culture for technology adaption.
   - **Talent scarcity:** There is real scarcity of persons talented with expertise in digital skills for AI and ML and other modern issues such as (say) DT.
   - **Use identification:** In the present time, AI solutions are often viewed as hyped and overestimated. As a result, often organizations may implement them to keep pace without clear strategy or utilization issue in mind.

   Apart from the above, additional challenges mainly include: leadership issue, data privacy issue, standard compliance issue, customer requirement and experience.
4. **AI and TSN application:** Time-Sensitive Networks (TSN) have brought about huge changes automation. As a set of Ethernet standards, TSNs guarantee precise timing and low latency for data transmission across a network. To ensure that Information technology (IT) and Operational technology (OT) do not affect each other a number of modifications must be made in the Ethernet structure. This is done through TSN, which can be conceived as tool box with several tools having specific purpose. AI in TSN can optimize network configurations, predict and prevent potential issues, and manage traffic flows in real-time, ensuring reliable low-latency data transmission. AI can analyze network data to adapt the changing conditions, especially for fluctuating traffic demands/network topology changes. Some of the key application areas of AI in TSN are: *Dynamic scheduling, Traffic prioritization, Bandwidth allocation, Fault prediction* and *diagnosis, Optimized network configuration, Enhanced reliability, Higher network efficiency* and *reduced operational overhead.* Robotic vision, *3D images with synchronized cameras,*

*Frequency synchronization in the feed-in and load management of alternative power generators* etc. are examples of some of the applications where there is scope to deploy TSN with AI. This also, assists in applying it in industrial automation. Industry 4.0 requires support for different application profile. Based on IEEE/IEC 60802 TSN with high data rates and low latency (delay) can assists in video inspection/ Augmented reality (AR) for supporting maintenance. As discussed above implementing AI in this combination could result in powerful most acceptable option.

5. **Comparison of ML and DL:** Both ML and DL are very much connected to each other. ML is a subset AI focused on algorithms development for learning. DL is a subset of ML used for ANN with many deep layers brain. The relation of AI, ML and DL has been clarified in Fig. III 4.1b. Even though both ML and DL have similar goal but methods they deploy vary significantly. Difference between ML and DL has been elaborated in Table III 4.1a.

## 4.1.1 Machine and deep learning

Data science comprises several steps such as Collection, Processing, Exploratory data analysis (EDA), Machine learning, and Visualization as shown Fig. III 4.1c.

So, it is a part of data science. It shows some similarity viz. regression, with statistics, ML is different from statistics. ML is based on Decision tree, Rule induction, Neural network etc. According to Tom Mitchell ML can be defined as "*A computer program is said to* ***learn*** *from experience E with respect to some class of tasks T and performance measure P, if its performance at tasks in T, as measured by P, improves with experience E*". ML is a technology that gives the computer the ability to make it to behave more similar to humans i.e., the ability to learn. As discussed, ML is a subset of AI whereas DL is a subset of ML for ANN as shown in Fig. III 4.1b. The overall goal of ML is to build models that imitate and generalize data. These models need to learn how to discriminate certain things to achieve a desired end result. ML can learn for themselves and also, with the emergence of the internet the huge amount of digital information generated can be stored, and made available for analysis for learning. ML algorithm refers to a set of rules/processes used by AI to carry out tasks encompassing mainly; discovering new data insights and patterns/prediction of values from a given set of variables. These algorithms enable ML to learn. Thus, underlying algorithms ML algorithms are essential for computers to learn from data, to perform the desired task. Deep learning is a type of ML that builds neural-like networks, having functional resemblance with human brain.

**FIGURE III 4.1B** AI, ML and DL interrelations.

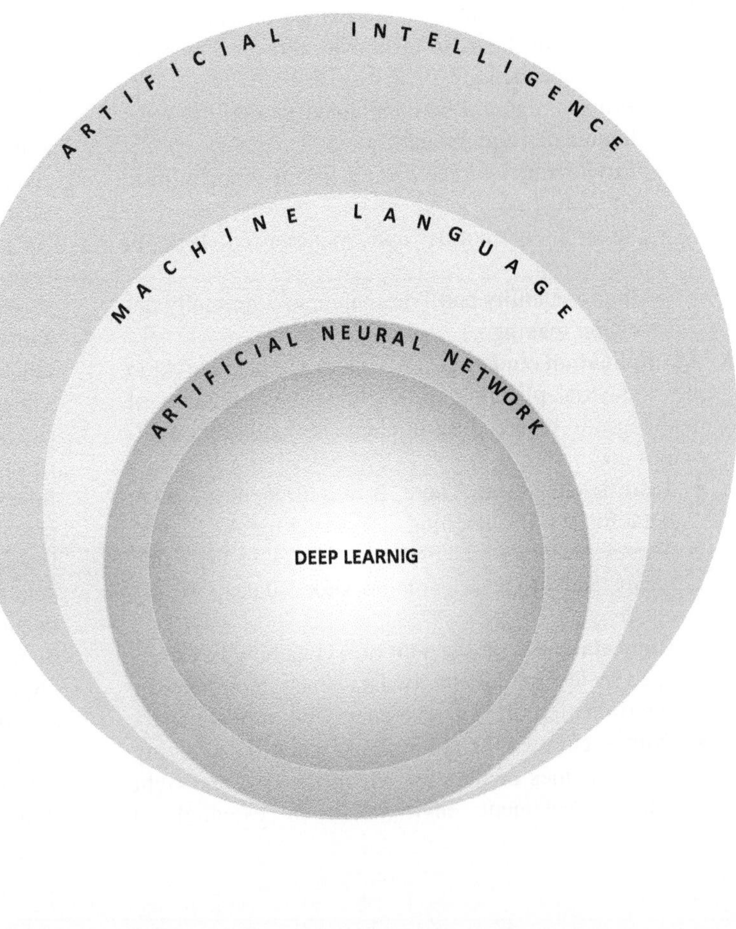

**TABLE III 4.1A** Differnce between machine learning and deep learning.

| Machine learning | Deep learning |
|---|---|
| A variety of algorithms are in ML | DL algorithm is mainly related to ANN |
| Comparatively, ML deals smaller data set than that for DL | Comparatively DL deals larger data set than that for ML |
| Comparatively simpler computation for lower-level tasks and easier to interpret | This is normally complex (black box) tasks. It is not easy to interpret (black box) |
| Apply statistical algorithm for hidden pattern in the system | Use ANN to learn hidden pattern and relationship |
| Lesser time in learning/training | More time in learning/training |
| Models created manually from extracted data | Models created automatically from extracted data |
| Can work with lesser computing power of CPU | This calls for high performance machine with GPU |

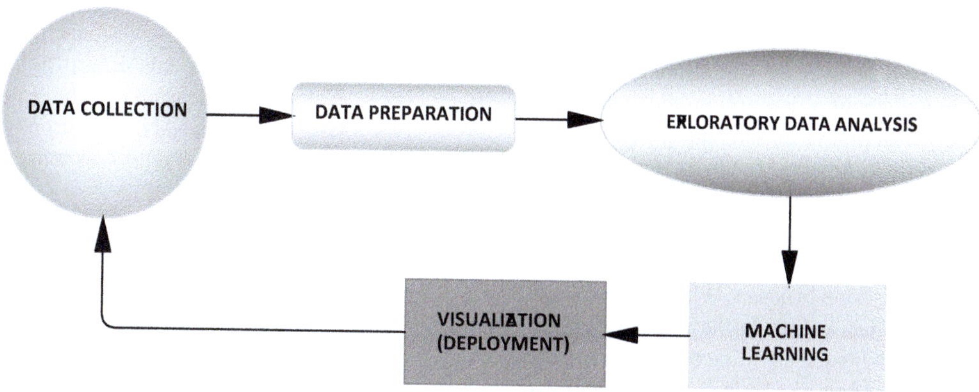


**FIGURE III 4.1C** Data science process.

Deep learning algorithms are the driving force behind many applications of AI especially those using ANN. The lack of pretrained data is what makes this type of machine learning so valuable. In DL, automating tasks, analyzing data, and making predictions etc. are done without human intervention. DL algorithms need to be able to make connections without always knowing what they're looking for. Deep learning can ingest unstructured data in its raw form (text or images). Also, DL automatically can determine the set of features that distinguish different categories of data from one another enabling use of larger data sets without human intervention.

1. **Basics of ML:** The goal of AI is to imitate and mimic human behavior, and machine learning gives us the mathematical tools that allow us to do that. In ML there will be different algorithms, different techniques (within those algorithms), are used to build a model that is application appropriate. ML a subset of AI focuses on the development of algorithms and models. Based on data, these algorithms allow computers to learn, to make predictions or to take decisions. ML helps to build systems that can automatically improve and adapt their performance with experience under different environments. As shown in Fig. III 4.1c, ML is data driven technology where large amount of data generated by enterprises on daily bases, are analyzed to establish notable relationships in data, so that accurate decisions can be made. There are a number of types of ML viz. Supervised/Unsupervised/Reinforced/Stochastic (*refers to a process where the outcome involves some randomness and has some uncertainty*)/Hebbian learning, etc. Let us now look into details of ML.

- **ML Explained:** ML algorithms are set of codes that helps for exploring, analyzing, and finding relations and meaning in complex data sets. Each algorithm is a finite set of unambiguous step-by-step instructions that a machine can follow to achieve a certain goal. ML model can establish or discover patterns to make predictions or categorize information. ML algorithms make use of training data (a subset of the larger set}, which can be expanded to represent the problem environment more realistically. Different ML algorithms analyze data in different ways but they can provide more accurate results. Common ML algorithms use regression and classification to predict target categories, find unusual data points, predict values, and discover similarities. Advanced AI are capable of providing more personalized support, reduce response times, provide speech recognition and improve customer satisfaction. The major benefits industries can gain from machine learning algorithms are: new content creation from vast amounts of data include supply chain management, transportation and logistics, retail and manufacturing etc. All these embrace generative AI, to automate tasks, enhance efficiency and provide valuable insights.

- **Working Principle:** Three major processes involved in working of ML algorithm are:
  - *Decision process:* In general, machine learning algorithms are used to make a Based on labeled or un-labeled input data, algorithm will produce an estimate about a pattern in the data in the form of prediction or classification.
  - *Error function:* An error function is used to evaluate the prediction of the model. For known examples, an error function makes a comparison to assess the accuracy of the model.
  - *Model optimization process:* When the model fits better to the data points in the training set, then weights are adjusted to reduce the discrepancy between the known example and the model estimate. The algorithm will repeat this "evaluate and optimize" process, until a threshold of accuracy has been met.
- **Features:** Various features of ML are:
  - It is a kind of data mining, as for functioning, ML deals with the huge amount of data.
  - With ML, systems can learn itself from past data for automatic improvisation.
  - From the given dataset, it detects various patterns on data.
  - Branding is an important issue for many, with ML it will is easier to target relatable customer base and meet customer demands.
- **Learning rules:** There are different learning laws and these are: *Hebb's*, *Delta*, *Wldrow and Hoff LMS*, *Perceptron*, *Correlation*, *lnstar* (Winner-take-all) and *Outstart* learning laws

2. **ML classification:** ML algorithms are broadly categorized into three types:
   - **Supervised learning**: Algorithms learn from labeled data with known relationship between, the Input/output (I/O). Here, training datasets are provided to the system. The algorithm *learns* a set of inputs along with corresponding correct outputs and learns by comparing its actual output with correct outputs to find errors. So, algorithms analyze the data and produce an inferred function. By analyzing and generalizing patterns in the labeled data, the model can make predictions or classify new, unseen data accurately. There is another type referred to as *Semisupervised* type, which can combine the best of both worlds i.e., *improved accuracy* associated with *supervised* ML and the *ability to make use of cost-effective* unlabeled data, as in *unsupervised* ML discussed below.
   - **Unsupervised learning**: Unsupervised learning algorithms are comparatively harder. The machines need to learn on its own without any supervision. These algorithms work with unlabeled data to discover inherent patterns/structures/groupings without knowing any prior data or information. Thus, there are no predefined outcomes or targets. The models identify relationships, clusters, or anomalies in the data, enabling insights and discoveries. *Competitive* learning is an example of this type ML where in ANN, neurons competes to "fire" or become active in response to a specific input..
   - **Reinforcement learning:** Reinforcement learning is defined by characterizing a learning problem and not by characterizing learning methods. Algorithm is developed with principles of behavioral psychology in mind. Algorithms use trial and error, within a specific context, receiving feedback in the form of rewards or penalties. An agent interacts with an environment and learns to make decisions to determine ideal outcome through performance feedback.

   It is worth noting that there exist subdivisions of various ML algorithm types mentioned above. These are further elaborated in Table III 4.1b.
3. **ML Application area: In modern days,** ML finds its applications in various application areas. Frankly speaking it has revolutionized by improving decision-making processes. In support of the statement Table III 4.1c has been presented.
4. **Discussions on DL:** Even when some information is unknown, industries can make complex decisions through model, based on extracting information and analyzing them from large amount of industrial data. DL computer models learn to perform tasks by considering examples rather than being explicitly programmed. The term "deep" in DL refers to the number of layers in the network. Naturally it transpires that DL is based on ANNs discussed before. As discussed before, DK is a subset of ML. So, DL can be defined as the branch of ML that is based on ANN architecture, which uses layers of interconnected nodes called neurons that work together to process and learn

**TABLE III 4.1B** Classification of ML algorithms.

| ML CLass | Description | Remarks |
|---|---|---|
| **Supervised learning: (Regression and classification)** | | |
| **Linear regression** | Best fit straight line between I/O | Predict continuous value |
| **Logistic regression** | Predicts probabilities and assigns data points for binary dependent variable and uses S-shaped logistic function for modeling relationship | Binary classification tasks (binary or multiclass) |
| **Decision tree** | Splits data into branches based on feature values, creating a tree-like structure with each node represents a feature leaf-nodes provide the final prediction. | Both classification and regression tasks i.e., predicting/classifying data into categories |
| **Support vector machines (SVMs)** | Finds the best boundary (a hyperplane) to separate data points into classes using support vectors for hyperplane. Handles linear/nonlinear problems | Classification/regression: Speech recognition text categorization, data classification, |
| **k-Nearest neighbors (k-NN)** | Nonparametric algorithm classifies data points based on their proximity and association to other available data. Classification: assigns class label, regression: predicts the value of its k nearest neighbors | Classification/regression:<br>Image video recognition, handwriting detection |
| **Naive bayes** | Principle of class conditional independence from the Bayes theorem i.e., all features are independent of each other (naive) for a given outcome with each predictor has equal effect on that result. | Spam filtering, text classification, |
| **Random forest** | Random forest is an ensemble process to combines multiple decision trees. "Forest" refers to uncorrelated decision trees reducing overfitting with final prediction on majority voting | High precision and reduced bias of results. Can handle missing values in features. Stock market, e-commerce |
| **Gradient boosting (G boosting)** | Boosts underperforming regression algorithm, by combining it with weaker ones to create a stronger one. Build models sequentially, new model correcting errors in previous ones. | Classification/regression: XGBoost, LightGBM, CatBoost; remote sensing |
| **Neural networks** | Labeled data to train/learn the relationship of I/O, error is minimized using back progression. NNs learn from adjustments based on the loss function through gradient descent. | Primarily for DL |
| **Artificial NN** | The learner algorithm recognizes patterns in input data using building blocks called neurons | Primarily for DL<br>Classification/regression image classification |
| **<u>Unsupervised learning</u>: (Clustering/Dimensionality reduction/Association rule mining)** | | |
| **Clustering** | Identification of natural groupings in the data. Multiple types are:<br>Centroid-based: *K-Means/K-Mode*; distribution-based: *Gaussian mixture models, Expectation-maximization*; connectivity-based: *Affinity propagation, Divisive clustering*; density-based: *DB scan*<br><u>Application:</u> *Market/image segmentation, Anomaly detection, Medical imaging* | |
| **Dimensionality reduction** | Simplification of datasets by feature nos. reduction retaining the important information. Various types are: Principal component analysis, t-distributed stochastic neighbor embedding (t-SNE), isomap, independent component analysis, autoencoder etc.<br><u>Application</u>: *Numerical, Categorical and ordinal* and *Time series data* | |
| **Association rule mining** | Find patterns between items in large dataset, frequent item sets by iterating through data and pruning. Various types are:<br>Frequent pattern-growth, equivalence class clustering and bottom-up lattice traversal. Apriori is an algorithm for frequent item set mining. With application of aprori, analysts can uncover valuable insights from transactional data<br><u>Application</u>: Customer segmentation, fraud detection etc. | |

*Continued*

**TABLE III 4.1B** Classification of ML algorithms.—cont'd

| ML CLass | Description | Remarks |
|---|---|---|
| **Reinforcement learning: (Model-based and model free)** | | |
| **Model-based** | Use an environmental model to predict outcomes and the agent can plan actions through simulating. | Bellman equation, markov decision process Monte Carlo tree |
| **Model free** | Agent learns directly from experience by interacting with the environment to adjust actions based on feedback. Two types are value-based method and policy-based method. | Value-based: SARASA, Q-learning.<br>Policy-based: Actor-critic algorithm, asynchronous advantage actor-critic |

**TABLE III 4.1C** ML application areas.

| Area | Details of application(s) |
|---|---|
| **Autonomous vehicle** | For perception, decision-making, and control. |
| **Finance** | ML used for credit scoring, fraud detection, and trading. *Credit scoring:* Assessment of credit risk avoiding human error<br>*Fraud detection*: ML analyzes enormous at a time to detect fraud based on a person's spending patterns.<br>*Trading*: With ability to efficiently assess data and patterns, for quick decision-making |
| **Health care** | Disease diagnosis, patient information, and personalized medicine.<br>*Diagnosis*: Analyzing data for trends/red flags for earlier diagnoses<br>*Patient information*: Data collection for patient information and real-time assessment.<br>*Medicine*: Data for probable side effect and personal impact of medicine |
| **Image and speech recognition** | Identification of objects in images,<br>facial recognition and transcribe speech |
| **Marketing and sales** | Online branding with suggestion and advertisement collected brand data to get customers a unique and personalized experience. |
| **Natural language** | ML enables chatbots, sentiment analysis, translation. |
| **Oil and gas** | *Source:* Analysis of ground data to discover potential new energy source.<br>*Distribution:* Streamline oil distribution system<br>*Reservoir modeling*: ML can optimization of hydraulic fracturing, reservoir simulation |
| **Retail** | ML empowers recommender systems, forecasting, inventory management. |
| **Transportation** | Data analysis for identification patterns and trends of traffic for public routes efficient etc. |
| **Utilities** | Here ML is used for uncovering hidden energy patterns, learning customer's energy behaviors etc. |
| **Area** | Details of application(s) |
| **Autonomous vehicle** | For perception, decision-making, and control. |
| **Finance** | ML used for credit scoring, fraud detection, and trading. *Credit scoring:* Assessment of credit risk avoiding human error<br>*Fraud detection*: ML analyzes enormous at a time to detect fraud based on a person's spending patterns.<br>*Trading*: With ability to efficiently assess data and patterns, for quick decision-making |
| **Health care** | Disease diagnosis, patient information, and personalized medicine.<br>*Diagnosis*: Analyzing data for trends/red flags for earlier diagnoses<br>*Patient information*: Data collection for patient information and real-time assessment.<br>*Medicine*: Data for probable side effect and personal impact of medicine |
| **Image and speech recognition** | Identification of objects in images,<br>facial recognition and transcribe speech |

*Continued*

**TABLE III 4.1C** ML application areas.—cont'd

| Area | Details of application(s) |
|---|---|
| **Marketing and sales** | Online branding with suggestion and advertisement collected brand data to get customers a unique and personalized experience. |
| **Natural language** | ML enables chatbots, sentiment analysis, translation. |
| **Oil and gas** | *Source:* Analysis of ground data to discover potential new energy source.<br>*Distribution:* Streamline oil distribution system<br>*Reservoir modeling*: ML can optimization of hydraulic fracturing, reservoir simulation |
| **Retail** | ML empowers recommender systems, forecasting, inventory management. |
| **Transportation** | Data analysis for identification patterns and trends of traffic for public routes efficient etc. |
| **Utilities** | Here ML is used for uncovering hidden energy patterns, learning customer's energy behaviors etc. |

from the input data. ANN architecture has been discussed at length in III 4.0. *Deep generative learning* is a DL type, which focuses on creating new output from learned input. *Deep generative learning* is the basis for *generative AI* and *foundation models* and it adds generation to pattern recognition. In a layered NN, the first layer extracts low-level basic features viz. edges and shapes from the data. Next layer combines these features into complex patterns, and so on until the final/output layer produces the desired result. This is forward propagation. Back propagation refers to supervised learning algorithm discussed begore. This calls for a dataset with known correct outputs. Backpropagation compares the network's output with the correct output to adjust the weights in layers in the network accordingly. This process repeats until the network converges on the correct output. So, back propagation is an important part of DL to train complex models quickly and accurately. Common DL models of this category include CNN, RNN, LSTM etc. discussed in III 4.0.1.4.

- **DL Features:** Apart from CNN, RNN discussed, a few DL features are:
  - Activation functions: These introduce nonlinearity to neural networks, to learn complex relationships e.g., Sigmoid, Tanh already discussed.
  - ANN Deep NN (DNN): DL involves ANN and DNN with multiple layers (deep architectures) allowing the model to learn hierarchical representations of data.
  - *Representation Learning:* DL models learn to represent data, extracting features *Autoencoders:* Autoencoders are unsupervised learning models that learn efficient representations of data by encoding and decoding it.
  - at different levels to facilitate better generalization.
  - *Backpropagation:* As already explained, backpropagation the training algorithm is used in DL to minimize the difference between predicted and actual outputs.
  - *Feature Learning*: As discussed, DL algorithms learn hierarchical representations of data, to form more complex features, to understand intricate patterns.
  - *Transfer Learning:* Transfer learning refers to DL model pretraining for a large dataset and fine-tuning it for a specific task.
  - *Dropout:* Dropout is a regularization technique of DL to prevent overfitting.
- **DL Challenges:** In spite of great achievement in various field, major challenges of DL are:
  - *Data availability:* It is a big concern to gather huge data for DL to learn from.
  - *Interpretability:* Complex DL models work as black box making interpretability dificult.
  - *Overfitting:* With again and again training of model, DL becomes too specialized leading to overfitting and results and poor performance on new data.
  - *Time-consuming*: Based on computational resource, DL can take very large time frame.
  - *Computation cost:* DL calls for costly hardware with high processing power such as GPU, hence high computational cost.
- **DL Pro and cons:** Advantages and disadvantages of DL has been tabled in Table III 4.1d.

5. **DL: Types and Applications:** After gathering some knowledge on DL, let us now concentrate on types of DL and their applications.
   - **Types of DL:** Deep learning AI can be used for supervised, unsupervised as well as reinforcement machine learning. it uses a variety of ways to process

**TABLE III 4.1D** DL: Pros and cons.

| Advantages | Disadvantages |
|---|---|
| DL algorithms can achieve *high accuracy* performance in certain tasks viz. Natural language processing | DL calls for *huge data* and *costly powerful* computational *resources.* |
| DL models possess high *scalability* for handling large and complex datasets, with *flexibility* to cater to a wide range of tasks | DL calls for large labeled data for training. This is time consuming and costly too |
| DL models are capable of *continuously improve* them with availability of data | Black box feature of DL make it not easily interpretable and understand. |
| DL algorithms can *automatically discover and learn* features from data | Overfitting discussed before are issues of concern |

these. However, DL applies mostly to unsupervised and reinforcement learning.

- *Supervised learning:* DL algorithms viz. CNNs, RNNs discussed above, are used for mainly in supervised tasks such as image classifications/recognition, language translations, sentiment analysis, to name a few.
- *Unsupervised learning:* Auto-encoder, generative models are examples of unsupervised DL algorithms, which can be used for anomaly detection, dimensionality reduction and clustering as discussed.
- *Reinforced learning:* Deep Q networks and Deep deterministic policy gradient (DDPG) used to reinforce tasks such as robotics and game playing etc. are examples of reinforced learning in DL.

- **DL applications:** Application of DL in various areas are as follows:
  - *Computer vision*: DL models enable machines to identify and understand visual data i.e., automatically extracts information and insights from images and videos. *Object detection and recognition, Image classification/segmentation* are some usage viz. Facial recognition, image detailing.
  - *Speech recognition:* DL models can analyze human speech despite varying speech patterns, pitch, tone, language, and accent. DL model can transcribe spoken words, making it possible to perform tasks such as speech-to-text conversion, voice search.
  - *Recommendation engines*: DL methods can be used to track user activity with personalized recommendations.
  - *Generative AI*: This creates new content and communicate with end users and can assist in automating complex workflows, brainstorming ideas, knowledge searches etc.
  - *Natural language processing (NLP)*: Computers use DL algorithms to gather insights and meaning from text data and documents, also can Translation, Automatic summarization, Business intelligence analysis, Indexing phrasing etc. Sentiment analysis is part of it
  - *Sentiment analysis:* Sentiment analysis process extracts and understands idea expressed in text. It is a kind of natural language processing used by enterprises to understand customers' experience/opinion about the product/service to identify the scope of improvement. It also helps to assess trends and patterns in the data.
  - *Business process*: DL can help to automate repetitive tasks, speed up decision making, and optimize operations and improve businesses. DL also can be used for improved predictive modeling using historical data.
  - *Marketing strategy*: DL can be used to connect the target audience and cater to their specific needs. DL in this way helps to strategize sales and marketing.

### 4.1.2 Artificial intelligence: application discussions

Technicality and specialization of AI, can be divided into subfields each with focus on the solutions to specific problems. AI system features include: *learning, reasoning, knowledge, planning, communication,* and *perception.* Mathematical optimization, logic, methods based on probability, and many others tools are used in AI. Accordingly, current approaches in control engineering cover various methods such as statistical methods, computational intelligence, and traditional symbolic AI making the interdisciplinary approaches. Associated with control and sensor systems, common AI tools are very useful in *automation applications* They assists in control system engineering with knowledge-base, fuzzy rule-based logic, automatic knowledge acquisition, NN, genetic algorithms, case-based reasoning, ambient-intelligence etc. With development of power and affordability of computers, applications of these tools are widespread. Control engineering tools tend to have less computational complexity and they can often be implemented with low-capability microcontrollers also. The appropriate deployment of new AI tools will contribute to the creation of more capable control systems and applications. Impact of AI on control engineering mainly include *data mining* techniques, *multiagent* systems, and distributed self-organizing systems. Machine intelligence can combine a wide variety of advanced technologies to give machines an ability to learn, adapt, make decisions, and display new behaviors. Some of major issues of AI in industrial control area are:

- **AI-boosting breakthroughs:** *Cheap parallel computation, Big data, Better algorithm, Recursive algorithm, Increased availability, Control systems – interdisciplinary decision making* and *Analog computing comeback*
- **Common applications include:** *Assembly, Building modeling, Computer vision, Sensing, Environmental engineering, Human-computer interaction, Internet use, Robotic manipulation, Robotic programming, Walking robot, Wheel chair assistance* and *Medical systems.* Apart from the above issues listed below are very effective in DT as well as Industrie 4.0.
- ***Customer interaction*:** AI can allow automate some repetitive tasks in the customer support journey viz. Chatbots (connect *customers with the answers from embedded knowledgebases*) replies to basic questions, Interactive voice response (IVR) or to route the enquiry to the agent with the right skills. So, these are useful to improve customer experience, agent performance and planning future needs.
- **Self-service:** AI and ML powered self-service applications are becoming very popular. Here, they continuously learn and improve based on interactions. For this, one has to get started with self-service tools for knowledge feed and training.
- **Artificial general intelligence (AGI):** Apart from robotics, pattern analysis and recognition, AGI can be used in industrial applications. AGI, which is comparable with human cognitive skills can be used for smart multimodal HMI communications. Major properties of AGI utilized in industries are: knowledge-based presentations, natural languages communication, the use of strategies, the evaluation of irregularities and acting.

As is clear from the above discussions that this subsection has been mainly dedicated for application of AI on industries including Industrie 4.0 and DT.

1. **Manufacturing: AI impact:** AI has direct impact on manufacturing. AI can *enhance efficiency, cost reduction, improved quality control,* and *predictive maintenance.* AI automates tasks to augment human capability, data analysis, refinement of product inspection and quality and optimize production processes. AI in manufacturing has been elaborated below:
   - **AI: logistics:** AI can be used to track production, demand forecasting, inventory control coupled with resource management.
   - **AI: Production process:** AI-powered process mining tools are used to automatically identify and eliminate bottlenecks in production processes and monitor performance. Major issues involved are: *Increase yields, Improve accuracy with workplace safety, Boost productivity* and *reduce cycle time.*
   - **AI: design and development:** AI-enabled software can optimize designs mainly in *Raw materials, Size and weight, Manufacturing methods, Cost* and *other resource constraints.* It also assists in product development by: *Reducing trial* and *error time and cost, Problem predictions, Streamline maintenance* and *debugging* and *Less time-to-market.*
   - **AI: Factory automation:** For this, operators rely on their experience and huge signals. Major issues here are: *Human-intensive system can be erroneous, leading to equipment malfunction, Automate several complex tasks in factories, Building central repository* for operational data, *Quick anomaly detection, Reduction in the number of resources, Scaling production* according to *demand fluctuations and manufacturing strategies.*
   - **AI: robotics:** AI automates decision-making and repetitive tasks at manufacturing plants, through robots and ML. On account of self-learning, it is possible to improve handle the tasks better. Also, robots do not have fatigue, hence no break and consistent output without human-like errors is possible.
   - **AI: Autonomous vehicle:** Use of AI in 24/7 autonomous vehicles on the production floor, can assist automate assembly lines.
   - **AI and IIoT:** IIoT refers to smart, connected sensors (viz. Industrial analytics, continuous performance monitoring) to generate huge operational data in real-time.
   - **AI: OT -IT Mixing:** AI can help in marrying OT-IT; such as: *Collection and integrating data, Real-time tracking and performance monitoring, Predictive analytics and capacity planning,* and *big data analytics* for improvized resource utilization.
   - **AI: Connected factory:** AI assists in building smart connected factory. This would facilitate: *Asset utilization monitoring, Remote touchless access, Real time intervention, Production scaling* etc.
   - **AI: Predictive maintenance**: AI can check and predict equipment health, so assists in predictive maintenance.

- **AI: warehouse management:** AI automate wire house operation.
- **AI: Supply chain management:** AI systems assist to improve last-mile deliveries by assessing various scenarios (time/cost/revenue/demand). Also, optimal delivery in terms of route, tracking, delivery performance assessment.

2. **AI in Process control**: AI uses learning algorithms embedded in the controller for development of process models and diagnostics for control loop to monitor performance with adaptive tuning to accurately identify problematic issues for recommending tuning improvement. Also, AI can modify and adapt process models based on process changes. For any change in the setpoint (or output when controller is in manual), AI can capture the process input and output data to efficiently identify the process dynamics. Application of multidisciplinary AI algorithm are characterized mainly by: *High reliability*, *Convenient operation of control*, *enhancement of control accuracy* and *improvement of the working efficiency and quality* of production. AI is applied specifically to build a reliable model. AI learns control laws/the model through ML out of historical data that describes the way a product has responded to different conditions or different instructions in the past. Fig. III 4.1d illustrates application of AI in process controls.
   Fig. III 4.1d1 shows the design philosophy of AI in Process control. Based on application, AI in process engineering and control can be classified into *Data collection (information extraction)*, *Predictive modeling*, *Process optimization*, Diagnostics and detection, and *Process control* to result in *AI process models*, as shown. Generalized architecture and implementation of AI in process control has been detailed in Fig. III 4.1d2. This shows how physical system in conjunction with digital twin can be used to develop hybrid model from historical and current data.
   Another area of application of AI in process control is predictive maintenance, which uses data from various sources to predict health of equipment. This enables to schedule maintenance proactively avoiding unplanned downtime reducing maintenance costs, and extend the lifespan of the equipment. There are many challenges AI needs to face in process controls. Major challenging issues are:
   - Design phase: Improvement of model identification from historical data
   - Implementation phase: Identification of models from real-time operating data
   - Online phase: Dynamic identification and manage changes in the process state
3. **AI in monitoring:** The basic purposes of development of plant monitoring solutions are: Increase the Plant availability, *Core functionality*, *Plant operational efficiency*, *Cost* and *Return on investment* (ROI). Some of the major issues in different industrial plants are:
   - **Process functionalities:** AI-based automation system also helps in continuous control performance monitoring for every loop for quick assessment of automation and control conditions. Overview displays of the plant identifying abnormal control conditions in the system, area and module level. After abnormal condition(s) is detected, it is easier to get detailed performance metrics and historical information for the specific control blocks. The control conditions monitored for the overview and Summary displays include: *Uncertain Input*, *Limited control*, *Incorrect mode*, *Large variability*, *Process oscillation*, *Tuning problem* (if any with) *recommended tuning* and *Device alert.* Apart from loop, It is possible to generate overview displays with summary status information for all equipment and devices monitored for maintenance state. In this connection, asset alert levels reported are: *Asset failed*, *Maintenance soon*, *Advisory* and *communications failed.* AI in plant monitoring can generate performance reports to assist activities such as: operations, maintenance and control performance tracking to identify opportunities for improvement.
   - **Production Process:** Major issues related involvement of AI-based monitoring are:
     - *Productivity Issues:* With application of AI in plant supervisions can effectively reduce issues such as *equipment failure*, *defective product*, *theft* and *security*, *poor inventory management* etc. to improve plant productivity.
     - *Quality control:* Application of AI in quality control, ensures precise quality controls. In many cases especially in food and beverage/pharmaceutical applications even the slightest variations in production processes can lead to defects, rejections, and increased operational costs.
     - *Operational efficiency:* In order to be competitive in a global market, it is essential that the production process must be optimized with AI so that there will be reduction of downtime wastage, and minimization of energy consumption.
     - *Cost management*: Threats from rising production costs, due to labor and raw materials would reduce profit margins and economic viability. Therefore, it is needless to emphasize that deployment of AI in different forms can keep cost under control.
   - **Safety issues:** AI-based systems monitor and identify unsafe conditions/practices industrial systems in real-time. These include identification of improper use of safety equipment/hazardous activities or use of

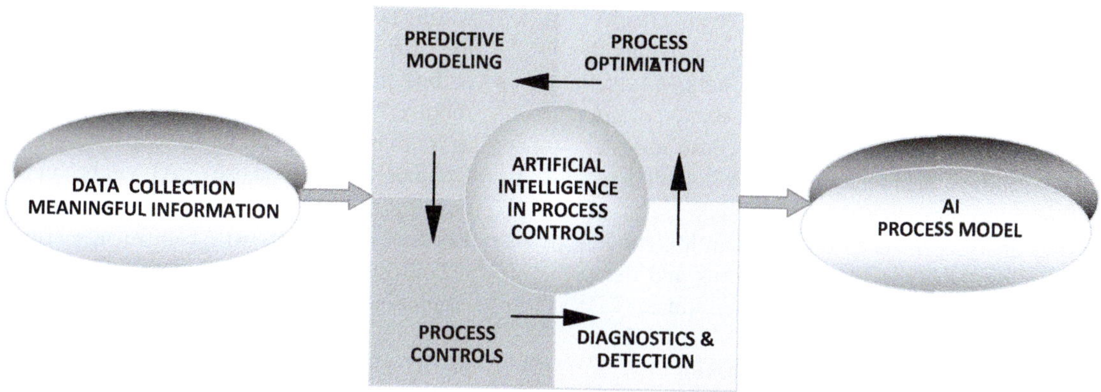


FIG III 4.1d1 DESIGN PHILOSOPHY FOR AI IN PROCESS CONTROL

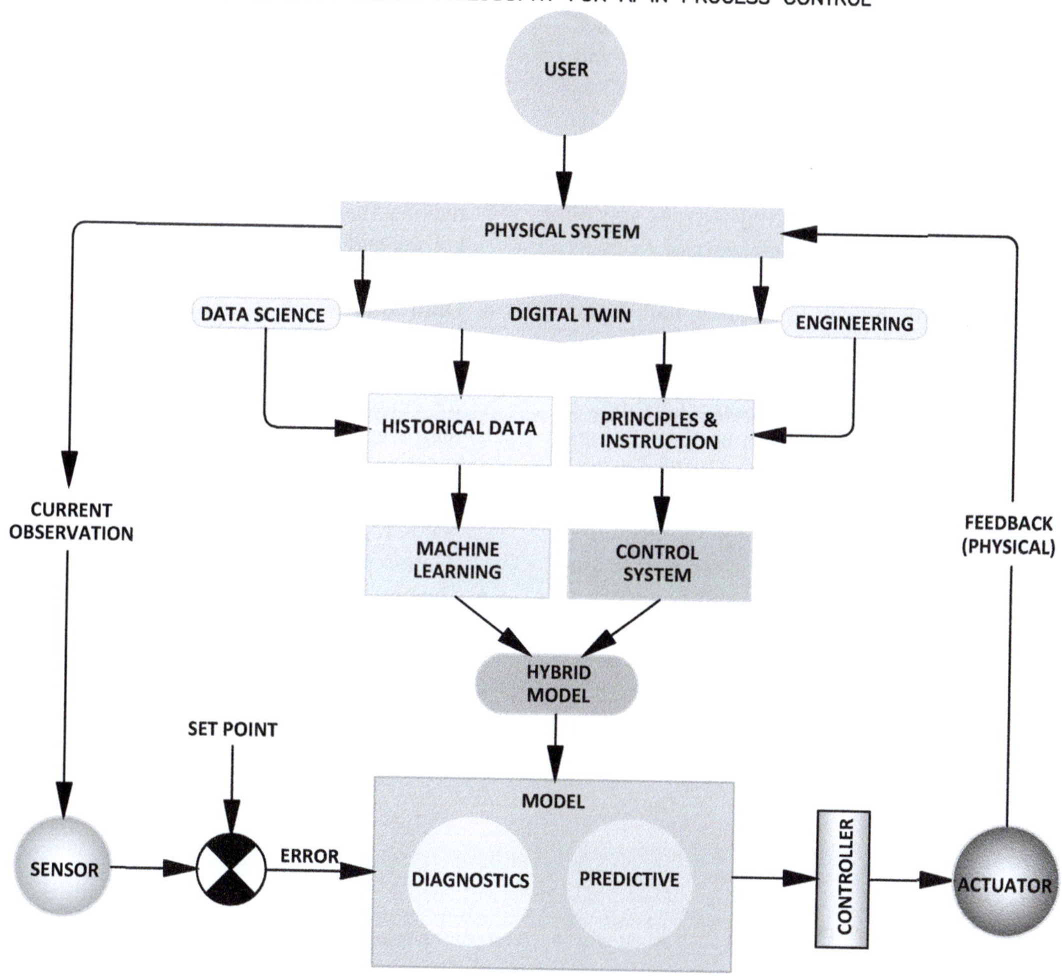


FIG III 4.1d2 ARCHITECTURE OF AI IN PROCESS CONTROL

**FIGURE III 4.1D** Application of AI in process control.

hazardous materials. These systems also take corrective actions to ensure functional safety.

- **Predictive maintenance:** ML algorithms can analyze data from sensors and historical data to predict machine failure in advance. So, predictive maintenance activities help not only to schedule maintenance but also minimize system down time. On account of AI it is possible to get to know wear and tear patterns in machines/systems. As a result there will be huge saving in cost and time.
- **AI: Fault detection and Diagnostics:** AI in fault detection and diagnosis brings together insights

from data science and ML to deliver an effective diagnostic system for industrial plants. This mainly covers: *various AI-based diagnostic methodologies*, Development of *intelligent programs to predict and prevent* fault occurrences to cause disaster. Fundamental AI techniques employed in fault detection and prediction are:

- After analysis of set of historic data ML models and algorithms can identify patterns and detect anomalies/fault in the system in real time.
- Anomalies are detected by identifying the deviations from the expected value/behavior
- By utilizing historical data, predictive models are Developed to predict future events or outcomes.
- For predictions, predictive models are trained with historical fault data and other associated variables and conditions for forecasting faults.
- AI with domain knowledge, can be utilized as an expert systems to detect and predict faults.

4. **AI in Industrie 4.0:** Advanced issues for Industrie 4.0, AI can be utilized in many ways. Major issues are:
   - **Quality 4.0:** With rise in the complexity of products to meet the changing market demand in very short time it is extremely difficult to maintain high levels of quality and to comply with quality regulations and standards. So, in Industrie 4.0, "Quality 4.0" explained in Fig. III 4.1e is necessary. Quality 4.0 takes the help of AI algorithms to notify emerging production faults impacting product quality viz. AI solution use high-resolution cameras in place of human visual inspection. In Quality 4.0, data are collected on the performance of the product in use/field. Predictive quality control is another important issue discussed.
   - **Predictive quality control***:* AI has revolutionized the entire quality control process, wherein manufacturers have moved from reactive to proactive quality management. Apart from reduction in wastages, it also assists in strengthening customer trust for the brand.
   - **Robot collaboration:** For Industrie 4.0, human-robot collaboration needs to be efficient and safe as more industrial robots participating in shop floor. Obviously, advanced use of AI will be central to this development.
   - **Product development:** AI shows new way of product design products. In view of Quality 4.0, now emphasis is on details on the issue defined by designers and engineers and considered as input into an AI algorithm. Technical issues covered would include data describing restrictions with various parameters such as material types, available production methods, budget limitations, and time constraints. AI after examining all possibilities to get optimal solutions.
   - **Market adaptability:** AI as a core element in Industrie 4.0, can be used to optimize production in various ways viz. improvement in supply chains, making them interactive to changes on the market beforehand. So, strategic plans can be implemented in line with AI recommendations.
   - **Data analysis:** It is possible to carry out complete data analysis from the collected data to get to know the instant insights into production process for continuous improvement by quickly identifying inefficiencies and issues.
   - **Comprehensive reporting:** Enterprise can be benefitted from long-term data storage and complete status reporting that gives a clearer picture of production performance.
   - **Dashboard support:** It is possible to develop own dashboard for displaying relevant metrics, improving visibility and decision-making.
   - **Fault diagnosis**: Digital solutions of Industry 4.0 can monitor the machine's condition and shop floor activities in real-time, to identify the defect or

**FIGURE III 4.1E** Quality 4.0 explained.

**Q lity 0 Qulity 0 refers to the future of qulity a well a orga zti ona excellence in cona ction with Industrie 0 This is more tha technology d involves *people, technology* & process. AI, bigdta dta science, DL, ML, Block cha n ae tools for it. Quality 0 witness observes quli ty shift from process opertor to process designers. This involves coti nuous qulity with rel -time dta ad IoT.**

failures in real-time and help diagnose the root cause of fault. DL-based approach toward this is common and found successful.

5. **Use of AI for DT:** From Vol II we understand that DT is adopting new technologies for reshaping business involving: *New concept of business models*, *Operation optimization* and *Improved customer experiences.* During discussions, improvement issues, AI driven DT and implementation issues will be discussed with little overlaps.
   - **Improvement area:** With deployment of AI in DT, improvements achievable are:
     - *Customer experience:* DT being customer-centric, implementation of proper strategy can help in improving customer experience. AI and ML can deliver meaningful, personalized experiences, tailoring products to better fit specific customer needs. Actionable practical issues include actionable insights of customers impression about the enterprise (sentiment analysis) and implement specific solution for serving customer better with application AI e. g., chatbots.
     - *Improved productivity:* Application of AI significantly help to optimize the process bringing business benefits. With AI-based automation, better allocation of resources, scaling can be achieved and repetitive, time-consuming tasks so-called "digital workforce" can be deployed for the routine processes.
     - *Security improvement:* With application of AI and ML, it is easier to detect malicious behavior and anomalies that might be a sign of a security breach violation. Therefore, security will be improved a lot and it will provide better surveillance and take predictive action by detecting fault early.
     - *Decision making:* Since AI decisions are taken based on data analytics, meaning better identification of specific patterns, deliver actionable insights, it is always better for making any decision. So, data-backed decisions are taken for improved KPIs. Such data-backed decisions also help in identifying the potential bottlenecks before they emerge, to minimize the business risks.
   - **AI driven DT:** Major issues of AI that could drive DT to new model concept are:
     - *AI Strategy:* Business strategies driven by AI change continuously with developments of advanced technologies such as ML, DL data analytics and different combinations of them. AI driven strategy use predictive analytics to forecast customer demands, automate processes to reduce costs, downtime supply chain problems and personalize services to enhance customer satisfaction. Thus, these changes are not mere improvement but total transformation.
     - *AI governance:* With integration of AI with DT, obviously, DT will be strongly governed by data analytics. This also provides relief to user from Operational inefficiencies, Security vulnerabilities, Ethical missteps (especially for algorithmic bias).
     - *Digital Infrastructure:* AIs drive innovation, streamline operations, and support transformation sustainability. So, robust digital infrastructure built around *Cloud-based platforms*, *Scalable AI models* is also essential to integrate AI. This calls for a well-designed digital architecture.
     - *AI and Culture:* Enterprise culture, mindsets are important components in DT. AI can play vital role in breaking down mindset barriers with data analytics by revealing real-time insights, AI helps teams collaborate and make decisions more agilely and effectively.
   - **Implementation issues:** Following are major implementation issues for AI in DT:
     - Assessment of present state of the business, including associated bottlenecks, and system integration capabilities for AI to improve the situation.
     - Identification specific reason and objectives for adopting AI for DT in hand.
     - Review of DT strategy planned to check readiness to integrate AI
     - Establishment of data governance practices necessary for AI-based DT after it is found that sufficient necessary data (with legal right to use the same).
     - For implementation of AI-based DT, sufficient resources, in terms of monetary and human are necessary for the implementation.

## 4.2 Futuristic digital control system

Trust, reliability etc. are important when there will be the need to share data to third parties, partners and other external organizations. This is mostly done through a distributed database (say) a cloud-based platform. Blockchain technology (primarily associated with bitcoin) is a technology that can functionally provide a service that's a bit like 'escrow for anything', achieved by all transactions being logged on a secure shared ledger that no single organization or group controls [10]. Blockchain technology provides secure and transparent supply chain management and by integrating blockchain into automation systems, reliable traceability, provenance, and trust in supply chains can be achieved. Big businesses houses such as IBM, Samsung, and Barclays are already working on blockchain-related projects and services. Blockchain technology will cause major disruption to many existing industries, including energy management, supply chain, banking, real estate, and voting.

Automation and controls are integral part of almost all industries for driving efficiency, productivity, and innovation. As technology continues to advance, the future of automation and controls hold exciting possibilities. In industries, IoT integration and data analysis have become crucial leveraging IoT/IIoT technologies to optimize processes, improve maintenance strategies, and enable predictive analytics. Advanced cognitive capabilities such as ML and AI in conjunction with advancement in Robotic process automation (RPA) will allow bots to more intelligently interpret the interfaces they work across, better handle errors, and manipulate unstructured data. Also, these will add more and more intelligence to the system making it easier for decision making and achieving the goal. Other trends in future automations include: Turning big data into smart data to support Industrie 4.0, Digitalization and digital support focusing on customer requirements, and sustainability (to take care of environment, society, and employees). In this section brief discussions on Blockchain technology with special reference to automation, future trends of control and automation especially toward Industrie 4.0 and Digital transformation (DT) will be covered.

### 4.2.0 Future control and automation trends

*Simplify production environments* and unleash *extraordinary innovation*, Improved system *lifecycle management* and *Risk mitigation* etc. are major goal based on which automation will proceed toward future. Unification, simplification and integration of automation are really challenging. However, based on this trend, many modern production systems have evolved from an automated process to a fully autonomous system. In order to explore digital capabilities, marrying IT with OT (i.e., collaboration between OT and IT) is essential, to ensure data, security, and industry-specific knowledge merge with analytics to drive outcomes. Current trend in modern software-defined design uses object-oriented programming to deliver context-rich data for defined objects – automatically. Use of Software as a service in cloud-based software model accelerates software development bringing profitability to the user. In modern days, much of the repetitive and mundane work have been automated with use of AI and Robotic process automation (RPA), which are the key contributor for entering into "Hyperautomation" era. "Hyperautomation" refers to the concept of automating everything in an enterprise to streamline entire business processes making use of AI, ROA and other technologies. Gather insights on the processes, Identify the structured and unstructured data, Predict outcomes in terms of efficiencies and ROI, Determination of automation technology and platform, Automate complex business process and technology and use of AI and RPA etc. are major components of "Hyperautomation". This is also seen as democratizer of automation. In this subsection brief discussions will be presented on future trends control and automation in various industrial plants.

1. **Control system Trends:** The future trend of control systems viz. Distributed control system (DCS)/Programable logic controller (PLC) with AI/ML, is driven by the integration of advanced technologies such as IoT, cloud/edge computing, virtualization types etc. to support data analytics, remote monitoring as necessary. Traditional controllers are replaced with AI-based Self-tuning types and sensors also possess AI with cognitive nature and memory function [10]. Advanced process controls use predictive model-based expert systems to handle multivariable interactive controls. Linguistic Fuzzy logic handling with vagueness may be utilized with smarter sensor to sense disturbances due to environment and take self-correcting actions. Workers will be connected to enterprise asset more logically than before. Major issues in future trends include but not limited to:
   - **Integration with other systems:** Control systems will integrate with other sub systems with associated controls viz. PLC, Enterprise resource planning (ERP), Manufacturing execution systems (MES), etc. for seamless integration of various plant systems and improving the overall efficiency of the plant.
   - **Remote monitoring and control:** Remote monitoring and control will be more important and easier to implement with advancements in IIoT. This would enable users to optimize operations and reduce costs by monitoring and controlling processes remotely.
   - **Virtualization:** As the name suggests, virtualization refers to a virtual version of an underlying service, so that multiple Operating system (OS)s and applications can be run with same hardware simultaneously to provide increased utilization and flexibility coupled with cost-effective, hardware-reducing, and energy-saving. These are mainly used for cloud computing. This also helps to get more flexible and scalable DCS architecture [11].
   - **Industry 4.0 and DT:** With integration of AI and ML in control systems to make them intelligent, it is possible to enable advanced extended capabilities such as predictive maintenance, real-time

optimization, autonomous decision-making, predictive diagnostics, customer experience details, making intelligent controls compatible to requirements for DT and Industry 4.0 technologies. Also, advanced analytics not only make intelligent control systems suitable for Industrie 4.0, but also make suitable for real-time *process modeling* and *optimized control* strategies.

2. **IoT and IIoT issues:** Integration of IoT with control and automation systems, will allow for real time data collection, remote monitoring, advanced analytics etc. These would improvize system efficiency and ensure better safety for the industrial plant. More and more connected devices to automation systems may require new communication technologies, as well as more powerful edge computing and data analytics capabilities. With introduction of 5G technology (Fig II 1.2b) and prevalent IIoT control and automation systems must be designed in such a way that systems are capable of handling huge volume of the increased data traffic.
3. **Cloud Edge computing:** Cloud computing refers to the delivery of computing service (including server, storage, database, networking software analytics and intelligence) over internet to offer greater innovations, faster actions lower operating cost for efficient business infrastructure. To avoid cyber security issue private virtual servers will also come into effect. These shared resources and system redundancies will have direct impact on future control and automation systems. Therefore, cloud-based control and automation solutions will allow not only remote access but also the ability to scale as needed, at reduced costs. Also, edge computing (i.e., processing of data closer to the source), would provide reliable data communication with reduced latency in industrial environments where real-time control and decision making is critical.
4. **Control Center:** In view of Industrie 4.0 and DT, future control rooms will have advanced Workstation (WS) with different kinds of visualization viz. AR and Virtual reality (VR) described in Fig II 16.0c. *Ambient intelligence, where electronic systems and environments are sensitive and responsive to the presence of people, will be given greater attention* [10]. Therefore, major trending issues mainly include: *Advanced HMI, Immersive visualization* –AR and *Digital twin, Data analytics, Cloud/edge computing, AI -ML support, Enhanced Reliability and Security Collaborative environment, Training arrangements* etc. So, AR digital twin and HMI technologies in conjunction with AI will provide operators with a more intuitive and immersive interface for monitoring, control and automation to improve safety, performance and efficiency of the industrial processes. Some important features are:
   - **Advanced WS:** Some of the significantly greater capabilities may be: , *Unlimited source management, control and interactivity, Advanced data analytics and immersive visualization, Secured management with scaling of control room applications* etc.
   - **Advanced security:** Control room platforms with stronger security strategies mainly are: *administrators to decide on the extent of network and internet access, Deployment of backup for failover, ensuring no downtime,* Keeping the *on-premise server physically secured with the option to run it completely air-gapped, choice to manage the platform through the cloud, Facility for cloud interaction with protection without additional software,*
   - **Collaborative approach:** To support factory, there shall be improved communication and collaboration breaking the boundary with security for the adoption of collaboration tools and technologies as a result.
   - **AR and VR:** As described in Vol II, AR and VR in conjunction with digital twins are quickly becoming more commonplace in operation centers and control rooms.
   - **Sustainability in design:** Sustainability is becoming a hot issue in control room and console design. Eco-friendly materials make up more of console construction. PVC may be replaced with more eco-friendly plastics such as ABS. Local manufacturing is preferred, because of low transportation emissions.
   - **Console design:** Some of the issues in console design would include: Adjustment of flexible monitor arm, adjustment of console height better comfortability and other details discussed in II 1.0 and II 1.1. The design shall also cover support for plug and play technology and latest display technology with ultra-wide and curved displays.
   - **AI Impact:** We understand that uses of AI in automation help operators to work more efficiently and assisting decision making and automate routine tasks. Major issues are: *Real time insight, Pattern recognition, Stream-lined operation* etc.
   - **Layout issues:** Control room layouts mainly feature consoles in clusters. Also, issues such as *Collaborative layout, Interactive surfaces, Integrated communication.*
5. **Safety and security:** With the increased integration and connectivity of control and automation systems to the internet and other networks, cybersecurity will become an increasingly important aspect of future control and automation systems. In place of *reactive* safety and security systems predictive safety and security systems will be used in future systems. These will utilize predictive data analytics and AI viz. expert systems and ANN for these. On account of AI it would be

possible to early detect system/equipment degradation to initiate safety action and anticipatory alert to operator through proper alarm management. AI systems would also help to prevent access, detect and block or eliminate intruders, learn from mistakes and find system vulnerabilities, and eliminate or place roadblocks to intruders [10].

### 4.2.1 Influencing factors in future automation

Intelligent control and automation brings about sea changes in the operations of industrial plants and processes giving rise to Industrie 4.0 and DT for which there are several factors to influence future of industrial control and automation. *AR/VR*, *IoT*, *Digital twin*, *Cloud/Edge computing* (data analytics) and *IIoT* directly impact on control and automation future trends. This subsection has been developed to focus on impact of these factors on future automation. As explained in Fig II 16.0c, AR is an interactive experience to obtain enhanced the real world with computerized digital information utilizing suitable software. AR is expected to play a crucial role in automation to perform tasks more safely, efficiently and accurately, to enhance productivity, training, maintenance, and quality control across various industrial sectors. Key issues for AR are: *Visual guidance and instructions*, *Remote assistance*, *Diagnostics and machine maintenance*, *Training and work force*, *Quality control*, *Asset identification with tracking* etc. Next important factor is IoT, which enables smart bridging of three spheres viz. human, digital and physical. This convergence is possible through connectivity and communication capabilities in a secured manner. In very simple term IoT can be conceived as bridges between the 'Internet world' and the 'real world'. The most fundamental issue in IoT is its *ubiquity* of access. In IoT is a *collection* of *communication, computing, connectivity, contents* and *convergence*. Digital twin refers to digital representations of people, processes and things and it is more connected, providing a rich *real-time* view industrial processes and assets. Digital Twin technologies have proven especially useful when integrated with AI/ML and IoT in Industrie 4.0 applications. On account of digital twin it is possible to program the robots offline for multiple scenarios and tested. In modern days engineering manufacturing and supply chain management use digital twin for better and accurate virtual representations of objects with simulations of operational processes. Edge computing and cloud computing boost computational efficiency and allows data analytics. However, they also differ significantly in architecture and operational focus as has been elaborated in II 16.0.1.2 and3. *On-demand self-service*, *Broad network access*, *Resource pooling*, *Remote computing and storage*, *Rapid elasticity*, *Measured service* etc. are essential characteristics of cloud computing. Cloud computing supports Web browsing - enhancing mobility and this is independent of device and location. This focuses on maximizing the effectiveness of the shared resources. However, it invites outside inherent risks. Cloud-based analytics on demand as a service is a new phenomenon. Cloud analytics allows for the simultaneous recording and processing of data and enterprises can track data from different locations. Edge computing handles data processing closer to its origin, i.e., at the "edge" of the network—edge servers, in contrast to cloud computing, which centralizes data processing at remote data centers. Edge: *Distributed data processing*, *Enhanced analytical capability*, *improved decision making* and *emerging and innovative applications* are a few advantages of edge computing. Edge computing that has great promise to empower automation engineers to perform analytics on the edge for a wide range of applications. Edge computing and IIoT are perfect match. When IoT is utilized in industries specifically it named as Industrial IoT or *IIoT*. IIoT is an enterprise-grade counterpart of IoT with some more impressive/additional features. The technology can bring together brilliant machines, advanced analytics, and the people involved at work. It can offer machine-to-machine communication with big data analytics in an industry to drive unprecedented levels of efficiency, productivity, and performance. Many people also call it as *Industries 4.0*, but this not true picture it enhances Industries 4.0. In order to cover the entire aspects of futuristic controls, any discussions without these influencing factors will be incomplete. So, briefly these are covered here with special reference to future automation. [*There may be some repetition with respect to other volumes, but to complete the discussions here same have been included from different angle.*] However, these are discussed at length in Vol IV of this book. Let the discussions start with discussions of AR &VR.

1. **AR and VR Discussions:** In industrial automation, robotics and CC, AR/VR technologies play very important role, for better performance improving productivity, efficiency and enhance user experience. Actually, there are AR, VR and Mixed reality (MR). As mentioned, *AR enhances,/augments, the real world with digital information. VR, fully immerses people in a virtual world with suitable hardware.* In virtual world there will be *imagery and sounds, users can move around in all directions, manipulate objects* etc. Mixed reality blends imagination and reality so that users can both see and interact with the real world in the virtual environment. Let us now look into usage of AR and VR in industries.
   - **AR Benefits:** A few benefits of AR are: *Quick identification of problem, Improved productivity, Operational efficiencies with lower downtime*, Simplified

process, Enhanced product development, Streamlined warehouse management, Reduced risk.

- **AR: Industrial application:** By incorporating the physical with the virtual to augment the way people work, augmented reality gives more information and context about the object through digital information and working with digital twin. Various usages are: *Design and product development, Maintenance, Operational control, Safety, Training and learning, Quality control* etc.
- **VR Applications:** VR can provide suitable interactive simulation allowing creation an immersive experience of virtual models, simulations of products and testing. Applying VR, a flexible and immersive environment can be created. Major VR applications are: *Prototyping, Manufacturing process simulation, Factory floor planning, Normal and safety training, Inspection, Quality control, Promotion* etc.

2. **Role of IoT:** IoT refers to the network of physical objects-devices *with electronics, software, sensors, and network connectivity* to *sense, collect, exchange data* and *control remotely across* existing network infrastructure based on a layered architecture. Each of these layers includes the application of a range of diverse technologies for the data transmission, processing and storage. Thus, there will be vulnerabilities and threats in IoT environment. So, protection methods must be implemented within environment. As discussed, major benefits shall include: *Enhanced operational efficiency, Predictive maintenance, Real tome monitoring and control, Support Cloud computing (CC) and Edge computing (EC), Data driven Decision making* etc. Data security and integration with legacy systems are major challenges. Apart from various applications already discussed, IoT-based future applications mainly include: *Smart factory, Smart city, Smart traffic control, Smart energy, Smart grid, Smart health care services, Educational applications, Agriculture application Supply chain management*, etc. Major issues here are:
   - **AI integration and data analytic**: Current trend in IoT is to integrate advanced data analytics as well as AI. Vast amounts of data generated by IoT devices, in conjunction with benefit of AI/ML are able to provide deeper insights and predictive capabilities..
   - **Data processing:** EC is an emerging trend to go hand to hand with IoT-based automation. With the growth in the number of IoT devices, the data generated volume is growing requires data processing.
   - **Security and Privacy:** With a large number of connected devices transmitting data over networks, the risk of cyber threats and data breaches is a growing concern. Therefore, suitable measure to ensure security and data privacy is very important for IoT adoption.
   - **IoT-based automation and services:** Enterprises in various industries are adopting IoT-based automation to enhance efficiency, reduce costs, and improve overall operational effectiveness. Service providers are also not far behind in this adoption.
3. **Digital Twin concept:** According to encyclopedia *"The Digital Twin is a representation of an active unique* "product" *which can be a real device, object, machine, service, intangible asset, or a system consisting of a product and its related services"* [12]. Digital Twin has been explained in Fig. III 4.2a.
   From Fig. III 4.2a1, it transpires that "Digital Twin" can be conceived as the convergence between physical and virtual products i.e., Digital Twin refers to a real-time digital representation of a physical object. Digital Twin is a digital model of a real-life object, and can be used as a proxy for a real machine, allowing users to test concepts and hypotheses before applying them to physical machines themselves. Let us look into the details.
   - **Technological issues:** Digital Twins work around digital data/information. Data– acquisition, modeling, and application together constitute Digital Twins. As shown in Fig. III 4.2a2, four major technologies directly support Digital Twins are: AI, CC, IoT and Extended reality, which is umbrella term used to describe immersive technologies AR, MR and VR already discussed. Out of these technologies, use of a particular technology, depends on the application type, to a greater or lesser extent. Also, it is possible to manipulate digital environments or modify to introduce additional resources to optimize Digital Twin performance. Combination of twin data with business rules, optimization algorithms data analytics etc. Digital Twins can support and automate decision making.
   - **Application issue:** With advent of the concept of Industrie 4.0 have enabling technological advancements in sensing, monitoring, and decision-making tools, industrial plants specially manufacturing industry have undergone a rapid transformation. This has resulted, deployment of Digital Twins in industries the manufacturing industry have increased many folds. This technology holds great potential for a range of activities in the industry and can radically change the face of it. So, with the support of other technologies discussed Digital Twin can be implements successfully for the real-time monitoring and optimization of the process. *Product design, Process design and optimization, Cross-functional collaboration, Preventive maintenance, Supply chain management* etc. are common application area of Digital Twin in industries.

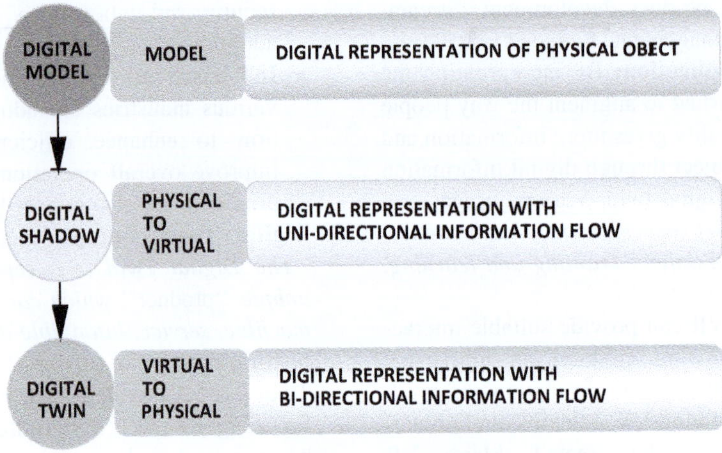


FIG III 4.2a1 DIGITAL TWIN INTEGRATION LEVELS

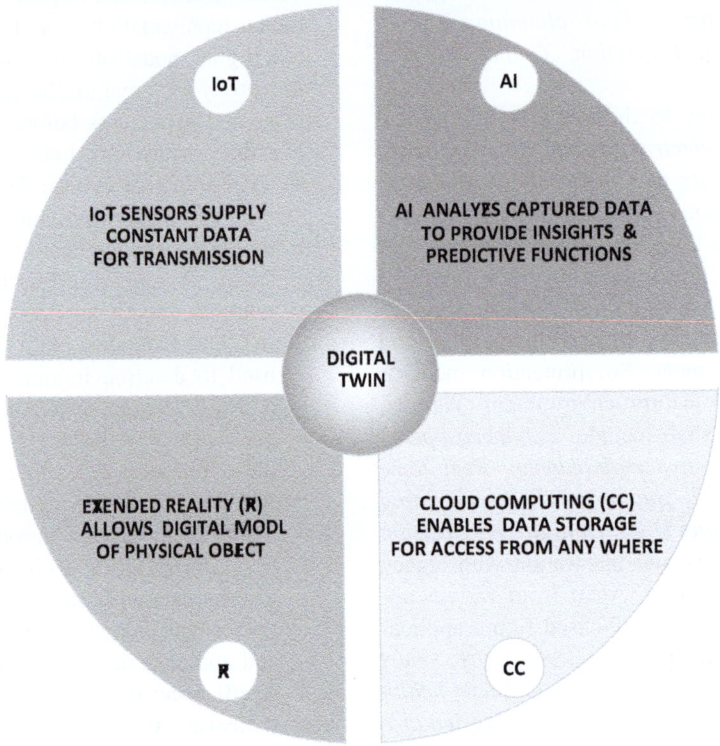


FIG III 4.2a2 DIGITAL TWIN TECHNOLOGICAL DETAILS

**FIGURE III 4.2A** Digital twin explained.

4. **CC and EC:** As already mentioned, in the initial discussions, that in spite of having the similar goal, there are differences between them. Latency is a key difference between EC and CC. EC is done at the edge of network; hence it slashes latency due to data travels. When compared with EC, CC can provide virtually unlimited scalability due to its vast, centralized infrastructure. However, EC can curtail cost in data storage and transmission. Let us look into both of them little closely.

- **CC in manufacturing:** CC mainly consists of servers, storage, databases, networking, software, and analytics. With CC in service manufacturers can skip own physical data centers but technologies such as AI/ML, big data analytics etc. accessible to organizations. CC assists to expedite production, optimize workflows through digitalized operations. CC-based MES simplifies deployment and lowers the cost of having a system that increases manufacturing efficiency. Therefore, CC provide innovative solutions, data analytics and ERP to modern smart factories. With emergence of AI/ML, IoT/IIoT and enormous benefits more and more enterprises are adopting DT integrating CC into it.

Importance of CC is not restricted to storing and retrieval of data but it increases the efficiency of manufacturing operations by analyzing real-time data, monitoring machinery health and potential failures, and providing seamless cross-communication collaborations for decision making. CC also assists in supply chain management by monitoring inventory levels for effective operations.

- **CC: Some benefits:** CC assists increasing efficiency, flexibility, and competitiveness. Also, CC offers flexibility and scalability, helping manufacturers to expand market i.e., exploring new market. Major issues here are: *Flexibility and scalability, Collaborations and information sharing*, *Improved operational efficiency*, *Enhancing customer satisfaction* and *Reduce costs.*
- **CC: Challenges for manufacturing:** Since CC deals with centralized cloud storage system hence data security and privacy are major challenges. So, major challenging issues are: *Security and privacy, Integration with legacy systems*, *Reliability and downtime*, *Skills gap* and Finding the right cloud service.
- **EC description:** As discussed, due to proximity to data source, EC has reduced latency. This proximity reduces latency and enhances the speed of data analysis, enabling real-time decision-making. It also offers enhanced data privacy and security especially crucial for industries dealing with sensitive information. It also possesses scalability with improved reliability and resilience in continuous industrial operations without any downtime. EC offers improved inventory management. Smart shelves and IoT sensors can monitor stock levels in real-time with local data processing. It goes well with IIoT to handle vast amounts of data generated by industrial equipment.
- **Benefits of EC:** *Security*, privacy are top advantages of EC. Apart from these issues such as *low latency*, *edge nodes* (Isolated, processing own data), *cost effectiveness, increased reliability and network efficiency, real time analytics, and easier to scaling* are major advantages of EC.
- **EC: challenges:** Some of the challenging issues include but not limited to: *Data management and storage, Network infrastructure requirement, Cost of network management and maintenance, Technological complexity with requirements for hardware* (manage and monitor than a centralized architecture) and *cyber security.*
- **Fog concept:** Fog computing is a form of distributed computing that brings computation and data storage closer to the network edge, where many IoT devices are located. It can be considered as an extension of cloud computation. Fog computing reduces the dependence on the cloud for these resource-intensive tasks, to improvize performance and reduce latency and providing protection to IIoT devices.

5. **IIoT: Future Automation:** As IIoT enhances productivity, dependability, and competitive performance by concentrating on profitability, often loosely IIoT is referred to Industrie 4.0. IIoT has been explained in Fig. III 4.2b with divisions in IIoT with IoT. It also creates a number of opportunities such as improved customer experience, data confidentiality, continuous data streaming etc. As discussed before, it goes well with edge/Fog computing for making real-time improvements to manufacturing productivity and product quality. Also, in many industrial applications, fog computation serves as a link between static and dynamic computation environments with protection for IIoT's security needs. Massive data produced by IIoT systems and devices demands for the use of extremely complex highly performing computing platforms for big data analysis [not traditional data processing]. This focuses on big data and AI for decision making.
   - **Major benefits:** Some of the common benefits of IIoT are:
     - *Business agility:* When IIoT devices share data in real-time in an intelligent network, data are analyzed for self-learning, so that enterprise can respond to available opportunity and risks in a speedy way with decisiveness.
     - *Productivity and efficiency:* Combination of advanced analytics and IIoT data leads to recommendations, strategies and decision-making help to attend increased productivity and efficiency.
     - *Machine health:* Because of huge data with analysis, machines get chances diagnostics, fault detection and predictive maintenance.
     - *Inventory management:* With data sharing of IIoT, enterprise gets a bigger network for smaller but distributed warehouses with a broader inventory range and more choices of individual items i. e., IIoT helps to maintain a geographically distributed warehouses, for better inventory management.
     - *Worker safety:* IIoT fitted with sensors can monitor the working environment to provide more safety to the work force.

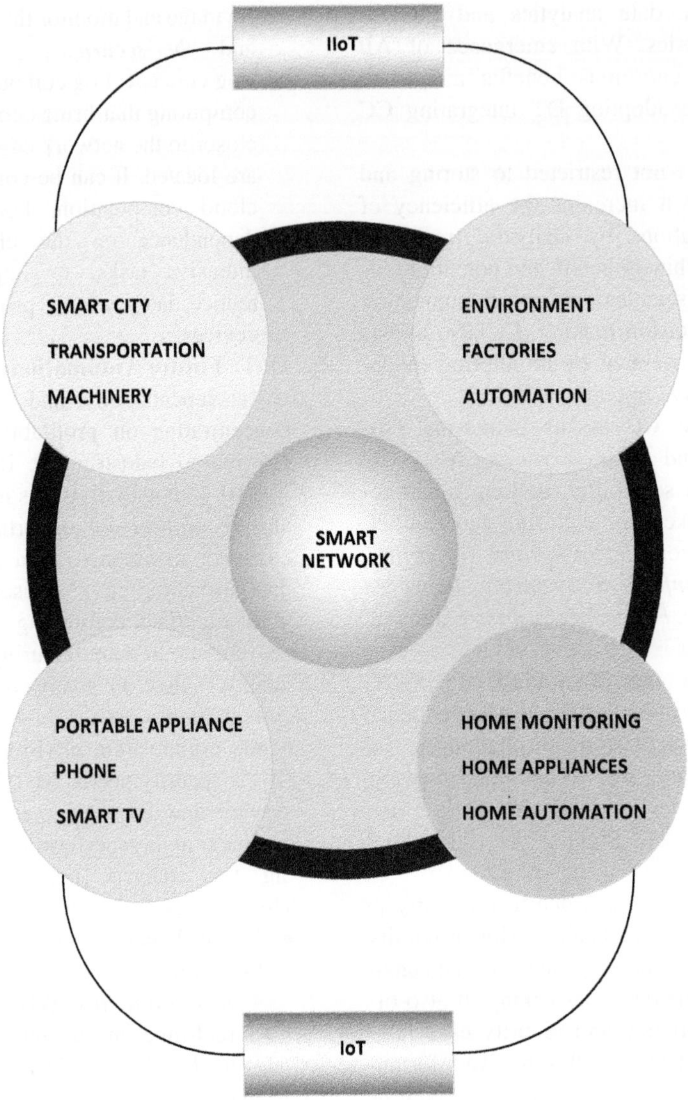


**FIGURE III 4.2B** IoT and IIoT in uses.

- *Customer experience:* IIoT not only include monitoring data from machine, devices and environment but also connects customers to get customer experience.
- **Major challenges:** IIoT services face many challenges and could become inconsistent due to heterogeneity, dynamicity, and data volume. All these are very crucial for industrial applications. As there are a lot of devices connected to each other in IIoT and a lot of data needs to be analyzed within a short period, any lack in proper maintenance, failure to ensure a secure network service or data confidentiality, a lack of individual access control policies, the absence of a solution for machine errors, and human errors can negatively impact the ability of industries to operate successfully. With big data and AI can make a significant impact in different application areas of Industry 4.0 [13]. So, it is important to know the challenging issues, which are: *Real time performance, Energy efficiencies, Safety and security, interoperability issue, coexistence, Initial investment* and *Limited research on OT to marry with IT.*
- **IIoT Future:** The future of IIoT will be deeply intertwined with DT and Industrie 5.0 builds upon the principles of Industry 4.0 with focus on human-machine collaboration, sustainability, green initiatives push for economic circulations etc. Major issues to be covered shall include but not limited to: *Human-centric design of IIoT systems, AR, Collaborative robots, Customer feedback loop, Mass customization, Monitoring of resource usage, Smart recycling* etc.

Let this part of discussions be completed with discussions on Block chain in next subsection.

### 4.2.2 Blockchain technology industrie 4.0 and digital transformation

Bitcoin a "cryptocurrency" are generated through a process called mining and users are referenced as "miners". Special software used is called "mine" blocks, or create the hash required to update the blockchain. Blockchain technology supports *decentralized method for recording transactions.* Blockchain, (the "distributed ledger") technology, has emerged as an object of intense interest in the industrial system and beyond as a way of recording transactions or any digital interaction in a way that is designed to be secure, transparent, highly resistant to outages, auditable, and efficient. Transactions are recorded in distributed ledger and grouped together in blocks. Also, transaction record is stored in thousands of computers worldwide They are secured using a form of cryptography called "hashing." Because the ledger is distributed and secured using hashing, it is theoretically impossible to make changes once something is recorded [14]. When any changes are made, it would generate a different result. So, Blockchain consists of three types of elements viz. *Blocks*. Blocks record (in correct sequence without being tampered with), *Chain* represented by Transactions (the actions) and *Network*. Therefore, tampering with or forging can easily be identified and prevented. This young and changing technology is spreading rapidly and will bring about a major disruption in the existing business. Blockchain technology primarily meant for bitcoin could also disrupt the instrumentation and controls of the plants as well. In order to avoid disruption, it is better to pay heed now. Apart from Blockchain, here brief recapitulating discussions have been covered on Industrie 4.0 and DT. Industrie 4.0 and DT have been discussed at length in Vol II of this book hence only important recapitulations have been covered here, with emphasis on Blockchain. Let the discussions start with Blockchain technology.

1. **Discussions on Blockchain:** Blockchain finds its usages in *real time IoT and IIoT,* Distributed control system (DCS)/Integrated control system (ICS) and access control etc. Blockchain technology has been applied in many fields such as Electric Management Information System (EMIS) to improve the management and data security of the information systems, which is necessary for. Certain unique and secured features of blockchain are also highly demanded by DCS/ICS. The industrial control system network needs to realize the safe and reliable protection mechanism. Blockchain (distributed data ledger – acting as a distributed database) replaces the dependence on the central server. Single point failure in ICS can be avoided in this process as it is distributed in nature. Therefore, principles of blockchain are well suited to DCS/ICS also. A block is a collection of data transmitted by each device node, and the relevant information and records are included, which is the basic unit for *forming* the block chain as detailed in Fig. III 4.2c. This shows both organization of blocks and associated "Proof of work."
   - **Working principles:** With reference to Fig III 4.2c, a blockchain is a scalable, highly decentralized peer-to-peer system. It works on the mechanism of consensus rather than the trust. The decentralized peer-to-peer system eliminates the monopoly to prevent the single user or the group of users from meddling or taking-over the control because all the participants possess equal status/right of abiding by the same protocol. The "block" part of a blockchain refers to a transaction or group of transactions. Each time a new block of data is generated, it's added to the previous ones. This linear sequence of blocks form the "chain" in a blockchain. Rules ("smart contracts") can be set to regulate under what condition(s) a transaction can take place and what must happen when it does. Thus, cryptographically secured, automated transactions between external organizations can be created. Once smart contracts are established, the rules for those are enforced automatically when the set conditions are met.
   - **Network:** The network is a collection, or a group of computers referred to as a node, which could be simply a personal computer/any other system operating on the network. Nodes are defined on the internet means anything with IP address. When any transaction occurs, or enter into the system, it is first verified by all the nodes, and once the consensus is made among most of the nodes, that transaction becomes the permanent part of the blockchain.
   - **Cyber security:** Blockchain technology provides an immutable ledger for transparent logging of compliance actions, while smart contracts ensure uniform application of security measures. Blockchain offers a new path toward greater security with strong encryption, more effective verification of data ownership and by reducing vulnerabilities. Here, the principal advantage is the use of a distributed ledger. Transactions are recorded across every node in the network, making it difficult for attackers to steal, compromise, or tamper with data, unless a vulnerability exists at the platform level. Another traditional weakness is eliminated through blockchain's collaborative consensus algorithm and

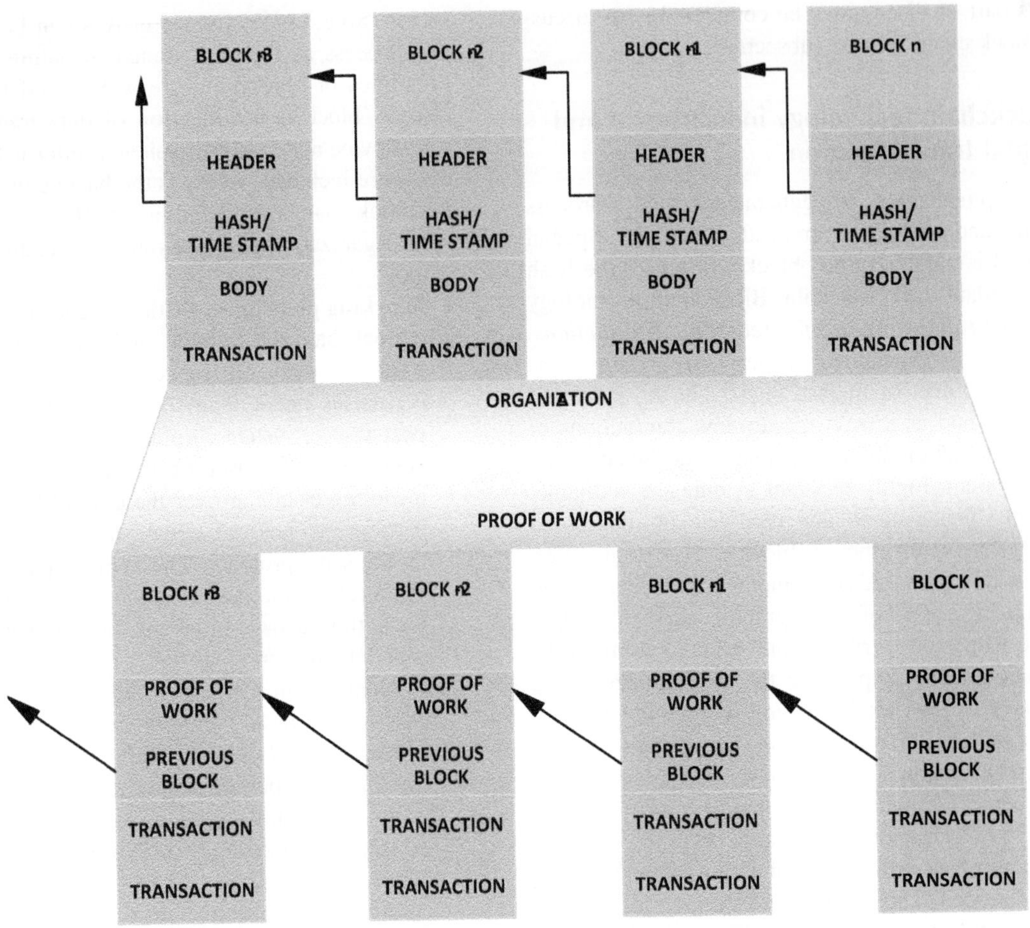


FIGURE III 4.2C Organization of the blockchain with proof of work.

strengthens authentication and secures data communications and record management.

2. **Pros and cons of Automation Blockchain:** System advantages and disadvantages are:
   - **Advantages:** Major advantages of blockchain are as follows:
     - *Distributed nature and transparency*: Blockchain data is recorded in distributed ledger comprising thousands of devices on a distributed network, the system and the data are highly resistant to technical failures. Each block is visible to every member ensuring trust between parties. As one block is linked to another block in a transaction process across the distributed ledger, it is impossible to delete or overwrite i.e., tamper with, malicious attacks on DCS)/ICS.
     - *Trust and lower cost:* In blockchain none has to trust a single party and it reduces the overall costs and transaction fees by cutting out intermediaries and third parties.
     - *Stability:* As blocks are very unlikely to be reversed and every change is tracked and permanently recorded on a distributed and public ledger. This makes the system stable, hence very good storing records or any other data where an audit trail.
     - *Consensus:* Only after all the parties in the blockchain agree to the process, transactions are carried out, hence the transactions are legitimate and consensus.
     - *Process Integrity and security:* Blockchain maintains process integrity and high levels of security. Every individual who enters a blockchain is provided with a unique identifier linked to its account with high encryption for high security.
   - **Disadvantages:** Major disadvantages of blockchain are:
     - *High Power consumption*: Black chain is faster in processing requiring high power for mining.
     - *Maintenance cost:* Maintenance cost of Blockchain is high to run the complex processes of transactions.
     - *Storage size and inefficiency:* Blockchain ledgers can grow very large over time and the ledger may be too large for individuals to download and store. Also, for maintaining proof of work, system is highly inefficient.,

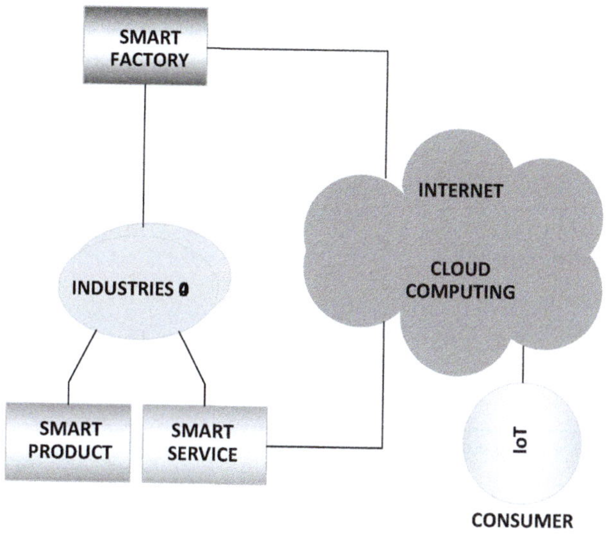


***[FOR DETAILS REFER CHAPTER 16 & 17 IN VOLUME II ALSO]***

**FIGURE III 4.2D** Basic structure of industrie 4.0.

3. **Industrie 4.0 Basic:** [These are already discussed in Chapter 16 and 17 in Volume II of the book. However, for completeness of the discussions same have been repeated some bits here].
   With reference to Fig. III 4.2d it is observed that there will be smart devices used to connect objects of the physical world with virtual networks for cloud computing. Here, smart devices stand to represent an intelligent device, which can exchange data from one place to another over internet (private/public).
   Now, let us go through various components of Industries 4.0 shown in Fig III 4.2d.
   - **Smart Factory:** Smart Factory basically refers to networked adaptive production systems. These networked production systems work with the help of software so as to get connected to networks, which can add value to the production process. Operators benefit above all from the prompt dissemination and utilization of data [15]. For secured and predictable exchanges, real time networking capability and scalability are very important. So, on the basis of cloud computing, decentralized intelligence and recommendations for operation/decision, operating personnel of smart factories can take decision and manage their shown in better way.
   - **Smart products:** Smart products refers to flexibly adapted product to meet the needs of the customer as well as retaining flexibilities for future adoptions, interactive configuration and commissioning.
   - **Smart services:** Smart Services go hand to hand with smart products. Smart services refer to as a combination of virtual and physical services that offer added value to the customer [15]. Smart services always ensure its current market orientations and market demand, taking the help of networked software as shown.
4. **IIOT and Industries 4.0:** Although IIoT often referred to as Industrie 4.0, yet are not same. Both the systems go in hand to hand as they have lot of similarities such as:
   - **Efficiency- and results-oriented:** Both are meant for improving plant efficiencies.
   - **Technology Support:** Success of both systems pivoted on High speed networking technology and digital controls.
   - **Human involvement:** Both the systems need human to implement new technology.
     IIot is a component of Industrie 4.0 and without it Industrie 4.0 could be nonexistent. In spite of all these, they have some differences too as noted below:
   - **Philosophy:** In Industrie 4.0 philosophy for *digitalization*; *automation*; *material*; *new technology*. IIoT is the enabling technology for *big data analytics*; *connected people*, *devices*, *machines*; *predictive maintenance*;
   - **Scope and Role:** Industrie 4.0 has much larger role in for analysis, action and long-term sustainability as well as for lean, efficient operations and for sustained and advanced processing.
   - IIoT is mainly meant for implementation for smart devices and differentiates itself from consumer products through internet.
   - **Action:** In addition to equipment it manifested in buying in management and leadership is for Industrie 4.0. Whereas IIoT, is meant for equipment/device.
5. **Digital Transformation:** Digital transformation requires more and more bridges to be built in an encompassing digital transformation strategy. For digital transformation both digitization and digitalization are necessary. Digital transformation encompasses all aspects of business, regardless of whether it concerns a digital business or not. According to iScoop definition in Google "Digital transformation is the profound and accelerating transformation of business activities, processes, competencies and models to fully leverage the changes and opportunities of digital technologies and their impact across society in a strategic and prioritized way". Digital transformation in broader perspective refers to the customer-driven strategic business transformation that requires cross-cutting organizational change as well as the implementation of digital technologies [16]. Digital transformation obviously may initiate a number of digitalization projects; however, one has to remember that it does not mean only digitalization but digital transformation requires the organization to deal better with overall change to meet the challenges of end-to-end customer-driven strategies. For detailed discussions on Digital transformation Chapter 1 of Volume II may also be referenced.

With this let us conclude the present discussions to move on to Process intelligence and Robotics in next chapter.

## List of abbreviations used

**AGI** Artificial general intelligence
**AI** Artificial intelligence
**ANN** Artificial neural network
**AR** Augmented reality
**ART** Adaptive resonance theory (NN)
**BRNN** Bayesian regularized NN
**CC** Cloud computing
**CNN** Convolutional neural networks
**DCS** Distributed control system
**DDM** Data-driven model
**DDPG** Deep deterministic policy gradient
**DL** Deep learning
**DSP** Digital signal processing
**EC** Edge computing
**EDA** Exploratory data analysis
**EMIS** Electric Management Information System
**ERP** Enterprise resource planning
**FLS** Fuzzy Logic Systems
**GAN** Generative adversarial networks
**GPU** Graphics process units
**HMI** Human machine interface
**I/O** Input/Output
**ICS** Integrated control system
**IIoT** Industrial IoT
**IoT** Internet of things
**ISS** Intelligent soft sensor
**IT** Information technology
**IVR** Interactive voice response
**LSTM** Long/short term memory
**MES** Manufacturing execution systems
**ML** Machine learning/Manual loader
**MLP** Multilayer perceptron
**NLP** Natural language processing
**NN** Neural network
**OS** Operating system
**OT** Operational technology
**PID** Proportional integral differential
**PLC** Programmable logic controller
**PLR** Perceptron learning rule
**RNN** Recurrent neural networks
**ROI** Return on investment
**RPA** Robotic process automation
**SAAS** Software as a service
**SOM** Self-organizing map (NN)
**SVM** Support vector machines
**TSN** Time sensitive network
**VR** Virtual reality
**WS** Work station

## TABLES (with Clause Reference)

| Cl. | Table no. | TABLE Title |
|---|---|---|
| 4.0 | III 4.0a | Difference between Artificial Intelligence & Machine Learning |
| 4.0 | III 4.0b | Artificial Neural Network Types with Applications |
| 4.1 | III 4.1a | Differnce between Machine Learning and Deep Learning |
| 4.1 | III 4.1b | Classification of ML algorithms |
| 4.1 | III 4.1c | ML Application Areas |
| 4.1 | III 4.1d | DL: Pros and Cons |

## LIST OF DRAWINGS (with Clause Reference)

| Cl. | Dwg. No. | TITLE | Dwg. Type |
|---|---|---|---|
| 4.0 | III 4.0a | Basic Classification of ANN | AUTOCAD |
| 4.0 | III 4.0b | ANN Activation Functions | AUTOCAD |
| 4.0 | III 4.0c | Neuro Fuzzy System: Conceptual Working | AUTOCAD |
| 4.1 | III 4.1a | Artificial intelligence (AI) Approaches | AUTOCAD |
| 4.1 | III 4.1b | AI, ML and DL interrelations | AUTOCAD |
| 4.1 | III 4.1c | Data Science Process | AUTOCAD |
| 4.1 | III 4.1d | Application of AI in Process Control | AUTOCAD |
| 4.1 | III 4.1e | Quality 4.0 Explained | MSWORD |
| 4.2 | III 4.2a | Digital Twin Explained | AUTOCAD |
| 4.2 | III 4.2b | IoT and IIoT in Uses | AUTOCAD |
| 4.2 | III 4.2c | Organization Of The Blockchain With Proof Of Work | AUTOCAD |
| 4.2 | III 4.2d | Basic Structure of Industrie 4.0 | AUTOCAD |

## References

[1] O.I. Abiodun, A. Jantan, A.E. Omolara, K.V. Dada, N.A. Mohamed, H. Arshad, State-of-the-art in artificial neural network applications: a survey, Heliyon 4 (11) (2018) e00938.

[2] Real-life and business applications of neural networks; smartsheet; internet document; https://www.smartsheet.com/neural-network-applications.

[3] Neural Networks for Process Modeling, B.A. Jensen, J. Abonyi, in: third ed., in: B.G. Liptak (Ed.), Instrumentation Engineers Handbook — Process Control and Optimization, vol. II, CRC Press, 1995.

[4] Activation Functions: Neural Networks; A. Ralhan; Internet Document, 2018. https://medium.com/@abhinavr8/activation-functions-neural-networks-66220238e1ff.

[5] 7 Types of Neural Network Activation Functions: How to Choose; Missinglink AI; Internet document; https://missinglink.ai/guides/neural-network-concepts/7-types-neuralnetwork-activation-functions-right/.

[6] K. Maladkar, 6 Types of Artificial Neural Networks Currently being Used in Machine Learning, Analytical India Magazine, Jan'2018. https://analyticsindiamag.com/6-types-of-artificial-neural-networks-currently-being-usedin-todays-technology/.

[7] A. Ahmad, M.K.A. Hamid, Neural Networks for Process Monitoring, Control and Fault Detection: Application to Tennessee Eastman Plant, Malaysian Science and Technology Congress, Melaka, 2001. Internet Document, https://www.researchgate.net/publication/255635179_Neural_Networks_for_Process_Monitoring_Control_and_Fault_Detection_Application_to_Tennessee_Eastman_Plant.

[8] B. Lipták, Artificial Intelligence in Process Automation, Control Global, November 06, 2006. https://www.controlglobal.com/articles/2006/221/.

[9] S. Malhotra, AI Explained: Rule-Based AI Vs Machine Learning for Enterprises, Oodles AI, June' 2020. https://artificialintelligence.oodles.io/blogs/rule-based-ai-vs-machine-learning/.

[10] W.L. Mostia, Control Systems of the future. https://www.controlglobal.com/articles/2015/controlsystems-of-the-future/, March' 2015.

[11] The future of DCS – emerging trends and technologies; automation community. https://automationcommunity.com/future-of-dcs/, Dec'2023.

[12] S. Chatti, T. Tolio, Digital twin. The international academy for production engineering, in: R. Stark, T. Damerau (Eds.), CIRP Encyclopedia of Production Engineering, Springer, 2019. https://www.researchgate.net/publication/334737259_Digital_Twin.

[13] Industrial Internet of Things enabled technologies, challenges, and future directions; S. F. Ahmed, M.S. B. Alam, M. Hoque, A. Lameesa, S. Afrin, T. Farah, M. Kabir , GM Shafiullah and S.M. Muyeen; Computers and electrical engineering; science direct; https://www.researchgate.net/publication/372235111_Industrial_Internet_of_Things_enabled_technologies_challenges_and_future_directions.

[14] Blockchain-based Technology for Industrial Control System Cyper Security; M. Mao & H.Xiao; International Conference on Network, Communication, Computer Engineering (NCCE '2018); Advances in Intelligent Systems Research, volume 147; published by Atlantis Press.; this is an open access article under the CC BY-NC license; https://www.atlantis-press.com/article/25896629.pdf.

[15] L. Waschbusch, A. Stark, SLM IoT basics: what does industry 4.0 mean?; SPOT LIGHT METAL. https://www.spotlightmetal.com/iot-basics-what-does-industry-40-mean-a-842216/, June'2019.

[16] J. Bloomberg, Digitization, Digitalization, and Digital Transformation: confuse them at your Peril; (Enterprise and Cloud); Forbes.com. https://www.forbes.com/sites/jasonbloomberg/2018/04/29/digitization-digitalization-and-digital-transformation-confuse-them-at-your-peril/?sh=2c53a8a02f2c, April'2018.

Chapter 5

# Process intelligence and robotics

Carrying on and maintenance of a competitive business edge in a changing global market is really a challenge to meet various issues such as: *Regulation, Deglobalization, Supply chain insecurity*. Obviously, enterprises need to have improved efficiency and effective process management to build an agile, future-ready enterprise that can keep costs down as well as adapt quickly in an ever-changing landscape. Technological advancement provides enterprises to get a complete picture about the insight of an enterprise. Also, these help in understanding actual business operation. A lack of process visibility combined with ever-changing Information technology (IT) and compliance environment remains one of the biggest barriers to process efficiency which can result in financial loss and reputational damage to the enterprise for *Inconsistent customer experience* and *Constant vulnerability*, etc. New technology also recommends way and means to improve operations. "*Workflow management*" with agile project management, the business process– the key to unlocking its power is Process intelligence (PI) which can combine business intelligence with process mining. Artificial Intelligence (AI)/Machine learning (ML) are applied in PI also. Often PI is considered as a kind of Business intelligence (BI). Process mining uses specialized software to "mine" all of business' end-to-end processes. PI recommends easy-to-follow suggestions for correcting business bottlenecks for increasing operational efficiencies. So, PI provides: *Full transparency, Reduced costs, Improved resource management, Smoother workflow, Data-driven decisions., Ongoing improvement.*

With the advancements in mechatronics, electrical, electronic computer engineering and data science development of robotics have gained acceleration providing the ability to the machines for adapting in their ever-changing environment. Unlike traditional machines able to tolerate minimal variations. The main idea of converging AI and Robotics is to try to optimize the autonomy level by learning the adjustment for adapting in the changed environment and the capacity of prediction is a kind of measure of autonomy. Stanford University, defined 18 topics of importance for AI and these are: *Technical trends and surprises, Key opportunities for AI, Delays with translating AI advances, Privacy and machine intelligence, Democracy and freedom, Law, Ethics, Economics, Safety and autonomy, AI and warfare, Loss of control of AI system, Criminal uses of AI, Psychology of people and smart machines, Neuroscience and AI, Collaboration with machines, AI and human cognition* and *AI and philosophy of mind*. Important ethical and legal issues pertinent to AI and robotics mainly include: *Threat to privacy, Threats to security and weaponization of AI, Economics and employment issues, Human bias in AI, Legal responsibility, Civil rights for AI and robots*. It is important to put into perspective the current development of AI and its specific limitations such as *Intelligence as a multi-component model, Large datasets and hard generalization, Black box and a lack of interpretation, Robustness of AI*. While AI can do much good, including by making products and processes safer, it can also do harm. The main concerns related to AI include: the application of rules designed to protect fundamental rights including personal data and privacy protection, etc. On account of falling costs of robotics with wages rise, populations age, and supply chain pressures, robotics solutions are preferred more and more.

By leveraging Internet of things (IoT)/Industrial IoT (IIoT) and cloud computing technologies, automated production through robotics can offer better efficiency, capability making robot economics more cost-effective.

As the name suggests, a collaborative robot is referred to as Cobot i.e., a robot designed to collaborate with human workers. Cobots, have become increasingly prominent in industry and research Development of Cobot calls for creating many complex systems for *sensing, communication* and *movement* alongside humans *safely and effectively*.

Cobots use advanced technologies that can mimic the basic senses in human being for carry out its tasks, understanding the environment and adjust accordingly. Also, Cobots must communicate, move, talk, understand for assisting human coworkers. In contrast to industrial robots, Cobots can interact directly and can have common workspaces and workflows for a shared task. They have the potential to make work more efficient and relieve people of dangerous tasks. Naturally, here worker safety is of prime importance. Industry 4.0 compatible Cobots have played a major role in automating various processes. Versatility, user-friendliness, small footprint and affordability, etc., are

**Plant Intelligent Automation and Digital Transformation. https://doi.org/10.1016/B978-0-323-89892-8.00017-X**

a few qualities of Cobot made it suitable for deployment in processes that were not previously automated.

So, in this chapter all these aspects will be covered. Let the discussions start with PI.

## 5.0 Discussions on process intelligence (PI)

As mentioned, PI can identify bottlenecks for improved operational efficiency. In order to address vulnerabilities and to get deep insights for improvement including customer experience, intelligent processes must be built from the ground up, to have a comprehensive view of the entire end-to-end business landscape across various operational processes and functions. This approach would help enterprise to transform, optimize and gain control of the processes to reach the goal. Processes must be streamlined to optimize resources, improve operational efficiency and reduce cost. With end to end mapping and redesigning processes all stakeholders can see, model and maintain business process in a central place. These also allow thorough analysis in identification of inefficiencies and bottlenecks to make changes to minimize impact and maximize Return on Investment (ROI)—hence implementation of the best practices for a better future performance. Thus, from the discussions it transpires that with a process-centric view, an integrated PI suit is able to cover:

- Business process analysis including the design and documentation of process information for monitoring and analyzing processes and their performance.
- Process mining for in-depth analysis, discovery, monitoring and enhancement of business processes.
- Automation for streamlining and enhancing business operations for increased efficiency with reduced costs, better accuracy and scalability.
- Effective risk and compliance management for adherence to relevant laws, regulations and standards.
- Increased accessibility, with the help of Artificial intelligence (AI)

In this connection, important issues to note are:

- Task mining software helps to look at the steps in a given task by accessing user logs and interfaces. So, it shows all the steps relative to that user, enabling user to look at task in larger context. It offers a few benefits such as: *Identify repetitive and time consuming jobs inefficiencies* and *patterns of disruption*, *Hard data metrics productivity optimization*, *Ability to act as driver for process automation.*
- Process mining such as task mining, uses process mapping and event logs to look at the totality of all business processes at the enterprise level.

As explained, with PI the user can start from scratch and the technology can bring the best of automation, helping user to understand the financial impact of the decisions based on verifiable data. Also, PI leverages artificial intelligence and cognitive algorithms to help deepen the benefits gained from Robotic process automation (RPA) and intelligent automation.

### 5.0.0 Important issues for PI and BI approach

In order to make PI useful, AI needs to be given proper business context with training for how and where things are happening as well how things *should* run, end-to-end. PI-based primarily on *Analysis*, *Improvement*, *Monitoring* and *Management*, can provide powerful, insightful information from the processes. Thus, it can offer Process orchestration, *Accelerate business initiatives* and Unlock possible values through AI/ML. PI harvests data by combining of AI, analytics, and process automation, for identification system inefficiencies to improvise work performance. The mechanics might be complex, but the business function is simple. With PI it is possible to get enterprise leadership with unprecedented insights for leading from front. Process intelligence tools employing ML algorithm for data collection also include: *Process mining and modelling*, *Task mining*, *Auto process discovery tool* and *Enterprise Digital Twin.* So importance of PI lies mainly on: *Data-driven decision*, *Identification and eliminates inefficiencies* for faster progress, *Cost optimization* and Establishment of the best practices. Process intelligence (PI) is a continuous exercise. Suitable consistent endeavor is necessary for regular review and change management at all the different phases such as: *Definition of goal and object*, *Selection of proper intelligence tool*, *Preparation and data integration*, *Data mining and analysis*, *Application of insight to optimize the process.* However, there will be number of challenges the systems need to face for implementation. As mentioned before, AI and ML also play great role in development of PI. It is worth noting that PI helps transform BI insights from "actionable" to "actioned". It is important to integrate BI with the system. However, there are difficulties such as: *Different data sets and formats*, *Poor data quality*, *Data Siloing*, *No BI stakeholder*, *No key metrics to focus*, *Lack of data security* and *Poor training*, etc., for such integration. Therefore, various challenges from *Ineffective data architecture*, *Unclear BI strategy*, *Weak content management*, *Low adoption rates*, *High fees* and *Hard to deliver mobile BI*, etc., to be considered. In this subsection, discussions on vary important issues on PI and BI shall be covered.

1. **PI Principle and Implementation:** PI is very important for any enterprise at it connects various business

process with advanced tools. Let the discussions divided in tow parts:

- **Working principle:** PI refers to a technological process that automatically captures and analyzes how work flows through an organization's different systems. As shown and explained in Fig III 5.0a, PI is a cyclical process comprises several parts viz. "Collection," "Aggregation," "Clean (Transformation)," "Analysis," "Automation" and "Action." So, works by collecting data from different business components/subsystems, then using advanced analytics (AI/ML) to automatically analyze, interpret the data for getting the process insights showing details on recommended path ways, identification of bottlenecks, scope of improvements by optimizing and automating various activities. Naturally based on these decisions are made for various actions. As it is cyclical, so that impact of decision making could be fine-tuned also.
- **PI Implementation**: Several steps involved in PI Implementation process are:
  - *Goals and objectives:* The PI implementing process starts with establishment of clear goals and objectives such as (say) customer satisfaction. Normally, goals are quantitative and measurable, so that success/failure could be assessed and measured.
  - *Selection of proper intelligence tools*: There are many choices available, but selection of tool could be simplified by categorizing options such as: *Process analytics* (e.g., PowerBI), *Task mining* software, etc.
  - *Mining and analysis*: There are a number of tools for process mining and analytics The workflow organizes and processes the raw data for analytics involving cleaning transformation, etc.
  - *Preparation and integration*: One of the major tasks of PI tool is to accommodate data from various sources in one platform for use. So, data after extraction should be cleansed, transformed, and put in server which then acts as messenger for data request and the data sources. The business insights can be obtained from contextualized data. These insights can lead to innovations that improve operational efficiency and leverage opportunities for expansion and achieve goal.
  - *Insight application*: In next step, the contextual data as completely as possible are captured for more useful view of the data. The insight that contextualized data is invaluable for developing the actions to improvise the processes

2. **PI discussions:** PI is developed based on *Process mining, Task mining, hybrid PI,* etc. PI are developed for *Visualization, Analytics, Real time monitoring*

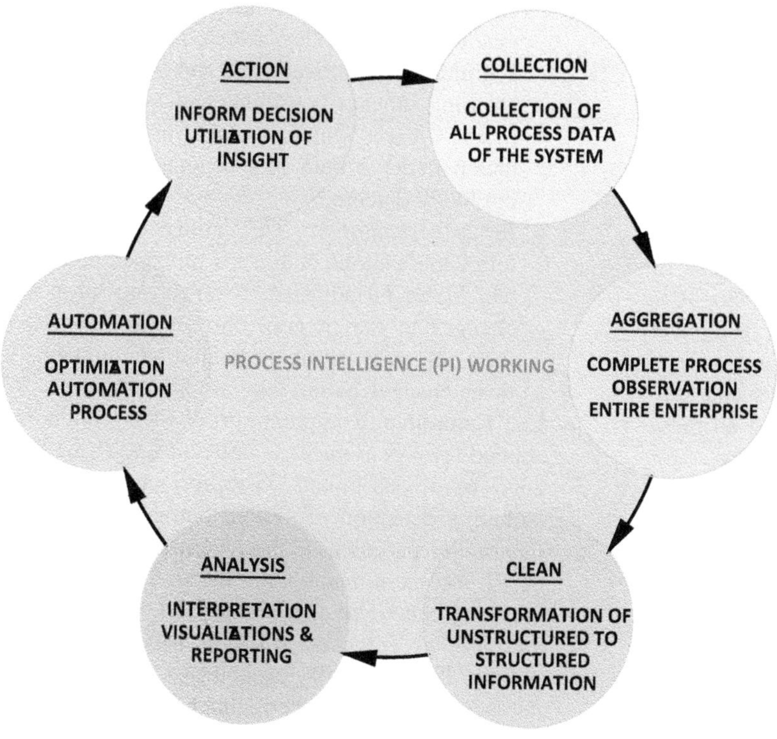


**FIGURE III 5.0A** Working of process intelligence.

Automated insights, *Predictions,* etc. Also, Process discovery, Operational excellence, Conformance checking can be achieved. Now little details about PI shall be explored for the following:

- **PI feature:** Let us know look into various issues mentioned as PI features:
  - *Process mining:* Extraction of process information for creation of process model with comprehensive picture of processes.
  - *Task mining:* Capturing of detailed user interactions within applications for understanding task activities and identification of inefficiencies
  - *Reduced costs:* Resource optimization, and implementation of changes efficiently after identification of areas for process improvement with scope of improvements for achieving reduction in cost.
  - *Monitoring and Alarm:* Real time monitoring and generation of alarm on certain deviation.
  - *Conformance checking:* Comparison of the actual process execution with respect to the defined standard for finding deviations and noncompliance
  - *Operational efficiency:* Identification and elimination bottlenecks to optimize workflows cycle times to attend better operational efficiency
  - *Decision making:* With data-driven insights, decision makings are done for better strategic decisions
- **PI Pros and cons:** PI offers a number of benefits as well as it has a few drawbacks also. These advantages as well as disadvantages have been tabled in Table III 5.0a.
- **PI challenges:** Employees opposition to changes, integration to existing system, etc., pose challenges to PI implementation, in addition to common challenges for implementing PI listed below:
  - *Personnel Issues:* For lack of in-house skills, PI implementations are often outsourced. So, it is important to ensure that concerned personnel have advanced knowledge in related areas.
  - *Strategies*: PI have many components and success depend on coordinate operation of all these components. So, it is complex for implementation.
  - *Prioritizing*: Prioritization of various process involved in any enterprise is difficult. This is related to fixing goal/objective, making implementation quite difficult.
  - *Data quality***:** As data-based decision making is at the heart of PI, so, in case of poor-quality data PI cannot get success. So, it is an important issue.
  - *Security and privacy***:** Since PI mainly deals with a lot of data and connected to outside so data security may at times pose problem. Also, data privacy is an important issue, especially for sensitive data, in implementing PI.

3. **Components and use example:** Here, discussions have been presented on major components used in PI and use of PI in enterprise.
   - **Major Components** [1]**:** Major components constituting PI are:
     - *Process Discovery*: Process discovery comprises *Process mining* helping developing process model from events recorded, while, *Task mining* task can be used for: filling process gaps with user interfaces, observation and analysis of the actions of administrative staff not captured by the event logs. With such comprehensive approach it is possible to identify bottlenecks, redundancies, deviations, etc., to improvise efficiency.
     - *Real time monitoring*: This allows user to define specific process patterns/standard, which shall be detected as they occur and when they are deviated from it, for issuance of alerts/triggering an action. This proactive approach helps minimize risks associated with noncompliance, reduce processing times, and ensure a consistent, high-quality customer experience.
     - *Process Analysis*: These specialized tools help to visualize and analyze process data. These tools are accessible all stakeholders/users so that they can participate in the analysis and decision-making process. This would foster culture of continuous improvement and accountability ensuring enhanced customer satisfaction.
     - *Process prediction*: This makes use of historical data patterns to train an AI engine for prediction the likely outcome.
     - *Process simulation*: This utilizes process histories to develop/configure a process simulation, which can be adjusted to reflect any proposed changes to the process or resource allocation, with the goal of understanding the impact of those changes before they are implemented.
   - **Use Example:** In any enterprise, PI is commonly applied for *procurement* department for getting benefits from end-to-end visibility into processes including detection of system bottlenecks. *Human resource* department of the enterprise can get help of PI for recruitment with the services such as: *handling candidate applications*, *interviews*, *offers*, and *offer acceptances*. All these are prone to data mining. PI can identify capability gaps and help *IT department* to reduce the maintenance costs for enterprise software. Also, it can also improve the chances of implementing a good solution through the use

**TABLE III 5.0A** Advantages and disadvantages of process intelligence.

| PI advantages | PI disadvantages |
|---|---|
| Increased productivity | High-cost implementation |
| Enhanced efficiency: Identification of bottle neck and initiate action | Data privacy and security becomes Critical for critical information |
| Data driven decision making | High technical expertise required |
| Improved customer experience | Potential data biasness |
| Process improvement and cost reduction | Success dependent on quality of collected data |
| Predictive actions including risk mitigation | For process improvements requires organizational change management |

of process mining tools that deliver data-driven insights.

4. **BI and Embedded BI:** Business intelligence (BI) refers to data and technology-driven process that involves the analysis and presentation of business information. Embedded BI refers to *Embedded analytics* – a data analysis technology. Details about them are:
   - **BI Goal**: The Main objective of BI is to assist enterprise take well informed decisions based on insights of business operations, customer behavior, market trends, and associate competitions their in. From analysist point of view important elements of BI are: *Data analysis, to identify patterns, trends, and anomalies* and *Data visualization, – presentation of the analyzed data in a visually accessible manner*, viz. dashboards/reports. As part of the BI process, relevant data is collected and prepared for analysis. BI data (both historical and real-time data gathered from a combination of internal IT and external sources) is *typically* stored in a data warehouse built for an entire enterprise. Data preparation, Analytical querying, Data visualizations with Key performance indicator (KPI) and performance metrics and Decision making based on analytics results. Thus, BI yields insights that inform business strategies and operations. Business analytic is sub set of BI to provide the prescriptive, *forward-looking* analysis.
   - **Embedded BI**: Embedded analytics in process intelligence refers to the integrated data analysis capabilities directly within the system. Key issues in embedded analytics in process intelligence are: *Seamless integration*, *Real-time insights*, *Contextual understanding*, *Improved user experience.* On account of *Ease of adoption*, *Productivity/speed to value*, *collaboration*, etc., this is found to be beneficial. Embedded analytics brings data analysis and visualization functionality to the applications or platforms workers use to perform processes and tasks. For utilizing embedded analytics, it is necessary to ensure, that correct amount of quality data is available, appropriate embedded analytics tool is selected and workers are trained. Embedded analytics can handle specific processes viz. inventory planning.
5. **BI Analytical trends:** This is also referred to as augmented analytics i.e., more AI applications in BI i.e., to *collate*, *enhance*, *automate*, *analyze*, *share* and *activate* BI data for data driven decisions. BI has been explained in Fig III 5.0b.
   The process used in BI has been shown in Fig III 5.0b1. BI life cycle phases have been detailed in Fig III 5.0b2. Some of the associated important issues are:
   - **Preparation and processing:** Organizing huge volumes of raw data for to transfer, cleanse, align, standardize, analyze and integrate in order to generate maximum value from their BI platform.
   - **Predictive analytics:** Predictive analytics is an advanced analytic in business intelligence. It uses historical data, live BI data and learned business context. Then AI forecasts future business outcomes.
   - **BI Insights:** AI-powered BI solutions can personalize insights based on individual user preferences and roles.

## 5.0.1 Discussions on process and business intelligence

PI and BI are two key technologies have emerged. It is important to know about them for understanding intricate business processes. BI platforms is popular for visualization of data analysis insights while PI is comparatively newer and popular for pinpointing near real-time process execution and providing insights helpful in improvising processes including RPA. Process intelligence allows users to define process rules aligning with an organization's compliance requirements. This way, process owners are alerted of any deviations so the business can avoid potential risks. Some important issues are:

- Process discovery is a combination of technologies, such as ML, computer vision to effectively map organizational processes.
- Process mining uses actual/real time data to create a model to reflect a process execution.
- Task mining involves capturing, analyzing user interactions from work station to perform tasks related to higher-level process execution.
- Process analytics tool are designed for interpreting multistep process data.
- Automation-driven process monitoring is aimed to monitor to quickly detect any deviations from expected behavior of the ongoing business process. Also, AI-based prediction of future outcomes by analyzing all possible variations of previous process instances.
- Process simulation can build "What if?" scenarios also.

Both PI and BI are interrelated to each other as shown in Fig III 5.0d2. Leveraging BI Insights process Improvement through guidance from BI. Also, use of PI can enhance BI.

Various associated issues are:

- Both PI and BI rely on accurate and consistent data duly managed by robust data management practices.
- For successful implementation, both call for strong coordination among different departments
- Both systems call for careful change management strategies upon detection of inefficiencies through process intelligence are needed to implement improvements effectively.

In previous subsection some detailing about PI have been established. In this subsection same will be further elaborated along with detailing of BI with application details.

1. **Detailed discussions on PI:** PI analyzes and visualizes data from various sources with a focus on the information about how processes run, interact, and depend on each other continuously. PI is a combination of several technologies and methods for both analysis and execution such as process mining (process modeling) detailed later, task mining. PI and task mining closely related under umbrella of business process optimization, with task mining focusing on granular details of individual user actions. PI takes a broader view, using data from various sources. In Fig III 5.0c PI has been explained in details.

   Fig III 5.0c1 [2], shows various components of PI, which is combination with process mining and task mining, PI offers several benefits such as: *Detailed process understanding*, *Root cause identification*, *Faster process discovery* (discussed below), *End-to-end monitoring and improvement*, *Link processes to work patterns*, etc., to help optimizing business processes and drive operational efficiency. The relation between processes and tasks have been explained in Fig III 5.0c2. Let us look into details on various issues related to PI.

   - **Process Mining:** As clear from the discussions that process mining is a subset of PI that focuses on analyzing event logs generated by information systems such as Enterprise resource planning (ERP) applications. Process mining numerical deep-dive analysis works well on all types of processes, no matter how many data sources are involved or how complex a process may be. Some process mining tools may not be able to connect with certain types of data sources or may struggle with complex processes. It can carry out real-world data to analyze processes. Task mining and process mining often go hand in hand. While both are pretty powerful, they bring better results and higher work coverage when used simultaneously.
   - **Process discovery:** Essentially, process discovery monitors, captures and document employees' digital footprints i.e., user interactions pertinent to various information and applications. Process discovery can work together with task mining for enhanced visibility into both quantitative and qualitative aspects of business processes. This is in addition to offering insights into process efficiencies and bottlenecks. This is a combinational techniques such as ML and computer vision for mapping organizational processes. In combination with process mining, this can significantly shorten the time to map and understand enterprise's processes. The insights gained are the main initiatives automation.
   - **Task mining:** Task mining issues are popular component in understanding PI and associated software to compliment process mining. Basically, task mining refers to capture task and workflow activity from workstation device in order to provide insights on work efficiency, team resourcing and automation opportunities. Task mining can provide a microscopic view of user interactions, hence can be used to identify opportunities for automation by understanding how users currently perform tasks, while process intelligence is utilized to identify broader process inefficiencies and design process improvements. It offers benefits such as: *Detailed process insights and Validation of process models.*

2. **PI RPA and IA:** PI can significantly enhance the automation capabilities of both RPA and AI by providing deep insights into the processes, allowing for more

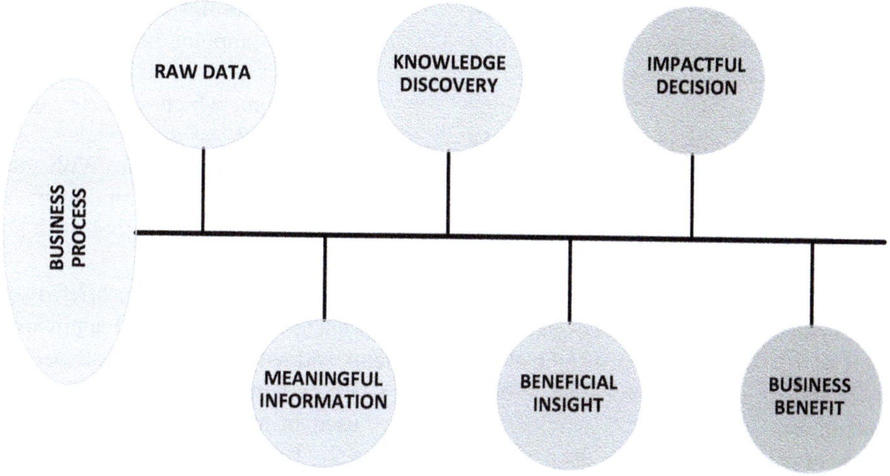


FIG III 5.0b1 BUSINESS INTELLIGENCE (BI) PROCESS

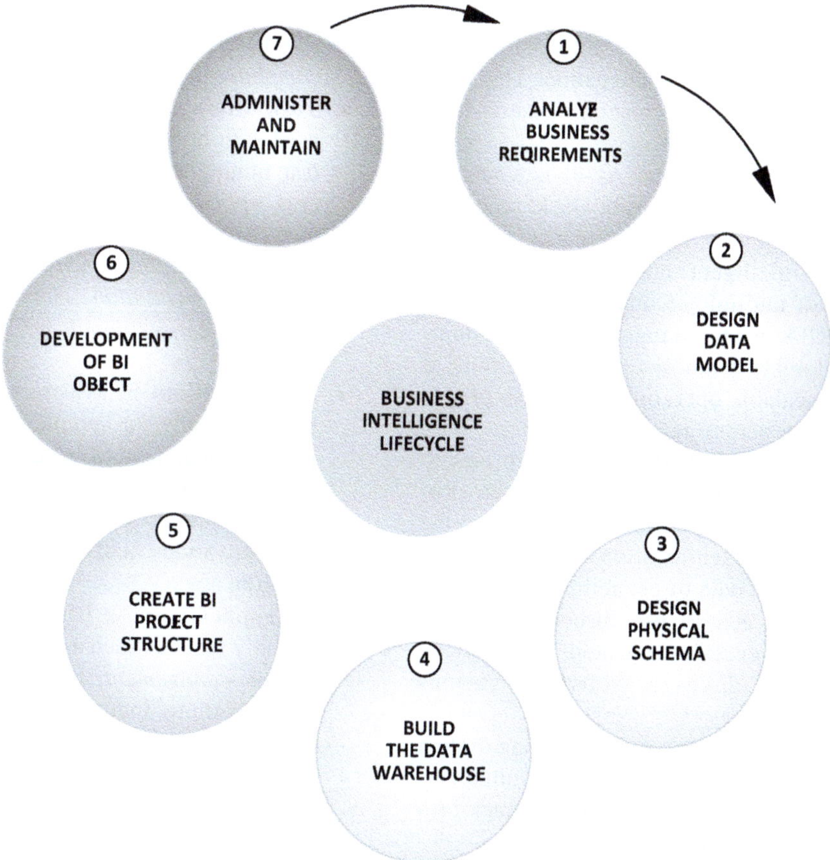


FIG III 5.0b2 BUSINESS INTELLIGENCE (BI) LIFECYCLE

**FIGURE III 5.0B** Business intelligence explained.

targeted and effective automation strategies for improved efficiency, accuracy. etc., to achieve greater business value. Key issues addressed are: *Optimized process identification*, *Improved automation*, *Unstructured data handling*, *Scalability and adaptability*, *Data analysis*, and *Enhanced decision-making*. Here, major issues are:

- **Deep insights:** PI provides deep insights into an actual process. PI can uncover delays, siloes or duplicates and finds scope of optimizing processes for both automatic system and operator. So, PI can guide in identifying the best possible application of automation for the best outcomes.

- **Processes visibility:** PI guides to find the optimal path to attend the highest efficiencies and achieve the best ROI through accurate visualization of highly variable processes in different dimensions. At the design stage, it is so as to visualize the imperfections in workflows to take care and decide on optimized strategy.
- **Optimized processes:** End-to-end processes monitoring is a part of PI. So, with deep insight of the process it is possible to improvise processes to optimize their operations.
- **Automations performance:** Through deep insight proper PI always help in running AI and RPA at peak performance.
- **Improved automation:** PI can reduce the time for process discovery and prioritize automation jobs, helping accelerating process automation. This also, assist in scaling and implementation of automation faster.
- **Governance and compliance:** As PI is a reliable data-driven process, it can easily track processes aligning with compliance and governance goals.
- **Default IA:** Intelligent automation comprises combination of multiple automation technologies such as AI, RPA. All these together these technologies can automate business processes helping in streamlined, scalable operation as well as proper decision-making across an enterprise.

3. **PI guidance:** As discussed, PI helps enterprises to discover, automate, analyze, and optimize their processes in an effective and faster way Some of the major key factors are identification of major inefficiencies, with process data to get actionable insights to improve process, etc. PI uses process analytics viz. Microsoft Power BI, dedicated process/task mining and process discovery software. Various pertinent issues here are:
   - **Target baseline:** Quick comparing existing system against best practices, then knowing the initial/current state of any process, it is possible to track process automation benefits with high accuracy.
   - **Discovery stage**: In this stage, PI creates history of processes using process and task mining to extract systems and interface data. Such data assist in monitoring and visualizing process performance in user-friendly displays in dashboard.
   - **Data analysis**: Dashboard could be developed to show all the entire process steps, variations bottlenecks and various outcomes. These data can be used for modeling and analysis viz. for a root cause analysis.
   - **Process automation:** After thorough analysis, it is easier to pinpoint bottleneck. Here, human-in-the-loop is important to decide the process requiring automation, which can be done by setting up, manage and test the models.
   - **Process improvement:** With introduction of automation, it is easier to optimize. Also, with business process management tool, it is possible to improve business process.
4. **Application areas:** Here, brief discussions have been presented applications and applicability of various systems and associated sub0systems.
   - **Process mining application:** Process mining can be used in many different fields for various purposes viz. Business process optimization, Compliance checking, identifying bottlenecks operation efficiency improvements, etc. Some common usages are: *Order management*, *Order-to-cash*, *Logistics*, *Purchase-to-pay*, *Lead-to-order*, *Auditing and compliance,* etc.
   - **Task Mining application:** It is common that PI are used with task mining. In such cases application of task mining are found for the purposes such as *Process optimization*, *Customer experience improvement*, *Compliance monitoring*, *Employee experience enhancement*, *Digital transformation (DT) implementation,* etc. Here, common usages are: *Data entries*, *Invoicing*, *Reporting*, *Reconciliation,* etc.
   - **PI application:** PI assists enterprises across industries especially for complex industries, to eliminate process unknowns to better plan and execute *business process automation.* Naturally, there will be different purposes in different field. So, application of PI in various fields have been presented as Table III 5.0b.
   - **BI application:** As BI comes with so many user-friendly features such as: *Data visualization*, *Performance management*, *BI reporting*, it is very useful across various industries. Some of the application areas are:
     - *Manufacturing*: For manufacturers across multiple locations operating in diverse product offerings, BI is an incredibly useful tool. It can be used to analyze real-time data to forecast market size and demand so that the manufacturing process can be adjusted accordingly.
     - *Finance*: With implementation of BI software financial institutions, can automate loan applications and business process and ensure compliance.
     - *Retail*: BI can collect all relevant data such as *customer demographics*, *product sales*, etc., to

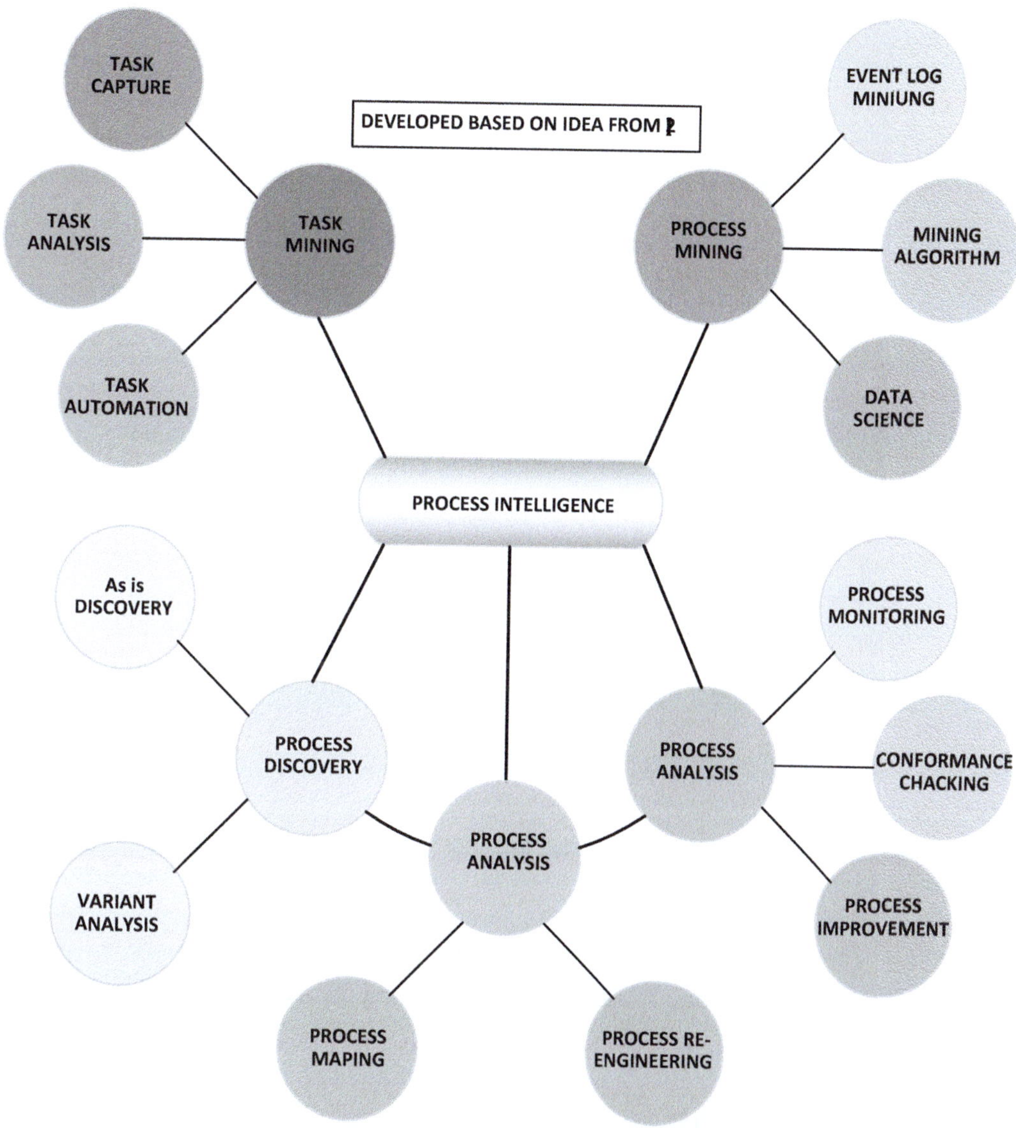


FIG III 5.0c1 FUNCTIONAL DETAILS FOR PROCESS INTELLIGENCE

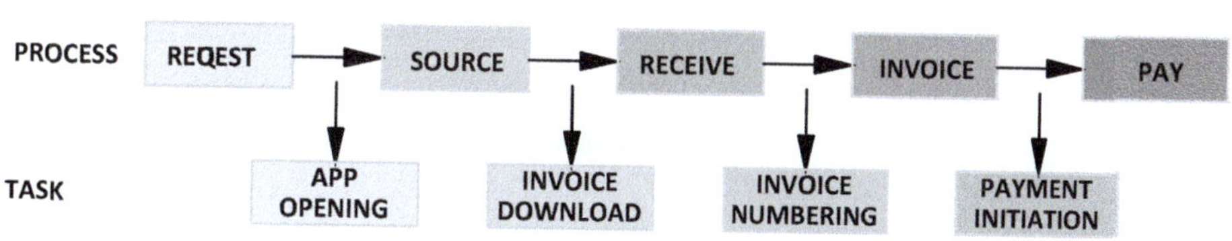


FIG III 5.0c2 TASK AND PROCESS RELATION

**FIGURE III 5.0C** Process intelligence explained.

provide process insight including customer behavior/experience.

- *Hospitality*: The hospitality/hotel business deals in a lot of receipts, particularly those with several locations. By implementing BI software, all these can be integrated this is more helpful for chain of hotels/restaurants.

**5.** BI details with PI comparison:
- **BI Conceptual discussion:** "BI" is an umbrella term referring to various tools, methods, and processes for collecting, evaluating, and data presentation about business. BI subsystem and interrelation between BI and PI has been detailed Fig III 5.0d.

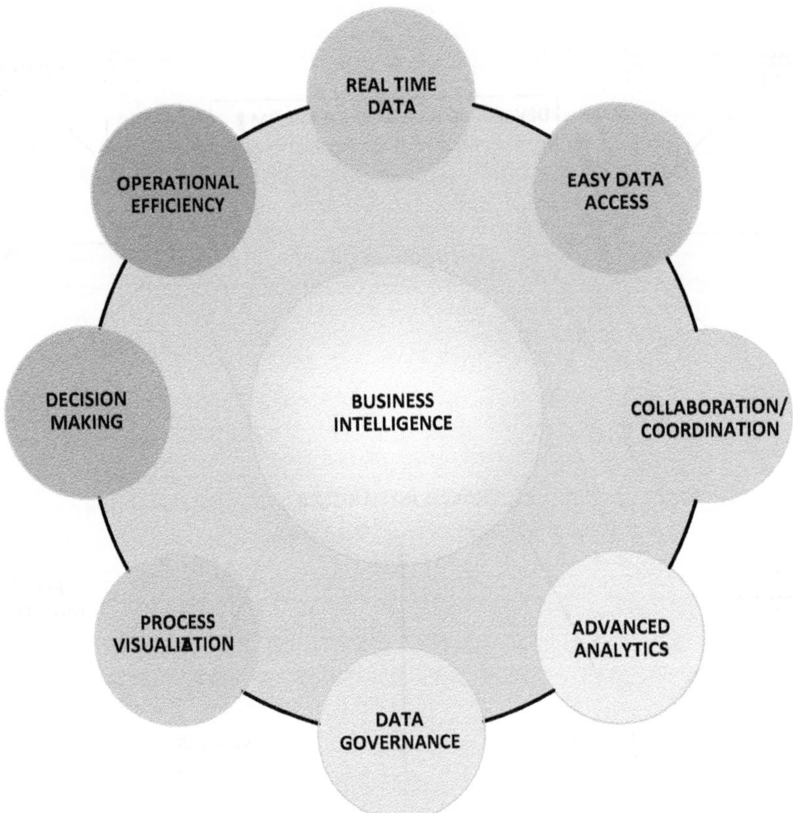


FIG III 5.0d1 BUSINESS INTELLIGENCE (BI) SUB-SYSTEMS

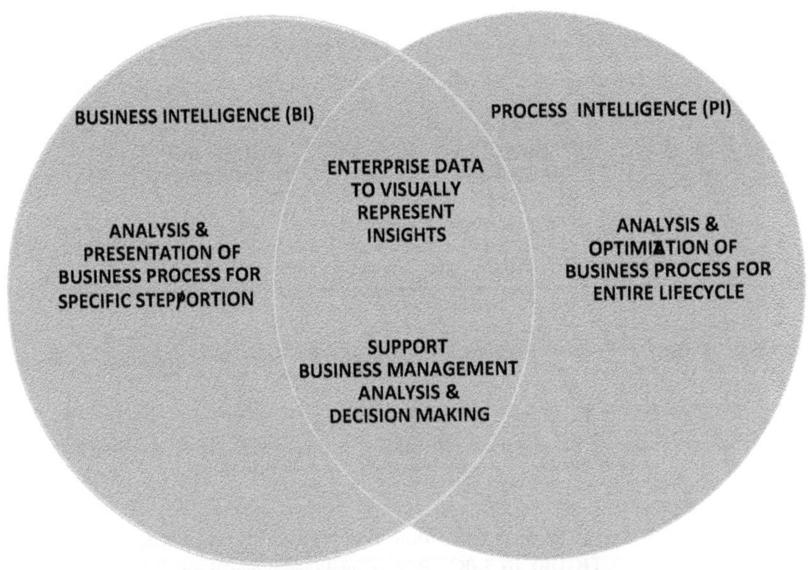


FIG III 5.0d2 RELATION OF BUSINESS INTELLIGENCE (BI) PROCESS INTELLIGENCE (PI)

**FIGURE III 5.0D** Business intelligence sub systems and relation with PI.

BI comprises various functional subsystems such as Real time data collection, Easy data access, Collaboration, Advance analytics, Visualization, Data guidance, Decision making and Operational efficiency as detailed out in Fig III 5.0d1. BI is utilized to understand and measure performance against their goals. Here, data from different sources are combined and presented in user-friendly visualizations viz. dashboards/graphs'/reports so that users through information analysis can gain insights about the patterns and discontinuities and answers to various questions about the enterprise, competitors, customers, etc. Modern solutions have a strong focus on self-service, enabling employees without a technical background, ask and answer questions as well as share data-driven insights with other stakeholders.

- **BI Working solutions:** BI tools utilize collected raw data from business systems, for its operation. The data is them processed and visualized for analysis. BI is developed for specific issue and provide snapshot analysis for making decision. However, these can be used for planning in conjunction with various other tools to understand business performance, follow trends, and make better decisions. So, it is designed with the focus on descriptive analytics – detailing. BI typically focuses on current happening inside business and related issues. This can indicate various variation of product both demographic and temporal variations also, but cannot explain *why* or *how* these things happen, nor it can suggest any action needed then.
- **BI data quality and security:** A BI solution totally depends on data quality to present correct and accurate insight for any decision making. Therefore, data quality management is very important. Poor quality data can lead to commercial disaster because in that case, there will be wasteful *financial investments, bad decision and corrupt strategy* (because of bad analysis due to data quality), *Poor and ineffective decision intelligence* (based training on poor data). Maintenance of the security, quality and integrity of data is extremely crucial for BI strategies (optimizing BI outputs and data privacy compliance). Now, as BI becomes more democratized through self-service, and is active multiple cloud domains, it is becoming vulnerable due to the risks of unauthorized access, misuse, and lack of trust in data increase. Therefore, suitable care must be paid towards this.
- **PI and BI Comparison:** PI and BI are interrelated. Both of them use enterprise data for getting insight, analysis and data driven decision making. When PI deals with entire process lifecycle, BI is related to specific step/portion as shown in Fig III 5.0d2. Comparison of PI with BI has been tabled in Table III 5.0c.

## 5.0.2 Process intelligence in industries

Even in modern days, manufacturing companies tend to approach the design and manufacturing process the same way they had for many years. There are many *constraints, which* can limit the potential performance of products. In this age of globalization, they face a dynamic competitive environment characterized by disruptive technologies with complex supply chains. All these put both challenges and opportunities. Process improvement efforts are quick to implement and would bring results in short time. However, enterprises normally take a long-term view and recognize the need for continual process improvement efforts. Another popular topic used in predictive analytics is Decision intelligence (DI). DI software uses ML algorithms to process a company's big data and creates a semantic layer of business rules and contexts to ground predictive analytics. Process/operation optimization, efficiency improvement, cost reduction, predictive analysis and data driven decision making are common expectation from PI by both process and manufacturing industries. Issues related process industries in connection with PI are:

- **Process optimization:** *Bottlenecks and inefficiencies identification, Streamlining operations*
- **Data-driven decision making:** *Real-time insights, Predictive analytics,*
- **Compliance and risk management:** *Early detection of Issues, Risk mitigation*
- **Cost savings and resource allocation:** *Optimized resource allocation* and *Reduced costs.*

Similarly, associated issues of PI for manufacturing industries are:

- **Process optimization:** *Real-time (RT) data analysis, Predictive maintenance, Digital Twin* and *Automation*
- **Quality control and product assurance:** *Defect detection, Predictive quality assurance*
- **Supply chain management:** *Demand forecasting, Logistics optimization*
- **Safety and Security:** *Risk assessment, Predictive analytic.*

This subsection deals with use of PI in different industries.

1. **Process improvement:** Some common approaches for process improvement are:
   - Detailed explanation and understanding of the problem and fixing target for improvement.
   - Root causes of the problem with ways and means to address root causes.
   - Measurement of performance levels.
   - Make it stick: Learning; retrieving from memory for learning.

   Data and its analysis involve assessing diverse data categories derived from organizational activities, customer needs, process efficiency metrics and system event logs. Analyzing these data sources helps in understanding patterns, optimizing processes and enhancing decision-making capabilities. Various intelligent

**TABLE III 5.0B Process intelligence application areas.**

| Application area | Purpose |
|---|---|
| Risk and compliance | *Identification and report, Ensure adherence to policies and procedures, Strengthen internal controls* |
| Retail logistics and supply chain | *Omni channel strategies, Increase hiring effectiveness, Optimization of logistics and deliveries, Optimize order management, Streamlining inventory* |
| Customer services | *Speed resolution times, Streamlining returns processing and Increased responsiveness to customer* |
| IT sector | *Max-technology investments, Match technologies to needs and Secure against cyber threats* |
| Transformation | *Improved project execution, Identification of scope for automation and innovative technologies* |
| Fund transactions | *Invigorate transaction, Efficient reporting, collateral and funds management and Establish operational controls* |
| Financial services | *Inform and management of risk, Timely regulatory filings, monitor and maintain process controls, Financial closure and Automatic report generation,* |
| Health services | *Optimize patient flow, improve quality of care, Faster claims processing and Streamline policy issuance and underwriting* |

analytical tools can provide richer insights in a process analysis platform to reconstruct the actual process and finding the process performance. Once inefficiencies are identified, PI provides would assist to streamline operations. As PI calls for continuous businesses monitoring and analysis, it also ensures the process generally are optimized and aligned with changing business needs. With actual process performance at a handoff level, it is easier to suggest improvements after identifying bad steps not adding value to the system (i.e., redundant step/slow unproductive issue, etc.) or acting as constraint/disturbance. Such issues could be eliminated and hard process improvement could be achieved. Instead of focusing on short term fixes, it will be wiser to identify and address the problems at their core. This will bring greater and longer lasting benefits. Another important issue is data driven decision-making, which is very close to actual process. As per some study it has been found that data driven decision making improves productivity also. As already discussed, AI and ML are now utilized PI processes. Therefore, training is essential for obtaining skills and knowledge, suitable mentorship model is also helpful and effective.

2. **Manufacturing Process Intelligence:** Manufacturing process intelligence (MPI) is an operational intelligence solution meant to provide broad visibility and decision support for achieving and sustain manufacturing excellence. MPI refers to the comprehensive integration of software, tools, and techniques used to collect, analyze, and present data from a manufacturers' operations by leveraging AI and data analytics. It can improve visibility across global operations, making the right information, with sufficient granularity and context available quickly, for faster and better decisions. In present time, preconfigured Intelligence packs are available with MPI for analytics, dashboards, and KPI, etc. MPI can simplify the gathering and aggregating of clean data quickly, resulting in valuable analytics and reporting that is delivered in near real-time. MPI reduces the complexity of performing mission-critical analysis on current and historical data [3]. So, MPI extracts detailed, granular data for a rich operational context to analytics and reports. MPI can modify, create, and share analyses across the enterprise and can provide global visibility. MPI also help in increasing flexibility and efficiency in data analysis, visualization, and contextualization. MPI plays an important role in Industrie 4.0. Let us now explore MPI little bit more:
   - **Components:** Major components of MPI are:
     - *Data collection*: IoT sensors, controller (say Programmable logic controller-PLC), and many other sources including manual data inputs.
     - *Data Processing*: This refers to cleaning, organizing, and analyzing this vast dataset
     - *Data Visualization*: After data processing comes data visualization to translate processed data into visual formats, such as graphs, charts, and dashboards.
     - *Predictive analytics*: This leverages historical data along with real-time inputs to anticipate future outcomes.

**TABLE III 5.0C Comparison of process intelligence and business intelligence.**

| Issue | Process intelligence | Business intelligence |
|---|---|---|
| **Focus** | Check s data in specific process for identification inefficiencies, variations etc. with aim to identify scope of improvement within the processes. | Analyzes aggregated data across the entire enterprise for market trends, customer behavior, performance and strategic decision-making. |
| **Focus range** | On entire process lifecycle | On individual process step |
| **Data source** | Event log, audit tables, user interactions, workstation activities | Various sources for analysis and visualization patterns, trends, and anomalies. |
| **Tools and techniques** | Process and task mining, process discovery, prediction and simulation | Software dashboard, data: Warehousing/visualization Reporting, |
| **Applications** | Order processing analyzing time, production bottleneck identification work flow optimization. | Creates dashboards with KPIs, for customer demographics, sales trends, demand forecast, |
| **Outcome** | Business process optimization, variations in process instances, understanding of steps in global context. | Insights for making informed, data-driven decisions |
| **Similarities** | Enterprise data for visual insights to support business management, analysis, and decision-making processes. The same data warehouses may be leveraged with outcome variations based on platform's capability. Both use AI and ML | |

- *AI ML*: Manufacturers now adopt ML and AI through Natural language processing (NLP).
- **Feature:** Quality control, Predictive maintenance with ML, Production data analysis and NLP in data analysis are some of the special features in MPI.
- **Challenges:** Following are major issues that can prevent to utilize full potential of MPI.
  - *Data silo*: A major challenge is the existence of data silos that are isolated datasets. These data sets cannot be easily integrated with other datasets and often offer hindrance in comprehensive analysis.
  - *Management of change(MOC)*: For digitalization and DT, MOC may be necessary and often this poses problem.
  - *System integration*: At times system integration pose challenge especially for merging legacy systems

3. **PI and DT:** AI, ML, IoT and cloud are game-changing technologies. However, the maximum benefits are achievable when these are used in conjunction with PI, which can identify where and how to use them. Combination of the power of process mining and discovery, PI provides actionable insights showing how to turn substandard into super standard workflows. As discussed, DT refers to the integration of digital technology into all areas of the functionalities and business process pertinent to an enterprise. This calls for entire work culture, operation as well as delivery value to the customers. Apart from many other issues (discussed in Vol II), Intelligent process automation (IPA), which incorporates AI in automation to create smart business processes, is the most crucial issue in DT. Therefore, process mining tool may be utilized to assess and analyze the current state of operation to identify various bottlenecks and inefficiencies in the process. Process insights, based on hard data that reflects real world conditions, enabling process optimizations by *Eliminating unnecessary steps*, *Reducing handoffs*, and *Standardizing the execution of tasks.* Only after this, IPA including RPA can be implemented for best results. Fig III 5.0e shows, the concept of data driven process management to assist DT for achieving process management centered around human. In the path of implementation of DT, it faces many challenges as tabled in Table III 5.0d.
4. **PI Improvement:** From the discussions it is clear that PI mainly depends on data collection, analysis and process optimization. So, for PI improvement major issues to be looked into are:
   - **Data Collection:** *Relevant data sources identification*, *Comprehensive collection of relevant data* and *Effective processing and management of data.*
   - **Analysis:** *Use of suitable tools to analyze, visualize processed data*, *Root cause analysis*, and *Identification and prioritize improvement area*
   - **Optimization and improvement:** Implementation *of process improvement*, *Continuous monitoring with KPI tracking*, *Leverage AI and ML for proper tuning.*

Apart from the above, focus should be also on proper training in using AI and ML, employees experience and enterprise culture, which is important for DT as well. Real-time data and insights empower for making informed decisions that drive efficiency and innovation. It helps mitigate risks, especially for high-risk domains. Also, common PI tools and techniques such as *Process audits*, *Root cause analysis*, *Process assessments*, *Predictive analytics*, *Process discovery* and *Process mapping* should also be suitably taken care. PI is a continuous practice enabling enterprise to identify patterns and trends, leading to a culture of ongoing improvement and innovation.

5. **Concluding Discussions:** in the concluding discussions on PI, brief account on BI reporting, PI software categories, PI software selection and major PI software in the market shall be covered.
   - **BI reporting:** BI reporting refers to the process of preparing, interpreting and presenting data in such ways that reports are easy to interpret, digest the takeaways data without knowhow of a professional data analyst. This should be easy for decision-makers to understand also. So, BI reporting data are in various formats for employees to easily interpret and act on that information in a timely manner without any data analyst. Major function of BI reporting are Data— *preparation*, *interpretation* and presentation. *Predictive analytics*, Flexible data presentations, *Enhanced data actionability*, *Operational efficiency*, Informed faster decision-making, etc., are some of the benefits offered by BI reporting.
   - **Categories PI software:** PI software can be broadly classified into the following:
     - *Process mining:* Tools to extract PI from event logs.
     - *Task mining*: For collection of process information from the workstations
     - *Process discovery*: Software tools using any of the above category for identification of "as-is state" of business processes.
     - Hybrid PI: Software to combine process mining, process discovery and/or task mining
     - *Self-built analytics*: – manually created tools and dashboards
   - **Selection of PI software:** For selection of the best PI software, following factors may be considered:
     - *Robust analytics and reporting*: As the name suggests, powerful analytics and reporting are main issues in unlocking the full potential of PI (advanced data analysis, real-time reporting, with customizable reports)
     - *Security and compliance*: Ensuring the security and compliance of PI software is critical for protecting sensitive business data. This can never be sacrificed.
     - *User-Friendliness*: For better adoption, user friendly interface cannot be overestimated. Intuitive software easy to use software tool such as drag-and-drop, customizable dashboards and clear visualizations are always a better choice.
     - *Scalability and flexibility*: PI software offering scalability and flexibility, for quick adaptability to changing business requirements without sacrificing functionality/performance are only welcome.
     - *Customizability with adaptability*: Selected PI software must be easily customizable and adaptable to fit specific needs of business processes and requirements of the enterprise, ensuring a tailored solution for the business.
     - *Total cost of ownership (TCO)*: TCO for PI software includes the initial purchase price, as well as ongoing costs such as: *maintenance*, *support*, *upgrades*. Also, costs for licensing fees, implementation and training expenses should be considered.
   - **Common PI software:** Some of the common PI software has been listed in Table III 5.0e.

## 5.1 AI and robotics with robot details

The robotics and AI theme is experiencing accelerating adoption, showing entrance into the "Early Majority stage". Robotics and AI are the engines behind an impending Industrie 4.0. The pinnacle of effective robotic and networking integration is the smart factory. Specialized AI is used in robot when the robot is to perform only specified special tasks. However, in general there are two kinds of AI. These are:

- **Weak (narrow) AI** can reproduce an observed behavior very accurately. It can carry out the task for which they have precise trained. These are extremely efficient in their own field but lack generalization ability viz. self-driving cars, and industrial robots.
- **Strong AI:** It refers to an intelligent system with consciousness and ability to think and reason like a human being. In addition to assimilating information as in weak AI, it also can modify its own functioning, i.e., autonomously reprogram the AI to perform general intelligent tasks with the help of human-like cognitive abilities including consciousness, sentience (*feeling* and *sensation*), sapience (*wisdom* and *intelligence*) and self-awareness.

An important application of AI in robotics is for the task of perception. Robots can sense the environment by means of integrated sensors or computer vision, which have improved the quality of both sensing and vision. Perception is important for planning and creating an artificial sense of self-awareness in the robot. This also allows Human-computer (robot) interactions (HCI) and cognitive robotics. Two main domains are robotic perception (understanding activities, emotions, etc.) and ability to navigate an environment. Cognitive robotics mainly focuses on acquiring knowledge and the autonomous capacity to learn mainly based on imitation and experience. Mimicking the human cognitive system can help to regulate the process of acquiring knowledge and understanding, with use of sensors. Machine vision integrates captured image with computer vision algorithms for automatic inspection and robot guidance. The development of AI is closely coupled with the understanding the human brain.

As discussed, any new technology, the use of AI brings both opportunities and risks. So these are guided with legal and ethical guidelines already mentioned. This section has been dedicated to the discussions on use of AI in robotics.

### 5.1.0 Discussions on AI-based robotics

Automation refers to carrying out a task using technology, in place of manual labor. It could be complete/partial manual labor replacement. Process automation types are:

- **Business process automation (BPA):** BPA is the automation of activities to achieve a certain business goal, by transforming the infrastructure of business processes.
- **IPA** stands for a combination of process redesign with automation and ML/AI. In addition to rules-based automation, IPA integrates decision-making capabilities with the help of cognitive technologies.
- **RPA** use computer programs in the same way as a human operator would do. RPA can be implemented by integrating RPA into existing business processes.

Let us now look into robotics. Robotics is a multidiscipline subject Involving Mechanical, Electrical, Electronics and communication and computer engineering knowledge. Here, our discussions is restricted to application of Robot in industries with special reference to application of Artificial intelligence (AI) for it. So, as a prat of it, a brief discussion will be put forward on basics of industrial robots will be covered here.

1. **Industrial automation:** Before going into details on industrial robot, let us start the discussions on industrial automation types. *Higher productivity*, *High-quality products*, *Increased safety*, *Greater flexibility*, *Low production costs*, *Accurate data collection*, etc., are some of the major benefits offered by industrial automation. There are mainly four different types of automation in industries. These are as follows:
   - **Fixed Automation:** Fixed automation, also called Hard automation is that type of automation in production line where the operational sequence production is fixed by the equipment configuration e. g., "machining transfer lines found in the automotive industry**".** Here, the machines are permanently set up to perform a specific task, which when is completed, the machine is reset for the next task. In this type, the initial cost may be high. In this type of automation, the programmed commands are contained in the machines in the form of (cams/wiring and other) hardware, which is not changed for product styles. When large volume production is necessary then fixed automation is helpful.
   - **Programmable Automation:** Programmable automation is more complex than fixed automation and the machines can be programmed to perform a variety of tasks. This type of automation is normally found in factories where large quantum of products is done in batches. For each new batch, reprogramming is necessary. In this there is a period of nonproductive time followed by a production run for each batch. Therefore, for reprogramming and change over there will be nonproductive time. Naturally rate of production is lower here but it can accommodate new products with different style and feature. A numerical-control machine is a good example of programmable automation.
   - **Flexible Automation:** Flexible automation is an extension of programmable automation, avoiding the changeover and reprogramming time. In this case number of varieties products are limited under this program for quick automatic change over. Here, reprogramming is done offline. These systems are used in industries where there are variations in products from one to the other viz. CNC machines.
   - **Integrated automation:** This is the most complex type of industrial automation in which all the machines are connected to a central computer—automation system. This configuration allows control of all aspects of production lines from a single control center. This type of automation suits well for the industries where tailor made products are produced in small quantities so that changes to the production line can be made quickly and easily. Here, various processes, tools such as CAD/CAM, or intelligent machines and operators/robots/cranes are seamlessly controlled through a single control center.
2. **Robotic Application: AI and ML:** Any form of automation, without AI/ML lack in *Flexibility and adaptability*, *Dexterity and complex task*, *Visual feedback*. Robots with AI can provide many advantages such as *Vision*, *Intelligent placement* scalability, etc. There

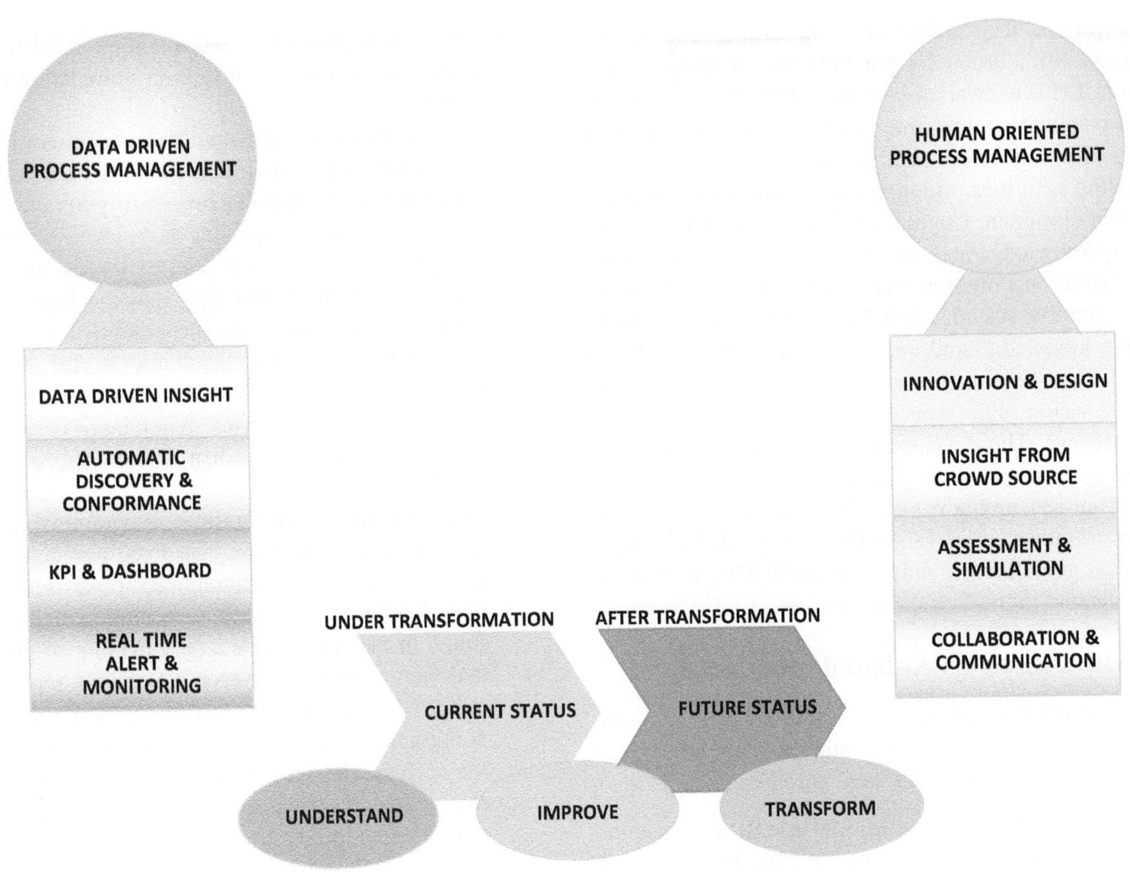


**FIGURE III 5.0E** Data driven process management approach to DT.

are many benefits of AI/ML integrated in industrial robotics for manufacturing production. Robots with integrated AI can monitor their own accuracy and performance, signaling when maintenance is required to avoid expensive downtime. Same is also applicable for, production engineering including a few batch process industries such as food beverage and pharmaceutical industries. Every year there is tendencies to increase market and deployment of robot to increase plant efficiency and profitability. Major industrial robot manufacturers shall include ABB, FANUC Robotics, KUKA, Yaskawa Electric and Kawasaki Heavy Industries, to name a few. Replacement of human with robot for monotonic jobs such as welding, assembly, disassembly, pick and place, packaging and labelling, palletizing have been introduced decades ago. Fresh Programming would be necessary for new product/tasks. Also, normal robots cannot track unidentified errors or remembering previously performed actions for long. So, people need to undertake painstaking and time-consuming reprogramming every time. At this point AI in robots brings about a revolution in manufacturing industries. Combination of AI and ML works together in industrial robots.

- By ML robots are trained to perform specific and exact task (s) robot has to perform. The knowledgebase of AI helps it to remember
- AI infrastructure helps the robot to understand various rules, and inference engine of AI provides the most optimal way of performing that task.

FANUC, KUKA, ABB – the leading global robot manufacturer have already greater emphasis on including ML, AI and suitable connectivity in robots to achieve automation in true sense. This would also help in utilizing robots on multiple fronts in industrial applications. According to ABB ("Innovation highlights 2020" ABB Review | 01/2020 –https://new.abb.com/news/detail/56162/innovation-highlights-2020) "The combination of artificial intelligence (AI) and robotics can substantially improve the flexibility of factory automation by replacing rigid applications with learning capabilities." This combination would bring about robot functionalities for improved productivity under safest condition. Assembly, Packaging, customer services are a few areas where AI has been introduced to perform various tasks better.

The enormous benefits of combing AI and ML into industrial robotics technologies cannot be overestimated

**TABLE III 5.0D** PI Solutions for challenges in DT implementation.

| Challenge for DT | PI solution | Remarks |
|---|---|---|
| **Legacy systems** | Monitoring, optimization for DT, comprehensive mapping, testing and simulation | Smooth transformation roadmap without losing continuity |
| **Cost and uncertain Resource allocation** | Identification high-impact area, Resource optimization, data-based ROI– short vs long term | Assured effective resource use/allocation |
| **Management of risk and compliance** | Risk assessment, audit trails, auto compliance reporting, mitigation proactively | Stable sustained environment for DT and regulations adherence, |
| **Lack of visibility** | Real time discovery, including inefficiencies, cross department transparency | Focused automation and optimization effect |
| **Inefficient decision making** | Real time, predictive data analyses, data driven decision making | Effective decision-making, quick response to changing condition |
| **Lack of cooperation, siloed departmental operation** | Real time data, strategic alignment, unified process view | Eliminating silos drives, cohesive decision-making, meaningful sustained DT |
| **Transformation success** | Baseline measurements, defined KPI, ROI proof | Dashboard analytics, fine tune strategies. |

for manufacturing industries: Some of these benefits shall include but not limited to the following:

- Increase production uptime, production rates/higher productivity.
- Capable of for learning and performing multiple tasks
- Ensure more safety for operations in hazardous environments.
- Reduced programming time for new products, and less overall production downtime
- Predictive Maintenance and preventive actions alert for maintenance.
- Self-corrections to improve performance (better accuracy) and optimum solution.

3. **AI in Robotics discussions:** AI robots can be explained better as intelligent automation applications where robotics provides the body and AI supplies the brain and sensory organs. However, robots do nor possess "sixth sense" like human. AI and robotics are closely related and at times they are used interchangeably. In reality, they are distinct and not complementary. With substantial development of AI, the combination of AI and robotics are gaining tremendous applications in different industrial applications for increased productivity and efficiency, improved safety and greater flexibility. Most common way of use of AI in robotics is through machine learning, enabling robots to learn and perform specific task from human actions. AI gives robots a computer vision to navigate, detect and determine their reactions, etc. Edge computing is another way AI can be used in robotics for application demanding massive amounts of data gathered by robot-based sensors for analysis and to make decisions like human quickly. AI assists in learning through sensors such as: Temperature and humidity sensors, Ultrasonic sensors, Vibration sensors, Time-of-flight optical sensors, Millimeter-wave sensors, etc. So, advanced sensor technology, camera modules, etc., make it possible to capture high-quality images that go in as input for AI use in robots.
   - **Current trend:** At present the focus on AI in robotics is task performing capability of robot. With application of AI, importance will be on the type of input a robot can provide for AI's "mind." This allows the exploration and experimenting with real objects for proper understanding.
   - **AI Advancements:** The revolutionizing impact of AI in robotics mainly include: making machines more intelligent, autonomous, and adaptable. As result of advancements in ML, there has been significant enhancement in the capabilities of robots. Deep learning (DL) especially in neural networks, allow robots to process vast amounts of data and recognize complex patterns for making well informed decisions viz. object detection, predictive maintenance. With enhanced computer vision, robot interaction with environment have been revolutionized. NLP is another example in advancement of AI in robotics in the extensive use of spoken languages.
   - **Future of AI in Robotics:** In future, AI may be referred to as Artificial general intelligence (AGI), which could hold the potential to reach levels of true human understanding, total integration of the

**TABLE III 5.0E Partial list of common PI software.**

| Software | Features | Main purpose |
|---|---|---|
| **Celonis:** | Enterprise businesses, event log mining, wide partner range | Business optimization, Process mining |
| **UiPath—PM**[a] | Automation suite, training, flexible accessible | Intelligent process automation, RPA |
| **IBM—PM**[a] | Process discovery and intelligence solution, easy to use, process simulation and visualizations | Alignment of business and design process with standard |
| **FortressIQ** | Process discovery and mining solution, automation integration, task capture and enterprise client list | Automation-focused enterprises |
| **ProcessMaker** | Integrated AI powered solution, easy implementation, automated insight | Does not Require extensive data science and process mining resources in-house |

[a]*PM: Process Mining.*

computational system of AI in robot and the mobility, sensing (such as touch, vision and hearing) as well as the ability to interact with physical objects with sensory feedback from the actions taken.

- **AI and robotic comparison:** As mentioned, there are several distinctions between AI and robotics and these are compared and tabled in Table III 5.1a.

4. **AI-Robotic: Application details:** On account evolvement of AI in robotics, applications of AI in robotics are in rise. AI integrated robotics have revolutionized work responsibilities across various industries including automating routine to introducing flexibility and learning capabilities into tedious applications even involving some cognitive actions at times. Many enterprises integrate AI robots into their routine procedures for increased productivity, efficiency, safety and better customer experience. These Ai integration in robots, transform the ways robots perceive, navigate, and interact with environments. Advanced ML models, computer vision, and NLP are extending the boundaries of robots' capabilities, making them more autonomous, adaptable, and efficient. AI in robot also, enables operation of Multimodal large language model (MLLM) described in Fig III 5.1a.

   Various common functions of AI in robotics are:

   - **Perception:** *Perception* refers to the ability robot, to interpret and understand sensory data from its surroundings. Integrated AI models process these inputs from sensors.
   - **Navigation:** AI-powered navigation systems use algorithms to plan and execute paths, avoid obstacles, collisions in adapting to the changing environment.
   - **Identification:** Object identification enables robots to recognize and classify items within their environment accurately. With introduction of AI, this is done accurately with several AI Training. Computer vision with DL play important role.
   - **Collision avoidance**: With integration of AI, robots can navigate complex environments safely and efficiently without colliding with obstacles with the help of sensor technologies, perception and decision-making.
   - Application details: Application of AI-powered robots in different area/sector has been enlisted in Table III 5.1b with details of impact discussed later.

**Features and application: Industrial robots:** There are a number of robots used in different purposes in modern times. Detailed classification of robots have been presented in Table III 5.1e. There are several advantages of industrial robots in manufacturing and production engineering including a few batch process industries such as food beverage and pharmaceutical industries. Every year there is tendencies to increase market and deployment of robot to increase plant efficiency and profitability. Major industrial robot manufacturers are: ABB, FANUC Robotics, KUKA, Yaskawa Electric and Kawasaki Heavy Industries, etc., to name a few. Major types of industrial robots with advantage disadvantages and uses have been tabled in Table III 5.1c.

### 5.1.1 AI and robotics detailing and discussions

For understanding the impact of AI, it is important to draw lessons from the past as well as to anticipate its future directions and potential legal, ethical and socio-economic implications. Robotics and AI are closely related and as a combination it is referred to as *robotic artificial intelligence*. For its *Enhanced capabilities, Increased efficiency and productivity,* robotic AI find applications in complex tasks in everchanging environment. Major issues here are:

- AI in robotics, allows learning from experience, adapt to new situations, making decisions based on sensor data.
- Machine learning (ML) algorithms are used for data analysis data, pattern recognitions, etc., the tasks in dynamic/unpredictable environment.
- AI enable robots to have better vision needed to interpret and understand visual information. This is very useful in object recognition, navigation, and manipulation.

So, this combination continues to redefine possibilities and elevate technological platforms. Combination of AI and robotics are at the forefront of technological revolution, pushing the boundaries of modern fulfillment into a new era. *Increased productivity*, *Enhanced accuracy*, *Scalability and flexibility* in operation, *Data-based decision*, *Energy efficiency*, *Real time* (RT) *control and monitoring*, *Cost reduction*, *Predictive actions* such as detection maintenance, *Reduced downtime* and Improved performance are major advantages of integrating AI in robotics. The system is not without challenges. Cost complexity in implementation are major ones. AI becomes more adept at learning and adapting to different scenarios for more intuitive systems capable of handling complex tasks. Also, robotics fleets can adjust proactively. Robotics advancements would include more agile and versatile robots to perform a wider range of tasks. In this sub section, discussions has been presented on AI and robotics starting with use of AI in robotics.

1. **AI use in Robot:** Combination of AI and robotics technology, takes the benefits of AI to the next level. This combination helps in ensuring operations flow continuously at optimum efficiency. helps ensure operations flow continuously at optimum efficiency. By this time some knowledge has been gathered about use of AI/ML in robot, robotics and automation. Fig III 5.1b shows typical use of AI in robotics.
   Here also, the discussions will be similar but with little bit more on details. ML/DL (ref: chapter 4) as a part of AI are used in robots. Reinforcement learning– a subset of ML, discussed (Chapter 4) are mainly used in robotics. *R*einforcement learning assists robots to learn and adopt motor skills and control strategies, allowing them to perform complex tasks such as *walking*, *running*, or *playing games*. As a continuance of III 5.1.0.4 various functionalities of AI in robotics are discussed here. AI assists robots with Intelligence and adaptability, so that robot can carry on various functionality assigned to it. Major ways AI utilized in robot shall include but not limited to the following functionalities (Ref: III 5.1.0.4 also):
   - **Computer vision**: On account of presence of AI in robots, they can visualize and recognize patterns, which robots have never encountered. AI smoothens detection as well as works accurately on the patterns (better than traditional robots).
     - *Object Recognition*: Power of AI in computer vision, allows robots to recognize, identify objects in their environment by helping robots understand the surroundings, create maps, and also navigate (discussed before) through complex environments. So, AI/ML provide direction to the robots to grasp an object.
     - *Visual serving:* Based on visual feedback, AI directs robots to track and precisely manipulate objects. These are extremely critical for many tasks such as welding, painting, filling/bottling/assembling delicate components.
     - *Surrounding Mapping:* With process camera and sensor data, AI algorithms can help to map surroundings, identify obstacles, and plan safe and efficient collision free paths for robots to navigate i.e., path planning.
   - **NLP technology:** The NLP empowers robots to perceive and interpret text written in human language. In combination with Automatic speech recognition (ASR), computer systems can recognize human speech. AI powered robots can understand direct commands, extract meaning from the variations of human expressions and can also perform sentiment analysis to come up with a relevant text/verbal response (speech synthesis). So, major issues are:
     - *Human interaction:* Robots can understand to respond to NLP commands, enabling more intuitive and collaborative interactions with humans.
     - *Generative AI*: One of the latest developments, by which AI in robotics brings machines closer to humans so that they can produce unique text or, express vocally (through text-to-speech conversion)
     - *Voice control:* AI powered robots can be controlled through voice commands, making them accessible for a wider range of users with ASR.
     - *Sentiment analysis:* As mentioned above, AI can analyze human text and speech to understand *emotions* and adjust robot behavior accordingly.\

- **Motion Control-** Proper control of locomotive parameters are extremely important to get a human-like robotic motion. ML helps robotics through dynamic interaction.
- **Warehouse automation:** Synergy created by combination of AI with robotics, could be used extensively in warehouse automation capabilities. AI is capable of predicting demand, accurate forecasting, optimize stock level while robots streamline the movement of inventory in most effective manner. Thus, improved workflows are achieved by eliminating bottlenecks. Also, there will be predicted inventory needs, inform reordering points, and improve replenishment processes. These help in enhancing labor planning, better labor efficiency and safeguard operational integrity. Apart from AI-based inventory, there will be use of Autonomous mobile robots (AMR) or Automated guided vehicle (AGV) s for warehouse automation.
- **ML Discussions:** From the above discussions it transpires that AI and ML help robotics to attend top of its assigned tasks. Apart from the various tasks discussed above ML (ref Chapter 4) helps robotics in the following areas also.
  - *Data use:* AI Powered robotics can adopt such data use and analysis through ML for bring success to robotic projects.
  - *Decision-making:* ML helps AI algorithms to learn from data and make decisions in real time, so that robots can adapt to dynamically changing environments and adjust in unexpected situations.

2. **AI Robot benefits:** From the above discussions, it is clear that AI has taken robots to a new height. Here, are some of the benefits of AI powered robots:
   - **Enhanced capabilities:** This is achieved though *capability to execute* highly detailed and complex tasks, improved ML, autonomously learning from data for improvising knowledge and for better adaption
   - **Increased accuracy productivity and efficiency:** AI in robotics offers all these benefits
     - *Automate task repetitions:* AI, can use robots to manage many activities that are boring and time consuming to relieve workers' burden.
     - *Better accuracy:* Powered with AI, robots function autonomously, so, proactivity of AI eliminate/reduces chances of errors especially human error during tasks such as picking, packing, and sorting. Also, intelligent sensors, advanced sensor data processing (in real-time, dexterous robots), DL and reinforced learning contribute greatly to increase the accuracy of intelligent machines.
     - *Increased productivity*: Ai powered robots are **in** advantageous position in terms of efficient and work accuracy, which brings faster results with fewer defects, losses—hence enhanced productivity. Thus, AI algorithms optimize robot tasks for speed and efficiency, resulting substantial increase in throughput. Also, capability of AI algorithms for some data analysis and precise calculations definitively increases productivity and product quality.
     - *Enhanced efficiency:* AI powered robots are autonomous robots and can work 24/7, without ensuring uninterrupted service and product delivery hence better efficiency in conjunction with other issued discussed above.
   - **Safety issues:** AI in robotics could have huge impact on domains and environments with increased risks for human health and life. However, AI methodologies are nondeterministic, hence may be challenging to get a consistently predictable outcome in all situations. So, there is lot of potential for AI-powered mixed reality in safety. A few issue to look into are:
     - *Hazardous environments:* Robots with AI can be used in risky areas to carry out important work without costing human lives; these robots can. Through cameras/sensors AI-robot can identify when workers may move into dangerous areas. In mining, AI-robots can handle toxic substances in chemical plant or can navigate underground tunnels and handle explosives, reducing the risk of cave-ins and explosions.
     - *Enhanced collaboration:* AI can bring robots closer to humans in safe and efficient manner. Some AI-based robots have cognitive quality in them also. Since robots avoid human fatigue hence in some cases may offer better results and safety.
   - **Various operational issues:** AI powered robots offer a number of operational benefits: such as *scalability*, *flexibility*, *creativity*, and data-driven decision-making.
     - *Scalable flexible operation*: AI and robotics can help in adapting quickly the changing demand. Based on demand robot functions could be augmented. So, AI in robot provides flexibility and scalability especially in terms of labor requirements
     - *Energy efficiency*: AI-driven robots can select the best paths and operational strategies to minimize energy usage, contributing to more sustainable operations.

- *Reduced downtime*: Predictive maintenance powered by AI can forecast robotics equipment failures before they occur, reducing downtime and maintenance costs.
- *Cost reduction*: AI-based robots not only reduces labor costs through auto operations but also ensures all forms of automation are working at maximum efficiency. This ensures cost reduction in all forms.
- *Monitoring and control:* AI enhances real-time tracking of operations with problem-solving issues making it easy to adjust and regulate operation based on monitoring. *Data-driven decision*: Because of AI there will be capability in handling huge data hence AI provides insights empowering to strategically make decision to upgrade operations, including best use of robot.

3. **AI in robot discussions:** *AI &ML, Cobots expanding to new applications, Mobile manipulators, Digital Twins, Cloud connectivity, Humanoid robots* are some of the important issues in current trends in AI in robot. Therefore, it is important that we look into the issues in depth. Having gathered some knowledge, here we discuss a few important issues pertinent to AI in robotics. Let the discussions start with AI functionality and components in robotics.
   - **Functionality and components:** AI/ML technology supplies brain and transforms robots from simple robot into intelligent robot allowing it to sense and perceive the environment by sensors, data collection, some analysis and then make decisions using algorithms. Intelligent robot can learn from experience (ML). As already mentioned, intelligent robots can autonomously navigate, recognize object, adapt the environment and go beyond preprogrammed actions. For these actions key components are:
     - *Sensors and Actuators*: Sensors and actuators are the physical interfaces of AI-powered robots for environment to perceive (collect data about the surroundings through sensors) and to respond (initiate actions to interact with and manipulate its environment through actuator) viz. Camera (sensor), motors (actuator). Noise, disturbances and energy efficiency are some issues here.
     - *Algorithms and software*: These form the "brain" AI-powered robots for various processing functions.The software includes low-level controland high-level AI algorithms to perform tasks from perception to decision-making.
     - *Human Interfaces*: Human-robot interfaces allow for seamless communication between humans and AI-powered robots. It covers wide range of interfaces from simple interaction to voice commands, gesture recognition, augmented reality, etc.
     - *Connectivity and data processing:* With AI-power, robots can access computational resources and large datasets through cloud connectivity enabling complex AI processing. Also, with cloud remote monitoring and control of robot fleets is possible. Edge computing for distributed databases can also be supported.
       a. **Applicational impact:** From the discussions above we have noted that there will be enhancement in *capability to handle complex task*, increase in *productivity*, *efficiency*, *precision* and *system safety* with sharp fall in *downtime* and *overall cost.* AI can be applied for *Automation, Analytics (Data/image/video), Predictive analytics, Computer vision, NLP, Risk assessment, Personalization, Optimization, Process improvement, Customer services, Quality control, R&D, Marketing, Decision making,* etc. Here, we discuss detailed account of impact of AI-powered robots across various domain of applications. Table III 5.1d shows Application and impact of AI- Robots across Domains
   - **AI and Robot: Challenges and negative impact:** There are several issues that pose challenges for implementing AI in robot. Major issues are:
     - *Technical issues:* Design and development of AI strong algorithms to work reliably across diverse environments and scenarios is a big problem. Robots needs to function effectively under different conditions, hence AI systems must be robust and reliable to handle different contexts safely. Management of power demands and battery life pose problem in implementing AI in robot.
     - *Ethical issue:* Concerns about job displacement, Privacy and Data security are some of the major ethical issues to deal with. AI robots are more autonomous; hence liability of failure becomes more obvious. It is challenging to find failure attribution (to manufacturer, software designer or operator) is legally challenging. This is a big ethical issue especially for critical AI application.
     - *Implementation cost*: High initial cost for development and implementation is quite high, making it difficult of small and medium-sized enterprises to implement. AI robots call for special training for operation and maintenance adding cost to the enterprise. Also, integrating AI robots into existing systems is not easy especially with existing legacy systems. To establish seamless communication after integration at times are difficult.
     - *Sustainability and climate change*: AI has direct impact on sustainability, climate change and

**TABLE III 5.1A Comparison between artificial intelligence and robotics.**

| Issue | Artificial intelligence | Robotics |
|---|---|---|
| **Conceptual (Purpose)** | Complex real-world problems solution; with ML and DL enable analytical capabilities also. | Deployed for design, development and construction of (programmable) robots interfacing with other devices/ humans. Perform autonomous or semi-autonomous tasks. |
| **Adaptability focus** | On learning and adapting from data and experience, | Creation of physical systems meant to adapt environment and tasks |
| **Automation degree** | IPAs use of AI for making applications more intuitive and "human-like." | RPA involves and works on the basis of programmed scripts. |
| **Hardware** | General purpose computer | Special purpose |
| **Input** | AI program is in symbols and rules | Electrical signal in the form of speech waveform or images |

environmental issues. In one hand through efficient supply chains, predictive maintenance it reduces carbon foot print. On the other hand the energy and resources required to create and maintain AI models would rise carbon foot print to a large extent. Even if AI is applied to climate-conscious project, could leave society in a bad environmental situation.

- *Deepfakes and misinformation*: The spread of deepfakes threatens to remove the lines of separation between fiction and reality.
- *Human biases*: AI is often tainted with a habit of human biasness in training algorithms.

  AI tools could reinforce these in the minds of users and perpetuate social inequalities.

4. **Mobility and security:** Application of AI robot in various area/sector has been mentioned in the above discussions. Computer vision, NLP, etc., also have already been discussed. Safety issue related to AI-robot also have been discussed above. Here, discussions have been put forward on mobility and security issues important in manufacturing sector.
   - **Autonomous mobility:** Autonomous mobile robots are quite popular in warehouse application by replacing forklifts, in production process they move products from one conveyer to another. *Autonomous mobile robots* (AMRs) are known for unique ability to navigate in an uncontrolled environment with a high level of understanding through sensors, blueprints, AI, and vision. AMRs are now in use in many environments. AI-enabled robots are capable of taking better decisions on alternative routing, by using all the data available to them, optimizing the path efficiency, leading to much more accurate and faster navigation without probable accident/collision. AI powered outdoor robots can deal with various varying temperatures and humid conditions.
   - **Cyber security:** The integration of AI in robotics systems in one hand has enhanced the quality of our lives, on the other hand it has also made these systems vulnerable to security attacks. The physical components, algorithms, and data that makeup AI-Robotics systems can be exploited by malicious actors, potentially leading to dire consequences. There are Two common but distinct types of cyber threats. Cyberattacks on robots could cause damage/harm robots by targeting a specific new vulnerability in the robot's software system. Other attacks might use AI technologies to conduct automated intrusions against industrial automation networks. These attacks can potentially even shut down entire plants, causing significant losses. AI systems can use anomaly detection approaches to continually monitor network communication traffic to identify any behavior that is significantly out of the ordinary.
5. **Human robot interaction (HRI):** HRI is an interdisciplinary (robotics, computer science, communication) field of study dedicated to understanding, designing, and evaluating (user friendly, safe and effective) how humans and robots interact. Interaction refers to communication, collaboration, and the development of robots with human in ways and means (with focus on communicate, cooperate, and share tasks). Communication between a human and a robot may take several forms viz. in close proximity or remote interaction (Mars Rovers separated in space and time).
   - **Basic discussions:** With ML algorithms robots can adapt to user preferences and behaviors, creating personalized and intuitive interactions. This is facilitated by NLP in seamless communication in HRI with voice commands and context-aware responses. Computer visions provide AI-powered robots with enhanced perception, to recognize and interpret human gestures, emotions, and facial expressions. With reinforcement learning AI-robots can learn

MLLM: ML models operate in one data mode – text/image / audio etc. AI activated Large language model (LLM) can understand and generate human-like text after trained on huge amounts of text data. AI activated Large multimodal model (LMM) can process and understand multiple types of data like text, images, audio, and video. Multimodal LLM (MLLMs) are AI models, which transform the capabilities of AI. MLLMs enable machines to process and understand diverse forms of input text, images, and audio etc. So, MLLMs are designed to integrate information across different modalities, enabling more complex and versatile.

**FIGURE III 5.1A** Multimodal large language model.

**TABLE III 5.1B** AI integrated robot applications.

| Area/Sector | Application note |
|---|---|
| **Assembly line** | With advanced vision, AI can help a robot to learn and real-time course correction in complex manufacturing. |
| **Packaging** | AI provides save certain motions with constant refinement. AI ensures frequently for quicker, faster action at lower cost with better accurate packaging. |
| **Customer experience** | Robots can leverage AI, NLP abilities to interact with customers like human in many ways it learnt. Customer: Greetings, reply to queries including recommendation, hotel robotic concierge etc. |
| **Manufacturing** | With AI in robots, in the manufacturing industry for decades, can be modified to assess situations in real time and perform dynamic actions to ensure better productivity, efficiency in safe environment. AI-robots can handle wide range of work from assembly to quality control with greater precision. |
| **Inspection and quality control** | AI powered robot can leverage computer vision in machines with sensors to inspect quality including defect detection in the products. |
| **Automation** | Integrating AI in robots, enable systems to learn, adapt, and take informed decisions—a big leap to encompass complex, context-aware workflows. AI in robot can automate tasks such as document reading and data analysis previously required human involvement. There can be single access point for process monitoring to track several manufacturing processes globally. |
| **Robot human collaboration** | Robots-human collaboration with AI assistance relieves human from monotonous work so that they can decide and solve problems. |
| **Rescue operation** | AI powered robots can used in disaster for rescue search for alive People, pets and also to survey the impact area. |
| **Agriculture** | AI in helps to understand weather conditions, advises for using fertilizer, water, and the time to harvest etc. to improve process efficiency and increase crop yields. AI can determine if the crop is ripened/ready for harvesting. They can plan path also to avoid collision. With ML, computer vision robot can be used to identify specific fruit/vegetable for pick up. |
| **Open source** | For AI, many robots can be taught to do custom specific tasks |
| **Autonomous vehicle** | With sensors, computer vision, and ML models, AI enables robots to independently navigate complex environments. |
| **Health care** | With adoption of AI in robots in healthcare, it us deployed for many complex functions such as: surgery. |
| **Predictive maintenance** | AI and ML algorithms can used to analyze data that anticipates mechanical breakdowns of equipment. |

**TABLE III 5.1C** Major industrial robot: features and applications [4,5].

| Type | Advantage | Disadvantage | Application |
|---|---|---|---|
| **ARTICULATED** | High speed, large work volume/type Multiplane alignment easily | Complex programming and kinematics, dedicated controller required | Arc/spot welding, machine tending, material/Glass handling, steel cutting assembling, foundry. Pick place, packaging, dispensing |
| **CARTESIAN** | Less costly, highly positional, simple operation and programming, heavy loading | Single direction movement requiring large work area. Complex to assemble | Assembling, pick place, material handling, loading unloading |
| **CYLINDRICAL** | Simple operation, assembly, installation. Less floor space, large payload. | Cannot go around obstacle, low directional accuracy in rotation, | Assembling, coating, diecasting |
| **DELTA** | High speed and accuracy | Complex operation with dedicated controller | Assembling, pick place, dispensing, inspection. Food pharmaceutical, electronic industries. Flight and automotive simulator, fiber optic alignment |
| **POLAR** | All round reachable, below and above obstacle, less floor space with large work volume | Low accuracy and repeatability in direction for rotary motion, complex control algorithm needed, short vertical reach | Diecasting, material handling, forging, welding, glass handling |
| **SCARA** | High speed and repeatability and large work volume and type | Difficult offline program, limited planer surface and needs dedicated controller | Machine loading, assembling, semiconductor wafer handling, palletizing and packaging |

**TABLE III 5.1D** Application and impact of AIs-powered robots across domains.

| Application | Impact |
|---|---|
| **Manufacturing** | |
| Quality control | No defect, high quality product, least manual intervention |
| Cobot/AI-robot | Improved safety, productivity and accuracy, least downtime |
| **Retail/E-commerce** | |
| Customer issue | Queries chatbots, engagement data, behavior &preferences categorization in distinct groups, |
| Stock and inventory | Identification trends and patterns demand forecast, optimize inventory levels, minimizes overstock and stockouts |
| **Aerospace** | |
| Auto rover | Better object identification, efficient space surface research, space ship repair, soil collection |
| Companion | Improved work experience and efficiency |
| **Supply chain and logistics** | |
| Warehouse space | Analysis product demand and inventory, strategic arrangement of storage areas |
| Shipment | Efficient real-time tracking and monitoring of shipments |
| Resourcing | Predictive analytics, optimized task allocation |
| Supply planning | AI-driven supply planning, greater control over expenses |
| **Application of advanced drones** | |
| Agriculture | Optimized yield, reduced waste, energy used and cost, weed removing |
| Military | Surveillance against enemy |
| Rescue operation | Prioritization, successful safe mission |
| Transportation | Improved navigation, planning and speedy delivery |
| **Military** | |
| Application | Used as armed systems to attack the opposing Forces or as medicare agents to help friendly forces. |
| **Underwater** | |
| Exploration | Robots can explore places that humans cannot reach. |
| **Health care** | |
| Surgery | Better precision, Reduced risk and complication |
| Medical images | Accurate analyzes, diagnostics, customized treatment plans |
| Assistance | Improved efficiency, better performance and patient care |
| Services | Personalized, improved emotion state of patients |
| Fraud detection | Overbilling/unnecessary procedures, extra claims |
| **Smart home** | |
| Service robots | Support people limited abilities, low effort house chores |
| Social robot | Increased personal efficiency, learning companion |
| **Food preparation** | |
| Chief | These chefs can create food using hundreds of different recipes. |

*Continued*

**TABLE III 5.1D Application and impact of AIs-powered robots across domains.—cont'd**

| Application | Impact |
|---|---|
| **Banking and financial services** | |
| Fraud prevention | Analysis of transaction patterns, identifying anomalies, and flagging fraud to detect and prevent |
| Credit scoring and risk assessment | Alternative data sources for assessing the creditworthiness, for proper risk assessment |
| **Security** | |
| Intelligent action | With cutting-edge features and actionable intelligence, these can assist human security/even replace them. |
| **Entertainment** | |
| Action | These can be used behind the sets in movies and serials to manage the camera, provide special effects, etc. |

from human feedback to optimize actions in real-time. Despite these advancements, challenges persist in the field of AI-driven HRI. Ethical considerations discussed, and the use of AI-robot with responsibility are major challenges also. Ensuring the safety and security of AI-driven robotic systems remains paramount, requiring robust measures against malicious attacks and unintended consequences. Human-robot trust remains a critical challenge, necessitating transparent AI algorithms and effective communication strategies [6].

- **AI-HRI: Advancement with challenges:** There has been significant advancements in AI and this has revolutionized AI-HRI. ML algorithms are used to analyze user data. Robots can use reinforcement learning to adapt their behavior based on user feedback, ensuring that interactions are tailored to individual preferences and needs [6]. Advancement in voice recognition have improved the accuracy and reliability of speech recognition in HRI. Robots can now understand a wider range of accents, languages, and speech patterns, allowing for more seamless communication with users [6]. Also, there has been advancements in object detection, recognition, and tracking algorithms have improved robots' ability to perceive and interact with objects in their environment. Reinforcement learning techniques enable robots to learn from experience and optimize their behavior in real-time based on feedback from users and the environment. Robots can now adapt their behavior in real-time based on feedback from users and the environment based on reinforce learning. So, AI-enhanced HRI have transformed the way humans and robots interact. However, there are many challenges for these viz. concerns are raised about privacy and data security, as users may be unaware of how their data is being used or shared. Also, the safety and security of AI-driven HRI systems is crucial. It is extremely important to prevent potential harm to human and protect against malicious use of AI technologies.

### 5.1.2 Robot details with specification

Robot may not resemble human beings in appearance nor may perform functions in a human-like manner. Robot refers to any automatically operated machine to replace human effort. According to Britannica "Robotics, design, construction, and use of machines (robots) to perform tasks done traditionally by human beings". Almost all sector of life there will be scope for application of robot. Now there are trends for AI-robot applications. Such robots are widely used in all industries such as automobile manufacturing to perform repetitive tasks, in industries where work must be performed in environments hazardous to humans and widely used in modern manufacturing in Industrie 4.0 already discussed in vol II of the book. On account of their efficiency, versatility, and production capabilities, industrial robots have become a mainstay of manufacturing worldwide. Industrial robots are valuable tools in industrial applications and are available in different types. It is important to make proper selection of the type best suited for the application and for this it is essential to understand the types of robots with their variations. Depending upon payload, reach, axes, and application capabilities, etc., there will be variations in each category. With so many variables to consider it can make the robot selection process seem overwhelming, but it does not have to be. Understanding each technical issue in robot specification will be helpful for selection. Collaborative robots commonly known as "Cobots" possess the innate ability to work in tandem with humans, instead of working as replacement of

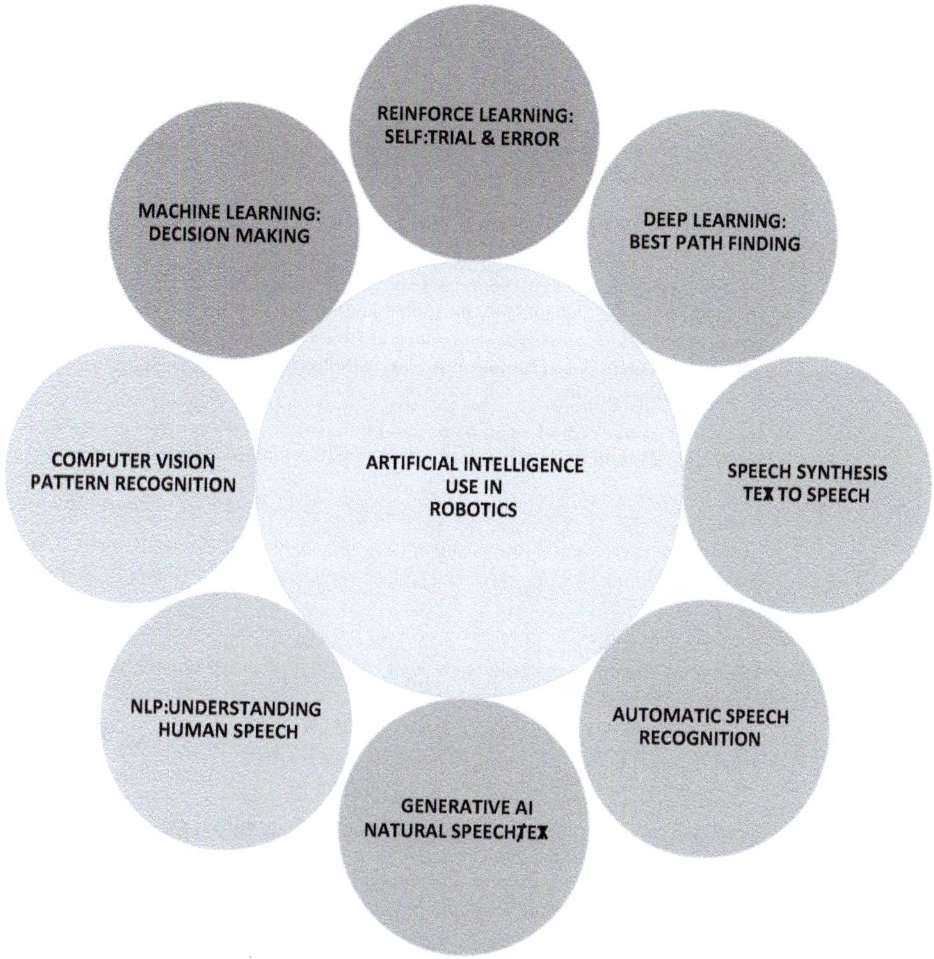


**FIGURE III 5.1B** Artificial intelligence use in robotics.

humans as seen in traditional industrial robots. Advanced AI, IIoT, machine vision, cognitive computing, mobile technology, and touch technology make it possible for lower-power robots to be aware of their surroundings and perform multiple types of tasks safely in close proximity to human workers as Cobots. Usually taking the form of a fixed robotic arm, Cobots aid workers in accomplishing strenuous or tedious tasks. Cobots being AI -powered robots, possess learning ability. Reprogramming ability of Cobots help industrial workers to simply readjust a Cobot along its desired track. Cobots will remember their new positioning and repeat the action.

1. **Discussions on types of Robots:** Classifying robots is as difficult as defining robots and differentiating robots from automation. Robots can be classified on the basis of application/movements and so on. Features and application of a number of industrial robots have been presented in Table III 5.1c. Classifying robots by application are: *Industrial robot* (designed for used in industries e.g., manufacturing), *Service robot* (a robot that performs useful tasks for humans – ISO). There will classifications based on locomotion/kinematics, which is more common and covers wide range of robots, which are enumerated in Table III 5.1e.
2. **Selection criteria:** Presently, there is a very large number of different industrial robots. It is important to have a knowledge about these selection principles. There are mainly three levels engineering tasks viz. *Macro planning* (higher level planning), *Micro planning* (operating level decisions) and *Operation executers* (fulfilling the production task). There are two different issues Selection requirement (*How to choose*) and Technical issues for selection criteria.
   - **Selection requirement:** Following are major selection requirements:
     - *Task definition:* Definition of task robot needs to carry out with reach payload, etc.
     - *Production issue:* This refers to both cycle time and production though put.
     - Flexibility *in operation*: This is related to the variety of tasks to be performed as well as the degree of freedom and no. of axes needed.

**TABLE III 5.1E** Classification of robots (locomotion/kinematics).

| Robot type | Descriptive details |
|---|---|
| **6 axis** | All in one robot. Used as human arm for pick up function. These can be from toy size to giant size, used for large assembly jobs. |
| **Airborne** | Robot capable of flying drone is an example. |
| **Aquatic** | These robots work on or under water and used for Underwater exploration of oil, gas or minerals. |
| **Articulate** | They offer wide range of movements such as forward, backward, upward, and downward motion. Articulated arm resembling human arm with joints such as elbow, shoulder, or wrist. They may have 2–10 joints. Higher the joints more degree of movement. Used for assembly, arc welding, material handling, machine tending, and packaging. |
| **Automated guided vehicle (AGV)** | AGVs rely on tracks/predefined paths and often require operator oversight. Normally they work in controlled environments such as warehouses and factory floors. |
| **Autonomous mobile robot (AMR)** | AMRs move throughout, make decisions in real-time. These are equipped with sensors and cameras and processing devices to analyze and ingest information to make an informed decision avoiding collision, picking precisely the right parcel etc. |
| **Cartesian** | They have three linear axes for straight line movements rather than rotating. They are mounted at right angles to each other offering high precision and repeatability due to rigid structure. These are less expensive; highly customizable and easy to use and program. Used in industrial/manufacturing sectors. |
| **Cobot** | Work along side human in collaboration. So, they share Spaces with workers to help them accomplish more. |
| **Cylindrical** | Here, the robotic arm can move up and down along a vertical member. Also, the arm can rotate about the vertical axis and can also extend or contract. Mostly they have rotary and linear actuators and can be placed in the middle of a workspace to work anywhere around it. Good for assembly operations, spot welding. |
| **Delta** | These are the fastest and most expensive. The end of arm tool (EOAT) of a delta robot can move along all cartesian axes and rotate around the vertical axis, resulting in a dome shaped work envelope. They come as complete solutions but for more complicated jobs. Expensive for speed and operational precision. There are at least three links connected to an three links connected to an EOAT and a common base. |
| **Humanoid** | A subset of AMR, and identified as robots that perform human-centric functions and often take human-like forms. These are used for directions or offering concierge services. |
| **Hybrids** | Combination of various types of robots discussed here. |
| **Parallel** | These are closed system to support a single platform where both accuracy and dynamic response are needed viz. Drilling and milling. |
| **SCARA** | It stands for selective compliance arm for robotic assembly. This type of robot has two parallel revolute joints. They are all-in-one robots, meaning equipped with x, y, z, and rotary motion; ready-to-go, apart from the EOAT. The work envelope of an SCARA is a torus. |
| **Spherical (polar)** | Robot works in a spherical system. It can move in a biangular and single linear direction. Polar robots, also known as spherical robots, they use the three-dimensional polar coordinate system r, θ, and φ for work envelope in spherical range. |
| **Wheel and legged** | Robots with wheel, which can be classified as bi/tri/quadra pedal robots. |

- *Speed-of operation*: This refers to a very high production speed and requirement.
- *Automation goal*: This not only refers to goal but also degree of automation envisioned. *Space and* footprint: This is vital as it is related to utilization of available space in selecting robot type.
- *Miscellaneous engineering issues*: This stands for cost consideration for design, assembly, and implementation as well as reliable operation, etc.

- **Technical issues:** Major selection factors for robot type from various available types are:
  - *Axes* – Axis of robot represents the degree of freedom i.e., an independent motion of the robot in different directions. So, more the axes the robot will be more movement capability of it. It is extremely important to ascertain the number of axes when selecting an industrial robot, which usually have three to seven axes.
  - *Payload*: Robot must be able to handle maximum weight required for the application. This is specified by *payload capacity* expressed in kilograms. This is found from the workpieces and the maximum weight of any end-effectors integrated with the robot.
  - *Reach*: This refers to all the required areas robot to reach. This reach can be broken down into (say) vertical and horizontal. Vertical reach determines the maximum vertical sweep a robot arm can obtain. Horizontal reach refers to the maximum distance obtained from the center of the robot base to its wrist.
  - *Speed and range*– The speed is related to the speed of each robotic axis. Range defines the scope of movement for each robotic axis in degree.
  - *Working environment*: This indicate the actual working condition and ensures that the robot to be selected is suitable for the specific conditions (temperature, dust, etc.)
  - *Robot mass* - Robot mass representing the weight of a robot (robotic manipulator only) expressed in Kg is important mounting the robot on a shelf, table, or overhead.
  - *Structure* - Structure refers to robot's work envelope and functionality. There are variation of this with the type of industrial robot.

3. **Performance specification:** Different applications call for different sets of performance parameters specifications/level. However, with betterment cost also shoots up at a high pace. Here, performance issues in general will be has been covered with some overlapping of discussions mentioned above.
   - **Axes: Degree of freedom:** As already noted it refers to the ability of a robot to move and orient the EOAT in 3D space. Possible six degrees of freedom as shown in Fig III 5.1c are: forward or backward, left or right, up or down, yaw, pitch, and roll. Base on a flexibility requirement and design, number of axes are determined. Simple pick and place type industrial robots may have 3-axes, while complicated/versatile applications may call for all six degrees of freedom or more.
   - **Accuracy and repeatability:** *Accuracy* refers to the ability of a robot to position itself or its load at a specific point, which is measured by determining how close its final state is to a set state defined by the user. So, it is a measure of a robot's ability to perform a movement and achieve the exact intended result without *online teaching*. *Repeatability* is the ability to return to the exact same position and is expressed in millimeters plus or minus the point of alteration to determine the robot's margin of error i.e., it is the measure of how the robot maintains its final position across several operating cycles. This is measured by getting the average of the distances between the final positions of the EOAT or a point on the robot after each cycle.
   - **Work Envelope:** The work envelope as spatial specification of a robot, indicates the swept area, reach, and stroke i.e., it shows the area the robot's EOAT can access. The size of the work envelope depends on the type of robotic arm (Cartesian, polar, SCARA, etc.) and the length of its linkages.
   - **Speed and acceleration:** Speed and acceleration are the influential the design factors for operating speed and throughput of a robot. These factors depend on the power rating and the kinematic design of the arm of the robot's driver, actuators, and transmission components.
   - **Controller specifications:** Every robot has its own controller, which contains necessary electronics and programming devices for its positions, speeds, and clamping or unclamping grippers. Different brands of robots have different programming methods. IP class, environmental condition and physical sizes are specified as mechanical issues. Also, no of Input output ports with signal types, communication standards, Power supply requirements are electrical specification for the control box. Also, programming type such as teach through/teach pendant types are specified.
4. **Industrial robot: Application detailing:** Apart from simple pick and place operations, industrial robots are deployed for more versatile and critical functions with *sophisticated sensors*, *advanced controls* and *powerful actuators*. Major common applicational details have been enumerated here.

- **Material handling:** Material handling robots are utilized to move, pack and select products. These functions are now automated for the transferring of parts from one piece of equipment to another. These result direct reduction of labor costs with increase in productivity. Loading and unloading are also, kinds of material handling where robotic systems with high load capacity reach and mechanical advantage are utilized.
- **Sorting:** Sorting processes refer to pick and place capability and high-speed monitoring of robotic systems. Visual sensors to detect variations in size, color, or shape reare utilized and odd items are rejected.
- **Product assembly:** Industrial robots are widely used as assembly machines. Robots routinely can assemble products, eliminating tedious tasks to increase output with reduced operational costs Their EOAT is usually mechanical grippers that pick, place, and orient small/large parts in quick succession. Modern end effectors are usually equipped with sensors to ensure that they have the correct part and the part is in the correct orientation for the assembly process. Also, robots can be programmed to assemble in a specific style and reprogrammed to produce a different product style saving time and money in the long run.
- **Various mechanical operation:** Common secondary processes such as, *grinding*, *polishing* are used to improve the product's final appearance and surface properties. These processes involve A robotic arm can easily mimic this simple movement of the tool to replace repetitive, oscillating motions of the abrasive or polishing material by human.

After building dexterity into robots, it is used for various manufacturing operations such as intricate cutting, grinding, machining that is otherwise very difficult to automate and time consuming viz. the production of orthopedic implants such as knee/hip joints. Robots are often used for nonconventional machining includes waterjet cutting, laser cutting, abrasive jet machining, etc.

- **Deburring:** Deburring process where robots are used, involves the robot to hold a rotating tool, usually a sanding drum, wire wheel, or carbide deburr tool, and follows a preprogrammed path to deburr and smooth parts from casting or injection molding. This relieve human from exposure to debris or dust during deburring.
- **Palletizing and Depalletizing:** Both palletizing and depalletizing processes are labor-intensive and may cause process bottlenecks. So, deployment of robot in these processes are common. Robotic palletizers offer better product handling and cost-efficiency. EOATs of robots could be mechanical, pneumatic, and vacuum grippers.
- **Welding:** Robotic welding systems are very common, especially in automotive manufacturing plants. Also these are widely used in many high-volume metal fabrication processes.
- **Painting and Coating:** Painting and coating demands for high accuracy and repeatability for a layer with uniform thickness of paint, which may be hazardous chemicals also. Naturally, here deployment of robot with suitable programming is very helpful.
- **Inspection task:** The use of robots with sensors for inspection is in a rise to meet quality specifications. Robotic inspection systems use measuring devices such as optical sensors, proximity sensors, ultrasonic probes for inspection of parts or assemblies.

5. **Cobot versus Industrial robot:** Cobots are designed to work alongside human employees to assist employees with work, which are too dangerous, strenuous, or tedious jobs. Cobots create a safer, more efficient workplace without eliminating factory jobs. Industrial robots are used to automate the production process, and work *in place of* those employees. As Cobots are capable of "learning" on the job these are normally more easily programmable and reprogrammable than their industrial counterparts. Usually, industrial robots are meant for handling larger materials in heavy manufacturing processes. Brief difference between Cobot and Industrial robots have been listed in Table III 5.1f.

## 5.2 Cobot and robotics: detailed discussions

When compared with Cobot, industrial robots are typically larger requiring more footprint, faster, more powerful with a higher payload capacity and greater range of motion. Also, for human safety, mostly industrial robots are enclosed in cages/barriers.

Cobots are extremely popular in manufacturing processes for safe and collaborative ways of working with human workers. It is important to decide the choices between *Cobots* and *robots*. These are done based on:

- **Need assessment:** it is important to assess specific needs and requirements based on. *type of tasks*, *human-robot interaction level*, *accuracy*, *speed*, and *budget available*.
- **Workspace:** Available workspace is important consideration. Cobots are compact and lightweight, easier to integrate but industrial robots call for more covered workspaces.
- **Future:** In addition to current needs it is important to consider future requirements as well as impact of technological advancements.

Cobots are normally equipped with powerful onboard computers, which are able to connect to IoT in any factory environment. Also, there are different types of possible interactions of Cobot with human for shared working environment. These are:

- **Coexistence:** Human and robot work alongside each other without a fence, without shared workspace.
- **Cooperation:** Cooperative robots refer to industrial robots with a virtual barrier between human operators and themselves. Cooperative robots are more complex requiring more advanced programming and are expensive.
- **Collaborative:** A collaborative robot (cobot) is a robot that can work alongside humans through direct interaction without conventional safeguarding fences.
- **Synchronization:** In addition to the above there is another type called synchronized, which is self-explanatory.

From the above discussions it is clear that in this section details will be presented on collaboration of robot with human Cobot and their programming, etc.

## 5.2.0 Discussions on cobot

Cobots are the newer development in robotics, which have significantly changed the landscape of the automation and robotics globally. On account of having sensors, they can feel the environmental conditions Tand can work with workers in a collaborative fashion. For any disturbance and an interruption in operation, Cobots are able to go to the safe mode unlike industrial robot, which would continue and potentially injure someone. While insisting on utility of robots, we underestimate/misunderstand the best uses of robotic solutions. Cobots are specifically designed to work in collaboration with human to perform task with minimal disruption to/increased safety and improved efficiency. Cobots should not be considered a low cost short cut for fully automated robot. Both Cobots and industrial robots can be customized. However, from previous discussions we noted that there will be difference in size, capacity and application between Cobot and industrial robots. Obviously, there will be difference in cost engineering and possible flexibility in customizing both types. Cobots are available in customizable preengineered designs, so that it would call for reduced the cost and time for customizing. As discussed, Cobot can assist employees with work in dangerous location or some, strenuous,/tedious jobs. Thus, Cobots can create a safer, more efficient workplace without eliminating factory jobs unlike industrial robot meant for automation replacing human. Cobots free up employees for more meaningful tasks that are less mundane and are less prone to repetitive motion injuries. With its user-friendly interface and ease of programming, it may be possible to get a Cobot for work in short time. Much less expensive than a industrial robot, Cobots are designed to meet the needs of small and medium businesses. Also, payback period for Coots are short. Therefore, it is possible to grow business quickly with Cobots. Cobots are much used in small and medium enterprise (SME) because Cobots bring efficiency and precision in production at an affordable cost, making them a worthy investment. In this subsection discussions have been presented on Cobots.

1. **Cobot tools:** Cobots use the same tools like other robots. Manufacturers are concentrating on manufacturing tools for Cobots allowing a robot to learn new operations to become more versatile. Some common Cobot tools are:
   - **Cobot vision:** With 2D/3D cameras Cobots can locate objects, recognize patterns, and scan barcodes, etc., making the integration process of a cobot much easier. This assists in the product's supply and needing organized supplies of objects for pick items up by Cobots. Computer vision allows optical inspections. Vision can assess the width and length (2D) and height also (3D) of an object especially for objects whose heights vary or if objects have to be stacked. With introduction of AI, the possibilities are increasing.
   - **EOAT:** EOAT stands for all types of tools that can be mounted on the end of a Cobot (arm). Various EOATs include: grippers, glue dispensers, screw machines, sanders, tool changers, sensors, welding and soldering irons and companies are developing new EOAT for specific uses. This tooling is easy to assemble and disassemble so that Cobots can perform different tasks. This facilitates flexibility in the production environments, and Cobots can do varieties of work instead of doing the same job for longer periods of time.
   - **Range extenders:** Range Extenders provide the robot with an extra axis and more range creating more automation possibilities and higher productivity. Normally, Cobots work only around the axis from where the foot is joined. So, by putting them on a range extender, the field of operation can be enlarged. Cobots can be attached to the Range extenders fast and easily as they have user-friendly programs and, communicating with the cobot is simple.
   - **Tool changers:** To make a Cobot versatile, there are different automatic tool changers making it conceivable for a Cobot to change tools entirely autonomously, to carry out multiple operations making it more productive and flexible with quicker ROI.
   - **Supply/Feeding systems:** These include machines that offer the items to be dealt with to the robot viz. a tool that always places screws in the same position ready for the cobot.

- **Sensors**: Cobots have built-in sensors by design to make the cobot collaborative and sensitive. Also, there are extra sensors for higher sensitivity.
- **Software**: Each Cobot has its own instinctive, user-friendly software to address a variety of applications. Also, simulations are available.
- **Safety:** Safety includes all tools contributing to the safety of the human and the robot i.e., collaborative robot cooperation's safety viz. Sensors/screens to detect someone gets too close to the robot.

2. **End of arm tooling (EOAT):** EOATs are the key to extract full potential of Cobots, allowing Cobots to adapt to different tasks. The right EOAT can turn a Cobot as a versatile member of the team, capable of handling various task with human. EOATs can be conceived somewhat as the hands of Cobot, which uses it to do various tasks assigned. their jobs. *Accurate placement*, *Cycle consistency*, *Mold protection*, *Improved quality*, *Reduced maintenance time* and *Additional versatility* are some of the facilities provided by EOAT. *Figure out the job*, *Type of task handling*, *Match tool*, *Work space*, *Safety issue* and *Future plan* are some of the issues need to be considered for selecting right EOAT tool for performing specific task type/task. Some EOAT types are:
   - **Grippers:** *Grippers* allow a cobot to pick up and move objects and they act like robotic fingers to grab, hold, and move items. There are many different types of grippers with their own characteristics. These include finger grippers, vacuum grippers, and magnetic grippers. The choice of gripper depends on the process being automated. Various types to choose from are mechanical grippers (solid objects), vacuum/pneumatic grippers (for delicate or irregular items) Each type has its own characteristics and power, for gripping strength, precision, or adaptability to shapes and sizes. Pneumatic grippers– well suited for less intricate tasks, are cheaper with high grip force but are less heavy than other grippers. When precision is needed, three- or two-finger grippers are the best solution to adjust object's shape, picking them and holding them firmly. On account of better regulation electric grippers are used for more intricate tasks. So, load capacity, gripping width and gripping force are major specifying parameter.
   - **Suction cups:** Suction cups use vacuum technology to stick to objects, lifting them without leaving a mark. Suction cups are perfect for gentle touch and working on smooth, nonporous surfaces, making them ideal for handling glass, metal sheets, etc.
   - **Screwing machine:** Screwing is a repetitive, monotonous task well suited for Cobots, which can attach products independently/with a worker with the help of screwing machines.
   - **Sanders:** With a sander tool as EOAT, Cobot can sand, polish or deburr surfaces with built-in sensors to ensure consistent quality.
   - **Soldering and welding tool:** These are precision task for which Cobots are perfect for the tasks with suitable soldering/welding tool.
   - **Glue dispenser:** A dispenser is utilized for glue or sealant syringe as EOAT, enabling Cobot to move in the desired pathway for distributing glue/sealant evenly and precisely.
   - **Tool changers:** Tool changers are meant for allowing Cobot to switch between different EOATs automatically.
   - **Specialized tools:** Sometimes Cobot needs to do some specialized tasks. These tools are tailored for specific tasks, turning Cobot fit to do the job.
3. **Cobot types:** Based on programming and safety features there are different kinds of Cobots. Each Cobot type deploys special technologies and methods to maintain safe operating environment. From safety standard, four types of Cobots are:
   - **Power and force limiting Cobots:** These Cobots are secured choice for human-robot teamwork–enabling direct collaboration with human worker without any extra safety barriers, external scanners, or vision systems (if *risk assessment finds safe*). For better human safety, force, and power limiting Cobots do not have any exposed motors, sharp corners, or pinch points. Advanced sensors are provided to prevent accidents and stop when they touch a person, to reduce the chances of harm. These Cobots can be slower, smaller, and less strong than the other (three) types. Heavy loads and precise tasks might be tricky for them. So, these are presently limited to small applications. These are cost-effective option that doesn't compromise safety. Power and force limiting Cobots have following subcategories:
     - *Joint sensing:* These Cobots are smart robots and they can feel the forces on their joints to help working smoothly with human. They have two versions: a) uses sensors in their joints for super-precise measurements and monitoring force and speed *b) the other uses motor* current to check forces. These are made very simple and easy to understand and it is easy to teach them to perform tricky tasks following specific paths. Based on force, it is possible to change what they're doing in real-time. Thus, these flexible and can perform delicate and precise jobs like human. Joint-sensing Cobots are a smart choice for making work more automated and safer.
     - *Skin sensing:* Skin-sensing Cobots can feel touch. They have special sensors to check if something

touches, if detect any touch, they stop to keep people safe. The technology may be new but offers highest safety. These robots call for good training to work well and ensure great safety. They are often used in industries such as electronics, pharmaceuticals, and food, where careful and precise tasks are required [7].

- *Inherently safe:* Inherently safe Cobots focus on safety through their light payloads and lack of power, making them relatively harmless. The tools they use on their "hands" could change how safe they are. These Cobots are best for jobs that are not very risky and involve working with people. but they may not be good for really tough tasks.
- *Force sensor base:* These Cobots feature a force-torque sensor at the base, allowing them to react to small impacts These Cobots are strong but also safe to work with people. They can lift heavy things and feel touch with special sensors.

- **Safety monitored stop Cobots:** Safety monitored stop Cobots are mostly used for applications that require the least human inter action. This can be conceived as a a traditional industrial robot attached with different safety sensors to sense when a human enters into the workspace of Cobo, to instantaneously stop working on detection. This ensures safety to human worker.
- **Speed and separation Cobots:** These Cobots are often referred to as "fenceless" Cobots work better when there is also less interaction with a human, but a human requires easy access to the Cobot. Safety laser range scanners are generally utilized to set up areas near the Cobot's workspace.
- **Hand guiding Cobots:** Hand guiding Cobot feature a safety rated part on the end of the Cobot's arm to enable an operator to manually move or guide the robot around. Hand guidance can be programmed easily and quickly.

4. **Selection Issues:** For Cobot selection suitable considerations to be taken into account are:
   - **Application details:** A clear understanding of the specific applications of the Cobot needs to be clearly defined. A few issues in this regard are:
     - *Task category*: *Specific task details* such as: Dirty, dull, hazardous, or physically challenging; assembly, packaging, welding, material handling, etc.
     - *Payload capacity*: Payload capacity: handling of the weight of the objects. The weight and reach requirements establish the size and number of joints necessary to properly control the payload.
     - *Movement and speed*: Proper evaluation of the Cobot's *movement* and *speed* is important to carry out tasks efficiently. Consideration for movement types such as: fixed, repetitive motion or a variable movement based on sensor data, movement space, direct human interaction, etc. The speed at which the movement needs to be completed is also important. The specific motion will decide number of axes or joints. The motion task and design also help assessing the torque and speed requirements of the joint motors and the length of the arms between joints.
     - *Work place*: It is important to ensure the Cobot's reach and workspace are adequate. The heavier the payload and the longer the reach data determines the more torque and power. The lengths of the elements between joints determine the moment loading. Also, other data such as offset from different axes, especially Z axis or length of common normal, etc. also, The available footprint must be considered for the entire reach of the robot.
     - *Performance and other issues:* Performance specifications of Cobot are important and it includes: accuracy (for tasks requiring precision) details, repeatability, etc. Reliability and sustainability are important to save money in the long run. Through scalability it is possible to accommodate the existing configuration modification or additional growth or change.
   - **Safety issues:** One needs to give due considerations to following issues:
     - *Risk assessment:* We know that Cobots have sensors that stop motions to avoid human injury. However, they can still injure a person even with advanced sophisticated sensors. So, a key factor when determining the use of Cobot is risk assessment.
     - *Features*: Major issues are: *Integrated safety sensors*, *Collision detection*, *Force limiting*, and *Emergency stop mechanisms.* Safe motion configurations are determined by the outcome of a safety risk assessment. Typically, safety functions include limits that prevent excessive force, torque, or speed based on sensors. As Cobot applications include humans in close proximity, designers must run a thorough analysis to select the appropriate *safe motion* and *safe machine* SIL recommendations must be incorporated (also ref: III 5.2.0.3). Based on level of collaboration necessary in the application, one needs to choose Cobot with appropriate safety features for that level including, design to limit force and speed to prevent injuries to human.
     - *Standards:* Relevant standards are: ISO 10218, ISO/TS 15066 Cobots need to comply.

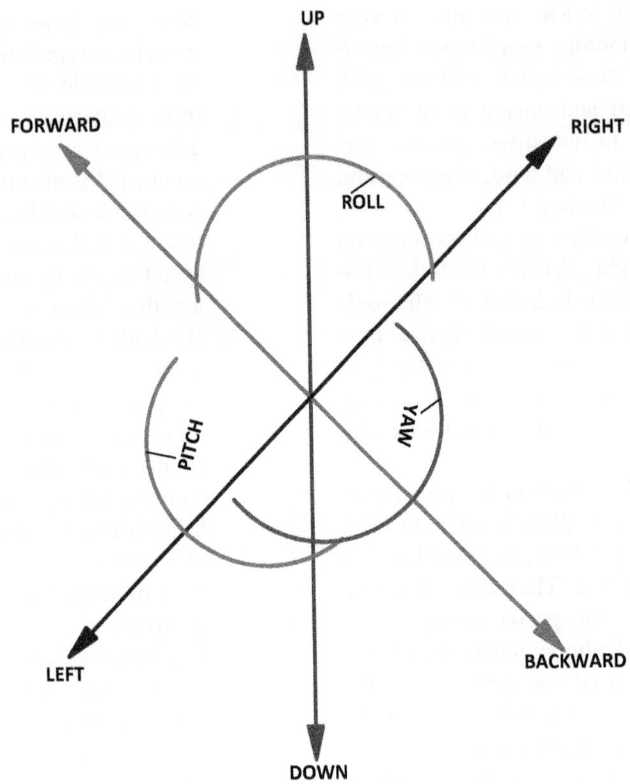


**FIGURE III 5.1C** Robot: degree of freedom.

- *Control modes:* Study and grasp various safety control modes available for selection of appropriate for the application.

- **Programming and integration:** Issues related to programming, system integration and compatibility include but not limited to the following:
  - *Programing:* Programming of Cobot needs to be intuitive and *user-friendly* and capability to carry out off/on-line programming quickly and easily. This important to shorten time for smooth implementation and adaptation of changed production needs.
  - *Ease to use:* Since Cobots, are designed to work in conjunction with human these must be easy to operate by workers with varied skill levels. An interface for a Cobot should accommodate graphical user interfaces for intuitive and easy learning.
  - *System integration:* Modern Cobots are equipped with advanced vision systems and sensors to detect and identify objects making them highly versatile. So, integration of these is necessary. Consider how the Cobot will integrate with existing production systems, machinery, and software. System integrators after performing all of setups for a number of customers before, can offer something that will cost less even when the system integration is being outsourced.
  - *Compatibility:* Cobots can be used within a network with other devices. Cobot control systems must have capability and the right to communicate with associated control system (PLC/DCS/HMI). Also, for complete automation, IoT/IIoT would call for Cobot communication remotely through the internet. While operating on network through common bus, Cobot would need to communicate and share the same communication protocol.
  - *Training and support:* It is essential to check and ensure that supplier provides adequate training and technical support for personnel responsible for programming and maintaining the Cobot. Also to ensure there is access to help when needed to resolve any technical issues.

- **Ergonomics and Environment:** Human factors along with technological aspects are important considerations for the design of modern Cobots and collaborative manufacturing workplaces in the context of Industrie 4.0. A few issues are:
  - *Ergonomics:* It is necessary to consider the ergonomic implications for the Cobot to ensure

minimum potential risks to human operators. Also, selected Cobot should have flexibility in operation.

- *Working conditions:* Due consideration in the physical working conditions at the work place are important to choose a Cobot.
- *Flexibility:* Choose a Cobot that is flexible and can be adapted to different tasks and environments.

- **Compliance and commercial issue:** following area few issues to be looked into:
  - *Compliance issue:* It is necessary to consider that all the regulations and standards applicable to the use of Cobots are well met by the selected Cobot both for design and implementation.
  - *Cost issue:* A thorough cost-benefit analysis before purchase decision is necessary. Here main issues are initial capital cost for purchase, recurring operational cost for maintenance, energy consumption, training as well as any modifications to the production line.
  - *ROI:* Expected ROI is calculated to assess how quickly the investment will be paid for Productivity, reduced costs, and improved safety are major factors under consideration.

5. **Robotic: Collaboration and Application:** With discovery of collaboration and programming, work can be performed in many ways. Let us look into these details of collaboration and applications of robots in different industries. Major issues here are:
   - **Collaboration:** Depending on the area of application, humans and Cobots can work together with varying degrees of proximity. Accordingly, there can be four terms defined for some predefined cases. Let us look into the details below:
     - *Collaboration:* Humans and robots interact in a *shared* workspace. In *Direct* collaboration, both humans and robots are working on the same *piece* in *simultaneous* actions. If humans and robots are working on the same *piece* but in *alternating* action it is referred to as *Indirect* collaboration
     - *Coexistence:* Humans and robots work in adjacent workspaces without safety fencing but without sharing a common workspace and work independently of one another on different tasks. Here, autonomous work on different parts (man/robot coexistence). Here, the sensitivity and safety of a Cobot can greatly ease. Here, also there will be speedy commissioning/programming, training and reuse.
     - *Cooperation:* In human-robot cooperation, humans and robots work in the same workspace. There is no direct interaction hence termed as fenceless operation also. Here, the Cobot either is inherently safe or the workspace is protected with sensors. So, there will be two types in one *physical/gesture/handling assistance*, robotic systems of physical assistance *guided by the operator*. In another *physical assistance/gesture/handling, physical assistance systems* attached to the operator
     - *Sequential operation:* Human and robot share the workspace but perform tasks at different times. Human lines up product in a tray for a Computer numerical control (CNC) machine tending robot.
   - **Application area:** On account of many tools, comprehensible software and adaptability, Cobots are capable of performing a number of tasks as enlisted in Table III 5.2a.

### 5.2.1 Robotics: various aspects and programming approach

From the discussions on Cobot in previous sub section it has been noted that there are a few benefits Cobots offer. These are: *Small* and *compact*, *Easy installing and programming*, Flexible, *Mobile*, *Consistent and precise*, *Reduction in production cost* and *Good impact on employees*. However, the Cobots may need a restricted working space for worker's safety. The initial cost of investment may be high and work speed may fall with safety processes put in place. The collaborative robot might have to halt every time it detects the presence of a person in its specified working space. There are a good number of companies focusing on the development and production of Cobots with each type has its own unique features and specifications. For an automation, it is important to look in details in to the requirements, for successful integration. With Industrie 4. 0, the goal is to make production lines more agile, reduce costs, reduce changeover times and increase throughput while customizing products. The use of new technologies in the frame of Industrie 4.0 is crucial. This eliminates all nonadded values and optimizing all processes. In the advanced robotic automation, there will be decentralized intelligence with defined characteristic allowing devices and equipment to make decisions and take actions autonomously, without human intervention. Advanced robotics is a key element in the movement toward advanced automation, which is helping to dramatically improve factory operations. Robotics are all around us and their uses are increasing every day. Whether you're ready to move your company into the future with robotics or are just beginning to consider automation possibilities. Plant leaders need to take a few things into consideration when planning where to implement robotic automation viz. for a robotic arm, one needs to consider the other factors

**TABLE III 5.1F** Difference between cobot and industrial robot.

| Issue | Cobot | Industrial robot |
|---|---|---|
| **Human** | Designed to work in partnership (including upgradation, maintenance) | Automated to complete automated task with no/little human interaction. Exception: Upgradation, maintenance |
| **Programming** | Because of on job learning, easier to program -more flexible and adaptable | Task automation with complex programing may be for one time or never changing |
| **Accessibility** | Stand out for accessibility, easy integration and cost especially for SME[a] units | Expensive requiring long and energy-intensive integration processes. |
| **Flexibility and adaptability** | Flexible, reconfigurable with light weight and foot print | Mainly fixed for specific pupose with larger foot print |
| **Safety** | Safe enough to function around humans viz.<br>Force/torque sensors, soft padding rounded edges | May call for safety infrastructure emergency push button |
| **Manufacturing** | Because of size and proximity to human, not meant for heavy material handling | Meant for large job with heavy material handling |
| **Functionality** | Production with low volume, high-mix/safely automate processes with human. | Fixed equipment for high–precision/volume, and speed of production. |
| **Sensors** | Rely on a variety of sensing technologies. | Mostly feature vision as perception capabilities |
| **Explosive area** | May not be possible | Good fit for application |
| **Cost and ROI** | Low costs, setup and<br>Training costs, faster ROI. | Higher cost additional cost for safety infrastructure. High ROI in high precision work |
| **High speed and accuracy** | Not a good fit for speed but mediocre for accuracy | Good fit for high speed and high precision work |

[a]*SME: Small medium enterprise.*

(say) EOAT for it to function properly. Robotics can automate the processes with continuous, 24/7 tasks to reduce labor expenses. The potential exists to reduce no of labors through automation. In case of Cobots, instead of replacing workers, Cobots frees them up to use their skills in more complex activities/human supervision. With this background let us start our journey in exploring various aspects of Robotics and programming issues.

1. **Robotic Model:** Robot models are used to represent the kinematic and dynamic properties of robots and other rigid body systems. The model is represents a rigid body tree containing rigid body and joint elements. Here, basic building blocks and improvement models aspects have been touched upon.
   - **Building blocks:** Basic robotic blocks for advance automation are:
     - *Data models*: This also referred to as Digital twin comprising a digital representation of products and the production system with life cycles. In such scenario is possible to carry out *virtual* equipment setup, integration and commissioning of the systems with the help of simulation. Also, production robots can access the models.
     - *Cloud-edge computation*: As the name suggests computation power and storage, etc., are shifted to network's edges, marrying Information technology (IT) with Operations technology (OT). All these approaches help overcoming various challenges such as data latency, limited bandwidth. In this arrangement collected data can be transferred to cloud services for processing. Computing at the edge help in addition, adjusting environmental changes autonomously.
     - *Data processing:* Different data types of various sizes, velocities, etc., from transactional objects to IoT/IIoT demands suitable processing. With advanced data processing, it is possible to continuously optimize processes.
     - *Workflow control*: With this it is possible to synchronize all tasks in manufacturing and logistics processes. This is extremely important. Efficient management of the entire workflow requires integration of both machines and human workers.
   - **Improved model:** In addition to the building blocks discussed, there are a few generalized functional building blocks, which combined with equipment-specific innovations, help unleash the potential of advanced robotic systems. These four types of functionalities are:
     - *Perception*: Major robotic improvements such as: NLP, computer vision, and different sensors support increased autonomy, dexterity, and precision. With these improved functionalities robots can perform complicated tasks without interruption.
     - *Integrability*: On account of integrability, it is easier to carry out the setup and robot teaching process. All these are p because of new service-oriented architectures, better connectivity, access to data models, and programming& interface improvements.
     - *Adaptability*: Advances in data-processing technologies and access to cloud services enable robots to learn and autonomously adapt to complex, changing environments. Improved workflow control systems discussed above.
     - *Mobility*: ML and computer vision along with other sensor as applicable help robots to do jobs autonomously i.e., self-guiding themselves viz. Self-driving robots and aerial drones.
2. **Safety: Cobot and robot:** During discussions on selection of Cobot (III 5.2.0.4), we have noticed that risk assessment and safety is an important issue in selection process. Every Cobot and robot must have emergency stops push button as a safety measure. Additionally, as Cobots are coming in close proximity of human, *hand detection sensors*, *speed* and *separation monitoring* systems are often in place. Before put any Cobot in operation, it is essential every possible unsafe situation should be evaluated to ensure foolproof safety controls. There are a few safety standard whose requirements must be complied with by Cobot and industrial robot based on applicability. Here, main priority is to assure that all human workers are safe. Major issues to deal with are:
   - **Cobot safety features:** Some of the common safety features include:
     - *Power and force limiting*: As, discussed, the idea behind this, is to set maximum level for the robot's force output, so that in case of accidental contact it won't cause injury.
     - *Speed and separation*: With no human within workspace, the robot can operate at desired speeds but as human enters the guarded area, it slows down to a safer speed.
     - *Hand-guiding*: If present in robot, this technique lets an operator manually guide the robot through programmed paths, allowing safe, close collaboration.
   - **Cobot safety Pitfalls:** A few probable safety pitfalls are:
     - *Irregular risk assessments*: Cobot environment changes over time, it is necessary to carry out regular risk assessment. Failure to reassess risks in changed condition would result new dangerous blind spots.

    - *Blurred boundaries:* It is possible for workers to inadvertently cross into off-limits areas so, . proper monitoring and safety measures are essential
    - *Latest standards*: It is essential to comply with the latest standard, which may be changing.
- **Cobot safety Guidelines:** A few common safety guidelines are as follows:
    - *Safe work place*: Safe design of workplace is the requirement. So, designated areas for Cobots, clearly marked and equipped *with safety features such as barriers, emergency stop buttons, sufficient lighting including warning lights, floor markings*, and *clear signage* to guide human movement.
    - *Training of team*: It is recommended that suitable training should be arrange for everyone in the team i.e., all receive thorough instruction, guideline on safety protocols, operation procedures, and emergency response measures.
    - *Clean system*: Like any other machine Cobots must be clean and in tip top conditions including inspecting functionality and wiring, checking for loose bolts, and ensuring all safety features are functioning as intended.
    - *Built-in safety*: Right from design, safety features should be considered i.e., selected model must have built-in safety measures such as force limitations collision detection/avoidance and speed restrictions.
- **Cobot safety validation:** RIA 15.06(discussed below) calls for all robot (industrial robot and Cobot) applications have a formal risk assessment performed. For Cobots, RIA 15.806 requires that specific functional safety validation testing is performed before placing the robotic system into production. Before releasing the Cobot application to production, an ANSI/RIA compliant risk assessment and specific documentation are required. Machine safety specialists (MSS) can guide in this respect.
- **Safety standard discussions:** The RIA ISO/TS 15066 technical specifications and the ISO 10218 safety standards explain the safety performance and functions of a collaborative robot. However, ISO 10218-1 and ISO 10218-2 provides safety framework in broader perspective for industrial robots. Under TS 15066, speed and force monitors of collaborative robots are configured based on human contact area, application data, and workspace hazards. The *ANSI/RIA R15.06* is US-based standard to work with ISO 10218 to define safety requirements for collaborative robots. RIA TR15.806-2018 Guidance to Testing force and pressure in Cobot application text gives guidelines on finding conditions of the measurement devices, test measurements, and accurate testing methods. It also provides metrics and test methods for measuring the forces and pressures associated with transient and quasi-static contact actions of collaborative applications.

3. **Some branded robots:** As discussed, there are many robot manufacturing companies, each having its own feature. Normally, each robot brand has its own specific language viz. Universal (URScript). ABB (RAPID), Fanuc (Karel). This is because, proprietary language mostly designed to optimize the performance. Modern robot programming mostly calls for proper training for associated complexities and the use of new technologies. Alos, there exists trend in the use of robotics open software viz. ROS, Matlab (Franka Emika) for sharing knowledge with the robotics community. Human machine interface (HMIs), based simple robot tasks programing, easy to be used by operators. Some of robot brands have been included here.
    - **Universal robots:** This is the largest cobot manufacturer in the world, and caters to almost half of the robotic market. In total, they released two different series of total different Cobots. "URScript is the custom programming language, developed by Universal Robots, that controls the robot arm. Any interactions with PolyScope get converted into URScript commands, and are subsequently sent to the robot to be executed."[https://www.universal-robots.com/developer/urscript/Courtesy: UNIVERSAL ROBOTS]. Differences include greater precision and sensitivity due to a built-in, tool-centric force/torque sensor in the E series viz. UR 30, UR 3e, UR 5e, UR 10e [https://www.universal-robots.com/Courtesy: UNIVERSAL ROBOTS]. The robots vary in payload and range.
    - **ABB Robotics:** ABB robotics offer wide ranges of robots including AMR controllers. IRB 1010, IRB 1520ID, IRB 2400, etc., are some of the example from the wide ranges of selection process based on payload, reach and controller types. ABB use high level programming language RAPID. They can support open programming languages.
    - **Techman robot:** Techman Robot's cobots are unique in that they have an integrated vision system. A camera is placed on the head of the robot and the accompanying software is equipped with smart vision programs. They offer smart simple safe Cobot solution. These include pattern matching, object localization, barcode scanning and color recognition. TMscript is the programming language used in this brand
    - **FANUC**: FANUC cobots stand out in the market by providing an extensive array of options, a higher

payload capacity, extended reach, and increased speed compared with other cobot series. CR and CRX are the common series of Cobots. These collaborative robots are certified for safety, enabling them to seamlessly collaborate with humans and enhance the efficiency of operations. FANUC has a wide variety of products to accommodate nearly all automation environment. There are two programming languages: Teach pendant is very similar to RAPID of ABB and Karel.

- **Franka Emika:** Franka Emika offers wide choices. Franka Emika's Panda is an advanced robot whose unique feature is its sensitivity and extreme precision. Sensors are placed in the seven joints of this robotic arm, allowing the robot to handle the most delicate objects.

4. **Flexibility and intelligence:** From precious section in this chapter, the close relationships between AI and robotics have been established. These relationships are reshaping many industries and sectors with automated systems to make them more adaptive, efficient and can act in smart/intelligent modes. As already discussed, in contrast traditional robots, AI-empowered robotics can learn, adapt, and operate with immediate and agile responsiveness in *dynamic* environments. The adaptability in robots the property by which they can efficiently learn to perform new tasks regardless of their specifics and external conditions. Therefore, AI-based robotics systems, through adaptability can perform *multiple and overlapping* tasks that integrate judgments and active responses to dynamic alterations in task. Let us discuss flexibility and intelligence in two parts:
   - **Flexibility:** "Flexibility" typically refers to the equipment's ability in adapting to new tasks, changing/or varied conditions/environments without bringing major structural or software changes. Cost reduction, increased productivity and the automation of a greater diversities for complex/variable applications with lesser functional blocks are direct benefit of robotic flexibility. So, in this frame the collaboration between robot and human could be increased to improve work performance. This also helps in an intelligent robotization and automation of manufacturing lines. The creation of a specific programming tool as an interface between programming needed for robot and easy operations expected.
   - **Intelligence:** AI-driven robots are designed to adapt in real-time to the *changes* in *operation*, conditions and environment making them most suitable for applications in.

   ML is the key factor in adapting AI in robotics. ML algorithms allow robots to learn from experience and adapt to the environment and managing exceptions seamlessly, to improve performance based on past data. There are two major areas where impact could be seen. In vision and perception it impacted in *Object detection and recognition*, *Environmental awareness*, *Human detection and interaction*, *NLP for*

**TABLE III 5.2A** Cobot application area.

| Task | Description |
|---|---|
| **Assembly line** | Parts selection and orientation, glue dispensing, welding, joining, surface treatments and painting, soldering, fastening: Screws/bolts/nuts deburring, grinding and polishing, etc. |
| **Welding** | A subset of assembly cobots provide solutions for industrial welding in supplementing welding staff. |
| **Machine tending** | It may be mundane and repetitive task; can tend to machines: *CNC, injection molds, 3D printers, bending machines* etc. |
| **Material handling** | When used with different types of end effectors, cobots can handle a variety of materials and can reduce both the time and cost viz. *Primary/secondary packaging, palletizing* etc. |
| **Inspection** | Able to perform quality testing and inspection tasks, cobots can load and unload parts into inspection machines and can also be outfitted with external sensors, to perform the inspection to decrease defects to a large extent. |
| **Pick-and-place** | This simple application can grow in complexity with added tools. Various tools used are: *Double gripper, Sensitive gripper,* materials with different sizes etc. |
| **Bin picking** | With vision technology and smart software, the cobot identifies materials in a bin, then picks them up one at a time often used for supplying a conveyor belt |

*communication and adaptability*, *Contextual awareness*, *Task adaption* and *Multi-language capability*. Predictive analytics such as: *Maintenance prediction*, *Task optimization*, *Environmental adjustment*, *Adaptive action*, *Proactive error correction* and *Dynamic task allocation* are also examples of AI impact on robotics. Typical architecture in Fig III 5.2a [8] shows how expert system can be used linking HMI. , as desired in the programing tool. This architecture is based on a knowledge engineering and management system [8].

5. **Programming approach:** Through robot programming robots are instructed to perform tasks by defining their movements, actions, and sensor interpretations. All these are done through Tech by showing methods (power/manual lead)/textual language programming. General programming approach comprises: *Lead -through* programming, *Supervised learning*, *Dynamic programming*, *C or Python* type language and *Robot operating system* (ROS). *Error handling*, *Safety*, *Robot programming* issues are: real time control, sensor integration and communication protocol. Cobot programming is calls for knowledge on the Cobot specific language by the operator. A typical process management philosophy has been depicted in Fig III 5.2b where, various the technical aspects related to the process optimization are combined to use with Cobot. This could result in new robotic support for much better performance and the sustainability. This robotic support system will allow to define for user friendly HMI for operation and this will cover: *data*, *instructions* and processes independent of the Cobot language and easy to learn and apply.

   Associated software is composed of:

   - A user interface for human interaction with the Cobot
   - Interface for exchanging data/instruction with the Cobots, and modifying behaviors with dynamic data
   - Interpretation module for Cobot or operator adaptations from dynamic data
   - A scheduler for Cobot response and comparing the robot reaction.
   - A tool manager to manage relation between modules in software and reference model system to manage generic data
   - An expert system for rule and fact basis, an expert knowledge module and an inference engine, etc., for continuously capitalizing expert knowledge to improve performance.
   - A learning system for capitalizing human expectations and Cobots behavior or Cobots expectations and human reaction.

With this in mind let us discuss *robot programming language*.

### 5.2.2 Discussions on robot programming

In robotics, textual programming languages such as Python, C++, Java, and MATLAB form the foundation for development and simulation. For easy integration of ROS, Python and C++ are favored for facilitating experimentation with new concepts. As have seen there are robot-specific proprietary programming languages viz. ABB's RAPID or KUKA's KRL. These languages help optimize the capabilities of their robots, offering specific solutions but with limited flexibility. Both C++ and Python are two versatile object-oriented programming languages, which are common and extremely popular in creating a wide variety of different programs/functional code. However, choice between the two is very tricky as both programming languages have very different programming approaches, syntaxes and usages. Generally, C++ is a better choice for large system developments and is also the hardest to learn and write. Python robot framework to develop control software, describe the control scheme. For this robotics programming, basic knowledge of mathematics (including trigonometric functions and vectors) and basics of Python to use of basic Python libraries and functions. A major issue in transition to robotic automation, is robot programming. The programming method can seriously impact ROI directly, so finding the optimal programming method for the application is critical. Two common programming methods are: Teach pendant programming and Offline programming (OLP) software. Teach pendant method uses, an electronic control box to program the motions of the robot, in OLP the user can program the robot on a computer, independent of the hardware. OLP optimization is possible and validation can be carried out in simulation environment that matching with the real-world robot cell to ensure the generated program is error- and collision-free. OLP, allows to generate robotic code and then upload it to the robots without any production downtime and loss of revenue. In addition to textual languages coding fundamentals through visual/graphical programming languages shall also be covered here. In this sub section brief discussions will be presented on robotic programming to conclude the discussions on process intelligence and robotic. Major issues here are:

1. **Robot: Programming basics:** Primarily robot programming is done to make the robot understand the tasks it needs to perform. Programming normally covers guidance to: The *path to be followed*, The *destination point to be reached*, *Interpretation of sensor data*, *Actuation details of end-effector*, *point to point movement details of various parts*, etc. Two basic issues in robot programming are Types of programming languages and Types of programming methods. Fig III 5.2c shows

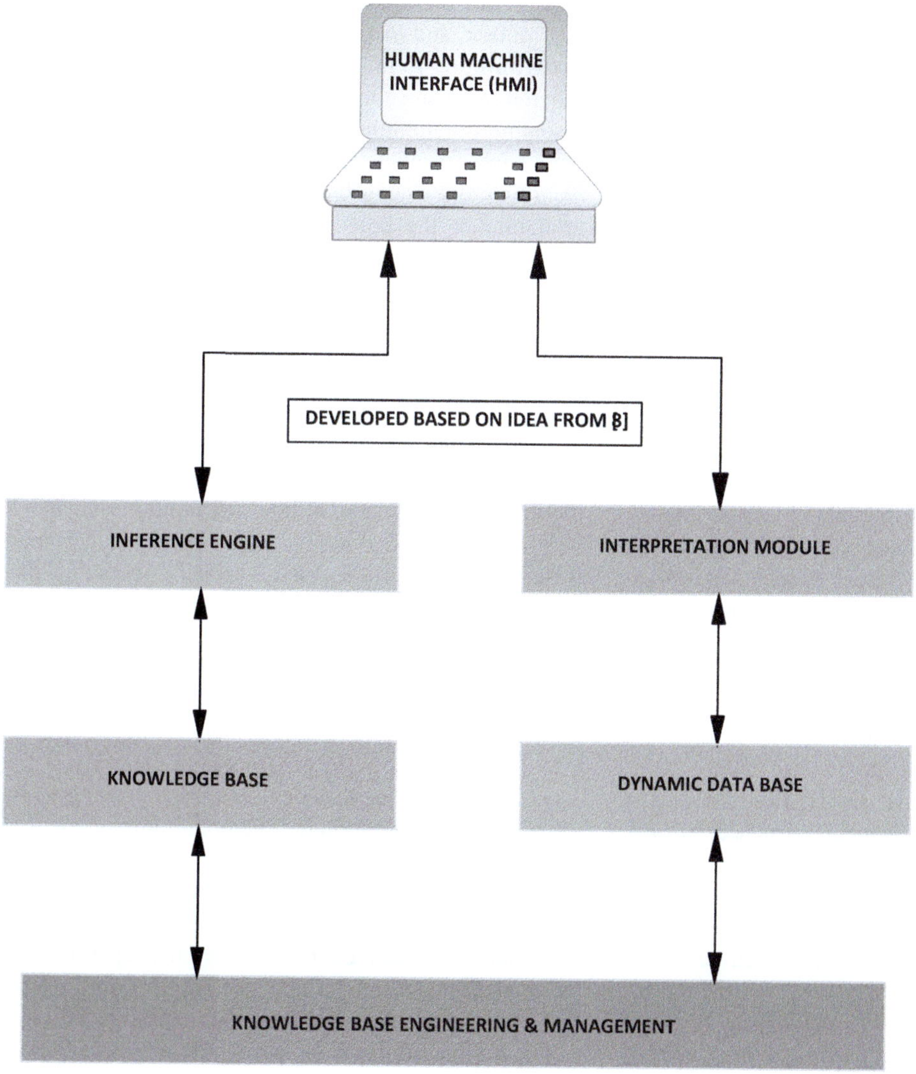


**FIGURE III 5.2A** Expert system and HMI as program tool.

the Basics of Robot Programming. Fig III 5.2c1 details out programming languages and programming methods.

- **Programming language types:** Robot programming language (RPL) can be divided in to three brought categories. These are: *Robot library general language* (are proprietary– e.g., RAPID for ABB robot as discussed in III 5.2.1.3), *Existing computer language* (C++) and *Specialized manipulation languages* (e.g., VAL II). Mostly, RPLs are implemented as a combination of textual and teach-pendant programming. Various RPL types are:
  - *Textual*: C/C++, Python, Java, LISP, MATLAB, etc., are textual and existing computer languages.

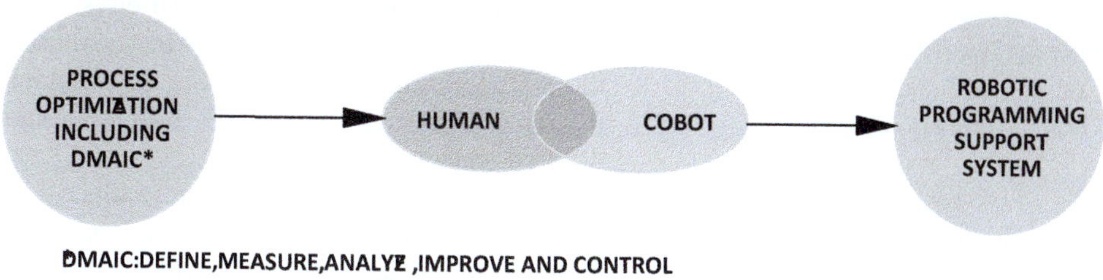


**FIGURE III 5.2B** Process management philosophy with cobot.

Languages. Here codes use keywords, operators, and other symbols with text-based specific syntax to be followed.

- *Scripting*: This is mainly used for task automation purposes, mainly to tackle repetitive tasks, manage dynamic web content, or support processes in larger applications. *PHP*, *Ruby*, *Python*, bash, *Perl* are example of this.
- *Algorithmic*: As the name suggests, this is used for mathematical or symbolic computations.
- *Assembly*: This low-level language is used to get fine-grained control over hardware for some high performing job.
- *Visual or graphical*: WAVE, VAL II, AML are examples of these types of language. These are specialized language category. Here code is represented visually, often through blocks or diagrams, with syntax is based on the arrangement and connection of these visual elements. These languages primarily use a "teach by showing" or "drag-and drop" for its programming.
- *Proprietary languages*: Already discussed at length.

- **Programming types:** There are various programming methods are there and these are mainly:
  - *Text language:* They use an English-like language to establish the logical sequence of a work cycle. The programmer writes code in a programming language to control the robot's actions i.e., human readable commands to instruct robots are used. They use OLP method for programming.
  - *Teach by showing:* Teach-by-showing, are of *two* types: *Powered lead through* (discrete point programming), which uses a teach pendant to instruct a robot to move in the working space. (*a small handled control box equipped with switches, dials, and buttons used to control the robot's movements)* in the space meant for point-to-point motions not continuous movement and complex geometric paths in space. Another type is *Manual lead through* (continuous path programming) in which the programmer simply moves the robot physically through the required motion cycle with robot controller recording the position and speed.
  - *Miscellaneous programming:* Apart from the two above there are a few other types such as: *Tech pendant* with control box – already described, *Offline programming*, *Visual programming*, *Task level programming* (Task-based programming is the most intuitive approach to distributing the computation of an application over a set of nodes with main focus on the concept of a task) and *Programming through demonstration.*

2. **Programming details:** Two main categories of robot programming are Online and offline detailed below. In online programming involves movement of the robot's arm through a sequence of positions, for recording and saving on robot's systems. Offline programming is the process of writing a program on a separate computer and then uploaded to the robot. As discussed, online teach pendants are typically handheld devices, which may be wired or wireless for teaching by showing. Major issues here are: *Joint coordinates*, *Global coordinates*, *Tool coordinates*, *Workpiece coordinates*, and *Lead through.* On the contrary, OLP does not call for use robot movements to create a program, instead it moves programming to a virtual environment. Despite its advantages, offline programming creates a new challenge: Real robots and working zones may differ from the virtual copy, The parts including fixtures/clamps may differ from the perfect 3D model, due to preassembly. Here, brief discussions on programming details shall be discussed.
   - **Programming issues:** Some common criteria when looking at the programming methods include:
     - *Learning curve*: If programming the robot is too difficult to learn, manufacturers are going to be reluctant to automate. The programming solution needs to be easy enough for current and future employees to grasp and understand.
     - *Learning process*: Loops and conditionals issues let robot learn and to do something repetitive and make decisions based on input. Loops allow robot to carry out the same action repetitively multiple times. "If then" provides conditional issue. If there is obstacle, take Y; otherwise X path. So, without loops and conditionals robot only *follows instructions blindly* without *reaction* to environment changes/*decision*-making.
     - *Complex shapes*: Complex part profiles can be hard to replicate. The programmer needs to have the confidence that they can reproduce the results consistently while maintaining accuracy.
     - *High-mix, low-volume production:* In high-mix, low-volume production runs, programming delays can have serious implications on delivery times and margins. If the production style is one that has a variety of parts in lower quantities, the programming method needs to be adaptable to rapidly change parts and cycle through different customer requests.
     - *Performance*: Achieving consistent results can be critical to meeting customer expectations, especially in industries with very stringent requirements. Repeatability can be an important factor

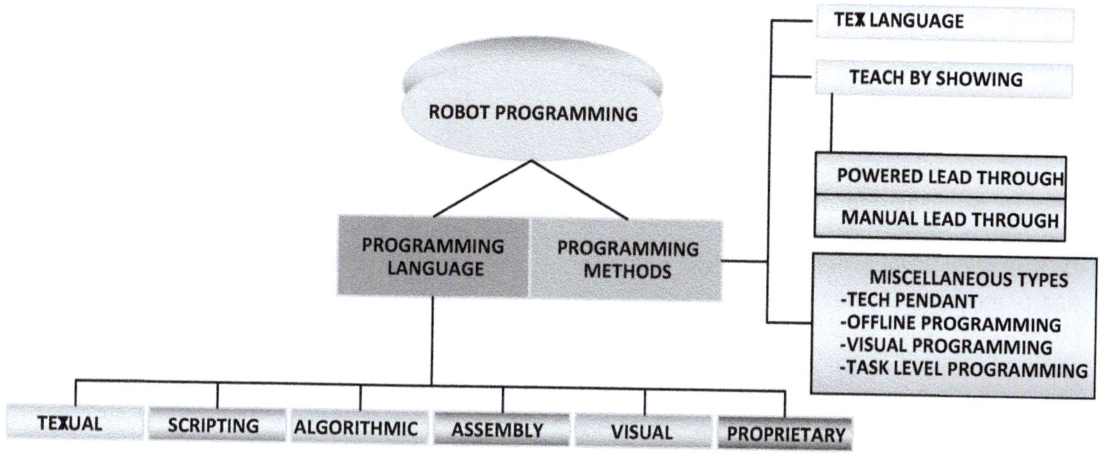


FIG III 5.2c1 PROGRAMMING LANGUAGES & METHODS

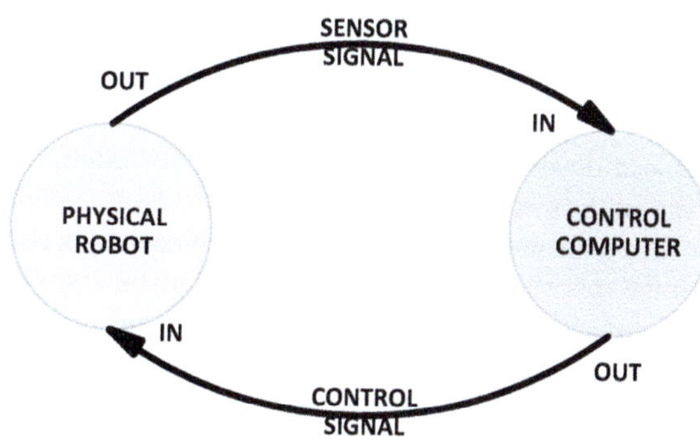


FIG III 5.2c2 ROBOT ADJUSTMENTS IN FEEDBACK

**FIGURE III 5.2C** Basics of robot programming.

to consider when selecting the programming method.

- *Challenge (models)*: Two major challenges are Perception versus Reality, and Control fragility. It is impossible to get to know the true state of the environment. Control software can guess the state of the real world based on sensor data. It can only attempt to adjust the state of the real world through the generation of control signals. So, at the beginning a representative real world the model is to be prepared and presented so that sensor signal could be interpreted to make decision. Fig III 5.2c2 has been developed to explain the concept. As long as the real world behaves according to the assumptions of the model, we can make good guesses and exert control. If the real world deviates from these assumptions, however, we will no longer be able to make good guesses, and control may be lost. It is quite possible that once control is lost, it may never be regained, unless some outside force restores it. This is one of the key reasons that robotics programming is at times very difficult. From this it transpires that good performance will be available as long as the environmental conditions remain within the narrow confines of its internal model. *One key to the advancement of robotics is the development of more complex, flexible, and robust models.*

- **Programming steps:** Starting with task definitions, there are several steps in robot programming. Major steps are:
  - *Goal/task:* The goal of software programming is to set a predetermined goal point i.e., defining the robot's task including interaction to the environment, which refers to robot's workspace understanding through sensors (as applicable) for analysis and interaction of robot with environment.
  - *Language selection:* As already discussed, there are different kinds of programming languages and selection of the same is extremely important. These languages could be textual, visual or robot

specific. Common choices are: Python (easy to learn and versatile) C++ for good speed, control and power. ROS are choices on advanced robotics.

- *Method selection:* Again, based on language and purpose there are variation of programming method. So, language and methods should be selected together.
- *Decision and actions:* In this step close coordination is important. In this step the robot's movements and paths reach, etc., needs to be finalized. Also, it is necessary to fix load lifting capability for motor powering up and movement. Also, it is necessary to integrate sensors so that robot can perceive the environment to react. All these are well connected to each other so, it is essential to confirm that robot hardware and software speaks in same language and frequency so that implementation of algorithms for decision making based on sensor data and programmed logic.
- *Program implementation*: Code development for programs and to control the robot's actions and interaction with environment to ensure the program communications with the robot's hardware. This also decides interface of robotic control system.
- *Simulation testing*: Simulation software can be used to test the program in a safe, virtual environment before actual implementation in robot. A good robot simulator is much more accessible and is a great place to start having wide choices of robot programming languages such as C++, Python based on speed development easiness, etc. Some of the common functional tests include: Physics of robot's movements, collisions with obstacles, testing of new sensors. So, simulation is a software testing means. Upon successful simulation in virtual environment, program are again tested on the physical robot and identify and fix errors.
- Based on the above actual program is deployed and then it can be fine-tuned as necessary under different environments.

3. **Programming language:** Java, MATLAB, LISP, Pascal, C++, Python, are some common textual programming languages used. Let us look into these from robot programming.
   - **Java:** IBM's Java is popular programming language for robots. As an interpretive programming language, Java commands are interpreted by the *Java virtual machine* allowing the same code to run on different computers. It is more popular to developers working with AI- robots.
   - **MATLAB:** MATLAB is very useful for development of robotic control systems with suitable interface as well as data analysis and simulations, etc. In conjunction with robotics toolbox for MATLAB, this can create systems for robotic arm research and simulations.
   - **Lisp:** Easy to use, LISP is one of the oldest robotic programming languages. It is seen more in AI and robotic applications, especially for open-source framework used to develop robotic applications. Tree, or nonlinear, data structures are some of the features, which made Lisp easy to use.
   - **Pascal:** Majority of proprietary robotic software are developed around Pascal language viz. ABB and Kuka, use Pascal as a base for their own in-house programming languages. However, direct use of this language in robot programming is very limited.
   - **C++:** C++ is now a general-purpose object oriented, cross-platform language. Naturally, it is used for writing code in a huge variety of different contexts. It supports scalability, procedural and generic programming, for supporting wide usages. The C++ compiler accepts multiple source files with each source file (treated as a separate compilation to be compiled. There are advantages to organizing an application with one procedure per file. C++ uses compiler to convert source code in to machine code i.e., into instructions that a computer can follow, which makes it a very fast and efficient language to run.

     Issues such as: portable and multi-paradigm, object-oriented programming language, low level manipulation capability, memory management capability, scalability for use in small to large projects, huge user base hence lot of free support, etc., are a few *advantages offered* by C++ as robot programming language. Inability to automatically filter out unnecessary data, no support for built-in code threads, difficult to understand for complex syntax, etc., are *a few disadvantages* associate with C++ language.
   - **Python:** Python is a cross-platform, object-oriented programming language. This high level, interactive language also has dynamic semantics. Features such as dynamic typing (defers the determination of the type i.e., variable type is determined during program execution not explicitly during interpretation) and binding (defers the determination of the actual method for invoking object i.e., the interpreter determines the appropriate method to invoke based on the object's type or class hierarchy at runtime) make Python so a very popular choice. It is well suited for scripting and automation language and for complex application development. This is very easy to learn and maintenance and debugging,

etc., are quite easy too. This portable language offer massive standard library. Python is not only good language but also can work quickly and effectively to improve productivity. It uses interpreter, which helps to use features of dynamic type and bind. Because of line byline execution, Python is quite slow to run and mostly run as server-side programming language with lot of memory usage. Python database is under developed and code needs to be thoroughly tested before use.

4. **Teach pendant:** During the teach phase, the user would guide the robot either by hand or operators can use the pendant to ""teach" the robot specific motions and sequences of movements. Teach pendants are *hand held* boxes that allow control of each manipulator joint or of each Cartesian degree of freedom. Some such controllers allow testing and branching, so that simple programs involving logic can be entered. They are devices equipped with switches and dials used to control a robot's movements within a determined space. Some teach pendants have alphanumeric displays and are approaching hand-held terminals in complexity. Out of various programming methods discussed, only *lead-through programming* uses teach pendants to program the robot with position and functional data. During lead-through programming, the operator works alongside the machine and uses the teach pendant to deliver data and process parameters. Activities of Tech pendant may from application to application as well as even the same line of work will vary in the steps involved. Articulated robots common in manufacturing would not perform wide range of jobs without tech pendant. With suitable software applications, it is possible to use mobile devices such as Personal digital assistants (PDAs) to act as teach pendants. Let us now look into:
   - **Features**: Various features are:
     - *Wired/wireless versions available*
     - *Emergency stop button for safety*
     - *Buttons and switches for controls, Display screen with touch screen feature for commands and displays*
     - *Command recall and editing function*
     - *Many can store a history of programming for easy reference.*
     - Operators communication with the robot to program on: *range of motion, cycle times, speed of operation, interaction with others,* etc.
   - **Applications:** Teach pendants are used for motion control of automated/robotic machinery in different applications including in medical area. Teach pendants are used in industrial applications to perform demanding, hazardous, or repetitive tasks. Major application areas are: *Material handling and assembly*, *Surface processing*, *Dispensing*, *Cutting*, *Welding Loading*, *Unloading*, tasks, etc.
5. **Offline programming (OLP):** OLP has evolved very fast in the recent years, and it has become more user-friendly, powerful, and intuitive. Robot OLP generates robot programs in computer software (virtual environment) based on 3D CAD data. Once the robot program is generated, optimized, and validated, the code is generated and uploaded to the robot controller/physical robot. OLP is helpful in ensuring the right robot selection. Some of the features of OLP are:
   - **Advantages:** OLP is common in almost all robotic applications for some of the benefits as listed:
     - *Quicker robot integration*: Creation, adjustment and program testing in a virtual environment. Then install in the robot caving time compared with Tech pendant.
     - *No downtime*: On line program uploading causes no delay/down time.
     - *Application universality*: OLP software are "robot agnostic" meaning it can generate code for a variety of robot brands and ability to communicate across many platforms.
     - *Safety*: OLP is done in a computer in a safe location then uploaded to the robot without the need to access. Hence reduced risk of accidents and injuries.
     - *Performance:* Robot programs are better optimized, resulting better productivity and performance.
     - *Quick change over*: OLP can program a Part of robot for, while it is completing the other part enabling quick changeover without time loss.
     - *Easy to use*: OLP requires process knowledge, but does not require advanced programming or robotics expertise. OLP is built around the user as it is designed to be a clean, easy to use and elegant experience.
     - *New configuration*: Intuitive user interface allows the user to design and test various configurations.
     - *No final changes*: Last-minute fixture/tooling modifications are avoided.
   - **Selection process:** Common issues to be looked into for software selection are:
     - *Right OLP*: The best way to select OLP is to engage programmers, shop floor owners, integrators, and other manufacturers to justify the selection reason.
     - *Cost issue*: To assess software cost one has to take into account proper understanding of the expected ROI, the maintenance costs, and other added costs.
     - *Provider and integrator*: Software provider with deep process knowledge in solving robotic

programming challenges is always a better choice. Reputed software integrator is always preferred.
- *OEM software*: OEM software at times are better choice as all applications do not fit the same standard size.
- *Test*: It is advised to ask for a test run of the software for checking easy to use and understanding of the software as well as it is fit for the application.

- **Application area:** OLP can be applied to almost all robotic applications. Some of the best use cases for OLP are: *Assembly*, *Material handling*, *Welding* (orientation access), *Cutting*, *Painting* (Orientation unified paint thickness and standoff distance), *Dispensing* (Adhesive), *Surface finish* (bead blasting and deburring).

Tables (with clause reference)



List of drawings (with clause reference)



## List of abbreviations used:

**AGI** Artificial general intelligence
**AGV** Automated guided vehicle
**AI** Artificial intelligence
**AMR** Autonomous mobile robots
**ASR** Automatic speech recognition
**BI** Business intelligence
**BPA** Business process automation
**CNC** Computer numerical control
**DI** Decision intelligence
**DL** Deep learning
**EOAT** End-of-arm tooling
**ERP** Enterprise resource planning
**HCI** Human-computer (robot) interactions
**HMI** Human machine interface
**IIoT** Industrial IoT
**IoT** Internet of things
**IPA** Intelligent process automation
**IT** Information technology
**KPI** Key performance index
**LLM** Large language model
**LMM** Large multimode model
**ML** Machine learning
**MLLM** Multimode LLM
**MOC** Management of change
**MPI** Manufacturing process intelligence
**MSS** Machine safety specialists
**NLP** Natural language processing
**OLP** Offline programming
**OT** Operational technology

**PDA** Personal digital assistants
**PI** Process intelligence
**PLC** Programmable logic controller
**ROI** Return on investment
**ROS** Robot operating system
**RPA** Robotic process automation
**RPL** Robot programming language
**RT** Real time
**SCARA** Selective compliance arm for robotic assembly
**TCO** Total cost of ownership

## References

[1] J. Knisley, What Is Process Intelligence? ABBYY, 2024, in: https://www.abbyy.com/blog/what-is-process-intelligence/.

[2] L. Numminen, Process Intelligence Vs Process Mining, Workfellow, 2023, in: https://www.processmaker.com/blog/process-intelligence-vs-process-mining/.

[3] Manufacturing process intelligence; delmia; Dassault Systems; catalog; internet document; https://www.inceptra.com/wp-content/uploads/2019/03/DELMIA-APRISO-Manufacturing-Process-Intelligence-DS.pdf.

[4] 6 Major types of industrial robots used in the global manufacturing 2018; technavio; Aug'2018; internet blog; https://blog.technavio.com/blog/major-types-of-industrial-robots#:~:text=areas of application.-,Articulated Robots,base with a twisting joint.

[5] 4 types of robots every manufacturer should know; manufacturing innovation blog; manufacturing extension partnership (https://www.nist.gov/mep); NIST US; https://www.nist.gov/blogs/manufacturing-innovation-blog/4-types-robots-every-manufacturer-should-know.

[6] A. Obaigbena, O. Augustine Lottu, E.d. Ugwuanyi, B.S. Jack, E. O. Sodiya and O.D. Daraojimba; AI and human-robot interaction: a review of recent advances and challenges; GSC Advanced Research and Reviews; internet document; https://gsconlinepress.com/journals/gscarr/sites/default/files/GSCARR-2024-0070.pdf.

[7] Types of cobots: a quick guide; qviro, Internet document, https://qviro.com/blog/Cobot-types/, 2023.

[8] C. Emeric, D. Geoffroy and D. Paul-Eric; Development of a new robotic programming support system for operators; 30th International Conference on Flexible Automation and Intelligent Manufacturing (FAIM2020) June' 2020; Science Direct; Elsevier; https://www.sciencedirect.com/science/article/pii/S2351978920318679.

Chapter 6

# Industrial data communication

Data refer to distinct pieces of information usually formatted in a special way which is agreed and accepted upon by senders and receivers. Data can exist in a variety of forms—as numbers or text, image, audio or video. In computer network, these are exchanged as bits and bytes through transmitting media which could be wired/wireless. In this context, it refers to exchange of information between two of more computerized devices/machines/systems. Communication has increased significantly in importance in recent years, especially with cloud computing, Iot/IIoT.

In data communication, encoded information is sent from one point to another by means of electronic transmission system with the help of transmission medium. In computer network, data communication is in digital form yet actual information (say voice) may be in analog form. Normally physical quantity is converted in to electrical signals varying with time with possible intermediate processing. Other than electrical signal via transmission media there can be sequence of pulses of light in case of optical fiber transmission or an electromagnetic wave in case of wireless transmission. Protocols follow certain rule to perform the sets of function. With advancement in use of internet, people found it convenient to integrate various networks, that is, integrating enterprise network with control or Integrated automation and control system (IACS) network, which would provide not only better over control but better business solutions. Now with further advancement, cloud computing Internet of things (IoT)/ Industrial IoT (IIoT) are now reality. People are now moving toward remote controls and Industrie 4.0 converging Information technology (IT) and Operational technology (OT) for better business efficiency in the world of competition. Therefore, importance of data communication in this era of digital transformations cannot be overestimated. In modern data communication, there is hardly any fundamental difference among (except security checks) in data communication in single home computer, multiprocessor computerized Local/Metropolitan/Wide area network (LAN/MAN/WAN). Both the technology and the technical standards organizations are driving toward integrated public systems that make virtually all data and information sources around the world easily and uniformly accessible [1]. It is worth noting that there is some overlap of networking and data communication so, here some network features have also been discussed.

## 6.0 Fundamentals of data communication

Data Communication process basically consists of hardware and software. Hardware parts include mainly the sender and receiver computerized devices, transmission medium as applicable and the intermediate devices, including communication devices through which the data transmission takes place. The software part mainly involves set of protocols which specify certain rules for communication. These protocols takes care of "what," "how" and "when" of data communications. Signals can be represented as a function of time/frequency domain. With the help of Fourier analysis, it is possible to show, that any signal consists of components at various frequencies where each component can be represented by a periodic sinusoidal signal. For understanding of data communication few important terms listed below are important.

- **Signal spectrum and energy spectrum:** Signal spectrum describes magnitude and phase characteristics of the signal in frequency domain. Fourier analysis is complex function having magnitude spectrum and phase spectrum. Squaring magnitude spectrum one can get *energy spectrum* which shows how energy in a signal is distributed in frequency domain. Energy spectrum is important for data communication.
- **Signal Bandwidth:** Bandwidth can be defined as part of the electromagnetic spectrum occupied by the signal, that is, the frequency range over which a signal is transmitted. When for analog signal bandwidth is measured in terms of frequency, for digital signal it is measured in terms of bit rate, that is, bits per second, bps. It is worth noting that signal bandwidth and channel bandwidth are different.
- **Bandwidth and Throughput:** Bandwidth corresponds to a higher maximum rate of data transfer. In networking the word "bandwidth" is used in a way to express on data transfer rate not on spectral characteristics. Therefore, it is expressed not in Hz but in bits per second.

**Plant Intelligent Automation and Digital Transformation. https://doi.org/10.1016/B978-0-323-89892-8.00004-1**

- **Absolute and Effective bandwidth:** Out of absolute spectrum signal energy is contained in a relative narrow band out of absolute spectrum. This narrow band where energy is contained constitutes *Effective* bandwidth. Higher the data rate higher will be requirement of effective bandwidth and lesser will be distortion.
- **Center frequency:** While defining bandwidth, it is assumed that it is centered around a frequency. This frequency is referenced as Center frequency. Higher the center frequency higher will be the bandwidth.
- **Channel Bandwidth:** Channel carrying signal, has bandwidth. For analog signal channel bandwidth is the maximum range of frequencies that the channel can carry. For digital signal, channel bandwidth represents the maximum bit rate supported by the channel.

**DC component:** In spectral analysis, if there a signal component with zero frequency then that component is DC component which is important in encoding.

## 6.0.0 Data communication: conceptual discussions

Automation may be conceived as "*The equipping of a device so that it completely or partially works as intended without the involvement of humans*" (DIN 19233). With increase in digitalization and integration of various systems, requirement and importance of high-performance industrial communication systems cannot be overestimated especially in cases of IIoT, cloud/edge computation, Industrie 4.0 and Digital transformation (DT). The fundamental purpose of any communication is the exchange of data between two or more parties. In this section short *recapitulations* of basics of data communication in computer network will be covered for better understanding of subsequent discussions. In this connection detailed discussion will be presented on industrial data communication from very basic concepts in this subsection. Let the discussions start with characteristics and components.

1. **Characteristics and Components**: Each data communication has fundamental characteristics and components discussed below:
   - **Characteristics of data communications:** Four fundamental characteristics of data communications are:
     - *Delivery*: The data must be delivered to the correct destination and user only.
     - *Accuracy*: The communication system must deliver the data accurately to the intended user, without any errors. During transmission, when the data get corrupted, it will directly affect the accuracy of the delivered data and uncorrected data are unstable.
     - *Timeliness*: The system must deliver data in a timely manner otherwise it could be useless. This is more so in real time data transmission. For audio and video data, timely delivery means delivering data as they are produced in the same order, without significant delay. This kind of delivery is called real-time transmission. In case of a video stream some delay in orderly manner even can be accepted but that in IACS data any delay could be a catastrophe.
     - *Jitter*: Jitter refers to the variation in the packet arrival time. Uneven Jitter may affect the timeliness of data being transmitted. If some of the packets arrive with 20 ms delay and others with 30 ms delay, then this would result uneven quality in data.
   - **Components of data communication:** In its simplest form, following components are essential for data communication.
     - *Source*: The source is the device that generates the data to be transmitted.
     - *Transmitter*: Data generated by a source cannot be, transmitted directly in the form data have been generated. So, transmitter is used. Transmitter transforms and encodes the information in suitable manner (based on destination requirement) so that converted electromagnetic signals can be transmitted across some sort of transmission system.
     - *Transmission system*: The transmission medium is the physical path by which data travels from transmitter to receiver. This physical path can be a channel or a complex network. Also, transmission channel could be copper wires, optical fibers and wireless electromagnetic waves.
     - *Receiver*: The receiver receives the signal from the transmission system and converts it into a form that is suitable to the destination device for handling data.
     - *Destination*: Destination device, for which source sent data, utilizes the incoming data from the receiver.

   Let us now look little deep in to the issue. It is important to ensure that transmission channel is most effectively utilized as well as suitable congestion control techniques are utilized to avoid channel jamming/congestion. In computer network communication, protocol the main Software (SW) plays great role for data transmission.
     - *Protocol concept*: Protocol, which governs data communication, is the agreed upon set or rules to be used by the source and destination for fruitful communication. Protocol is necessary not only for interfacing as well as for signal generation to make data communication fruitful over the channel. Addressing and routing with the help of header

source/destination address etc. is responsibility of protocol to ensure only destined receiver(s) get the message. Also, a specific route through this network must be chosen. Timing issue, that is, some sort of synchronization is important and well taken care of by protocol through encoding. Apart from these, there are a few issues like flow control and error handling are also important task of Protocol. It is important to ensure that source does not sends data at a rate faster than that can be processed and absorbed at destination. This is flow control job in protocol. In any communication there is potential for error (e.g., channel distortion) before transmitted signal reaches destination. So, error detection and corrections are important issue in communication protocol. Recovery techniques are needed in situations in which an information exchange, such as a database transaction or file transfer, is interrupted due to a fault somewhere in the system [1]. From the above it transpires that there are three major elements are syntax, semantics and timing issues.

2. **Common communication issues:** There exists some differences two terms which normally are used interchangeably viz. data and signal/communication and transmission.
   - **Data and Signal:** Data and signals are not same thing. Data are entities that convey information, whereas signals are electromagnetic representation of information or data. Data and signals both are two types analog/digital. *Analog data* are continuous in nature (normally varying with time), for example, audio, video, output from process transmitter. *Digital data* are discrete in nature, for example, text, integer. Analog signal is a continuous (normally varying in nature) electromagnetic wave/electrical signal that may be propagated through a variety of communication channels. A digital signal is a sequence of voltage pulses capable of transmission through channels. Digital signal has inherent advantage. Theoretically any data can be represented by $2^n$ digital signal where n can be infinity hence very high frequency can be achieved (normally value of n is limited to 3, 4, 5 or 6). Various combinations have been enlisted in Table III 6.0a.
   - **Communication/Transmission:** Transmissions could also be both analog as well as digital. At the first place now people are going mostly for digital transmission. Also, practically transmission of digital signal through analog transmission does not take place. So, discussions will be limited to digital communication/transmission only. Following are a few major issues for which digital communication is preferred;

**TABLE III 6.0A** Combinations of different data and signals.

| Data | Analog signal | Digital signal |
|---|---|---|
| Analog | Same spectrum as the data or different spectrum by carrier. Modulations (AM/FM etc.) | Encoded to produce digital bits, for example, PCM, delta modulation |
| Digital | Data are encoded to produce analog signal viz. ASK,PSK, FSK | Different voltage level or data encoded (line coding) |

   - *Technological*: With advent of Very large-scale integration (VLSI) especially with introduction of embedded systems/microcontroller there has been severe reduction in size and cost of device coupled with more functionality. Analog devices could not cope up with digital.
   - *System integration*: It is not only easier but also economical to integrate various networks together with digital signal and transmissions. This is not an easy task in analog system, so now analog data are converted to digital bit stream.
   - *Integrity*: Data integrity in digital transmissions are well maintained through repeaters.
   - *Security and privacy*: Security and privacy are easily available through (say) encryption techniques). This is hardly possible in analog counterpart.
   - *Utilization*: Utilization of the system is more and economical.

   From the above discussions it is clear that there could be both analog and digital type data as well as signals. However, data communication is digital. With reference to Table III 6.0a we need to address four possible combinations. Let us now start with an important issue of analog data (say audio/video) to digital signal. Naturally we need to convert analog data to digital signal via Analog to digital converter (ADC). Similarly, in some cases like in controls, we may have to convert digital data into analog signal via Digital to analog converter (DAC). The simplest technique is Pulse code modulation (PCM), which involves sampling the analog data periodically and quantizing the samples. Pulse amplitude modulation (PAM), Pulse width modulation (PWM)_ and PCM techniques are used to convert analog to digital signals also.
   - **Sampling and Quantization:** Fig. III 6.0a details out various issues related to sampling and quantization. Sampling including sampling theory and

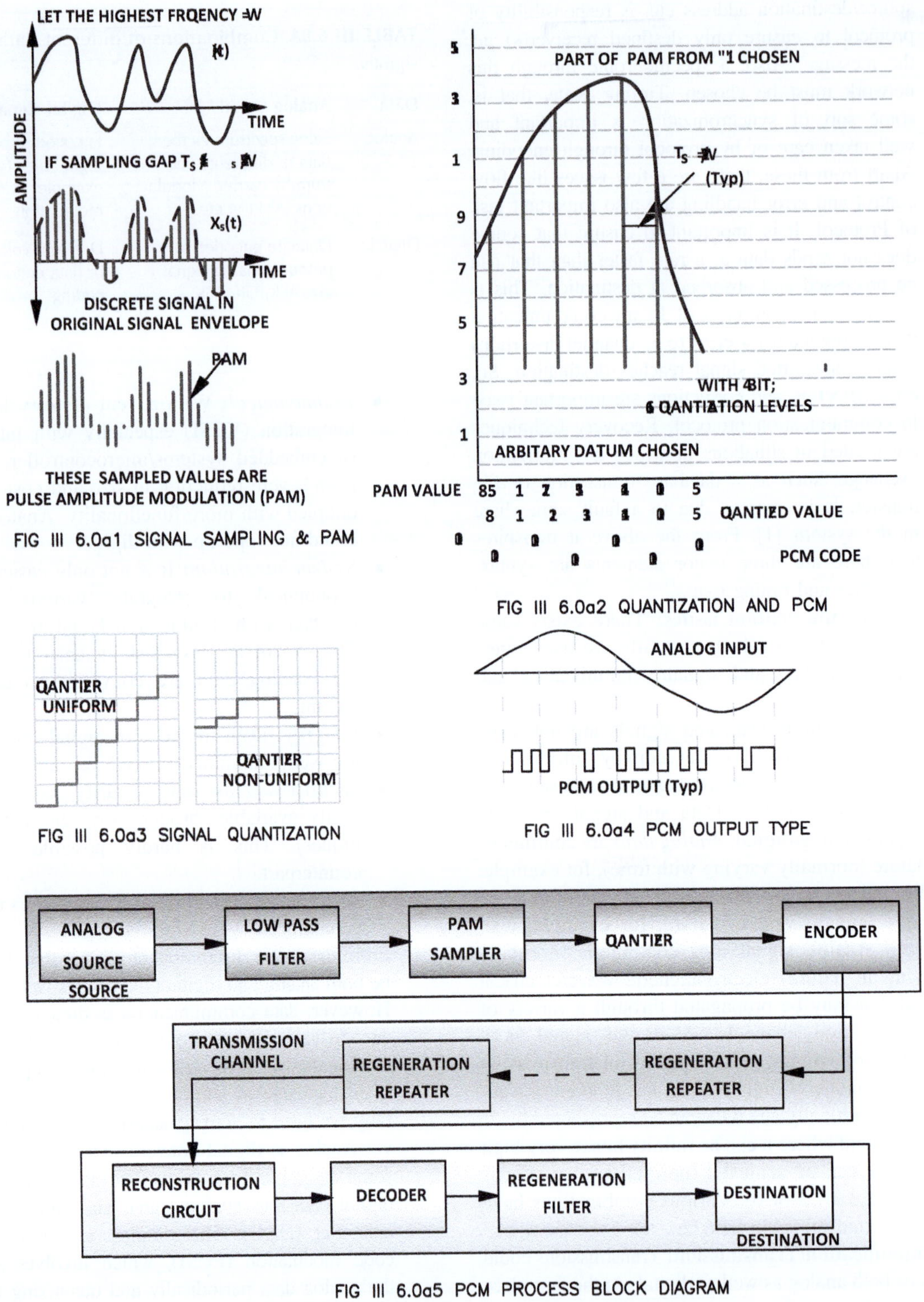


**FIGURE III 6.0A** Sampling and quantization: PAM PCM.

sample value quantization are very important issue in digital communication as detailed below:

- **Sampling:** Sampling represents the process of measuring the instantaneous values of continuous-time signal in a discrete form. Analog signal from source has to be digitized in 1 or 0 form. So sampling converts analog signals into discrete signals.
- **Sampling Rate:** To generate discrete signals analog signal is sampled at a fixed gap called sampling period. If sampling gap in time scale is $T_S$ then

sampling frequency $f_s$, often called sampling rate is given by

$$fs = 1/T_s \qquad \text{(Eq. III 6.0a)}$$

Nyquist sampling theorem tells minimum sampling rate at which a signal can be sampled to get back the original signal envelope.

- **Nyquist Sampling Theorem:** "*If a signal X(t) is sampled at regular intervals of time and at a rate higher than twice the highest signal frequency, then the samples contain all the information of the original signal.*" Let a continuous signal has band limited with no frequency components higher than W Hz. Then, if we sample this signal at sampling rate

$$f_S = 2W \qquad \text{(Eq.III 6.0b)}$$

Then, original signal can be reproduced. With reference to Fig. III 6.0a1 let X(t) is one analog signal, it is sampled at a rate $f_s$ by multiplying it with a impulse train to obtain sample signal $X_s(t)$. We see that original enveloped is produced as long as $f_S = 2W$ if the function X(t) has highest frequency W Hz.

- **Quantization:** Quantization is a process for representing the sampled values of the amplitude by a finite set of levels, which means converting a continuous-amplitude sample into a discrete-time signal with amplitude of a finite set of levels. The quality of a quantization depends upon the number of quantization levels. The discrete amplitudes of the quantized output are representation levels and the spacing between the two adjacent representation levels is step-size. On account of quantization there could be loss of information due to signal distortion. The difference between an input value and its quantized value is called a *Quantization error.* There are two quantization levels viz. uniform quantization (uniformly spaced quantization levels) and nonuniform quantization (unequal quantization levels—may be logarithmic relation). If there are 16 quantization levels, then for uniform quantization there will be equal 16 signal levels. On the contrary for nonuniform quantization spacing near base line (DC line say), spacing may be closure whereas as they go up beyond base lines gap/level increase. So, uniform quantization gives equal absolute error for all signals meaning for low level signals will produce more error than high level signals. Fig. III 6.0a3 may be referenced for arbitrary signal quantization for both types.

3. **Pulse code modulation:** Pulse code modulation is also a method for ADC taking the help of sampling and quantization for transmission through the digital communication network. PCM is in binary form, so there will be only two possible states high and low(0 and 1).In PCM non uniform quantization (i.e., are not equally placed quantization) levels are used. Companding (compression and expansion both) is used in PCM. Now when we use 16 ($=2^4$) different levels, then there will be 4 bit, for quantization of PAM signal in PCM encoding. This is shown in Fig. III 6.0a2. The figure also shows quantized values are converted to binary in PCM. More the number of levels in quantization, there will be more bits in PCM and better will be the recovery of signal and noise will be less. Typical analog signal and encoded PCM wave form has been depicted in Fig. III 6.0a4 whereas PCM process has been illustrated in Fig. III 6.0a5 which is self-explanatory.
4. **Pulse shaping and ISI:** There is a form of distortion of a signal, in which one or more symbols interfere with subsequent signals and delivered output is unreliable a poor in quality. This is Inter-symbol interference (ISI). This is more so for RF communication where analog carrier signal is used. This calls for pulse shaping which has been illustrated in Fig. III 6.0b. ISI is important contributor for requirement of pulse shaping. Unless distorted, we know from Fourier analysis that any time limited pulse like rectangular pulse has infinite bandwidth. So, rectangular pulses/symbols are not suitable for transmitting data at the highest possible bit-rates through channels with band limits. As a result, there could be ISI. ISI, which may be caused due to multipath Propagation, Nonlinear frequency in channels, is unwanted and must be eliminated to get quality output. When signals pass through various media over long distances characteristic of the medium and physical environment causes some symbols to be spread beyond given time interval. Thus, there can be interfere with the following or preceding transmitted symbols. For studying ISI in great details Eye Pattern (a graphical representation) resembling human eye is used. When the effect of ISI increases, traces from the upper portion to the lower portion of the eye opening increases and the eye gets completely closed, if ISI is very high. Actual eye patterns are used to estimate the bit error rate and the signal-to-noise ratio [2]. It is common in any communication to send information on a carrier. In digital communication, digital data are sent on a carrier by changing phase or frequency or amplitude of the digital information. To address above issues like ISI, to smoothen out any transition in channel, filters are used. Filters play important part in communication channel in effectively eliminating spectral leakage, reducing channel width, and eliminating ISI [2]. Generating band limited channels,

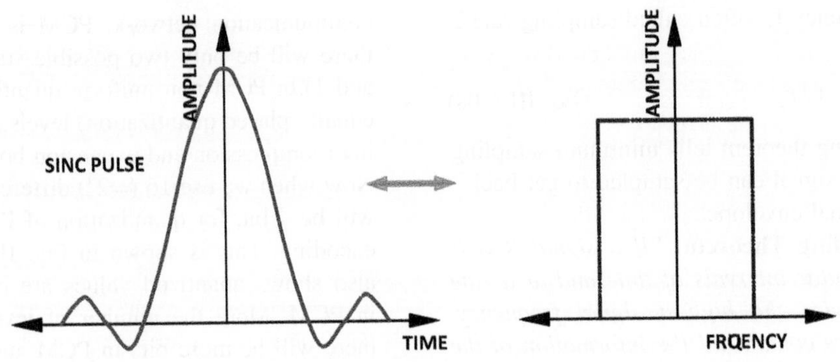


FIG III 6.0b1 SINC. SIGNAL FOR PHASE SHAPING FILTER

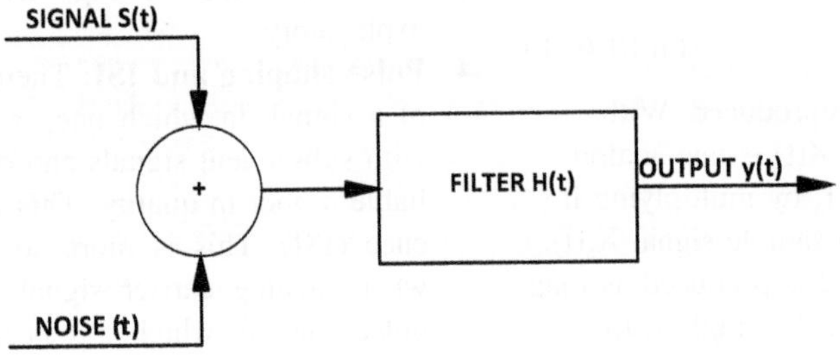


FIG III 6.0b2 CONCEPT OF MATCH FILTER

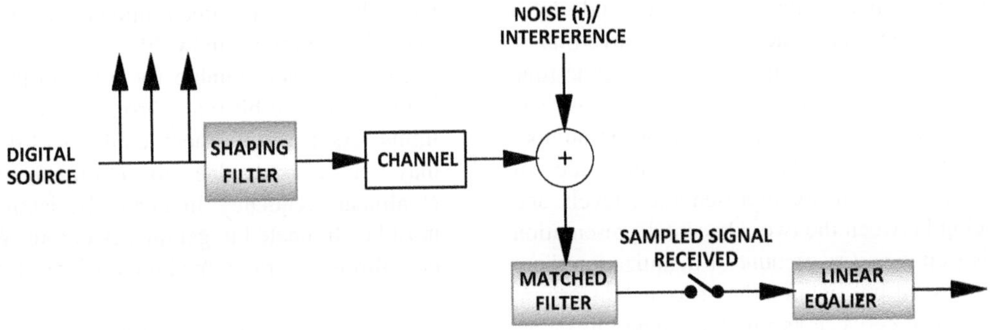


FIG III 6.0b3 DETECTION PROCESS AT RECEIVER (SHOWING FILTERS & EQUALIZER)

**FIGURE III 6.0B** Pulse shaping.

reduction in ISI are two important issues in digital data communication. These requirements are well met by pulse shaping filters. A convenient way of generating symbols with the time-domain shape, an impulse of the appropriate strength for each symbol may be generated [3].

- **Pulse shaping filter concept:** A Sinc pulse along with an FFT spectrum of Sinc pulse has been shown in Fig. III 6.0b. Sinc pulse is periodic, with a maximum amplitude in the middle and it appears as a square wave in the frequency domain (a smaller portion) to effectively limit a communications channel within specific frequency range as in Fig. III 6.0b1. So, it can serve both the purposes stated above. In Sinc is a pulse shaping filter which eliminates any signal beyond given cutoff frequency. In any modulation carrier sinusoid carrier causes transitions in its phase and amplitude. Thus, when pulse shaping filter is applied to the modulated signal, it is limited to a specific frequency band and high frequencies are eliminated. So, major part of the channel power is limited within to specified bandwidth. The required bandwidth for a channel is directly related to the symbol (baud) rate and is centered at the carrier frequency [4]. Therefore, pulse shaping filter would give better solutions. Apart from Sinc filter Raised cosine and Gaussian filters are also deployed for the similar purposes.

- **Match filter concept:** Matched filtering is used to detect a known piece of signal embedded in noise. Typical match filter has been shown in Fig. III 6.0b2. The filter will maximize the signal to noise ratio (SNR) of the signal being detected with respect to the noise. Signal to noise ratio will be maximized, when the impulse response of the filter is exact reverse and time delayed copy of the transmitted signal. Here, the input signal is *s(t)* and the noise is *n(t)*. The objective is to design a filter, *h(t)*, that maximizes the SNR of the output, *y(t)*. It is quite possible for the reflected signal to overlap with a subsequent symbol period. Match filter is very useful to filter out signal reflections in transmission process.
- **Equalizer and equalization:** Equalizer is used to treat the transmission losses which would affect the quality of the signal. Equalizer is helpful in eliminating ISI and in maximizing signal to noise ratio. In fact, it attempts to reverse distortions in the signal during transmission. This process is referenced as equalization. This is located at receiver. There are several types of equalizers with broad classification as linear and non linear type. Fig. III 6.0b3 shows an equalizer in the receiver.

### 6.0.1 Digital modulations and signal transmissions

Digital communication uses digital modulation techniques which may have some similarity, yet are different. Various digital techniques are discussed here as a part of recapitulations of communication basics including encoding and conversion techniques. Digital modulation is the process of encoding a digital information/data onto a carrier signal by varying its characteristics (amplitude, phase, or frequency) of the transmitted signal. The encoding process affects the transmitted signal's robustness and bandwidth. Digital modulation offers a number of benefits like *High capacity for data transmission with excellent bandwidth efficiency*, *Signal to noise ratio (S/N R) improvement* with increase in bit/sample, *Less susceptibility* to crosstalk, waveform distortion, *Enhanced data security* etc. are a few advantages of digital modulations. Here, some characteristics of the carrier analog signal viz. amplitude, frequency, or phase will be changed as per digital data present. Since digital modulation provides *more information capacity*, *higher system availability* coupled with better *communication quality and data security*, digital modulation finds much higher demand when compared with its analog counterpart. Several techniques for digital modulation are as follows: Amplitude shift keying (ASK), Frequency shift Keying (FSK) and Phase shift keying (PSK) of Differential PSK (DPSK). Apart from these there are Delta and adaptive delta modulation techniques. All these will be covered in this section briefly. Digital signals are transmitted by using one of two different approaches viz. baseband transmission or broadband transmission (using modulation). Each of these types uses either Time division multiplexing (TDM) or Frequency division multiplexing (FDM) so, discussions on these as well as discussions on base band have been provided here.

1. **Common digital modulation:** As indicated above, here three common characteristics will be deployed for modulations as shown in Fig. III 6.0c, and listed below:
   - **Amplitude Shift Keying (ASK):** In this type of modulation, the amplitude of carrier wave (which is a periodic sine wave) is varied as per digital data 0 or 1 as shown in Fig. III 6.0c1 (ASK wave form). ASK modulated, gives a zero value for Low input (0), while it gives the carrier output for High input (1). Major components are:
     - *ASK Modulator*: Typical block diagram of ASK modulator has been shown below the modulated waveform. The carrier signal is a high-frequency sinusoidal which is fed to a switching unit. The binary sequence from the message signal makes the unipolar switch either to close or open based on value of binary 1 or 0 in digital data. The high signal closes the switch to produce output as carrier signal at high input, otherwise output will be low (0) as there is low input causes switch to open. There is a band-limiting filter, to shape the final output pulse.
     - *ASK Demodulator*: Two types of demodulations in ASK viz. Asynchronous and synchronous demodulations has been shown below ASK modulator in Fig. III 6.0c1. *Asynchronous ASK demodulator* consists of *rectifier* (half waves), Low pass filter (*LPF*) and a *comparator*. Positive half output from the rectifier is passed through LPF to eliminate high-frequency components and filter out the envelope from which the comparator delivers a digital output. *Synchronous ASK demodulator* consists of a *square law detector*, *LPF*, a *comparator*, and a *voltage limiter*. Square law detector provides output voltage is proportional to the square of the amplitude-modulated input voltage. LPF eliminates the higher frequencies and as above, the comparator delivers a digital output. Voltage limiter helps to get a clean digital output at desired voltage level.
   - **FSK Modulation:** In FSK, the frequency of the carrier signal varies according to the digital signal changes. Two different frequencies near carrier

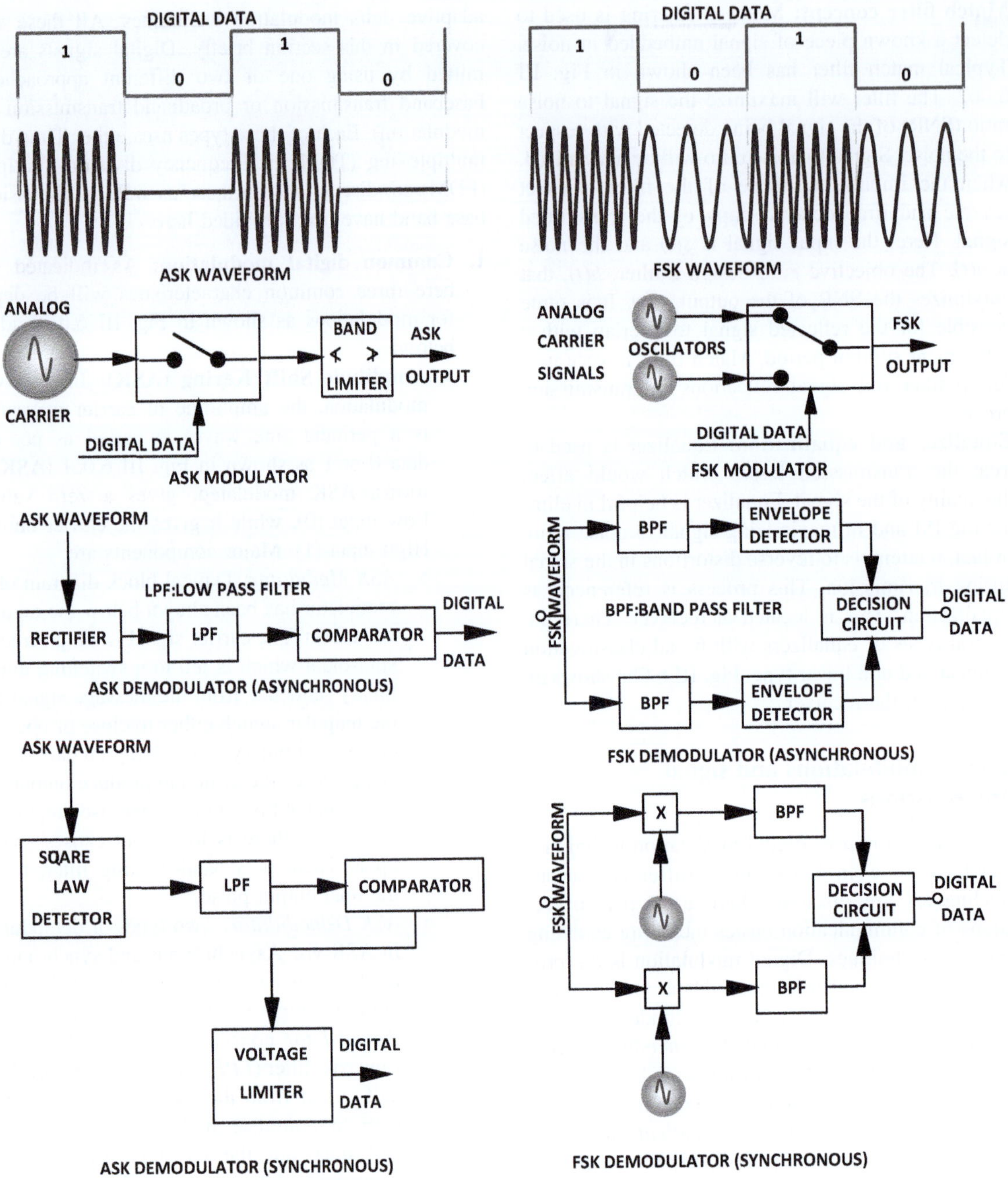


FIG III 6.0c1 AMPLITUDE SHIFT KEYING METHOD

FIG III 6.0c2 FREQUENCY SHIFT KEYING METHOD

**FIGURE III 6.0C** Digital modulation technique 1: ASK FSK.

frequency represent "0," "1," that is, high in frequency for high binary data (1) and low frequency for low binary data (0). This has been depicted in Fig. III 6.0c2 FSK waveform.

- *FSK modulator*: There are two *oscillators* producing high and low frequency carrier wave. Both of them are connected to a switching unit with internal clock which is applied to both the oscillator to avoid abrupt phase discontinuity. Based on digital data high (1) and low (0) switch chooses input frequencies to produce required FSK wave form.
- *FSK demodulator*: Here, also there are two types of demodulation methods viz. *asynchronous* and *synchronous* demodulation. *Asynchronous FSK demodulator* consists of two sets of each of *Band pass filters* (BPF) and *envelope detectors*, and a *decision circuit.* BPFs are tuned two

carrier frequencies. The output from these two BPFs goes to the envelope detector. The decision circuit chooses which output from envelope detectors is more likely and selects it from any one of the envelope detectors. It also re-shapes the waveform to a rectangular one. *Synchronous FSK demodulator:* It consists of two sets of each of mixer, BPF and local oscillator. As shown, FSK signal input is given to the two mixers with local oscillator circuits. The output of mixers passes through two BPFs. These circuits act as detector whose output is fed to the decision maker circuit to select most likely signals as digital data.

Fig. III 6.0d details out digital modulation technique 2 for various techniques below.

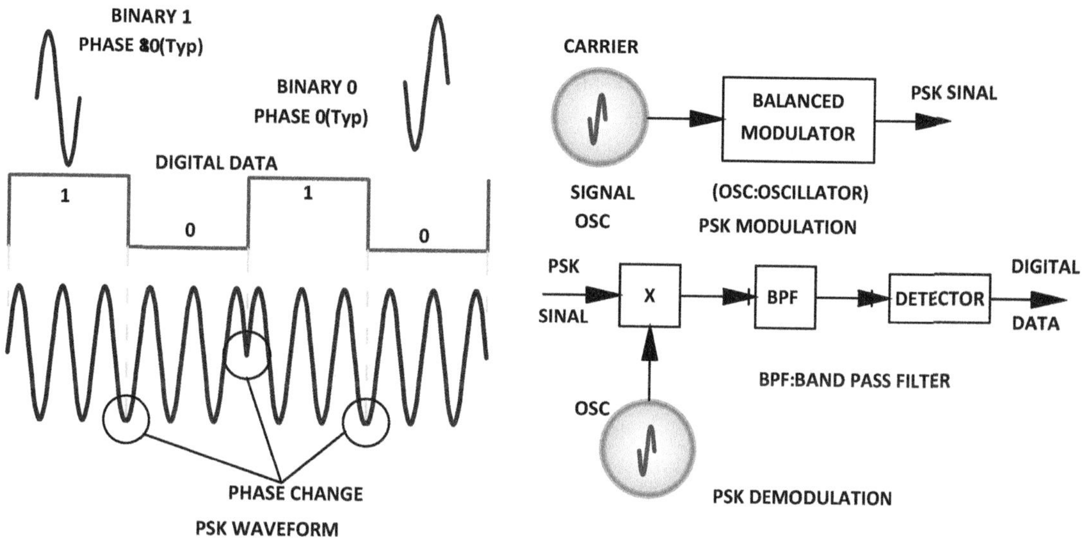


FIG III 6.0d1 PSK MODULATION

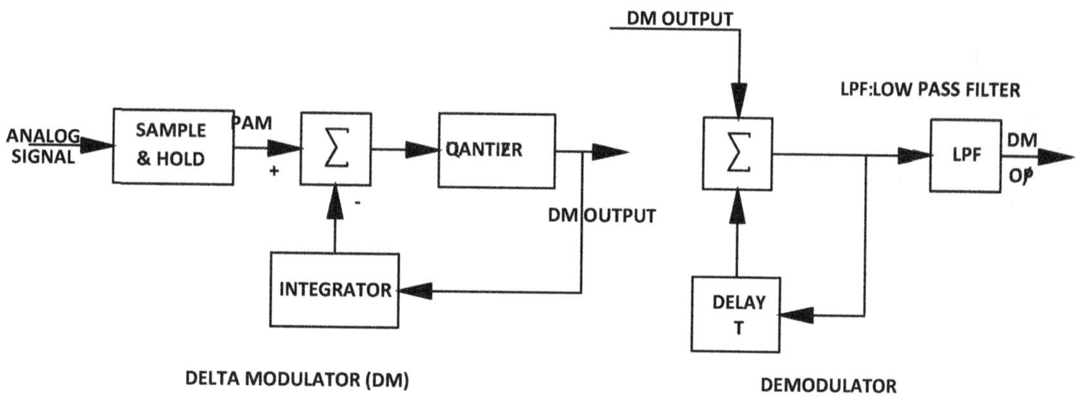


FIG III 6.0d2 DELTA MODULATION (DM)

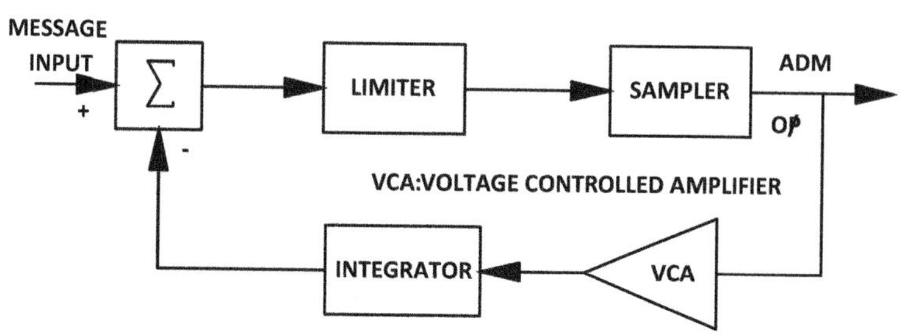


FIG III 6.0d3 ADAPTIVE DELTA MODULATION (ADM)

**FIGURE III 6.0D** Digital modulation technique 2: PSK, DM ADM.

- **Phase shift keying (PSK) and Differential PSK:** In this digital modulation process, data are conveyed by modulating the phase of a constant frequency the carrier sine/cosine wave. The phase of the carrier is discretely varied in relation either to a reference phase or to the phase of the immediately preceding signal element, in accordance with digital data to be transmitted. PSK is of two types viz. *Binary Phase Shift Keying (BPSK)* and *Quadrature Phase Shift Keying (QPSK). BPSK* is a two-phase modulation scheme, where the 0's and 1's in a binary message are represented by two different phase states in the carrier signal: $\vartheta = 0°$ for binary 1 and $\vartheta = 180$ degrees for binary 0 as detailed in Fig. III 6.0d1. In *QPSK* carrier sine wave takes four phase (eight or 16 also) reversals viz. 0°, 90 degrees, 180 degrees, and 270 degrees. Normally in QPSK each signal element can represent 2 bits of data, for example, 00, 01, 10, and 11 can be represented by 45 degrees, 135 degrees, 225 degrees and 315 degrees (with 90 degrees phase separation). This provides higher data rate. There is another kind called Differential PSK (DPSK). Here, once the data-bit is "0," the signal's phase will not be inverted and continued. Once the data-bit is a "1," then the signal's phase will be inverted.
  - *BPSK modulator*: BPSK modulator consists of the *balance modulator* which multiplies two inputs viz. carrier sine wave and the binary data sequence to produce PSK modulated output.
  - *BPSK demodulator*: It consists of a mixer with *local oscillator circuit, BPF* and *two-input detector circuit.* Signal recovery is done by mixer passes through BPF to complete basic demodulation process. Next, a detector with bit clock rate is needed to produce the original binary message signal.

2. **Discussions on Delta modulation:** Delta (and adaptive delta) modulation is also a kind of digital modulation. In this type of modulation, the sampling rate is much higher and the step-size after quantization is very small value delta (δ). So, this type is called delta modulation. Adaptive delta modulation is a refinement of delta modulation, which could solve the granular noise and slope overload error caused in delta modulation. Similar to delta modulation except, that here the step size is variable according to the input signal (unlike fixed in delta modulation). Here both types have been recapitulated.
   - **Delta Modulation**: This is one digital modulation technique has its root in PCM and can be conceived as a differential PCM. It provides a staircase approximation over-sampled base signal. The difference between the present sample and previous approximated sample is quantized into two levels, that is, ±δ. If the previous approximation is below the present sampled analog value, then the error is quantized in positive way, that is, by +δ, and when it is above the present sample value then the error is quantized by -δ. As seen the step size δ is very small but sequence rate is much higher than Nyquist rate. It has very simple modulator and demodulator implementation with user decides the bit rate.
     - *Delta modulator*: As shown in Fig. III 6.0d2, analog signal is first passed through on sample and hold circuit to get PAM. Oversampling is used to get better signal to noise ratio. As shown basic components are *adder, quantizer, integrator* and *encoder.* Within integrator a delay $T_s$ (not shown) is introduced so that the output of the integrator is a staircase approximation delayed by $T_S$. As stated earlier the staircase approximation is compared with the present sampled input signal at the adder to generate difference as the error signal which is passed on to quantizer. Here based on error signal sign, error is quantized into either of *two values* **±δ.** The output is coded to produce desired delta modulator output signal.
     - *Delta demodulator*: Delta demodulator consists of a low pass filter, adder and a delay circuit as shown in Fig. III 6.0d2. A binary sequence is given as an input to the demodulator. The staircase approximated output is given to the LPF to eliminate noise. The adder and delay circuit generates the staircase approximated signal output. Delay of sampling period T is added to the input signal, when binary signal is 1, and subtracted if binary signal is 0 to approximate suitable output.
   - **Adaptive Delta Modulation:** Analog signals are continuous with varying slope. Naturally for proper reproduction of the signal, it is necessary to control the adjustment of step-size, according to the requirement in *analog* signal. In adaptive delta modulation where step size is regulated through one voltage-controlled amplifier. The gain of the voltage-controlled amplifier which determines the step size is adjusted by the output signal from the sampler (in proportional manner) as shown in Fig. III 6.0d3. ADM quantizes the difference between the value of the current sample and the predicted value of the next sample, that is, a variable step height is used to predict the next values, for the faithful reproduction.

Apart from these there are other methods such as *M-ary encoding* methods are deployed for transmission of 2 bits utilizing $2^2 = 4$ conditions viz. voltage level, amplitude, phase and frequency. Here M stands to represent different conditions. There can be *M-ary ASK, M-AryFSK and M-AryPSK.*

3. **Multiplexer and de-multiplexer:** Multiplexing (MUX) is a technological method by which it is possible to transmit several signals is concurrently over a single data link or communication channel. So there has to be a device where this mixed signal is transformed back several signals. This is done at de-multiplexer (DEMUX) in receiver. Multiplexing enables the transport layer to support multiple simultaneous connections between hosts; each identified by a unique combination of source and destination IP addresses as detailed in Fig II 8.0a3. This shows importance of multiplexing in transport layer of ISO OSI model to be discussed later in connection with networking. So, multiplexer takes many inputs to send signal single output and de multiplexer takes single input to produce multiple outputs. These are two popular ways to achieve this. One dividing the channel time into divisions so that each input has a time slot giving rise to Time division multiplexing (TDM). In Frequency division multiplexing (FDM) a number of signals are transmitted at the same time but each source transfers its signals in the allotted frequency range. To avoid overlapping there should be sufficient frequency gap between two adjacent signals to avoid overlapping. From the discussions in II 8.0.1 we understand there is another kind, that is, Wavelength-division multiplexing (WDM) used in optical communication systems. In WDM, to allocate each input signal to a wavelength band, we split the possible bandwidth of an optical fiber into a number of wavelength bands.
   - **Multiplexer:** Let us look into pros and cons of multiplexer:
     - *Mux* advantages: Allows multiple data streams to be transmitted over a single channel, more data in given bandwidth (TDM) reducing cost of transmission, time (FDM).
     - Mux disadvantages: Complexity, less flexibility, higher latency are major disadvantages in addition to loss of all signal in case of failure of mux.
   - **Demultiplexer:** Let us look into pros and cons of multiplexer:
     - *Demux advantages*: Separation and sending to respective destinations, Enhanced security, easy scaling, probability of error during transmission.
     - Demux disadvantages: Implementation complexity, Costly affair, Limited overall scalability, Different data type may call for different security,
4. **Mux Types:** As discussed in above, and in II 8.0.1. two common mux types are:
   - **TDM: Time division multiplexing (TDM):** TDM is possible when data transmission rate of media is *greater* than that of the source. Total available channel time is divided among willing users, so that each of them is allotted with definite defined time slot which may be very small. "n" no. of TDM the signals may have same frequency, but they are transmitted one-by-one in time scale. Two types of TDMs are:
     - *Synchronous TDM (STDM):* In STDM each user has preassigned fixed and same time slot of possession given turn by turn. Thus, if an one input is not *ready or does not have* data to send then its time slot remains empty. In STDM, total speed of various inputs should *never* exceed the capacity of path. Time slots are organized into frames each of which consists of one or more time slots dedicated to each sending device.
     - *Asynchronous TDM:* This is also called statistical TDM *(STDM)* where the time slots are flexible and are allocated dynamically depending on the speed of source or their ready state to save channel capacity. In asynchronous TDM, if there are "*n*" input lines then the frame contains not more than *m* slots (statistical) where $\underline{m} < \underline{n}$. Here the total speed of input lines can be greater than the capacity of the path.
   - **FDM:** In this analog multiplexing technique a number of input signals, each of which is allotted *different* frequency range are transmitted at the same time. So, FDM requires that the bandwidth of a link should be greater than the combined bandwidths of the various signals to be transmitted. There would suitable frequency gap between adjacent signals to avoid over-lapping. Signals are transmitted in allotted time to decrease the probability of collision. Guard bands are used between two successive channels to avoid interference.
   - **Pros and cons of TDM and FDM:** Pros and cons of both the systems has been enlisted in Table III 6.0b.
5. **Baseband and broadband transmission:** Baseband refers to the unmodulated signal, while broadband refers to a transmission system using multiple frequency bands to transmit data, enabling higher speeds and simultaneous transmissions. Baseband signals often sent over copper/fiber-optic cables referred to as baseband wires. We modulate or convert the baseband signal to a different frequency. As a broadband signal has a wider bandwidth, it can carry more information and often used to analog voice and video signals with data.
   - **Baseband transmission (line coding):** As discussed, baseband technology transmits a single digital data signal/stream/channel at a time. Baseband transmission requires a low-pass channel. Baseband technique uses digital signals. For long distance

**TABLE III 6.0B** Pros and cons of TDM and FDM.

| Multiplexer | Advantages | Disadvantages |
|---|---|---|
| TDM | **1.** Full utilization of available channel width<br>**2.** No inter modulation distortion and less cross talk<br>**3.** Simpler TDM circuitry | **1.** Synchronization is essential for operation.<br>**2.** Slow band fading it can wipeout |
| FDM | **1.** Simultaneous transmission of a large number of signals.<br>**2.** Synchronization is not necessary between sender and receiver<br>**3.** Easier demodulation<br>**4.** Due to slow band fading only single channel is affected | **1.** Large bandwidth for channel is essential<br>**2.** Large nos. of modulator and filters required<br>**3.** Inter modulation distortion and high cross talk<br>**4.** Affected by wideband fading |

transmission, repeaters are used for regeneration of digital signal on account of attenuation of signal over distance. As baseband transmission uses whole frequency spectrum of the medium for the transmission, FDM cannot be used. Baseband supports *bidirectional communication*, meaning send and receive can be done simultaneously, (but) using two separate electric circuits together. By using TDM, baseband can transmit signals of multiple nodes simultaneously, that is, Utilization of whole bandwidth for a given time slot. Baseband technology is mainly used in Ethernet networks to exchange data between nodes. Baseband is used over three cable media types viz. co axial, twisted-pair, fiber-optic.

- **Broadband Transmission (Using signal modulation):** Broadband technology uses analog signals (electrical signal or optical) as the carrier wave to modulate digital signals and transmits the carrier wave across the channel/medium. Band pass channel is used in broadband transmission. In broadband technology, multiple data/signals can be transmitted simultaneously at the same time using FDM. Broadband is used for long distance transmission. Since analog carrier wave is used so, these can be regenerated using amplifiers, for long distance transmissions. Broadband supports only unidirectional communication meaning nodes connected at both ends of a medium can send or receive data but cannot perform both actions at the same time simultaneously. Broadband finds its applications transmissions of audio, video, and data simultaneously. Radio waves, coaxial, fiber-optic cables are used for this.
- **Difference between Baseband and Broadband digital transmission:** Difference between two types has been enlisted in Table III 6.0c.

## 6.0.2 Error and flow control in data communication

Both error and flow controls are extremely important in data communication. Error control ensures reliable data transmission by detecting and correcting errors, that is, detection and correction of errors as they happen. Thus, error control refers to error-free and reliable data frames data the transmission between source and destination with basic focus on the detection and correction of errors in various approaches like parity check, check sum etc. Flow control manages the rate of data transfer as well as ensures the integrity of the received data. Flow control refers to the transmission of data frames from source to destination with focus on the proper flow of data without loss of data.

**TABLE III 6.0C** Difference between Baseband broadband transmission.

| Issue | Baseband | Broadband |
|---|---|---|
| Signal type | Digital | Analog carrier |
| Signal transmission | A single data stream at a time | Multiple signal waves at a time |
| Topology | Works nice with bus topology | Works well with bus and tree topology |
| Encoding | Line coding manchester/ diff manchester type | PSK is well fitted |
| Multiplexing | Supports time division multiplexing | Supports frequency division multiplexing |
| Transmission | Bidirectional | Unidirectional |
| Signal range | Short distance | Long distance possible without attenuation |
| Signal boost | Repeater used to boost | Use amplifier to boost signal |
| Network | Mainly LAN | Mainly used in cable TV, telephony |
| Media | Guided media Coaxial, twisted pair and fiber optic cables | Radio wave, coaxial, fiber optic |

Different approaches include feedback rate base etc. Flow control at data link layer, coordinates that amount of data from the source, that can be sent before it is well received and acknowledged at destination point so that there is no loss of data. It is one of the most important duties of the data link layer. On account of flow control source point needs to wait for some sort of an acknowledgment (ACK) before continuing to send more data. There are two kinds of flow control techniques viz. *Stop and Wait* Protocol, *Sliding Window* Protocol. In network error and flow controls are important. The Data link layer (DLL) in ISO OSI model, is responsible to ensure that the error-free data are transferred between the nodes in the network. Here, framing is done to take care of two main functionalities viz. *Reliable data transfer* and *flow control. Character count, Starting and ending characters, with character stuffing, Starting and ending flags*, with bit stuffing and *Physical layer coding violations* are major issues in framing method. A few important issues are:

- **Character count:** A header field to specify the number of characters in the frame.
- **Character stuffing:** In this method, frame starts and ends with the ASCII character sequence DLE STX/ DLE ETX.
- **Bit Stuffing:** Data frames to contain an arbitrary number of bits allowing character codes with an arbitrary number of bits per character.
- **Physical layer coding violation:** This is applicable to networks in which the encoding on the physical medium contains some redundancy.

With these in mind let us start discussions on error and error detection.

1. **Error and Error Detection:** Any mismatch between data transmitted and data received is referred to as an error in communication. In communication, data transmitted are verified for correctness. Error detection stands for this verification process to check whether data have been corrupted during transmission or storage. After error is detected, attempts are made to fix them, ensuring accurate data delivery. This is error correction.
   - **Errors and error types:** After resolving the issue of start and end of the frame, it is necessary to ensure error-free transmission. Error types and error detection methods have been shown in Fig. III 6.0e. During transmission changing of bit 0 to 1 or bit 1 to 0 during transmission causes error. Error occurrences are generally of two types viz. *single-bit* error and *burst error* both types have been depicted in Fig. III 6.0e1.
     - *Single-bit Error*: As the name signifies, here one bit is altered without affecting nearby bits, it is more often found in parallel transmission in presence of white noise, than in serial transmission.
     - *Burst error*: A burst error generally refers to a contiguous sequence of bits in which the first and last bits and any number of intermediate bits are received in error. However, it is not necessary that only consecutive bits are changed. The length of burst error is measured from first changed bit to last changed bit as shown in Fig. III 6.0e1, it is 8bits in the given case. Burst error happens more in serial transmissions due to noise occurring for a longer duration affects multiple bits. So, the data rate and duration of noise are main responsible factors.

   Standard method to ensure proper delivery is to provide sender with some feedback. Positive feedback for safe receipt at the other end while negative one is to signify something wrong. However, it has some problem if the acknowledgment is lost or associated timer is off. So, let us now look for error detection methods and control techniques.
   - **Error detections:** For error-free transmission the first issue to address is error detection. Basic approach for error detection is the use of redundant bits. These additional bits are added to facilitate error detection. *Parity check, Two-dimensional Parity check, Checksum* and *Cyclic redundancy check* are the most commonly used methods which will be discussed here.
   - **Parity check:** Blocks of data from the source are subjected to a parity bit generator form. One extra bit is sent along with the original bit frame to make number of 1s either even in case of even parity, or odd in case of odd parity. At source number of 1s in the frame is counted. For even parity if the number of 1s is even then additional bit will be 0 to keep the number of 1s even. If the number of 1s is odd, to make it even a bit with value 1 is added as shown in Fig. III 6.0e2.
     - *Two-dimensional Parity check:* As shown in Fig. III 6.0e3 Parity check bits are calculated by placing the frame one below other and calculate parity for each row, in the similar way described above. Parity check bits are also calculated for all columns, then both are sent along with the data as shown. At the destination, these are compared with the parity bits calculated on the received data. Often it is referred to as block error control. It can detect two-bit errors. However, though probability is less but simultaneous occurrence of two-bit errors in two characters at the same column positions may go unnoticed at the destination point.

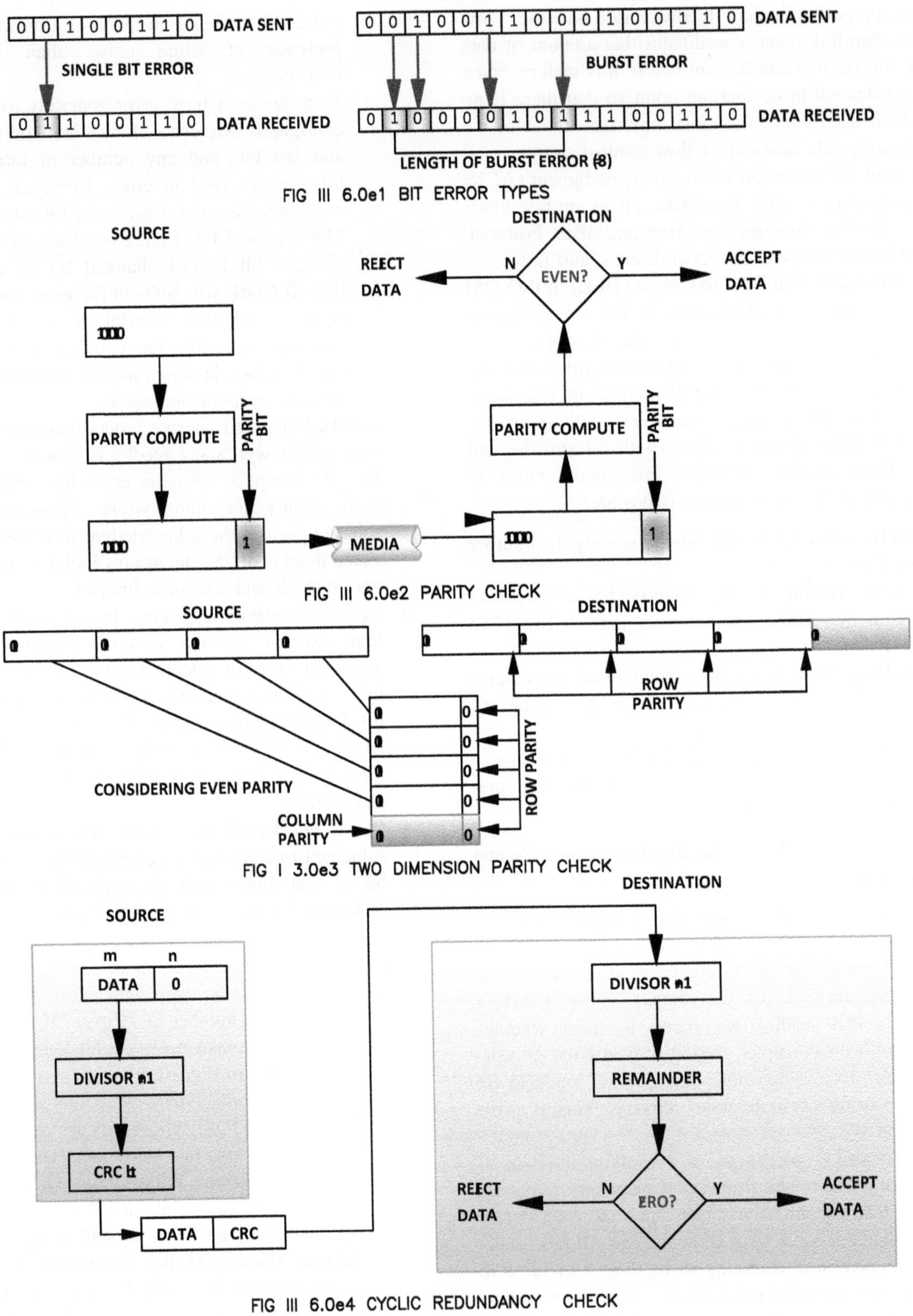


**FIGURE III 6.0E** Error types and error detections.

- *Checksum*: In this scheme, the data are divided into k segments each of m bits and at the source end segments are added for obtaining complement of check sum. Check sum is obtained by complementing (the sum of segments). Along with data segments, the checksum segment is also sent. At the destination, all received segments are added to get the sum which is complemented. Received data are accepted if compared result is zero, otherwise rejected.
- *Cyclic redundancy check*: After parity check, CRC is the most widely used for error—detection. CRC is based on binary division. The divisor is generated using polynomials. In CRC a sequence of "extra" redundant bits known as *CRC bits* are added at the end of data. At the end of bitstream n numbers of 0s are added. It is then divided by a predetermined divisor which has" n+1" bits as shown in Fig. III 6.0e4. Remainder thus obtained is the CRC. Now CRC replaces "n" 0s. The combined sequence of data plus CRC is transmitted by the sender. At destination, the number is divided by same (n+1) divisor to check if ZERO is the remainder for accepting the data or else reject it.

2. **Error Control: Error Control:** Basically, there are two types of error control mechanisms. One is based on error correction and hence called Forward error control, while the other is based on error detection method referenced as Feedback error control.
   - **Feedback error control:** Here with each character, some additional information is added for error detection. At the destination end if error is detected the information is retransmitted but no correcting action takes place. Every time when an error is detected, specified frames are retransmitted. This process is called ARQ (Automatic Repeat ReQuest) *is a group of error—control protocols in data communication over noisy or unreliable communication network.*
   - **Forward error control:** Here redundant information is transmitted with the data enabling destination point to detect the erroneous bit as well as location of the error in the data. On account of number of additional information, to be sent it is not that much popular.
   - **Error correcting codes:** In error-correcting codes not only the error bits will be identified but also could be corrected. Only additional issue is that here more redundant bits are needed calling for improper utilization of channel bandwidth for information.
     - *Single-bit error correction*: Main aim of this is to identify the position of erroneous bit to correct it. For 7-bit ASCII code, at least three additional bits will be required to identify the possibility of error in the seven positions. Also, if an error occurs at the redundant bits themselves, to identify it, additional bits are required. Hence the total number 4 bits are required. So, for single bit correction, 2K <: M + *K*+ 1 will be necessary, where M = message bits, and K represents redundant bits. Thus, single bit error in 7-bit ASCII code, at least 4 redundant bits are needed. Hence, the transmitted data contains 11bits for each data units [5].
     - *Hamming code and distance:* Hamming (1950) developed a code for *single* bit error correction by inserting multiple parity check bits at *selected* suitable positions of data before transmission to enable destination point to recalculate the check—bits, compares it with the received check—bits, for determining the error bit. The number of bit positions in which two code words differ is Hamming distance. In this system, a matrix is developed to calculate the values of redundant bit $k_1$, $k_2$, $k_3$ from the message codes.

3. **Delays and efficiency:** Delays in communications systems directly influence the overall performance and reliability of data transmission. Delays mainly arise from sources, for example, signal propagation, processing time, protocol handling etc. Delays in transmission directly impact efficiency, affecting data transfer speed and reliability.
   - **Transmission Delay ($T_D$):** Time taken by the source to send all the bits of a frame onto the medium is transmission delay $T_D$ very clearly depicted in Fig. III 6.0f1. $T_D$, shown very clearly depicted in Fig. III 6.0f1 can be obtained by dividing the frame data size(D) by the bandwidth(B) of the link.
   - **Propagation Delay ($T_P$):** It is the time taken by a bit of the frame to reach from one side to the other side is called propagation delay. It is the ratio between the link length and the propagation speed over the specific medium. So, if the distance between source and destination is d and s be the speed of electromagnetic wave in the specific medium, then

$$T_P = d/s \qquad \text{(Eq. III 6.0 c)}$$

So, as shown total time for sending the data frame and the acknowledgment will be:

$$\text{Total time} = T_D + T_P \text{ (data frame)} + T_P \text{ (acknowledgment)} \qquad \text{(Eq. III 6.0d)}$$

   - **Efficiency η:** From the above, it is clear, that source is active only for $\mathbf{T_D}$ time and the rest $2T_P$ time it is waiting for the acknowledgment. So,

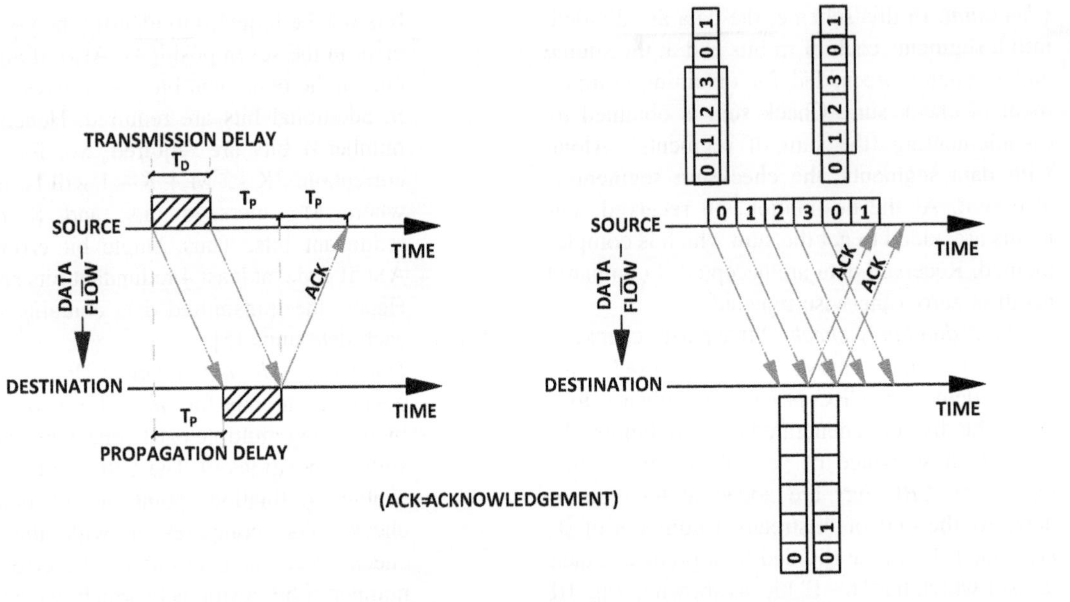


FIG III 6.0f1 STOP AND WAIT DATA FLOW CONTROL FIG III 6.0f2 SLIDING WINDOW DATA FLOW CONTROL

**FIGURE III 6.0F** Data flow control types.

$$\text{Efficiency } \eta = \text{Useful Time / Total Time} = T_D \;/\; (T_D + 2T_P) \qquad \text{(Eq. III 6.0e)}$$

4. **Flow control techniques:** For reliable data transfer in data communication a few techniques like "Stop-and-wait," "Sliding window" protocol are used. These are direct fall out of error control discusses. The error control mechanism is divided into two categories, that is, Stop and Wait ARQ and sliding window. Basically "Stop-and-wait" is a simpler version of "Sliding window." Let us study them individually here. Flow control processes have been depicted in Fig. III 6.0f.
   - **Stop and Wait Protocol:** This method is very simple and straightforward. Source point will the send one frame at a time to destination. After this source will *stop and wait* for the acknowledgment from the destination. When the source gets the acknowledgment then only it starts sending the next data frame and *stop and wait* for the acknowledgment. The process continues as shown in Fig. III 6.0f1, source needs to wait $2T_P$ times before next frame is sent. Here there could be a deadlock situation when any frame is lost before it could reach destination. In such case destination point cannot send any acknowledgment as it has not received any frame. Again, the source point will not send the next frame as it is waiting for the acknowledgment. To avoid any such situation there is need for a timeout timer, so that the source will wait for fixed amount of time pertinent to timer, for the acknowledgment and if the acknowledgment is not received then it will send the frame again. Referring to Fig. III 6.0f, it is seen that there are two types of delays (defined below), involved in this process. Stop and wait protocol is very simple to implement but one frame can be transmitted at one time. It has a few other disadvantages also. For long distance $T_P$ may be more than $T_D$, hence more idle time for the source as well as more time to send a frame. Naturally, efficiency will be poor.
   - **Sliding Window Protocol:** In this protocol the waiting time is reduced by utilizing the same as shown Fig. III 6.0f2. Here, window refers to a buffer to store the frames with numbering. For n sized window, numbering will be 0 through n-1. Here it is possible to send n frames at a time. When the destination point acknowledges the frame, then that frame need not be stored as it is already received. Therefore, as shown, window at source point slides to the next frame and this window now contains a new frame plus other unacknowledged frames of the window, that is, at any time window contains unacknowledged frames. It is clear from Fig. III 6.0f2, When the acknowledgment is received at the source, then the new window contains the frame 1, 2, 3, 0. (in place of 0,1,2,3) In this way, the window slides hence the name sliding window protocol. In sliding window protocol, the efficiency can be made maximum (i.e., 1). From the above discussions and figure, it is clear that the source can send the data

frame all the time, that is, even in $T_D+2T_P$ time. $T_D$ is the time, for one packet. So, in one unit of time, we can send $1/T_D$ packets. Therefore, in $T_D+2T_P$ time one can send $(T_D+2T_P)/T_D$ packets, that is, if one sends 1+2a (assuming $a = T_P/T_D$) packets then the efficiency is 1 [6].

5. **Automatic Retransmission ReQuest (ARQ):** ARQ referenced in previous section is related to both error control and flow control. ARQ reside in data link and transport layer of OSI model.
    - **ARQ working principle and types:** As discussed earlier, on receipt of data correctly, destination point sends acknowledgment to the source. When source does not receive the acknowledgment within a specified time, (i.e., timeout), the source understands that the frame has been corrupted or lost during transit and it retransmits the frame. This process is repeated until the correct frame is transmitted. There are three types of ARQ viz. *Stop-and-Wait ARQ, Go-Back-N ARQ* and *Selective Repeat ARQ*
    - **Stop-and-Wait ARQ:** It provides unidirectional data transmission with flow control and error control mechanisms, suitable for noisy channels. The source retains a copy of the sent frame. If within specified time it does not receive positive acknowledgment, the copied frame is retransmitted before sending next frame.
    - **Go-Back-N ARQ:** This is applicable for sliding window protocol whose principle already discussed above. If the acknowledgment of a frame is not received within the time period, all frames starting from that frame are retransmitted.
    - **Selective Repeat ARQ**: This is similar as above, only here only the erroneous or lost frames are retransmitted and the good frames are buffered.

Let us now explore some connections and other protocol types in next sections.

## 6.1 Miscellaneous data communication issues

There are a few miscellaneous issues like *Signal attenuation, noise, distortion* etc. to impair data transmission, or security threats, data quality concerns, scalability etc. All these can seriously affect the communication performance and data delivery. There are a few physical layer issues like *Bit error/Data corruption, Packet loss* (nondelivery of data packet to intended destination), *transmission medium/line loss.* There could be issues related to software such as *Software errors* (software specific problems), *Device specific* issues (viz. driver missing), *Data update* issue, *Server related* issue. Apart from all these issue like *Reliability, Addressing, Error and Flow Control* are very much pertinent. Therefore, these issues must be suitably be addressed. It is worth noting that attenuation and distortion in data communication are two important aspects. As the signal travels in transmission medium, part is absorbed by the medium, so the signal gets weaker the further it travels. This effect is referenced as *Attenuation.* On the contrary *Distortion* refers to degradation and alteration of the signal. Signal gets distorted by *external influences* as it travels along the communication path. Noise and disturbances have direct impact on these. *Noise, Distortion* and *Attenuation* are major reasons behind, hence these are discussed in a separate subsection. The discussions put forward in this section is a kind of recapitulation of pertinent issues in data communication. To restrict the book volume, details on these cannot be covered at length and these are available in any standard book on communication at bachelor level. In this section, appropriate details on all these as necessary, have been covered so that it is possible to understand and master advanced level of data communication in industrial plants and systems.

### 6.1.0 Data communication basic issues

Data refers to the *raw unrecognized facts* which are collected for processing. Information refers to the data that has been processed, organized, and given context so that decisions can be made based on this information. Data Communication is a process of exchanging data or information between two or more devices through transmission medium which may be guided (wire. Fiber optics) or unguided (wireless). Data communication involves both Hardware (HW) and Software (SW). Normally, this exchange takes places in is presented in format and rules that is agreed and accepted upon by its creators and users. The software part responsible for to maintain certain formats& rules is also called as a *Protocol.* Three important parameters protocols are:

- **Syntax:** the *data structure* or format and *particular order arrangement* of data.
- **Semantics:** meaning and interpretation of each bit section and associated action/decision based on interpretation,
- **Timing:** Meant for sender about the readiness of the receiver to receive the data.

Another important data communications consideration is *Data rate limit* which refers to how fast in bits per second data can be sent over a given communication channel. This depends on *Bandwidth available, The level of the signals* and T*he quality of the channel* (the level of noise; S/NR). Nyquist (noiseless channel) and Shannon (noisy channel) theoretical formula are deployed for this purpose. Data transmission and networking are backbone of modern technology, for easy information sharing among devices. However, rapid technology breakthroughs and exponential data growth pose substantial concerns. In this subsection, a few important issues which could pose challenges will be addressed. Let us start the discussions on data communication issues with data quality parameters.

1. **Quality parameter:** Transmission quality depends very much on *Bandwidth*, *Spectral efficiency*, and *Speed*. These are three different inter-related things. Higher the bandwidth means higher the channel capacity. Naturally, higher spectral efficiency indicates better utilization of the given capacity. Speed, refers to the amount of information transmitted in unit time hence influencing factor for the system efficiency. Apart from these three issues, *Incomplete data*, *Inaccurate data*, *Duplicate data*, and *Inconsistent data* are also crucial to maintain data quality. Let us now look into details on three things discussed:
   - **Bandwidth:** Bandwidth refers to the amount of frequency space allocated to transmit a signal. It is a measurement indicating the maximum capacity of a transmission medium to transmit data over a network connection in a given amount of time expressed in megabits or gigabits in 1 s. Similar to pipeline concept; the more bandwidth a transmitting medium has, the more data it can send and receive at one time. Wider the bandwidth faster will be data transmission. Often *bandwidth* and *speed* are used interchangeably but it is not proper. Essentially, *speed* refers to the rate at which data can be transmitted, while the definition of *bandwidth* is the capacity for that speed.
   - **Spectral efficiency:** The amount of data transmitted over a given spectrum/bandwidth with minimum transmission errors is referred to as *spectral (/bandwidth) efficiency*. So, this is a measure of how effectively a limited frequency spectrum is used to transmit data and can be expressed as bits per second per Hertz, that is, (bps/Hz). Obviously, higher the spectral efficiency more will be the data transmission within the same bandwidth.
   - **Data Rate:** Speed/The data rate is the amount of digital data that's moved from one place to another in a given time. This refers to the amount of data transmitted per unit of time, in bps/Mega bps (Mbps). Speed is a key factor in transmission quality and is influenced by bandwidth and spectral efficiency.
   - **Relationships:** A *wide* bandwidth can support high data rates, which can lead to smoother and faster transmission of content. Apart from bandwidth, S/NR and the efficiency of the modulation scheme are also important. As, stated, speed is a direct indicator of transmission quality. Lower speeds can lead to delays, buffering, and other problems, negatively impacting the quality. Again, with higher spectral efficiency more data can be squeezed into a given bandwidth, to improve overall system capacity and potentially reduce costs by making better use of the available spectrum. The transmitted signal should occupy smallest bandwidth of the allotted bandwidth, with better spectral efficiency.
2. **High level issues:** There are two parts of this high-level issues. These are:
   - **Security issue**: An important issue is that the communication signal should never be intercepted/ or it should never be attacked by virus or malware. Vulnerabilities in protocols, like IP spoofing, can increase vulnerability in the network. For this suitable security arrangements, encryption etc. are under taken. Major security threats are:
     - *Network sniffing*: Intercepting and analyzing data in a network.
     - *Denial-of-Service (DoS)*: Overloading network traffic to deny services to legitimate users.
     - *IP spoofing*: Deceiving a system by with a false IP address
   - **Networking issues:** Some issues related to networking are:
     - *Congestion*: Heavy traffic on a network can create congestion to cause delays, packet loss, and performance degradation.
     - *Address*: With increase in number of internet-connected devices increases there could be constraints on IP addresses.
     - *Protocol issue*: Incompatibility between different networking protocols/standards leading to failure of communication.
3. **Implementation:** Data communication faces various challenges like impairments (noise, interference, attenuation-discussed later), security and, scalability issue (Sniffing, spoofing, DoS, *Congestion*, *Bandwidth limits* and Address depletion—already discussed), service quality issue and implementation challenges. Major issues related to quality and implantation issues are:
   - **Quality of Service (QoS):** This refers to meeting of certain specific performance requirements like higher speed, better reliability, and lower latency with seamless nice connectivity.
   - **Implementation Issues:** A few implementation issues are:
     - *Network design*: major issues here are selection of network topologies protocols and data traffic.
     - *Redundancy*: Implementing redundancy and making the network fault tolerance in case of single component failure.

- *Interoperability:* To ensure seamless communication across different systems used in network.
- *Security:* Use of encryption, implementation of firewall, and other suitable antivirus to ensure protection of unauthorized access as well as intrusion detection and protection system.
- *User expectations:* management and balancing performance requirements, detection of performance bottle neck and get solution. This also must include monitoring network performance to maintain the quality services.

4. **Disturbances:** Disturbances in data communication and control systems refers to unwanted signals and events that interrupt the normal intended operation of the system. Disturbances are often called transmission impairments, which responsible for the degradation of the signal quality resulting errors/loss of data. A few factors are:
   - **Delay distortion:** The signal propagation velocity in a guided medium varies with frequencies it is fastest at the center of the frequency, but falls off at the two edges of frequencies. This delay distortion gives rise to jittering problem. Therefore, any delay distortion should be avoided.
   - **Noise:** Noise causes signal impairment and is caused when an unwanted signal gets inserted between transmission and reception. Various noise types are as follows: *Thermal* noise, *Impulse* noise, *Cross talk*, *Flicker* noise, *Shot* noise etc. The system should be free from all possible noise as far as achievable.
   - **Interference:** Interference in data communication refers to Unwanted signals that disrupt or degrade the transmission of data over a communication channel is referenced as interference viz. Electromagnetic interference (EMI), Radio-frequency interference (RFI). Interferences in network may be caused by parallel communication on similar/contiguous frequencies, data collision, throughput loss etc. So, communication should be free from any interference both in terms other signal operating at same frequency/from neighboring channels.
5. **Attenuation:** Attenuation refers to the loss of signal strength, caused by various internal/external factors. In data communications, cables and connections are responsible. Distance, frequency, weather are factors responsible for this also. Signal attenuation should be minimum. These are detailed out in next subsection.

Let us have a detailed look on *signal attenuation and distortion* in next subsection,

### 6.1.1 Signal attenuation and signal distortion

*Attenuation*, *Distortion* and *Noise* are main contributors for signal impairment. Signal strength decreases with increasing distance as it has to overcome resistance of medium. So, there will be loss of signal energy. This is signal attenuation. Attenuation represents the loss of energy-per unit distance travel. It is expressed as ratio of out power and input power per unit distance. Attenuation mainly depends on *noise*, *physical surrounding* and *travel distance*. For guided media, this reduction in strength, or attenuation, is generally exponential. So, expressed in logarithmic form or in decibel (dB)/distance. Often it is expressed in terms of attenuation at a particular frequency relative attenuation at a reference frequency. Influencing factor for attenuation are as follows: *Distance range* (gradually dissipate in strength over longer distances), Noise/*Interference and wire size* for guided media. Since in fiber optics signals travel on high-frequency light waves, and less susceptible to electrical noise, attenuation in fiber optics will be lesser than that in copper. As signal in unguided medium signals pass through air, naturally there would be many other issues to cause attenuation of signal, for example, atmospheric condition. Naturally, for unguided media, attenuation is a more complex function of distance and the makeup of the atmosphere. We had seen that in attenuation signal strength decreases with distance, but it also varies with signal frequencies. Attenuation being inversely connected to Channel capacity. Channel capacity is detailed out in Fig. III 6.1a wherefrom we see, as attenuation lowers S/NR, so, channel capacity will also be lowered. Distortion refers to the signal changes of its form or shape. Distortion can occur in a composite signal made of different frequencies. Each signal component has its

Cha nel Capait y: The maimum rte a which dta can be reliably tramit ted over a given communiction pth, or channel, under given conditions, is referred to as the *channel capacity* i.e. limit for the dta rte. According to Shannon formula Channel capait y C= BI og2(1+S/N R) where B is the bdwi dth and S/NR in is given by $S/NR_{dB} = 10\log_{10}\frac{Signal\ power}{Noise\ power}$. Therefore, influencing factors ae: Bd width, S/NR (o ise level), Error rte (Codig techique for correctio) Ba ndwidth is cotried by nture of medium & tra smitter etc.

**FIGURE III 6.1A** Channel capacity.

own propagation speed through a medium hence its own delay in arriving at the final destination. Differences in delay may create a difference in phase. Data signals must be transmitted over a medium with a defined frequency and speed. Digital data are vulnerable to delay distortion, because of spillover of data packet bit to destabilize the overall packet transmission pattern. Details on Signal attenuation and distortion will be dealt in this subsection.

1. **Attenuation issues:** We know a few issues responsible for attenuation. Signal -noise is a major issue. Since it affects many other systems and very important, it is treated separately. *Distance, Frequency, Medium, Material properties* and *Noise (interference)* are major causes of attenuation. Apart from *Signal loss*, it causes *Reduced data* rate and *Signal degradation.* Other factors associated with attenuation are:
   - **Detection level:** For extracting information and establishing communication between source and destination, detection level is important. Signal detection refers to the process of identifying and extracting meaningful information from a received signal. So, the signal must have enough strength so that it could be detected at the receiving end by receiver electronics. Detection level is important also for: *Data integrity, Proper data reception* and *Efficient communication.* Major challenges are too weak signal due to attenuation, noise/interference etc. For signal recovery amplification (may use difference amplifier), *repeaters*, suitable filter, adaptation of error detection and correction techniques. Often *Thresholding,* that is, setting a minimum signal strength level to differentiate between valid signals and noise.
   - **Variation with frequency:** Attenuation co-efficient (*signal weakens per unit of distance*) increases with increase in frequency. Also, the absorption of by objects, media is more for higher frequency signal than lower frequency signal. Apart from these high-frequency signals are more susceptible to scattering and interference. For all these factors, there will be attenuation variation with frequency, that is, more for high-frequency signals. This is more associated with analog signal transmission. It is dealt by many ways and means one of them is to use signal equalizer etc. However, this is really not much pertinent to data communication we are interested in.
   - **Distance range: (between Transmitters and receivers):** In case of point-to-point communication signal strength should be optimum so that it is strong enough to be recognized but is not that strong to cause overload to the receiver. As attenuation varies with distance, beyond certain distance would become very high, thus it is necessary to use repeaters at regular intervals. Attenuation variation with distance, is complex in case broadcasting/multicasting modes, where there will be variation of distance of different receivers from the transmitters. In case of wired communication, mismatch in impedance could enhance attenuation whereas for wireless communication, obstacle, weather can interfere, causing signals to weaken.
2. **Signal and Noise:** The signal represents the meaningful information to be detected. On the contrary, the noise is the random, unwanted variation or fluctuation that interferes with the signal. As discussed, *Thermal, Cross talk, Impulse* etc. are various sources of noise. There are many ways like external interference, internal signal distortion, channel limitations are major reasons behind introduction of noise in to the system. Alias signal and S/NR are major issues covered here.
   - **Signal to noise ratio:** Signal-to-noise ratio, S/NR, is a measure of the strength of the desired signal relative to background noise (undesired signal). S/NR can be determined by using a fixed formula of comparison, of the two levels. The ratio is typically expressed as a single numeric value in decibels (dB), that is, S/NR in dB = = 10 log10($P_s/P_n$) [Ref: Fig. III 6.1a]. Here, $P_s$ and $P_n$ represents signal and noise power in Watts respectively. This ratio can be negative, zero or positive. A signal-to-noise ratio over 0 dB indicates that the signal level is greater than the noise level. The higher the ratio, the better the signal quality. S/NR plays an important role in communication. If the noise is very near the actual signal, then, received signal cannot be used fruitfully. It is impossible to detect a signal when the S/NR becomes $<2$. Therefore, it is important that signal received is at much higher level than associated noise, that is, S/N is high.
   - **Figure of Merit:** The ratio of output SNR to the input SNR is called Figure of merit. *Shannon–Hartley theorem* relates bandwidth, SNR and channel capacity and is very important for data communication.
   - **Alias signal:** We know that Nyquist sampling rate must be *at least 2 times greater than the highest frequency* component in the complex signal. When the analogue voltage is not sampled frequently (< Nyquist rate) then this results in a shambles because high-frequency signals will appear as low frequency signals at the output, which will be very confusing. This phenomenon is called *aliasing.* Alias signal causes a false lower frequency component to appear in the sampled data of a signal hence a poor representation of the analog signal. For this antialiasing filter which is basically a low pass filter is used.

3. **Degradation and Distortion:** Degradation and distortion are two related important parameters in communication have been detailed out below:
   - **Degradation:** Any change in a signal which alters the basic waveform or the relationship between various frequency components; it is a *degradation* of the signal. So, it refers to the loss of quality and integrity of a signal while traveling through a transmission medium and can manifest *attenuation* (reduction in signal strength), distortion (changes in the shape or characteristics of the signal) of the signal's shape, or the introduction of unwanted noise (Unwanted signals interference). Signal degradation can lead to *errors* in data transmission, *reduced data rates*, and *unreliable* communication. This degradation can affect both analog and digital signals. In analog signals, degradation leads to a gradual loss of signal quality, making it unclear/Fuzzy. In digital signal it can introduce bit errors/cause blocks of data to become corrupted. Signal degradation can occur due to: physical distance, change in noise margin (digital signal), electromagnetic interference to change quality in analog signal. *Reduced data rates*, *Increased error rate*, *Limited range* and *Communication failure* are major impact of degradation in data communication. To mitigate the situation, use of *Amplifier/Repeater*, Use of *Twisted pair cable with better shield* (its grounding), *Signal equalization*, *Proper selection of transmission medium*, *Error detection and correction* are some of the methods are applied.
   - **Distortion:** Signal distortion in data communication refers to the changes/alteration change in the characteristics (shape, frequency and amplitude) of a signal/data representation during transmission, resulting in a change at the receiver compared with the original signal at the sender. Noise of various forms, S/NR and attenuations are major contributing factor for this. When signal quality is compromised due to imperfections in the transmission medium it is referred to as transmission/communication impairment (Shape distortion, noise and attenuation). Environmental parameters, distance between sender receiver and medium characteristics are major contribution for this. Distortion could be linear such as distortion in amplitude, phase (delay distortion) or distortion can be nonlinear, for example, signal clipping, generation of new frequency. *Reduced signal quality*, *data error*, *difficulty in decoding*, *reduced transmission speed*, *unreliable communication* etc. are major impact on communication. *Signal regeneration*, *Use of Equalizers*, *error correcting code* are a few mitigation techniques can be used.
4. **Delay distortion issues:** Delay distortion is a major reason for the timing jitter problem, where the receiver clock deviates from the incoming signal in a random fashion so that an incoming signal might arrive earlier or late. Delay distortion, is a major contributor for signal impairments. From III 6.1.0.4, we know that delay distortion occurs because the velocity of propagation of a signal through a guided medium varies with frequency, that is, the highest velocity near the center frequency and fall off toward the two edges of the band. Thus, various frequency components of a signal will arrive at the receiver at different times, resulting in phase shifts between the different frequencies. So, the received signal is distorted due to varying delays experienced at its constituent frequencies. This kind of distortion is mainly appearing in case of composite signals comprising various frequency components in it. While transmitting this composite signal, if a certain delay happens between the frequencies components, then there may be the chance that the frequency component will reach the receiver end with a different *delay constraint* from its original which leads to the change in shape of the signal. The delay happens due to environmental parameters or from the distance between transmitter and receiver etc. Delay distortion is particularly critical for digital data. For a sequence of bits is being transmitted, because of delay distortion, some of the signal components of one bit position will spill over into other bit positions, causing *intersymbol interference*, which is a major limitation to maximum bit rate over a transmission channel. For a given a bandwidth of $B$, the highest signal rate is $2B$. This limitation is due to the effect of inter-symbol interference, such as is produced by delay distortion. Equalizing techniques can also be used for delay distortion.
5. **Delay distortion: Result and remedy:** As discussed above, this would result in phase shifts among the different frequencies. Delay distortion is a major reason for the timing jitter problem, where the receiver clock deviates from the incoming signal in a random fashion so that an incoming signal might arrive earlier or late [7]. Digital data are highly vulnerable to delay distortion. This is because when any transmitted data packet bit spills it can destabilize the overall packet transmission pattern. On account of delay distortion, some of the signal components of one bit position will spill

over into other bit positions, causing inter-symbol interference, which is a major limitation to maximum bit rate over a transmission channel [1]. Normally Equalization techniques can be used to smooth out the delay distortion.

### 6.1.2 Noise and interference in data communication

In data transmission, signal received, consists of the transmitted signal, and some distortions injected by the transmission system, plus unwanted signals injected in to it from transmission medium and or somewhere between transmission and reception. Noise/interference, is as undesirable electrical signals that distort an original (or desired) signal, or interfere with it. Signal noise, is any unwanted interference which would degrade communication signal. Signal noise affects both analog and digital signals. However, as digital "bits" are involved in digital communication, the strength (power) of noise necessary to affect a digital signal is much higher than that necessary to affect analog communication. Analog signals represent an infinite range of possible values with established range. So, it gets affected easily. Industrial plants vital processes are automated and controlled based on the measured signals. Therefore, any signal variation due to noise can lead to unpredictable and potentially damaging results. In many cases the noise will be unpredictable, like transients caused by lightning, whereas 50 or 60 Hz ac "hum" from power circuits close to the data communications cable are predictable. Various sources of noise, Type of noise, Impact of noise, mitigation methods are major issues addressed here. In this connection discussions in II 18.1.0 and III 18.1.1 in Vol II may be repeated. There may be some overlap/repetitions on various issues. These have been included here briefly to complete the meaningful discussions. Shielding, grounding of cables etc. have been covered in II 18.2 hence not repeated again. When signal travels from source to destination in a channel or transmitting media, an interference modifies the original signal in disruptive way. Interference may come from another signal in other circuit or from the same circuit. Since interference in the communication networks comes as undesirable nuisance, it is a kind of noise but an artificial noise. The discussions here also will cover interferences in wireless communication to complement the discussions in II 18.1.1.

1. **Noise issues:** Noise is an *unwanted signal* that interferes with the original signal and corrupts the parameters of the signal to a certain extent. The discussions here shall be read in conjunction with discussions in II 18.1.0. In wireless communication, any energy interference which affects the quality of original wireless signal is referenced as noise. Noise can be natural or manmade. Noise is random signal with no defined pattern, no defined frequency/frequency band and any fixed amplitude. Noise injection can take place at any location where the network is exposed to and can happen on account of many factors. On account of these sort of unpredictability it is meaningless to trouble shoot noise. Some pertinent issues are:
   - **Frequency analysis:** Effect of noise can be evaluated by examining its frequency spectrum. Noise can be seen to fall into three categories: *Wideband*, *Impulse* and *Frequency specific* noise. Wideband noise contains numerous frequency components with varying amplitudes. Impulse noise is a burst of noise which might last up to 20 ms and Frequency specific noise may have fixed frequency (range) with varying amplitude.
   - **Common sources:** There exist three contributing factors for an electrical noise and associated problem. These are as follows: *Electrical noise source*, *Coupling mechanism* and *Communication circuit.* Apart from Lightning strikes, High voltage surges (electrical faults) Fluorescent lighting tubes there are a number of industrial noise sources like common industrial large electrical motors, welding equipment as described in II 18.1.0.
   - **Noise coupling:** These are already discussed at length in Vol II of the book. In this connection section II 1.1.0, II 18.1.1 and Fig II 8.0e may be referenced. As discussed therein there will be four forms of coupling of electrical noise into the sensitive data communications circuits. These are as follows: *Impedance* coupling, *Electrostatic*/capacitive coupling, Electromagnetic/inductive coupling and *Radio frequency* (RF) radiation.
2. **Results of Noise:** *Improper grounding*, *Ground loop*, *Poor wiring* and *proximity* to *electrical equipment* etc. could cause electrical noise. Noise pick up in industrial automation could result in
   - **Process signal distortion:** This would cause improper interpretation of signal monitoring plant operation may be jeopardized and or would result spurious alarms.
   - **False process control:** This comes as a corollary from the above. Naturally for an automated plant process control may be highly affected and as a consequence quality product of the plant could be affected quite seriously.
   - **Loss of signal:** there could be apparent loss of signal. Although majority of present-day plant devices use noise filter, yet at times this may not be sufficient and any such filtering could cause delay hence delay distortion!

- **Receiver effect:** Noise in indirect manner puts limit to the weakest signal to be received by the receiver and limits the sensitivity of the device as well.

3. **Industrial plants Common causes:** Common causes of noise in industrial plants are:
   - **Ground loop:** Improper grounding often form a ground loop to inject noise signal.
   - **Poor wiring:** Poor wiring practices, use of unshielded cables and or grounding shield at both ends are common cause to inject noise.
   - **Cheap and poor circuitry:** It is not uncommon that these kind of products often cause easy pick up of external and internal noise signals.
   - **Nonterminated wire:** Often long extra cable left unterminated would act as antenna to pick up noise. It is always better to terminated extra wires.
   - **Close proximity to electrical devices:** When sensors are very near to electrical equipment (especially of high electrical rating) it may pick up noise from these electrical machines especially when such machines start/stop. Suitable care must be taken to avoid these.
   - **Defective electrical machines:** Many times, large motors/commutators requiring maintenance left unattended. Naturally associated sparks etc. from them could cause noise into the systems.
   - **Cable Routing:** It is always recommended to use separate tray/tier for instrumentation and control cable from (*even*) *low voltage* electrical cables with gap as specified in standards in practice at that place. These cable mixing or close proximity could cause noise interference.

   Majority of the above issues discussed are related wiring practices which have been detailed out in II 18.2 of vol II of the book.

4. **Noise: Classification and Types:** From discussions in Chapter 18 of Vol II reader is aware of various types of common noise in industrial system. [*The discussions pot forward here complements the discussions in II 18.1 and in some cases some repetition to make self-sufficient document*]. Noise can be classified as:
   - **External noise**: Noise generation external, that is, outside the system are external noise. Electrostatic interference (ESI)/Electrical storms, EMI, RFI and Cross talk are examples of external noise. This may occur in the communication medium. It is not possible to completely eliminate external noise.
   - **Internal noise:** When noise is generated from within the system itself (internal noise). Thermal noise, shot noise and Flicker noise are example of internal noise. All these are discussed in II 18.1.0 and II 18.1.1 also.

   Various noise types are:
   - **EMI:** Electromagnetic Interference is unwanted noise or interference in an electrical path or circuit caused by an outside source. Electromagnetic Interference causes degradation in the performance of electrical devices, in the form of poor operation, malfunction or stop working completely. EMI can be caused by natural or human-made sources. Variations are as follows: Radiated EMI, Narrow/broad band EMI, Coupled (Inductive/capacitive) EMI, conducted EMI.
   - **RFI:** RFI refers to the disruption of electronic devices caused by external radio frequency signals. Unwanted radio waves interfere with the normal operation of communication systems, leading to degraded performance or complete signal loss. *Electromagnetic devices*, *Industrial equipment*, *Communication overlapping*, *Weather conditions*, *Solar activity* etc. are major causes of RFI
   - **Cross-talk:** Cross-talk refers to signals interfering with each other electromagnetically, for example, two cables running parallel one carrying signal and other also signal or low voltage power could cause cross talk. For details II 18.1.1.5 may be referenced.
   - **Electronic and common noise:** Electronic noise is created in network transmitter or added through the communication media such as air/wire. Other common noise are Flicker, Hiss and Buzz.
   - **Thermal noise:** Thermal noise can occur in all transmission media and communication equipment, including passive devices due to thermal agitation of electrons. It is characterized by a uniform distribution of energy over the frequency spectrum with a Gaussian distribution of levels. From II 18.0 we get: the noise power, P, in watts, is given by P ¼ kT D f, where, k is Boltzmann's constant in joules per Kelvin, T is the conductor temperature in Kelvin, and D f is the bandwidth in hertz. It is a function of temperature and often is referred to as white noise.
   - **Shot noise:** Shot noise (quantum noise) refers to the variation in a signal caused by the quantized nature of the light and electricity making up the signal. When the electrons and holes cross the barrier, shot noise is produced. This is found in a diode, a transistor, but not in a resistor because of the absence of a potential barrier. Ref: II 18.1.0.
   - **Transit-time noise:** This is a kind of random internal noise. As it occurs in transistors is called

transit time noise. Transit time is the duration of time that it takes for a current carrier such as a hole or electron to move from the input to the output. At low frequencies this time is negligible. But when the frequency of operation is high and the signal being processed is the magnitude as the transit time, then problem can occur [8].

- **Intermodulation noise:** When signals at different frequencies share the same transmission medium or pass through a nonlinear device, the result may be intermodulation noise. The effect of intermodulation (IM) which could produce signals at a frequency that is the sum or difference of the two original frequencies or harmonics and subharmonics of input signal frequencies. Naturally, these resulting IM components may be inside or outside the frequency band of interest of the device. Only when it is produced signals are *inside* the band of interest then IM become *IM noise* which is very significant in RF communication. IM noise can result from either design characteristics of a system, or from manufacturing defects. However, later it becomes a big issue.

5. **Interference issues:** As already discussed that major Interference source are as follows: EMI, Cross talk, Conducted interference, Common-mode interference (CMI), Inter-symbol interference (ISI). Also wireless interferences are as follows: Adjacent-channel interference (ACI),

Co-channel interference (CCI), Intercarrier interference (ICI) etc.

Interference is of particular concern for unguided media, but is also a problem with guided media. In guided media such interference an come from nearby cable already pointed out at the beginning of previous section. Interference from competing signals in overlapping frequency bands can distort or wipe out a signal. RF interference can be generated by almost any device that emits an electromagnetic signal. However, it is most important to note that the single biggest source of Wi-Fi interference is their own Wi-Fi network. This is because Wi-Fi is a shared medium that operates in the unlicensed RFs within the 2.4 and 5 GHz range. IEEE 802.11n allows multiple radios in in single AP could be vulnerable from interference point of view. There three approaches to address interference issue in wireless data communications. These are [9]:

- Lowering the physical (PHY) data rate.
- Reducing the transmit power of the affected AP.
- Changing the AP's channel assignment.

None of these approaches are foolproof, so it is ever challenging to mitigate RF interference, as a flood of new Wi-Fi-enabled devices hit enterprise networks [9].

## 6.2 Data communication approaches

There are mainly two kinds of switching in network. These are *Packet* and *Circuit* switching. In *packet switching* instead of sending data in one piece, it is broken into smaller so called "packets" each of which can travel independently through the network, taking different paths to reach the same destination. *Circuit switching* is a traditional method may be used in traditional telephone networks using a dedicated path. Difference between the two systems have also been elaborated. *Efficiency*, Fault tolerance, Reliability and Scalability are major benefits of packet switching. However, there also exist some disadvantages like *Delay*, *Overhead*, *Complexity* etc.

Data link layer second layer in the ISO-OSI model, is primarily responsible to move the datagram across a specific link. The IEEE has defined a number of LAN technologies in the data link layer, including Ethernet, Fast Ethernet, Gigabit Ethernet, and token ring. The actual data link layer is split into two sublayers, called the Medium access control (MAC) sublayer and the Logical link control (LLC) sublayer. Datalink layer protocols include: *Synchronous data link protocol* (SDLC), *High-level data link protocol* (HDLC), *Serial line interface protocol* (SLIP), *Point to Point Protocol* (PPP), *Link access procedure* (LAP), *Network control protocol* (NCP), *Address resolution protocol* (ARP). They offer services like *Framing and link access*, *Reliable delivery*, *Flow control*, *Session setup* and termination *Error* detection *and correction*, *Half-duplex* and *Full-duplex*.

In the present days with smart devices network are also developed at the field level. Fieldbus systems are used for field level communication, local loop and connecting field devices to the control level. Legacy implementations that used Ethernet-only for enterprise and supervisory level combined with non-Ethernet Fieldbuses. However, Industrial Ethernet (IE) available now (e.g., EtherCAT) is expected to continue to displace traditional fieldbuses, homogenizing all levels of industrial networks. Till date there is no well suited (by protocol TCP/IP) protocol in real time applications, supervisory level and enterprise automation, that is, serving all levels. Also, industrial Ethernet *must* be compatible with existing fieldbus systems to take care about the existing investments [10]. There exists a number of different standards exist in the market; they do not append each other for different field communication standard but have gone in their own way. In this section short discussions will be presented on IEC 61158 and IEC 61784 which play great role in fieldbus and field level communication protocols. Table III 6.2a has been presented below for applicable IEC standards necessary for cabling for industrial network data communication.

**TABLE III 6.2A** List IEC standards for cabling in Industrial communication.

| Purpose | Standard | Title | Connected to |
|---|---|---|---|
| Design | IEC 24702 | -Generic cabling- industrial premises | IEC61918 |
| Design | IEC61158-2 | -Fieldbus specification-Physical layer specification and services definition | IEC 61784-1, 2 and 61,918 |
| Design | IEC 61784–1 | -Profiles-part 1: Fieldbus profiles | IEC61158-2 |
| Design | IEC 61784–2 | -Additional fieldbus profiles for real time networks based on ISO/ IEC/IEEE 8802–3 | IEC61158-2 |
| Implementation | IEC 61784–5 | -Profile-part 5-3: Installation of fieldbuses-installation profiles- | IEC61918 |
| Implementation | IEC 61918 | Installation of communication networks in industrial premises | IEC 61702, IEC61158-2 and IEC61784-5 |

## 6.2.0 Packet switched network—X.25 and frame relay

In packet switching, total message is divided and grouped in to packets and are sent via various routes/nodes routes to reach the destination. Packet switching are two types viz. connectionless—datagram switching and connection-oriented—virtual circuit switching. Ethernet and Internet Protocol are examples of connectionless packet switching. X25 Frame relay are examples of virtual circuits. A short discussion will be presented on two protocol suites which use packet switching and virtual circuits. Two terms packet switching and virtual circuits needs to be clarified prior to going for the main discussions. Let the discussion star with Fig. III 6.2a.

1. **Packet Switching:** Packet switching is a process in which data are grouped and grouped data are transmitted from source to destination, over a digital network into packets. So, in this process the message is divided and grouped into a number of units called packets. Packet switching uses the *store* and *forward* technique; when forwarding the packet each hop stores the packet to forward allowing packets can be discarded for suitable reason. Each packet in a packet switching technique has two parts: a header and a payload. The header contains the addressing information of the packet and a sequence number. The addressing information is used by the intermediate routers to direct it toward its destination. The payload carries the actual data. These packets hop from one router to another across the network and at each hop, the router looks at the packet's destination address and decides the best route to send it along and provides high flexibility. The process is shown in Fig. III 6.2a1. There is no need to establish a dedicated circuit for communication. Let us look into details:
   - **Packet transmission steps:** Various steps in packet switching are:
     - *Creation*: Data are divided into packets containing part information with headers.
     - *Routing*: Routers find the best path to reach the destination.
     - *Forwarding*: Packets hopping through multiple network nodes, to avoid congestion*Reassembly*: After arriving at the destination, packets are reassembled into the original.
   - **Delay types;** As discussed, there are several kinds of delays are associated with packet switching. These are as follows: *Transmission delay*, *Processing delay*, *Queuing delay* and *Propagation delay.*
   - Packet versus Circuit: switching Basic differences between Packet and Circuiting switching have been enumerated in Table III 6.2b.
2. **Packet switching: Discussion:** Here, various aspects of packet switching networks have been discussed briefly. Major issues are:
   - **Packet loss:** In packet-switching, packets take different routes and may encounter congestion, or unable to reach destination—this is known as *"packet loss."*This can be caused due to various reasons like Congestion (too much traffic packets dropped), Hardware fault (e.g., Router failure), Transmission loss (Poor signal interfered), Security threat (DoS). These packet losses can be handled in different network differently. These are as follows: Detection of acknowledgment from receive, resending missing packets, prioritizing *speed over reliability/buffering.*
   - **Application and redundancy:** Packet switching technology is used in real-time applications. However, on account of different paths taken by packets, there will be various arriving timing introducing

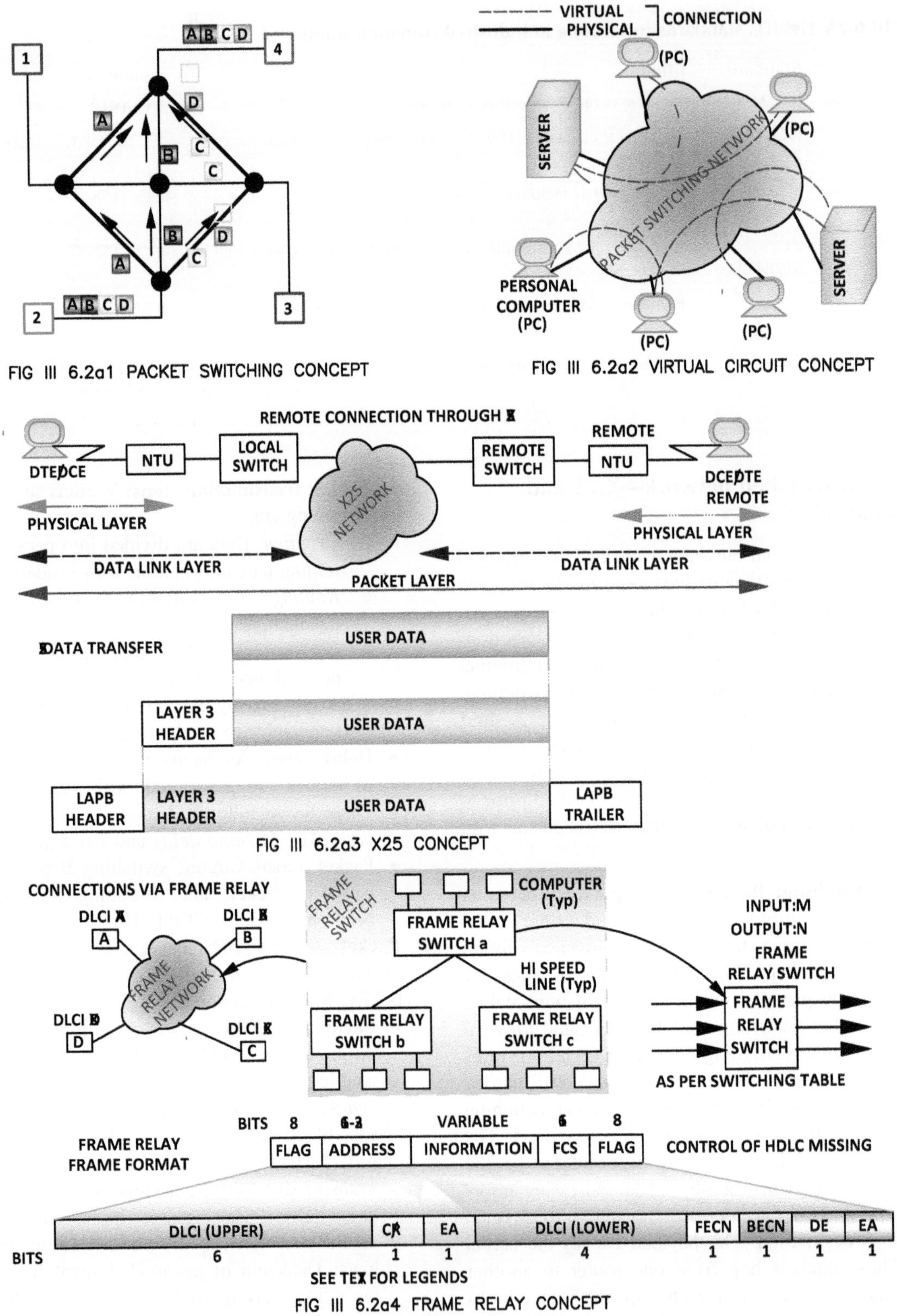


**FIGURE III 6.2A** Packet switch network—X25 and frame relay.

**TABLE III 6.2B** Difference between packet switching and circuit switching.

| Issue | Packet switching | Circuit switching |
|---|---|---|
| Transmission | Divided into packet, then sent independently | Dedicated path for each packet |
| Transmission and processing | Both source and intermediate nodes | At source only |
| Physical path | No specific physical path | Specific physical path |
| Protocol | Complex protocol handling | For simple protocol handling |
| Call set up | No call set up is necessary | Call set up is necessary |
| Delay | Non uniform | Uniform |
| Reliability | Comparatively less reliable | Comparatively more reliable |
| Resource | Resource wastage more | Resource wastage less |
| Latency | Comparatively more latency | Comparatively less latency |
| Overhead | Comparatively more overhead | Comparatively less overhead |

latency and jitter in the system. In some case like User datagram protocol (UDP) are preferred in scenarios where speed matters more than accuracy, as they prioritize fast transmission over packet loss recovery. Network redundancy in a packet switching network ensures that if one route experiences congestion or failure, packets can automatically reroute through an alternative path enhancing fault tolerance and reliability, preventing disruptions [11].

- **Switching types:** Two types of packet switching are common. These are:
  - *Datagram packet switching*: Here, each packet is treated independently and each one can take a different route and arrive at different times, but they all eventually get at the desired destination. During transmission, when each packet reaches a node, it decides which path is the best to follow the next. This dynamic decision making, improves the performance of data transmission.
  - *Virtual circuit packet switching*: A predetermined path is set up before any packets are sent and all the packets follow this path. Here, packets arrive in order without the need for reordering at the destination.

3. **Virtual circuit/Virtual connection:** Virtual circuit/connection refers to the transportation of data over a packet-switched network in such a way that it appears as though there is a dedicated physical link between the source and destination. It first establishes a logical path/virtual connection with a signaling protocol, between sender and receiver and all packets in the information will follow this predefined route as shown in Fig. III 6.2a2. Virtual circuit identity is provided by switches/routers to uniquely identify this virtual connection. Data are divided into small units and all these small units are appended with help of sequence numbers and packets arrive in order at the destination. So, there are three phases viz. *Setup* (All address transferred), *Data transfer* (packet header containing information length, timestamp, sequence number, etc.) and *Tear-down phases* (established connection between the source and destination is terminated). There are 2 types of Virtual Circuits viz. *Switched Virtual Circuit* (*SVC*) and *Permanent Virtual Circuit* (*PVC*).
4. **X25 Protocol: X25 Protocol:** Defined by International Telecommunication union (ITU) -T in 1970s X.25 was a packet switched communications over WAN. It is later replaced by Frame relay which like X25, is also *connection-oriented* network but without error and flow control. Here same physical line can be utilized by multiple logical channels. X25 protocol does not have any transport layer. With the help of X25 network one Data terminal equipment (DTE)—host/terminal, can establish connection and exchange data with one or more other X.25 Host(s) through X25 network cloud. Each packet contains address and payload. This address provides the routing information to be used by the network in carrying packet to establish the connection. So, X.25 network switching nodes is well aware of each connection. In reliable X25 protocol, it is possible to make several virtual circuits simultaneously through multiplexing. However, X.25 networks cannot forward the packets until they have been completely received, resulting in transit delays [12]. X25 connection and data format have been elaborated in Fig. III 6.2a 3.
   - **X25 Addressing:** X.25 uses digital string (up to 15 digits typical 12digit) as Network user address (NUA). Format of addressing is guided by X121 in public network with first 4 digits are known as the Data network identification code (DNIC), allocated by ITU to identify the network and the remaining digits are assigned at the discretion of the network operator.
   - **X25 Virtual circuits:** There exists much similarities between SVC of X25 and telephone call. Calling party initiates the connection and the called party receives the connection. The calling party or client supplies the address of the server.

- **X25 Layers:** X.25 has three layers are:
  - *Physical Layer:* Commonly used interface of ITU (EIA) are as follows: V24 (EIA RS 232), V28 (EIA RS 423), V11 (EIA RS 449), X21 (X24/V11), V35 RS422 etc. X.21 physical implementer is commonly used for the linking.
  - *Datalink Layer:* Link access protocol—Balanced (LAPB) is concerned with link access procedures for exchanging data over the link. Control information for transmission over the link is attached to the packets from the packet layer to form the LAPB frame.
  - *Network layer:* Packet layer protocol (PLP); This layer is concerned in defining the format of data packets and their control and transmissions.
- **X25-Packetization and Data Transfer:** In X25 it is possible to chain data packets together by "M-bit," so that at destination point they can be linked together again. It is also possible to use X.25 as a simple character stream but x25 is more popular for its ability to preserve the boundaries of blocks of data. Carrying TCP/IP data over X25 follows rules of RFC1356. In such case entire X25 would be layer2 for TCP/IP.
- **X.25-Point-to-Point:** X25 can be used for Point-to-point communication also. In such case there shall be one DTE and one Data communication equipment (DCE). PPP treats X.25 framing as a bit synchronous link. The link must be full-duplex. RFC Provides the rule for PPP in X25.
- **X.25: PLP:** X.25 Packet Layer Protocol is in network layer to handle the virtual circuits with multiplexing capability. Normally, logical channel number *Zero* is reserved for link control traffic which is concerned with restart and diagnostic packets.
- **X.25 Packet Format:** The X.25 Packet contains a *header* and *payload* or packet body. Within header top 4 bits of first byte of header there is General format identifier (GFI) to specify the format of the rest of the header (varies according to sequence number). It also contains Q-bit and D-bit.
  - *Sequence Numbers:* Data packets contain both a send sequence number P(S) and a receive sequence number P(R).
  - *M-bit—More Data:* The M-bit only appears in data packets (within the Packet Type field), and is used to link data packets together.
  - *D-bit—Delivery Confirmation:* The D-bit applies to Data packets.
  - *Q-bit—Qualified Data:* Qualified data enables a parallel channel of data to be transferred at the same time as normal data, typically used for control information.
  - *Interrupt Packets:* They provide a mechanism whereby a DTE can send a message which can overtake normal data flow—perhaps, useful if the virtual circuit is flow-controlled and the window is blocked.

5. **Frame Relay: Frame Relay:** Frame Relay is a high-performance WAN protocol developed for use across Integrated services digital network (ISDN). This is LAN-to-WAN interconnection technology for routers and bridges, with high bandwidth and low-delay characteristics. It can effectively optimize bandwidth, consolidate networks, and do multiprotocol bridging and routing. A Frame Relay node can start switching traffic out onto a line as soon as it has read the first 2 bytes of addressing information without caring if the frame is error-free or not. Frame relay supports variable length packets and statistical multiplexing—the characteristic features of packet switching and delivers frame of data from source to destination, passing through several switches, within only a few bytes' delay. As, no error checking is done at any intermediate but done at the ends, it can offer as high speed 2044 Mbps compared with about 64 kbps in X.25. In terms of memory capacity requirement, handling a large number of small packages and high speeds large files transferred Frame relay is more efficient when compared with X25. While X.25 guarantees data integrity and network managed flow control at the cost of some network delays. Frame Relay switches packets end-to-end much faster with no guarantee of data integrity at all. Frame relay networks are unable to correct or recover errors in frames or retransmit discarded frames due to transmission error, illegal frames, or network congestion.
   - **Functioning:** Frame relay supports both PVC and SVC and based on transmission requirement. Frame relay functions on virtual circuit created and used by data link layer. Data link connection identifier (DLCI) identifies virtual circuits in Frame relay protocol. After establishment of virtual circuit DTE is provided with one DLCI by which it starts communicates with another remote node. Frame relay switches, discussed later, play important role here. The connection details have been depicted in Fig. III 6.2a4 (leftmost sketch).
   - **Features:** A few features of Frame relay are as follows: a *connectionless service*, Wide *variation in speed56kbps to 25 Mbps*, *variable long frame* length (4kBytes), allows *bursty data*, *no delivery guarantee*, and *no error control.*
   - **Layers of Frame relay:** There are two layers in Frame relay, so, packet switching, fragments the message into transmission units called frames.

- *Physical layer:* Any ANSI supported protocol can be used by user.
- *Data link layer:* Modified version of HDLC frame format is employed by the frame relay at the data link layer.

- **Equipment and architecture:** Basically, DTE DCE and Frame relay switches are basic equipment in Frame relay. Virtual circuit SVC/PVC and switches form the architecture of frame relay.
- **Switching:** As packet switching protocol Frame relay protocol uses frame switches to route packet from the source to destination. Basically, each switch is a special type of computer which forwards frames based on the address contained in it. Typical frame switches with interconnections have been shown in Fig. III 6.2a4. As shown, each switch is connected to several computers (in our case 4). A Frame switch has "M" input and "N" outputs as shown in Fig. III 6.2a4 (extreme right). For each Frame relay switch, there will be table a switching table and Frame relay switches act according to the table.
- **Format:** As already discussed, frame format with the exception that since in Frame relay there is no flow/error control hence "control" field is missing. Typical such frame format along with address details of Frame relay and bit size in each case has been shown in Fig. III 6.2a4 (Two most popular octets). As seen from the figure that Address of Frame relay could be 16–32 bit (actually in octets, that is, 2 octets, 3 octets or 4 octets. There are some changes in format of address with different octets. 10-bit DLCI refers to virtual connection of Frame relay switch and DTE. This 10 bits DLCI is represented in two parts upper DLCI (6bits) and lower DLCI (4bits). Various details (Ref: Fig. III 6.2a4) are:
  - *C/R (command/Response):* This is distinguishing bit for command/response for upper layer to understand and not used by Frame relay.
  - *Ea—Extended Address:* With this bit, the frame relay address can be extended from the original 2-byte address to 3 or 4- byte 1 in this field indicates it is the final byte, and 0 indicates that there is another address byte to follow.
  - *FECN-Forward explicit congestion notification:* This bit is set to indicate that traffic is congested in the direction of travel also, the destination is informed about the congestion via this bit.
  - *BECN—Backward Explicit Congestion Notification:* This bit is set to indicate the congestion in the direction opposite to the direction of frame travel also it informs the source about it.

### 6.2.1 Discussion on data link protocols

The data link layer manages the delivery of local frames between nodes on the same level of the network. These are also referred to as data-link frames. Major functions of data link protocols are as follows: *Framing* (DLL encapsulates data received), *Flow control* (regulation of data to prevent overwhelming), *Error detection and correction* (with mechanism like parity checks, checksums, or Cyclic redundancy check (CRC)), *Media access control* (management of devices access to the physical medium), *Addressing* (physical addresses to identify and locate devices) and *Reliable* and efficient *delivery with* between nodes. Various data link protocols are required for Wide area network (WAN) and modem connections. Logical link control (LLC) is a Local Area Network (LAN) data link protocol. As part of interface, there are four basic things happen in circular manner. These are set up, transfer, disconnect and idle (after disconnection). Link set-up phase and disconnection phase are collectively referred to as link management (not to be confused with Link management protocol), that is, except data transfer all other three form link management. Data link protocols include but not limited to High-Level Data Link Control (HDLC), Synchronous Data Link Control (SDLC), Link access procedure/protocol (LAP)—LAP-Balanced (LAPB), LAP-D channel (LAPD), LAP for modem (LAPM), Serial line interface protocol (SLIP), Link control protocol (LCP)/ Point to point protocol (PPP), Network control Protocol (NCP) etc. Here we shall discuss various data link protocol types. Frame formats for various protocols have been depicted in Fig. III 6.2b. Let the discussions start with HDLC.

1. **HDLC Protocol:** General-purpose bit-oriented data link layer protocol HDLC (ITU-T: 1979) has different modes of operation. Depending on HDLC mode used it provides different types of services. Data are encapsulated with header (comprising address and control frame) and a trailer containing CRC. The frames are separated by HDLC flag sequences. Flag sequences are sent between each frame and for *no data* to be transmitted. It is used with various serial interface defined in the physical layer viz. EIA/TIA-232, V.24, V.35 etc. HDLC protocol uses sliding window protocol with 3-bit sequence number. Now let us look into details about it briefly.
   - **Type of stations:** There are three types of stations and these are:
     - *Primary station (P):* Primary station controls the operation of the link through commands. It can hold separate sessions with different stations.
     - *Secondary station (S):* Secondary station operates under the control of the P. This can only

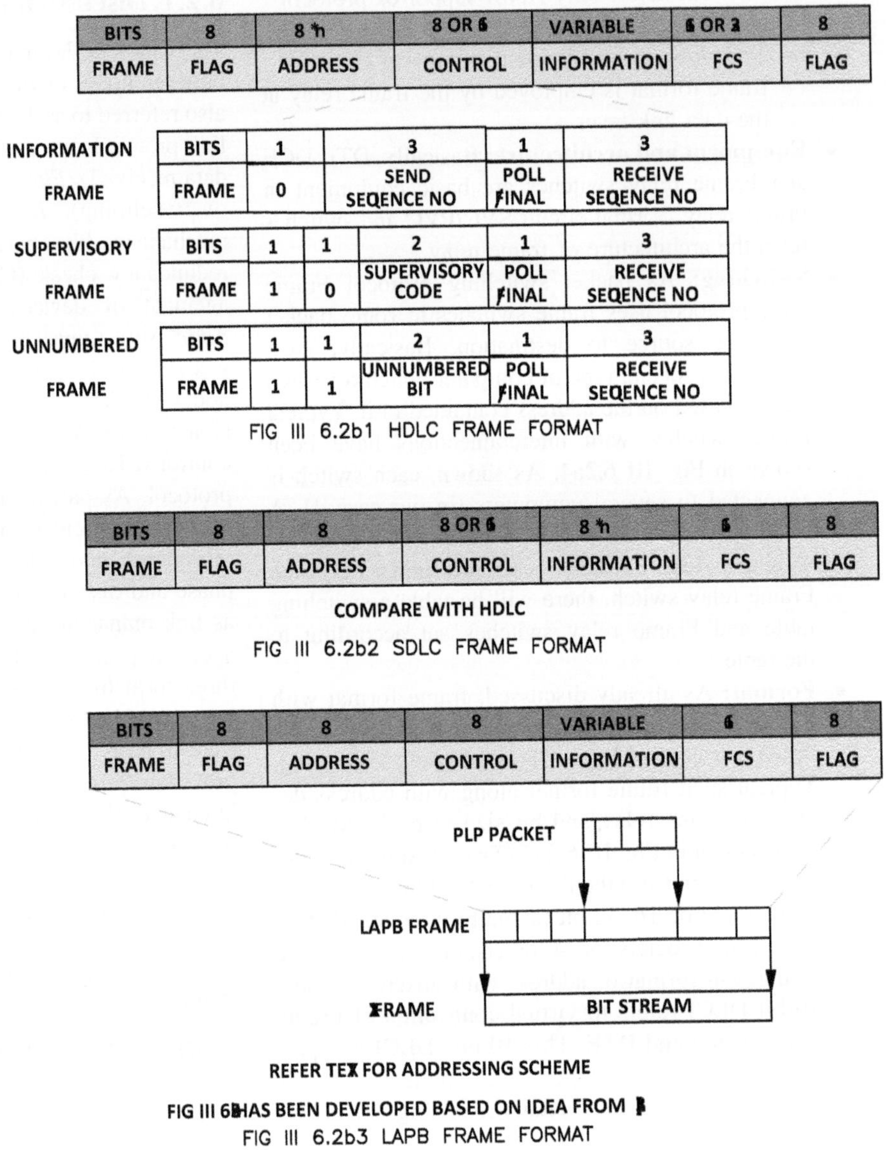


**FIGURE III 6.2B** Frame format for HDLC, SDLC and LAPB.

send responses to one primary station and they can talk to each other via P only.

- *Combined Station (C):* Combined station (C) is a combination of P and S with associated features hence this can transmit and receive commands and responses from one other station.

- **Link configurations:** Link configurations support full-duplex and half-duplex. There are three types of configuration viz. *Unbalanced* (one P to talk to several S over switched, unswitched, point-to-point or multipoint paths), *Balanced* (*two C* stations communicate *point-to-point* link—switch/unswitched) and *Symmetrical* type (two independent point to point unbalanced station).
- **Data Transfer modes:** Three data transfer modes types are:
  - *Normal response mode* (NRM)*:* It is used for unbalanced configuration, for example, data transfer of P to S (S needs permission from the P before it can transmit data) P- multidrop line, point-to-point. It is mainly used on multipoint lines.
  - *Asynchronous response mode* (ARM)*:* It is used for unbalanced configuration for transmission by S without permission of the P (it is hardly used). It is also used by P for initialization, error recovery, and disconnection, hub polling by P
  - *Asynchronous balanced mode* (ABM)*:* This is used for balanced configuration, for example, C initiated transmission or P to P transmission. Here either station can initiate transmission without permission from the other. This is commonly used on point-to-point links.

- **Frame Structure/Format:** Frame format with associated bit sizes has been depicted in Fig. III 6.2b1. The format consists of Flags at the beginning and end of the format used to differentiate frames. Address field consists of number of octets. Control fields are 8/16bit long. Changes here causes changes in frame class as discussed above. Frame check sequence (FCS), that is, checksum may be 16/32 bit and CRC is used. Information/data field can be arbitrarily long.
- **Classes of frames:** There are three classes of frames Viz. Information Frame, Supervisory frame and Unnumbered frame. For three classes there are variation in control field of HDLC frame.
  - *Information Frames*: Information frames are used for the data transfer between stations. The send sequence (next), and the receive (next) sequence.
  - *Supervisory Frames*: As the name suggests it supervises. These frames are used to acknowledge frames, request for retransmissions or to ask for suspension of transmission. The Supervisory code denotes the type of supervisory frame being sent. There are various types of supervisory frame viz. Type 0: acknowledgment, type 1: negative acknowledgment, Type 2 Receive not ready and Type 3 Selective Reject.
  - *Unnumbered Frames:* Unnumbered frames are used for link initialization or link disconnection. The Unnumbered bits indicate the type of Unnumbered frame being used. Another important issue is Poll/Final applicable to all the three classes of frames are important.
- **Poll/Final:** It is used when a computer/node is polling a group of terminals [13], that is, when it is set at poll p, it is used by primary- "P" station to get the bit is called Poll when used by the primary station to obtain a response from Secondary-S station, and Final f is set when used by the Secondary "S" station to indicate final or the end of transmission.

2. **SDLC protocol:** SDLC can be conceived as a subset of HDLC from its inception. It was meant to connect remote devices to central computers in either PPP/point-to-multipoint connections. Also, it ensures that data units move correctly and in the right order from one point to the next. Two types of links of SDLC: are as follows: I/O and non-I/O channel communications in various configurations like PPP, Multipoint, Loop and Hub types. In SDLC, a link station is a logical connection between adjacent nodes. Only one Primary Link Station is allowed on an SDLC line. A device can be set up as a Primary or a Secondary link station. SDLC supports line speed of 64kbps. Here all communication occurs via the single primary device. Typical SDLC frame format has been presented in Fig. III 6.2b2. Hardly there is much difference between them. Various field are as follows:
   - **Flag:** 8 bit at the beginning and end.
   - **Address:** Secondary address of 8bits
   - **Control:** Like HDLC, the variations in this field give rise to three kinds of frame classes: *Information* (for send receive sequence), *Supervisory:*(for acknowledgment, status, stop/termination) and *Unnumbered* (sequence numbers, for set up).
   - **Information/Data:** Path information unit (PIU)/Exchange identification (XID).
   - **FCS:** This is standard checks carried out on sending and receiving.
3. **LAP–LAPB:** Link access protocol LAP when used in balanced it is referred to as LAPB. Similarly, when used for channel of ISDN it is referred to as LAPD. Similarly, LAP Frame-Mode bearer services (LAPF) provides the data link for frame relay networks. Let us concentrate now on LAPB. LAPB is bit oriented data link-layer protocol to manage packet framing communication between DTE and DCE. So, LAPB circuits can be established by DTE or DCE. It is mostly used with packet switching X.25 protocol. Since LAPB is a subset of HDLC, it has the same frame format, frame types, and field functions in limited way. Primary (P) station initiates the call and the Secondary (s) station responds. So, LAPB is restricted to the Asynchronous balanced mode (ABM) transfer mode only as it is appropriate only for combined stations (C).
   - **Frame structure:** Frame structure of LAPB has been shown in Fig. III 6.2b3 courtesy [14]. Here bit details of various fields are as follows:
     - *Flag:* At the beginning and end 8 bit (bit stuffed)
     - *Address:* 8 bit has basically two values Viz.: For binary value: **00000001** command is from DTE and response from DCE and Again, for binary value: **00000011** command is from DCE and response from DTE. These two values are not actual address because communication is in balanced mode, in full duplex mode and DTE or DCE may initiate or terminate communication, that is, Primary P or Secondary S stations. Thus, two addresses indicate whether DTE or DCE is in control of the communication. Since it is used with X.25 where address is handled by X121 (see X.25 discussed) hence this address is not much of importance.
     - *Control:* The control field distinguishes the frame types like HDLC. The types are is an I-frame, an S-frame, or a U frame discussed below.
     - *FCS:* As usual it is meant for error checking for transmission.

- **Frame Types:** There are three types of frames like HDLC. These are as follows: *I-Frames (*Information for upper-layer information like sequencing, flow control, and error detection and recovery etc.), *S-Frames (*Supervisory for control details viz. status reporting etc.) and *U-Frames (*Unnumbered carries no sequence number; used for link management).

4. **SLIP:** SLIP is a data link control facility (with limitations like lack of any error-detection and correction mechanisms) can be used for transmitting IP packets, between an Internet service provider (ISP) and a home user over a dial-up link. Major issues here are:
   - **Characteristics:** It has two new special characters viz. END (decimal 192) and ESC (decimal 129) used for framing and escape sequence. These characters work together to ensure that the END character is used correctly to mark the end of a packet and that the data within the packet is not misinterpreted necessary for reliable communication over serial lines using SLIP. Data are transferred synchronously. A payload (data) and a flag serving as an end delimiter make up an SLIP frame. Because of absence of any standard specification, there is no maximum packet size. However, the standard value is 1006 bytes of datagram for both sending and receiving.
   - **Features:** Some important features are:
     - *Configuration:* It can support a number of network configurations.
     - Overhead: Low overhead makes it suitable for use in microcontrollers.
     - *Implementation*: It is easier to implement.
     - *Authentication*: It doesn't perform any data authentication and dynamic IP allocation
     - *Error handling*: This lacks in error detecting or fixing errors during data transmission.

5. **Other protocols:** Apart from the above there are a few more data link protocols. PPP and LCP frames are depicted in Fig. III 6.2c. Some popular of them are:
   - **PPP:** PPP provides same functionality as SLIP in dial-up and leased router-router lines. It offers a framing technique to describe frames. However, this character-oriented protocol also has an error-detection function. As very reliable protocol, it is used to transport not only IP packets but also other kinds of packets. It is used similar to serial interfaces as HDLC Major features are as follows: *Packet framing*, *Bit transparency*, *Error detection* and *Multi-protocol.* PPP Frame has been elaborated in Fig. III 6.2c1
   - **LCP** It was developed by IEEE 802.2. It is employed to offer HDLC-style services on LAN to establish, configure, test, maintain and terminate links for the transmission of data frames, LCP is essentially a PPP protocol. LPP Frame has been depicted in Fig. III 6.2c2
   - **NCP:** Old NCP enables users to access computers and a few other devices at distant locations and transfer files between two or more computers. PPP is typically made up of a collection of protocols. NCP is always available. TCP/IP took over from NCP.

**FIGURE III 6.2C** Frame format for PPP and LCP.

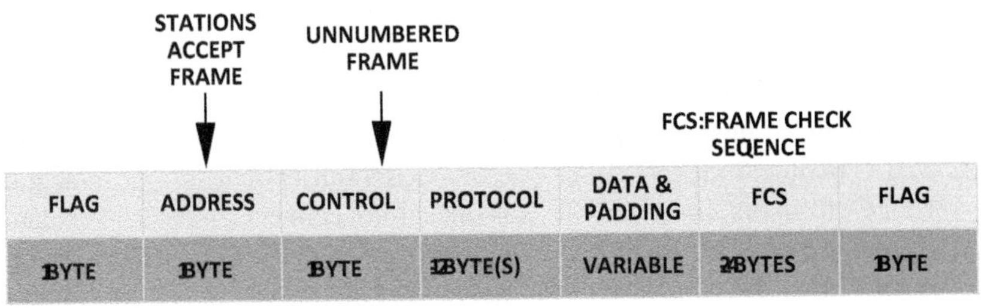


FIG III 6.2c1 FRAME FORMAT FOR PPP

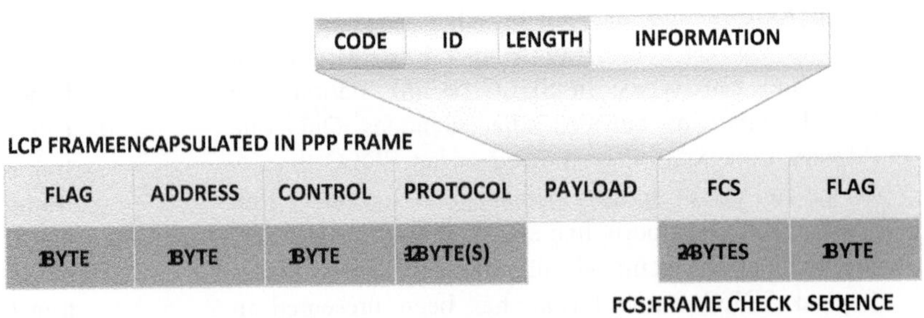


FIG III 6.2c2 FRAME FORMAT FOR LCP

### 6.2.2 Field level communication standards —IEC 61158 and 61784

With official completed IEC 61158 in 2009, resulted in many protocols commonly referred to as "fieldbus" and provided a quantum leap in technology. Enhancements to Electronic device description language (EDDL) have made it easy to use field level network "fieldbus." With more rigorous testing of devices and systems has improved interoperability between the fieldbus products from different manufacturers and provided simpler solutions for hazardous areas as well. Also, troubleshooting tools have evolved, and engineering guidelines have been established. Many enterprise giants began to implement networks at field level and referred to as *fieldbus*. So, Field level communication (FLC) standards have been developed with the aim to establish a uniform secured way for the field level devices (sensors, actuators, and controllers) to communicate and share data independent of vendor. The standard needs to enable secure, reliable, and interoperable communication between these devices, regardless of the manufacturer or platform. From Table III 6.2a, it is clear, that two standards viz. IEC 61158 and IEC 61784 play major role for field level communications. In this sub section, brief discussions will be presented on various aspects of these standards to understand the issue.

1. **Importance of FLC Standard:** Robust FLC standards are very important for various reasons, including associated challenges as well. Some issues are:
   - **Benefits:** Benefits of FLC standards includes but not limited to: *Interoperability* (devices from different manufacturers can communicate seamlessly), *Secured* (prevention of authorized access and protection of sensitive data), Reliability (reliable data transmission even under harshest condition) and Digital platform (common digital platform to connect field devices with higher level devices and cloud.
   - **Physical layer:** At physical layer a number of challenges mentioned below are well taken care of by FLC standard practices which otherwise could pose problems like
     - *Cable type*: In absence of suitable cable specification may jeopardize signal integrity.
     - *Wiring*: Incorrect wiring/improper grounding (loop) creates communication problems.
     - *Termination*: Less/excessive termination impacts signal level and network health.
     - *Signal level*: Non uniform/unspecified signal levels can cause problems with terminators/devices/communication network.
     - *Attenuation*: Unspecified, excess cable lengths can lead to attenuation issue.
   - **Security Issue:** Unprotected fieldbus networks are vulnerable to data protection and unauthorized access. Suitable standard protects against same.
   - **Interoperability:** Interoperability is important for FLC standard. Major issues are:
     - *Fieldbus standards*: This aims for open and interoperable standards except a few specific limitations of some protocols viz. IEC 61158/ IEC 61784
     - *Addressing and communication*: Proper device address configuration selecting appropriate communication mode is crucial for optimal performance. Also, selection of proper fieldbus is extremely important.
   - **Troubleshooting:** This is an important issue in FLC standard. Here pertinent issues include but not limited to: *Signal measurement*, *Interoperability test*, *Device type specific issue* and *pattern recognition.*
2. **Discussions on FLC standards:** IEC 61158 and IEC 61784 are two important standards in industrial data communication. IEC 61158 focuses on the specifications for fieldbuses and associated protocols. IEC 61784 defines communication profiles to be utilized by fieldbus protocols. So, IEC 61158 provides the "*building blocks*" for fieldbus whereas IEC 61784 establishes how these "building blocks" can be *connected* to create communication networks. It is structured into various parts, each addressing specific aspects of fieldbus, like physical layer specifications, data link layer services, and application layer protocols. In IEC 61158 there are three layers defined for fieldbus specification. These are Physical layer, Data link layer and Application layer. Other than physical layer, other layers together in fieldbus specification is called communication stack as shown in Fig. III 6.2d. In IEC 61158, other parts are arranged according to the layers defined for Fieldbus specification, that is, by physical layer, data link layer, application layer protocols, and services. IEC 61158-2: Fieldbus for use in industrial control systems—Part 2: Physical Layer specification" that covers non-Ethernet Fieldbus cabling details. It is advisable that IEC 61158 should be used in conjunction with the Fieldbus profiles documents IEC 61784-1, IEC 61784-2 and IEC 61784-5. According to IEC61158, a fieldbus is a digital, serial, multidrop, data bus for communication with industrial control and instrumentation devices such as—but not limited to—transducers, actuators, and local

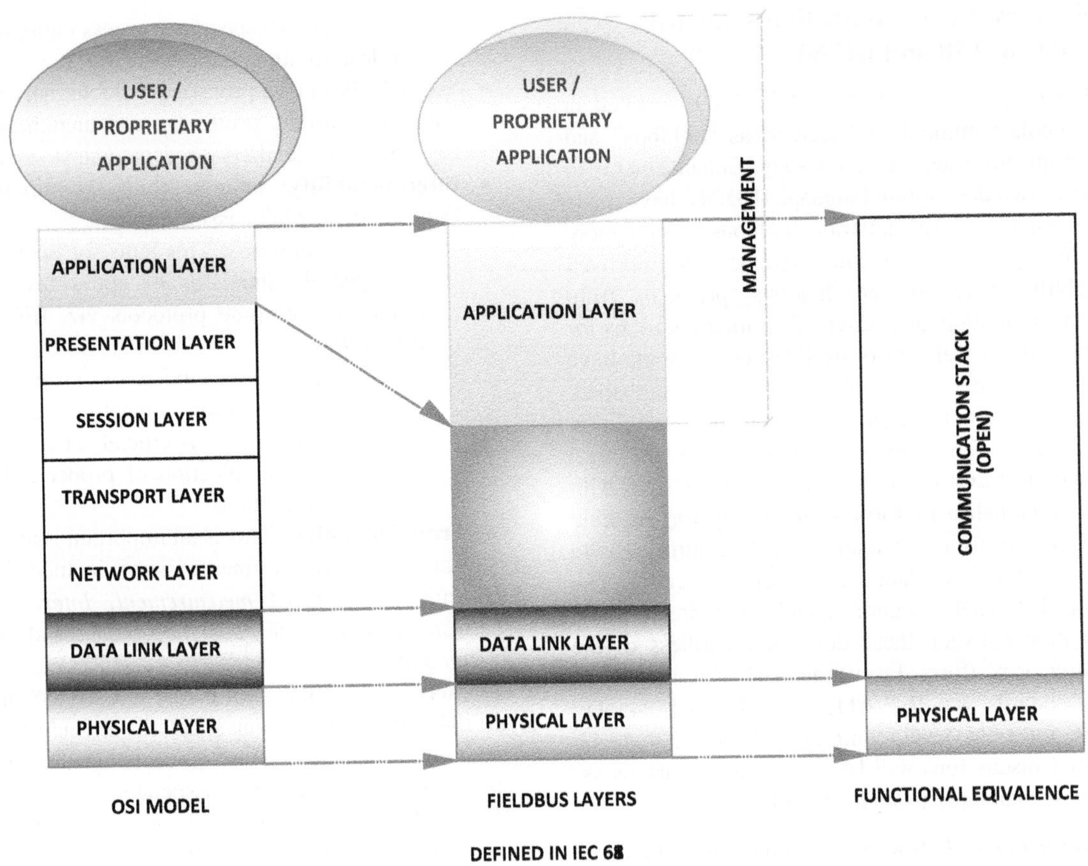


**FIGURE III 6.2D** Fieldbus and OSI model.

controllers. The Physical Layer specified in this international standard provides for transparent transmission of data units between Data link layer entities across physical connections. The physical layer provides services used by data link protocol and systems management. Basic structure of Fieldbus has been elaborated in Fig. III 6.2d, where it has been compared with respect to OSI model as well. The standard specifies type 1,2,3,4 6 and 8 physical media covering twisted pair wire, optical (single and dual fiber mode), co axial cable and wireless media. It also includes transmission through RS -232 and RS- 485. It also covers power via bus as well as intrinsic safety issues, full duplex mode of transmission and nonreturn to zero encoding. Now let us look into the structure of IEC61158. It is worth noting that in IEC 61158, abbreviations have been used as shown here: Physical Layer (PhL), Data link Layer (DLL) and Application Layer (AL). As discussed, IEC 61784 defines communication profiles. IEC 61784–3 deals with functional safety fieldbuses. Process control, Discrete manufacturing, Real time Ethernet (RTE), Building automation and standardized field communication including fieldbus are major application areas of these standards.

3. **IEC61158 structure:** IEC 61158 provides a standardized framework not only for fieldbus but for communication in various automation domains, enabling interoperability between different devices and systems. IEC 61158 has *various parts* as listed below:
   - IEC 61158-1: Fieldbus for use in industrial control systems—Part 1: Introductory guide.
   - IEC 61158-2: Fieldbus for use in industrial control systems—Part 2: Physical Layer specification (and service definition).
   - IEC 61158-3: Fieldbus for use in industrial control systems—Part 3: Data link layer service definition.
   - IEC 61158-4: Fieldbus for use in industrial control systems—Part 4: Data link layer protocol specification.
   - IEC 61158-5: Fieldbus for use in industrial control systems—Part 5: Application layer service definition.

**TABLE III 6.2C** List of common fieldbus features (Ref: IEC 61784 and 61158).

| Bus | CPF | IEC 61784–1 | IEC61784-2 | PhL | DLL | AL |
|---|---|---|---|---|---|---|
| Foundation FieldbusH1 | 1 | Profile 1/1 | | Type1 | Type1 | Type9 |
| FF HSE | 1 | Profile 1/2 | | 8802–3 | TCP/UDP/IP | Type5 |
| F F H2 | 1 | Profile 1/3 | | Type1 | Type1 | Type9 |
| CIP<br>ControlNet<br>DeviceNet<br>EthernetIP | <br>2<br>2<br>2 | <br>Profile 2/1<br>Profile 3/3<br>Profile 2/2 | Profile 2/2, 2/2.1[b] | Type 2<br>Type 2<br>8802–3 | Type 2<br>Type 2<br>TCP/UDP/IP | Type 2<br>Type 2<br>Type 2 |
| Profibus PA | 3 | Profile 3/2 | | Type 1 | Type 3 | Type 3 |
| Profibus DP | 3 | Profile 3/1 | | Type 3 | Type 3 | Type 3 |
| Profinet (4 types) | 3 | Profile 3/3 (Type CBA) | Profile 3/4 to 5 (I/O A, B, C) | 8802–3 | TCP/UDP/IP, Type 10 | Type 10 |
| World FIP (3 types) | 5 | Profile 5/1 to 3 | | Type 1 | Type 7 | Type 7 |
| Interbus | 6 | Profile 6/1 to 3 | | Type 8 | Type 8 | Type 8 |
| HART[a] | 9 | Profile 9/1 | | | | Type 20 |
| Ether CAT | 12 | | Profile 12/1, 2 | Type 12 | Type 12 | Type 12 |
| Modbus | 15 | | Profile 15/1, 2 | 8802–3 | TCP/IP | Type 15 |

*PhL*, DLL and AL are defined in IEC 61158.
[A]*HART is a protocol.*
[B]*For Ethernet Time synchronized.*

- IEC 61158-6: Fieldbus for use in industrial control systems—Part 6: Application layer protocol specification.
- IEC 61158-7: Network Management

4. **IEC61784 structure:** Ethernet is not part of the Fieldbus specification, however, standard IEC 61784 (RTE) relates it to IEC 61158, providing guidelines for using Ethernet in automation environments. So, IEC 61784 defines protocols for RT networking using *Fieldbus*, *Ethernet* and *IEEE 802.3*. IEC 61784 has *approved* total 5 parts as listed below:
   - IEC 61784-1, Industrial communication networks—Profiles—Part 1: Fieldbus profiles—Communication profile (CP)s for continuous and discrete manufacturing on Fieldbus.
   - IEC 61784-2, Industrial communication networks—-Profiles—Part 2: Additional fieldbus profiles for real-time networks based on ISO/IEC 8802-3
   - IEC 61784-3, Industrial communication networks—Profiles—Part 3: Functional safety fieldbuses. It has several parts to include additional specification for different CPFs.
   - IEC 61784-4, Industrial communication networks--Profiles—Part 4: Profile for secure communications in industrial networks (security—profile specific).
   - IEC 61784-5 (several parts), Industrial communication networks—Profiles—Part 5: Installation of fieldbuses.
5. **CP Families (CPF):** The most of the available and recognized fieldbuses (and HART protocol) has been assigned one communication profile number. Table III 6.2c enlist most commonly used fieldbuses.

Standard individually specified various layer characteristics. Based on these the buses can be compared. Tables III 6.2d–6.2f have been presented below for comparison of various popular bus types to conclude the discussions on Standards on Fieldbus systems.

CPs can be many, they are all organized into families. 16 different families originate. Some important ones are elaborated in Table III 6.2g.

With this let us conclude the discussions on Field level communication system as well as discussions on Industrial data communication in this chapter. Let us now move on to Fundamentals of Fieldbus system in Chapter 7.

**TABLE III 6.2D** Comparison of some Fieldbuses on Physical layer (PhL).

| Bus | Standard | Pwr-C[a] | Data rate | IS[b] | Distance -max | Device (nos) |
|---|---|---|---|---|---|---|
| Foundation FieldbusH1 | IEC61158/ ISA SP 50 | yes | 31.25 Kbps | Yes | 1.9/9.5 Km | 32/Segment |
| FF HSE | IEC 8802/ IEEE 802.3 | No | 100Mbps/ 1000 Mbps | No | 100m | No limit |
| Profibus PA | IEC61158 | Yes | 31.25 Kbps | Yes | 1.9/9.5 Km | 32/Segment |
| Profibus DP | IEEE1451.2, TIA 485 | No | 9.6 Kbps-12 Mbps | No | 1512m | 247/Segment |
| Profinet | IEC 8802/ IEEE 802.3 | No | 100Mbps/ 1000 Mbps | No | 100m | No limit |
| Modbus | IEEE1451.2, TIA 485 | No | 9.6 Kbps-12 Mbps | No | 1512m | 247/Segment |
| HART* | Bell 202, 4–20 mA | Yes | 1.2–9.6 Kbps | Yes | 3 Km | Analog 64 |

All buses have *digital* data communication except HART: where it is *digital over analog*
[a]*Pwr-C power with communication.*
[b]*Intrinsic safety.*

**TABLE III 6.2E** Comparison of some Fieldbuses on Data link layer (DLL).

| Bus | Standard | Link type | Error detect | Relation |
|---|---|---|---|---|
| FF H1[a] | IEC61158/ISA SP 50 | Token passing | 16bit CRC | Client/server, Pub/Sub, Sink/source |
| FF HSE | IEC 8802 | Token passing | 16bit CRC | Same as FF H1 |
| Profibus PA[a] | IEC 61158 | Token passing | 16bit CRC | Master/slave |
| Profibus DP | IEC 61158 | Master slave addressing | 1 bit | Master/slave,Pub/Sub |
| Profinet | IEC 8802 | Token passing | 16bit CRC | Master/slave |
| Modbus | | Master slave addressing | 1 bit | Master/slave |
| HART[b] | | Flat addressing | CRC | Master/Slave |

[a]*Deterministic communication for.*
[b]*HART not fieldbus but protocol.*

**TABLE III 6.2F** Comparison of some Fieldbuses on Application layer (AL).

| Bus | Data type | Field control | Peer to peer[b] | Alert and trends |
|---|---|---|---|---|
| FF H1 | A-I/O, D-I/O, PID, PD, CS, MI/O etc. | Yes | Yes | Yes |
| FF HSE | As above | Yes | Yes | Yes |
| Profibus PA | A-I/O, D-I/O | No | No | yes |
| Profibus DP | A-I/O, D-I/O | No | No | No |
| Profinet | A-I/O, D-I/O | No | No | Yes |
| Modbus | Registers | No | No | No |
| HART[a] | Command | Yes | No | No |

Standard IEC61158 and ISA SP50, *A*, Analog; *CS*, control sector; *D*, digital/derivative; *I*, input/integral; *M*, multiple; *O*, output; *P*, proportional.
[a]*HART not fieldbus but protocol.*
[b]*Communication.*

**TABLE III 6.2G** Some important communication profile families.

| CPF No. | Design basis | Brand |
|---|---|---|
| CPF 1 | Fieldbus | Foundation fieldbus |
| CPF 3 | Fieldbus | Profibus |
| CPF 4 | Industrial ethernet | Profinet |
| CPF 5 | Fieldbus | World FIP |
| CPF 13 | Ethernet IEEE 802.3 | Ethernet Powerlink |
| CPF 15 | Ethernet IEEE 802.3 | MODBUS |
| CPF 16 | Ethernet IEEE 802.3 | SERCOS |

## List of abbreviations used

**ABM** Asynchronous balanced mode
**ACI** Adjacent-channel interference
**ADC** Analog to digital converter
**ARP** Address resolution protocol
**ASK** Amplitude shift keying
**BPF** Band pass filters
**bps** Bits per second
**CCI** Co-channel interference
**CMI** Common mode interference
**CP/CPF** Communication profile/CP families
**CRC** Cyclic Redundancy Check
**DAC** Digital to analog converter
**DLCI** Data link connection identifier
**DLL** Data link layer
**DNIC** Data network identification code
**DoS** Denial-of-service
**DPSK** Differential PSK
**DTE** Data terminal equipment
**EDD-L** Electronic Device Description- Language
**EMI** Electromagnetic interference
**ESI** Electrostatic interference
**FCS** Frame check sequence
**FDM** Frequency division multiplexing
**FLC** Field level communication
**FSK** Frequency shift keying
**GFI** General format identifier
**HDLC** High-level data link control
**HW** Hardware
**IACS** Integrated automation and control system
**ICI** Intercarrier interference
**IE** Industrial Ethernet
**IIoT** Industrial IoT
**IoT** Internet of things
**ISDN** Integrated services digital network
**ISI** Inter-symbol interference
**ISP** Internet service provider
**IT** Information technology
**ITU** International Telecommunication union
**LAN** Local area network
**LAP** Link access procedure
**LAPB (D) (F)** Link Access Protocol Balanced (D channel) (Frame)
**LCP** Link control protocol
**LLC** Logical link control
**LPF** Low pass filter LPF
**MAC** Media access controller
**MAN** Metropolitan area network
**NCP** Network control protocol
**NUA** Network user address
**OT** Operational technology
**PCM** Pulse code modulation
**PIU** Path information unit
**PLP** Packet layer protocol
**PPP** Point to point protocol
**PSK** Phase shift keying
**PVC** Permanent virtual circuit
**PWM** Pulse width modulation
**RF** Radio frequency
**RFI** RF interference
**RTE** Real time Ethernet
**S/N R** Signal to noise ratio
**SDLC** Synchronous data link control
**SLIP** *Serial line interface protocol*
**STDM** Synchronous/Statistical TDM
**SVC** Switched virtual circuit
**SW** Software
**TDM** Time division Multiplexing
**VLSI** Very large-scale integration
**WAN** Wide area network
**WDM** Wavelength-division multiplexing
**XID** Exchange identification

## References

[1] W. Stallings, Data and Computer Communications, 8th ed., Prentice Hall, Pearson, 2007.

[2] Digital Communication - Pulse Shaping; Tutorialspoint; Internet Document; https://www.tutorialspoint.com/digital_communication/digital_communication_pulse_shaping.htm.

[3] Inter-Symbol Interference and Pulse Shaping; School of Computer Science; University of Manchester; CS3282: Digital Communications '05-06; http://www.cs.man.ac.uk/~barry/mydocs/CS3282/Notes/DC06_6.pdf.

[4] Pulse-Shape Filtering in Communications Systems; National Instruments; 2019; https://www.ni.com/en-in/innovations/white-papers/06/pulse-shape-filtering-in-communications-systems.html.

[5] Error Control in Communication Networks; D. Thakur; Communication Network; e-computer Notes; Internet Document; https://ecomputernotes.com/computernetworkingnotes/communication-networks/errorcontrol.

[6] What is Flow-Control in Networking?; Admin; After Academy; January'2020; internet document; https://afteracademy.com/blog/what-is-flow-control-in-networking.

[7] D. Communication Concepts, V.K. Garg, Y.C. Wang, The Electrical Engineering Handbook, Elsevier, 2005. https://www.sciencedirect.com/science/article/pii/B9780121709600500736.

[8] S.C. Mohontal, M.F. Ali, M. G. Sadeque3 Dept of Eee, PUST, Bangladesh, Study of different types of noise and Its effects in

communication systems, International Journal of Engineering and Management Research 5 (2) (2015). www.ijemr.net/DOC/StudyOfDifferentTypesOfNoiseAndItsEffectsInCommunicationSystems(410-413).pdf.

[9] Coping with Wi-Fi's biggest problem: interference; D. Callisch;, Ruckus Wireless; Network World; 2010; www.networkworld.com/article/2215287/coping-with-wi-fi-s-biggest-problem-interference.html#:~:text=To%20minimize%20co%2Dchannel%20interference,travel%20beyond%20these%20architected%20limits.

[10] Finding your way in the maze of industrial Ethernet standards; Fluke networks; Internet document; https://www.newark.com/pdfs/techarticles/fluke/3433787_6513_ENG_A_W.PDF.

[11] Packet Switching; M. Hakimi; IO river; https://www.ioriver.io/terms/packet-switching.

[12] X.25 Networking Guide; Farsite Communications; Internet Document; https://www.farsite.com/X.25/X.25_info/X.25.htm#WhatIsAnX25Network.

[13] A.S. Tanenebaum, Computer Networks, 4th ed., Pretice Hall, Pearso, 2009.

[14] Link Access Procedure Balanced (LAPB); InetDaemon.Com; Internet document; www.inetdaemon.com/tutorials/telecom/x25/lapb.shtml#:~:text=Link Access Procedure%2C Balanced (LAPB,25%2C Frame Relay and ATM.

# Chapter 7

# Fundamentals of fieldbus systems

Even though it was first installed in the early 1970s, standardization work on Fieldbuses started in the mid1980s. Later in the second half of the 1980s IEC started working on it.

The most promising results were the French FIP and the German PROFIBUS and were standardized on the respective national level and finally proposed to the IEC for international standardization. Unfortunately, the approaches of the two systems were completely different. In view of the same, 1990–94 witnessed German-French Fieldbus war. In modern industries, different kinds of embedded systems coexist and are utilized in the most of the industrial areas which are growing number of areas. Each requires adjustments to operate correctly, and to integrate with eventual larger ecosystems. These embedded systems are interconnected through fieldbuses. They are optimized to serve the specific requirements put on by critical and sensitive tasks emerging in industrial environments. Field level communication includes wireless systems, Real time Ethernet (RTE)/Industrial Ethernet (IE), and Fieldbus (both hardwired and wireless) also and each are having different market share. Fieldbus systems are digital communication networks that connect field sensors, devices actuators, and controllers in automation network in industrial plants. They not only offer reduced wiring but also offer improved diagnostics, enhanced control capabilities utilizing multidrop, serial bus architecture for bidirectional communication.

Major features offered by fieldbus/field-level communication include the following: *Digital communication, Serial bus architecture, Bidirectional communication, Open architecture, Distributed control, Real-time control, Power over the network, Diagnostics and Self-monitoring, Interoperability and standardization, Reduced wiring and costs, Hazardous area with reduced design complexity and simplified verification, Faster Factory/Site acceptance test (FAT/SAT)*, and *Reduced downtime with predictive maintenance strategy*. Another big advantage is that power and communication are transmitted via the same shielded twisted-pair cable is another important advantage. Selection of an open, digital fieldbus system for allows end users to select best-in-class instrumentation and integrated plant automation for both green and brown field projects. Many criteria must be considered when selecting a fieldbus system. Cycle time, Determinism, Protocol efficiency, Expansions, Diagnostics, Security are a few issues normally considered for selecting fieldbus system.

In this chapter basics of field-level communication with special reference to fieldbus shall be discussed.

## 7.0 General discussions on fieldbus system

As discussed, field-level communication encompasses IE, Wireless and fieldbus technologies also. Table III 7.0a enumerates the, market share of all these globally.

Here, discussions are mainly concentrated around Fieldbus infrastructure which brings a common frame of reference to the technology; choices are guided by specifications, registration, and application pertinent to field sensors/devices/actuators. Fieldbuses in automation, digitalize signals to communicate. This distinguishes fieldbus from protocols like HART. It offers less requirements for instrumentation as well as lesser space, Remote parameterization, predictive maintenance, etc. It also provides simple prove of explosion protection—Fieldbus intrinsically safe concept (FISCO), etc., have been discussed at length in Chapter 5 of Volume II of this book and the same may be referred to for safe fieldbus. Plant asset management is about getting the greatest benefit out of investment, and is well achievable in fieldbus technology. Fieldbus technology allows many smart devices to share one communication network. Two of the benchmarks issues that drive Foundation infrastructure application are the *freedom to choose and the power to integrate* during fieldbus implementation/installation. The Fieldbus infrastructure brings a common frame of reference to the technology; choices are guided by specifications, registration, and application. Fieldbus cables are defined by several standards. However, in this connection, ISA Standard SP 50–02 may also be referred to. It is worth noting that as safe fieldbus has been discussed at length in Chapter 5 of Volume II of this book and the same may be referred to for safe fieldbus. In section general discussions on

Plant Intelligent Automation and Digital Transformation. https://doi.org/10.1016/B978-0-323-89892-8.00003-X

**TABLE III 7.0A Field-level Communication Technologies: Market share.**

| Technology | % share | Brand name | % share | Remarks |
|---|---|---|---|---|
| Fieldbus[a] | 22–27 | [a]Foundation fieldbus takes 80%–90% share in process control along with HART** | | |
| Fieldbus | | PROFIBUS | 7 | |
| Fieldbus | | MODBUS RTU | 5 | |
| Fieldbus | | CC link | 4 | |
| Fieldbus | | Others FBs | ~6 | |
| RTE/IE | 66–71 | | | |
| RTE/IE | | PROFINET | 23 | |
| RTE/IE | | Ethernet I/P | 21 | |
| RTE/IE | | EtherCAT | 16 | |
| Wireless | 7 | | | |

[a]*Figures mentioned here are approximate subject to upgrade with time; **HART: Highway addressable remote transducer.*

fieldbus shall be presented. Let the discussions start with conceptual discussions on Fieldbus.

## 7.0.0 Fieldbus conceptual discussions

Generally, fieldbus infrastructure refers to a control system that uses a digital, two-way, multidrop communication link among smart digital measurement and control devices. Fieldbus network is like a Local area network (LAN) for process control, remote Input/Output (I/O), and high-speed factory automation applications. Fieldbus system Specification details out the physical medium/the bus or network, voltage levels for communication, timing, and connections with each function defined by a protocol standard. There exist similarities between different fieldbus types, for example, both FOUNDATION Fieldbus H1 and PROFIBUS PA transmit data and power. With the growth of plant automation, also increases number of monitoring and control points. This means there would be huge increase in cabling/wiring. Here comes the utility of digital data transfer through fieldbus which is meant to replace the bundle of parallel cables by a single bus cable to connect various devices at different levels from the field to the control level. Naturally, this reduction in cable also saves cabinet spaces as well as time for planning, designing, and installation. So, utility of fieldbus in modern plant cannot be overestimated. There exists large number of Fieldbus types approved by international standards, it is not an easy task to select one fieldbus type, Therefore, it is better to spend some time on selection process of fieldbus systems. Issues like *Process type* (process/manufacturing plant), *Topology*, *Hierarchical structure*, *Media support Protocol efficiency*, *Security*, etc., are a few important issues needs consideration. Let the conceptual discussions start with fieldbus concept.

1. **Fieldbus what and why:** Fieldbus can be conceived as a simple means of communicating with I/O devices (sensors, switches valves, drives, indication lamps, etc.) without the need to connect each individual device back to the controller. Thus, a fieldbus is an industrial network system for real-time distributed control. A fieldbus is a serial bus system used in machines and systems to connect I/Os to each other and to a number of masters/controllers making it possible to exchange data between different system components over long distances in master slave mode and under high external load. Power can also be transmitted over the bus as well. Implementation of fieldbus system architectures is growing at a rapid pace in diverse industries around the world. So, in present time, through its usage, fieldbus technology has achieved industry standard status. This is more so, for process control systems with a wide variety of instruments exists to accomplish specific types of process control. Fieldbus specifications also include Safety instrumented functions (SIF), which make fieldbus technology suitable for Safety instrumented systems (SIS) applications. Higher initial costs as each node must have communication equipment for transmitting and receiving the data. As against that there are a number of advantages. Major reason to use Fieldbus communication includes but not limited to:
   - Flexibility in addition, modify and deletion of devices on the bus.
   - Higher range of coverage than that in point-to-point communication.

- Huge reduction in cabling/wiring and the total costs.
- Extensive diagnostics for fault detection and detailed information of devices.
- Data and power transmission over same bus
- Interoperability to connect products from different manufacturers.
- Easy installation and operation of the fieldbus with associative devices.
- Reduced design complexity with simplified verification for hazardous areas.
- Reduced time for FAT/SAT and downtime for predictive maintenance.

2. **Common issues:** There are a few common issues for fieldbus. These are as follows:
   - **Fieldbus Architecture**: Fieldbus follows ISO—Open system interconnect (OSI) model with limited layers. Instead of seven layers it uses physical layer, data link layer, and application layer. As the communication takes place over a *fixed* network *routing* and *transport layers* are in true sense redundant. Also, in an industrial control environment, the network software processes are generally *invariant*. So, requirements of the session and the presentation layers are insignificant. Therefore, third to sixth layers of OSI models have been omitted in the Field bus protocol as depicted in Fig III 6.2d. It is worth noting that fieldbus Data link layer (DLL) protocol is a hybrid protocol that is capable of supporting both scheduled and asynchronous transfers. It defines three types of data link layer entities, a Link master (LM), a Basic device (BD), and a bridge. Link master devices can assume the role of the bus master, referred to as Link active scheduler (LAS) which decides which device transmits next and for how long.
   - **Cyclic and Acyclic communication:** There are two types of communications. These are as follows:
     - *Cyclic communication*: In cyclic communication, data are communicated in every fixed interval, without any lag or latency. This means data are updated continuously and at a predetermined schedule. A control loop requires a measured variable to be updated on a cyclic basis.
     - *Acyclic communication*: Alarm, Trend data, Operator data update, set point modification, or Controller tuning are examples of acyclic communication. In acyclic communication, data are communicated in every fixed interval, without any lag or latency, then it is cyclic communication. It means data are updated.
   - **Management functions:** In fieldbus, there are two management functions. These are as follows:
     - *Network management*: Network management function is to manage various parameters for fieldbus communication. Generally, communication software executes the communication function. The parameters for determining actual operation are called the Network management information base (NMIB).
     - *System management (SM)*: SM performs the management of the parameters needed for the construction of a functional control system, rather than communication.
   - **Cabling Challenges:** There are a few cabling challenges some discussed are in III 7.0.0.4. However, two common issues are as follows:
     - *Environmental factor*: Hersh environment and exposure to corrosive chemicals can all impact the performance and lifespan of Fieldbus cables and connectors. For this reason, selection of associated components like connector and cable is important.
     - *Mismatch issue*: Mismatch in communication protocols/cables/connectors will result in communication problems. So, compatibility testing prior to implementation for any changes or upgrades to a Fieldbus system, it's essential.
3. **Working of Fieldbus:** The availability of low-cost, powerful processors suitable for field instrumentation made it possible to replace the bulk of these cables with a single pair of cable called bus enables enhanced fast data transmission in the network. Fieldbus—a serial bi-directional digital communication protocol, allows smart field sensors/devices/actuator communicate with each other and with host systems in real-time control. Fieldbus utilizes packetized communication, that is, exchanges information/data in small packets sent sequentially and multiplexed over time. Major issues are as follows:
   - **Network communication**: Bus-connected network allows for multiple devices to connect to a single connection point, which then connects to the controller and provides real-time error-free data and DC power communication with faster response times.
   - **Packaged data:** In fieldbus, data/information are transferred in *small data packets* that are transmitted sequentially and multiplexed over time.
   - **Topology:** Fieldbus supports various kinds of topologies based on associated fieldbus specification.
   - **Standardized:** Fieldbus protocols are standardized, and structurally they follow IEC 61,158 and profile as per IEC 61,784.
   - **Power and Device quantity**: A fieldbus through a single twisted pair wire carrying both a digital signal and DC power (*through a fieldbus power supply*) that can connect up to 32 fieldbus devices to any control system.

- **Voltage and length:** Voltage range 9–32VDC is the allowed voltage range, and it can support a distance range about 1900m with Shielded twisted

4. **EDDL discussions:** EDDL (Electronic Device Description Language) and Field device tool/Device type management (FDT/DTM) are more advanced ways to describe devices. As discussed in I 12.3.0, [*To complete discussions on Fieldbus, discussions on EDDL has been included here also*] Electronic device description language (EDDL) is a programming language used to create Device descriptions (DDs) to describe the attributes of field devices in an *Operating system (OS) and Human machine interface (HMI)-neutral environment.* EDDL—a text-based description of the variable contained in field devices, is a universal, proven, and state-of-the-art method of accessing diagnostic, real-time, and asset management *information contained* in huge numbers of field instruments from a host of manufacturers. This provides optimum data and device interoperability. It defines associated variables and ways to access and carry out various operations like. The description defines each variable, and describes the way to access it. With EDDL technology, a user can *calibrate instruments*, *diagnostic issue*, and get data through Graphical u*ser interface* (GUI)/User interface (UI) displays, *process alarms.* It also helps to get high-level software Manufacturing execution systems (MES), UI/ Supervisory control and data acquisition (SCADA), plant historians, asset management, and Enterprise resource planning (ERP). As mentioned, this is supported by virtually every vendor. So, end users can get the associated benefits.
5. **Component and termination:** Following are basic requirements for components and termination based of IEC 61,158 standard for fieldbus.
   - **Component:** Fieldbus standard IEC 61,158–2 identifies the components required to create a complete segment like a host; fieldbus power supply, cable distribution, termination, and field devices. Apart from fieldbus component parts, standard also specifies the media and network configuration requirements to ensure proper data integrity before data-link layer error checking and interoperability between devices at the physical layer.
   - **Host:** This being first component of a segment is responsible for control of the segment and ensuring communication.
   - **Trunk:** The trunk refers to the main cable running along the entire segment to connect all other components for the segment.
   - **Power Supply:** Fieldbus power supply, different from the standard bulk power supply, to offer a matching impedance to the segment, which prevents this signal from being short circuited by the bulk power supply. Voltage level already discussed in III 7.0.0.3.
   - **Device Coupler:** The device coupler allows various numbers of devices to be parallel connected to the trunk. There are several types of junction boxes available on the market for this purpose.
   - **Other issues:** Some common important issues are as follows:
     - *Cable damage*: Being exposed to harsh industrial environments, Fieldbus cables are often get damage due to physical wear and tear. Through suitable cable management like cable routing and strain relief, situation may be combatted.
     - *Connector damage*: Connectors are critical components and failure connector can cause communication breakdowns. This can be circumvented by regular maintenance.
     - *Signal Interference*: Any signal interference can jeopardize fieldbus communication. To circumvent the situation Shielded twisted pair (STP) cable can be used.
   - **Segment Termination:** Every fieldbus segment must be terminated at both ends for proper communication and to avoid signal reflections. The primarily terminators like a current shunt for the control network. These are shown in Fig III 7.0a1and3. Most device couplers use manual on/off DIP switches to terminate couplers. In a segment, the last device coupler should contain the terminator, and all couplers between the last coupler and the H1 card should have their terminator switches also.
   - **Instrument connection**: Fieldbus connection details have been detailed out in Fig III 7.0a. A typical fieldbus connection with host and power supply has been shown in Fig III 7.0a1 which details out connect devices via junction boxes. Also, each fieldbus device connects to the segment in parallel, via a "drop" on the fieldbus segment called a spur. Fig III 7.0a2 shows the simplest spur connection as "T." Connection through device couple has been detailed in Fig III 7.0a3 with short circuit protection.
   - **Cables:** IEC 61,158–2 recommends cable details, for example, *length*, *gauge*, *inductance*, *capacitance,* etc. Some cable types used have been enlisted in Table III 7.0b.

### 7.0.1 Discussions on fieldbus with standards and field-level protocol

When the desire for serial cabling continued growing and the Fieldbus technology was introduced at the time and gained immense popularity. Fieldbus can communicate via just one cable and transferred information—in the form of bits—one after the other, rather than in parallel. Fieldbus

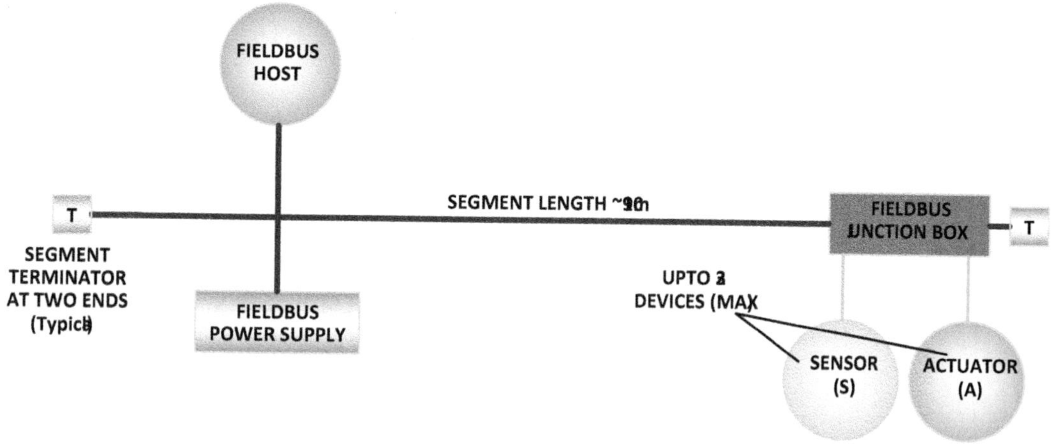


FIG III 7.0a1 FIELDBUS CONNECTION WITH POWER SUPPLY,HOST AND JUNCTION BOX

FIG III 7.0a2 FIELDBUS CONNECTION: SPUR AS "T'CONNECTION

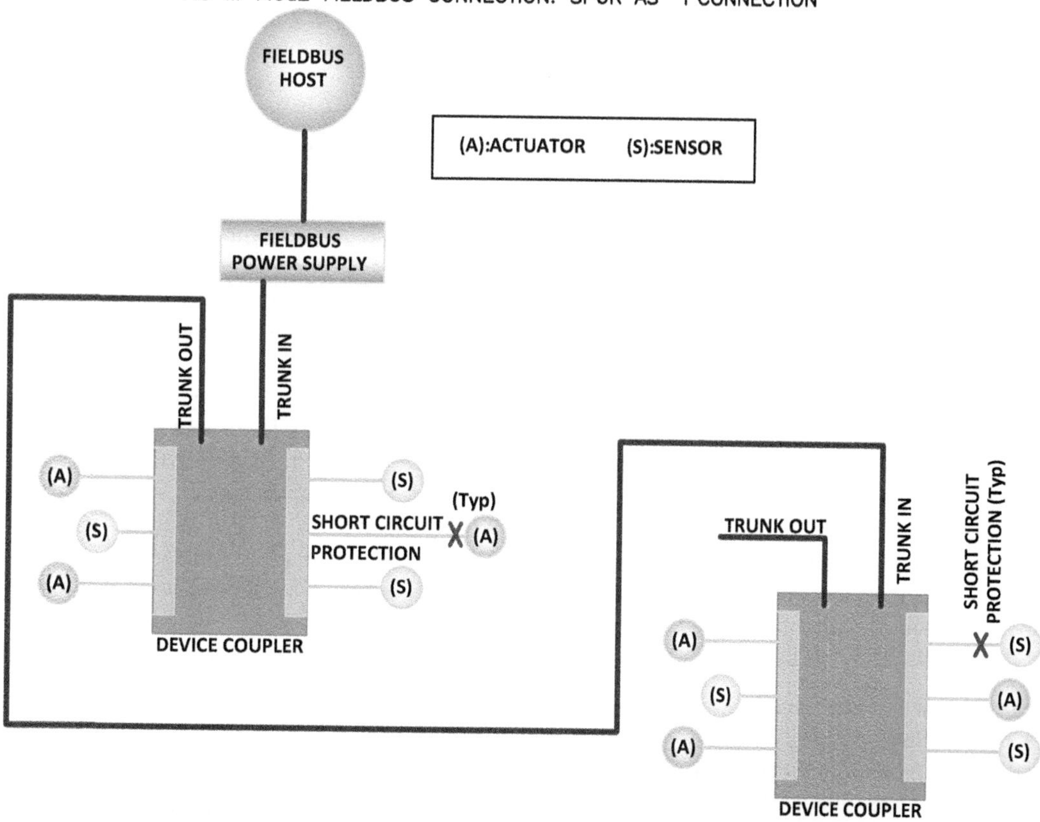


FIG III 7.0a3 FIELDBUS CONNECTION: DEVICE COUPLER & POWER SUPPLY

**FIGURE III 7.0A** Fieldbus connection details.

**TABLE III 7.0B** IEC 61,158–2 cable recommendations (Partial).

| Parameter | Type A | Type B | Type C | Type D |
|---|---|---|---|---|
| Description | Single STP | Multiple STP | Multiple UTP | Multicore + overall shield |
| Cross section | 0.8 mm$^2$ | 0.32 mm$^2$ | 0.13 mm$^2$ | 1.25 mm$^2$ |
| Trunk length | 1900m | 1200m | 400m | 200m |
| Impedance @ f = 31.25 kHz | 100 + 20 Ω | 100 + 30 Ω | Not specified | Not specified |
| Attenuation at f = 39 kHz | <3.0 dB/km | <5.0 dB/km | <8.0 dB/km | <8.0 dB/km |

UTP: Unshielded twisted pair

replaces analog systems by reliable accurate digital communication. Fieldbus can utilize various network topologies for efficient data transmission. The topology of a fieldbus describes the structure of the network. The transmission media used for data transmission are another distinguishing feature of fieldbuses. In this context, fieldbus standards are crucial for modern industrial automation. So, it is obvious that there will be standards for Fieldbus to establish reliable communication between smart field instruments and host systems in industrial automation, for real-time control. These standards support wide range of fieldbus types and the protocol choice depends on the specific application which can range from simple and cost-effective solutions to high-performance and complex systems. As backbone of communication in process automation/manufacturing, fieldbuses connect sensors/devices/actuators and controllers, for efficient and reliable data transmission between participants to establish simple but reliable and efficient real-time communication in complex systems with huge reducing wiring cost and complexities. HART technology is in operation for last over 40 years. while Fieldbus technology has been available for the last 2 decades, offering significant advantages over HART.

Both HART and fieldbus like Foundation fieldbus (FF) are both communication protocols used in process control systems, but they have distinct characteristics and capabilities. In this subsection discussions have been presented on application areas of various standards on fieldbus systems. Also, short comparative discussions have been presented on HART protocol with fieldbus.

1. **Fieldbus Requirements:** Fieldbus technology in industrial automation ensures reliable, efficient, and deterministic communication between among various participating devices like sensors/devices/actuators/controllers, etc. Normally, fieldbus systems should be developed in plug-and-play mode offering low maintenance requirements. Basic requirements for fieldbus technology have been detailed below:
   - **Communication and Topology:** Low latency, deterministic serial real-time communication. Fieldbus technology must support various network topologies.
   - **Interoperability:** Devices from different manufacturers must communicate using standardized protocols in line with the standards (IEC 61,158/61,784)
   - **Reliability:** Extremely robust reliable communication under harsh industrial conditions. This is ensured with High error detection and correction capabilities.
   - **System Integration:** Fieldbus technology must have capability pf system integration especially to higher-level systems like MES. SCADA, ERP, etc., via gateways/other means.
   - **Redundancy and Safety:** In line with IEC 61,784–3, must support redundancy and safety systems in fieldbus technology.
   - **Diagnostic issue:** Fieldbus technology needs to support status monitoring, diagnostics/remote diagnostics, as well as tools for system configuration.
   - **Physical layer:** In line with IEC 61,184–2 fieldbus technology should have physical layer which can tolerate noise, Electromagnetic interference (EMI)
   - **Addressing:** There will be unique address for each device in the network through automatic address generation in fieldbus.
   - **Scalability:** This is scalable in nature to support nodes from a few to hundreds.
   - **Bandwidth:** Efficient use of network bandwidth must be adopted for both cyclic (control) and acyclic (configuration) data.
2. **Discussions on IEC 61,158:** Fieldbus with associated standards facilitate connection between various devices, reducing wiring and simplifying communication in complex systems. During the discussions in III 6.2.2, we came across various fieldbus standards. Let us now discuss features of the standard and comparison between them.
   - **IEC 61,158 Features:** International standard IEC 61,158 refers mainly to structure of serial digital communication in fieldbus technology for industrial

automation. As discussed, it is structured into seven parts, each addressing different aspect of the communication stack mentioned in Fig III 6.2d. Table III 7.0c briefly elaborates various parts of IEC61158.

- **IEC 61,158–2 and FISCO:** The IEC 61,158–2 standard was developed for interconnection of various automation system components. So, some of the salient features of IEC 1158–2 related to transmission technology are presented in Table III 7.0d. Also, a few features for FISCO have been enlisted below [1]:
- Each segment should only have one power supply.
- No power fed to the bus when a station is sending.
- In steady state all field devices consume constant current
- Field devices are passive current sinks.
- Passive line termination at both ends of the main bus line.
- Linear, Tree, and Star topology allowed.

3 **Discussions on IEC 61,784:** IEC 61,784 focuses on fieldbuses and Real time Ethernet (RTE)/Industrial Ethernet (IE) for automation systems. The standard consists of several parts, each detailing different aspects of industrial Communication profiles (CP) and CP family (CPF). Let us also compare this standard with IEC61158.

- **IEC 61,784 Parts:** International standard IEC 61,784 entitled Digital data communications for measurement and control-communication profiles has several parts detailed out in Table III 7.0e. This standard details out mainly information on various communication profiles based on the protocols defined in IEC 61,158.
- **IEC 61,158 Vs IEC 61,784:** Table III 7.0f details out the comparison of two important standards for fieldbus IEC 61,158 and IEC 61,784.

4. **Modes and Relations** There are several ways in which various devices connected to the fieldbus. These also give rise to various modes of operation and relationships.

- **Communication Mode:** There are several modes in which fieldbus operates such as *Master slave* common in MODBUS, PROFIBUS DP/PA, etc. Another mode is bus *Arbitration* as in Foundation fieldbus (FF), WorldFIP. *Token passing (token ring)* is a deterministic way as found in FF-H1. *Stochastic/*

**TABLE III 7.0C** IEC 61,158: Parts details [PLs: Physical layer; DL: Data link; AL: Application].

| Standard | Title | Description |
|---|---|---|
| IEC 61158–1 | Overview and guidance | General overview and structure; relation of various parts to the OSI model; guidance for role and selection of different CP; guidance for the application of the standard. |
| IEC 61158–2 | PL specification and service definition | PL specification; defining: Signal level, transmission media, topology, bit timing; interoperability of devices |
| IEC 61158–3x | DL service | DLL services: Communication mechanisms, error detection and handling, message prioritization; interface between application and physical transmission |
| IEC61158–4x | DL protocol | DLL protocol definition: Frame, header, cyclic redundancy checks (CRCs); reliable data transmission between nodes |
| IEC61158–5x | AL service | Services offered by the AL layer; model: Client server/publisher-subscriber; communication object; standard accessibility to field devices |
| IEC61158–6x | AL Protocol spec. | Definition, syntax details: Structure; syntax, timing; state machine definition and interoperability of communication |
| IEC61158–7x | Network management | CP definition for PL, DL, AL; each profile corresponds to a specific protocol type; selection of a consistent protocol stack; to use with IEC 61784 |

"x" representing a specific fieldbus type as applicable

**TABLE III 7.0D** Features of IEC 61,158–2 [1].

| Features | Requirements | Features | Requirements |
|---|---|---|---|
| Transmission | Manchester bit Synchronized | Data security | Preamble, error<br>Free start and End delimiter |
| Speed | 31.25 kbps | Cable | 2-Wire STP |
| Explosion protection | IS EExia and ib; EEx d/m/p/q | No. of station | 32/Segment; max 126 address available |
| Topology | Line/tree/combination | Repeater | 4 |

**TABLE III 7.0E IEC 61,784: Parts details. [ FB: Fieldbus; ICS: Industrial control system].**

| Standard | Title | Description |
|---|---|---|
| IEC 61784–1 | Profiles for continuous and discrete manufacturing relative to fieldbus use in ICS | CP for IEC 61158: FB protocol; Guide to use a specific type of FB protocol in manufacturing |
| IEC 61784–2 | Additional profiles: Continuous and discrete manufacturing relative to fieldbus use in ICS | Extended part of IEC 61784–1; Additional communication profiles for specific or emerging applications |
| IEC 61784–3 | Functional safety fieldbuses – General rules and profile definitions | CP for functional safety applications; CP based on FB protocols suitable for safety-critical applications. |
| IEC 61784–4 | IEC 61784–4: Ethernet-APL (advanced physical layer) profiles | Profile for Ethernet-APL; ethernet in hazardous area for robust communication |

**TABLE III 7.0F Comparison of IEC 61,158 and 61,784.**

| Topic | IEC 61158 | IEC 61784 |
|---|---|---|
| Purpose | Fieldbus design, implementation, maintenance framework; ensure device interoperability | Selection and implementation of networks; choice of the appropriate protocol and configuration for specific use |
| Scope | Specification and definitions of protocol and layers viz. Physical, data link and application | Definition and specification of CPF for protocols in IEC 61158; |
| Structure | Divided into several parts, describing different aspects of the fieldbus including multiple protocol types | Description of CP and categorized in to CPF |
| Focus | Definition of protocols | Application of protocols |
| Level | Use in lower level | Important in higher level |
| Usage | Useful for developers/ engineers | Useful for system integrators and engineers |

*random bus access is* another way for communication mode as found in CSMA/CD.

- **Device relationship:** From IEC 61,158–5 we get that there are several device operational relationship these are as follows: *Client-Server*, that is, request and response relationship. In *Publisher–subscriber* (also called the producer–consumer), the publisher broadcasts data at its allotted time slot, the subscribers recognize data that they are of interest to update their local copies. In *Source-sink*, source transmits when it detects a necessity to report. One of the sinks receives the (multicast) message ensuring that the event is detected.

5. **Fieldbus and Protocol:** Fieldbus allows access to a wider range of device data and functionality, including complex diagnostics and control capabilities. HART, while providing valuable data access, is typically limited to the primary process variable and a few associated parameters, and is designed to be used with existing 4-20mADC instrumentation. Table III 7.0g shows comparison between Fieldbus and HART.

**TABLE III 7.0G Comparison between fieldbus and HART.**

| Issues | Fieldbus | HART |
|---|---|---|
| Communication | Fully digital in dedicated bus | 4-20mADC on digital overlay |
| Data rate | Faster 31.25 Kbps | Slower 1.2 Kbps |
| Topology | Multipoint | Point to point, multi point |
| Real time | Suitable for real time control | Limited real time control |
| Control | Can support local control | No support for control |
| Cost | Higher cost | Lower initial cost |
| Cable | Both STP and UTP | STP for short distance UTP |
| Cable length | Limited | Normally not restrictive |
| Push and poll | FF devices push data with events latched and time stamped | Polled periodically |

### 7.0.2 Fieldbus: Important issues

The operational characteristic of process plants and manufacturing units are not same, naturally various fieldbuses have been developed keeping an eye on these varied requirements. Also, there are certain industries slightly mixed in nature such as food beverage, pharmaceutical industries, or drilling operation (including mud engineering), etc. Naturally, it is very important to select fieldbus best suited for the application. Here, matching of CPF with application requirements is essential. On account of two-way multidrop communication, in place of analog signal, there has been huge difference in engineering. Even bigger differences in engineering have emerged from assigning the control functions to various field devices scattered throughout the plant, and allowing each of these devices to perform distributed control independently. In view of these following issues are raised [2]:

- Designing of control schemes inside fieldbus devices which incorporate some function blocks in each fieldbus device, and communication links between different devices.
- Startup, operation, and maintenance of the fieldbus system.
- The scope of fieldbus engineering also covers the development to maintenance of fieldbus devices, with different activities performed by users of different background. New engineering issues mostly include the following: *Fieldbus device selections and control-loop design, Design of communication and scheduling information,* and *Setting of engineering data in fieldbus devices.* Various stages are as follows:
- Manufacturer determines the function of the device with the parameters to be communicated over the fieldbus.
- System builder determines the required control function, and selects the devices, control-loop architecture, and parameter values to be used to realize this control function.

Engineering of the fieldbus system is performed in three steps: *Registration of devices, Control loop configuration,* and *Configuration data generation.* Engineering tools available are also used. Predictive diagnostic made it possible to shift from reactive and preventive maintenance practices to predictive maintenance practices to significantly reduce maintenance workload and cost while improving overall plant availability. Fieldbus provides a framework for implementing a vast array of device and process diagnostics. Major diagnostics are as follows: *Equipment diagnostics* (detection and alarm problems with field devices and process equipment) and Loop diagnostics. This would result in improvement in process availability, enhancing safety and environmental compliances as well as alarm monitoring, etc. In this subsection discussions on implementation steps including hardware and software issues including hardware safety have been covered.

1. **Plant and fieldbus:** Based on applications, there will be variations in best use i = of fieldbus type. Let us discuss the same starting with plant type.
   - **Plant Types**: It is needless to tell that there are several types of plants each having own requirement for fieldbus. Let us look into two important types of plants:
     - *Process Plants:* Process plants normally are continuous processed (or batch process) and are dominated by continuous regulatory control. Measurements and controls are continuous and modulating type [3]. Therefore, HART (included in IEC 61,158), Foundation fieldbus, PROFIBUS PA, are more popular in process industries. In fact, first two take major market share in process plants.
     - *Manufacturing industries*: Manufacturing units like automotive, bottling, and machinery industries, are mostly controlled using discrete logic and sensors. At times robots are used calling for special controls. Obviously, the network types are developed simple discrete I/O focus on low overhead and small data packets, but they are unsuitable for larger messages like configuration download [3]. In these cases, Interbus and AS-I and/or more advanced protocols oriented toward discrete logic include CIPs, and PROFIBUS (DP), etc., are suitable.
   - **Selection Issues:** A few criteria needing attention during selection process are as follows: *Media support, Variety of devices/integrating* all devices (industry/device specific), Cycle time (Deterministic or nondeterministic), *Transmission speed, style/type, Protocol efficiency, Different data and devices support* (viz. transmitters, encoder, Robotic, drive), *Easiness for device addition/deletion, Security features, Diagnostic features,* Product *availability and* product support. For convenience, comparison of various fieldbus types has been presented in Table III 7.0h
2. **Planning and Engineering** [4]: It is very important to plan fieldbus system implementation in plant especially for the first time in process control application. Small guidelines have been prescribed here for planning and engineering of fieldbus in a project. Its worth noting that cable selection is an important issue. Since this has already been covered in Table III 7.0b, cable discussions have not been included in the following discussions:
   - **Segment planning:** IEC 61,158–2 elaborates all requirements to be followed for proper operation of

**TABLE III 7.0H Comparison of various fieldbuses**

| Fieldbus | Physical media | Nodes | Distance[1] | Major area |
|---|---|---|---|---|
| AS-I<br>1993 | 2 wire cable | 31 | 300m | Assembly, packing material handling |
| CAN open<br>1995 | Twisted pair, (signal and power) | 30 | 1000m | Sensors, actuators, automotive |
| ControlNet<br>1996 | Coaxial, fiber optic | 99 | 1000m | Mission-critical, plant-wide networking of PCs, PLCs |
| DeviceNet<br>1994 | Twisted pair for signal and power | 64 | 500m | Assembly, materials handling, welding machines |
| Foundation fieldbus<br>1996<br>H1<br>HSE | Twisted pair/fiber<br>As above | 240<br>IP address | 1900m<br>2000m | Process sensors<br>Higher level communication |
| Industrial ethernet<br>1976 | Thin coax, twisted pair fiber, thick coax | 1024+ | ~200m | Miscellaneous |
| Interbus<br>1984 | Twisted pair, fiber, slip ring | 256 | 400m | Assembly, material handling/welding machine |
| MODBUS<br>RTU/ASCII<br>Modbus + | Twisted pair<br>Twisted pair | 256<br>32perseg. Total; 64 | 350m<br>500 m/segment | |
| Profibus<br>1995<br>DP/PA | Twisted pair/Fiber optic | 32/Segment | 200m | Process sensor/Inter PLC communication |
| World FIP<br>1988 | Twisted pair/Fiber optic | 64 | 2 Km | Real time controls: Process/machine |

fieldbus network. This calls for careful planning, design engineering especially for first time projects. Otherwise also such meticulous approach allows significant time savings during commissioning compared with classic interface technology. Here major issues are as follows: *Specification of environment* and *cycle time*, *Loops design* for *control and monitoring*, *Identification of the longest cable lengths*, *Selection of segment topology*, *Cable selection*, *Determination of number of devices connectable*, *Power supply and wiring interface design* and *Verification of the required trunk and spur length* (based on number of devices).

- **Environmental issues:** Environmental issues include but not limited to: *Shock/vibration*, *Corrosion ratings/requirements*, *Type of enclosure*, and *IP class/NEMA ratings of enclosures.* These are in addition to environmental conditions for device selections.
- **Loop design:** Designers need to develop typical loop with the highest demands or constraints. Major issues are as follows: *Longest cable length from the control room*, *Field devices current consumption*, *Control loops with demand for shortest response times* and *Highest bus cycle time acceptable and monitoring loops with acceptable longer bus cycle.*
- **Cycle time:** Bus cycle time is the time, taken for a master (host) to communicate with *all* slave devices on the bus. For faster responsiveness, a shorter bus cycle time is desired. for faster system responsiveness. However, it depends on protocol and hardware used. The bus cycle time of monitoring and control loops influences the overall system responsiveness. Many factors that influence the bus cycle time are as follows: *Fieldbus system* and *DCS system used*, *Number of devices on the segment*, *Amount of data to be transmitted*, and *Time required for asynchronous data exchange.*
- **Power supply:** The maximum current provided by the power supply determines the maximum number of field devices and wiring interfaces. As per standard a minimum prescribed current is 10 mA per field device (most devices have normally 10 ... 20 mA) with a 20% current reserve minimum per

segment. At least one short circuit condition must be considered for adding up current consumption to take care of intermittent short circuits. Power supply maximum voltage plus load current in *conjunction* with voltage drop on the cable is determining factor for maximum cable lengths. This is again influenced by cable type, load, and topology. Planner needs to keep 10% reserve to keep 10V as minimum to meet standard requirement of 9 V available at the field device.

- **Topology issue:** Typical Topology with power supply and barrier has been depicted in Fig III 7.0b. Even though theoretically, any topology can be selected, yet the trunk-and-spur and tree topologies are recommended because: *Usable for all segments of the plant, Easy to understand, work with as well as easy to maintain, Ensure limiting short circuit*

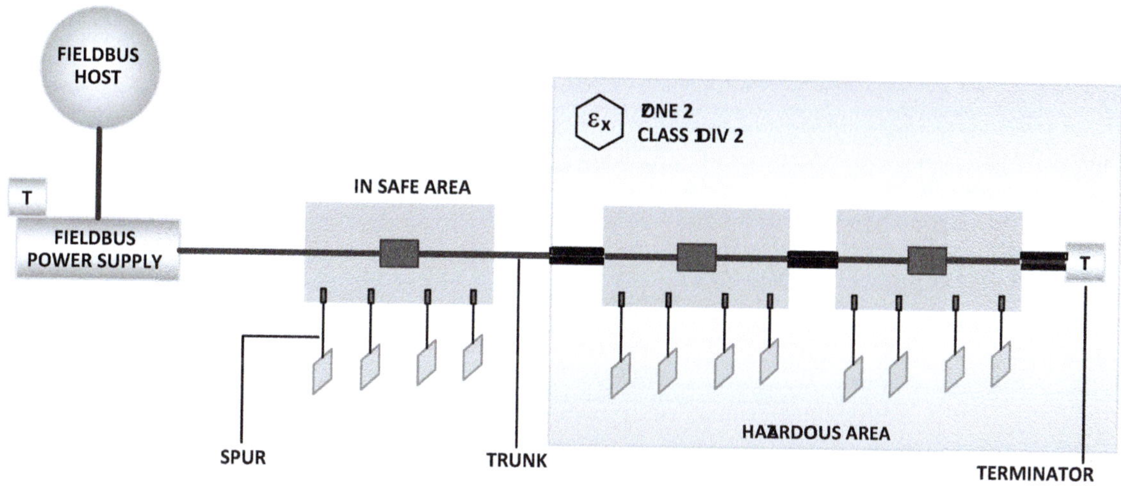


FIG III 7.0b1 FIELDBUS IN SAFE AND HAZARDOUS AREA WITH TRUNK & SPUR SHOWN

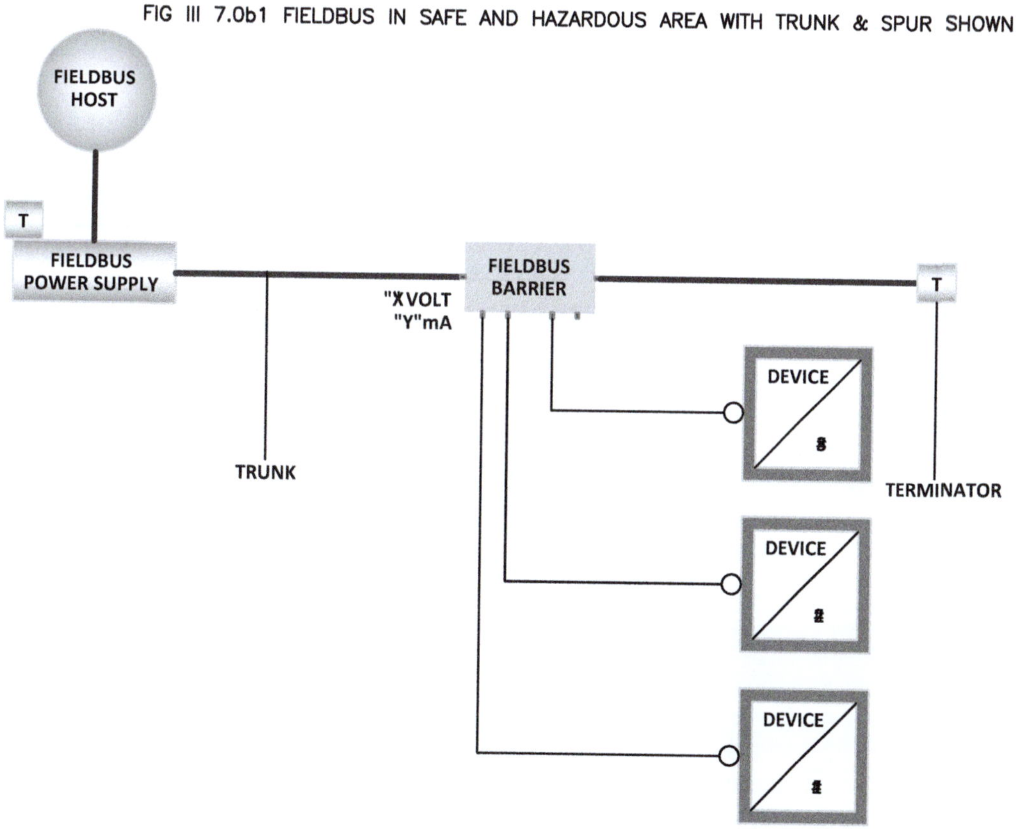


FIG III 7.0b2 FIELDBUS IN HAZARDOUS AREA WITH FIELDBUS BARRIER

**FIGURE III 7.0B** Fieldbus segment planning in various area.

*current for the trunk*, etc. As shown in Fig III 7.0b1 typically, the power supply is installed at one end of the trunk (in a cabinet) which is the longest part of the segment.

- **Junction Box and trunk:** For the safe area, distribution units with simple terminators and connectors are permissible. Simple junction boxes, are also installed near the field instruments based on the selected topology layout. Typical spur length based on no of participants is elaborated here (the figure inside bracket indicates the max no. of devices upto: 120m (1-12 nos), 90 m (13–14), 60m (13–18), 30m (19–24), and 1m (25–32). Fig III 7.0b2 depicts typical segment layout with power systems, barrier, etc.
- **Engineering tools:** Fig III 7.0c shows use of engineering tools for fieldbus design and engineering. A few important issues here are [2]:
  - *Builder*: Generates data compliant to fieldbus—to be set in the fieldbus devices
  - *Network startup function*: Downloads data to fieldbus devices.
  - *Maintenance function*: Uploads parameter values of blocks.

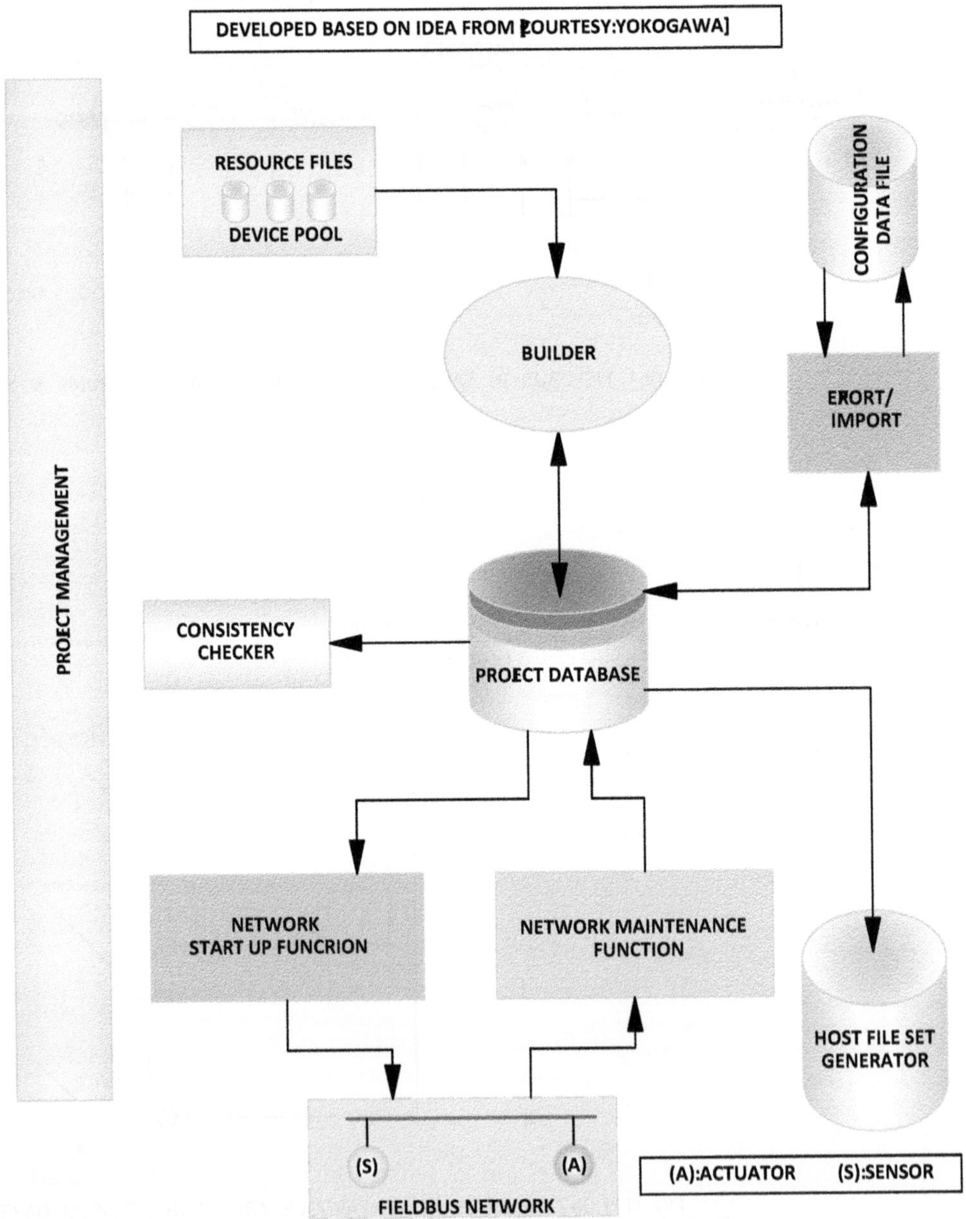


**FIGURE III 7.0C** Functional details of fieldbus engineering tools [2].

   - *Host-file set generator*: Generates engineering data about the fieldbus system for host applications.
   - *Project data base*: It can generate and download set values for host interface devices. It provides an external interface to import resource files and imports and exports value files.

3. **Diagnostic issue:** In fieldbus design and implementation, diagnostics are crucial for identification and resolution of communication issues to ensure optimal performance, with enhanced system reliability. A well-designed and maintained segment can operate without any problems for many years. Even to decrease cost further, online physical layer diagnostic module could be able to monitor several segments at the same time.
   Important diagnostic issues pertinent to fieldbus are as follows:
   - **Diagnostics and diagnostic tools:** Diagnostic tools and techniques help in pinpointing problems viz. faulty wiring, impedance mismatches, signal degradation, and even detect anomalies in field devices themselves. So, major issues diagnostics address are as follows: *Proactive monitoring*, *Real-time fault detection*, *Device-specific diagnostics*, *Troubleshooting*, *Online monitoring* and *analysis*, *System documentation* and *configuration*, and *Expert system.* Also, it is cost-effective to opt for attaching, and retaining, advanced automatic diagnostic test equipment on every trunk. This would actually reduce construction and commissioning expenditure and operational expenditure. This automatic diagnostic test device would offer cost reductions in every phase like construction, commissioning, and operational phase. Also, advanced diagnostics offer a variety of diagnostic functions leading to a more comprehensive reporting structure to reveal potential issues which may arise during plant operation. Various physical layer diagnostic tools include the following: handheld device, basic online device, advanced online device, and advanced mobile device.
   - **Diagnostic infrastructure:** The information from the diagnostic hardware can be transferred through the fieldbus itself/a dedicated diagnostic network. Latter one will be the most effective. Fieldbus failures can occur at any time without warning and it can range from an insignificant change to a fatal crash. Some of these are as follows: Pole-to-shield faults (unbalance due to a low-resistance contact to the shield by a cable pole), pole-to-pole fault, power supply drift, power supply failure (/bad health), device failure, polarity inverse, termination fault, water ingress fault, cross talk, noise, and interference.

4. **Implementation Steps:** Fieldbus, covering various aspects like physical layer installation, network topology, device configuration, and troubleshooting, can be implemented in a guided manner in step-by-step way. Following are key steps/issues:
   - **Planning and Design:** Major issues here are as follows: *Topology selection* (Based on requirements of the system, *selection* of the appropriate fieldbus network topology), *Segment design* (as discussed, segment planning on the basis of the *maximum number of devices* and the *maximum cable length*), and *Configuration* (Defining *addresses*, communication *parameters*, along with other settings).
   - **Physical layer:** As the name suggests, it deals with, physical media such as:
     - *Cable selection*: To use *twisted pair,* (STP preferred) cables designed for the specific fieldbus protocol.
     - *Termination*: Suitable cable termination at both ends to ensure no signal reflections.
     - *Wiring*: The wiring guidelines for the specific fieldbus protocol must be followed.
     - *Topology*: Mainly bus with spur but other types also for specific fieldbus.
     - *Power supply*: As discussed above, suitable and adequate power to the fieldbus network with the selected power supply with conditioner.
   - **Device Integration:** Major issues here are as follows: *Device addressing* (Assignment of unique address to each device), *Block configuration*, that is, configuration of the Function block (FB)s within devices for performing desired functions. Other functions include *Data link services* for communication between devices and host.
   - **Network Management:** This refers to *interoperability* (among different fieldbus devices and host systems) and *system management* (related to network health monitoring, troubleshooting, and device updates).
   - **Commissioning issue:** Device identification, configuration, and verification, etc., as per guidelines of IEC 61,158 are part of, major commissioning activities. Key steps include creating placeholders, reconciling differences, confirming commissioning, and calibrating/ranging the devices, configuration files (CFF), host configuration, and device description (DD). It is essential to confirm that the terminators are correctly located. It is also necessary to ensure that the DIP termination switches sometimes are set correctly, to create proper start-up of the fieldbus system. In case above recommendations are not followed, there could be problems during commissioning, and these could be: The

erratic behavior instruments, mysterious drop off the segment, etc. Following are some testing issues:
- *Network verification:* Testing of the network for proper communication and data integrity.
- *Device verification*: Verification of correct device configuration and communication.
- *Integration*: Fieldbus commissioning with control system for seamless integration.
- *Protocol and safety*: To follow specific implementation guidelines for different fieldbus protocols as well as safety guidelines for wiring in hazardous areas.

- **Redundancy issue:** In fieldbus, all communications and power are dependent upon a single twisted pair trunk cable. If the trunk cable fails, the fieldbus segment lost to the control system, and devices on the segment can no longer talk to each other. Although fieldbus instruments can continue to operate if the *control system* fails, any cable fault (open or short-circuit) could render the entire segment inoperable. Various fieldbus vendors, including major process control companies, have developed redundant fieldbus schemes as shown in Fig III 7.0d. Here, a "voting scheme" is needed if control systems cannot tell when a fieldbus segment fails.

5. **Grounding and Surge:** Proper grounding/earthing for noise protection and surge protection from lightening is an important issue in fieldbus design and implementation.

   Major issues are as follows:
   - **Grounding:** It is obvious that proper grounding is essential and important for immunity to noise of the segment as well as for proper explosion protection. This grounding of fieldbus system should be chosen in line with the grounding philosophy of the plant. Therefore, at design and engineering, these issues should be checked. Fig III 7.0e. Single and multipoint grounding have been shown in Fig III 7.0e 1 and 2. Capacitive grounding types have been depicted in Fig Fig III 7.0e 3.
   - **Surge protection:** Protection against the effects of a lightning strike is referred to as surge protection. A risk analysis based on the location of the fieldbus systems and system topology can be used. A surge can affect the fieldbus cables travels either toward the control system or toward the field devices through the field junction box. Surge protection should be used at various stages to prevent damage to the system. Surge protection in the control room to protect the control system against surges come

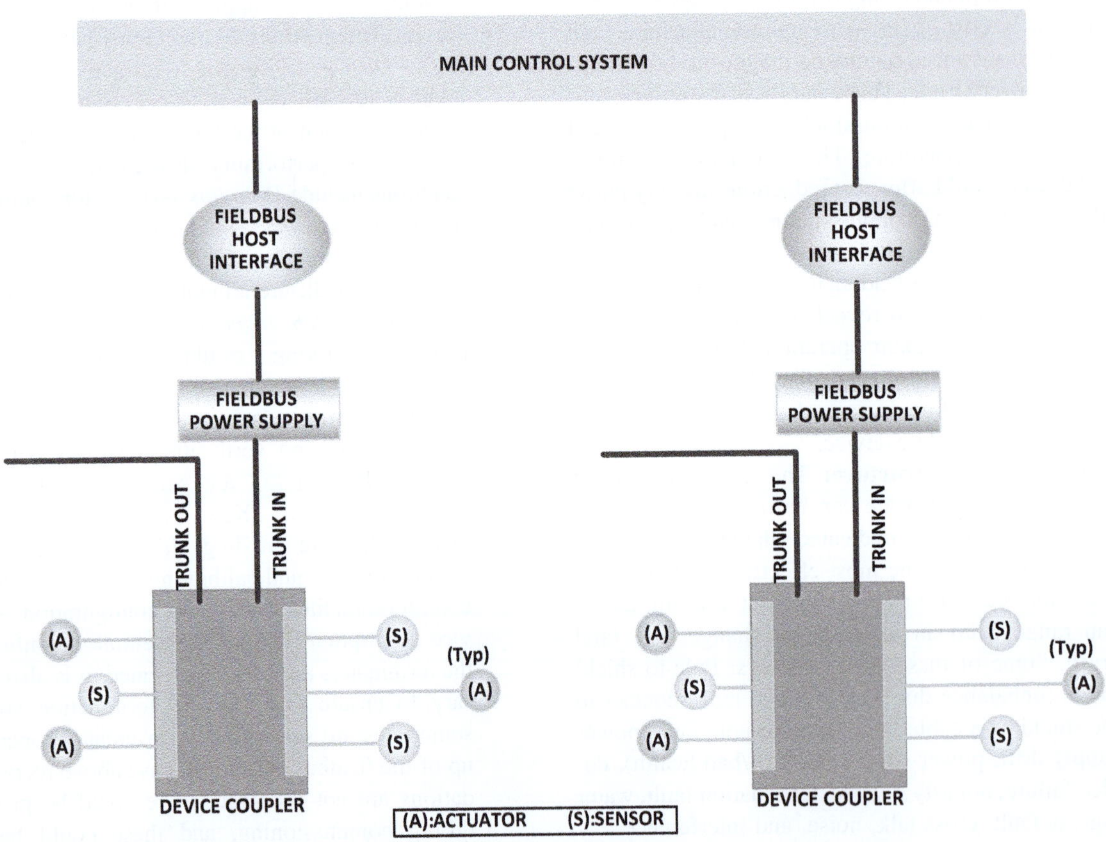


FIGURE III 7.0D Fieldbus redundancy system.

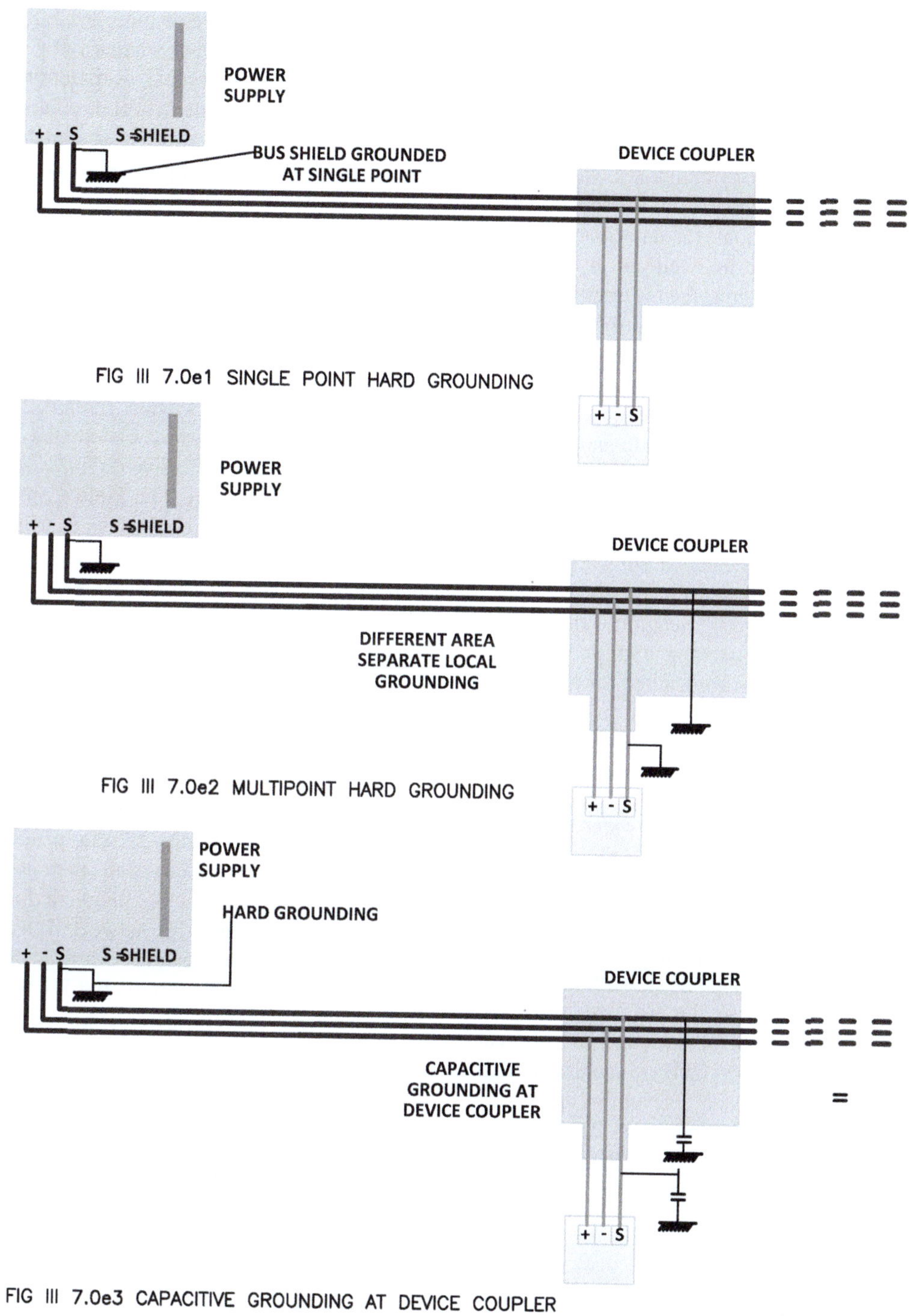


**FIGURE III 7.0E** Ground details of Fieldbus system.

from the field. Surge protection at the trunk connection of the field junction box protects the field side against surges. Surge protection at the spur side of the field junction box protects the electronics against surges. Surge protection at the field devices protects the field devices against surges.

## 7.1 Foundation fieldbus details

Proposed as a concept in 1984, Foundation fieldbus (FF) is a standard for digital field instrumentation enabling field instruments to communicate with each other and also to execute continuous control algorithms. That is, FF extends the concept of Distributed control system (DCS) at field level. Because of FF design & development for control-in-the-field enable strategies across the plant it is possible for field instruments can communicate directly with one another to maintain continued safe operations, during a host system failure. The technology comprises wide choice of physical layers and software platform. FF supports the NAMUR NE107 recommendations for data management for intelligent instrumentation as well as it supports exhaustive diagnostic system managed effectively. FF provides deterministic process control: control communications happen on schedule, without delays caused by other traffic on the bus. If a message doesn't get through, it tries again. If fieldbus devices lose their connection to the host system, they are capable of maintaining safe and effective control across the bus. Keeping the user informed. FF H1 was developed specifically to meet the needs of the process industry such as: *Delivery of power and communications over the same pair of wires*, *Withstand capability of the harsh and hazardous environment of process plants*, *Application in hazardous area including supporting intrinsic safety* and supporting existing wiring. Thus, H1 technology in FF is designed and well suited for field-level interface and device integration within the process plant environment for which this information-driven systems offer enhanced connectivity, real-time data, and advanced analytics. FF is well covered by a number of standards like ANSI/ISA 50.02, IEC 61,158, CENELEC EN50170:1996/A1, etc. A good number of companies are member of Fieldbus foundation.

For configuration and operation *Actual physical, DD, Common File Format CFF*) are important. These DD and CFF are furnished by the device supplier or Host. Parameters and capabilities are defined in these files and used for off line configuration. DD help operation of devices from different suppliers to work on the same bus. CFF are standard ASCII Text files to describe capabilities and functions of the device.

In this section foundation fieldbus shall be discussed briefly to cover all salient issues at length enabling to understand why this in conjunction with HART take nearly 85% of market share.

## 7.1.0 Foundation fieldbus overview

Foundation Fieldbus (FF) is basically meant/well suited for process plant applications. FF can implement independently distributed control. Honestly, FF can be conceived as a LAN for process sensors, actuators, and control devices compatible to FF. According to Fieldbus foundation "*Foundation Field is an integrated total architecture for information Integration—FOUNDATION field bus is an all-digital, serial, two-way communication system.*" It consists of H1, which interconnects field equipment and High-speed Ethernet (HSE 100 MB PS). Link Device is an HSE device to connect one or more H1 Links. It has been incorporated in international standard IEC 61,158–2. As usual, FF H1 is an interoperable, bi-directional, digital (I/O conversion unit is eliminated), serial, publisher-subscriber communications network, which is suitable for use in hazardous and potentially explosive areas (Ex zones 0 and 1), as well as intrinsic safety (IS) applications also. Based on the individual environment, each H1 segment supports a length of 1900 m and can connect max. 32 field devices. Using bridges, these limits can be increased. FF provides deterministic process control meaning that its control communications happen on schedule, without delays caused by other traffic on the bus. In case a message doesn't get through, it tries again. Even if fieldbus devices lose their connection to the host system, they are capable of maintaining safe and effective control across the bus keeping the user informed. The Field Comm Group tests and registers FF (H1 and HSE) devices to ensure interoperability of registered products from multiple vendors This provides end user choices for the best instruments for the application regardless of the host system supplier. In this subsection brief discussions on FF have been presented to cover overview of FF system.

1. **FF Family:** After withdrawal of H2 bus, Foundation fieldbus family consists of two types of buses viz.
   - **HI bus:** H1 Bus based on IEC 61,158–2 standard used filed devices like Sensors, Transmitters, and I/O, etc. Conceptually, H1 comprises six parts viz. links, devices, blocks and parameters, linkages, loops, and schedules. So, there are three types of devices on a Fieldbus H1 network: Link masters (controlling communication traffic), basic devices (Incapable of scheduling), and H1 bridges.
   - **HSE Bus:** HSE bus: it is High speed Ethernet Bus provides integration of high-speed controllers, subsystems via Link Device, Data Servers, and Workstations. HSE Device is any FF device-type connected directly to HSE Media. HSE devices contain a field device access agent, system management kernel and network management agent.

Fig III 7.1a depicts fundamental issues related to Foundation Field bus. Before going into details about the fieldbus it is better various benefits it can offer.

2. **FF Benefits:** Major benefits of FF are in two parts:
   - **H1 BUS:** At the field level H1 bus offers following benefits:
     - *Space and wiring reduction:* Reduced wiring, Marshalling, cabinets, hence reduced space room size due to less wiring.
     - *Power Supply:* Common Power supplies, IS.
     - *Control in field and device addition:* It supports control in field and easy device addition.

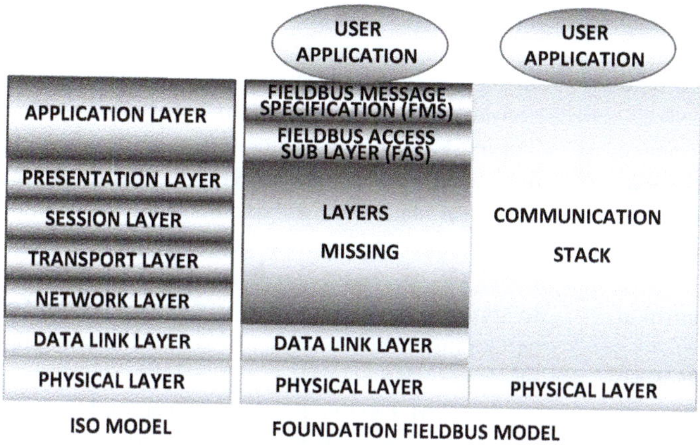


FIG III 7.1a1 FOUNDATION FIELDBUS STRUCTURAL MODEL

SPUR LENGTH TABLE (LENGTH IN METER)
TABLE FOR FOUNDATION FB CABLING

| DEVICE NUMBERS | [illegible] | [illegible] | [illegible] | [illegible] | [illegible] (MAX) |
|---|---|---|---|---|---|
| [illegible]DEVICE [illegible]PUR | [illegible] | 90 | 60 | [illegible] | 1 |
| [illegible]DEVICE [illegible]PUR | 90 | 60 | [illegible] | 1 | 1 |
| [illegible]DEVICE [illegible]PUR | 60 | [illegible] | 1 | 1 | 1 |
| [illegible]DEVICE [illegible]PUR | [illegible] | 1 | 1 | 1 | 1 |

PART OF FIG III 7.1a2

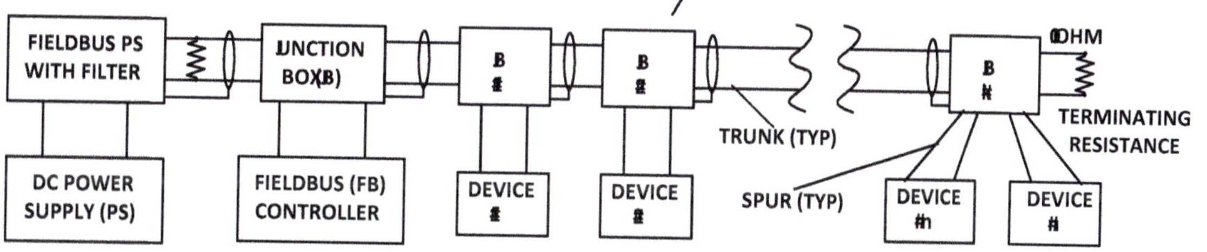


PREFERRED CABLE:SHIELDED TWISTED PAIR AWG [illegible] ([illegible] mmSq) MAXLENGTH :[illegible]ntr.

FIG III 7.1a2 FOUNDATION FIELDBUS CABLING SCHEME

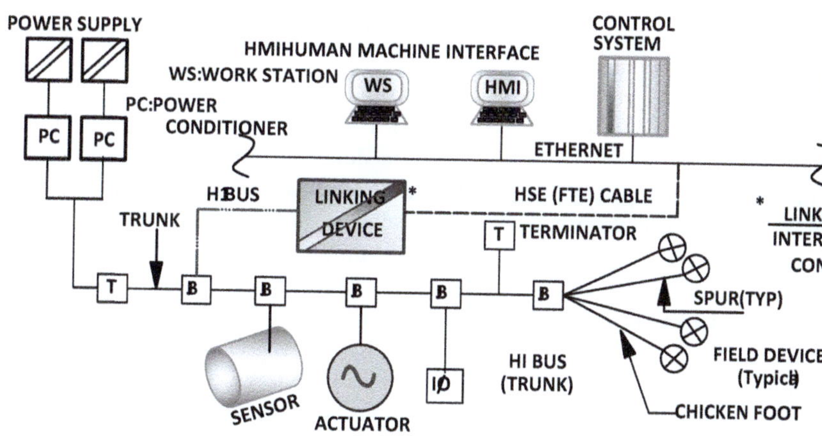


DEVELOPED BASED ON GUIDANCE OF FIELDBUS FOUNDATION

* LINKING DEVICE IS FB INTERFACE MODULE TO CONNECT H[illegible] HSE

FIG III 7.1a3 FOUNDATION FIELDBUS CONFIGURATION

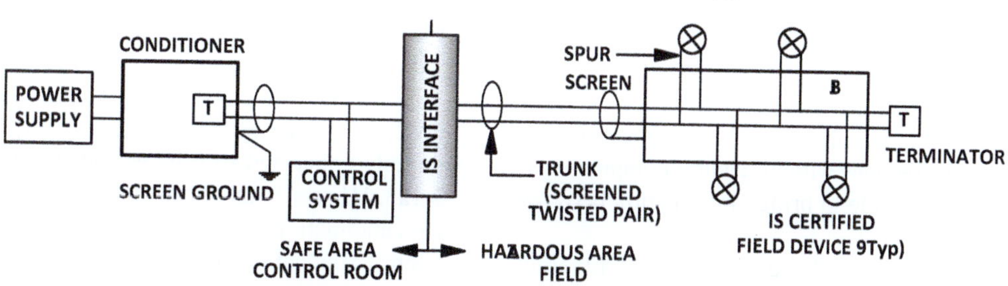


FIG III 7.1a4 FOUNDATION FIELDBUS INTRINSIC SAFETY APPLICATION

**FIGURE III 7.1A** Foundation fieldbus fundamentals.

- *Signal quality and functional capability:* It offers distortion-free quality digital communication signal and enables control capability at the field level.
- *Performance and accuracy:* Better performance for Self-diagnostic and communication capability. This improves plant safety and reduces plant downtime. Also, it offers better measurement accuracies [4].
- *Hardware requirement:* For distributed control at field, less HW requirement.
- *Flexibility:* Higher Flexibility and sophistication.
- *Additional information:* More data/multi variables from each device, that is, More Information for operation and maintenance helps Plant personnel to get notified early so that corrective actions taken quickly and safely.
- *Diagnostics and upgrade:* It offers enhanced diagnostics and online upgrade.
- *I/O mismatch and failure:* No I/O mismatch with very fewer point of failure.
- *Intrinsic safety:* It offers intrinsic safety for operation in hazardous area. The beauty is that here, common intrinsic safety can be used for number of field devices hence reduction in number of intrinsic safety barriers [4].

- **HSE Bus:** HSE bus has the following benefits:
  - *High-end functions:* It offers high performance in asset management functions like diagnostics, calibration, Identification, etc., help to take proactive management actions
  - *Interoperability:* Since the system is an open system so, it is interoperable enabling user to mix various subsystems to form an integrated system.
  - *Functional blocks:* Easier evolution is possible due to standardized function blocks. Both the *buses* use preengineered functional blocks to ease set up. Same Functional blocks of H1 can be used here also eliminating the need for proprietary programming languages. Same language system can be used over the entire system.
  - *Remote configuration:* Remote configuration of devices is possible.
  - *Peer To Peer communication*: Peer to peer communication enables to communication between two devices without central computer system. It is possible to bridge information between HI networks.
  - *Ethernet standard:* Since it is standard Ethernet so there is no need for any special tools and cables.

3. **FF Model and layers:** FF specification is based on layered communication model as shown Fig III 7.1a1. it also shows comparison of OSI model (Ref: III 13.1) discussed with FF model. FF H1 consists of three layers viz. physical layer, data link layer, and application layer as shown in Fig III 7.1a1 which also shows that in FF, OSI layers three to six are *missing.* Fieldbus message specification (FMS) is in OSI model application layer. There is a subdivision OSI application layer to accommodate Fieldbus access sublayer (FAS) along with FMS described above. In FF model, data link layer to application layer together of OSI model is referenced as Communication stack as shown. Functionally, *FAS maps* the FMS into DLL.
   - **Physical Layer:** Physical Layer of H1 bus which complies with IEC 61,158–2, provides for transparent transmission of data between DLL entities across physical connections. Upon receiving *message* from communication stack physical layer converts the same to physical signal for transmission in the medium and vice versa. The conversion includes addition removal of start/end Delimiters, preambles, etc. [1].
   - **Communication Stack:** Communication stack is layered software supporting communication between devices. It is meant to provide encoding and decoding of user-layer messages, deterministic control of message transmission, and message transfer [5]. Communication Stack comprises of Layer 2 to 7. Data Link Layer (DLL) regulates the message on to the bus through deterministic LAS (detailed later). FAS uses *scheduled* time-restricted data and Unscheduled data (on request, alarm) to provide services to FMS whose services allow user applications to exchange messages.
   - **User Application area:** It consists of a no. of blocks as per Foundation Field Bus. Each of these blocks represents different functions such RESOURCE BLOCK, TRANSDUCER BLOCK, and FUNCTION BLOCK. While first two blocks are used to configure various devices, Function block is used to develop control strategy.
4. **Configuration discussions:** Two communication protocols of FF are H1 and HSE. H1 is meant for use with field devices and is a low-speed bus. HSE, uses Ethernet as the physical layer where Work station (WS), HMI can be connected. Also, HSE bus can be directly connected to Ethernet bus as shown in Fig III 7.1a3. H1 is a LAN for field devices used in both process automation with built-in capability to distribute the control application across the network. Although H1 is basically meant for process automation yet, it finds its application in manufacturing automation too. As seen in Fig III 7.1a3, there is one Fieldbus interface module to link H1 bus to HSE bus. This FB interface module is

referred to as *Linking module.* HSE bus can be directly connected to any Ethernet as shown. While H1 is optimized for traditional field control applications, while HSE, employing low-cost, Ethernet device is designed for high-performance control applications and plant information system. The combined H1/HSE Fieldbus provides solution for full integration of basic and advanced control for continuous process/hybrid/batch/discrete control systems, as well as high-level, supervisory applications. Various features are as follows:

- **H1 Bus Features:** Major features of FF H1 are as follows:
  - *Common power and IS* (as applicable): Min Voltage for communication is *9V.* Instead of direct feeding usually power is fed through suitable power conditioner as shown in Fig III 7.1a3 and III 7.1a4.
  - *Speed and encoding:* H1 bus communicate @ 31.25 KBPS with Manchester Encoding.
  - *Standard:* IEC 61,158–2; Token Passing is the communication mode deployed in H1 bus.
  - *Device in H1 bus:* Max 32 devices per H1 Bus in Line, Tree, star topologies are allowed. The Max. Number of units in Hazardous units is reduced on account of power supply limitation (<< 32). Device types in H1, have already been described in III 7.1.0.1.
  - *Spur and Trunk:* Though Daisy chain is allowed, but normally/preferably devices including link master, are connected through JB. The connection between devices to JB is called *Spur.* JB to JB connection in H1 bus is termed as *Trunk.* A number of devices connected to a common junction box form *Chicken Foot* as shown in Fig III 7.1a3.
  - *Bus length and Distance:* Max length of H1 bus (sum of trunk and spur): *1900m* without repeater with max possible spur length is 120m. Max Four repeater allowed. So, maximum possible distance is *9500m* (with 4 repeaters). 1900m connection length is possible with Shielded twisted pair cable with 0.8 $mm^2$ conductor. Screen of the shielded cable shall be made ground at one end, and both ends shall be properly terminated by resistors as detailed out in Fig III 7.1a2.
- **High Speed Ethernet (HSE):** Following are major features HSE, the backbone of FF:
  - *Speed and functionality:* HSE which is based on 10/100 Mbps standard Ethernet/IP/TCP/UDP protocols supports functions of H1, but at a much higher bandwidth. HSE data handling capability is quite large and has all the functionalities of FF. This makes it suitable for plant-wide integration in the process industries.
  - *COTS:* HSE use low cost, Commercial Off the Shelf (COTS) Ethernet devices. A variety of commercially available interconnection devices such as hubs, switches, bridges, routers, and firewalls are deployed.
  - *Device connection (direct):* Devices designed for HSE can be connected to HSE directly, for example, HSE linking devices, HSE field devices running function blocks, or host computers.

5. **Cabling concept:** Typical cabling has been depicted in Fig III 7.1a2. As discussed, power conditioners must be used with power supplies to prevent the power supply from shorting out the communications signal and shutting down a segment. Junction boxes shown with one or more connected field devices. Based on above connections, there are two different topologies that exist viz. Tree and spur.
   - **Tree topology:** it is commonly referred to as Chicken shown in Fig III 7.1a3. It consists of a single fieldbus segment connected to a common junction box to form a network. This is possible when the devices on the same segment are well separated but within the general area of the junction box.
   - **Spur Topology:** This topology consists of fieldbus devices that are connected to the bus segment through a length of cable called a spur as detailed out in Fig III 7.1a2.

The relation of spur length with numbers of devices in a spur has been detailed out in the table associated with Fig III 7.1a2. There are different types of cables used as Foundation fieldbus. Only one preferred type has been illustrated in Fig I 3.2i2. Various types of cables used in FF are listed below (sizes are conductor size):

- **Shielded Twisted pair:** Size: 0.8mm2 (AWG 18) used for length up to 1900m.
- **Single/Multi twisted pair with overall shield:**
- **Multi twisted pair without shield:** Size: 0.13mm2 (AWG 26) used for length up to 400m.
- **Multicore (not twisted pair) without shield:** Size: 1.25 mm2 (AWG 16) used for length up to 200m.

### 7.1.1 Foundation fieldbus technology discussions

By this time is established that FF is a digital, bidirectional, multidrop communication protocol widely used in industrial automation. FF has various features like; *Digital communication, Distributed control, Multinode capability, Bidirectional communication, Open standards, Interoperability, Dependent on specialized knowledge, High initial cost, Complex configuration, Easy access for device setup and diagnostics (including NAMUR NE 107 diagnostic*

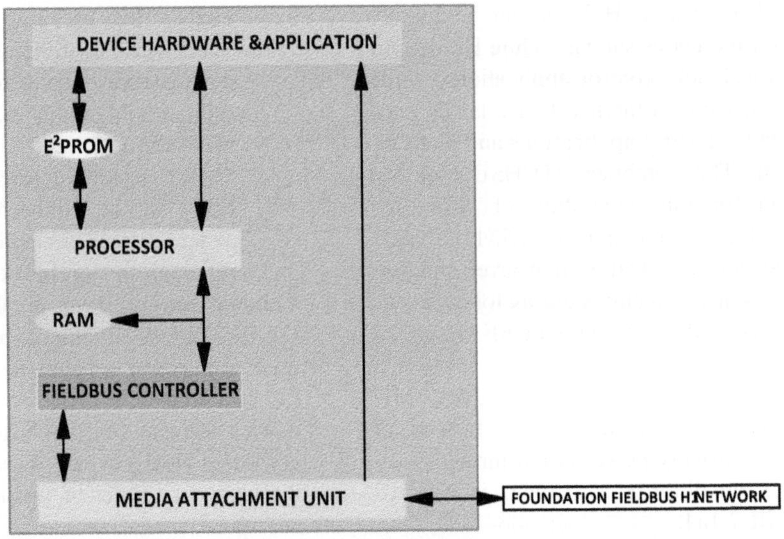


FIG III 7.1b1 FOUNDATION FIELDBUS H1: HARDWARE BLOCK

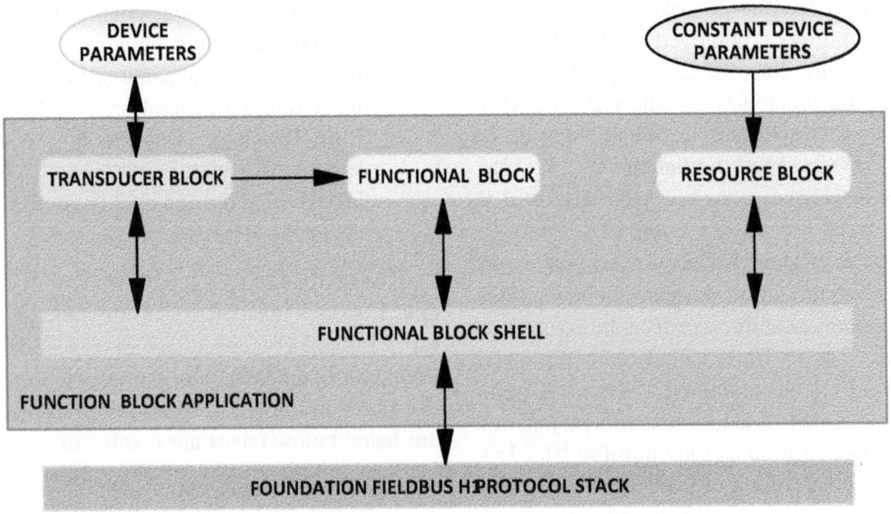


FIG III 7.1b2 FUNCTION BLOCK APPLICATION

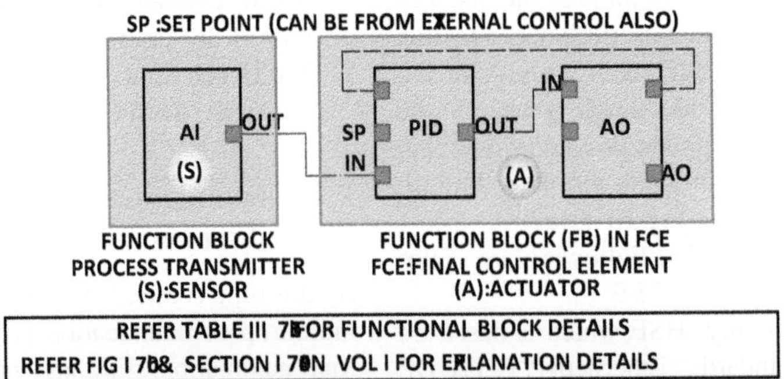


FIG III 7.1b3 FOUNDATION FIELDBUS: FIELD CONTOL LOOP IMPLEMENTATION

**FIGURE III 7.1B** Foundation fieldbus: Parts and functioning.

*classification*) and *FDD frame with DD/EDD operating/management tool.* FF uses H1 Spur cable: "Type A," as per IEC-61,158–2 and Trunk cables: Multiple twisted-pair, with individual and overall shields. "Type A" as per IEC-61,158–2. Various parts and functioning of FF have been elaborated in Fig III 7.1b. The hardware of an FF H1 has been detailed in Fig III 7.1b1. Field device has to support a physical layer as required for Manchester coded bus powered (MBP) communication (Fig III 7.2a). FF H1 field device comprises different hardware components required for connecting a *device* to the FF H1 network, for performing low-level operations within the FF communication, and for executing various functions associated with H1 stack and allied function block application. The Medium attachment unit (MaU) the hardware component of FF H1 *field device* that provides the direct connection of the device to the fieldbus. Functional block application has been detailed in Fig III 7.1b2. An FF H1 implementation does not necessarily require the use of an operating system. However, this communication protocol deals with different tasks, which are more or less time-critical and thus are associated with various priorities [6]. As already discussed distributed local control is supported by the system. Local control capability of fieldbus has been detailed out in Fig I 7.0b Clause I 7.0.3.4. However, for completeness of discussions same has been detailed in Fig III 7.1b3.

Various FF characteristics of FF make it unique. Device integration in FF impacts maintenance management as it enables technicians to verify the status of field devices. FF guarantees interoperability by strict mandatory interoperability test regime and this creates competitiveness among manufacturers. Depending on the type of installation FF can impact on capital expenditure with lots of savings in operating and maintenance expenses. As field devices communicate information on device status, maintenance can be addressed before problems become a reality, and a number of costly shutdowns can be reduced substantially. FF enables proactive maintenance especially in the process industry, and this would result considerable cost savings both in terms of routine and preventing maintenance. Because of extensive diagnostics, checking, and unnecessary inspections are reduced at the same time proactive maintenance can be initiated. Thus, key to predictive maintenance strategy is the extensive diagnostics. It also provides a wealth of information from the field about both the devices and the process. However, it is worth noting that FF is not suitable for applications requiring extremely fast response <200 mS viz. Machine condition monitoring systems. Some major issues of FF have been discussed in this subsection as a part of FF technology.

1. **FFB Component:** FF can be divided into three parts H1 Bus and its components, Linking device, and HSE Bus and its components. There could be another component gateway.
   - **H1 bus components:** Three types of components in H1 bus are as follows:
     - *Basic Devices:* Basic devices are those devices which are incapable of scheduling communication. Basic devices cannot become the LAS. Field devices, I/Os are commonly examples of basic devices. Field devices consist of physical device, DD, and CFF (by the devices supplier).
     - *Link Master:* Any H1 bus device which is capable of controlling the communication traffic on a link by scheduling the communication on the network could be a link master. On account of link master only, H1 fieldbus can operate independently. An interface board in control system/Personal computer (PC) can act as link master or a simple basic device (say) a field transmitter with link master capability interface card can be a link master. So, link masters need not be separate devices. There can be more than one in the bus but only one can be active at any instant of time. Active link master controlling the bus is referenced as *LAS.*
     - *H1 bridge:* Bridge devices are always link master devices, and they must be the LAS. One H1 bridge is a device connected to multiple H1 links whose data link layer performs forwarding and republishing between and among the links [7].
   - **HSE devices:** Any device directly connected to HSE bus is HSE device. HSE devices must contain an Fieldbus data access (FDA) Agent, an HSE SMK (System management kernel), and an HSE NMA (Network management agent) Virtual field device (VFD). Link devices, Gateway devices, and HSE field devices (which additionally contains functional block Application Process.
     - *Linking Device*: Linking devices are HSE devices to connect H1 bus to HSE network, so that access to HSE devices and H1 devices is established. A linking device may also contain an H1 bridge to provide H1 to H1 communications between bridged H1 links. It is possible to opt for redundant Linking devices for connecting H1 buses, for example, LD 800HSE of ABB (*This is qualified for redundant use, where the corresponding H1-ports of two Linking Devices are connected to one H1 link. A separate redundancy link between primary and secondary Linking Devices ensures reliable failover and monitoring of redundancy From ABB LD800HSE document*).

- *HSE gateway*: It is used to get access to non-Foundation fieldbus, for example, ControlNet, via functional blocks.
- *LAS*: LAS is not any component but controlling function of Link master. Any discussion on FF will be incomplete without it. Any link master can become LAS. There can be > 1 link master in a network, and it is possible to operate the system with redundant LAS, but only one is active. If one LAS fails other Link Master will become LAS. LAS controls so many functions with the help of a series of commands.

2. **Communication stack:** The communication stack comprises FMS, FAS, and DLL. FAS services are described by Virtual communication relationships (VCR). Three types of VCR are as follows: *Publisher–subscriber* (data publication), *Client–server* (Operator message), and *Report distribution/Source–sink* (Event notification). The DLL manages access to the FF through LAS, by splitting data into frames to send on the physical layer, receiving acknowledgment frames, and re-transmission for incorrect reception. It also performs error checking to maintain reliable virtual channel to the next layer. FAS provides an interface between the DLL and FMS layer. FAS provides communication services such as client/server, publisher/subscriber, event distribution, etc. FMS defines a model for applications to interact over the Fieldbus. FMS allows read and write information about the object dictionary, read and write the data variables described in the object dictionary, and perform other activities such as uploading/downloading data and invoking programs inside a device [7]. Within FMS layer, there is System management (SM) and Network management. System management is responsible for addressing assignments/device tags, maintenance of the function block schedule of device, and distributes application time. Network management contains objects that other layers of the communication stack use, like data link, FAS. In H1 bus, there are a few services available such as: Devices in the H1 Bus can *exchange data*, Devices *are served in definite time*, No two devices can access the bus *simultaneously*, Token management for Scheduled communication (SC) and *Unscheduled communication* (USC). Time-critical tasks such as process controls, etc., fall under SC. On the contrary, functions like *Parameterization, Diagnostics,* etc., are US which are executed in break time between 2 SCs. All services are controlled by LAS responsible for control and scheduling of communication over the bus. There can be > 1 link master, and any link master can become LAS. It is possible to operate the system with redundant LAS. If one LAS fails other link master will become LAS. LAS controls so many functions with the help of a series of commands it broadcast over the bus, for example, it broadcast *time distribution* signal to synchronize various operation Major functions of LAS have been tabulated in Table III 7.1a.
LAS continuously polls—the unassigned device so it is possible to include any device at any time by special command *Probe-node* and *Probe response* Before any operation, LAS refers to the list to check any SC. Before any operation LAS refers to the list to check any scheduled tasks. Depending on the device, there are separate schedule times for different actions so devices know precisely the task to be performed. LAS waits for a precise time to send Compel data (CD) to activate the schedule operation. It waits for the idle

**TABLE III 7.1A LAS functions.**

| Activity | Discussions | Remarks |
|---|---|---|
| Scheduled communication (SC) | 1) All time critical tasks have strict schedule created during configuration.<br>2) a compel data are sent to the device. Cyclic data are sent according to a list.<br>3) each device has different schedule based on its function and task. | 1) Device(say sensor) publish specific data upon receipt of compel data and it can be received by others (say CV).<br>2) In SC point of time and sequence is well defined |
| Unscheduled communication | 1) On request device parameters, diagnostics data are sent.<br>2) LAS passes token to all in the bus | 1) sent between break of SC.<br>2) LAS grant permission to access the bus when token is given to it.<br>3) each device can send data till the time is expired or it finishes the data transmission. |
| Live list | Maintains a live list to recognize the device sending data upon token passing. | |
| Synchronization | All devices have exactly same time as in DLL, due to time synchronization. | |
| Redundancy | In case of more than one link master, if LAS fails other takes over. | |

**TABLE III 7.1B** Some standard FF functional blocks.

| Symbol | Functional block name |
|---|---|
| AI | Analog input |
| DI | Discrete (digital) input |
| AO | Analog output |
| DO | Discrete (digital) output |
| BG | Bias/Gain |
| CS | Control sector (over ride) |
| ML | Manual loader |
| RA | Ratio |
| PID | Proportional integral derivative |
| PD | Proportional derivative |

time to start SC as shown Fig III 7.1c2. Deterministic SC uses a *publisher/subscriber* method, that is, data are "published" once the bus and all devices that need the data "subscribe" to the same transmission. Diagnostic, configuration information, alarm event, etc., are example of USC. Some standard FF functional blocks have been detailed in Table III 7.1b. There are many advanced continuous/discrete function blocks as well viz. Lead/lag, count up/down timer.

Important issues related to communication stack have been detailed in Fig III 7.1c. Associated interfaces and issues are as follows: The *Virtual field device* (VFD) is a model for remotely viewing data described in the object dictionary, *User layer*, *Device description*, *Various blocks* like *Resource*, *Transducers*, *Function* as detailed in Fig III 7.1c1. Communication schedule has been detailed in Fig III 7.1c2. Depending on device there are separate schedule times for action, enabling devices to know precisely the task to be performed. In that case, it waits for the idle time to start SC.

Major FMS services are as follows: Context management, Object directory, Variable access, Program Invocation, Upload/Download services, and Event services.

3. **Miscellaneous Issues: discussions:** Fig III 7.1d details out pertinent miscellaneous issues, those and a few other details have been elaborated below:
   - **LAS Functions:** Link schedule optimization has been detailed out in Fig III 7.1d1. LAS detailed in Table III 7.1a and Fig III 7.1c2 performs following major functions
     - LAS provides scheduled communication (sends a *CD message to a device* allowing it to *publish specific data receiving CD message*) control on the H1 network.
     - LAS allows unscheduled communication (with *pass token allowing the device to send message until it has finished*/token time expire) on H1 network.
     - A live list (devices that response to the pass token) is maintained to recognize devices on each H1 link. This is a part of network management.
     - LAS provides time synchronization, for all the devices in the data link have exactly the same time.
     - Probe node message is used by LAS to detect a new device in the network
     - LAS ensures backup and LAS Redundancy for other to take over the control.
   - **Application Layer Functions:** Various application functional issues are as follows:
     - *FAS and VCR*: FAS: FAS provides interface to its service users and the DLL. This specification defines the FAS model, the FAS services, and its protocol. It creates the VCR referenced above, like speed dialing system in telephone. After configuration such VCR no. is to be entered, to communicate with another field device on the bus. Different types of VCR are listed in Table III 7.1c
     - *FMS and VFD*: As discussed, FMS establishes the communication services for data in the fieldbus network by specifying *how data types are assigned* to specific communication services, using *object descriptions* to organize these services, that is, it provides standard message format as "object description" [where *Index 0* stands for *dictionary header*, and 1–255 define standard data types]. FMS also defines *VFD* [*a mechanism to make the object descriptions of a field device and its associated data*] used to make the object or *DD* [*comprehensive description*] of a field device as well as the associated data available to the network. These two are used to access from remote.
   - **Telegram:** The telegram is structured according to the Fieldbus DLL and FMS. This includes *Start delimiter* (SD), *Frame control* (Frame/message code), *Source and Destination address*, *Data*, *Frame Check Sequence* (FCS- CRC error detection), and *End delimiter* (ED). In this connection, Fig III7.1d2 may be referenced.
4. **Functional Block and Device Description:** Both these two issues are very much important for FF communications. Let us look into these d with little details:
   - **Functional block:** Three primary functional blocks are as follows: *Resource block* used for *device characteristic* (Device name, Manufacturer, Serial No), *Transducer bock* for device *configuration* (read sensor, and command output value), and *Function block* meant for *control strategy* building with input and

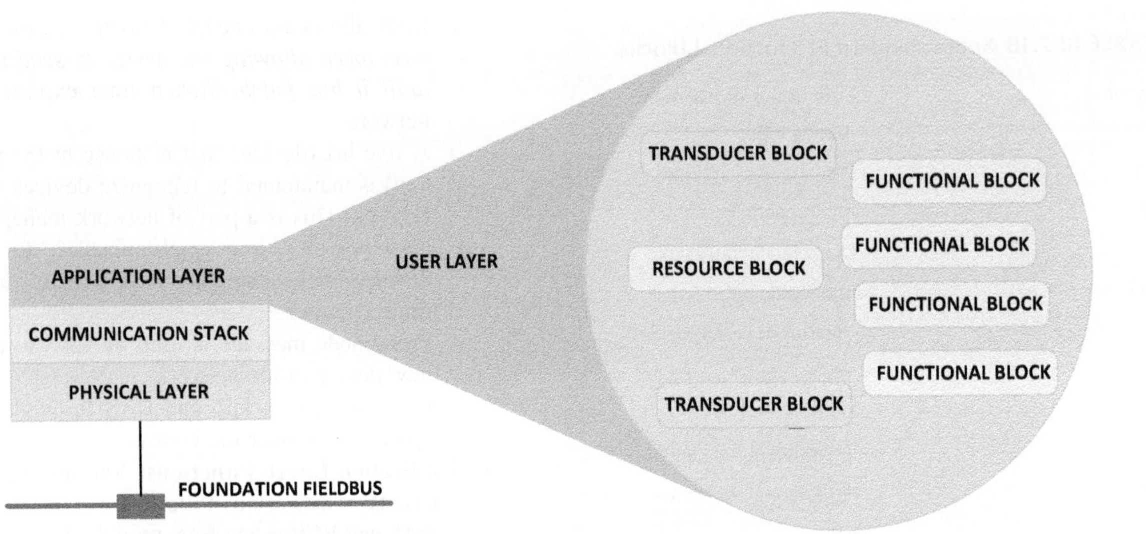


FIG III 7.1c1 USER LAYER DETAILING

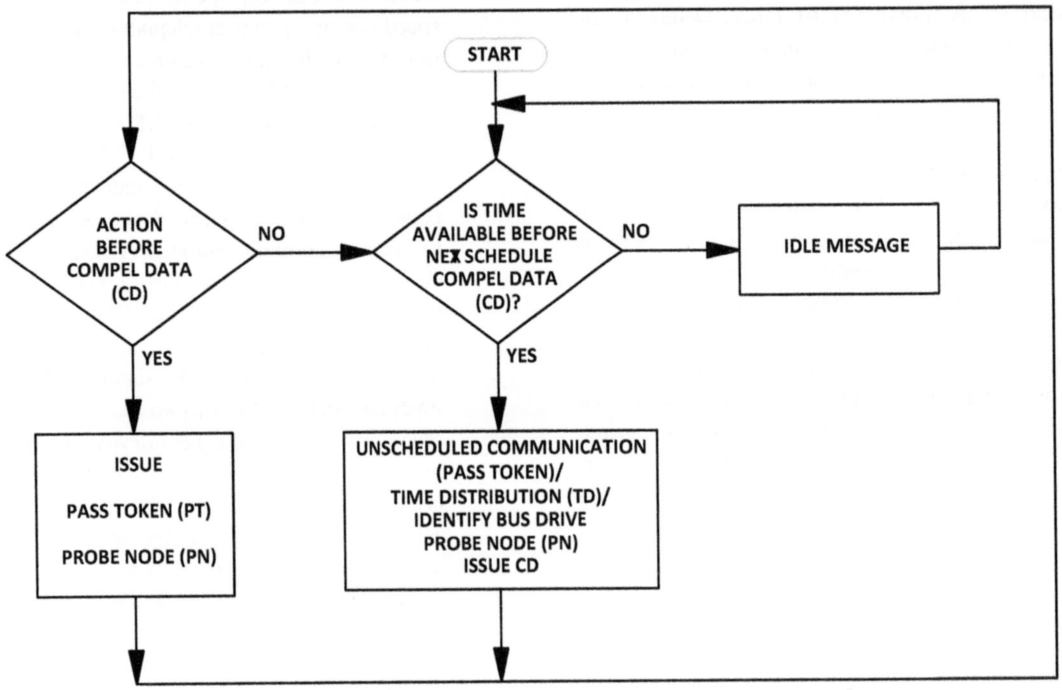


FIG III 7.1c2 BRIEF COMMUNICATION SCHEDULE

**FIGURE III 7.1C** Communication stack: Important issues.

output of function blocks linked over Field Bus. Fig III 7.1e elaborates functional blocks with associated objects.

Standard FF Functional blocks have been discussed in Table III 7.1b. Various blocks with description have been detailed in Table III 7.1d.

- **Device description:** DD is a software file. Any control system or host can operate with the device when it has the associated DD which provides the information a host system needs both in terms of the *meaning of the data* and the device capabilities. It is unique to device type, and it serves as a "driver" for a device similar to that in a printer. A device driver contains an extended description of each object contained in VFD such as human interface for functions (calibration and diagnostics). Some of the information DD

**TABLE III 7.1C** VCR types.

| Cleint/server | Report/ distribution | Publisher/ subscriber |
|---|---|---|
| Operator communication | Event notification, alarm, trend | Data publication |
| Changes for set point, device data remote diagnostics, display access. | Operator alarm, trend to historian | Actual value or PID block to OP console |

provides is as follows: *The label for a parameter*, *Engineering units*, *Help text*, *Diagnostics* and *Methods* for device-related tasks, that is, a predefined sequence of operations executed in a device. These predefined task sequences are useful in setting ups/commissioning. FF makes the standard device drivers available on a CD-ROM. The user can obtain the incremental device driver from the device supplier. Device suppliers add supplier-specific features such as calibration and diagnostic procedures to their device's incremental description, on the top of standard DD. Before a DD is accepted by the FF, it has to pass

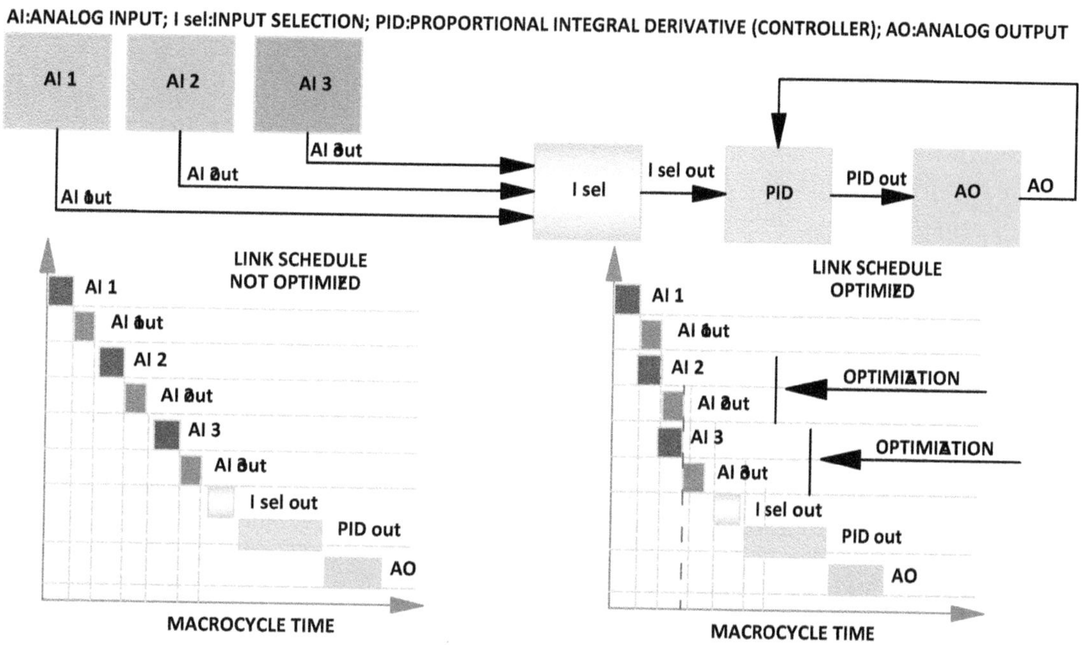


FIG III 7.1d1 LINK SCHEDULE OPTIMIZATION

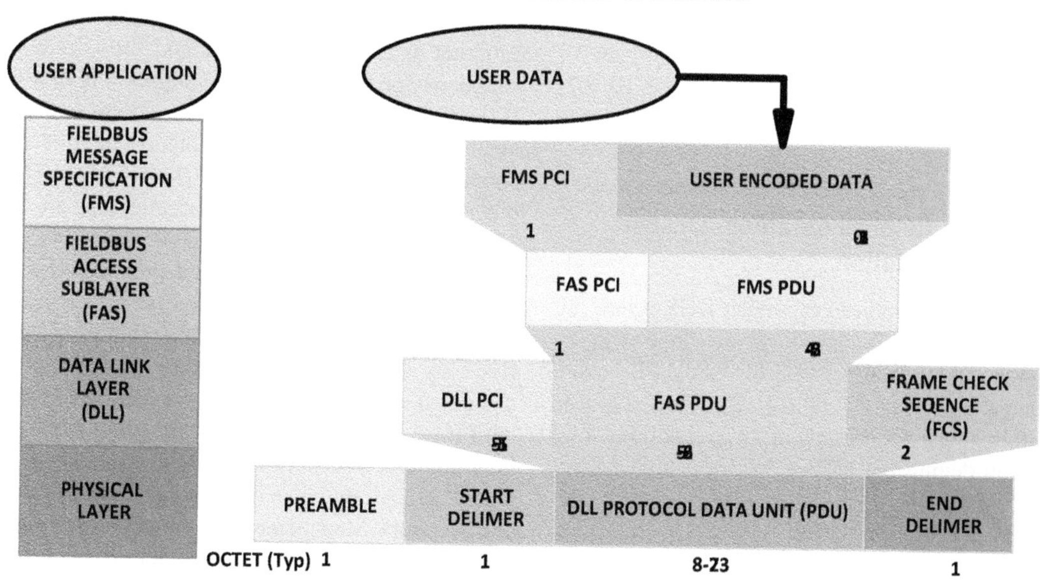


FIG III 7.1d2 FOUNDATION FIELDBUS TELEGRAM

**FIGURE III 7.1D** Link schedule and foundation fieldbus telegram.

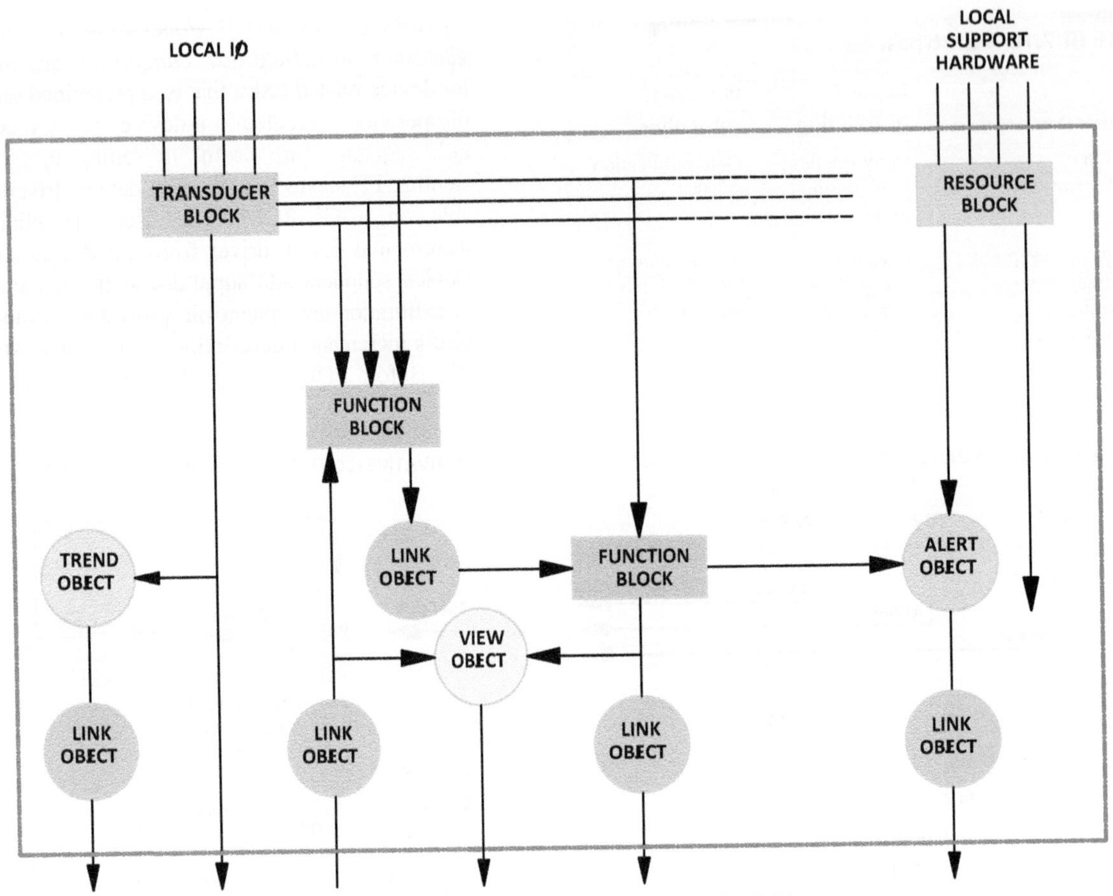


**FIGURE III 7.1E** FF functional blocks and associated objects.

FF interoperability test to guarantee that it will operate without problem with other FF devices.

5. **Modes and parameters:** Variations in mode types determine how device operates, that is, in which mode the device operates. Similarly parameter refers to input/output/controller, etc. Here, let us discuss them separately.
   - **Modes of operation:** Various modes of FF devices are enumerated in Table III 7.1e
   - **Parameters and mode parameters**: Table III 7.1f defines the parameters and mode parameters for FF devices.

## 7.1.2 Foundation fieldbus detailing: Design and implementation

Foundation fieldbus resembles very much DCS where, all the calculations and logical "decisions" are done in specified *controllers*, taking from various I/O cards. FF system is one step ahead allowing all control algorithms could be executed within the field instruments (i.e., includes field controllers unburdening the DCS to perform higher-level tasks as necessary) to make automatic control "decisions." In fact, DCS may not even be necessary if needed in certain mode of operations. So, FF can be conceived as an all-digital, two-way, multidrop communication system that brings the control algorithms into instrumentation, that is, FF is a LAN comprising FF devices like sensors, actuators, and control devices especially for process control applications. As noted, FF has two communication protocols in it. These are H1 transmitting at 31.25 Kb/s to connect field devices, and HSE, uses 10 or 100 Mbps Ethernet as the physical layer and provides a high-speed backbone for the network. FF supports both a high-speed HSE and low-speed H1 network, there exist two architectures with same basic functionality but difference in the communications stack. Apart from FAS, FMS, etc, the other key components in FF architecture are as follows: *Leyer management entity, Network management information base, Object directory, System management information base, etc.* The allowed number of devices in H1 in hazardous area depends upon the safety concept adopted (Ex i/Ex d) with their individual properties. FMS layer discussed, defines a model for applications to interact over the FF where the object dictionary (describes data that can be communicated) and the virtual field device are important. There are two management layers in FMS viz. System and Network Management. Former one assigns addresses for device tags, maintains the function block schedule for

**TABLE III 7.1D** Various block and object description.

| Block/Object | Description |
|---|---|
| **Description of blocks:** | |
| Resource block | The base block needed for each function block/fieldbus device and only one is defined for each fieldbus device. It contains the associated hardware-specific characteristics containing set of parameters NOT input or output, i.e., data specific to the device as already detailed. It also contains an algorithm to control and monitor the overall health and operational status of the device hardware. The algorithm may also generate events. Execution is dependent upon the physical characteristics. |
| Transducer block | Transducer blocks decouple local I/O functions from the function blocks, i.e., at least one transducer block that allows the I/O blocks to access data. They can read from sensor to write into actuator. Thus, the transducer block could be executed to obtain good data from the sensor/to ensure valid writes to the actuators without burdening the function block(s). For transducer block, running calibration and diagnostics for field devices are feasible. With standardized signal, they also isolate the function block from vendor-specific characteristics of the sensor/actuator. Main purposes are as follows: Identify the device and software, provisions for a standardized signal for analog I/O. |
| Function block | This represents the automation functions in application having input parameters as per specified algorithm and internal control parameters. Thus, it comprises a complete set of parameters, an algorithm and events to produce output parameters for use within the function block applications (same or other). Function blocks are *executed* to a schedule/on the occurrence of an event. For I/O functions, they are are linked to the transducer blocks. |
| **Description of objects:** | |
| Alert objects | On detection of alarms and events, alert objects messages are generated by sending an event notification with specified waiting period for an acknowledgment. |
| Link object | Link objects contain information on I/O of function blocks hence used for external access to trend, view and alert objects. They identify the parameter or object to be linked. |
| View and trend objects | To view objects aid the management and control of function blocks through the configuration and operational parameters visible to the user and to group operator data. |

the function blocks in that device, and distributes application time. On the contrary, network management contains objects that other layers of the communication stack, for example, DLL FAS. In this last subsection on FF, will be on FF design engineering and implementation issues.

1. **Design and development:** Fieldbus base technology in common comprises a transceiver circuit known referred

**TABLE III 7.1E** Modes of operation.

| Type | Description |
|---|---|
| Auto mode | The block is using a local setpoint value (by an operator<br>Through an interface device) in the normal block algorithm to determine the output value. |
| Manual mode | The block output is not calculated by the normal block algorithm but the output is written by the operator directly. |
| Out of service mode | The block is not running at all and used during block configuration. Also, some devices may require out of service when changing certain parameters. The output is maintained at last value or, at an assigned value and setpoint is at last value. |
| Cascade mode | The block receives setpoint value from another function block for use in the normal block algorithm to determine the output value, i.e., an upstream controlling block, and a<br>Downstream controlled block (AO block) or upstream PID feeds a setpoint into a second PID that is acting as the downstream block. The output parameter of the upstream block is connected to the cascade input parameter of the downstream block. Also, the backward calculation output parameter of the downstream block is connected with the backward calculation input of the upstream block |

**TABLE III 7.1F Parameters and Mode parameters.**

| Type | Description |
|---|---|
| **Parameter Types** | |
| Input | An input parameter obtains its value from a source external to the block |
| Control (Contained) Parameter | A control (/contained) parameter is a parameter whose value is configured, set by an operator, higher level device, or calculated. |
| Output | An output parameter may be linked to an input parameter of another function block. Output parameters contain status. The output status indicates the quality of the parameter value. |
| **Mode-Parameter Types** | |
| Target | This is the mode requested by the operator. |
| Normal | This is a mode during normal operating conditions. It does not affect the algorithm calculation. |
| Actual | Current mode of the block, may be different from target |
| Permitted | The permitted modes are configured based on the application requirement. |

to as *MAU* mentioned, a *communication interface integrated circuit* (IC), and *communication software.* The basic developmental issues have been detailed in Fig III 7.1f with basic building block connections detailed in Fig III 7.1f1.

Let us now look into the details of them in the following discussions:

- **MAU discussions:** It is responsibility of the transceiver MAU to physically connect the IC to the fieldbus network, that is, IC and MAU together ensure reliable and efficient communication. This basically converts the IC's signals into bus signals suitable for transmission over the physical medium Associated functionality includes but not limited to: *Data transmission* and *reception*, *Signal conditioning*, *Protocol implementation*, *Power management* (bus signal), Jabber inhibit and low-power signal (by monitoring the signal using a timer), etc.
- **Interface IC:** This is the heart of the Fieldbus node to implement the protocol, handle data formatting and offering interfaces for connecting to other system components. Major objective of interface IC are as follows: The objectives for the development of the communication purpose ICs, were: Interface with various types of microprocessors (viz.Intel, Motorola), *To minimize the frequency of interruption*, *Providing buffer* in First-in/first-out (FIFO) through Direct memory access (DMA).
- **Communication software:** Communication software for use in field devices is structured for high performance. The performance of communication processing depends on: the processing time for immediate response ($V_{MRT}$) and the response time at the application layer FMS level. $V_{MRT}$ affects the performance of the whole fieldbus segment. The response time at the FMS level affects performance for the device and not the whole segment.
- **Control system:** Configuration using common files has been detailed in Fig III 7.1f2. It is necessary to enable the user to establish an engineering scheme that can be applied to multivendor field devices with no field devices actually connected. In fieldbus communication, a system is constructed by connecting devices from multiple vendors. Interoperability is therefore required in every aspect of system operation [8].

2. **Engineering effort:** The engineering effort is more complicated with the increased availability of multichannel and multivariable fieldbus devices. Selecting the fieldbus-centric instrumentation and control systems involves considerable research work because greater capabilities are available in fieldbus but also because all instrument types are not available in fieldbus version and may have to opt for 4–20 mA and HART devices and I/O for safety instrumentation systems and certain extremely fast-response special applications. Let us look into the issues with some details staring with general considerations.
   - **General considerations:** Two major tasks in system design are as follows: *assignment of devices to fieldbus segments* and *system configuration.* It is important to optimize segment through proper design issues so that segment does not exceed physical and functional constraints. After two to three iterations of checking of following issues can optimize segment design:

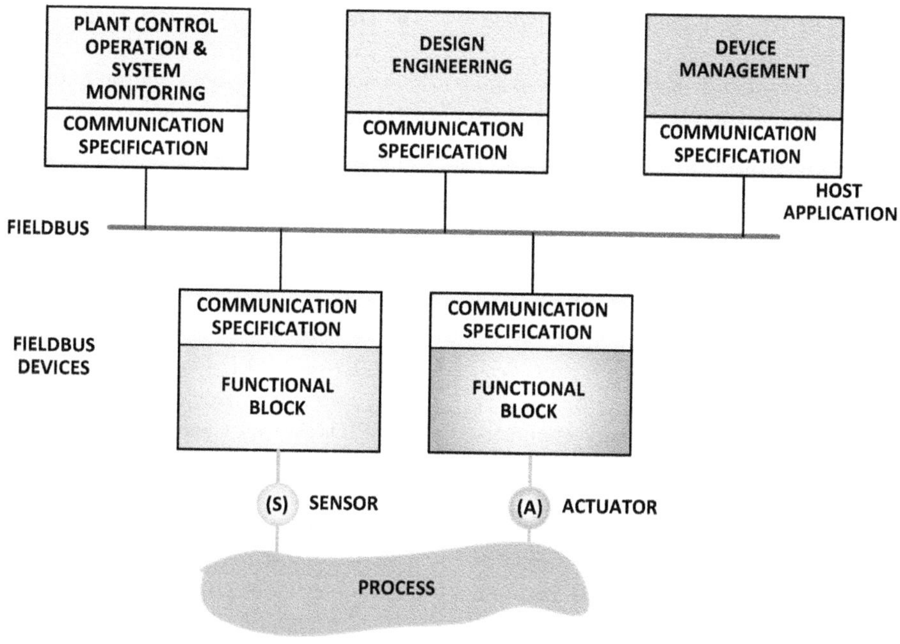


FIG III 7.1f1BASIC BUILDING BLOCKS

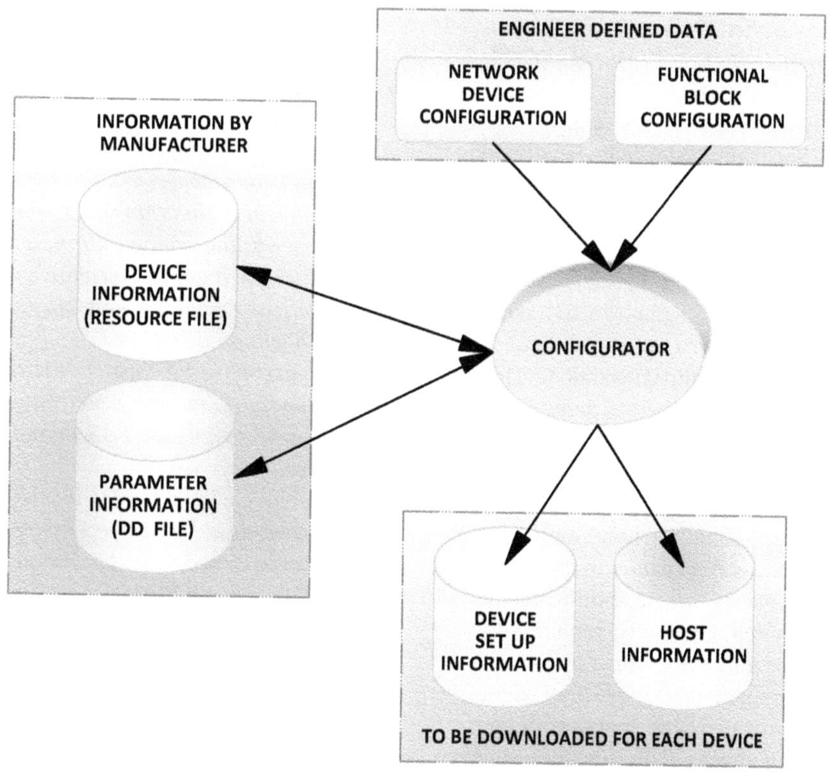


FIG III 7.1f2 FF CONFIGURATION

DEVELOPED BASED ON IDEA FROM 9[COURTESY:YOKOGAWA]

**FIGURE III 7.1F** FF building blocks and configuration [8]. *[Courtesy: Yokogawa].*

- *Assessment of device type and its associated limits*
- *Ensure that all components of the control strategy are on the same segment.*
- *Trunk and spur cable lengths are within limits.*
- *Voltage drop calculations for each segment are done.*
- *The macrocycle time* (Fig III 7.1d1) *is within specified limits.*
- *At least one instrument has backup LAS capability.*
- *Number of VCR and functional block is exceeded.*

Let us now divide entire engineering into different parts to discuss starting with preliminary engineering considerations.

- **Preliminary Engineering:** Major issues to be considered at this stage shall include the following: *Technical specification, Fieldbus* (general/Hazardous area) *type, Engineering format* and *detailed documentation, Segment topology selection, Segment design issues* (discussed), *Loop definition* and *required response time, Device/hardware planning with requirements, Redundancy issues, Naming convention, Host requirements, FF media* (Cable) *consideration, Control assessment* and *assignment, Spare issues, FF instrument interface,* etc.
- **Segment design:** Segment design can be divided into:
  - *Preliminary segment design*: For basic segment design a number of issues need due considerations: Major documentation needed includes: Pipe and instrumentation diagram (P&ID), equipment layout, Hazard area classification, FF approved guidelines for implementation. These are necessary to arrive at various other issues like selection of FF (general/Hazardous area) type, voltage drop and attenuation calculation, Segment execution time and criticality rating, Spare capacity, etc. Detailed flow diagram is available from Fieldbus Foundation in this regard.
  - *Final segment design*: At this stage little, detailed information is required. Generally, required documents include the following: Instrument location drawing, Instrumentation cable tray layout, I/O lists, List of critical control loops, with identification of loops requiring loop execution times faster than 1 Sec, Control details with more details on complex loops.
- **Hardware planning:** This constitutes listing of major hardware requirements like Numbers of H1 interface card, Quantity of power conditioning units, Numbers of junction boxes with type classifications, Barriers and Termination box-type & quantity, and Surge protection.
- **Device requirements:** FF Hi connected devices are expected to perform digital, two-way communication network with required power (*voltage range of 9V to 32V, current consumption between 10 mA to 30 mA*), as per IEC 61,158–1, for operation. So, they must be certified and interoperable with host systems, and their DD must be available. Major pertinent requirements shall include the following: *Certified latest interoperability tested device, Fieldbus Foundation approved device carrying the "Check Mark" logo, Device having availability of DD and CFF files, Device possessing link master capabilities,* Devices with *required functions* ( AI, PID, AO etc) and *suitable execution time, Device having operating voltage, current, and power consumption as* per *IEC 61,158. Insensitivity to polarity,* Device for hazardous area, should be *certified hazardous area with certification for safety parameter of device/safety parameter of I.S. barrier, Device capable of performing advanced diagnostics, self-test function and can provide necessary diagnostic information, Online calibration capability* and *certified by host control system manufacturer* as FF device.
- **Engineering issues** [9]: Following are a few engineering issues to be considered:
  - *Close co-ordination between engineer, designer, and programmer*
  - *Segment validation in system vendor's scope, also off-line configuration data should be made available by device vendor.*
  - *Critical complex loops may be implemented by conventional I/O*
  - *To use more FF-type multivariable transmitter to economize on pipe penetration.*
  - *Use of redundant LAS in redundant H1 card if available*
  - *Devices with different macro cycle time should not be mixed on same segment*
  - *Simple closed-loop devices on same segment with PID controller in valve positioner, avoid control in Transmitter*
  - *Complex loop controllers/control strategy in Host system controller*
  - *Package supplied FF instruments wire to plant FF junction box.*

3. **HSE Detailing:** Let the discussions start with reference to Fig III 7.1g. FF HSE comprises:
The physical layer, The communication "stack," and the user application.
Even though FF HSE is based on ISO–OSI model, yet fifth and sixth layers are missing. There exists the FMS layer (OSI seventh layer) mapped onto the transport layer (OSI fourth layer) by the FDA agent. Thus, apart

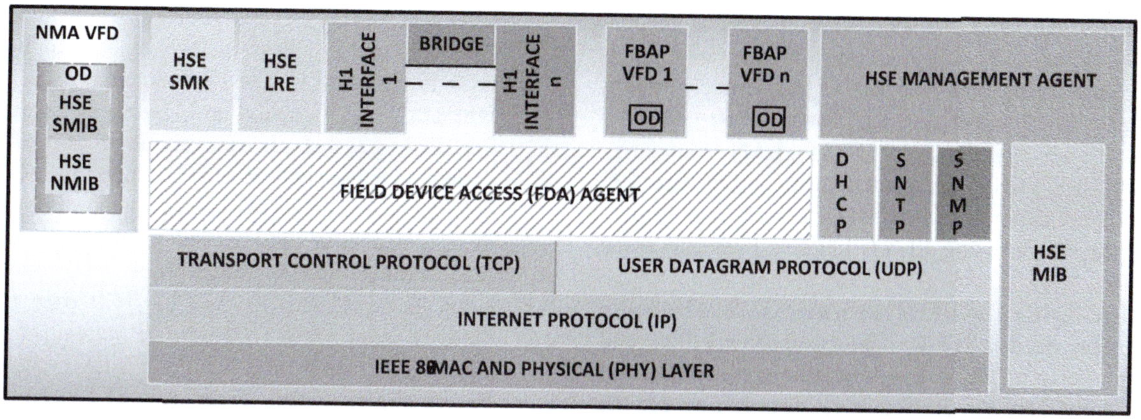


FIG III 7.1g1 HSE ARCHITECTURE

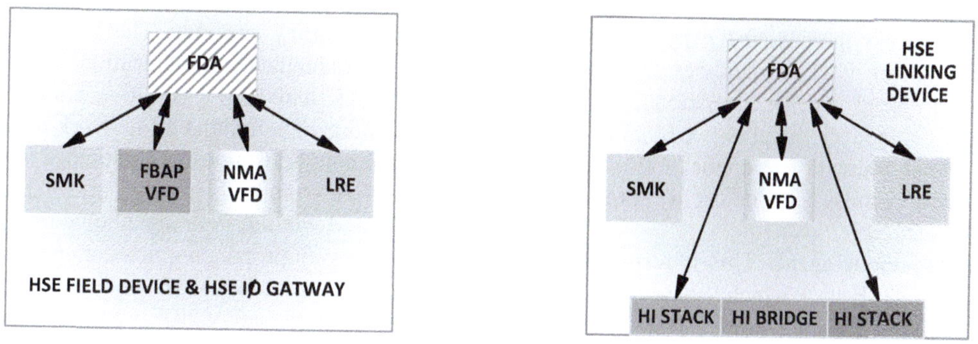


FIG III 7.1g2 FUNCTION OF FDA AGENT IN HSE DEVICES

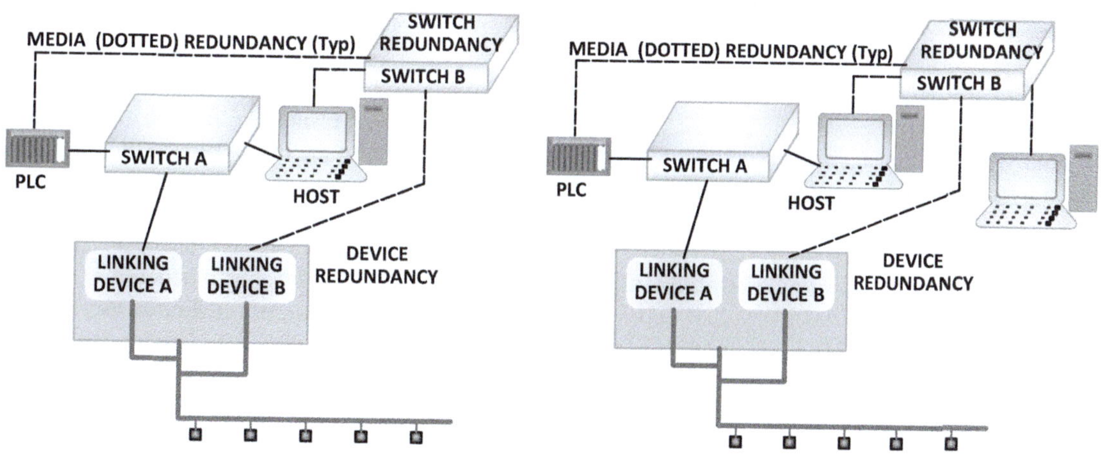


FIG III 7.1g3 HSE REDUNDANCIES

ABBREVIATIONS USED:

NMA:NETWORK MANAGEMENT AGENT; VFD:VIRTUAL FIELD DEVICE; OD:OBECT DIRECTORY; HSE:HIGH SPEED ETHERNET;

NMIB:NETWORK MANAGEMENT INFORMATION BASE; SMIB:SYSTEM MIB; SMK: SYSTEM MANAGEMENT KERNEL;

LRE:LAN (LOCAL AREA NETWORK) REDUNDANT ENTITY; FBAP:FUNCTION BLOCK APPLICATION PROCESS;

DHCPI DISTRIBUTED HOST CONTROL PROTOCOL; SNTP:SIMPLE NTEWORK TIME PROTOCOL; SNMP:SN MANAGEMENT PROTOCOL

**FIGURE III 7.1G** HSE Explained in details.

from physical layer, the communications stack encompasses layers 2, 3, four and seven in the OSI model. User application part (not part of ODI model) mainly comprises: System management kernel (SMK), Network management agent (NMA). LAN redundancy entity (LRE), HSE management agent and the function block application process, H1 interface, and bridge. With reference to Fig III 7.1g1, let us look into some details:

- **HSE SMK:** The HSE system management kernel in linking devices to HSE subnetworks to H1 links, and also performs the task of forwarding requests and responses between HSE and H1. This is used for the UDP connectionless services to transfer protocol messages. When a device first powers-up on network, it checks the device ID (by the manufacturer) for each of its HSE interfaces. This check is necessary to send receive operational messages. Then, device announces its presence by multicasting its operational address. The application monitoring the multicast address, can detect the presence of the new device.
- **HSE Management agent:** This is responsible for the part of the HSE stack which is based on Ethernet standards comprising components like:
  - Dynamic host configuration protocol (DHCP) client to manage the IP addressing of the devices on a network.
  - Simple network time protocol (SNTP) client responsible for maintaining time synchronization with on the network.
  - Simple network management protocol (SNMP) agent responsible for managing the HSE presence layer protocols by sending messages to different parts of a network.
- **HSE NMA:** This is adapted to HSE-specific management and addresses various issues like *Information for support VCRs* for H1 bridge objects, *Time synchronization* information, *Device identification*, *List of VFDs accessible* from the device, and List of function block schedules for the device, etc.
- **HSE LRE:** LRE is used only in HSE network. This refers to the use of a secondary backup device to take over operation for a primary device failure. LRE utilizes multiple network interfaces to maximize so that HSE device can communication with the other devices on the HSE network. LRE in a device is responsible diagnostic messages and maintenance.
- **HSE-Ethernet: Communication stack:** As discussed, HSE is a Ethernet based hence support Ethernet physical layer viz. *100BASE-TX*, *100BASE-FX* and *100BASE-T4*. So, communications stack comprises: Application layer, *Transport layer* (TCP/UDP), *Network layer* (IP), and DLL (IEEE 802.3). Application layer is The application layer is managed by the FDA Agent. All hi-Speed Ethernet devices contain an FDA agent, detailed in Fig III 7.1g2, that provides access to network and system management data. HSE supports the same field messages as H1 with different format to optimize transport using TCP and UDP. HSE can use both connectionless services of UDP and the connection-oriented services by TCP over the IP network layer [*TCP/UDP and IP explained in III 13.1*]. Application modes support UDP connection, for example, Client/server (or TCP connection *also*), Publisher/subscriber, and Report distribution support.
- **Redundancy support:** The HSE redundancy provides redundant communication paths and redundant devices, and allows selective connectivity through all ports. The redundant network could be either a dual network/single fault tolerance as shown in Fig III 7.1g3 with redundancies at different levels.

4. **Cable and wiring:** Poor signal quality can cause intermittent loss of proper communication to devices, unreasonably long downloads, and lost data. Electrical scheme for IS safety has been illustrated in Fig III 7.0b2. Some suggestive critical components are as follows:
   - STP, specifically for fieldbus is preferable. The type of cable, shielding decides overall length of the trunks and drops. Cable Type A [Table III 7.1b] gives the best results irrespective of the total permissible length of the bus. Also, the bus cable shielding must exhibit continuity throughout the length.
   - Proper terminators reduce noise on segments due to signal reflections. One terminator on each end of the H1 trunk is necessary as practice for proper performance. 7/8″ connector (Minifast)/screw terminals and cable gland/entry are common terminations.
   - Bus powering and IS operation as shown in Fig III 7.0b2 is very much possible. Essentially, power conditioners discussed below, are used between a supply and the H1 trunk delivering 9 … 32V DC consistently from conditioner at the appropriate total load including cable resistance. Single/redundant power supplies made for fieldbus, can be used. Built in short-circuit protection is necessary for power supply and conditioner.
   - To use lightning arresters wherever necessary
   - With multiple devices sharing a cable, only one device should transmit at a time to prevent signals from collision/interference. The LAS directs this traffic, use gap of silence for this.
   - The length of a fieldbus wiring system, the number of devices on a network/segment is limited by the

power distribution, attenuation, and signal distortion. For this, ISA 50.02 may be referenced.

5. **Power and grounding:** In this part, we discuss power supply, power conditioner, and grounding details, etc. Fig III 7.1h elaborated electrical scheme for power and grounding in FF system.
   - **Power systems:** Two major parts of FF power system are as follows:
     - *Power supplies*: Power supplies shall comply with IEC 61,158–2. Based on design, FF devices can be either powered from the segment (bus) or locally powered. Bulk power supplies provide direct current in segment voltage can range from 9 … 32V (common: 24 ±2V DC), according to FF specification. Bus-powered devices usually require 10 … 30 mA. Obviously, it is essential to calculate power to assess the number of devices in a bus segment. Here, major important guiding issues are as follows: *Power supply output voltage*, Each *device current consumption* and *minimum operating voltage*, *Device* and *power supply location*, *Sectional cable resistance*, and *Short circuit current* (of 10 mADC).
     - *Power conditioning*: FF systems, power conditioning is utilized to ensure a clean and reliable power supply for fieldbus devices and preventing from signal disturbances from the power supply by trying to maintain a constant voltage level. Power supply disturbance issue, signal integrity, and device compatibility necessitate use of conditioner with power supply. Power conditioners comprise resistors and inductors to create a high impedance network at communication frequencies. The inductor lets in the DC power on the wiring, but it prevents signals from going into the power supply. One fieldbus power supply conditioner is required for each fieldbus network segment. Typical power conditioner scheme has been detailed in Fig III 7.1h1 It also ensures matching impedance for maximum power transfer. It also provides some form of signal isolation for the fieldbus segment, to prevent the communication from power supply noise/disturbances. The combination signal/power wires have a plus (+) and minus (−) polarity associated to the power conditioner outputs, which must be wired to the appropriate terminals on the devices. Some devices are polarity insensitive, meaning they still work with reverse connection.
   - **Grounding Issues:** Fieldbus segment must be grounded at only one point to prevent ground loops. By grounding the cable shield at the control room

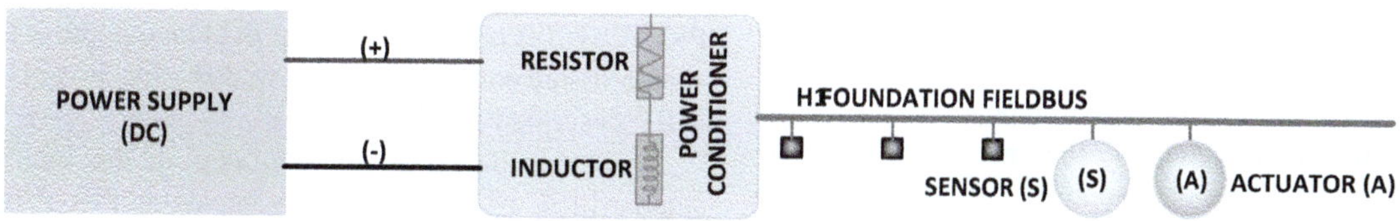


FIG III 7.1h1 POWER SUPPLY AND POWER CONDITIONER

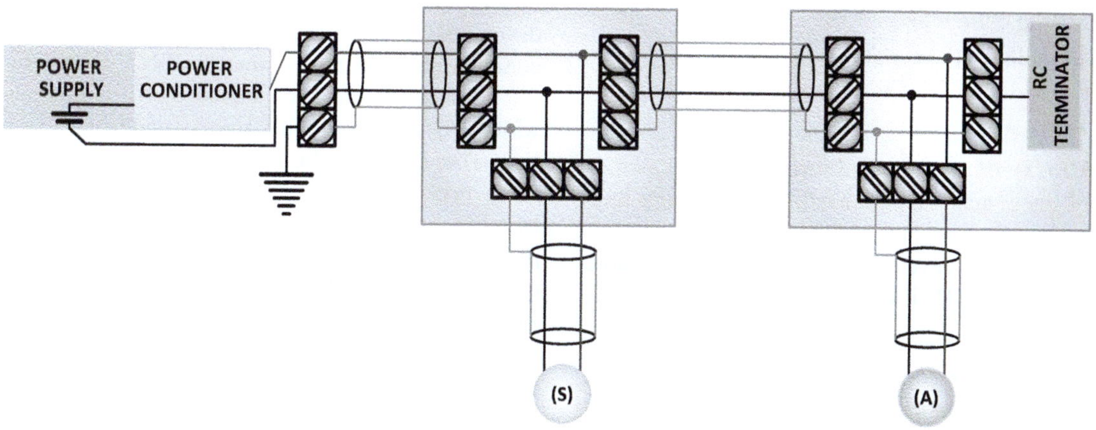


FIG III 7.1h2 FOUNDATION FIELDBUS GROUNDING SCHEME

**FIGURE III 7.1H** Foundation fieldbus electrical scheme.

end of the segment same is normally achieved. Let us look into details of grounding of various items:

- *Shield grounding*: The shield of the Fieldbus cable should be grounded at only one point as shown in Fig III 7.1h2. Normally, the shield is grounded at the control cabinet, and the shield is left open at the device end.
- *Device grounding*: As stated, ground loop must be avoided to prevent from interference in communication. Each device on the system is to be grounded individually, to avoid any ground potential. Also, the grounds for a group of devices in a common physical location can be connected done with proper grounding.
- *Miscellaneous other issues*: Ground loops causing communication problems must be carefully avoided. Only power supply with conditioner should be used and prescribed wiring guidance must be followed. This includes proper routing and termination. Segment design tools discussed should be used.

With this, let us conclude the discussions on FF to proceed to discuss PROFIBUS.

## 7.2 PROFIBUS discussions

PROFIBUS and FF have two different origins. FF was specifically designed to address the needs of process automation for offering safe reliable and effective real-time process control operation over bus. The PROFIBUS family was designed for hierarchy in communications for a PLC system, primarily io address the needs of discrete manufacturing building automation, etc. PROFIBUS (Process FieldBus) was originally created in 1989. Three different protocols in PROFIBUS are as follows:

- *PROFIBUS FMS* offers backbone for PLC systems to connect user interfaces, PLCs,

  CNC machines. It is a complex protocol and is no longer used today. This is mainly replaced by PROFIBUS DP in 1993
- *PROFIBUS DP* is used for discrete processes like assembly lines, material/pipe handling, packaging lines, building automation, and motor control.
- *PROFIBUS PA* is developed to use this in process automation. Like FF H1, it is also MBP.

  During the discussions in this section, we will find that the greatest differences between PROFIBUS PA and DP are their physical layer. So, when integrated PROFIBUS DP can act as backbone network to connect the fieldbus segments, that is, PROFIBUS DP can serve as a interface for control system, which can transport layer in transmitting data between the control system and the fieldbus. PROFIBUS DP can be used in an architecture to accommodate both discrete and process automation. PROFIBUS DP protocol was created after PROFIBUS FMS lost its expected potential. In 1998 PROFIBUS PA (Process Automation) was launched. PROFIBUS PA is very much the similar (same) as FF H1 based on the same standard and meant for process automation. In PROFIBUS family another member is PROFINET. PROFINET is an application of Industrial Ethernet (PROFINET has been discussed at length in chapter 17of this volume). PROFINET can be applied from control to communication in the field for a uniform vertical network. It also helps in easy IT integration. Based on functionality there are variations in PROFIBUS DP. These are DPvo, DPv1, and DPv2. DPv0 and DPv2 handle cyclic data with DPv2 is meant deterministic exchange. DPv1 is meant for acyclic data exchange. Fail-safe an optional feature available in PROFIBUS slave device, allowing user to specify the states of slave outputs in the case of a master/bus failure. Twisted, shielded copper pair RS-485 cables are the standard for PROFIBUS DP in nonintrinsically safe applications, with DB9 connector.

In this section, discussions on PROFIBUS system have been presented [for PROFINET refer Ch17 of this volume].

### 7.2.0 PROFIBUS overview

PROFIBUS is one of the most proven and popular field bus standards introduced in Germany. Like FF, it is also based MBP described in Fig III 7.2a.

Now this has been incorporated in the international standard IEC 61,158–2. Frankly speaking, PROFIBUS is a family of fieldbus and protocol system. PROFIBUS family offers a digital network responsible for providing the communication between the field devices and the control system (s). Even though PROFIBUS FMS was first developed. Yet as discussed, FMS is no longer in use. To complete the structure, shown in PROFIBUS structure in Fig III 7.2b [1] has been included. All plants have tasks which are associated with process and/or factory automation. These are as follows: *process automation*: *measurement, actuation, control*, and *factory automation*: *filling, storage, conveyance, drives, and other manufacturing processes*. PROFIBUS family cater to both these requirements with different PROFIBUS types. Both PROFIBUS and PROFINET have two different message types viz. cyclic and acyclic. PROFIBUS provides wide address space to accommodate 126 devices on a single network. PROFIBUS supports longer wire lengths without the need of repeaters and digital signals maintain integrity over long distances. PROFIBUS possesses an interesting feature. The protocol is the same irrespective of the transmission

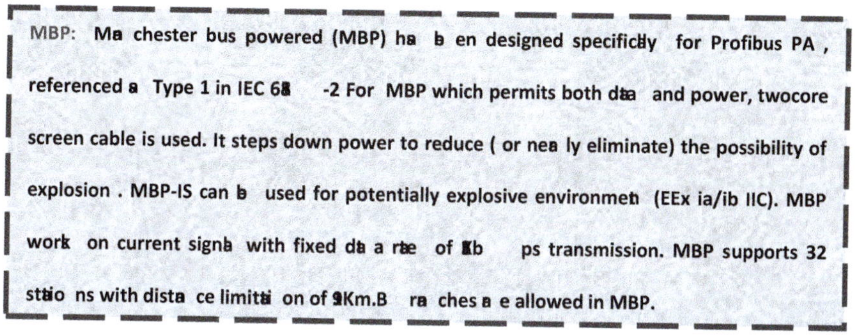


**FIGURE III 7.2A** Manchester bus power (MBP).

medium/technology used. So, PROFIBUS DP (using RS-485 copper/fiber optics/infrared, etc.) and PROFIBUS PA (utilizing MBP-IS) can use the same protocol, and the devices can exchange cyclic I/O data with a standard DP master. PROFIsafe is an integrated safety technology, providing maximum safety for the applications with Flexibility in integration of all types of safety components. PROFIsafe is Compliant to the most stringent safety requirements up to Safety integrity level 3 (SIL3) as per IEC 61,508, 61,511, and IEC 62,061. It can be applied for highly integrated safety applications like Emergency shutdown (ESD), Fire and gas (F & G) systems, and Burner management systems (BMS). In this subsection, brief discussions on PROFIBUS overview have been presented.

1. **PROFIBUS Family with applications:** PROFIBUS family consists of three compatible versions which can operate all together seamlessly. Fig III 7.2b1 shows all three types of PROFIBUS. As discussed FMS is no longer in use. In DP, there are three versions DPv0 refers for DP basic shown and DPv1 and DP v2 are used for two different signal exchanges already discussed.
   Complete PROFIBUS architecture and communication profile (OSI model) have been detailed in Fig III 7.2c which is self-explanatory. PROFIBUS architecture details out both members (DP and PA) along with process automation and asset management system in Fig III 7.2c1. Communication specification along with comparison with OSI model (ref: III 13.1) has been detailed in Fig III 7.2c2 [see Table III 7.2a also].
   PROFIBUS family provides solutions for a wide range of application areas including manufacturing and process automation. All these details along with coupling with PROFIBUS DP have been detailed in Fig III 7.2d. PROFIBUS family is standardized in IEC 61,158 type 3. As shown in Fig III 7.2d1, PROFIBUS is mainly meant for discrete system hence various family members like PROFIBUS DP for factory automation, there will be PROFIBUS drive for motion control and PROFIsafe for safety applications (use both RS 485 and MBP-IS). Even with different they all fall under PROFIBUS DP, that is, is based on RS 485. On the contrary, PROFIBUS PA (similar to FF H1) is dedicated for process automation, that is, it is based on MBP. PROFIBUS PA is based on the PROFIBUS DP-v1 version. Trunk and Spur used in PROFIBUS PA and its segments coupling to PROFIBUS DP have been depicted in Fig III 7.2d2. Apart from trunk and spur, this also details out field distribution with barrier.
2. **FMS and DP discussions:** In this part discussions on FMS and DP will be discussed
   - **PROFIBUS FMS (Field Message Specification):** This is a high-end application-level communication. With Extension of other family member PROFIBUS DP and suitable integration of the same with Ethernet gave rise to PROFINET. As discussed at the beginning, PROFINET made *FMS practically obsolete.*
   - **PROFIBUS DP:** PROFIBUS DP—an extension of PROFIBUS, is low-cost field and controller level communication. Majority of PROFIBUS devices belong to this category. It contains a subset of the functionality of PROFIBUS standard and was primarily developed for the fast processes in factory automation. In conjunction with PROFIBUS-PA systems for process automation, a PROFIBUS-DP system is used at the control level for quick transmission of the data. DPV0 is the foundation for PROFIBUS (PROFIBUS Basic in Fig III 7.2b1), and it is mainly responsible for cyclic data exchange between master slave and also does diagnostic reporting. DPv1 and2 are enhancement and further enhancements. There is a variant of PROFIBUS-DP—DPv1. In addition to the cyclic exchange of data with a PLC, this allows the field devices to be configured via acyclic services like alarm handling. The extended version DPv1 allows up to 127 participants including up to 32 masters. The control system (say PLC) serves as a Class 1 master. It uses

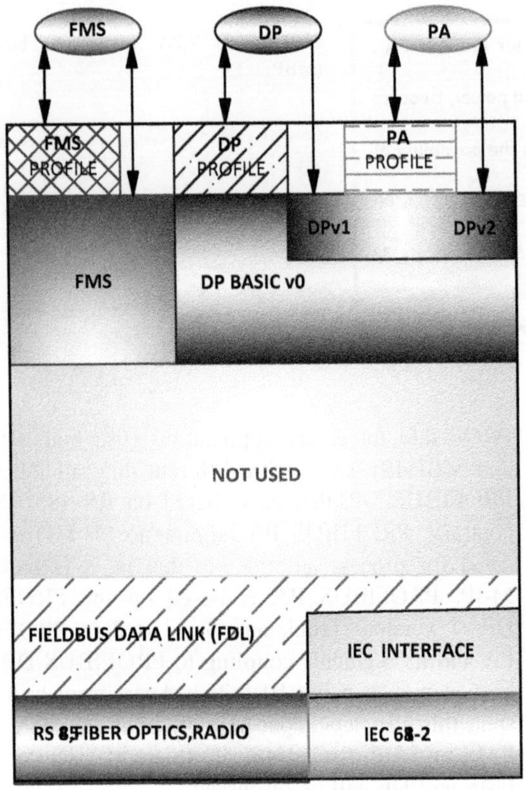


FIG III 7.2b1 PROFIBUS LAYERED STRUCTURE

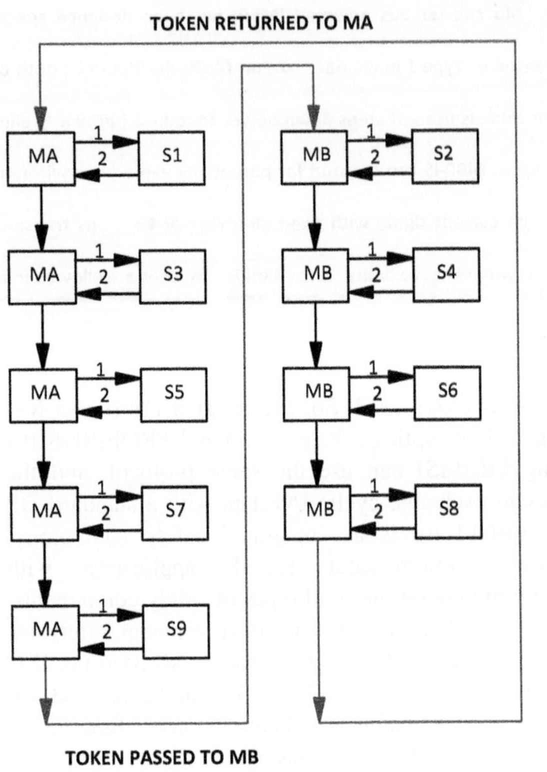


FIG III 7.2b2 PROFI TOKEN PASSING COMMUNICATION

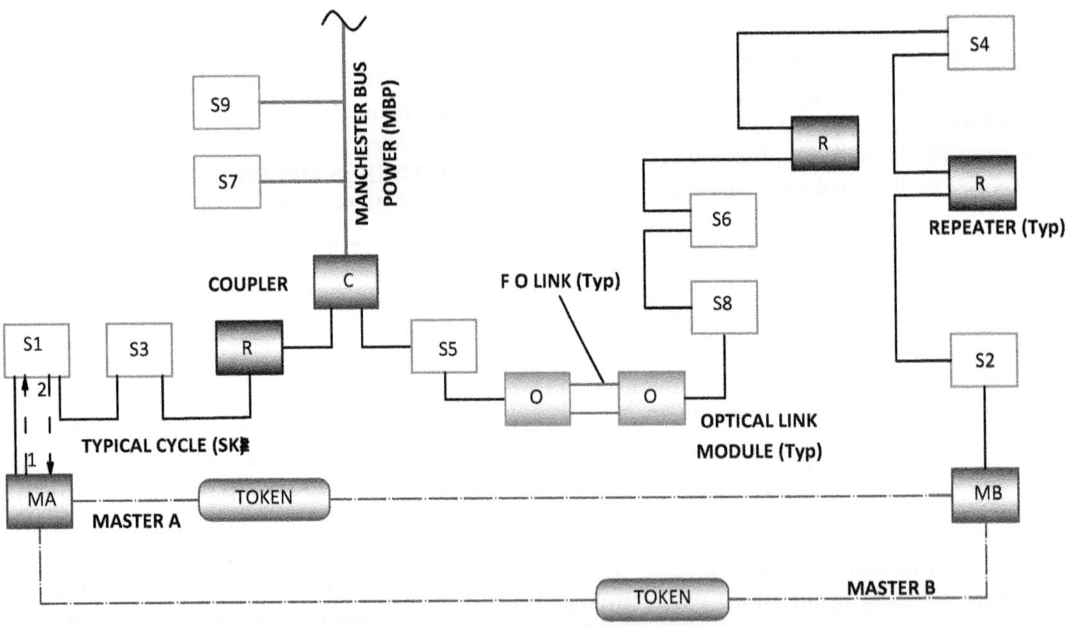


FIG III 7.2b3 PROFIBUS CONFIGURATION AND COMMUNICATION (PA)

ADAPTED FROM 1

**FIGURE III 7.2B** Profibus fundamentals [1].

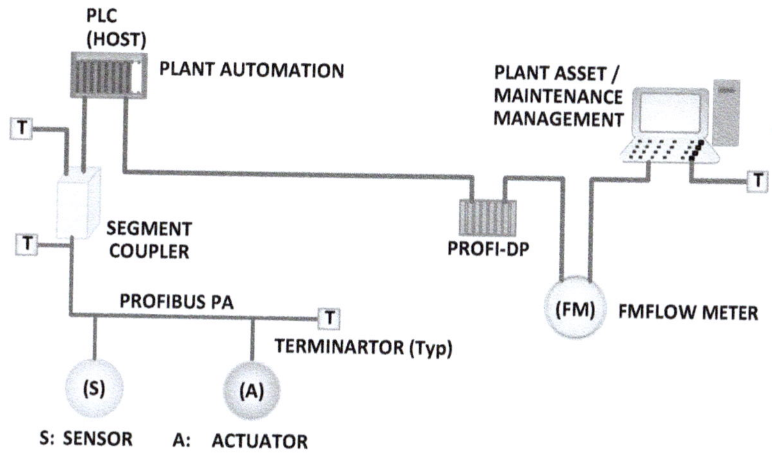


FIG III 7.2c1 PROFIBUS ARCHITECTURAL DETAILS

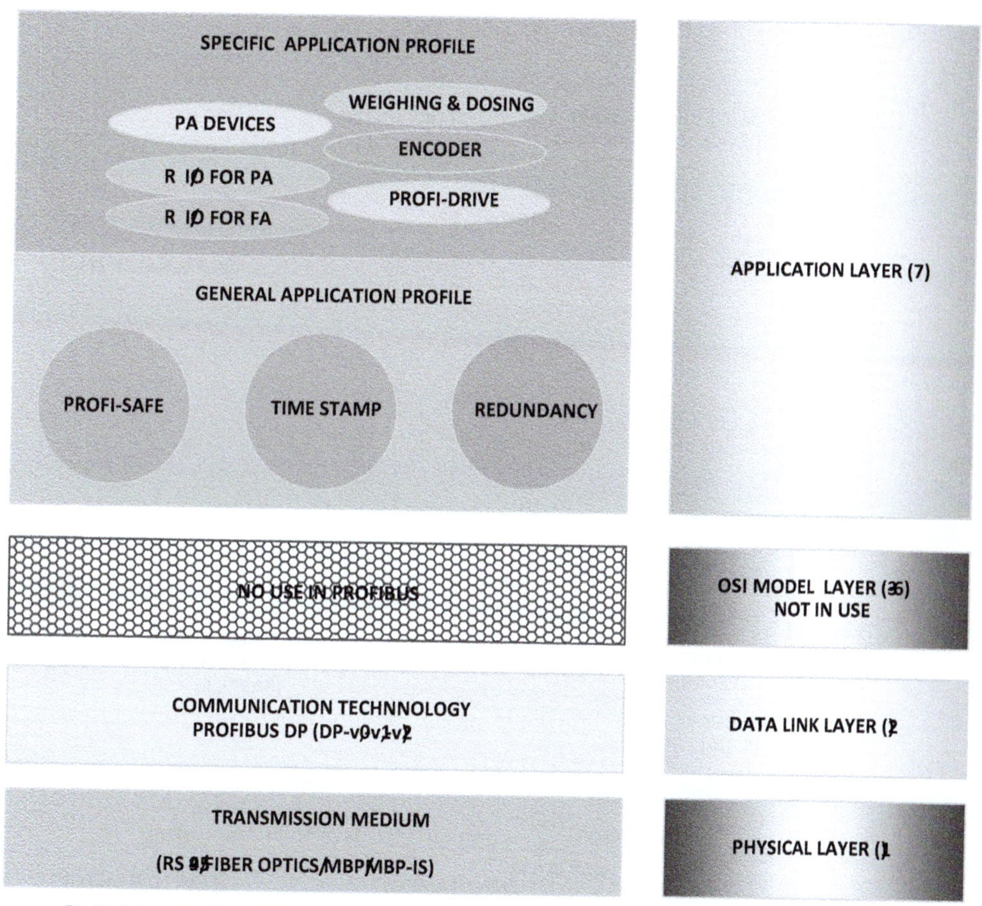


FA:FACTORY AUTOMATION; MBP:MANCHESTER BUS POWERED; R I/O:REMOTE INPUT/OUTPUT;

FIG III 7.2c2 PROFIBUS COMMUNICATION SPECIFICATION

**FIGURE III 7.2C** Profibus architecture and communication specification.

the cyclic services to acquire measurements and output control commands. DPv2 is responsible for deterministic cycle time mode, slave to slave communication Synchronization, Up and downloads. PROFIBUS DP in manufacturing and PROFIsafe use DPv0 to DPv2. PROFIBUS DP in process automation and motion control (PROFI-drive) use DPv1 and DPv2, respectively. Some of

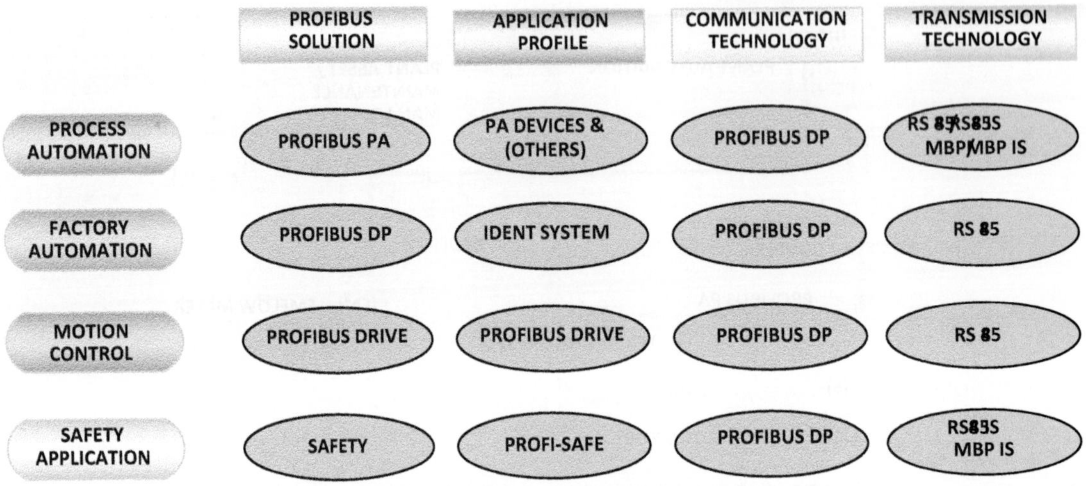


FIG III 7.2d1 PROFIBUS FAMILY PROFILE

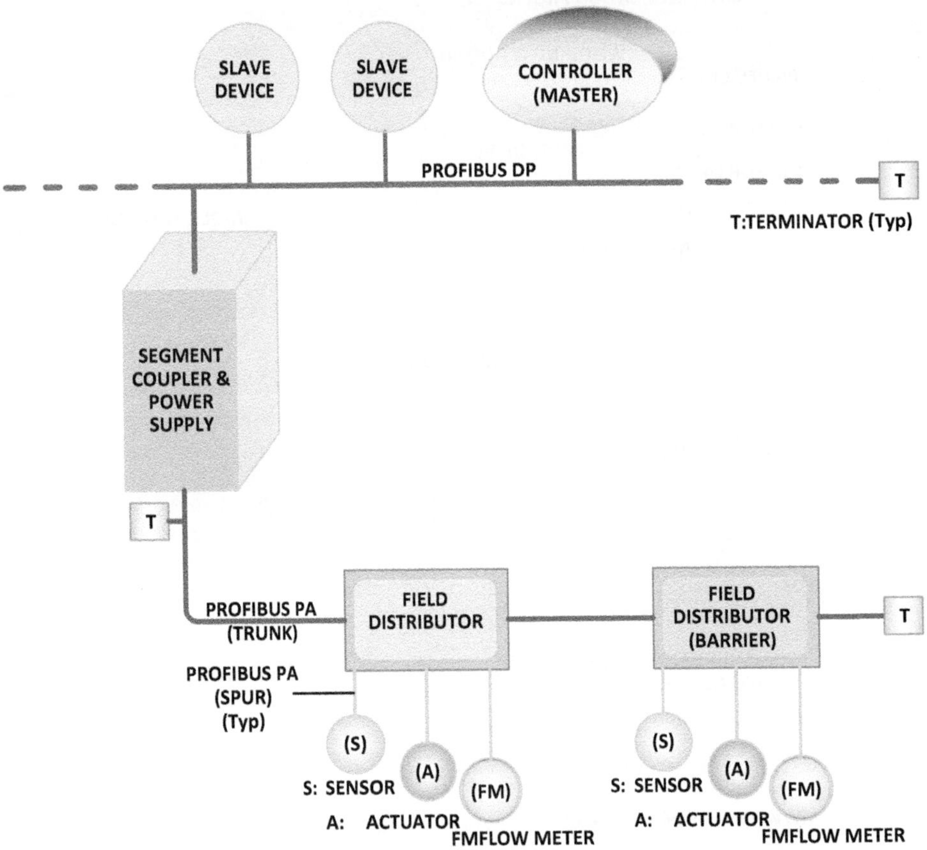


FIG III 7.2d1 PROFIBUS PA & DP WITH BARRIER

**FIGURE III 7.2D** Profibus family profile and profibus PA and DP coupling.

the key features of PROFIBUS DP are as follows: Based on open standard EN 50,170, Plug and play operation, Up to 244 bytes of data per message, Up to 126 stations to the bus, Up to 32 stations per bus segment, Class 1 Master (Central controller) and Class 2 Master Diagnostic, configuration, or startup tool. The GSD file defines the slave for the master.

3. **PROFIBUS PA:** PROFIBUS PA (process automation) is an extension of PROFIBUS-DP for process plant field devices. This is a cost effective Two wire connection for field devices (e.g., $T_X$) to communicate. It can offer to draw intrinsically safe power from the bus to the connected devices and can handle data according to the international standard IEC 61,158–2 based on MBP as described in Fig III 7.2a. So, as intrinsic safe, it can be used in hazardous area. A maximum of 32 participants can be connected to a PROFIBUS-PA segment and is governed by the master/slave method. The profile specifies how the various services of the PROFIBUS communication protocol are used. The PA profile defines the functions and parameters to adapt the devices to the respective application and process conditions, for example, transmitters, actuators, valves, etc. Like FFH1, PROFIBUS PA can be used with a variety of field devices, including transmitters, actuators, valves, and analyzers. DP/PA coupler connects DP controllers to PA devices, allowing for communication between the two systems. There have been improvements in PA profile with main focus on the optimization of lifecycle management of the devices.
4. **Special feature and functionality:** In addition to fail-safe operation PROFIBUS has a few optional operations. These are discussed here.
   - **Fail-safe operation:** Fail-safe operation in PROFIBUS systems is a safety-critical feature to ensure handling of system failures gracefully for minimizing the risk of harm to people, equipment, and the environment. Fail-safe is a special state allowing the function block in PROFIBUS to act when an abnormal condition is detected. When the I/O function blocks detect an abnormal condition in the event of a failure, such as communication loss or sensor malfunction, that is, the abnormal situations are detected through different ways and the block enters in *Fail-safe* mode. A master in PROFIBUS system, can switch to "*clear mode*" or may fail. These issues can cause communication failure when a specific output data message is sent. Fail-safe feature is an optional feature slave device which can transition output to specified value configured as a fail-safe state. On such an occurrence, nonfailsafe devices usually clear their outputs to zero. Fail-safe systems comply with Process technology ordinance (PTO) requirements, by which slaves must have the option to clear outputs upon master failure or clear mode switch.
   - **Optional functions:** PROFIBUS systems include several optional device functionalities to enhance network capabilities and management. Majority of them are as follows: *Simple device replacement, Device-specific diagnostics, Dynamic reconfiguration, I-Device, Shared device, Device,* controller, network *and media redundancy, Time stamping, Dynamic reconfiguration, I&M Records, Fast startup, Application profiles, the output SYNC feature* (motion control), *the input FREEZE feature,* etc. Normally, vendors place keywords in the GSD file.
5. **Service area:** PROFIBUS systems are widely used in industrial automation for communication between field devices and control systems Major areas are as follows:
   - **Factory automation and robotics:** Simple to large low-cost distributed control and communication, connecting devices including remote I/O.
   - **Process automation:** High-speed time-critical application with intrinsic safety features—as applicable,
   - **Building management systems:** Complex communication tasks; Building automation to control and monitor various systems viz. HVAC
   - **Transportation:** Utilized in transportation systems, such as railways and airports
   - **Other industries and safety application:** Many other applications including safety applications through PROFIsafe.

For smoother interaction between the bus nodes of an automation solution, *the basic functions and services of the nodes must match*, that is, they need to "speak the same language" and use the same concepts and data formats. Based on this idea we get from Probus.com [10] "General application profile" and "Specification application profile." Partial list has been presented in Table III 7.2a.

### 7.2.1 PROFIBUS: Communication characteristics

Fieldbus systems are characterized by two-way digital communication networks for connecting and controlling devices at the field level in industrial automation. There are different fieldbus protocols exist, each having own specific characteristics adding strengths and weaknesses to the system. PROFIBUS is a bi-directional digital network supporting multiple devices on a single cable, using a master-slave communication mode of communication, that is, the master device initiates communication for a slave device to respond. Apart from bidirectional digital communication, few distinct characteristics of PROFIBUS mainly include the following: *Master-slave, Multidrop network* (i.e., multiple devices in a multidrop network in a single cable, forming), *Encoding, High speed transmission, Interoperability, Flexibility and scalability,* Support various topologies (like bus, star, ring configuration), *Addressability, Cyclic and acyclic, ISO-synchronous slave-to-slave communication, Diagnostics* and *Configuration.* PROFIBUS DP is characterized as an open industrial standard for integrated serial communication. PROFIBUS

**TABLE III 7.2A** Partial application profile [Complete list ref [10].

| Profile name | Profile content |
|---|---|
| **Specification application profile** | |
| Dosing/Weighing | The use of dosing and weighing systems on PROFIBUS |
| Encoder | The connection of rotary, angular, and linear encoders with single/multiturn resolution |
| PA devices | The properties of process automation devices |
| PROFIdrive | The device behavior and access behavior to data for Variable speed electric drives |
| Remote I/O | The interchangeability of remote I/O devices in process/factory automation. |
| **General application profile** | |
| Data type | Dealing with data type definitions of profile variables and parameters, programming languages of field devices and controllers as well as platform aspects for field devices |
| Host application | Each application profile and the functional Requirements for the host and the engineering system |
| PROFIsafe | This is safety-related devices communication with safety controllers via PROFIBUS |
| Diagnostics | Dealing with the different aspects of diagnosis with information on field devices faults |
| Time stamping | Time assignment of certain events and actions with time stamping. |
| Redundancy | The mechanism for field devices with redundant communication behavior. |

DP supports multiple primary systems with several secondaries. *Master-slave*, *Token passing*, *Linear bus topology*, *Support for upto 125 devices* are architectural beauty noticed in PROFIBUS DP. As already discussed, it can support wide varieties of physical layer medium both for normal as well as hazardous areas through intrinsic safety (IS). DP-v0, DP-v1, DP-v2 are various PROFIBUS DP to facilitate different communication needs. GSD files, error handling are important characterized issues in PROFIBUS. The GSD from device manufacturer is the electronic data sheet for the communication properties of each and every PROFIBUS device. A particular advantage of PROFIBUS is openness, which in turn brings with it compatibility with a large number of device and system manufacturers. Devices are usually integrated by means of mapping their functionality to operator software. In this subsection, brief discussions on PROFIBUS characteristics.

1. **Physical layer: Transmission medium:** As seen from Fig III 7.2b1, Copper, FO, and Radio can be used for in PROFIBUS. The transmission media used include either twisted pair shielded cables for FMS/DP, even fiber optics or radio waves [11]. Communication technology uses RS 485 standard, in compliance with IEC 61,158–2. Also, power over the bus (ref Fig III 7.2a) is used in PROFIBUS PA ((especially for explosion-hazardous areas). Table III 7.2b shows transmission media for PROFIBUS realization.

**TABLE III 7.2B** Transmission media in profibus.

| Media | FMS/DP | PA |
|---|---|---|
| RS485 | Twisted pair or with four wire with power supply (IEC61158 half duplex) speed: 9.6 (1200m) to 12000 KBPS (100m) | |
| Optical | Hi speed communication with electrical isolation | Not applicable |
| Radio wave | Possible | Not applicable |
| IEC 61158–2 | Not applicable | MBP @ 31.5 KBPS ref: Fig I 3.2f |

**TABLE III 7.2C** Speed, segment length and branches in profibus DP.

| Speed (kbps) | Segment length (meter) | Branch length/ segment |
|---|---|---|
| 9.6 | 1200 | 32 × 3m |
| 19.2 | 1200 | 32 × 3m |
| 45.45 | 1200 | 32 × 3m |
| 93.75 | 1200 | 32 × 3m |
| 187.5 | 1200 | 32 × 2m |
| 500 | 400 | 32 × 1m |
| 1500 | 200 | 32 × 0.3m |
| 3,000–12,000 | 100 | Not allowed |

For PROFIBUS DP there is correlation among speed of Transmission length and allowed branching a detailed out in Table III 7.2c.

2. **Communication:** Following are issues related to communication in PROFIBUS:
   - **Topology and Communication** [1]: The circuits can include *Repeaters*, *Couplers* and *Optical link modules* (for transforming optical signals to electrical signals) as shown in Fig III 7.2b3. Function of signal (segment) coupler is to convert MBP signal to compatible for RS 485 connection.
   - **Signal (segment) coupler:** Following are a few characteristics of Signal Coupler
     - *Isolation:* Electrical Isolation between safe and IS Bus
     - *Slave:* It is like a slave to PROFIBUS DP
     - *Baud rate:* Baud rate adaption (PA; 31 kB PS) faster (12,000 kB PS) also.
     - *Power supply:* Powering of PA Bus
     - *Adaption:* Adaption of IEC 61,158–2 technique for in PA when other side is RS 485
   - **Operational aspect** [1]: PROFIBUS operates on Master-Slave mode with even more than one master. Master stations control the network communication, and slave responds to the call from master. Each of the masters can communicate with a number of slaves. The system is highly democratic meaning that all the masters have equal right while all the slaves connected to any master have equal right. A simplified method of communication and token passing has been shown in Fig III 7.2b2. Assuming arbitrarily MA has the token, then it one by one communicated with slaves S1, S3, S5, S7, and S9. After that, MA passes the token to MB. Then, MB communicates one by one from S2, S4, S6, and S8. Then passes on token back to MA. In case of multimaster system, communication is regulated by token passing method as shown by token passing (red [light gray in print version]) lines in the figure. Master having the token controls the bus.
3. **PROFIBUS (DP) characteristics** [13] [1]: Extended version DPv1 allows up to 127 participants including up to 32 masters. Out of that, 126 is reserved for Generic device and 127 is used for Broadcast message. However, from IEC 61,158–2 and RS 485, only 32 per segment is allowed. Following are major features of DP (DPv1):
   - Physical layer: RS 485; Fiber Optic (FO); Radio wave
   - Max. Length: 1200 m for copper; F.O. Several Km
   - Total participants: 126 including 32 masters in segments
   - Speed: 9.6 kbps to 12,000 Kbps; Ref: Table III 7.2c
   - Access method: Token passing Master slave
   - I/O per station: Up to 244 Bytes
   - Communication: In cyclic operation, only one master can write on Slave.
   - Alarm: Alarm acknowledgment possible.
   - HW Address setting: Binary Dip switch or Rotary switch
   - SW Address setting: Over PROFIBUS, special SW and link by hand-held device/controls.
   - Topology: It can be Linear in RS 485 in daisy chain or tree as per IEC 61158-2.
4. **PROFIBUS (PA) characteristics** [1]: Following are major features of PROFIBUS PA:
   - Transmission Technology: IEC 61,158-2—MBP (Fig III 7.2a)
   - Maintenance and diagnostics: Yes, from available instruments
   - Distance: <1900m max four repeaters allowed
   - Coding: Manchester code –mean value @+9 mA
   - Devices: Max 126 in 32 segment max.
   - Remote Voltage: 32VDC max.
5. **PROFIBUS DP and PA Discussions:** In this part specification of both Profi DP and PA shall be covered. Also, a comparison of Profi DP and PA has been presented. Let it start with specification.
   - **PROFIBUS DP and PA: Specification:** Major specifying issues for both PROFIBUS DP and PA are as follows:
     - The physical layer used in PROFIBUS DP is RS485, IEC 61,158–2 (physical layer—PA) RS 485. RS 485IS/optical for high speed in DP and MBP & MBP-IS for PA shown in II Fig III 7.2d1.
     - STP cabling is used for signals.
     - Cable length is dependent on the Baud rate as detailed in Table III 7.2c for segments

**TABLE III 7.2D** Comparison between profibus PA and profibus DP.

| Issue | Profibus PA | Profibus DP |
|---|---|---|
| Purpose | This is used to enhance and replace traditional hardware signal/HART | Interface and activate sensors and Actuators to host/controller within industrial automation |
| Technology | It uses MBP/MVP IS | Use RS 485 R S 485 IS with different physical medium. |
| Network | Alternative network to HART/Traditional 4–20 mA loop (bus, start, tree) | High performing bus network. Can transmit 1000s of I/O in a few milliseconds. |
| Baud rate | 31.25 kbps only | <12 Mbps |
| Cable | STP for data and power <1900m | STP for data only <1200m |

- The number of devices is 32 for each segment with max. 126 including four repeaters.
- Bus power is 24 VDC supply; Bus power is up to 0/5A for every segment.
- Transmission rate from 9.6k to 12M bits per second.
- The size of data transfer mainly depends on the type of device like the input is up to 246 bytes and output is up to 246 bytes.
- The device identification possible, based on the particular ID number for every device.
- The topology used is a trunk with branching.

- **Comparison of PA, DP and PROFIBUS and PROFINET:** The comparison of these has been presented in Table III 7.2d and Table III 7.2e respectively.

With this, *let us conclude discussions on PROFIBUS characteristics to start detailed discussions* on PROFIBUS.

### 7.2.2 PROFIBUS: Detailed discussions

Let the discussions on PROFIBUS, start with reference to Fig III 7.2c depicting PROFIBUS architecture and communication specification. PROFIBUS is a standardized protocol based on IEC 61,158 and IEC 61,784, providing a robust communication infrastructure for a variety of industrial applications. From the discussions made so far that various key aspects of PROFIBUS include the following: *Reliability*, *Standardization*, *Master-slave topology*, *Token ring*, *Flexibility*, *Communication profiles*, and Defined network structure. Even though it was mainly developed

**TABLE III 7.2E** Comparison between profibus and profinet.

| Issue | Profibus | Profinet |
|---|---|---|
| Concept | Engineering GSD files | Engineering GSD files |
| Protocol | Fieldbus within automation | Ethernet based |
| Physical layer | RS 485 based | Ethernet based |
| Baud rate | 9.6 to 12,000 Kbps | Up tp 100 Mbps |
| Main types | Profibus PA and DP | Profinet I/O, profinet CBA |
| Communication | Master slave | Provider consumer |
| Cable length | <1000m | <100m |
| Wireless | Possible | IEEE 802.11, 15.1 |
| Files used | ASCII file | XML |
| Address space | 126 | Unlimited |
| Telegram | 244 bytes | 1440 bytes (cyclic) |
| M2M comm[b] | No | Yes |
| V-integration [c] | No | Yes |

*[b]Machine to Machine communication.*
*[c]Vertical Integration.*

for manufacturing and factory automation, it can support process automation through PROFIBUS PA. *Cost-effectiveness*, *Interoperability*, *Scalability*, and *Flexibility* are a few major benefits. In addition to connecting field devices, (arranged in a multidrop fashion, requiring a single STP cable) in a digital, serial, two-way communication protocol enables advanced diagnostics from all fieldbus components including the physical layer for better plant availability. A segment coupler converting establishes, interconnection between PROFIBUS PA and DP. This segment coupler converts the PA physics based on the IEC 61,158–2 of PA to physics based on RS 485 for DP. Also, converter converts the synchronous data transmission to asynchronous data transmission. Based on different related application profiles, it provides solutions for a wide range of application areas including manufacturing and process automation—this is detailed in Fig III 7.2d1. Different versions of DP have been defined. DP-v0 provides then basic functionality of the communication, addressing cyclic communication as well as device-specific, module-specific, and channel-specific diagnostics, in particular for quick fault localization. DP-v1 version supplements DP-v0 with functionality for acyclic communication. In this concluding subsection, discussions on PROFIBUS in details shall be covered to complete the discussions.

1. **Working and start up:** Here, two important aspects of PROFIBUS have been discussed in different parts, with depictions in Fig III 7.2e
   - **Working Principle:** PROFIBUS working principle is based on a number of simple concepts viz. *Master slave* and *Token passing ring.* Basic concept of the same has been discussed in III 7.2.1.2 with reference to Fig III 7.2b2 and three in PROFIBUS PA. In view of the above, here emphasis will be given on various device types (in PROFIBUS DP). In a master-slave protocol, one device on the network is defined as the *master,* and it is the device that is in command. All other devices on the network are called *slaves.* As already discussed, token passing is like a relay race where a baton is passed from one device to the next. This means the device having token can only can communicate. Slave device in the network can only talk only when it is asked for by master. Various device types in PROFIBUS DP have been detailed out in Table III 7.2f
   - **Configuration tool:** When a vendor develops a slave device, it must develop a device description (GSD) file to completely describe the functionality of the device, for example, baud rates support, startup parameter choices, etc. The installation process for different configuration tools, varies but are very simple viz. in some case installation means copying the GSD file into a "GSD subdirectory," or simply "imports" the new GSD file into the hardware catalog by selecting that option from a menu. After all the appropriate GSD files are installed into the configuration tool, a bus configuration can be defined in a straightforward process. This configuration database file can be loaded into the master device as shown in Fig III 7.2e1. Configuration tool for cyclic I/O can perform the following:
     - Process *GSD* (device description) files, maintain a hardware devices catalog for configuration on the bus
     - Allow the specification of PROFIBUS device address, the I/O data to be exchanged
     - For specific address mode, allow selection of certain startup parameters
     - Selection of the system baud rate
     - Generation of the database file for use by the master
   - **Start-up sequence:** The PROFIBUS master continually scans the network looking for nodes which it is not already in a data exchange. The master issues a *check configuration* telling the slave Its identity, number of modules the slave has, and how much data it will send and receive. Again, the slave will just respond with a short *acknowledge.* The master next issues a second *get diagnostic* message to verify that there were no problems with the parameters and the configuration that it sent to the slave. This has been detailed out in Fig III 7.2e2. Diagnostics are a fundamental function of a fieldbus protocol and provide complete diagnostic functionality to the network. In addition to the cyclic diagnostic message, a device written to DP-v1, PROFIBUS PA profile specifications will also provide diagnostics viz. cyclic status byte and acyclic diagnostic messages.
2. **Field device and PA issues:** Let us discuss various PA devices including segment coupler and PROFIBUS PA devices
   - **Field device issue:** General but important issue is device selection based on interoperability, that is, all devices must be interoperable. Based on the above objective, first DCS PA devices then other PA devices are selected. PA certification. Field device management (Device's fieldbus address and physical device tag are set before installation), Field device power and Device upgrading (PA device firmware changes can be made), are various issues to be considered.
   - **Segment coupler:** The segment coupler is the core of every PROFIBUS PA. It consists of a (redundant) gateway and separate fieldbus power supply (could be redundant also). Signal coupler could be

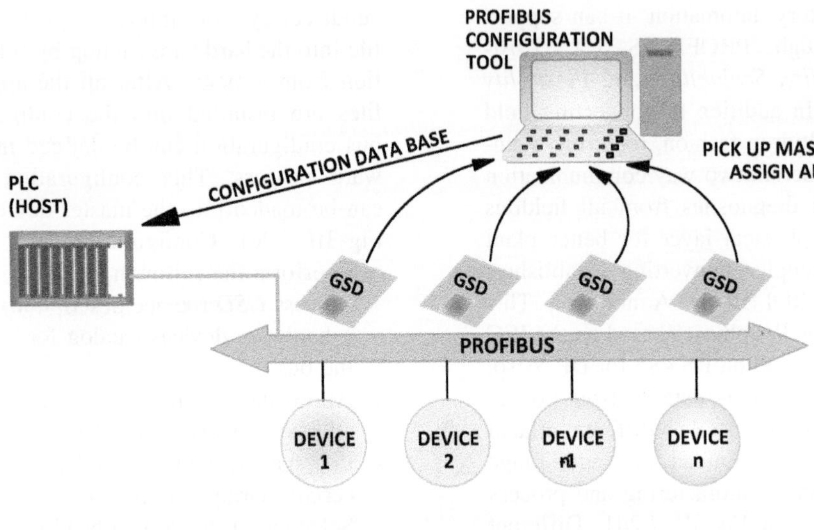


FIG III 7.2e1 PROFIBUS CONFIGURATION TOOL

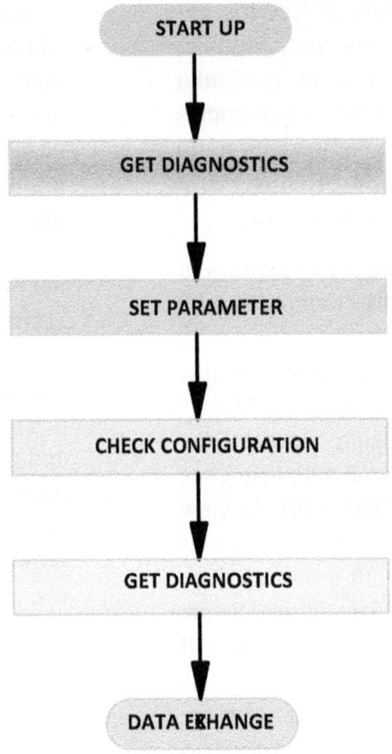


FIG III 7.2e2 PROFIBUS START UP SEQUENCE

**FIGURE III 7.2E** Configuration tool and start up sequence.

redundant to avoid any problems in terms of *availability/PROFIBUS PA cycle time limitation.* Various issues related to segment coupler:

- *Functionality*: *Gateway*, *Power supply,* and *Transparent communication* between DPM 1and2 with PA, Up to 70% of cycle time (dependent on: *nos. of instrument* in a loop, *process* and *loop type*) should be used for cyclic data exchange and 30% for acyclic communication, etc., are major functionality
- *Grouping*: For maximizing the benefits of the PA, all devices of a particular loop should be connected to the same PROFIBUS DP master class 1 [not mandatory].
- *Fieldbus power supply*: Part of segment couple. V less than 25VDC and current < 350 mA. It also provides basic layer diagnostics (trunk over current, under voltage, bulk power under voltage) as well as advanced layer diagnostics (noise termination fault etc).

**TABLE III 7.2F** Profibus DP device types.

| Device type | Description |
|---|---|
| Master class I | This master (DPM1) –central controller, uses cyclical communication command to exchange process data with its associated slave devices during the configured timeslots. Devices of this type are often integrated into a memory PLC or an automation station of the process control system. It can support upto 125 slaves. Determines the baud rate (slaves to auto-detect), detects another master during the gap time and manages the token transfer between masters. |
| Master class 2 | This (DPM2)- a tool for commissioning), accesses the fieldbus by using acyclic services allowing the configuration and parameterization of intelligent field devices. With developments in DP v1 and DP v2, DPM2 has been specifically used to set device parameters via acyclic communication. DPM2 is part of operator/engineering stations which need not be permanently connected to the bus system |
| Slave device | A profibus DP slave device is a passive communication node that reacts to prompts from the master by sending a response message only (without having right to control the bus). So, this is a peripheral device to exchange I/O information with master. GSD file defines the slave for the master. Based device, the maximum I/O capacity can be max 246 byte and 244 byte respectively. By combining different modules, modular devices can be adapted flexibly to respond to prevailing input and output data requirements. |

- **PROFIBUS PA Device Loading:** PROFIBUS PA supports up to 32 devices including PA Master (integrated with segment coupler) per segment. There exist two other limitations such as physical layer limitation and DCS/PLC/controller limitations.
  - *Physical layer limitation*: Power to field devices are supplied through the fieldbus cable in the same segment. Depending on devices, the average current, drawn by the each device is between 12 and 18 mA DC. Also, IEC 61,158–2 provides limitation by defining the max *allowed spur length* as a function of connected nodes to the segment. As per the standard max allowed length are specified in the bracket for numbers of address in use such as addresses: *1–12 (120m), 13–14 (90m), 15–18 (60m), 19–24 (30m) 25–32 (<1m).*
  - *Controller limitation*: The Nos. of addressable nodes typically PROFIBUS supports is in the address range from 0 to 127. Address 127 is a broadcast message (not usable as a device address), hence address 126 is the default address of every PROFIBUS PA device.

3. **Miscellaneous Issues:** Here, a few important aspects of PROFIBUS have been discussed in assorted form. Let the discussions start with PROFIBUS concept.
   - **PROFIBUS concept:** Following are some of the key concepts:
     - *Open standard based on EN 50,170 with data rates up to 12 MB.*
     - *Plug-and-play operation.*
     - *Up to 244 bytes of input/output data* per *message.*
     - *Up to 126 stations may connect to the bus.*
     - Up to 32 stations per bus seg
   - **Watch dog:** In PROFIBUS slave devices, the watchdog is a safety mechanism to prevent the slave device from entering an uncontrolled state. Watchdog monitors the communication between the slave and the master. A slave may assume a defined state when the associated watchdog time expires without having received a message from its assigned master. This timer is reset by the slave every time the master talks to the slave.

     So, when the time indicates that the master has not communicated hence the slave is not being controlled. The timeout state is clear by default, or it may be set via user-defined parameters of the parameterization telegram and GSD file (or by hardware switches on the slave).
   - **GSD Files:** GSD is a human-readable ASCII text file specifying keywords as mandatory/optional with the corresponding data type and their border values, that is, GSD files contain information about the basic capabilities of a device for system integrators to determine the communications options/available diagnostics. GSD files have a standard file format as per the PROFIBUS guidelines including the editor for syntax and semantics. The configuration tool uses GSD files for data testing, entering limits and validity related to the performance of the individual device. All PROFIBUS devices (class 1 masters and slaves) have their own GSD files. The GSD file is the fundamental building block for the master parameter record to be used by a PROFIBUS configuration tool permits plug and play interoperability among different devices from different manufacturers.

| FRAME | SD | LE | LEr | SD | DA | SA | FC | DSAP | SSAP | DU | FCS | ED |
|---|---|---|---|---|---|---|---|---|---|---|---|---|
| BYTE (B) | B | B | B | B | B | B | B | B | B | VAR | B | B |

DU:VAR=VARIABLE LENGTH
DU-- DATAUNIT /NET DATA ---- ITO BYTES

NET DATA LENGTH (LE) =DU + FC + DA+ SA+ DSAP + SSAP

ABBREVIATIONS

SD:START DELIMITER; LE:NET DATA LENGTH
LEr:LE REPEAT; DA:DESTINATION ADDRESS
SA:SOURCE ADDRESS; FC:FUNCTION CODE
SAP:SERVICE ACCESS POINT
DSAP:DESTINATION SAP; SSAP:SOURCE SAP
DU:DATA UNITS; FCS:FRAME CHECKING SEQENCE
ED:END DELIMITER

FIG III 7.2f1 TELEGRAM FRAME STRUCTURE WITH EXPLANATION

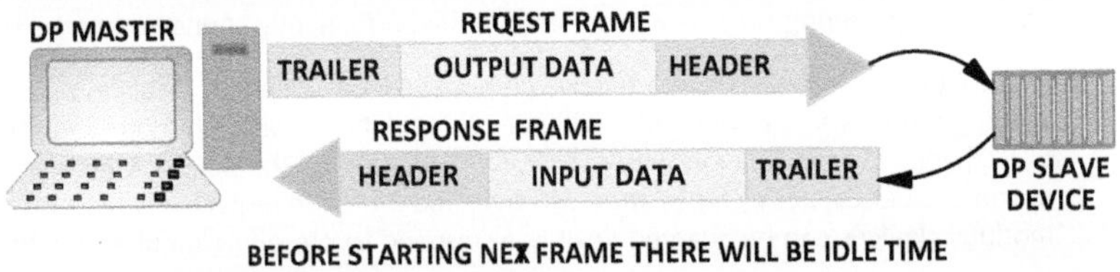


FIG III 7.2f2 TELEGRAM EXCHANGE BETWEEN dp MASTER AND SLAVE DEVICE

**FIGURE III 7.2F** Profibus telegram structure and telegram exchange.

- **PROFIBUS Telegram:** A PROFIBUS telegram can be as long as 256 bytes—up to accommodate upto 244 bytes of data per node per message, plus 11 bytes of overhead referred to as "*Header.*" As detailed in Fig III 7.2f. For specifying *data type and functions* to be performed, all data exchanges between master and slave are handled by using Service access points (SAP) which are two types Source SAP (SSAP) and Destination SAP (DSAP). All telegram headers are 11 bytes, except for data exchange telegrams which have 9 bytes of header, that is, the DSAP and SSAP bytes are dropped. Details of data and overhead along with explanations are detailed in Fig III 7.2f1. While method of data exchange has been depicted in Fig III 7.2f2
- **Bus timing**: PROFIBUS is designed for deterministic response (ability to precisely predict the system behavior of over time) signifying that a reliable system reaction time can be calculated. Here, important issues are as follows:
  - *Bit time* (Tbit): A bit-time is the time taken to transmit 1 bit, that is, it is the reciprocal of the transmission rate (baud rate). One Tbit at 12 MB baud rate = 1/1 MB = 83ns/bit.
  - *Sync Time*: The synchronization time refers to the minimum time a station must remain in the idle state before it can accept another request.
  - *Slave reaction time*: The reaction time is the time it takes a slave to respond to a message—expressed in min/max values of delay.

  Apart from these there will be initiator delay and ideal time, etc.
- **Fail-safe mode**: A PROFIBUS master runs in either "*operate*" or "*clear*" (class 1 master sends 0 length data, i.e., data set to 0) mode. In a DP master-based slave device, failsafe refers to whether the *slave will process output telegrams with zero length* data, or *not.* Whether the combined master/slave system is considered fail-safe depends on the *actions taken by the slaves* if the master *fails, or if the master switches to clear mode.* The failure of a master should not cause errors in any of its slaves. Using a fail-safe mode, the slave outputs can automatically switch to a fail-safe state in the event of master failure/when the master switches to clear mode. A slave may assume a fail-safe state if its watchdog time expires without having received a message from its assigned master. Also, it can be in fail-safe mode when master switches from operate to clear mode. Masters may go to clear mode if a slave is disabled, as specified in associated GSD file.

4. **Profile and design:** Here, two important issues will be touched upon. Profiles refer to various features and aspects to indicate total coverage by PROFIBUS will be discussed. The design of PROFIBUS system is standardized and follow certain rules. Under design same will be covered.
   - **PROFIBUS Profile:** PROFIBUS profile clarifies and defines many details like The *data* with *associated format* presented on the network, *Standard functions* available for manipulation of data, Device *set up parameter, Procedural details for communication* of parameters over the communication network, etc. Device profiles are a wonderful

PROFIBUS feature, and the protocol's usability. PROFIBUS has profiles for a wide selection of equipment viz. drive systems, process automation, etc., as detailed in Table III 7.2a. Those are partial list and not a complete list and it is ever expanding.

- **PROFIBUS design rules**: Following are basic design rules followed in PROFIBUS design:
  - *PA design*: Basic major PA design rules include the following: *Maximum spur lengths* must correlate to number of spur lines, Use of *Class A, PROFIBUS PA cable* (from same manufacturer throughout), *Total cable length* cannot exceed 1900 m, actual length depends on coupler used, *Ensure > 9 V* at each slave,
    *Number of devices limitation* is dependent on the current drawn, *Termination* at the extreme the *segment ends*, *Recommended grounding* practices to be followed,
    *IS design affects* maximum spur lengths and total length, etc.
  - *DP design*: PROFIBUS DP running on copper RS-485 must follow basic design rules like *Max 32 nodes* in one segment, with nodes to be arranged in *daisy chained style*, *Termination* at the extreme *ends*, *Maximum directly addressable* devices 125 per *PROFIBUS DP interface* card, *Max 6 repeater* between master and devices, Maximum length per segment depends on baud rate, *Recommended grounding* practices to be followed, Design for *maximum baud rate*, but *run* it at the *lowest* possible *baud rate,* etc.

5. **Diagnostics:** Like other fieldbus, in PROFIBUS protocol, diagnostics are a fundamental function to provide complete diagnostic functionality to the network. Diagnostics ensure that the instruments can report any conditions to the master and that a variety of actions can be implemented to deal with the matter. Here, diagnostics in different views and applications have been enumerated. Diagnostics have importance in asset management from maintenance point of view, whereas same diagnostics is important as process/equipment status in automation. From maintenance view point diagnostics can tell "*Failure,*" "*Maintenance required,*" "*Functional check*" and "*Out of Specification.*" In case of automation, PROFIBUS diagnostics provide status like "*Bad,*" "*Uncertain*" and "*Good.*" Some associated issues here are as follows:
   - **Cyclic diagnostics:** Two-way communication is crucial that the slave device can alert the master of an event. In all PROFIBUS devices, a slave can request the master to read diagnostics during the next cycle. For this, the slave sets a diagnostics request flag in its data exchange response, as a diagnostic request. After the slave issues a diagnostic request, the master issues the command in the next bus cycle. The diagnostic response message is made up of the diagnostic message plus an extended diagnostic. In this connection Fig III 7.2e2 may be referenced.
   - **Diagnostics in DP-V1:** PROFIBUS DP diagnostics refer to real-time monitoring and fault detection on the network, for quick identification and resolution of issues. Major PROFIBUS DP diagnostic capabilities are as follows: *Diagnostic telegrams, Cyclic data exchange, Physical layer check, Network monitoring Quality index,* etc. So, in addition to the cyclic diagnostic message, a device written to DP-v1 PROFIBUS PA profile specifications will also provide two additional sources of diagnostics: cyclic status byte and acyclic diagnostic messages.
   - **Diagnostic software:** The PROFIBUS diagnostics software is thus an essential component of an overall PROFIBUS diagnostics concept. A network diagnostic tool uses in-depth packet analysis to collect and inspect the data to send over a computer network. Based on the logged analysis data, the PROFIBUS diagnostics software provides the various signal and logical analysis capabilities for testing PROFIBUS networks.

With this, let us conclude the discussions on *fundamentals of fieldbus systems* to look forward to discuss other protocols.

## List of abbreviations used

**BD** Basic device
**BMS** Burner management system
**CF/CFF** Configuration files/Common file format
**CP/CPF** Communication profile/CP families
**CRC** Cyclic Redundancy Check
**DCS** Distributed control system
**DD** Device description
**DHCP** Dynamic host configuration protocol (
**DLL** Data link layer
**DMA** Direct memory access
**DSAP** Destination SAP
**DTM** Device Type Manager
**EDD-L** Electronic device description-language
**EMI** Electromagnetic interference
**ERP** Enterprise resource planning
**ESD** Emergency shutdown
**F&G** Fire and Gas
**FAS** Fieldbus access sublayer
**FAT** Factory acceptance test

**FB** Functional block
**FDT** Field device tool
**FF** Foundation fieldbus
**FIFO** First in first out
**FISCO** Fieldbus Intrinsic safety concept
**FMS** Field message specification
**GUI** Graphical user Interface
**HART** Highway addressable remote transducer
**HMI** Human machine interface
**I/O** Input/Output
**IC** Integrated circuit
**IE** Industrial Ethernet
**LAN** Local area network
**LAS** Link active scheduler
**LM** Link master
**LRE** LAN redundancy entity
**MaU** Medium attachment unit
**MBP** Manchester coded bus powered
**MES** Manufacturing execution systems
**NMA** Network management agent
**NMIB** Network management information base
**OS** Operating system
**OSI** Open systems interconnect
**P&ID** Process and instrumentation diagram
**PC** Personal computer
**PTO** Process technology ordinance
**RTE** Real time Ethernet
**SAP** Service access point
**SAT** Site acceptance test
**SC** Scheduled communication
**SCADA** Supervisory control and data acquisition
**SIF** Safety instrumentation function
**SMK** System management kernel
**SNMP** Simple network management protocol
**SNTP** Simple network time protocol
**SSAP** Source SAP
**STP** Shielded twisted pair
**UI** User interface
**USC** Un-Scheduled communication
**UTP** Unshielded twisted pair
**VCR** Virtual communication relationships
**VFD** Virtual field device
**WS** Work station

## References

[1] Power plant Instrumentation and control handbook, S. Basu, A. K. Debnath, 2nd rev, Elsevier; Academic Press, 2019. https://www.elsevier.com/books/power-plant-instrumentation-and-control-handbook/basu/978-0-08-102804-9.

[2] T. Shoji & M. Hiroshi; *Fieldbus System Engineering*; Yokogawa; Internet document; https://web-material3.yokogawa.com/rd-tr-r00028-005.pdf.

[3] Fieldbuses for Process Control: Engineering, Operation and maintenance; Jonas Berge, ISA Publication, 2001.

[4] C. Stroehle; Fieldbus Foundation: *Fieldbus Basics*; Pepperl + Fuchs pte ltd; For Fieldbus foundation marketing Society Singapore;.

[5] *Plantpax process automation system: Foundation fieldbus design considerations; catalog numbers 1757-FFLDx, 1757-FFLDCx*; Allen Bradley; Reference manual; https://literature.rockwellautomation.com/idc/groups/literature/documents/rm/rsfbus-rm001_-en-p.pdf.

[6] *Guide to Implementing FOUNDATION™ Fieldbus H1 Field Devices; Industrial Softing*; White paper; Internet document; https://industrial.softing.com/fileadmin/secure/Industrial/White_Papers/White_Papers_English/Guide_to_Implementing_FOUNDATION_fieldbus_H1_Field_Devices.pdf.

[7] Foundation Fieldbus Overview; 370729-01D, National Instruments, Jan;2014. http://www.ni.com/pdf/manuals/370729d.pdf.

[8] M. Yoshitsugu; *Design concepts of foundation fieldbus products*; Yokogawa; internet document; https://web-material3.yokogawa.com/rd-tr-r00027-001.pdf.

[9] Foundation Fieldbus Engineering Consultant Perspective, Fieldbus Foundation – India marketing committee; H.G. Mistry, 2008.

[10] *Profibus system description technology and application; PI profibus profinet; internet document*; https://www.profibus.com/download/profibus-technology-and-application-system-description.

[11] PROFIBUS-PA; *Technical information;part4: Communication; Samson AG*; https://www.samsongroup.com/document/l453en.pdf.

## Further reading

[1] *Fieldbus engineer's guide; process automation; Field connex; Pepperl +Fuchs; Catalog*; https://files.pepperl-fuchs.com/webcat/navi/productInfo/doct/tdoct3032__eng.pdf.

[2] Field Communication PROFIBUS-DP/PA: *Guidelines for planning and commissioning; Metso; E+H*;https://valveproducts.neles.com/documents/softwarepackages/Profibus/BA198Fen.pdf.

Chapter 8

# Various protocols and fieldbus systems

Fieldbus systems and protocols are essential for modern industrial automation, providing a reliable and efficient way to connect and control various devices in an industrial environment. These protocol and fieldbus systems use digital communication networks at the field level to exchange data between field devices and control systems. The Highway Addressable Remote Transducer (HART) and Broadcast Rapid Access Intelligent Network (BRAIN) are examples these protocols and fieldbus systems. Major key concepts mainly include *interoperability*, *digital communication*, *real-time control*, *reliability*, *low cost*, *lower wiring*, and *improved diagnostics*, especially in fieldbus systems. Fieldbus protocols come in a myriad of types, each developed to cater to specific application requirements and operational environments. Therefore, the landscape of protocols is so diverse that they can serve huge variations in industrial needs. The manufacturers of automation devices had to develop fieldbuses with differing features and functions that best suited the industrial need and their market niche and to compete on a technical level. Apart from digital communication, there exists hybrid communication that transmits field device information by a combination of conventional analog DC and digital signal. In addition to the transmission function, the field device's span and zero-point adjustments can be remotely performed. Maintenance information such as self-diagnostics of the field device can also be obtained using a dedicated terminal. These hybrid communication protocols have been developed independently by vendors, and devices from different vendors are unable to exchange total information with each other. The digital data communication speed of hybrid communication is slower than that of fieldbus communication. Fieldbus transmits data in the form of digital signals. Various types of data can be transmitted via fieldbus. Real-time controls can be implemented in fieldbus systems. Signal transmission errors rarely occur in digital signal transmission. Fieldbus international standards have been unified by the International Electrotechnical Commission (IEC)/ IEC Technical Committee/ TC65/SC65C WG6 and International Society of Automation SP50 Committee. We have gathered some information on the fundamentals of fieldbus systems and, starting from there, we will explore some protocols like HART/BRAIN, which are basically hybrid communication systems. The discussions will also cover various fieldbus systems including special fieldbus systems.

## 8.0 Discussion on HART, BRAIN, and can-bus-can open protocol

This section deals with various kinds of protocols, some of which use hybrid communication and some of which belong to fieldbus systems. The most popular field device, the HART protocol, is a backward-compatible enhancement to 4–20 mA DC transmission, allowing two-way communication with smart field devices. The HART signal is superimposed on a 4–20 mA current loop and is modulated for two-way digital communications and, with backwards compatibility, it can be used with already-existing infrastructure, making it a cost-effective proposition. A frequency-shift keyed (FSK) signal superimposed on the 4–20 mA signal constitutes the standard HART transmission at 1200 bps. Two different frequencies are used for digital 1 and digital 0. An alternative to the standard FSK signal is a coherent 8-way phase-shift keyed signal, which increases the digital transmission rate. This version uses a 3200-Hz carrier that has eight different phases for information [1]. Based on the Institute of Electrical and Electronics Engineers (IEEE) 802.15.4 standard, 2.4 GHz time division multiple access (TDMA) communication is utilized by wireless HART. Wireless HART is a wireless mesh network that connects field devices to a wireless HART gateway. Like HART, the hybrid-type transmission protocol BRAIN also use superimposition of an analog signal on a digital FSK signal. The device is equipped with BRAIN communications capabilities, so that range changes, tag number setup, monitoring of self-diagnostic results, and zero-point adjustment can be handled remotely via the hand-held terminal. Personal computer (PC)-based configuration tools are available for BRAIN to perform numerous tasks like initial setup, daily maintenance, troubleshooting, and configuration backup for device replacement. This incorporates the open field device tool (FDT)/device-type manager (DTM) standard and is compliant with DTMs per the FDT1.2 and FDT2.0 simultaneous standard. A controller area network (CAN) is basically a vehicle bus standard that allows devices like electronic control units (ECUs), sensors, and actuators to communicate through the bus and with other systems. On the other hand, CANopen is a standardized high-level protocol in which CAN bus is utilized as an underlying

Plant Intelligent Automation and Digital Transformation. https://doi.org/10.1016/B978-0-323-89892-8.00022-3

physical layer. CANopen can provide standardized communication, configuration, and device profiles, so that various devices can be easily integrated and managed in a CAN network. Widely used in vehicles and industrial automation, the compact, cost-effective CAN bus is the basic protocol that allows devices to send and receive messages. CANopen is developed on top of the CAN bus as a high-level protocol for standardized communication, network management (NMT), data transfer, and device profiles. CANopen and CAN bus gained importance because of issues like *transmission security*, *speed*, and *reliability* (the uniqueness of the data transmission of the CAN network). In a CAN network message transmissions are prioritized. Transmitter uses broadcasting mode to send a message to all nodes. Basic issues related to the HART/BRAIN protocols and CAN bus and CANopen have been enumerated so that readers are familiarized with them.

## 8.0.0 Fundamentals of the Highway Addressable Remote Transducer protocol

Both HART and BRAIN are not fieldbuses but are two protocols that provide data access between intelligent field instruments and host systems. Out of the two, HART is the most popular and has been included in the IEC 61158 and 61784 standards. HART paved the way for digital communication at the field level. As mentioned, HART communication uses the 4–20 mA loop and superimposes a digital signal to the loop using a HART modem. The modem signaling sends two different frequencies that act as the 0 or 1 in the digital communications utilizing FSK modulation. The modem also capacitively couples the voltage signal to receive the digital (HART) signal as a primary variable and is shown as receiver (RX) inside the remote transmitter. This is detailed in Fig. III 8.0a.

Typical signal superimposition and multidrop configuration of the protocols are depicted in Fig. III 8.0b.

As shown in Fig. III 8.0b, in both cases (FSK technique for digital signal transmission) the 1200 bps frequency represents 1; however, in case of 0 representation, HART uses 2200 bps (BRAIN uses 2400 bps) frequency. Also, HART can be implemented wirelessly. There are two basic methods of connecting a HART device to the control system visualization (viz.): point-to-point connection (HART host and a HART device) and host in multidrop connection. The HART protocol communication is half duplex from host to device and vice-versa. Two modes of HART communication between host and field device are:

- **Request-response mode**, in which the field device responds based on the host's command initiation. This is also known as master-slave mode.
- **Burst mode communication,** in which the host initiates a command for a burst mode response. Once started, the field transmitter responds with a repeated HART response until the host sends a command to stop burst mode.

With these in mind let us elaborate discussions on the HART protocol in this *subsection.*

1. **HART communication concept:** HART is an open, bidirectional, *master-slave* communication protocol that utilizes the Bell 202 standard FSK technique to exchange data between intelligent field devices with the host system [2]. There can be two masters to each HART loop viz. The primary master is the control system, say a *distributed control system (DCS)/PC/programmable logic controller (PLC)*, and the secondary master can be a *handheld terminal* or *another PC.* Field devices such as transmitters and actuators can respond to commands from the primary or secondary master. In HART communication a digital signal gives access to secondary and other useful auxiliary data for operations, commissioning, maintenance, and diagnostic purposes. HART communication methods have been detailed in Figs. III 8.0a and b. The HART *command set* provides uniform and consistent communication for all field devices. Three classes of command sets are *universal* (mandatary for all), *common practice*, and *device specific.* In addition to these, special commands for wireless HART include nonpublic commands. Host applications may implement any of the necessary commands for a particular application.
2. **HART: major features:** Some of the major features of HART are [2] *system and plant availability*, *two-way communication*, *superimposition of an FSK digital signal on a current loop*, *master-slave protocol*, *support of both point-to-point and multidrop communication*, *regulatory compliance*, *standardized command*, *advanced diagnostics with lower downtime*, *better process optimization solutions*, *remote configuration*, *enhanced data access*, *cost effectiveness*, *lower cost for maintenance*, and *support of safety integrity level* (SIL).
3. **HART operating modes:** HART devices can operate in point-to-point-to-point or multidrop mode. There is another way to look at the modes of operation of HART. These are master-slave (request-response type) and burst mode.
   - **Point to point:** In this mode, a single device can be connected to either of the masters/hosts discussed. Here, a single device and a handheld calibrator can be connected to the host without affecting conventional 4–20 mA DC transmission, that is, superimposition of the signal as shown in Fig. III 8.0b1 for point-to-point mode.
   - **Multidrop:** As shown in Fig. III 8.0b2, in multidrop mode a single pair of wires can connect a number of devices including primary and secondary masters as

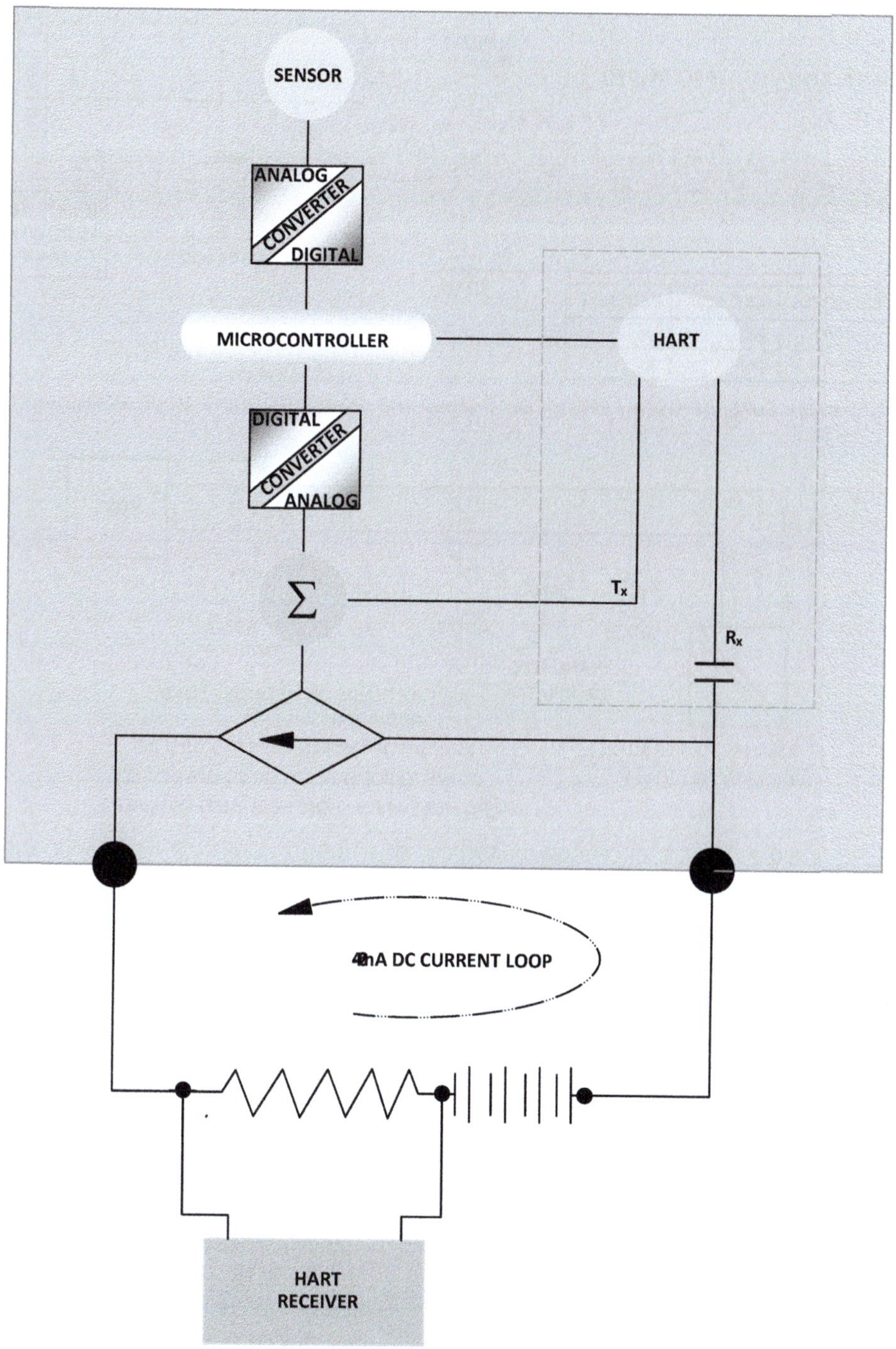


**FIGURE III 8.0A** Highway Addressable Remote Transducer (HART) superimposed on a 4–20 mA current loop.

required. Depending on applicability, an external power supply and safety barriers can be included in the loop. In multidrop mode, all field device polling addresses are >0, and the current through each device is fixed to a minimum value (typically 4 mA).

- **Master-slave mode:** In this type of mode, slave devices simply transmit data once a request from the master device is initiated as a response.
- **Burst mode:** HART burst-mode publishes cyclical process data using "burst" messaging, that is, a device is instructed to publish the response to a command continuously without host action. This mode allows faster communication. In this mode the master device instructs (HART reply message continuously) the slave device to transmit a typical HART reply message continuously (until master orders the slave to stop bursting).

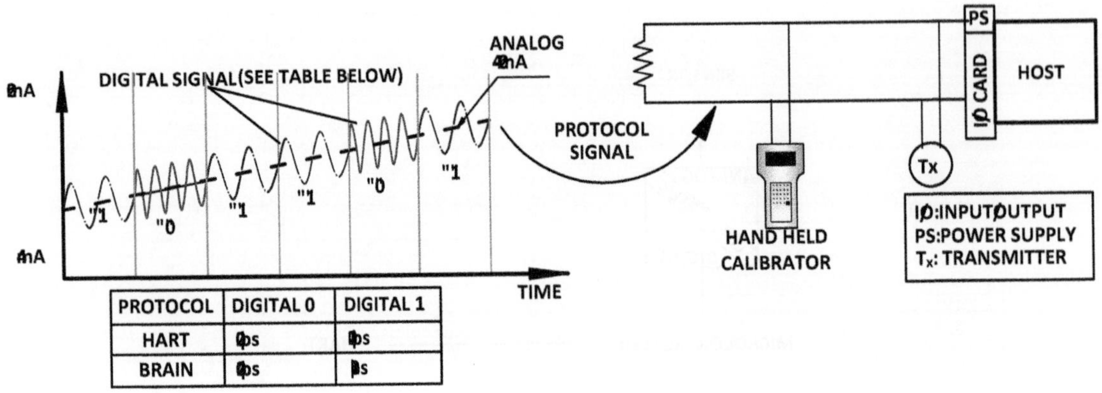


FIG III 8.0b1 SUPERIMPOSED PROTOCOL SIGNAL AND POINT TO POINT MODE COMMUNICATION.

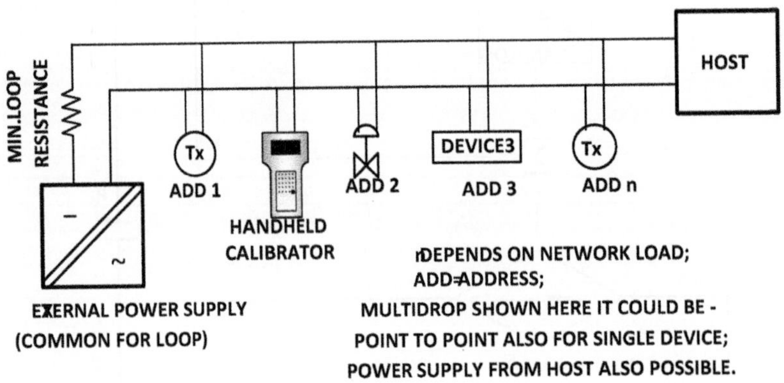


FIG III 8.0b2 MULTI DROP COMMUNICATION WITH HOST, CALIBRATOR & POWER SUPPLY

**FIGURE III 8.0B** Highway Addressable Remote Transducer (HART) and Broadcast Rapid Access Intelligent Network (BRAIN) protocol fundamentals.

4. **Discussions on HART:** In this section we discuss important aspects of HART as follows:
   - **HART command types:** With reference to III 8.0.0.1, we understand that there are mainly three kinds of commands, which are:
     - *Universal commands*: Universal commands are well supported by all devices with HART protocol.
     - *Common practice commands*: These functions are implemented by many, but not all, HART communication devices.
     - *Device-specific commands*: Device-specific commands represent functions that are unique to each field device. These are *available* from device manufacturers.
     - Command table: Table III 8.0a elaborates the brief command types mentioned above.
   - **HART device description (DD)**: HART uses electronic device description language (EDDL) as per the IEC 61804-2 standard. Host applications use DDs to obtain information about the variables and functions contained in a HART field device. EDDL is used to write a DD to combine all the information needed by the host application into a single structured file. The DD includes all the information needed by a host application to fully communicate with the field device. When HART suppliers supply the DD for their product, the DD provides information for a DD-enabled host application to read and write data according to the device's procedures. As discussed, a HART protocol uses the Bell 202 FSK standard to superimpose digital signals with two different frequencies: 1200 Hz and 2200 Hz (representing bit 1 and bit 0, respectively). When sine waves with these frequencies are placed over DC analog signal cables then the data transfer occurs without affecting (due to the standard value of a frequency-shift keying signal equivalent to zero) the 4–20 mA signal. Thus, the 4–20 mA DC current loop communicates the primary measured value and the additional device data are communicated through a digital signal that is overlaid on it. Therefore, jointly, the two communication channels offer a very robust and low-cost communication solution that is very simple to utilize and configure.

**TABLE III 8.0A** Highway Addressable Remote Transducer command details.

| Universal | Common practices | Device specific |
|---|---|---|
| Read manufacturer and device type | Read selection of up to four dynamic variables | Read or write low-flow cut-off |
| Read primary variable (PV) and units | Write damping time constant | Start/stop clear totalizer |
| Read current output and percent range | Write damping time constant | Read/write density calibration factor |
| Read <4 predefined dynamic variables | Calibrate (set zero, set span) | Read/write details on materials/ construction |
| Read/write 8-character tag, 16-character descriptor, date | Perform self-test | Trim sensor calibration |
| Read/write 32-character message and final assembly number | Perform master reset | PID enable |
| Read device range values, units, and damping time constant | Trim PV zero and Write PV unit | Write PV unit and valve characterization |

- **HART characteristics:** The following are major characteristics of the HART protocol [2]:
  - *Communication speed*: This is 1200 bps.
  - *Number of devices support*: Theoretically, in a loop there could be as many as 100. However, based on allowed load impedance 15 are typically accommodated.
  - *Noise immunity and error detection*: Online devices can be added or removed without much degradation of performance. Noise immunity conforms to IEC 801-3 (*radiated, radio frequency (RF), electromagnetic field immunity)* and IEC 801-4 (*electrical fast transient/burst immunity*) class 3. It can detect up to three corrupt bits in a telegram as error detection.
  - *Hazardous application*: It is intrinsically safe (IS) compatible and Exi.
  - *Addressing*: Special long-format addressing is used. During configuration mode in bus the addresses are set.
  - *Cable*: Normal instrumentation twisted pair cable is sufficient.
  - *Data exchange*: Status, diagnostics, calculated value, and others. Some of the generic parameters are process variable (PV) reading, analog output reading, secondary variable reading, change (tag, data, PV unit, range, damping, output transfer function, and polling address), output trimming, loop checking data, zero and range change, and modem action.
  - *Load limit*: This is 1100 ohms for the entire loop.

5. **Handheld calibrator:** Normally, HART-compliant modems referred to as "*handheld calibrator/communicators*" are complete with *display*, *keyboard*, *memory*, *operating system*, *rechargeable battery*, and others. These are available in a rugged, handheld design with varying enclosure ratings.
   - **Features:** Standard features available in handheld communicators mainly include:
     - Quick startup and connect: 15 s
     - Update new device profiles and firmware
     - Dongle coordinates AC power adapter and serial communication
     - Clone/upload/download configurations
     - Off-line mode for developing/reviewing configurations
     - Document and store >200 configurations
     - As-found/as-left history with device management software
     - A few lines of backlit display graphic and full alpha-numeric keypad
     - Edit directly from Review/Edit mode
     - Menu access key including home key, dedicated text edit keys, contrast key for mode adjustment, thumb keys for menu, up, down, and select
     - Programmable lock-out settings
     - Elastic hand strap and adjustable shoulder strap
   - **Specification:** Common specification details are:
     - *Processor and memory*: fast processor with RAM/E$^2$Prom
     - *Storage*: large number of HART devices >600
     - *Language*: often multilingual with DD support
     - *User interface*: high-resolution pixel line display, with back-lit display
     - *Power*: lithium/alkaline battery back-up (about 60 h battery life)
     - *Housing*: plastic case
     - *Software updates*: facility for downloading software updates and DDs

- Device maintenance: troubleshooting, devices, view device error messages
- *IS/certification options*: available for use in hazardous environments with certification from CE/ATEX, and others
- *Accessories*: lead sets, carrying cases, and others

With this, let us conclude the basic discussion on HART and look into an advanced discussion, including a discussion on the BRAIN protocol.

### 8.0.1 Discussion on HART and BRAIN protocols

Other than fieldbus systems, the majority of smart field devices are HART enabled. HART is a bidirectional communication protocol that provides data access between intelligent field instruments and host systems. Similar (*bidirectional digital signal over analog signal utilizing FSK in half duplex mode*) to HART, there exists another protocol called BRAIN, which is often referred to as the HART of Japan because of its popularity with instrument suppliers from Japan—viz. Yokogawa. So, these are international standard for sending and receiving digital information superimposed on analog signals between smart devices and the host system. People seeking the benefits of intelligent devices with digital communication with extensive diagnostics can opt for HART technology, which offers long-term solutions for plant operators. It is claimed that smart transmitters using BRAIN can offer enhanced diagnostic capabilities and possess features like zero and span adjustment, instrument loop identification, and access to multiple PVs. *Increased diagnostics*, *reduced maintenance*, and *enhanced control* are some of the additional features claimed to be included in BRAIN. Like HART, it also has an independent handheld communicator.

Both HART and BRAIN technologies are easy to use and very reliable when used for commissioning and calibrating smart devices and for continuous online diagnostics. It is better to use a host with these protocols. Its major utilities include d*evice configuration/reconfiguration, device diagnostics, device troubleshooting, reading the additional parameter values for the devices*, and *device health status*. In this subsection, short discussions on both HART and BRAIN protocols are presented.

1. **Benefits of HART:** Table III 8.0b elaborates the benefits of HART communication.
2. **HART specification:** Table III 8.0c briefly elaborates HART specifications.
3. **HART discussion:** We will conclude the discussion on HART with its telegram details and HART dynamics.
   - **HART telegram details:** Normally, the HART protocol operates according to the master-slave method with initiation of any communication by the master (control station/operating device). HART accepts two master viz. control systems: the primary master and a PC laptop/handheld terminal, the secondary master. The other type is broadcast command: a HART command received by all devices. Telegram structure is detailed in Fig. III 8.0c. Each individual byte is sent as an 11-bit universal asynchronous receiver/transmitter character with a start, a parity, and a stop bit. The basic

**TABLE III 8.0B** Benefits of Highway Addressable Remote Transducer communication.

| Issue | Description |
|---|---|
| Highway Addressable Remote Transducer technology | Leverage the capabilities of intelligent device for operation<br>Increase asset productivity and system availability<br>Early warning for performance variation in device/process<br>Quick troubleshooting: identification and resolutions<br>Continuously validate the integrity of control/automation |
| Plant availability | Devices and systems integration for quick problem detection<br>Real-time detection of device/process connection problems<br>With new, early warnings to minimize deviation impact avoidance of unscheduled shutdowns/process disruptions. |
| Maintenance cost | Quick verify and validate loop and device configuration<br>Remote diagnostics to reduce unnecessary field checks<br>Capture performance trend data for predictive maintenance<br>Reduce spare inventory with device management |
| Regulatory compliance | Automated safety shutdown testing and record keeping of compliance data<br>Improved safety integrity level with better diagnostics<br>Intelligent multivariable devices for accurate reporting |
| Interoperability | Supports interoperability among smart devices and is widely adopted industry standard |

**TABLE III 8.0C** Highway Addressable Remote Transducer specification.

| Feature | Description |
|---|---|
| Technology | Physical layer: Bell 202 standard, using FSK at 1200 bps speed |
| Data link layer | Normal master-slave—request-response; two masters allowed |
| Network | Routing, end-to-end security, and transport services |
| Transport | Ensure end-end communication is successful |
| Application | The commands, responses, data types and status reporting supported by the protocol. Four command types discussed |
| Digital signal | Signal 0: 1200 Hz; signal 1:1200 Hz |
| Slave | Communication about 15 slave devices in multidrop pair cable |
| HART interface | HART enabled 4–20 mA input cards, HART multiplexer (RS-422, RS-485, or RS-232 serial connection), slide-in PLC gateway cards, custom-coded software interfaces |

*FSK*, Frequency-shift keyed; *HART*, Highway Addressable Remote Transducer; *PLC*, programmable logic controller.

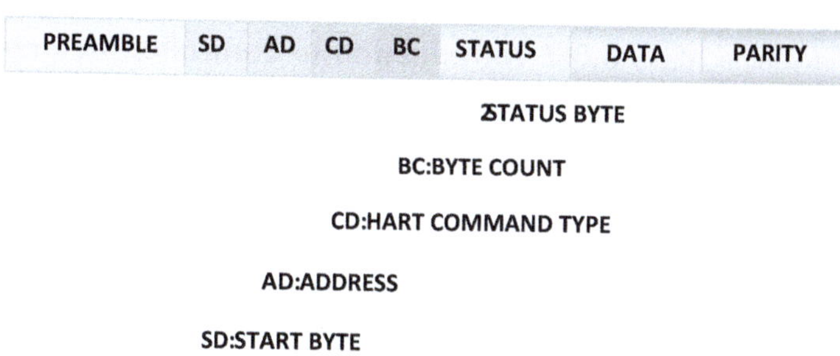


FIG III 8.0c1 HART TELEGRAM STRUCTURE

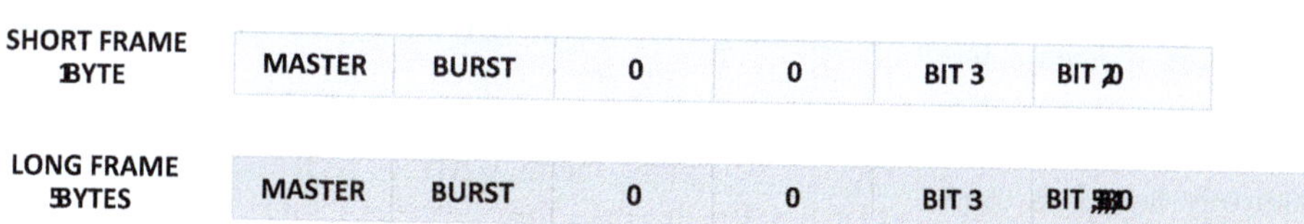


FIG III 8.0c2 HART ADDRESS FORMAT

**FIGURE III 8.0C** Highway Addressable Remote Transducer telegram details.

structure of a HART telegram is shown in Fig. III 8.0c1. Some tasks of HART telegram are:

- The preamble: >3 hexadecimal FF characters for synchronization
- The start byte: indicates master-slave/slave-in-burst mode. Also short/long frame discussed below
- Command byte: encodes the three categories of master commands discussed
- Byte count: indicates the message length, necessary since numbers of data bytes/telegram can be from 0 to 25
- Status bytes (two): are only in reply messages from slaves

The HART protocol has two telegram formats for different forms of addressing, as depicted in Fig. III 8.0c2. In addition to the short frame (1 byte with 1 bit serving to distinguish the two masters and 1 bit to indicate burst-mode telegrams) containing 4 bits, slave address format, a long-frame address format with 5 bytes (the field device is identified using 38 bits), has been introduced as an alternative. HART transactional details have been enumerated in Table III 8.0d.

**TABLE III 8.0D** Typical Highway Addressable Remote Transducer telegram timings [3].

| Parameter | Calculations | Values |
|---|---|---|
| Byte/telegram (character) | | 25 message character + 10 control character |
| Telegram size | 35-character × 11 bits | 385 bits |
| User data rate | 25 × 8 bits/385 bits | 52% |
| Time/bit | 1/1200 bit/sec | 0.83 ms |
| Transaction time | 385 × 0.83 ms | 0.32 s |
| Time per user data byte | 0.32 s/25 bytes | 13 ms |

- **Dynamic and device variable:** Superimposed HART in its devices can provide a lot of additional data on the primary variable. In addition to diagnostics, two important types of HART variables are dynamic and device variables. These variables support *IEEE 754 floating point* value. These could be retrieved by HART hosts/interface devices from the field device with HART commands. Dynamic variables consist of the *primary*, *secondary*, *tertiary*, and *quaternary* variables. Device variables may also be used in more sophisticated or multivariable field devices to provide additional process, diagnostic, or status-related information. Revision 6 and above have *device variables* in HART. Each field device can define up to 240 *device variables* [4].

4. **BRAIN discussion:** BRAIN is similar to HART for field instruments in remote accessing. In this case, there shall be terminating resistance of a minimum of 250 ohms as loop resistance. Typically, BRAIN communication uses a point-to-point connection, meaning one transmitter communicates with one host device at a time. When compared with fieldbus, the bandwidth of the digital communication is relatively low. It is used by Yokogawa for their field instruments. Similar to a HART handheld terminal, they have a BT 200 handheld terminal. However, there is a USB modem to be connected to one PC. This modem is compatible with both HART and BRAIN devices. Major features described for HART are applicable for BRAIN too.
5. **BRAIN calibrator:** The handheld configurator offers a few features like:
   - Low-voltage alarm
   - Security code to set up for protection of set-up parameters
   - Upload/download to set up parameters for future reference
   - Support printing with print density control
   - Dual language—English and Japanese
   - Basic functions of setup, alteration, and display of parameters
   - BRAIN communication, automatic power off
   - Built-in battery

   With this, we conclude the discussion on hybrid-type digital communication through protocols like HART and BRAIN and concentrate on CAN bus and CANopen.

## 8.0.2 Introduction to controller area network bus and can open protocol

A CAN bus is a high-integrity serial bus system for networking intelligent devices. CAN busses and devices are common components in automotive and industrial systems. Developed by Bosch in 1986, the CAN protocol was adopted as International Standards Organization (ISO) 11898 in 1993 and by then CAN 2.0 was established. The CAN bus system enables ECUs to communicate amongst them without complex wiring. An ECU can prepare and broadcast information (e.g., sensor data) via the CAN bus (comprising two wires: CAN low and CAN high). The broadcasted data are accepted by all other ECUs on the CAN network to check the data and decide whether to receive or ignore them. CAN is well described in ISO 11898-1 for the data link layer and ISO 11898-2 for the physical layer, which describes:

- **Baud rate:** 1 Mbit/s (classical CAN)/5 Mbit/s (CAN flexible data-rate [FD])
- **Cable length:** <500 m (125 kbit/s) or 40 m (1 Mbit/s)
- **Termination:** CAN bus terminated using 120-ohm resistance at both ends

With a need for *increased advanced functionality* in vehicles, growth in the internet of things (*IoT*), utilization of *cloud computation*, and use of artificial intelligence (*AI*) to make *autonomous vehicles* are a few trends noted in the CAN bus system. Based on CAN, *CANopen* is used widely in the automation of embedded control applications. CANopen is key in many applications like machine diagnostics/production optimization. CANopen can be conceived as the communication language where CAN bus is used as a transmission medium. In this subsection, CAN bus and CANopen are briefly discussed, including differences between the two.

1. **Description and benefits:** Fig. III 8.0d depicts the conceptual details of CAN bus.
   The discussions parts have been enumerated in two parts.
   - **Description:** CANopen is a CAN-based communication system. It comprises higher-layer protocols and profile specifications. CANopen has been developed as a standardized embedded network with highly flexible configuration capabilities. It was originally designed for motion-oriented machine control systems, such as handling systems. In terms of the open systems interconnection (OSI) communication systems model, CAN bus covers the first two levels: the physical layer and the data link layer. CANopen covers the top five layers: network (addressing and routing), transport (end-to-end reliability), session (synchronization), presentation (data encoded in a standard way, data representation), and application. The application layer describes how to configure, transfer, and synchronize CANopen devices. CAN and CANopen in the OSI model have been detailed in Fig. III 8.0d1. CAN bus is bit stream–oriented (serial data) transmission linear bus. A two-wire twisted pair cable is used as the transmission medium, and connected units are referred to as "nodes" in parallel via stubs. As discussed, on both ends, the bus must be terminated with a resistor, as detailed in Fig. III 8.0d2, which also shows that a three-wire cable (viz. CAN-high, CAN-low, and CAN-ground) is required for the transmission of CAN signals. The nine-pin D-sub connector is the connector for the CAN bus. The bus length and length of the stubs depend on the bit rate (max 1 Mbit/s). If the bit rate is > 250 kbit/s, it is a high-speed CAN, whereas when it is < 250 kbit/s it is a low-speed CAN.
   - **Benefits of CAN:** It is *simple and low cost, easy access* (one-point entry), with an extremely robust and efficient CAN bus network, which offers many advantages like:
     - *Simple, low-cost, lightweight network*: CAN is an inexpensive, durable network that can assist communication among multiple CAN devices/ECUs. A major benefit is that ECUs will have a single CAN interface. This decreases overall cost and weight in automobiles.
     - *Broadcast communication*: Each of the devices on the network is intelligent, having a controller chip. As messages are broadcasted by the ECU, all devices on the network see all transmitted messages, and based on the need can decide if a message is relevant for receiving. Naturally, such a structure allows modifications to CAN networks with minimal impact; in fact, additional nontransmitting nodes can be added without modification.
     - Error capabilities: The CAN specification includes a cyclic redundancy check (CRC) to perform error checking on each frame's contents. Frames with errors are disregarded by all nodes. Global and local errors are differentiated by the controller. In the event of too many errors, individual nodes can stop transmitting errors or disconnect themselves.
     - *Priority*: Every message has a priority assigned, so if two nodes try to send messages simultaneously, the one with the higher priority gets transmitted, postponing the other. This arbitration

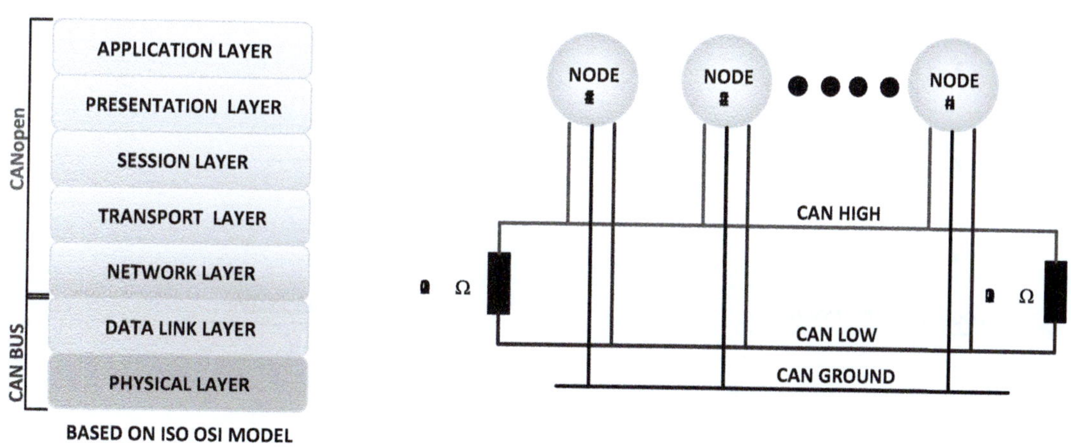


FIG III 8.0d1 CAN BUS AND CANopen

FIG III 8.0d2 CAN BUS: 3 WIRES RESITANCE TERMINATION

**FIGURE III 8.0D** Detailed description of controller area network bus.

of nondestructive and noninterrupted transmission allows networks to meet deterministic timing constraints.

- *Wiring*: This enhanced performing network has less wiring since it is a distributed control.

  *Software* development: Since at times CAN chip manufactures provide the data link layer and the physical layer interfaced to the chip, the software developer needs to develop the application coding.
- *Noise*: CAN provides the ability to work in different electrical environments, ensuring noise-free transmission.

2. **Working and characteristics**: Fig. III 8.0e shows a CAN network with voltage levels.
   Whereas the above figure details the working principle and bus voltage levels, Fig. III 8.0f shows an arbitration flow chart and node access methods.
   With all these in mind, let us look into the following:
   - **Working overview:** CAN is a peer-to-peer network without any master. Individual nodes have access to read and write data on the CAN bus. The CAN frames do not contain addresses of either transmitting/intended receiving node(s), but contain a unique arbitration ID in the frame. For transmitting data, the node checks to see if the bus is busy and then simply

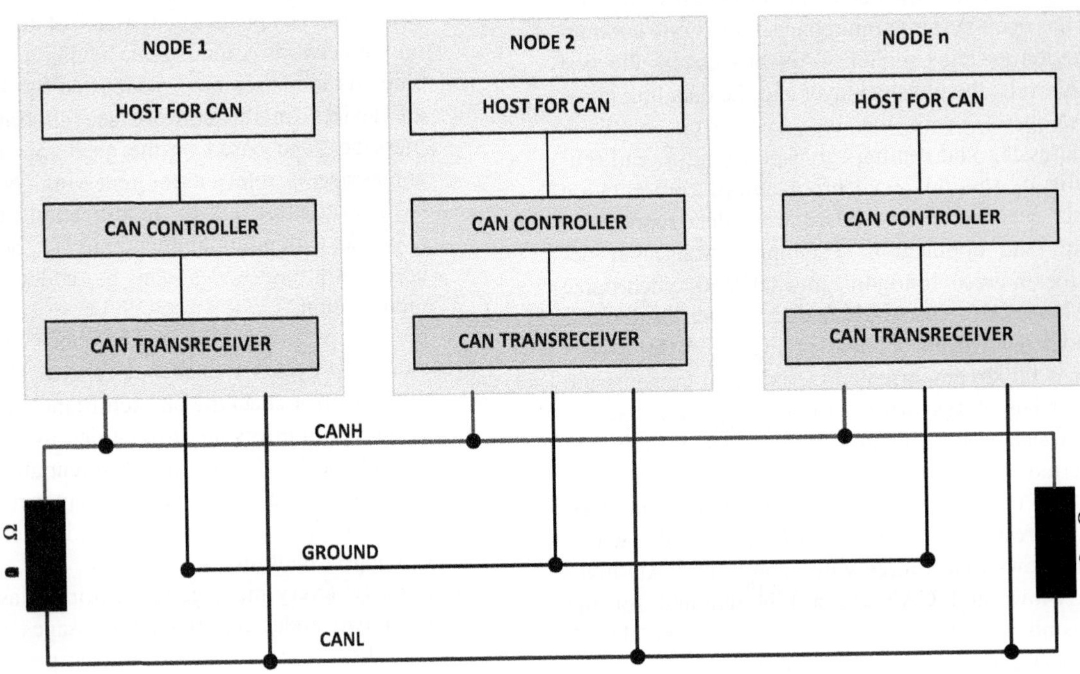


FIG III 8.0e1 CAN NETWORK WITH BASIC ELEMENTS

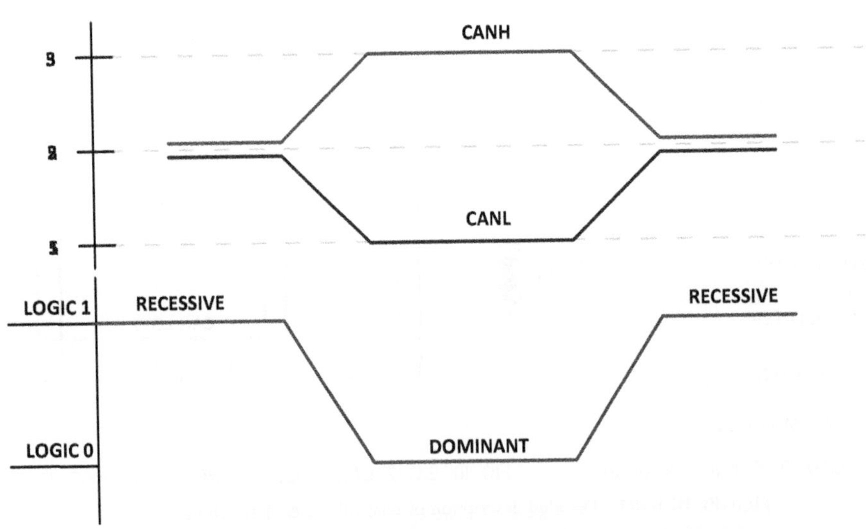


FIG III 8.0e2 CAN RECESSIVE AND DOMINANT CHARATERISTICS

**FIGURE III 8.0E** Controller area network and voltage characteristics.

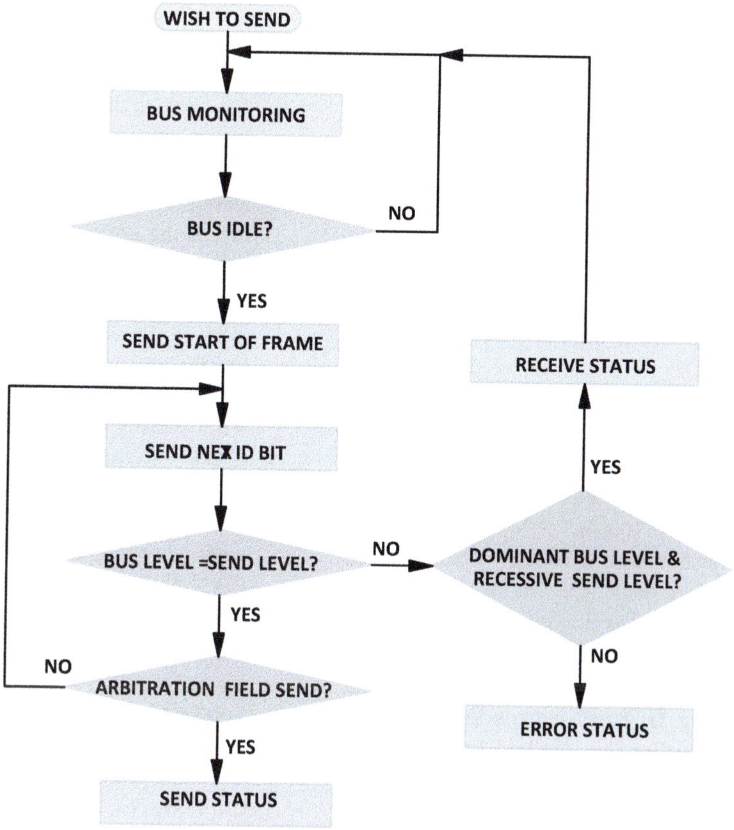


FIG III 8.0f1 ARBITRATION FLOW CHART

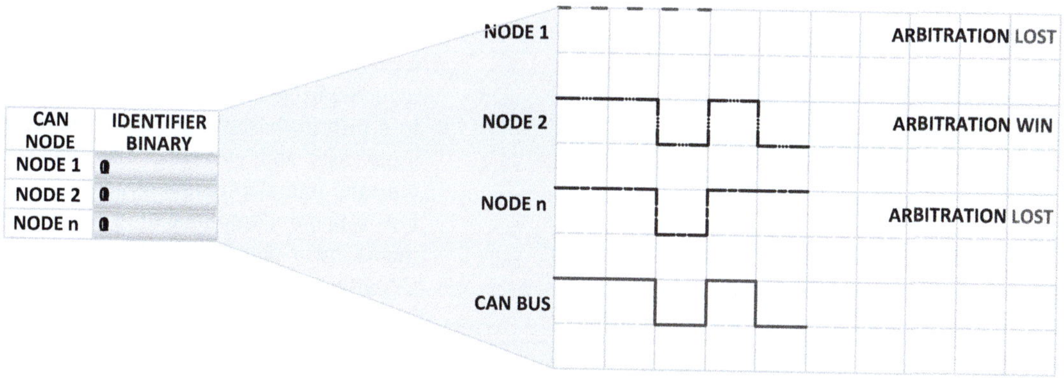


FIG III 8.0f2 NODE ACCESS METHOD

**FIGURE III 8.0F** Arbitration flow chart with node access method.

writes a CAN frame onto the network. All nodes on the network receive the CAN frame and decide whether to accept the frame based on the arbitration ID of the transmitted frame. If multiple nodes try to transmit a message onto the CAN bus simultaneously, the node with the highest priority (*lowest arbitration ID*) will get preference, as explained in Fig. III 8.0f2. In this way, deterministic communication among CAN nodes is ensured.

- **CAN explained**: A CAN network comprising the basic CAN bus described before and multiple CAN nodes, each of which has three basic elements, is shown in Fig. III 8.0e1.
  - *Host*: It is basically a microcontroller meant to carry out some applications for a specific job. The host decides the action on the received message and the action on the next message to be sent.

- *CAN controller*: The CAN controller is responsible for the communication functions pertinent to the CAN protocol. It triggers the transmission/reception of the CAN messages.
- *CAN transceiver*: It is responsible for the transmission/reception of the data on the CAN bus by converting the data signal into the stream of data collected from the CAN bus for understanding of the CAN controller.

- **Signal transmission:** As discussed, there are two lines, CAN low and CAN high, referred to as CANL and CANH. The transmission occurs when the differential voltage applied to these lines. As shown in Fig. III 8.0d2, the bus is terminated with 120-Ω resistance at each end. When CANH and CANL are applied with 2.5 V, then the actual differential voltage would be 0 volt at CAN bus. This is an ideal state of the bus and is read by the CAN transceiver as a recessive or logic 1. On the contrary, when CANH is pulled up to 3.5 V and CANL is pulled down to 1.5 V, then the bus's actual differential voltage would be 2 V. It is referred to as a dominant bit or logic 0. Fig. III 8.0e2 depicts the same.
- **CAN communication:** For the standard frame, the message identifier is 11 bit (special extended-type 29 bit). Fig. III 8.0f1 shows a typical flow diagram for arbitration, which is self-explanatory. Node access to the network has been depicted in Fig. III 8.0f2. As shown in this case, we consider three nodes, viz. node 1, node 2, and node 3, with message identifiers 0 × 7F3, 0 × 6B3, and 0x6D9, respectively. As we see, the *11th and 10th* bits of the 3 bits of nodes are recessive, so the bus bit will also remain recessive; *9th* bit: node 1 has a recessive bit whereas *nodes 2 and 3* have a dominant bit, so the bus will also remain dominant (node 1 lost the arbitration); *8th bit*: both nodes 2 and 3 are sending a recessive bit, so that the bus state will remain recessive; but in the *7th bit*, node *2 is sending a dominant bit* whereas node *3 has sent a recessive bit*, so that the bus state will remain dominant. In this case, *node 3 has lost the arbitration.*
- **Characteristics:** From the above discussion we understand that the recessive state is logical 1 (both CANH and CANL are at 2.5 V) and the dominant state is logical 0 and (voltage difference between CANH @ 3.5 V and CANL @1.5 V is 2 V). Based on the above, major characteristic issues are:
  - To transmit logic 1 (recessive state) on CAN bus, both CANH and CANL will be applied at 2.5 V.
  - To transmit 0 (dominant state) on CAN bus, CANH will be applied at 3.5 V and CANL will be applied at 1.5 V.
  - The ideal state of the bus is recessive.
  - If the node reaches the dominant state, it cannot move back to the recessive state by any other node.
  - All stations are equal and independent in sending and receiving data and have equal responsibility for bus access, error handling, and failure monitoring. Failure of one node does not call for failure of the entire system.
  - Communication between the stations takes place via broadcast, having no individual station address. Thus, new stations are integrated into the system without a new configuration.
  - The messages can be assigned and prioritized with identifiers in the CAN frame.
  - CAN provides the messages with content-oriented addressing (not node identifiers). All CAN frames are available to each CAN node, each receiver is responsible for selecting the CAN frames, and receiver-selective addressing is very flexible.
  - Security procedures: A CRC minimizes data transmission errors to ensure network-wide data consistency.

3. **Features and applications:** Here, basic features and associated applications shall be discussed separately.
   - **CAN features:** The following are major features based on the above discussion:
     - Being a multimaster, message-oriented system, in CAN bus any node can send and receive messages, so it has distributed control with high flexibility. As a physical medium it uses two wires to reduce wiring complexity.
     - In a priority-based arbitration, the winning node broadcasts and everyone else can receive, but only the intended recipient acts.
     - For reliable communication in noisy environments, error detection and correction are part of CAN bus communication.
     - It offers a reliable, efficient, and cost-effective communication system for all environments demanding real-time data exchange.
   - **CAN application:** Major application areas are:
     - Automotive: most modern vehicles, for almost everything from engine management to power windows, lights, and safety features like airbags
     - Industrial automation: used in manufacturing and factory automation systems for real-time monitoring and control
     - Process automation: food processing and robotic applications
     - Medical equipment viz.: patient monitors and infusion pumps
     - Electronic equipment: in aviation, navigation, and marine electronics

- Building automation: elevator and escalators
- Security monitoring and management

4. **Discussion on CAN protocol:** Here, the CAN protocol will be briefly discussed, including structural details, architecture, and other topics.
   - **CANopen structure:** The CANopen device structure is detailed in Table III 8.0e.
   - **CANopen protocol:** The CANopen protocol stack implements several CANopen communication objectives (COBs), which are communicated with any CANopen standard bit rate. Process data objects (PDOs) and service data objects (SDOs) are two mechanisms used to access an object directory.

     From the above it transpires that it offers high flexibility in configurations such as *definition of the internal device structure, influence and control of network behavior and implementation of desired network behavior into the device, communication and data processing as well as device participation in point-to-point communication, adjustment of data rates from 10 kbps to 1000 kbps*, and *device internal error detection and viewing.*
     - *SDO*: With SDO services, CANopen nodes can read, edit, change, or access the values of another node's object dictionary over the CAN network. This is a confirmed communication service with two CAN data frames with different CAN-identifiers, as detailed in Fig. III 8.0g1. With peer-to-peer in client (device to access object dictionary) server (the owner of the accessed object dictionary) mode, communication between two CANopen devices can be established. The presence of overhead deteriorates the flexibility of SDOs, making it less ideal for real-time operational data when a PDO protocol can be used.
     - *PDO*: Essential PDO in CANopen carries the bulk of information. PDOs are used for broadcasting high-priority control and status information in real-time operational data across CANopen nodes. As shown in Fig. III 8.0g1, a PDO consists of a single CAN frame that communicates $<8$ bytes of application data (object parameter values in a single frame). Communication takes place in *producer* (produces data for master) and *consumer* (as master receives data from produce) modes. So there are two types of PDOs. A transfer PDO is the data coming from the node (produced) and receive PDO is the data coming to the node (consumed). The configuration parameters specify the COB-ID, *transmission type*, *inhibit time*, *event timer*, and others. There are various types of these, including time-driven transmission (at a fixed time interval), individual polling (by remote request; not very common), and synchronized polling (initiated using an SYNC signal).
     - *NMT protocol*: This refers to the communication behavior of a CANopen device. The NMT enables a CANopen network to control the communication state of individual nodes. NMT services include the ability to change the state of a slave between initializing, preoperational, operational, and stopped. On receipt of the NMT, the CANopen device is forced to transit to the commanded NMT state viz. *initialization state* (after power-on or reset), *preoperational state*, *operational state*, and *stopped state.*
     - *Special function protocols*: Some special function protocols are the *synchronization protocol* (to synchronize several devices), emergency protocol (internal device errors trigger the emergency protocol), *time-stamp protocol*, and *error control protocols* viz. heartbeat protocol and boot-up protocol.
   - **Frame details:** Fig. III 8.0g shows the CANopen protocol, frame formats, and CAN bus message frame formats.

**TABLE III 8.0E** Controller area network open device structure.

| Part | Function | Remarks |
|---|---|---|
| Protocol stack | Communication handling over network | Managements: message framing, addressing, and error handling |
| Application software | Internal control functionality<br>Provides interface for hardware | Control logic and interaction with the process hardware |
| Object directory | Device configuration and diagnostics<br>Interface for protocol stack and application software<br>Used data type references<br>Storage of communication and application parameters | Acts as central database for communication and application information. Devices implement an object dictionary and standard calls for a 16-bit bit index and an 8-bit subindex |

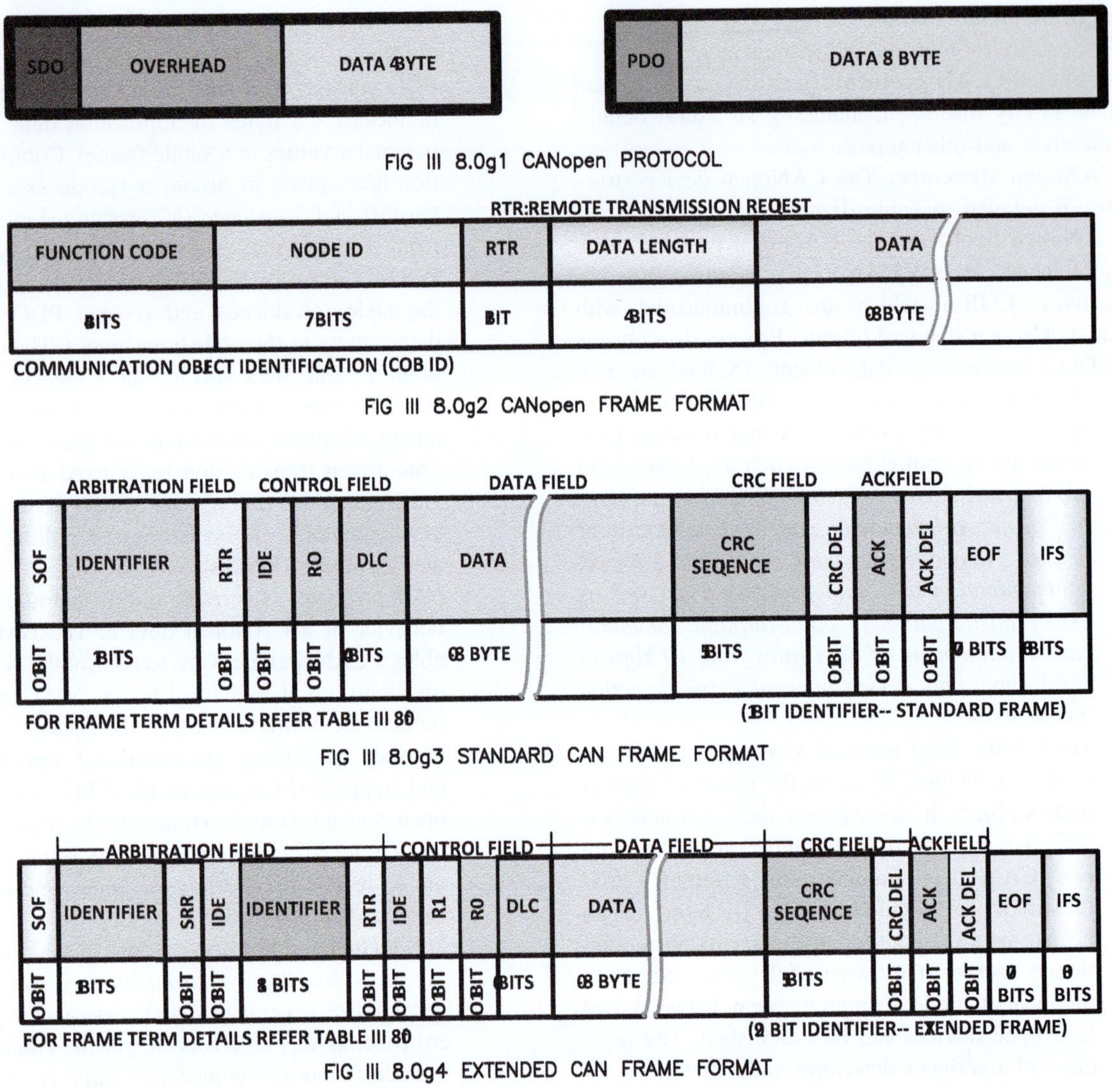


**FIGURE III 8.0G** Controller area network open protocol, frame, and controller area network bus frame formats.

- *CANopen frame format*: We have gained some knowledge on the CANopen protocol from the above discussions. The CANopen message frame format has been shown in Fig. III 8.0g2. As shown, it consists of the COB-ID, CAN-ID, and control bits. This is similar to that of the CAN bus detailed below. All COB-IDs (comprising functional code and actual node ID) used are different to avoid any conflict with the bus. There will be a remote transmission request as discussed above. There will also be a 4-bit–long sequence.
- *CAN bus frame fields*: CAN bus communication takes place via CAN frames. Normally, a standard CAN frame with an 11-bit identifier (CAN 2.0A) is the type used in most cars. Apart from the data field there are four other field types: viz. *arbitration*, *data*, *CRC*, and *acknowledge*.
- *CAN frame details*: There are two types of message length, viz. *standard* and *extended*, as detailed in Fig. III 8.0g3 and 4, respectively. Comparing the two figures, it is clear that the difference is the additional 18-bit identifier in the arbitration field. The CAN bus message structure is further elaborated in Table III 8.0f.

- **Architecture with layers:** The CAN bus layer architecture and special CAN FD frame format has been detailed in Fig. III 8.0h.

  CAN bus layered architecture has been detailed in Fig. III 8.0h1. It consists of two layers of OSI seven-layer model viz. physical and data link layer, whose specification is defined in CAN standard ISO 11898-1. Brief details include:

  - *Physical layer*: The physical layer is responsible for the transmission of raw data.

**TABLE III 8.0F Controller area network bus message structural details.**

| Slot | Bits | Field | Description |
|---|---|---|---|
| SOF (start of frame) | 1 | | SOF is the dominant bit to mark message start and synchronize the nodes after an idle period. |
| Identifier* | 11 | Arbitration | The identifier sets the message priority; the lower the value the higher the priority.* |
| RTR (remote transmission request)/SRR (substitute remote request) | 1 | Arbitration | RTR bit is dominant when information is requested. The identifier determines the node, though all receive the request.<br>SRR: in extended controller area network (CAN) message frame. |
| IDE (identifier extension) | 1 | Control | Dominant IDE defines standard CAN 2A, recessive IDE refers to extended CAN 2B. |
| R0 (R1) | 1 | Control | Reserved for future use (CAN2B). |
| DLC(data length code) | 4 | Control | DLC specifies the data length in the data field. |
| Data | 0–64 | Data | The actual data being transmitted. |
| CRC (cyclic redundancy check) | 15 | CRC | Used to detect the corruption during the transmission time comparing CRC. |
| (CRC) DEL | 1 | CRC | Delimiter; for form check and synchronization. |
| ACK | 1 | Acknowledge | Receiver's acknowledgment by overwriting the bit with a dominant bit (no separate packet) an error, bit to remain recessive. |
| (ACK) DEL | 1 | Acknowledge | Recessive; delimiter is used to separate the ACK from the subsequent CRC. |
| EOF (end of frame) | 7 | | EOF: The end of the CAN frame |
| IFS/(IMF) | 3 (1) | | Interframe space (then bus idle)/Intermission field (1 bit IMF for CAN FD Fig. III 8.0h2) |

*[The standard identifier is 11 bit* (standard CAN 2A)*; 2048 different message identifiers. For extended, it is 29 bit* (extended CAN 2B)*; 537 million different message identifiers.*

- *Data link layer*: It is responsible for node-to-node data transfer, error detection, and correction at the physical layer, allowing establishment and termination of the connection. It has two sublayers: media access control (*MAC*; for access to the medium through encapsulation and decapsulation, error detection, and signaling) and logical link control (for frame acceptance filtering, overload notification, and recovery management). *These will be elaborated in connection with a network later.*
- **CAN bus types:** As per the ISO 11898 standard, the three types of CAN bus are:
  - *Low-speed CAN* (ISO 11898-2): Simple and economic wiring low-speed CAN has a maximum data transfer rate of 125 kbps. Fault-tolerant systems, diagnostics, dashboard controls, and displays use this.
  - *High-speed CAN* (ISO 11898-3): Critical communications between subsystems requiring high update rates and high data accuracy use this. The data transfer rate can be from 1 kbit to 1 Mbit per second. The bandwidth requirement of new automotive applications is increasing every year, so automobile original equipment manufacturers are now installing CAN FD.
  - CAN FD: CAN FD is essentially an extension of the original CAN standard as specified in ISO 11898-1 and is fully compatible with classical CAN systems. CAN FD allows ECUs to dynamically change their transmission rates and select larger or smaller message sizes based on real-time requirements. Backwardly compatible CAN FD supports the CAN 2.0 special protocols like SAE J1939. The CAN FD data format is depicted in Fig. III 8.0h2

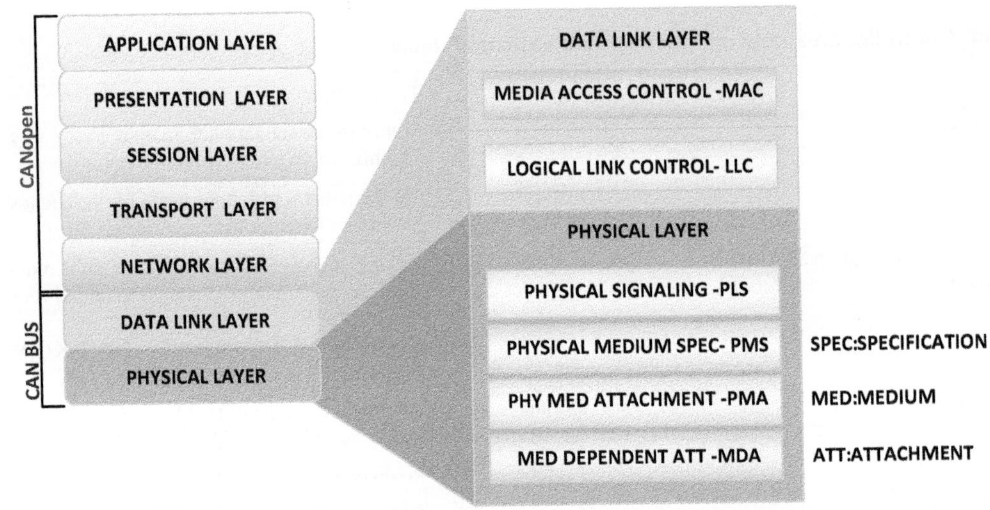


FIG III 8.0h1 CAN BUS LAYERED ARCHITECTURE

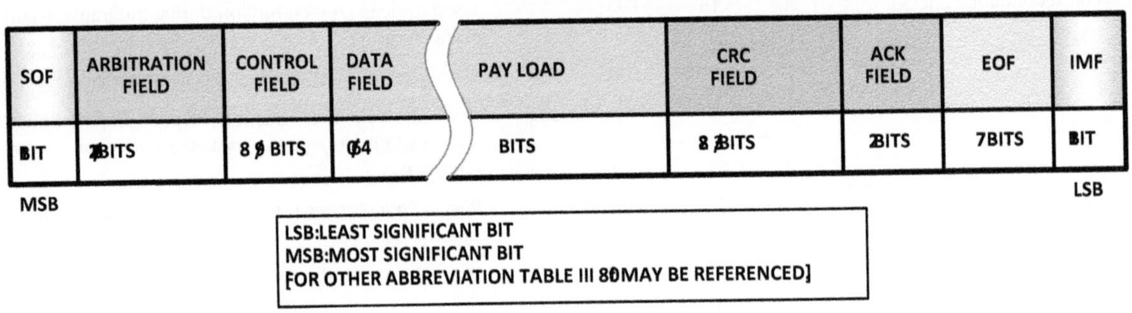


FIG III 8.0h2 CAN HD FRAME

**FIGURE III 8.0H** Controller area network bus layered architecture and controller area network flexible data-rate frame.

5. **Difference between CAN bus and CANopen:** The interrelation of CAN bus and CANopen has been depicted clearly in Fig. III 8.0d1. Table III 8.0g elaborates the differences between CAN bus and CANopen. With this, let us *conclude the discussion on CAN* bus and move on to other fieldbus types.

## 8.1 Modbus, Interbus, and Control and Communication Link

In this section, three more fieldbuses that are also popularly used in industries will be discussed, including Modbus, Interbus, and Control and Communication (CC)-Link.

**TABLE III 8.0G** Differences between Controller area network bus and controller area network open.

| Features | Controller area network (CAN) bus | CANopen |
|---|---|---|
| Definition | It is twisted pair with high and low-level difference for real-time data transmission | It is an application layer protocol based on CAN bus |
| Standard | ISO 11898 (−1,2,3) | CIA 301 (application layer and communication profile) |
| Level | Low: physical and data link level | Higher level (Fig II 8.0d1) |
| Purpose | Basic communication (transmission medium) | Configuration, standard communication (language) |
| Interoperability | Devices need to implement specific profiles | Standard communication and device profiles |

Modbus devices communicate using a master-slave (client-server) technique. Only one device (the master/client) can initiate a transactional query. Masters can address individual slaves or can initiate a broadcast message to all slaves. Slaves return a response to all queries addressed to them but never respond to broadcast queries.

The slaves (servers) never initiate messages on their own; they only respond to the master by supplying the requested data or by taking the action requested in the query. A master's query will consist of a slave address (or broadcast address), a function code defining the requested action, any required data, and an error checking field. A slave's response consists of fields confirming the action taken, any data to be returned, and an error checking field. A slave could be a server or any peripheral device (input/output [I/O] transducer, valve, network drive, or other measuring device) that processes information and sends its output to the master, the bus.

Interbus is an open, nonproprietary standard that is promoted by the INTERBUS Club and currently supported by over 700 device manufacturers with new technical developments and products. This is an open, master-slave fieldbus system known for its deterministic, cyclic, full-duplex data transfer that operates in an active ring topology. It allows for both process and parameter data exchange between a central master and up to 511 slave devices. Interbus supports a distance of up to 400 meters and is extendable with repeaters. Interbus is especially suitable for use in the field of industrial sensors and actuators and is laid out for consistent networking from the control level down to the last limit switch. The availability and reliability of Interbus, as well as its application stability, is documented by more than 1.5 million networked field devices.

Developed by Mitsubishi Electric, CC-Link is an open network, allowing multiple automation equipment manufacturers to incorporate CC-Link compatibility into their products. There are several versions, including the original CC-Link (operating at 10 Mbit/s), viz. CC-Link/LT (for reduced I/O applications), and CC-Link Industrial Ethernet (IE) with options like CC-Link IE Field and CC-Link IE time-sensitive network (TSN). CC-Link incorporates many different field devices that can be configured into a wire-saving communications network. Using the remote device net mode, it is possible to connect up to 64 remote devices. RS 485 network-based CC-Link features include: Transmission speed: 156k/625k/2.5M/5M/10M, Overall distance: 100 (10 Mbps).1200 (156 kbps), with 64 nos. of maximum connected stations etc.

This section will begin with a discussion on Modbus.

## 8.1.0 Discussion on Modbus protocol

In 1979, Modicon developed the Modbus Protocol for communication amongst Modicon PLCs, which later became a de facto standard for connecting various devices with control systems as well as communication between networks [2]. This is an open protocol and anybody can use it. "Modbus" is a registered trademark of Schneider Electric USA, Inc (current owner of the Modicon brand). It is basically an application layer communication protocol for transmitting information between electronic devices over serial lines/Ethernet/transmission control protocol (TCP)/internet protocol (IP) (detailed in *III 16.1.2*). In Modbus serial architecture, there is one master and up to 247 slaves, each with a unique slave address. The Modbus protocol is the grandfather of modern fieldbuses and is very popular for its simplicity, openness, and ubiquitous nature—it is used everywhere. In addition to its popularity in factory automation, it is also popular in process automation. Typical Modbus connections, transactions, and data formats [2] are depicted in Fig. III 8.1a [2].

Typical master-slave connections are depicted in Fig. III 8.1a1. A Modbus serial protocol single master can control several (up to 247) slaves, as shown in Fig. III 8.1a2. Physical connections may be through *RS 232/RS 422/285*. As shown, connections of slaves with the master can be of two types: daisy chain or a direct serial link connecting several slaves. Fig. III 8.1a3 shows that TCP/IP (detailed in *III 16.1.2*) is used over Modbus. So, Modbus comes in several varieties including remote terminal unit (RTU), American serial code for information interchange (ASCII), TCP/IP, and User Datagram Protocol (UDP)/IP. The basic data structure of Modbus and typical Modbus transaction details are detailed in Fig. III 8.1a4 and 5 Modbus can handle floating-point data, long-integer data, and other data requirements. Reading the Modbus interface and slave documentation is key to understanding and implementing Modbus. Different manufacturer's devices can be used in the same Modbus network [5].

1. **Architecture overview:** Modbus (the original version) is a serial master-slave protocol with data transactions that include multiple slaves that respond to the master's requests to read from or write data, up to a maximum of 247 slaves (each with a unique slave address). As shown in Fig. III 8.1a, Modbus comes in several varieties. Fig. III 8.1b elaborates this in more detail; types of Modbus and a combination of TCP with serial connections have also been detailed.
   Modbus is an application layer protocol that is independent of the data transmission medium. As discussed, data transactions are on a master-slave basis and are controlled by the master (client). Data are based on 16-bit registers that can contain discrete on/off or 16-bit integer values. In some cases, two or more integer registers that represent floating data or long integer values are used. Diagnostic data can be requested by a serial master from the slave, and the slave can send

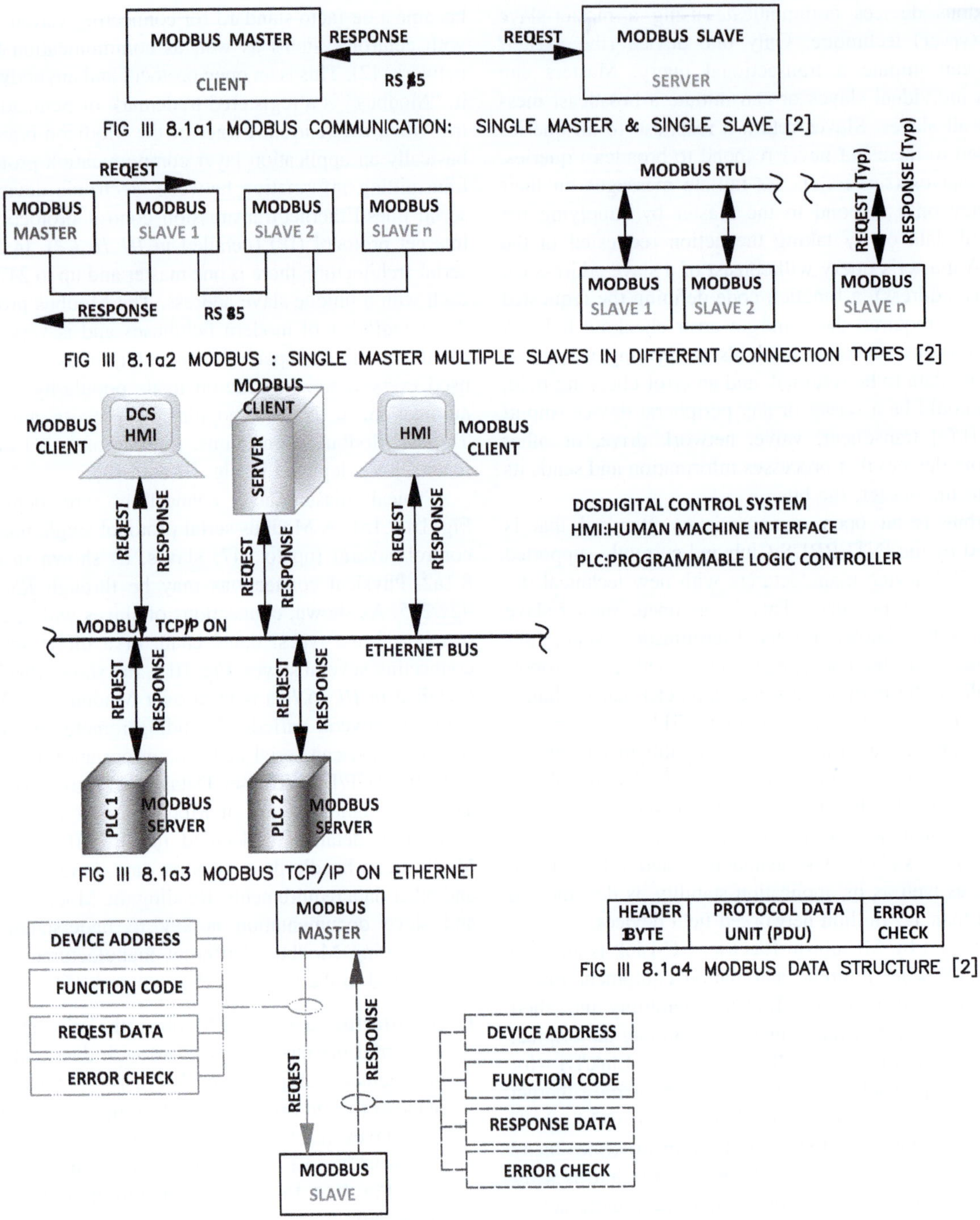


**FIGURE III 8.1A** Modbus protocol basics.

error codes to the master if they perceive there is something wrong with the request they received. Fig. III 8.1b1 shows that Modbus can be divided into several types based on architecture. These types are:

- **Modbus plus**: As shown in Fig. III 8.1b1, Modbus Plus is a proprietary high-speed, deterministic, peer-to-peer network protocol developed by Schneider Electric for industrial automation. At the physical layer it uses proprietary cabling and terminators. High-level data link control can be used as a data link layer, as shown. For fast token passing and high-level data link control processing, one dedicated coprocessor is used. It is distinct from the standard Modbus protocol, using token passing for

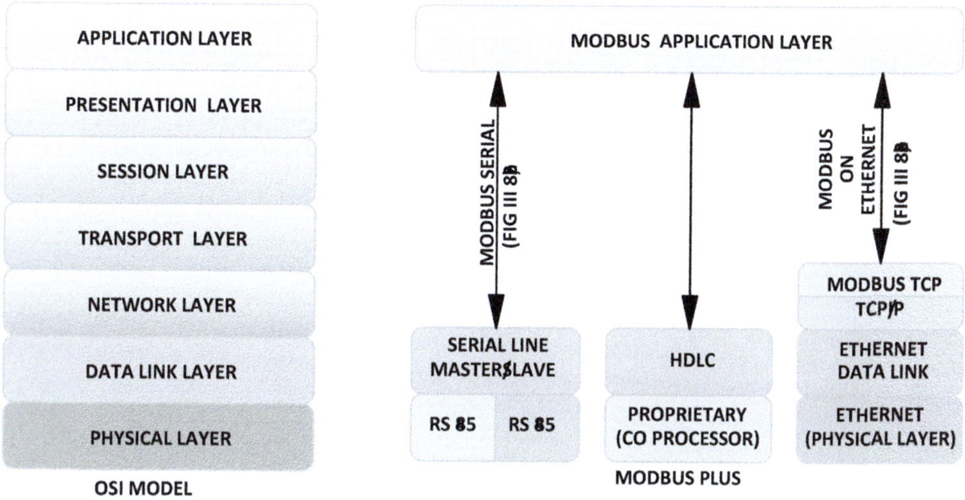


FIG III 8.1b1MODBUS TYPE DETAILS WITH MOD PLUS

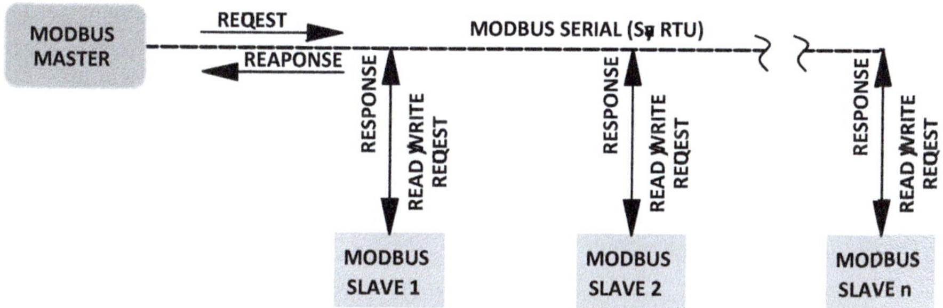


FIG III 8.1b2 SERIAL MODBUS CONNECTION WITH MULTIPLE SLAVES

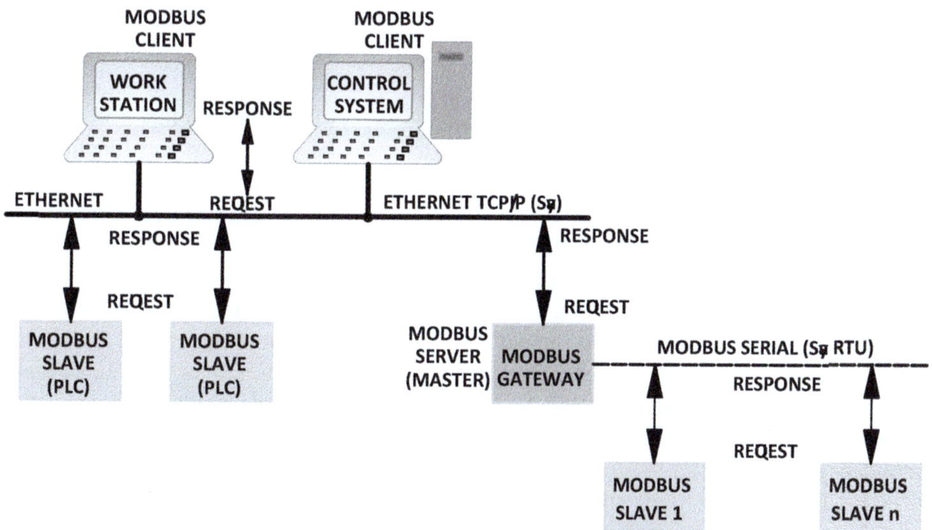


FIG III 8.1b3 MODBUS ON ETHERNET DETAILS CONNECTIONS

**FIGURE III 8.1B** Modbus types and connections.

communication and supporting faster data exchange. In contrast to master-slave, it uses peer-to-peer communication and devices in the network that can communicate with each other at high speed.

- **Serial Modbus:** Fig. III 8.1b2 shows serial Modbus. Based on encoded messages, there are two types of serial Modbus viz. RTU and ASCII, as detailed in Table III 8.1a. Fig III16.1c also shows a Modbus serial data frame.
  - *RTU:* Modbus RTU is simply Modbus over serial communication media like RS485, RS422, and RS232. In RS 485, the master communicates

**TABLE III 8.1A Details of format for serial transmission modes.**

| Issue | Modbus ASCII | Modbus RTU |
|---|---|---|
| Coding | Hexadecimal, ASCII characters 0–9, A–F one hexadecimal character contained in each ASCII character of the message | 8–bit binary, hexadecimal 0–9, A–F *Two hexadecimal characters* contained in each 8–bit field of the message |
| Bit/Byte | 1 start bit, 7 data bits, *LSB sent first,* 1 bit for even/oddparity; no bit for no parity *1 stop* bit if *parity* is used; 2 stop bits if *no parity.* | 1 start bit, 8 data bits, *LSB* sent first, 1 bit for even/oddparity; no bit for no parity *1 stop* bit if *parity* is used; 2 stop bits if *no parity.* |
| Error check | LRC | CRC |

*ASCII,* American serial code for information interchange; *CRC,* cyclic redundancy check; *LRC,* longitudinal redundancy check; *RTU,* remote terminal unit.

with multiple slaves (supports up to 32 nodes) over a range of 4000 feet. The speed, that is, the baud rate or bits per second, is typically 9600–19200. RS232 is used for point-to-point communication over short distances. In Modbus RTU, bytes are sent consecutively without space in between them, with a 3.5-character space between messages as a delimiter. For each *eight–bit byte*, one start bit, eight data bits, 1 bit for parity, and one stop bit are sent, for a total of 11 bits per byte. RTU messages are smaller sized, which allows for more data exchange in an identical time period [5].

- *ASCII:* Modbus ASCII uses RS-485/RS-232 as the physical layer, in which RS-485 is preferred for longer distances and multidrop networks and RS-232 for simple point-to-point connections in shorter distances. Modbus ASCII marks the start of each message with an ASCII colon character ": " and the end of each message is terminated with ASCII carriage return/line feed (*CR/LF*) characters. This allows the spacing between bytes in the message (variable), making it suitable for transmission through some modems. The data in a Modbus ASCII message uses ASCII characters. Here, the messages are terminated with an error checksum called a longitudinal redundancy check (LRC).

- **Modbus TCP-UDP:** Fig. III 8.1b3 shows Modbus (TCP) typically implemented on an Ethernet network. Modbus TCP/IP is a reliable (guaranteed delivery) and connection-oriented protocol. It also shows that connection to an RTU is well suited for local area networks and wide area networks. So, they can be used for high bandwidth and long-distance viz. connecting PLCs, human-machine interfaces (HMIs), and others, as shown. Modbus UDP/IP uses the faster, connectionless UDP protocol. This is less reliable (no guaranteed delivery) but suitable for broadcast and multicast. This calls for application-level error handling (as against built-in error handling in TCP) and can be used in situations where speed is paramount and some data loss is acceptable.

2. **Descriptions of coding details:** There is no data by-exception transmitted in standard Modbus. Some important issues pertinent to serial Modbus are as follows:
   - **Function code:** Let the discussion start with functional codes for request/response.
     - *Request function code*: The function code advises the *addressed slave* device on the type/kind of action to be performed and any additional byte for additional information that the slave will require to perform the requested function.
     - *Response function code*: For the normal response of a slave, the function code in the response will be an echo of the function code in the request. The data bytes contain the data collected by the slave, like register values/status. In case of an error, the function code is modified to indicate that the response is an error response, and the data bytes contain a code that describes the error. The error check field allows the master to confirm that the message contents are valid.
     - *Function field:* In framing discussed later, there will be various functional codes in the functional field. The function code field tells the addressed slave what function to perform. Some of the common functions include, but are not limited to 01 (0 × 01) *read coils*, 02 (0 × 02) *read discrete inputs*, 03 (0 × 03) *read holding registers*, 04 (0 × 04) *read input registers*, 05 (0 × 05) *write single coil*, 06 (0 × 06) *write single register*, 15 (0 × 0F) *write multiple coils*, and 16 (0 × 10) *write multiple registers*.
   - **Serial coding:** ASCII coding: hexadecimal ASCII printable characters 0 … 9, A … F. It is eight-bit binary: hexadecimal 0 … 9, A … F. RTU coding: two

hexadecimal characters contained in each eight-bit field of the message. Refer to Table III 8.1a for details.

**ASCII error:** ASCII error checking is an LRC. The LRC field contains two ASCII characters as a result of an LRC calculation that is performed on the message contents (exclusive of the beginning colon and terminating carriage return and line feed [CRLF] characters).

- **RTU error:** In RTU mode, the error-checking field contains a 16-bit value implemented as two eight-bit bytes. The error check value is the result of a CRC used here. A CRC calculation is performed on the message contents.

3. **Serial transmission:** Here, the discussion will be a little elaboration of the previous discussions, including the difference between two serial Modbus communications. Basic coding errors of the two systems are shown in Table III 8.1a.
Two serial Modbus systems are distinct and cannot directly communicate, that is, nodes in one Modbus network must be of the same transmission mode. They have some similarities as well as differences, which are detailed in Table III 8.1b.
4. **Addressing and framing:** In this part, we discuss the addressing scheme and message frame details. Let us look into the addressing scheme.
   - **Addressing scheme:** In general, Modbus addressing is organized around 16-bit (0/1) states. It can also have 16-bit integer values in I/O or holding registers. A typical addressing scheme in a Modbus system is shown in Fig. III 8.1a5. On account of encapsulation by the Ethernet, there will be some additional overhead and the error checksum is different. Modbus messaging is based on two important terms: application data unit (ADU) and protocol data unit (PDU). In the general Modbus frame address, function code, data, and error check comprise the ADU and function code and data together are called the PDU. In case of Modbus TCP/IP ADU, there is no separate error check and the Modbus Application Protocol (MBAP) header, function code, and data form the ADU [6]. The PDU is the same. For a detailed discussion, vol III may be referenced. The Modbus message includes the address, a function code, data start addresses, and the data being sent to (written) or to be sent back (read) to the master/client, with an error checksum at the end (CRC/checksum/LRC). A typical format has been depicted in Fig. III 8.1a4. The size of the serial Modbus PDU is limited to 256 bytes. Modbus slave addresses are limited to 1-255. Addresses 248-255 are reserved; only 1-247 are available for the user. Modbus data can include starting data addresses, data quantity or count, and actual data that are read or to be written. If the Modbus slave/server has a problem with the master/client request, the slave/server will issue an error response back to the master/client [5].
   - **Message framing:** The Modbus message frame is shown in Table III 8.1c.
5. **Various Modbus issues:** Here, discussions will be presented on various issues related to Modbus before concluding the subsection on Modbus. Various issues include:
   - **Error checking:** The master is configured to wait for a predetermined timeout interval, which is long enough for any slave to respond normally. On the

**TABLE III 8.1B** Comparison of Modbus ASCII and Modbus RTU.

| Issue | Modbus ASCII | Modbus RTU |
|---|---|---|
| Readability | Messages are readable | Binary code; not readable |
| Data transmission | Lower transmission efficiency | Smaller size, more information/time; higher transmission efficiency |
| Frame start | Character ':' | 3.5-character silence (Table III 8.1c) |
| Frame end | CR/LF | 3.5-character silence (Table III 8.1c) |
| Message gap | 1 sec | 1.5-times character length |
| Start bit | 1 | 1 |
| Data bit | 7 | 8 |
| Error check | Longitudinal redundancy check | Cyclical redundancy check |
| Parity | Even/odd OR none | Even/odd OR none |
| Stop bit | 1 bit OR 2 bits for none | 1 bit OR 2 bits for none |

*ASCII*, American serial code for information interchange; *RTU*, remote terminal unit.

**TABLE III 8.1C** Serial Modbus frame details.

| Modbus | Start | Address | Function | Data | Error check | End |
|---|---|---|---|---|---|---|
| ASCII | 1CHAR | 2CHARS | 2CHARS | n CHARS | 2CHARS | 2CHARS CRLF |
| RTU | T1-T2-T3-T4* | 8 bits | 8 bits | nX8bits | 16bits | T1-T2-T3-T4* |

*ASCII*, American serial code for information interchange; *CRLF*, carriage return and line feed; *RTU*, remote terminal unit.
**T1-T2-T3-T4 3.5-character times at no transmission.*

other hand, MBAP and Modbus Plus use frame checking at a level above the Modbus contents of the message. There are two kinds of error checking in Modbus serial networks. These are:

- *Parity checking* (even or odd for character): Users can configure controllers for "*No*" parity or for *even* or *odd* parity checking. Based on this, the parity bit will be set in character. While transmitting the message, the parity bit is calculated and applied to the frame of each character. Naturally, the receiving device counts the quantity of 1 bits and sets *an error* if they are not the same as configured for that device.
- *Frame checking* (LRC or CRC) for the entire message frame: As discussed, in *ASCII mode*, an error-checking field that is based on an *LRC* is applied regardless of any parity to check the contents of the message, excluding the 'colon' at the beginning and ending CRLF pair. In RTU mode, the error-checking field is based on the CRC that is applied, regardless of any parity to checks; the contents of the entire message to check method are used for the individual characters of the message. The CRC field contains a 16-bit binary value that is calculated by the transmitting device to append the CRC to the message. The receiving device recalculates a CRC during receipt of the message and compares the calculated value to the actual value it received in the CRC field to detect an error if they do not match.

- **Wireless Modbus:** Wireless Modbus enables Modbus devices to transmit information wirelessly. This can be designed for any type, for both Modbus TCP/IP and Modbus RTU. Modbus via wireless is transparent to the control system or host and the slave devices, that is, the host does not know the existence of the wireless Modbus network. When a Modbus master makes a request to a slave and the packets arrive at the transmitting radio, that radio will usually reorder the packets and encrypt them before transmission. Once the RF packets are received by the "slave" radio, it deencrypts them and puts them back in order to represent a valid Modbus packet. Assuming that the packet has not been damaged or corrupted, it will then be sent to the destined slave. The slave will respond back to the master and the process starts again. The cost savings from wiring infrastructure is the main benefit for using wireless Modbus devices in industrial applications. With wireless technology, it is possible to quickly and cost effectively obtain real-time data from anywhere, even remote locations.
- **HART and Modbus:** HART-to-Modbus conversion is done in a HART interface module (HIM), which can be utilized to integrate HART-enabled field devices with Modbus-based control systems. In HIM, the HART data are digitally mapped to the HIM's Modbus memory, where it can then be polled by a PLC or DCS that is acting as the Modbus RTU Master. Thus, with HIM supporting Modbus RTU communication, all the HART data can be brought to the control system in a simple and cost-effective manner.
- **Comparison: Modbus RTU and TCP:** Table III 8.1d shows a brief comparison of the two.

Let us now *conclude the discussions on Modbus to explore the Interbus system.*

### 8.1.1 Interbus communication

In 1984, Pheonix contact developed the open INTERBUS fieldbus system, which is covered by the international IEC 61158 standard as CPF 6. INTERBUS is one of the world's most popular fieldbus systems and can be used to network sensors, actuators, I/O, and field devices commonly used in control systems to system parts. It also connects higher level control systems. A pure master-slave–based Interbus system has very few protocol overheads. The widespread use of Interbus for sensor/actuator–level applications is mainly due to the relatively simple interfacing requirements, which are supported by protocol driver chips. These reduce the number of external components required for direct input or output interfacing to a minimum. As shown in Fig. III 8.1c, the Interbus system is designed as a ring-type network with a central master-slave access

**TABLE III 8.1D** Comparison of Modbus remote terminal unit and Modbus transmission control protocol.

| Issue | Modbus RTU | Modbus TCP |
|---|---|---|
| Medium | Serial RS 485/RS 232/RS 422 | Ethernet (TCP/IP) |
| Format | Binary | Binary |
| Technical base | Mainly device communication point to point | TCP/IP, local area network/internet (Ethernet switch) |
| Distance | Limited by serial cable length versus baud rate | Can span across huge distances via Ethernet |
| Data rate | Limited baud rate: 9600/19,200 bps | Much faster (10/100 Mbps or higher) |
| Performance | Slower due to serial communication overhead | Faster and more responsive due to Ethernet |
| Frame | Address, function, data, CRC | Header, function code, data |
| Header | No distinct message header | Header carrying message information: length and transaction identifiers |
| Cost | Lower due to simplicity | High for Ethernet |
| Applications | Local automation | Large projects: small grid, building automation |

*CRC*, Cyclic redundancy check; *IP*, internet protocol; *RTU*, remote terminal unit; *TCP*, transmission control protocol.

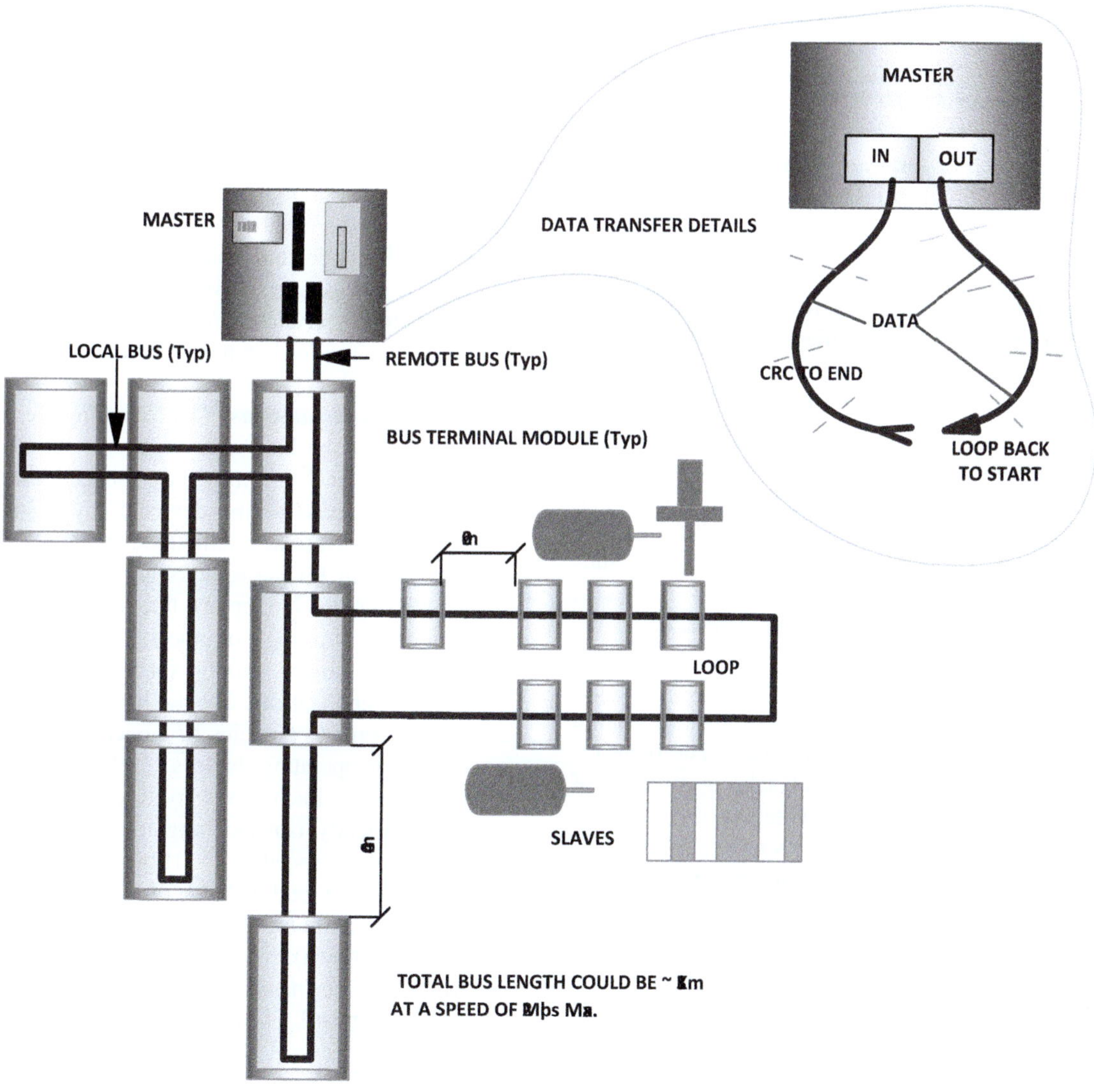


**FIGURE III 8.1C** Interbus protocol basics.

procedure as detailed therein. It has the structure of a distributed shift register. The different registers of the devices connected to the ring are a portion of this shift register. Fast and easy handled communication Interbus is characterized by low susceptibility to electromagnetic interference, noise, and data loss. As many as 512 subscribers can participate in a network, with a provision for modification of the number of subscribers in the loop, which will be closed by the last subscriber. The loop is closed by the *loop-back bit* combination and sent to the master, signifying that everything functions suitably in a closed circle (i.e., everything functions correctly). As each participant is connected by two bus terminals, a tree structure physically results, permitting incoming and outgoing signals processed by the master. Interbus runs on a transmission rate of 500 kbps. Branch bus terminals enable a subnetwork, and coupling elements prevent faults from spreading from one network to another, ensuring stability. The local buses branching from the bus terminal connected to the remote bus would divide the network. These two can then be switched on and off independently of each other. Addressing is done automatically, so that time-consuming calibration is not required; also, this promotes better maintenance and smooth switching to the serial bus system. Interbus allows both cyclic and acyclic message communications. Cyclic communication is used for the real-time exchange of process data between devices. Acyclic communication is used for non–time-critical tasks like configuration or diagnostics. In this subsection, a brief *discussion on Interbus* has been presented.

1. **Network topology:** Interbus is in a ring topology. This means that each device amplifies the incoming signal and retransmits it so that a high transmission rate can be maintained over longer distances. In physical appearance, Interbus is an "*open*" tree structure, since the data forward and return lines in the Interbus system lead to all devices via a single cable. As shown in Fig. III 8.1c, there can be a main (remote) line with the bus master. This can be used to form seamless subnetworks up to 16 levels deep with up to 512 devices, allowing quick adaption of bus. The ring is automatically closed by the last device. Known for low configuration expenditure and countless topology creation, Interbus is flexible in design, allowing quick expansion to cover larger distances with repeaters. The point-to-point structure of the main and subring systems of Interbus is ideal for the implementation of completely different physical transmission technologies, that is, any part of the system can be converted from copper to fiber optics or both. The ring structure permits data to be sent and received simultaneously in full duplex mode and well implementation of the system's self-diagnostic feature. A ring system with active device coupling provides a segmentation of the entire complex into electrically independent subsystems, including a connection and disconnection of devices without interference; it also enables monitoring of every single transmission path and unambiguous fault location. The local buses branch from the bus terminal, which is connected to the remote bus, and in doing so, they divide the network and a subnetwork is formed. Major topological issues are:
   - **Segmentation:** The point-to-point connection eliminates the need for termination resistors. Adaptive flexibility in design allows the user's requirement of addition removal of devices to be met. The coupling elements also referred to as bus terminals (ref Fig. III 8.1c) between the bus segments (remote and local) enable the connection and disconnection of a subsystem and thus make it possible to work on the subsystem without problems. Also, coupling elements prevent faults from spreading from one network to another, ensuring the stability of the system. As shown in Fig. III 8.1c, with *branch terminals*, at a local bus new branches can be created, enabling the connection and disconnection of devices.
   - **Addressing:** In the Interbus system, data are automatically assigned to devices using their *physical location* in the system. So, addressing is different from other systems. This plug and play feature highly reduces the installation effort and makes the system easily understandable and service friendly. The ability to assign "*easy-to-understand*" software names to the *physical addresses* allows device addition and removal without readdressing existing devices.
2. **Components of Interbus:** The following are major of components of the Interbus system:
   - **Remote bus:** The remote bus connects the master/controller board to the devices; any branch from this is referred to as a remote bus branch. The physical medium for the remote bus could be a copper cable using RS-485, optical fibers, infrared transmission paths, slip rings, or others. The remote bus provides the power supply for the connected I/O modules and sensors. Special bus terminal modules and certain I/O modules or devices such as robots, drives, or operating devices can be used as remote bus devices. Each has a local voltage supply and an electrically isolated outgoing segment [7]. The transmission speed is 500 kbps or 2 Mbps, with coverage of a total distance of up to 12.8 km. There can be a maximum of 255 subsegments (each 400 m apart, as shown).
   - **Master:** The master is also referred to as the controller board. The master is responsible for traffic

control. The zoomed-in panel of Fig. III 8.1c shows a master with one in and another out part. It sends output data (from out) to the corresponding modules and receives input data (at in). The master also monitors data transfer and carries out diagnostic functions. In case of an error it displays the same and sends out error messages to the host system.

- **Bus terminal:** The bus terminal modules have an embedded bus to connect to the remote bus. The bus terminal amplifies the data signal (repeater function) and electrically isolates the bus segments The distributed local buses branch out of the bus terminal module with I/O modules (with power supply from it), which establish the connection between Interbus and the sensors/actuators [7]. With the branch terminal, it is possible to switch branches on/off separately during operation. Bus terminal modules connect a branch network with the main network. It can be switched off separately, but does not permit any further branching.
- **Local bus and loop:** Both local bus and loop are identified in Fig. III 8.1c. The local bus branches from the remote bus through a bus interface module to connect local bus devices. Here, the power supply is arranged by bus terminal module. Local bus devices are typically I/O modules. Distributed sensors and actuators on machines or systems are networked as a loop for Interbus. An untwisted, twisted pair cable is used for the transportation of data and power supply to the connected devices. In an Interbus loop, two participants can be placed at a distance of up to 20 m. All elements function as slaves with regard to a master connected to a remote bus.
- **Media issue:** In many cases, data from an Interbus system is transmitted via a copper cable using a differential signal according to RS-485, which calls for a separate twisted-pair cable for the forward and return lines, and a fifth wire provides equipotential bonding between the modules. When using optical fiber technology, the Interbus structure is the same, with data transmitted by two fibers. The current must simply be converted into light and vice versa. Thus, both copper and optical fibers can be mixed between individual devices, making retrofitting easier. With a suitable converter, the combined operation of different media in a single system does not pose any problem.

3. **Transmission protocol:** By this time, we know that apart from the application layer for the application interface, there will be a physical layer, which is important for the timing condition (viz. baud rate, permissible jitter, etc.) and cabling/cable coding, among others. Another important layer, the data link layer (DLL), is responsible for the integrity of the data. This layer takes care of the support of both data types occurring in sensor/actuator technology. The DLL of Interbus ensures a deterministic approach for the cyclic data transport. Interbus is based on the summation frame procedure, which uses collision-free time division multiple access (TDMA) transmission procedures. With the summation frame, Interbus offers an excellent, high rate of efficiency for protocols with smaller overhead. As a result, there will be economy in cabling and improved electromagnetic compatibility.
   - **User interface:** The Interbus system can integrate both cyclic I/O data and noncyclic message transport in a single system for the entire field of industrial sensors/actuators. The Interbus application interface has two parts based on the two data types. Direct memory access helps to avoid lengthy service access procedures that have a direct impact on the real-time characteristics of the process data protocol. It supports standardized software interfaces, dynamic data exchange, open platform communication (OPC), and others to get direct access. For accessing process data, users are not required to familiarize themselves with complex communication structures.
   - **Peripheral message specification (PMS) services:** Based on the structure of the multimedia messaging service (MMS), a compatible subset of the defined communication services created for the Interbus system is referred to as PMS (about 25 PMS services).
     The number of services can easily be extended for certain devices.
   - **Interbus loop:** The standardized Interbus protocol enables the amount of data per device to be reduced as desired, without reducing the efficiency and therefore the data throughput. Interbus acts as the crucial link between industrial sensors and actuators, enabling their seamless integration with the control system. The cost-effective connection of sensors/actuators in the Interbus transmission technique is known as the Interbus loop. A special bus terminal module enables connection to the Interbus remote bus, whereby the Interbus loop, together with all other Interbus products, can be used on a common bus.
4. **Protocol structure:** Interbus supports both cyclic and acyclic messages. Interbus uses a master-slave procedure for data transmission and the client-server paradigm for the parameter channel. It is possible to transmit parameter data between two slaves (peer to peer), that is, both slaves can adopt both the client and server functions. Unlike other fieldbuses, Interbus works according to the summation frame method, which uses only one protocol frame for messages from all devices, as shown in Fig. III 8.1d.
   In Interbus, there are two telegram formats with start and stop bits. These are *the 5-bit status telegram* and

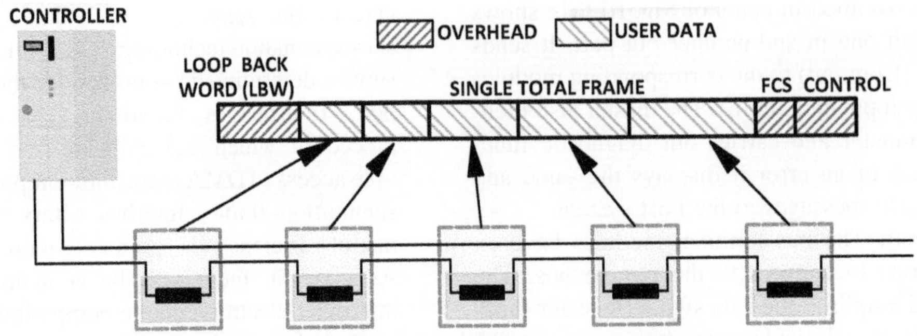


FIG III 8.1d1 INTERBUS SUMMATION FRAME METHOD

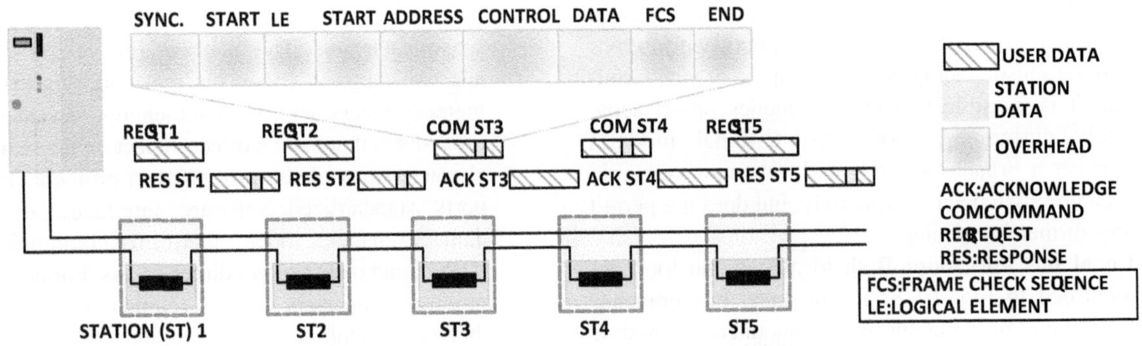


FIG III 8.1d2 INTERBUS MESSAGE BASED TRANSMISSION

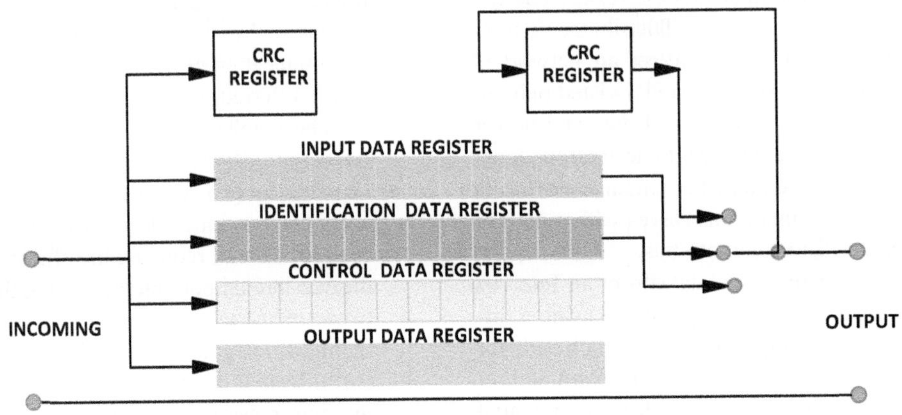


FIG III 8.1d3 INTERBUS SLAVE NODE MODEL

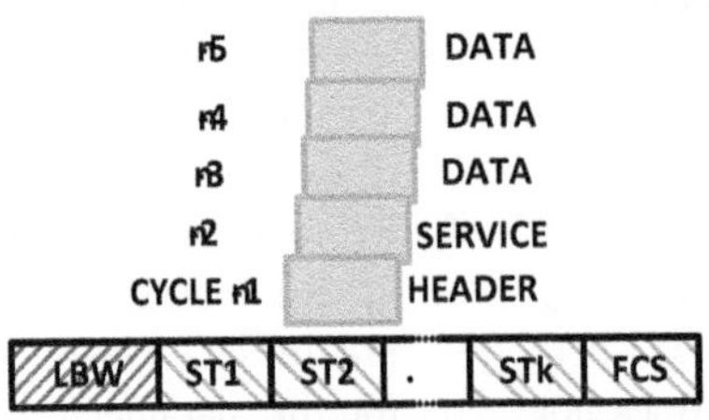


FIG III 8.1d4 INTERBUS SUMMATION FRAME: SEGMENTATION AND RECOMBINATION

**FIGURE III 8.1D** Interbus protocol details.

*the 13-bit data telegram.* The status telegram defines the activity on the medium during pauses. The slave nodes use the status telegram to reset their internal watchdogs for a fail-safe state. The data telegram is used to transmit a byte of the layer 2 payload. The remaining bits of both telegrams are used to distinguish between data and ID cycles and the phase data transfer and frame check sequence (FCS) within a cycle. Fig. III

8.1d1 shows that the summation frame structure of Interbus supports two cycle types: the *identification* cycle for system configuration and *error management* and a data transfer cycle for the transmission of user data. The summation frame consists of a special 16-bit loop-back word (preamble), the user data of all devices, and a terminating 32-bit FCS, as detailed in Fig. III 8.1d2. Summation frames consist of the header, loop-back word, and data save and end information. Data from all connected I/O devices are grouped together in a block. The additional information that is required is transmitted only once per cycle. Interbus ensures constant and predictable sampling intervals for setpoints and real-time control values. In highly efficient protocol Interbus, data can be simultaneously sent and received by the ring structure of the Interbus system in full-duplex mode. The logical method of operation of an Interbus slave can be configured between its incoming and outgoing interfaces by the register. Interbus contains a number of binary memory cells to push digital information from cell to cell to clock pulses. Each device has a certain number of buffers assigned to a preset number of cells for different tasks. An Interbus device contains three registers that are connected in parallel. I/O data are transferred using the data register. The type of Interbus device is defined in the identification register. Fig. III 8.1d3 shows the method as a register, which is formed by the devices that are connected in a ring system. For simultaneous parameter data and time-critical process data, the data format of the summation frame *must* be extended by a specific time slot. In several consecutive bus cycles, a different part of the data is inserted in the time slot provided for the addressed devices. The protocol architecture of Interbus provides the cyclic process data and an acyclic parameter data channel using the services of the PMS as well as a peripheral network management channel.

- **Data transfer:** The master needs to know which devices are connected to the bus so that it can assign the right I/O data to the right device. Once the bus system has been switched on, the master starts a series of identification cycles, which enable it to detect how many and which devices are connected. The identification cycle is also used to find the cause of a transmission error and check the integrity of the shift register ring. Each slave has a 16-bit identification in the data register. The master can use this ID code to assign a slave node to a defined device class. The control data registers are switched in parallel to the identification data registers, whereby the individual devices can be managed by the master. The individual registers are switched in the different phases of the Interbus protocol via the selector in the ring, as shown. Data to be transmitted to the master (input data) are loaded to the input data register during data transfer. The output data registers and the CRC register are switched to the input data register *in parallel*, as shown in Fig. III 8.1d3. Data to be transferred to the periphery are stored in the output buffer of the master in sequence following the sequence of the output stations' connections to the bus. The master first shifts the "loop-back word" through the ring, and following the loop-back word, all the output data are shifted through the shift register, that is, data are placed on the bus by the master. After finishing a valid data transfer cycle, output data from the output data register are written to *memory* and then accepted by the local application. The *CRC registers* shown are used during the FCS to check for correct data transmission.
- **Peripheral communication protocol (PCP):** The PCP software performs the task of inserting a different part of the data in several consecutive cycles. It inserts one part of the telegram in each Interbus cycle and recombines it at its destination. This is detailed in Fig. III 8.1d4.
- **Cycle time and summation frame:** The cycle time refers to the regular time interval at which data is exchanged/updated, that is, the time required for I/O data to be exchanged once with all the connected modules depends on the amount of user data in an Interbus system. Because the summation frame has a set length, the cycle time also remains constant. Interbus, through the summation frame method, ensures a deterministic approach for operation, which is essential for fast controllers. The amount of user data for the summation frame method is >60%, and for the master-slave structure, there will be no access conflicts. Therefore, it avoids potential error sources [7].
- **Determinism:** The summation frame method ensures that the process image for all devices is consistently updated, because all the input data originates from the same point of scan time and all the output data are accepted by the devices simultaneously. This deterministic approach, that is, the guaranteed time for cyclic data transfer between spatially distributed devices, is an important feature of Interbus.

5. **CMD and Interbus discussion:** We discuss configuration, monitoring, and diagnostics (CMD) and Interbus separately, as follows:
   - **CMD discussion:** CMD-type software is an integrated tool for use in all control systems. CMD clearly facilitates the work with Interbus for smooth operation. In the case of a fault with a subscriber or a short circuit, the network functions up to the fault location.

In cases of *transmission pauses* and *the subscribers' fault*, the subscribers are placed into a safe "idle state" condition so that a fault does not cause permanent damage. However, the master can maintain the ring so that the network continues normally while the problem of the failed subscriber is removed. The important features are:

- *Independence and flexibility:* It is independent of the control system and is highly flexible for program extensions, new functions, and add-on programs.
- *Lifecycle*: The CMD tool can be used for the entire life cycle of a system, from planning, configuration, and startup through monitoring to evaluating diagnostic data during servicing.
- *Planning*: Using CMD software, the user can specify the Interbus system configuration.
- *Diagnosis*: Diagnostic data from the controller board can be read and evaluated while the system is running.
- *Data preprocessing*: Using the function block diagram as per the IEC 61131 programming language, it is possible to program limited process data preprocessing directly on the controller board.
- *Operating data* [7]: CMD can be operated with the latest Windows operating system. This enables the operation of different controller boards and control systems with a single software tool. CMD data are available anywhere by using a serial interface.

- **Diagnostics discussions:** System diagnostics is an important issue and diagnostics errors must be located quickly in increasingly complex systems. Interbus uses the CRC in each device to monitor the transmission paths and, in the event of CRC errors, it is possible to determine the segment where the error occurred.

  Common errors are a *cable break, device failure, line short circuit,* and diagnostics of *temporary interference*. Interbus is similar to a linear system, but unlike linear systems, the individual devices in the Interbus system are networked so that each one behaves as a separate bus segment. In case of a fatal error, the outgoing interfaces of all devices are fed back internally via *a bypass switch.* For a *line interrupt* between the devices, the master activates each separate device in turn to clearly identify the inaccessible device. In an Interbus ring system, the termination resistor is opened and closed together with a bypass switch to ensure the condition of the closed ring is always met. Therefore, Interbus diagnostic features are essentially based on the physical segmentation of the network into numerous point-to-point connections. This feature is very useful for use with fiber optics.
- **Interbus features:** Some of the major features of Interbus are:
  - *Topology*: active ring with remote bus devices that include repeater functionality
  - *Communication*: master-slave, fixed telegram length, deterministic
  - *I/O and speed*: maximum 4096 I/O points and 500 kbps
  - Bus length: as detailed in III 8.1c
  - *Interbus diagnostics*: full diagnostics through an external microcontroller
  - *Application area*: sensor/actuator, machines, process, and system production
- **Interbus discussion:** To operate Interbus, centralized intelligence and control are required, which guide the operation of Interbus according to the master-slave method. In master-slave, lower-level subsystems can also be created. The submaster behaves like a slave in the entire network and transfers its control system functions to "its" devices.

Interbus master functions according to the automation task assigned, and all of them offer the basic diagnostics discussed. The master is only responsible for bus control and monitoring the telegrams sent and received by different automation components within the bus configuration.

The hardware overhead for Interbus devices increases in proportion with the width of the data, which is restricted to 20-byte input data and 20-byte output data. As discussed, there will be two cycles: an *ID cycle and data cycle*. An ID cycle is issued when the Interbus system is being initialized and upon request, and, during this cycle, the master reads the ID register of every module connected to the bus to generate the process image.

A data cycle refers to the actual transfer of data. During this cycle, the input data from the registers of all devices are transferred to the master and the output data are transferred from the master to the devices in full duplex mode.

### 8.1.2 Control and Communication Link in industrial data communication

CC-Link was developed by Mitsubishi in 1996 and finally launched in 1999 as an open network fieldbus system mainly used by automation systems made in Asia. CC-Link is the leading open networking technology in Japan and Asia and is growing strongly in North America and Europe. The CC-Link partner association was founded later. Depending on the size and complexity of the automation system, CC-Link comes in several variants, each with its own physical medium, data rate, and typical use cases. The four standards within the CC-Link family are CC-Link, CC-Link/LT, CC-Link safety and CC-Link IE, and CC-Link IE TSN. These standards offer different functionalities and performance characteristics for industrial automation.

- Based on RS-485, standard CC-Link has a data rate of 10 Mbps/ for both control data and information.
- CC-Link/LT is a bit-oriented network for actuators and sensors and offers a transmission speed of 2.5 Mbps. It also offers simplified and minimum efforts for connecting field devices to the switch cabinet.
- Though CC-Link safety is based on the CC-Link standard, this network has an additional level with the security function in the top layer of the protocol. With this, it possible to detect communication errors that could lead to a malfunction of the emergency stop function.
- CC-Link IE (and IE TSN) is based on IE and is designed for seamless data exchange at all levels of a production plant, that is, it is designed from the topmost control level to the production level.

CC-Link also supports various network topologies, for example., Star, Line, and Ring. There are different bus cables for different bus versions. The standard CC-Link bus requires a three-core cable (cross-section 0.5 $mm^2$) with an overall shield. CC-Link IE (and IE TSN) uses conventional category 5 enhanced (CAT 5e) Ethernet cables. CAT6 cables can also be used.

CC-Link provides high speed, deterministic communication to link a wide range of multivendor automation devices over a single cable. It is designed for ease of use, simplicity of system design, speedy installation, and reliability in operation. CC-Link is mainly used in factory automation, with strong representation in automotive manufacturing, electronics and semiconductor fabrication, food and beverage packaging, and machine tooling and Computer numerical control (CNC) systems, among others. Conceptual details pertinent to CC-Link covering a number of issues have been detailed in Fig. III 8.1e.

In this subsection, *CC-Link with its associated issues* will be briefly discussed.

1. **Basics of CC-Link:** Let the discussion start with a conceptual discussion on CC-Link along with its general features, as detailed below:
   - **CC-Link conceptual discussion:** Conceptually, CC-Link is a type of fieldbus that connects devices like sensors, actuators, and controllers for industrial automation. CC-Link covers a few issues for its successful implementation. These are:
     - *Open network*: CC-Link is an open network, that is, specification is available openly and adopted by different manufacturers in their devices for integration.
     - *Architecture*: CC-Link is based on a master-slave configuration, with a master device to manage communication in the network with slave devices.
     - *Reliability and availability*: Standby managers, automatic return mechanisms, and hot swapping are a few features that make CC-Link a highly reliable and available system with quick communication failure, which reduces the associated downtime.
     - *Flexibility*: CC-Link offers high flexibility and can be used for a wide range of applications, such as motion and I/O control and safety functions, and it can handle both cyclic and transient communication. With seamless message protocol (SLMP), which is detailed in Fig. III 8.1e2, Ethernet-enabled devices can be used in a CC-Link network.
   - **CC-Link general features:** The general features of CC-Link are:
     - *High-speed communication*: CC-Link offers a communication speed of 10 Mbps, allowing for efficient data transfer. It also offers relatively fast communication speeds viz. refresh rate <3.9 ms for about 64 I/O stations.
     - *Station support:* CC-Link can support a large number of stations, as many as 64.
     - *Application versatility*: CC-Link is suitable for a wide range of automation tasks, including discrete manufacturing, machine controls, and process controls.
     - *Scalability:* With flexible configurations, users can build systems to cater to specific needs.
     - *Open network:* Being an open network, CC-Link allows devices from different manufacturers to communicate with each other.
     - *Cost effectiveness*: The open nature and widespread availability of compatible devices make CC-Link a cost-effective solution.
   - **Comparisons of different variants:** As discussed, there are several variants of CC-Link, which are compared in Table III 8.1e.
   - **Common application area:** Common industrial application areas of CC-Link are:
     - *Machine tools*: servo drive integration, compact motion controllers, HMI networks
     - *Automotive industry:* robotic arms, conveyor control, safety I/O, PLC network
     - *Electronic industry:* wafer handling, cleanroom automation, high-speed device networking
     - *Food and beverage:* packing lines, motion control, and others
2. **General specification:** Let us discuss the general specification for CC-Link, as detailed in Table III 8.1f. For CC-Link IE TSN, the synchronous accuracy of a

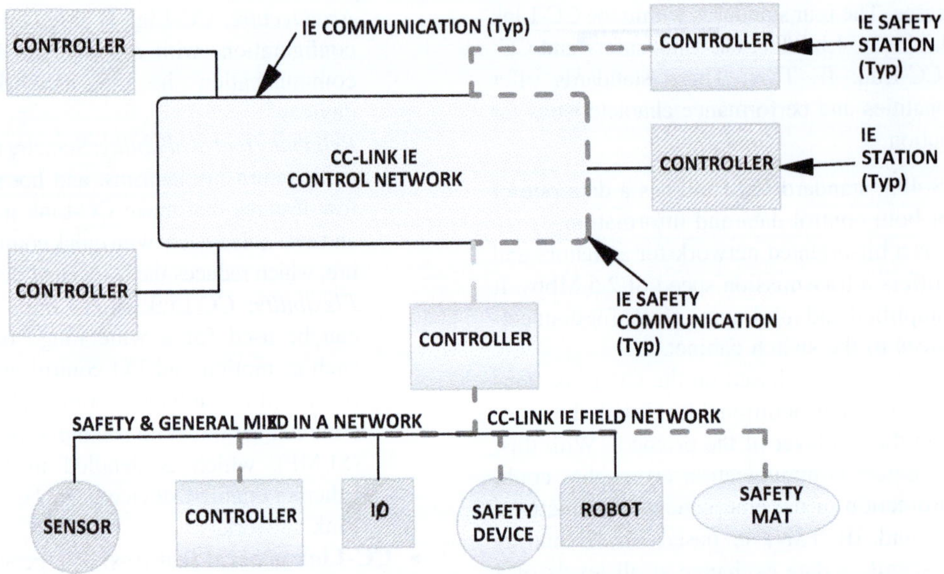


FIG III 8.1e1 CC–LINK IE WITH SAFETY CONFIGURATION

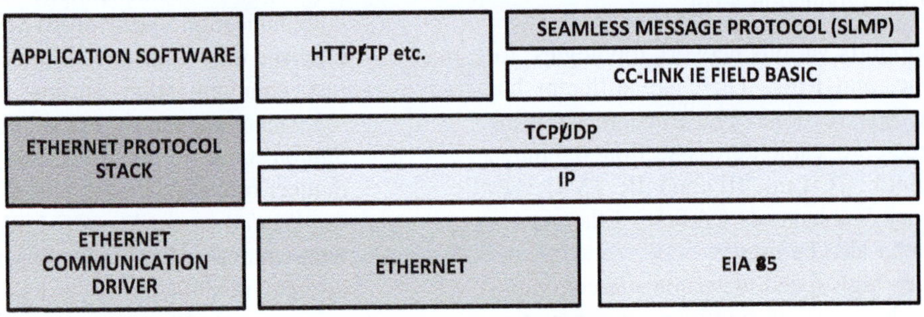


FIG III 8.1e2 SLMP DETAILING

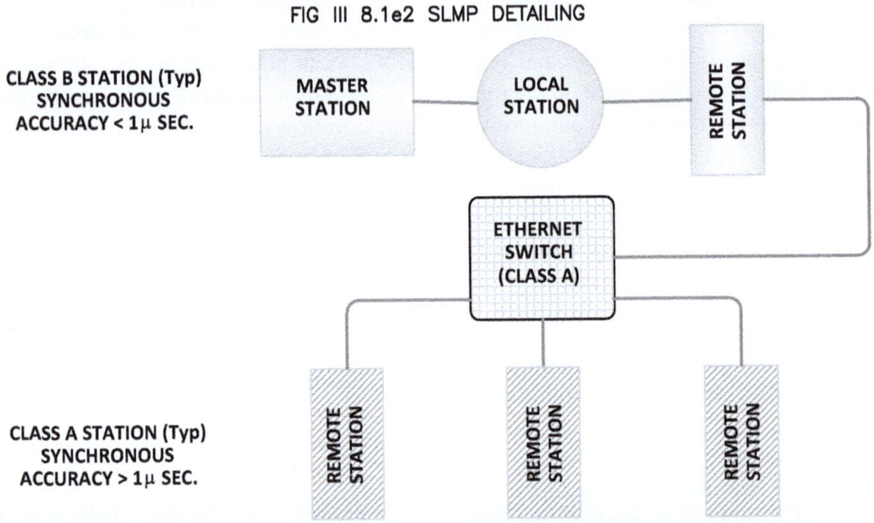


FIG III 8.1e3 CC–LINK STATION CLASSIFICATION

**FIGURE III 8.1E** Control and Communication Link conceptual details.

system varies based on the combination of connected devices and switches. Based on synchronization accuracy there will be two classes: class A, which has a synchronization accuracy $> \pm 1$ μs and high accuracy and class B, which is a system within $\pm 1$ μs of high-accuracy synchronization, as detailed in Fig. III 8.1e3. *Now we shall touch upon variants of Control and Communication Link.*

**TABLE III 8.1E** Comparison of Control and Communication Link variants.

| Items | Classic | LT type | IE *control | IE* field | IE*field -B* |
|---|---|---|---|---|---|
| Medium | Twisted pair RS 485 | 3-Wire cable | Ethernet CAT 5e/6 | | |
| Topology | Daisy chain | | Star/ring | Star/ring/line | Star/line |
| Speed | 10 Mbps | 156 Kbps | 1 Gbps | | 100 Mbps |
| Stations | 64 | | 120 | | 64 |
| Timing | Deterministic timing | | | | Based on load, limited deterministic |
| RT sync | Moderate | Low | Very high | High | Moderate |
| Ideal use | High-speed M/ C* | Small automation | Controller network | Large Field I/O network | Low-cost Ethernet |

**B, Basic; IE, industrial ethernet; I/O, input/output; LT, Link terminal (light weight); M/C, machine; RTL, Reat time.*

**TABLE III 8.1F** Control and Communication Link general specification.

| Feature | Specification *(with note in the bracket as applicable)* |
|---|---|
| Speed | 1 Gbps/100 Mbps (variants see Table III 8.1e) |
| Cable type | Twisted pair/Ethernet CAT 5e/6 (variants see Table III 8.1e) |
| Maximum stations | 64/120 (variants see Table III 8.1e) |
| Typical distance | 100m station to station — typical |
| Topology | Common line/star (variants see Table III 8.1e) |
| Communication | Time shared |
| Max safety/station | 120 |
| Temperature | *Operating* (–)20–55°C/*Storage* (–)25–75°C nonfreezing |
| Humidity | 5%–95% noncondensing |

3. **CC-Link IE (all) discussion:** The features and characteristics of CC-Link IE, including IE field and IE TSN are discussed below.
   - **CC-Link IE (all) features:** A few features of CC-Link IE, including IE field and IE TSN are:
     - *Open network:* CC-Link IE is an open standard 1 Gbps Ethernet that includes a CC-Link IE field network as a fieldbus manufacturing network.
     - *Speed and cabling:* It supports 1 Gbps networks using a standard CAT 5e cable.
     - *Wide bandwidth*: CC-Link IE provides safe, high-speed control with simultaneous quality and traceability data exchange.
     - *Integration:* It integrates various functions like *safety*, *motion control*, and *energy functions* into a single network.
     - *Communication programming*: Memory: map-based communications are adopted.

       *Implementation and maintenance*: A standard Ethernet physical layer with diagnostic functions. So, it is simpler to implement and maintain.
     - *Flexible topologies*: It supports flexible topologies viz. star, line, mixed star and line, and ring setups.
     - *Flexible industrial IoT systems*: A TSN is employed as a time-sharing communication method. This allows integration of real-time control and non–real-time information on the same network.
     - *Deterministic communication*: The CC-Link IE field network provides deterministic control communications without requiring Ethernet switches, which is *a key difference* from other IE solutions.
   - **CC-Link IE (all) characteristics:** CC-Link IE is a high-speed, deterministic network designed for automation applications. These are characterized by an

*open network* with support for *redundancy* (redundant configuration), *high-speed communication* (1 Gbps) bandwidth, *deterministic control* (predictable and reliable data exchange with fixed scan times), *seamless integration*, and *comprehensive diagnostics* (efficient network diagnosis and troubleshooting to ensure high availability) to name a few important issues. CC-Link IE TSN further enhances performance, synchronization, and determinism, enabling more advanced motion control and other real-time applications. It also allows integration of real time and non–real time information in a single network. As discussed, with SLMP (Fig. III 8.1e2), Ethernet-enabled devices can be used in a CC-Link network. With CC-Link IE it is possible to communicate data seamlessly from the plant-level enterprise network to the production floor network. This integrated approach provides cost reduction for the total system engineering. CC-Link IE field networking allows seamless communication of data from field devices to controllers (controllers to other controllers) at gigabit speed. The CC-Link IE field supports both "cyclic" and "transient" methods of data exchange based on "*token passing*" technology and an optimized protocol. Unlike other IE solutions, the CC-Link IE field network delivers deterministic control communications without Ethernet switches. There can be a total of 120 stations with up to 100 m of CAT 5E cable between each station in a single network, and there can be as many as 239 networks.

4. **CC-Link LT details:** CC-Link LT is designed for sensors and actuators and other reduced I/O applications to simplify and minimize wiring for field devices and within control panels. In a way it eliminates wiring errors. This caters to *small point dispersion*, a *smaller I/O*, and an *open network with reduced wiring and sensor network.* Some major features include, but are not limited to an *open network, high-speed response, simpler programming, no need for setting in device stations, reduced unused I/O addresses, high noise immunity*, and an *easy* and *efficient network connection.* Table III 8.1g details the CC-Link LT specification.
5. **CC-Link safety:** CC-Link IE safety is a safety field network of CC-Link that realizes safe system architecture. Here, the high-speed communication of CC-Link IE is realized, and the same system can be mixed by safe and general controls, as shown in Fig. III 8.1e1. So, a CC-Link IE network can easily be expanded into the safe control system. A safety station connected to CC-Link IE safety has an IE safety communication layer. The IE safety master protocol is responsible for establishing a connection with the IES slave (IESSLP) protocol machine, which is specified as its safety parameter, and performing safety communication with the IESSLP. Some features are:
   - Safe communication can be at a transmission rate of 1 Gbps in CC-Link IE.
   - Both safe and general control, including motion control, can be mixed in a single network.
   - CC-Link IE safety can be implemented based on an international safety standard.
   - CC-Link IE safety communication satisfies the requirements of SIL3 as per IEC 61508.

   CC-Link IE safety details are shown in Table III 8.1h.

## 8.2 Fieldbus discussion: actuator-sensor interface, world factory instrumentation protocol, and serial real-time communications system

AS-i is an acronym for actuator-sensor interface, which is one of the most high-tech buses for industrial automation. Basically, this is the universal interface between the higher

**TABLE III 8.1G** Control and Communication Link LT specification.

| Feature | Specification |
|---|---|
| Input/output points update | All in 1.2 ms @ 2.5 Mbps |
| Communication speed | 156 kbps, 625 kbps and 2.5 Mbps |
| Error control | Cyclic redundancy check |
| Number of station support | Max 64 |
| Communication type | Broadcast polling and interval timed response |
| Trunk line length (m) | 35 (2.5 Mbps)/100m (625 kbps)/500 (156 kbps) |
| Branch line length (m) | 4 (2.5 Mbps)/16 (625 kbps)/60 (156 kbps) |
| Diagnostics | Network diagnosis, internal loopback diagnostics, etc. |

**TABLE III 8.1H Control and Communication Link Industrial Ethernet safety details.**

| Feature | Specification/description |
|---|---|
| Standard<br>IEC 61508:2010 | Functional safety standards for E/E/PE safety-related systems. Safety integration level: safety integrity level 3 |
| IEC 61784-3: 2010 | Industrial communication networks—Profiles—Part 3: Functional safety fieldbuses |
| Communication | Connection type communication method |
| Safety connection | Maximum 2 connections/station |
| Maximum number of safety links refresh | Input: 4-100 octets/output: 4-100 octets |

control level and binary actuators and sensors. Naturally, this saves huge cabling costs in connecting the field level to automation equipment to field devices. An individual sensor and actuator connection not only has a high cost but also gives rise to a considerable source for errors.

World factory instrumentation protocol (WorldFIP) is a fieldbus protocol meant to provide a link between sensors/actuators at level 0 and field controllers (say PLCs) at level 1 in industrial automation systems. Remote connection of the WorldFIP is important for monitoring, maintaining, and controlling devices in a factory calling for asynchronous transfer mode as a backbone network. However, WorldFIP was proposed and developed for use with all types of automation, including centralized, synchronous, and asynchronous applications. This fieldbus protocol was not only dedicated for real-time requirements but also for transferring the monitoring information from the plant and along the network to the supervisory equipment. This is done without interfering with basic real-time tasks. The WorldFIP protocol is relatively simple to learn in comparison to other protocols. It maximizes the use of international standards.

SERCOS is an acronym for serial real-time communications system, which is an internationally approved communication standard for motion control (IEC 61491). This communication network is designed for high-speed serial communication of standardized closed-loop data, in real-time, over a noise-immune fiber-optic cable. SERCOS III is a universal communication for all applications. The vision of a real-time, nonproprietary communication solution permeates the world of industrial automation.

From the above, it is clear that this section is dedicated to covering these three fieldbus systems.

## 8.2.0 Actuator-sensor interface

AS-i is designed primarily for binary (on-off) sensors and actuators, although analog elements are also easily incorporated. In a complex automation system, numerous control devices may be networked on different data networks and with different protocols with differences in their time response, degree of protection, type and use of data to be transmitted, and many other aspects. The actuator sensor level is the lowest zero level in the field area meant to form a network. So, here, the data volume may be small, but with very high-speed data transmission. AS-i can form sensor/actuator networks and link them to higher control levels in a very simple, inexpensive way, meeting all industrial requirements including compliance with the standards. AS-i enjoys the support of the industrial automation community in Europe and is quickly becoming a US standard too. Thus, AS-i is becoming a worldwide de facto standard for simple I/O networking; users are attracted to its high speed, power on bus, simplicity of operation, and insulation displacement wiring system. AS-i can be used to select the network topology from a linear, star, or tree structure. The AS-i protocol thus allows for modular expansion of an industrial communication, enabling individual adaptation to the topology of existing systems. On account of a robust, balanced signal with parity redundancy, cabling becomes simple. The signal segment can be laid out in a different topology, without drop limitations as long as the signal segment cabling length is <100 m. There are no terminating resistors. Currently in AS-i, users are switching to round cable with M12 connectors, although flat cable is still in use. Fig. III 8.2a shows the architecture and connection details of AS-i.

The disposition of the slave and use of repeaters in AS-i are detailed in Fig. III 8.2b.

In this subsection, the AS-i fieldbus system has been discussed.

1. **Main features:** Some of the major features including safety are:
   - As many as 31 devices can be connected onto the AS-i cable, each with 4 bits of input and 4 bits of output with a cycle time of <5 ms.
   - AS-i uses low-cost, two-core unscreened 1.5 $mm^2$ stranded cable with a length <100m.
   - Power and data are combined in one cable—nominally 30 V at <8A over the bus cable.
   - Both special flat profile and round profile cable are in use.

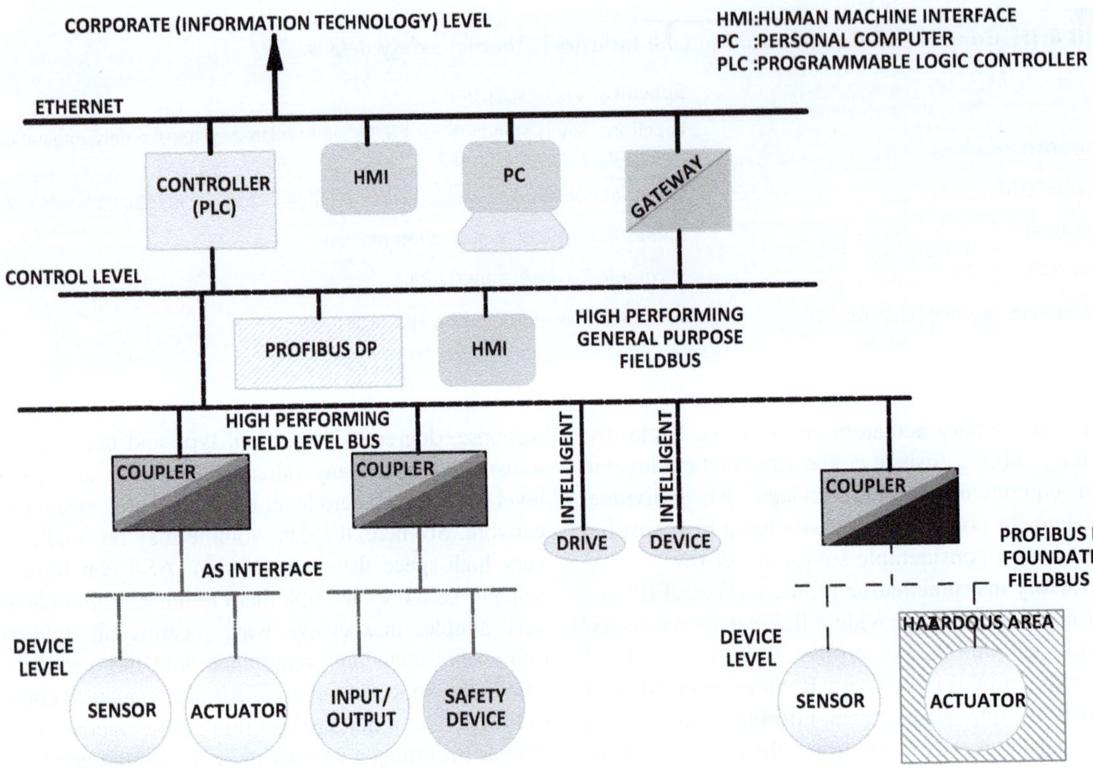


FIG III 8.2a1 AS INTERFACE ARCHITECTURE AND APPLICATIONS

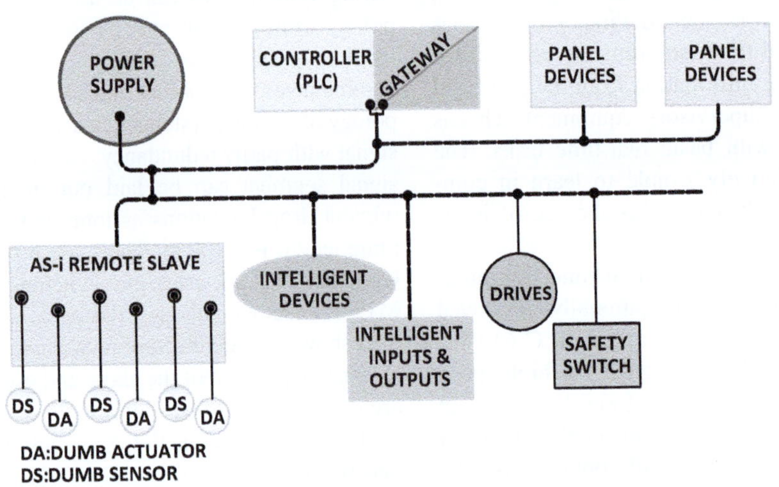


FIG III 8.2a2 AS DUMB AND INTELLIGENT DEVICES WITH CONNECTION DETAILS

**FIGURE III 8.2A** Actuator-sensor interface (AS-i): architecture and connection details.

- It is extremely low cost to implement based on a self-contained chip (can be embedded or built directly into the sensors or actuators). So, it can be implemented even for the simplest of devices (e.g., proximity switches, panel indicators).
- Safety protection (for man, machine, and environment) systems are part of AS-i.
- It does not call for a safe PLC.
- Using standard AS-I, safety at work can be used in the networking of safety devices viz. safety door switches, e-stops, and others. The following features are important:
  - AS-i is also widely used in functional safety systems with (safety protection systems) high integrity operation and a low failure rate.

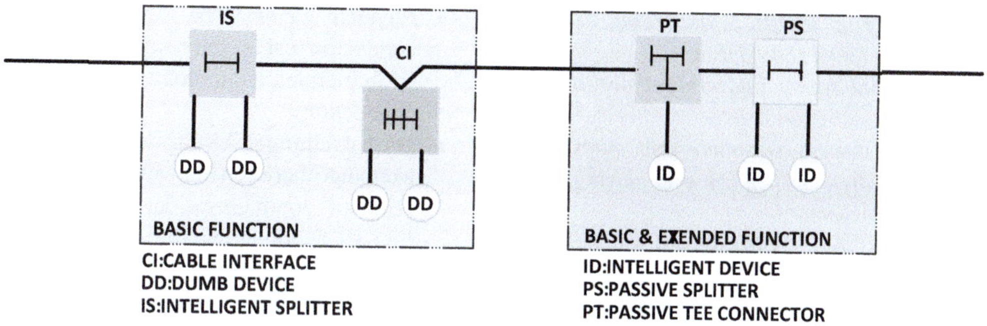


FIG III 8.2b1 INTELLIGENT AND STANDARD SLAVE DETAILS

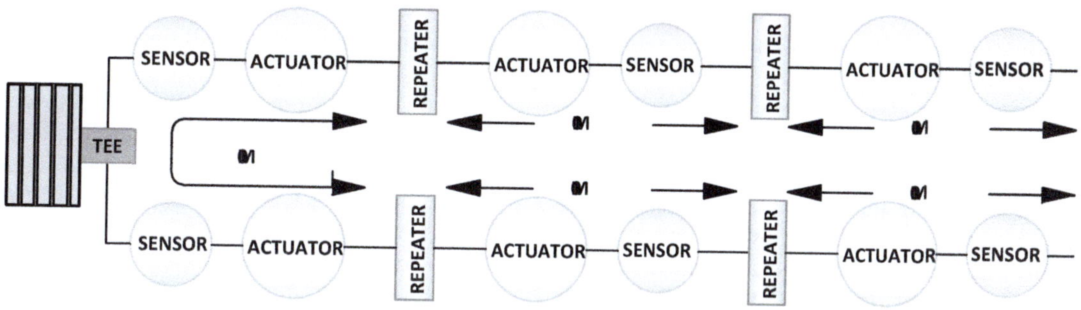


FIG III 8.2b1 INTELLIGENT AND STANDARD SLAVE DETAILS

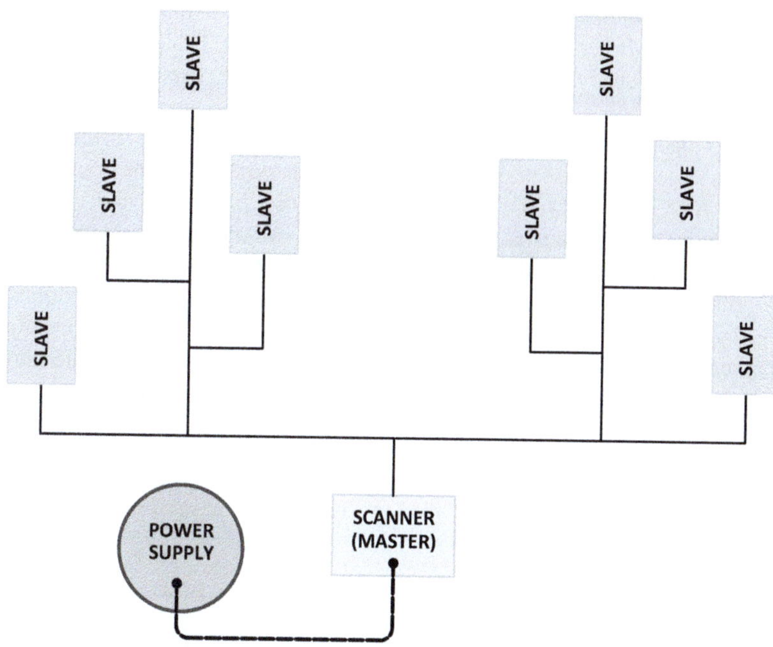


FIG III 8.2b3 TREE TOPOLOGY INAS INTERFACE

**FIGURE III 8.2B** Actuator-sensor interface: slave, repeater, and topology details.

- *Special "safe slaves" are required to provide the required safety functionality.*
- *A standard AS-i master can be used with safe slaves, and the safety function is provided by a "safety monitor".*
- *The special "safety monitor" provides the safe shutdown or stop functionality.*
- *The safety monitor allows implementation of both simple and powerful safety procedures.*

There can be 31 safety modules per network.

- *Safety category up to SIL 3 according to IEC 61508* and EN *ISO 13849-1 Ple.*
- *Control I/O and safety information are in the same network.*
- *Adding safety devices is simple and quick and can be added wherever needed, even during final phases of the project.*

2. **Architectural issues:** AS-i is a standardized fieldbus system for connecting sensors/actuators to a higher-level controller via a common data transmission medium. AS-i is also implemented as a subfieldbus system in established, high-performance fieldbus systems and IE for communication purposes. At the field device level, AS-i is governed by the master-slave principle; communication between the AS-i master and its slaves (field devices) occurs as a cyclic request. AS-i topology is completely open with no termination and supports bus, star, ring, and tree structures. AS-i bus topology is unrestricted, so the most direct connections could be made between the bus and various sensors/actuators, making the installation layout fit each application (the maximum length restriction of the AS-i bus is 100 m). For a maximum of 31 digital slaves the cycle time is 5 ms. Power supply/modules can be placed anywhere in the segment. AS-i technological applications are detailed in Fig. III 8.2a1. Master-slave principles and key AS-i architectural issues are enumerated below.
   - **AS-i operation:** AS-i operates on a master-slave principle.
   - **Master-slave bus:** In AS-i is a master-slave bus; all the slaves are managed by a single master that polls each of the slaves present on the bus and waits for a reply. The cycle time is 5 ms maximum. For a maximum of 31 digital slaves the cycle time is 5 ms. Communications are always initiated by the bus master.
   - **Physical layer:** The AS-i cable is a flat, two-conductor cable carrying both power and data. With piercing technology, it allows easy and secure connections.
   - **Data exchange:** AS-i uses a simple protocol for data exchange between the master and slaves, enabling efficient communication for basic I/O functions. Fig. III 8.2a2 shows the *power supply* and *master-slave* (both dump and intelligent devices) connection details for better clarity and understanding.
3. **Components:** AS-i comprises a variety of components, as listed below:
   - **Master AS-i:** AS-i is a self-configuring system that automatically performs all functions required for its correct functioning, including *self-diagnostic* functions. The AS-i master is meant to provide the connections to the higher-level controller. Its duty is to organize data transmission so that the signals of the sensors and actuators are available to higher-level bus systems. Major duties of the master include *network monitoring, polling signals, transferring parameter settings* to the nodes, and *performing diagnostics.* The AS-i master specification clearly defines different master categories with different ranges of functions known as a "profile," as elaborated in Table III 8.2a.
   - **Master interface types**: AS-i accepts different types of masters like the *PLC, master module (incorporating communication port to directly transfer the data,* transparently, *to the PLC application program),* and *gateway master (to convert the AS-i bus from a simple communication node to a higher-level bus* from a third-party vendor like MODBUS Plus or Profibus DP ).
   - **Slave AS i:** Slaves are actually distributed I/O modules that recognize the data bits transmitted by the master and return their own data. Up to four binary

**TABLE III 8.2A** Actuator sensor interface master profiles specification.

| Master profile | Functional specification |
|---|---|
| Master M0 | Configured with the station configuration in cable called "expected configuration", it can exchange input/output (I/O) data with the individual slaves. |
| Master M1 | All the functions as specified for the actuator sensor interface (AS-i) master. |
| Master M2 | Functionally same as that of master M0, additionally can assign parameters to the AS-i slaves. |
| Master M3 | Functionally same as that of master M2, additionally can assign parameters to the AS-i complex slaves and 16-bit I/O. |
| Master M4 | Functionally same as that of master M2, additionally can assign parameters to the combi slaves, transfer of serial data, extended address mode (8 I/O), 16-bit inputs in the extended address mode, and fast/superfast analog value transfer. |

sensors/actuators can be connected to one standard AS-i module. AS-i slaves are available as digital, analog, and pneumatic modules, and as intelligent nodes viz. motor starters, membrane keyboards. Fig. III 8.2b1 shows slaves as intelligent and standard components in basic and extended function forms, respectively.

- **Power supply:** As an essential component, the AS-i power supply unit produces a controlled DC voltage of 30 V with a high degree of stability and low residual ripple. They work on the principle of a primary switched-mode regulator.
- **Cable and links:** The common cable for AS-i is the yellow profiled cable. In complex automation network configuration, AS-i can also be connected to a higher level fieldbus (e.g., PROFIBUS). For this, it requires a gateway (e.g., Profi DP/AS-I link), often called a coupler, as shown in Fig. III 8.2a1. This is used as an AS-i *master in the AS-i network* but functions as a *slave* in the higher-level fieldbus (e.g., PROFIBUS). Gateways for all standard buses are available: *MODBUS, Modbus TCP, PROFIBUS, PROFINET, CC-Link, CAN Bus, EtherNet I/P, DeviceNet*, and others.
- **Repeaters and extenders:** Up to two repeaters can be connected in series, in which all components can be distributed to the segments in any way. Much larger extents are possible because parallel arrangement of several lines in a star configuration is possible. With extenders, the AS-i cable can be lengthened by 100 m. This is detailed in Fig. III 8.2b2. In the design of the network, a "tree topology" can be used to run cable anywhere required by the application, as shown in Fig. III 8.2b3.

4. **Technical requirements:** Let us look at various issues related to various *technical requirements*: the *master-slave principle, data transmission*, and specification issues have been elaborated here. Fig. III 8.2c was developed based on [8] Courtesy Siemens]] and elaborates master-slave and data transmission Principles.
   - **Important technical issues:** As listed below, here, we deal with various technical issues and requirements for AS-i.
     - *Slave configuration:* The AS-i configuration basically refers to drawing up a list of the configured slaves and storing it in the master. Configuring is usually performed at the master, mainly by automatically reading in the network configuration. The slave address and type (ID code), the I/O configuration (I/O code), and the parameters (for intelligent sensors) are defined during configuration. The master uses the list to check that the programmed configuration matches the actual configuration.
     - *Parameterization and operation*: Normally, slaves are not parameterized. Only intelligent slaves *with* required options are parameterized as defined in the data sheet in terms of the requirement and functions. After installation of all the components of the AS-i system, the slaves are addressed and, as necessary, parameterized, and configuration is complete, that is, after the start-up phase, it is time to start the system switches to normal operation, and the master runs in protected mode. The system is continuously monitored not only during start up but also during normal operation.
     - *Response time and layout*: A network of 31 nodes and one master can do a complete read-write in 5 msec. The actual time per discrete node to use is 0.165 msec. The passive parameters of cable attenuate the signal to reduce the amplitude and distort the shape. The layout of the AS-i signal segment mainly follows rules like: The total distance of the bus cable *cannot exceed 100 m* per signal segment and the bus cable *cannot be laid out with closed loops.* AS-i–approved masters allow for two repeaters. Essentially, the repeater retransmits the signal. The strength and shape are brought back to the original condition so the signal can travel another 100 m. Fig. III 8.2b2 shows the distance limits for AS-i with repeaters.
   - **Master-slave principle:** The AS-i master connected to the AS-i cable controls the data exchange with the slaves via the interface to the AS-i cable. Fig. III 8.2c1 elaborates master-slave principles, with the two interfaces of the *AS-i master communication profile (CP)*. The process data and parameter assignment commands are transferred via the interface between the master central processing unit and master CP. The user programs have suitable function *calls and mechanisms* available for reading and writing via this interface. Information is exchanged with the AS-i slaves via the interface between the *master CP and AS-i cable*. The AS-i slave has an integrated circuit to provide the attachment of an AS-i device (comprising eight configurable data I/O, four parameter outputs, and two analog channels of 16 bits) for the sensor/actuator to the common bus cable to the AS-i master. The operating parameters, configuration data with I/O assignment, identification code, and slave address are stored in additional memory ($E^2$PROM).
   - **Data transmission principle:** Fig. III 8.2c2 details data transmission principles. The above figure elaborates various system components in which the data

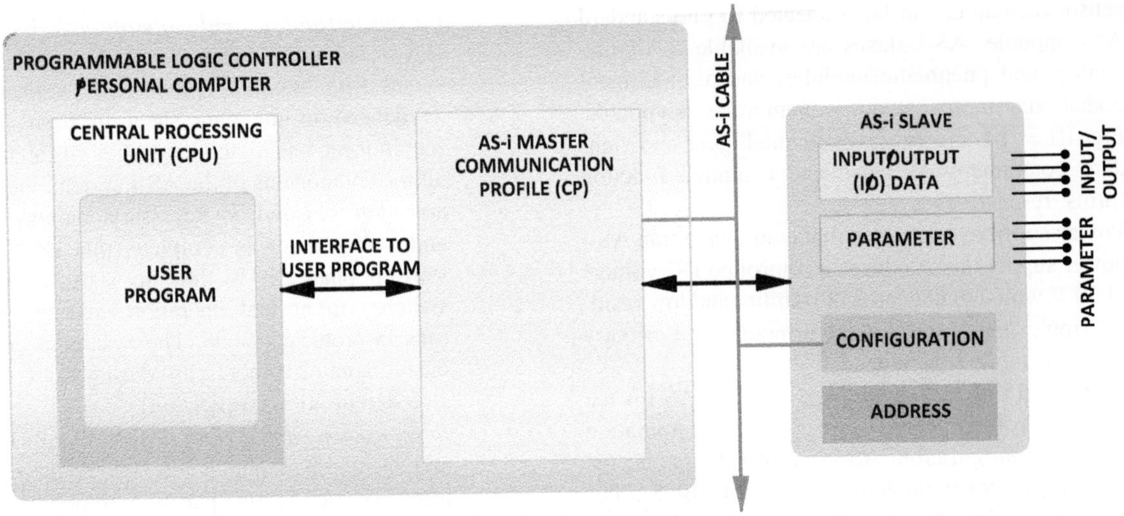


FIG III 8.2c1 MASTER SLAVE PRINCIPLES

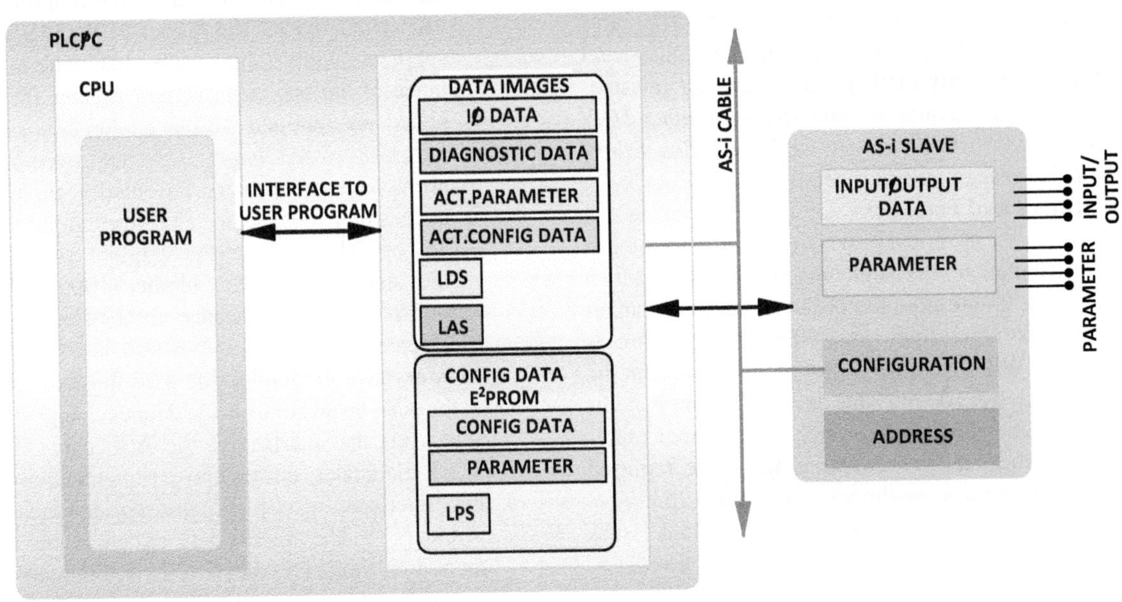


ABBREVIATIONS USED:
ACT:ACTIVE; CONFIG:CONFIGURATION; E $^2$PROM:ELECTRICALLY ERASABLE PROGRAMMABLE READ ONLY MEMORY
LAS:LIST OF ACTIVE SLAVE; LDS:LIST OF DETECTED SLAVE; LPS LIST PERMANENT SLAVE

FIG III 8.2c2 DATA TRANSMISSION PRINCIPLES

DEVELOPED BASED ON IDEA FROM 8[COURTESY:SIEMENS]

**FIGURE III 8.2C** Master-slave and data transmission principles [8].

fields and system lists are stored. Various data pertinent to masters and slaves are:

- *AS-i master*: *data images, I/O data*, configuration *data*
- *AS-i slave*: *I/O data, parameters, actual* configuration *data*, and *address*
- **Technical specifications:** Table III 8.2b elaborates the technical specifications of AS-i.

5. **AS-i discussion:** Let us discuss a few issues related to AS-i. This concluding discussion on AS-i will cover the following topics:
   - **AS -i benefits:** Some major benefits are:
     - *Less cabling and cost:* Both the data and power supply are transmitted by a standard, unshielded pair of nontwisted cables (1.5/2.5 mm2) installed

**TABLE III 8.2B** Technical specification: actuator sensor interface.

| Features | Specification |
|---|---|
| Max node numbers | 31/62 in version 2.1 |
| Max input/output (I/O) points | 248 (4 I, 4 O/node—V1.0), 434 (8 I, 3 O/node—V2.1) |
| Max cable length | 100 m without repeaters, 300 m with 2 repeaters |
| Cable | Unshielded two-wire cable for both power and data |
| Power supply | 24 VDC |
| Device support | Discrete device supported. Analog device: Limited support in V1 and V2.1; much more in version 5 |
| Topology | Tree, star, line |
| Addressing | Each slave has specific address set by master/programming |
| Hot swap: slave | Yes supported |
| Multimaster | Not supported |
| Error detection | Master can automatically detect invalid slave transmissions and schedules repeats |

directly on the machine. So, there will be less and simplified wiring. This results in cost savings of around 15%–40% compared with other bus systems.

- *System efficiency*: The reduced cabling and lesser components are more compact, resulting in space saving. These reductions in material and labor costs make AS-i a more resource-efficient solution.
- *Simplicity*: AS-i is popular for its simplicity, having device power right on the network cable, and not needing master or slave programming.
- *Open standard*: AS-i is based on open standard and hence can be integrated with high-performance fieldbus systems like Profibus and Foundation Fieldbus through bus coupler.
- *Response time*: AS-i Master can poll up to 31 connected slave devices and get a response within 5 ms. So, the cycle time is very short. It is much shorter in version 5, as discussed below.
- *Reliability:* AS-i is highly reliable, with high availability, expandability, and adaptability.
- *Deterministic*: AS-i is inherently deterministic and repeatable. Transmissions between the master and slave are predefined and cyclic.
- *Diagnostics*: The system offers built-in diagnostics, allowing for quick identification and troubleshooting of issues.

- **Cable issue:** A standard two-wire round cable can be used, but the *famous yellow, flat* cable is always preferred for AS-i. It not only provides an efficient installation but also ensures correct polarity due to the mechanical keying, The yellow AS-i cable ensures peak electrical performance regardless of the network length (<100 m/segment) and topology. In addition to using the yellow AS-i network cable, a black, mechanically keyed flat cable supplies auxiliary power for output devices such as lights, valves, or actuators. AS-i uses unshielded cable. If shielded cable is used, it is important to connect the *shield wire* to a *solid machine ground*, especially for the exposed data/power leads.
- **Implementation issue:** AS-i network implementation refers to connecting field devices (sensors/actuators) to one central AS-i master using a two-wire cable (flat yellow preferred). Installation is done through piercing technology (method of connecting field devices to the network using a flat, two-conductor cable with piercing pins). The major implementation steps are:
  - *Cable installation*: It is installed using a yellow flat cable connection.
  - *Device connection*: Slaves are connected to the cable using piercing technology.
  - *Addressing*: Each slave is assigned a unique address using a programming device.
  - *Configuration*: The AS-i master is configured.
  - *Special printed circuit board (PCB)*: AS-i PCBs are used to integrate mechanical switches.
  - *Serial gateway*: Use a serial gateway to integrate higher-level devices. This is the least economical approach with the fastest time to market.
- **Latest versions:** At this time (2025), version 5.0 is available on the market. AS-i version 5 can handle

larger data volumes at much higher transmission speeds. AS-i5 is capable of integrating intelligent sensors such as IO-Link at ease. This supports up to 32 bytes of IO-Link process data. ASi-5 is backwards compatible with ASi-3 (also compatible to version 2.1), allowing for the coexistence of both technologies on the same cable. Utilizing ASi-5 orthogonal frequency-division multiplexing (OFDM) for data transmission, more data can be sent with shorter cycle times, comparable to Ethernet systems. ASi-5 deploys a multicarrier modulation scheme with OFDM, with each carrier transporting 20 bits in a 1.2-ms cycle. It can support up to 96 devices (rather than 62 in version 2.1/3.0). It uses CRC for error detection.

### 8.2.1 Fundamentals of WorldFIP and factory instrumentation protocol input/output

On account of the increased system robustness associated with the distribution of system intelligence, high manageability, and fewer expenses, the use of fieldbuses in distributed industrial control applications has become widespread. Real-time operation is an essential part of this to ensure timely transmission of information. In addition to having the common features of fieldbuses such as low cabling and component costs, high noise impunity, and modest speed of transmission over distances, WorldFIP has the ability for real-time synchronization. WorldFIP is peculiar because one bus caters to three different automation levels. It also caters to different applications differently, for example, for processing WorldFIP offers 31.25 Kbps and 1 Mbps, for manufacturing and drive it also offers a 1 Mbps data rate, but for drive it can also offer 2.5 Mbps. The first international industrial communication standard WorldFIP, a three-layered fieldbus architecture, is described in volume 3 of EN50170. The WorldFIP provides data transmission at S1 (31.25 kbit/s), S2 (1 Mbit/s), and S3 (2.5 Mbit/s) on shielded twisted pair and at 5 Mbit/s on fiber optics. A bus arbiter (BA) controls the bus and other stations with medium allocation including periodic transfer, request for aperiodic transfer, and unacknowledged and acknowledged aperiodic message transfer requests. All network schedules are provided by BA table (BAT), which includes all defined variables and their timings. The BA uses an orderly sequence of microcycles and macrocycles (MCs) for them. Manchester coding is used here. WorldFIP also contains producers and consumers. Producers give their own consumer variables to consumers through messages, which include the data to subsequently activate their own actuators. The chips used for these functions are FULLFIP2, MICROFIP, FIELDDRIVE, and FIELDDUAL. Globally, there are many transport systems like buses, tramways, subways, locomotives, cars, and high-speed trains that run in a network utilizing the WorldFIP protocol. Major application coverages include traction and brake control, air conditioning systems, lighting systems, radio, communication systems, high-voltage equipment, and pantographs.

FIPIO is an acronym for factory instrumentation protocol I/O, which is a subset of WorldFIP. The FIPIO fieldbus is a standard means of communication between different control system components. As high as 256 devices can be connected at the connection point integrated in the processor. This follows producer/consumer mechanisms in line with the WorldFIP standard. It is designed for a remote location of I/O up to 15 km away. The FIPIO protocol supports both deterministic data transfer and random messaging.

1. **Overview of WorldFIP:** A basic overview of WorldFIP has been provided here. Let the overview start with *layered architecture* and *elementary transactions.* Fig. III 8.2d details layered architecture elementary transactions.
   - **Layered architecture**: The WorldFIP protocol comprises three communications layers, as shown in Fig. III 8.2d1. Like other fieldbuses, the three layers are the *physical layer, data link layer,* and *application layer.* The different elements in the application layer are manufacturing periodic (/aperiodic) services, which are *responsible for* managing *the exchange of data (variables) between devices on the fieldbus*; a subset of MMS; and a message control system for application of the service element, which offers a high-quality MMS service exchange as a part of the management of data flow between the application layer and physical communication medium. MMS is an ISO 9506 protocol for uniform communication handling regardless of network type and connected device. Some of the important services are *reading and writing, application downloads, alarm handling, safe peer-to-peer communication,* and *creating, deleting, starting, and stopping of a program.*
   - **Transactions and frames:** "Elementary transaction" refers to the basic unit of data exchange involving the transmission of a producer-consumer pair of *frames* "ID_DAT" followed by "RP_DAT," including details for *turnaround and silence time.* All these are shown in Fig. III 8.2d2. The buffer transfer mechanism uses these two types of frames, which are crucial for real-time communication. For scheduled transactions, these are analyzed to ensure timely data delivery. Every transmitted WorldFIP frame is sandwiched between information from the physical layer. The frame is placed *between* the beginning and end of a frame. WorldFIP frames begin with a control byte used by subscribers to

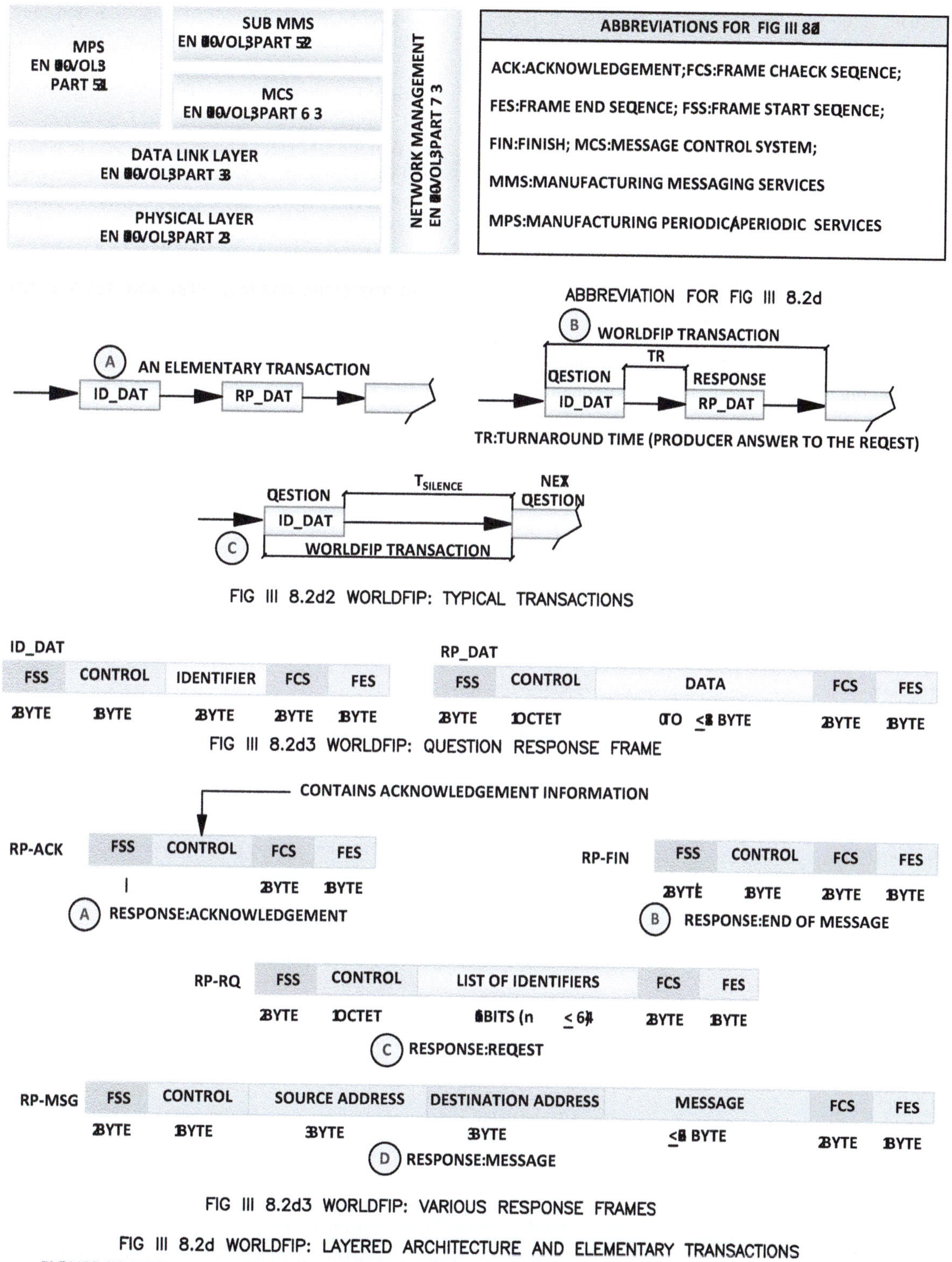


**FIGURE III 8.2D** World factory instrumentation protocol (WorldFIP): layered architecture and elementary transaction.

recognize the type of frame received, that is, "requests", "acknowledgment frames", and others. All WorldFIP frames end with a 2-byte FCS used by the frame receiver to verify the integrity of the frame received. All question frames, be they ID-RQ (request) or ID-MSG (message) frames, are ID_DAT type, as detailed in Fig. III 8.2d3. When there will be any aperiodic variable or message transfer requests these will be response request transfer (RP_RQ), response message transfer (RP_MSG), and response acknowledgment transfer (RP_ACK), including the end-of-message transaction response frame (RP_FIN). All these are detailed in Fig. III 8.2d4.

- **Important transaction issues:** Important issues/key features of WorldFIP transactional issues have been elaborated in Table III 8.2c.
- **Variable exchanges:** The MAC for WorldFIP is based on centrally controlled access and with producer-distributor-consumer (PDC) as the underlying communication model. A master node, that is, BA, starts all the transactions, sequentially, by transmitting a frame with the identification of a given variable (distributor role), as already elaborated. The BAT is organized as a sequence of elementary cycles (ECs) to guide and support the orderly release of the transactions. In WorldFIP there can be one or several BA(s). The traffic cyclic pattern is the periodic nature of the exchanges and is equal to the mathematical least common multiple of the variable periods, referred to as a MC as discussed. The WorldFIP has several modes of operation in what concerns the synchronization of the ECs and MC.
- **WorldFIP cycle:** The WorldFIP cycle is detailed in Fig. III 8.2e. In a WorldFIP there can be three types of elements viz. message, periodic, and aperiodic variables. The above figure explains BAT, MC, and EC in a lucid way and does not need any further explanation, since it is self-explanatory.

  Let there be a 20-ms MC with a number of ECs. Fig. III 8.2e1 shows all three elements viz. message, periodic, and aperiodic. It also shows an MC with 4 ECs in which there are three PVs (A, B, C) with different periodicity. The BA concept has been detailed in Fig. III 8.2e2. Here, both EC and MC are also defined and marked.
- **Network configuration**: As shown in Fig. III 8.2e3, WorldFIP network topology is of a bus type in which a number of stations or nodes are connected

**TABLE III 8.2C Important issues related to World factory instrumentation protocol transactions.**

| Feature | Definition/Explanation |
|---|---|
| Producer-distributor-consumer (PDC) | World factory instrumentation protocol (WorldFIP) utilizes PDC model where single producer provides data for one or more consumers. Viz. *one-to-one, one-to-many, one-to-all (broadcast).* |
| Bus arbiter (BA) | As a centralized component, BA manages access to the communication bus and controls all data transfers between different stations. |
| BA table (BAT) | BAT is a configuration table that dictates the scheduling of data transfers, specifically for buffer transfers. |
| Elementary cycle (EC) | A fundamental unit of time in the communication cycle. It represents a specific time slot on the bus where data exchanges can take place. |
| Macrocycle (MC) | The overall cyclic pattern of traffic on the WorldFIP network. It is basically the repetition of a sequence of ECs. |
| Three data types | *Cyclic variables* are always transmitted for time-critical closed loop control, *event variables* are sent on demand/change of state, and *Messages* **are** nontime critical, viz. diagnostics. |
| Schedulability | A guarantee that real-time data exchanges occur within specified deadlines is schedulability. Determined through characteristic timing analysis. |
| Schedulability analysis | Schedulability analysis helps to determine the network capability to handle the required data exchanges without violating real-time constraints. |
| Timing | The duration of an elementary transaction is determined by the transmission timing of both frames including turnaround time. |
| Synchronous mode | In synchronous mode, elementary cycles have a fixed duration, and transactions may be delayed for idle time. |
| Asynchronous mode | A BAT scan can be triggered by an external event, independent of the termination of the current scan. |
| Synchronous transmission | This is characterized by the transmission of individual bytes/characters with start and stop bits to delineate each unit of data. |

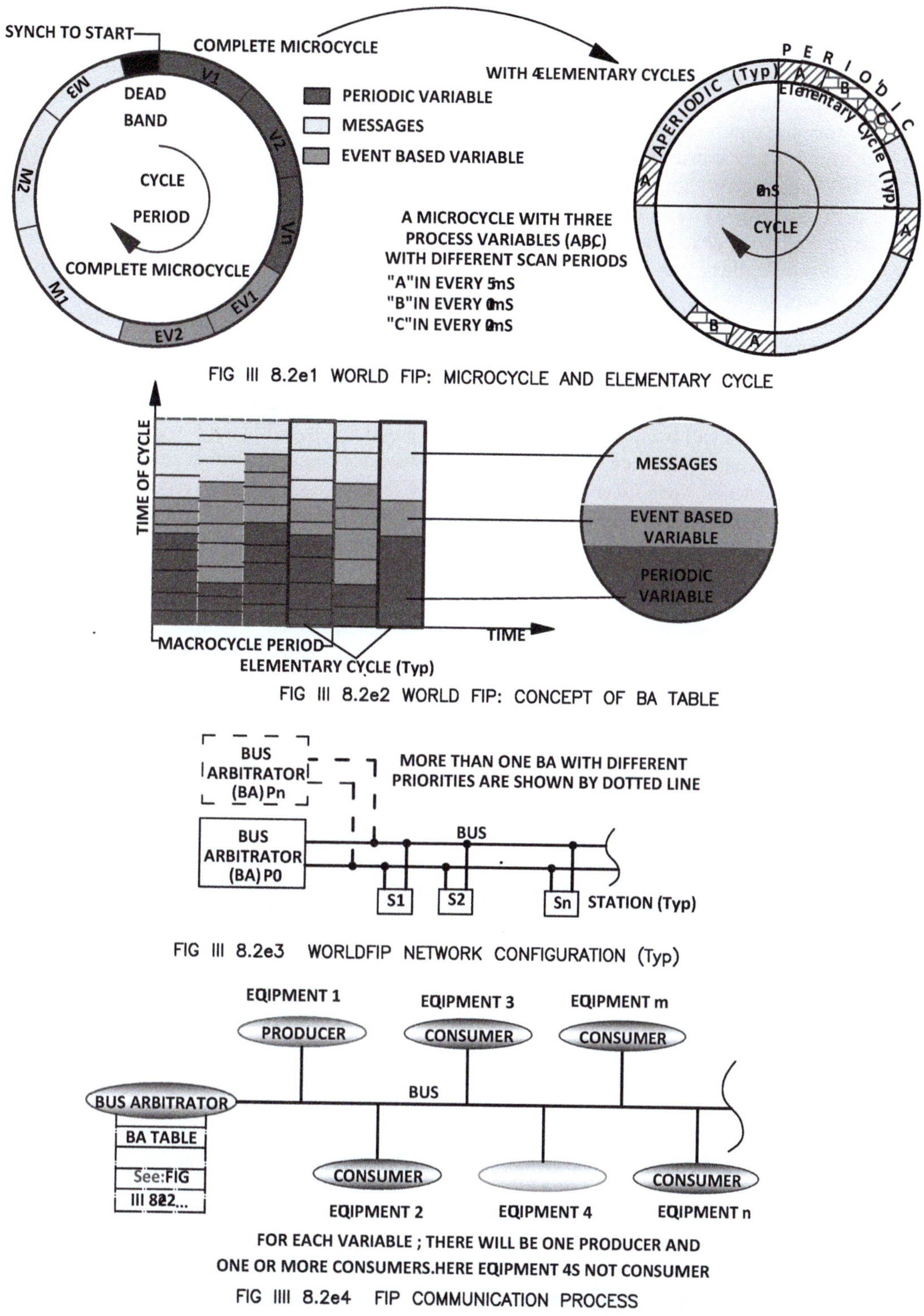


FIG III 8.2e WORLDFIP CYCLE EXPLANATION

**FIGURE III 8.2E** World factory instrumentation (WorldFIP) protocol cycle explanation.

via the network bus. In WorldFIP there are two types of nodes viz. BA (also referred to as the centralized scheduler) and the others are stations/ nodes. The BA manages and controls the accessibility to the transmission medium including the decision on which node transmits at any specific time. The BAT residing in the BA (Table III 8.2c) scans the periodic variables at exact time intervals, as detailed in Fig. III 8.2e2. Normally, there will be one BA; however, there can be more than one BA with different priorities, as shown in Fig. III 8.2e3 by dotted lines. Other than BA, all other nodes are producer or consumer(s) nodes. At any point of time, any of these remaining nodes can be *either producer or consumer nodes.* At times a node can be both types. Based on a call from BA, when a producer sends data consumers (s) listen to it.

- **Communication process: producer consumer model:** In WorldFIP, the PDC model relates producers and consumers in the distributed network system, which is also shown in Fig. III 8.2f3. Corresponding to each PV there is one and only one producer and one or several consumers. For a process *sensor*, the station that provides the variable value will act as the *producer* and its value will be provided to all the concerned consumers of the variable. This is shown in Fig. III 8.2e4, where equipment 1 is the producer and all others are consumers except equipment "4". For managing transactions associated with a single variable, a unique identifier is associated with each variable [9].

2. **Physical layer:** WorldFIP follows bus topology in its network architecture, that is, all devices are attached to the same bus. In this protocol, the transmission is balanced bipolar with Manchester encoding. Originally, there were three classes of speed: S1 (31.25 Kbps), S2 (1.0 Mbps), and S3 (2.5 Mbps). Accordingly, there were three conformance classes viz. hi (*CH*), medium (*CM*), and low (*CL*). Thus, *CH: S2 (1.0Mbits) or S3 (2.5Mbps); CM: S2 (1.0Mbits); CL: S1 (31.25 Kbps), or S2 (1.0Mbits).*
   However, as per FIP 98, there is another speed of 5.0 Mbps allowed. As physical medium twisted pairs (C46-604), optical fibers (C46-607) are used. The maximum lengths in copper and optical lines are as per Table III 8.2d.
   Each cable segment can accommodate a maximum of 64 nodes. With a repeater, which can connect up to four segments at the same time, more nodes can be accommodated. The cable segment length depends on the data bit rate, as shown in Table III 8.2d. Bus topology with media redundancy as well as repeaters has been detailed in Fig. III 8.2f, which also details media access.
   Basic bus topology with different media types is detailed in Fig. III 8.2f1. Media redundancy combing copper and fiber optics is detailed in Fig. III 8.2f2.
3. **Data link layer:** The WorldFIP data link layer is made up of a set of produced and consumed buffers, which can be locally (through application layer services) or remotely accessed (through network services) [9]. The WorldFIP protocol supports two types of transmission services. These services are *exchange of identified variables* and *message transfers.* Both time critical and non–time-critical data are handled here. WorldFIP mainly makes use of cyclic nature of exchange for real-time data, yet it could be acyclic or event based also. In this connection Figs. III 8.2d2–4 and III 8.2e1,2 may be referenced. Cyclic transfers are triggered by BA. Message transfers are of an aperiodic broadcasting type, which could be either an *unacknowledged* or *acknowledged* type. Variable types have been elaborated in Fig. III 8.2e1. These are:
   - **Cyclic variable:** These are always transmitted periodically for time-critical variables (e.g., closed-loop control).
   - **Event-based variable:** Event-based variables are transmitted on demand or change of state (e.g., for alarms or display).
   - **Message:** Messages are nontime critical (e.g., for maintenance, configuration, and diagnostics).

     The media access method is detailed in Fig. III 8.2f3, where the step-by-step accessing details have been detailed.
4. **Brief addressing types:** There are two types of exchanges, hence two types of addressing also viz. variable and message addressing.
   - **Variable addressing:** Each variable in the system is associated with a unique identifier to characterize that variable. Since there are 16 bits to encode this address, $2^{16} = 65536$ variables can be identified. These are used for both periodic and aperiodic variables. Addresses are related to physical location (producer and consumer[s]) and used to know the identifier of the variables.
   - **Message addressing:** Message exchanges could be point to point and/or multipoint on a single segment. Each message contains a 24-bit address including the address of the source and destination station. So, $2^{24} = 16777216$ addresses are possible for the messages. Addresses are divided into a segment number (8 bits) and 16 bits for local service access point.
5. **FIPIO details:** FIPIO is the broader version of WorldFIP standard and is popular for deterministic, real-time data exchange capabilities. Thus, FIPIO is suitable for industrial automation and process control applications calling for reliable real-time communication with precise timing. Like WorldFIP, FIPIO is based on bus

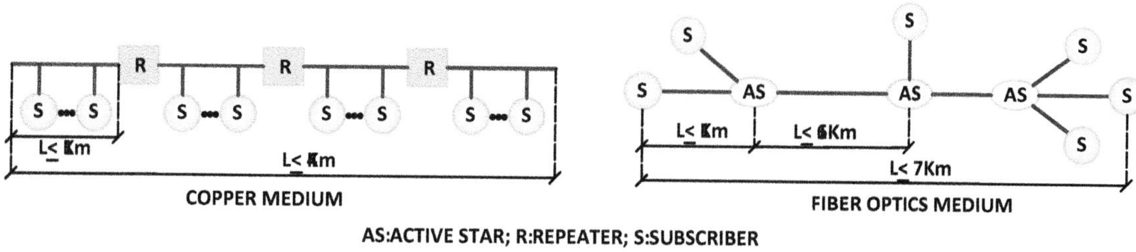


FIG III 8.2f1 BUS TOPOLOGY IN DIFFERENT MEDIA

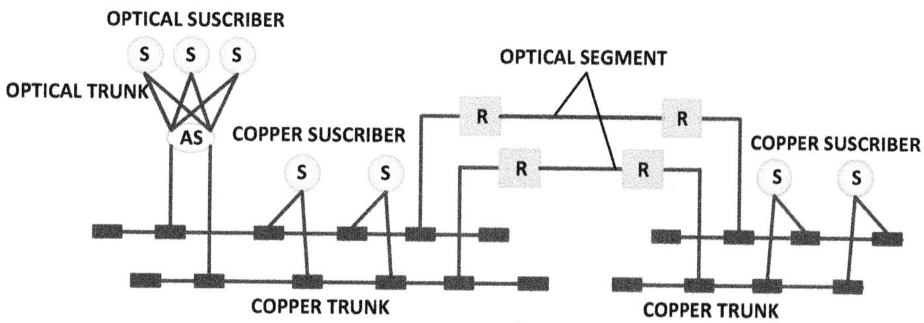


FIG III 8.2f2 BUS TOPOLOGY WITH MEDIA REDUMDANCY

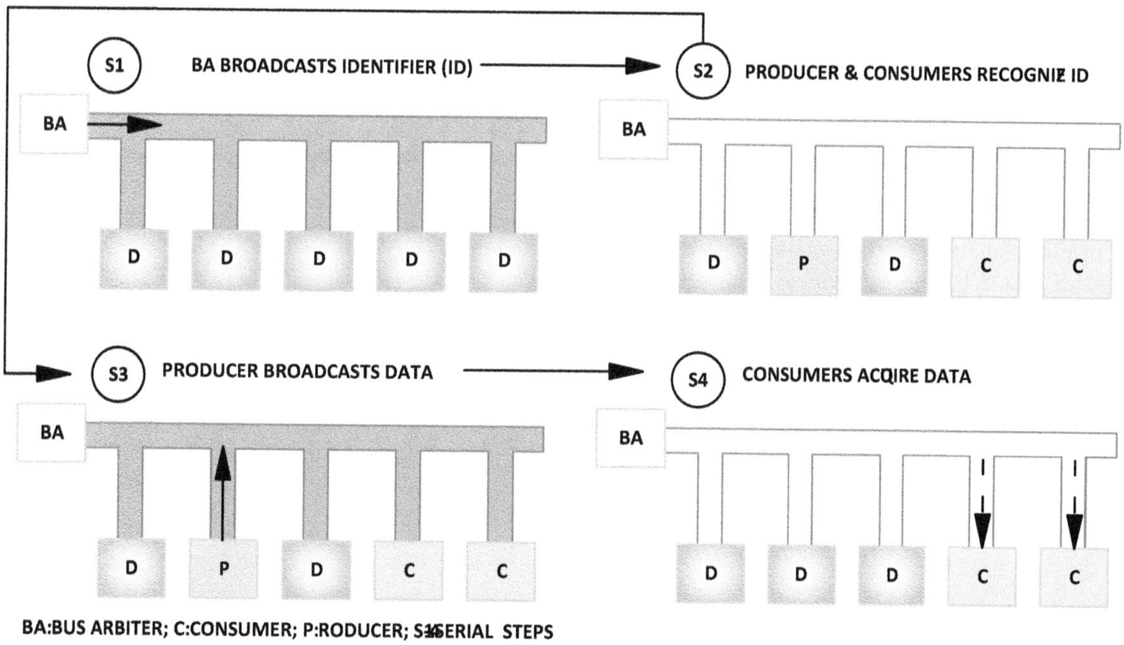


FIG III 8.2f3 WORLDFIP: MEDIA ACCESS DETAILS

**FIGURE III 8.2F** Bus topology and media access.

**TABLE III 8.2D** Maximum allowed length in World factory instrumentation protocol for different media.

| Medium | 31.25 Kbps | 1.0 Mbps | 2.5 Mbps | 5 Mbps |
|---|---|---|---|---|
| Copper | 5-20 km | 1-4 km | 0.5–1.5 km | 0.3–0.7 km |
| Optical fiber | 40 km | 40 km | 40 km | 40 km |

topology and deploys a producer/consumer model for efficient data exchanges. This protocol is not only compatible with other WorldFIP devices but also the protocol can handle errors and maintain communication integrity in challenging and harsh industrial environments. FIPIO can be connected to I/O devices located at up to 15 km. Some technical details are shown in Table III 8.2e.

*Let us conclude the discussion on WorldFIP here to move on to discuss* SERCOS.

### 8.2.2 Discussion on communication over serial real-time communications system

SERCOS has been a leading bus system for industrial automation for over 2 decades. It is an IEC-standardized open digital interface for high-speed, real-time communications between industrial controls, motion devices, I/O, other peripheral devices, and standard Ethernet nodes. SERCOS can be characterized by *easy*, *fast*, *open*, and proven technology. The concept of the SERCOS communication fieldbus system is depicted in Fig. III 8.2g.

SERCOS is a real-time technology with proven quality, universal application possibilities, and high security of investments. All these make the SERCOS automation bus the first choice in *mechanical engineering* and *construction*. An *efficient* and *deterministic* communication protocol based on an optical transmission system for high noise immunity is the foundation for SERCOS' success. SERCOS is in use in the most varied market sectors and applications. SERCOS has become the de facto standard for challenging applications demanding dynamics and precision. Also, SERCOS specifies as many as 700 standardized parameters that describe the interaction between control systems, drives, and other peripheral devices using universal semantics. Without additional hardware, SERCOS devices can be integrated with other Ethernet devices. A logic system integrated into each SERCOS device allows the direct connection of Ethernet devices, as shown in Fig. III 8.2g1 (LHS). The integrated switch monitors compliance with a defined period of time so that the real-time behavior of the SERCOS network is not negatively hindered. This process also allows an external switch, as shown in Fig. III 8.2g1 RHS. Fig. III 8.2g2 demonstrates the use of motion control, drives, and the easy combination of these with other Ethernet devices. The basic communication cycle comprises a real-time (SERCOS) cycle (RTC) and combination with a unified communication channel (UCC), as shown in Fig. III 8.2g3, which has been detailed to clearly show the communication cycle. In this subsection, details on SERCOS (III) have been discussed, including technology and applications.

1. **Benefits, features, and application:** Three important issues are:
   - **Benefit of SERCOS III:** SERCOS III offers a number of benefits, as listed below:
     - *Fast Ethernet:* A recognized cost-effective, fast Ethernet protocol supports direct communication between the peripherals and controller.
     - *Versatility*: With an open Ethernet protocol using any IP, the SERCOS system shows versatility for consistent peripheral, drive, and safety communication. It also offers improved information technology (IT) integration.
     - *Safety:* Common Industrial Protocol (CIP) safety on SERCOS supports IEC 61508-compliant safety functions up to SIL 3.
     - *High performance:* SERCOS III provides high bandwidth and synchronization, enabling faster update rates and precise control.
     - *Energy efficiency*: SERCOS supports intelligent energy consumption control, leading to significant energy savings.
     - *Extensive diagnostics*: The synchronized data and cross-communication (CC) offers efficient network diagnostics and troubleshooting.

**TABLE III 8.2E Important technical details of factory instrumentation protocol input/output.**

| Features | Specification |
|---|---|
| Network size | Up to 256 nodes in max 15 segments |
| Data rate | 31.25 kbps, 1 and 2.5 Mbps, 6 Mbps (optical fiber bus) |
| Max network length | <40 km |
| Media | Shielded twisted pair/fiber optics |
| Addressing | Remote I/O |
| Remote input/output (I/O) | Discrete and analog I/O in IP20 and IP65/67 module |
| Communication | Bus topology; peer to peer |

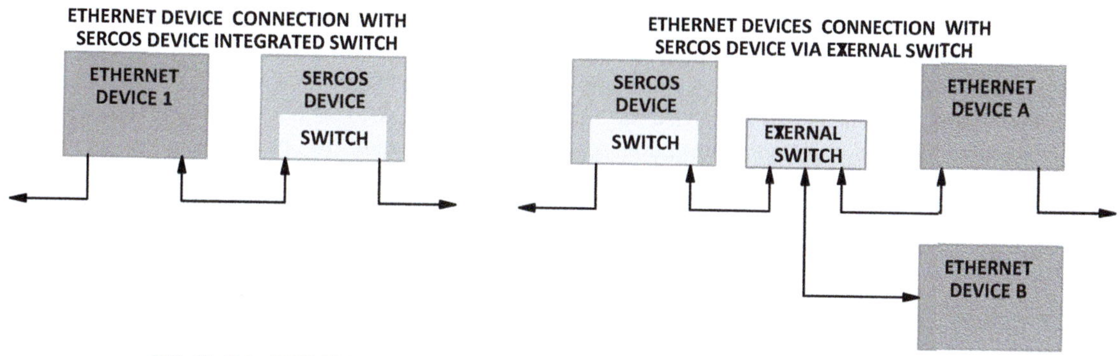


FIG III 8.2g1SERCOS DEVICE CONNECTION TYPES WITH ETHRNET DEVICE(S)

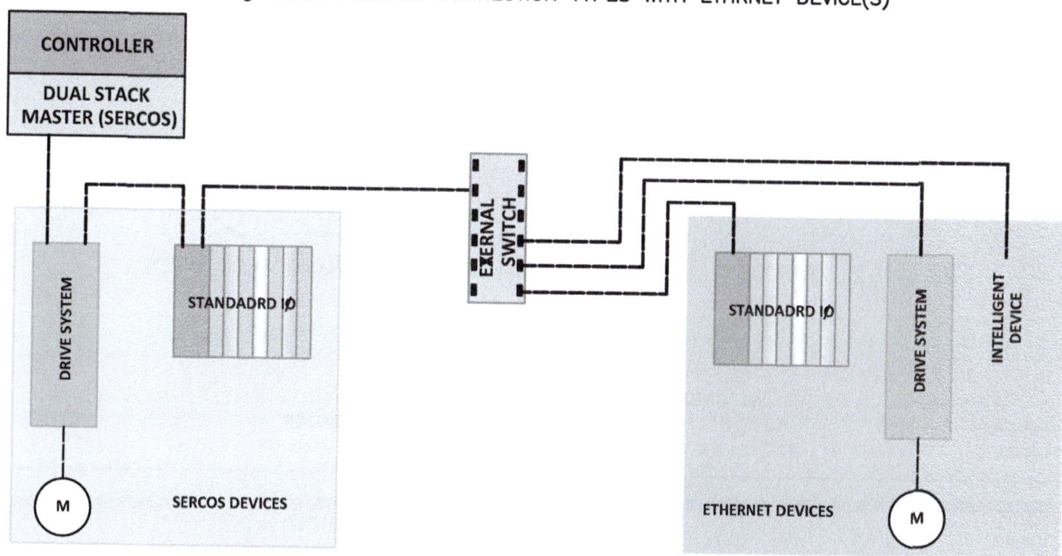


FIG III 8.2g2 SERCOS DEVICE TYPES CONNECTION WITH ETHERNET DEVICE TYPES

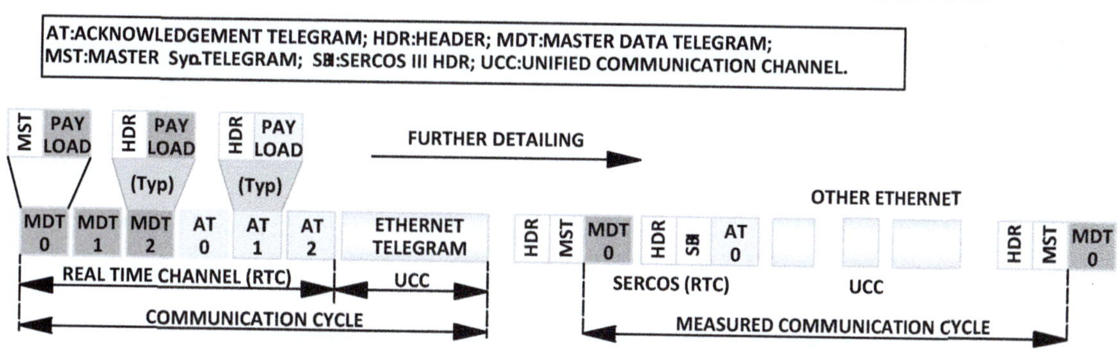


FIG III 8.2g3 BASIC COMMUNICATION CYCLE CONCEPT

**FIGURE III 8.2G** Serial real-time communications system (SERCOS) fieldbus operation concept.

- *Fool proof:* Support to open international standards and backward compatibility makes it fool proof.
- *Vendor neutrality*: It is supported by numerous manufacturers, offering flexibility and interoperability.

- **Main features:** A few pertinent features are as follows:
  - *Standardized Ethernet:* Easy to use, flexible, cost-effective IE technology, it is a standardized solution defined in IEC 61158, 61784, and 61800-7
  - *IE with media redundancy*: SERCOS III utilizes IE, offering high-speed deterministic communication crucial for motion control applications. Also, it offers media redundancy for reliable communication.

- *Universal automation bus*: In the SERCOS III universal automation bus, communications among standardized profiles for drive technology, I/O, controllers, and others can be carried out. It also supports point-to-point communication for both controller-to-controller (C2C) and direct cross-communication (DCC) among slave devices (Fig. III 8.2h1).

Also, it allows the parallel transmission of standard IP protocols over a non–real-time

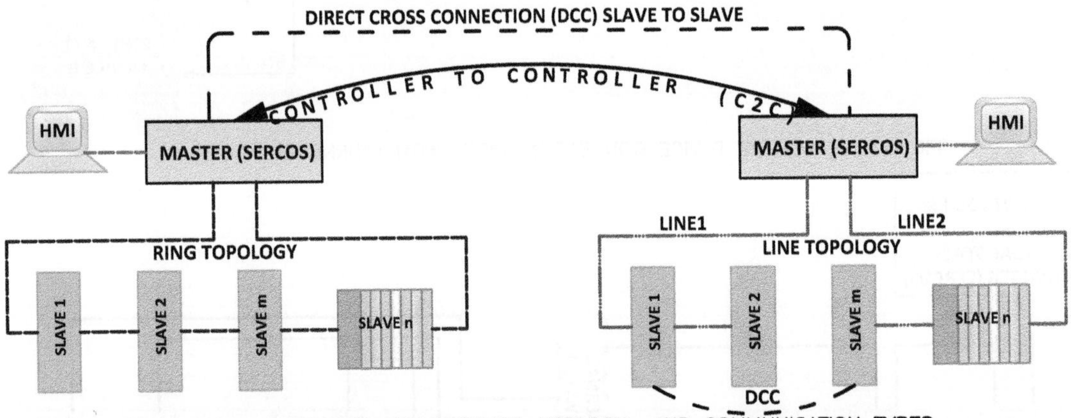


FIG III 8.2h1 SERCOS III ARCHITECTURE: TOPLOGY AND COMMUNICATION TYPES

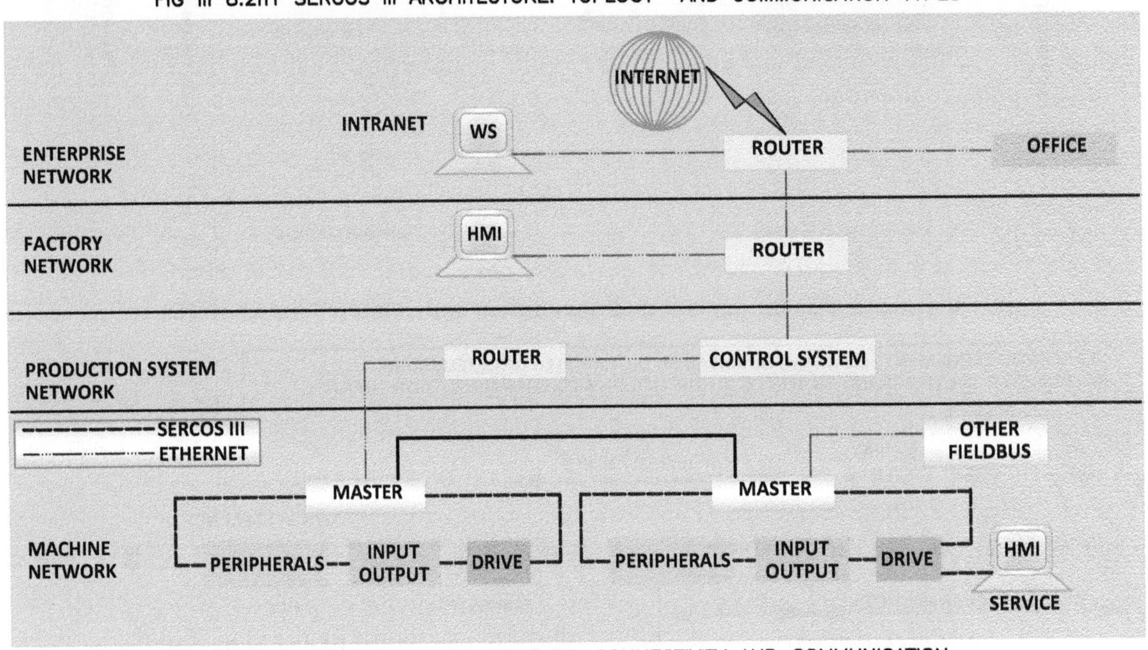


FIG III 8.2h2 SERCOS III ARCHITECTURE: CONNECTIVITY AND COMMUNICATION

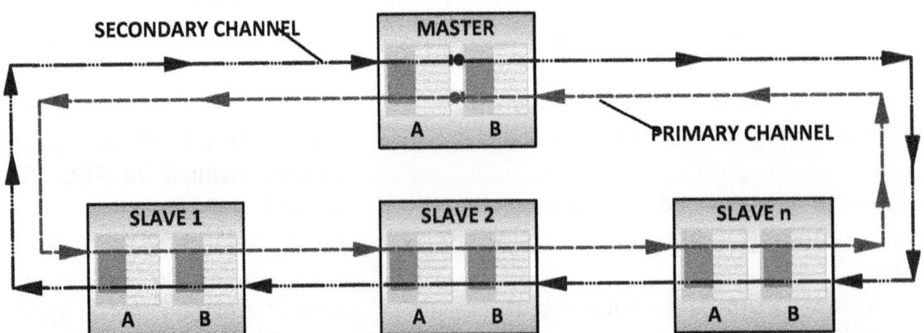


FIG III 8.2h3 SERCOS III ARCHITECTURE: RING TOPOLOGY WITH REDUNDANCY

**FIGURE III 8.2H** Serial real-time communication system (SERCOS) automation architecture and communication.

channel, alongside its real-time communication. This also facilitates the seamless integration of safety automation functions.

- *Optimal bandwidth utilization*: All device user's data up to 1494 bytes is bundled together with 44 bytes of overhead data. With this packet size, the available bandwidth for *productive data* increases over 90%, and deterministic data traffic is obtained without collisions.
- *Online plugin*: Devices can be added and configured during active cyclic communication without disrupting the network.
- *Increased precision*: Real-time precision demands accurate signal timing and execution, which is well ensured by SERCOS III cyclic communication. Unique hardware synchronization combined with least jitter (±10 ns) ensures maximum real-time precision.
- *Telegram optimization*: The number and length of real-time channels are adjusted automatically to cope with the data volume and number of nodes registered. Thus, automatic optimization of telegram optimizes efficiencies. For further details, refer to [10].

- **Applications:** Common application areas mainly include, but are not limited to factory automation, building automation, process control, sensors, actuator and other field devices communication, digital/analog I/O modules, drive controls, and PLC.

2. **Operational details:** In this part, both operational and transmission principles are covered. Let us start with operation principles.
   - **Operating principles:** SERCOS is a high-performance protocol that operates on a master-slave principle utilizing a cyclic communication (a time slot–based approach) philosophy with reserved time for real-time communication. This is a combination of the openness of standard Ethernet with the real-time accuracy required to cater to the demands of industrial automation. SERCOS III uses a dual channel. One channel *is dedicated to real-time communication* for SERCOS-defined telegrams. The other is the UCC for handling other Ethernet traffic (i.e., IP-based protocols). For details, refer to Fig. III 8.2h.

     SERCOS supports decentralized intelligence and an unlimited capacity to work in real time through direct communication between devices. SERCOS slaves can communicate via CC directly and with a minimum of communication dead times to ensure unlimited real-time communication and intelligent automation structures. When the data are directly communicated between the slaves in DCC (/CC) within an individual communication cycle, there will be minimum delay, even with increased cycle times. Controllers also use the same principle for direct communication. These are detailed in Fig. III 8.2h1. All real-time data are synchronized (i.e., in relation to the general communication cycle). This makes data processing extremely easy, efficient, and flexible at individual network nodes. Thus, the decentralized automation solutions can be installed with minimal reaction times and a great amount of flexibility.
   - **Transmission principles:** SERCOS communication is based on a time slot process with a cyclical transmission of telegrams based on a master-slave principle. The communication cycle with varied *cycle times* (viz. 31.25 μs, 62.5 μs, 125 μs, and multiples of 250 μs up to a maximum value of 65 ms) is divided into two channels. These are *a real-time channel for SERCOS telegrams* and a UCC for other Ethernet traffic (IP-based protocols such as TCP/IP and UDP/IP can be transmitted). With this, deterministic, real-time communication and non–real-time data occur within the same network. Key issues associated with SERCOS transmission are *cyclic communication with real-time channels, UCC, time slots, full duplex communication,* and *redundancies.* Decentralized automation solutions are also created as a result of this bandwidth of cycle times.
3. **Architectural discussions:** Master and slave devices can be realized either in line or ring topology and the Ethernet cable can be connected to any port on a master or slave device. *Network redundancy* is only supported in ring topology, where the master sends out each frame *twice*, once each over the *primary* and *secondary* channels. SERCOS III combines a deterministic RTC and UCC, as discussed and shown in Fig. III 8.2g3. The overall processing delay in a slave from incoming port to outgoing port is ~1 μs constant. So, the frame round-trip delay in a network with 10 slaves is 10 μs in ring topology and 20 μs in line topology [11]. Its *flexible automation system, 10 to 100 times faster data throughput, less expensive cost, proprietary technology, standard components, global support,* and *consistent IT implementation* are some of the major reasons that fast Ethernet technology is considered a future-proof technology. This can be based around SERCOS III, as detailed in Fig. III 8.2h2. Major issues here are:
   - **Service channel (SVC):** This noncyclic channel is used by the master to exchange non–real-time data with slaves. It is primarily used for reading and writing SERCOS parameters (identification numbers) of a slave and for executing procedure commands. The master data telegram (MDT) and AT frames also embed an asynchronous communication channel used by the master to exchange diagnostic data.

- **Topology and redundancy**: As discussed, network redundancy is supported in ring topology, as detailed in Fig. III 8.2h3 with the channel identified. The primary port of the master is connected to the first slave and the secondary port is connected to the last slave. The master transmits SERCOS III frames simultaneously on both ports. In case of an Ethernet cable break (ring break), the slave that detects the break immediately starts the loopback mode. The slave sends the MDT/AT frames back on the same port where the frames were received. The master detects the ring break scenario in the status information of the AT frames. After the ring break is physically resolved, the master issues a ring heal command to the slaves to restore the ring topology connection [11].
- **RTC**: There are a number of data exchanges, as detailed in SERCOS III. *Master-slave* data exchange represents functional data between the master and slave. As discussed and shown in Fig. III 8.2h1, *CC/DCC* refers to direct peer-to-peer communication between devices (slave) via the master. C2C represents communication between controllers or drives and peripheral slaves under CC. This also accounts for the exchange of safety-related data on a *master-slave*/CC device channel.

4. **Communication protocol:** With SERCOS, real-time data are sent according to the IEEE 802.3 Ethernet protocol (type 0 × 88CD) standard in *cyclical telegrams* in master-slave, DCC, SVC, and safety modes. Fig. III 8.2i shows SERCOS frame and communication details. Let the discussion address a few pertinent issues related to protocol, which are listed below.

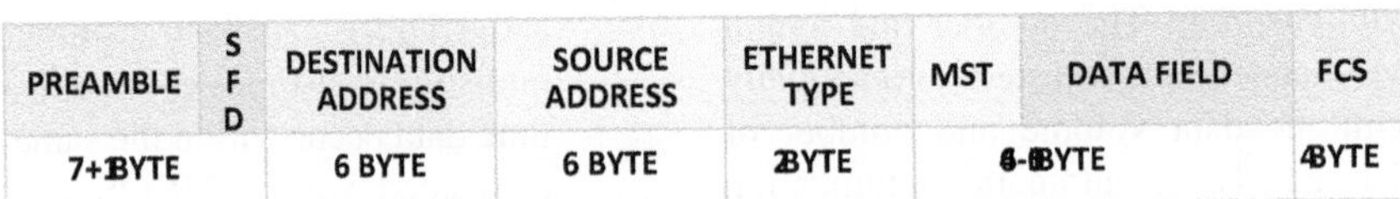


FIG III 8.2i1 SERCOS GENERAL FRAME STRUCTURE

FCS:FRAME CHECK SEQENCE; MST:MASTER SYNC.TELEGRAM; SFD:START OF FRAME DELIMITER

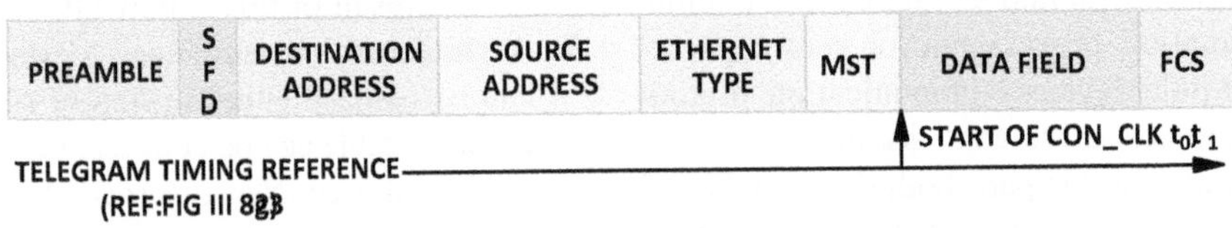


FIG III 8.2i2 SERCOS FRAME SYNCHRONIZATION WITH MDTO

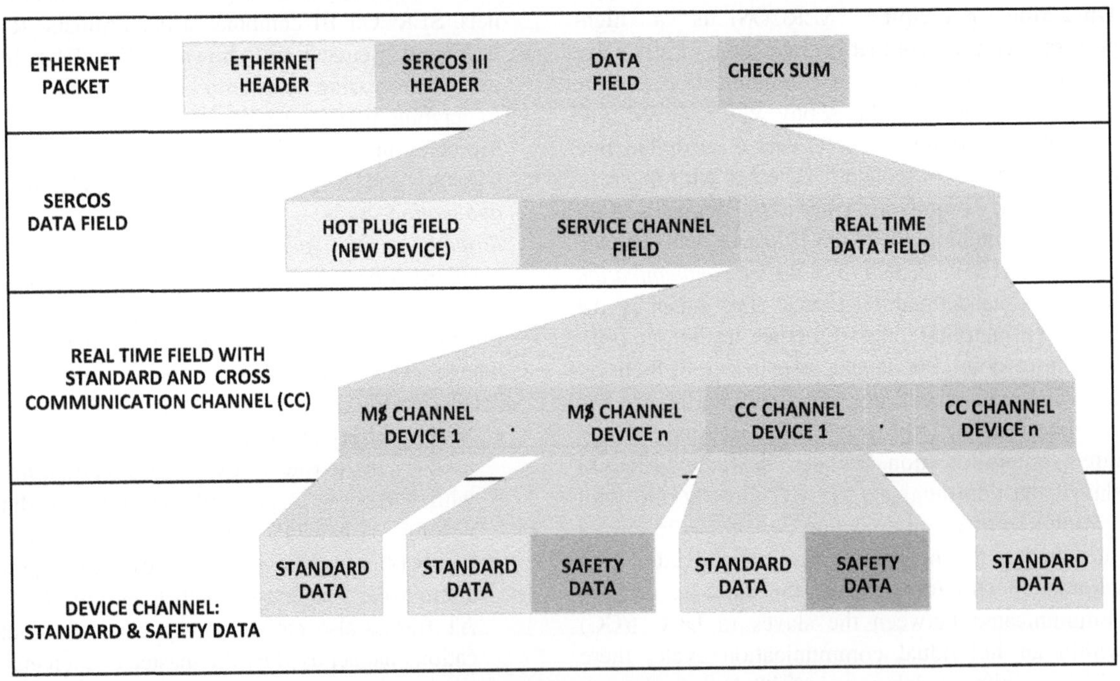


FIG III 8.2i3 SERCOS DATA COMMUNICATION DETAILS

**FIGURE III 8.21** Serial real-time communications system (SERCOS) frame and communication details.

- **General frame**: The master can only generate and handle SERCOS III MDT and AT Ethernet frames. A general frame structure has been shown in Fig. III 8.2i1. The MDT frame transfers data from the master to the slave while the AT frame transfers data from the slaves to the master and other slaves. SERCOS III frames are broadcast frames. Each slave processes the frame by *taking or placing data* from the data field while forwarding the modified/unmodified content to the secondary port. The master *receives back* the modified frame; in line topology, the *last slave loops back the frame* and in ring topology *the frame is received on the master's secondary port*.
- **Frame synchronization**: This procedure has been detailed in Fig. III 8.2i2. The *MST field* in the MDT0 frame is used by the master for slave synchronization. The MST field has its own FCS. Each slave validates the MST FCS and uses the MST time reference as an internal synchronization event.
- **Protocol structure:** SERCOS telegrams contain a SERCOS header and a data field embedded in the Ethernet frame, as detailed in Fig. III 8.2i3. The SERCOS header describes which *phase the network* is in and the position of the MDT and AT telegrams in the communication cycle. As shown, the three areas are *hot plug (*exchanges data with slaves that have been added to the network while the operation is running or will have to be plugged in again), *SVC* (total number of communication channels that exchange acyclic data between master and slaves), and *real-time data* (to create acyclic, cyclic, or clock-synchronous connections, plus real-time communication between any devices in the SERCOS network). This clear, robust structure helps in application development.
- **Connectivity:** As shown in Fig. III 8.2i3, the vertical communication in the context of industry 4.0 and IoT captures more and more at the sensor/actuator level. With this, the system will become more and more intelligent for easy implementation of decentralized automation concepts.
- **SERCOS with** OPC UA: OPC UA plays a very important role in advancing the convergence of IT with operation technology, thus enabling the consistent exchange of information between the different control levels. OPC UA provides a secure, platform-independent communication framework for industrial automation, whereas SERCOS is a fieldbus standard optimized for motion control and real-time applications. The OPC UA companion specification describes details on a domain-specific SERCOS information model for OPC UA, making the functions and data provided by SERCOS devices accessible via OPC UA.

5. **Miscellaneous issues:** SERCOS III is the open, IEC-compliant system used for providing real-time IE. It overcomes the wasted bandwidth of some other TCP/IP-based Ethernet solutions. It is based on Ethernet frames, defining a new type for SERCOS. A few pertinent issues here are as follows:
   - **Safety issue:** SERCOS encompasses CIP safety. CIP safety on SERCOS means *the CIP safety standard is used as the functional safety protocol, CIP safety is adapted to SERCOS* as shown in Fig. III 8.2i3, and there are *SERCOS-specific safety profiles.*
   - **Implementation:** Some of the important issues related to implementation are:
     - *Controller and communication modules*: A large range of hardware options are available including Field-programmable gate array (FPGA) components, application-specific integrated circuits and microprocessors with integrated SERCOS functionality for controller and communication modules, and SERCOS fieldbus implementation on FPGA for real-time control of metal laser 3D printing. Dedicated hardware based on a Xilinx FPGA was developed in house for the real-time control of melt pool temperature and track width.
     - *Slaves*: On the slave side, specific hardware is always used to handle the fast telegram processing, synchronization, topology switching, direct linking of any Ethernet node, hot plugging, and DCC.
     - *Driver software*: Driver software is used to connect the controller to the device logic. Some of the open-source software available are *common SERCOS master API (CoSeMa), SERCOS IP services, SERCOS UCC Ethernet network driver,* and the *SERCOS III Softmaster.*

## List of abbreviations used

**ADU** Application data unit
**ASCII** American serial code for information interchange
**AT** Acknowledge telegram
**ATM** Asynchronous transfer mode
**BA** Bus arbiter
**BAT** BA table
**BRAIN** Broadcast Rapid Access Intelligent Network
**CAN** Controller area network
**CC** Cross-communication
**CIP** Common Industrial Protocol
**CMD** Configuration, monitoring, and diagnostics
**COB** Communication objective
**CRC** Cyclic redundancy check
**CRLF** Carriage return and line feed

**DCC** Direct cross-communication
**DD** Device description
**DLC** Data length code
**DLL** Data link layer
**EC** Elementary cycle
**ECU** Electronic control unit
**EDDL** Electronic device description language
**EMC** Electromagnetic compatibility
**EOF** End of frame
**FB (D)** Functional block (diagram)
**FCS** Frame check sequence
**FES** Frame end sequence
**FSK** Frequency shift keying
**FSS** Frame start sequence
**HART** Highway Addressable Remote Transducer
**HDLC** High-level data link control
**HDR** Header
**HIM** HART interface module
**I/O** Input/output
**IE** Industrial Ethernet
**IESMAP** IE safety master protocol
**IESSLP** IES slave protocol
**IoT** Internet of things
**IP** Internet protocol
**IT** Information technology
**LAN** Local area network
**LRC** Longitudinal redundancy check
**LSAP** Local service access point
**LSB** Least significant bit
**M/S** Master-slave
**MAC** Media access controller
**MBAP** Modbus Application Protocol
**MC** Macrocycle
**MCS** Message control system
**MDT** Master data telegram
**MMS** Multimedia messaging service
**MPS** Manufacturing periodic/aperiodic services
**MST** Master sync telegram
**NMT (protocol)** Network management (protocol)
**OFDM** Orthogonal frequency-division multiplexing
**PCP** Peripheral communication protocol
**PDC** Producer-distributor-consumer
**PDO** Process data object
**PDU** Protocol data unit
**PID** Proportional integral and derivative
**PLC** Programmable logic controller
**PMS** Peripheral message specification
**PNM** Peripheral network management
**PV** Process variable
**RF** Radio frequency
**RTC** Real-time (Ethernet) channel
**RTR** Remote transmission request
**RTU** Remote transmission unit
**SDO** Service data object
**SIL** Safety integrity level
**SLMP** Seamless message protocol
**SOF** Start of frame
**STP** Shielded twisted pair
**SVC** Service channel
**TCP** Transmission control protocol
**TDMA** Time division multiple access
**TSN** Time-sensitive network
**UART** Universal asynchronous receiver/transmitter
**UCC** Unified communication channel
**UDP** User Datagram Protocol
**UTP** Untwisted twisted pair

## References


1. Joseph Wu; *A Basic Guide to the HART Protocol— Application Note*; Texas Instruments; Internet Document; https://www.ti.com/lit/an/slaaeh0/slaaeh0.pdf?ts=1749532554022&ref_url=https%253A%252F%252Fwww.google.com%252F.
2. S. Basu, A.K. Debnath, in: Power Plant Instrumentation and Control Handbook, 2nd rev, Elsevier; Academic Press, 2019. https://www.elsevier.com/books/power-plant-instrumentation-and-control-handbook/basu/978-0-08-102804-9.
3. Samsons AG; *Technical Information; HART Communications; Part 4*; Internet document; https://www.samsongroup.com/document/l452en.pdf.
4. *Bridging the Gap between HART Devices and IIoT, the Industrial Internet of Things; Smart Industry- Technology Report*; Moor Industries; White paper; https://www.miinet.com/images/pdf/articles/smartindustry-2019-02-techreport-bridging-hart-iiot.pdf.
5. W.L. Mostia, in: Introduction to Modbus, Control Global, 2019. www.controlglobal.com/articles/2019/introduction-to-modbus?utm_source=hs_email&utm_medium=email&utm_content=68965126&_hsenc=p.
6. *MODBUS Messaging on TCP/IP Implementation V1.0b; Standard Guide*; Modbus Organization; https://modbus.org/docs/Modbus_Messaging_Implementation_Guide_V1_0b.pdf.
7. *Interbus Basics; Interbus Club*; Internet document; http://www.interbus.de/dl/Dok_interbus_basics_en.pdf.
8. *AS-Interface - Introduction and Basics; C79000-G8976-C089-04 ; Simatic Net*; Siemens. Com; Internet document; https://cache.industry.siemens.com/dl/files/856/1171856/att_961903/v1/SYH_asi-grundlagen_76_201805231008141616.pdf.
9. E. ovar, F. Vasquest, L.M. Pinho, in: Engineering Real Time Applications with World FIP: Analysis and Tools, IFAC Intelligent Components and Instruments for Control Applications, Buenos Aires, Argentina, 2000. https://www.sciencedirect.com/science/article/pii/S1474667017393461.
10. *SERCOS III Real-time Communication with Ethernet*; DBR Automation: Rexroth Bosch Group; Internet document; https://www.dbrautomation.com/catalogs/Control/catalogo_sercosIII.pdf.
11. *Sercos III Communication Development Platform; TI Designs: White Paper*; Texas Instruments; https://www.ti.com/lit/ug/tidu534a/tidu534a.pdf?ts=1720279005321&ref_url=https%253A%252F%252Fwww.google.com%252F.

Chapter 9

# Miscellaneous communication protocols and issues

The International Standards Organization's (ISO) Open Systems Interconnection (OSI) model was introduced in 1979. The very next year, Digital Equipment Corporation (DEC), Intel, and Xerox (DIX) released the first Ethernet specification, version 1.0, under the name *Ethernet Blue Book*/DIX standard, which mainly defined *Thick Ethernet* in the case of 10 Mbps carrier sense multiple access with collision detection (CSMA/CD). Ethernet is only a specification of layers **1** and **2** in the OSI model. It is not a complete network protocol but a subnet on which other protocols such as the Transmission Control Protocol (TCP)/Internet Protocol (IP) suite can work. The most important functions of Ethernet are filling in of the physical layer, *sending and receiving* the serial bit streams over the medium, and *detecting* collisions.

Filling in of the data link layer: Here, there are two parts: the media access control (MAC) sublayer for the *access mechanism* to the network (CSMA/CD) and building of the data frames and the logical link control (LLC) sublayer for data reliability and supply data channels for higher-level applications. Ethernet implementation at the physical layer can be either two-wire or four-wire. The two-wire implementation can be (say) 100BASE-TX for copper wire and 100BASE-FX for fiber optic cable. There can also be thick and thin coaxial cable. These are all discussed in detail in Chapter 16 of this volume. Ethernet Powerlink is a real-time protocol for standard Ethernet. This open protocol is managed by the Australian company B&R. It extends Ethernet (Institute of Electrical and Electronics Engineers [IEEE] 802.3) with mechanisms to transfer data with predictable timing and precise synchronization. Powerlink offers suitable mechanisms, including data exchange, which is based on a publish/subscribe relationship that is completed in precise isochronous cycles that are effective for the automation industry, and device synchronization, which is highly accurate and transmits less time-critical data asynchronously on IP-based protocols like TCP or User Datagram Protocol (UDP) and higher layer protocols like Hypertext Transfer Protocol (HTTP) and File Transfer Protocol. Dynamic Data Exchange (DDE) is a Microsoft software meant to exchange data between two Microsoft products. Open Platform Communications (OPC) is a software that implements the OPC standard and thus provides the standardized OPC interfaces to the outside world. The OPC server is the basis of OPC communication. ControlNet utilizes the Control and Information Protocol (CIP) for the top three upper layers of the ISO-OSI model. The designers of ControlNet adapted the CIP layers on transport, network, data link, and physical to fit the needs of the ControlNet network. DeviceNet is a combination of the CIP protocol and the controller area network (CAN) physical layer. In CIP, every network device represents itself as a series of objects, each of which is simply a grouping of the related data values in a device. Fiber Distributed Data Interface (FDDI) is a network standard that uses fiber optic connections in a local area network (LAN) that can extend the range up to 200 km. Two token rings in an FDDI network are the primary and secondary rings (redundant backup), one operating in a clockwise and the other in an anticlockwise direction. Individually they operate at up to 100 Mbps capacity; hence the effective capacity is raised up to 200 Mbps. The American National Standards Institute (ANSI) product FDDI operates on an OSI model with a physical and MAC of layer 2. FDDI technology can also be used as a backbone for a wide area network (WAN). In the 1990s, with advancements in manufacturing systems, a sophisticated remote monitoring approach became attractive. With industrial remote monitoring, machine-connected sensors can share large volumes of real-time data, which are necessary for data analysis, plus operators enjoy the convenience of getting performance insights remotely from any device. Remote monitoring also helps maximize uptime while minimizing maintenance costs. Monitoring in industrial systems can be for used for manufacturing processes, maintenance, emission monitoring, and others. Field device integration (FDI) provides solution for the integration of field devices connected to a multitude of networks, operating systems (OSs), and industrial control systems. Today's field devices are more like embedded intelligent microcontroller-based devices deployed across many networks offering a broad range of data and are required for data analysis as part of the Industrie 4.0 approach. Electronic Device Description Language (EDDL) is a standardized language used to

**Plant Intelligent Automation and Digital Transformation. https://doi.org/10.1016/B978-0-323-89892-8.00002-8**

describe the capabilities and behavior of automation devices in a consistent way to represent device information regardless of the specific communication protocol or host system, allowing them to be integrated into various automation systems.

Industrial Ethernet (IE) is a specialized version of standard Ethernet that is designed to withstand harsh industrial environments. Although both use similar networking principles, IE features more ruggedized hardware, enhanced reliability, and real-time communication capabilities to cater to the needs of industrial automation.

*Confidentiality*, *integrity, reliability*, *availability*, *authenticity*, and *access control* are major issues in the *security* of industrial data communication. Therefore, data communication security refers to the measures and practices employed to protect sensitive information during its transmission. Cybersecurity, system integration, and migration are major challenging issues, and naturally the focus will be on these issues to address this situation in the future. Based on the above discussion, readers can see that this chapter is dedicated to miscellaneous communication protocols (details on Ethernet have been covered at length in Chapter 16 of this volume). Let the discussion begin with an introduction to the Ethernet family.

## 9.0 Miscellaneous protocol families

In the OSI model, Ethernet describes the bottom two layers: the physical layer and the data link layer (DLL). IP is the third network layer, and the TCP is the fourth transport layer, as shown in Fig. III 9.0a.

The major standards are:

IEEE802.1 Bridging (networking) and Network Management; IEEE802.2 LLC.

IEEE802.3 CSMA CD (Ethernet); IEEE802.5 Token Ring; IEEE 802.6 DQDB.

IEEE802.11 Wireless LAN and Mesh (Wi-Fi certification); IEEE802.15 Wireless PAN.

A request for comments (RFC) starts as a recommendation; after successful discussion, it is then approved by the internet community. The most important protocols are *RFC 768* User Data Protocol, *RFC 791* IP, *RFC 792* Internet Control Message Protocol, *RFC 793* TCP, and *RFC 826* Address Resolution Protocol.

RFCs are one of the main ingredients that contribute to the success story of Ethernet 802.3. Over years, Ethernet has evolved from standard Ethernet (10 Mbps), to Fast Ethernet (100 Mbps), to Gigabit Ethernet (1 Gbps), to Ten-Gigabit Ethernet (10 Gbps). Ethernet uses Manchester phase encoding for 10BASE-T systems. Fast/Giga Ethernet uses block encoding such as 8B/10B ('block coding' is used for processing groups of bits rather than outputting a signal for an individual bit). The radio connections for a wireless LAN take place in the 2.4 GHz frequency band, the so-called ISM band (industrial, scientific, and medical), or in the 5 GHz band. For hard real-time data transmission, standard Ethernet may be unsuitable because data traffic can be delayed. To develop Ethernet-based, but real-time capable fieldbuses, manufacturers have pursued various approaches in their efforts to eliminate such delays. These solutions are commonly referred to as "real-time IE" technologies. Powerlink is designed to give users a single, consistent, and integrated means for handling all

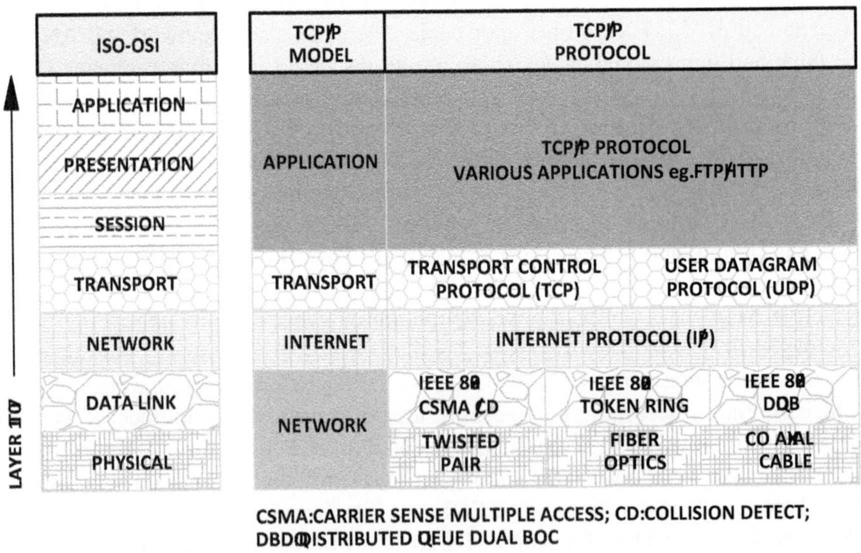


**FIGURE III 9.0A** Ethernet: International Standards Organization's (ISO) Open Systems Interconnection (OSI) and Transmission Control Protocol/Internet Protocol (TCP/IP) model.

communication tasks in modern automation. Fast Ethernet has proven its tough real-time characteristics in thousands of applications. The Ethernet Powerlink Standardization Group ensures openness and continuous advancement. In this section, various IE family networks and Ethernet will be briefly discussed. DDE, OPC, ControlNet, and DeviceNet all are communication technologies for industrial automation with varied purposes and services. Microsoft's DDE is a legacy technology for exchanging data between applications (viz. RSView32, a human-machine interface [HMI] software). There are different kinds of OPC applications. ControlNet is designed for high-speed, deterministic control applications. ControlNet provides reliable, high-speed control and input/output (I/O) data transfer by setting a specific time for data transfer. Also, ControlNet can handle both scheduled and unscheduled messaging. DeviceNet focuses on connecting field devices, is widely accepted, and supports a multitude of hardware. So, in this section all these will be discussed. Let the discussion start with the Ethernet family.

## 9.0.0 Introduction to Ethernet family protocols

Ethernet is a well-recognized technology in the home and office environment. Recently, there has been trend in moving data in industrial applications utilizing Ethernet on account of its speed and physical media simplicity. The fundamental problem in using standard Ethernet in industrial applications is that the TCP/IP protocol, which is used for routing and delivering data, does not inherently provide the real-time, deterministic (guaranteed) performance required in most of automation and industrial applications. Described here are a number of variations on Ethernet, more specifically IE. "Ethernet family" refers to a collection of (wired) networking technologies, mainly standardized under the IEEE 802.3 standard. This family encompasses various specifications for different data rates, cable types, and physical layer implementations. Therefore, one may conclude that the Ethernet family is related technologies that define how devices connect and communicate over a network and how it meets application requirements with determinism in a varied environment. Apart from standard Ethernet, Ethernet I/P and Ethernet for Control Automation Technology (EtherCAT) are examples of family members. Here, the basics will be covered in the section on communication, with details provided in Chapter 16 of this volume.

1. **IE:** The requirements and challenges for IEs have been briefly discussed. Industrial automation can be divided broadly into:
   - **Supervision/routing level:** Data transfer between the individual systems may not be very time critical. Standard Ethernet has been used here for years.
   - **Control level:** Here, high-speed, real-time deterministic data transfer is essential.
   - **Field level:** The real-time characteristics of IE, as discussed above, have special importance.
   - **Determinism**: Industrial applications demand a real-time protocol designed to utilize the physical layers of Ethernet and provide deterministic (guaranteed) communication at all levels. Such a real-time deterministic protocol should also offer a reasonably high speed of communication so that a higher bandwidth will be available for the transfer of a large quantum of process data and high-speed requirements in shop floor applications. There exist a number of challenges for the use of standard Ethernet in industrial environments, as listed below. Therefore, standard Ethernet needs a number of modifications before it is fit to be used in industries. With these modifications, IEs were developed. Several IEs exist on the market.
   - **Challenges:** A few common challenges are *communication*, *process concerns*, *environmental challenges* including *temperature* and *humidity*, *noise*, *vibration*, *hardware*, *chemicals/corrosion*, *fire and explosion hazards*, *topology*, *cable/media of communication*, *cost of down time*, and *diagnostics and security*.
2. **Common IEs:** There are around 15–20 different protocols for IE that conform to the IEEE 802.3 standard for Ethernet. Control and Communication Link IE, EtherCAT, EtherNet/IP, Modbus TCP, Powerlink, PROFINET, and Sercos III are a few examples. There have been modifications in the MAC layer in many IE protocols to achieve low latency and jitter coupled with determinism. There exist a number of challenges for standard Ethernet when it must be used in an industrial environment. A brief comparison among various IEs covered in the book is presented below in Table III 9.0a.

   Using a real-time approach, modifications have been made in software for EtherCAT and PROFINET using Ethernet/IP and Modbus TCP standard Ethernet software. Whereas EtherCAT is based on CANopen, Ethernet/IP is based on CIP and PROFINET is based on Profibus.
3. **EtherCAT and Ethernet/IP comparison:** Although both are types of IE, Ethernet/IP and EtherCAT are two different types with distinctive characteristics and different ways of implementation. They are not comparable, but each can be the right choice for many specific applications. Therefore, depending on the application, one has to make one's own choice. To facilitate this, the *pros and cons* of each system are discussed below. Based on CANopen, EtherCat (developed by Beckhoff Germany) is popularly known as "on-the-fly

**TABLE III 9.0A** Comparison of various Industrial Ethernets [2] [*Courtesy:* RTA].

| Industrial Ethernets | Automation technology | IEEE 802.3 MAC | High-performance motion |
|---|---|---|---|
| **EtherCAT** | Shared frame | Master Yes; ASIC for slaves | Shared frame |
| **Ethernet/IP** | CIP | Yes, requires 1588 hardware in some applications | CIP Sync. 1588 |
| **Modbus TCP** | Real time publish subscribe (RTPS) | Yes | None |
| **PROFINET** | PROFINET I/O | Yes, requires Ertec ASIC | PROFINET IRT (1588) |

processing". Developed by Rockwell automation, Ethernet/IP is based on CIP. Both systems conform to the IEEE802.3 protocol and inexpensive CAT5 cable can be used as physical medium for either. In EtherCAT the physical layer is *converted in the coupler to E-bus* to meet the requirements of the electronic terminal block. EtherCAT is highly synchronized to meet IEEE1588 and is well suited for multiaxis motion control. Ethernet/IP (like any other CIP network) additionally uses CIP sync to meet IEEE1588.

- **EtherCAT advantages:** The major advantages of EtherCAT are as follows:
  - *Fast:* It uses extremely fast "on-the-fly processing" that is facilitated by special Application-specific integrated circuit (ASIC) chips present in each slave. The time wasted processing on the fly is in the order of nanoseconds.
  - *Flexibility:* It offers good flexibility, including flexibility in topology selection, with very highly synchronized operation.
  - *Single cable and less wiring:* EtherCAT uses only one 4-wire cable with $+2 \times 24$ V DC/3 A over it. It also has uses much less wiring, hence has fewer chances of error. Space and cable track usage is highly optimized.
  - *Power supply:* It has power supply forwarding to all connected devices. It supports scalable connectors for different voltages from 24 V DC to 600 V DC with a load of 64 A.
  - *Low cost:* It is low cost with high performance; for use of standard hardware, there will be a reduction in hardware and installation costs.
  - *Safety protocol:* It has a safety protocol and multiple device profiles.
  - *Slave connection:* Slave devices have an I/O port. No managing switch is necessary; a simple daisy chain connection is used.
  - *Safety black channel (*safety network hiding):* Black supports integrated black channel safety. Black channel*
  - Redundancy: It has full redundant operation with an extremely fast switch over.
  - *Telegram size:* EtherCAT message telegrams are up to 60 KB, so there is huge node data movement to the master.
  - *Master action:* Master devices simply issue the message and receive the response.
  - *Slave numbers:* There is no limit to the number of slave nodes on a network.
  - *TCP/IP support:* The EtherCAT master processes TCP/IP and UDP/TCP packet data passed to the slave device. The slave can support TCP/IP applications (e.g., HTTP, Hypertext Transfer Protocol Secure, or Hypertext Markup Language [HTML]).
  - *Bus:* EtherCAT requires no underlying bus system to manage nodes and connections.
  - *Working with other protocols:* EtherCAT can work as the transport protocol for other protocols like BACnet, Ethernet/IP, PROFINET IO, or Modbus.
  - *Network quality:* EtherCAT offers very low jitter and latency.
- **Ethernet/IP advantages:** The major advantages of EtherCAT are as follows:
  - *Wide use:* It is a widely used protocol in manufacturing units in North America; currently, it has 15% of the market share compared with 7% for EtherCAT.
  - *Flexibility:* Ethernet/IP with managed switches provides a lot of flexibility.
  - *Ethernet transport and control:* It retains the standard TCP/IP of standard Ethernet, so access and signaling are available off the shelf.
  - *CIP standard:* Since it is based on CIP, network devices are a collection of objects with defined access, object behavior, and extensions. Major supporting tools and training are available.
  - *Message transfer:* No special hardware is required. Cyclic and acyclic messages are transferred using UDP and TCP, respectively.

- **EtherCAT disadvantages:** It is comparatively less simple and rather complex in understanding and implementation.
- **Ethernet/IP disadvantages:** It has limited bandwidth and speed. Also, processing of UDP and TCP eats away a lot of central processing unit (CPU) time.

4. **Discussion on EtherCAT and Ethernet I/P:** Let us discuss the fundamental details of EtherCAT and Ethernet I/P as follows:
   - **EtherCAT basics:** EtherCAT was developed specifically for real-time automation. It is based on the CANopen protocol and on Ethernet. However, since it was specifically developed and optimized for industrial automation, there have been modifications in traditional Ethernet software controls. EtherCAT is a very fast and deterministic network and processes data using dedicated hardware and software. "On-the-fly processing" is the main theme behind this protocol. This allows fast communication. EtherCAT is a master-slave system. The master controls the network and communication can easily be implemented using an industrial personal computer (IPC) with a standard network card. It uses a full duplex, master-slave configuration and accommodates any topology. It can process 1000 I/O points in less than 50 μs and communicate with 100 servo axes in 100 μs. Let the discussions start with Fig. III 9.0b, which shows the basic concept and EtherCAT fieldbus configurations. This has been developed based on an idea from [1].
     - *Bus integration*: EtherCAT has high bandwidth, making it possible to embed conventional fieldbus networks as an underlying system through an EtherCAT gateway. This is very helpful for migration from other fieldbuses (viz. PROFINET) to EtherCAT. Therefore, it is possible to integrate decentralized gateways and reduce the physical size of the connected IPC, that is, in EtherCAT, a single Ethernet port is needed to connect these devices.
     - *System description*: As shown in Fig. III 9.0b1, when the telegram continues to pass through the device, it receives data messages with its address header and also inserts data into the telegram. This is specifically mentioned in the figure, since it is typically applicable to all slaves. It is somewhat like loading and unloading in a moving train. So, it is referenced as "on-the-fly processing". For this reason, as the telegram passes through devices its size changes due to insertion and receiving. This is clearly seen when one compares the size of two telegrams. Another interesting part is that the frame path and cable routings are different. When cablings are done, say in daisy chain, the frame path forms a logical ring, irrespective of the actual topology. Cable paths are indicated and marked by brown (dark gray in print version) dotted lines, whereas green (gray in print version) lines with dots indicate the frame path, which is also indicated with arrows to mark the logical ring formation. Fig. III 9.0b2 shows EtherCAT fieldbus connections, which support field sensors and can work with Profibus, DeviceNet, and CANopen.
     - *EtherCAT protocol*: The EtherCAT protocol is optimized for processing data by transporting it directly within the Ethernet frame as several subtelegrams, each of which serves a particular memory area (up to 4 gigabytes in size) of the logical ring (i.e., process images). EtherCAT is based on CANopen, meaning the parameters specified via CANopen device profiles. In addition to master-slave data exchange, EtherCAT is also used for communication between controllers (master/master). The EtherCAT protocol has also been discussed at length in III 17.0.
     - "On-the-fly processing" *and fieldbus memory management unit (FMMU)*: Unlike traditional Ethernet, here, packets are not received, interpreted, and copied as data at every connection. In EtherCAT, there is a newly developed FMMU in each I/O terminal. FMMU reads the data *addressed to it*, and the telegram continues to pass through the device. Similarly, input data are inserted while the telegram passes through. This process is shown in Fig. III 9.0b1 with a difference in frame sizes. On account of this data reading and insertion there will be some delay in the telegram. Delays in the telegrams are only a *few nanoseconds.* Since an Ethernet frame reaches the data of many devices in both the send and receive directions, the usable data rate increases to approximately 80% [30]. As indicated, this is fully duplex and the same features are fully utilized, so effective data rates of >100 Mbps (up to 80% of 2 × 100 Mbps) [1] can be achieved.
     - *Change of physics:* The Ethernet protocol conforms to IEEE 802.3 and remains intact right up to the individual terminals without any subbus. The physical layer is converted from twisted pair or optical fiber physics to E-bus in the coupler to meet the requirements of the electronic terminal block. E-bus uses low-voltage differential signaling, as detailed in Fig. III 9.0b3, for internal device communication, which is cost effective. Again, at the end of the terminal block, on the outside, the physical bus characteristics are converted back to the Ethernet physical layer. On

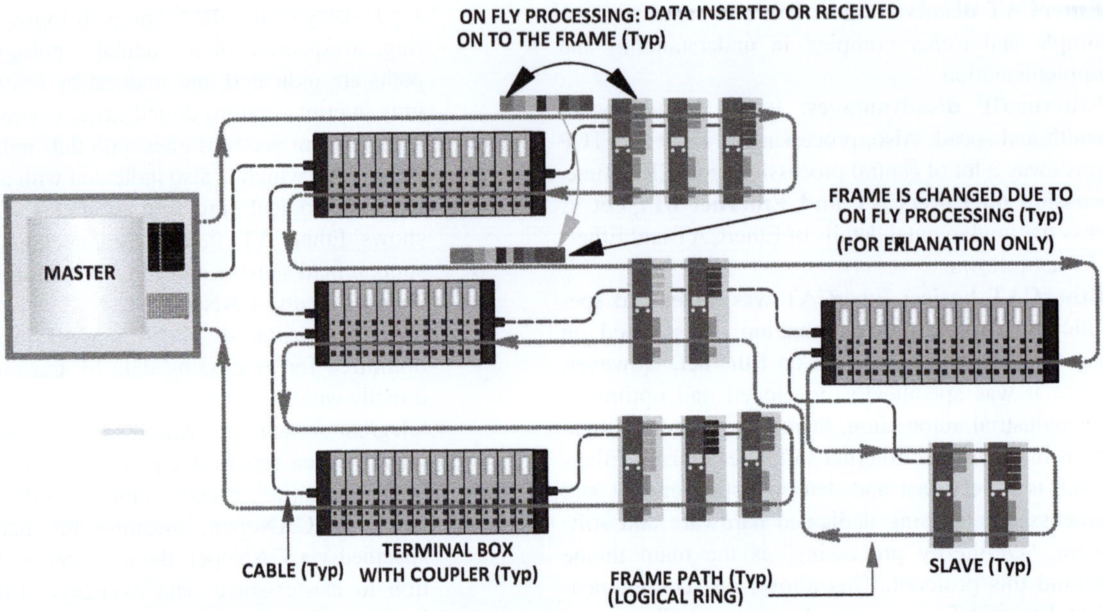


FIG III 9.0b1 ETHERCAT CABLE AND FRAME PATH (LOGICAL RING) WITH FLY ON PROCESSING

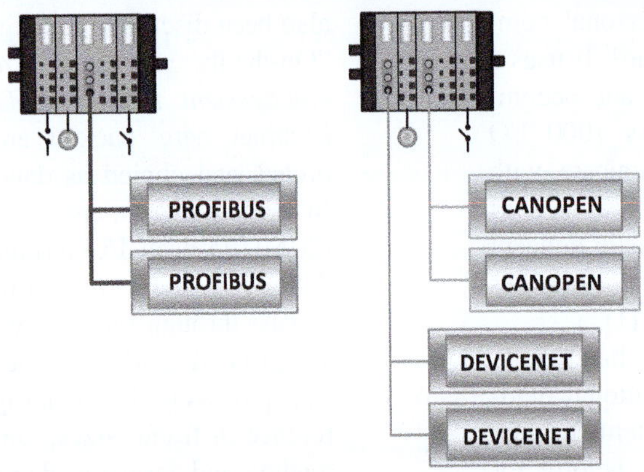


FIG III 9.0b2 DECENTRALIZED ETHERCAT FIELDBUS AT INTERFACE TERMINAL [2]

Low voltge differetiaSigalig (LVDS) is ahigh-speed digitalterfae,stadadized uder ANSI/TIA/EIA 644rMach 1996.

As the nme sigifies,it work ondifferetiasigalig. Two sigls ae geerted with opposite polaities d datramission references the two signals to one another. At the receiver with high input impedance 3.5 mA flow across 100Ω to create a differential of 3mV.Out of my advtges,its bility to reject commonmode oise d its low power cosumptionare worth otig.

It fids its pplicationinEtherCAT.

FIG III 9.0b3 LOW VOLTAGE DIFFERENTIAL SIGNALLING DETAILS

**FIGURE III 9.0B** EtherCAT basics and low-voltage differential signaling details.

the control side, the TwinCAT Y driver for Ethernet complements the FMMU technology to integrate transparently into the system, so that it appears as an OS-compatible network driver [1]. Twin CAT can also act as a fieldbus card, as shown in Fig. III 9.0b2. There exists internal prioritization and a buffer at the transmitter end for a free transmission channel for Ethernet frames. The OS's Ethernet frames are only transmitted in the gaps if sufficient time is available. At the receiving end, all the Ethernet frames received are examined by the TwinCAT

I/O system to filter out those with real-time relevance, and other frames outside the context of the real-time system are passed on to the OS.

- *Implementation*: EtherCAT technology adapts a specially optimized low-cost design to restrict an increase in device cost when adding an EtherCAT interface. In addition to hardware and software requirements, development support, the availability of communication stacks, and the interoperability of EtherCAT devices from various manufacturers (required to perform a *conformance test*) are important when developing an interface. The interface for an EtherCAT master device is very simple and uses direct memory access, requiring no additional CPU capacity. EtherCAT slave devices use inexpensive EtherCAT slave controllers (ESCs) in ASIC, Field-programmable gate array, or integrated in a standard microcontroller. ESCs are available with internal dual-ported RAM and a number of FMMUs.

- **Ethernet I/P:** The Control and Information Protocol (CIP) provides both real-time and informational message structures and is designed to be "*wire independent*" in that it can work with any data link and physical layer. As CIP is freely available, accessible to anyone, easy to understand, and widely supported by many manufacturers of industrial equipment, it is a natural candidate for use in building other industrial communication standards. ***"IP"* in** Ethernet/IP protocol stands for industrial protocol, meaning it has associated determinism. Ethernet I/P is an application layer protocol and a part of CIP. This means that Ethernet I/P is simply the way data are organized in TCP/UDP. Here, CIP covers the top three layers of the OSI model (discussed later), covering the *device profile*, *message,* and *connection profile.* In UDP and TCP, IP represents the network and transport layers whereas it is based on Ethernet infrastructure in the Ethernet MAC (data link) and physical layers. Ethernet/IP is a certifiable standard supported by DeviceNet and ControlNet to ensure the consistency and quality of field devices. Each CIP device is required to make an identity object available to the network. The identity object contains related identity data values called attributes. Attributes for the identity object include the vendor ID, date of manufacture, device serial number, and other identity data. There are three types of objects found in CIP viz. *required object*, *application object*, and *vendor-specific object.* This is covered in a later section on networks.

5. **MODBUS TCP:** MODBUS is a widely used, open industrial protocol. MODBUS TCP combines the MODBUS remote terminal unit with standard Ethernet and the universal TCP standard. As MODBUS does not have any particular network topology, all topologies implementable in standard Ethernet including switched Ethernet can be configured. Speeds of 10/100 Mbps in half and full duplex mode are achievable. Normally, there can be around eight masters and a number (60–70) of slaves. Usually, twisted pair Cat5 cable as per IEEE802.3 with an RJ45 connector are used for 100BASE-T Ethernet.

### 9.0.1 Ethernet Powerlink protocol

Standard Ethernet–based Powerlink is a second-generation fieldbus that applies the full power of information technology (IT) to the automation area. Ethernet Powerlink (often referred to as Powerlink) is especially suited for drives, I/Os, visualization, and data exchange between programmable logic controller (PLC) systems as well as for motion control applications. Fig. III 9.0c shows the basics of Ethernet Powerlink.

Ethernet Powerlink provides IE with a single, consistent, and integrated means for handling all communication tasks in modern automation. It is suitable for machine applications, plant engineering, and process industry applications. Total coverage is detailed in Fig. III 9.0c1. This reflects that Ethernet Powerlink can integrate all components in industrial automation including HMI systems and motion and safety controls. The Ethernet Powerlink cycle not only accommodates transmissions of actual payload data for applications, but, in its asynchronous phase, it also provides a fixed timeslot reserved for transfers of miscellaneous user data. Ethernet Powerlink is characterized by the freedom it provides to users to choose any type of network topology. Ethernet Powerlink's speed is very much suitable for motion control for less than microsecond level synchronization and a few hundred microseconds cycle time. The Ethernet Powerlink communication profile is well-suited for real-time Ethernet applications by extending Ethernet (IEEE 802.3) with mechanisms to transfer data with predictable timing and precise synchronization. Live modifications can be made to the network at any time and in any way without compromising the application. Ethernet Powerlink has good future prospects as a fieldbus because it is *open and independent, proven and robust, market oriented with no license fees, available now,* and *future proof.* Let us exploring it starting with its operating principle.

1. **Operating principle:** Ethernet Powerlink is a software-based solution complying 100% with Ethernet standard IEEE 802.3. Ethernet Powerlink ensures that all features, benefits, and the flexibility of Ethernet technology are available in the real-time protocol. For

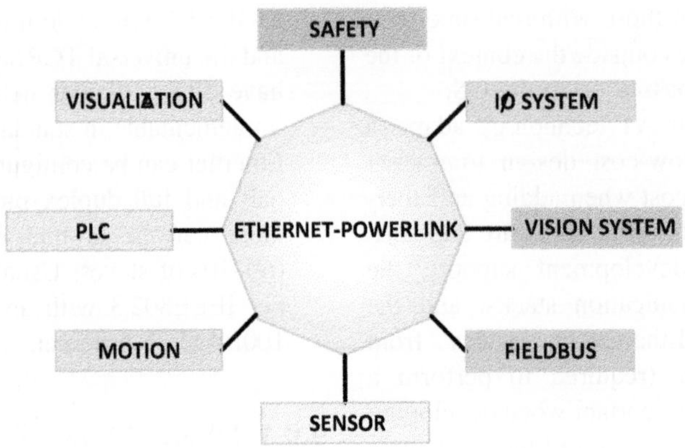


FIG III 9.0c1 ETHERNET–POWERLINK COVERAGE

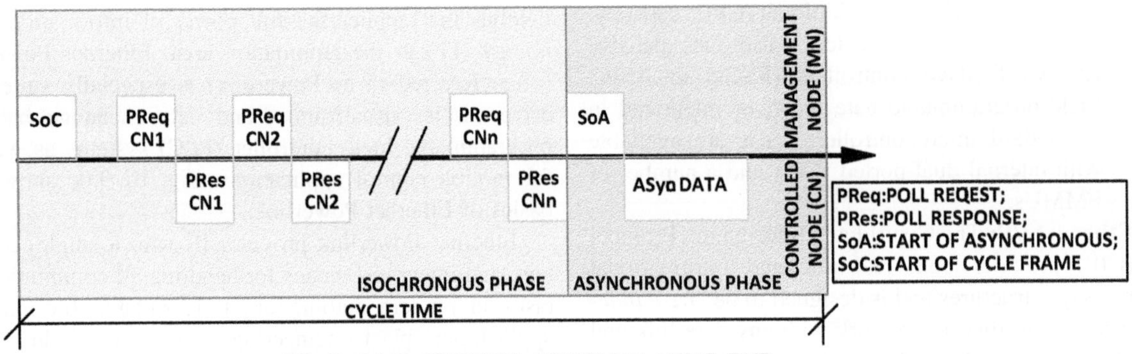


FIG III 9.0c2 ETHERNET–POWERLINK CYCLE TIMR

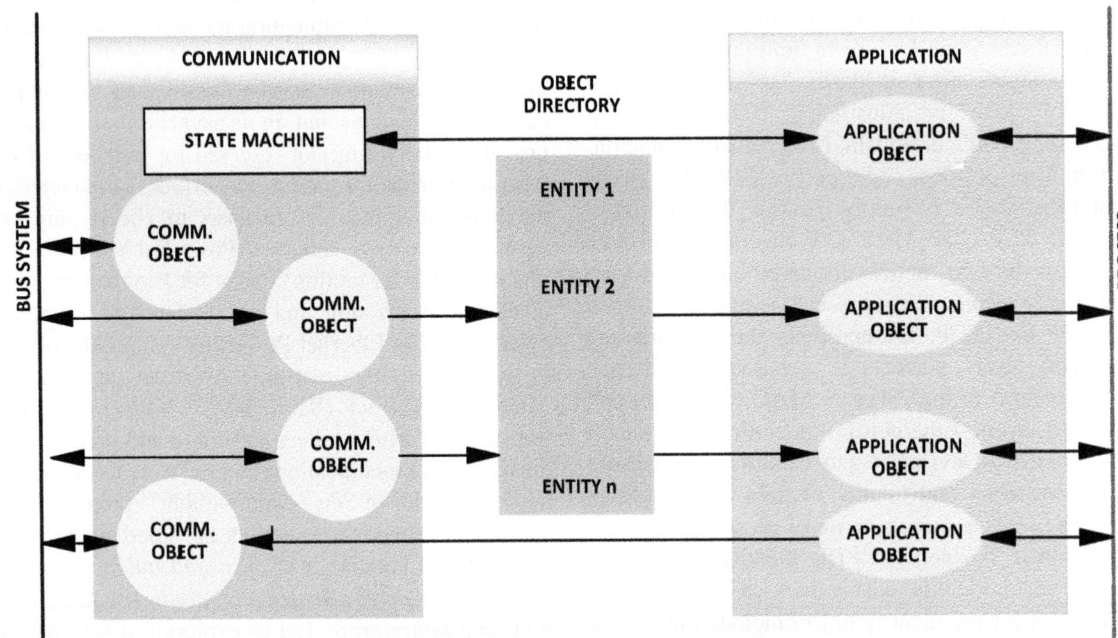


FIG III 9.0c3 ETHERNET–POWERLINK DEVICE MODEL

**FIGURE III 9.0C** Basics of Ethernet Powerlink.

real-time capability, Ethernet Powerlink resorts to a mixed polling and timeslot procedure, allowing only one node at a time to send data. In contrast to standard Ethernet, this procedure ensures that nodes cannot "speak" at the same time (no arbitration), as is seen in an organized roundtable conversation. In Ethernet Powerlink networks, one node (viz. PLC/IPC/motion controller) is arbitrarily designated to function as the managing node (MN), like a moderator. All other devices operate as controlled nodes (CNs). Responsibilities of the MN include defining *the clock pulse for synchronization* and *managing the data communication cycle.* Thus, the MN successively polls each CN in one cycle via a "poll request" message, conveying additional data to CNs. Via the "poll response" message, each CN then sends its own data to all other nodes. Ethernet Powerlink has a mixed polling and time slicing mechanism *to*:

- **Ensure** transfer of time-critical data within very short and precise isochronous cycles with configurable timing.
- **Synchronize** networked nodes with high precision in the microsecond range.
- **Transmit** less time-critical data in a reserved asynchronous channel.

  An Ethernet Powerlink–capable device can handle the following operating states:
- **Basic Ethernet mode**: The device can operate in Ethernet networks when real-time data transfer is not necessary.
- **Preoperational mode**: During system start/after connection, the initialization and configuration data are loaded via the asynchronous channel.
- **Ethernet Powerlink mode:** After the boot procedure, the device enters real-time operation. The MN checks the timing.

  As shown in Fig. III 9.0c2, the three periods in Ethernet Powerlink *cycle-time* are:
- **Start** phase: This starts with the MN sending the start of cycle frame to all CNs. All networked nodes synchronize themselves to the MN's clock
- **Isochronous** phase: The MN assigns a fixed time window to each node to transfer time-critical data (payload). All other nodes can always listen to all data during this phase (publish/subscribe).
- **Asynchronous** phase: The MN grants the right to send ad hoc data to one particular node (noncritical data packet). Standard IP-based protocols and addressing are used during this phase.

  The "object directory," which serves as an interface between the communication and application, is another important item in Ethernet Powerlink. The complete description of a device's application with respect to the data items in the Ethernet Powerlink *object dictionary* is called the device profile (Figs. III 9.0c3 and III 9.0d3). Thus, the *object directory* is a group of objects that can be accessed through the network. This group of objects is basically a collection of data items that have an influence on the behavior of the application objects (device functionality), communication objects, and state machine used on this device.

2. **Architecture and performance:** Let the discussions start with reference to Fig. III 9.0d.

   Fig. III 9.0d1 shows the network structure of Ethernet Powerlink. This shows that Ethernet Powerlink is well suited for both centralized and decentralized automation concepts due to its brief cycle times, flexibility, and high-bandwidth performance. As Ethernet Powerlink technology closely adheres to the Ethernet standard, it is well suited to these requirements. Ethernet Powerlink distinguishes between real-time domains and non–real-time domains. This separation makes it suitable for machine and plant concepts. It also satisfies the increasing security demands to prevent hacker attacks at different levels. A few related issues are:
   - **Topology:** A free choice of topology in Ethernet Powerlink is extremely helpful for modular system extensions so that machine upgrades, plant expansions, and ongoing additions are allowed due to excellent scalability without any negative impact on the network's real-time capability and performance.
   - **Cross-traffic:** Direct cross connection (DCC) enables communication between components bypassing the master. All devices send data in broadcast mode so that all other nodes can receive these transmissions.
   - **Miscellaneous issues:** As discussed, it allows devices to be added or removed without disrupting the network. It has two periods viz. isochronous period (dedicated to time-critical data exchange, with the MN sequentially polling each CN) and asynchronous period (to handle non–time-critical data and allow for communication between CNs).
   - **Performance:** Ethernet Powerlink covers all performance classes as a software solution. Due to the use of standard hardware, various manufacturers provide cost-optimized solutions for hard real-time applications. General performance data are:
     - 0.1 μs system synchronization
     - *100 μs cycle time with* Jitter <1 μs
     - *100 Mbps bandwidth*
     - *High data throughput even for* brief *cycles*
     - *240 nodes*
3. **Description and features:** IE through system on a chip (SoC) can provide improved methodology to manage I/Os, sensors, and actuators for motion control. The SoC

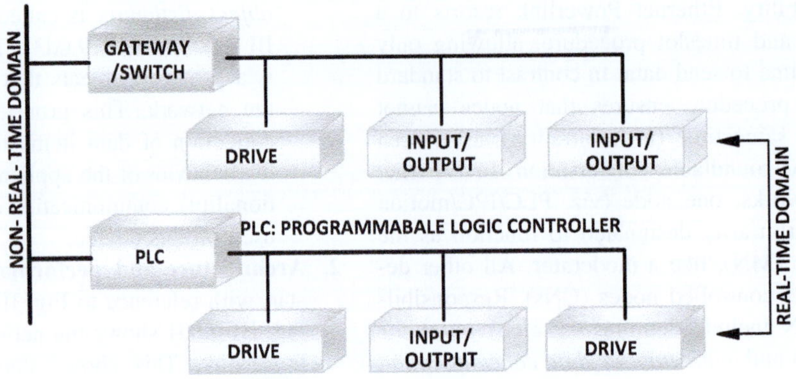


FIG III 9.0d1 ETHERNET POWERLINK: NETWORK STRUCTURE

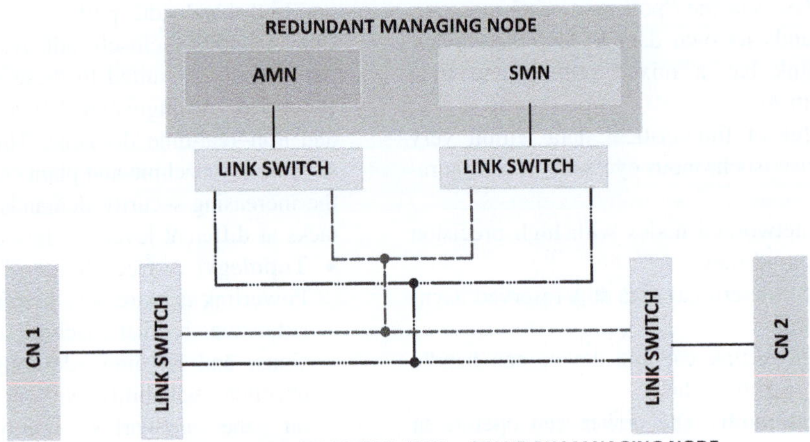


AMN: ACTIVE MANAGING NODE; SMN: STAND BY MANAGING NODE

FIG III 9.0d2 REDUNDANCIES IN ETHERNET POWERLINK

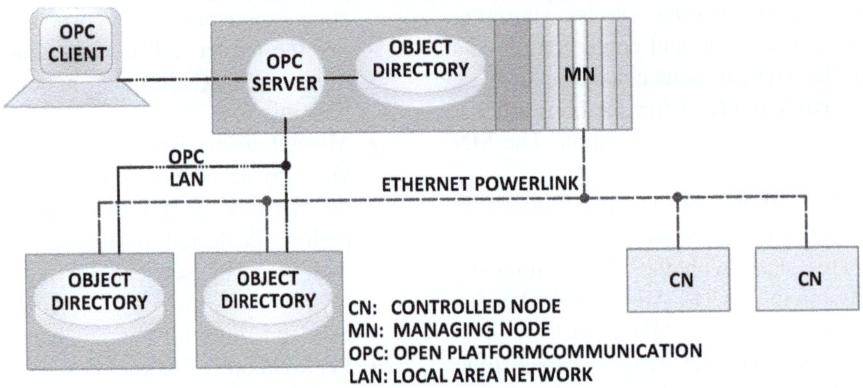


FIG III 9.0d3 ETHERNET POWERLINK: OPC UA INTERFACE

**FIGURE III 9.0D** Ethernet Powerlink network details.

can provide a protocol for real-time, deterministic, jitter-free Ethernet communications with reliable deterministic synchronization capabilities over large distances. In addition to being implementable in SoC, Ethernet Powerlink also supports Ethernet standard features like cross-traffic, hot-plugging, and wide selections of network configurations and topologies. CANopen is the basis for the communication profile of Ethernet Powerlink. Process data object for process variables and service data objects (SDOs) for configuration and remote objects are reused. For safety applications, Ethernet Powerlink can be used with the open-source SAFETY stack. Key features of Ethernet Powerlink include:

- **Standard compliance:** IEEE 802.3 Fast Ethernet IP-based protocols are supported (TCP, UDP, etc.) and it integrates well with CANopen profiles EN 50325-4. IEEE 802.3 and IEC 61784-2 (real-time capable, Ethernet-based fieldbuses) are the main basis of Ethernet Powerlink. Other supported standards are IEC 61158 and ISO 15745.
- **Unique addressing:** A unique addressing method allows for simple software configuration/setting up of a network and reliable service. This makes it the first choice for machines with a modular design devised for quick and easy extensions.
- **Multiplexing**: Multiplexing (Fig. III 9.0c2) optimizes bandwidth use in the isochronous phase. As requirements of data sampling intervals are application specification, not all CNs need be polled in every cycle.
- **DCC:** When polled, each node is free to broadcast into the network, and every other node is generally able to receive these broadcasts. Nodes can tell by the target addresses on data packets whether any such data are intended for them. In this way DCC is established and without the MN communication is established.
- **Asynchronous data:** The isochronous phase (time-critical payload) is followed by the asynchronous phase for the rest of the cycle, as shown in Fig. III 9.0c2. The asynchronous phase is for non–time critical data packets. This also helps to integrate components and plant segments. SDOs are for device configuration and diagnostics, application data such as surveillance camera feeds, and protocols like TCP/IP for device configuration or maintenance from a web browser.
- **Security:** Ethernet Powerlink real-time domains are *strictly separated* from non–real-time domains via gateways, which are often integrated into the controllers. The Ethernet Powerlink network constitutes its own domain, represented to the outside through one single IP address. The gateways operate like a firewall in this set up and internally resolve the public IP address via network address translation.
- **Response chaining:** As seen in Fig. III 9.0c2, Ethernet Powerlink includes a "Poll Response Chaining" feature, that is, instead of requesting the CNs sequentially through PReq frames, the CNs are requested together by the PRes frame in multicast to improve performance.
- **Hot plugging:** This enables adding to and disconnecting from a network line without impacting/compromising network functionality and rebooting for added and replaced nodes to become operational.
- **Redundancy:** Ethernet Powerlink supports the implementation of redundancy, both in media and the master, as detailed in Fig. III 9.0d2. As an example, ring redundancy, the economic option in machine engineering, has been shown. Here, there are two active MNs and a standby MN. Similarly, two links are connected via link switches (LSs) to connect MNs and CNs with redundant links.

4. **Safety and security:** Ethernet Powerlink provides clear dividing lines and access controls on the machine level. On the one hand it is important to guarantee that external access to the machine network is only possible *by authorized persons.* Also, it is important to guarantee that the timing of the real-time domain cannot be *influenced* by malicious attacks on the higher-level network. Even small-time deviations in the microsecond range could cause reduced production quality and could even damage machine parts. For this reason, there will be separation between real-time and non–real-time domains in Ethernet Powerlink to ensure security in all aspects. The Ethernet Powerlink Standardization Group's safety working group is currently busy developing EPLsafety, the safety-related extension of Ethernet Powerlink. The main goal is to develop an open system with the highest performance, absolute independence from nonsafe transport protocols, and transparent data exchange between safe and nonsafe areas. For data management, EPLsafety uses an object dictionary (Fig. III 9.0c3) whose structure and format borrow from the mechanisms present in the CANopen object dictionary well appreciated by users. EPLsafety will cover applications in all areas like machine automation, process control, and transport systems.
5. **Miscellaneous issues:** Here, we discuss two important miscellaneous issues: OPC interface and benefits and applications.
   - **OPC–Extensible Markup Language (XML) interface:** OPC Unified Architecture (UA) is a vendor-independent communication protocol for industrial applications to close the gap between operations technology (OT) and IT using semantic data descriptions with integrated security features, as shown in Fig. III 9.0d3 (OPC UA interface). The OPC UA Ethernet Powerlink companion specification describes how machine data are mapped from the OPC UA server to the machine controller. The transport of OPC UA packages in Ethernet Powerlink uses its asynchronous phase, where it can transport all types of Ethernet data. The only IEEE-standardized IE protocol, Ethernet Powerlink truly lives up to its role as an open, standardized, and future-proof protocol for industrial communication in the converging worlds of IT and OT. Ethernet Powerlink also uses XML standard for its device description files. An XML-based description (Ref: Volume I) format gives manufacturers the major

advantage of using a consistent, standardized description for different fieldbuses or real-time Ethernet systems for their devices. Moreover, XML provides a range of options for device descriptions that other languages have failed to implement to date, including multilanguage capability, an option for textual descriptions, and parameter and value characterizations by physical units and scaling functions [3].

- **Benefits and applications:** Due to being low cost and high performance, Ethernet Powerlink fieldbus systems are almost inevitable in machine control applications. The major benefits of Ethernet Powerlink are:
  - Ethernet Powerlink provides a low-cost, high-performance solution through the use of Ethernet hardware and functions as per open international standard.
  - It has detailed diagnostic information and guaranteed bandwidth for troubleshooting.
  - Easier operation, greater flexibility, and robust security make Ethernet Powerlink a great choice, especially in machine controls.
  - Ethernet Powerlink can connect and synchronize up to 240 nodes in a single network
- In addition to real-time capabilities, additional features like *hot plugging*, *direct cross-traffic*, and *multiplexing* are available.
  - It has advanced motion control capabilities with a cycle time up to 400 μs.

On account of these benefits, Ethernet Powerlink has applications in many sectors. Applications are mainly found in sectors like *automotive*, *mining*, *chemical*, *paper industries*, *power and energy*, *food and beverage*, *semiconductor*, *timber*, *plastics*, *logistics*, *maritime*, *metals*, *pharmaceutical*, *process*, *textile*, *transport*, and *packaging*.

### 9.0.2 Introduction to Dynamic Data Exchange–Open Platform Communications, ControlNet, and DeviceNet

OPC (originally Object Linking and Embedding for Process Control) is an interface specification for data communications that is popular in industrial environments. It is faster than DDE with more features meant mainly for industrial applications. On account of its complex programming requirements with specialized use, it is also popular in nonindustrial applications. OPC connections are always *client-server*. Setting up the Cascade Data-Hub (a software component designed to facilitate data exchange and integration between OPC Data-Hub and other systems) to use OPC is simply a matter of configuring it to act as a client or as a server. Details on OPC are discussed at length in Ch.15 of this volume. DDE is a form of interprocess communication in Windows that allows two applications to exchange data in client-server mode, initiated by the client. ControlNet is a state-of-the-art, high-speed, deterministic open network used for the transmission of time-critical application information. It provides suitable I/O bandwidth, real-time interlocking, peer-to-peer messaging, and programming on the same link. It is well suited for repeatable performance for both discrete and process applications. It utilizes CIP for the upper layers of the OSI mode. ControlNet was designed to provide reliable, high-speed control and I/O data transfer by using programming that sets the logic to specific timing over the network. The ControlNet network mainly communicates with PLCs, I/O chassis, HMIs, personal computers, drives, and robots. Rockwell Automation created DeviceNet as an application layer protocol on top of CAN in the 1990s. DeviceNet is also part of the CIP protocol family and is designed for the direct connection of sensors and actuators in the field. Although DeviceNet is based on CAN specifications (physical communications standard), it is restricted in functionality for easier implementation. DeviceNet, a low-level application layer protocol in industrial applications, simply connects industrial devices, like sensors and actuators, with higher-level devices such as PLCs. It is developed mainly for tasks like configuring, accessing, and controlling industrial automation devices. DeviceNet primarily deals with application data rather than nonapplication layer protocols. A DeviceNet explicit message has specific information in specific bytes of a message. CAN has been chosen as the DeviceNet physical layer because of its *extremely robust nature*, *open technology*, *small processor footprint* (memory needs), and *inexpensive components*. As per the CAN and OSI model, DLL can be conceived as a composition of MAC and physical signaling. Similarly, the physical layer is composed of a medium attachment and transmission medium. The basics of ControlNet and DeviceNet are introduced here; for a detailed discussion, chapter18 in this volume may be referred to. Let us begin the discussions on Windows-based data exchanges.

1. **DDE-OPC Basics:** The basics of DDE and OPC are elaborated in Fig. III 9.0e.
   DDE is a standardized concept of interprocess communication available to applications running in Windows. Applications can use DDE for one-time data transfers and ongoing exchanges. DDE always takes place between a client and a server application. The client initiates the exchange by establishing a conversation with the server so that it can request data or services from the server. Various levels and services are:
   - **Service name (level 1):** This refers to the server name, which is unique among all servers.

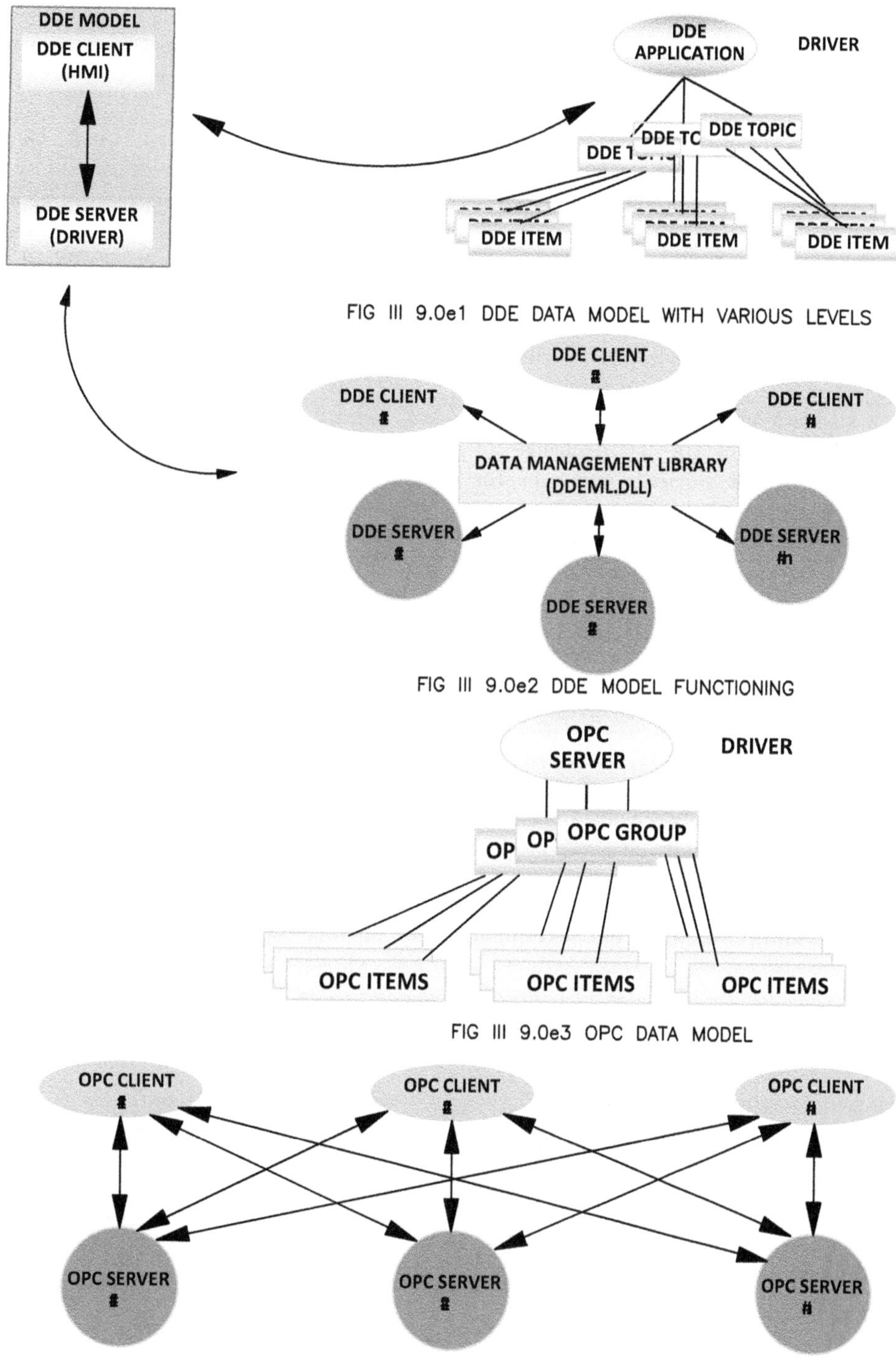


**FIGURE III 9.0E** Basic data exchanges in Dynamic Data Exchange and Open Platform Communications.

- **Topic name (level 2):** This identifies a logical data context in the server. These two levels are used by the client for a conversation with a server.
- **Item name (level 3):** This is used to name the data items for a conversation.
- **Request (service):** This allows the client to read data items in the server.
- **Execute (service):** This service of the client makes the server execute a command.
- **Poke (service):** This is used by the client to modify server data.

- **Advice loop (service):** This is a client service allowing them to subscribe to data items in the server. For active subscriptions, the client will be notified of any changes/modification.

DDE methods of data exchange between applications with details on levels are elaborated in Fig. III 9.0e1 and 2. Data exchange in OPC, which is based on Microsoft's Component Object Model (COM) and Distributed COM (DCOM), and OPC specifications for software are elaborated in Fig. III 9.0e3 and 4. Readers may compare Fig. III 9.0e1 with III 9.0e3. Based on Microsoft's COM/DCOM technologies, OPC facilitates data exchange between industrial automation systems by providing a standard interface. OPC is an open, standards-based infrastructure for the exchange of data and specifies a set of software interfaces for several software objects as well as the methods (functions) of those objects. This software standard is well supported by all major hardware vendors and most OPC products (clients and servers) operate very much in plug-and-play fashion. Users can develop their own clients and servers by means of developers' toolkits. It is even possible to develop simple HMI systems without any programming knowledge at all. The scope of OPC is to focus primarily (though not exclusively) on the exchange of "raw" data, that is, the efficient reading and writing of data between an application and process control device. Details on OPC are provided in Chapter 15 of this volume.

2. **ControlNet features and models:** Major features are:
   - **Networking capabilities:** Some of the capabilities are:
     - Deterministic, reliable, and repeatable with high throughput (5Mbps)
     - Same physical media for both I/O control and programming
     - Easy configuration and maintenance with increased diagnostics
     - Multiple controllers controlling independent I/Os on the same link with scheduled peer-to-peer interlocking between controllers
     - Multicast capabilities: multiple controllers sharing input data and multiple controllers and reduced programming requirements
   - **Low installation costs:** Widely used *RG-6 quad shield coaxial cable* with connectors and tap with *integrated drop cables* for node connections
   - **Programming and networking:** Some of these features are:
     - User-selectable I/O and PLC interlocking update times
     - Online feedback of network bandwidth use
     - Configure I/O and PLC interlocking without ladder logic
   - **Flexible architecture:** Some of the pertinent issues are distances up to 30 km or more using repeaters in various topologies, simplified design, and various cable options for a wide range of applications, environments, and media redundancies.
   - **Model:** ControlNet models are explained in Fig. III 9.0f.

     There can be two types of models viz. Source/Destination Model and Producer/Consumer Network Model. These are explained in Fig. III 9.0f1 and 2.

**FIGURE III 9.0F** ControlNet model explanation.

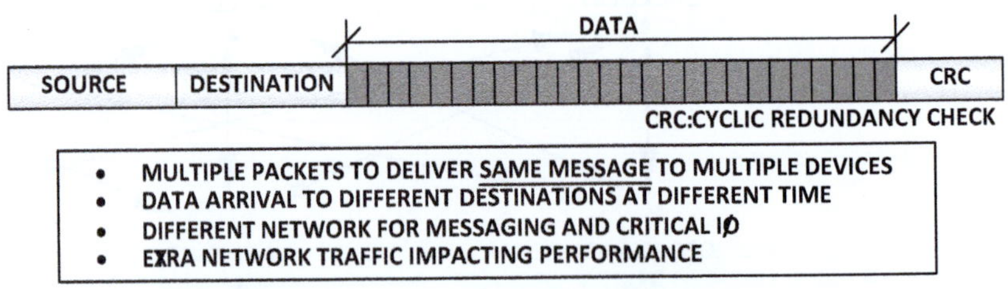


FIG III 9.0f1 CONTROLNET SOURCE DESTINATION NETWORK MODEL

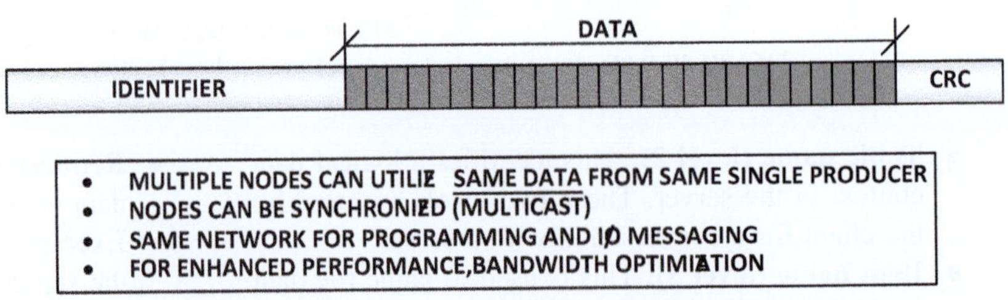


FIG III 9.0f2 CONTROLNET PRODUCER CONSUMER NETWORK MODEL

3. **ControlNet layers:** Various layer details of ControlNet are:
   - **Physical layer:** RG 6 coaxial cable (BNC connectors) is common; other cable types like "flexible", "direct burial", or "plenum", which are rated based on the application and environment, are also used. For longer distances, optical fiber is often used. In a normal case without a repeater a maximum distance of 1 km can be covered. There can be a maximum of 5 repeaters with no impact on the number of nodes (maximum 99 nodes). BNC connectors can be a "twist-on"/"locking type" or a more rugged design that is a "screw threaded" type for more harsh environments.
   - **DLL:** ControlNet uses a "token-passing" ring through a coaxial cable. A "token" grants a device the right to transmit. The node, having possession of the token, transmits data frames until it reaches the end of its transfer/the token time limit. Then, the token is passed sequentially to other devices (next logical device). In ControlNet, the nodes are assigned a MAC ID address from 1 to 99, and each node also knows the address of its predecessor and successor. To maintain integrity, LLC performs error detection and correction. In the DLL, the timing for ControlNet is performed through network update time (NUT), which is based on a fixed and repetitive time cycle. The NUT can be set from 2 to 100 ms for the overall update rate of the network. The request packet interval determines how often devices transmit during the NUT.
   - **Network and transport layer:** It adapts CIP in these layers. It has connected and unconnected messaging. Unconnected messages are used for low-priority messages. Connected messaging uses settings in each node viz. "real-time I/O data".
4. **DeviceNet technical issues:** In this part, let us address some important issues pertinent to CIP and DeviceNet. These issues are:
   - **CIP object:** Three types of objects defined in the CIP protocol are:
     - *Required objects*: These are required by the specification for inclusion in every CIP device. Various required objects are the *identity* object, *message router* object, *network* object, and *connection* object.
     - *Application object*: Application objects define the data encapsulated by the device and hence are specific to the device type and function.
     - *Vendor-specific object:* Objects not found in the profile for a device class are termed "vendor specific" to include additional features of the device and extension objects
   - **Connections with IDs:** In DeviceNet messaging (messaging between a *master* and a *slave* device), the *Connection IDs* are predefined, enabling low-resource devices to optimize processing of these messages. Using the filters in CAN controllers, these messages are easily identified and processed while all other messages are quickly discarded. The two types of messages are:
     - *Unconnected messages*: This refers to a special virtual access port, which is in every DeviceNet device, to send a few predefined messages to a DeviceNet device without first making a connection.
     - *Connected messages*: These are messages produced and consumed across a connection between two devices. Connected messages can be either *explicit request/response*-type messages or I/O messages containing formatted data. Master-slave, offline, and I/O are some connected-type messages.
   - **Messaging:** Three kinds of messages transmitted between DeviceNet nodes are *peer messages* (unformatted messages between any two nodes), *explicit messages* (request/response-type messages from a master to a slave), and *I/O messages* (to transfer predefined I/O data between a master and a slave).
   - **Node configuration**: The individual participants are configured via electronic data sheet files provided by the manufacturer for each slave.
   - **Addressing:** Addressing is implemented via rotary switches/software via addressing tools or the master. Suitable checking is done to eliminate and correct.
   - **Diagnostic system:** In DeviceNet, the diagnostic message can be transported via additional input bytes (status bytes).
5. **DeviceNet communication:** For communication, it is important to look into various layers of DeviceNet and communication modes.
   - **Physical layer specification:** DeviceNet uses a network topology of the trunk/derivation type, allowing the signal and power wirings to be present in the same cable. Power is supplied by the power supply connected to the network, feeding the CAN transceivers of the nodes, and has the following characteristics: 24 Vdc, with the DC output isolated from the AC input, and current capacity compatible with the installed equipment. The four standardized types of cable used are *thick* (trunk) *cable, mid cable, thin* (drop) cable, and *type V cable*. The Baud rate used depends on the cable length of the network, as shown in Table III 9.0b below.
   - **DLL details:** The DeviceNet DLL is defined by the CAN specification already discussed [ref: III 8.0.2]. As discussed, it has two possible states: dominant (logic level 0) and recessive (logic level 1). A node can bring the network to the dominant state to transmit information. For data frames, Fig. III 8.0h2 may be referenced.
   - **Network and transport layer:** In DeviceNet, a connection has to be established prior to data

**TABLE III 9.0B** DeviceNet: network size and baud rate.

| Baud | Network size | Max derivation | Total derivation |
|---|---|---|---|
| **125 kbps** | 500 m | 6 m | 156 m |
| **250 kbps** | 250 m | 6 m | 78 m |
| **500 kbps** | 100 m | 6 m | 39 m |

exchange. The DeviceNet telegrams are classified into groups that define specific functions and priorities. A DeviceNet node can be a client, a server, or both. Furthermore, clients and servers can also be producers and/or consumers of messages. The protocol also has a resource for the detection of nodes with duplicated addresses (Mac ID) to avoid this issue.

- **CIP application:** In the application layer, it uses the object-orientated CIP for the *transport of I/O device control data* and *transport of configuration and diagnosis information of the system being controlled.* Any DeviceNet node (master/slave) is molded by a set of CIP objects to encapsulate data and services for determining its behavior.
- **Bus cycle time:** The bus cycle time depends on many factors like number of participants, the relevant amount of data of the individual slaves, transmission rate, and type of communication.
- **Communication mode:** The DeviceNet protocol presents two basic types of messages: I/O and explicit. Each one is adequate to a specific kind of data.
  - I/O: It is a kind of synchronous telegram dedicated to the movement of priority data between one producer and one or more consumers. Types of I/Os are *cyclic*, *change of state* (slave automatically sends data to master for any change in input data), *polled* (polled by master), and *bit strobe* (master sends to all slaves).
  - Explicit: This is a general-purpose telegram without priority and is used for asynchronous tasks (e.g., configuration of the equipment).

  Let us now move on to discussions on FDDI, remote control EDDL etc.

## 9.1 Data communication: Fiber Distributed Data Interface, remote controls, field device integration, and Electronic Device Description Language

In the mid-1980s ANSI standardized FDDI as a group of networking specifications. FDDI is an optical data communication standard for long-distance networks that provides communication with fiber optic lines over <200 km at a speed of 100 Mbps. FDDI has dual primary and secondary communication rings, where the primary ring works and the secondary ring remains as backup. FDDI Full Duplex Technology (FFDT) uses the secondary ring as an additional primary channel; hence, FFDT is without redundancy, supporting data transfer up to 200 Mbps. FDDI is also used both by single and multimode fiber optics. Multimode (larger core) fiber optics use a light-emitting diode as a source, whereas single-mode fiber optics use a *laser* for data transmission only.

In many industries like manufacturing, energy, aerospace, defense, and oil and gas, instead of multiple systems for different business functions and locations, the trend in these industries is to collate as much information as possible into fewer operation centers that can be located remotely from the site. Now, implementing a remote manufacturing strategy has become critical to ensure resilient operations. Early diagnosis of equipment problems and optimization of maintenance and operations are essential elements for efficient operation of plants in the power and process industries. Remote monitoring systems can collect data accurately, track assets, and continuously monitor a facility or equipment condition. So, massive amounts of data are essential for proper operation, and can be collected using a remote monitoring system without being present 24/7. Therefore, it helps detect problems that cause downtime and gives information about the performance of the system and hardware involved. The hardware and software required for remote monitoring vary from vendor to vendor, but they mostly use the same intranet- or internet-based technologies. These remote monitoring connections are usually made to controllers/PLCs and HMIs via internal intranets or the internet, often via a virtual private network (VPN) router.

FDI specifications help instrument and automation manufacturers develop compatible products and host systems for managing field devices. For the user, the specification combines the attributes of competing device integration technologies: EDDL and field device tool (FDT)/device-type manager (DTM). A key objective of the specification is to combine the simplicity of the text-based DD technology with the flexibility of specialized Windows-based FDT features and complex graphical representation. By this time, the reader is familiar with FDT and EDDL from discussions in I 12.3.0 and I 12.3.1. In this section, discussions will mainly be on device integration. As a result, there will be some repetition in the form of recapitulation to complete the discussions.

Let this section begin with a discussion on FDDI.

### 9.1.0 Fiber Distributed Data Interface

Fig. III 9.1a shows details on FDDI. In an FDDI network, data transfer speeds of 100 Mbps over a fiber optic cable are supported using the token passing technique. As discussed, there are two physical paths in the form of "rings" (primary and secondary rings). So, there are counter-rotating rings with the *primary* one carrying data. Should there be a break in the ring, the stations can "wrap" the ring through themselves, as detailed in Fig. III 9.1a3. Normally, in an FDDI ring, the token passes by each network device fast. If the token is not seen within a given time (maximum time that it takes to circulate the largest ring), it indicates a network problem. It can fall back on the other ring. FDDI stations incorporate secondary rings into the data path to route traffic around the fault. As discussed, in FFDT the secondary ring is also used as a redundant primary channel to support data transfer speeds up to 200 Mbps. On account of optical data communication, it can be used to transfer data up to 200 km at a speed of 100 Mbps. Although FDDI was developed for fiber optic cables, copper cable was later included and is called a Copper-Distributed Data Interface (CDDI), which has been standardized with twisted-pair physical medium-dependent. Another important issue worth noting is that although the physical network is a ring it uses IEEE 802.4, that is, token bus protocol *not* IEEE 802.5 (token passing ring). FDDI also has error-detection and correction capabilities. An FDDI network requires a special network card and fiber optic cable, both of which are more expensive than copper interface cards and twisted pair cables. It is expensive and somewhat complex in nature. However, a major reason in favor of FDDI is its high speed, at almost 10 times increased speed compared with IEEE 802.3, that is, it offers increased bandwidth over Ethernet, is consistent with other IEEE 802 LANs, and is standardized by ANSI. Also, FDDI could be accommodated in several Very Large Scale Integration chips.

1. **FDDI network features (with use):** The major features of FDDI are:
   - **Media:** FDDI uses optical fiber as its physical medium (exception CDDI).
   - **OSI model:** FDDI mainly operates in the physical and DLLs of the OSI model. In the IEEE protocol, the data link has two parts comprising the LLC MAC layer. As shown in Fig. III 9.1a1, In FDDI, the physical layer has two parts viz. physical protocol (PHY) and physical media dependent (PMD). FDDI operates on PHY, PMD, and MAC.
   - **Speed:** FDDI operates on a data rate of 100 Mbps and can support thousands of users. For FFDT it can be 200 Mbps. So, it has increased bandwidth.
   - **Distance coverage:** FDDI can be used in LANs up to 100 km (200 Km wrapped) for long-distance communication with a maximum length between nodes of 2 km. Voice and multimedia communication can be achieved through FDDI.
   - **Traffic class:** It supports synchronous and asynchronous traffic classes.
   - **Protocol:** It uses a ring-based network and IEEE 802.4 token bus standard protocol.
   - **Packet size:** It is <4500 byte.
   - **Redundancy:** An FDDI network comprises two token rings viz. a primary ring for data and token transmission and a secondary ring to fall back on, with the exception of FFDT.
   - **Uses:** It is used for an LAN (with a maximum of 500 nodes) and WAN.
2. **FDDI structural details and operation:** An FDDI network consists of several FDDI nodes and connecting medium optical fiber. FDDI network nodes are referred to as station/concentrator nodes. *Station nodes* provide end-user device connections, whereas *concentrator nodes* can provide connections to end-user devices or other concentrators. The basic FDDI structure is illustrated in Fig. III 9.1a1. At the lowest level is a PMD layer, which is responsible for the fiberoptic interface, data driver, and receiver operation for FDDI stations (i.e., the physical medium interface). The next level is PHY, which specifies the encoding and framing of data and control information exchanges. There are various types of physical layer protocol viz. PHY-A, PHY-B, and PHY-M or PHY-S to take care of variations in control information exchanges. PHY-M is used in a concentrator, whereas PHY-A and PHY-B are used in a dual station and PHY-S in a single station, as discussed below. PMDs and PHYs form physical layers. Next is DLL, starting with MAC. The MAC sublayer specifies the protocols for logical ring formation and control, transmission and reception of packets at a station, and repetition and stripping of packets on the ring [28]. Station management (SMT) is meant for *n*-layer management. It also manages local interfaces to PMD, PHY, and MAC layers, as shown, to cater to the requirements of IEEE 802.2 for LLC. As shown, above LLC lies client protocols such as the Digital Networking (DECnet) Protocol, OSI, TCP/IP, and local area transport (LAT). LAT is a type of DLL used in networks and protocols developed by DEC for connecting various DEC computers (e.g., VAX, alpha). DECnet is a network protocol family developed

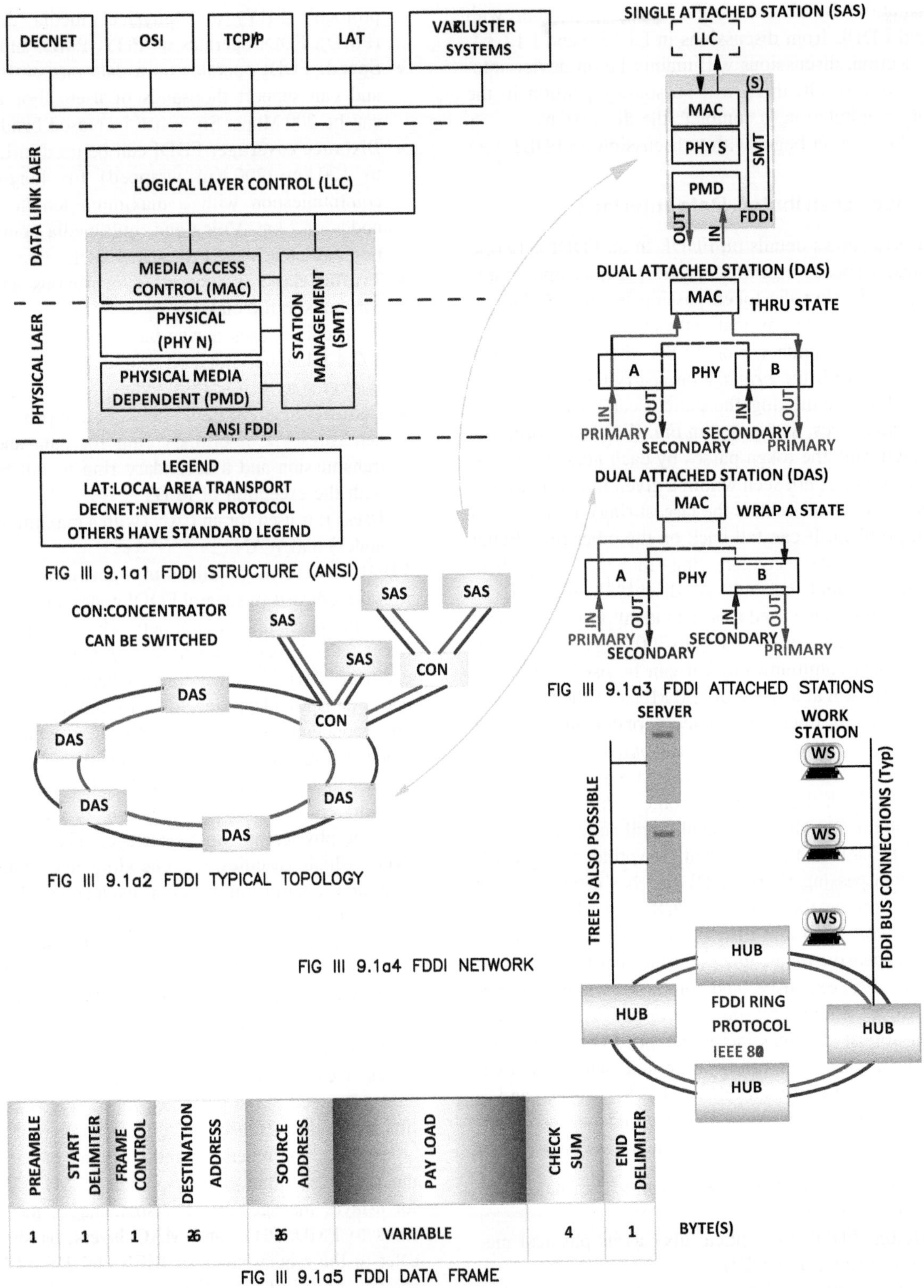


**FIGURE III 9.1A** Fiber Distributed Data Interface (FDDI) details.

by DEC for connecting their range of products. Three types of stations viz. single attachment stations (SAS), dual attachment stations (DASs), and concentrator are detailed in Fig. III 9.1a2.

- **SAS:** Fig. III 9.1a3 (top part) shows an SAS, which is a class B station with the simplest configuration and the basic components of FDDI. This is connected to a primary ring without the ability to wrap. There will be a single incoming and outgoing data path connected to the PMD for data exchange through the link layer. These are class B stations.
- **DAS:** Fig. III 9.1a3 (middle and bottom sketches) shows DASs, in which there will be a second physical layer (i.e., PHY A, PHY B). Later, one provides connections to a secondary ring for use in the event of breakage on the primary ring. DASs are class A stations connected to both rings with the ability to wrap. Naturally, there will be two states viz. fault free/THRU state and WRAP An STATE. Under fault-free operating conditions, THRU STATE is shown in the middle sketch of Fig. III 9.1a3, where the primary data path enters through PHY-A, travels through the MAC, and exits through PHY-B. The secondary data path enters through PHY-B and exits directly through PHY-A. When a problem is detected in the primary data path, either within the station or on one of the PHYs, the station wraps the two data paths, thereby providing an alternate route through the secondary data path, as shown. Here, the primary input is passed on to the secondary path via MAC (i.e., the secondary path is active).
- **Concentrator:** There will be additional PHY names as PHY master or PHY-Ms in the primary path. Here, a number of SASs can be joined. As shown in Fig. III 9.1a2, SASs are joined to each PHY-M.

3. **Networking:** An FDDI LAN can be formed by joining multiple stations to form a logical ring topology, as shown in Fig. III 9.1a2. From the figure, it is clear that there can be two types of topologies, such as the logical ring with dual trunk ring and a tree cluster, in which to several SASs are connected a concentrator. As shown in Fig. III 9.1a4, hubs can be put on the rings to connect workstations (WSs) and servers in bus or tree topology.
4. **Data frame:** As FDDI naturally use the IEEE 802.4 protocol, the frame format of FDDI is similar to that of a token bus, as shown in Fig. III 9.1a5, where all fields along with associated byte sizes have been mentioned. Single byte preamble, start delimiter, and end delimiter are used for synchronization to mark the beginning and end of the frame, respectively. Six-byte addresses are used for source and destination addresses. A single byte frame control specifies whether the frame is a data frame or control frame. As stated at the beginning, there will be a provision for error detection and correction, and for this purpose a 4-byte check sum is used. Before the check sum lies the variable byte payload.
5. **Standards:** Various standards used are:
   - **PMD:** ANSI X3.166-1990 and ISO 9314-3:1990
   - **PHY:** ANSI X3.148-1988 and ISO 9314-1:1989
   - **SMT:** ANSI X3T9.5/84-49, Revision 6.2, May 1990
   - **MAC:** ANSI X3.139-1987 and ISO 9314-2:1989

*With this, the discussion on FDDI comes to an end to move on to remote controls.*

### 9.1.1 Discussion on remote controls

Remote controls in manufacturing and production systems allow for monitoring and controlling equipment and processes from a distance, including expert services. Remote control services can provide expertise services to the customer in a cost effective and efficient way that is available 24/7. This is especially important for customers suffering from the dual challenge of a *lack of expertise* and *rapidly advancing technology*.

The key aspects here are remote access, monitoring and control, increased efficiency, better safety, real-time data analytics, diagnostics and troubleshooting, and others. The software and hardware for remote controls and the internet of things (IoT)/industrial IoT (IIoT) may vary by vendor. With the use of these technologies, it is easier and more economic to implement remote monitoring and controls to machines and processes. These remote monitoring and control connections are usually made through a VPN router. *Efficient and immediate problem solving, access* and *data recovery*, and *capacity and vision for growth* are some major benefits of remote controls. However, remote controls also have some challenges. Various industry challenges are:

- **Interoperability:** There is a lack of interoperability, especially for IoT solution. So, adoption of open standards and protocols is crucial for the smooth data exchanges necessary for remote control.
- **Employee and environmental safety:** The industry needs to establish robust connectivity channels with remote production assets and environmental parameters to improve the safety of operations and reduce risks and exposure.
- **Cybersecurity:** This is a big challenge for remote operations and controls, This may come from chances of attack, integration of OT and IT systems (especially if they are old ones), and any lack of robust security arrangements.
- **Data management:** Managing big data from IoT devices and real-time data processing, including data consistency, are major data management issues.

- **Cost of ownership:** A siloed view of operations causes several challenges and increases costs of ownership.
- **Other issues:** Other challenging issues include *supply chain resilience*, *remote work force management*, and *product quality*.

With reference to Fig. III 9.1b, let the discussions on remote monitoring and control begin.

1. **Objectives, benefits, and features:** The objectives, benefits, and features of remote monitoring and controls are as follows:
   - **Objectives:** Remote control objectives mainly include:
     - Central assets management with a detailed view of asset properties and operating parameters
     - Automated alarm generation in case irregular conditions are detected
     - Remote access to diagnose and troubleshoot resolution of issues
     - Critical machine health for connected equipment
     - Unified factory performance with key performance index (KPI); measuring and monitoring production KPI
     - Detailed view of the line performance and KPIs
   - **Benefits of remote controls**: Remote controls offer the following advantages:
     - They improve safety for people and the environment.
     - They provide automated in place of manual data collection and data can be stored in the cloud. This also provides immediate action for data analysis.
     - Direct data collection from machines helps to achieve a higher level of performance for better data accuracy.

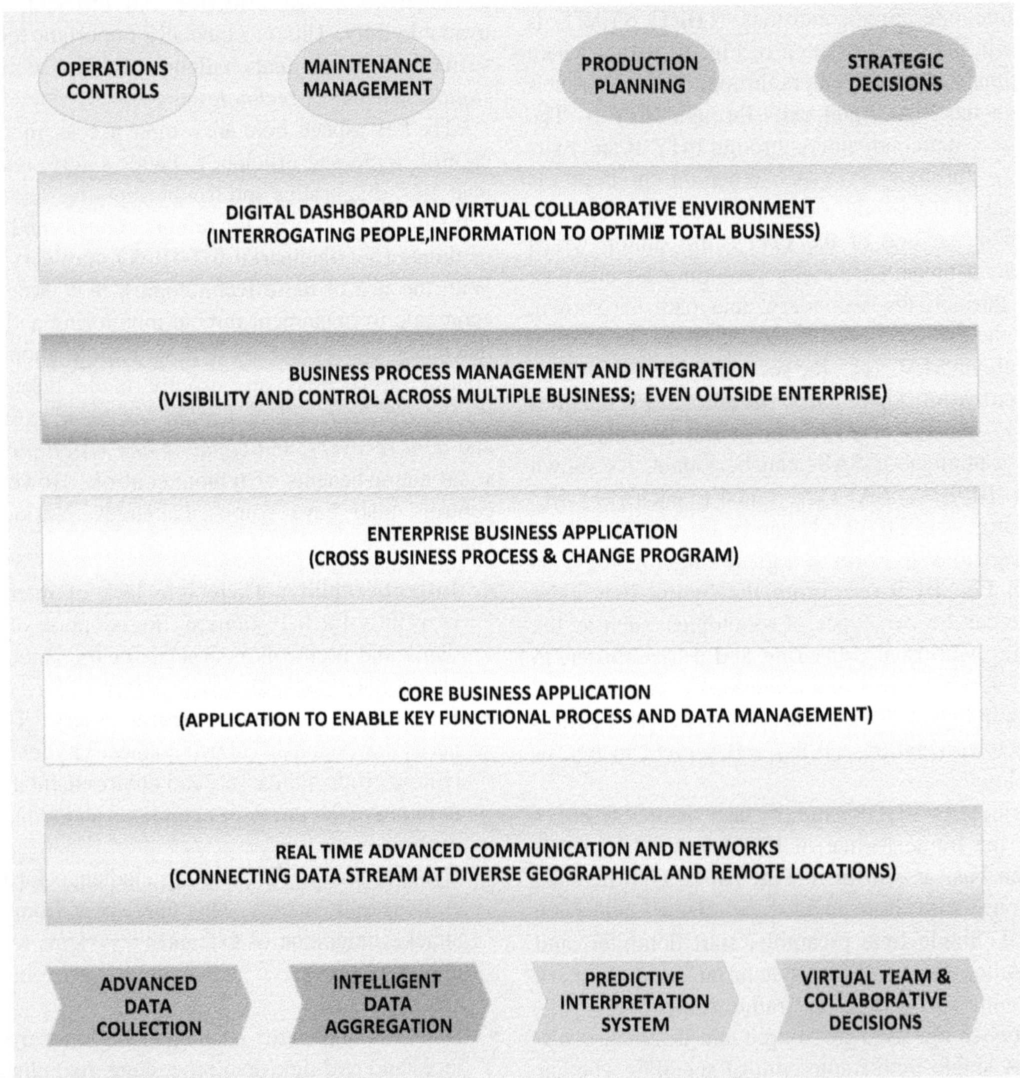


**FIGURE III 9.1B** Remote monitoring and control details.

- There is increased work efficiency with a remote system, with proper decision making enhancing productivity and efficiency.
- Remote control systems provide reliable complete visibility of the operations.
- With embedded software it is possible to receive timely alerts/updates.
- Experts are brought closer to the problems (and solutions).
- Technical collaboration shortens the time to resolution.
- Solutions can be deployed via remote connections.
- In remote systems, backing up the data is available to cater to situations when the server goes down.
- Deployment of remote Factory acceptance test (FAT) and simulator. This allows the customer to access and review ongoing projects that are in the FAT phase via a dedicated and secure website.
- Plant engineers minimize the number of trips to the facility, reducing chances of contamination in contamination-sensitive processes.

- **Features of remote controls:** Some important features are:
  - *Realtime monitoring*: instant decision making, proactive problem solving, and optimized resource allocation
  - *Scalability and flexibility*: future proofing, customization
  - *Advanced analytics and visualization*: performance optimization and predictive maintenance regular validation
  - *Robust security protocol*: multiple authentication, end-to-end encryption
  - *Seamless integration*: unified operation and enhanced efficiency
  - *Implementation*: investment in high-speed, low-latency communication infrastructure, training, regular review, and update monitoring
  - *Mobile accessibility*: mobile features
  - *Automated reporting and compliance management*
  - *Energy management and sustainability:* renewal energy and less energy consumption
  - *Collaboration and knowledge management*

2. **Secured access:** *Secure data transmission*, *secure user access at the service center*, and *audit and traceability* are major features of remote controls. The Remote Access Platform (RAP) provides remote connection security between the service center and the customer site. RAP enables secure, real-time remote monitoring and control of devices located at customer sites. RAP components, with security and redundancy (as applicable), can be used for around-the-clock operation. With consent of the plant's IT, the first step is to connect remotely to access the VPN by authorized personnel. There will be a list of authorized technicians allowed for access to provide service and support. Necessary safeguards for secured connection mainly include a VPN for the plant's control system connection, an isolated network plant IT network for the VPN, routine system backup, and use of virtual machines. With available remote access, authorized technicians are able to support plants. With prior remote access established, a technician will be connected to the plant in a short time to begin monitoring the system live and troubleshoot for lesser plant downtime. Also, control systems can be remotely accessed, monitored, and controlled (as allowed) for a specific issue. Remote access for control also provides a flexible time frame for system maintenance without hampering production schedules, since remote access can occur at any time during the day. Saving on time of travel to the site and cost of some control system modification (including addition/deletion) work, it is beneficial to start the modification from a remote access. It saves resources and provides other benefits, as discussed above.
3. **Security issue:** As discussed, RAP provides remote connection security. Various security details are:
   - **Secure data transmission:** The RAP and service center carry out a two-way authentication prior to initiating communication. RAP connects to the specific IP address(es) of its service center to initiate the connection, allowing the customer to control and limit outbound communication.
   - **Security and control:** The customer can set permissions on each remote activity.
   - **Data collection**: Secure data transmission begins at the source, with control over the types of data being collected for transmission. Based on the asset owner's security policy, configurable RAP can enable/disable data access.
   - **User access at the service center:** Access control is granular to the level of privileges and the actual scope of privileges for any given site is restricted. User accounts are governed by strict account management procedures, so customers can be assured that the principles of least authority and separation of duties are applied.
   - **VPNs and security layers:** Cybersecurity is more important than ever as threats continue to rise and more systems are monitored and supported remotely. In addition to a firewall and VPN connection, the encryption used in a VPN ensures that data cannot be intercepted, and that only authorized users can access the HMI, PLC, or other networked

devices. VPNs are offered in two main configurations: traditional and hosted. A traditional VPN is best administrated by an IT professional. A hosted VPN starts with the connected devices, controller, and HMI connected to a VPN router at the plant.

- **Cybersecurity:** Depending on the task, many regular site assessments can be complemented by remote assessments to reduce complexity and labor costs through remote automation of manual tasks. This, in turn, brings uniformity to security, compliance, and change management. Security fingerprints analyze computer- and network-based assets to determine deviations from accepted security standards. Such fingerprints give a comprehensive site cybersecurity status and identify strengths and weaknesses.

4. **Service details:** With the introduction of upcoming technologies like 5G, IIoT, sensors, data communication networks, and software applications, industrial remote monitoring and control services can offer real-time supervision and management of industrial equipment/processes, diagnostics, and maintenance services from a remote location. *Real-time data acquisition control and automation, data transmission, cloud storage and analysis, alerting and notifications, remote access and visualization, predictive maintenance,* and *enhanced safety* are some of the services commonly utilized. Here are some common services:
   - **Remote monitoring:** This is meant for real-time tracking of equipment and assets, including insights like status and performance from a remote location for early identification of potential issues.
   - **Remote maintenance:** Industrial remote control systems facilitate remote access to machinery diagnostics, troubleshooting, and maintenance, reducing the need for on-site visits.
   - **Access and control:** Industrial remote access enables service teams to connect to machines remotely, allowing for troubleshooting, programming, and software updates, enabling operators to control machinery from a distance, enhancing safety in hazardous environments and improving operational efficiency.
   - **Diagnose:** Fingerprints for automation systems including measurement and analysis of system performance, network communication, and controller loading. These are used to generate performance reports, compare performances, and diagnose fault. Based on the fingerprints, improvement recommendations are determined and scheduled for implementation.
   - **Remote FAT:** Remote FAT saves time and money for customers using limited resources.
5. **Remote functioning:** Internet-based data monitoring brings new capabilities and unprecedented access to process measurement and control. With internet-enabled instrumentation, remote access can be anywhere, including utilizing mobile phone technology. To make the system compatible with Industrie 4.0, as a first step, it is necessary to establish the industrial connectivity. Typically, IoT/IIoT sensors (to measure various parameters viz. temperature and pressure) are deployed for collecting data. These sensor data are transmitted over a network to a central system. These collected data are processed, stored, and analyzed using an edge (/cloud) computing platform. These processed data are available for display and monitoring in a central place in an HMI, including anomaly detection alarm generations. Based on predefined rules or operator input, the system can trigger automated actions viz. equipment shutdown, parameter adjustment, and others. This approach makes things easier, more secure, and more cost effective. The main issues are:
   - **Easier**: For connectivity, OPC UA and MQTT platform can be used.
   - **Secure**: It uses modern encryption.
   - **Cost effective**: It preprocesses data; only changes are sent, decreasing bandwidth usage.
   - **Health monitoring:** Real-time visibility into the health and status of critical factory assets through continuous monitoring of essential parameters.
   - **Predictive maintenance:** It provides a maintenance action by identifying potential issues before they happen.
   - **KPI:** It rapidly develops dashboards for KPI with sources of operational data to provide real-time visibility into production status and critical KPIs (e. g., performance availability).

Let us conclude the brief discussion on remote controls and move on to discuss FDI EDDL.

### 9.1.2 Field device integration and Electronic Device Description Language in device integration and communication

Readers are already familiar with FDI, FDT EDDL, and others as a part of the discussion in I 12.3. To complete the discussion and provided meaning these are recapitulated.

FDI refers to the integration of field devices into a process automation system with the help of the combined strengths of technologies like FDT/DTM and EDDL, providing a common platform for device integration across different systems and protocols. Here, FDT/DTM provides a framework for integrating various field devices through a common interface and EDDL describes device parameters and functionality. Both FDT and EDDL address interoperability and the seamless integration of diverse field devices into automation systems with simplified configuration, diagnostics, and maintenance, but with different approaches and capabilities. So, in short, FDI has been developed by taking the best features of EDDL and FDT and combining them in one common standard. FDI can provide increased security, prepare for future devices, and support multiple protocols.

FDI and EDDL are detailed in Fig. III 9.1c.

Modern field devices are almost like embedded computers connected to control systems through networks. They generate large amounts of data: both process data and device data. These data require processing and converting into the required usable format. Also, software and systems upgrades, configuration, and maintenance are ongoing processes to manage. Thus, using these data to leverage digital connectivity, scalability, analytics, and others for improved operational performance and safety is a challenge. In this environment, it is necessary to have a standardized and interoperable model to comprehensively describe and manage various automation components. The tools for commissioning, calibrating, diagnosing, and maintaining these instruments must be capable of taking full advantage of the fieldbus communication protocols. Further, the tools must use these advantages in the simplest possible way, keeping the end user in mind. FDI technology is a device integration technology that plays a key role in addressing device management in an operating plant, as illustrated in Fig. III 9.1c1. EDDL is the engine in device configuration and maintenance software that allows the use of a single tool for all the devices they support, and EDDL can be linked to HTML for internet access. EDDL is based on *device identification*, *device description* (parameter attribute), *functionality* (scaling and mode switching), and *structural information* (parameters and functions). EDD is a text file written with EDDL that contains the description of a device for copying the file to the necessary location in a distributed control system (DCS).

1. **FDI features and benefits:** Some important features and benefits are:
   - **Features**: Various feature are as follows:
     - *Unified platform*: FDI offers a single, integrated platform for the management of various field devices, irrespective of their communication protocols/manufacturer.
     - *Interoperability*: By eliminating compatibility issues, it ensures seamless communication among different devices and systems.
     - *Security*: FDI incorporates enhanced security features to protect against unauthorized access and cyber threats.
     - *FDT tool*: This is a common way to enable an FDT host to support FDI devices through FDI-DTM, as detailed in Fig. III 9.1c2 and discussed in I 12.3.0.
     - *Diagnostics*: It supports device-specific diagnostic information across the entire enterprise for troubleshooting and maintenance.
     - *FDI host support*: Existing EDDs may be processed by FDI hosts, as shown.
     - *New standards*: EDDL has been standardized and harmonized; however, new innovations and market requirements would bring changes to the EDDL standard.
     - *Future proofing*: FDI is designed to accommodate new devices and technologies as they emerge, ensuring long-term compatibility and adaptability.
   - **Benefits:** The major benefits are as follows:
     - *Integration complexity*: FDI reduces the complexity of integration for new devices and systems, saving time and resources. It also helps different groups collaborate.
     - *Improved efficiency*: The standardized, streamlined device management enhances diagnostics, increasing operational efficiency.
     - *Productivity*: It empowers plant operating personnel with access to process and device information with diagnostic data to enhance productivity.
     - *Cost savings*: With simplified integration and improving efficiency, FDI can lead to significant cost savings in the long run.
2. **FDI device package:** The FDI device package includes the *device definitions/drivers*, *user interface plug-ins (UIPs)*, *certificates*, and *useful documentation* to simplify integration with field devices and FDI host-enabled plant, enterprise, and cloud-based systems. *A unified container*, *standardization*, *comprehensive information*, *plug-and-play integration*, and *reduced economic integration efforts* are some main features of the FDI device. The device package is the core FDI technology and common host components. The FDI device package services all devices and tools. The device comes with an FDI package used by all the tools or systems, including standalone personal computers, field instruments, and process control and automation systems. The FDI device package makes sure that all device management tools function with devices with no issues, regardless of the manufacturer. A *device package* is a file (with a user-friendly file name and driver design for one preferred tool) created by the device manufacturer that allows the device to be integrated in a system. With FDI it is very easy to identify the right package for the right device. The base technology of FDI is EDDL, and a device package contains an EDDL file [4]. Optionally, the device package may have a UIP

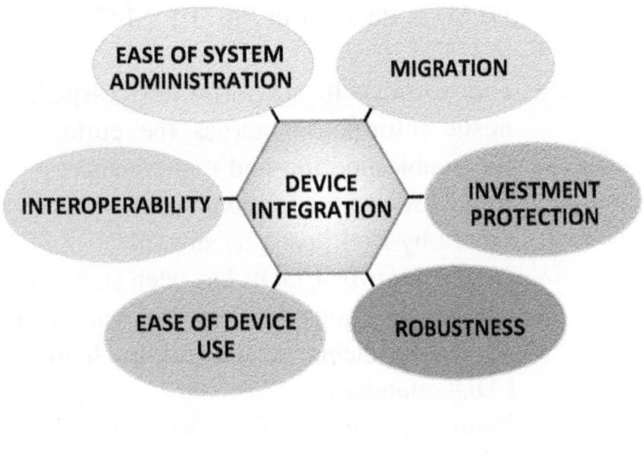


FIG III 9.1c1 FIELD DEVICE INTEGRATION CONCEPT

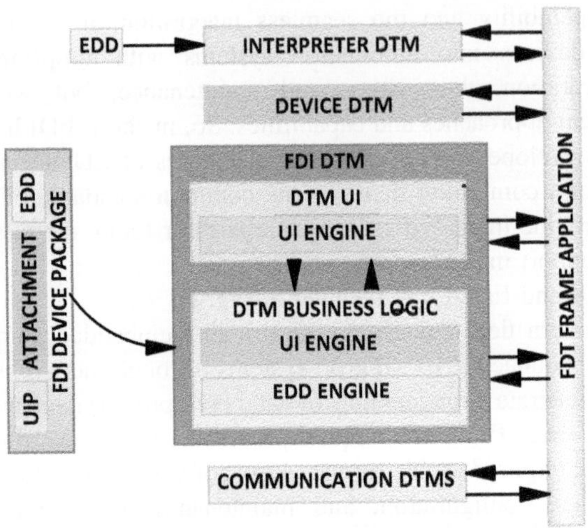


FIG III 9.1c2 FDT SUPPORTING TOOL FOR FDI (DTM)

**ABBREVIATIONS:**
CFF:COMMON FILE FORMAT; DTM:DEVICE TYPE MANAGER; EDD (L):ELECTRONIC DEVICE DESCRIPTION (LANGUAGE); FDI:FIELD - DEVICE INTEGRATION; FDTFILED DEVICE TOOL; GSD:GENERAL STATION DESCRIPTION; UI:USER INTERFACE; UIP UI PLUG IN

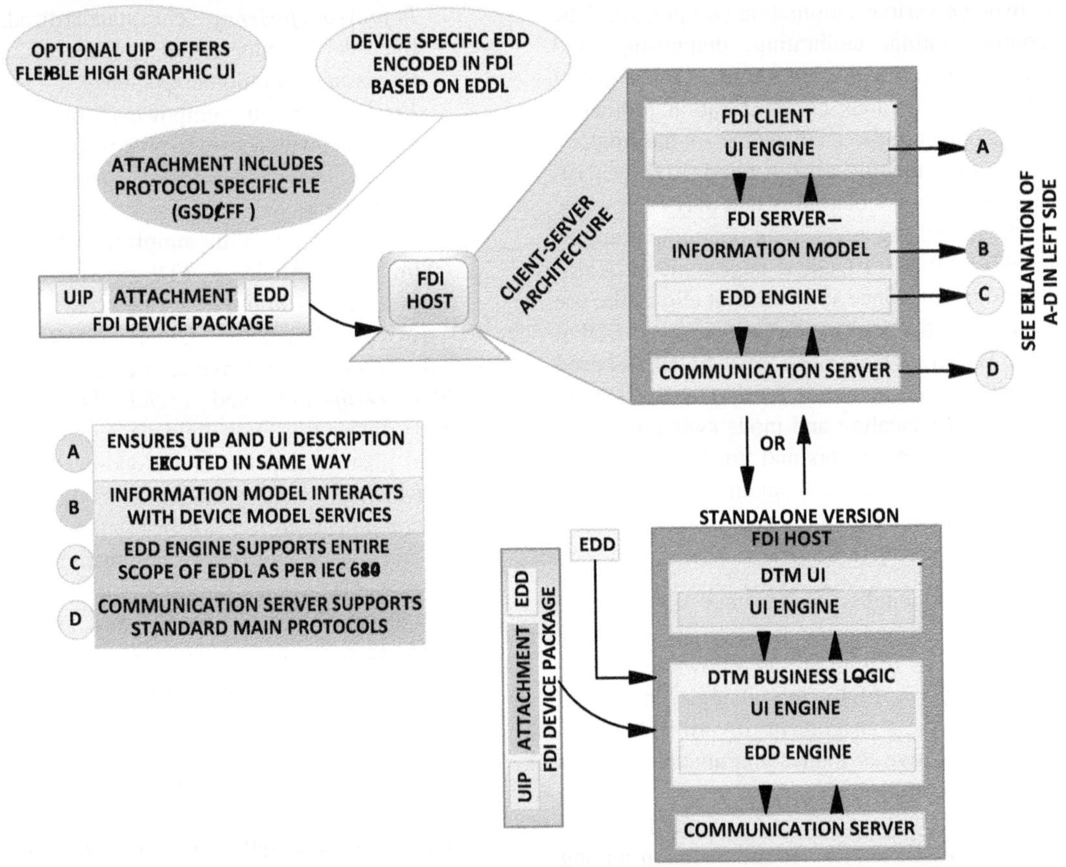


FIG III 9.1c3 FDI HOST TYPES AND CLIENT DETAILS

**FIGURE III 9.1C** Field device integration and Electronic Device Description Language details.

and other files, viz. the Profibus general station description file or common file format. This information is provided in Table III 9.1a.

Let us now discuss important components here viz. FDI host, clients, and others.

- **FDI hosts:** FDI device packages are imported by FDI hosts for users to operate the device. An FDI host can be a standalone software component (integral part of controller) or a software component with a client server architecture. Both options are detailed in Fig. III 9.1c3. An FDI host typically consists of an FDI client, an FDI server, and one or more FDI communication servers, mainly including some products like handheld field communicators, standard DCS software components for the device/asset management historians, and analytics packages. FDI host components are:
- **FDI client:** As shown in Fig. III 9.1c3, this refers to the user interface (UI) engine, which ensures that UI elements of the FDI device package (e.g., UI description and UIP) are executed in the same way in various host systems.
- **FDI server:** This consists of two elements.
  - *Information model*: The information model of an FDI server interacts with device model services (included in FDI Common Host Components), which retrieve EDD information from the EDD engine.
  - *EDD engine*: In line with standard IEC 61804, the EDD engine supports the entire scope of EDDL in a multiprotocol manner with backward compatibility.
- **Communication server:** This supports standard protocols like HART, PROFIBUS, PROFINET, and FOUNDATION Fieldbus in addition to communication supports like the OPC UA communication server.

3. **Common host components:** The common host component software, which is in the host system, ensures that the representation of the FDI *device package* is similar in various tools. The common host components consist of the EDD engine and UI engine. All FDI device packages will be tested and approved with reference to the FDI reference. The usage of common host components ensures that the representation of the FDI device package is similar in various tools. Various issues related to the common host components are:

- **UI design:** The guide provides examples of source code or sketches of the graphical representation of controls or frames and standardizes labels viz. device settings, diagnostics, and others. In addition to guide translation in different languages, these will be based on the harmonized EDDL applicable for all three major protocols—HART, PROFIBUS, and FOUNDATION Fieldbus—and all new EDDs must use the updated and optimized IEC 61804-3 standard. FDI-compliant systems have standard FDI host components. Also, manufacturers provide an easy-to-use interface and associated step-by-step guidance for use. Device manufacturers use EDDL to create FDI packages for definition, content, and structure. This creates a consistent look and feel and makes it easy to use.
- **Full device functionality:** With powerful graphics, including trend charts and table grids, it is possible to set up advanced diagnostics of sophisticated (complex) devices. Various software and tools based on FDI technology are suitable for all phases of the life cycle, including DCS operator consoles and DCS engineering consoles. The FDI packages make devices fully interoperable.
- **Device database:** FDI is a single common technology for all systems and devices inclusive of the *codes* for device manufacturer, *device type*, and *version*, so that the control system and intelligent device management software can automatically pick the correct FDI package.
- **OPC UA interface:** With use of the standard OPC UA interface in FDI hosts, easy access from other

**TABLE III 9.1A** Field device integration device package details [4].

| Item | Description |
|---|---|
| **Device definition** | Description of all device parameters including label, data type, and local help |
| **Business logic** | Wizards that steps a user through the processes. Conditional to hide internal dependencies |
| **User interface description device package** | Visualization graphics and task/role-based hierarchical menu system |
| **User interface plug-in** | Optional software component for set up and diagnostics not available in the device |
| **Attachments** | Other optional integration files, manuals, and others |

applications is possible and applications can be developed without the supplier's FDI host support. OPC UA services in the FDI server provide safe and secure access to the device. Generic OPC UA clients can be *maintenance tools/manufacturing execution systems* (MESs) and others.

- **Robustness and DCS integration:** The FDI package keeps a dedicated folder with version name on the system for a device type. FDI packages do not overwrite system files or make registry entries. Hence, the robustness is well maintained. FDI packages can be an integrated part of DCS for intelligent device management within the DCS WSs.
  - *Package implementation*: The FDI package for a registered device is downloadable and can be copied onto the system easily because the EDDL portion of FDI packages is encrypted text. While loading the device package, there is an option to integrate the device with both device pages and plug-in or only the EDDL text-based device pages.
  - *Operation and maintenance*: Using predictive diagnostics therefore becomes a part of operations and maintenance.
  - *Engineering WS*: By loading FDI packages in a DCS engineering WS, it is possible to carry out system and device configuration from the WS, providing easy checks for ranges and other settings, and system administration/FDI packages ensure the system can be kept up to date with these new devices.
- **Device navigation:** For easier navigation, device manufacturers separate common features and parameters from special ones.
- **License fee:** The EDDL portion of FDI packages requires no license key, so system administration is easier and less costly.
- **Investment protection:** The FDI technology was designed to extend the life of the system, and since this is text based, it is independent of Windows updates.

4. **Discussion on EDDL:** Here, characteristic issues of EDDL as well as functional support will be covered to complete the discussion on EDDL. The major issues are:
   - **EDDL characteristics:** EDDL makes it possible to use the same software for many different versions of devices. Using the EDDL standard, device vendors can describe all the information on their device and optimal means of display to the technician using the device management systems or handheld EDDL, eliminating driver software. A few characteristic functions are a *unified UI*, *universal language*, *interoperability independent of platform*, efficient diagnostics, *easy updates*, *low complexity*, *low cost*, and *multilanguage support*. EDDL is a markup language. Through interactive scripts, EDDL helps technicians through a series of steps to complete complex tasks viz. diagnostics, calibration, and others. Characteristically, EDDL files are small and easily transferable across the internet. Files for many types and versions of devices can be stored in handhelds.
   - **EDDL: functional support:** Robust EDDL is supportive of external secured accessibility, investment protection, international standards, single universal solutions, and uniform consistent displays so that it is easier for technicians to deal with. Major supportive functions are [5]:
     - *Identification*: tag, device ID, manufacturer, serial numbers, versions, and others
     - *Dynamic monitoring*: device internal parameters and logic
     - *Parameterization*: ranging, advanced setup, reconciliation, and others
     - *Diagnostics*: sensor failures, loss of supply air, memory failures, and others
     - *Information*: materials of constructions and primary/final elements
     - *Performance analysis*: step response, hysteresis, and others
     - *Device security*: write protection
     - *Documentation*: datasheets, manuals, procedures, loop diagrams, piping and instrumentation diagram, and others
5. **Benefits of EDDL:** The major benefits of EDDL are listed in Table III 9.1b.

Now let us explore Industrial communication: important issues.

## 9.2 Industrial communication: Important issues

Ethernet is a physical layer connection, and is widely considered the standard. It can work together with the TCP/IP protocol, which is based on a client/server model in which a server computer provides services to a client computer. Based on IEEE standards, Ethernet technology provides data connection and power over a single Ether-net port, which can be of different types. Standard Ethernet consists of a power device, Ethernet port, and Ethernet switch. The Ethernet switch plays a fundamental role in transmitting the data over Ethernet ports. The power device supplies power to the Ethernet switch with the Ethernet port. Based on its application there will be different versions of Ethernet. The office version of Ethernet is not suitable for industrial purposes. Similarly, IE is different from standard Ethernet. IE can withstand the harsh

**TABLE III 9.1B** Major benefits of Electronic Device Description Language.

| Issue | Explanation |
|---|---|
| **Independent of** | Operating system, distributed control system platforms and versions (no drive), and communication paths—interoperability |
| **Interoperability** | Interoperability provides streamlined integration with less development time and lower costs |
| **Easy usability** | Unified user interface (UI)d, single UI to all devices, common look and feel, built-in state-of-the-art graphics |
| **Full functionality** | Full functionality and diagnostics of their devices, including detailed parameter sets, configuration methods, status information, trends, data visualization analysis, and optimization. Reduced translational errors |
| **Diagnostics** | Signature diagnostics and predictive maintenance |
| **Safe operation** | No influence on runtime stability, easy updates, and device additions during operation |
| **Lower cost for** | Education and training, development, and maintenance |
| **No invest risk** | Controlled lifecycle management and transparent backward compatibility |
| **High scalability** | **Scalable**: From handheld configurator to manufacturing execution systems, from simple to highly complex devices |

environment of industrial plants. It sustains higher temperatures and is more durable, enabling the connection without interruption. Although IE works on the same basic principle as standard Ethernet, there are multiple factors that differentiate them, including hardened switches or IE switches, rugged cables, and connectors. Fast Ethernet and Gigabit Ethernet in the industrial zone can be installed to get gigabytes of data speed.

Security in communication is an important design issue that should be considered right from the beginning. Security has been briefly discussed in connection with "remote controls" in this chapter. Here, three different types of malicious software (malware) are briefly explained: worms, viruses, and Trojan horses. Worms are self-replicating programs that use security flaws to spread, often without the user of the machine being aware of this. Worms need not harm the machines; however, they quite often consume bandwidth and cause harm due to increased network traffic hindering important information exchange. Worms with payload (i.e., code designed to perform actions on the infected system) can cause damage to the infected machine and its system. Viruses and Trojans can both significantly impact communication security systems in different ways. Viruses are malicious programs that replicate themselves like worms and spread by attaching to other files, potentially corrupting/deleting data. Trojans are disguised as legitimate software but can be used to steal sensitive information, disrupt systems, or grant attackers remote access. The growing demand of extremely high data rate services, together with emerging advanced IoT/IIoT Industrie 4.0, including digital transformation applications, bring new technical challenges to network communication, especially wireless communications. The full integration of computing, communication, and control with artificial intelligence (AI) and machine learning (ML) is essential for the realization of these. In this section, major challenges and future trends in industrial data communication have been covered. Future wireless communication systems must support a universal and ubiquitous cyber physical infrastructure for a huge variety of applications with novel network structure, spectrum access schemes, and resource allocation solutions, while taking into account energy efficiency and security/privacy considerations [6].

To complete the discussion, all three main issues have been included in this section. Let the discussions start with IE.

### 9.2.0 Difference between Ethernet and Industrial Ethernet

Mainly developed in the 1970s, Ethernet was standardized as a group of LANs following the IEEE 802.3 standard. Ethernet allows computers to connect over one network, which is the heart of modern world communication amongst devices. As stated before, based on its application, Ethernet technology can be divided into two primary categories: *standard Ethernet* and *IE*. IE essentially refers to Ethernet that operates efficiently in a rugged industrial environment. "Standard" Ethernet, on the other hand, is Ethernet that is designed for home, business, and other uses besides industrial plants with harsh environments in real time. IE refers to Ethernet in industrial settings. IE protocols like EtherCAT, EtherNet I/P, and PROFINET are meant to provide *real-time control and determinism*. IE is a technology that connects multiple devices in an industrial set up/facility. IE offers the ability to integrate various devices and systems that were previously separate. Obviously, this makes the factory environment more efficient and productive, offering the most efficient and profitable automation solutions. Some basic characteristics of IE include the use of industrial-grade Ethernet hardware to withstand harsh environmental conditions and sharing of physical layers and DLLs. In terms of speed, data transmission rates range from 10 Mbps to 1 Gbps, with 100 Mbps being the most popular. Compared with commercial Ethernet, the speed needed for IE is much lower because the control and automation data bandwidth remain relatively low. When using Ethernet, streams of data are separated into shorter pieces or frames, each containing specific information such as the source and destination of the data. These data allow the network to accept and send data. Important issues include:

- **Medium:** In Ethernet technology, the medium is twisted pair, fiber optic, or wireless. IE requires more rugged connectors, cables, and switches.
- **Segment:** This is a single shared medium.
- **Node:** These are devices that attach to a segment.

In this subsection we discuss a comparative study between Ethernet and IE.

1. **Main IE components:** The primary tools to establish an IE system are:
   - **Hardened switch:** Hardened switches are complex switches. They can be the power over Ethernet (PoE) switches (Ref: II 18.0), with unique capabilities for IE that are not present in standard Ethernet switches viz. endurance of high temperatures and resistance to damage. Thus, better functionality and high performance are expected. The industrial environment, especially the shop floor, always has a high noise level due to the continued working of machinery. Therefore, they must have higher noise tolerance capability. There are multiple varieties of these switches, including *layer 2 switches*, *layer 3 switches*, and *PoE switches*.
   - **Ethernet ports:** These are different from standard Ethernet ports, having features like *rugged connectors*, *power supply*, and *data transmission*.
   - **Power devices:** Normally, power devices are usually high-power devices. Therefore, the industrial sector uses PoE (Ref: II 18.0) to get the maximum power supply of 100 W.
   - **M12 Ethernet cable and connector:** For harsh environments with extreme temperatures, vibration, moisture, and corrosive chemicals, M12 Ethernet cables designed for such environment are suitable. Solid copper conductors and a tinned copper drain (ground) wire are used in an M12 cable with a suitable M12 cable jacket. M12 connectors are circular nickel-plated or plastic connectors with a 12-mm locking thread and different coding with different pins for various applications. These cables are available with various Ingress protection (IP) ratings viz. IP67 and IP68. The rugged M12 connectors are designed to ensure a secure, reliable connection between field devices and controllers and between controllers and network switches. Also, M12 cables are well suited for washdown environments as in the food, beverage, and pharmaceutical industries.
2. **IE characteristics:** Ethernet has now made significant inroads into the fieldbus installed base, accounting for 65% of new installations. IE is distinguished by several key characteristics that make it suitable for use in demanding environments.
   - **Determinism:** Standard Ethernet is probabilistic based on the assumption that devices will probably not communicate at the same time to cause a collision. In industrial settings, timing is critical. IE supports deterministic communication to ensure with certainty that data are delivered within a predictable timeframe. In critical applications like motion control, delays in data transmission can lead to errors, and even to fatal accidents. IE uses real-time protocols, prioritization mechanisms, and quality of service (QoS) techniques for determinism.
   - **Latency and jitter:** Network latency refers to the time taken for data to travel from one point to another on a network. It can also mean the time taken to complete a round trip. In factory automation, especially in a time sensitive network like motion control, there has to be predictable and repeatable network latency, that is, devices must sense and take action in "real time" to ensure very low and guaranteed latency. Also, variation in latency, referred to as jitter, must also be kept to a minimum level in IE.
   - **Integration issues:** Open, industry-standard protocols make it easier to integrate devices from different manufacturers and can be implemented using off-the-shelf industrial-grade Ethernet switches and cabling. Many industrial protocols use Ethernet such as Ethernet PHY and DLLs for automation and control systems. So, Ethernet data handling must be modified to meet low latency and jitter requirements.
   - **Real-time data exchange:** Industrial processes call for real-time data exchange to maintain operational efficiency, determinism, and safety. IE supports high-speed communication, which allows for real-time monitoring and control.
   - **Ruggedness:** IE components, including switches, cables, and connectors, are built to withstand physical stresses. So, they should feature the various

ruggedness properties discussed before viz. metal enclosures, IP ratings, and resistance to chemicals and moisture.

- **Scalability:** IE networks must offer easy scalability to accommodate more devices or expanded coverage areas. This flexibility is important in growing industrial environments where additional machinery and sensors may be added over time.
- **Redundancy and reliability:** For continuous operation, IE networks often include redundancy features in devices, media, and power supplies to prevent network failure and improve reliability. These systems are designed to maintain network integrity.
- **PoE:** As already discussed in II 18.0, PoE allows data and power to be delivered over an Ethernet cable at the same time. PoE-capable IE switches can supply high power at up to 100 W per port.

3. **IE protocol types:** Time-sensitive networking is also an Ethernet extension and is defined by IEEE as an LAN solution with a more deterministic approach that will have a guaranteed maximum latency or coexistence with background or other traffic. The most common types of IE are:
   - **EtherCAT:** EtherCAT, developed by Beckhoff, is an open, real-time IE network. EtherCAT typically implements a ring topology in which one device is the master (standard Ethernet controller) and all others are slaves. Instead of TCP/IP, it deploys a unique method of on-the-fly packet processing and Ethernet communication. A telegram traversing through the network can build upon itself after each node. Each slave node processes its datagram and inserts the new data into the frame while each frame is passing through. It can provide scalable connectivity for entire automation systems. It delivers a flexible, low-cost, and network-compatible IE solution.
   - **Ethernet/IP:** Ethernet/IP ("industrial protocol"), developed by Rockwell, is an application-layer protocol on top of *TCP/IP* that utilizes standard physical, data link, network, and transport layers with CIP over TCP/IP. CIP provides a common set of messages and services for industrial automation control systems, and it can be used in multiple physical media. Ethernet/IP is the more flexible and compatible protocol thanks to the use of off the-shelf components and its object-oriented basis. This is more cost effective (but slower than PROFINET) and leans toward star topologies. Ethernet/IP offers complete producer-consumer service and enables very efficient slave peer-to-peer communications.
   - **Modbus TCP/IP:** Modbus TCP/IP is an IE control and supervisory control and data acquisition (SCADA). This is designed to address the problems of slow data transmission rates, incompatible communication protocols and media, and short distances of traditional data acquisition equipment. Data are processed in the Modbus format, but are wrapped and transmitted in Ethernet mode. A Modbus TCP network is normally programmed directly in the code of the controller using function blocks or other libraries.
   - **Powerlink:** Powerlink was developed by B&R. It is implemented on top of IEEE 802.3, allowing a free selection of network topology, cross connection, and hot plug. It uses a polling and time slicing mechanism for real-time data exchange. A Powerlink master controls the time synchronization through packet jitter in the range of tens of nanoseconds. Barriers to implement Powerlink are quite low due to the availability of open-source stack software.
   - **PROFINET:** PROFINET networks, developed by Siemens, are generally faster and costlier than their competitors. PROFINET networks are primarily deployed using line topology, which reduces the amount of cable needed, although tree and star topologies are also supported. PROFINET uses standard, unmodified Ethernet. It can use TCP/IP and UDP/IP communication for non–time-critical tasks such as parameterization and device configuration. For time-critical communication, PROFINET uses a real-time channel (PROFINET RT/PROFINET Class B), bypassing normal Ethernet packet processing. It uses a software-based real-time approach and has reduced the cycle time to around 10 ms. Class B is typically used in factory and process automation. PROFINET Class C (PROFINET "Isochronous Real Time" [IRT]) is isochronous and real time, requiring special hardware to reduce the cycle time to less than 1 ms to deliver sufficient performance on the real-time IE for motion control operations. The PROFINET protocol makes it easy to exchange between devices without worrying about user interactions.
   - **Sercos III:** Sercos III combines on-the-fly packet processing for delivering real-time Ethernet and standard TCP/IP communication to deliver low latency IE. It is very much like EtherCAT. A slave processes the packet by extracting and inserting data to the Ethernet frame on the fly to achieve low latency. Sercos III supports ring or line topology. One key advantage to using ring topology is communication redundancy. If the ring breaks due to failure of one slave, all remaining slaves still get the Sercos III frames with I/O data. Sercos III separates input and output data into two frames. It is comparable with EtherCAT and PROFINET IRT in speed (Fig. III 9.2a).

**FIGURE III 9.2A** Spanning Tree Protocol (STP) and Rapid STP in networking.

**Spa ning tree protocol (STP) & Rpid STP (RSTP): STP ensures a loop-free logica topology for Ethernet network by dynamica ly blocking redud a t path to preven broadca t storms and data duplica ions that can occur in network redud ant link. STP is defined in IEEE 8[illegible]. This suited where recovery time in the tune of minutes are a cceptable a in stad a d Ethere t. In case of IE, this is not a ceptle and RSTP is used. RSTP can quickly ada pt to link failure and recover through traffic rerouting with least down time and a twork resilieae . Here recovery time may b sub ms. RSTP is developed a per IEEE 8[illegible]v. Unlike STP (which ha 5 port sta es) it is 3port sta es viz. *Discarding, Learning* a d *Forwarding*.**

4. **IE challenges:** Major challenges IE faces are:
   - **Industrial environmental conditions:** IE technology is increasingly being deployed at remote field locations, hence the IE switch, controller, machines, sensors, and cabling will need to be robust enough to handle harsh field sites like high temperatures, corrosion, dust, electromagnetic interference (EMI), high-voltage transients, severe shock, and vibration. Fire and hazardous conditions may prevail in many industrial set up viz. oil and gas application. So, these call for additional//intrinsic safety requirements for IE. Hardware for IE must be capable of withstanding these harsh conditions. Therefore, withstanding these conditions and working with *reliability* is a big challenge.
   - **Bandwidth-hungry devices:** As remote field applications are covered by a single network, especially for bandwidth-hungry applications, bandwidth and network availability will be a challenge and critical factors to consider during network design.
   - **Scaling issues:** IE network scaling is a very challenging issue. As networks grow, managing increased data traffic and latency become critical. A major challenge is to ensure network security during scaling. With expansion of the network, there will be potential vulnerabilities, making it crucial to implement robust security measures. Major issues include *managing increased data traffic and latency*, *compatibility issues with legacy systems*, *network security during scaling*, and *balancing cost and performance*.
   - **Redundancy and recovery time:** For high network availability, network redundancy is deployed. For IE, a very quick recovery in the tune of milliseconds is called for. STP and RSTP are explained in Fig. III 9.2a. As stated there, for IE, RSTP is acceptable because recovery is in the tune of milliseconds. A longer recovery period may jeopardize plant safety. This is compounded when the number of switches increases. As field-level applications are aggregated onto the network, network redundancy becomes even more critical in ensuring network resilience.
   - **Security of IE:** Security vulnerabilities and loopholes in most industrial networks are, unfortunately, discovered only after a security breach has occurred. Therefore, maintaining the security of an IE network is of prime importance, that is, there is a stringent need for security. Implementing advances in security protocols to protect a high-capacity network is essential.
   - **Manageability for easy operation:** Switch manageability is an important aspect of maintaining large-scale networks. Operators and engineers will also need tools to efficiently manage switches from onsite and remote locations.
5. **Ethernet and IE comparison:** Based on the above discussion, it is clear that there are distinct features that are characteristic of standard Ethernet and IE. These are elaborated in Table III 9.2a.

Now let us move on to security issues in industrial communication. This will supplement the detailed discussion on "Security" in the next volume.

### 9.2.1 Security issues in data communication

In addition to communication between devices for real-time monitoring and control of the target system, industrial communication also performs auxiliary functions like parameterization, configuration, asset management, and firmware upgrading. At a higher level it acts as a coordinator for systems such as SCADA. The use of a fieldbus decreases cabling and costs, making the installation manageable. Modern industrial communication networks are increasingly based on open protocols and platforms for integration in office, IT, and internet environments.

**TABLE III 9.2A** Comparison of standard Ethernet and Industrial Ethernet.

| Issue | Standard ethernet | Industrial ethernet (IE) |
|---|---|---|
| **Purpose** | Home office data center, commercial and other nonindustrial application | For industrial applications and harsh and stringent industrial environments |
| **Real time and network** | Consumes time due to network bottlenecks CSMA/CD and needs more efficiency and speed | Industries demand real-time communication, so network must recognize this and data exchange must be fast for action. They handle time and even sensitive data in real time. |
| **Protocol** | TCP/IP and other standard networking protocols | Some real-time protocols are PROFINET, EtherCAT, Modbus TCP, and Ethernet/IP. |
| **Topology** | Typically star | Star, ring bus |
| **Speed and performance** | Typical range of speeds: i10/100/1000 Mbps (Fast and Gigabit Ethernet) | Typical range: 100 Mbps and 1 Gbps–10 Gbps. So, more bandwidth |
| **Operation** | Any downtime should be avoided | Downtime makes for a costly affair in industry and IE |
| **Determinism latency and jitter** | It is not deterministic on its own | For determinism IE use real-time protocols, prioritization, and QoS. Low latency and jitter (through switches) |
| **Reliability-stability** | Comparatively less reliable with less capability to eliminate electromagnetic interference (EMI) | IE uses different protocols capable of withstanding harsh environment and fulfilling the networking options at the industrial level to eliminate EMI |
| **Security and safety** | Emphasis is on cyberattacks and hacker penetration | In addition to cyberattacks and hacker penetration, IE must also take care of safety coming from fire and explosion hazards. So, hardware of IE should be explosive proof/meet intrinsic safety requirements |
| **Bus power supply and PoE** | Not called for | The fieldbus power is quite common in IE. They use PoE |
| **Device compatibility** | Standard and compatible | Rugged for more reliability and compatibility, depends specific use |
| **Guided media** | Unshielded twisted pair (UTP): Cat5e, Cat6, Cat6a; coaxial/fiber optics | UTP, STP, and M12—special-type fiber optics |
| **Management tool** | General network tool | Remote diagnostics, device monitoring, and central managing tools |

*CSMA/CD*, Carrier sense multiple access with collision detection; *IE*, Industrial Ethernet; *IP*, Industrial Protocol; *PoE*, Power over Ethernet; *STP*, shielded twisted pair; *TCP*, Transmission Control Protocol.

Particularly with the TCP/IP protocol suite, a wide range of cryptography-based secure communication protocols are available. We focus on industrial communication systems, which have a number of security-relevant characteristics that are distinct from office IT systems. Earlier, automation systems were not linked to each other and were not connected to public networks like the internet. Now, for accurate and up-to-date information about the plant and process, the status must be available not only on the plant floor, but also at the management level in the enterprise and even to supply chain partners. This results in increasing interconnection between different automation systems and between automation and office systems. Today, open and standardized internet technologies and the internet itself are increasingly used for that purpose. So, the available digital communication interfaces in both systems and devices pose security risks unless they are correctly addressed. This is especially true for advanced cyberattacks. A cyberattack can transpire through a *passive* attack or an *active* attack. In a passive attack, an attacker observes a system to identify open ports or vulnerabilities with the intention to collect information about the target. So, detecting passive attacks is challenging. In active attacks, cybercriminals manipulate a system or disrupt its standard operation to collect sensitive information. To protect a system from such a cyberattack it is necessary to follow security standards for industrial communication. Relevant ISA/IEC standards include:

ISA/IEC 62443-1-1: Technical specification defines according to its name definitions around the topic.

ISA/IEC 62443-2-1: This standard has a concept for a cybersecurity management system (CSMS) for industrial automation and control systems (IACS) including risk analysis, addressing risks with the CSMS, and monitoring and improving the CSMS.

ISA/IEC 62443-3: This part is published as a publicly available specification/prestandard for industrial control system security policy.

Here, security in industrial communication will be briefly discussed, which supplements detailed discussions in *Volume IV*.

1. **Common threat sources:** Common sources of security threats in industrial communication include malware, ransomware, supply chain attacks, denial of service (DoS)/distributed DoS, eavesdropping, and advanced persistent threats. These threats can disrupt operations, steal sensitive data, and cause significant financial and reputational damage. Network security threats are constantly evolving, and the variety of attack methods presents significant challenges for security management against a wide array of cybersecurity threats and vulnerabilities. Common industrial communication security threats with mitigation methods have been elaborated in Table III 9.2b. Apart from these cybersecurity issues, there can be other threats like *information leakage/tampering/open-source system.*
   - **Information leakage:** This can occur through an *IT leak* (leak through programming and configuration of field devices), *human leak* (social engineering attacks can give an attacker the opportunity to access confidential information), or *physical leak* (happens on the manufacturer side, but is more probable when the device is employed in the field). This is for field devices.

**TABLE III 9.2B** Common industrial communication security threats.

| Threat | Description | Mitigation |
|---|---|---|
| **Ransomware** | Infiltration with malware; a variety of ransomware attack vectors. Holds the network hostage until the victim pays. Impact: *reputational damage, revenue loss, long-term operational disruption.* | *Robust cybersecurity hygiene, regular updates and patches, employee training, endpoint detection and response,* and others |
| **Distributed denial-of-service attack** | A network or website is flooded with an immense volume of traffic by attacker to cause outages and significant operational disruption. | *Constant vigilance, threat detection tool deployment,* and *continuous monitoring* |
| **Supply chain attack** | Supply chain attacks exploit vulnerabilities in third-party vendors, for downstream effects. *Challenges*: lack of visibility, increased attack surfaces. | *Continuous monitoring, monitor fourth-party vendors* and *prioritize risk assessment during vendor onboarding* |
| **Global business expansion** | This is applicable for global businesses—the risk of cyberattacks for complex and large digital ecosystems. | Security policy, continuous mapping, risk management |
| **Unpatched application and server** | Overlooking maintaining a regular patching cadence due to the frequency of updates, the scale of enterprise systems, and a lack of resources. | *Patch management cadence,* vulnerability tool, address high-risk asset in time |

- **Tampering:** This occurs with field devices, which can be manipulated in different ways. *Spoofed firmware updates or change device parameters* can be inserted through the PC software controlling the device/the device's integrated UI, which alters the device firmware. These will cause the device to malfunction, creating damage through deployment of the field device.
- **Open source:** Open-source systems can be considered a risk/threat since the source code is readily available to anyone, including malware creators. This falls under OS vulnerabilities.

2. **Threat vector:** Threat vectors represent the method through which malicious actors can gain access; the system can only be stopped by thoroughly understanding it. Threat vectors are closely linked with attack surfaces, which represent the amalgamation of possible pathways an attacker may use to exploit. The motivations for using cybersecurity threat vectors vary by the type of attacker. Cybercriminals that effectively leverage cybersecurity threat vectors include *individual hackers*, *disgruntled former employees*, *politically motivated groups*, *hacktivists*, *cybercrime syndicates*, and *state-sponsored groups*. Threat vectors impact activities by *successfully infiltrating with a threat vector*, *stealing valuable information*, *launching ransomware for extortion*, *damaging systems*, *causing system failures*, and *taking control of systems*. Some of the major players are:
   - **Malware:** Malware describes many strains of software that are purpose-built for attacks. Malware gains unauthorized access to systems to cause damage from stealing sensitive data to disrupting operations. These attacks can be combatted using sandboxing and firewalls.
   - **Replay attack**: A replay attack is initiated with malicious access to communications, so that these communications can be reused.
   - **Brute force attack**: Passwords and encrypted data can be targeted by a brute force attack. The encrypted data can be accessed if the password is decoded using all the attempts of the possible key combinations.
   - **Misconfiguration**: Many threat vectors can be associated with misconfiguration, which can facilitate threat vectors when setup pages are enabled, a user uses default usernames and passwords, or errors occur when setting up cloud services. Automation, monitoring applications, and device settings are ways to protect against this.
   - **Dictionary attack:** The password may also be broken using a dictionary attack. This method can break the password only if all passwords in a predefined list are tried [7].
   - **Eavesdropping:** Eavesdropping in several forms can be used by cyber attackers to intercept communication and steal sensitive data in transit.
   - **DoS attack**: This attack refers to a cyberattack intended to make the system resource unavailable to the intended user. This is common in electric power grid systems.
   - **Poor encryption:** When encryption is implemented poorly or not at all, cybersecurity threat vectors multiply (i.e., the sensitive information is open to attacks). Data can be stolen during transmission/from data storage.
   - **Phishing:** Phishing is a social engineering attack method that uses email, text messages, or telephone calls.
   - **Remote access services:** Remote access services are examples of cybersecurity threat vectors that turn an important productivity tool into a potential point of entry for cybercriminals.
   - **Insider threats:** Insider threats are either malicious or negligent. A malicious insider is a user with internal access privileges (e.g., an employee or former employee uses their access to attack an organization).
   - **Drive-by download attacks:** Drive-by download attacks infect a device while users are simply browsing websites (even trusted websites). The malware is downloaded when users clicks on a link, pop-up window, or advertisement.
   - **Weak, compromised, or stolen credentials:** Credential exposure is a common cybersecurity threat vector. This arises out of *weak passwords* and password reuse, making users' login credentials easy targets for cybercriminals.
3. **Common attacks:** It is important to protect the master devices. If a master device gets compromised, the attacker can issue commands that appear "legitimate" to the slaves. Like a man-in-the-middle attack, a communication attack may occur silently, avoiding detection while gathering information about the system. In the case of DoS, detection could happen after it is executed. In addition to these, there may be spoofing of information like source/destination addresses. Protocols and systems unable to authenticate the source or destination address are vulnerable to spoofing. Some common forms of attacks are:
   - **Man in the middle:** The attacker acts as a legitimate partner and acts at both endpoints of the communication as if the attacker were the expected recipient.
   - **DoS:** As the name suggests, the user is deprived of the expected service from the server, that is, it decreases the *availability* of the system/service for its intended purpose.

- **Spoofing:** Using a technique called "spoofing," unauthorized parties gain access to restricted information. Culprits disguise their identities by modifying the address of the computer from which the scheme has been launched.
- **Eavesdropping:** In eavesdropping, the aim is to violate the *confidentiality* of the communication.
- **Virus:** In a virus-based attack, a legitimate user is manipulated into bypassing the *authentication* and *access control* mechanisms and is lured into executing the malicious injected code.
- **Trojan:** A Trojan is a special virus in which the malicious functionality is *hidden* behind the desired functionality (posing as some application) of the user. Trojans are typically employed to circumvent *confidentiality* or *access control* objectives.
- **Worm:** A worm is malicious code whose propagation mechanisms rely on automatic exploration and exploitation of vulnerabilities in the targeted system.
- **Malicious programs:** When detected, they can be combated with antivirus programs that are installed, but in the meantime they can corrupt the system to bring down computers or entire systems.
- **Piracy:** This is a kind of security threat as well. It is easier to create and sell intellectual property by stealing it.
- **Breaking into a system:** By violating the *authentication* and *access control* objectives, the attacker obtains the ability to control aspects of the behavior of the communication system to overcome the *confidentiality* and *integrity* objectives [8].

4. **Protocol and network security:** Here, security issues related to communication protocols and networks have been briefly discussed. The following are major issues:
   - **Communication protocols:** Security objectives are achieved by security protocols at different layers, as listed below:
     - *Link layer*: These are extensions of the point-to point protocol viz. the *password authentication protocol.*
     - *Internet layer*: Security at the internet layer is provided by IP security, which is an *optional* extension of version 4 (IPv4) and mandatory for IPv6.
     - *Transport layer*: Applications running on top of TCP (not UDP) can be secured by the secure sockets layer or the transport layer security (TLS). TLS refers to a cryptographic protocol for secure communication over a network.
     - *Application layer*: Typical examples of application layer security are *HTTP digest authentication (DA), Pretty Good Privacy,* and *XML Web Services.*
   - **Network security:** The following are some common network security systems:
     - *Firewall*: Firewalls are built to protect their computer/network systems from outside intruders. A firewall is a software module deployed at the border of a network or network zone to protect it against unauthorized access from neighboring zones. Basically, it is a gateway meant to examine the protocol-specific header fields of incoming data packets to decide the admission of incoming data. Firewall types and products vary based on the protocol layers.
     - *Intrusion detection*: An intrusion detection system is used to discover attacks based on known attack profiles and/or unusual system behavior.
   - **Wireless network:** Many current wireless industrial automations rely on IEEE 802.15.4.
     Wireless links can be targeted by jamming attacks by disrupting authorized communication in a system with a device that deliberately transmits different kinds of signals into a busy medium, decreasing the signal-to-noise ratio of the ongoing communication. There are variations in security *for short-range wireless links*; Bluetooth authentication and encryption rely on a shared "link key" between each pair of devices. IEEE 802.11 wireless LANs have higher transmission power but are more vulnerable to an eavesdropping attack.
5. **Security: objective and threat mitigation:** Here, let us elaborate separately the main objective of security in industrial communication and how various threats in industrial communication are mitigated.
   - **Security objective:** The three major perspectives of security systems are how the attacker reaches the target information system, the fundamental types of threats to the system (i.e., security objectives), and the confidence with which one can achieve the relevant security objectives. So, fundamentally, the security objective is the protection of information and deals with:
     - *Confidentiality*: This refers to preventing the disclosure of information to unauthorized persons or systems.
     - *Integrity*: The integrity objective refers to preventing *undetected modifications* of information by unauthorized persons or systems viz. product recipes/sensor values/control commands.
     - *Availability*: Availability ensures that unauthorized persons/systems cannot deny access/use to authorized users.
     - *Authentication*: Authentication is concerned with determining the true identity of a system user and

mapping the same to a system-internal principal with the identity of the user in the system.
- *Authorization*: The authorization objective (access control) is concerned with preventing access to the system by persons or systems without permission to do so.

- **Threat mitigation:** In industrial communication security threats, vulnerabilities should be addressed right from the security assessment and can be mitigated through *assessment*, *preventative measures*, *robust security controls*, and *proactive strategies* with *protective measures*. *Evaluations* of cybersecurity for vulnerabilities, *monitoring* for suspicious behavior, and detection coupled with implementation of *corrective measures* are the major steps. The main purpose of the mitigation technique is to reduce the risk to the network system. Each type of attack is addressed in the mitigation technique. Various mitigation issues are described in Table III 9.2c.

### 9.2.2 Industrial communication: Challenges and future trends

Industrial networking and communication systems act like the backbone of industrial processes, encompassing the interconnected web of devices, systems, and applications that facilitate data exchange and communication within production environments. In industrial communication, more than one computer transfers information and instructions to each other and shares resources. So, we can define data communication as the transfer of data in the form of a digitized analog signal over a point-to-point or point-to multipoint communication channel. This enables real-time monitoring, control, and optimization of operations, thereby driving productivity and efficiency and enabling informed decision making. Therefore, in addition to its several benefits, data communication also poses significant challenges that can hinder efficiency. Therefore, facilitating communication in harsh and expansive industrial environments poses several significant challenges. The

**TABLE III 9.2C Various threat mitigation techniques in communication.**

| Mitigation means | Detailed description |
|---|---|
| *Preventive techniques* | |
| **Network segmentation** | Isolate critical systems from noncritical networks and the internet. Dividing the network into smaller, isolated segments (e.g., firewalls) to limit the impact of a breach and the network is shielded from attackers' lateral movement. |
| **Access controls** | Implementing strict access controls viz. role-based multifactor authentication and access to restrict/limit access to operations technology/critical systems with regular review permissions. |
| **Vulnerability and patch** | Regularly assessing systems for vulnerabilities, update software and firmware to apply applicable patches, and configuration management. |
| **Data encryption** | For unauthorized access encryption of sensitive data (all time). |
| **Password** | Use of strong, unique passwords, and regular changes. |
| **Remote access** | Secured remote access: strong authentication and encryption. |
| **Protections** | There will be endpoint protection with a firewall, antivirus, and others. |
| *Proactive strategy for mitigation* | |
| **Response planning** | Incident response strategy to detect, contain, and recover from a cyberattack |
| **Assessment and monitoring** | Regular cybersecurity assessments for vulnerabilities/weaknesses and monitoring for suspicious activity and intrusions. |
| **Training** | Educating and training staff members on cybersecurity best practices. |
| **Zero-trust approach** | With zero trust, no user or device is automatically trusted prior to verification. |
| *Compensating measures* | |
| **Redundancy and back up data** | Redundant systems with regular backup of critical data to ensure business continuity. |
| **Physical protection** | Protecting physical access to industrial equipment and other infrastructure with suitable measures viz. surveillance and access control. |
| **Third-party risk** | Assessment of third-party risks to ensure security standards. |

integration of 5G technology into industrial networking heralds a paradigm shift in connectivity. The new facility can offer ultralow latency, high bandwidth, and support for massive device connections. In addition, 5G opens doors to real-time data exchange, remote operations, and augmented reality applications. Future wireless communication systems should effectively support a universal and ubiquitous cyber physical infrastructure, taking into account the energy efficiency and security/privacy considerations for a huge variety of applications with network structure, spectrum access schemes, and resource allocation solutions. So, there are several challenges inherent to future wireless systems. The industrial domain is characterized by the coexistence of legacy systems and cutting-edge technologies. Organizations often find themselves balancing the optimization of existing infrastructure with the benefits of embracing digital transformation. Industrial networking and communication stand at the crossroads of new innovation and operational efficiency, playing a pivotal role in the modern industrial ecosystem. With the introduction of IIoT and approach of Industrie 4.0 coupled with digital transformation, the challenges and trends within this domain have become more pronounced than ever before. In this subsection, future challenges and trends in industrial communication are briefly discussed. Challenges in different communication types have been discussed separately, yet there may be some commonality, and hence some repetition.

1. **General challenges:** Some of the common challenges are:
   - **Operating condition:** Industrial environments typically experience extreme temperatures, corrosion, humidity levels, and others. These are one of the biggest challenges in industrial communications. Communication devices used in such environments must be extremely durable so that they can operate under these conditions. In addition to these, the explosive environment in some plants, especially chemical and process plants, pose serious safety issues and are a challenge to industrial communication.
   - **EMI:** EMI/environments under which the communication technology must operate in also pose a major challenge in industrial communications. Most communication devices are susceptible to EMI. As the use of the electromagnetic spectrum continues to increase and electronic devices become more complex, the issue of EMI becomes progressively challenging.
   - **Reliability and resilience:** Industrial operations spanning over vast areas with harsh environments coupled with EMI create big challenges, making reliable connectivity crucial in these diverse settings. EMI, signal attenuation, and equipment failure can lead to network outages, impacting production efficiency and data flow.
   - **Real-time monitoring:** A key challenge in industrial communication is real-time monitoring. It is important for those working in industrial communication to know everything about their network in real time.
   - **Long-distance transmission:** Industrial communications often take place across huge geographical areas. Facilitating reliable, authentic, secured, and effective communication across such long distances is a challenge for industrial communication applications. High-speed communication across long distances is managed by IE switches with fiber optic connections.
   - **Multiple media:** Often, industrial applications require the use of various communication media. Since the differing media also have different communication standards, maintaining a communication standard between the many devices is also very challenging.
   - **Industrial cybersecurity:** With the proliferation of connected industrial environments, the risk of network attacks on critical infrastructure has grown exponentially, escalating vulnerability and cybersecurity threats. Protecting industrial networks from malicious actors requires robust cybersecurity to protect both IT and OT realms.
2. **Networking challenges:** A few issues related to network challenges are:
   - **Cybersecurity safeguard:** As discussed, with modernization, that is, with the proliferation of an interconnected industrial environment, a big challenge is ever-increasing vulnerability and cybersecurity threats. This calls for protecting industrial networks from malicious actors, which requires robust cybersecurity measures encompassing both the IT and OT realms. Therefore, organizations must implement stringent protocols, regular vulnerability assessments, network monitoring, and advanced intrusion detection systems to fortify their defense mechanisms. *Encryption*, *access control*, *network segmentation*, and *data backup* are some of the common security actions that form part of the security measure.
   - **Scaling issues:** With the increase in size of industrial systems, the number of connected devices and data exchanges have increased immensely, leading to *network congestion*, *increased latency*, and loss of data packets. All these directly impact real-time industrial communication. Some scaling solutions are network segmentation, addressing QoS, and upgrading network hardware and bandwidth.

- **Bandwidth and latency:** With an increasing number of connected devices and exponential increase in data generation, concerns around network bandwidth and latency become more pronounced and challenging. This is especially so in industrial scenarios requiring real-time, quick decision making; low network latency may lead to inefficiencies or safety concerns.
- **Comprehensive assessment:** It is necessary to make a journey of a network upgrade after a thorough evaluation of the existing infrastructure. This includes assessing *network architecture*, *identifying bottlenecks*, and *pinpointing areas susceptible* to security vulnerabilities.
- **Edge computing:** At times, edge computing has emerged as a potent solution to enhance industrial network *efficiency* and *responsiveness*. Edge computation with high-performance computing and predictive analytics holds the promise of revolutionizing industrial operations.
- **Software-defined networking (SDN):** SDN is capable of modifying the networking landscape by transferring network control from hardware infrastructure to a software-based system. SDN enables dynamic configuration, centralized management, and simplified provisioning of network resources.

3. **Wireless challenges:** There are a number of challenges for wireless industrial communication. These are often called grand challenges. These include:
   - **Security and privacy**: The broadcasting nature of wireless transmissions is very much susceptible to eavesdropping, creating opportunities for breaches of highly secretive information and privacy. Novel transmission technologies should be adapted.
   - **Communication infrastructure:** The growing demand of wireless services calls for the limitations of static resource allocation to be overcome.
   - **Interference and fault:** High device densities or devices with high frequency can cause interference, impacting signal quality and reliability, even faults. So, networks should be designed to withstand interferences and failures and maintain communication.
   - **Access techniques:** Effective wireless access techniques are essential for future generations of wireless communication systems to fundamentally meet the stringent requirements for very high spectral efficiency, very low latency, massive device connectivity, very high achievable data rate, ultrahigh reliability, excellent user fairness, high throughput, support to diverse QoS, energy efficiency, and cost reduction [9].
   - **Resource and spectrum use**: The spectrum availability for suitable wireless communication has become scarce. This calls for the exploration of a new spectrum.
   - **Resource and interference:** For emerging applications, new resource and interference management schemes are expected in the future.
   - **Efficiency enhancement:** Due to the enormous energy consumption expected by the massive number of connected nodes in future wireless networks, several strategies need to be implemented to make self-sustainable communication systems. Improving the channel condition during transmission is necessary for reducing transmission energy consumption and improving efficiency.
   - **Architecture and functional analysis:** Future wireless networks are envisioned to have a dynamic topology in nature because of network densification, wherein users will have many connections.
   - **Coding and modulation:** New coding and modulation schemes will be crucial in future wireless networks to ensure the achievable gains predicted.
4. **SCADA communication:** There are many challenges in SCADA, which are discussed here. There are cybersecurity vulnerabilities, especially when new technologies like IoT/cloud are integrated with legacy protocols. This also gives rise to huge complexity. Naturally, there will be a growing focus on enhanced security measures, viz. *encryption*, *multifactor authentication*, and utilization of AI for threat detection. Future trends in SCADA technology are focused on various issues like *increasing security*; *enhancing connectivity*, including integration of *IoT/IIoT*; *data analysis*; incorporating *AI* with ML; connecting to *cloud computing* (flexible, scalable, and cost-effective solution); and *edge computing*. Let us look into these issues in the following section.
   - **Selection issue:** Certain SCADA systems offer support for time-stamped data. In these systems, the data samples are time stamped with the real-time clock and the message is presented to the SCADA system. This approach effectively addresses network latency issues and helps mitigate challenges that can occur when a field device experiences temporary disconnection from the system.
   - **Scalability and legacy systems:** Expanding SCADA systems to accommodate growing needs can be complex, especially when integrating new technologies and ensuring adequate resources. Now, legacy system not equipped with modern cybersecurity features may cause such integrations (with new technologies) to be very difficult and may also enhance vulnerabilities.
   - **Cost effectiveness:** Managing budgetary and resource constraints in SCADA is a big challenge. At times, SCADA system costs can be prohibitive.

Making it cost effective without quality compromises is a big challenge.

- **Connectivity:** With limited or inconsistent network coverage, it is challenging to ensure reliable communication with remote assets.
- **Cybersecurity:** On account of open connections and integration, SCADA systems are increasingly vulnerable to cyberattacks from malware, DoS, man in the middle, and others. Future SCADA systems needs to prioritize robust cybersecurity measures, including encryption, *multifactor authentication*, *anomaly detection*, and *intrusion detection and prevention systems.*
- **Access control:** Because of the unique nature of industrial environments and the need for operational continuity, it is challenging to maintain access control in SCADA.
- **Data management:** Tackling and managing the vast amounts of data generated extracting meaningful insights can be a challenge.
- **Change management and training:** Managing change and providing adequate training during SCADA adoption are crucial for ensuring the success of the implementation.
- **Cloud/edge computation:** Cloud-based SCADA systems offer more *scalability*, *accessibility*, and cost savings. In edge computing, data processing occurs closer to the source, hence there are faster response times and reduced bandwidth requirements. Therefore, selection between the two as well as security and data integrity are vital issues to be taken into account and are challenging.
- **AI and ML:** As discussed, AI and ML are integrated into modern SCADA systems to facilitate *threat detection*, *predictive maintenance*, *optimization*, and others.
- **Remote monitoring:** Remote monitoring and control capabilities are becoming increasingly prevalent in SCADA, allowing monitoring and control from anywhere.
- **Digitalization:** Modern SCADA systems are integrated with many other digital technologies like IIoT/IoT devices and enterprise networks for more comprehensive and efficient industrial operations. Therefore, suitable integration to support digitalization efforts is a real challenge.
- **Standardization:** Efforts are being made to standardize communication protocols for interoperability between different SCADA systems.
- **Predictive maintenance:** With the introduction of AI and ML, efforts are being made to enable predictive maintenance strategies.

5. **Countermeasures:** To mitigate/provide countermeasures to challenges in industrial communication systems, a multilayered approach is crucial. Some major issues are:
   - **Security:** *Secure network design (network segmentation* and *custom protocol), robust authentication and access control design,* and *secured communication (encryption, incident response,* and *anomaly detection)* are some of the major issues.
   - **IT and OT convergence:** This convergence as well as integration with IIoT/IoT may cause security concerns, which can be overcome by a *unified security strategy*, *zero trust*, and *central monitoring*, as discussed before.
   - **Compatibility:** There are issues pertinent to hardware/software incompatibility. These will prevent interoperability. Major countermeasures toward these include *standardization*, *upgrading* hardware/software, *testing and validation*, *translation and conversion*, and other interoperability solutions (as applicable).
   - **Bandwidth limitation:** During periods of high traffic, there may be network congestion and other issues. These are mainly overcome by *bandwidth expansion*, *QoS control*, and *compression.*
   - **Human error:** Human errors are big challenges that can be addressed with *standardization*, *double checking*, *training and education*, and other methods.

After gathering some knowledge on miscellaneous communication protocols and issues, let us proceed to discuss wireless communication.

## List of abbreviations used

**AI** Artificial intelligence
**CAN** Controller area network
**CDDI** Copper-distributed data interface
**CFF** Common file format
**CIP** Common Industrial Protocol
**CN** Controlled nodes
**CSMA/CD** Carrier sense multiple access with collision detection
**CSMS** Cybersecurity management system
**DAS** Dual attachment station
**DCC** Direct cross connection
**DCOM** Distributed COM (Microsoft)
**DCS** Distributed control system
**DDE** Dynamic Data Exchange
**DDoS** Distributed DoS
**DECnet** Digital networking
**DLL** Data link layer
**DoS** Denial of service
**DT** Digital transformation
**DTM** Device-type manager
**EDDL** Electronic device description language
**EDR** Endpoint detection and response
**EDS** Electronic data sheet

**EMI** Electromagnetic interference
**FDDI** Fiber Distributed Data Interface
**FDI** Field device integration
**FDT** Field device tool
**FFDT** FDDI Full Duplex Technology
**GSD** General station descriptions
**HMI** Human-machine interface
**HTML** Hypertext markup language
**IACS** Integrated automation and control system
**IIoT** Industrial IoT
**I/O** Input/Output
**IoT** Internet of things
**IP** Internet protocol/ingress protection
**IPC** Industrial personal computer
**IPSec** IP security
**IT** Information technology
**KPI** Key performance index
**LAN** Local area network
**LAT** Local area transport
**LED** Light-emitting diode
**LLC** Logical link control
**LVDS** Low-voltage differential signal
**MAC** Media access controller
**MES** Manufacturing execution system
**ML** Machine learning
**MN** Managing node
**NAT** Network address translation
**NUT** Network update time
**OPC** Open Platform Communications
**OS** Operating system
**OT** Operational technology
**PDO** Process data object
**PLC** Programmable logic controller
**PMD** Physical media dependent
**PoE** Power over Ethernet
**PPP** Point-to-point protocol
**QoS** Quality of service
**RAP** Remote access platform
**RPI** Request packet interval
**RT** Real time
**RTPS** Real time publish subscribe
**SAS** Single attachment station
**SCADA** Supervisory control and data acquisition
**SDN** Software-defined networking
**SDO** Service data object
**SMT** Station management
**SoC** System on a chip
**SSL** Secure sockets layer
**STP** Shielded twisted pair/Spanning Tree Protocol
**TLS** Transport layer security
**UDP** User datagram protocol
**UI** User interface
**UID** UI description
**UIP** UI plugin
**UTP** Unshielded twisted pair
**VM** Virtual machines
**VPN** Virtual private network
**WS** Workstation

## References

[1] EtherCAT System Documentation; Version 5.4, Beckhoff, 2020. https://download.beckhoff.com/download/document/io/ethercat-terminals/ethercatsystem_en.pdf.

[2] *What is Industrial Ethernet: How Does It Differ from Commercial Ethernet?*; Real Time Automation; Internet document; https://www.rtautomation.com/industrial-library/what-is-industrial-ethernet/.

[3] *General Information About Ethernet POWERLINK and OPC UA*; OPC Foundation; Write up; Internet document; https://reference.opcfoundation.org/POWERLINK/v100/docs/4.

[4] J. Berge, T. Blevins, S. Mitschke and L. Winkel; *How FDI Technology Meets the Device Integrations Needs of Your Plant/Facility*; ISA Interchange; ISA document; https://blog.isa.org/field-device-integration-new-technologies-process-industries-fdi.

[5] *EDDL: Unlocking Device Information*; Emerson.com; Internet document; https://www.emerson.com/documents/automation/article-eddl-unlocking-device-information-en-37612.pdf.

[6] D.Benevides da Costa & H.C. Yang; *Grand Challenges in Wireless Communication*; Frontiers; https://www.frontiersin.org/journals/communications-andnetworks/articles/10.3389/frcmn.2020.00001/full.

[7] K. Imtiaz, M.J. Arshad, Security challenges of industrial communication protocols: threats, vulnerabilities and solutions, International Journal of Computer Science and Telecommunications 10 (4) (2019).

[8] D. Dzung, M. Naedele, T. Von Hoff, Security for industrial communication systems, in: Article in Proceedings of the IEEE, 2005.

[9] D. Benevides da Costa & H.C. Yang; *Grand Challenges in Wireless Communication*; Frontiers; https://www.frontiersin.org/journals/communications-andnetworks/articles/10.3389/frcmn.2020.00001/full.

Chapter 10

# Wireless communication

Wireless communication is a broad term that encompasses a variety of technologies to Transmit and exchange information without the use of any guided medium i.e., without physical connections. There are several types of wireless communication technologies such as Radio frequency (RF), Infrared (IR), Microwave: and light wave (optical fiber—a guided medium or free space optical communication). A computer network making use of Radio frequency (RF) is normally referred to as wireless network with RF connections between nodes in the network. Bluetooth, ZigBee, LTE, 5G etc. are also part of wireless communication. Wi-Fi acronym for "*Wireless Fidelity*" is specific to the wireless protocol IEEE 802.11 (IEEE 802.11—a,b, g, n, ac ax) specification. Internet, Cloud computing (Public cloud, private cloud, hybrid cloud and community cloud) are also part of wireless communication. A wireless network keeps devices connected to a network while they are free to move about. This is made possible by wireless communication technologies such as Wi-Fi, Bluetooth, and cellular networks. Now with the advent of 4G/LTE and especially 5G, not only several issues in wireless communication have been resolved, but it is possible to go for remote monitoring and control of industrial plants easily.

Wi-Fi communication takes place in most parts of the world these are the 2.4 and 5 GHz frequency *bands (Some 6 GHz* frequencies also). Bluetooth, ZigBee, etc. also operate in 2.4 GHz band. Wi-Fi network connection modes are: *infrastructure*, ad hoc, *Wifi direct. Wireless communications* are taking over the majority of wired connections in many industries, radically reimagining of the industrial landscape. The digital evolution brings multiple industrial wireless communications options that ensure seamless data transfer, remote monitoring and control and streamline the production process. This technology harnesses most of the industrial communication protocols, such as: Highway addressable remote transducer (*HART*), Open platform communication universal architecture (*OPC UA*), *Ethernet/IP*, *PROFINET*, and *EtherCAT* etc. to name a few. By eliminating the limitations of wired networks, industrial wireless communication enhances operational flexibility and reduces installation and maintenance costs. Wireless navigate the complexities of industrial settings, allowing for device-to device connectivity that can hold up under extreme conditions. Essentially, it enables a much more agile and decentralized system layout. The core of industrial wireless communication is reliability and latency. These are the primary concerns when the technology needs to support critical systems. Industrial communication protocols for wireless connectivity have made tremendous strides in reducing system complexity, downtime, latency and improving reliability. As there are no wires to encumber the system, maintenance, and troubleshooting occur with greater efficiency. Wireless systems are inherently adaptable to novel technologies and smarter industrial environments, which is a crucial advantage in an era characterized by rapid technological evolution. The adoption of industrial wireless communication, however, isn't without its challenges. Interference from other wireless systems operating within the same frequency range can disturb communication. Frequency hopping is used to ensure adequate performance and to reduce the impact of interferences. It alternates frequency channels for each transmission as a measure against interference. With these ideas in mind, we shall discuss wireless industrial communication systems in this chapter.

## 10.0 Basics of wireless network communication

From small area, wireless can be spread over to wide area in different network such as: Local area network (LAN), Personal-area network (PAN), Metropolitan-area network (MAN), and Wide-area network (WAN). Major components make up a wireless network's topology are client (user device) and Access point (AP) also called Wireless AP (WAP). In Wi-Fi network, the AP will advertise the specific network referred to Service set identifier (SSID), offers connectivity to users (client) see when they look at the list of available networks on their devices. A Wi-Fi-based wireless network sends signals using radio waves. So there will be a transmitter, which sends out the electromagnetic waves. Receiver is the device that picks up the electromagnetic waves and converts them back into a signal. There will be antenna that helps to transmit and receive the electromagnetic waves. Channel: This is the medium through which the electromagnetic waves travel. Thus, a client device receives the beacon transmitted by the

Plant Intelligent Automation and Digital Transformation. https://doi.org/10.1016/B978-0-323-89892-8.00010-7

AP and converts the RF signal into digital data, then that data is passed along to the device for interpretation. If the user wants to connect to the network, it can send messages to the AP trying to join and (when security is enabled) providing the proper credentials to prove *Association* and *Authentication.* If everything is order, data from the client (or from the AP to the client) is converted from digital data into an RF modulated signal and transmitted over the air. When received, this is demodulated, converted back to digital data, and then forwarded along to its destination. The Internet of things (IoT)/Industrial IoT, a network of interconnected devices that communicate and exchange data with each other. Wi-Fi systems more or less support of data rate of 54 Mbps for an indoor distance of around 30.5m (100 ft). Wi-Fi can function as Time division duplex (TDD) form, hence on account of channel sharing these are half duplex. TDD is a mode of data communication that transmits information both ways using wireless technology whereby the time cycle is divided into a number of transmission periods followed by reception periods all within the given frequency band. time is partitioned into pulses that would allow a transmission of signal within the available time intervals, the received signals in between the time frames of the next succeeding pulse. In this manner that can support both the bidirectional methods of data transfer at the same speed. Before we proceed on details on the topic, it is better to gather general idea about wireless in Network communication. In this section discussions will be concentrated on general discussions, commonly used terms and methodology as well as overall view of wireless communication in industrial plant network.

## 10.0.0 General discussions on wireless network communication

In Industrial communication-based around wireless communication is rapidly evolving to meet the demands of modern facilities to cater to the requirements toward connected factories, data analysis, cloud computing remote control etc. Wireless communication offers many advantages such as: reduced cabling and installation costs, increased flexibility, and suitability for harsh environments. However, it is not devoid of challenges also! Reliability, security, latency and determinism are major challenges. With developments of modern technologies such as 5G and Wi-Fi six (increased capacity and efficiency– IEEE 802.11ax) and Software-defined radio (SDR). SDR has been detailed in Fig III 10.0a. for customized wireless solutions for specific industrial application are promising significant advancements in this field, particularly for real-time control applications. Wireless HART, ISA 100 standard are examples of some industrial applications of wireless communication. In this subsection, short discussions will be put forward on pros and cons of the system, operating radio wave spectrum along with focus on ISA 100 standard. Let the discussions start with Advantages and disadvantages of wireless network communication.

1. **Advantage and Disadvantages:** Following are a few advantages and disadvantages of Wireless network communication.
   - **Advantages of Wireless network:** Listed below are a few advantages:
     - *Ease of Use:* Wireless network is rather easier to use, without prior information on it.
     - *Clutter free Layout:* Wireless network provides clutter free networked places.
     - *Planning and installation:* Wireless ad-hoc networks allow for communication without previous planning, unlike wired networks. Naturally installation and set ups are easier than wired counterparts.
     - *Portability:* Wireless networks allow portable devices to be used within the network coverage and access to the network is not bounded by the length of the cables.
     - *Disaster impact:* Wireless networks can handle disasters, namely earthquakes, flood much better than wired counterpart, which may break down completely in disasters. Here “much better” is used because, under disaster if antenna tower collapses then the situation could be different.
     - *Flexibility:* Since radio waves can penetrate walls, senders and receivers can be placed anywhere (even nonvisible), so, within radio coverage, nodes can communicate without restriction.
     - *Scalability:* Wireless LANs are scalable by nature, hence addition and removal of devices in network is easier compared with wired network.
     - *Cost Impact:* Since there is no direct cable cost and once one user is given access through access point subsequent user addition will not cost, so on an average cost of wireless network is less costlier than its wired counterpart.
   - **Disadvantages of Wireless network:** Listed below are a few disadvantages:
     - *Noise interference:* On account of radio waves there could be interference from nearby sources hence susceptible to noise interference.
     - *Quality of Services (QoS):* On account of bandwidth limitations QoS of wireless is typically lower than wired counterpart.
     - *Speed:* Wireless networks inherently slower than wired network
     - *Security against error and high bandwidth:* For security reasons it is necessary to encrypt the

SDR: "Radio in which some or all of the physical layer functions are software defined"-SDR forum and IEEE definition. SDR technology provides an efficient and less expensive solutions to function modifications allowing multimode, multiband multi-functional wireless device which can be further enhanced using software upgrades. This also provides more flexibilities than traditional radio devices.

**FIGURE III 10.0A** Software defined radio (SDR).

information. Thus, the bandwidth requirements will be more than the wired LANs

- *Restrictions:* Several govt./nongovernment institutions world-wide regulate the operation and restrict frequencies to minimize interference. Therefore, national and international frequency regulations have to be considered [46]. Also for license free operation 2.4 GHz ISM band has to be used.
- *Power Supply design:* For proper communication devices in wireless network need required power supply and the devices (especially in remote place) may operate on battery. Naturally, wireless network design must take into consideration power saving and power management. This may not be applicable in most of the wired network.

2. **RF Spectrum and Network issues: RF Spectrum and RF network issues:** As discussed before, wireless LANs operate mainly operate in RF range and some Infrared region. Our focus is mainly on LAN operating in RF communication region, which lies between the frequencies of 9 kHz and 300 GHz electromagnetic spectrum with division. Table III 10.0a Beyond Extra high frequency (EHF) starts infrared region, which also used in wireless communication. In RF region there is one unlicensed frequency band, referred to as Instrument, scientific and medical band (ISM) is mostly used in LAN. Mostly used in LAN (IEEE 802.11a) is 5.8 GHz. ISM 2.4 GB is used by IEEE 802.11b, g, blue tooth and Zigbee. Zigbee also uses 915/868MHzISM. There a few challenges RF networking faces. This has been enlisted below [1] [**courtesy:** ***S. Rackley's book***]:
    - **Link Reliability:** Signal propagation, interference, equipment siting, link budget.
    - **Media Access:** Sensing other users (hidden station and exposed station problems), Quality of service requirements.
    - **Security:** Wired equivalent privacy (WEP), Wi-Fi Protected Access (WPA), 802.11i, directional antennas.
3. **ARP Routing and NAT:** Short explanation of these terms have been presented below:
    - **Address resolution protocol (ARP)**: Each Physical layer data transmission is addressed to the Layer 2 Media access controller (MAC) address of the

**TABLE III 10.0A** Division of radio frequency spectrum.

| Frequency range | Description (transmission type) | Wave length |
|---|---|---|
| Up to 300Hz | Extremely low frequency (ELF) | |
| 300–3000Hz | Voice frequency (VF) | 1000-100 Km |
| 3-30 KHz | Very low frequency (VLF) | 100-10 Km |
| 30-300 KHz | Low frequency (LF) | 10-1 Km |
| 300–3000 KHz | Medium frequency (MF) | 1000–100m |
| 3–30 MHz | High frequency (HF) | 100-10m |
| 30–300 MHz | Very high frequency (VHF) | 10-1m |
| 300–3000 MHz | Ultrahigh frequency (UHF) | 1000–100 mm |
| 3–30 GHz | Super high frequency (SHF) | 100–10 mm |
| 30–300 GHz | Extremely high frequency (EHF) | 10-1 mm |

network interface card of the receiving device, not really its third layer Internet protocol (IP). In order to address a data packet, the sender first finds the MAC address that *corresponds to the immediate destination* IP address with a label of MAC address on the data packet. This is done using ARP. Sender broadcasts a message that requests the device with a certain IP address to *respond* with its MAC address. Therefore, *ARP converts/translates IP addresses into MAC addresses* within a local area network (LAN).

- **Routing**: Routing mechanism enables a data packet to find its way to the destination. Router compares the destination address of each data packet with router table, which is a table of addresses held in memory.
- **Network address translation (NAT)**: NAT refers to a process allowing multiple devices on a private network to access the internet using a single public IP address i.e., it translates the private IP addresses of devices within a network into a single public IP address, for communication with external networks such as the internet. When a computer sends a data packet to an IP address outside a private network, the gateway that connects the private network to the Internet will replace the *private IP source address* (say 192.168.0.1), *by a public IP address* (Say–. 205.55.55.1). The receiving server and Internet routers will recognize this as a valid destination address and route the data packet correctly. When the originating gateway receives a returning data packet it will replace the destination address in the data packet with the original private IP address of the initiating computer.

4. **IEEE Standards: IEEE Standards:** There are a number of IEEE standards used for networking especially for wireless networking. At this point it is important to note the purposes of various IEEE standards. Table III 10.0b enlists IEEE 802 standards with special reference wireless networking. List below is not total instead major ones (inactive ones are star marked) have been listed.
5. **ISA 100 Standard:** IEEE 802.11/15/16 etc. are developed for general purpose wireless network communications. ISA 100 is a wireless protocol especially meant for plant *controls* and *automation.* It covers both factory automation as well as process automation. It can be adapted from machine health control and monitoring, to basic process control including Monitoring of well heads or remote process, condition and environmental monitoring. Since it takes the help of wireless back haul ISA 100.15 and ISA 99 with IPv6 in its architecture, it is ready for IoT integration in a secured manner.

Fig III 10.0b depicts some features of ISO -100 wireless protocol. Fig III 10.0b1 accounts for structure of protocol along with various benefits ISA-100 wireless protocols offer. ISA-100 wireless protocol also possesses some unique features (which are not available with most of the industrial protocol) have been elaborated in Fig III 10.0b 2.

*With these basics in mind let us make familiar with terms and techniques in wireless digital communication techniques next.*

### 10.1.1 Common terms and techniques in wireless digital communication

Wireless communication is a broad term that encompasses a variety of technologies and methods used to transmit information without the use of physical connections. This form of communication has revolutionized the way we interact with the world, enabling devices to communicate with each other over distances without the need for guided media. Its application ranges from simple mobile phone to Ship Aviation, Automation, SCADA etc. to name a few. Naturally, there will be several terms and technique used for these. In this section a few terms used commonly in wireless digital communications and a few common wireless digital communication techniques have been discussed.

1. **Backhaul:** Back haul is a commonly used term in communication to mean a high-speed channel capable of transmission of high bandwidth at very fast speed. Wireless backhaul is the use of wireless communication systems to transport data between the internet and subnetwork i.e., the connection that flows from the wireless cell site to the internet is called the backhaul. Backhaul connection could include cables, fiber optics, or microwave/wireless components, depending on the circumstances. Also, wireless backhaul refers to the use of wireless communication links to connect base stations or access points to the core network, typically in different frequency bands than the access links. It's a cost-effective way to connect new sites, especially where wired connections are impractical or expensive.
2. **Wireless Ethernet: Media Access:** This is an interesting part of wireless LAN (Ethernet). Ethernet works on Carrier sense multiple access/"Collision detection" (CSMA/"CD"). However, "CD" is only possible if the PHY layer transceiver enables the device to listen to the medium while transmitting. This is possible in a guided medium network, where invalid voltages arising out of collisions can be detected, but is not possible for a radio transceiver since the transmitted signal would overload any attempt to receive at the same time [1]. So, as explained in *III 10.0.2.1*, in

**TABLE III 10.0B** Major IEEE standards for wireless network.

| IEEE 802 | Overview | Conceptual details |
|---|---|---|
| 802.1 | Bridging | MAC-based LAN/MAN bridging and management. |
| 802.2* | Logical link | Logical link control (LLC) specification |
| 802.3 | Ethernet | CSMA/CD over guided media |
| 802.5 | Token ring | The original token-passing standard. Also called "IBM token-ring." |
| 802.11 | Wi-fi | Wireless LAN media access control and physical layer specification. |
| *802.11a* | | High speed WLAN standard, supports data speed to 54 Mbps; use OFDM in the 5 GHz ISM band. Ratified after 802.11b |
| *802.11b* | | DSSS defined in the original 802.11, with boosted data speed 11 Mbps on 2.4 GHz ISM band |
| *802.11d* | | Enables MAC level configuration of allowed frequencies, power levels and signal bandwidth; global roaming |
| *802.11e* | | Quality of service (QoS); TDMA to prioritize and error-correction for enhanced performance |
| *802.11f* | | Interaccess point and interoperability of access points of multiple vendors. |
| *802.11g* | | Use OFDM modulation; enhanced speed 54 Mbps on the 2.4 GHz ISM band; interoperability same as 11.b |
| *802.11h* | | Dynamic frequency selection (DFC) and transmit power control (TPC); management in 5 GHz |
| *802.11i* | | Additional security for WLAN applications with<br>Advanced encryption standard (AES), authentication, and key exchange. |
| *802.11j* | | Japanese regulation for 11a; frequency range 4.9–5.0 GHz |
| *802.11k* | | Radio resource measurements for networks |
| *802.11m* | | Maintenance of802.11 family |
| *802.11n* | | High data rates of 150, 350 and up to 600 Mbps using multiple input and multiple output (MIMO); |
| *802.11r* | | Enables fast basic service set (BSS) to BSS transitions for mobile devices |
| *802.11s* | | Extended service set (ESS) support by extending 802.11 MAC. Suited for mesh network |
| *802.11u* | | Generic standardized approach for working with non 802.11 network, namely Zigbee |
| *802.11v* | | Enhancements of reliability through put with reduced interference. |
| *802.11w* | | Enhancements of network security by protection and management of data frame. |
| 802.11ac | WiFi 5 | Operates on 5 GHz with higher speed |
| 802.11ax | WiFi6 | Operates on 2.4 and 5 GHz with higher speed in improved crowded environment |
| 802.12* | Demand priority | Raises ethernet data rate to 100 Mbps with media utilization. |
| 802.15 | Wireless PAN | Communication specification |
| *802.15.1* | Blue tooth | Original 2.4 GHz frequency hopping spread spectrum (FHSS) specification |
| *802.15.2* | LAN PAN | Coexistence of LAN with PAN |
| *802.15.3* | UWB** | Short range high bandwidth ultra-wide band** link |
| *802.15.4#* | Zigbee<br>LR WPAN | Short range sensor network<br>Low rate wireless PAN (wireless sensors) |
| *802.15.5* | Mesh network | Develop MAC and PHY mesh network with PAN; enhanced reliability, extension of network, better battery life and easy configuration. |
| *802.16* | Broad and wireless | Total eight subdivisions 802.16 a to h. Fixed and mobile broadband wireless access methods used to create wireless MAN (WMAN) |
| *802.22* | | Wireless regional area network |

(Disbanded standards not listed; *marked are Inactive standards; *Wireless standards* numbers are indicated in italics).
# 802.15. Four is for point to point and low energy communications. ZigBee offers extra services, namely start topology routing, encryption, application services

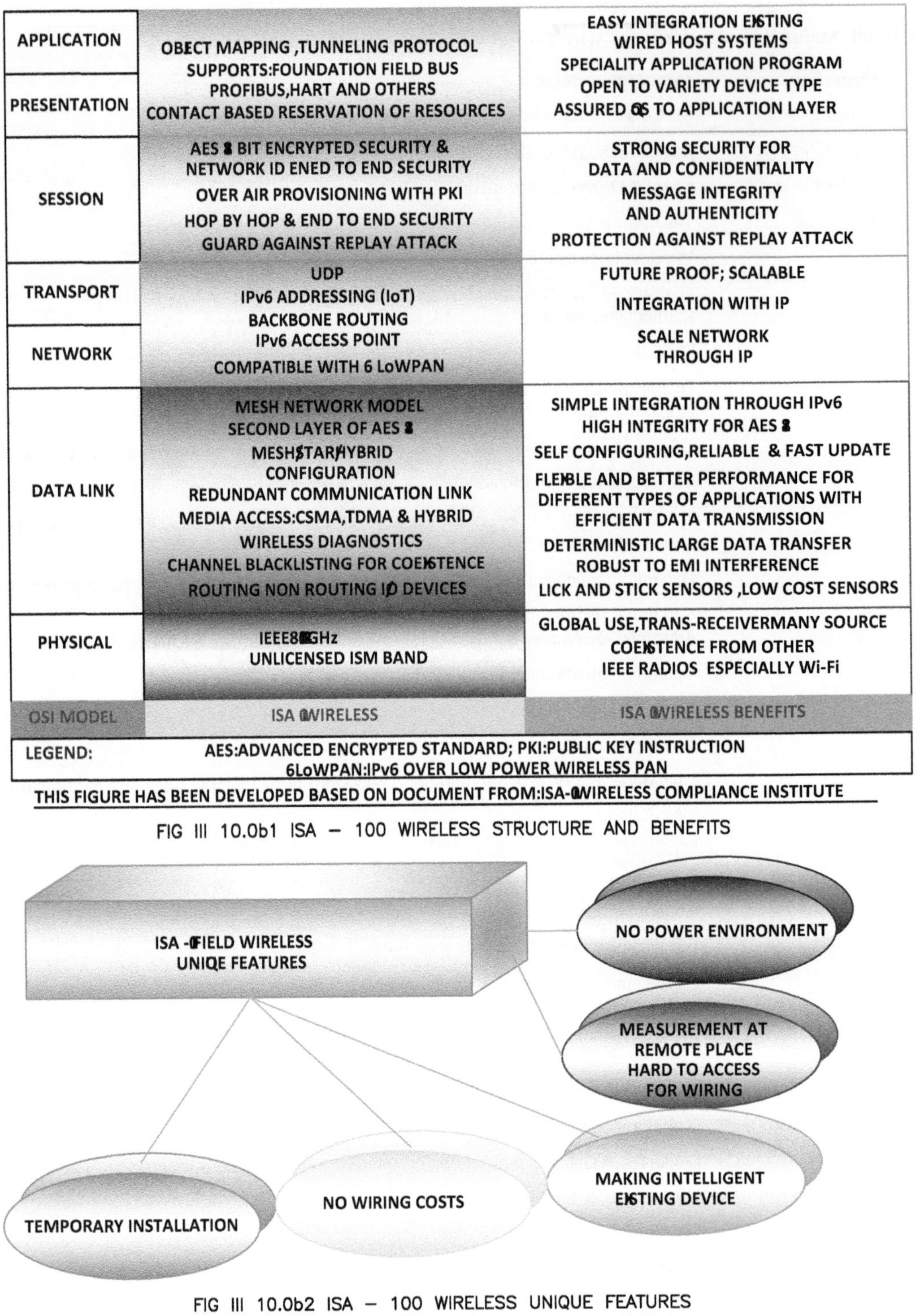


**FIGURE III 10.0B** ISA-100 wireless technology.

wireless networks 802.11, a variant of CSMA/CD known as CSMA/CA is used, where the CA stands for Collision Avoidance. CSMA/CA will first transmit the intent to send for data transmission to avoid collision. So, as the name suggests CA as per IEEE 802.11 standard, is effective before a collision, and is mainly used in wireless network. CSMA/CA helps in saving energy efficiently as the protocol waits until the channel is clear to avoid unnecessary energy wastage. CSMA/CA is efficient in channel use.

3. **Spread Spectrum (SS):** Narrowband signals are easily jammed (such as narrow road during rush hours due to heavy traffic) by other signals, and also there could be interference from other signals in the same band. Also, on account of fixed narrow band signal can be intercepted causing security threats. Spread spectrum is a *radio frequency data transmission technique that spreads the signal over a wide band of frequencies.* Basic philosophy behind spread spectrum is to use more bandwidth than the original data/message keeping signal power same. As in SS signal does not have any distinguishable peak in the spectrum, the signal is very difficult to distinguish from noise.
Wireless digital communication techniques to discuss next are shown in Fig III 10.0c.
Thus, on account of this modulation technique, the original data/message signal is less susceptible to noise and interference and interception. In narrow band more power is necessary to maneuvering jamming, whereas for the same power, in spread spectrum same can be done and power density will be less. Typical scheme for spread spectrum has been depicted in Fig III 10.0c1. Here, encoded signal is modulated with spreading generator (spreading signal generator) to convert narrow band data input into modulated spread band signal for transmission over the channel. At the demodulator the transmitted signal is again multiplied with spreading signal so as to extract the encoded data. Finally encoded data is decoded to get the information. Depending on the technique used spreading signals types can be different. Two important issues are:
   - **Pseudo-Noise (PN) Sequence:** A Pseudo (random) Noise (PN) sequence is a sequence of binary numbers (e.g., $\pm1$, 0,1) with certain autocorrelation properties. PN data sequence can be generated using a linear feedback shift-register circuit. Pseudo noise coding sequence is used in spread spectrum techniques. It is a type of cyclic code.
   - **Frequency Hopping (FH):** Frequency hopping systems is a system in which the transmitter changes the carrier frequency according to certain "hopping" pattern. Therefore, the signal sees a different channel and a different set of interfering signals during each hop over in spread band of frequencies and could also avoid the chances of interception at fixed frequency and the problems of interferences and jamming discussed earlier. Two kinds of frequency hopping are: *slow and fast* frequency hopping.

4. **Various SS:** Let us discuss various spread spectrum types in this part starting with DSSS. There are two major techniques to spread the spectrum, namely Direct sequence spread spectrum (DSSS) and Frequency hopping spread spectrum (FHSS). Spread spectrum methods and FHSS and DSSS have been explained in Fig III 10.0c. Apart from these, High rate DSSS (HR-DSSS) is also another technique
   - **DSSS:** DSSS use Differential phase shift keying (DPSK) (III 6.0.1), to modulate information on to carrier after encoding. In DSSS there will be rapid phase transition of PN code hence, to the data making it larger in bandwidth. As period $T$ of a signal gets shorter (i.e., R rate is increased) in time, the bandwidth $B$ of the signal increases: $R = 1/T = 2B$ (Nyquist Rate). In DSSS technique, each bit of the data is mixed (multiplied by) with the spreading code, which is pseudo noise sequence. This multiplied signal is transmitted. The receiver uses the same code to retrieve the original message. This is more explained in Fig III 10.0c2. Here each bit of the message will contain the entire PN sequence. Here, data spread spectrum as shown is different (but has the spreading of PN code) because in order to generate the output data ExOR logic has been used between PN code and signal data as detailed out in Fig III 10.0c2. Another feature is on account of signal spreading power density is much lower shown in Fig III 10.0c2. Here each bit is transmitted using *Barker sequence* (01,110,111,000).
   - **FHSS:** As the name implies, here there will be frequent change of carrier frequency. In this method as shown in Fig III 10.0c3, entire frequency spectrum is used and is divided into 79 frequency bands of width 1 MHz. Similarly, time has been divided in number of division equal to hop time. During each hop time carrier stays and uses any of the frequency slot then jumps up or down to another slot as shown. This frequency hopping in a pattern *known* to sender and receiver. Thus, because of this hopping, interception is practically impossible. In the said figure these hopping patters are shown for a definite limited time and frequency slots only for explanation. In the figure also the hopping pattern is marked.
   - **HR-DSSS:** In HRDSSS (IEEE 802.11b) Differential quadrature phase shift keying (DQPSK) is used. HRDSSS uses a combination of DQPSK and Complementary code keying (CCK) for modulation. Either four or 8 bits are encoded in each symbol period. Four bits are used for 5.5 Mbps (4 bit chunk) communications, and 8 bits are used for 11 Mbps (8bit chunk). In either case, 2 bits are always modulated using DQPSK, and the remaining bits are modulated using CCK (*a modulation technique used in IEEE 802.11b Wi-Fi to achieve higher data rates of 5.5 and 11 Mbps in the 2.4 GHz*

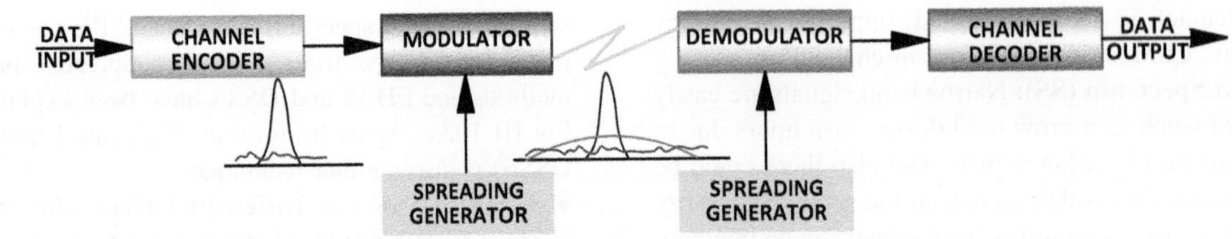


FIG III 10.0c1 SPREAD SPECTRUM BASICS

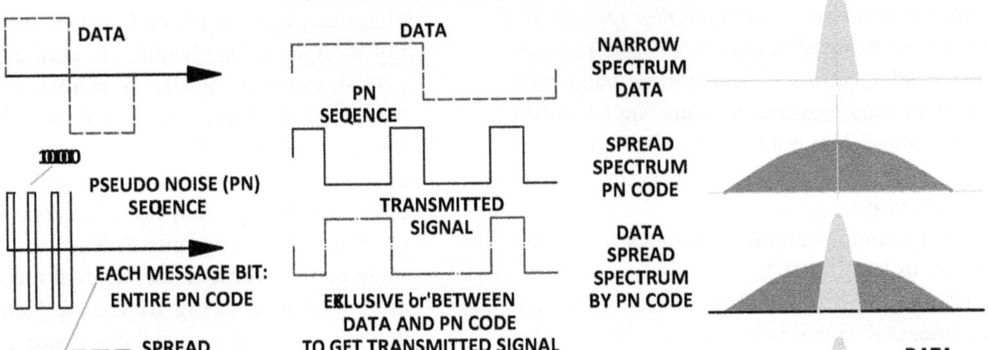


| I/P1 | I/P2 | O/P |
|---|---|---|
| 0 | 0 | 0 |
| 0 | 1 | 1 |
| 1 | 0 | 1 |
| 1 | 1 | 0 |

FIG III 10.0c2 DIRECT SEQUENCE SPREAD SPECTRUM EXPLANED

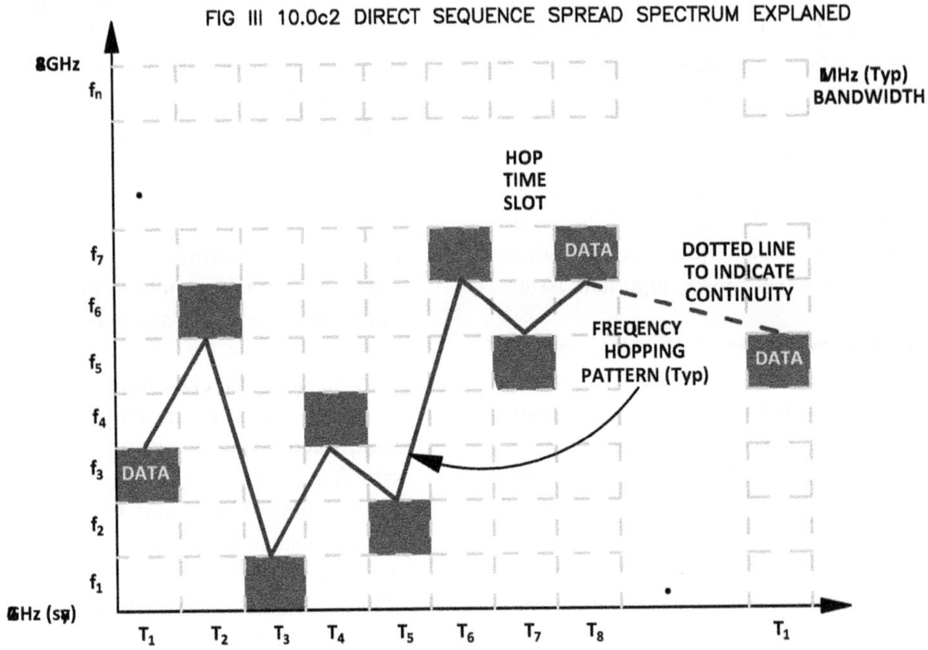


FIG III 10.0c3 FREQUENCY HOPPING SPREAD SPECTRUM EXPLANED

**FIGURE III 10.0C** Wireless digital communication techniques.

*band*). The coding method is much complicated. Here different pin codes are used for different bit sequences.

- **Difference between FHSS and DSSS:** Brief difference between two have been detailed in Table III 10.0c below.

5. **OFDM:** Orthogonal frequency-division multiplexing (OFDM) is a digital modulation technique used for wireless data communication over multiple narrowband subcarriers to enhance transmission efficiency, especially in *challenging wireless environments*, by mitigating the effects of multipath fading and

**TABLE III 10.0C** Difference between DSSS and FHSS.

| Feature | DSSS | FHSS |
|---|---|---|
| Frequency band | It transmits data using a wider frequency band. | It transmits data using a Narrowband carrier that hops among different frequency Channels |
| Pattern | Data transmission is Encoded and decoded using a Pseudo-random binary Sequence or *chip code* | Data transmission is Encoded and decoded using a Specific pattern called *hopset.* |
| Susceptibility | More susceptibility to multipath fading | Less susceptibility to multipath fading |
| Interference | Less resistant to interference | More resistant to interference |

interference. This significantly differs from SS type modulation. As the name signifies, this, is similar to frequency division multiplexing. In frequency division multiplexing total bandwidth is divided into different slots to transmit different data/information at different frequency band with some gap in between the bands so that there can be spread out sidebands on either side (not overlapping, for the receiver to separate them out). This is not the case with OFDM, here signals can overlap, saving bandwidth. Here, frequencies are chosen to ensure minimum interference between adjacent subcarriers by controlling the spectral *width* of the individual subcarriers (tones) so that the frequencies of subcarriers *coincide with minima* in the spectra of adjacent subcarriers i.e., the peak of one band will always have zeros of neighboring bands as detailed in Fig III 10.0f2. An advantage of OFDM is that, the symbol rate is much low when spread across multiple carriers so, the wireless link is less susceptible to intersymbol interference (ISI).

*Let us now direct our discussions toward LAN overview in next subsection.*

## 10.0.2 Wireless LAN over view (IEEE 802.11)

Wireless local area network (WLAN) is probably the most widespread, commercially significant and well-developed wireless networking technologies dominated by the IEEE 802.11 (*Wi-Fi*) standards including various specifications and amendments such as 802.11a, 802.11b, 802.11e, 802.11g, 802.11n etc. The standard outlines/defines the *Architecture, the MAC* and *physical layer* specifications WLAN accommodating mobile stations also. Major components in Wi Fi IEEE 802.11 architecture are: *Stations* (STA) (comprise all devices, which can be either AP or client) connected to WLAN, *Basic service set* (BSS) – a group of stations communicating at physical layer level, *Extended service set* (ESS) and *Distribution system* (DS) to connect APs in ESS. Depending upon the mode of operation, BSS can be *Infrastructure*, and *Independent types.* In this section brief discussions on overview of IEEE 802.11 wireless LAN shall be covered. Let the discussion start with wireless media access touched upon in previous subsection.

1. **Wireless: Media Access Explained:** Sharing media access among many nodes in a wireless network is quite complex because a wireless network station is unable to detect a collision between the transmissions between stations. This is because, radio transceiver is unable both to transmit and to listen for other stations transmitting simultaneously. The 802.11 standard defines a number of MAC layer coordination functions to coordinate media access among multiple stations. In standard 802.11, media access can be either contention-based, as in Distributed coordination function (DCF), or as in the optional Point coordination function (PCF). In DCF all stations essentially compete for access to the media, where as in PCF stations can be allocated specific periods during which they will have sole use of the media. Out of a number of MAC protocols that utilize carrier sensing, CSMA is the most common. Node desires to send data tests the channel activity. If channel is busy, node waits for predetermined period. Once the medium is free, the station waits a predetermined period called Distributed interframe spacing (DIFS) to send Request to send data (RTS) data to check if AP is free. If not, wait again. Now when AP confirms clear to send then it sends data i.e., transmission is complete. This is depicted in details in Fig III 10.0d where both logical and temporal views have been explained.
2. **Basic Service Set (BSS):** The 802.11 standards cover the PHY and MAC layer and Logical Link Control (LLC) services specified in the 802.2 standard, which also links too higher layers. In the 802.11 WLAN is a cellular structure. Each cell, under the control of an AP is Basic service set (BSS). This is detailed in network discussions. There are two modes of

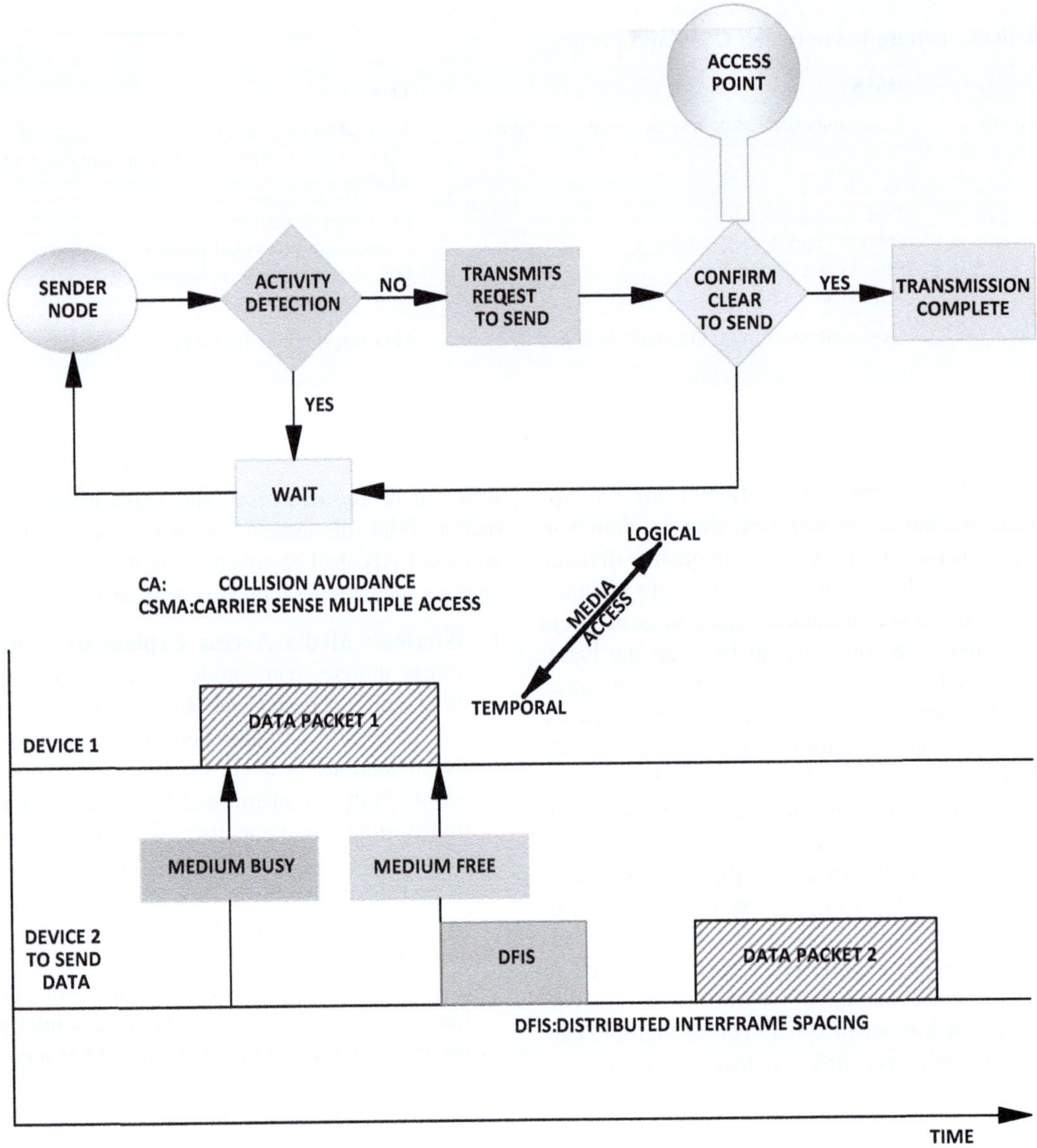


**FIGURE III 10.0D** Wireless media access: Csma CA

operational topologies for a BSS as depicted in Fig III 10.0e. Two topologies are:

- **Ad-hoc (independent)mode:** As shown in Fig III 10.0e1, an ad-hoc network is formed when a group of two or more 802.11 stations communicate directly with each other without AP or connection to a wired network. This peer-to-peer operating mode allows quick wireless connections for data sharing among a group of wireless enabled devices. Under ad-hoc mode the service set is referred to as Independent basic service set (IBSS). In IBSS all stations broadcast beacon packets.
- **Infrastructure mode:** As shown in Fig III 10.0e2, in this mode, stations communicate through AP even if two wireless station belongs to same cell. A WLAN with an access point and several wired devices connected through an Ethernet hub or switch is an example of infrastructure mode.

3. **IEEE 802.11 Protocol stack:** Fig III 10.0f shows the protocol stack for IEEE 802.11.
Part view of IEEE 802.11 protocol has been depicted in Fig III 10.0f1, where physical layer is quite similar to OSI model. HRDSSS (11 Mbps) and OFDM (54 Mbps) explained above in III 10.0.1, were introduced later to get higher bandwidths. Each PHY has associated MAC sublayer, which decides channel allocation. Above MAC layer is LLC, which follows IEEE 802 detailed out during network discussions later. In 2001, 802.11g was introduced with OFDM. Fig III 10.0f2 shows details of OFDM.

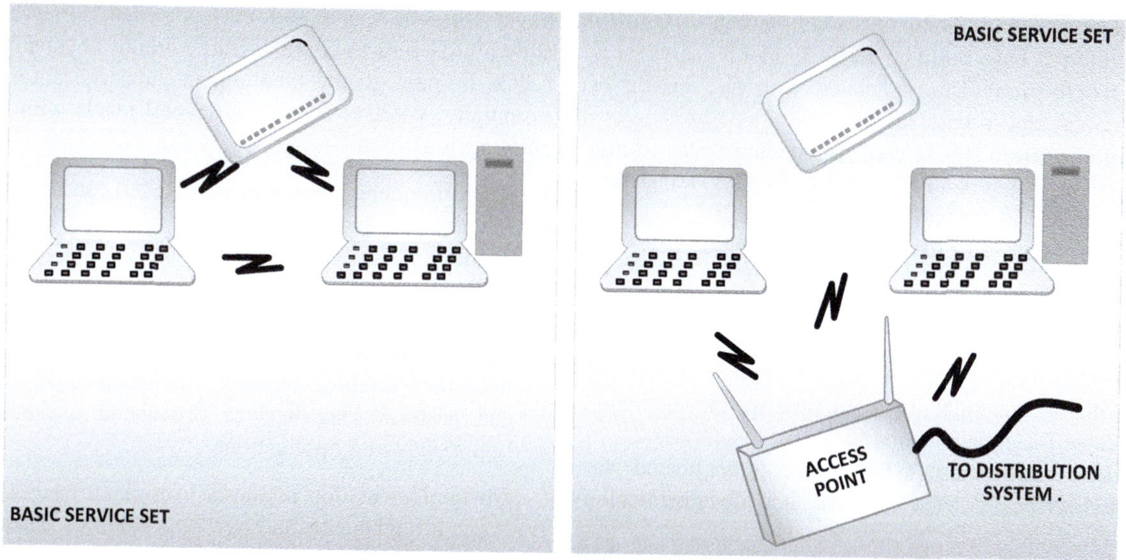


FIG III 10.0e1 AD–HOC TOPOLOGY FOR BSS FIG III 10.0e2 INFRASTRUCTURE TOPOLOGY FOR BSS

**FIGURE III 10.0E** Basic service set operation types.

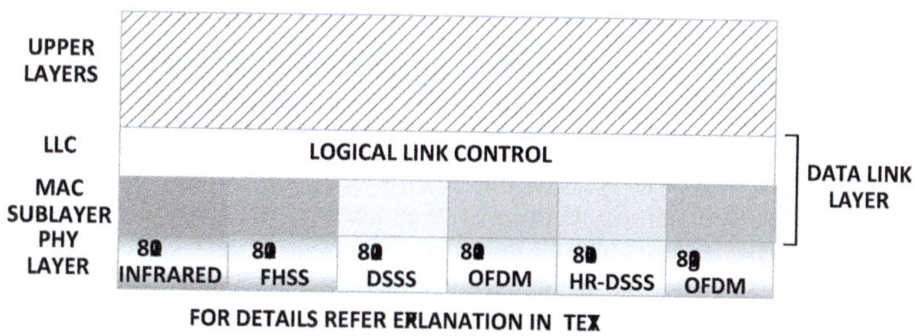


III 10.0f1 PARTIAL PROTOCOL STACK (IEEE 802.11)

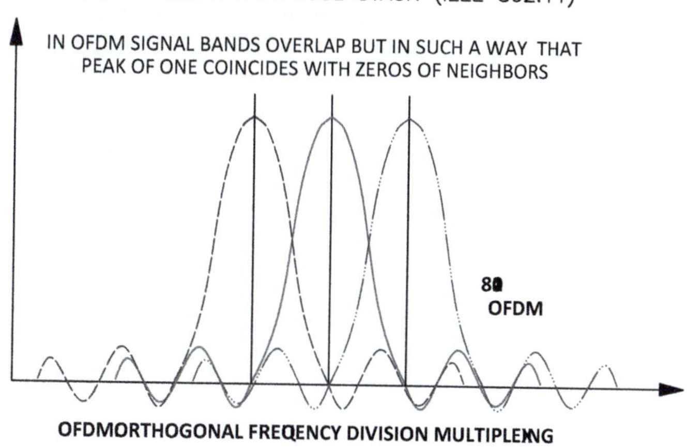


III 10.0f2 OFDM SIGNAL TRANSMISSION TECHNIQUE

**FIGURE III 10.0F** Partial protocol stack (IEEE802.11) and OFDM.

4. **Services offered:** According to IEEE802.11 there are nine services. Out of these, five distribution services offered by base stations are:
   - **Association:** This is related with mobile services to connect to base station. AP can only receive or deliver any data when a station is associated i.e., the information necessary for delivery of data are available with distribution system.
   - **Disassociation:** It is for breaking of relationship by either station or base station.

- **Reassociation:** With this a station can change its preferred base station—useful in mobile services.
- **Distribution:** This is related with the routing of frames sent to base station.
- **Integration:** This is used for sending frame to non 802.11 network. This service handles translation from 802.11 format to another format.

  Four services offered by station within *cell* shall include:
- **Authentication:** In order to avoid unauthorized communication, a station needs to authenticate itself before sending data. After a station is associated with base station, it has to face special challenge frame for authentication [2].
- **Deauthentication:** When above authenticated station wishes to leave network, it is deauthenticated and cannot be in the network.
- **Privacy:** This is related with encryption and decryption services.
- **Data delivery:** Ways and means to transmit and receive data. As it follows Ethernet Pattern so delivery is not guaranteed [2].

5. **Wireless sensor network** Wireless sensor networks (WSN) also offer enhanced plant asset management through the freeing up of cable resources and are currently drawing the most attention because of benefits directly relate to application area. The majority license-free wireless networks use 2.4 GHz bandwidth and are based on the IEEE 802.15.4 standard based on which also ISA-100.11a is built utilizing IPv6 for connectivity toward this. Low cost WSNs eliminate the needs of power and signal wiring and makes it possible to have measurements at sites that are hard to access and/or for temporary points where power supply is hardly possible to reach. WSN architecture is structured into three main layers: *Physical*, *Data Link* and *Application* layers. WSNs have the power of distributed communication, computing, and sensing features and are characterized as infrastructure less, fault tolerant and self-organizing networks. Various characteristic features are *Resource constraints*, *Communication paradigm*, *Application-based design*, *Scalability and design*, *Error-prone* (Node failure and unreliable communication), *Dynamic topology* etc. Without disturbing user benefits of both Wireless HART and ISA100.11a, efforts are on to a phased converged specification for both in ISA 100.12.

## 10.1 Wireless communication functionality

In wireless communication, information is transmitted through the air using electromagnetic waves, which can travel through space and even through some objects, allowing devices to communicate without physically connected through guided medium. In industrial automation, essentially, wireless can act at several levels within a facility such as:

- Radio-frequency identification (RFID) can be deployed for asset inspection, tracking, safety, security, and location.
- WSN are mainly used for condition monitoring, wireless instruments and measurements.
- WLAN covering areas such as mobile operator terminals, data logging, security, maintenance and IT etc. can perform long-distance broadband backhaul and high-bandwidth applications.

On implementation of these, in a plant, each of these areas can bring benefits and new opportunities. There are a number of challenges for industrial automation wireless networks. There are concerns about coexistence and interference leading to reliability and latency problems and about multiple protocols sharing the same bandwidth. Also, security is a great concern to the users especially for the potential for jamming, sabotage and the compromising of network privacy. Two most common standards used in industrial automation are WirelessHART and ISA100.11a. Although the two have some features in common, in reality they are very different, namely the scope of ISA100.11a is much wider. WirelessHART focuses on monitoring from HART-enabled field instruments, whereas ISA100.11a offers the scope to cover everything from field instruments to control room integration. Here detailed discussions have been presented on different ways to access the media in transmitting information through wireless means. Great concern on signal propagation concerning characteristics of various wireless components have been covered. propagation technology. Emphasis also have been put on various wireless technologies adapted in industrial automation. Detailed explanation of each of them have been elaborated. In view of current trend emphasis also have been put on IoTs/IIoTs, which have become part and parcel for Industrie 4.0. Wireless communication is the basis of IIoT/IoT, connecting billions of sensors/devices. Choosing the right protocols and technologies ensures seamless operation, reliability, and scalability. Looking at the current trend one can conclude that in future the trend would be to integrate Information technology (IT) and Operational technology (OT), to utilize more and more data for data analysis, edge/cloud computing etc. essential for Industrie 4.0and Digital transformation (DT). For all these requirement of 5G and other improved wireless technologies play very important role. Obviously, security issue can never be overestimated. Apart from all these detailed discussions have also been presented to cover various factors affecting wireless communication propagation. Now, let the discussions start with various wireless network propagation and access method in next section.

### 10.1.0 Wireless multiplexing and multiple access technologies

From initial discussions, we understand that in multiplexing in data communications multiple signals or data streams are combined into one signal over a shared medium to allow efficient use of resources and can significantly increase the amount of data exchange over a network. So, multiplexing is the sharing of a medium or bandwidth. Time division, Frequency division and wave division multiplexing with different variations within them, are commonly used.

- In frequency division multiplexing the bandwidth of a single physical medium is divided into a number of smaller, independent frequency channels.
- In time division multiplexing, instead of sharing a portion of the bandwidth, time is shared in the bandwidth.
- In Wavelength division multiplexing (WDM), multiple optical signals, each with a different wavelength are transmitted simultaneously over a single optical fiber to increase the capacity of optical fiber.

Having knowledge on the above we see that in Frequency division multiple access (FDMA) users are separated by frequency bands. Users have specific frequencies to transmit but fixed frequency allocation may lead to low bandwidth efficiency. Similarly in Time division multiple access (TDMA) users are separated by in time slots. It has moderate bandwidth efficiency but calls for guard time to avoid probable interference. Code division multiple access (CDMA) separates users by unique codes, all allowing multiple users to share the same channel hence shows greater bandwidth efficiency. Space division multiple access (SDMA) use utilizes location to separate users by focusing signals in specific spatial directions. It gives more bandwidth efficiency and range but inflexible. We see in all cases the philosophy is to separate users to avoid interference as well as get best benefit and efficiency. The key differences in utilization of different dimensions would results in variations in bandwidth efficiency, complexity, data rates and applicability. In this subsection various multiplexing and multiple accesses shall be covered. Wireless Multiplexing and multiple accesses have been detailed in Fig III 10.1a. Let the discussions start with WDM.

1. **Wavelength division multiplexing (WDM):** WDM is a fiber-optic transmission technique allowing multiple light wavelengths (Two or more colors) to send data over the same medium on one fiber i.e., multiple high-bit-rate data streams of (say) from 10 Gb/s upto recently, 800 Gb/s, each carrying distinct throughputs, can be multiplexed over a single fiber. Fig III 10.1a1 depicts WDM details. Based on number of channel accommodation, two WDM categories are:
   - **Coarse wavelength division multiplexing (CWDM):** CWDM is defined by WDM for lower-capacity applications, with typically up to <u>*eight channels*</u> with a typical spacing of 20 nm between the channels. Compact and cost-effective CWDM is used for short-range communications where spectral efficiency is not an important requirement.
   - **Dense wavelength division multiplexing (DWDM):** DWDM is used for multiplexing a large number of optical signals $\leq$80 channels onto a single fiber, with a spacing of 0.8 nm or less between the channels. Naturally, in DWDM the number of multiplexed channels is much denser as compared with CWDM, and is defined by frequency instead of wave length as in CWDM. Cooling LASER is used here for temperature and better performance and higher lifespan.
2. **Time division multiple access (TDMA):** As discussed, TDD, the simplest TDMA, uses half duplex form for uplink and downlink duplex communication. TDMA is used in Bluetooth (with TDD in Chord less phone). Fig III 10.1a2. shows that the time axis is divided into time slots that are assigned to users according to a slot allocation algorithm. The master device provides the system clock to determine the timing of slots. In each time slot, the master polls slave devices to identify devices in need to transmit and allocates transmission time slots. So, here nodes are active for short period on the same frequency but transmission is discontinuous in nature. Synchronization in time domain, limited capacity and requirements for guard time are characteristics features of TDMA.
3. **Frequency division multiple access (FDMA):** As discussed before, FDMA provides user, continuous channel with fraction of total available bandwidth. Fig III 10.1a3 shows FDMA explanation. Each user has allocated specific frequency to carry on continuous transmission. So, there will be filtering in frequency domain. Frequency division duplex (FDD) is simple form of FDMA where for the available bandwidth is divided into two channels to provide duplex communication. By nature, this is inflexible as frequency is scares resource.
4. **Code division multiple access (CDMA):** CDMA is a multiple access method in wireless communication and is based on key concepts such as *Spread spectrum:*, *Pseudorandom codes*, and *Orthogonality*

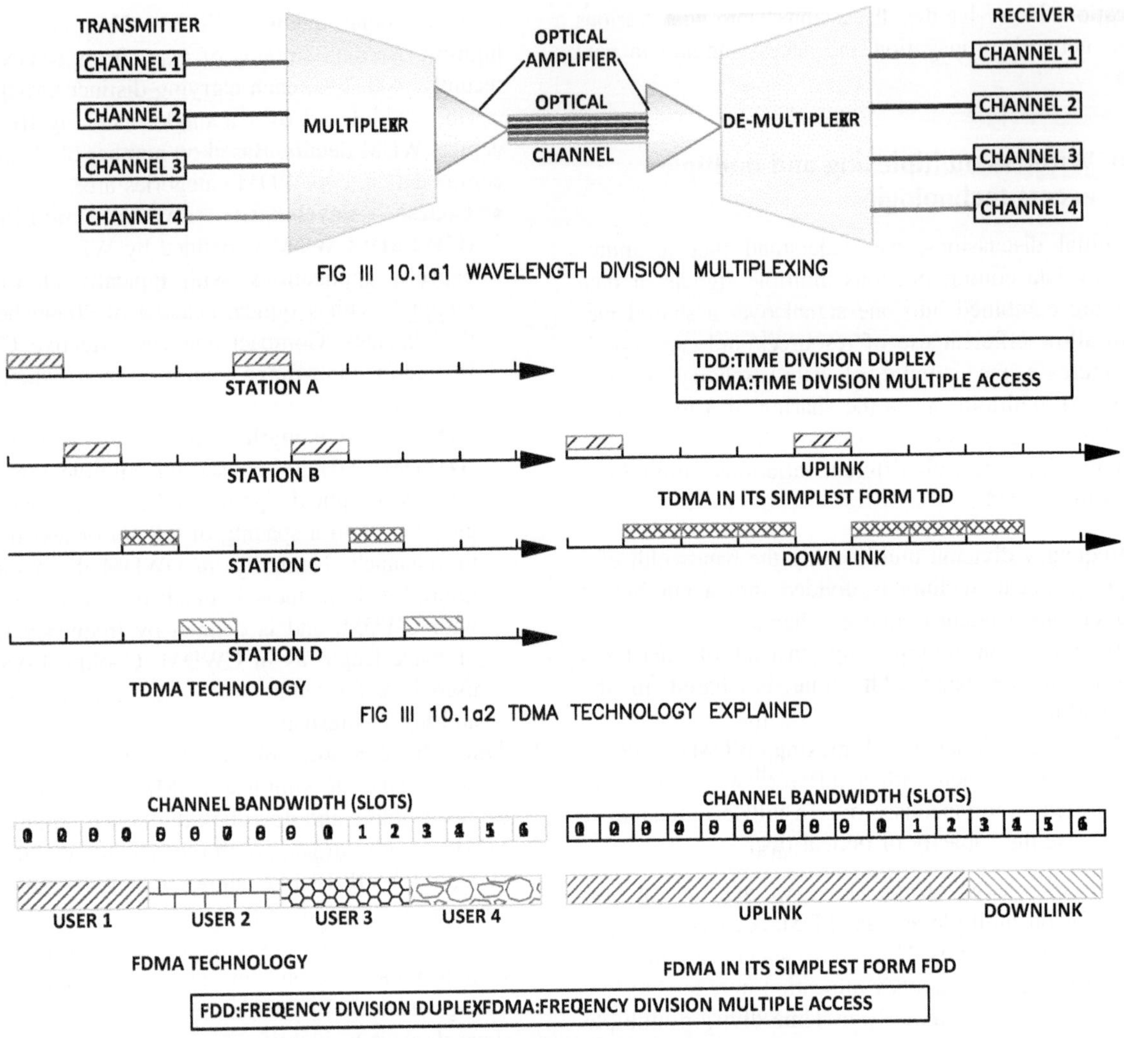


**FIGURE III 10.1A** Wireless multiplexing and multiple access technologies.

*(unrelated)*. Each user's data is spread over a wider bandwidth by multiplying it with their unique code. CDMA is closely related to DSSS, where a pseudo-noise code is used to spread a data signal over a wide bandwidth in order to increase its immunity to interference. As noted above, if two or more transmitters use different, orthogonal pseudo-noise codes in DSSS. Nodes/users can be active at the same place at the same moment uninterrupted. This is continuous, interference limited transmission with limitless channel—very flexible in nature.

5. **Space division multiple access (SDMA):** SDMA aims to multiply the data throughput of a wireless network by using spatial position as an additional parameter to control user access to the transmission medium. This is very simple with increased capacity but inflexible with antenna typically fixed. Only one terminal can be active in a sector.

### 10.1.1 Wireless: Propagation and technology discussions

No discussion of wireless communication is complete without some discussion on the propagation of wireless signals. Wireless propagation covers radio/electromagnetic wave travel and how this travel is affected by the environment and other factors. These are crucial for design and development of efficient wireless communication systems. Transmitted wireless signal to lose power as it propagates from the transmitter to the receiver. Reflections, diffraction, and scattering create multiple propagation paths between the transmitter and the receiver, each with a different delay and impact. Key factors include free space loss (signal strength decreases with distance), various environmental factors and various interferences. As result of these, the wireless propagation leads to a loss of received signal power as well as the presence of multipath. The signal can reach the receiver by straight, reflection, scattering or diffraction. When a signal reaches the receiver from the transmitter in a single path, without suffering any reflections, diffractions, or scattering, this is known as propagation along the *Line-of-sight* (LOS). When *Reflection* occurs when a wave impinges on an object that is smooth. *Scattering* is what happens when a wave impinges on an object that is rough or has irregularities with dimensions on the order of the wavelength. *Diffraction* is the "bending" of waves around sharp corners. The main characteristics of wireless communication channel are: *Path loss, Fading and shadowing, Interference Doppler shift* etc. Thus, we find that communication path for wireless transmission cannot be directly observed when transmitting radio waves, especially, for a problem, it difficult to ascertain where to look. This may result the incorrect impression that wireless communication is difficult to use or unreliable. Such confusion can be clarified with a greater understanding of the mechanisms and characteristics of wireless communication. In view of this, brief discussions have been covered on wireless signal propagation. Wireless technology is another important issue, which has also been touched upon. In modern industrial automation, wireless technologies are increasingly playing vital role with enabling flexible, scalable, and cost-effective communication between equipment/machines, sensors, controllers, and remote monitoring and control systems. On account of wireless technology, integration of IT with OT, IoT/IIoT, edge and cloud computing, data analysis with automation system is possible to attend Industrie 4.0 and digital transformation. As industrial systems grow in complexity and scale, traditional wired infrastructure often becomes expensive and difficult to maintain. Wireless technologies offer a practical solution by providing mobility, ease of installation, and the ability to connect to hard-to-reach locations. Here, various wireless type of technologies used in industrial automation shall be covered.

1. **The Process:** Reliable wireless communication calls for the radio waves to travel from the transmitter to the receiver with no loss in signal strength. Unfortunately, this is impossible in real world. As soon as the RF signal leaves the transmitting antenna, the signal begins to degrade. Any object will variously reflect, refract, diffract, or absorb the signal's energy, which is a common propagation characteristic. In order to understand the process, it is to get familiarized with commonly used terms enlisted in Table III 10.1a.
   Let us now, look into simple wireless communication process in Fig III 10.1b where there will be wireless communication from Device one to Device 2.
   Radio waves are amplified and attenuated at various points including *at the devices' wireless module* and *at the devices' antenna* as shown. During propagation of the signal through space, with repeated amplification and attenuation, the power level of the radio waves finally received by the receiving side wireless module

**TABLE III 10.1A** Commonly used terms in wireless propagation.

| Wireless term | Explanation |
|---|---|
| Reflection (Smooth surface) Scattering (Rough surface) | Reflection: The property of a wave/beam to bounce off of a smooth surface and proceed in a direction different from that which was intended. RF with frequency < 1 GHz reflects from ionosphere. RF with frequency > 1 GHz RF will reflect off of just about anything smooth, for example, doors, filing cabinets, buildings, even roads. Scattering: Can be conceived as multiple reflections where, reflection occur in several ways when a wave strikes an uneven surface. |
| Refraction | Refraction: a Property of a wave to change direction when traveling between different mediums. |
| Diffraction | RF wave encountering an obstacle, it tends to travel around it, like water. This is most pronounced at long distances (shorter distances– RF shadow) and at lower wavelengths. Diffraction occurs at a greater degree as the edge of an obstacle becomes sharper, as in the ridge. |
| Fade margin | The fade margin calculated in dB, refers to the signal level above what is required for a reliable signal. |
| Multipath | Multipath is a transmitted signal arriving at a receiver at different times through different paths this can cause signals to arrive in corruption or total loss. The time differential between these signals is measured in nanoseconds and is called the "delay spread." four multipaths are: *Upgrade* (increase in signal strength), *Downgrade* (decrease in signal strength), *Nulling* (cancellation) and *Data corruption* (signals exhibit different delay spreads) |
| Fresnel zone | Normally, the antenna is connected to the transmitter by a transmission line having characteristic loss per foot of cable. At the antenna, the signal strength is the sum of all the gains and losses from the transmitter to the antenna output called *Equivalent isotropically radiated power.* |

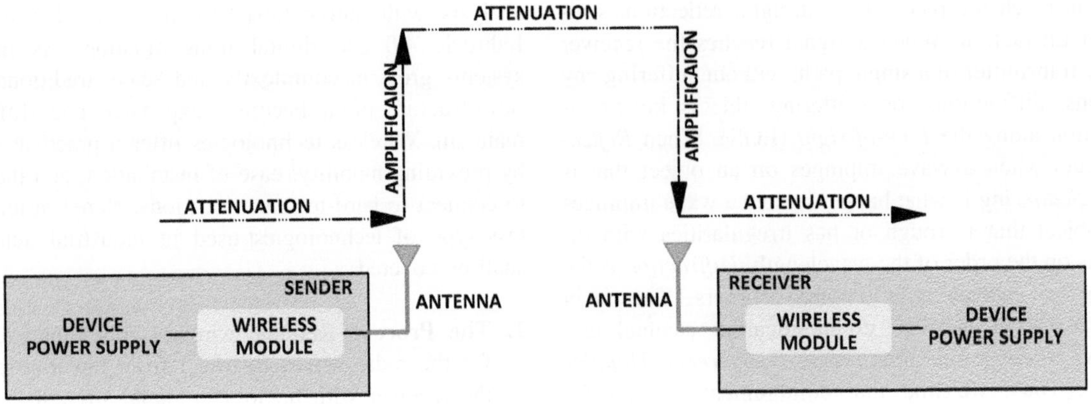


DEVELOPED BASED ON IDEA FROM ISA GUIDE [COURTESY:ISA]

**FIGURE III 10.1B** Wireless propagation process.

determines the quality of the wireless communication. Obviously, received signals should have a power level sufficient to ensure healthy wireless communication. With reference to Fig III 10.1b, let us look into attenuation and amplification at various stages in a simple manner:

- **Attenuation 1:** Loss of signal intensity within the circuits and cables as discussed.
- **Amplification 1:** The signal is amplified at the transmitting antenna for RF signal.
- Then the radio waves are propagated through space, from the antenna of Device one to the antenna of Device two
- **Attenuation 2:** Attenuation occurs due to propagation: reflection, scattering diffraction etc. and the influence of obstacles in the signal pathway.
- **Amplification 2:** The receiving antenna of Device two amplifies the signal. Then the signal is transmitted from Device 2's antenna to its wireless module.
- **Attenuation 3:** Transmission loss of signal intensity occurs along the cables and circuits of Device 2.

2. **Transmitter receiver characteristics:** In wireless communication, the purpose of transmitter is to convert information into electromagnetic waves, amplifying and filtering the signal for transmission, and regulating power consumption. On the other hand major functions of receiver include accurately capturing the signal, having high sensitivity to detect weak signals, offering good noise immunity (reject interference), and selecting the correct signal frequency. Major characteristic features are:
   - **Transmitter Characteristics:** A few characteristics expected of transmitter are:
     - *Power level*: The power level at which a sender can transmit is specified by law in the country of operation. Also, there will be power management for power consumption
     - *Conversion and Modulation*: Input information signal conversion suitable for wireless transmission, mixing information signal with a carrier signal for transmission
     - *Filtering and amplification*: Removal of unwanted signal, prevent interference as well as increasing power of the modulated signal to make sure it can travel desired distance. *Frequency and Bandwidth*: Operation within a specific frequency band and bandwidth to match the receiver.
     - *Antenna issue*: This is crucial for the electromagnetic waves transmission with the correct gain and polarization.
   - **Receiver Characteristics:** Following are typical characteristics of receiver
     - *Sensitivity*: The ability to detect and process very weak desired wireless signals. As, transmitters are limited by law on their transmission power, which also detects the reception sensitivity of the receiver
     - *Range and Power Management*: The range of signal strengths the receiver is capable of handling and it should be efficient in power management
     - *Noise interference*: A measure of the receiver's internal components degradation the signal-to-noise ratio of the received signal. Also, the ability to withstand and ignore interfering signals
     - *Selectivity*: The ability *to select* the desired signal from a multitude of received signals. *Demodulation*: Extraction of the original information signal from the modulated wave.
     - *Antenna Gain*: The receiver's antenna's gain influences the signal strength received.
3. **Antenna:** Antennas are essential hardware for signal conversions as well as transmitting and receiving radio

waves for wireless communication. Effectiveness of antenna depends on issues such as directivity, gain, and robustness to withstand industrial interference obstacles. Functionally, antenna performs functions such as: *Signal conversion* (electrical signal into electromagnetic wave and vice versa), *Radiate/absorb electromagnetic* (RF) waves, *Data transmission.* Antennas are characterized by certain specific issues such as: *Frequency band* (antennas are designed for a frequency band), *Robustness*, *Gain* and *Directional pattern*, *Connectivity*, *Efficiency* and *Productivity.* Frequency range is extremely important in selecting antenna for specific application. Frequency band decides distance coverage range and type of directivity. Also, specifying details on application helps in narrowing down antenna selection/Two important issues for antenna are:

- **Gain**: Gain is the degree of amplification of the radio wave. So, higher the power of the radio wave due to high gain of the antenna, better will be the quality of RF communication.
- **Directivity:** This measures the degree to which a radio wave is concentrated in direction. Based on directivity there are two types, namely *Omnidirectional*— having equal coverage in all direction and *Directional* — pointing RF concentration within a targeted area in a specific direction.

4. **Industrial wireless Technologies:** There are following major wireless technologies deployed for wireless communication in industrial plants and facilities. In each case, Definition/Description, Strength, Weakness and Application details have been enumerated.
   - **WiFi (IEEE 80,211):** Set of specification defined by IEEE, for WLAN using RF signal to exchange data/ information.
     - *Definition*: Wi-Fi is a wireless networking technology that uses RF signal to provide high-speed wireless communication. It refers to IEEE 802.11 standard as per Table III 10.1b
     - *Strengths*: High data rates suitable for large data images/video streams. High availability and supported by many devices.
     - *Weaknesses*: Susceptible to interference, not optimized for real-time, low-latency control.
     - *Application*: Industrial, control room terminals, portable operator panels.
   - **ZigBee (IEEE 802.15.4):** This stands for low power low data rate wireless communication as defined in IEEE standard mentioned.
     - *Description*: ZigBee and IEEE 802.15.4 These protocols that provide the network infrastructure required for wireless sensor network applications. The standard defines the physical and MAC layers, while ZigBee defines the network and application layers. Major characteristics revolve around long battery life, low cost, small footprint, and mesh networking to support communication between large numbers of devices.
     - *Strengths*: Low-power sensor networks with mesh capabilities. Self-healing networks increase reliability, support network with large no. of devices.
     - *Weaknesses*: Limited data bandwidth, Slow data rate Susceptibility to interference and security threats
     - *Application*: Distributed large sensor networks, temperature/humidity monitoring.
   - **Bluetooth:** Details discussed here covers Bluetooth low energy (BLE) also. This stands for short range wireless interconnections.
     - *Definition*: This is a wireless communication standard that allows devices to exchange data over short distances. It has many versions such as Bluetooth 1.0 through Bluetooth 5.4.
     - *Strengths*: Low power consumption, point-to-point/short-range mesh networks, Suited for handheld or wearable industrial tools.
     - *Weaknesses*: Short distance and limited device connections.
     - *Application*: Wireless configuration of instruments, Machine diagnostics, some operator interface, mobiles and consumer electronic devices.
   - **Long range (LoRa):** The discussions includes both LoRa and LoRa WAN. These refer to low power consuming technologies designed for transmitting data over extended distances, (in Km). They operate in ISM (Industrial, scientific, and medical).

**TABLE III 10.1B** Common wireless standards: IEEE 502.11 [Ref: III 10.0b also].

| IEEE standard | 802.11a | 802.11b | 802.11g | 802.11n | 802.11ac |
|---|---|---|---|---|---|
| Frequency | 5 GHz | 2.4 GHz | 2.4 GHz | 2.4/5 GHz | 5 GHz |
| Max data rate | 54 Mbps | 11 Mbps | 54 Mbps | 600 Mbps | 1 Gbps |
| Indoor range | 100 ft | 100 ft | 125 ft | 225 ft | 90 ft |
| Outdoor range | 400 ft | 450 ft | 450 ft | 825 ft | 1000 ft |

- *Description*: Based on spread spectrum, LoRa is a simple means of achieving long-distance, low-power wireless communication. LoRa is a physical layer used to establish long-distance communication links. LoRa technology has revolutionized IoT and IIoT landscape, providing a means to connect devices over long distances without the need for extensive power consumption. The ability to maximize the range of LoRa WAN enables the deployment of wireless sensor networks across vast geographical area making things economic and logistically reliable.
- *Strengths*: Very long-range communication (<10 Km open areas), extremely low power consumption, ideal for remote sensor IoT/IIoT.
- *Weaknesses*: Very low data rates; best for periodic or small data packets may require additional protocol for two-way communication.
- Application: Water and gas metering, remote oil/gas site monitoring, agriculture automation.

- **NB-IoT/LTE-M:** Narrowband-IoT, and Long term evolution for machines use cellular technology for IoT/IIoT in industrial scenario.
  - *Definition*: These new form of connectivity are Low power wide area networks (LPWAN) developed for IoT. These are narrow band technology. As the name suggests, Narrowband-IoT (NB-IoT) is a low power wide area technology developed for IoT. Long term evolution for machines (LTE-M) is also the simplified also form of LPWAN.
  - *Strengths*: Use of mobile networks, supports roaming and deep indoor penetration, efficient use of bandwidth, long battery life, reliable connectivity, low cost, and security specially suited for IIoT/IoT applications.
  - *Weaknesses*: Dependency on mobile operators, not ideal for mission-critical control tasks, does not support VoLTE.
  - *Application*: Smart meters, IIoT/IoT, remote alarms.
- **Wireless HART:** This is wireless version of popular HART. This is IEC62591-1.
  - *Description*: Built upon the existing HART, wireless HART is a wireless protocol designed for industrial process measurement, control, and asset management applications. WirelessHART has the ability to self-organize and self-healing, into a "full-mesh" network, so that every device has multiple redundant communication paths with a line of site range of ~250m, installed in the range of 50–100m. Network formation is a secure process to authenticate devices before they are allowed to join the network.
  - *Strengths*: Developed specifically for industrial process control; deterministic communication; mesh topology; robust against interference.
  - *Weaknesses*: High cost and complexity with limited bandwidth.
  - *Application*: Process plants and production lines.
- **ISA 100.11:** IEC62591-1 is meant for HART, whereas ISA 100.11 is an open and supports multiple protocols.
  - *Description*: ISA100.11a is designed as a family of standards to support a wide range of wireless industrial plant needs including process automation, factory automation etc. Major design issues include: *Support for multiple protocols*, *Open standards*, *Support for multiple applications*, *Flexibility*, *Reliability*, *Security* etc. ISA100.11a defines the protocol stack, system management and security functions for use over low-power, low-rate wireless networks.
  - *Strengths*: Designed for process automation with support for complex traffic, high security, and interoperability with other protocols.
  - *Weaknesses*: Niche technology with fewer vendor options than WirelessHART.
  - *Application*: Industrial Process and Factory automation for reliable wireless control.
- **5GURLLC:** This stands for fifth generation Ultra reliable low latency communication (URLLC).
  - *Definition*: This is a 5G networks critical service designed for mission-critical applications requiring extremely high reliability and near-instantaneous response times.
  - *Strengths*: Unprecedented low latency (~1 ms), extremely high bandwidth, and ability to connect thousands of devices.
  - *Weaknesses*: Requires significant infrastructure investment.
  - *Application*: Factory automation, Automated guided vehicles (AGVs), remote robotics, augmented reality maintenance.

Above discussions have been summarized along with application details in Table III 10.1c.

5. **Wireless in Automation:** Various issues related to wireless communication industrial automation encompass the following:
   - **Industrial Interferences:** Industrial communication systems must comply with the stringent requirements for reliability, availability, and determinism in order to serve real time automation applications. Therefore, the development of wireless communication systems, for the extreme time critical area of factory automation, is a big challenge. Apart from interference on industrial radio channels from ones of mobile radio channels there are certain physical

**TABLE III 10.1C** Overview of wireless technology [3].

| Technology | Band[A] | Speed[B] | Range[C] | Feature | Application |
|---|---|---|---|---|---|
| **WiFi** | 2.4/5 | <1000 | 30–100 | High throughput | High speed data |
| **Zig bee** | 2.4 | 0.25 | 10–100 | Mesh support, low power, simple sensor | Environmental monitoring |
| **Bluetooth/ BLE** | 2.4 | 1–2 | 5–30 | Point to point; low energy Short range | Equipment configuration/ diagnostics |
| LoRaWAN | 0.433 /0.868 | <1 | <1000 | Long range low power and bandwidth | Remote and environmental monitoring |
| NB-IoT/ LTE M | Licensed cellular | National scale | Low-medium | Remote monitoring | Remote monitoring |
| WirelessHART | 2.4 | 0.25 | 30–100 | Low latency, high reliability | Safety-critical control |
| **ISA100.11** | 2.4 | 0.25 | 30–100 | Interoperability | Advanced and safety-critical control |
| 5G URLLC | <6 | In gbps | 100–500 | High reliability, Low latency; | High speed data; Safety-critical and real time control |

A: Frequency Band in GHz; B: Speed (Data rate) in Mbps; C: Range (Typical) in meters

phenomena occur in industrial facilities to create interferences. These phenomena are:

- *Reflections:* This occurs, when EM-waves encounter reflecting objects, whose dimensions are much larger than the wavelength.
- *Scattering*: Scattering can happen because of one of the two reasons/both. Either when the dimensions of the encountered object are much lower than the wavelength of the EM-wave or when the surface structure is classified very rough in comparison to the wavelength.
- *Diffraction:* As discussed, diffraction occurs when EM-waves encounter sharp edges.
- *Shadowing*: Shadowing is caused by obstacles, which completely block the propagation paths of EM-waves.
- Doppler effec ts: This would arise, either when there is a relative movement between transmitter and receiver, or a mobile obstacle in the propagation field causing reflection, scattering, diffraction or shadowing.
- *Large scale fading*: This results from widespread movements. Large-scale fading can be caused by path loss of signal as a function of *distance* and *shadowing* by large objects.

  *Small scale fading*: Small scale fading characterizes the fast fluctuations of radio channels over short distances. This could be due to the constructive and destructive interference of the multiple signal paths between the transmitter and receiver and may occur at the spatial scale of the order of the carrier wavelength (frequency dependent).
- *Delay spread*: The delay spread describes the mean spread in time of transmitted impulse.

  Based on the implementation, three interferences in CSMA/CA are:
- *Type-1*: A weak signal, incapable of inducing interferences at the receiver, is detected at the transmitter, can cause the medium to be classified busy, and delaying the transmission
- *Type-2*: Interferences caused by multiple radios accessing the medium simultaneously.
- *Type-3*: The source of interference is out of the detection range of the transmitter,

  Broad categories of interferences have been detailed in Fig III 10.1c.

- **Discussions: Performance issues:** Wireless technologies may be applied, as a supplement to existing wired communication networks, across all levels of the model. Basically, the control level consists of Programmable logic controllers (PLCs)/Distributed control system (DCS) and (intelligent) decentralized I/O-Terminals, which are interconnected by one or more field buses with variations in the number of devices and distances as per application. Current field buses aim at convergence to the well-established Ethernet IT-networks with data rates of up to 100 Mbps. Depending on the area of application typical cycle times varies from 10 mS … 100 mS and in some case such as motion control <1 mS. At field or the lowest level of automation, sensors,

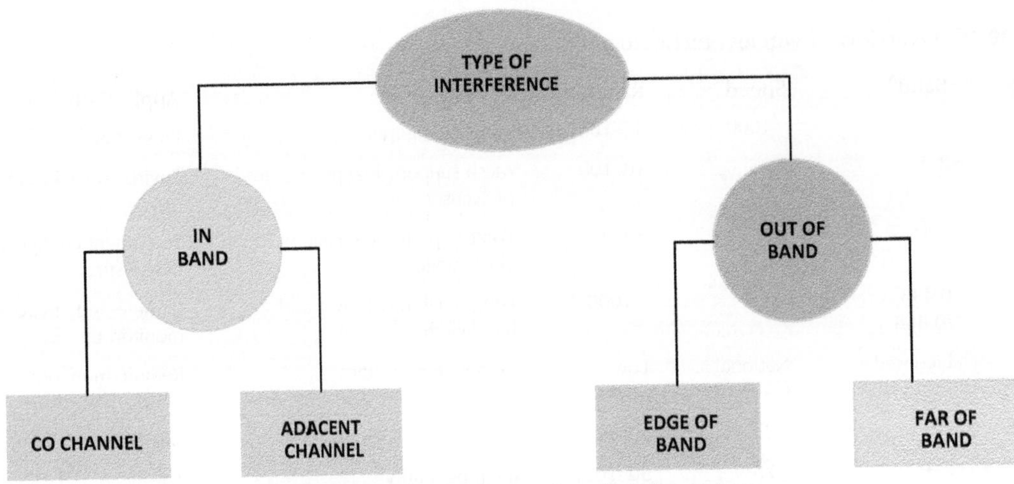


FIGURE III 10.1C Broad categories of interference types.

actuators, drives, and manufacturing robots are equipped. The number of I/O-points is very high. However, on the basis of a master/slave protocol, additional information, such as diagnosis or parameters for configuration, can be exchanged with PLC/controller besides the process data. IO-Link specifies data rates of at least 4.8 KBaud and typical 38.4 KBaud. So, corresponding to the amount of process data, cycle times of 2 ms can be achieved at data rates of 38.4 KBaud. Having understanding the performance requirements mentioned above, wireless technologies can be applied and would be useful on the control as well as at the sensor/actuator level. However, besides the stringent requirements concerning latency, determinism, reliability, and data integrity, the requirements with respect to costs, energy consumption, distances, bandwidth, node density, and network topologies differ. Here, the bandwidth and data rates are of major importance. The network topologies for wireless solutions range from simple cable replacement point-to-point and point-to multipoint connections up to cellular networks. However, the development is focused on high-speed wireless sensor/actuator networks supporting large numbers of devices. Star topology consisting of wireless sensors/actuators, wireless I/O-concentrators, and a master base station are very common. Master station acts as the interface to a super ordinate control system.

*Error detection* (retransmissions) or *Error correction codes* are classical methods to improve the performance of radio channels [effectively adding redundancy to the transmitted data]. However, their effectiveness mostly depends on the properties of the channel. Based on criticality and importance, automation have been categorized in different classes as in ISA SP100 and have been tabled in Table III 10.1d.

## 10.1.2 Wireless communication: miscellaneous issues

When a radio wave propagates through space, its power attenuates and cannot be amplified during transmission through space. The loss of radio wave intensity (L) during propagation through space is proportional the square of the distance (I) and and inversely proportional to the square of the wavelength (λ). So, the loss can be expressed as

$$L = (4\pi l/\lambda)^2 \qquad \text{(Eq. III 10.1a)}$$

The equation describes the case when propagating through ideal space. In actual space, the effects vary due to factors such as attenuation by obstacles, reflection, and diffraction, which are not incorporated in the above equation.

There are many factors that affect wireless communication. These factors can be categorized as:

- **Obvious:** *Physical obstructions, Network range* and *distance between devices, Wireless network interference, Signal sharing,*
- **Well known:** *Network usage and load, Poorly deployed antennas, Local environment characteristics, Spectrum channel limitations, Signal reflection and signal restriction*
- **Less known:** *Transmitter power limitations, Backwards compatibility, Polarization of signal, Speed loss for wireless overheads* etc.

Other than environmental factors other factors mainly include: *Distance and free space loss, Frequency, Mobility, LoS, Ground/Sky wave, Network planning* and Emerging technologies. Wireless communication plays a great role in the IoT/IIoT ecosystem by by providing flexibility, scalability, and lower infrastructure costs (obvious due to savings in wiring cost). An industrial sensor with wireless communication connects IIoT devices

**TABLE III 10.1D** Usage classes of ISA SP100.

| Category | Class | Application | Description |
|---|---|---|---|
| Safety | 0 | Emergency action | (Always) critical |
| Control | 1 | Closed loop regulatory control | (Often) critical |
| Control | 2 | Closed loop supervisory control | (Usually) noncritical |
| Control | 3 | Open loop | Human in the loop |
| Monitoring | 4 | Alerting | Short-term operational consequences—even based/maintenance |
| Monitoring | 5 | Logging and downloading/ uploading | No immediate operational consequence (history data, sequence-of-events, preventive maintenance) |

to create smarter and more efficient systems. So, wireless IoT technologies is growing rapidly across industries such as manufacturing and transportation to optimize processes, enhance customer experiences, and gain a competitive edge in the market. Therefore, in this subsection brief account on these factors as well as IIoT/IoT ecosystems have been covered.

1. **Common factors:** Here we discuss various common and obvious factors as listed below with special reference to signal path loss.
   - **Physical obstructions:** Normally, wireless signals are capable of overcoming obstacles. However, at times, wireless signals can have trouble penetrating certain solid objects such as hills, some buildings, single walls, or even people. Therefore, more obstructions between the transmitter and receiver, can affect the signal strength with more chances.
   - **Range and distance:** There will be more drop of signal strength when the two communicating devices set further apart. This is because with increase in networking range and distance, there will more signal spreads covering wider area making the signal weaker with spreads.
   - **Network interference**: Wireless network interference is common. Signals operating at similar frequencies can cause interference with each other having significant negative impact on the network performance.
   - **Signal sharing:** In hared network connection, channel resources are shared amongst the user affecting quality of communication, especially, for nonduplex communication.
   - **Path loss:** The signal path loss is essentially the reduction in power density of an electromagnetic wave or signal as it propagates through the environment in which it is traveling. Major reasons for signal loss are: *Free space loss*, *Absorption losses*, *Diffraction*, *Terrain*, *Multipath*, *Atmosphere*, *Building*, *Vegetation* and *Vessel.*
2. **Environmental and design factors** From the above discussions, we understand that signal path loss is an important issues, This can be predicted by *statistical method* or *through deterministic approach.* All these factors influence system design also. In this connection some environmental and other factors shall include but limited to:
   - **Local Environment:** This is more prominent for indoor networks Wall thickness and construction properties (*wire meshing* to cripple signal and *the materials* used have different effect on the performance in indoor networks) can be one of the biggest inhibiters of wireless signals.
   - **Spectrum limitations:** This severely affects wireless networks operating at popular frequency bands such as 2.4 GHz but may start to affect the 5 GHz band also later. The channels have to overlap each other to be able to fit within the overall operating range. This overlapping causes interference if wireless devices are using neighboring channels.
   - **Signal reflection:** Signal reflection, referred to as Multipath fade, occurs typically in buildings that have complex and intricate structural layouts. Reflected signals cause differences in overall path distance lengths on their way to the receiver. As discussed, these can cause unpredictable signal values at receiver. Also, the times at which the reflected signals reach the receiver are different due to the varying distances and the spread in the delay between the signals would create ISI.
   - **Poor antennas:** It is obvious that the highly directional antenna type that must be pointed directly at the receiver for proper signal pick up. Other types of antenna must be orientated correctly to ensure that they are transmitting signal on the correct plane.
   - **Signal restriction:** For security reasons it may be required to restrict the transmission of wireless signal to only the areas that are desired to provide network access for. This can be troublesome as controlling signal propagation is not easy.

- **Usage and Load:** More clients that are utilizing the network bandwidth, the less that there is to share between them.
- **Link budget:** Like others wireless communication, link budget is applicable to industrial wireless communications system design, namely LoRaWAN, LTE-M and NB-IoT. Link budget refers to the accounting of all the gains and losses, transmission power level, receiver sensitivity etc. in any wireless transmission system. The link budget mainly covers: *Transmitter power, Both side antenna gains, Attenuations at both side antenna feeders, Path losses, Receiver sensitivity* etc.

3. **Less Known factors:** Less known factors affecting wireless communication include *Signal polarization, Fresnel zone obstruction, Overhead of the wireless protocol, Atmospheric conditions, Backwards compatibility* etc. can impact wireless communication. Some of these less known affecting factors are:
   - **Signal Polarization:** This refers to the orientation of RF wave affecting reception
   - **Fresnel Zone:** This is related to The Earth's curvature affecting reception. The central portion of the signal path, known as the Fresnel zone, which when faces obstruction this can happen.
   - **Atmospheric conditions:** Affecting factors are: heavy rain, humidity, snow, and storms.
   - **Time of the day:** Some atmospheric phenomena can influence signal strength, at different times of the day.
   - **Channel overcrowding:** The number of wireless networks in a dense area can have impact on Wi-Fi channels quality.
   - **Backwards compatibility:** To support older devices, newer wireless standards may need to operate with reduced performance, slowing down—current IEEE 802.11 is an example.
   - **Power limitations:** Regulations set by OFCOM (UK) dictate that the devices operating on certain frequency ranges must adhere to maximum transmitter output power levels.
   - **Wireless overheads:** The wireless protocols themselves require a certain amount of "overhead" to manage connections (encryption, packet translation), channel access, and error correction consuming bandwidth to reduce the effective data speed.
4. **IoT/IIoT Protocols**: In order IoTs/IIoTs function efficiently, wireless communication protocols deployed must meet specific requirements such as: *Low power consumption, Adequate range and coverage, Security, Scalability* and *High transmission rate* (high bandwidth). Major protocols are with associated features and applications have been tabled. In Table III 10.1e
5. **IoT/IIoT**: **Protocol and Challenges:** Let us now discuss, factors to be considered for selecting suitable wireless protocol for IoT/IIoT and various challenges wireless communication in meeting the requirements. Let us start discussions with selection process
   - **Protocol selection**: In order to select the right wireless protocol for an IoT/IIoT important issues to consider shall mainly include: Application details, Range requirement, Power consumption, Bandwidth, Latency, Security, and Devices numbers per unit area as well as cost consideration. Then comes the question of various protocols discussed to match. Major issues are:
     - *Power efficiency*: Power consumption is Critical for battery-powered devices. Lo RaWAN, is excellent. NB-IoT is also good
     - Range versus Data: To ascertain the range requirement and match with the same available from various protocol discussed. Data through put and range are well balanced in NB-IoT, LoRaWAN is very good for long range but lower data than NB=IoT. NB-IoT is a good choice for bulk data/frequent updates.

**TABLE III 10.1E** IoT/IIoT protocol details.

| Protocol | Benefits | Limitation | Application |
|---|---|---|---|
| Wi Fi | Hi speed use of existing facility cost saving | Hi power consumption | Video surveillance sensors, standard devices, |
| Wi Fi 6 | Multiple devices in congested area, Spee10 Gbps | High cost and power consumption, range limitation | Public network |
| Zigbee | Large mesh, reliable low band | Shorter than cellular | Smart lighting, industrial IoT |
| Bluetooth | Less power, ideal for short range | Limited range | Proximity sensor |
| LoRA WAN | Low power, long range, | Small data packages | Environment sensor, logistics |
| NB-IoT | Low power, wide coverage, 4G/5G | Limited bandwidth | Smart metering, building automation |
| 5G | 10 Gbps speed, 1M sensors/Sq km | Cybersecurity, network slicing | Autonomous vehicle, aotomation |

- *Security*: Support for strong encryption and secure authentication mechanisms to protect against cyber threats is an important issue is selection.
- *Safety issues*: Safety issues mainly cover: *Avoid interference, Proper grounding, Training*
- *Bandwidth*: Some applications such as video surveillance, systems call for protocol with high bandwidth.
- *Environment:* Remote, urban, or dynamic settings require different strengths.
- Scalability: Protocols should be capable of handling a growth of devices in the network without performance issues.

- **Wireless Challenges** [4]: Major challenges for IoT/IIoT wireless communication are:
  - *Interference*: Crowded frequency bands can reduce reliability and signal quality.
    *Compatibility*: It is quite complex to integrate devices using different protocols.
  - *Power constraints*: For battery-operated IoT devices protocols optimized for low power consumption is a big requirement.
  - *Vulnerabilities*: Unsecured IoT devices are targets for hackers, for jeopardizing data integrity.

*Having gathered some knowledge on various wireless protocol types used in industrial applications, let us now start discussing wireless in Industrial applications.*

## 10.2 Wireless communication in industrial applications

Industrial wireless technologies are mainly installed inside wireless instrumentation field devices used for production/process automation. Purpose-built for the rigors of industrial use. Wireless communication network for these must satisfy the radio performance requirements of the industry in a well-balanced manner. ISA100 wireless standard enumerate necessary requirements for industrial wireless technology suitable for a variety of plant applications such as control and monitoring, asset management, and safety alarms. ISA100 Wireless devices can run on battery power for years before being recharged, can communicate over long distances, can have reporting intervals of seconds, and can be explosion-proof. Process Automation (PA) refers to the technology used to automate industrial processes, often found in power and process plants. Process variables are continuously and automatically manipulated to produce the final product. The wireless device communication system needs to ensure the reliability of communication meets the application requirements. There may be numerous sources of radio interference. Wi-Fi. ISA100 Wireless offers a variety of reliability, robustness, and availability features to provide a reliable path for data transmissions. Also, there exists wireless fieldbus to cater to the requirements of industrial automation. Wireless transmission at the fieldbus level can be utilized in at least two ways: All nodes use wireless transmission-referred to as "pure wireless fieldbus." In other option in a network only a few nodes are without wires. This approach is often selected for cost and performance reasons when a number of nodes can be easily connected through wires. Compared with wired, radio transmissions suffer from Bit error rates (BER). The signaling rate is limited in kilobits per second and seldom exceeds 10 Mbit/s, and can be jammed easily. On account of low spectrum, spatial reuse is limited, with small transmission distances. High cost, undetected collisions, hidden terminal effect, limited radio band made wireless transmission less popular. WirelessHART is an open, industrial wireless protocol built on the HART standard, developed for process monitoring and control. HART utilizes a self-organizing, self-healing mesh network topology for reliable communication through multiple redundant pathways, allowing any device to act as a router. Some key features of wireless HART are: *Open standard, Mesh network, Backward compatibility* and *Security*. Each WirelessHART network includes three main elements: *Wireless field devices connected to process or plant equipment, Gateways allowing communication between devices and host applications* and *Network manager is responsible for configuring the network, scheduling communications*. WirelessHART technology for process automation and other industrial sector are quite popular compared with wireless fieldbus. WirelessHART is already incorporated into the products of several major suppliers in this sector. WirelessHART was released as an open and interoperable wireless communication standard designed to address the needs of industry for reliable, strong and secure wireless communication in real-time industrial measurement and control applications. Also, approval from IEC is obtained and associated standard would come in the market very soon. In this section discussions will be presented on ISA standard, wireless fieldbus and wireless HART.

### 10.2.0 Wireless communication: ISA 100 standard for plant automation

Wireless communication has now been utilized in plant automation area especially with introduction of Industrie 4.0, IoT/IIoT and Digital transformation ideas wireless communication gains tremendous momentum. ISA 100 standard provides wireless communication protocol for plant automation including machine/factory floor automation as well as process control automation too. It also

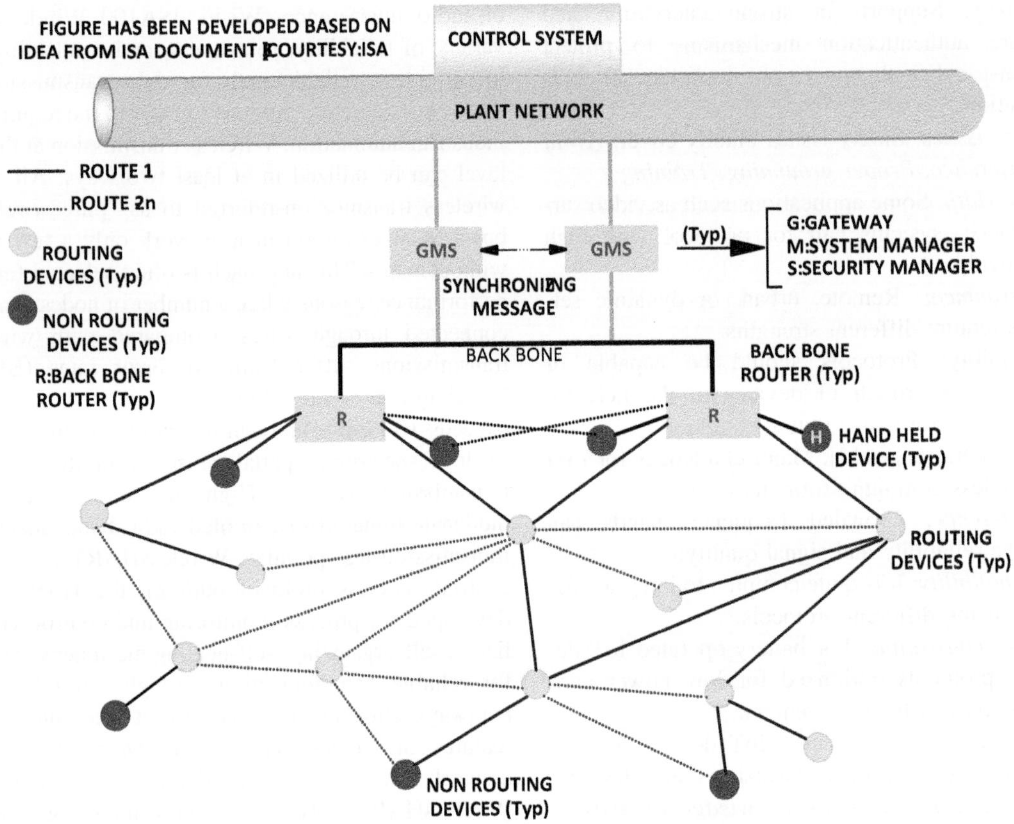


FIG III 10.2a1 ISA 100 ARCHITECTURE

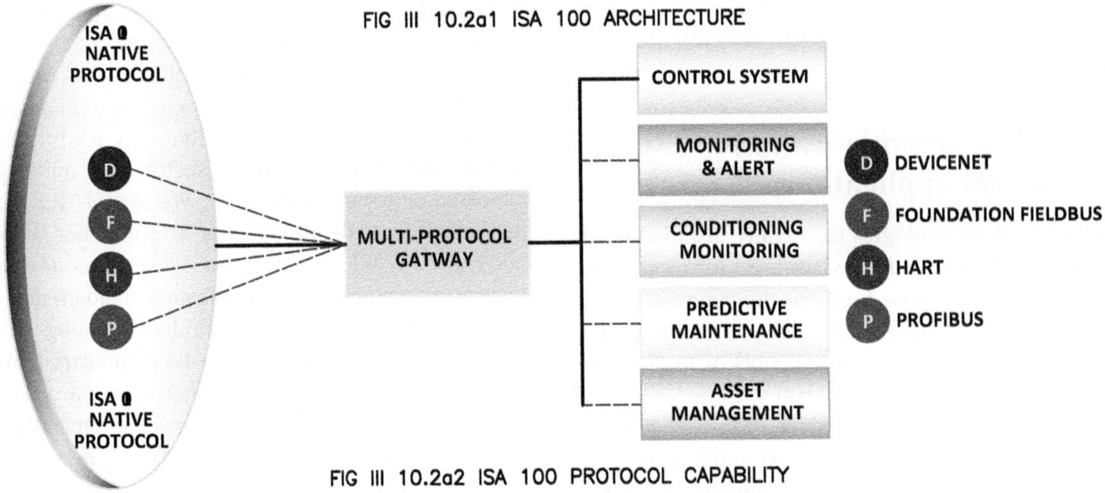


FIG III 10.2a2 ISA 100 PROTOCOL CAPABILITY

**FIGURE III 10.2A** Isa 100 protocol overview.

supports multiple other protocols to enhance automation protocol support to integrate entire network; field to control room including total asset management. ISA100 Wireless is primarily focused on process automation applications. Also, various wireless technologies used for varieties of wireless technologies as detailed here. Various wireless technology supports include: *Remote operation support: 4G LTE, On site maintenance support: Wi FI* or *IIoT/IoT: LoRaWAN and Process control ISA 100/Wireless HART*. To avoid interference, ISA100 Wireless has a feature called *channel blacklisting* that avoids Wi-Fi frequencies to minimize such interference. For noise avoidance channel hopping are used. For Communication path avoidance: mesh networks. In this subsection brief discussions on.

ISA 100. Let the discussion start with Fig III 10.2a, which have been developed based on idea from Ref. [5] [Courtesy: ISA].

1. **Overview of ISA 100:** This standard was developed for create open standardized, reliable, and secure wireless communication especially meant for industrial automation and control systems, with interoperability between different vendors' devices. ISA 100 is international standard (IEC 62,734) with IPv6 wireless network protocol. It is highly scalability and flexibility unattainable so that it could be applied for existing systems as well. In order to ensure high reliability, ISA 100 Utilizes mesh networking so that devices can act as repeaters, creating multiple communication paths. For providing security, it incorporates 128bit encryption. It optimizes long battery life. This standard is based on universal object-oriented application to eliminate protocol barrier and easy to add new applications. IPv6 addressability makes ISA100 to integrate IoT. It offers QoS for application layers and covers wireless back haul 100.15. It is well supported by ISA 99 for strong security in every message for data protection and confidentiality. It can be used in hazardous area with explosion proof and safety applications. As major sensor networks operate in unlicensed ISM 2.4 GHz band so there will be possibilities of interference due to coexistence causing reliability and latency problems. These are well handled in ISA 100. Even though there exists some similarity between ISA 100.11a and wireless HART yet there are differences. One thing to be noted that wireless HART covers field devices yet ISA 100.11a covers everything from field instruments to control room. Typical architecture of ISA 100.11a has been detailed out in Fig III 10.2a1. ISA 100 supports a number of other protocol suites paving the way complete plant system integration including enterprise network. This is well depicted in Fig III 10.2a2. ISA 100 protocol suite consists of several standards as listed here; *ISA 100.11a*: Process Application; *ISA 100.12*: HART and ISA 100.11a convergence network; *ISA 100.14*: Trust worthy wireless; *ISA 100.15*: Wireless Back haul Backbone network; *ISA 100.21*: Tracking and identification of personnel and asset.
2. **Important ISA 100 parts:** There are several parts of ISA 100, namely ISA100.11a is an international standard with multiprotocol capability, ISA 100.12 for HART convergence, ISA 100.15 for wireless back bone Backhaul and ISA 100.21 for various asset tracking. Here, we briefly, two important parts of ISA 100 covered are:
   - **ISA100.11a:** As shown in Fig III 10.2a2, apart from control and monitoring, ISA 100 cover entire plant automation system including asset management, condition monitoring and predictive maintenance. ISA 100.11a covers multivendor device interoperability. Security, which is considered throughout the whole system not just at the Physical layer/MAC sublayer, is a major design facet of ISA100.11a (leverages security aspects of the IEEE 802.15.4–2006). ISA 100.11a the most important part of ISA 100 standards, finds its usage in total control systems, namely Closed-loop supervisory and regulatory (critical) controls as well as in open loop control systems i.e., it covers all classes except class 0 in Table III 10.1d. In this connection for architecture, Fig III 10.1a1 may be referenced. It can participate monitoring, alarm and logging functions also.
   - **ISA100.15:** Wireless Backhaul Backbone network mainly focuses on definition of standard network common interfaces for interoperability between the backhaul backbone wireless networks and different field wireless networks. This is to sit above field wireless networks such as ISA 110.11a, wireless HART and other PAN devices including ZigBee, RFID, UWB, 802.11a/b/g/n. This is positioned below the control application protocols to provide the clear common mechanisms to support variety of application protocol translation. This ensures, Quality of service (QoS), security etc.
   - **ISA 100.12 and 100.21:** ISA 100.12 evaluates the specifications of WirelessHART specification and the ISA100.11a to provide a technical differential analysis with associated benefits. Also, It can suggests successful coexistence options for both networks and draw up recommended vendor profiling. ISA 100.21 is meant for asset tracking. There are many techniques for asset tracking e.g., GPS/Wi Fi-based asset tracking. ISA 100.21 examines various mathematical principles to implement this.
3. **ISA 100 Control features:** ISA 100 meant for plant automation can support full wireless control as well as any combination of wired/wireless sensor with wired/wireless actuators. General requirements for wireless communication mainly include: *Cyber security, Data integrity* and *Operational reliability & robustness.* It can support SIL2 in hazardous applications. All these requirements are well satisfied by ISA100. ISA 100 is versatile in nature and supports *Reliable mesh networking, Interoperability, Designed from the bottom up* etc. It offers better latency around 100 ms with publication rate of 1–2 s. On account of wireless "*Backhaul Backbone*" network, ISA can be engineered with scalability network flexibility, redundancy as well as interoperability. ISA 100 also offers multiple access points, field routers, and meshing/

nonmeshing devices. Features such as deterministic battery life discussed earlier, and interoperable peer to peer connection make it highly suitable for control and monitoring application. Highly reliable, deterministic wireless communication along with dual communication up to process control network makes the system well suited for plant automation applications. In this connection Fig III 10.0b1 may be referred to also.

4. **Security feature:** ISA100 Wireless provides robust, multilayered, policy-based security. As mentioned in Fig III 10.0b1, ISA 100 offers multiple-layered 128-bit AES block cipher. Only devices that have legitimate and unique secret join keys are admitted to the network. ISA100 wireless' security features are designed with scalable encryption in two security layers, namely Data Link Layer and Transport Layer. Data Link Layer security protects the network by providing hop-by-hop assurances so that each message is flawlessly transmitted to the next hop, with detailed performance and security diagnostics.Transport Layer security protects data and provides end-to-end assurances that mission critical messages received are secret and authentic [6].
5. **IPv6 in ISA 100:** As mentioned earlier, IoT and IIoT applications are call of the day! ISA 100 with IPv6 makes the things much easier for this. ISA100 Wireless incorporates IPv6 in its network layer and transport layer. This seamless end-to end routing with IoT/IIoT in a single environment. Features of IPv6 not only enhance IP addresses but also offers an array of major other features such as: *Support for newer services*, *Security enhancement*, *Directed data flow*, *Easier network configurations*, *efficient routing* and *efficient packet processing*. These features would result improved network and service performance and reduced down time with lower life cycle cost.

### 10.2.1 Wireless communication: HART

The lack of an open standard to both fulfills the industrial requirements and ensure that there exist many vendors was mainly responsible for very slow progress in adoption of wireless technology in the industrial automation. Even though HART is quite old in industries, yet its popularity is still going strong even after version 7, which can offer open standard for wireless interface referenced as WirelessHART. Other similar ones are Zigbee, ISA 100 etc. By the ways of technological means such as mesh networking, channel hopping, and time-synchronized messaging, WirelessHART proves reliability even in the presence of interference. This strengthens coexistence with other wireless networks. For security and privacy for network communications, wireless HART uses encryption, verification, authentication, key management etc. Smart data publishing and other ways and means make batteries, solar and other low-power options practical for wireless devices working at inaccessible locations. All these features make WirelessHART the best approach for wireless process automation network. Both standards, namely WirelessHART, and ISA100 contain many similarities, they also contain differences. WirelessHART extends HART by introducing device types, ISA100.11a introduces the concept of roles and allows these roles to be applied in various combinations. ISA100.11a includes backbone routers for bridging subnets. WirelessHART, which uses AP (which can be used in parallel to create a mesh of thousands of devices), Backbone routers limit the throughput into and out of a single subnet to the throughput of one radio. In this subsection brief discussions on WirelessHART have been presented.

1. **General discussion:** WirelessHART (IEC62591) is open standard protocol providing cost-effective, simple and reliable way to deploy new measurement and control without the wiring costs. WirelessHART can be used with existing wired instruments also. WirelessHART offers reliable solution with the help of technologies such as mesh networking, channel hopping, and time-synchronized messaging and can coexist with other wireless networks. Network security and privacy issues are ensured by use of encryption, verification, authentication, key management, and open industry-standards best practice to achieve this. Smart Data Publishing, solar and other low-power options for wireless devices etc. make it as one of the best approaches for wireless approach in plant automation application. Release 7.0 HART protocol included WirelessHART with features such as powerful unsolicited data transfers, event notifications and advanced diagnostics. WirelessHART also find applications in rotating equipment such as kiln dryers, environmental health and safety applications. As state earlier that there are some similarities of ISA 100 with WIrelessHART yet there are differences. ISA100 does not specify *process automation protocol application layer* to include existing devices such as WirelessHART. Fragmentation and reassembly are supported by both but in ISA 100 it is in Network level whereas in WirelessHART it is application layer. WirelessHART devices, which must support routing.
2. **Working principle:** WirelessHART has been detailed in Fig III 10.2b.
   Fig III 10.2b1 depicts a WirelessHART network. It comprises a group of *network devices* including a field

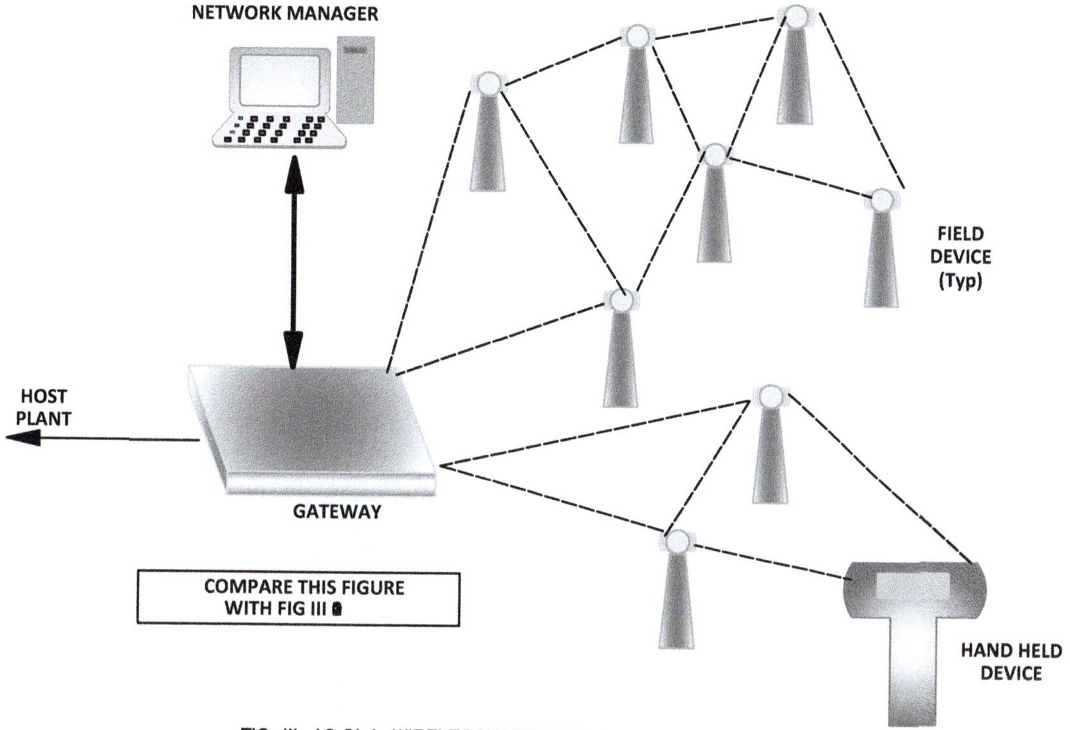


FIG III 10.2b1 WIRELESSHART NETWORK ARCHITECTURE

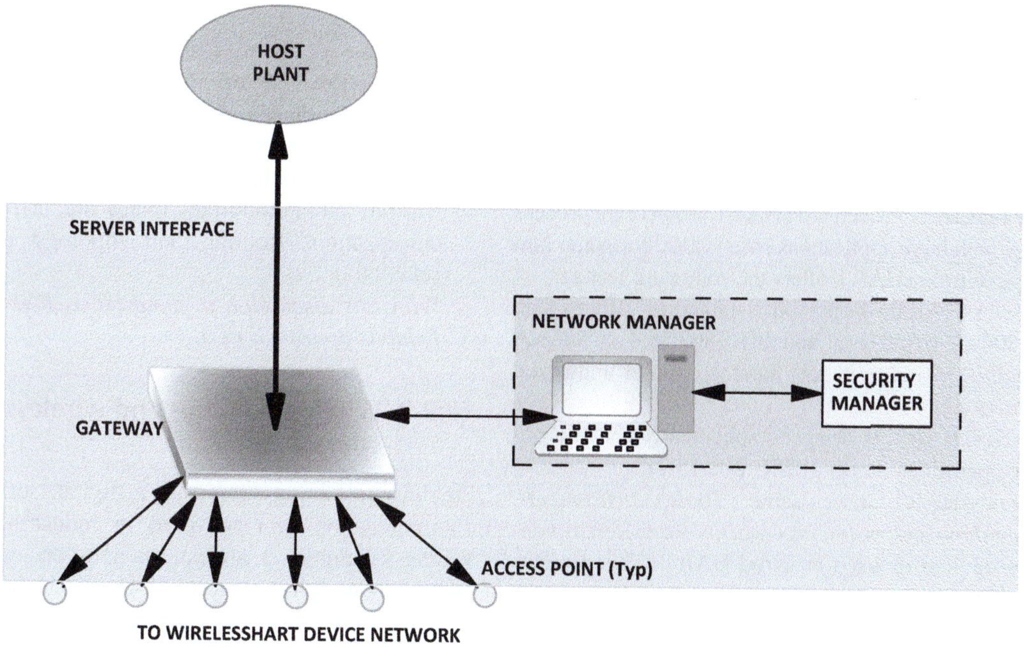


FIG III 10.2b2 WIRELESSHART DEVICE NETWORKING DETAILS

**FIGURE III 10.2B** Wireless HART protocol overview.

device (directly connected to the process plant), handheld devices. As shown in Fig III 10.2b1, the network supports both direct connection between device and gateway (star topology) and connections over multiple hops (mesh topology). Each network device therefore must be able to function as *Source*, *Sink* and *Router*. The WirelessHART gateway acts as the bridge connecting the WirelessHART network to the process plant. In this connection Fig III 10:2a may be referenced for comparison. The basic schematics are illustrated in Fig III 10.2b2. It consists of a *virtual gateway* and one or more network access points. Host applications access the network devices through the service interface, which can have single or multiple ports. Operators can also monitor or configure particular field device through the process plant backbone.

3. **Wireless HART components:** As shown Fig III 10.2b, three essential components in each WirelessHART network are:
   - **Field Device**: Wireless field devices deployed for process measurement
   - **Gateway:** Gateways build network and that enable communication between various field devices and host applications connected to plant communications network.
   - **Network Manager:** Major responsibilities of network manager mainly include configuration of the network, communications scheduling between devices, managing message routes, and monitoring network health. This also includes security manager as shown in Fig III 10.2b2. The Network Manager can be integrated into the gateway, host application, or process automation controller.

   Apart from the above, there are a few other components in WirelessHART. These are:

- **Adapter:** This is used to add wireless capability to existing HART device
- Rep**eater:** As the name suggests, it extend the network.
- **Handheld terminal:** This is used to check various parameters of connected devices by connecting it to the network. Also, it serves diagnostics and other maintenance purposes as a tool for maintenance personnels.

4. **Major features:** Proven used HART protocol for process industry with huge applications base is ideal for going into wireless. WirelessHART offers the following features:
   - WirelessHART built on standards IEC61158, 61,804–3, 61,591e-1, and IEEE802.15.4 is *Simple, Reliable, Secure.* It can meet all major industrial requirements.
   - WirelessHART is designed specifically for low power operations as per (IEEE 802.15.4)
   - WirelessHART uses same *Tools, Instruments, Know-how* and as *Interoperability* for different vendors devices as used in wired HART. This protects investments.
   - WirelessHART uses mesh topology with self-organizing and self-healing characteristics
   - WirelessHART uses DSSS technology and adjustable transmission power to provide reliable communication
   - WirelessHART uses Hopps" across channels to avoid radio interference, coexistence with other wireless networks with black listing function
   - WirelessHART can adjust communication paths (prescheduled time window) for optimal performance, enabling reliable, power-efficient, and scalable communication

5. **Implementation issues:** The Wireless HART network specification enables a reliable, secure, and scalable architecture and it gets more robust as more devices are added to an existing network. Design guidelines support the deployment of small networks with <10 Wireless HART devices as well as installations supporting thousands of devices. WirelessHART implementation is not a simple task. On one hand, the standard has to meet the basic requirement for the various protocols and devices. On the other hand it has to leave sufficient room for different realization schemes. WirelessHART needs to support compliant products from various vendors covering diverse array of applications such as process monitoring, asset management, process control, health safety and environmental monitoring. To simplify this task the majority of protocols are designed in a layered fashion. Each of the layers are independently implementable and addresses different part of the communication problem to the level of complexity per layer. For effective implementation of WirelessHART, the following design goals are essential:
   - **Computational complexity:** WirelessHART has very strict timing requirements for which the computational complexity should be kept as low as possible.
   - **Memory consumption:** The WirelessHART stack maintains a set of communication tables for storing information about routing, end-to-end transactions etc. The stack should leave as much memory as possible for the running application.

   Various other important issues are: MAC layer synchronization, Routing and topology control, Link scheduling etc.

   *With this discussion in mind let us look into wireless fieldbus approach next.*

### 10.2.2 Wireless fieldbus and wireless systems comparison

Fieldbus technology refers to a digital, serial, multidrop communication protocol used in industrial automation. Wireless fieldbus is an extension of the concept implemented over wireless technologies to enjoy advantages such as: Flexibility, Easy installation, Cable cost saving, Mobility, Avoidance of harsh environments etc. There are a god numbers of suppliers that supply wireless fieldbus systems each with their "type" of wireless fieldbus system. This means they have their own modem application, such as: 802.11 Wi-Fi etc. Major differences in these wireless fieldbus systems come from Bandwidth and operational frequency, Standards, Frequency modulation, Network topologies etc. In general, the industrial environment is error prone and reliability of wireless technology is lower than wired. Therefore, it is difficult to support real-time transmission and locating many node in industrial environment. In view of these, it is necessary to design the wireless Fieldbus can support many nodes and guarantee real-time transmission on mixed traffic in industrial environment. Various approaches for fieldbus devices are discussed in this subsection. In this connection Fig III 10.2c may be referenced. In general, wireless technologies (IEEE 802.11, IEEE802.15.1(Bluetooth), and IEEE 802.15.4(LR-WPAN)) can be applied to the wireless fieldbus or hybrid industrial network. Bluetooth has low cost and fast transmission rate, but it should be composed of few node and has short range. Hence, it is difficult to apply in wide factory area for a large number of nodes. Another ways is IEEE 802.15.4. A each node of LR-WPAN has low price and a coordinator node can support many I/O nodes. However, it can't handle mixed real-time traffic efficiently. Therefore, efficient transmission of mixed real-time traffic has been required.

In this subsection brief discussions on wireless fieldbus with coverage of wirless Foundation fieldbus and Profibus will be covered. This subsection will also cover comparison of various wireless standards in industrial automation.

1. **General discussions:** In fieldbus there are three types of data transmissions, namely real-time periodic data, sporadic alert data and messages for network maintenance (non-real-time data). On account of these most existing fieldbuses are primarily developed for wired technology. Timing and reliability catered to by wired fieldbus systems are well established. Apart from the above issues, it is also true that in spite of many benefits of wireless system (mainly lower cost and time), in industrial environment wireless technology is more error prone and less reliable than wired counterpart. In view of inherent benefit of wireless transmission, several technologies such as wireless LANs (WLANs), wireless, personal area networks (Zigbee, LR WPAN) and Wireless sensor networks (WSN) have been tried for developing wireless fieldbus systems. On account of short distance and limited number of nodes, Bluetooth has not been mentioned here. WLAN has high cost and has reliability limitation, IEEE 802.15.4 LR-WPAN can't handle mixed real-time traffic efficiently. Major challenges for wireless fieldbus systems shall include but not limited to the following:

    - **Transmission errors:** These may be caused by either channel outages (signal strength below a critical threshold values) and/or interference and jamming.
    - **Security issue:** The wireless medium very easily fall prey to interception and an attacker to eavesdrop, to insert malicious packets,
    - **Power supply:** In most of the fieldbus systems, the same cable can be used for communication and power supply purposes. So, power supply is a pertinent issue. Batteries and energy scavenging could be a solution.
    - **Path loss:** Signal strength of a radio signal decreases with the distance, so, signal may not be received properly when distance is more. This path loss is associated with distance limit in wireless fieldbus.
    - **Half-Duplex operation:** The wireless transceivers may not be able to transmit and receive simultaneously on the same channel as their own signals would drown all signals from any other stations.
    - **Physical layer:** Most wireless channels use preamble for synchronization. Naturally this would cause increase in PHY overhead.
    - **Channel Errors:** A wireless transmitter propagates waveforms in several spatial directions. Naturally, waveforms are subject to reflections, diffraction, or scattering resulting multiple waveforms could appear at receiver especially in industrial environment. These would cause multipath fading symbol interferences discussed earlier.

    Apart from these there could be issues such as lack of consistency (world FIP), token passing issue (in Profibus) and Problems for Carrier-Sense Multiple Access (CSMA—as in CAN) [7]. Hybrid wired and wireless fieldbus in master slave modes is possible as shown in Fig. III 10.2c1. Here, Profibus in hybrid mode where both wired as well as wireless Profibus is connected [8,9]. Wireless Profibus extenders are available in both Profibus *master* and *slave* functionality, and will connect to a Profibus DP LAN at full bus speed of 12 Mb/s. Also, Wireless and remote I/O (WIO) technology, including a suite of solutions to assist automation for development WIO devices are available from Foundation Fieldbus. "WIO will enable the utilization of an open, interoperable fieldbus automation infrastructure incorporating both Foundation fieldbus HSE and industrial wireless applications" [Courtesy Foundation Fieldbus].

2. **IEEE 802.15.4 FB:** There are various wireless fieldbus approaches. Some of them could be: *WirelessHART* (IEC 62,591), *ISA100.11a (IEC 62,734)*, *Wireless Profibus/Profinet*, *miscellaneous wireless options*. First two have already discussed, In Profibus approach only

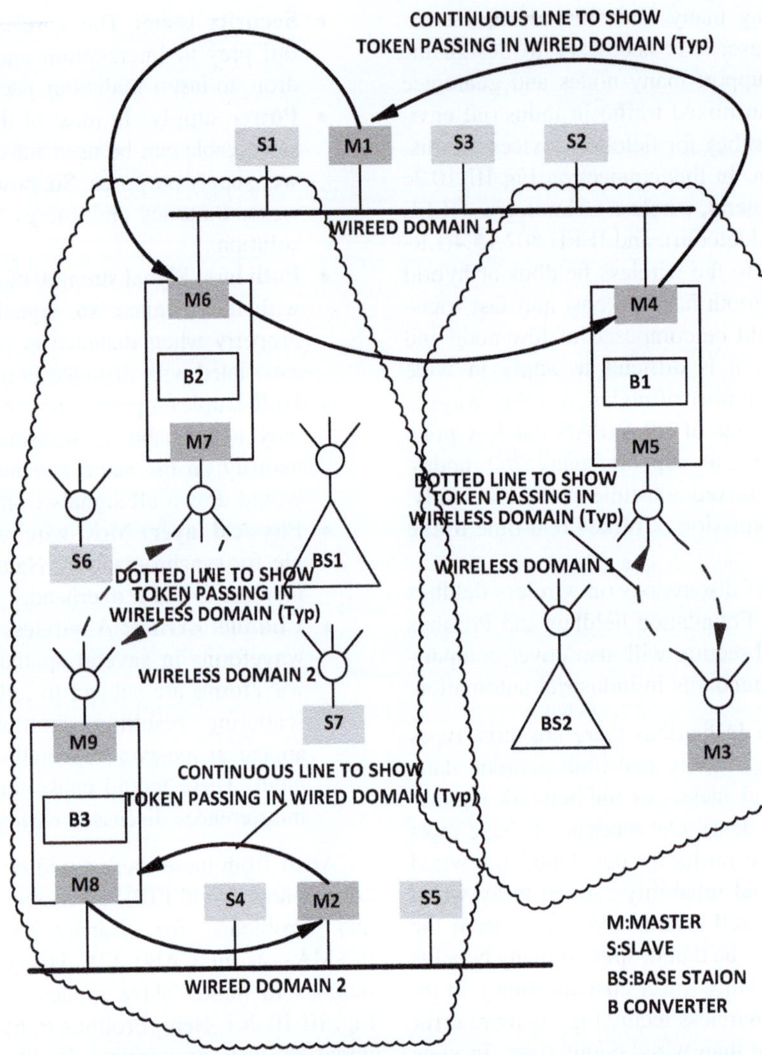


FIG III 10.2c1 WIRED AND WIRELESS HYBRID PROFIBUS [8][9]

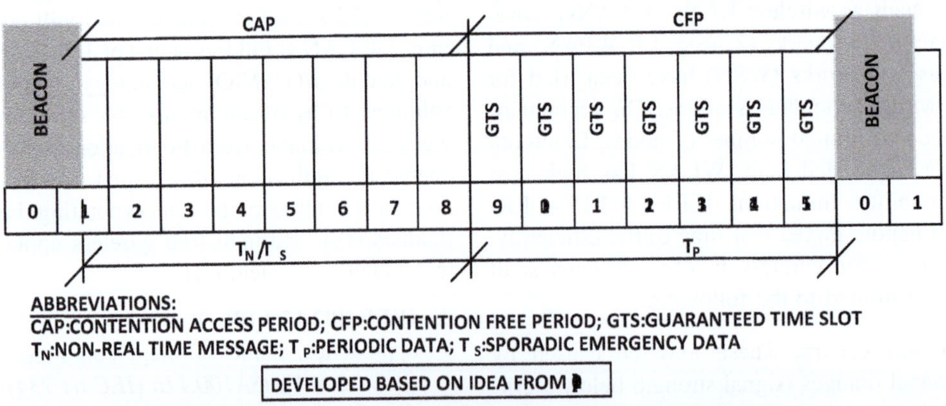


FIG III 10.2c2 MODIFIED SUPER FRAME STRUCTURE: IEEE 802.15.4

**FIGURE III 10.2C** Hybrid profibus and IEEE 502.15.4 super-frame.

through gateways to WLAN/Bluetooth could be done as shown. Profinet is similar to Ethernet hence standardized through WLAN discussed or modern 5G approach. Modbus over Wi-Fi/Bluetooth/LoRa etc. Apart from these, another way could be IEEE802.15.4, which let us discuss here now.

IEEE 802.15.4 can be divided into two periods: *active* period and *inactive* period. Active period is ranged from the start of beacon frame to the end of Contention free period (CFP). In this period, every node can transmit their data. From the end of CFP to the start of next beacon frame, *inactive* period, every node turn off its radio, and sets a timer to awake itself later. The active period is composed of beacon broadcast, Contention access period (CAP), and CFP. For fieldbus approach, in order to transmit three types of data effectively, the super-frame structure of IEEE 802.15.4 has to be modified super-frame structure as shown in Fig III 10.2c2. The real-time periodic data is transmitted in the CFP and real-time sporadic data and non-real-time message can be transmitted using CAP. The inactive period is defined to zero to reduce delay. The real-time performance depends on a length of the basic period. The basic period is used to schedule periodic node and determine transmission period of periodic data [10].

3. **FF Remote I/O:** FOUNDATION Fieldbus is a wired digital protocol. However, it can be integrated with wireless communication through Wireless Foundation fieldbus (WFF). WFF uses mainly the IEEE 802.11 physical layer in combination with Foundation fieldbus (FF) technology user and application layers to provide wireless services. This is a hybrid approach allows for wireless integration into existing wired FOUNDATION Fieldbus systems. The WIO development is part of the Fieldbus Foundation's initiative to design and deploy an infrastructure that will accommodate evolving wireless standards inclusive of WirelessHART and ISA100.11a. The foundation's WIO gateway provides an interface to both technologies. It uses Electronic device description language (EDD-L) and Function blocks (FB) such as wired fieldbus to provide interoperability with the other WIO gateways. WIO continues to broaden the automation infrastructure capabilities of Foundation technology by providing interoperable gateways for wired and wireless HSE devices.

*Combined with the ability to handle remote I/O, this greatly extends the range and capabilities of Foundation technology to encompass many more devices throughout the entire plant or facility, regardless of their communications technology. The business value lies in taking the large amounts data from these devices through HSE and turning it into useful information in an open environment. You will be able to enjoy the benefits of wireless networking and remote I/O along with the powerful infrastructure of Foundation technology [11].*

4. **Wireless Profibus and Profinet:** It is important to note that Profibus and Profinet both in terms of technical approach as well as application so, these will be discussed separately.
   - **Wireless Profibus:** Basically, Profibus is natively wired deterministic, fieldbus. Major difficulties in realizing wireless Profibus are: *Maintenance of determinism and low latency* and *Standardization.* For these reasons they are not normally deployed for critical controls but for monitoring. In order to enable wireless Profibus, vendors developed gateways and adapters: for RS-485 to Wireless bridge (namely, WLAN, Bluetooth). These are mainly used for remote I/O, mobile machinery, cranes and AGVs. Usually, two converters are needed to realize one wireless path: cable-wireless to wireless-cable. The cabled devices that use the same master also need to operate on this speed. This means that the user needs to put in some work to create a reliable wireless application. The cabled devices that use the same master also need to operate on this speed. The wireless Profibus is a special fieldbus, for servicing wirelessly. Three different approaches for this are: *Wireless modem as gateway*, Wireless transfer with change of the timing (slow approach) and Transparent wireless transfer without change of the timing. In last approach instead of cable wireless link is used.
   - **Wireless Profinet:** Being similar to Ethernet protocols, it can ride on wireless technologies. Profinet over WLAN (IEEE 802.11) and now over 5G are supported. Profinet Real time (RT) possible over WLAN for noncritical loops and Profinet Isochronous RT (IRT) requires very low jitter and is difficult wirelessly. So, when there is no standardization of Wireless Profibus, Profinet is over WLAN and 5G.
5. **Comparison of Wireless systems:** Let us conclude the discussions on wireless communication with comparative studies of various wireless protocols in Table III 10.2a.

*Let us now explore Utility Protocol and Smart Grid Communication in next chapter.*

**TABLE III 10.2A** Comparison of wireless protocols.

| Feature | WirelessHART | WLAN | Profinet | Industrial 5G |
|---|---|---|---|---|
| Usage | Monitoring and slow loop | Monitoring and control | FA[a], motion, AGV | Hi speed automation. IIoT |
| Topology | Mesh | Mesh/start/Hy[b] | Star | Star |
| Medium | 2.4 GHz ISM | 2.4 GHz ISM | Wi Fi/5G | Licensed 5G |
| Determinism | TDMA | TDMA with QoS | RT + IRT | Possible |
| Data rate | 250 Kbps | 250 Kbps | Mbps/Gbps | Gbps |
| Latency | 100-500 mS | 50-200 mS | 10 mS | 1 mS |
| Standard | IEC 62591 | IEC 62734 | PI guide | 3 GPP[c] |
| Security | AES-128 encryption | AES-128/256, flexible | WPA3, 5G SIM-based | In built |
| Energy | Highly efficiency | | Low efficiency | |

a: Factory automation; b: Hybrid; c: third Generation partnership project

## List of abbreviations used

**AGV** Automated guided vehicle
**ARP** Address resolution protocol
**BER** Bit error rates
**BLE** Bluetooth low energy
**BSS** Basic service set
**CAP** Contention access period
**CCK** Complementary code keying
**CDMA** Code division multiple access
**CFP** Contention free period
**CSMA/CD** Carrier sensed multiple access/Collision detect
**CWDM** Coarse wavelength division multiplexing
**DCF** Distributed coordination function
**DCS** Distributed control system
**DIFS** Distributed interframe spacing
**DPSK** Differential phase shift keying
**DQPSK** Differential quadrature phase shift keying
**DS** Distribution system (IEEE 802.11)
**DSSS** Direct sequence SS
**DT** Digital transformation
**EDD-L** Electronic device description-language
**EHF** Extra HF
**ESS** Extended service set
**FB** Functional block
**FDD** Frequency division duplex
**FDMA** Frequency division multiple access
**FF** Foundation fieldbus
**FH** Frequency hopping
**FHSS** FH spread spectrum
**GTS** Guaranteed time slot
**HART** Highway addressable remote transducer
**HF** High frequency
**HRDSSS** High rate- DSSS
**IIoT** Industrial IoT
**IoT** Internet of things
**IP** Internet protocol
**ISI** Inter symbol interference
**ISM** Instrument, scientific and medical band
**IT** Information technology
**LAN** Local area network
**LoRa** Long Range
**LOS** Line-of-sight
**LPWAN** Low power WAN
**MAC** Media access controller
**MAN** Metropolitan area network
**NAT** Network address translation
**NB-IoT** Narrowband-IoT
**OFDM** Orthogonal frequency-division multiplexing
**OPC** Open Platform communication
**OT** Operational technology
**PAN** Personal area network
**PCF** Point coordination function
**PLC** Programmable logic controller
**QoS** Quality of service
**RF** Radio frequency
**RFID** Radio-frequency identification
**RT** Real time
**SDMA** Space division multiple access
**SDR** Software-defined radio
**SS** Spread spectrum
**STA** Station
**TDD** Time division duplex
**TDMA** Time division multiple access
**UA** Universal architecture (OPC)
**URLLC** Ultra reliable low latency communication
**WAN** Wide area network
**WAP** Wireless access point
**WDM** Wavelength-division multiplexing
**WFF** Wireless Foundation fieldbus
**WIO** Wireless and remote I/O
**WLAN** Wireless local area network
**WSN** Wireless sensor network

## References


[1] S. Rackley; *Wireless Network Technology—From Principles to Successful Implementation*; Internet document; http://home.ustc.edu.cn/~wfsun/lab/course/wireless/Steve%20Rackley%20-%20Wireless%20Networking%20Technology.pdf.

[2] A.S. Tanenbaum, Computer Networks, fourth ed., Pearson Prentice hall, 2009.

[3] *Types of Wireless Technologies in Industrial Automation; Zero Instrument; Internet Document*; https://zeroinstrument.com/types-of-wireless-technologies-in-industrial-automation/.

[4] S. Rathore; *Iot in Wireless Communication: Exploring the Best Protocols and Technologies*; Hashstudioz; 2024 https://www.hashstudioz.com/blog/iot-in-wireless-communication/.

[5] *The ISA 100 standards—Overview and Status*; The ISA 100 roadshow presentation; ISA; http://isa100wci.org/Documents/PDF/ISA100_Standards-Overview_WCI_Webinar_V3_12Apr10.aspx.

[6] Architecture for Industrial Internet of Things (IEC 62734); *ISA 100 Wireless; ISA Document; Internet Document*; https://isa100wci.org/hubfs/2022%20ISA%20Website%20Redesigns/ISA100/PDFs/Documents%20and%20Forms/EP06-3405-ISA100-Wireless-Systems-for-the-Future-brochure-WEB_4-7-25.pdf.

[7] Wireless Technology in Industrial Networks; A. Willig, K. Matheus and A. Wolisz; Invited paper; *Proceedings of the IEEE*, 93, 6, 2005; https://www.dca.ufrn.br/~affonso/PET0351/papers/wireless2.pdf.

[8] L.L. Ferreira, E.Tover and M.Alves; Enabling inter-domain transactions in bridge-based hybrid wired/wireless profibus networks; *Conference: Emerging Technologies and Factory Automation, 2003. Proceedings. ETFA '03. IEEE Conference* 1; 2003; 10.1109/ETFA.2003.1247682; IEEE Xplore; https://www.researchgate.net/publication/4044841_Enabling_inter-domain_transactions_in_bridge-based_hybrid_wiredwireless_PROFIBUS_networks/figures?lo=1.

[9] S. Basu and A. K. Debnath; *Power Plant Instrumentation and Control handbook*; 2nd rev; Elsevier; Academic Press; 2019; https://www.elsevier.com/books/power-plant-instrumentation-and-control-handbook/basu/978-0-08-102804-9.

[10] D.K. Choi, J. I. Lee, D. S. Kim, and W. C. Park; Design and implementation of wireless fieldbus for networked control systems; *SICE-ICASE International Joint Conference 2006*; 2006; Korea;.

[11] *Fieldbus Foundation Brings More Information Through HSE with Wireless and Remote I/O; Control Engineering*; 2011; Internet document; https://www.controleng.com/fieldbus-foundation-brings-more-information-through-hse-with-wireless-and-remote-i-o/.

Chapter 11

# Utility protocols and smart grid communication

Dynamic data exchange (DDE) is a Microsoft Windows interprocess communication (IPC) protocol that allows running applications to exchange data through a client—server conversation. DDE is a messaging system based on shared memory to facilitate both one-time data transfers and continuous updates for various applications. Interapplication communication, real-time (RT) updates, industrial automation, and remote controls are some of the application areas of interest here. However, some limitations of this such as *complexity*, *security* (lack of), *legacy status,* etc. paved the way for advanced protocols such as open platform communication (OPC), ease of integration, connectivity, interoperability, freedom of choice, etc., which makes OPC more popular in open platform operation. OPC is the interoperability standard for the secure and reliable exchange of data in the industrial automation and other application. It is an independent platform that ensures the seamless flow of information among devices from multiple vendors. The OPC foundation is responsible for the development and maintenance of this standard. OPC provides a standard interface for industrial automation systems to exchange data securely and reliably across different vendors and platforms, enabling communication between factory floor devices, control systems, and enterprise IT. Architecture, platform, reliability, security, configuration, etc. are a few issues where DDE and OPC differ.

Developed by Harris, Distributed Automation Products in November 1993, is an open, object-oriented protocol. This finds its major usage is the electric, water, and oil/gas utility industries for SCADA communication systems, i.e., the communication between master stations and remote devices such as remote terminal unit (RTU)s and intelligent electronic devices (IED)s. This is an open and public protocol that ensures interoperability, longevity, and upgradeability. DNP3 is a nonproprietary protocol that is available to anyone by visiting the website. DNP3 provides the rules for remotely located computers and master station computers to communicate data and control commands. It is not a general-purpose protocol. DNP3 was designed to optimize the transmission of data acquisition information and control commands from one computer to another. Many vendors offer products that operate using transmission control protocol/Internet protocol (TCP/IP) to transport DNP3 messages. Link layer frames are embedded into TCP/IP packets. This approach has enabled DNP3 to take advantage of Internet technology and permitted economic data collection and control between widely separated devices. Utility communication architecture (UCA) is standard protocol common in supervisory control and data acquisition (SCADA) applications for utility communications, designed to enable multivendor interoperability for devices. This is a standardized object and ISO model for defining communication protocols and data formats. This plays a great role in manufacturing message specifications (MMS), generic object models for substation and feeder equipment (GOMSFE), and the common application service model (CASM), and an explanation of UCA interoperability. Various standards used are also detailed. IEC 60870 refers to coverage and requirements for telecontrol systems pertinent to power industry, covering data transmission, message formats, and application functions for between two systems in SCADA, i.e., master–slave communication in substations and SCADA systems. It defines various requirements including communication over low-bandwidth, bit-serial communication links, protection to send information using standard application server data unit (ASDU, operation of telecontrol protocol over Ethernet networks using TCP/IP, etc.). IEC 61850 also refers pertains to standardization in communication in power utility automation systems, including substation automation and smart grid operations. It defines the requirements for standard data models and protocols for interoperability, reduced wiring, and unified communication between IEDs from different manufacturers. It also provides requirements for generic object-oriented substation events (GOOSE) messaging for high-speed communication, and MMS protocol for client–server interactions. Smart grid refers to an electrical network based on *digital technology* used to supply electricity to consumers via two-way digital communication.

Smart grid communication uses information and communication technologies (ICT) for RT information sharing and control between components of the power grid, allowing for improved efficiency, reliability, and

**Plant Intelligent Automation and Digital Transformation. https://doi.org/10.1016/B978-0-323-89892-8.00020-X**

sustainability. Reliability, performance (IEEE 1646), security, scalability, network latency, data delivery criticality, etc. are major considerations in smart grid communication.

The chapter covers mainly some utility protocols and smart grid system and associated standards. Under utility program discussions on DDE, OPC, UCA, and DNP3 have been covered. Here it is important to note that since OPC is treated in details in III 15.0, here only essential details have been presented to complete the discussions. *Let us start discussions with DDE-OPC.*

## 11.0 Dynamic data exchange and open platform communication

Out of several methods, Windows provides for transferring data between applications: one method is the DDE protocol, which stands for a set of *messages* and *guidelines*. DDE was first introduced by Microsoft back in 1987. OLE's predecessor, DDE, is a standardized concept of interprocess communication, where data move dynamically among applications in the Microsoft Win32 application programming interface (API). DDE can be used by applications for one-time data transfers and for ongoing exchanges when one application sends updates to another as new data becomes available. This protocol can be conceived as a communication bridge between two software devices for data exchange. Although it is quite old and have lost popularity after introduction of OPC, yet it is still in use for enterprise programming, statistical analysis system (SAS) as well as in industrial application area, e.g., developing programmable logic control (PLC) ladder logic in Excel sheet. Data in a DDE is identified by a three-level hierarchy: *Service*, *Topic,* and *Item*. An application that processes the DDE protocol communicates with another DDE application via standardized messages. One of the applications is defined as the client, the other as the server. Client and server are holding a DDE conversation. DDE has a major use in transferring information in conversations (which is the communication between two types of applications named client and server). This message-driven protocol conducts its communications in the background, i.e., user can do whatever they are doing even during the execution of a DDE request. DDE messages can be of different types such as *Request*, *Advice*, *Poke*, and *Execute* messages. Many applications also use a DDE management library (DDEML) that provides functions to manage DDE conversations. *System topic* tells about the general interest of a DDE client. It gets associated with a data *Item* that helps users to locate information about the system. A *topic name* is a string that helps in the identification of a data context.

OPC is an interoperability standard for secure data exchange in industrial automation and other areas. It is designed to be platform-independent so devices from different vendors can exchange information. The original standard was limited to Microsoft technologies, so its application was restricted to Windows systems, but the standard is now more platform-independent, scalable, and extensible. The OPC standard has come to play a vital role in Industry 4.0, a movement toward smart manufacturing that embraces digital transformation and cutting-edge technologies, including artificial intelligence (AI) and Internet of thing (IoT)/industrial IoT (IIoT). Industry 4.0 emphasizes communication and interoperability: The OPC standard enables systems to communicate more efficiently and effectively, making it possible to collect and analyze information from every part of the manufacturing process. This helps to break down data silos, improve efficiency and responsiveness, and facilitate better business decisions. In this section, discussions shall be presented on DDE and OPC. OPC is discussed at length in III 15.0; hence, OPC is discussed here mainly as comparison purpose with DDE.

### 11.0.0 DDE—Conceptual discussions

DDE sends messages between applications that *share* data and *uses shared memory* to exchange data between applications. The DDE protocol can be used for one-time data transfers and for continuous exchanges in which applications send updates to one another as new data becomes available. Windows also supports the DDE management library (DDEML) for applications to share data. The DDEML provides functions and messages that simplify the task of adding DDE capability to an application. For message-based architecture of Windows, passing messages is the most appropriate method. Messages contain *WParam* (16-bit word parameter) and *lParam* (32-bit long parameter) for passing data. The DDE protocol defines exact use of *wParam* and *lParam*. The DDE protocol has specific rules for allocating and deleting global atoms and shared memory objects. DDE is most appropriate for data exchanges that do not require ongoing user interaction. DDE can be used to implement a broad range of application features such as *linking to real-time* (RT) data, *performing data queries* between applications, etc. Important concepts and terminology for understanding DDE are *client, server, and conversation*, *application, topic and item names*, *the system topic*, *permanent data links*, and *atoms and shared memory objects*. DDE has been conceptually presented in Fig III 11.0a to help in fully understanding the concept of DDE.

1. **Client–server conversation**: DDE is client–server-based technique. Two applications participating in DDE are said to be engaged in a DDE conversation. Client needs to initiate the exchange by establishing a conversation with the server. Client server network

FIG III 11.0a1 DDE NETWORK COMMUNICATION

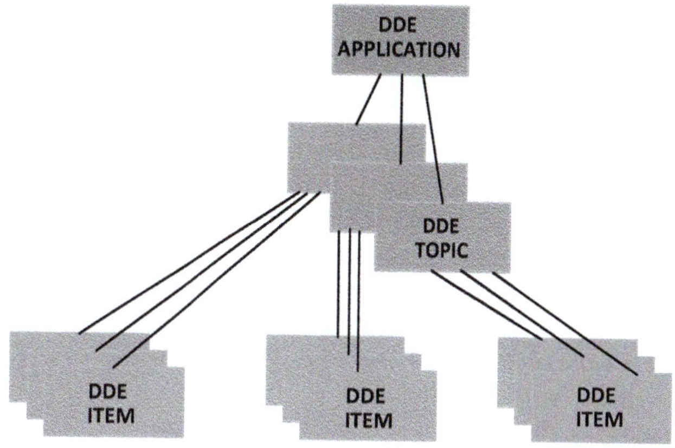


Fig III 11.0a2 DDE DATA MODEL

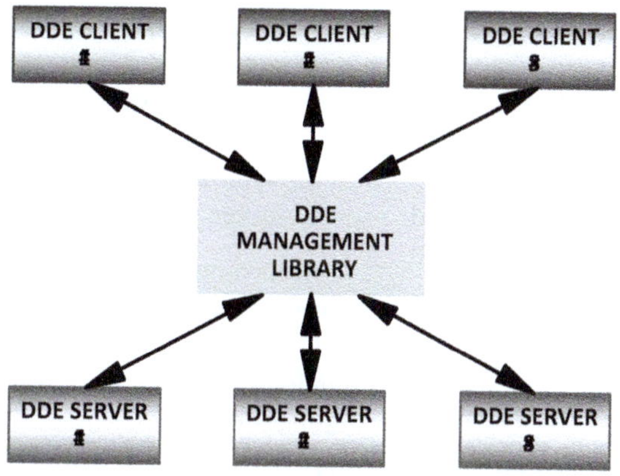


FIG III 11.0a3 DDE MANAGEMENT LIBRARY

**FIGURE III 11.0A** Dynamic data exchange (DDE)—concept.

has been clearly shown in Fig III 11.0a1. Client sends request for data and service from the server A server would oblige client with required data/service. To initiate a conversation with another application, it is necessary to know the name of the other executable program, and both programs must be running. One server can serve many clients at the same time. Similarly, a client can request data/services from several servers. Since a DDE conversation is identified by the *pair* of handles to the windows in the conversation, *no* window should be engaged in *more than one* conversation with another window. The client application/the server application must provide a different window for each of its conversations with a particular server or client application. It is client who terminates the conversation when it no longer needs the data or services from the server. An application can be both a client and a server.

2. **Application topic items**: Each DDE conversation is *uniquely* defined by the application name and topic determined at the beginning by the client and server. Normally, the application name is the name of the server application. Fig III 11.0a2 shows three levels of hierarchy:

- **DDE application:** This is unique server name among all the servers in the system. A string expression identifies an application that can participate in a DDE conversation [MS support].
- **DDE topic:** There could be several DDE topics under each DDE application. DDE topic is used to identify a logical data context in the server. According to MS Support, *topic* refers to "a string expression that is the name of a topic recognized by *application.*" The DDE topic is a general classification of data within which multiple data item. To establish a conversation, DDE application and DDE topic names are used by the client.
- **DDE item:** This refers to the name of data items available for conversation. There will be a number of DDE item under each DDE topic as shown in Fig III 11.0a2, e.g., boiler BMS programming (in Excel sheet). So, topic name refers to topic names loaded into the spreadsheet, e.g., monthly sales or ladder logic coal firing, whereas item name refers

to one or more cells in the spreadsheet. According to MS Support, item refers to "a string expression that is the name of a data item recognized by *application.*"

3. **Miscellaneous issues** [1]: Apart from the above, other issues behind DDE concept are as follows:
   - **System topic:** Applications should support the system topic at all times. This topic provides a context for information that may be of general interest to another application. Some items under system topics are *Format, Help, Return message, Status, System item* (list), *Topic item* (list), etc.
   - **Permanent data link:** Once conversation begins, client needs to establish one/more permanent data link, which is a communications mechanism by which the server notifies the client whenever the *value of a specified data item changes.* This data link is permanent till the data link/the DDE conversation is terminated. There are two kinds of permanent DDE data links: *warm* (only notification by server would be sent only on client request) and *hot* (the server immediately sends the changed value).
   - **Atoms and shared memory objects:** Certain arguments of DDE messages are global atoms or shared memory objects. Applications using these arguments must follow explicit rules about when to allocate and delete them. DDE uses shared memory objects for the following: T*o carry a data item value to be exchanged, To carry options in a message,* and *To carry a command execution string.*
4. **Key characteristics**: Some typical characteristics of DDE include the following:
   - **RT data transfer:** DDE allows for continuous data exchange between applications when available and hence may stay synchronized. DDE commands, such as DDE-REQUEST or DDE-EXECUTE, may wait for the command to complete before the client continues.
   - **Interprocess communication:** DDE is a mechanism for *different* applications to communicate and exchange data.
   - **Background operation:** DDE can transfer data in the background, i.e., while data is being exchanged, it allows users to continue working in other applications.
   - **Security risks:** DDE can be a security risk, as it can be exploited by malicious actors to launch unintended processes.
   - **Legacy technology:** DDE is a legacy method for interprocess communication in Windows; often alternatives are preferred for security and functionality.
5. **The DDE message:** Because DDE is a message-based protocol, it employs no functions or libraries. All DDE transactions are conducted by passing certain defined DDE messages between the client and server windows. Major DDE services shall include the following:
   - **Request:** This is used by client to read data items in the server.
   - **Poke:** With this service, client can change data items in the server.
   - **Advice loop***:* This allows the client to subscribe to data items in the server. During active period of subscription, server needs to notify client in case there is any change in the said data item in the server. An advice could be a message about notification and/or the new data.
   - **Execute:** By this client, make server to execute the commands. Major execute commands include *Update requests* (the client asks the server to notify for a change in data item for continuous updates) and *Data pushing* (the client can send data to the server without being requested).

Even though the name is client server, yet client and server communicating through DDE *never directly talk to each other.* In reality, they actually send messages to Windows, which passes the information on to the correct destination for carrying out the communication. Windows support DDEML to facilitate such exchange of data.

### 11.0.1 DDE protocol functionalities and applications

The client application starts a DDE conversation with the server application by specifying the application program and topic names and defines the exchange data by specifying the item name. As primary function, DDE allows Windows applications to share data. As discussed, the data communication is established by a simple, three-segment model, viz., each program as "application" name, in each application information by groups known as "topic," and each topic could serve up individual pieces of data as an "item." A common use of DDE was for custom-developed applications to control off-the-shelf software, viz., Excel spreadsheet. There are different uses of DDE, and it is said to be the most appropriate system for data exchanges where there is no requirement of ongoing interaction of the user. Usually, the user gets a *method* from the application for establishing a link between the applications in which data is exchanging, and hence, because of this link establishment, there is no further user involvement in data exchange between the applications viz. *Real-time data linking, Data queries between two applications, Helps in the creation of compound documents*, etc. In this subsection, brief discussions will be provided on functionalities of the protocol and associated issues. Let the discussions start with protocol itself with features.

1. **The protocol:** Being message-based architecture, it is very important to adapt appropriate method for passing information. DDE came into use when Windows were 16 bit; hence, messages contain only two parameters *WPARAM* and *LPARM* discussed at the beginning, for passing data. The DDE protocol defines use of WPARAM and LPARAM for passing larger pieces of data. The use of Windows DDE is most appropriate for data exchanges that do not require ongoing user interaction [1]. As discussed, client application provides a method for the user to establish the link between the applications exchanging data. Once it is done, then applications can exchange data automatically. There is a set of rules applied to decide when and how data is sent. DDE has its own protocol to monitor the communication between applications. Protocol details out the procedures for starting and ending of DDE sessions, as well as exchanging data with another application. It also specifies instruction for another application to execute a macro or command. DDE conversation always takes place between two windows. Since a DDE conversation is identified by the pair of handles to the windows in the conversation, no window should be engaged in more than *one conversation* with another window [1]. So, an application makes sure that a pair of client and server windows is never involved in more than one conversation even through a hidden window. As DDE is based on only a few parameters, it is rather elementary in nature. As DDE does not use any specific functions or libraries, it is easier to use this protocol. However, with introduction of OPC and its loses popularity.
2. **Working of DDE:** DDE mainly works under the following steps:
   - **Client–server:** In a conversation between two windows, one application acts as the DDE client, requesting data, and the other as the DDE server, servicing the data.
   - **Conversation:** Conversation begins when the client initiates a DDE conversation by sending an "INITIATE" message to the server for its service.
   - **Messaging:** Applications communicate by posting messages to each other, in agreed standardized format.
   - **Data hierarchy:** Data in a DDE is identified by a three-level hierarchy: *Service*, *Topic*, and *Item*.
   - **Data transfer:** Data is transferred via a shared memory buffer between the client and server.
3. **DDEML:** The DDEML is a dynamic-link library (DLL) used by applications to share data. The DDEML provides functions and messages that simplify the task of adding DDE capability to an application [1]. Applications use DDEML to manage DDE conversation as it helps to avoid entire processing time involving sending, posting, and processing functions. DDEML also provides a facility for managing the strings and data that DDE applications share [1]. For various connections of DDEML, Fig III 11.0a3 may be referenced. In DDEML, supported server names can be registered and names can be broadcasted for other applications in the system so that system can use it to connect the server. To ensure compatibility among DDE applications, DDEML forces everyone to implement the DDE protocol in a consistent manner. Existing applications using message-based DDE protocol are fully compatible with those that use the DDEML so that message-based DDE can establish conversations and perform transactions with applications that use the DDEML. Also, either the original DDE protocol or the DDEML can be used by a client application to communicate with content manager on demand. *To use the application programming interface (API) elements of the DDEML, it is necessary to include the DDEML header file in the source files, link with the DDEML library, and ensure that the DDEML dynamic-link library is in the system's search path* [1]. For details, recent document from Microsoft [45] may be referenced.
4. **Security issue:** Here, security issues with "DDE" can refer to **DDE protocol** in Windows applications. Vulnerability comes from embed malicious DDE queries in documents, embedded by attackers. In such cases in open file, information may be passed on to other location. Concurrent version system (CSV) *files are used for version control and management.* Vulnerability comes from CSV Injection. Thus, attacker can control data that is exported to an interjected CSV file. Mitigation methods mainly include the following: *Not to open suspicious file, Disable DDE Options, Stay and keep updated.* On account of vulnerability, people are opting for newer alternatives: (say) OPC.
5. **Application:** As already discussed at the beginning of this subsection, there are many application areas of DDE. As it is rather simpler, DDE can be used for a broad range of applications including linking to real-time data, viz., stock market updates or plant automation. It also allows for generation of compound document, e.g., a chart produced by a graphics application in word processing document. There are many automation software that support DDE protocol for communicating with other software, viz., Wonderware Intouch SCADA or RS Linx (Rockwell automation).

### 11.0.2 OPC basics and comparison of DDE with OPC

This message-driven DDE protocol has a major use in transferring information in conversations. DDE is an old standard, initially used by Microsoft. DDE shows poor performance and less reliability in interprocess communication, suitable for low-demand applications. DDE lacks in robust configuration and platform independency. OPC is a standardized interface for accessing process data. Object linking and embedding (OLE), component object model (COM), and distributed component object model (DCOM) were developed. When this is applied to the process control, OPC (OLE for process control) is developed. OPC is based on the Microsoft COM/DCOM standard and has been expanded according to the manufacturer's requirements when they need access to the data in the area of automation. OPC is a robust, vendor-independent, high-performance industrial automation standard for real-time data exchange, developed by the OPC Foundation for cross-platform compatibility and data security. Industrial automation product manufacturers can write OPC data access servers and software and could become an OPC client. Typically, OPC clients are programs for the acquisition of operating data. OPC servers are provided for PLCs and Fieldbus cards. An OPC server is an executable program, which is started when the connection between the server and client is made. Because of DCOM characteristics, it is possible to access the OPC server even if it is another computer. Windows-specific DDE is meant for sharing and updating data between applications running on the same computer. This is suitable for low-demand cases being slow process not suited for RT operations. This "plug and play" communication, developed by the OPC Foundation, is based on COM/DCOM technologies, which now can support TCP/IP the broader platform. In performance, OPC is significantly faster and can handle larger volume of data in more robust data management, compared to DDE counterpart. OPC supports vendor-independent and cross-platform, promoting easy integration of systems. So, it is very popular standard for use in industrial automation, SCADA systems, and real-time applications involving PLCs, DCS, and other control devices.

In this subsection, brief discussions on OPC and comparison of DDE and OPC have been presented. It is worth noting that OPC is discussed at length in III 15.0 so brief details have been presented to complete the discussions.

1. **OPC: Conceptual idea:** The OPC standard refers to a series of specifications developed by industry vendors and end-users. These specifications are meant for the interface between *client–server* and *server–server*, including access to *RT data, alarms and events, historical data*, and other applications. During initial days, it acted like "middle-man"—a standard interface for data exchange between PLC and human–machine interface (HMI)/SCADA. As a result, a number of products emerged allow end-users to implement systems using best-of-breed products with seamless interaction via OPC. For Industrie 4.0 approach, manufacturing industries adopted service-oriented architecture (SOA). SOA refers to a method of making software components reusable and interoperable through service interfaces, so that they can be rapidly incorporated into new applications. Therefore, with the introduction of SOA in production systems came new challenges in security and data modeling. For these in mind, the OPC Foundation developed the OPC unified architecture (UA) specifications to address these needs and at the same time provided a *feature-rich technology* open-platform architecture that was future-proof, scalable, and extensible. These made OPC a universal standard interface in automation.
2. **Purpose of OPC:** Major purposes of using OPC in industrial automation include the following:
   - **Interoperability:** Core purpose of OPC is to allow *diverse devices and applications* of different manufacturers to communicate seamlessly, breaking down data silos in industrial environments. So, OPC is an interoperability standard for secure data exchange in industrial automation and other areas. It is designed to be platform-independent so devices from different vendors can exchange information.
   - **Universal translator:** OPC provides common "language" to abstract complex, proprietary device protocols into a standardized interface.
   - **Standardization:** OPC specifies a consistent method for accessing field data from various devices, to ensure that a client can get data in the same way regardless of the underlying hardware or software.
   - **Legality:** By registering as an OPC, entrepreneurs can enjoy the benefits of a separate legal entity while maintaining complete control over their business operations.
3. **Functions of OPC:** Functional details of OPC include but not limited to:
   - **Client–server:** OPC servers act as a bridge between the industrial devices and the client applications that need to access their data and allows the data to be shared and monitored by providing a standardized interface for data access and can perform tasks. Functionally, OPC operates as a client–server model. An OPC server converts (translates) device-specific data into OPC format. An OPC client requests and receives the required data. The OPC server converts the hardware communication protocol, as well as any software that needs to

connect to the hardware becomes the OPC client. OPC Instructions are used by OPC clients and servers to identify, send, and control commands that are executed in a controller or I/O module.

- **Data access**: Data access facilitates real-time data exchange, so that the clients can read and write current data values, which include the value, a timestamp, and a quality indicator (validity of the data).
- **Execution method**: Clients can also execute methods on the server, enabling control functions beyond simple data access.
- **Security:** As a part of security functionality, OPC UA includes built-in security features e.g., encryption for secure data transmission, crucial for IIoT and Industrie 4.0.
- **Alarm and event:** OPC servers are capable of monitoring and provide access to alarms and events generated by devices.
- **Historical data:** The standard allows access to historical data for analysis.

4. **OPC types:** Basics of classification of OPC has been enumerated below to complete the discussions. Detailing on these is available in III 15.0
   - **OPC classic:** As discussed during initial discussions of this subsection, this first version of the OPC standard was based on COM and DCOM (a Microsoft technology) for exchanging data between software components across a network. Originally, the OPC acronym stood for *OLE for Process Control.* The specifications provide separate definitions for accessing process data, alarms, and historical data.
   - **OPC UA:** OPC Foundation released version 1.0 of OPC UA in 2006. This was done, for shift toward SOA (reusable services capable of communicating across different platforms and languages) discussed before. Also, Industrie 4.0 approach platform has thoroughly examined OPC UA and validated it as a *key technology* for implementing the "Communication Layer" of the "Reference Architecture Model Industrie 4.0.'" (RAMI4.0). The OPC UA protocol has been standardized as part of the IEC 62,541 series also. Recent version of OPC UA is not based on Microsoft COM technology and so enables cross-platform compatibility. For details on RAMI, Fig II 16.2a, in Volume II, may be referenced.
   - **OPC: AE:** Released in 1999, OPC AE describes the ways OPC server software can monitor systems and send alarms and events to client software.
   - **OPC:DA:** Released in 1996, it deals with OPC client–server architecture for exchanging data detailing out software read and write data, the data types, and structures available.
   - **OPC: HDA:** Released in 2001, the specification defines the queries and analytics related to the historical time-stamped data collected from devices.
5. **Comparison of DDE and OPC:** Let us conclude the section dedicated for DDE and OPC by tabling the comparison between DDE and OPC in Table III 11.0a

**TABLE III 11.0A** Comparison between DDE and OPC.

| Features | DDE | OPC |
|---|---|---|
| Technology | Microsoft; general purpose; Windows message based | Initially, Microsoft COM/DCOM technologies. Now regulated by OPC foundation—cross-platform |
| Configuration | Limited | Robust configurable |
| Architecture | Windows based | COM/DCOM and TCP/IP Cross-platform |
| Reliability | Low—for demanding application | High and robust |
| Passing | One value per request | Multiple values per request |
| Interoperability | Not tested | High interoperability |
| Time stamp | No supported | Yes, as standard response |
| Scalability | Not much | Scalable |
| Security | Limited features | Enhanced features |
| Quality flag | No quality flag | Yes, as standard response |
| Resource | High resource utilization | Low utilization |
| Supportability | Becoming obsolete; people opt for OPC. | Heavily supported |
| Performance | Slow in adequate for large volume of data | High-performance, capable of handling Large data loads quickly |

## 11.1 Discussions on UCA DNP3, and IEC 60870 standard

Utilities spend an ever-increasing amount for real-time information exchange; costs for data integration and maintenance are exploding. UCA defines a "suite" of protocols to address utility communication requirements with focus mainly on substation. UCA has also been developed for applications in different areas such as the electric, gas, and water utilities of SCADA. It is important to note that UCA is an architecture, rather than a simple, protocol [2]. UCA provides standards to dramatically improve device data integration into the information and automation technology, reducing engineering, commissioning, operation, monitoring, diagnostics, and maintenance costs. UCA also helps to increase the agility of the whole life cycle of utility automation systems. The UCA 2.0 really is a family of basic communications protocols to meet the challenges for a wide range of utility services and SCADA. Developed in 1993, DNP3 (IEEE Std 1815) is a comprehensive protocol standard that defines the rules by which computers communicate with one another. The DNP3 protocol specifically defined the interaction between *utility computer systems* with remote communication in mind with special focus on providing a lightweight means of transporting simple data values with a high degree of integrity. DNP3 is based on an object model. This model reduces the bit mapping of data that is traditionally required by other less object-oriented protocols. The DNP3 defines two types of endpoints that communicate with one another—a master and an outstation/remote device. The master provides the interface between the human network manager and the monitoring system. The remote (RTUs and IEDs) provides the interface between the master and the physical device(s) being monitored and/or controlled. Signals to be sent over serial links, multidrop serial links, radio links, dial-up connections, and dedicated networks using TCP/IP or UDP. Communication in the area of SCADA, open standard distributed network protocol 3 (DNP3), is quite popular. Although it draws many of its elements from IEC 60870, yet it has many features different from IEC 60870. Not only in electrical power distribution, it is very widely used in water sector. Use of DNP3 in the gas and oil sectors has also been gaining popularity. In this section, brief discussions on both of them will be presented to understand the utility service of SCADA. IEC standard 60870 has six parts, defining general information related to the standard *operating conditions*, *electrical interfaces*, *performance requirements*, and *data transmission protocols*. IEC 60870–5 is a protocol standard for telecontrol, teleprotection, and other telecommunication functions for electric power systems. Some important issues are as follows:

- IEC 60870–5-101 (IEC101) is a standard for power system ***monitoring***, ***control***, and other *communications* for electric power systems automation.
- IEC 60870–5-104 (IEC104) is an extension of the IEC 101 to include *transport, network, link, and physical layer extensions* to enable a *full network access*. IEC 104 also contains various types of mechanisms to synchronize data.
- For communications within Ethernet local area networks (LANs) from the data through *TCP/IP*.
- Two separate link layers enable data transfer over of both *Ethernet and serial media*.

This section has been dedicated to discuss UCA, DNP3, and IEC 60870 standard.

### 11.1.0 General discussions on UCA DNP3 and important terms

UCA supports data exchange between control centers based on IEC 60870-6-503: TASE.2 Services and Protocol, IEC 60870-6-802: TASE.2 Object Models, and IEC 60870-6-702: TASE.2 Application Profile. For UCA, common parts and the parts for modeling and communication for intelligent devices are based on IEC 61850. It is possible to use UCA to interconnect data sources within and between utilities using WAN with TCP/IP and Ethernet, which is an obvious choice for its popularity, low HW and SW cost, and scalability in terms of bandwidth. Apart from TCP/IP configuration, three-layered model, it can be configured in full ISO/OSI seven-layered solutions. Like other protocol discussed earlier, the application layer is built on the services of the MMS.

UCA is object model–based system comprising CASM and GOMSFE. There are two other interfaces types in pipeline but not yet published. Typical three-level structure with CASM for UCA has been depicted in Fig III 11.1a (developed based on idea from 3).

UCA is the umbrella for UCA distribution energy resource (DER) discussed later. Since UCA is based on object model, as seen in Fig III 11.1a1, data object model and service interface sit on communication profile. Data object model is at the highest level. Service interfaces to data model for various services such as definition, reporting, retrieving processed data controlling devices including file transfer as we had seen in previous occasions, i.e., it describes the data and devices and their access methods. Possible communication profiles have already been discussed before. Field controller (server) CASM exchange data with CASM of user (client) based on GOMSFE. So, it is better to look into CASM and GOMSFE. DNP3, specifically, was developed to facilitate communications between various types of data acquisition and control equipment. Therefore, it plays a crucial role in SCADA

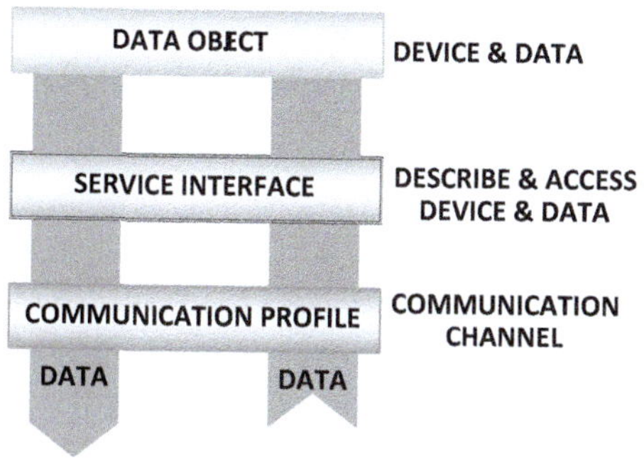


FIG III 11.1a1 THREE LEVEL UCA

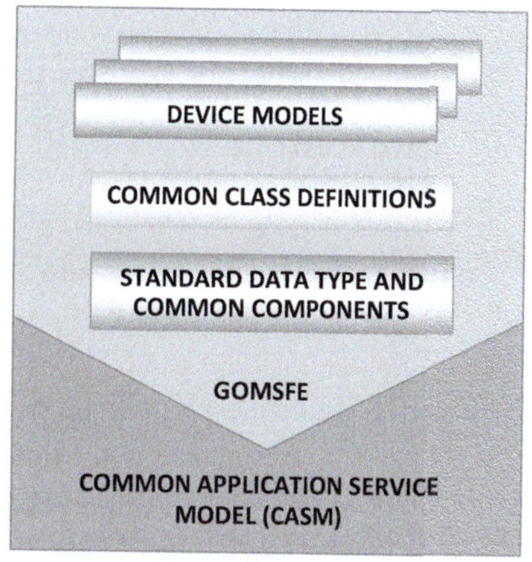


FIG III 11.1a2 UCA LAYERING (GOMSFE & CASM)

THIS FIG I 8 HAS BEEN DEVELOPED BASED ON IDEA FROM 8

COURTESY:SCC AND RESEARCH GATE.

**FIGURE III 11.1A** Overview of utility communication architecture [2].

systems, where it is used by SCADA master stations (control centers), RTUs, and IEDs as well as communication between RTUs. DNP3 is object model–based system where it uses IEC 60870-5 as its basis. IEEE and the Electric Power Research Institute (EPRI, Palo Alto, USA) originally initiated the development of an open communication architecture so that it will be easier and better for exchange of information between various elements, using different technologies, of power and other utility service. Standards Coordinating Committee (SCC) of IEEE finally came out with IEEE TR 1550 in July 1999 on UCA version 2.0.

1. **UCA: Discussions**: UCA protocol has been developed to standardize communication among various devices and systems within utility environments to enable seamless integration, foster multivendor interoperability, and achieve reduced costs for engineering, operation, and maintenance. UCA provides a common framework, including object models for data and standardized interfaces, to connect diverse utility equipment with SCADA. UCA has been designed and developed to address interoperability issues using a decentralized, object-oriented philosophy. UCA is object-oriented architecture rather than a protocol. So, UCA breaks down large problems into smaller blocks with defined interfaces that are easier to solve. In UCA, *a system is designed in such a way that the information is organized as a collection of devices containing objects, which can be independently accessed by multiple systems* [3]. The database is decomposed into a collection of smaller and simpler blocks of information for flexible information system. Devices compliant with UCA are able to give complete description of all the data it has over the medium, on request from SCADA system. It comes out with the full list of data points. UCA point names are a logical device model based on object-oriented techniques. So, to access a data point, the functional names will be used and nothing to remember during configuration as these are easily recognized. UCA does not subscribe new, utility-specific protocols. MMS is an internationally standardized (International Organization of Standardization 9506) messaging system for real-time networked data exchange and supervisory control [3]. The UCA Forum has been formed to precisely define ways and means for MMS implementations in UCA. CASM, GOMSFE, and Inter-control Center Communications Protocol (ICCP) specify these requirements. CASM, GOMSFE, and ICCP together form UCA.
2. **Common Application Service Model (CASM):** CASM: Typical structure of CASM has been detailed out in Fig III 11.1a2. The UCA-CASM is a generic communications services for data access, data (and exception) reporting, logging, device control, tagging, and self-describing devices along with related services. CASM also provides detailed data objects mapping to MMS and mapping of generic services to MMS services. It not only helps to isolate models from services but also helps in application interoperability. The CASM services are abstract and may be mapped into any number of existing communication application

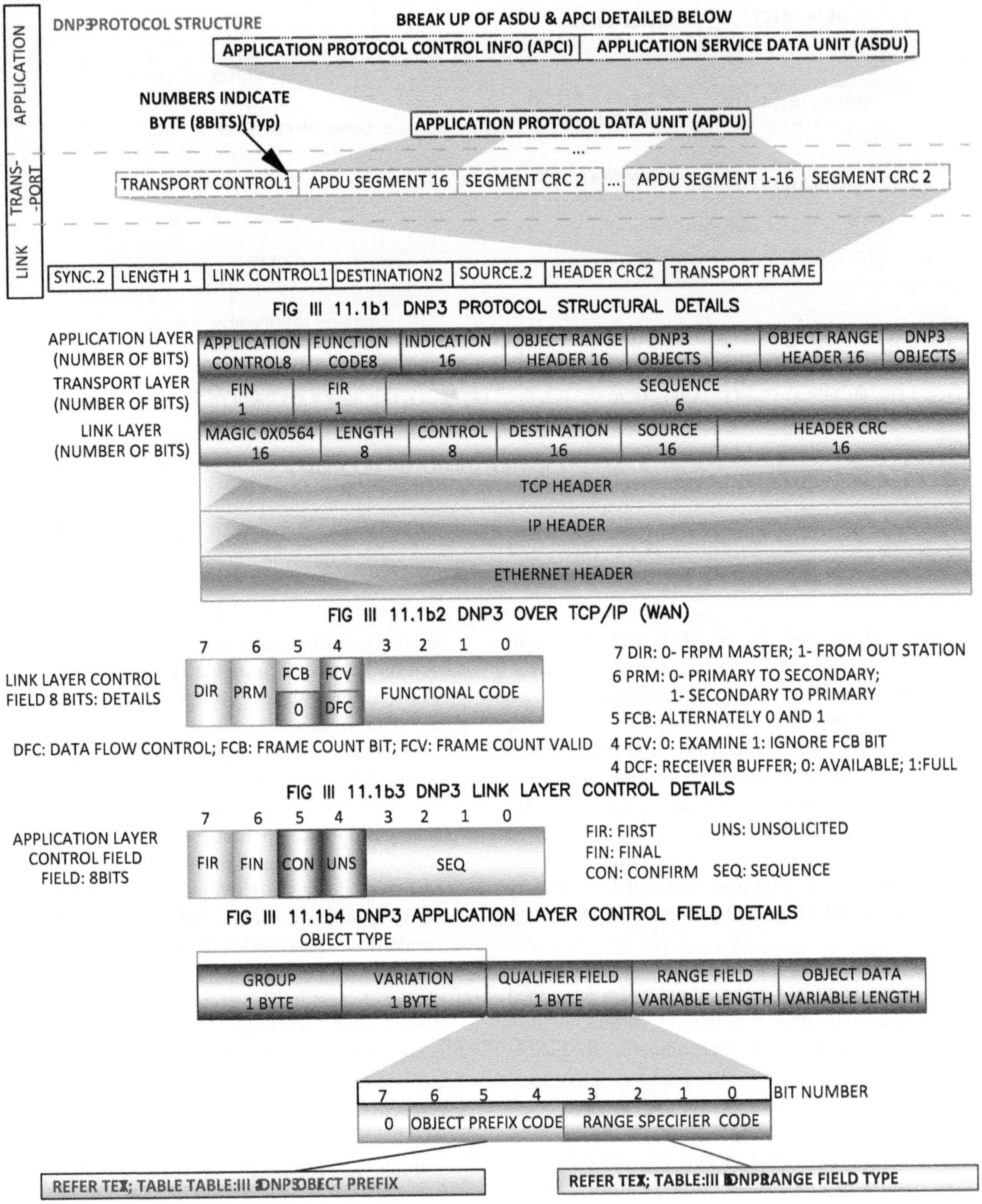


**FIGURE III 11.1B** Overview of DNP3 protocol.

standards [2]. CASM is a document, which specifies the step-by-step methodology, for UCA 2.0.

3. **Generic Object Models for Substation and Feeder Equipment (GOMSFE):** GOMSFE is responsible for the development of protective relay/IED models. Object modeling is a technique used for identification of data elements with their behavior, attributes, and standards. It also lays foundation for all the required information needed for self-description. Starting with a base set of models, each of the relay vendors has added draft models for an additional one or two functions, which brought the total to 13 models [2]. Based

on these models, two building blocks, viz., *Basic Relay Object* and *Basic Time Curve Object*, have been developed. Based on these, several building blocks have been developed. There are several such building blocks, viz., *generic input/output*, *measurement functions*, *transformer functions*, *switch functions*. GOMSFE document is like a dictionary for standard object modes with associated names to describe device and their functions for substation IED. Every UCA 2.0-compliant device uses the same naming conventions. So far, the discussions were mainly restricted to use of UCA in electrical systems. On account of enhanced capability for development of integrated business/enterprise model and simplicity for implementation, UCA also finds its application in gas industry SCADA.

4. **DNP3 Basics:** DNP3 can be conceived as a language than a communication between different devices especially in SCADA application. As DNP3 is based on an object model, bit mapping of data is greatly reduced. DNP3 has a few predefined objects, which makes it easier a somewhat more comfortable design and deployment for SCADA framework. As it is open, any manufacturer can develop own DNP3 equipment that is compatible with other DNP3 equipment—making it very popular. DNP3 maintains a library of common object, and both the master and remote use the same for exchanging data. As open intelligent SCADA protocol, DNP3 offers the following major features:
   - **Layer 2 protocol:** As layer 2 protocol, it allows multiplexing, data fragmentation, and link control.
   - **Interoperability:** It offers interoperability between different vendor's equipment.
   - **Request and respond:** It can use multiple data in single message, and only changed data sent as response.
   - **Error detection:** It uses multiple frames to offer excellent error detection by cyclic redundancy check (CRC).
   - **Priority and unsolicited response:** Priority assignment and unsolicited response (alarm) are possible.
   - **Optimization:** It is designed to optimize the transmission of data acquisition information and control commands from one computer to another.
   - **Time synchronization:** It supports time synchronization.
   - **Peer-to-peer communication:** DNP3 supports it with multiple masters.
   - **File transfer:** File transfer is possible.
   - **Polled and quiescent operation:** It supports these operational modes.
5. **Important terms:** Listed below are a few terms which are key to the standard and will be discussed in subsequent discussions:
   - **Abstract communication service interface (ACSI)** [4]: This powerful abstraction technique isolates services and information models from their underlying protocols. ACSI defines naming conventions in the communication of substations, a virtual substation image, real-time data access, device control, event reporting, data typing and discovery of data types, publisher/subscriber, and file transfer from the physical devices.
   - **Manufacturing messaging specifications (MMS)** [4]: MMS [5] is a kind of messaging system for communicating real-time data and control information between devices in the grid. MMS is independent of *application function* and *device manufacturer*. MMS supports the set of services and named objects. Part 8–1 specifies the mapping of this ACSI to the existing MMS. With MMS, the server can send data without an explicit request provided open TCP socket connection is available. The MMS server represents the IEDs that need to be controlled like those located in smart grid distribution networks. An MMS server can represent an individual DER, a microgrid, or a home grid [36].
   - **Process bus** [6]: The process layer or process bus of the substation is responsible for gathering information, such as voltage, current, and status information, from equipment (transformer) and sensors connected to power system. As IEC 61850, there are two different protocol types, viz., Part 9.1: Unidirectional multidrop point-to-point fixed link carrying a fixed dataset and Part 9.2: A "configurable" dataset that can be transmitted on a multicast basis from one publisher to multiple subscribers.

### 11.1.1 Detailed discussions on DNP3 protocol

Enhanced performance architecture (EPA) refers to a simplified, three-layer communication model based on the OSI model. This is the foundational framework for the DNP3 protocol. So, it has its root with IEC 60870-5. Layers of DNP3 protocol are *physical layer*, *data link layer* (DLL), pseudo-transport layer, and *application layer*.

DNP3 is an open, intelligent, robust, and efficient modern SCADA protocol. It can do the following:

- Request and respond (unsolicited also) with *multiple data* types in *single* messages
- Message segmentation into multiple frames for error *detection* and *recovery*
- Include only *changed data* in response messages
- Assign priorities to data items and request data items periodically
- Support time synchronization in standard time format

- Allow multiple masters and peer-to-peer operations and user-definable objects including file transfer

With these issues in mind, let us now look into little details about the protocol and later would compare the same with other protocols such as T101 and Modbus. Let the discussions start with reference to Fig III 11.1b and associated Table III 11.1a and Table III 11.1b. So, the figures should be read in conjunction with the tables referenced.

1. **Protocol structure with layers:** DNP3 supports a variety of physical layers such as RS-232, RS-422, and RS-485. It can be encapsulated and can be sent over TCP/IP also. Apart from physics, it has three main layers, viz., link layer (data link layer), a pseudo-transport layer, and application layer. Transport layer is referred to as pseudo because it functions only to build application data. This has been clearly depicted both in Fig III 11.1b1 and 2. As clearly shown in Fig III 11.1b1, application protocol data unit (APDU) consisting of application protocol control information (APCI) and application service data unit (ASDU) in application is fragmented into 16-bit segments each with a two-byte CRC field appended to it. This is passed on to link layer where it adds header to it.
   - **Data link layer:** DNP3 uses FT3 frame format. It starts with magic bytes to identify it as DNP3 frame. Next field is length, which gives the length of all the fields following it, except the CRCs length. It also contains destination/source address and CRC each with 2-byte (16bits) length. Before destination address, there lies control field of 8-bit size. Details of control field have been elaborated in Fig III 11.1b3, which is self-explanatory. Link layer has a number of functional responsibilities, which include but not limited to synchronizing and handling the data flow control (DFC) bit based on buffer availability and the frame count bit (FCB) in the control field. It also controls physical layer, unpacking the data link frame from the physical layer into user data and checking and removing CRC. This layer is also responsible for error notification, service data unit (SDU) exchanges, and SDU delivery quality between peer link layers.
   - **Transport layer (pseudo):** The transport layer is only 1 byte (8 bits) long as it is basically used for fragmentation of large DNP3 packets into smaller segments. Out of 8 bits in transport layer, there are 2 bits for FIN and FIR signifying final and first, i.e., to indicate that this is the final and/or first fragment in the sequence. Six bit sequence number is used for segment reassembly. Thus, a total of 255 segments is possible. Naturally for a single segment payload, both final (FIN) and first (FIR) will be 0. Fragmentation of data into one or more data segments and reassembling data link frames, controlling all aspects of the data link excluding data link configuration, and sequencing SDUs are major

**TABLE III 11.1A** DNP3: Object prefix code (meant for Fig III 11.1b5).

| No. | Description | No. | Description |
|---|---|---|---|
| 0 | Object packed without prefix | 4 | Object prefixed with 1-octet object size |
| 1 | Object prefixed with 1-octet index | 5 | Object prefixed with 2-octet object size |
| 2 | Object prefixed with 2-octet index | 6 | Object prefixed with 4-octet object size |
| 3 | Object prefixed with 4-octet index | 7 | Reserved |

**TABLE III 11.1B** DNP3: Range field type (meant for Fig III 11.1b5).

| No. | Description | No. | Description |
|---|---|---|---|
| 0 | 1-Octet start stop indexes | 6 | No range field used |
| 1 | 2-Octet start stop indexes | 7 | 1-Octet count of objects |
| 2 | 4-Octet start stop indexes | 8 | 2-Octet count of objects |
| 3 | 1-Octet start stop virtual addresses | 9 | 4-Octet count of objects |
| 4 | 2-Octet start stop virtual addresses | A-C | Reserved |
| 5 | 4-Octet start stop virtual addresses | D-F | Reserved |

functions of transport layer. In addition to these, like link layer, the layer is also responsible for layer error notification, and peer-to-peer SDU exchange at transport layer.

- **Application layer**: Other than DNP3 objects, there are four types of field in application layer.
  - *Control field:* This is 1-byte field with detailed structure as shown in Fig III 11.1b4. This is meant to accommodate the functionality of fragmentation in the application layer. As shown, it also indicates whether message is unsolicited or is a confirmation. FINAL/FIRST and sequence for different DNP3 objects are similar to that discussed earlier.
  - *Function code:* This 1-byte code is used to identify the function of the targeted device, e.g., Read/Write, Confirm, Select, Operate, Cold Restart, Response, etc.
  - *Indication code:* This 2-byte code will be present in packets with a "response function" code to provide the details about the status of the request.
  - *Object ranges*: This is detailed out in Fig III 11.1b5 in association with Table III 11.1a & Table III 11.1b. The range field is used to determine the number of DNP3 objects within the range. The object type sent in a DNP3 object range is determined by the object group and variation within that group. The object types supported vary by each DNP3 capable device, and each manufacturer generally documents its supported types [7]. The qualifier field is for indexing of the objects as detailed out in Table III 11.1a & Table III 11.1b, which are self-explanatory. Range specifier code specifies the range field type.
  - *Functions*: Quiescent operation, unsolicited report, polled report-by-exception operation, and polled static operation are major functions of the layer [8].

2. **Element communication and layered communication:** There are 27 basic function codes used for data exchange between masters (control center) and remotes such as remote terminal unit(RTU)s/IEDs. Some of those function codes are used for a master's request and receive response of status from remote. Other function codes enable a master to change a remote's settings, or master to control the remote. There is one unsolicited response of remote meant for process or system alarms. From the discussions on protocol structure, it is clear that DNP3 uses the layered communication model. There are ASDU and the packaged objects. APCI block is used with ASDU to make an APDU as shown in Fig III 11.1b1, to pass on to pseudo-transport layer. At the transport layer, APDU is broken into several segments with a maximum size of 16 bytes appended with 2-byte CRC. Also, at this layer, these segments are added with 8-bit transport control header for sequence control. After that, it goes to link layer, which also adds a header to the control and address information. The packet is now ready for delivery. If the serial transport is used, the packet assembly is completed and placed on the transport media for delivery. For wide area network (WAN) communication, these layers would be mapped to the four-layer model developed by the Department of Defense with IP layer omitted. Fig III 11.1b2 shows DNP3 message over TCP/IP. The three DNP3 layers are rolled up into the application layer in TCP/IP model. The assembled packet is wrapped in the TCP by the transport layer, which in turn is wrapped in the IP. User datagram protocol (UDP) can be used for reliable delivery. The fourth layer is the network interface layer (of TCP/IP model) where the assembled packet is actually interfaced to some kind of transport media.
3. **TCP/IP traversing:** From the aforementioned, an idea about communication is gathered. Now when looked at TCP/IP traversing, the following steps are followed there:
   - **TCP/IP transmission:** After message is put over TCP/IP, the transport layer adds ID data block on to it to identify the ports for master and that of remote DNP3. The packet is then passed to the IP layer where a data block holding the IP and *media access* addresses of the master and the remote is added and the entire packet is passed to the network interface layer. The network interface layer checks media access and availability to place the packet on it for transmission.
   - **TCP/IP receipt of message:** After message is receipt, same process in reverse order takes place. Network interface layer pulls off the message, checks packet totality and validity so that it can pass to IP layer. The IP layer checks the media access and IP address before passing on to TCP/UDP layer where the target port is checked for connected applications. If an app is listening at the target port, the packet is passed to the application layer. If the listening app is the remote DNP3 process, the read request is passed [7]. Message then moves through the three layers of DNP3 in reverse direction to check the request.
4. **Object library and classes:** DNP3 protocol contains data points with a variety of data types. Data points are grouped as per data types. These groups are referred to as data *objects*, viz., binary input object, analog input object, counterobject, etc. The collection of these data object groups is called *data object library*, which is free for DNP user group members for use in SCADA

systems. DNP3 data objects can be further classified based on object their variations, e.g., 16-bit analog input, 32-bit floating-point analog input, etc. This library not only makes it easy for designing the DNP3 remote responder to use these common objects to report but also makes it easier for masters to integrate the data from remotes. Again, data types can be divided as *static* objects (current value of the object) or *event* objects (result of data change). There are four classes, viz., class 0 through class 3. Class 0 is static data, whereas classes 1, 2, and 3 objects are event objects. The highest to lowest priority are assigned to class 1 through class 3, respectively. Common data groups include the following:

- Binary inputs (single-bit read-only values)
- Binary outputs (single-bit values whose status may be read, or that may be pulsed or latched directly or through SBO type operations)
- Analog inputs (multiple-bit read-only values)
- Analog outputs (multiple-bit values whose status may be read, or that may be controlled directly or through SBO type operations)
- Counters, time, and date
- File transfer objects and others
- For each object group, or data type, one or more data points exist. A data point is a single data value of the type specified by its object group.

5. **Comparison of DNP3 with other similar protocols:** As discussed earlier, DNP3 has its root with IEC 60780 yet they are not same. Also, Modbus can be used for master to RTU communication. So, it is interesting to compare the features of these protocols. Table III 11.1c enlists such comparison.

### 11.1.2 Brief discussions on overview of telecontrol protocol: IEC 60870

International standard IEC defined series of IEC 60870 standards meant for telecontrol SCADA and power system automation applications. Part 5 of the standard is concerned with communication profile for sending basic telecontrol messages between a central telecontrol station and telecontrol outstations. Basic purpose and scope of IEC 60870 mainly include *telecontrol SCADA* and *interoperability*. Within IEC 60870, the IEC 60870-5 family of standards defines telecontrol protocols, primarily IEC 60870-5-101 and IEC 60870-5-104, for SCADA. The IEC 60870-5 protocol stack is based on the reduced reference model called EPA (comprising *physical* layer, *link* layer, and *application* layer),

**TABLE III 11.1C** Comparison of DNP3, T101, and Modbus.

| Features | DNP3 | T101 | Modbus |
|---|---|---|---|
| Standard/year | Open ind./1993 | IEC 1995/2000,01 | NA/1979 |
| Architecture | 4 layer (pseudo-transport) | 3 layered EPA | Application: Messaging protocol |
| Physical layer | Balanced mode<br>Supports multiple<br>Masters and slave<br>Peer-to-peer comm<br>RS 232/485 and<br>TCP/IP over IEEE 802.3/<br>X.21 | Balanced mode:<br>Point to point/<br>Multipoint: Using X.24/X.27<br>Unbalanced mode<br>Point to point/multipoint: Using V.24/<br>V.28 | Balanced mode:<br>RS 232<br>Peer to peer<br>Comm. TCP/IP over IEEE 802.3 |
| Data link | Frame: FT3 | Frame: FT 1.2 | Mode: ASCII/RTU |
| Application layer | Both support time synchronization, time-stamped events, select before Operate, polled report by exception<br>Unsolicited responses and data group/classes | | No time stamp; SOE without time nor event list with time. No polled report by exception. Checksum for end-to-end communication |
| Application layer (contd) | Remote start stop, poll by priority and internal indication | No internal indication/application confirms for events | |
| Device addressing | Pairs of devices may swap master and slave roles | Pairs of devices will not swap master and slave roles | Addresses: ASCII:2 character RTU: 8 bits |
| Major configuration parameter | Baud rate, device addresses, fragment size | Baud rate, device addresses, frame length, size of link and ASDU address. Point number size and structure | Baud rate, mode: ASCII/RTU, parity mode |
| Market | North America | Europe | |

which is discussed here also. The companion standards IEC 60870-5-101 and IEC 60870-5-104 are derived from the IEC 60870-5 protocol standard definition. IEC 60870-5-101 allows two alternative transmission procedures, unbalanced and balanced one. In application areas, it covers data and event acquisition, interrogation, clock synchronization, etc. Telecontrol protocol mainly focuses on key features such as *reliability*, *time synchronization*, *efficiency*, and *security*. The main focus of discussions is Part 5 along with associated companion standards for exchange of telecontrol messages. IEC 60870-6-502 is concerned with ICCP. So, short discussions on ICCP have also been covered later. To understand various issues of Part 5 of the standard, Fig III 11.1c has been presented in the following.

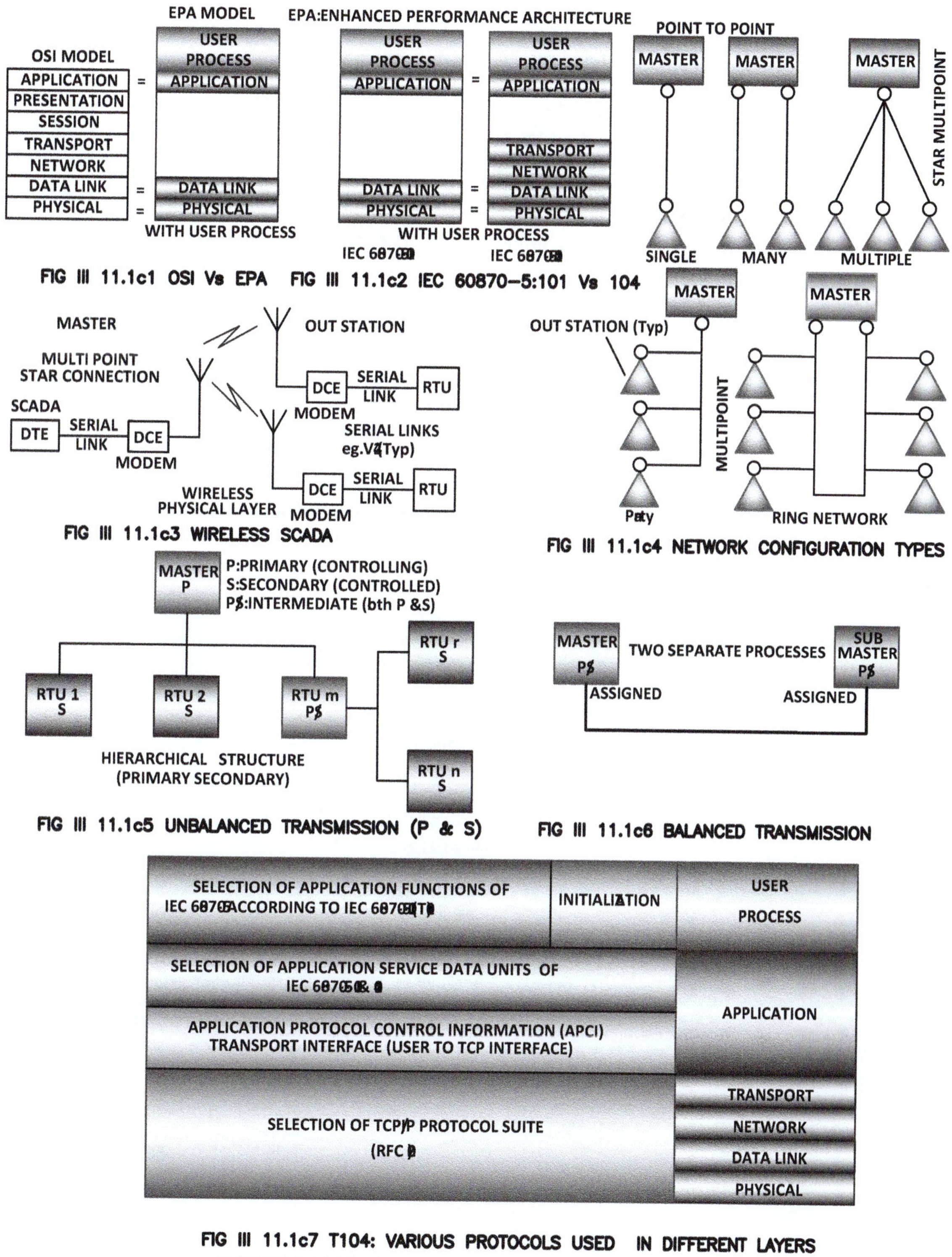


FIGURE III 11.1C Overview of IEC 60870-5 transmission protocol.

1. **Structure of IEC 60870 standard:** Structure of IEC 60870 standard has been enlisted in Table III 11.1d. Although IEC 60870-5-1 to IEC 60870-5-5 are the core specifications, yet companion standards, IEC 60870-5-101 to IEC 60870-5-104, are also important as each of these is intended for specific application level, data objects, and functions to completely define the protocol. Often 101–104 are referenced as T101–T104 where "T" represents telecontrol. *T101* covers general communications with remote terminal units (RTUs), including data (generic) types and services for electrical substation systems. The data types are generic and suitable for wider SCADA applications. *T102 and T103* are mainly concerned with electrical *protection* systems, e.g., line/transformer differential protection. *T104* deals with the transmission protocol over networks using TCP and IP protocols. IEC 6-870-6-502 covers ICCP by which data exchange over WAN between utility control centers, regional transmission operators (RTOs), and other generators is possible.
2. **Discussions on IEC 60870-5:** This part of the standard is related to transmission protocol. It is important to look little deep in this standard. Listed in the following are short discussions on various parts of IEC 60870-5.
   - **IEC 60870-5-1: Transmission frame formats:** This standard deals with frame format types for data link layer. It gives four choices, viz., FT1.1, FT1.2, FT2, and FT3. Each of them has different security level toward data error. Both fixed and variable length versions for frames are available with associated set of rules for transmission.
   - **IEC 60870-5-2:** Link transmission procedures: It deals with the internal processes for the service primitives (communication between adjacent layer) and transmission procedures. The service primitives could cover communication between the link layers and its higher-level user also. Transmission procedure regulates the sequence of events in physical communication with the help of a control field in each side.
   - **IEC 60870-5-3: General structure of application data:** It describes general structure of the application data and rule for forming application data (say) by encapsulating it within an application data unit and passing it to underlying layer—similar to that described in OSI model.
   - **iEC 60870-5-4: Definition and coding of application information elements:** This is meant to define information elements, and a common set of information elements for transmission of information, viz., signed and unsigned integers, fixed and floating-point numbers, time elements, so that a set of information building blocks are made for use by companion standard.
   - **IEC 60870-5-5: Basic application functions:** It deals with the highest-level functions of the transmission protocol, i.e., those application functions above layer 7 of the OSI model as detailed in

**TABLE III 11.1D** Structure of IEC 60870.

| Standard | Description |
|---|---|
| IEC60870-1 | General considerations |
| IEC60870-2 | Operating conditions |
| IEC60870-3 | Interfaces (electrical characteristics) |
| IEC60870-4 | Performance requirements |
| IEC60870-5[a] | Transmission protocols |
| IEC60870-5–1 | Transmission frame format |
| IEC60870-5–2 | Link transmission procedure |
| IEC60870-5–3 | General structure of application data |
| IEC60870-5–4 | Definition and coding of application information elements |
| IEC60870-5–5 | Basic application functions |
| IEC60870-5–6 | Guidelines for conformance testing for companion standards |
| IEC60780-5–7[a] | Security extensions for T101 and T 104 (*IEC 62351*) |
| Companion | **Standards for IEC 60870-5-101 to 104 (T101-104)** |
| IEC60870-5–101 | Companion standard for basic telecontrol tasks |
| IEC60870-5–102 | Companion standard for transmission of integrated totals |
| IEC60870-5–103 | Companion standard for protection communication |
| IEC60870-5–104 | Network access using standard transport profiles |
| IEC60870-6 | Telecontrol protocols compatible with ISO and ITU-T recommendations |

[a]*IEC 60870-5–7 is for security extension for T101 and T104 ("T" explained in the following).*

Fig III 11.1 c1. It refers to application service primitives, i.e., the request and response indications passed between the application layer and the application user. A set of basic application functions and associated service primitives, or requests and indications, are presented. Basic application functions supported are *data acquisition, event acquisition, command transmission, interrogation, clock synchronization, transmission of integrated total, changes in protocol and link parameter*, and *acquisition of transmission delay.*

3. **Enhanced performance architecture:** IEC 60870-5 standard protocol has been developed based on reduced OSI model. It is called enhanced performance architecture (EPA), which (like field bus model) includes physical layer, data link layer, and application layer. Selected application functions of IEC 60870-5-5 described above have been added as user process as shown in Fig III 11.1c1. User process has been included to define telecontrol system operations with various application functions and is important interoperability among devices to make the protocol completely operable. Further EPA details have been enumerated in Table III 11.1e.
   - **Physical layer:** The physical interface is defined in terms of the electrical characteristics, and individual signals passing over the interface. It deals with hardware-dependent specifications for IEC 60870-5-101/IEC 60870-5-104 communication interfaces and configuration types.
     - *Communication interfaces:* It includes communication interfaces such as V.24/V.28 FSK with maximum speed of 1200 bps, V.24/V.28 modem (9.6 Kbps), X.24/X.27synchronous (64 Kbps), etc. A typical wireless communication interface utilizing DTE and DCE has been illustrated in Fig III 11.1c3. These provide both unbalanced and balanced full-duplex serial data transmission between master and RTUs.
     - *Configurations:* There can be different configurations as per IEC 60870-1-1, viz., point-to-point, point-to-point (many/multiple), multipoint star, multipoint-party line, multipoint-ring as shown in Fig III 11.1c4. Here there are two types of devices such as controlling device (marked as master) and controlled device (outstation device). Here it is worth noting that in point-to-point, one master is connected to one outstation device (even in multipoint to point). In multipoint, a master station is connected to a number of outstations. When a full-duplex channel is used, in the point-to-point configurations either the master or the outstation(s) can transmit messages and there is no question of collision. In the multipoint configurations, the master communicates in parallel to all connected outstations. The outstations share a return communications channel; one may transmit at a time. Ring arrangement, master can offer redundancy, should the first port fail it can use second port.
   - **Data link layer**: By this time, we are aware that data link layer specifies data frame formats, bit order of information from least significant bit (LSB) to most significant bit (MSB). It also specifies the transmission procedures (e.g., primary/secondary, balanced/unbalanced, etc.) to govern its transmission and reception.
   - *Primary and secondary*: Based on ability to initiate communications in a communication, devices/stations can be classified as primary and secondary. Primary stations can initiate communications, whereas secondary stations have to wait till they are polled by the primary station, so that they can transmit data. Based on hierarchical structure of the system, an intermediate station can be both a controlled and a controlling station, i.e., primary as well as secondary as shown in Fig III 11.1c5. In the figure, the hierarchical network is shown, with primary and secondary ports marked.

**TABLE III 11.1E Enhanced performance architecture (EPA) Details.**

| Stack details | Description |
|---|---|
| Physical layer | Selected ITU-T recommendations |
| Data link layer | Link transmission procedures of IEC 60870-5-2[a] |
| | Transmission frame formats of IEC 60870-5-1[a] |
| Application layer | Application information elements of IEC 60870-5-4[a] |
| | Application service data units of IEC 60870-5-3[a] |
| User process | Application function of IEC 60870–5[a] |

[a]*Selected part of the standard used, as applicable.*

- *Unbalanced and balanced transmission*: "Balanced transmission" and "unbalanced transmission" are related to "primary" and "secondary." In unbalanced transmission controlling station (master) acts as a primary on the link, and there will be a number of controlled stations acting as secondary stations without being peer-to-peer at the link level. So, communication shown in Fig III 11.1c5 is an example of unbalanced communication. Here there can be two sets of unbalanced communication one between P/S station to S RTUs and other from master to any of RTUs (including P/S). So, here, controlling station acquires data by polling station in sequence, leaving no chance for data collision. Balanced transmission refers to the configuration where any station on a link may act as a primary, to initiate communications. Since stations are in peer-to-peer level, it is also called peer-to-peer communications. Fig III 11.1c6 shows a balanced communication. It is worth noting that each station can act as a primary and a secondary at the ports for this link. These may in fact be thought of as two separate processes within each station, which in fact is how they are logically within the stations. Station A has a primary process and a secondary process operating simultaneously for that link, and station B has the same. As per IEC 60870-5-101, only point-to-point links can be balanced and multipoint links are unbalanced. This is in contrast to DNP3 (next section), which supports balanced transmission only.
  - *Control field*: The control field of the data frame is central to the operation of the transmission procedures. This field is almost identical to that used by the DNP3 protocol as it was derived from the same source document, IEC 60870-5-2 1992.
  - *Service procedures:* There are three main types of services in the data link layer. These are detailed out in Table III 11.1f for unbalanced transmission. For balanced transmission, only "SEND/CONFIRM" and SEND/NO REPLY" (initiated by controlling station only) are supported.
- *Link initialization:* This data link service is used after a station has been off-line and first becomes available again. Station/link initialization in *unbalanced* mode are as follows: *Master sends link status request until status of link received*, *Master sends link reset*, *Link is active on receipt of ACK*, and *Slave generates station initialization complete event.* Station/link initialization in *balanced* mode are as follows: *Each station sends link status request until status of link received*, *Each station sends link reset*, *Link is active on receipt of ACK at each station*, and *One or both may generate station initialization complete event.*
- **Application layer**: Application layer defines the information elements for structuring application data and the communication service functions. It has different parts:
  - *Overall message structure.* It is meant for application service data unit (ASDU) to show how this is carried as link user data by the data link layer for T101. For T104, the ASDU is carried by the TCP/IP protocols.
  - *ASDU structure*: The structure of the ASDU is in two main sections, viz., data unit identifier, and data itself, made up of one or more information objects. The data unit identifier specifies the data type, addressing to identify the data type and additional information on transmission field.
  - *Common address*: The common address of the ASDU is either one or two octets in length, fixed on a per system basis. The address is called a common address because it is common to all of the data contained within the ASDU. Common types are as follows: *Interrogation command*, *Reset process command*, or *Clock synchronization command.*
  - *Message addressing and routing*: This deals with important concept in understanding addressing under IEC 60870-5 is the difference between control and monitor directions. With the assumption of hierarchical structure, every

**TABLE III 11.1F** EPA Data link support services for transmission.

| Service type | Purpose |
|---|---|
| SEND/NO REPLY | Global messages and for cyclic set-point commands from the controlling station. Used for broadcasting and where response is unimportant |
| SEND/CONFIRM | Control commands and set-point commands from the controlling station |
| REQUEST/RESPONSE | Data polling from the controlled outstations. Response is with data not confirmation |

station is either a controlling station or a controlled station. Naturally there may be multiple controlled stations, with one or more controlling stations. In such a system, control messages such as commands or interrogations are transmitted by the controlling station, and these result in actions and return information transmitted by the controlled station.

- *Information elements*: Depending on the variable structure flag, there may be multiple information objects within the ASDU each containing a defined set of one or more information elements, or there may be just one information object containing a number of identical information elements. The information element is the fundamental component used to convey information under the protocol. The information elements are used as building blocks in the definition of the set of ASDUs under the protocol.
- *Set of ASDUs*: ASDUs are the application level units of data used to exchange SCADA information between controlling stations and controlled stations. There are ASDUs for each type of information carrying, e.g., control parameters, setpoints, commands to set or clear particular status points, and for measured analog values or countervalues.

4. **Companion protocol T101 and T 104:** Let the discussions start with Fig III 11.1c2, where both T101 and T104 models have been shown. T1O1 profile is a four-layer model as defined for EPA model in Fig III 11.1c1. On the other hand, in T104 profile, there are additional layers such as network and transport layers. These are added on account of TCP/IP layers; naturally, nonnetworked version T101 is different from networked version T104. T101 deals with communication for basic telecontrol messages between central station and outstations. The IEC 104 specification combines the application layer of IEC 60870-5-101 and the transport functions provided by a TCP/IP in networked communication. An important concept in understanding addressing under IEC 60870-5 is the difference between control and monitor directions [9] applicable for both T101 and T104. Communication exchange between controlling station and control station has been already discussed. *Monitor Direction* is a direction of transmission from RTU to controlling station (PC), Whereas control direction is a direction of transmission from controlling station to outstation RTU. *Reversed Direction* is a direction when monitored station is sending commands and controlling station is sending data in monitor direction [9].
   - **T101:** T101 has a number of subsections, and these have been discussed in the following:
     - *General rules*: As general introduction, it provides the selections for physical layer, link layer, application layer, and user process.
     - *Physical layer*: In this subsection, the requirements for the interface with external data communication have been described to select physical layers from standard ITU-T V.24A/V.28 (RS-232)/balanced X.24/X.27 (RS-485) signals. It also encompasses network configurations, e.g., point-to-point, multipoint, etc.
     - *Link layer*: Under this, selections of the frame format and frame lengths (fixed/variable) are covered. Also, detailed operation of the link layer is specified with the choices for unbalanced and balanced operation. Amendment 2 contains significant additional information clarifying the procedures with the use of state transition diagrams.
     - *Application layer and user process*: Here overall structure of the application-level data, ASDU, is defined to select from available ADSU sets. From the available information elements, the full set of ASDUs are constructed. It also describes application functions of IEC 60870-5-5. The user process subsection makes selections from IEC 60870-5-5 and includes additional detail for some of the functions.
   - **T104:** T104 is a standard for telecontrol equipment and systems with coded bit serial data transmission in TCP/IP-based networks for monitoring and controlling geographically widespread processes. Protocol standard defines the transferred data entities in the station object as equal to T101. Also, the implementation of the T104 protocol uses the same as station objects (STA) as T101 implementation. IEC 104 is designated according to a selection of transport functions given in the TCP/IP Protocol Suite (RFC 2000). Within TCP/IP, various network types can be utilized including X.25, Frame Relay, ATM, ISDN, Ethernet, and serial point-to-point (X.21). Fig III 11.1c7 shows protocols used in different layers [40]. IEC 60870-5 has information on a set of information objects. Only one type of data is included in any one ASDU. The information object types are grouped by direction, as discussed before, by the type of information (process information, e.g., measured value or a control command, or system information, e.g., interrogation command, parameter, file transfer). Thus, application data is carried within the ASDU with one or more information objects. The information element is the fundamental component used to convey information under the protocol [9]. Each ADSU consists of several information object containing data. Each APCI, shown in Fig III 11.1c7, starts with a start byte followed by the 8-bit length of APDU and four 8-bit

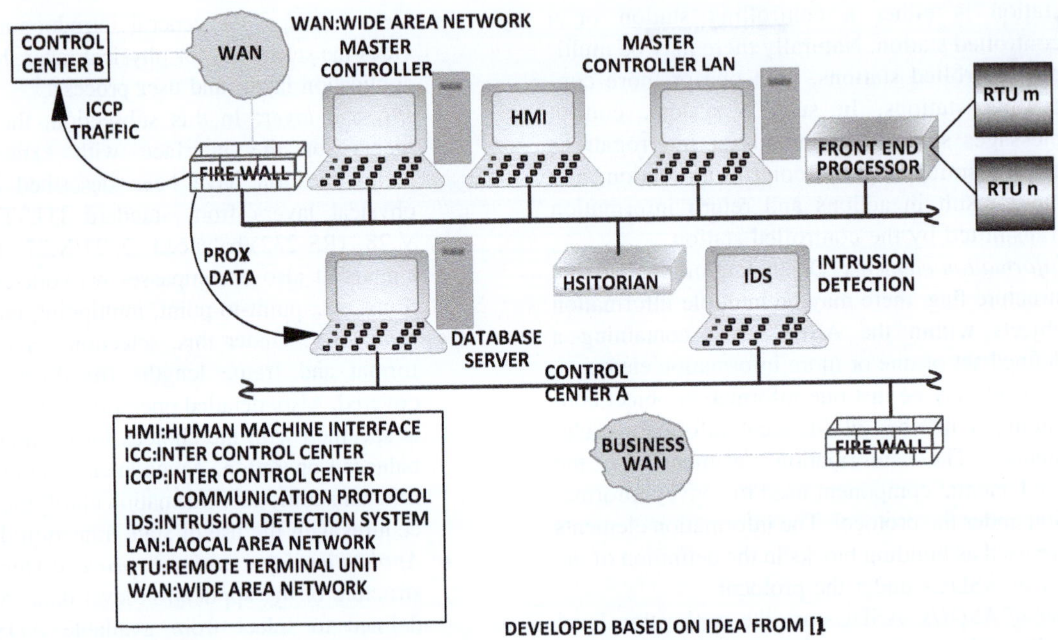


**FIGURE III 11.1D** ICC protocol (IEC 60870-6).

control fields (CF). APDU contains an APCI or an APCI with ASDU.

5. **Inter-control Center Protocol (ICCP—IEC 60870-6):** IEC Technical Committee 57 (Work Group 03) has developed IEC 60870 part 6 (compatible with ISO standards and ITU-T recommendations). The ICCP is also known as the Telecontrol Application Service Element (TASE.2) protocol. This is meant for communications over WAN between control centers in the electrical power sector/independent system operators (ISOs), RTOs, and other generating stations. Based on a request from a control center (client) to another control center (server), data transfers result in client/server principles. ICCP provides access control via bilateral tables maintained on an ICCP server. A bilateral agreement is between two control centers for data access. A bilateral table is a digital representation of the *Agreement.* The ICCP protocol designers assumed that additional data security measures would be implemented in various layers of the open systems interconnection (OSI) 7-layer stack. Secure ICCP is an extension of the existing standard ICCP. A typical secured ICCP network has been depicted in Fig III 11.1d developed based on idea from Ref. [10]. Transport layer security (TLS) is inserted into the appropriate layer of the standard communications profile. Here the network architecture that is to support the ICCP applications basically comprises an Ethernet LAN across a WAN. Because ICCP allows the sharing of control and status information between control centers, the actual database that will be used to exchange near real-time information should be in the form of a proxy to prevent direct access to the Master Controller LAN from outside users.

So, in the figure, proxy server has been used. The purpose of a proxy server is to prevent direct access to information hosted on a critical asset within a control systems operations network. This standard approach to security has been implemented on information technology (IT) network systems also [10]. For this reason, suitable firewalls have been implemented at main network to prevent direct access as well as in the proxy network to prevent from business LAN, which is integrated in the network through WAN.

Let us now explore new subject, i.e., smart grid and associated standard.

## 11.2 Fundamentals on smart grid and associated IEC standards

In electric energy transmission and distribution world now, “smart grid” is a buzz word. So, it is better to first define *smart grid*. There are many definitions; some are as follows:

- **IEEE Definition:** According to IEEE (https://ieeexplore.ieee.org/document/5768100), “The Smart Grid can be defined as an electric system that uses information, two-way, cyber-secure communication technologies, and computational intelligence in an integrated fashion across the entire spectrum of the energy system from the generation to the end points of consumption of the electricity."

- **Technopedia definition:** A smart grid is an electricity network based on digital technology that is used to supply electricity to consumers via two-way digital communication. This system allows for monitoring, analysis, control, and communication within the supply chain to help improve efficiency, reduce energy consumption and cost, and maximize the transparency and reliability of the energy supply chain. The smart grid was introduced with the aim of overcoming the weaknesses of conventional electrical grids by using smart net meters.
- **NIST definition:** Comprehensive definition used at NIST is "a modernized grid that enables bidirectional flows of energy and uses two-way communication and control capabilities that will lead to an array of new functionalities and applications."

These are definitions but it does not tell why there is need for smart grid. Energy needs are continually expanding, e.g., extensive use of electric cars, buses, and trucks; remote control of in-home appliances to promote energy conservation. There are four dominant factors, which are going to impact in electrical systems of the future. These are *government energy policy*, *environmental concern* (demanding for conservation, high system efficiency, and use of renewable energy), *customer efficiency needs*, and new developments in *intelligent control systems.* As such, bulk powers generated through nuclear, thermal, hydel, etc. on an average can offer overall conversion efficiency less than 40% (though in ultra super-critical power plants can offer better but still average conversion efficiency is less than 40%). Again, for a good electrical network, transmission and distribution (TD) loss will be around 7%–10%. Where these losses go? They impact environment. It is therefore clear that there are two driving issues here: one is environment concern and the other is high-quality reliable power availability with improved efficiency. Smart grid development tends to be driven by these two factors. While in Europe smart grid is envisioned primarily by environmental concern, in the United States smart grid has been motivated primarily by a desire for reliability improvements. In the United States, desirable characteristics of the smart grid include self-healing transmission and distribution power architectures that will be resistant to intentional attacks and natural disasters and very-high-quality power along a broad range of metrics that go well beyond outage statistics [11]. By this time, we got some idea about advancement in digital networking and communication. So, the availability of new technologies, e.g., distributed intelligent sensors, two-way secure communication, and advanced software support for control and data management, has opened up new avenue for the energy management system. So, smart grid is a reality. Major challenges for smart grid are in terms of reliability and security for both wired and wireless communication. Here reliability is defined in terms of the quality of accessible power and information. Loads may be viewed as an active participant in the network. Therefore, a system controller with bidirectional interaction (cyber and physical) between the utility-side network and the customer-side is necessary for development of smart grid system.

## 11.2.0 Conventional grid system and smart grid systems

The conventional grid, the electrical grid, is a *unidirectional* network, i.e., electricity flows unidirectionally from generators to substations, over transmissions lines, and eventually to consumers' outlets. Also, most of the equipment and lines of the conventional power grid are old; hence, grid elements are outdated and need to be maintained and supervised frequently.

With introduction of nonconventional energy sources such as solar panels, wind turbines, battery storage systems discussed in I 18.0, now distributed energy resources (DERs) are part of electrical grid. DER refers to a network of small-scale, modular, decentralized technologies such as solar panels, wind turbines, electrical vehicle (EV) battery storage systems, etc. located at or near the point of electricity consumption. Therefore, because of integration of DER, in smart grid, energy flow is bidirectional, i.e., DERs, such as solar panels, wind turbine, etc. are both consumer and producer of electricity, causing bidirectional flow of electricity. In contrast to traditional grid, smart grid enables bidirectional *flows of energy*, with *two-way communication and control* capabilities leading to an array of new functionalities and applications. Smart grids are also referred to as "modernized grid," which conveys the idea that smart grid is an enhancement to the conventional grid; A smarter grid impacts multiple aspects of energy production and usage: *the power production, the grid infrastructure*, demand response, *bidirectional flows of electricity, and information* for the *smart power management.* More integrated power products, DERs, substations, and lines are built and integrated. These are equipped with large numbers of sensors and actuators to enable the localization of failure, the rerouting of electricity, and to avoid outages. On account of the new smart grid infrastructure with a two-way transfer of both electricity and information in a decentralized architecture, the consumers also get the opportunity to participate in the processes and get many options to choose from. There may be a need for new businesses responsible for energy storage and supply as a backup for existing grid capabilities. These are more so with various data centers for computational clouds. Advanced information metering, monitoring, and management equipment can be embedded in the infrastructure. Customers may appreciate having the option to track their energy consumption and know when less costly energy is available so customers can manage their own energy usage. The smart grid is capable to monitor activities of the grid-connected system and consumer preferences of using electricity and provides real-time information of all the events. The key components of smart grid include smart appliances, smart substations, smart meters, and advanced synchro-phasor technologies. In this subsection, some details on smart grid (SG) have been discussed, starting with traditional grid systems along with differences between the two. Let the starting discussions throw some light on old conventional grid system.

1. **Conventional grid management system:** Although there exists regional, state load dispatch centers, for monitoring the grid system, yet conventional systems are not thoroughly integrated system and instead can be conceived as mostly isolated components of generation, transmission, substation, distribution, and the customer. Major bottlenecks associated with conventional systems can be summarized as follows:
   - **Generation:** Centralized sources of power generation. There is hardly any chance to augment power generation at distribution level, i.e., distribution emergency generation.
   - **Energy and information flow:** Unidirectional flow of energy from the generation to transmission, then to distribution and customer without interaction at downstream levels.
   - **Participation:** Passive participation by the customers; customer knowledge of electrical energy usage is limited to a monthly electricity bill at the end of the month—no smart metering systems.
   - **Real-time control coverage:** Real-time monitoring and control is mainly limited to generation and transmission, and only at some utilities, does it extend to the distribution system.
   - **Flexibility and renewable source of energy:** Conventional system is hardly flexible, to add on at any point along the grid, electricity from renewable/alternative sources, which can have large seasonal and time of the day variations
   - **Load management:** It is very difficult to efficiently manage new services desired by the users of electricity.
   - **Scheduling of consumer demand:** Depending on peak period, it is difficult for customer to schedule own demand.
   - **Alternative pathway:** In case of any weather or any other disasters, if there is breakage in network, energy management suffers till it is restored due to lack of many alternative paths.
2. **Smart grid systems with difference:** Smart grid systems have been conceived not only to remove major drawbacks of the conventional system but also to add facilities such as connecting to IoT/IIoT—demand of the modern time. One needs to bear in mind that the smart grid actually does not mean to replace the existing electric system but augments on the available infrastructure for much better utilization of existing assets and to empower the implementation of the new functionality. Basic concepts of smart grid system have been elaborated in Fig III 11.2a.

   Modern power systems for smart grid should possess the following characteristics/requirements. Difference between conventional and smart grids has been presented in Table III 11.2a to conclude the discussions on *smart grid*.
   - **Networking and central control system:** Suitable networking must be built to include all stakeholders such as bulk power generation, transmission, distribution, consumers, renewable energy sources, electric vehicle, distributed generation, and storage units under central control system forming a real-time wide area network with a large number of interconnections, for energy and information flow. On account of such large networking, it could be possible to restore energy flow network disruption due to any sort of disaster.
   - **Integration of control and communication:** To promote interoperability and open systems and to increase safety and operational flexibility, smart grid should allow for the integration of communication and control across the energy system to promote in line with requirements specified in international standards, e.g., IEC 61850, IEC 60870.
   - **Integration of renewal energy source:** To address environmental concern, people now go for renewable energy systems. In most of the cases, these are not suitable for direct grid connection, e.g., photovoltaic cell output is in small VDC. So, in such cases, suitable power electronics are used so that they can be connected to grid. Smart grid must allow for the integration of renewable energy resources, which also have temporal and seasonal variations.
   - **Electric vehicle integration:** To reduce $CO_2$ emission, electric vehicles are encouraged, so there will be more charging stations. Smart grid should allow for the integration of these charging stations, i.e., integration of electric vehicles in smart grid.
   - **Customer participation:** As stated earlier that loads are active participants, smart grid should allow for active customer participation through the use of

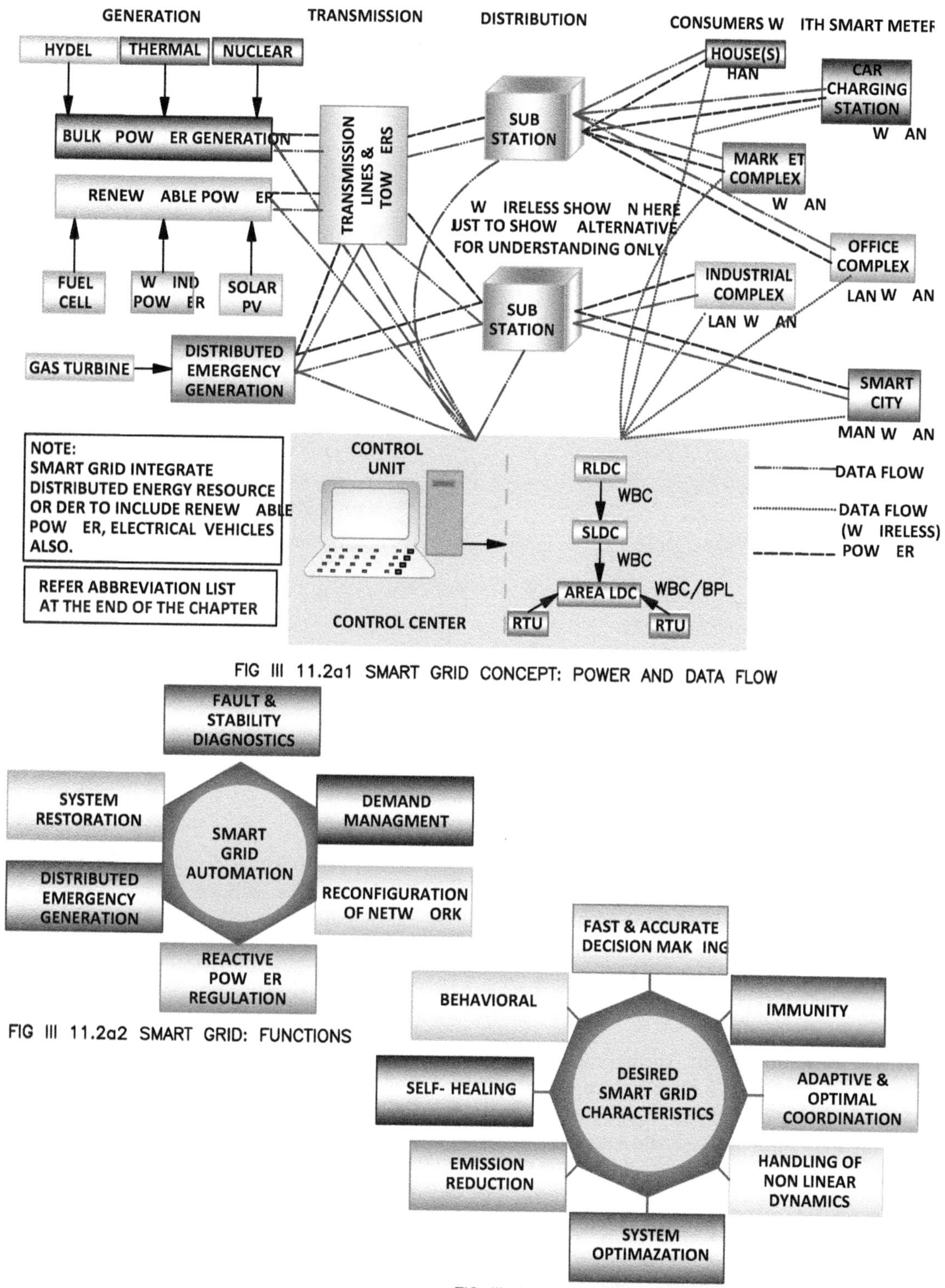


FIG III 11.2a1 SMART GRID CONCEPT: POWER AND DATA FLOW

FIG III 11.2a2 SMART GRID: FUNCTIONS

FIG III 11.2a3 SMART GRID: CHARACTERISTICS

**FIGURE III 11.2A** Smart grid control basics.

**TABLE III 11.2A Difference between conventional grid and smart grid (SG).**

| Issue | Conventional grid | Smart grid |
|---|---|---|
| Setup type | Electromechanical relay, meter, and switches | Smart digital electronic devices |
| Powe source | Centralized power generation; no alternative power generating sources | Both centralized and DER allowing alternative power generating sources |
| Energy flow | Unidirectional | Bidirectional |
| Information | Unidirectional information flow | Bidirectional information flow |
| Communication | Dumb communication, i.e., no data communication between various devices of the system | Smart communication, i.e., data communication between various devices of the system |
| Monitoring | Infrastructure and conventional equipment allow manual monitoring mainly | Infrastructure and equipment allow smart digital automatic monitoring |
| Protection | Semiautomatic | Fully automatic (automation) |
| Control impact | Limited control | Pervasive control |
| Sensor quantity | Infrastructure allows a few only | Completely sensor based |
| Customer participation | No customer participation | There is active participation of consumers in case of smart grids (can track) |
| Customer choice | Limited customer choices | Many customer choices |
| Sudden failure | Sudden failure may call for complete failure and customer may not het power | In sudden failure, power rerouting is possible so that customer gets power |
| Restoration | Manual restoration; upon failure | Smart grid has self-healing property with huge sensor actuators without intervention also |
| Environment | Since it does not support alternative power sources, it cannot reduce pollution | It can support pollution reduction by supporting DER |

smart meters so that there will be far better energy conservation and energy scheduling at consumer level. This could also help to balance peak demands.

- **Energy flow optimization:** One of the major goals of smart grid is to minimize TD losses so that there will be optimized energy flow to reduce lower the cost of energy;
- **Both-way communication:** There shall be both-way communication, and it must be secured through the use of suitable firewalls at required levels.
- **Sustainability:** Smart grid should allow for better utilization of existing assets to address long-term sustainability;
- **Distributed generation and management:** Smart grid should allow for the management of distributed generation and energy storage—DER (could be at local level also).

3. **Description: Smart grid:** With reference to Fig III 11.2a1, we find that there will be four major sections in smart grid like conventional grid system also. Only difference is that here all sections are not only interconnected to each other but also to control center (*central control center* better to call it) for information flow. Similarly, there will be interconnections among the transmission lines and substations (*not explicitly shown here as our main goal is to focus on control networking and not power flow*). In Fig III 11.2a1, power flow and information flow has been distinguished in color and byline type. It is clear that all data flow lines are connected to control center. Also, through various substations, generating stations can also exchange information. Depending on the type of network configuration deployed, there can be interconnections among them also to exchange information under supervision from *central control center* through open bus system also. In this diagram for simplicity, voltage/reactive power regulation and service providers are not shown. Various bulk power stations, which could be *nuclear*, thermal (fossil fuel), or hydel, have been clustered under bulk power generation. Similarly, renewable energy sources along with the associated power electronics have been clustered under renewable energy sources including storage units, to show its participation in smart grid. In case of high demand and/or failure of

bulk power generation, distributed emergency generating stations (say gas turbine—lower capital cost but high operating cost with lower startup time) have been shown. In smart grid, these could be added to substations also. It is worth noting that here all consumers with smart meters, including car charging stations, are participating in the network as discussed. Control centers may be in numbers but in hierarchy as shown in Fig III 11.2a1 separately, these may be regional load dispatch center (RLDC), state load dispatch center (SLDC), and/or area load dispatch center. They in turn connected via wide band communication (say Internet TCP/IP).

4. **Smart grid: Features:** Various features of SG are as follows and shown in Fig III 11.2a2 and Fig III 11.2a3.
   - **Secured communication:** Secured open network communication is the backbone of smart grid.
   - **Protective management:** Protective management of electrical network during emergency.
   - **Power restoration:** In smart grid, power is out and usually restores it automatically.
   - **Peak demand handling:** Utility suppresses demand at peak with lower cost.
   - **Distributed emergency generation:** Well-coordinated and safe management of distributed emergency generation.
   - **Loss reduction:** TD loss is reduced by $> 2\%$ and hence lowers emissions and customer bills.
   - **Optimization:** Optimization of asset, demand, transmission, distribution, and engineering.
   - **Customer participation:** Usage of smart meters for consumer participation.
   - **Home area network (HAN):** In individual house develops HAN so that consumer can schedule his consumption and avoid peak hours to get energy at lower cost. It also addresses conservation and storage issues.
   - **Reduced emission:** It supports electric vehicle and reduces losses to cause less emission.
   - **Renewable energy integration:** It helps to integrate renewable energy sources.
5. **Open communication infrastructure:** Open secured communication infrastructure is essential for reliable, highly secure, and manageable smart grid design, operation, and facing various eventualities during operation. This open network must be highly secured, self-corrective, capable of reconfiguring and restoration to handle load variations, variable participant types in real time. Thus, intelligent devices with smart controlling, routing, and communication algorithms developed as international standards are essential for dealing with very complex demands of power systems.

### 11.2.1 Discussions on smart grid

The smart grid represents an unprecedented opportunity to move the energy industry into a new era of reliability, availability, and efficiency that will contribute to our economic and environmental health. Primarily, SG is meant to deliver the optimal amount of information and load control for customers, distributors, and grid operations so that there will be high energy efficiency coupled with reduced system demands and costs. The SG is therefore associated with the production of energy through renewable energy sources, i.e., integration of DER, promoting reduced emissions, lower energy costs, and greater flexibility to accommodate new renewable distributed energy sources. Major contextual issues shall include but not limited to the following: *active network management (ANM) and SG, design and planning of SG, application of power electronics in SG, control and automation in SG, application of computational intelligence and data analysis, DER and SG, smart metering with demand management, economic aspects and liberalized market related to SG*, etc. *Efficient transmission of electricity, quick restoration of electricity after disturbances, managed reduced peak demand, reduced operations and management costs for utilities* to customer, *total integration of large-scale renewable energy systems and DER, better integration of customer–owner power generation systems, improved security, intelligent computations, improved reliability with higher asset utilization, better customer choices*, etc. are some of the major benefits of SG. Apart from coexistence of various types of renewable and nonrenewable energy sources, in SG management, it is necessary for the operator to coordinate the operation of the generation plants, and ensure the stable and secure operation of the system. Wide-area measurement system enabled by communication technologies and advanced communication infrastructure are part and parcel of smart grid operations. Energy-efficient transmission network carries the power from the bulk generation facilities to the power distribution systems as well as power exchange with DER. There exists communication interface among the transmission network, the various power sources, system operator, power market, and the distribution system. An important task of the system operator is to ensure optimal utilization of the transmission network, by minimizing the losses. Also, an important issue for the operator is to control the DERs in a coordinated way to ensure stability and power quality of the distribution system. In this subsection, brief discussions on various aspects of smart grid with special emphasis on control and automation issues related to SG shall be covered.

1. **Control and automation of SG:** For the proper functioning of control unit—the brain of SG systems, advanced integration of secured communication,

sensing, and control with optimization functions is an absolute necessity. Also, the integrated control system should be interactive and adaptive for making the system self-healing, efficient, and reliable for SG operation. System needs to have built-in intelligence and modeling capability so as to provide real-time forecasts for system operation. Typical such control and automation functional model has been depicted in Fig III 11.2a2 with its characteristic features in Fig III 11.2a3. Intelligent field sensors would communicate with main control system in a wide area monitoring and control system. The communication should not only be secured and accurate that it has to be quite fast. Fast Internet could be used but with necessary firewalls at every required stages. Major characteristics of communication shall include but not limited to *easy deployment, high data carrying capacity, high QoS with low latency* and *standard communication means, secured, wide network coverage.* The increasing complexity and nonlinear nature of future SG will require wide area monitor (WAM) and effective control systems such as adaptive wide area controller (WAC) [12], for SG to become self-healing and fault tolerant. Control system must also provide for dynamic optimization facility discussed earlier. It is possible to deploy artificial intelligence and/or fuzzy logic for dynamic optimization and overcome nonlinear nature of power systems. Control unit must coordinate with various controllers to be deployed at local levels to get the best results for SG operation under various situations. Two-way communication and automated energy management, benefiting both providers and users, demand open standards such as IEC 61850 (discussed later) to ensure this interoperability of all SG systems. Major issues under consideration include but not limited to the following:

- Internet-based open standards allow integrated communications.
- Accurate smart sensors/measurement devices and actuators are available.
- Control and monitoring methods are familiar.
- Today's operator interfaces allow detailed and secure interaction with the devices and systems that need managing.
- Development of WAN comprising WAM and WAC (explained in the following).

2. **SG: WAM, WAC, and power electronics:** These three different topics have been explained separately in the following:
   - **WAM:** It is called WAM as it functions over a wide geographical area. Technologically advanced WAM system upgrades the traditional grid to the smart grid by modernizing the delivery system of electricity and avoids severe outages/blackouts. Based on synchrophasor technology, WAM employs synchronous measurement/sensing technology, phasor measurement unit (PMU) to monitor and evaluate the state of the electric grid. Major reasons of deployment of WAM are as follows:
     - It is *fast and efficient* in detecting and responding to faults and disturbances. Thus, it results in improved reliability.
     - It provides *RT data about grid operations* and helps in better management of the grid.
     - It *reduces the impact of faults and disturbances* on the system by rapidly detecting and isolating them.
     - It *improves the sustainability* of the grid by integrating renewable energy and DERs.
   - WAC: Critical component of WAC in SG enhances *grid stability, reliability*, and *efficiency* by providing RT high-resolution data and automated control actions across a large geographical area. WAC systems also respond to the RT control requirements of an SG. The control messages in smart grid can be of different levels of criticality. Therefore, priority-based routing, depending on criticality of messages, is very essential in control message communication.
   - **Power electronics:** As stated earlier, output produced by renewable power sources and distributed energy sources are unsuitable for direct connection to grid. Thus, energy conversion systems, based on power electronic devices, are required. These inverters have to be flexible and intelligent to support any local conditions. Power electronics also helpful in smart grid in supporting flexible AC transmission system (FACTS) and high-voltage direct current (HVDC) are beyond the scope of this book, and for details on "FACT" and "HVDC," any standard book on power systems may be referenced.
3. **Control unit (center):** It is the brain of smart grid. Control unit comprising various other control units in hierarchical structure, e.g., RLDC/SLDC, etc. as shown in Fig III 11.2a1 together forms control unit establishing communication interface with all stakeholders such as bulk-generating facilities, renewable energy source, DER, transmission system, substation/distribution automation, demand management system (DMS), car charging units, consumers with smart meters and HAN, service providers (business houses to be engaged in building HAN, renewal energy prediction, etc.), and the market (expected to grow with introduction of smart grid with regulatory policy pricing, etc.). It will be responsibility of this control center with its infrastructure to meter, record, and control smart grid operation with secured communication for reliable and quality power delivery. Also, in emergencies as well as in disaster

situation (environmental or any other), it will be responsibility of control unit to find alternative path for power flow (as well as for information exchange) to restore the situation as quick as possible. In all these, role of international standards cannot be overestimated. To develop new power market, it is essential to have real-time information exchange in secured manner and interact with various service providers in this business model.

4. **Other important smart grid components:** A few major components involving various technologies such as *advanced metering infrastructure* (AMI), *demand response* (DR), *distribution automation* (DA), *self-healing*, *cyber security*, *system efficiency improvement*, etc. listed in the following are necessary to make smart grid a success.
   - **Intelligent appliances:** Intelligent appliances are important both as consumer and as sensing element. As sensing element, it has to be smart/intelligent so that it can be integrated with the network easily for data exchange with minimum wiring. On the other hand, intelligent appliances can decide when to consume energy based on customer preset preferences to avoid peak time power consumption. The smart appliance typically reports energy consumption to the home energy controller (HEC) and could also be controlled by the HEC according to the DR signals
   - **Smart meters:** The smart meter is an advanced energy meter to obtain information from the end users' load devices and measures the energy consumption of the consumer. Smart meters provide additional information to the utility company and/or system operator through several sensors, control devices, and dedicated communication infrastructure. The smart meters are enabled for automatic bill generation and collection, indication of consumption pattern to consumer, detect failures so that repair crews can reach exact location very quickly with no or minimum waste of time. Guiding standard is IEC 62056 with two versions, viz., device language message specification (DLMS) and companion specification for energy metering (COSEM).
   - **Smart substations:** In addition to usual functions, with smart substations, it is possible to carry out necessary splitting the path of electricity flow into many directions and find alternative routes during restoration process. Usually all protection and other relay types are intelligent and addressable type, so that health monitoring is also possible.
   - **Superconducting cables:** Superconducting cables are special cables usually consisting of the superconducting compound (in the form of coating) and conduction stabilizer. High-temperature superconductor (HTS) can carry three to five times the capacity of conventional AC cable and nearly 10 times capacity of DC cables. It has fault management capability. These are special cable used for long-distance power transmission and automated monitoring. It can predict fault also.
   - **Phasor measurement units** (PMUs) are high-speed sensors to estimate/measure the magnitude and phase angle of an electrical phasor quantity, e.g., voltage and current with a common time source for synchronization. This synchronization is provided by IEEE 1588 standard protocol. It offers accuracy in the order of 1 microsecond, much faster than the speed of existing SCADA technologies.

5. **Role of smart grid in different areas:** Smart grid plays a role in almost all sections of power system management: Major ones are listed in the following:
   - **Power generation:** Major roles of smart grid in generation are as follows:
     - Handling of coexistence of various types of power sources, viz., bulk power and renewable power (Fig III 11.2a1).
     - *Coordination:* Well-coordinated effort by system operator with various generating stations can only bring about good success and secure operation of SG.
     - *WAMS and WAC:* WAM/WAC enabled by communication technologies is to be used to control the operation of the generating stations.
     - *Communication infrastructure:* Communication infrastructure must be secured for communication among generating facilities, control center, transmission and distribution system electricity market, etc.
   - **Transmissions:** SG plays a role in transmission as follows:
     - *Communication interface:* As shown in Fig III 11.2a1, there exists communication interface between the transmission network and the bulk-generating stations, control center for SG power market, and the distribution system.
     - *Control and monitoring:* For secured and stability operation of SG, power flow and voltage on transmission lines must be controlled, and transmission network is monitored in real time to protect against any disturbance.
     - *Optimal utilization:* An important task of control center is to assure optimization of the transmission network, with the aim to minimize the losses/voltage deviations and maximize the reliability of the supply.

- **Distribution:** SG plays a role in distribution as follows:
  - *Substation automation:* Automation of substation (s) and distribution automation is a key issue for SG. So, it should be part of control network.
  - *DER:* Increased use of distributed energy resources (DERs) is important ref Fig III 11.2a1. So, well-coordinated control of DERs with central control unit is very important to ensure power system stability and quality of distribution network.
  - *Substation information exchange:* For SG, communication infrastructure enables information exchange the substations and a central control unit as shown.
  - *Customer participation:* Information exchange between distribution system operator and the customers for better operation of the distribution system is a new feature of the smart distribution systems [13].
- **Customer participation:** There are three main classes of customers, viz., house, commercial (including office), and industrial (including all plant types). SG plays a role in customer participations as follows:
  - *Customer role:* By peak-load shaving, valley-filling, and emergency response, customers are going to play an active role in better operation of the distribution system [32].
  - *Smart meter and HAN:* HAN will monitor and control the power consumption at the consumer premises through smart meters and would help customer to schedule power consumption pattern in a suitable way to avoid peak load, e.g., may schedule to run washing machine and dish washer at very late night or very early morning.
  - *Communication inclusion:* Necessary secured communication medium is necessary for customer to participate and interact with control systems as well as with market. will be required for the consumers to interact with the operators, distribution systems, and the market.

Communication protocols used in power systems normally follow a series of international standards. Important ones have been enlisted in Table III 11.2b.

With this, let us conclude discussions on smart grid and move on to discuss IEC 61850.

### 11.2.2 Brief discussions on overview of IEC 61850

IEC 61850 is an Ethernet-based international standard for communications and system architecture in substation automation and power distribution systems. It is a standard defined by the common work of both ANSI (North America) and IEC (Europe) together with the main vendors. It defines communication protocols for intelligent electronic devices at electrical substations. These protocols can run over TCP/IP networks or substation LANs using high-speed switched Ethernet to obtain the necessary response times <4 mS for protective relaying. IEC 61850 has a standard data modeling and naming convention of the IEDs, and a common language to configure the devices, providing interoperability between devices as well as engineering tools. The standard has a flexible and open architecture, enabling freedom in configuration of the IEDs according to the application needs. With the present form of SCADA inside the substation, it would not be possible to support next-generation intelligence and the demands for interconnecting many systems. IEC 61850 standard approach would help overcome the current limitations.

*Electronic device (IED)* with huge data points in a single device play a great role in substation and power system automation. Another important issue is communication system [14]. The following are a few points that form the basis for power system communication systems.

- Well-defined and describable in terms of data and services

**TABLE III 11.2B** Power systems communication protocols.

| Standard | Function—purpose | Remarks |
|---|---|---|
| IEC 60870–6 | Inter-control Center Communications Protocol (ICCP) | Control center to center |
| IEEE 1815 (DNP3*) | Control center to field devices—NorthAmerica | *Covered in next section |
| IEC 60870–5<br>–101<br>–103<br>–104 | Control center to field devices—European<br>Serial communications<br>Protection devices<br>TCP/IP network communication | |
| IEC 61850 | Substation automation and protection | Field |
| IEEE 1815 | Substation and feeder device automation | DNP3* |

- High-speed IED to IED communication
- Standard-based with multivendor interoperability
- Highly secured and reliable with high availability
- Support open connect networking over enter utility
- Guaranteed delivery time
- Well configuration support and autoconfigurable
- File transfer support

The aforementioned issues are basis for development of IEC 61850 for field communication and substation automation. Important communication and naming issues have been depicted in Fig III 11.2b.

1. **Structure of IEC 61850:** Structure of IEC 61850 has been enlisted in Table III 11.2c.
   Although IEC 61850 was developed with main focus on "within" the substations automation, with advantage of a comprehensive object-oriented data model and the Ethernet technology, now its usages have been extended for wide area substation-to-substation communication. Part 1 to Part 3 deal mainly on general ideas about the standard. Part 4 focuses on the project and management issues when substation is IEC 61850 enabled. Part 5 deals with implementations. So, Parts 3–5 together take care of general and specific functional requirements for communications in a substation. Part 6 is related to XML-based language for IEC configuration [15]. Part 7 is elaborative, to take care about logical concept. The definition of the abstract services could be found in Part 7.2, whereas abstractions of data objects are defined in Part 7.4. Part 8 deals with mapping objects (internal) into presentation and Ethernet link layers [15]. Part 9 deals with mapping also but on sampled values (SVs) to point to point. Finally, Part 10 deals with the testing methodologies.
2. **Substation automation system architecture:** Fig III 11.2b1 shows the typical system architecture of substation automation. IEC 61850 implements merging units (Mus) to convert analog current transformer (CT)/ voltage transformer (VT) signals into IEC 61850 sampled values. Mus, which could be located at the field or control room, are connected to the "process" bus, to collect data from voltage and current sensors (optical/electronic) and digitize data to pass on to Ethernet bus (—process bus). The collection points will be redundant Ethernet switches with 1 GB internal data buses and 1 GB uplinks that support Ethernet priority and Ethernet virtual LAN (VLAN) [14]. For sampling synchronization, clock needs to be adapted. At the substation level, an Ethernet station bus (redundant), as shown, will provide primary communications between the various logical nodes, for protection, control, monitoring, and logging functions. Communications will be either connection-oriented basis or a connection-less basis IEC generic object-oriented substation event (GOOSE). This is a publisher subscriber network where relays are subscribers and Mus are publishers. As shown, typical clients would include local HMI. There is remote gateway for remote communication through Internet with implemented suitable security, so that IEDs are not burdened with encryption work.
3. **IED modeling:** The IEC 61850 helps to develop a virtualized model for a device. It begins with an abstract view of the device and its objects and is defined in IEC 61850 Part 7. *IEC 61850 has 92 predefined logical nodes that can be extensible by manufacturers and 355 different data classes, which can be divided into seven categories: system information, physical device information, measurements, metered values, controllable data, status information, and settings.* An IED can be defined by one or more logical devices to differentiate functions in a device, which is connected to the network by one network address. On account of virtualization stated before, it is easier to manage the functions when they are classified hierarchically as shown in Fig III 11.2b2. With reference to Fig III 11.2b2, the information can be understood just by inspecting its data point reference. The first part is Relay1. The second part shows the logical node; here the first letter in the logical node shows the group of the node, in this case, X, to signify switchgear, and the remaining letters CBR1 mean circuit breaker 1. The abstract model is mapped to a specific protocol stack in section IEC 61850-8-1 based on MMS (ISO9506), TCP/IP, and Ethernet. IEC 61850-8-1 specifies method for transforming the model information into a named MMS variable object. Accordingly, the separation mark is "$" used for the basic communication mapped over MMS protocol. Functional constraint is used to specify which services can be used to access the following data. Because of this abstract data and object models, all IEDs can present data using identical structures to relate to respective power system function. The Abstract Communication Service Interface (ACSI—discussed in the following) models of IEC 61850 define a set of services and the responses to those services that enable all IEDs to behave in an identical manner from the network behavior perspective [14]. Logical node group symbols are as follows: **C:** control; **G:** generic; **I:** interfacing and archiving; **L:** system LN; **M:** metering and measurement; **P:** protection; **R:** protection related; **S:** sensor and monitoring **T:** instrument transformer; **X:** switchgear; **Y:** power transformer; **Z:** further power system equipment.
4. **Communication profile**: Part 8.1 defines profiles for the various layers of the communication stack. These are dependent on the service provided. Typical IEC 61850 communication profile with legends has been

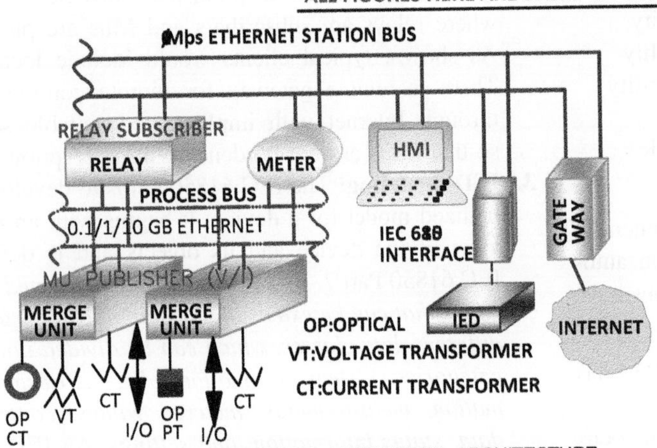


FIG III 11.2b1 SUBSTATION AUTOMATION ARCHITECTURE

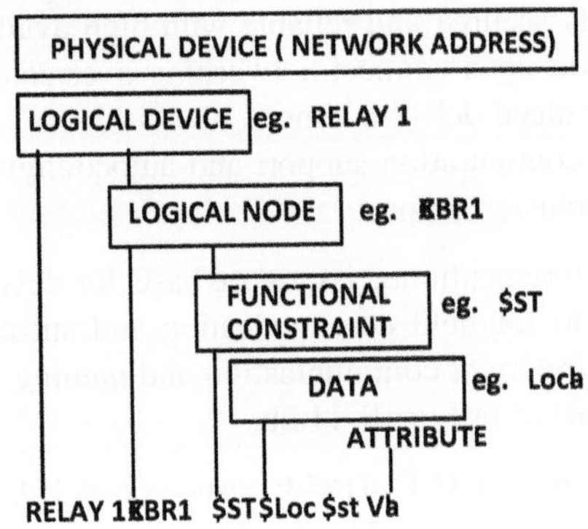


FIG III 11.2b2 ANATOMY OF OBJECT NAME

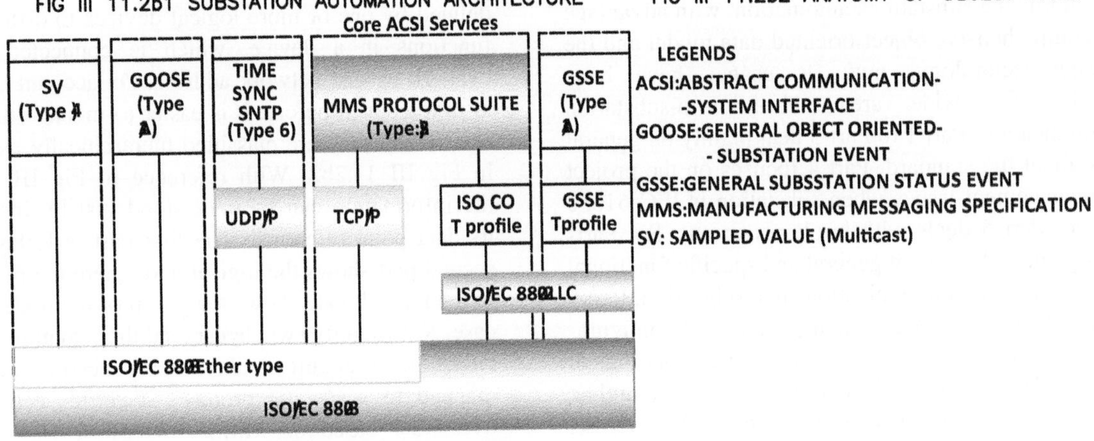


FIG III 11.2b3 COMMUNICATION PROFILE

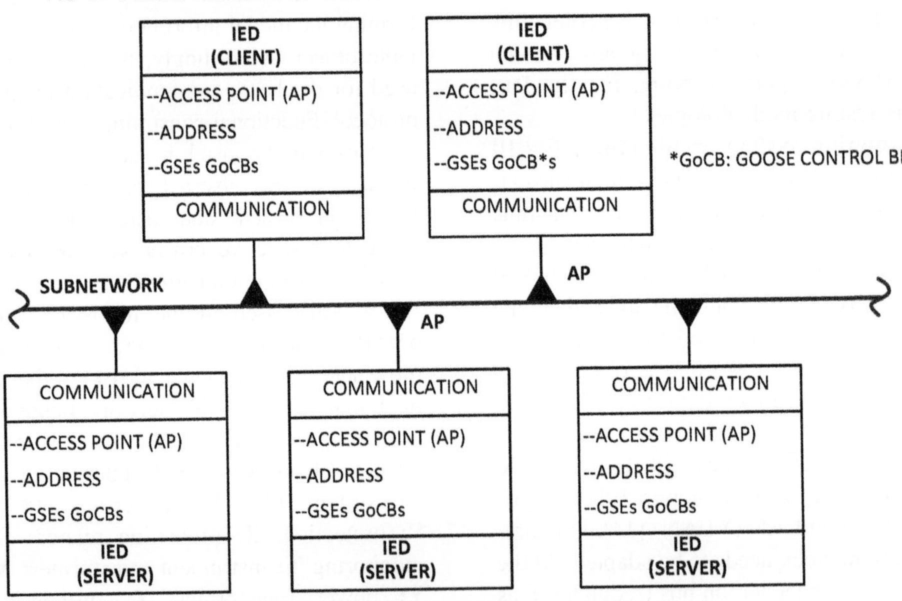


FIG III 11.2b4 COMMUNICATION NETWORK

**FIGURE III 11.2B** Overview of IEC 61850 protocol.

**TABLE III 11.2C** IEC 61850 structure (parts).

| Part No. | Title |
|---|---|
| 1 | Introduction and overview |
| 2 | Glossary of terms |
| 3 | General requirements |
| 4 | System and project management |
| 5 | Communication requirements for functions and device models |
| 6 | Configuration description language for communication in electrical substations related to IEDs |
| 7<br>7.1<br>7.2<br>7.3<br>7.4 | Basic communication structure for substation and feeder equipment<br>Principles and models<br>Abstract communication service interface (ACSI)<br>Common data classes<br>Compatible logic node classes and data classes |
| 8<br>8.1 | Specific communication service mapping (SCSM)<br>Mappings to MMS (ISO/IEC 9506—Part 1 and Part 2) and to ISO/IEC 8802-3 |
| 9<br>9.1<br>9.2 | Specific communication service mapping (SCSM)<br>Sampled values over serial unidirectional multidrop point-to-point link<br>Sampled values over ISO/IEC 8802-3 |
| 10 | Conformance testing |

depicted in Fig III 11.2b3. It is worth noting that time-critical (low delivery time <4 mS) messages such as sampled value (SV) and GOOSE applications handling raw data (types 1 and 4) are directly mapped into Ethernet data frame eliminating intermediate processing layer and time of processing. As shown, there are seven types of messages identified by type numbers, mapped into different communication stacks. The medium and low-speed messages, file transfer functions, and the command message with access control (types 2, 3, and 5) are mapped to MMS protocol operating over TCP/IP stack for connection-oriented layer. The time synchronization messages (type 6) are broadcasted to all IEDs in substation using UDP/IP. Generic substation status event (GSSE) has transport profile or t profile also in addition to type 1 data and operates over connectionless ISO services.

5. **IEC 61850: Communication:** The IEC 61850-8-1 standard defines the communication network with no relation to an existing media or protocol. As already discussed, the IEC 61850 standard describes in Part 7–2 the ACSI in a media and protocol-independent form. The communication section describes information routing among the IEDs. As shown in Fig III 11.2b4, client–server network contains the following parts:
   - Subnetworks
   - IEDs connected to different subnetworks
   - Access points per IED to subnetworks and address
   - IP address of LAN network
   - Link to GOOSE control block (GoCB) message in transmission direction.

With this, we *conclude the discussions* not only on smart grid. Now, let us proceed to discuss various issues related to *communication and networking* in plant automation.

## List of abbreviations used

**AC** Alternative current
**ACSI** Abstract communication service interface
**AI** Artificial intelligence
**AMI** Advanced metering infrastructure
**ANM** Active network management
**APCI** Application protocol control information
**APDU** Application protocol data unit
**API** Application programming interface
**ASDU** Application service data unit
**CASM** Common application service model
**COM** Component object model (Microsoft)
**CRC** Cyclic redundancy check
**CT** Current transformer
**CVS** Concurrent version system
**DA** Distribution automation
**DCOM** Distributed COM (Microsoft)
**DDE** Dynamic data exchange
**DDEML** DDE management library
**DER** Distributed energy resources
**DFC** Data flow control
**DLL** Dynamic-link library/data link layer
**DMS** Demand management system
**DNP** Distributed network protocol

**DR** Demand response
**EPA** Enhanced performance architecture
**EV** Electrical vehicle
**FACTS** Flexible AC transmission system
**FCB** Frame count bit
**GoCB** GOOSE control block
**GOMSFE** Generic object models for substation and feeder equipment
**GOOSE** Generic object-oriented substation events
**GSSE** Generic substation status event
**HAN** Home area network
**HEC** Home energy controller
**HTS** High-temperature superconductor
**HVDC** High-voltage direct current
**IC** Integrated circuit
**ICCP** Inter-control Center Communications Protocol
**ICT** Information and communication technologies
**IED** Intelligent electronic device
**IIoT** Industrial IoT
**IoT** Internet of things
**IP** Internet protocol
**IPC** Interprocess communication
**ISO** Independent system operators
**IT** Information technology
**LAN** Local area network
**LSB** Least significant bit
**MAN** Metropolitan area network
**MMS** Manufacturing message specifications
**MSB** Most significant bit
**MU** Merging unit
**OLE** Object linking and embedding
**OSI** Open systems interconnect
**PLC** Programmable logic controller
**PMU** Phasor measurement unit
**RAMI 4.0** Reference architecture model Industrie 4.0
**RLDC** Regional load dispatch center
**RT** Real time
**RTO** Regional transmission operators
**RTU** Remote transmission unit
**SAS** Statistical analysis system
**SCADA** Supervisory control and data acquisition
**SDU** Service data unit
**SG** Smart grid
**SLDC** State load dispatch center
**SOA** Service-oriented architecture
**TCP** Transmission control protocol
**TD** Transmission and distribution
**TLS** Transport layer security
**UCA** Utility communication architecture
**UDP** User datagram protocol
**VT** Voltage transformer
**WAC** Wide area controller
**WAM** Wide area monitor
**WAN** Wide area network

# References


[1] About dynamic data exchange; microsoft learning; Online0 microsoft topic; internet document; 2018; https://learn.microsoft.com/en-us/windows/win32/dataxchg/about-dynamic-data-exchange.

[2] K. Schwarz; *IEEE Utility Communications architecture (UCA) applies mainstream standard ethernet*; Schwarz Consulting Company (SCC); Research Gate; https://www.researchgate.net/publication/228805495_IEEE_Utility_Communications_Architecture_UCA_applies_mainstream_standard_Ethernet.

[3] *UCA 2.0 for Dummies; Issue 2 and volume 5; power grid international*; https://www.power-grid.com/2000/03/01/uca-20-for-dummies/#gref.

[4] O. Khaled, A.Marín, F. Almenares, P. Arias & D.Díaz; Analysis of secure TCP/IP profile in 61850 based substation automation system for smart grids; *International Journal of Distributed Sensor Networks*; First Published April '2016 Research Article' https://doi.org/10.1155/2016/5793183 .

[5] S. Basu and A. K. Debnath; *Power Plant Instrumentation and Control handbook*; 2nd rev; Elsevier; Academic Press; 2019; https://www.elsevier.com/books/power-plant-instrumentation-and-control-handbook/basu/978-0-08-102804-9.

[6] R. Mackiewicz, *Rechnical overview and benefits of the IEC 61850 standard for substation automation*; SISCO'; https://scadahacker.com/library/Documents/ICS_Protocols/SISCO%20-%20Technical%20Overview%20and%20Benefits%20of%20the%20IEC%2061850.pdf.

[7] *SCADA Distributed network protocol (DNP3); blog; internet document; Ixia Com*; https://www.ixiacom.com/company/blog/scada-distributed-network-protocol-dnp3.

[8] DNP, *650 Series Version 1.1 IEC Communication protocol manual (Section2); relion® 650 series*; ABB manual; ABB limited.

[9] Description and analysis of IEC 104 Protocol; Technical report; report no: FIT-TR-2017-12; P. Matousek; faculty of information technology; brno university of technology; czech republic; Dec' 2017; www.fit.vut.cz/research/publication-file/11570/TR-IEC104.pdf#:~:text=IEC%2060870%2D5%2D104%20protocol,engineering%20and%20power%20system%20automation.

[10] J. T. Michalski, A. Lanzone, J. Trent, and S. Smith; *Secure ICCP Integration considerations and recommendations*; SANDIA REPORT; SAND2007-3345; Sandia National Laboratories; 2007; https://www.energy.gov/sites/prod/files/oeprod/DocumentsandMedia/19-Secure_ICCP_Integration.pdf.

[11] C. It; B. Fahimi, A. Kwasinski, A. Davoudi, R. S. Balog, and M. Kiani; *Re printof IEEE power and energy magazine*; 2011; https://www.ieee-pes.org/images/files/pdf/2012-pe-smart-grid-compendium.pdf.

[12] An Overview on the Smart Grid Concept; C. Cecati, G. Mokryani, A.Piccolo; P.Siano; A conference Paper; Dec'2010; Research gate; https://www.researchgate.net/publication/224207844_An_overview_on_the_Smart_Grid_concept.

[13] Smart Grid: Concepts and Deployment; D.S. Chakrabarti; Indian Institute of Technology Kanpur; Internet Document; ttps://www.iitk.ac.in/ime/anoops/for_15/photos/PPTs/Day-2%20IITK/Smart%20Grid%20Concept%20&%20Deployment-%20Dr.%20Saikat%20Chakrabarty.pdf.

[14] R. Mackiewicz; *Technical overview and benefits of the IEC 61850 standard for substation automation*; SISCO, Inc; Control Global; https://www.controlglobal.com/assets/knowledge_centers/abb/assets/IEC61850_Overview_and_Benefits_Paper_General.pdf.

[15] *A QIP Course on smart grid technology: Smart grid protocols; Prof. A. Sharma*; IIT Kanpur India; http://iitk.ac.in/smartcity/qip/download/ppt/Day-4/Smart%20Grid%20Protocols.pdf.

Chapter 12

# Discussions on communication and networking issues

Modern industrial automation requires connected and integrated systems rather than isolated plant-level solutions. As part of Industrie 4.0 and Digital Transformation (DT), network integration occurs both horizontally and vertically. Horizontal integration connects multiple plants to support operations through data analysis, while vertical integration combines information technology (IT) and operational technology (OT) across different systems. Consequently, industrial communication and networking face challenges related to economics, real-time performance, integration of diverse devices and subsystems using different protocols, and the reliable and secure implementation of wireless solutions. Harsh industrial environments, loss of physical media, protocol incompatibilities, and environmental interference further complicate plant automation systems. Addressing these challenges requires careful protocol selection, robust physical infrastructure, and proper engineering to develop scalable and flexible system architectures. Key considerations in industrial communication and networking include performance, reliability, security, environmental challenges, interoperability, determinism, flexibility, scalability, heterogeneity, integration of wired and wireless media, protocol compatibility, and economics. Information and Communication Technology (ICT) provides native support for industrial automation and offers advantages in control, monitoring, and supervision, along with high performance, enhanced controllability, flexibility, and scalability. A network interface enables an intelligent node to connect to a network and exchange data. It acts as a communication bridge by managing data packets, ensuring proper formatting, and controlling network traffic to support seamless connectivity and resource sharing across networks and the internet. Major functions of network interface devices include data conversion, packet transmission and reception, protocol support, and physical- and data-link-layer access. Thus, a network interface is a dedicated hardware or software component that allows a computer or similar device to connect to a network and communicate with other devices. In hardware implementations, a networking interface enables communication among nodes by serving as the point through which data packets are transmitted and received.

A wireless device is a computer-based device that connects to the internet and other devices without physical cables, enabling wireless communication and access to various applications and services. Transmission media in intelligent computer-based networks are the physical channels that connect network nodes and facilitate data transmission. These media allow intelligent devices, telecommunications equipment, and other communication systems to send and receive data through electromagnetic signals. Depending on the type, transmission media may be guided, such as twisted-pair cables, coaxial cables, and fiber-optic cables, or unguided, such as Wi-Fi, microwave, and infrared communication systems.

Various industrial networking devices are used in automation systems, including media connectors, industrial routers, firewalls, endpoints, network servers, network switches, and Power over Ethernet (PoE) devices. Brief discussions of these devices, except for PoE (covered in Section II 18.0.2), are presented in this chapter. Accordingly, this chapter serves as a bridge between communication systems and networking in plant automation systems and introduces the fundamentals of communication and networking, including network interface devices, wireless devices, transmission media, and network hardware devices.

## 12.0 Fundamentals of network transmission/communication

A computer network is a collection of interconnected intelligent devices capable of exchanging data through communication links. A computer network should ensure reliability of the data communication and security of the data.

A few fundamental issues related to networking need to be defined before discussing network communication in detail. Network topology refers to the physical or logical arrangement of components within a network. For network communication, specific protocols are required. A protocol

Plant Intelligent Automation and Digital Transformation. https://doi.org/10.1016/B978-0-323-89892-8.00005-3

is a set of rules that governs communication within a network and determines how data are transmitted, received, and processed.

Since a network typically consists of multiple interconnected nodes, proper communication requires that each node have a unique address. Addressing methods are therefore used to identify individual devices and enable communication among them. Another important consideration is network security, which refers to measures such as encryption, authentication, and firewalls that protect a network from unauthorized access, hacking, and other security threats.

The ultimate objective is to achieve optimum network performance, which refers to the speed, efficiency, and reliability of a network. Factors affecting network performance include bandwidth, latency, and packet loss. Transmission impairments are also important considerations in network communication. The three major types of transmission impairments are attenuation, distortion, and noise.

Another important aspect of networking is the transmission mode used for data transfer. Depending on the communication system, the transmission medium may be capable of transmitting either a single bit or multiple bits at a time. When bits are transmitted one after another, the method is known as serial transmission; when multiple bits are transmitted simultaneously, it is known as parallel transmission. Serial transmission may further be classified as synchronous or asynchronous transmission.

Data encoding is another key concept in network communication. It refers to the conversion of analog or digital data into analog or digital signals suitable for transmission through a communication channel. In general, encoding involves converting information from one form to another using a defined set of codes. Data encoding ensures reliable and efficient transmission and enables correct interpretation of information at the receiving device. The encoded data may be represented as standardized digital symbols or analog signals. Encoding is performed at the transmitting end, while decoding is carried out at the receiving end.

This section discusses these fundamental networking concepts, including network topology, protocols, addressing methods, network security, transmission impairments, transmission modes, data encoding and decoding, and UART/USART communication techniques.

## 12.0.0 Network transmission issues

There are various network types and their applications including Network Control Systems (NCSs). Brief discussions on data transmissions in network will be presented in this subsection. Let us take the simplest form of example. When a key is pressed, an equivalent binary coded pattern, using standard coding schemes for keyed element is generated within the keyboard. This is used as information. Naturally to represent all characters of the keyboard, a unique pattern of 7 or 8 bits in size is used. So, by a key stroke information is sent. Similarly, but reverse procedure is followed at the receiver that decodes every received binary pattern into the corresponding character (say).

This simplest example has been cited here to indicate that for a computer system or network to function there shall be some means to exchange encoded data over a transmitting medium. In computer terms data is a digital bit or digitized analog signals. Data transmission is a means of transmitting digital or analog data over a communication medium to one or more devices in point-to-point, point-to-multipoint, or multipoint-to-multipoint transmission modes. Therefore, precisely one can say that data transmission describes the transmission of records or information having structured format which are computerized and generally unintelligible. The digital data comes from the source device in the form of digital bit streams which are positioned over a communication medium for transmission to the destination device. Data transmission, or data communication, refers to the transfer of encoded information from one point to another. It involves the exchange of data between two or more devices through a transmission medium, which may be guided (e.g., wire or optical fiber) or unguided (i.e., wireless). In digital communication systems, data are typically represented in digital form at both the source and destination, while the transmitted information may exist as either digital or analog signals. A signal is a physical quantity that varies with time and serves as the carrier of information between communicating devices. As discussed earlier, data communication is essential for the exchange of information among computers, terminals, and other computerized devices to support the execution of application programs and services. Effective data communication enables users to share data, access resources, and perform coordinated tasks across interconnected systems. The major reasons for the necessity of data communication are as follows:

- Reduces time and effort required to perform the required tasks
- Capturing necessary data/information and dissemination of same over network
- Keeping control over the tasks to attend desired objective without any adverse effect on environment
- Cost reduction and system optimizations with flexibility to expand or modify
- Develop integrated network and help in digital transformation approach to enhance productivity
- Supports secured and improved management control of an organization

Fig. 12.1 shows network data transmissions.

FIG III 12.0a1 COMMUNICATION COMPONENTS

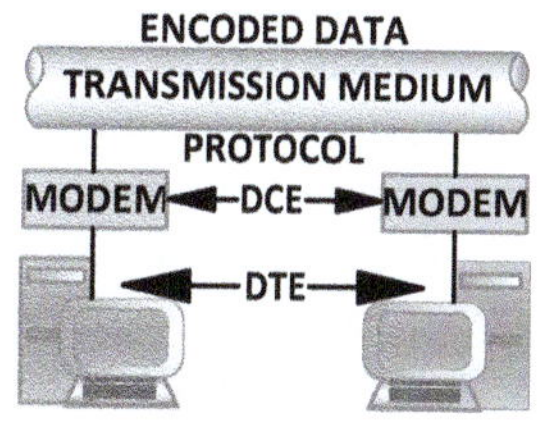


FIG III 12.0a2 DCE & DTE IN TRANSMISSION

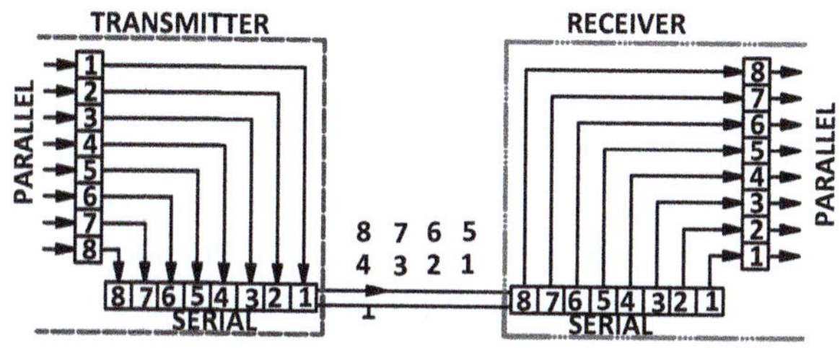


FIG III 12.0a3 SERIAL TRANSMISSION

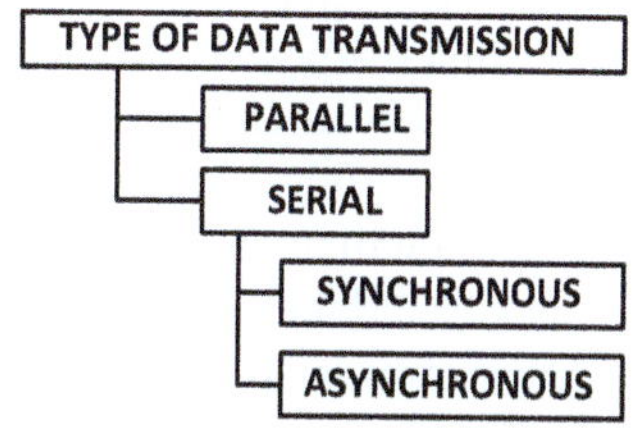


FIG III 12.0al4 SERIAL TRANSMISSION

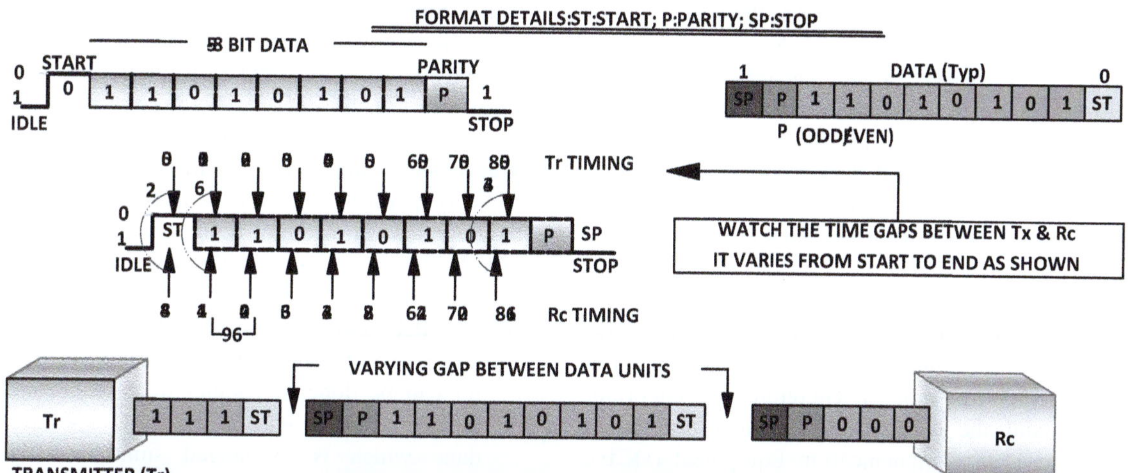


FIG III ■ASYNCHRONOUS SERIAL TRANSMISSION

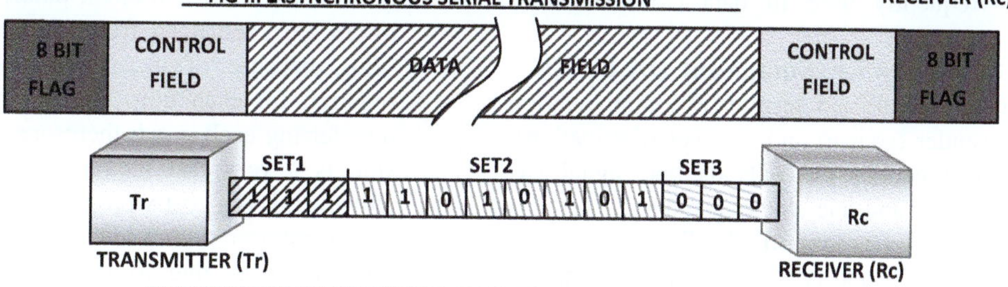


FOR BETTER UNDERSTANDING DATA SETS ARE DIFFERENTIATED WITH SHADES

FIG III 12.0a6 SYNCHRONOUS SERIAL TRANSMISSION

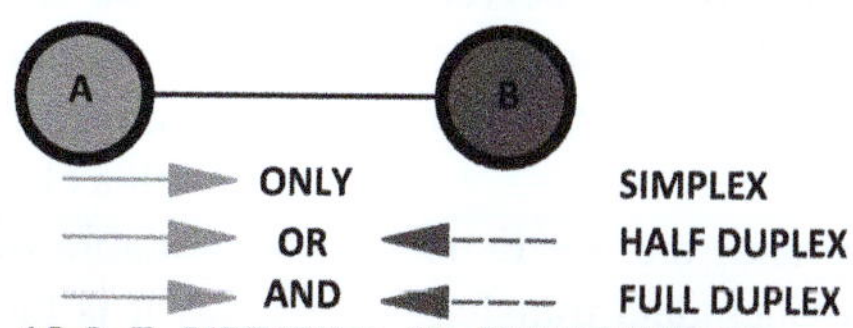


FIG III 12.0a7 DIRECTION OF TRANSMISSION

**FIGURE III 12.0A** Basic issues of network data transmission.

1. **Data Transmission Components:** A data transmission system consists of five fundamental components: (1) sender (source), (2) receiver (destination), (3) transmission medium, (4) information (data), and (5) communication protocol. These components work together to enable the reliable exchange of information between devices. This is shown in Fig. III 12.0a1. While the first four points are not difficult to understand the fifth (protocol) needs some explanation. Suppose a Bengali-speaking person calls a Chinese-speaking person and each speaks in his or her own language. Even if a communication link is successfully established, the purpose of the communication will not be served unless one or both parties can understand the other's language. Similarly, in data communication, a protocol is a set of established rules that enables the sender and receiver to understand and correctly interpret the exchanged information. During data communication, the sender transmits data in the form of packets. A packet generally consists of a header, source address, destination address, data type, actual data, information required for reassembly, and error-control information such as a parity check. The actual packet format varies with the protocol being used. The effectiveness of data communication depends on several factors, including proper delivery to the intended destination, accuracy of transmission, timeliness, and jitter control. When computerized devices are connected to a network, they are often connected through communication equipment such as a modem. Therefore, it is important to understand the terms Data Terminal Equipment (DTE) and Data Communication Equipment (DCE), which are frequently used in data communication systems.
   - **Data Communication Equipment (DCE):** DCE establish, maintain, and terminate communication between sender (source) and receiver (destination). The major functions of DCE include signal exchange, coding, and line clocking tasks.
   - **Data Terminal Equipment (DTE):** DTE is the interface for DCE and is a source or a destination in communication. DTE as source converts information to signal and as destination reconverts received signal into information. No two DTEs communicate directly; communication is carried out through DCEs.

   The interface between the DTE and the DCE is extremely important because it provides data transfer, a host of controls, status indication in both directions, as well as grounding. Even though a DTE can control a DCE through hardware and software means, it cannot provide the clocking function; this function is provided by the DCE. Typical DTE and DCE connection with transmitting medium is shown in Fig. III 12.0a2.
2. **Types of Data Transmission:** Because of mutual capacitance effects and the high cost of implementation, it is not practical to transmit data in parallel over long distances. Therefore, data are generally transmitted in serial form over communication links. However, processors within computing devices typically process data in parallel-bit mode. Consequently, the DTE interfaces with an I/O port that can accept data from the high-speed internal parallel bus of the user's terminal. The I/O port then converts the parallel data into serial form at a lower speed and in the proper format for transmission through a modem or other communication device. At the receiving end, the reverse process takes place. The I/O port accepts serial data from the modem and converts it into parallel form for transfer over the internal data bus. Thus, at the transmitting side, data are converted from parallel to serial format for transmission, whereas at the receiving side, the serial data are converted back into parallel format before being supplied to the processor. This is shown in Fig. III 12.0a3. Therefore, we see that there are two types of data transmission, namely, parallel transmission and serial transmission. Serial transmission is mainly used for data transmission over a communication medium. Serial transmission is further divided into two types: synchronous and asynchronous. This is depicted in Fig. III 12.0a4.
   - **Parallel Data Transmission:** Within very short distances (i.e., between different subunits), data transfer can be done in parallel transfer mode. In this mode a separate wire is used to carry each bit of data, which is exchanged simultaneously. This means that with each clock pulse, all n bits of one group are exchanged from one device to another. Thus, because of the multiple wire connections required for each unit, there is minimal delay in transferring each word; therefore, parallel transmission is a very fast method of data transmission. However, it is a costly proposition because it requires n transmission lines and may not be feasible over long distances due to increased susceptibility to distortion and signal degradation.
   - **Serial Transmission:** In serial transmission, data are sent one bit at a time from one intelligent device to another. Each bit is transmitted in synchronization with a clock pulse. Serial transmission is particularly suitable when the distance between devices is large, as it is more economical and requires only a single communication channel, thereby reducing wiring complexity and minimizing signal distortion over long distances. Unlike parallel transmission, serial transmission uses a single set of communication lines through which data bits are transmitted sequentially, one after another, with each clock pulse. In

many serial communication systems, the least significant bit (LSB) is transmitted first, followed by the remaining bits in sequence, with the most significant bit (MSB) transmitted last. As shown in Fig. III 12.0a3 signal type conversion at DTE is required. Therefore, the savings achieved through reduced cabling costs in serial transmission may be offset, to some extent, by the cost of serial-to-parallel and parallel-to-serial data conversion. Nevertheless, serial transmission remains the preferred method for long-distance communication because of its simplicity and lower wiring requirements. It is evident that serial transmission is generally slower than parallel transmission because data are transmitted one bit at a time rather than simultaneously over multiple channels. The two principal types of serial transmission are synchronous transmission and asynchronous transmission. From the preceding discussion, it is clear that there are several differences between parallel and serial transmission, particularly with respect to transmission speed, wiring requirements, transmission distance, and implementation cost. These differences are presented in Table III 12.0a.

3. **Serial Data Transmission Types:** Serial data transmission can be asynchronous or synchronous:
   - **Pros and cons of asynchronous transmission:** Asynchronous transmission is used for short bit sequences, and only one character is sent at a time. In asynchronous transmission, two devices are synchronized using start and stop bits. A start bit (0) at the beginning indicates the start of data transmission and alerts the receiver to the arrival of a new group of bits. A stop bit at the end indicates the completion of the character transmission. There are normally 5–8 data bits as shown in Fig. III 12.0a5. At the end, one stop bit (1) is sent to let the receiver know that the byte transmission is complete. An optional parity bit for odd or even parity checking may be inserted between the data bits and the stop bit. The stop bit may be 1, 1.5, or 2 bits in length. The idle time between the transmission of successive data bytes is known as the gap, which may be variable, with the minimum gap determined by the stop-bit duration of 1, 1.5, or 2 bits as shown in Fig. III 12.0a5. On account of additional start/stop bits asynchronous transmission calls for higher bandwidth. Details of data frame structure are also shown in Fig. III 12.0a5. Asynchronous transmission is well suited for keyboard-type terminals, Internet traffic, and similar applications. For short communication lines, it is more economical than synchronous transmission. This method does not require clock transmission and is comparatively easier to implement. Since each character is transmitted independently, corruption of one character does not affect the others. Proper recognition of the start bit is essential for successful transmission; otherwise, framing errors may occur. However, asynchronous transmission is slower and less efficient (approximately 73%) than synchronous transmission.
   - **Pros and cons of synchronous transmission:** It is necessary to determine when transmission starts and ends; therefore, timing control is required. Synchronous transmission is used for long bit sequences and requires clock transmission, often with Manchester coding. It is essential that both the transmitter (Tr) and receiver (Rc) use the same clock frequency, as even a small difference can cause errors. Synchronous transmission does not use start and stop bits. Instead, the bit stream is organized into longer frames containing multiple bytes, with no gaps between bytes in the data stream as shown in Fig. III 12.0a6. As Fig. III 12.0a6 shows, a typical data frame for synchronous transmission contains a preamble bit (flag) and postamble bit and control field. As various bytes are placed on the link without gap, it is the responsibility of the receiver to separate the bit stream into bytes for reconstruction of

**TABLE III 12.0A** Difference between parallel and serial transmission.

| Point of comparison | Parallel | Serial |
|---|---|---|
| Data flow | Multiple direction 8 bit | Two direction, bit by bit |
| Bit transfer rate[a] | 8 bit or 1 byte | One bit |
| Speed | Fast | Slow |
| Transmission distance | Only short distance | Long distance |
| Cost | High cost | Economical |

[a]*per clock pulse.*

information. High-level data link control (HDLC) uses bit-oriented synchronous transmission, which is well suited for high-speed communication between two computers (i.e., computerized devices). This method is much faster and without start/stop bit requires less bandwidth. However, it is more costly than asynchronous transmission and calls for accurately synchronized clocks at both ends. This makes the system somewhat complicated. Efficiency is calculated as data bit/(data bit + preamble + postamble).

There is another type of transmission called isochronous transmission, which is commonly used for audio and video applications, such as television broadcasting and real-time multimedia communication. Table III 12.0b compares asynchronous and synchronous transmission. This discussion would be incomplete without considering USART and UART, which are responsible not only for parallel-to-serial (and serial-to-parallel) conversion but also for supporting asynchronous and synchronous data transmission.

- **USART and UART:** Universal Synchronous Asynchronous Receiver/Transmitter (USART) and UART are very similar devices. In this discussion, the USART is considered because it provides the additional capability of synchronous communication, whereas both USART and UART support asynchronous communication. A USART acts as an interface between the processor and peripheral devices. It receives serial data from peripheral devices and converts them into parallel data for processing by the processor. Similarly, it receives parallel data from the processor and converts them into serial data for transmission to peripheral devices. The basic process is illustrated in Fig. III 12.0a3.
- **Differences between USART and UART:** The differences (i.e., the differences between synchronous and asynchronous modes) are as follows:
  - Synchronous mode is possible only with USART, but asynchronous mode is possible with either UART or USART.
  - Synchronous mode requires both data and a clock. Asynchronous mode requires only data. In synchronous mode, the data is transmitted at a *fixed* rate, which is not the case in asynchronous mode (i.e., no fixed rate data).
  - In synchronous mode, data are transmitted in block form, whereas in asynchronous mode, data are transmitted one byte at a time.
  - Synchronous mode allows a higher data transfer rate (DTR) than asynchronous mode when all other factors remain constant.
- **USART in Microprocessors:** The 28-pin 8251A chip is an example of a USART used with the 8085 processor, as shown in Fig. III 12.0b. As Fig. III 12.0b1 shows, USART typically consists of five major sections: read/write logic, transmitter, receiver, data bus buffer, and modem control. Fig. III 12.0b2 shows the interface between the USART and processor. Table III 12.0c describes USART pins and their functions.
- **USART in Microcontrollers:** The above description is mainly based on microprocessors. In microcontrollers, the USART/UART is a peripheral used for data conversion between serial and parallel formats. Most modern microcontrollers, such as PIC and AVR devices, include a USART; however, the synchronous function is not often used because other synchronous communication interfaces, such as I2C and SPI, are available. USART and UART peripherals

**TABLE III 12.0B** Difference between synchronous and asynchronous transmission.

| Point of comparison | Asynchronous | Synchronous |
|---|---|---|
| Data frame | One byte at a time | Chunk or data frames |
| Speed | Slow | Fast |
| Additional bit | Start/Stop | Not used |
| Use of clock pulse | No | Used |
| Time interval | Random | Fixed/constant |
| Gap between bytes | Variable gap | No gap |
| Implementation | Comparatively easier | Bit complicated |
| Cost | Economic | Bit costly |

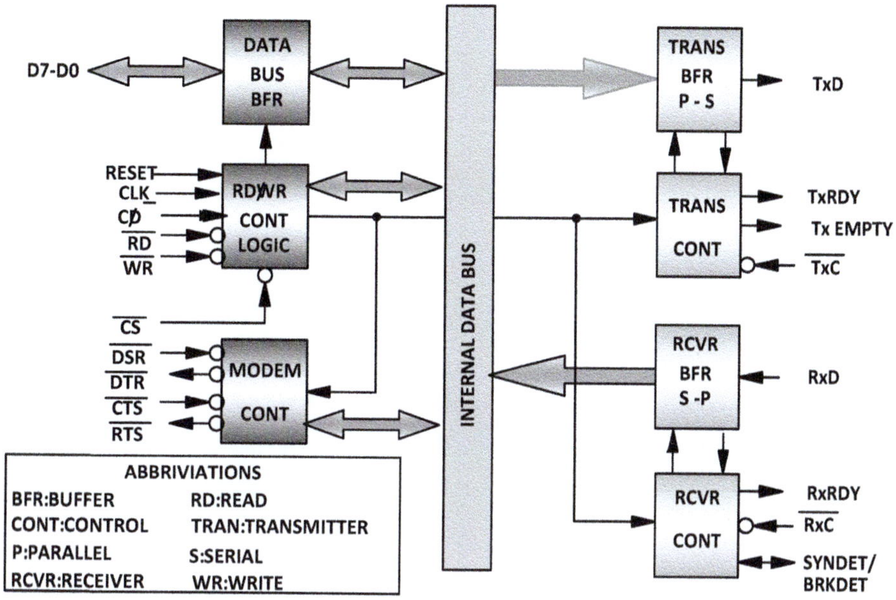


FIG III 12.0b1 USART BLOCK DIAGRAM

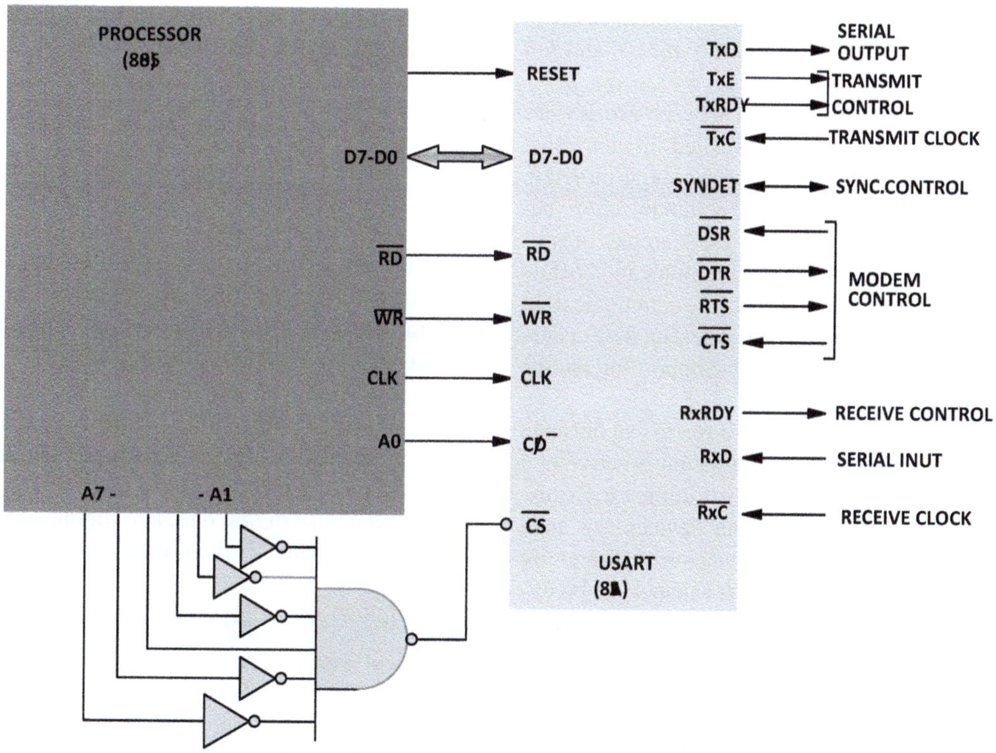


FIG III 12.0b2 USART PROCESSOR INTERFACE

**FIGURE III 12.0B** USART functional details.

have different capabilities and are used in different situations. This can be illustrated by an example. STM32 devices may include both USART and UART peripherals on the same chip. The USART is intended to handle most serial communication tasks during normal operation. When the microcontroller is operating in a low-power or sleep mode, the UART peripheral can handle low-speed communications while consuming less power [1] (. In this connection

**TABLE III 12.0C** USART pin details.

| Pin | Description | Pin | Description |
|---|---|---|---|
| $D_0$-$D_7$ | Parallel data | RxRDY | Receiver ready |
| C/$\overline{D}$ | Data buffer select | TXRDY | Transmitter ready |
| $\overline{RD}$ | Read control | $\overline{DSR}$ | Data set ready |
| $\overline{WR}$ | Write control | $\overline{DTR}$ | Data terminal ready |
| $\overline{CS}$ | Chip select | SYNDET/ | Synchronous detect/ |
| CLK | Clock pulse | BRKDET | Break detect |
| RESET | Reset | $\overline{RTS}$ | Request to send data |
| $\overline{TxC}$ | Transmitter clock | $\overline{CTS}$ | Clear to send data |
| $\overline{TxD}$ | Transmitter data | TxEMPTY | Transmitter empty |
| $\overline{RxC}$ | Receiver clock | $V_{CC}$ | $V_{CC}$ (5 V) (not shown) |

(at *Volume II*); Fig II 6.2b2 in II 6.2.0.3 may be referenced.

4. **Direction of Data Flow:** In any network direction of data flow is extremely important. There are three types of direction of data flow: *simplex, half duplex,* and *full duplex* (Fig. III 12.0a7).
   - **Simplex:** When communication between two devices occurs in only one direction, it is called simplex transmission, i.e., data flow takes place in a single direction only. Radio broadcasting and communication between a keyboard and a monitor are examples of simplex communication. Simplex communication offers the advantage of full utilization of the communication channel capacity. However, because data flow occurs in only one direction, it does not support interactive communication.
   - **Half duplex:** In this case, transmission can occur in both directions, but only one direction is active at a time. This mode is known as half-duplex transmission. In half-duplex mode, both devices can send and receive data, but not simultaneously. The entire bandwidth of the communication channel can be utilized during transmission, although communication can occur in only one direction at a time. Error detection is possible, and the receiver can request retransmission if necessary. However, when one device is transmitting data, the other device must wait, which may introduce delays in communication. Walkie-talkies and communication in field networks are examples of half-duplex communication.
   - **Full duplex:** When transmission occurs in both directions simultaneously, it is referred to as full-duplex communication, i.e., both stations can send and receive messages at the same time. Therefore, it is the fastest mode of communication. However, in the absence of a dedicated communication path between the devices, the capacity of the communication channel must be divided between the two directions. The telephone network is a common example of full-duplex communication.

## 12.0.1 Basic data encoding techniques

When signals travel through a communication medium, they become attenuated and distorted as the transmission distance increases. Therefore, techniques are required to match the properties of the transmitted signal to the characteristics of the communication channel to ensure efficient data transmission. In digital communication systems, data are encoded into digital signals using various encoding techniques. The source data may be either digital or analog. When the source data are analog, an additional analog-to-digital conversion step is required before encoding. Various encoding techniques are discussed in this section. The conversion of digital data into digital signals is known as line coding, which is the primary focus of the following discussion.

The following concepts need to be understood first:

- **Logical status:** Based on how zeros and ones are assigned to logical states i.e. Hi or Lo, there can be two types of logic viz. *Positive logic:* 0 for low voltage (say 0 V) and 1 for high voltage (say 5 V); *Negative logic:* 0 for high voltage and 1 for low voltage.
- **Line code:** The code used for data transmission of a digital signal over a transmission line to use bandwidth economically and to avoid overlap and distortion of signal, etc., is called line code.
- **Purpose of coding:** The primary purpose of line coding is to improve the efficiency and reliability of data transmission. By encoding data appropriately, bandwidth

can be utilized more economically, allowing more information to be transmitted within a given bandwidth. Line coding also promotes efficient power utilization and favorable power spectral density characteristics. In addition, it helps reduce the probability of transmission errors and may facilitate error detection. Coding schemes are generally designed to avoid long sequences of consecutive 1s or 0s and to ensure adequate timing information for proper synchronization between the transmitter and receiver.

- **Features of line coding**: Features of line coding mainly include the following:
  - *Signal levels:* Signal levels represent the number of distinct voltage values used in a signal. For example, a signal that uses zero and either a positive or a negative voltage (e.g., 0 and +5 V, or 0 and −5 V) has two signal levels. In contrast, a signal that uses zero, a positive voltage, and a negative voltage (e.g., −5 V, 0, and +5 V) has three signal levels.
  - *Bit rate versus baud rate:* Although bit rate and baud rate may appear to represent the same quantity, they are different. Bit rate represents the number of bits transmitted per second, whereas baud rate represents the number of signal elements transmitted per second. Therefore, depending on the encoding technique used, the baud rate may differ from the bit rate (data rate).
  - *Signal spectrum:* Different data-encoding techniques produce different signal spectra. Therefore, it is necessary to select an appropriate encoding technique that is compatible with the transmission medium in order to minimize attenuation and distortion of the signal during transmission.
  - *DC components:* After line coding, a DC component (i.e., a zero-frequency component) may be present in the signal spectrum. This DC component is undesirable because it can introduce distortion and errors in the received signal. In addition, it cannot pass through transformers used in communication systems and may result in unnecessary energy loss.
  - *Synchronization:* For proper reception and interpretation of a signal, it is necessary that the bit interval at the receiver matches that of the transmitter within the permissible limits of variation. Otherwise, the received signal may be interpreted incorrectly. Typically, the clock signal is recovered and synchronized from the received signal using a special hardware circuit known as a phase-locked loop (PLL). However, this process is effective only when the received signal is self-synchronizing and contains frequent transitions, preferably at least one transition per bit interval.
  - *Cost of implementation:* Simple encoding with low-cost implementation is always preferable [2].

Let the discussion start with reference to Fig. III 12.0c which will be referenced in next subsection dealing with encoding.

1. **Unipolar:** Unipolar signaling is also known as on-off keying (OOK). In this technique, the presence of a pulse represents a binary 1, whereas the absence of a pulse represents a binary 0. There are two variations of unipolar signaling: (a) non-return-to-zero (NRZ), in which a binary 1 is represented by a positive pulse (mark) having a duration equal to the bit interval, and (b) return-to-zero (RZ), in which a binary 1 is represented by a positive pulse (mark) having a duration equal to half of the bit interval before returning to zero. In both cases, a binary 0 is represented by the absence of a pulse. Between the two, NRZ requires less bandwidth than RZ. In unipolar encoding, only two voltage levels are used, and the signal employs only one polarity of voltage with respect to the time axis, as shown in Fig. III 12.0c1. Power in normalized condition is achieved by

$$\text{Normalized power} = \tfrac{1}{2}\left(V^2 + 0^2\right) = \tfrac{1}{2}V^2 \quad \text{(III 12.0a)}$$

   The greatest advantage of unipolar encoding is its simplicity and ease of implementation. However, it has several disadvantages, including the presence of a DC component, lack of self-clocking capability, the possibility of loss of synchronization, and the absence of built-in error detection or correction capabilities.

2. **Polar Non-Return to Zero (NRZ):** This is one of the most common and simplest methods for transmitting digital signals using two different voltage levels to represent the two binary digits. In NRZ encoding, the signal level may be represented by either a positive voltage or a negative voltage, depending on the encoding convention used. As shown in Fig. III 12.0c3 the two most common forms of NRZ polar coding are NRZ-L and NRZ-I. From the voltage characteristics of the signal, it can be observed that most of the signal energy is concentrated between 0 and one-half of the bit rate. One major drawback of this encoding scheme is the presence of a DC component. In addition, when the transmitter and receiver clocks are not synchronized, the receiver may be unable to determine where one bit ends and the next bit begins. This problem becomes more significant during long sequences of consecutive 1s or 0s. This synchronization problem associated with NRZ encoding is addressed by RZ encoding. Baseline wandering (discussed below) is a problem for both NRZ-L and NRZ-I; however, it is more severe in NRZ-L than in NRZ-I because of the signal transitions present in NRZ-I. NRZ and inverted NRZ encoding schemes are used in RS-232-based

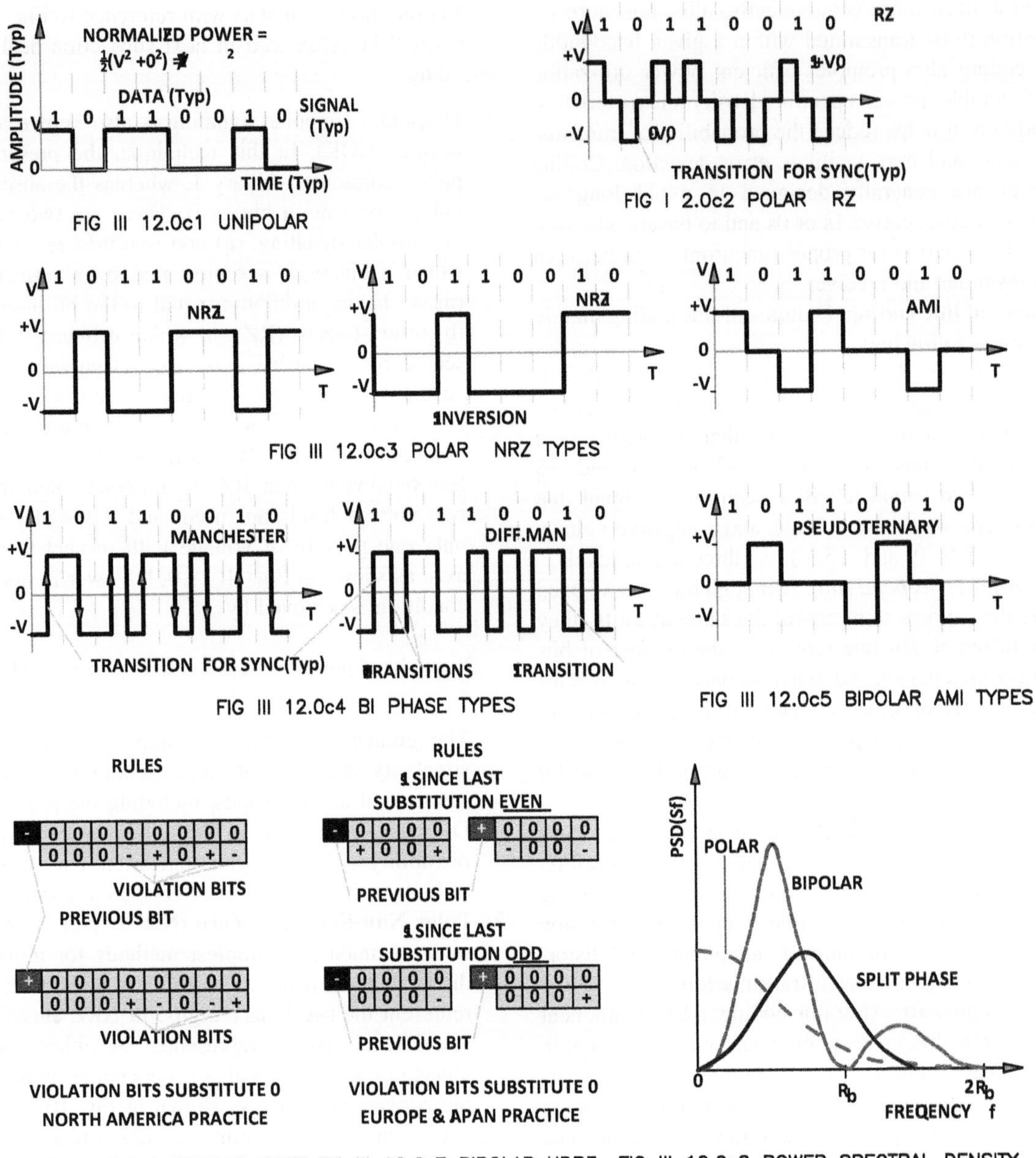


**FIGURE III 12.0C** Line coding techniques.

protocols and the Fiber Distributed Data Interface (FDDI), respectively.

- **Non-Return-to-Zero-Level (NRZ-L):** In NRZ-L encoding, one binary value is represented by a positive voltage level and the other by a negative voltage level. For example, a binary 0 may be represented by a positive voltage and a binary 1 by a negative voltage. Thus, the voltage level of the signal depends directly on the state of the bit being transmitted.
  The voltage pattern is depicted in Fig I 2.0c3.
- **Non-Return-to-Zero Inverted (NRZ-I):** As shown in Fig. III 12.0c3, in NRZ-I encoding, a binary 1 is represented by a transition (inversion) of the voltage level between positive and negative states at the beginning of the bit interval. In contrast, a binary 0 is represented by no change in the voltage level. Because a binary 1 causes inversion of the signal level, this encoding scheme is known as NRZ-I.
- **Advantages of the NRZ polar technique include:**
  - Easy and simple to implement.
  - *No low frequency component.*

- *In presence of noise, it is easier to detect* transition *than threshold.*
- Efficient use of bandwidth.

- **Disadvantages of the NRZ polar technique include:**
  - *No clock.*
  - No error correction possible.
  - Signal droop is caused when the signal is nonzero at 0 Hz.

3. **Polar Return-to-Zero (RZ):** As the name suggests, RZ encoding returns to zero in the middle of each bit interval. In polar RZ signaling, a binary 1 (high data) is represented by a positive pulse (mark) whose duration is half of the bit interval. The signal then returns to zero for the remainder of the bit interval. A binary 0 (low data) is represented by a negative pulse having a duration equal to half of the bit interval, after which the signal returns to zero for the remainder of the bit interval, as shown in Fig. III 12.0c2. Unlike NRZ encoding, RZ uses three voltage levels: positive, zero, and negative. The synchronization problem associated with NRZ encoding is absent in RZ because the signal changes state within each bit interval, allowing the receiver to synchronize using these transitions. However, the bandwidth requirement is higher, and the signal bit rate is twice the data rate. In addition, RZ encoding does not contain a DC component. The advantages of RZ encoding include simple operation and the absence of DC or low-frequency components. Its disadvantages include a larger bandwidth requirement, the need for three voltage levels, which increases encoding and decoding complexity, and its replacement in many applications by Manchester and Differential Manchester encoding schemes.
4. **Biphase Polar Encoding**: In biphase polar encoding, the signal changes state in the middle of each bit interval; however, unlike RZ encoding, it does not return to zero. Thus, it overcomes the limitations of NRZ encoding. However, the bandwidth required for biphase techniques is greater than that required for NRZ techniques. The predictable transition within each bit interval enables the receiver to maintain synchronization with the transmitted signal. The characteristic features of biphase encoding techniques are shown in Fig. III 12.0c4. In the figure, the normalized power spectrum for biphase encoding is also shown to provide an understanding of its spectral characteristics [2]. The major advantages of biphase encoding include the use of only two signal levels, the absence of a DC component, and good synchronization properties. It is used in the IEEE 802.3 standard for CSMA/CD systems and forms the basis of Manchester and Differential Manchester coding schemes. However, it requires a higher bandwidth because the signal bit rate is twice the data rate.
5. **Manchester Encoding Types:** Here both Manchester and differential Manchester coding are discussed.
   - **Biphase Manchester Encoding:** This is illustrated in Fig. III 12.0c4. As in RZ encoding, the signal changes state in the middle of the bit interval; however, it does not return to zero. Thus, Manchester encoding provides synchronization capabilities similar to those of RZ encoding while using only two amplitude levels. The mid-bit transition serves both as a clocking mechanism and as a means of representing data. In Manchester encoding, a negative-to-positive transition represents a binary 1, whereas a positive-to-negative transition represents a binary 0. Manchester encoding combines the synchronization feature of RZ encoding, namely the mid-bit transition, with the data representation characteristics of NRZ-L encoding. Manchester encoding is widely used in Ethernet networks.
   - **Biphase Differential Manchester Encoding:** Differential Manchester encoding is similar to Manchester encoding, with certain modifications. Like Manchester encoding, it includes a transition in the middle of each bit interval to provide synchronization. In addition to this mid-bit transition, a binary 0 is represented by a transition at the beginning of the bit interval. In contrast, a binary 1 is represented by the absence of a transition at the beginning of the bit interval, as shown in Fig. III 12.0c4. Thus, in this scheme, two signal transitions represent a binary 0, whereas one signal transition represents a binary 1. Therefore, Differential Manchester encoding combines the concepts of RZ and NRZ-I encoding. Differential Manchester encoding is widely used in Token Ring networks. Its advantages include the absence of a DC component, self-clocking through the mid-bit transition, and the possibility of error detection. However, it also has certain disadvantages, including higher bandwidth requirements, at least one transition per bit interval (and sometimes two), and a maximum modulation rate that is twice that of NRZ encoding.

## 12.0.2 Other digital data encoding methods

In the following sections, various encoding techniques are discussed, along with a few important related issues listed below:

- **Coding technique categories:** There are several line-coding techniques. However, the basic coding techniques are based on amplitude values, frequencies, phase relationships, and edges (changes in signal level). These form the basis for categorizing line-coding schemes, as discussed.

- **Types of line coding**: There are three types of line coding: unipolar, polar, and bipolar. Line coding types and their subdivisions are listed in Table III 12.0d.
These three types along with their variations are depicted in Fig. III 12.0c.
  - *Unipolar:* Unipolar encoding uses only one polarity of voltage, i.e., the signal exists on only one side of the time axis in a voltage-versus-time waveform. Because of this characteristic, unipolar encoding contains a DC component in its signal spectrum.
  - *Polar:* Polar encoding utilizes two voltage levels: one positive and the other one negative. By using two voltage levels, the average voltage level is reduced, thereby alleviating the DC-component problem associated with unipolar encoding schemes.
  - *Bipolar:* Bipolar encoding utilizes three voltage levels: positive, zero, and negative. In a bipolar encoding scheme, the zero voltage level represents a binary 0, whereas a binary 1 is represented by alternating positive and negative voltage pulses. The reverse arrangement is also possible in the pseudoternary bipolar encoding scheme, where binary 1 is represented by the zero voltage level and binary 0 is represented by alternating positive and negative voltage pulses.
  - *Multilevel:* In this method, various voltage levels are used to generate different signal patterns for encoding data bit patterns. The number of available signal patterns is greater than the number of bit patterns, allowing the additional (redundant) signal patterns to be used for synchronization, error detection, and DC balancing.
  - *Multiline:* Multiline encoding utilizes three voltage levels, +V, 0, and −V, to encode data. It offers lower electromagnetic emissions and reduced bandwidth requirements and is therefore used in high-speed Ethernet communication over copper media.
  - *Block coding:* This method is implemented by first dividing the data stream into groups of m bits, then substituting each group with an n-bit block (n > m), and finally combining the resulting blocks into a single data stream. This is done to utilize the redundant signal patterns for synchronization, error detection, and other control functions.

Other types are shown in Fig. III 12.0d as reference for further discussion.

1. **Alternate Mark Inversion (AMI):** There are two related schemes:
   - **Bipolar AMI/Pseudoternary:** Being bipolar schemes, both AMI and pseudoternary use three voltage levels, namely negative, zero, and positive. These schemes have the same signal rate as NRZ but without a DC component. For long sequences of zeros (in AMI) or ones (in pseudoternary), the voltage remains constant at the zero level. Thus, a sequence that produces a constant zero voltage does not create a DC component; this can also be demonstrated using Fourier transform analysis. In NRZ encoding, most of the signal energy is concentrated around zero frequency. In contrast, in AMI and pseudoternary encoding, the energy is concentrated around a positive frequency greater than zero. Therefore, these schemes are more suitable than NRZ encoding for many communication applications.
   General characteristic features include no DC component, three voltage levels (−V, 0, and +V), possible loss of synchronization for long sequences of zeros (AMI) or ones (pseudoternary), and lower bandwidth requirements.
   The signal streams of both schemes are illustrated in Fig. III 12.0c5.
   - **AMI Scheme:** In the AMI scheme, the zero voltage level represents a binary 0, whereas a binary 1 is represented by alternating positive and negative voltage pulses. If the first binary 1 is represented by a positive voltage, the second binary 1 is represented by a negative voltage, the third binary 1 by a positive voltage, and so on. Thus, successive

**TABLE III 12.0D** Line coding types (non-multilevel schemes).

| Categories | Unipolar | Polar | Bipolar |
|---|---|---|---|
| Category subdivisions | | Non return to zero (NRZ)<br>-NTZ L<br>-NRZ I | AMI (& Pseudoternary) |
| | Return to zero (RZ) | B8ZS | |
| | Biphase (2 types)<br>- Manchester<br>- Differential Manchester | HDB3 | |

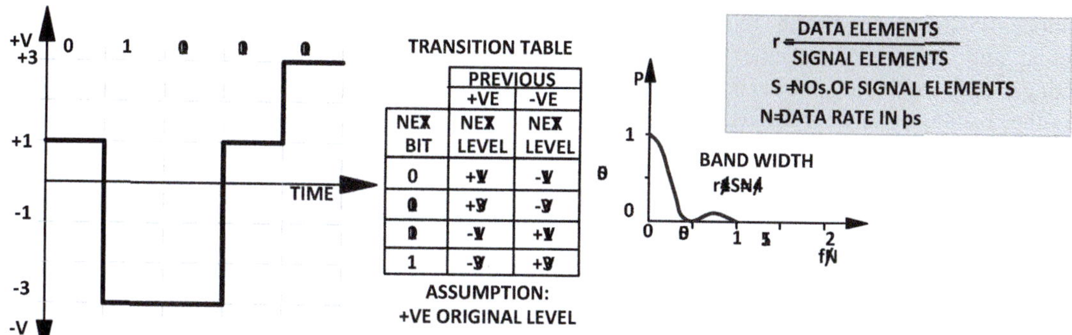


FIG III 12.0d1 B21Q(4) SCHEME (MULTILEVEL)

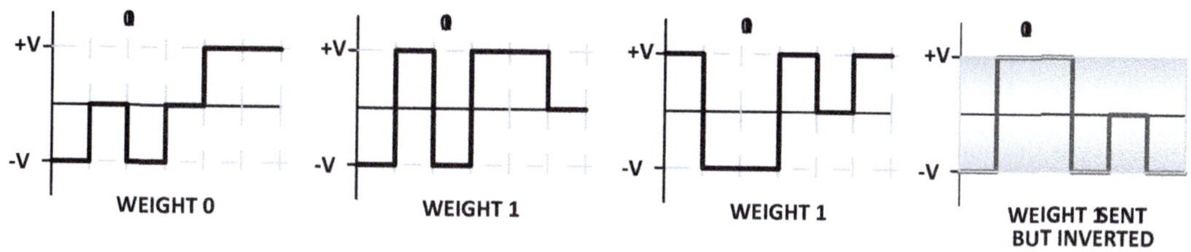


FIG III 12.0d2 B86T(3) SCHEME (MULTILEVEL)

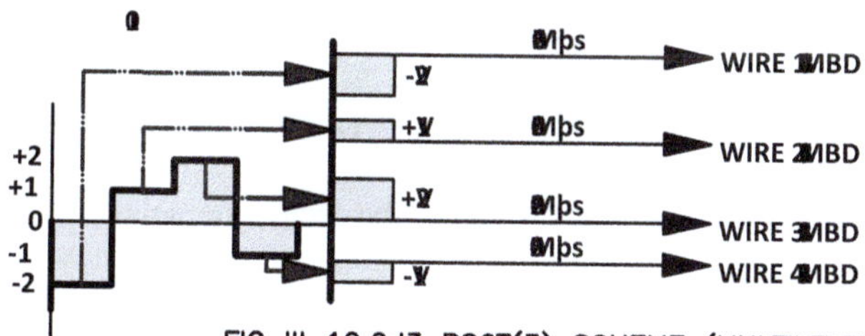

FIG III 12.0d3 B86T(3) SCHEME (MULTILEVEL)

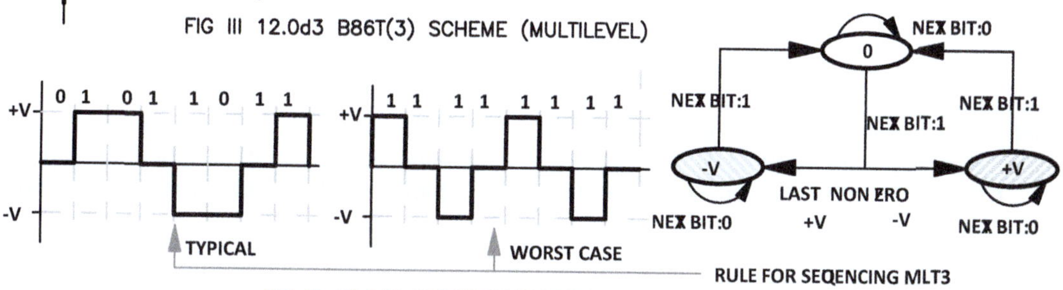


FIG III 12.0d4 MULTITRANSMISSION MLT3

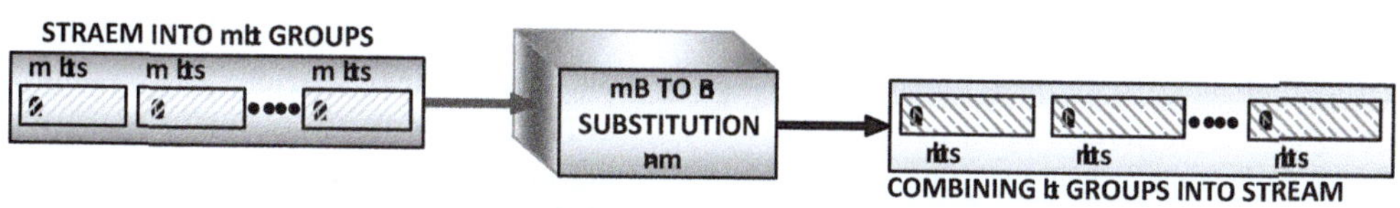


FIG III 12.0d5 BLOCK CODING METHOD CONCEPT

**FIGURE III 12.0D** Multilevel multiline and block coding transmissions.

binary 1s are represented by alternating positive and negative amplitudes.

2. **Pesudoternary Bipolar and Multilevel Scheme:** These are two different coding types:
   - **Pesudoternary Bipolar:** It is a type of AMI in which the roles of binary 0 and binary 1 are reversed. Here, logical 1 is represented by the neutral level (0 V), whereas logical 0 is represented by alternating positive and negative voltage levels. Thus, successive 0 bits alternate between positive and negative amplitudes. Like AMI, this scheme can experience synchronization problems with long sequences of identical bits, and no clock information is inherently present in the signal. However, it offers several advantages, including simple implementation, the absence of DC and low-frequency

components, lower bandwidth requirements than unipolar and polar NRZ schemes, the capability for single-error detection, and the absence of signal drooping, making it suitable for AC-coupled transmission lines.

- **Multilevel Scheme:** The goal of this scheme is to increase the number of bits transmitted per baud by encoding a pattern of m data elements into a pattern of n signal elements. In digital communication, only two binary values (0 and 1) are available. Therefore, grouping m bits produces $2^m$ possible data patterns. Similarly, if n signal elements are encoded using L voltage levels, the number of possible signal patterns is $L^n$. If $2^m = L^n$, each data pattern can be represented by one signal pattern. However, if $2^m > L^n$, the number of data patterns exceeds the number of available signal patterns, and some data patterns cannot be represented. Therefore, the condition $2^m \le L^n$ is generally used, with a larger number of signal patterns often chosen to provide redundancy and help prevent baseline wandering. This scheme is known as the mBnL - scheme. The first two symbols, mB, refer to the data elements, where m represents the number of binary (B) data elements. The last two symbols, nL, refer to the signal elements, where n represents the number of signal elements and L represents the number of voltage levels. The value of L is commonly specified as B (binary, 2 levels), T (ternary, 3 levels), or Q (quaternary, 4 levels). Thus, a 2B1Q scheme represents 2-bit binary data encoded into 1 signal element having four voltage levels, as illustrated in Table III 12.0e.
  Here S=N/4 (Eq. (III 12.0c), so data two times faster than by using N R Z-L. The typical 2B1Q scheme is illustrated in Fig. III 12.0d1, and has been already discussed in connection with Table III 12.0e. Similarly, 8B6T means that an 8-bit data pattern is encoded using six signal elements with three voltage levels. Six signal elements with three voltage levels can generate $3^6 = 729$ different signal patterns. On the other hand, 8-bit binary data can generate $2^8 = 256$ data patterns. Therefore, the remaining $729 - 256 = 473$ signal patterns can be used for synchronization, error detection, and related purposes. Part of this redundancy is also used to provide DC balance. Each signal pattern is assigned a weight of 0 or +1 in terms of DC value, with no pattern having a weight of −1, thereby helping to maintain overall DC balance in the transmitted signal stream. This concept can be understood with the help of Fig. III 12.0d2.
  As shown in the figure, let the first data pattern be represented by the signal pattern (−) 0 (−) 0 (+) (+), which has a weight of 0. The second 8-bit pattern, 01010011, is encoded as (−) (+) (−) (+) (+) 0, which has a weight of +1. The third data pattern would normally be encoded as (+) (−) (−) (+) 0 (+), which also has a weight of +1. To maintain DC balance, however, the sender transmits the inverted signal pattern, shown in the shaded portion as (−) (+) (+) (−) 0 (−). Since this pattern has a weight of −1, the receiver can recognize that the signal pattern has been inverted.

**TABLE III 12.0E** Voltage-level encoding.

| Bits | Voltage level (V) |
|---|---|
| 0 0 | 0 |
| 0 1 | 5 |
| 1 0 | 10 |
| 1 1 | 15 |

3. **Scrambling: Biploar B8Z4 and HDB3:** From the above discussion, it is evident that synchronization poses a significant problem in long-distance transmission. Another constraint arises from bandwidth limitations; therefore, the technique should not increase the number of transmitted bits. Let us now examine how AMI can be extended to support long-distance transmission without synchronization problems. One such approach is scrambling, which is based on a set of predefined rules.
   - **Characteristics:** Major characteristics of scrambling include the absence of DC components, elimination of long sequences of 0s, error-detection capability, and the use of predefined substitution rules, as implemented in B8ZS and HDB3 encoding schemes.
   - **B8Z4 Bipolar:** B8ZS stands for Bipolar 8-Zero Substitution. As stated earlier, it is an extension of the bipolar AMI scheme. This coding rule is used in North America to maintain synchronization during long sequences of 0 bits. When a long sequence of 0s occurs, some of the 0s are replaced with specific signal levels to introduce transitions for synchronization. B8ZS achieves this by inserting artificial signal changes, known as violation bits, within the string of 0s. When eight consecutive 0s occur, B8ZS substitutes the 0-string pattern according to the polarity of the preceding 1 bit. If the polarity of the previous 1 bit is positive, the eight 0s are encoded as 0, 0, 0, +, −, 0, −, +. Similarly, if the polarity of the previous 1 bit is negative, the eight 0s are encoded as 0, 0, 0, −, +, 0, +, −.

Both substitution rules are illustrated Fig. III 12.0c6 where violation bits are identified.

- **HDB3 Bipolar**: HDB3 stands for High-Density Bipolar 3. This coding rule is primarily used in Europe and Japan. Like B8ZS, it is intended to provide synchronization when long sequences of 0s occur in the bit stream. Here also, the pattern of violation bits depends on the polarity of the previous pulse. When four consecutive 0s occur, HDB3 examines the number of binary 1s transmitted since the last substitution and then applies the appropriate substitution rule to maintain synchronization and DC balance. Fig. III 12.0c7 explains the following rules:
  - EVEN: If the number of 1s is EVEN, then substitution is made on the FIRST and FOURTH 0s. The polarity of the substitution depends on the polarity of the previous 1. If the polarity of the previous bit is positive, then the violation is negative. If the polarity of the previous bit is negative, then the violation is positive.
  - ODD: If the number of 1s is ODD, then substitution is made on the FOURTH 0. The polarity of the substitution depends on the polarity of the previous 1. If the polarity of the previous bit is positive, then the violation is positive. If the polarity of the previous bit is negative, then the violation is negative.

From the above discussion, it is evident that certain characteristics make one type of line coding superior to another under specific conditions. It is well known that a higher signal rate requires a higher bandwidth. Therefore, from the standpoint of bandwidth limitations, coding schemes with lower high-frequency content are generally preferred because they require less bandwidth. It is also essential to have a synchronization mechanism to determine the beginning and end of each bit position. This can be achieved either through signal transitions within the encoded data or by using an external clock. Consequently, line-coding schemes that provide inherent synchronization, such as biphase encoding, are generally preferred. Similarly, the absence of a DC component is desirable because it allows AC coupling and electrical isolation.

4. Various issues in line coding: There are a few terms that need to be defined and explained so that the concepts of digital transmission and line coding can be better understood.
   - **Voltage-level encoding**: When bandwidth is limited and a high data rate is required, multiple bits can be grouped together and encoded as a single signal element. For example, consider the transmission of 2-bit data patterns. Different voltage levels may be assigned to represent the various bit combinations, as shown in Table III 12.0e.

     Here, we see that four different states (voltage levels) can be used to transmit 2 bits at a time. This method reduces the signal transmission rate by half and, therefore, reduces the required transmission bandwidth. Thus, the greater the number of signal levels, the greater the number of bits that can be transmitted per signal element. This principle is used in multilevel transmission schemes. In this example, 2 bits of information are transmitted using four signal levels.
   - **Modulation rate:** We often loosely use baud as a unit to express bits per second, but this is not strictly correct. When encoding techniques combine multiple bits into a single signal element, the baud rate and the bit rate (data rate) are no longer the same. When only one bit is transmitted per signal element, the baud rate (baud) is identical to the data rate in bits per second (bps). The data rate, or bit rate (R), is expressed as 1/Tb, where Tb represents the bit duration. In contrast, the modulation rate (D), expressed in baud, represents the rate at which signal elements are generated, i.e., the number of signal elements transmitted per second.

     So, one is related to data and the other to signal. The general relation between D & R is as given by Eq. (III 12.0b).

$$D = R/b = R/\log_2 L \qquad \text{(III 12.0b)}$$

     where $L$ represents the number of different signal elements and $b$ represents the number of bits per signal elements.
   - **Power spectral density (PSD):** As the information signals that generate a line code are stochastic (nondeterministic), the Fourier transform cannot be applied directly. Therefore, to study the performance of a line code, it is necessary to consider its power spectral density (PSD). The function that describes how the power of a signal is distributed over different frequencies in the frequency domain is known as the power spectral density (PSD). PSD is obtained as the Fourier transform of the autocorrelation function. It is generally represented by the symbol S(f) and is plotted against frequency to illustrate how signal energy is distributed over the available bandwidth. Typical PSD plots for different line-coding schemes are shown in Fig. III 12.0c8. Important PSD issues include:
     - The output distortion of a communication channel depends on the power spectral density (PSD) of the input signal, which in turn depends on the pulse rate, pulse shape, and pulse distribution.
     - Distortion in a communication channel may cause the output pulses to become wider than the input pulses.
     - Inter-symbol interference (ISI): A received pulse may be affected by one or more previously transmitted symbols.

- In NRZ-L and NRZ-I encoding, most of the signal energy is concentrated between DC and one-half of the bit rate.
- Manchester and Differential Manchester encoding have most of their energy concentrated between one-half and one times the bit rate. Thus, the bandwidth is reasonably narrow, and there is no DC component.
- AMI requires less bandwidth than Manchester and Differential Manchester encoding schemes.
- Baseline wandering: At the receiver, while decoding a digital signal, the incoming signal power is evaluated against a baseline (a running average of the received signal power). Long sequences of 0s or 1s may cause the baseline to drift, making correct decoding difficult. This drift of the baseline is referred to as baseline wandering. A well-designed line-coding scheme minimizes baseline wandering.
- Relation among S, N, and r: When the number of signal elements is represented by S and the data bit rate by N (bps), then:

$$S = N/r \qquad \text{(III 12.0c)}$$

And ratio $r = \text{Data element/signal element}$ (III 12.0d)

From Table III 12.0e we get r = 2/4 = ½. Here, a data element represents the smallest unit that contains a piece of information, whereas a signal element is the shortest unit of a digital signal in the time domain. The ratio r is an important factor in digital transmission. Increasing the data rate increases the transmission speed, whereas decreasing the signal rate reduces the bandwidth requirement. These concepts are utilized in various encoding schemes, such as multilevel coding, multiline coding, and block coding, as depicted in Fig. III 1ṅ2.0d and discussed below.

5. **4D-PAM5, MLT-3, and Block Coding:** Three important different types of coding include:
   - Four-Dimensional Five-Level Pulse Amplitude Modulation (**4D-PAM5**): As shown in Fig. III 12.0d3, there are four signal levels: (−)2,(−)1, +1, +2. There is also a 0-V level, which is used for forward error detection. Here, 4D-PAM5 represents the transmission of 4 data streams over four separate wires, using Pulse Amplitude Modulation (PAM) with five signal levels, of which four are used for data transmission and one (0 V) is reserved for error detection. In terms of a single-wire system, it is equivalent to 8B4Q encoding for 8-bit data. In other words, an 8-bit word is translated into a signal element having four different signal levels. The worst-case signal rate for this one-dimensional equivalent is N × 4/8, or N/2. Since four channels (four wires) are used simultaneously, the signal rate S is further reduced to (N/2) × (1/4) = N/8. This represents a significant reduction in the signal rate. As shown in Fig. III 12.0d3, all 8 bits are fed into a wire simultaneously and sent by using one signal element. Gigabit LANs use this technique to transmit 1-Gbps data over four copper cables, each capable of handling 125 Mbaud. This scheme provides considerable redundancy in the signal patterns because $2^8 = 256$ data patterns are mapped to $5^4 = 625$ possible signal patterns. The additional signal patterns can be utilized for purposes such as error detection and other control functions [3].
   - Multilevel Transmission, Three-Level (**MLT-3**): This scheme was introduced by Crescendo Communications for FDDI copper interconnects. Later, MLT-3 found widespread application in 100-Mbps data transmission over unshielded twisted-pair (UTP) cables. In MLT-3, the base waveform is a three-state alternating waveform, unlike the two-state alternation used in line-coding schemes such as Manchester or NRZ-I. MLT-3 alternates from −1 to 0 to +1, back to 0, and then back to −1, repeating this sequence indefinitely. A binary 0 is encoded as no change in the progression. Thus, MLT-3 can be viewed as a stop-and-go waveform, where a binary 0 represents "stop" and a binary 1 represents "go." Therefore, MLT-3 can represent a complete periodic waveform using the sequence 0, +1, 0, and −1 over four or more bits, as shown in Fig. I 2.0o4. MLT-3 uses three voltage levels, namely +V, 0, and −V, and follows three transition rules: if the next bit is 0, no transition occurs; if the next bit is 1 and the current level is not 0, the next level is 0; and if the next bit is 1 and the current level is 0, the next level is the opposite of the last nonzero level.

     With reference to Fig. III 12.0d4, we see that there are the three voltage levels, (−)V, 0, and +V, for three (ovals) states. The transition from one state (level) to another is shown by the connecting lines. Fig. III 12.0d4 also cites two examples of MLT3 signals.

     Although the signal rate in MLT-3 is the same as that in NRZ-I, the transition rules of MLT-3 shape the signal in a manner that reduces the required bandwidth. In the worst case, the input data stream is a nonperiodic function. However, by applying the MLT-3 transition rules, the resulting signal is shaped into a periodic waveform with a period equal to four times the bit duration. Consequently, the fundamental frequency of the MLT-3 signal is one-fourth of the bit rate. This makes MLT-3 a suitable choice for transmitting 100 Mbps over copper wire with a maximum frequency limit of approximately 32 MHz. In addition, MLT-3 interfaces emit less electromagnetic interference and require less bandwidth than most other interfaces operating at the same bit rate.

- **Block coding:** From the discussion so far, it is clear that redundancy in signal patterns can be used for synchronization and error detection. The same principle is utilized in block coding to improve the performance of line-coding schemes. In block coding, a block of m bits is converted into a block of n bits, where n is greater than m. Because block coding converts m-bit blocks into n-bit blocks, it is referred to as an mB/nB encoding technique (e.g., 4B/5B). The slash notation in block coding distinguishes it from multilevel encoding schemes. Block coding typically involves three steps: division, substitution, and combination, as shown in Fig. III 12.0d5.
  In the division step, a sequence of bits is divided into groups of m bits. The next step is substitution, in which each m-bit group is replaced by an n-bit group. Finally, in the combination step, the n-bit groups are combined to form a bit stream containing more bits than the original data stream, as shown in Fig. III 12.0d5.
  Let us now conclude the discussion on line coding by summarizing the characteristic features of the various coding techniques (Table III 12.0f). Next, we discuss network interface and wireless devices.

## 12.1 Network interface hardware and wireless devices

In this section, network interfaces, both wired and wireless [ref: Chapter 10 of this volume], are discussed. Various wireless devices are also included to complete the discussion. Wired interfaces may be broadly classified into two categories. The first category consists of devices responsible for serial communication, such as RS-232, RS-422, and RS-485. The second category includes wired networking devices such as hubs, bridges, gateways, and USB interfaces (covered in Volume II). Similarly, wireless networking devices include WAPs, VPN appliances, and related technologies.

In simple terms, a network device is a physical device used to connect other devices within a network. In some cases, its role is simply to forward packets of information to their destination. In other cases, it may act as a translator or block suspicious network traffic. Industrial plant automation networks are primarily concerned with interconnecting intelligent devices to create an integrated system. For these devices to exchange information, they must share a common communication protocol.

RS-232, RS-422, and RS-485 are serial communication standards that define the hardware interface rather than the software protocol. They often operate in conjunction with higher-level software protocols. For example, a device may communicate with a computer through an RS-232 port using the MODBUS protocol, whereas PROFIBUS may use RS-485 together with its associated software protocol. Each serial communication standard has limitations with respect to communication mode (simplex, half-duplex, or full-duplex), data transfer rate, and transmission distance.

In addition to serial communication interfaces, intermediary devices are required to facilitate the transmission of data between networked devices and therefore play an important role in network operation. A hub, now largely obsolete and replaced by switches, is a device that connects multiple network nodes into a single network segment. A bridge allows multiple network segments to be combined into a single network. In contrast, routing enables multiple networks to communicate while remaining separate. Routers connect packet-switched networks or subnetworks and manage traffic by forwarding data packets to their intended destinations, while allowing multiple devices to

**TABLE III 12.0F Characteristics of various line coding techniques [3].**

| Category | Scheme | Bandwidth B | Characteristics |
|---|---|---|---|
| Polar | NRZ I | N/2 | No self-synchronization, DC for long 0s |
| Polar | NRZ L | N/2 | No self-synchronization, DC for long 0s/1s |
| Polar | Bi phase | N | Self -synchronization, no DC, hi B |
| Bipolar | AMI | N/2 | No self-synchronization, D C for long 0s |
| Multilevel | 2B1Q | N/4 | No self-synchronization for long same double bit |
| Multilevel | 8B6T | 3N/4 | Self synchronization, No DC |
| Multilevel | 4DPAM5 | N/8 | Self synchronization, No DC |
| Multiline | MLT-3 | N/3 | No self-synchronization for long 0 |

share the same Internet connection. A gateway serves as an interface between different networks by translating data formats to enable communication.

As the name implies, a repeater receives, amplifies, and retransmits signals so that they can travel greater distances. Network security devices are also important. Intrusion Detection Systems (IDSs) monitor network traffic to identify potential threats, exploit attempts, or other incidents that may compromise network security. Firewalls provide a first line of defense by acting as a barrier between trusted internal networks and untrusted external networks.

A Network Interface Card (NIC) enables a computer to communicate with other devices on a network by providing a dedicated network connection and the necessary circuitry to convert digital data into transmission signals. A wireless NIC enables a computer to operate as a wireless station. Firewalls may be implemented as hardware-based or software-based solutions.

A modem is a network device that can act as an interface between wired and wireless networks. Functionally, it converts digital data into a form suitable for transmission over analog media and performs modulation and demodulation functions as required. An Access Point (AP), or Wireless Access Point (WAP), enables wireless devices to connect to a wired network. A VPN appliance is a dedicated device that provides secure remote access to a private network through Virtual Private Network (VPN) technology. Many devices used in wired networks, such as gateways, hubs, and switches, are also used in wireless networks.

Bluetooth is a Personal Area Network (PAN) standard that operates in the 2.4-GHz ISM band with a physical-layer data rate of 1 Mbps and an effective full-duplex data rate of approximately 432 kbps. ZigBee is another wireless networking technology that requires only a small fraction of the power consumed by an equivalent Bluetooth PAN, allowing battery life to extend from months to years. These wireless devices are also discussed in this section.

Let the discussion begin with network interfaces for serial communication.

Hardware required for a network, other than the main controller and peripherals, can be categorized as either transmission media or network interface devices. Connections between nodes/computers on a network may be wired or wireless. In wireless networks, nodes/computers communicate through wireless NICs. Devices available in the market may be broadly classified as commercial-grade or industrial-grade. These topics are briefly discussed in this subsection.

This subsection is dedicated to serial communication used in computer networks. The discussion begins with the current loop, which is now rarely used. Introduced in 1962 for communication between teleprinters and teletypes, it began to lose popularity following the introduction of the EIA-422 standard in December 1978 and the EIA-485 standard in April 1983. Powered by a constant-current source, the most common 20-mA current loop was capable of providing serial communication.

EIA-232 and CCITT V.24 are among the most common duplex serial interface standards. RS-232/EIA-232/V.24 is defined as the interface between DTE and DCE using serial binary data exchange. The interface exchanges data between two devices in a point-to-point connection, typically over distances up to 15 m. The transmission distance of EIA-232/RS-232 and CCITT V.24 generally cannot exceed about 30 m (100 ft) because the difference between transmitted voltage levels (±5 V) and received voltage levels (±3 V) provides only 2 V of common-mode noise rejection. In addition, increasing distance results in degradation of the slew rate because of distributed cable capacitance. Communication rates below 20 kbps are commonly used, although faster drivers are now available (e.g., MAX3225E) [4].

Differential data transmission (balanced differential signaling) offers superior performance in many applications because differential signals help cancel the effects of ground shifts and induced noise that appear as common-mode voltages on a network. Since logic states are represented by a differential voltage between two associated lines, the RS-422 interface is balanced with respect to ground. As a result, externally coupled noise produces nearly identical interference on both lines, leaving the useful differential signal largely unaffected.

RS-485 supports true multipoint communication, allowing up to 32 drivers and receivers on a single two-wire bus segment. This limit can be extended through the use of repeaters and high-impedance drivers/receivers. In addition, RS-485 drivers can tolerate data-collision (bus-contention) conditions and bus faults. Two forms of EIA/RS-485 are commonly used: a two-wire half-duplex configuration and a four-wire full-duplex configuration. In the two-wire half-duplex version, data transmission and reception occur alternately and must be controlled through appropriate software. RS-485 commonly uses a 9-pin D-sub connector (refer to the relevant standard for pin details). Pin assignments differ between the two-wire and four-wire implementations.

With this background, the discussion of network interface hardware begins with network hardware grading.

1. **Network hardware grading:** Network hardware is available in two grades: commercial and industrial. Industrial networks are designed to meet the demanding requirements of industrial environments, including real-time control and protection applications. Therefore, proper selection of network hardware is essential to ensure high availability and reliability. Industrial network components are often required to operate

under harsh environmental conditions, such as inadequate cooling, dust, moisture, corrosive atmospheres, and elevated ambient temperatures (e.g., near furnaces). In addition, they are expected to provide longer service life and greater reliability than commercial-grade equipment. Thus, environmental challenges and the need to maximize system uptime are two key factors that must be considered when selecting network hardware for industrial applications. Furthermore, industrial systems typically demand high availability, failover capability, and extended life expectancy. Table III 12.1a compares the two grades.

2. **Current loop:** Typically, a 20-mA current loop is suitable for distances around 610 m (2000feet) at data rates up to 19.2k baud. Logical 1 is represented by a loop current of 20 mA. If the circuit is interrupted, resulting in the absence of current, it represents a logical 0. Simplex and duplex TTY current-loop configurations are shown in Fig. III 12.1a. Devices used in 20-mA TTY current-loop systems may be classified as active TTY, passive TTY, and semiactive TTY.

- **Current loop scheme:** For implementation of a 20-mA TTY current loop, three major elements are required: a constant-current source, a current detector, and a switching element that operates within the current loop (as detailed out in Fig. III 12.1a). The current switch acts as the transmitter, whereas the current detector functions as the receiver. There are two modes of transmission: simplex and duplex. In simplex transmission, a single current source is provided at the active side, which operates as a semiactive device, as shown in Fig. III 12.1a2. Simultaneous two-way communication, i.e., duplex transmission, is also possible with a 20-mA TTY current-loop system. In duplex transmission, separate current loops are used for transmitting and receiving data, allowing both devices to communicate simultaneously. As shown in Fig. III 12.1a1, two 20-mA current generators are provided on the fully active side of the circuit to support duplex communication between the two devices, namely the active and passive devices.
- **Current loop specification:** A brief specification of the 20-mA TTY current-loop interface is presented in Table III 12.1b.

3. **RS-232 interface:** A typical RS-232/EIA-232/V.24 interface is illustrated in Fig. III 12.1bfig12_6. The figure shows a typical D-sub 9-pin configuration. Equivalent pin numbers for the D-sub 25-pin connector are also provided in parentheses for comparison. The typical D-sub 9-pin pinout is shown as well. In its simplest form, only three connections are required: TxD (Transmit Data), RxD (Receive Data), and GND (common signal ground). Additional pins are available for control, signaling, and clock functions when hardware handshaking is required.

The connection is straightforward and point-to-point in nature. For example, the DTE sends data through TxD, and the DCE receives it through the corresponding connection. Typical connections between DTE and DCE are illustrated in Fig. III 12.1b. In this figure, all nine pins of the D-sub 9-pin connector, together with their designations and functions, are described. Equivalent pin numbers for the D-sub 25-pin connector are also shown in parentheses.

Two types of grounding are used. The protective ground (not shown) ensures that both pieces of equipment are at the same earth potential, whereas the signal ground provides the return path for digital signals.

- **Connection details:** Although the EIA-232/RS-232 and V.24 interfaces were designed as point-to-point rather than multidrop interfaces, and their drivers are specified for a single load, it is possible to connect multiple slave devices to a master device. For example, multiple Single Loop Controllers (SLCs) may be connected to a single PC in a PC-based control system using a daisy-chain configuration, as shown in Fig. III 12.1b1.

**TABLE III 12.1A** Comparison of commercial and industrial network hardware.

| Issue | Commercial grade | Industrial grade |
|---|---|---|
| Availability | 5-7 years; less EMI effect. | Mil grade; 24–30 years w/o moving part like fans. It can withstand high EMI and function properly |
| Environment | Working environment is not very severe | Working environment may be quite harsh especially from temperature, corrosion, EMI effects etc. Also in some cases may call for high voltages e.g., Ethernet cable in MCC. |
| Cost | Initial cost with cost of replacement higher | Total cost is not high when all factors considered |
| Maintenance | More replacement & new installation cost | Overall maintenance cost is less hence total cost of ownership |

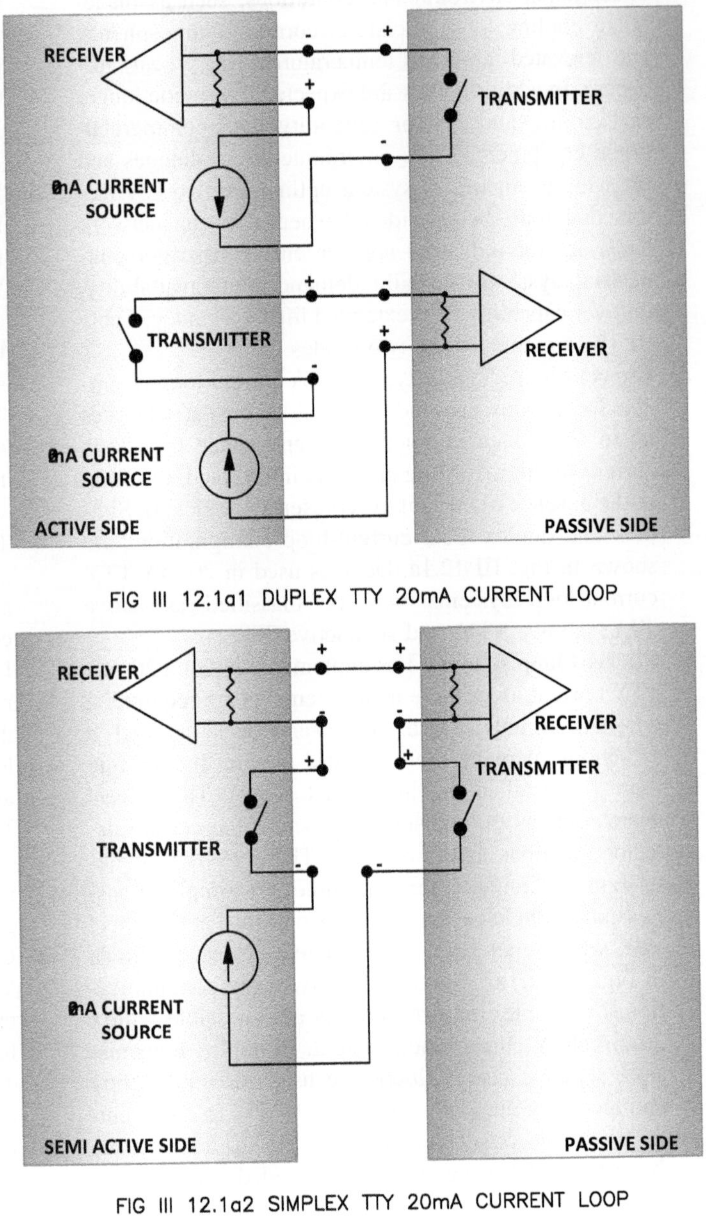


**FIGURE III 12.1A** TTY 20 mA current loop serial link.

In a daisy-chain configuration, the RS-232 signal enters through one receiver and is then looped through to a transmitter. This arrangement is repeated for each subsequent device along the communication line. One major disadvantage of this technique is its vulnerability to cable breaks. If a break occurs in the chain (as indicated by the red (light gray in print version) dotted line), all downstream devices beyond the break point will be disconnected from the network [4]. Since there are no fixed guidelines for cable selection, the choice of cable depends on the devices being connected and the required pin connections. Therefore, cable selection must be based on the type of equipment to be interconnected and the corresponding pin assignments of the interface [4]. RS-232 interfaces are commonly available with 4-, 9-, or 25-pin connectors and cabling arrangements. Four-pin cables provide the minimum required connections and often use jumpers to provide the necessary handshaking signals for devices that require them [5]. The 25-pin cable connects every pin, whereas the 9-pin cables do not include many of the uncommonly used connections.

- **Characteristics features** [4]: The RS-232 interface is characterized by parameters such as voltage levels,

**TABLE III 12.1B** Brief specification of TTY 20 mA current loop interface.

| Specifying point | Standard/Available data | To be specified |
|---|---|---|
| Data transfer channel | 2 Tx/Rx full duplex* | *Specify if simplex |
| Data transfer rate | Max 19,200 default: 9600 baud | |
| Data size | 8 data bits, one stop bit | Specify parity |
| Bit transfer | 2 × 20 mA | |
| Data buffer | 128 bytes receive buffer, 16 bytes transmit buffer | To specify |
| Load impedance | 500 Ω | |
| Cable type and max length | Twisted pair (1000m max) | Specify length |
| Voltage isolation (power to-signal) | 500V | |
| Power supply | Manufacturer's standard | To specify |
| Protection class | IP22 | |
| Environmental condition | Operating: Temp(max): 0–55°C Humidity: 95% noncondensing | To specify |
| Special feature | To specify as applicable | |

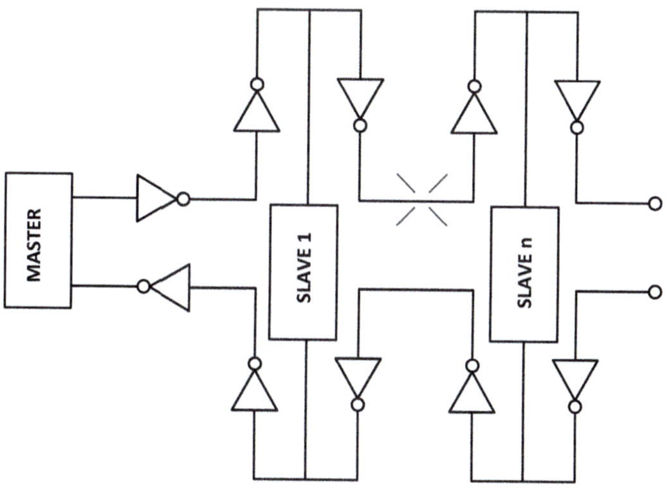


Fig III 12.1b1 CONNECTION DETAILS

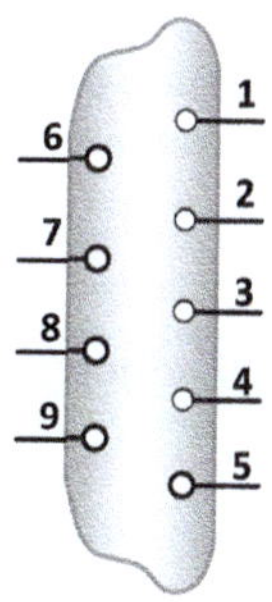


Fig III 12.1b2 D SUB 9 PIN: PIN DISPOSITIONS

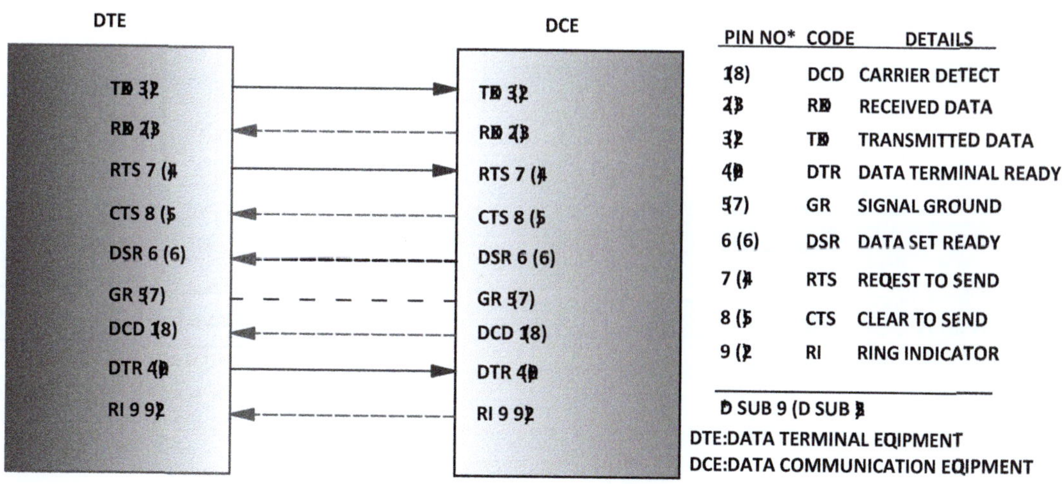


FIG III 12.1b3 RS–232 PIN CONNECTION DETAILS [4]

**FIGURE III 12.1B** RS 232 V24 interface.

signaling rate, timing, slew rate, voltage-withstand capability, short-circuit behavior, and maximum load. It uses pluggable connectors with defined pin assignments and specified functions for each circuit. However, the standard does not define the external power supply, character encoding (e.g., ASCII), communication protocol, framing characters, or error-checking methods.

- **Specifications:** Applicable standards for RS-232/V.24 include ITU-T V.24/DIN 66020-1, ITU-T V.28/DIN 66259-1, and TIA/EIA-232. A brief specification of the EIA/RS-232 interface is presented in Table III 12.1c. The data presented here have been compiled from reputable manufacturers.

4. **RS-422 interface:** Although RS-422 was primarily defined for communication between DTE and DCE devices, it is also widely used for point-to-point interconnection of digital equipment. In its simplest form, RS-422 can be used as an extension of RS-232, allowing conversion from RS-232 to RS-422 and vice versa. In EIA/RS-422, serial data transmission is performed in full-duplex mode at data rates of up to 10 Mbps and over distances of up to 1200 m. However, the achievable data rate decreases as the transmission distance increases; for example, at a distance of approximately 1.2 km, the data rate is typically around 100 kbps. Thus, the allowable interconnection length depends on the required data rate. The potential difference between the generator and receiver with respect to ground constitutes the common-mode voltage, which should be less than 7 V. A 37-pin D-sub connector is commonly used for RS-422 interfaces. A typical RS-232/EIA-232/V.24 interface is illustrated in Fig. III 12.1c.
The balanced connection configuration of RS-422 is illustrated in Fig. III 12.1c1. A single master can support up to 10 slave devices through an RS-422 interface, as shown in Fig. III 12.1c2 [4]. For operation, there must be at least one transmit-data channel (T) and one receive-data channel (R). Communication is generally coordinated by software. Alternatively, the transmission process may be managed by hardware using the control channel (I) and the signaling channel (C). For both RS-422 and RS-485, proper cable termination is essential in highly reliable applications to minimize signal reflections in the transmission line. Because RS-422 permits only one driver on the bus,

**TABLE III 12.1C** Short specification of EIA/RS-232.

| Specifying point | Technical data |
|---|---|
| Transmitted signal | |
| Binary 0/space/On | +5 to +15VDC |
| Binary 1/Mark/Off | −5 to −15VDC |
| Received signal | |
| Binary 0 | +3 to +13VDC |
| Binary 1 | −3 to −13VDC |
| Data format | |
| Start bit | Binary 0 |
| Data format | 5,6,7,8 bits |
| Parity | Odd/Even or mark/space |
| Stop bit | 1 (one/2 bits) |
| Driver open circuit voltage (max) | 25 VDC |
| Load @ Min ± 5V; max ±15V | 3–7 KΩ |
| Slew rate (v/μS) | 4–30 |
| Total load capacitance max | 2500 pF |

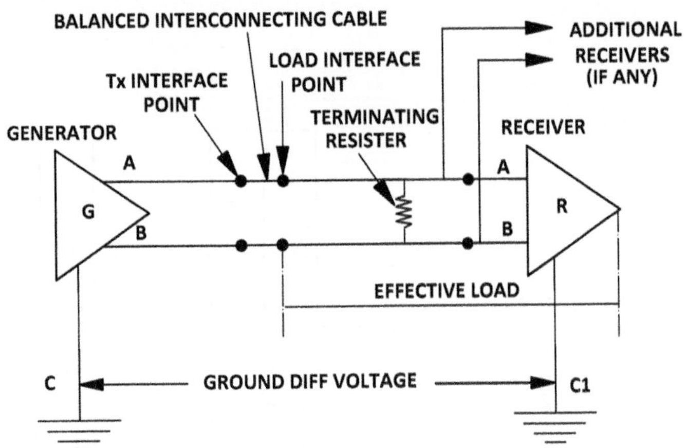


FIG III 12.1c1 BALANCED INTERFACE RS 422

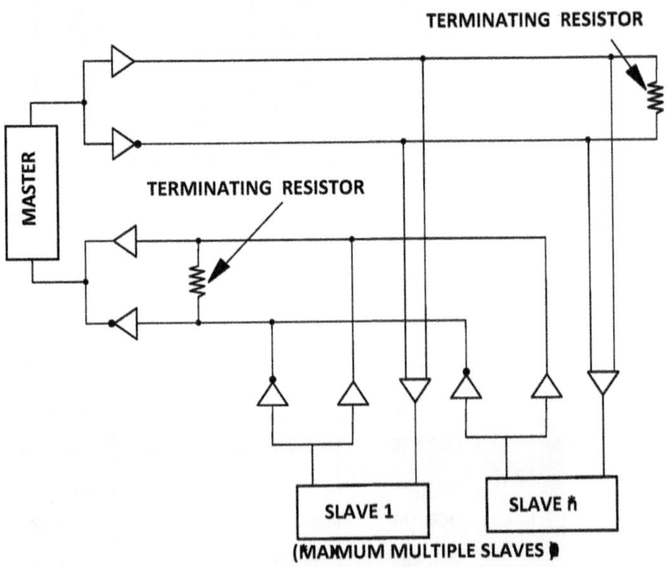


FIG III 12.1c2 RS 422 IN MULTIPLE SLAVE CONNECTION

**FIGURE III 12.1C** RS 422 interface.

the termination resistor, when used, is placed at the end of the cable near the last receiver. Reliable transmission is achieved by evaluating the differential-mode voltage between each pair of twisted wires. The governing standards for RS-422 are ITU-T V.11/DIN 66259-3 and TIA/EIA-422. A brief specification of the EIA/RS-422 interface is presented in Table III 12.1d. Data presented here are from reputed manufacturers.

5. **RS-485 interface:** RS-485 is a very important interface and is widely used in industrial communication systems, including fieldbus networks such as PROFIBUS. The following sections explore its key features, operating principles, and implementation details.
   - **Two-wire RS-485:** Like RS-422, RS-485 also uses a balanced differential transmission line. Although communication is carried out over two signal wires, a third wire is often used as a reference voltage (ground). Proper grounding is important to avoid data loss. The electrical signal levels and their logical assignments are the same as those defined for the RS-422 standard. However, unlike point-to-point communication, RS-485 supports multipoint communication and therefore requires an appropriate software protocol to address and identify individual devices on the network. At any given time, only one device is permitted to transmit, while all other devices remain in the listening mode. Typically, a two-wire RS-485 bus can extend up to 1200 m and should be terminated at both ends with termination resistors of 100–200 Ω. Individual devices may be connected to the main bus through branch lines up to 5 m in length. Because RS-485 uses twisted-pair cabling, it is relatively inexpensive and less susceptible to noise. Shielded twisted-pair data cables enable bidirectional asynchronous serial communication over long distances, for example up to 1.2 km at a data rate of approximately 100 kbps [4,6]. The highest transmission speed of RS-485 is typically 10 Mbps, although data rates of up to 35 Mbps may be achieved over short distances (e.g., 10 m) [4]. A typical RS-485 connection using a linear topology with twisted-pair cable and a ground wire is illustrated in Fig. III 12.1d. It is worth noting that RS-485 defines only the physical-layer characteristics of the communication medium. RS-485 forms the basis of communication standards such as PROFIBUS and MODBUS. Therefore, parameters such as transmission speed, data format, addressing, and data encoding are specified by these higher-level standards rather than by RS-485 itself. The governing standards for two-wire RS-485 are ISO/IEC 8482/DIN 66259-4 and TIA/EIA-485.
   - **Four-wire RS-485:** As stated earlier, this configuration supports full-duplex transmission. In the two-wire half-duplex system, the transmission and receiving paths are not separate. In contrast, in the four-wire configuration, the transmit and receive paths are completely separate and can therefore operate independently. The transmit path of a slave device has a third switching state (tristate), which places it in a high-impedance condition. Only the station that has been prompted to transmit data activates its transmitter on the bus line. The electrical signal levels and their logical assignments closely resemble those of RS-422. The maximum transmission speed and distance are similar to those of two-wire RS-485. As with the two-wire configuration, the bus cable must be terminated with resistors, and the cable should consist of shielded twisted pairs. The governing standards for four-wire RS-485 are TIA/EIA-485 and DIN 66348-2.
   - **RS-485 specification:** A brief specification of EIA/RS-485 is given in Table III 12.1e. The data presented here have been compiled from reputable manufacturers.

**TABLE III 12.1D** Short specification of EIA/RS-422.

| Specifying point | Technical data |
|---|---|
| Transmitted signal | |
| Binary 0 | +0.3 to + 6 V DC |
| Binary 1 | −0.3 to −6 V DC |
| Measuring point in relation measuring B, measuring A | |
| Negative: data line logical 1, Control line logical 0<br>Positive: data line logical 0, Control line logical 1 | UA - UB ≤ 0.3 V<br>UA- UB ≥ +0.3 V |
| Termination resistor | 100–200Ω |
| Driver open circuit voltage (max) | ±10 VVDC |
| Load @ Min ± 2 V; | 100 Ω |
| Driver output rise time (R\load 100 Ω | 10% bit width |
| Maximum output current | 150 mA |
| Differential receiver operating/withstand voltage | ±10 V/±12 V |

## 12.1.0 Network interface cards and other communication devices

This subsection is devoted to the discussion of NICs and other network communication devices used to connect not only individual devices but also entire networks. A variety

**FIGURE III 12.1D** RS-485 interface.

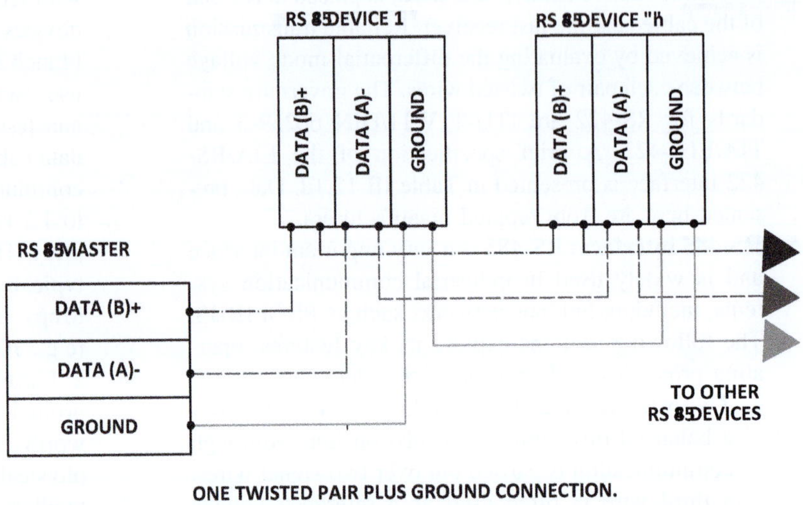


of devices, including bridges, gateways, hubs, modems, repeaters, routers, and switches, are used for this purpose. In addition, the NIC is responsible for connecting computers to a network. There are two main types of NICs: Ethernet NICs, which use 8P8C connectors for Ethernet cables, and Wi-Fi NICs, which enable direct connection to wireless networks. Several types of network bridging exist, including simple bridging, multiport bridging, and transparent bridging. Bridges may be configured for local bridging, where LAN segments are connected through local cables, or remote bridging, where the connected segments are part of a WAN. Similarly, hubs may be classified as simple hubs or multiport hubs. Routers

**TABLE III 12.1E Short specification of EIA/RS-485.**

| Specifying point | Technical data |
|---|---|
| Mode of operation | Differential |
| Allowed Tx Rx | 32 Tx & RX |
| Maximum cable length (twisted pair) | 1200m (~4000 ft) |
| Open circuit O/P voltage | +1.5 to 6VDC |
| Driver O/P voltage range | ±1.5 to 5VDC |
| Min/Max (SC) driving capability | 55/−250 mA (SC)/OP |
| Tx load impedance | 54Ω |
| Receiver sensitivity (–) 7 < V < 12 | +200 mV |
| Rx input voltage range | (–)7 to +12V |
| Rx input resistance (max) | 12 kΩ |
| Rx logic hi/Lo | >200 mV/<200 mV |
| Max common mode voltage | (–)7 to +12 V |

SC: Short Circuit.

provide connectivity between LANs and WANs and may be configured as either static or dynamic routers. Static routers retain fixed routing information until manually modified, whereas dynamic routers maintain and update their own routing tables automatically. Gateways are generally deployed at the network layer of the Open Systems Interconnection (OSI) model, which is discussed later. They are also used in various security applications, including firewall and proxy-server functions for scanning and filtering network traffic. In wired communication networks, repeaters are used to extend signal propagation distances. In wireless networks, repeaters are employed to increase the coverage area of a cell. All of these network devices are briefly discussed in this subsection. The discussion begins with the NIC. Figure III 16.0d may be referenced for connection details of common Ethernet network elements.

1. Network Interface Cards (NICs): All computers and computerized devices (nodes) are connected to a network through a NIC. In most computers, the NIC is integrated into the system and connected to the motherboard, although external NICs are also available, particularly for desktop computers. The NIC acts as a transceiver and uses the standard OSI model to support network communication. A NIC enables both wired and wireless communication within a LAN or larger networks such as the Internet through IP-based communication. It provides the necessary hardware for the physical layer and supports data-link-layer functions. There are two primary types of NICs. An Ethernet NIC provides wired communication, with one end connected to an Ethernet cable and the other connected to a modem, switch, or router. A wireless NIC includes an integrated antenna and communicates wirelessly

through a router using appropriate network protocols. Fiber Distributed Data Interface (FDDI) is a network technology that uses optical fiber for communication. FDDI converts data into digital pulses for transmission over optical fiber and offers the advantage of long-distance communication. A NIC typically includes memory for temporary data storage, connectors for physical-layer connections, processing circuitry for converting data into appropriate signal formats, and interfaces to standard system buses and connectors. In wireless applications, communication is achieved through routers and wireless networking protocols. Each NIC is assigned a unique MAC address, as discussed earlier.

2. **Bridges:** A bridge is used to separate a network into sections each with a collision domain and with separate bandwidth. It is used to reduce waiting time (for collision due to traffic) and improve network performance as well as increase network distance. A bridge operates at the data-link layer and examines incoming traffic. Based on the destination MAC address, it either forwards or blocks the data packet. As stated earlier, a bridge separates a LAN into different segments that are typically owned and managed by the same organization. The MAC addresses of connected devices are stored in a table. After building this table, the bridge listens to the data packets traveling on each segment and examines them before forwarding. Through a process known as filtering, the bridge ignores packets when both the source and destination devices belong to the same segment. However, if the source and destination belong to different segments, the bridge forwards the packet to the segment containing the destination device. A bridge connects network segments that use the same communication protocol. Different types of bridges include transparent bridges, translation bridges, and source-route bridges. Wireless bridges can also connect wired and wireless networks. A wireless Ethernet bridge is a device that enables non-wireless equipment to communicate over a Wi-Fi network.
3. **Gateways and modems:** Let us now examine gateways and modems, which are widely used in industrial communication networks.
   - Gateway: A gateway serves as the entry and exit point of a network and connects networks that use different protocols or networking models. It acts as a messenger or translator, receiving data from one system, interpreting it, and transferring it to another system in a format that can be understood. Gateways are similar to routers but generally provide additional functionality and are therefore more complex and costly. Typically, a gateway operates at layers 4–7 of the OSI model. A gateway may be implemented as a computer with software capable of converting data between two different computing environments. As a protocol converter, it can translate data from one proprietary LAN technology to another, for example, from Ethernet to Token Ring. Because of the additional processing required for protocol conversion, gateways are generally slower than routers and bridges.
   - Modem: A modem (modulator–demodulator) is a hardware device used to connect a computer or other network device, such as a router or switch, to the Internet. A modem may be installed in a Peripheral Component Interconnect (PCI) slot on the motherboard or provided as an external device. The basic function of a modem is to modulate digital data from a computer into analog signals suitable for transmission over telephone or cable lines and, conversely, to demodulate incoming analog signals into digital data that can be processed by the computer. The earliest type of modem was the dial-up modem, which is now obsolete. Modern modems typically include DSL modems, which operate over telephone lines using a wider frequency range, and cable modems, which communicate over cable television networks using coaxial cables.
4. **Repeaters and routers:** Both repeaters and routers - play critical but different roles in networking.
   - **Repeater:** A repeater operates at the physical layer to regenerate the signal over the same network (before the signal is too weak or gets corrupted) to extend the area of coverage of the network. As two-port devices, repeaters do not amplify signals; rather, they copy the signal bit by bit and regenerate it when the signal becomes weak. Depending on the type of signal being transmitted, repeaters may be either analog or digital. Repeaters can also be classified as wired or wireless devices, with wireless repeaters commonly used in wireless LANs and cellular networks. Based on their application, repeaters may be categorized as local repeaters, used within a limited LAN area, or remote repeaters, used to extend communication over longer distances.
   - **Router:** A router is a network-layer device that routes data packets based on their IP addresses. Routers provide connectivity between different IP networks and are responsible for forwarding IP packets according to their destination IP addresses. A router typically provides multiple Ethernet NICs, referred to as interfaces. Each interface must be assigned an IP address belonging to the network to which it is connected. Routers generally maintain dynamically updated routing tables, which they use to make forwarding decisions. Based on the

available routes, a router can determine the most appropriate path for transmitting data packets. It is the role of a router to interconnect multiple networks, including LANs and WANs. Therefore, routers must be capable of supporting multiple network-access protocols. Routers offer advantages such as improved security, reliability, and network performance.

5. **Hub and switch:** Both of them play crucial roles in connecting network nodes:
   - Hub: Network hardware is used to establish connections among multiple devices. A hub broadcasts incoming data to all devices connected to the network, even though all devices may not require the transmitted information. As a result, hubs consume more bandwidth and are now largely obsolete, having been replaced by switches and, in some applications, repeaters. As discussed earlier, there are two types of hubs. Active hubs have their own power supply and can regenerate, clean, boost, and retransmit signals, functioning in a manner similar to repeaters. Passive hubs, on the other hand, do not have their own power source and simply relay incoming signals without amplification or regeneration.
   - Switch: A switch (or Ethernet switch) provides connectivity points within a network using Ethernet cables, typically supporting cable lengths of up to approximately 100 m. It enables multiple devices to be connected in a star topology. The number of connected devices is limited by the number of switch ports, although this can be expanded by interconnecting additional switches to form an extended-star topology. A switch provides multiple Ethernet NIC interfaces, referred to as ports, through which it receives data. Using packet-switching techniques, the switch determines the destination device and forwards the data only to the appropriate destination port. Most Ethernet switches support multiple Ethernet standards on each port. Although individual ports can be configured to operate at specific data rates, modern switches typically negotiate automatically with connected devices and operate at the highest mutually supported data rate. Switches forward data based on destination MAC addresses and do not normally examine the IP addresses contained within packets. Therefore, switches operate primarily at the data-link layer and cannot be used to interconnect different IP networks. Such interconnection requires a router or gateway.

Table III 12.1f compares modems, routers, and switches.

Next, we discuss wireless network devices. For detailed discussions on wireless communication refer chapter 10 of this volume.

**TABLE III 12.1F** Comparison of modem router and switch.

| Modem | Router | Switch |
|---|---|---|
| Access the internet to connect computer | Access internet without modem | Not applicable |
| At data link layer. | At the network layer data link layer, and physical layer. | At layer data Link layer |
| Do not examine the data packet, so security threat | Router examines data packets hence secured | Not applicable |
| It is between a telephone/cable line and a router/computer. | It is between a computer network and modem. | It is between a computer network and computer |
| NA | Services like NAT, NetFlow and QoS available | No such service |
| NA | Store IP address in the routing table and an address on its own. | Store MAC address in a lookup table and maintain an address on its own. |
| NA | Networking device 2/4/8 ports | A switch is a multiport bridge. 24/48 ports. |
| At highest speed it can be 57000bps | The speed limit is 1-10 Mbps for wireless and 100 Mbps for wired connection. | The speed limit for the switch is 10/100 Mbps. |
| Not applicable | Faster routing decision | Complicated routing decision |
| WAN to computer/LAN connection | In various types of network environments (MAN/WAN), the router works faster compares to Switch | In a LAN environment, a switch is faster than router. |
| Both wired/wireless | Wired/Wireless | Wired |

NAT: Network address translation.

### 12.1.1 Wireless LAN & PAN devices

Advanced wireless networks can support roaming users who require continuous, uninterrupted connections for real-time applications. This capability makes remote operation of industrial plants possible. Easy setup, improved productivity, scalability, ubiquitous access, increased mobility, remote operation, cost savings, and enhanced security are among the major advantages of wireless networks. Two solutions that have historically led the industry are HomeRF and Wi-Fi. Consequently, Personal Area Networks (PANs) and their associated devices play an important role. A Wireless Network Interface Card (WNIC) connects to wireless networks, such as Bluetooth or Wi-Fi, using an antenna to communicate through radio waves. In large wireless networks, particularly in corporate environments with tens or even hundreds of access points, Wireless LAN (WLAN) switches are used for centralized management. Lower cost, simplified access-point management, and improved roaming performance are among the advantages provided by WLAN switches. An Access Point (AP), or Wireless Access Point (WAP), is a central device in a WLAN. By connecting wired and wireless networks, it functions as a bridge. In addition, it may provide functions such as Internet gateway services, switching, hub functionality, and network storage. Bluetooth is an industry specification for short-range RF-based connectivity among portable personal devices, with its functional specification first released in 1999. Bluetooth operates at 2.45 GHz within the Industrial, Scientific, and Medical (ISM) band. ZigBee is a standard based on low-power wireless technology that is widely used for remote monitoring, control, and sensor networks. ZigBee offers ultra-low power consumption but supports relatively low data rates over limited distances. Wireless USB refers to a standard for high-bandwidth, short-range wireless device connectivity. In addition, wireless USB adapters may use Wi-Fi to provide Internet access through a computer's USB port. Wireless USB devices share bandwidth using a host-scheduled Time Division Multiple Access (TDMA)-based media-access protocol. Wireless USB Wi-Fi adapters are commonly used to add or restore Wi-Fi functionality to devices, enabling wireless network connectivity and Internet access. This subsection provides brief discussions of various devices used in wireless LANs and PANs. Wireless USB technology is also briefly covered. To complete the discussion presented in Chapter 10 of this volume, Wireless NICs, APs/WAPs, WLAN switches, Radio-Frequency Identification (RFID), Bluetooth, Wireless USB, ZigBee, and related technologies are discussed.

1. **Network LAN Devices:** The following devices are similar to the wired LAN hardware discussed earlier, with a few differences. The main wireless LAN devices include:
    - **Wireless Network Interface Cards (WNICs):** These are used to communicate with other stations or an access point in a peer-to-peer network, enabling devices such as Personal Digital Assistants (PDAs) and computers to act as wireless stations. WNICs consist of a small antenna integrated onto the card and communicate using routers and various network protocols. They are also available as external USB devices and USB dongles. Some WNICs are Wi-Fi certified for IEEE 802.11 standards. WNICs are also available with external antenna connections to attach high-gain antennas when operating near the limits of wireless range.
    - **Access Points (APs):** These are the central devices in a WLAN, providing the hub for wireless communication among stations in the network. When connected to a wired network, an AP acts as a bridge between wired and wireless devices. Major functions may include serving as an Internet gateway (providing DHCP and NAT functions), a switching hub, a bridge, and a network storage server. APs may also support special features such as enhanced security and Simple Network Management Protocol (SNMP).
    - **Wireless LAN Switches (Controllers):** A wireless LAN switch, often called a controller, is a networking device designed to handle a variety of functions on behalf of a number of dependents connected to it.
2. **Bluetooth (IEEE802.15.1):** Bluetooth technology uses radio waves. Bluetooth, a PAN standard (IEEE 802.15.1), operates in the 2.4-GHz ISM band using a frequency-hopping Time Division Duplex (TDD) method for each radio channel. Bluetooth 2.0 can offer data rates of up to 3 Mbps (effective data rates of 721/56 kbps for asymmetric communication or 432 kbps for full-duplex communication) and ranges of up to 100 m, with relatively low power consumption. Low power consumption and guaranteed interoperability have made Bluetooth widely accepted in the mobile-phone community. There are three classes of Bluetooth devices: Class 1 (100 mW for up to 91 m or 300 ft), Class 2 (2.5 mW for up to 9.2 m or 30 ft), and Class 3 (1 mW for less than 1 m or 3 ft). Radio, baseband (positioned above the PHY layer), and link manager (used to manage baseband connections) are the three major layers of Bluetooth. Together, these form the transport layer, which is roughly equivalent to the OSI PHY and MAC layers. To achieve a high degree of interoperability, a middleware layer is built on top of the transport layer, with the application layer residing above it. The Bluetooth pairing process is used to establish authentication and encryption between two devices as a security measure. During pairing, the slave device receives a

frequency-hop synchronization data packet based on the master device's 48-bit MAC address in order to follow the frequency-hopping pattern [7]. The pairing process uses a series of keys before a connection is established. A Bluetooth PAN can support multiple devices, forming either a piconet or a scatternet, as illustrated in Fig. III 12.1e. The master in a piconet communicates at a data rate of around 720 kbps. Bluetooth supports a wide variety of devices, including mobile phones, PDAs, laptops, and desktop computers. Major Bluetooth application profiles include personal area networking, headset profile, synchronization profile, file-transfer profile, basic printing profile, and dial-up networking profile. Robustness, ease of use, reliability, low power consumption, fast connection establishment, ease of configuration, and seamless coexistence with Wireless LANs make Bluetooth an attractive solution for many industrial applications. Classic Bluetooth is used for Human–Machine Interfaces (HMIs), programming, service and maintenance, and real-time control tasks in industrial environments such as automotive manufacturing, energy plants, and warehouses. Bluetooth Low Energy (BLE) is increasingly used in metering, sensors, actuators, and other small devices that require interconnection [8].

3. **Radio-Frequency Identification (RFID):** RFID is a technology by which digital data encoded in RFID tags or smart labels are captured by a reader through radio waves. It is a part of Automatic Identification and Data Capture (AIDC) technology, which can collect data and pass it to a computer without human intervention. An RFID system is configured through reader/writers and host devices. RFID tags are rewritable, and the system writes data sent from the host device to the RF tag via the reader/writer. RFID systems consist of three basic components: an RFID tag, an RFID reader, and an antenna (in the form of an integrated circuit). The antenna transmits data to the RFID reader. The system uses tags and a two-way radio transmitter-receiver (reader), which sends a signal to the tag and then reads its response. The reader converts the received radio waves into usable data for the host computer to process and/or store for future analysis. RFID uses a wide range of RF frequencies and is categorized as low frequency (30–300 kHz), medium frequency (300–3000 kHz), high frequency (3–30 MHz), very high frequency (30–300 MHz), and ultra-high frequency (300–3000 MHz). Electromagnetic coupling, magnetic induction (near-field communication), and radio waves for UHF are the technologies used in RFID systems, each having its own advantages and disadvantages that must be considered during system selection. Memory type and power-supply method also classify RFID systems into different categories. Major industrial applications of RFID include inventory management, asset tracking, personnel tracking, access control to restricted areas, identification, and counterfeit prevention. Increased adoption of Wi-Fi has led to RFID-over-Wi-Fi applications, opening new areas such as the management of assets, people, vehicles, computer equipment, mobile phones, airport baggage, and even pets.
4. **ZigBee (IEEE 802.15.4):** The ZigBee 3.0 protocol is designed to communicate data in the noisy RF environments commonly found in commercial and industrial applications. Version 3.0 is a unified version that supports a wider range of devices. Like the Bluetooth master, every ZigBee network has a unique Full-Function Personal Area Network (PAN) Coordinator for network management. ZigBee networks require very little power for operation. In star and mesh topologies, devices normally communicate through the coordinator,

**FIGURE III 12.1E** Piconet and scatternet.

*Piconet:* As referenced in Bluetooth description, up to eight devices can b con ected in a n twork using Bluetooth technology. Such n twork is cā led Piconet. Here one Bluetooth device will be master and others will act as slaves i.e. seven sl es maim um. When two or more devices detect eah other and communicte through Bluetooth then Pi coet is formed.

*Scatternet:* When two or more piconets e connected by through a bridge node then a Scatternet is formed. In addition to seven slaves there can b pe d nodes o.

whereas in peer-to-peer mode, Full-Function Devices (FFDs) can communicate directly with each other. There are four distinct layers in the ZigBee protocol: PHY, data-link, network, and application-support layers, on top of which the ZigBee application framework resides. Thus, ZigBee consists of IEEE 802.15.4 PHY and MAC layers arranged one above the other. Above the MAC layer are the LLC 802 and LLC 802.15.4 layers, which interface with the network layer. Finally, the application-support layer, comprising functions such as security management and message formatting, supports ZigBee devices and related applications. The ZigBee application framework resides at the top. The ZigBee standard operates on the IEEE 802.15.4 specification and uses unlicensed frequency bands, including 2.4 GHz, 900 MHz, and 868 MHz. ZigBee supports three classes of devices: Full-Function Devices (FFDs), which support all IEEE 802.15.4 specifications; Network Coordinators, which are more complex devices with higher memory requirements; and Reduced-Function Devices (RFDs), which provide limited functionality. Unique features of ZigBee include low latency, low power consumption, support for up to 65,000 nodes in a network, 128-bit AES encryption, and support for multiple network topologies. It uses Direct Sequence Spread Spectrum (DSSS) and collision-avoidance techniques. This low-cost, low-power PAN technology finds extensive use in wireless Machine-to-Machine (M2M), Internet of Things (IoT), and Industrial Internet of Things (IIoT) networks. It is also widely used in smart-home applications, as well as in healthcare, asset management, industrial process control, and energy management.

5. **Wireless USB:** With Ultra-Wideband (UWB) radio technology, Wireless USB can deliver data at 480 Mbps within a range of 10 m while maintaining low power consumption. Ultra-Wideband technology is used for short-distance radio communication. Wireless USB devices share bandwidth through a host-scheduled TDMA-based media-access protocol. It supports a hub-and-spoke logical topology, with each host supporting up to 127 devices. Like wired USB, devices can be connected and disconnected at any time. Wireless USB also offers a "turn on and use" capability similar to the plug-and-play feature of wired USB. There are three types of Wireless USB devices: self-beaconing, directed-beaconing, and nonbeaconing devices.

In addition to the above, there are two other types of PAN technologies: IrDA, a low-cost, low-power, half-duplex, serial point-to-point communication standard that supports a range of at least 1 m with data rates of up to 115 kbps; and Near-Field Communication (NFC), which provides ultra-short-range wireless communication using magnetic-field induction to enable connectivity between devices that are in physical contact or separated by only a few centimeters.

## 12.2 Transmission medium and industrial network hardware

The physical medium of connectivity specifies the physical and electrical characteristics of network connections. Such media may be either guided or unguided. Bandwidth is an important consideration in data transmission. No transmission medium can transmit a signal without some loss of power. The range of frequencies that can be transmitted without significant attenuation is called the bandwidth. Bandwidth is a physical property of the transmission medium. However, the term bandwidth conveys different meanings in different contexts. For analog signals, it refers to a frequency range measured in hertz (Hz). In computer networks, it refers to the maximum data rate that a channel can support, measured in bits per second (bps) [9]. Bandwidth in guided media mainly depends on construction thickness, length of media, etc. For unguided media, factors such as signal interference, physical obstructions, distance from the router or access point, the number of connected devices, and network congestion are responsible for determining the available bandwidth. For a channel with bandwidth (B), according to Nyquist theory, the data rate is given by

$$\text{Max Data rate} = 2B * \log_2 D \text{bit/sec} \quad (D\text{: } \textit{discrete levels} \text{ in signal}) \qquad \text{(III 12.2a)}$$

With introduction of noise, the situation deteriorates. If $S$ is the signal and $N$ is the noise, then signal to noise ratio in $dB$ is given by

$$S/N = 10 \log_{10} * (S/N) \qquad \text{(III 12.2b)}$$

Combining the two equations, for $S/N$ in $dB$, according to Shannon theory we get

$$\text{Maximum bit rate (bit/sec)} = B * \log 2\,(1 + S/N) \qquad \text{(III 12.2c)}$$

So, this provides an estimate of the maximum data rate that can be achieved in a noisy channel.

As the name suggests, the physical layer in the OSI network model deals with the physical aspects of the communication system, such as the transmission medium, connectors, and signal encoding. Thus, the physical layer comprises the physical hardware and does not concern itself with routing information, addresses, or other message-packet contents, as these functions are handled by the upper layers of the OSI model. The physical layer therefore serves as the foundation for the remaining layers and provides the framework for data-signal conversion and transmission. It exchanges data (bits) over physical media, such as guided media (e.g., copper cables and fiber-optic cables) or unguided media (e.g., wireless radio-frequency signals and other communication channels). The functional characteristics of the physical layer in the OSI network model can be categorized into several areas. The standard attributes of the physical layer are:

- **Data signal conversion** (e.g., analog-to-digital conversion)
- **Data transmission:** For data communication, a physical medium is utilized, whether guided or unguided, with different modes of transmission, such as half-duplex and full-duplex, being employed as required.
- **Data modulation:** Mapping of digital data onto the carrier signal for transmission.

Therefore, as the physical layer and transmission medium form the foundation of all higher layers, they are responsible for providing a number of essential services, including physical-signal processing, bit-rate control, maintenance of data-flow rates, notification of physical-layer limitations, bit synchronization, direction of data transfer, determination of network topology, provision of interdevice interfaces, data modulation, support for error correction, and connection of computer systems to external networks [10]. In this section we discuss both guided and unguided media used in plant automation networks, along with various industrial grade hardware. Fig. III 12.2a shows the physical media for data transmission. Fig. III 12.2a1 details various types of physical media used for data transmission. The two main categories of transmission media are guided media and unguided media. For industrial-grade network hardware, major issues that must be addressed include environmental resilience and resistance, characteristic features, robustness, ruggedness, durability, reliability, scalability, and performance.

## 12.2.0 Guided media

Various physical media can be used for data transmission, each having its own advantages in terms of bandwidth, delay, cost, ease of installation, and maintenance. Transmission media are broadly classified into guided media and unguided media. Guided media provide a physical conduit from one device to another. Signals traveling through these media are directed and confined by the physical boundaries of the medium. Thus, signals are transmitted through a defined pathway, offering high speed, enhanced security, and reliable communication over comparatively shorter distances. Guided media are mainly categorized into metallic (copper) conductors and optical fiber. Metallic conductors, such as twisted-pair and coaxial cables, carry signals in the form of electrical current, whereas optical fiber transmits signals in the form of light. Guided media are often referred to as wired communication media and form the backbone of both LANs and WANs.

Major advantages of guided media include:

- Fiber optics, as a guided transmission medium, offer high stability with minimum signal attenuation and electromagnetic interference.
- Fiber-optic cables offer high bandwidth, enabling faster data-transfer rates and more efficient network performance.
- Guided media are less susceptible to external interference and therefore provide better security, signal integrity, and overall transmission performance.
- Guided media provide consistent performance with manageable signal behavior, making them ideal for high-performance networking.

There are also certain shortcomings of guided media, including, but not limited to, physical connectivity that limits mobility in networks, susceptibility to accidental and environmental damage, higher installation costs and hardware requirements, significant infrastructure investments in terms of both effort and cost, and disposal concerns. Twisted-pair cable is one of the oldest and most widely used transmission media. It consists of two insulated copper conductors twisted together in a helical form to minimize radiation and electromagnetic interference. Twisted-pair cables may be either Shielded Twisted Pair (STP) or Unshielded Twisted Pair (UTP). Another common transmission medium is the coaxial cable, which provides better shielding and greater bandwidth than UTP. Fiber optics are widely used for long-haul transmission and high-speed LANs. The transmission medium consists of ultra-thin glass fibers that carry information in the form of light pulses. A detector generates an electrical signal when light falls on it. By placing a light source at one end of an optical fiber and a detector at the other, a unidirectional transmission system is created that converts electrical signals into light pulses for transmission and then reconverts them into electrical signals at the receiving end. The protective coating of optical fiber shields it from heat, cold, and electromagnetic interference. Fiber-optic cables transmit data at much higher rates than copper cables. Currently, two modes of light propagation are supported: multimode and single-mode transmission, each requiring optical fibers with different physical characteristics. This subsection presents a discussion of guided media.

1. **Twisted-pair cable (guided metallic cable):** The least expensive and most commonly used transmission medium is twisted-pair copper wire (e.g., used in telephony for many years). The insulated copper wires are twisted together in pairs, although multipair configurations are also available. To minimize noise and crosstalk, the copper conductors are twisted together;

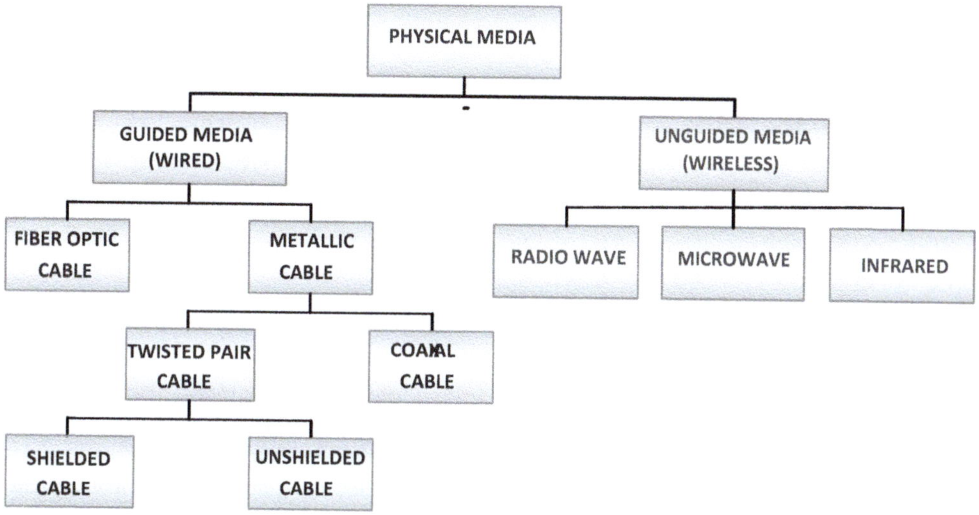


FIG III 12.2a1 TYPES OF PHYSICAL MEDIA

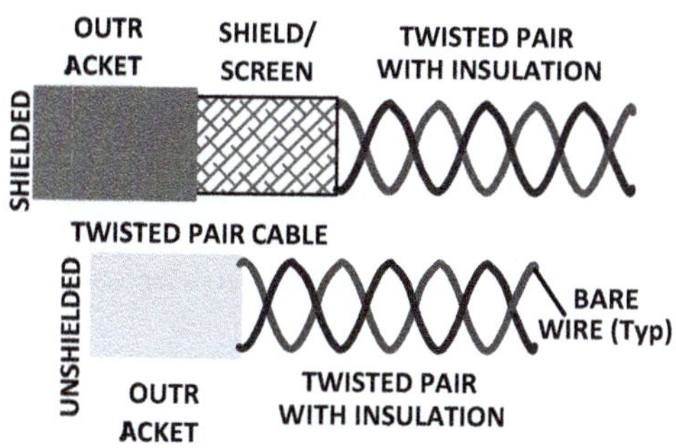


FIG III 12.2a2 TWISTED PAIR CABLE TYPES

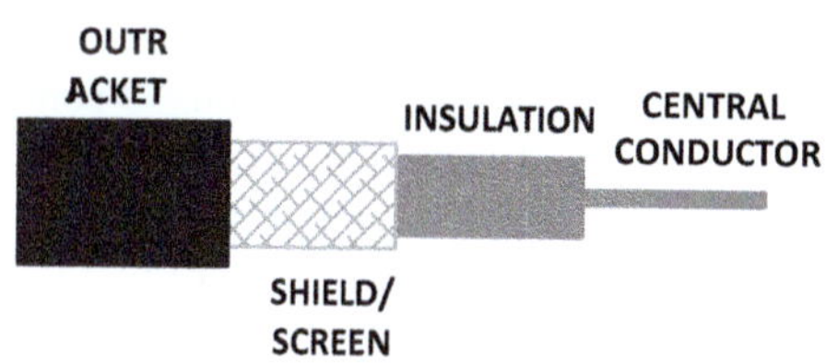


FIG III 12.2a3 COAXIAL CABLE

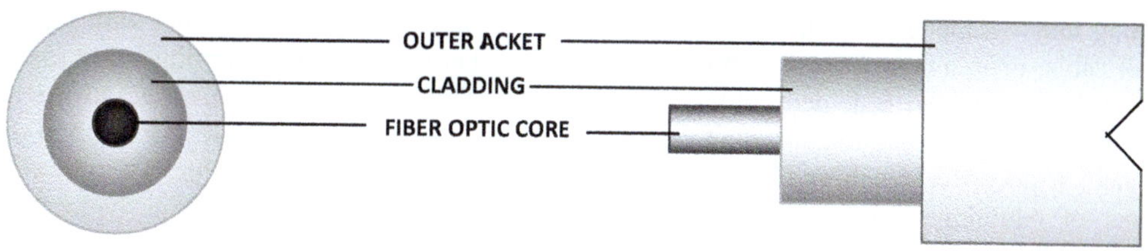


FIG III 12.2a4 FIBER OPTIC CABLE

**FIGURE III 12.2A** Physical media for data transmission.

hence, they are referred to as twisted-pair cables. As the insulated wires are twisted at 180° out of phase as shown in Fig. III 12.2a2, any noise or electrical interference induced on both wires tends to cancel itself out at the receiving end, thereby reducing the effects of noise and improving signal integrity. Twisted-pair cables are most effectively used in a system that uses a balanced line method of transmission. Normally twisted-pair cable consists of four insulated twisted copper pairs within a protective outer jacket. Twisted-pair cables may be either Shielded Twisted Pair (STP) or Unshielded Twisted Pair (UTP), as shown in Fig. III 12.2a2. In STP a twisted pair is confined in a foil or mesh shield that guards the cable against electromagnetic interference (with a drain wire). Naturally STPS offer a few advantages over UTP such as high data rate with less noise and cross-talk. Depending on the distance between source and destination and wire thickness, a data rate of 100 Mbps can be achieved through twisted-pair cables. However, STP is moderately expensive, requires grounding, and is comparatively more complex to install. There is another category of cable, FTP, which uses an overall foil shield (F) beneath the main cable jacket of a UTP cable. It is very similar to UTP but is less flexible because of the foil shielding. F/UTP is commonly used in

**TABLE III 12.2A** Popular UTP category copper cables.

| Category | Data rate max | Max. length | Application |
|---|---|---|---|
| Cat 3 | 10 Mbps | <100 m | Token ring, 10 BaseT ethernet |
| Cat 5 | 100 Mbps | 100 m | Token ring ether net fast ethernet |
| Cat 5e | 1 Gbps | 100 M | Ethernet, fast & gigabit ethernet |
| Cat 7 | 10 Gbps | 100 m | Gigabit ethernet, 10G ethernet (<100 m) |

10GBASE-T Ethernet applications. Category 3 (Cat 3), Category 5 (Cat 5), Category 5e (Cat 5e), and Category 7 (Cat 7) are twisted-pair Ethernet copper-cable categories defined by the Electronics Industries Alliance (EIA) and the Telecommunications Industry Association (TIA). Table III 12.2a compares UTP cables. Correct wiring of UTP cables is very important and is accomplished using the standard RJ-45 plug. A Registered Jack (RJ) is a standardized physical network interface used for telecommunications and data-communication equipment. The most common twisted-pair connector is the 8-position, 8-contact (8P8C) modular plug and jack, commonly referred to as RJ-45.

2. **STP & UTP types:** As discussed, there are two types of twisted pair cables:
   - **STP:** When twisted-pair cables are provided with a shield (or screen) to improve immunity to noise and interference, they are referred to as Shielded Twisted Pair (STP) cables. In multipair cables, shielding may be applied to individual pairs, to the overall cable, or to a combination of both. Refer to Volume II for details. Shielded cables may use either foil shielding or braided shielding.
     - Foiled: A polyester sheathing bonded with a thin sheet of Aluminum with drain copper wire is used in this cable. The cost of installation of foil-shielded cables is lower than for braided cables. These are used in places with high noise and cost is a factor.
     - Braided shield: Braided shielded cables consist of woven copper wires that provide high durability and excellent resistance to harsh environmental conditions, making them suitable for use in demanding industrial environments. The cost of braided shielded cables and their installation is high. These types of cables are mainly used in large production centers.
   - **UTP:** Cables without shielding are called Unshielded Twisted Pair (UTP) cables and are commonly used for computer networks within buildings and Local Area Networks (LANs). For Ethernet cabling, UTP cables typically contain four twisted pairs of conductors.
   - **Comparison of STP & UTP:** The additional metallic shielding in STP cables provides superior protection against electromagnetic interference (EMI), crosstalk, and external noise. However, the lower cost and greater flexibility of UTP cables give them an advantage in certain applications. A brief comparison between the two is given in Table III 12.2b.
3. **Coaxial cable (guided metallic cable):** As the name suggests in coaxial cable there are two conductors coaxially (concentric) placed as shown in Fig. III 12.2a3. The inner insulated conductor is weaved around by *metallic noise shield,* which also acts as a second conductor. The total assembly is wrapped in an insulating protective coating, called a jacket, which covers the outer conductor. The second (outer) conductor, or outer shield, protects the inner conductor from external electrical signals. The distance between the inner and outer conductors, together with the insulating material between them, determines the cable characteristics. This specialized insulation and shielding construction enables coaxial cables to support higher data rates than twisted-pair cables. The power rating of coaxial cable used for Ethernet is typically 50 Ω. There are

**TABLE III 12.2B** Comparison of STP versus UTP.

| Feature | STP | UTP |
|---|---|---|
| Shielding | Extra foil or braided shielding | No shielding |
| Protection | High EMI & interference | Low |
| Size | Higher in diameter | Lower in diameter |
| Cost | Higher | Lower |
| Flexibility | Lower | Higher |
| Application | High-speed/noisy environments | Standard LAN |

two types of coaxial cabling: ThinNet (flexible, with a maximum distance of about 200 m; RG-58 type) and ThickNet (less flexible, with distances up to 500 m; RG-11 type). 10BASE2 and 10BASE5 are Ethernet standards defined for coaxial cable. Baseband coaxial cable, which is lighter and easier to bend, was commonly used in LANs, whereas broadband coaxial cable, which is thicker and heavier, is commonly used in cable television systems.

4. **Fiber-optic cable (guided media):** Fiber optic cable uses total internal reflection of light through a core made up of glass or plastic. The core of fiber-optic cable is a thin glass fiber, flexible medium that conducts light pulses to represent a bit. In a fiber-optic cable, there is a very narrow strand of glass called the core, typically having a diameter of 62.5 μm. Surrounding the core is another layer of glass, called the cladding, which typically has a diameter of 125 μm. The cladding covers the fiber-optic core and helps confine the light signals within the core during transmission, as shown in Fig. III 12.2a4. Other parts include silicone coating, buffer jacket, strengthening material, and outer jacket. Fiber-optic cables are used for the transmission of large volumes of data with negligible crosstalk, electromagnetic interference (EMI), and attenuation over distances of up to 100 km. Compared with twisted-pair and coaxial cables, fiber-optic cables offer higher transmission speeds, greater bandwidth, and longer transmission distances. Because data are transmitted using light, fiber-optic cables provide very high bandwidth. Their small size and light weight allow a large number of optical fibers to be accommodated within a cable of relatively small diameter. Fiber-optic cables are widely available; however, although they are the most secure guided transmission medium, they are relatively fragile and can be more difficult to install and maintain. Retrofitting is also challenging. In practice, the additional optical components required for fiber-optic communication make the overall system more expensive than other guided media. Three types of fiber-optic cables are commonly used: single-mode fiber, multimode fiber, and Plastic Optical Fiber (POF). Light sources may be LEDs or lasers. There are two principal modes of operation for optical fibers: single-mode and multimode. Single-mode fiber provides higher transmission rates and longer transmission distances than multimode fiber, but at a higher cost.
5. **Guided media:** The various guided media discussed so far are compared in Table III 12.2c.

### 12.2.1 Unguided media

As the name suggests, unguided media transport electromagnetic waves without using a physical conductor. Unguided/unbounded/wireless transmission media transfer electromagnetic waves without using a physical medium or conductor. Unguided transmission media are used to transmit data over the air. Unguided media refer to wireless communication systems that use signals such as radio waves, microwaves, and infrared waves, offering flexibility and mobility in communication. The efficiency of wireless communication can be improved through techniques such as modulation, coding, multiple access, and smart-antenna systems. However, wireless communication can suffer from interference and limited bandwidth. The use of wireless communication is increasing because of the growing popularity of mobile devices, the Internet of Things (IoT), and cloud computing. Unguided media are mainly classified into three types: radio waves, microwaves, and infrared waves. Microwave communication can be further divided into two categories: terrestrial microwave and satellite microwave communication. Unguided media are characterized by:

- Air is the medium for electromagnetic energy to flow easily without any hindrance or intrusion.

**TABLE III 12.2C Comparison of various guided media.**

| Feature | Twisted cable | Co axial cable | Fiber optics |
|---|---|---|---|
| Signal | Signal transmission in electrical form in both wires | Signal transmission in electrical form in inner wire | Signal transmission in optical form through glass fiber |
| Noise | Lesser immunity | Higher immunity | No electrical noise |
| EMI | More affected | Less affected | Unaffected |
| Short circuit | Possible | Possible | Not applicable |
| Cost | Cheapest | Moderate | High |
| Installation | Easy low cost | Easy higher cost | Difficult costly |

- Propagation types of unguided signals include sky, ground, and line-of-sight.
- The electromagnetic spectrum ranges for wireless communication: 3 kHz to 900 THz.
- The unguided media makes use of an antenna for transmitting and receiving wave.

Radio-frequency (RF) waves are omnidirectional, easy to generate, and capable of traveling long distances while penetrating buildings relatively easily. Therefore, they are widely used for both indoor and outdoor communication. The properties of radio waves are frequency dependent. At lower frequencies, radio waves can pass through obstacles effectively; however, signal power decreases sharply with distance. This attenuation is known as path loss. RF communication can support higher transmission rates and is used in WAN and multicasting applications. Microwave transmission is also based on electromagnetic waves, typically operating in the frequency range of approximately 1–300 GHz. Microwaves are directional and propagate along a line-of-sight path, reducing the possibility of interference. When the transmitting and receiving antennas are properly aligned, communication between the two endpoints becomes easier to establish. Terrestrial microwave systems use high-frequency radio waves and require a clear line of sight between antennas, while providing high-speed communication over long distances. However, their performance can be affected by weather conditions such as rain, snow, and fog. Satellite microwave communication, on the other hand, offers wide-area coverage, making it suitable for global communication and remote locations where other communication infrastructure may not be available. In addition, the directional nature of microwave transmission allows multiple transmitters and receivers to operate with minimal interference. Unlike lower-frequency RF waves, microwaves do not penetrate buildings effectively. Infrared waves, operating in the frequency range of approximately 300 GHz to 400 THz, are used for short-range communication. Because of their high frequency, infrared waves cannot penetrate walls. This characteristic helps prevent interference between adjacent systems. Infrared systems are relatively directional, inexpensive, and easy to implement; however, they have the major limitation that they cannot pass through solid objects.

In this subsection we discuss unguided transmission media in industrial network communication.

1. **Unguided physical medium characteristics**: Unguided or wireless type physical media carry signals as electromagnetic waves. Wireless transmission types are attractive as they do not need any physical wire, can penetrate walls, and are suitable for use for long distance communication. This type of wire can also be used in mobile communication. However, such communication is highly dependent on the propagation environment and the distance over which a signal is to be transmitted. The propagation environment determines the path loss and fading characteristics of the signal. Depending on the communication range, different types of radio communication may be employed. Local radio networks can support communication over a few hundred meters and are therefore suitable for Local Area Networks (LANs). However, Wide Area Networks (WANs) and mobile wide-area communication require longer-range radio communication systems. From the above discussion, it is evident that unguided media are characterized by potential signal attenuation and interference, as well as limited bandwidth compared with guided media. Thus, the major characteristics of unguided media include the use of electromagnetic waves for transmission through air, mobility and flexibility, low installation cost, signal attenuation with distance, susceptibility to interference, and lower security with a greater possibility of interruption.
2. **Radio waves (unguided media):** These signals operate in the frequency range of 3 kHz to 1 GHz and can be transmitted in all directions through free space. Therefore, there is no requirement for the transmitting and receiving antennas to be aligned. It also means that a signal transmitted by one antenna can be received by all receiving antennas within its coverage area. Radio waves are easy to generate and can penetrate buildings without requiring antenna alignment. Radio-wave transmission is well suited for multicasting applications, such as FM radio broadcasting, where a single transmitter communicates with multiple receivers. Consequently, radio waves are widely used in WANs and mobile cellular communication systems.
3. **Microwave (unguided media):** Microwave signals, operating in the frequency range of 1–300 GHz, are line-of-sight transmissions, meaning that the transmitting and receiving antennas must be properly aligned. Microwaves radiated from transmitting antennas are narrowly focused; therefore, receiving antennas must be positioned in the signal path to receive the transmission. This directionality offers an advantage in that multiple pairs of antennas can operate simultaneously without causing significant interference to one another. Parabolic dish antennas and horn antennas are commonly used for microwave communication. Typically, microwave communication systems support bandwidths of around 10 Mbps. The two types of microwave transmissions are:
   - **Terrestrial microwave transmission:** It works on line-of-sight propagation. Therefore, the towers and antennas must be positioned within each other's line of sight and are often required to be very tall to

overcome the curvature of the Earth and obstructions caused by terrain and other obstacles. Microwave communication is generally better suited for short- to medium-distance communication; for very long distances, it can become expensive. However, it is particularly useful in difficult terrain and for communication across large bodies of water, such as oceans.

- **Satellite microwave transmission:** This type of communication is particularly suitable when wide-area coverage is required. Mobile communication systems make extensive use of this form of wireless communication because it can provide connectivity over large geographical areas.

4. **Infrared (unguided media):** Infrared communication supports high bandwidth and, consequently, very high data-transmission rates. Infrared signals, with frequencies ranging from 300 GHz to 400 THz, are used for short-range communication. Because infrared signals cannot penetrate walls, communication in one room does not interfere with communication in another, thereby reducing interference between systems. This characteristic also enhances security. A wide variety of devices, including computers, printers, keyboards, mice, and television remote controls, use infrared signals for communication.
5. **Comparison of guided and unguided media:** Guided media and unguided media are compared in Table III 12.2d.

### 12.2.2 Industrial network hardware (wired type)

Network functionality and network type depend on the purpose of the network. The functioning of any network depends heavily on the infrastructure supporting it. Naturally, the importance of understanding industrial automation networks cannot be overestimated. Their performance depends on the quality of the industrial hardware used. Primarily, network infrastructure facilitates effective communication and services among users, applications, devices, and processes within and beyond an organization. Network infrastructure refers to the collective hardware and software resources that make networking, management, business operations, and communication possible. It includes components such as servers, routers, switches, wireless access points, firewalls, cables, and other devices and systems. This infrastructure serves as the backbone for data transfer, resource sharing, and seamless interaction among various components, as well as smooth integration with the IT ecosystem. Important design considerations include, but are not limited to, robustness, proper network design, environmental conditions, scalability, and performance to ensure reliable and secure operation. The absence of a robust network infrastructure can expose services to severe consequences, directly affecting productivity and profitability. Moreover, in today's digital environment, which demands seamless connectivity, collaboration, and access to critical resources across the enterprise, a robust network infrastructure is essential. At its core, such an infrastructure ensures uninterrupted access to mission-critical applications and data repositories. This uninterrupted access translates directly into improved productivity, as users can access the tools they need without disruption. Furthermore, a well-designed infrastructure facilitates real-time communication and collaboration, enabling teams to work effectively regardless of geographical location. Beyond operational efficiency, a robust network infrastructure is also essential for maintaining a strong security posture [11]. Networking devices serve several purposes, including facilitating data transmission, supporting network configuration and management, ensuring efficient and secure connectivity with threat prevention, optimizing traffic flow to enhance performance, and extending network coverage while overcoming signal limitations. With the continuous development of industrial production and rapid technological advancement, intelligent manufacturing in the form of Industrie 4.0 has

**TABLE III 12.2D** Comparison of guided and unguided media.

| Features | Guided media | Unguided media |
|---|---|---|
| Signal energy | Signal energy is contained & guided within a solid medium | Signal energy propagates in unguided medium |
| Communication | Used for fixed point-to point type communication | Broadcasting in all directions |
| Media types | Twisted pair, coaxial cable and fiber optics | RF, microwave & infrared |
| Attenuation | Increases exponentially with distance | Increases proportional to square of distance |
| Capacity | Capacity increment possible | Capacity increment not possible |
| Installation | Complex, more time and costly | Simpler, less time and money |
| Network type | Discrete network type | Continuous network |

become a necessity of modern industry. As key enablers of intelligent manufacturing, industrial networks and networking products such as switches and routers play a vital role. This subsection presents a brief discussion of these topics.

1. **Industrial networking hardware:** Industrial-grade networking hardware is built for harsh environments due to its enhanced durability, wide temperature ranges, and ruggedized casings for vibration, shock, and electromagnetic interference resistance. Major features include but are not limited to:
   - **Power requirements:** Normally use DC input power.
   - **Ruggedized design:** Devices are hardened enough to withstand harsh industrial environmental conditions like high temperatures, moisture, dust, and strong vibrations. They should be selected with an appropriate life cycle to ensure long-term operation and reliability under severe environmental conditions.
   - **Fewer moving parts:** Devices incorporate fan-less designs to avoid moving parts and prevent dust and moisture ingress.
   - **Mounting:** Options are available for compact and convenient mounting, such as DIN-rail mounting.
   - **Enclosure:** Devices should be inside suitable enclosure for the environment it is exposed to. It may be enclosure like IP 67
   - **Supportive protocol:** Industrial devices must support required automation protocols for specific applications as well as to integrate seamlessly various control and automation systems.
2. **Industrial transmission media and connector:** Network transmission media provide seamless connectivity among integrated architecture components. Therefore, transmission media should offer excellent reliability, flexibility, and noise immunity. These specially designed cables, together with rugged connectors, must be compatible with the automation system to ensure network integrity and performance. In automation systems, nonstandard transmission media are sometimes used as part of proprietary solutions; however, such practices should be avoided whenever possible. Safety-wired and quick-disconnect systems (e.g., safety T-ports and patch cords) are often employed to reduce installation cost and time and to simplify troubleshooting. These features also improve system modularity and facilitate future system expansion. It is important to ensure that the transmission media are compatible with the specific version of the control system being used. Some transmission media and connectors may not be standard products and may need to be procured directly from the automation-system supplier. Therefore, availability of these materials should be assured with short delivery times. These products should be internally tested and certified by recognized agencies such as CSA and UL to ensure compliance with current standards. In many applications, such as machine-level installations, shielded Ethernet cables may be required to have a 600-V rating (e.g., Rockwell 600-V Ethernet Cat 5e cable). Another important consideration is that cables and connectors should comply with the appropriate ISO/IEC 11801 channel class requirements; for example, Class C, D, and E are intended for frequencies of 100, 250, and 500 MHz, respectively. A guide (from Rockwell—Industrial Ethernet Media) has been reproduced in Table I 2.2d (Table III 12.2e).
   STP cables should be used in high-noise environments, such as near drives and Motor Control Centers (MCCs), because the shielding helps reduce noise ingress and electromagnetic interference. These cables require proper equipotential grounding systems to prevent ground loops. Although STP cable installation is more complex and costly, the shielding helps reduce noise coupled into the balanced cable pairs. In other applications, an appropriate category of UTP cable may be used. Ethernet connectors must be properly certified and should have suitable IP ratings for the intended environment. Connectors for UTP and STP cables (including Gigabit Ethernet cables with 360° shielding) should be selected carefully. Such connectors are available in both straight and angular configurations, including 90° bent versions.
3. **Industrial routers/firewalls**: To obtain more diagnostic information while reducing cable and installation costs, Ethernet-based fieldbus networks are increasingly being used to communicate with field equipment. Fieldbus protocols such as IEC 61850 and EtherNet/IP are used to communicate with lower-level PLCs, field devices, and electrical systems, including MCCs. At the same time, automation industries and plant networks are moving toward increased digitalization, resulting in a corresponding rise in cyber threats and

**TABLE III 12.2E** Channel bandwidth & ethernet cable selection (courtesy: Rockwell inc.).

| Data rate | ISO/IEC 11801 channel class | UTP cable category |
|---|---|---|
| 10 Mbps | Class C | Cat 5 |
| 100, 1000 Mbps | Class D | Cat 5e & Cat 6 |
| 1, 10 Gbps | Class E | Cat 6 & Cat 6a |

system vulnerabilities. Consequently, security has become a critical concern. Therefore, modern systems are designed with enhanced security, confidentiality, and resilience to combat such threats. In addition, complex plant systems require rapid access to control-system data. The combination of these requirements makes system configuration and setup increasingly complex, creating the risk of incorrect configurations. For example, ABB's Ability System 800xA provides status monitoring and alarm functions to identify potential risks and future trouble spots at an early stage. The system includes a set of rack-mounted and DIN-rail-mounted network switches. Rack-mounted switches are primarily used for the client/server network, with redundancy achieved through the Redundant Network Routing Protocol (RNRP) and two completely separate networks. DIN-rail-mounted switches are used for field networks requiring multiple ports or optical interfaces. These systems employ two DIN-rail-mounted RNRP router/firewalls. According to ABB, the use of routers with longer service lives instead of servers reduces cost and effort. Maintenance costs are also lower. Because these are preconfigured, tailor-made networks, they can be installed easily, improving engineering and commissioning efficiency while reducing the risk of incorrect configuration (courtesy: ABB Limited).

4. **Industrial Ethernet switches:** Control networks offered by most automation manufacturers use dedicated network architectures that simplify design, implementation, and scalability through plug-and-play devices. These networks typically employ devices that comply with international standards. To manage the network, manufacturers often provide dedicated industrial-grade network switches. These switches are frequently equipped with built-in diagnostic software for monitoring communication-line health and supporting network redundancy.
   - **Ethernet switch with Internet operating system (IOS):** Industrial Ethernet switches provide dependable, rugged networking solutions for operation in harsh environments and are often available with IP67 protection. They offer capabilities such as network monitoring and port control. These switches facilitate integration with enterprise networks using dedicated operating systems (OS). They may support up to 24 ports, PoE, and NAT capabilities. Many models also include built-in SD-card support to simplify device replacement and configuration recovery. Such devices are specifically optimized for industrial automation applications, for example, the Allen–Bradley ArmorStratix 5700 [courtesy: Allen–Bradley].
   - **Smart switches:** Some of the features of industrial rated fan-less smart switches such as the Delta V are as follows (courtesy: Emersons):
     - *Network alerts and diagnostics*: Network alerts and diagnostics automatically report to control system workstations. Smart switches automatically report the status of switch health and network communications problems, excessive packet collisions, excessive dropped packets, etc.
     - *Plug-and-play installation:* No configuration necessary.
     - *Auto security lockdown:* This purpose-built switch enables the system to provide an advanced auto-port-lockdown security feature that helps prevent unauthorized access.
     - *Network diagnostic information:* The network device command center provides detailed network statistics to help in diagnosing network problems and security event reporting.
5. **Other network hardware:** The other types of common industrial networking hardware are listed below:
   - **Power over Ethernet (PoE):** PoE is a popular way of supplying power over Ethernet. In PoE technology, both power and data can be transmitted to network components over the same Ethernet cable. With PoE, power-sourcing equipment can deliver electrical power to PoE-enabled devices without the need for separate power cables, as discussed in detail in Section II 18.0.2.
   - **Network Server:** These systems provide high-performance computing, data storage, and robust connectivity for industrial automation systems, as well as for integrated systems that combine Information Technology (IT) and critical business applications while supporting high levels of virtualization. They are designed in a ruggedized manner, with components capable of withstanding harsh industrial conditions such as extreme temperatures, vibration, shock, and other environmental hazards. These systems are typically highly reliable, offering high availability with minimal downtime, which is particularly important for mission-critical operations. They also provide extended life-cycle management and enhanced security features for industrial applications, such as open-source Linux servers.
   - **Endpoint:** Endpoints are network hardware devices connected to the user ports of a network switch to serve devices located at the edge of the network, including sensors, controllers, HMIs, and other field devices. These devices send and receive data for end users while providing robust connectivity, including fiber-optic connections for noise-free,

long-distance communication. For industrial applications, endpoints are designed to operate reliably in harsh plant environments. They may also support PoE and are available in DIN-rail-mounted configurations. In integrated systems, endpoints can perform functions such as sending emails, sharing files, browsing the Internet, and supporting other network-based services.

Chapter 13 discusses other networking fundamentals.

## List of abbreviations used

**AIDC** Automatic identification and data capture
**AP** Access point
**DCE** Data communication equipment
**DSSS** Direct sequence spread spectrum
**DT** Digital transformation
**DTE** Data terminal equipment
**EMI** Electromagnetic interference
**HDLC** High-level data link control
**HMI** Human machine interface
**ICT** Information and communication technologies
**IDS** Intrusion detection systems
**IIoT** Industrial IoT
**IOS** Internet operating system
**IoT** Internet of things
**ISM** Industrial, scientific, and medical devices
**IT** Information technology
**M2M** Machine to machine
**NCS** Network control system
**NFC** Near field communication
**NIC** Network interface card
**NRZ** Non-return to zero
**OT** Operational technology
**PAM** Pulse amplitude modulation
**PAN** Personal area network
**PCI** Peripheral component interconnect
**PDA** Personal digital assistants
**PoE** Power over Ethernet
**POF** Plastic optical fiber
**PSD** Power spectral density
**RF** Radio frequency
**RFD** Reduced function device
**RNRP** Redundant network routing protocol
**RZ** Return to zero
**SNMP** Simple network management protocol
**STP** Shielded twisted pair
**TDD** Time division duplex
**TDMA** Time division multiple access
**UTP** Unshielded twisted pair
**UWB** Ultra-wideband
**WAP** Wireless access point
**WNIC** Wireless NIC

## References


[1] J. Beningo, in: USART vs UART: Know the Difference, 2015. https://www.edn.com/usart-vs-uart-know-the-difference/.

[2] *Data Communication Fundamentals (Module 2); Transmission of Digital Signal (Lesson 4); Version 2* CSE IIT, Kharagpur; Internet document; https://nptel.ac.in/content/storage2/courses/106105080/pdf/M2L4.pdf.

[3] Data Communication and Networking, McGraw-Hill Education, 2013.

[4] S. Basu, A.K. Debnath, in: Power Plant Instrumentation and Control Handbook; 2nd Rev, Elsevier; Academic Press, 2019. https://www.elsevier.com/books/power-plant-instrumentation-and-control-handbook/basu/978-0-08-102804-9.

[5] RS232 Protocol: Introduction to the Standard and the Interface, DwyerOmega.com, 2019. https://www.dwyeromega.com/en-us/resources/rs232-serial-communication.

[6] *Interface Serial; Pheonix Contact*; Steven Engineering; Internet document; https://www.stevenengineering.com/manufacturer/phoenix-contact/.

[7] S. Rackley; *Wireless Network Technology—From Principles to Successful Implementation*; Internet document; http://home.ustc.edu.cn/~wfsun/lab/course/wireless/Steve%20Rackley%20-%20Wireless%20Networking%20Technology.pdf.

[8] R. Nilsson, in: Industrial Wireless: Bluetooth can be Robust, Easy to Use, Control Engineering, 2013. https://www.controleng.com/articles/industrial-wireless-bluetooth-can-be-robust-easy-to-use/.

[9] A.S. Tanenbaum, D.J. Weatherall, in: Computer Network, fifth ed., Prentice Hall, 2011.

[10] What Is the Physical Layer: A Beginner's Guide, Coursera, 2024. https://www.coursera.org/articles/physical-layer.

[11] Core Components of Network Infrastructure and Why It Matters, Tata Communication, 2024. https://www.tatacommunications.com/knowledge-base/network/core-components-network-infrastructure.

Chapter 13

# Fundamentals of industrial networking

A system of interconnected computers, intelligent nodes/devices (including peripherals), to facilitate sharing of resources, controls and information among them, can be termed as conceived as computer network. The distributed system refers to a collection of independent computers appears to its user as a single coherent system [1]. Important issues such as Application programing interface (API) and Middleware in networking has been covered in *II 15.0.1*, hence repeated here and Volume II may be referred to. For computer networks that can be considered as distributed data bases, it is necessary to know more about network elements, network structures with their interrelation. In its simplest form one can consider there is one host computer as server and a few terminals as client interconnected mutually through a data communication subsystem of the network. There could be **N layered network.** N layer networks is mainly three layered network but with many levels. Unlike client server network, in a distributed network system, there will be computer elements/node and communication subsystem, which makes it possible to store, share and retrieve information through online access. It is needless to tell that there should be some transmission protocol to regulate system performance and Quality of service (QoS). Also, another term achieved through advanced control system is Quality of performance (QoP). Before we move on to specific details it is better to get familiarized with a few terms, which will be referenced during subsequent discussions. A distributed computer system consists of a set of autonomous computational resources, connected and communicated through a computer network. These networks may be Homogeneous or heterogeneous depending on *Variation in types of computational resources*, *Type of network* and *Type of network topology*. A few types include: Advanced research projects agency network (ARPANET), Simple object access protocol (*SOAP*), Peer to Peer (P2P), Message passing interface (MPI) etc. In case we look back one can find Mainframe-based systems in 1960. 1970 witnessed Client -server network, Two/three tier digital control networks were available in 1980. Distributed computing were available in 1990 and Post 2000 are era of Broadband and internet. The discussions started with though details about Network control system (NCS) to expose quality performance of network. Depending on applications there could be varieties of network topologies and network area types. Apart from these, brief discussions will be on network mode. The Open systems interconnection (OSI) model of ISO, is a *reference tool* for understanding data communications between any two networked systems. It divides the communications processes into seven layers each of which performs *specific functions* to support the *layers above* it and offers *services to the layers below* it. Understanding the function of each layer is instrumental in understanding data communication within Local/Metropolitan/Wide area networks (LAN/MAN/WAN). The discussions in this chapter also includes various kinds of transmission methods as well as vulnerability of network. Let the discussions of the chapter start with basics of network controls and types.

## 13.0 Network controls and types

It is needless to explain when intelligent devices/computers are connected together there will be easy access for data. Manual intervention is minimized and it is easier extraction and corelation of information and data. In its simplest form this network can be conceived as a central database with a number of users. The users sending request can be termed as client. So, servers, in conjunction with clients form the simplest client server network. The requirements and types of data handling in such network are not uniform and same e.g., plant control data must be reliable and real time data, whereas commercial data may not be real time one. For arriving at Industrie 4.0 state, it is necessary to have huge exchange of data through Information technology (IT) for implementing Operating technology (OT) for the plant by marrying IT with OT. NCS refers to the management and regulation of data flow within computer networks, ensuring that communication between sensors, actuators, and controllers occurs efficiently and reliably through shared, band-limited digital communication networks. The primary function of network control is to coordinate the operation of spatially distributed network elements, which may operate asynchronously, to achieve overall system objectives. Again, networked control systems refers to control systems in which the system to be controlled and the different components such as *Actuators*, *Controllers*, and *Sensors* are spatially distributed and communication

Plant Intelligent Automation and Digital Transformation. https://doi.org/10.1016/B978-0-323-89892-8.00006-5

between these components is achieved through the use of a digital communication network (shared by other applications). The network control and networking protocols go hand in hand. Protocols define the conventions for orderly coordination, and network control activities such as resource management, message control, and error handling. It is possible to integrate and coordinate multiple independent processes or plants in distributed systems.

A network topology is the physical and logical arrangement of nodes and connections in a network. Network topologies describe the arrangement of networks and the relative location of traffic flows. Two categories of topology are: *Physical topology* (physical layout of cables and devices) and *Logical topology (*data flow across the network, regardless of physical layout). Both physical and logical topologies are essential parts of network infrastructure. The type of network topology differs depending on the purpose/need of the network arrangement. As proper selection of network *affects* the performance, cost, reliability, and security of the network, it is important to choose the right topology. Properly designed network topology makes it easier for network admins to locate faults, troubleshoot issues, and allocate network resources. Hence it is important in scaling the network. According to the communication requirements, multiple types of network connections are available. Based on area coverage *Networks* can be divided as Personal area network (PAN) Local area (LAN), Campus area (CAN),MAN, WAN etc. In a plant there could be several controllers in a Distributed control system (DCS) connected for *extraction and corelation of information and data.* A number of Real time (RT) systems (controllers) in plants form LAN. A typical distance for LAN could be *1–2 Km.* In case, of a completely integrated system like in power plant, the network may stretch over Kilometers. When LAN is developed through wireless means (as discussed in Chapter 10 in this volume) it is known as Wireless LAN (*WLAN*). When this length limit goes beyond ~100 km, it will form Wide area network (WAN). In this section brief discussions will be presented on NCS, network topologies and network area types.

## 13.0.0 Network control system (NCS)

With advancement in computer and communication technology makes it possible for dynamic systems to interact through network. NCS is a distributed control system, comprising sensor actuator and other devices, and loops are *closed* through real time *communication network (packet-based shared and/or wireless communication network).* So, in NCS the different components are spatially distributed and communication among them is achieved through digital communication network. Major advantages of NCS is its improved reliability and capability of system integrations on account of reduce wiring, distributed processing, lower cost (especially for redundancy), more flexibility easier trouble shooting and maintenance and system integration indicated earlier. Major pertinent issues are: limited: *bandwidth, packet loss, synchronization* issue *delay etc.* Communication network shall mainly include, fieldbus and other network systems including Ethernet, Internets, wireless networking technologies etc. Two distinct types of communication networks applications are: *Data networks (–* large data packets with high data rate, infrequent burst transmission without real time constraints mainly for IT applications) and *Control networks (mainly in control application)***:** Control network supports *real time and time critical applications*. This must shuttle countless small but frequent packets among a relatively large set of nodes to meet the time-critical requirements [2]. NCSs have been used widely in industry/process control systems, micro-grids, aerospace systems, tele-operation, and intelligent systems [3]. Major issues of NCS arises from the sharing of the communication medium or the extra complexity associated with data transmission. In NCS encoded discrete signals in a packet are sent across the shared network. At the destination end naturally these are decoded. Let us first address network issues.

1. **Network Issues: Network issues:** Let us now look into a few issues listed below for designing NCSs.
   - **Driving Mechanism:** In computer network, nodes are either time or event driven i.e., triggered or initiated either by time or event. In the time-driven mechanism, the control components are trigged to work at regular intervals e.g., sensors reading. In event-driven mechanism, based on complex program the control components are trigged by predefined "events" viz. alarm.
   - **Data packet and constrains:** Packet-based data transmission is the distinguishing characteristic of NCS. As seen in DeviceNet, ControlNet etc. both computation and traffic related delays are significant implying that to fully utilize the packet communication structure one can decrease the transmission frequency or the packet size or both [4].
   - **Data Packet:** Encoded data packets are transmitted over the communication network. Fig. III 13.0a shows typical data packet consists of header payload and trailer. As shown it has two parts. On the top part a typical data packet where NSC and other applications share the same package. Let us assume that this is a sensor data for NCS and same is being used by other nodes also. Packet data consists of two parts viz. overhead shown by hatched part and actual data. Overhead has two parts, header in the front and

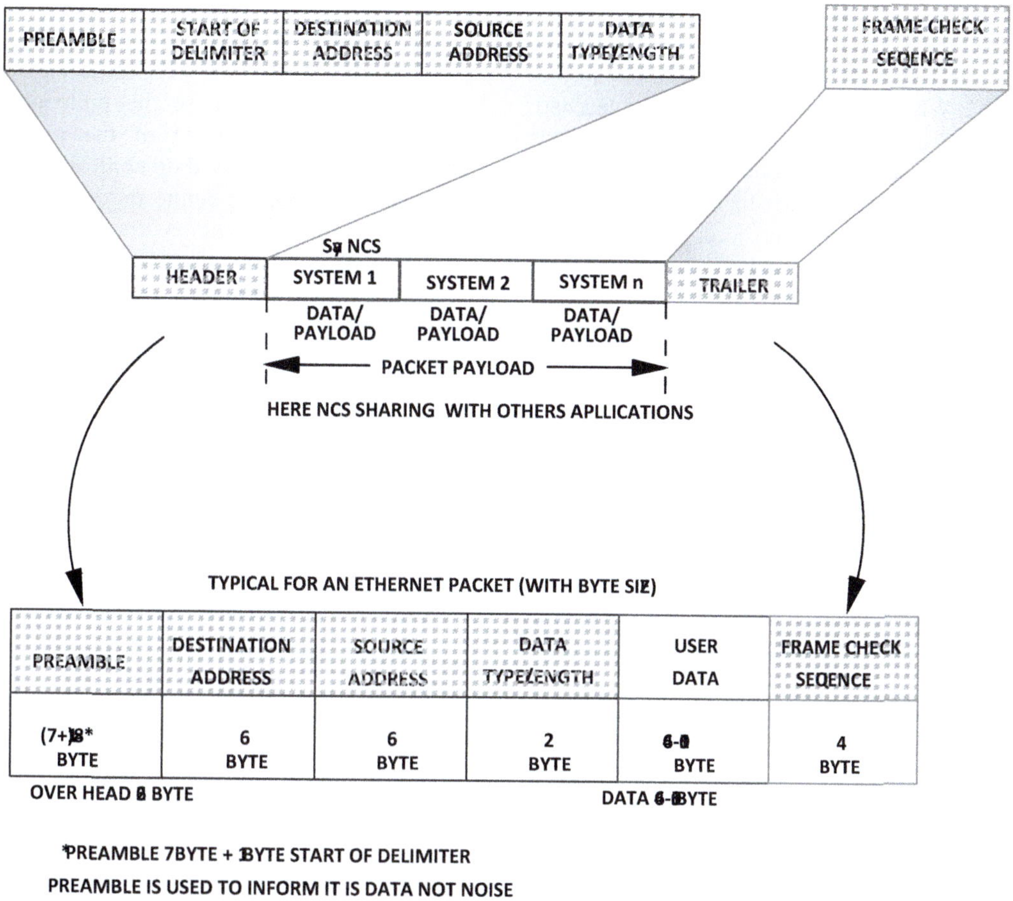


**FIGURE III 13.0A** Typical data packet details.

trailer at the end. Sandwiched between them is variable actual data as detailed out in the bottom part of Fig. III 13.0a. In that it is seen that total overload sum of header and trailer is 26 byte, whereas data could be between 46 and 1500bytes.

- **Network induced delay in Data packet**: Network-induced delay (e.g., sensor-to-controller delay) occurs when exchanging data among devices in a shared network This constant/time varying delays can degrade the performance of control systems and can even destabilize the system. Two types of network induced delays are: Backward channel delay $\tau sc$ (e.g., induced delay from sensor to controller) and Forward channel delay $\tau ca$ (e.g., delay from controller to actuator). In most cases two delays are not treated separately, instead the round-trip delay is of interest. The delay characteristics of an NCS primarily depend on the type of network protocol used with different impact on various network types:
- **Cyclic system**: Token bus, Token ring are cyclic systems where delays are bounded and considered as constant in most of the cases. In most of these cases it is assumed that total delay will be less than sampling period.
- **Random network**: Network such as Ethernet (CSMA/CD) or CAN, delays are unbounded. These are often dealt with different models [5]. These models are: Constant delay model, Stochastic delay model (deployed when the network delays are affected by network load, nodes competition, and network congestion etc.), Robust control: (to guarantee the robust stability and robust performance of the NCS), Predictive control (for complex dynamic systems). Apart from these there could be as Markov Chain Model, Hidden Markov Model [5].
- **Priority order network**: As in DeviceNet here, delays are bounded for data packets with higher priorities and unbounded delays for data packets with lower priority.
- **Data Packet Disorder**: Network induced delay affects different data packet differently, as a result a data packet sent earlier may arrive at the destination later or vice versa [6]. This obviously degrades control performance.

- **Data Packet Drop out**: On account of transmissions error it is quite possible some packets can be lost during transmission. This is called "data packet drop out". When network links become congested, data packet drop out can happen. Two methods are: either to resend the data till it reaches the destination or to simply ignore. In real-time systems, packet loss either should not occur at all or it has to be resolved through transmission protocol within the real-time limits [7].
- **Jitter**: Jitter refers to fluctuations/variations in arrival time and responsible for degradation in quality of service. This may be due to network congestion, causing timing drift and route change.
- **Latency**: It refers to average time, data packet takes to travel to destination. It should be as low as possible (close to zero).
- **Bandwidth**: It represents network capacity to transmit the maximum amount of data from one point to another for a given amount of time. This varies with network types and in Quality of service (QoS) this is optimized.
- **Multipacket and Single packet**: There may be multiple data packets for single purpose viz. for critical measurements redundant sensors are deployed and to enhance availability they are connected to NSC way. Here, if only one sensing data packet is lost, all the other sensing data packets may have to be discarded due to incompleteness. As shown in Fig. III 13.0a packet size used in NCS is quite large when compared with data size required to encode a single command. In such cases sensing data or the multiple control command are sent through a single data packet over NCS. This is referred to as single packet.

2. **Sampling and Traffic**: In NCS, sampling period plays an important role. It is quite important to go for most suitable network protocols and network scheduling strategies for a desired control performance. For NCS, a minimum transmission frequency is necessary to guarantee the network QoS. Again, with increase in transmission frequency, the network load also increases. So, it is necessary to consider the bandwidth limitation of the network. When transmission frequency reaches the maximum limit, the performance will decrease as too much data (due to high sampling frequency) will be produced to cause overloading and congestion in the network. Also, there may be more data packet dropouts. Naturally, a tradeoff has to be sought. Two ways to reduce the network traffic in NCS are: Compression of the size of data transferred at each transaction through sophisticated coding or, Minimize the frequency of transfer of information between the sensor and the controller/actuator. The characteristics of network traffic generated by NSC determines that the latter approach seems more promising with respect to the resulting traffic reduction, because the protocol overhead is the dominant component of the induced network traffic volume [8]. Based on control system specification it is possible to define the worst, unacceptable, acceptable, and best regions for fixed control systems as typically depicted in Fig. III 13.0b –developed based on idea from Ref. [9]. As shown there, for continuous control involving no sampling it is constant. For direct digital controls there will be sampling but the performance only depends on the sampling period, not any other constraints such as networked induced delay, etc. The performance degradation point A of sampling period PA in digital control could be estimated based on the relation between control system bandwidth and sampling rate. For NSC, point B has to the right of point A in x-axis, because of additional network-induced processing delay etc. As discussed earlier, as the sampling period gets smaller, the network traffic load becomes heavier, the possibility of more contention time or data loss increases, and longer time delays result [9], degrading the performance. Hence there will be a point C to get acceptable performance.
   This will be clearer from Fig. III 13.0c, where selection of best sampling point has been shown. From the above discussions it transpired that as sampling period is lowered system performance improves up to a certain maximum (say at B), beyond which it deteriorates (as seen in point C) on account of traffic congestion and packet loss etc. Again traffic/load increase with decrease in sampling period as shown by load curve. So, the load curve, which intersects system performance curve at B will give best sampling pint for the given network.
3. **QoS and QoS implementation**: Here discussions have been presented explaining QoS in network and its implementation:
   - **QoS in network (NCS)**: There does not exist any comprehensive definition of QoS and depends on network application. However, three parameters such as packet loss, latency and jitter in network are important parameters for QoS. Also, data throughput/bit rate and network delay also play crucial role in determining QoS. So, QoS can be seen as a technology that manages data traffic to reduce parameters such as packet drop, jitter, and latency on the network. Various models in implementing QoS are: Best Effort (NCS without suitable policy planning), Integrated Services (–IntServ with limited scalability but higher consumption of network resources), RSVP (– described in RFC

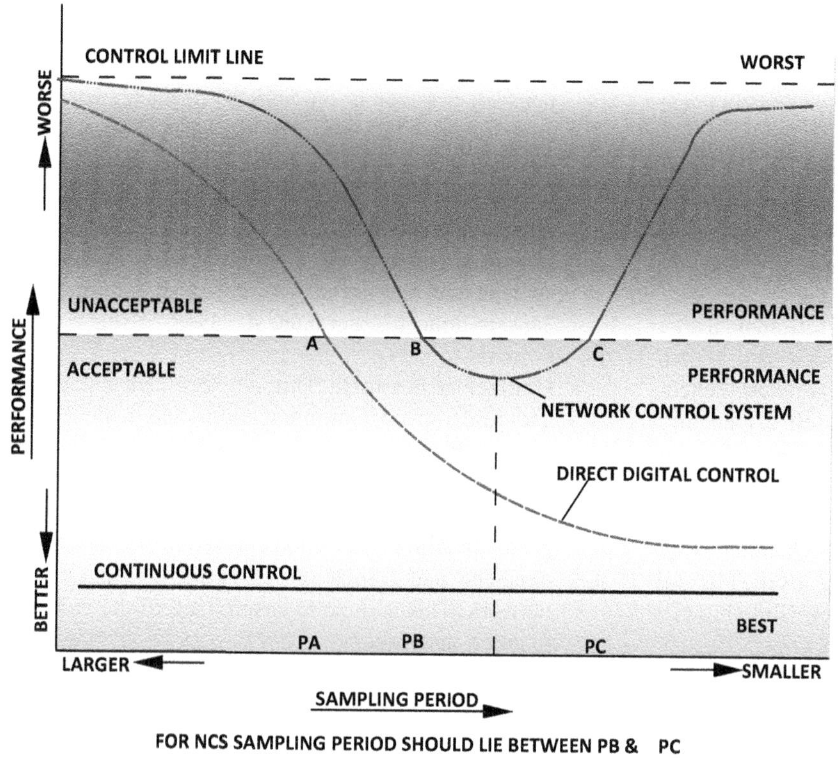


FIGURE III 13.0B Performance of NCS and other control types.

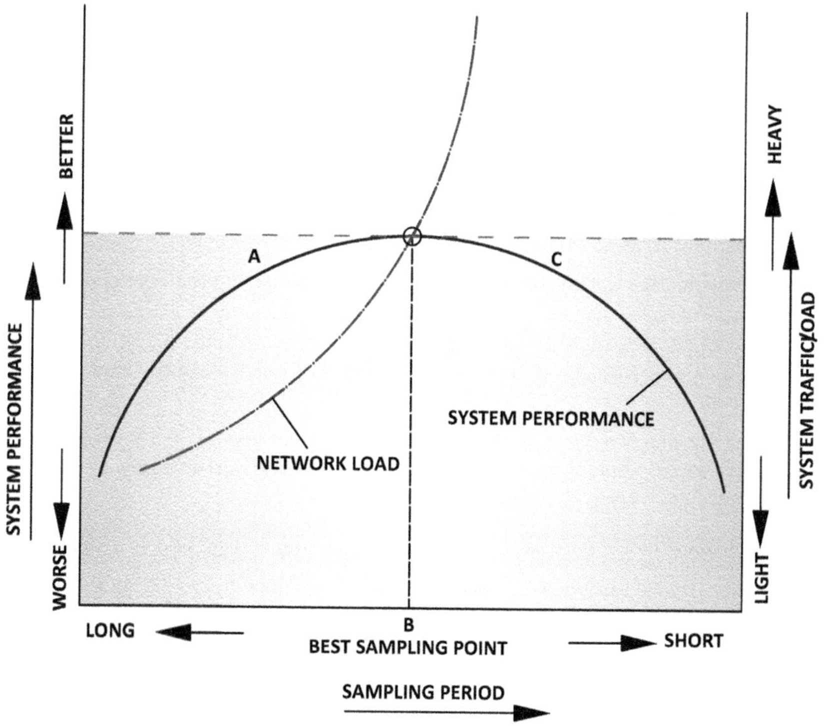


FIGURE III 13.0C Selection of best sampling points in NCS.

2205 is a transport layer protocol used to reserve resource in a computer network) and Differentiated Services (–DiffServ: comparatively simpler but scalable QoS model used to classify and manage network traffic).

- **QoS implementation**: Several mechanisms for QoS can be classified broadly categorized in several heading as shown in Table III 13.0a.

4. **QoS Technique**: QoS Techniques: Following are major QoS Techniques
   - **Scheduling**: Data packets from different traffic flows arrive at network elements such as a switch or router for processing. Suitable software scheduling techniques as shown in self-explanatory Fig. III 13.0d, can be adapted for principles of congestion management and congestion avoidance of implementation of QoS improvement.
     Three scheduling types are: FIFO Queuing (Ref: Fig. III 13.0d1), Priority Queuing (Ref: Fig. III 13.0d2) and Weighted Fair Queuing (Ref: Fig. III 13.0d3)
   - **Traffic shaping**: In scheduling we have seen how to handle priorities, in Traffic shaping, handles with providing smooth i.e., constant output data packets flow rate, in spite of having burst and/or variations in inflow. Two types of traffic shaping methods are Leaky bucket and Token bucket as detailed in self-explanatory Fig. III 13.0e.
5. **QoP and QoS**: NSC can be looked into from two angles: These are:
   - **Control over network**: This area refers to control system using employing the communication network as the medium of data transmission, i.e., control strategies and control systems design over the network trying to minimize the effect of adverse network parameters on NCS performance. It influences of inherent characteristics of the communication network on control system. Here main focus is on the issues of analysis, synthesis, and design pertinent to control system with the objectives to minimize effects of communication constraints such as data rate limitations and quantization effect. This is done to guarantee the QoP of the control system.
   - **Control of network**: This is concerned with study and research on communications and networks to make them suitable for real-time NCS, e. g., routing control, congestion reduction and management, efficient data communication, networking protocol. Aim of this is to achieve the effective management and allocation of limited resources of network. This is done to ensure QoS of the communication network.

Even though there exists difference in themes and contents between these two yet there are overlapping. In fact, the performance of control system and communication network in NCS restricts and influences each other. A common view is that the overall performance of NCS is associated with not only control algorithms but also with network resources allocation [10]. Naturally, for successful control systems through NCS one needs to QoP and QoS simultaneously giving rise to co design of the systems.

With this knowledge on NCS let us now explore network topologies.

**TABLE III 13.0A** List of QoS implementation Mechanisms.

| Method | Description |
|---|---|
| Classification and marking: | Sorting and differentiating packet; Marking each packet as network class according to priority. Classification and marking in Network device (router) |
| Congestion management | Based on classification and marking assist making queue for placement of packets to ensure QoS. |
| Congestion avoidance | Monitoring of network traffic for congestion to drop low-priority packets when congestion occurs e.g., Weighted random early detection and random early detection. |
| Shaping | Manipulation of traffic, prioritization real-time applications over less time-sensitive applications. Traffic shaping tools are buffers, generic traffic shaping and frame relay traffic shaping etc. |
| Admission Control | Used by network elements such as router with main aim to estimate the expected bandwidth for the incoming data stream and determine whether it is possible to assign needed bandwidth and effectively use of resources, to avoid congestion. It is kind o f congestion avoidance. |
| Link efficiency | Not exclusive for QoS; mechanisms are used to reduce delay, jitter, and IP overhead as well as maximize bandwidth e.g., Real transport protocol, transmission protocol. |

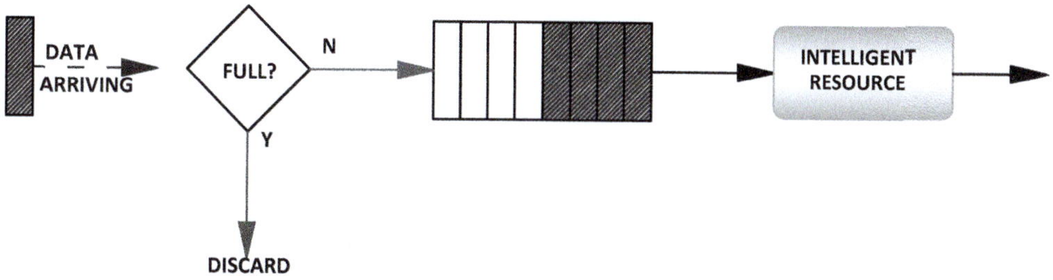


FIG III 13.0d1 FIFO QUEUE (SCHEDULING)

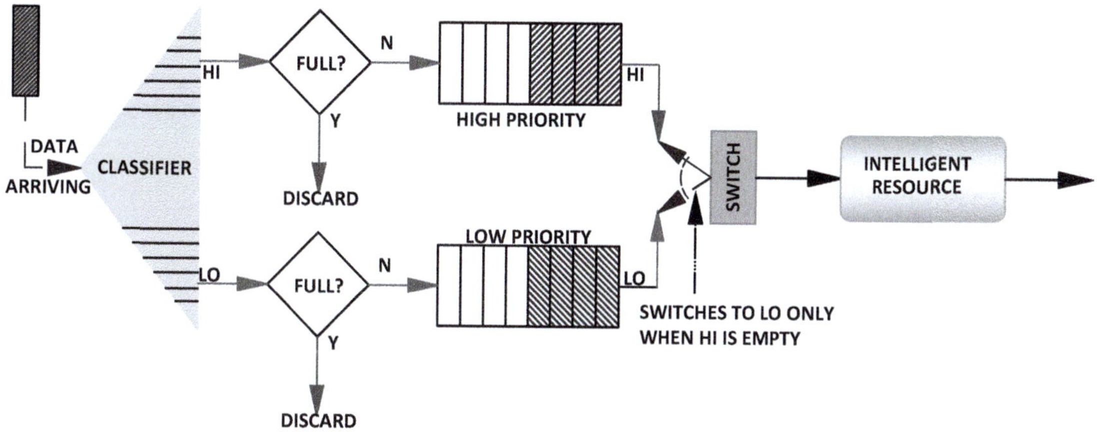


FIG III 13.0d2 PRIORITY QUEUE (SCHEDULING)

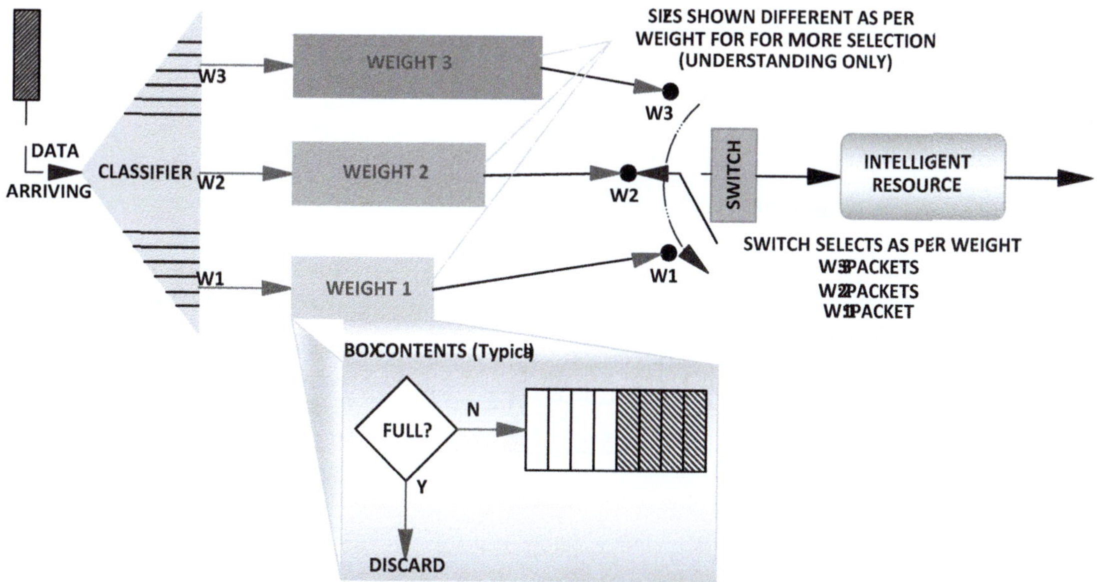


FIG III 13.0d3 WEIGHTED FAIR QUEUING (SCHEDULING)

**FIGURE III 13.0D** QoS improvement technique scheduling.

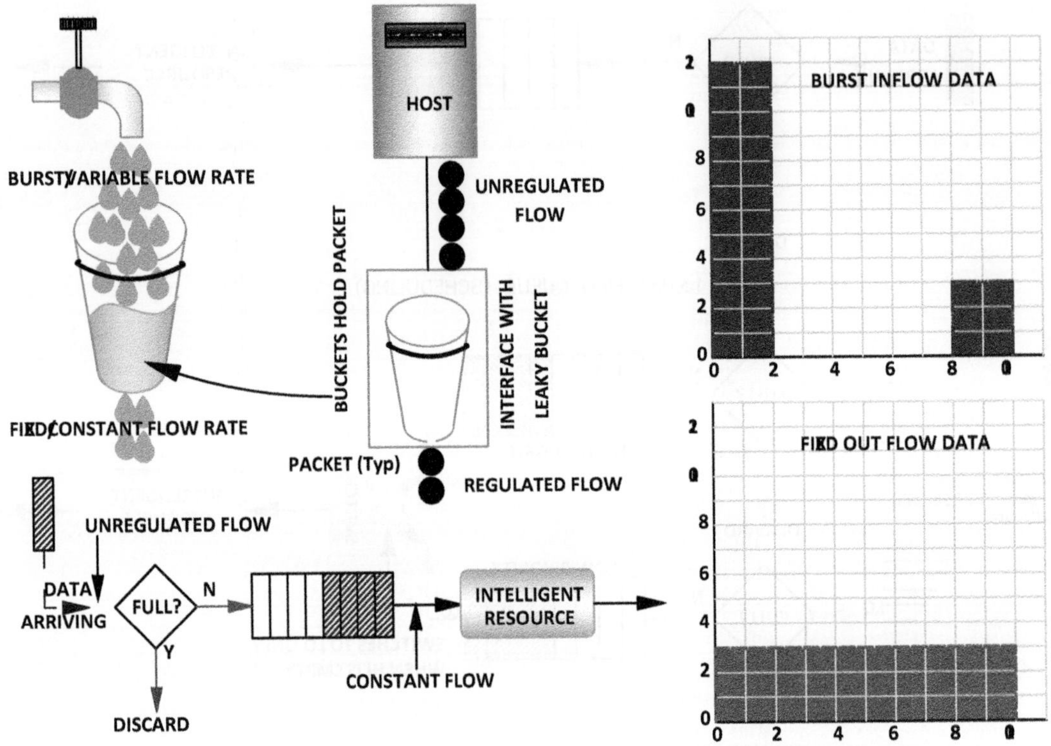


GRAPH SHOWS UNREGULATED INPUT DATA SMOOTHENED AT OUTPUT AT CONSTANT RATE

FIG III 13.0e1 LEAKY BUCKET METHOD

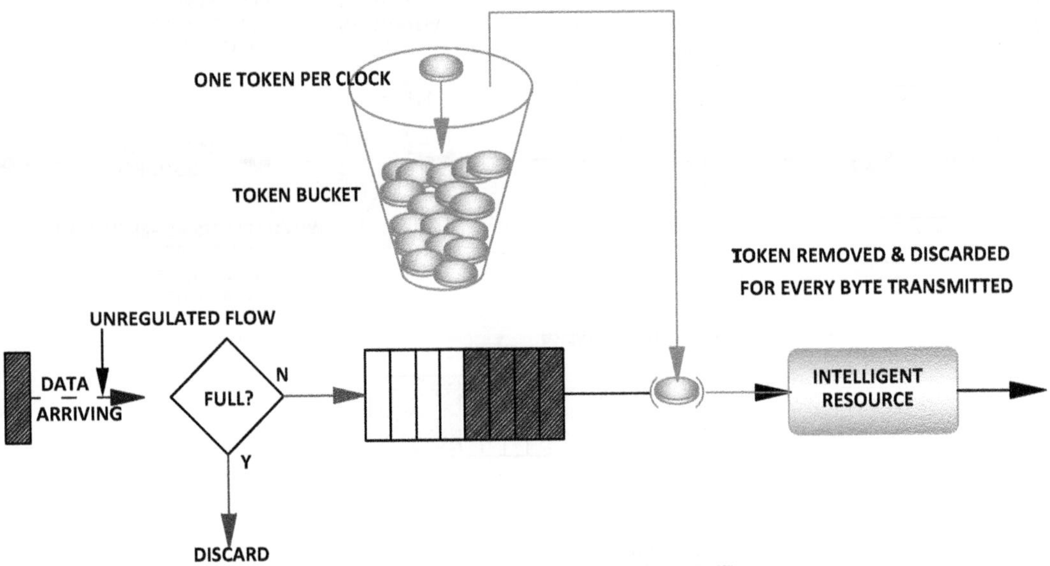


FIG III 13.0e2 TOKEN BUCKET TECHNIQUE

**FIGURE III 13.0E** QoS improvement Technique traffic shaping.

### 13.0.1 Types of network topology

Word "Topology" is derived from two Greek words topo and logy, where topo means 'place' and logy means 'study'. A network can be conceived as a layout or arrangements of nodes and communication link in an automation network. In computer networks, a topology is used to explain how a network is physically connected and how does information logically flow in the network. So, Network topology defines the physical layout, structure of the network both in terms of physical and logical also. Physical topology refers to an actual layout of the computer cables and other network devices/elements, describing connection details of nodes in a computer network. Logical topology represents the relationship between nodes and their logical connections. A logical connection will differ from a physical path when information can take an invisible hop at intermediate points. Logical topology provides the insight about the actual physical design of the computer network providing details about the data flow. Bound by network protocol, logical topology determines the way in which the devices communicate throughout the network. In network topology a node refers to any device or point in the network, such as computers, servers, routers, switches or any other intelligent device. Nodes are crucial for transmitting and receiving data within the network. A link (wired or wireless) signifies a physical or logical connection between two nodes, allowing data to flow between them. Each of these topologies has associated pros and cons. Choosing the correct topology for automation network is crucial to optimize its functioning i.e., it is important to determine which topology is best suited. Factors such as desired performance, cost, and fault tolerance are considered while selecting a network topology. Let the discussions start with reference to Fig. III 13.0f.

1. **Point-to point and Bus Topology**: As detailed in Fig. III 13.0f1, it is the simplest and easiest topology comprising of a direct link between two hosts computers. It is direct, fast with reliable connections. This is not based on any expensive Operating system (OS), server or expert for the network. However, it is possible between two computers in close proximity without any back up. Also, it less secured, log on may be avoided.
2. **Bus and Ring Topology**: There are some similarity between them bus and ring topology communication. These two are described separately for different pros and cons.
   - **Bus Topology**: Bus topology is simple kind of topology with common bus or channel (connecting all nodes) for communication in smaller local area network (LAN). This linear bus topology is very flexible, for adding/deleting any node without affecting other nodes. Over a few dozen computers can be added in the network in bus topology with terminators at both ends as shown in Fig. III 13.0f2. Through suitable medium access control technique, one station is allowed to send data at a time and all other stations listen to it, as it can work in a broadcast mode (CSMA/CD or bus master). A transmission from any station propagates the length of the medium in both directions and can be received by all other stations. However, concerned receiver(s) only accepts it verifying Media access control (MAC) address. As limited size LAN, It offers moderate reliability and performance. When data reaches the end, the terminator removes the data from the communication line (to prevent signal bounce and data flow disruption). The devices share the responsibility for the flow of data from one point to the other in the network. This single cable segment lends itself to being a single point of failure. Pros and cons have been enumerated in Table III 13.0b
   - **Ring Topology**: Nodes/stations in a ring topology are joined in a closed loop or ring as shown in Fig. III 13.0f3, which shows data flow in single and double direction. In normal ring topology if a device fails (or ring is snapped) entire network fails. In order to circumvent this situation, another secondary data path is provided in dual ring topology. Secondary ring provides redundancy in case of a failure in first ring and also can serve as an additional data path. The repeater is a function of Network interface card (NIC) attached to LAN node/station. The repeater receives data on one link and transmits them, bit by bit, on the other link (as fast as they receive) without any buffering. On account of NIC each node/station will repeat any signal irrespective of its destination and if any NIC fails the NIC controller can handle the defective repeater to pull off the ring so that ring stabilizes and function properly. Also, by using wiring concentrator, a faulty node can be bypassed. Ring topology is well described by Token Ring (IEEE 802.5) or FDDI. Pros and cons of the topology has been enumerated in Table III 13.0c
3. **Daisy chain Topology**: A daisy chain is an interconnection of controllers., computer devices, peripherals, or network nodes in a linear fashion somewhat like in series connection as shown in Fig. III 13.0f4. This is less popular. This is similar to ring topology where all nodes (exception: two at the ends) are connected to two hosts only. On account of linear configuration, one outage affects other connected items. Every intermediate host works as relay for its immediate hosts.

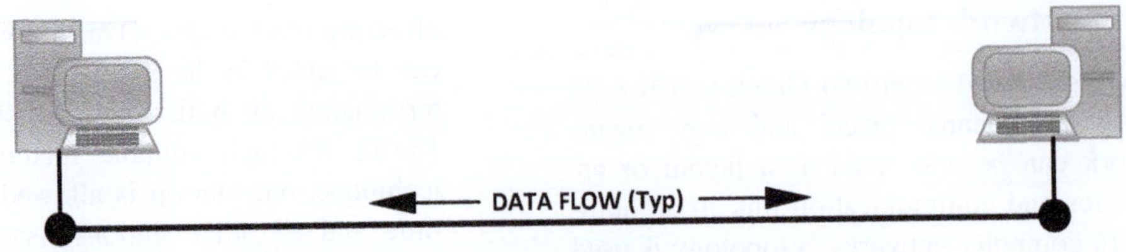


FIG III 13.0f1 POINT TO POINT NETWORK TOPOLOGY

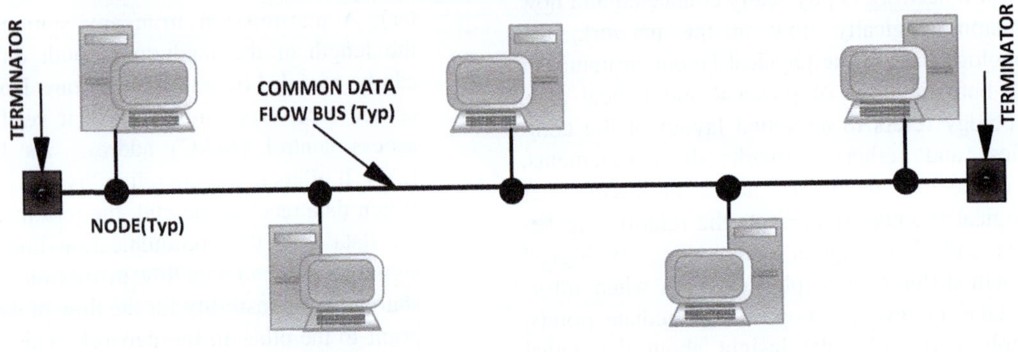


FIG III 13.0f2 BUS NETWORK TOPOLOGY

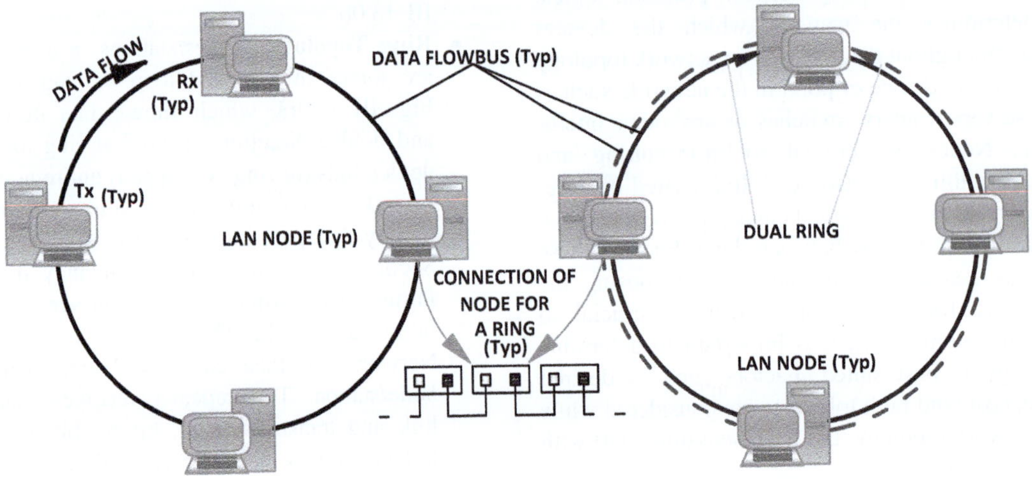


FIG III 13.0f3 RING NETWORK TOPOLOGY (SINGLE& DUAL)

FIG III 13.0f4 DAISY CHAIN NETWORK TOPOLOGY

**FIGURE III 13.0F** Network topology: PP, bus, ring and DC.

**TABLE III 13.0B** Pros and cons of Bus Topology.

| Advantages | Disadvantages |
|---|---|
| Simple inexpensive with less cable to accommodate N nodes/ drop lines | Broadcast mode is not secured; performance is poor with more nodes. |
| Small simple network, not affected for any node failure. Installation is simple and less expensive | Possibility of collision in heavy traffic, traffic congestion during communication. Higher bus length causes attenuation |

**TABLE III 13.0C** Pros and Cons of Ring topology.

| Advantage | Disadvantage |
|---|---|
| Cheaper to install and expand with less cable and least chance of collisions | Single failure can jeopardize network |
| Nodes have equal access, easy and faster token failure and error detection, | More power consumption for continuous token pass |
| Node addition/deletion requires two connection changes | Node addition/deletion is complex and calls for interruption |

Each link in daisy chain topology represents single point of failure (sometimes, bypass switch option available).

Various other network topologies have been detailed in Fig. III 13.0g

4. **Star and Mesh Topology**: These two topologies have similarities hence clustered here:
   - **Star Topology**: In star topology, hosts/nodes in star topology are connected to a central device, with P-to-P connection between node and central device as shown in Fig. III 13.0g1. Several different cable types such as shielded twisted-pair (STP), unshielded twisted-pair (UTP) can be used for this point-to-point link. Every communication between hosts, takes place through the central device only. Central device could be hub or repeater/switch or bridge/router or gateway. Central hub/switch acts as single point of failure but any node failure does not affect network. With a repeater as central device, operation is in broadcast fashion. A transmission of a frame from one station to the node is retransmitted on all of the outgoing links, hence is received by all other stations. Only one station can transmit at a time. When switch/router is central device, then it performs the switching or routing function. Incoming frame is buffered in the node to retransmit on an outgoing link to the destination station. This topology is very flexible and comparatively easier to maintain and it provides good reliability. With switches as central device high speed communication is possible. Pros and cons of star Topology have been described in Table III 13.0d
   - **Mesh Topology**: In mesh topology each node/station is connected to every other node/station as shown in Fig. III 13.0g2. Internet connection is an example. When connected (by twisted pair, optical fiber or coaxial cable), any two nodes are connected by dedicated point–point links between them. If N be the number of nodes/stations and they are connected with each other in mesh topology, then total number of ports that is required by each node/station is N-1 (with total connections = . N.(N-1)/2). As two nodes have P-2-P connection, there may not requirement of additional information/data packets for communication. This is not only inflexible but costly to add one node. Two kinds of mesh topology are: Full mesh (with all nodes connected to any other) and Partial mesh (same as full mesh, only a few devices connected with just 2/3 devices). Advantages and disadvantages of Mesh topology have been tabled in Table III 13.0e
5. **Tree and Hybrid Topology**: As the name suggest hybrid stands for combination of two topologies. Here this along with Tree Topology have been discussed.
   - **Hybrid Topology**: Hybrid topology combines two or more topologies inheriting merits and demerits of topologies. As shown in Fig. III 13.0g3, hybrid topology is costly but at the same time popular, scalable and reliable too. Even though in many cases these are costly and complex to implement yet, better flexibility, scalability, quicker error detection etc. are possible.
   - **Tree Topology**: Tree Topology, without any loop, is also known as hierarchical topology. The tree begins at the root, the pinnacle point, and expands to the network endpoints. It is used in cascading nodes in hierarchical manner with multiple levels and layers as shown in Fig. III 13.0g4. This topology copies star topology with inherited property of bus topology. The highest layer i.e., the root of the tree is known as core layer. From this highest layer middle layer is branched out to act as distribution layer. In the lowest level main hosts/nodes are connected for communication. From this it is clear that like bus topology, if core level fails entire network fails. The physical cable is known as a branching tree with all stations attached to it. When the connection point fails then that segment becomes unworkable. Scalability, flexibility and expandability are major characteristics of this topology. In view of the above one can consider that in tree topology a network can expand dynamically with only one active data path

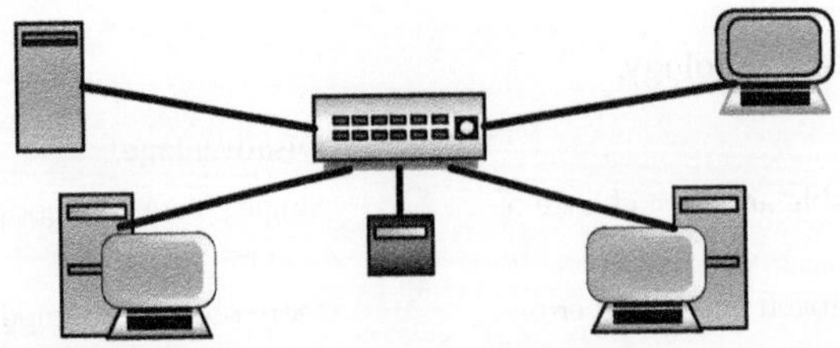

FIG III 13.0g1 STAR TOPOLOGY

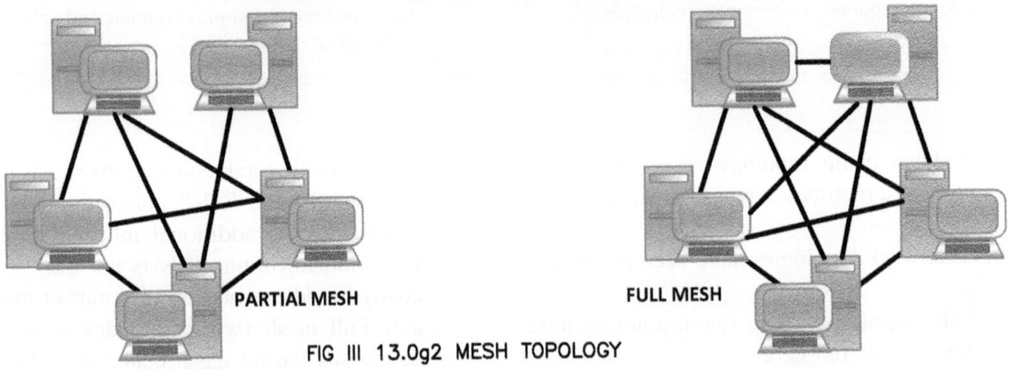


FIG III 13.0g2 MESH TOPOLOGY

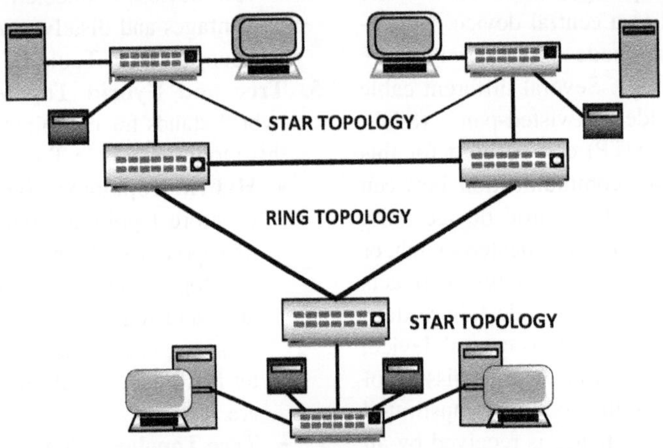


FIG III 13.0g3 HYBRID TOPOLOGY

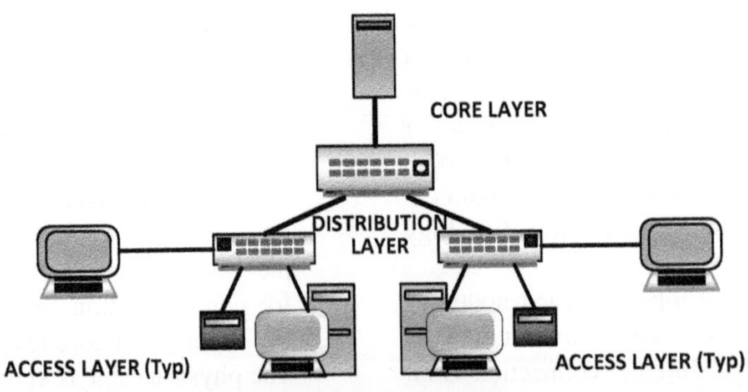


FIG III 13.0g4 TREE TOPOLOGY

**FIGURE III 13.0G** Network topology: Star, Mess, hybrid and tree.

**TABLE III 13.0D** Pros and cons for Star topology.

| Advantage | Disadvantage |
|---|---|
| Individual port in less expensive central device hence N cables for N device easy installation | Higher cable cost due higher number |
| For limited traffic fast performance possible and for node/cable failure other nodes can still work | Failure of central device means whole system will crash down; central device dependent |
| Node addition/deletion no disruption, easier scaling/upgrade. Reconfigure and troubleshooting | Limited no of nodes based on central device |

**TABLE III 13.0E** Pros and cons of Mesh Topology.

| Advantages | Disadvantages |
|---|---|
| Robust, reliable topology with P2P connection data can reach destination in different routes without traffic issue easy fault detection | More space requirement for links and cabling with difficult and costly maintenance |
| Node has privacy and security with expansion without interruption easy fault detection | Implementation is costly with high cabling requirement |
| Fast communication with fault tolerance and no total failure | Complex installation for dedicate connection |

between any two network endpoints. So, it is very useful in adding nodes but in hierarchical mode. Pros and cons of Tree topology have been enumerated in Table III 13.0f

Let us now move on to discuss area networks.

## 13.0.2 Types of area network

Based on geographical area coverage difference, network can be classified as different area network types. Apart from that there will be some classification of network based on media use e.g., wired LAN and Wireless LAN. Based on use also area network can be classified e.g., Storage/System area network (SAN). This means that there exists several issues based on which network can be classified. These factors influencing network classification are detailed in Table III 13.0g.

In this subsection we mainly concentrate on

- Private Area Network (PAN)
- Local Area Network (LAN) Types: wired LAN, wireless LAN (WLAN)
- Metropolitan Area Network (MAN)
- Wide Area Network (WAN)
- Other Network types e.g., Storage/System area network (SAN), CAN, Enterprise Private Network (EPN), Passive optical local area network (POLAN) etc.

Let the discussion start with reference to major area networks illustrated in Fig. III 13.0h.

1. **PAN Details**: As shown in Fig. III 13.0h1, PAN is arranged within typically within a range of 10 m (30′). Use of PAN for connecting the computer devices such as laptop, mobile phones, media player, and play stations is quite common. Common PAN using wired USB and wireless Wifi/Bluetooth are: Wireless modem, Bluetooth enabled headphones, wireless printers, and TV remotes etc. PAN networks are relatively secure and safe with distance limitations of 10 m. PAN may establish a bad connection to other networks at the same radio bands.
2. **LAN types**: As already aware during discussions in Chapter 10, Two LAN types are:

**TABLE III 13.0F** Pros and cons of Tree Topology.

| Advantage | Disadvantage |
|---|---|
| Failure of node/connection will cause affect associated segment only. | Heavy cabling, has high cable cost and if core element fails, network collapses |
| Expansion, error detection and maintenance is easy. | With more nodes maintenance is difficult and with hub failure attached nodes fail |

**TABLE III 13.0G** Network classification methods.

| Factors | Associated issues |
|---|---|
| Geographical | This represents coverage of geographical area that could be very short distance such as a few meters e.g., PAN using Bluetooth, it could be small area LAN such as control network within a plant. It could be global enterprise network WAN or internet. |
| Topological | Nodes/stations of a network can be connected to each other differently in some fashion. Here connection means physical/logical and/or both. This refers to topological difference in network already discussed in I 2.0.3. So not repeated again. |
| Architecture | Based on architecture computer networks can be discriminated e.g., Client-Server, peer-to-peer or hybrid.<br>Depending on architecture there could be one or more servers with several clients. Server takes and processes request on behalf of clients.<br>Peer to peer communication at same level say between servers.<br>Hybrid combination of both |
| Administration | Network can be private inaccessible from outside e.g., private network and enterprise, or it could be public (say) online shopping network, and or combination of both through suitable safeguard/firewall, e.g., enterprise network including control network where control network part is inaccessible but ordering can be done on line. |
| Application | Based on application there can be divisions such as<br>Resource sharing common in control network as in NCS or simple peripheral sharing<br>Information exchange common in modern Industrie4.0 and/or digital transformation<br>Interaction in dynamic web page<br>Other usage such as instant messaging, parallel computing etc. |

- **Wired LAN**: In Local Area Network (original area network conceived) a group of computers/computerized devices (often called nodes), are connected to each other in comparatively smaller area (within a few Km). In LAN nodes are connected through a communication medium viz. twisted pair/coaxial cable, optical fiber/wireless medium. Typical set of LANs (both wired and wireless) has been depicted in Fig. III 13.0h2. For making LAN a switch (or stack of switches) and data bus systems are used. LANs are privately owned and they use a private addressing scheme (unique in relation to node) as defined by the Transmission control protocol/ Internet protocol (TCP/IP). Here LAN definition given is slightly different from the definition available in the text books to take care of usage of LAN in plant control systems. In almost all industrial plants, various controllers such as Programmable logic control (PLC)/DCS of plants are connected in LAN where intelligent (computerized) nodes are connected through communication network to which field devices such as sensors/Final control element (FCE) may be connected through fieldbus. As LANs can be built with inexpensive network hardware such as hubs, switches, ordinary twisted pair cables. Building up LAN is less costly. Mostly wired LANs are used but wireless LANS are not far behind. Depending on configuration, high speed data transmission, easy deployment of suitable security/firewall, access controls are possible. However, with network integration of modern days for Industrie 4.0, security is a big concerned. LANs are also widely used to share resources starting from common controller simple devices such as printers, shared hard-drive. Table III 13.0h enumerates various advantages and disadvantages of LAN.
- **Wireless LAN**: This works mainly as per IEEE802.11 standard, which uses CSMA/CA (unlike CD, here, collision is prevented before they could occur) for media access control. It uses encryption for privacy also. Some of these already have been discussed in Chapter 10 Of this volume IEEE802.11 architecture comprises mainly of:
  - Access point: As discussed before, it is used to transmit and receive data over WLAN.
  - Stations (STA): Stations comprise all devices connected to the wireless LAN. Two kinds of work stations viz. wireless routers Access point (AP) and client (e.g., Laptop smart phone or interface controller).

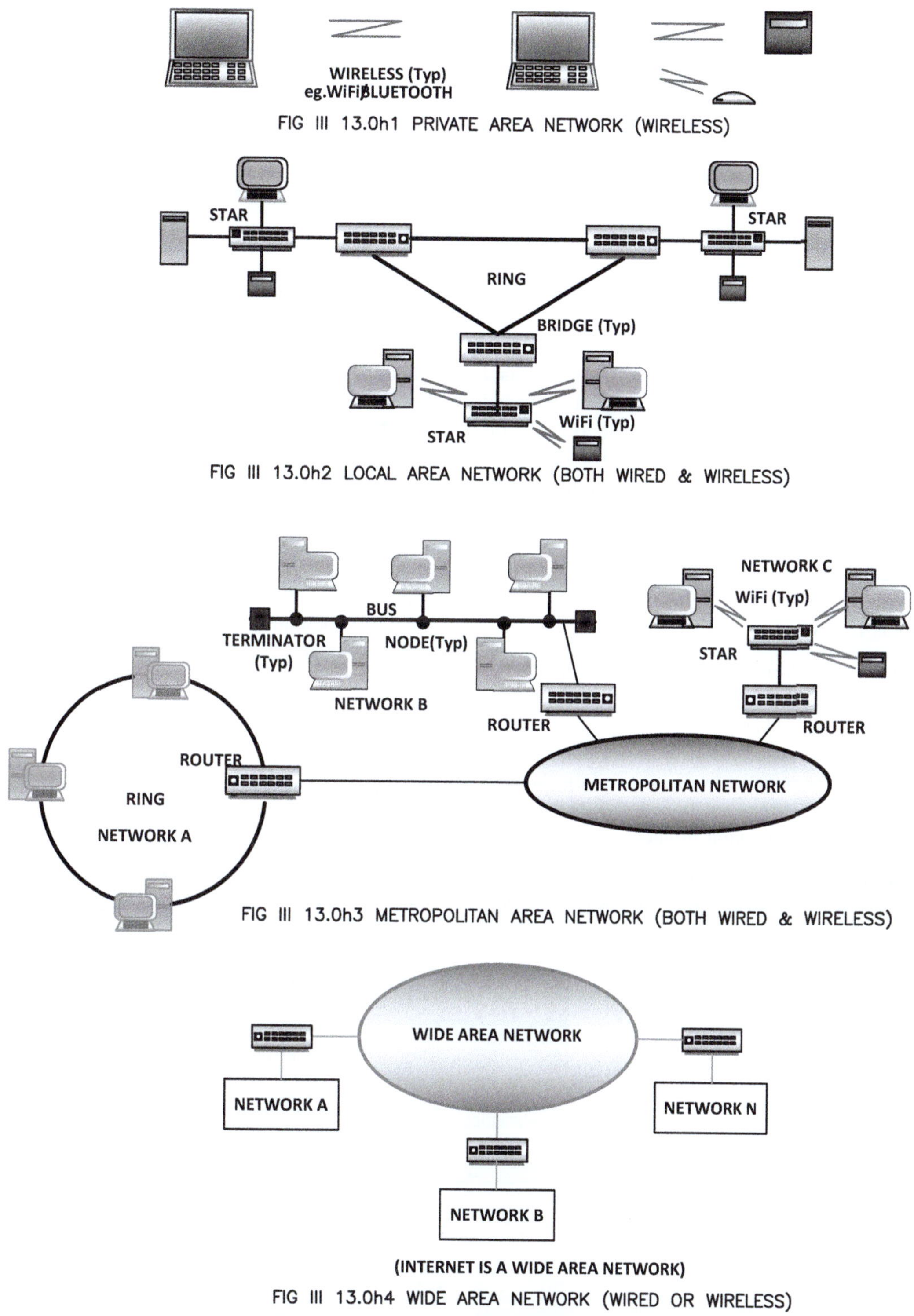


**FIGURE III 13.0H** Area network PAN LAN man WAN.

**TABLE III 13.0H** Advantages and disadvantages of wired LAN.

| Advantages of wired LAN | Disadvantages of wired LAN |
|---|---|
| Resource sharing hence hardware cost | High initial setup cost for LAN in automation with moderate users |
| Possibility of sharing application software, storage, internet to reduce overall cost | Administrator can check files hence privacy may be at stake |
| Security is easier to provide but integration is complex and elaborate | Security must be properly addressed especially for control system |
| Data/information can be transmitted easily to many users | Maintenance is complex and time-consuming |

- Basic Service Set (BSS): A basic service set is a group of stations communicating at physical layer level. BSS can be of two categories Infrastructure BSS (communication through access points) and Independent BSS (communication in ad hoc peer-to-peer basis)
- Extended Service Set (ESS): A set of all connected BSS.
- Distribution System (DS): It connects access points in ESS.

Typical arrangement of IEEE802.11 architecture has been elaborated in Fig. III 13.0i, which has been developed based on idea from Ref. [11].

Wireless LAN often used for cost saving due to cable but at times to make say internet available to a large number of persons. New devices are typically added and configured using Dynamic host configuration protocol (DHCP).

3. **MAN Discussions**: MAN consists of a computer network across a limited area of around 50 Km maximum, and shown in Fig. III 13.0h3. Area coverage of MAN is larger than LAN and may include several LANs. In MAN cables or optical fiber are used for moderate rate of transmission in distributed computations. The most widely used links and protocols in MAN are RS-232, ATM, ISDN, ADSL etc. Normally MAN services are provided by Internet service providers (ISP). MAN provides uplink for LANs to WANs or internet and expand. A few advantages mainly include: fast communication with high efficiency, with dual bus both way communication is possible. However, it calls for huge cabling and associated cable cost.
4. **WAN Discussions**: WAN covers a wide area, which can be even cover entire country. In WAN several networks may be connected together as shown in Fig. III 13.0h4. Internet for that matter is also a basic form of WAN. WAN can be private or it can be public leased network. However, normally WANs are not owned by any one organization but a collective or distributed

**FIGURE III 13.0I** Typical IEEE802.11 architecture.

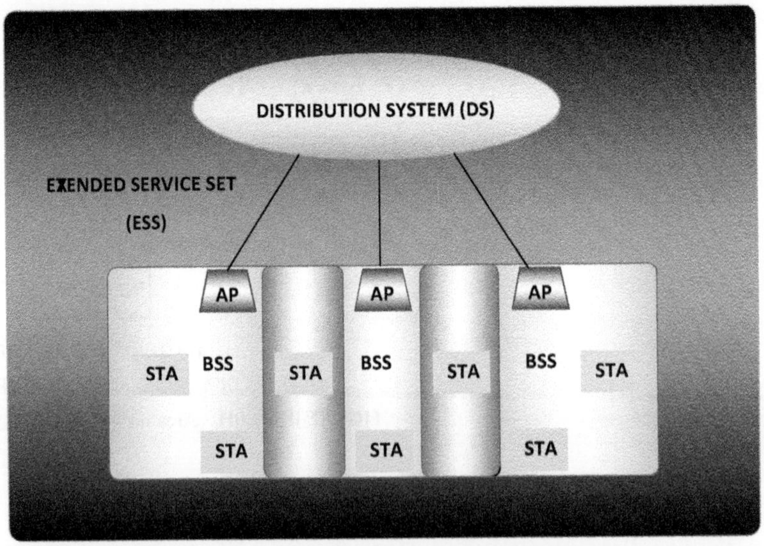


AP:ACCESS POINT; BSS:BASIC SERVICE STATION; STA:STATION

DEVELOPED BASED ON IDEA FROM 1

ownership. This is the point where WAN differ from LAN. WAN uses communication medium such as Public switch telephone network (PSTN) or Satellite links. WAN operates on low data rates. Major advantages mainly include: Communication coverage over large area, sharing software and resources including expensive devices, receiving updates (say– Security patch), high bandwidth hence high speed data transfer, working on updated data. However, it has a few disadvantages such as: security is more prominent (affected by virus and/or malware), Installation is expensive, complex and time consuming, maintenance and fault detection is complex and costly.

5. **Virtual private network (VPN)**: A VPN, is an encrypted "virtual" connection over the Internet from a device to a network. With VPN users can send and receive data through a virtual point-to point as if their devices were connected to the private network. Users can access a private network remotely. The encrypted connection helps ensure that sensitive data is safely transmitted. On account of encrypted traffic between the devices, traffic remains private as it travels. VPNs are not just restricted anymore, individuals interested in securing their communications over unsecured public WiFi networks and remaining anonymous during their online transactions have begun subscribing to paid VPN. Paid VPN services function very much like business VPNs but go through a VPN provider to reach the internet, rather than through a private business. These services are incredibly easy to use. All you have to do is download the software, install it on your device, and connect to the server of your choice. As long as VPN is connected, no one (not even your ISP) can know user and user's location as well as user's activity [12].

Let us now concentrate on ISO OSI model, which is the basis of modern networking.

## 13.1 Discussions on reference model

In 1977, ISO recognized the special and urgent need standards for heterogeneous network. Initial development result was ARPANET. Later Open systems interconnection (OSI) had been developed. The term "open" was chosen to emphasize that by conforming to those international standards, a system will be "open" to all other systems obeying the same standards. ISO determined a number of principles to be considered for defining the specific set of layers in the OSI architecture, and applied those principles to come up with the seven layers of the OSI Architecture. The OSI model is a reference tool to study the data communications between any two networked systems by dividing the communications processes into seven layers. Structured layering permits the network of OSI to be viewed as logically composed of a succession of layers, each wrapping the lower layers and isolating them from the higher layers. Thus, each layer both performs specific functions to support the layers above it and offers services to the layers below it. The three lowest layers focus on passing traffic through the network to an end system hence mostly hardware-related issues (responsibility of the network). The top four layers come into play in the end system to complete the process (responsibility of the host), hence mostly related to software. OSI networking model offers a generic approach through these networking functional multiple layers. Except the lowest layer, each layer relies on the layers below it to provide supporting capabilities and performs support to the layers above it till top application layer is reached. This model of layered functionality is also called a "protocol stack" or "protocol suite". Protocols, or rules, can do their work in either hardware or software or, as with most protocol stacks, in a combination of the two. The major benefits of the OSI model are.

- Easy user understanding about the big picture of networking
- Functioning of hardware and software elements together and understanding the same
- Easy troubleshooting by separating networks into manageable pieces
- Comparison of functional relationships on different networks
- Easy interpretation of vendor explanations of product functionality
- Understand new network technologies as they are developed.

The TCP/IP suite has become the industry-standard method of interconnecting hosts, networks, and the Internet. It is seen as the engine behind the Internet and networks worldwide.

TCP/IP supports both standard and nonstandard, with the help of a set of core protocols. One of its main goals of TCP/IP is to make sure that the data sent by the sender arrives safely and correctly at the receiver's end. To do this, the data is broken down into smaller parts called packets before being sent. These packets travel separately and are reassembled in the correct order when they reach the destination. The TCP/IP framework is used to model the communication in a network and is mainly a collection of network protocols and organization of these protocols in different layers for modeling. Unlike seven layers of OSI model it has four viz. Application, Transport, Network/Internet and Network access dealing with physical hardware and MAC address. TCP/IP is used over the OSI model because it is Simpler open standard with robustness and flexibility, Protocol driven design, practical, and widely adopted for real-world networking and the internet. It offers benefits such as: Interoperability, scalability, flexibility, standardization. However, security, data overhead etc. are matters of concern. The socket interface is one of several application programming interfaces to the communication protocols. A socket (address is a combination of IP address and port number) is a special type of file handle, which is used by a process to request network services from the operating system. Socket programming is a technique used to write applications that communicate over a network using TCP/IP protocols. So, this is referred to as TCP/IP Socket programming.

In this section OSI reference model, TCP/IP model as well as TCP/IP Socket programming will be discussed at length.

## 13.1.0 ISO–OSI reference model

In the year 1983–84, The International Organization for Standardization (ISO) developed a reference model called the Open system interconnect (OSI). An ISO OSI reference model consists of seven layers viz. Application, Presentation, Session, Transport, Network, Data link, and Physical (to remember: phrasing: "All People Seem To Need Data Processing"). There is another reference model TCP/IP. Both the models have been detailed in Fig. III 13.1a after this subsection.

The arrangement and data flow using OSI model is depicted in Fig. III 13.1a1 and with reference to above figure, following are details of various protocols.

- APDU: Application protocol data unit.
- PPDU: Presentation protocol data unit.
- SPDU: Session protocol data unit.
- TPDU: Transport protocol data unit (Segment).
- Packet: Network layer host-router protocol.
- Frame: Data-link layer host-router protocol.
- Bits: Physical layer host-router protocol

Functional details of each layer have been provided in Table III 13.1a.

Now, all seven layers, individually, will be analyzed for the purposes and functionalities.

All the layers have been described with reference to Fig. III 13.1a1.

1. **Physical Layer (Layer 1)**: The physical layer is the first and bottom-most layer of the OSI Reference Model and is responsible for transmitting Raw BIT over communication channel, and to make sure if a node sends one receiving node gets ONE not ZERO [13]. It provides transmission interface between the devices and the transmission media. So, it specifies mechanical, electrical and procedural network interface details including media type, connector type etc. As this layer is concerned with Signal, binary transmission and Media – issues etc. so associated protocol issues are physical interface, Bit rate, distance etc. Data encapsulation and synchronizations are other function of the layer. As a part of line configuration and data transmissions it defines physical connections between two devices and transmission modes (simplex/half duplex/full duplex). It is also concerned with associated network topology.
2. **Data link Layer (2)**: This layer converts the raw data in to byte, then byte to a Data Frame for transferring to Network layer without any transmission error [13]. Data link adds header and trailer at the two sides of the frame. It defines the format of the data on the network. Header to the frame contains destination address for physical addressing. Again, at receiving end, on receiving the signal, it decodes it into frames for delivery into the hardware.
   - **Error control**: Error control is achieved by adding a calculated value CRC at trailer of the frame. For any error then the receiver sends request for the retransmission of the corrupted frames.
   - **Access control**: In case of more than one in the same communication channel, then the data link layer protocols are used to determine which device has control over the link at a given time.
   - **Logical link control (LLC)**: It is responsible for transferring the packets to the network layer of the receiver. From the header, it identifies the address of the network layer protocol and also provides flow control.
   - **Flow Control**: Transmission speed is a great concern to ensure that constant data rate is maintained on both the sides so that no data get corrupted. There are two types of flow control process:
     - Stop and Wait for flow control [14]: Here it pushes the transmitter after the data is transmitted to stop and wait for the receiver's end acknowledgment of the frame. Naturally, the second data frame is sent over the medium, after this acknowledgment.
     - Sliding window: In this process, both the sender and the receiver will decide the number of frames after which the acknowledgment should be exchanged. So, for slow receivers, it creates is buffering space in the receiver. This process is time-saving as fewer resources are used in the flow control process.
   - **MAC**: Data-link layer supervises the physical addressing system called the MAC address for the networks and handles the access of the assorted network components to the physical medium. MAC address is responsible for the unique identification of each device residing on network.
   - **Flow control and Multiple Accesses**: Utilizing CSMA/CD this layer can provide access to multiple devices with proper synchronization between two devices. It supports layer two switches and bridges.

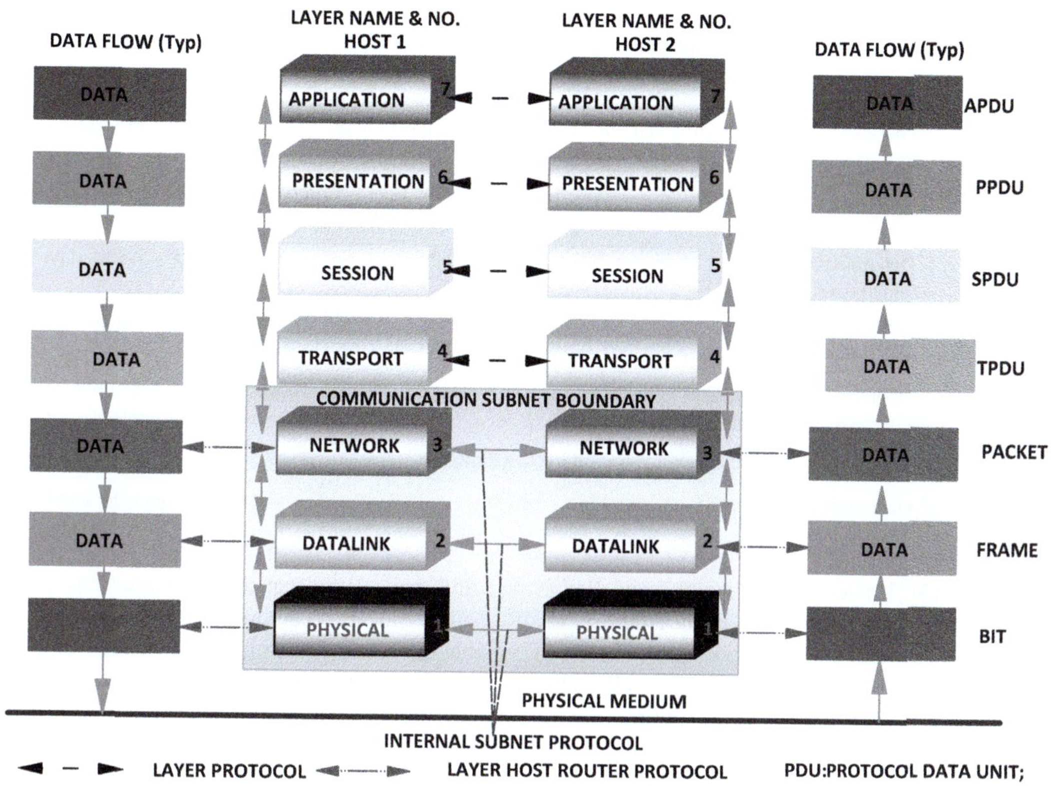


FIG III 13.1a1 ISO OSI REFERENCE MODEL

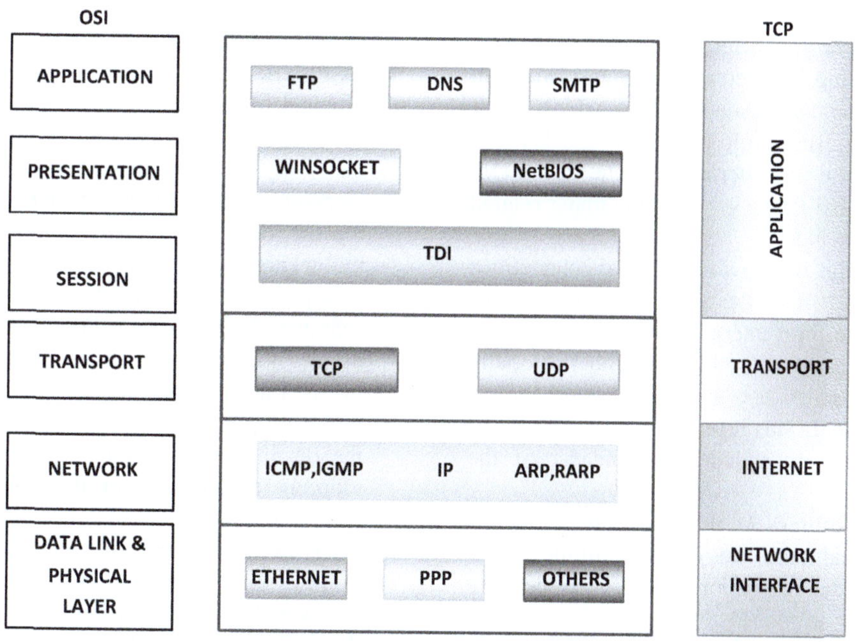


FIG III 13.1a2 TCP/IP REFERENCE MODEL MODEL

FOR INFORMATION FLOW REFER FIG III ▮

**FIGURE III 13.1A** ISO OSI and TCP/IP model.

**TABLE III 13.1A Functional details of different Layers in ISO OSI model [13].**

| No. | Name | Function | Remarks |
|---|---|---|---|
| 1 | Physical | Physical medium through which bits at specified signal level and rate moves from device to device | e.g., EIA/TIA 232, V.35 |
| 2 | Data link | Combines data byte in to frame, carries out detection and correction of error (checksum) for data transmission and provides access to media through MAC | IEEE802.2, HDLC |
| 3 | Network | Determination of data path by providing logical addressing of router through which data packet can be sent in shortest path | IP and IPX; path selection |
| 4 | Transport | Reliable end to end message delivery both connection oriented/connection less and error checking | TCP. UDP |
| 5 | Session | Opening, managing, and closing of session and provide synchronization. | Keeps application data separate. |
| 6 | Presentation | Building block of data compression and encryption | Translation function also |
| 7 | Application | Data meaning – purpose; user service | User interface FTP/HTTP |

3. **Network Layer (3)**: It is concerned mainly with Path/route finding and determination of Logical Addressing for data transfer. In a complex network with too many nodes, the finding an easy sure route in time-efficient way is a major task [13]. It uses using routing protocols, switching, error detection and addressing techniques [14]. It regulates the subnet and looks for the path how packet data can reach from source to destination for different networks working on the same or different protocol or for different topologies. It uses logical IP addressing and routers for communication to this function. Too many Data Packet transfer without suitable route will result traffic jam, causing delay. So, Network layer has to find alternative route. Issues are: Packet header, virtual Path e.g., IP. Some major related issues here shall include:
   - **Internetworking**: By providing logical connections between devices, it is the main responsibility of network layer to form internetworking.
   - **IP Addressing**: It is a 32-bit number globally unique for each network host It refers to logical network address. In this layer the source and destination addresses are added to the header of the frame.
   - **Routing and Router**: As discussed above it finds the best optimal path out of the multiple paths from source to the destination. Routers are specified in this layer.
   - **Packetizing**: With the help of IP, network layer receives the packets from the upper layer and converts them into packets. The process is packetizing.
4. **Transportation Layer (4)**: Transfer of data, in order correctly without any duplication, is the main responsibility of this layer. It is a layer for fragmentation and reassembly. It receives data from layer three and splits to many data packets (to facilitate network layer action) and makes sure that these are correctly received at the other end. At receiving end this layer needs to reassemble the same. It uses message headers and control messages to make sure that the end user gets the message in the order it was sent. Since it provides point to point connection between the devices it is also called an end-to-end layer. It also decides the type of service to be provided to the session layer. It shields the upper layers from changes in Hardware technologies. It ensures End to End Connection. Two protocols here are:
   - **TCP**: It is standard, connection-oriented and reliable protocol for communicate over the internet. At first the connection is established between the two hosts then only the data is sent over the network. The receiver always sends an acknowledgment once the first data packet is transmitted. After receiving the acknowledgment from the receiver, the second data packet is sent over the medium. When data is sent over the TCP connection, then the TCP protocol divides the data into smaller units known as segments. Each segment travels over the internet using multiple routes, and they arrive in different orders at the destination. The transmission control protocol reorders the packets in the correct order at the receiving end.
   - **User datagram protocol (UDP)**: UDP is a connectionless and unreliable protocol as in this case receiver does not send any acknowledgment when the packet is received, the sender does not wait for any acknowledgment. However, this makes it very easy to process any network requirement with loss of no time. This type of protocol is widely used in video streaming, online games some data packets losses do not have much significance.
5. **Session Layer (5)**: In session layer, process to process communication takes place between the nodes in a

network. Here Dialog Control (tracking who is to speak and who is to listen), TOKEN Management to prevent two sets of parties from doing the same operation at the same time! Synchronization of check points etc. takes place in this layer i.e., interhost communication e.g., NetBIOS (a protocol for some functions such as browsing, communication between servers) [13]. So, this layer is used to establish, maintain and synchronizes the interaction between communicating devices. Major functions shall include:

- **Dialog control**: As a dialog controller it allows the communication between two processes in half-duplex or full-duplex mode.
- **Synchronization**: Session layer adds some checkpoints in transmission for synchronization and recovery. It is very helpful for efficient delivery of data without any loss at the receiver end. If any error occurs in the middle of the transmission of data, then the transmission will take place again from the checkpoint.

6. **Presentation Layer (6)**: When communication between the two nodes takes place there shall be some defined Syntax and Semantics of information exchange. This layer looks after the same. The processors communicating may have different data representation, so someone has to define the same. It is responsibility of this layer to carry out the same, as a translator. Major functions are:
   - **Translator**: As translator, presentation layer handles the interoperability between the different encoding methods used by different hosts.
   - **Encryption**: It does encryption is needed to maintain privacy.
   - **Compression**: It also performs data compression for reduced bit transmission—important for multimedia.
7. **Application Layer (7)**: This is the topmost and seventh layer of the OSI reference model. This layer will communicate with the end users and user applications. Email, FTP etc. is the examples of application layer i.e., this layer is to offer service to the user (user interface Layer).

### 13.1.1 Transmission control protocol/internet protocol model

TCP/IP, is a set of protocol commonly used to communicate over internet. TCP/IP can also be used as a communications protocol in a private computer network also. TCP and IP are the two main protocols, though a few others are included in TCP/IP protocol suite (a set of rules and procedures).

- **TCP explained**: In TCP/IP Protocol suite, TCP is a reliable connection –oriented delivery service where data is transmitted in segments. As discussed, it can divide message into segments before transmission over the internet and reassembled in the right order at the destination address.
- **IP explained**: IP on the contrary is a connectionless unreliable (i.e., delivery is not guaranteed) datagram mainly responsible for routing and addressing Packets between the hosts/nodes. Here connectionless means that no session is established. IP defines how to address and route each packet to ensure right destination. Each gateway computer on the network checks this IP address to determine where to forward the message.
- **TCP/IP Functioning and characteristics**: TCP/IP uses the client-server model of communication. TCP/IP protocol suite is stateless, meaning that each client request is considered (new as it is unrelated to previous requests) and being stateless frees up network paths. However, transport layer itself, is stateful. It transmits a single message, and its connection remains in place until all the packets in a message have been received and reassembled at the destination. The TCP/IP model differs from ISO OSI model and it shows certain characteristics such as: Support for a flexible architecture (modification is easier), TCP is a connection-oriented protocol (and network remains intact as source and destination machines are functioning properly), TCP offers reliability and ensures data receipt in order.

Also, it allows flow control, so that sender can never overpower a receiver with data. With reference to Fig. III 13.1a2, let us begin discussion various TCP/IP layers.

1. **Network interface**: When compared with OSI model comprise both physical and data link layer and often referred to as network access layer. Some describes this as residing in layer 3, being encapsulated by layer two protocols. It defines how data should be sent physically through the network by using it. This layer is transparent to the user. Basic responsibilities shall include:
   - Error checking of incoming data packets
   - Computer network interface and tagging outgoing data packets for error checking
   - Acknowledgment of data—resend if not Acknowledged.
   - It supports several support protocols such as: Ethernet, Address Resolution Protocol (ARP), Serial line internet protocol (SLIP) used to send data across serial line), Compressed SLIP (CSLIP), Point-to-Point Protocol (PPP).
2. **Internet/network layer**: Internet (or Network) layer deals with packets for sending the packets from any network/any computer still they reach the destination irrespective of the route i.e., it is responsible for the functional and procedural method for transferring variable length data sequences from one node to another

with the help of various networks. IP address and address resolution takes place within internet layer. It consists of the following Protocols.

- **Internet protocol**: It is a connectionless unreliable data gram protocol. IP provides mechanism software to address and manage data packets being sent to computers. There are several headers associated with IP such as Source IP address, Destination IP address, Identification (inform destination host whether to be passed on to the transport layer protocol TCP/UDP), Protocol check sum (to check integrity of data). There are two versions of internet protocol (IP) are: IPV4 is major in use now with 32 bit addresses (Networking industries are slowly phasing IPV4 out) and IPV6 is replacement for IPV4, with 128bit addresses for future growth.
- **ARP**: ARP is used to convert logical address to Physical address. It is used to find specific address (hardware) from specific IP no discussed below.
- **Reverse ARP (RARP)**: RARP is used to resolve IP address without having own storage.
- **Internet control message protocol (ICMP)**: ICMP is (encapsulated within IP) used to report connection status back to the computer and for retry the connection. It also does error reporting.
- **Internet group management protocol (IGMP)**: IGMP is used by multicasting router for tracking group memberships.

3. **Transport layer**: Transport layer is meant for network connectivity to ensure that required quantum of data sent, is received by right node/machine at desired rate as well as for right application. This layer builds on the message, which it receives from the application layer. It also shields application layer from complexity of data. Error checking, flow control, verification of integrity and completeness are other functions. Two main protocols here are:
   - **Transport layer: TCP**: TCP is a reliable connection –oriented delivery service where data is transmitted in segments. It is used by transport by some application. Reliability is achieved by assignment of sequence number to each segment sent. Several sequence no. of TCP headers are Source port, Destination Port, Sequence number (sequence no of first byte of data), TCP check sum etc. There are separate port numbers for several applications e.g., TCP port no 20 and 21: FTP (data and control), 80: HTTP and 139: NetBIOS.
   - **Transport layer: UDP**: Connectionless unreliable (not guaranteed) protocol UDP is fast and cost-effective. Itis told by the application layer to which machine is supposed to transmit. Three headers of UDP are: Source port, Destination port, and UDP Check sum. Port functions are multiplexed message and give the idea where the message to be sent e.g., UDP Port 53: Domain Name System (DNS) and port 137: NetBIOS.
4. **Application layer**: This layer contains network applications and services that user interfaces in order to use network (internet). Application layer also have utilities such as printer. Some of these protocols here are:
   - **File transfer protocol (FTP)**: Used fr file transfer from one machine to other
   - **Hypertext transfer protocol (HTTP)/HTTPS ("S" Secured)**: It is used by the World Wide Web to manage communications between web browsers and servers.
   - **Simple mail transfer protocol (SMTP)**: This support email and sending data in email.
5. **TCP/IP and OSI**: After assessing pros and cons of TCP/IP, TCP/IP has been compared with OSI.
   - **TCP/IP: Pros and cons**: Pros and cons of TCP/IP has been elaborated in Table III13.1b.
   - **Comparison of TCP/IP and OSI**: Comparison of TCP/IP with OSI model has been presented in Table III 13.1c

### 13.1.2 TCP/IP, socket program and MAC address

Internet service is a client/server architecture, for exchange of distributed information systems. In client/server environment, one or more servers manage the centralized database and clients gain access to the data through the server. First, the client establishes a connection with one server by sending a segment with SYN and informs the server about the client should start communication, and asks for sequence number. Server (can either accept them

**TABLE III 13.1B** Pros and cons of TCP/IP.

| Advantages | Disadvantages |
|---|---|
| Independent of the operating system, supports routing-protocols connection between two computers of different types | Complex model to set up, manage and replacement and cannot guarantee delivery for high overhead. |
| Scalable client-server architecture TCP/IP model can establish internetworking between the organizations independently. | No separation between services interface and protocols |

**TABLE III 13.1C Comparison between two reference models: TCP/IP and OSI.**

| TCP/IP model | OSI model |
|---|---|
| It consists of 4 layers; presentation, session and application layers have been put together to form application layer. Also, physical layer and data link layers fuse to create Network interface layer. | It has 7 layers; it has separate application, presentation and session layers. Also it has separate physical and data link layers |
| This model uses the internet layer to define the routing standards and protocols. It uses header with 20bytes | This model utilizes Network layer for it. It uses header with 5bytes |
| It does not have strict boundaries, yet it is more reliable compared with OSI model. | It has strict boundaries but not that reliable. |
| It is protocol oriented, hence protocols were made before model with horizontal approach | In generic OSI model, protocols were made after model is made with vertical approach. |
| This model provides connectionless services but packet delivery is not guaranteed. It is internet oriented | OSI use connection-oriented and connectionless services and packet delivery is assured. |
| Complicated to set up, manage and replace | Protocols are well covered and it is rather easier to replace with new technology |
| It is associated with different computers | It standardizes switches, routers. |

or reject –if accepted) responds with SYN-ACK signal to signify the response and SYN signifies the sequence number it should able to start. The client acknowledges the response of the server. Then both create a stable connection for the actual data transfer process and the connection remains open until either the client or the server terminates. To ensure that each communication reaches the intended destination intact, TCP/IP model breaks down data into packets then these packets may travel along different routes (if one route is congested). All the packets are reassembled into the complete message at the destination. Sending the data in small packets makes it easier to maintain accuracy versus sending all the data at once. on the other end. To share the resources, support from some programs are necessary. These programs are essential for developing any software for a networked system (and Internet). Such programs create socket to send and receive data to/from the network. So, these programs are commonly known as Socket Programs. Socket programming provides a support mechanism for two programs (on different machines), to communicate over a network. The sockets in computer networking can be thought as interfaces that can "plug into" each other over a network. The sockets, created by some application program, are uniquely identified by the socket addresses. All the communication between two processes in the network (client and server) are carried out by the socket address. Network applications use sockets to communicate over a TCP/IP network.Every network interface has 6-byte long physical or MAC (Media Access Control) address. This is the hardware address that the lowest layer of the network architecture uses to communicate. Hence, to forward data using a physical link, the MAC address is used. In this subsystem brief discussions have been included on these.

1. **Socket**: A socket is a software concept for a connection, enabling applications to connect to a TCP/IP network. So, socket is a software endpoint for network communication pertinent to two applications. A socket is a concept that represents an endpoint of a connection with interfaces comprising, library of routines to help developers for building desired applications to communicate with other applications over Internet. An application running on a host creates a socket or doorway to connect with an application on another host. Messages pass through this socket or doorway. Sockets enable virtual TCP/UDP communication channels between hosts.A socket has three attributes viz. the IP address identifies a network node, the port number to identify a process on the node (viz. FTP: port 20/21) and the socket type determines a connection-oriented or connectionless protocol (discussed below).
2. **Socket programming**: Client program creates a socket and tries to connect server so that the server can create a socket object on its end of the communication, so that the client and the server can communicate. Major steps are [15]:
   - Server initiates server socket object, with port number. Also, server invokes the accept method of server socket class and waits for client to connect at the given port.
   - A client initiates a socket object, specifying the server name with the port number.

- The constructor of the socket class attempts to connect the client to the desired server and the port number. If communication is established, the client now has a socket object capable of communicating with the server.
- On the server side, the accept method returns a reference to a new socket on the server that is connected to the client's socket. Finally close socket when done.

3. **Socket Type**: The socket type determines the protocol being used to communicate through the socket viz. connection oriented or connectionless:
   - **Connection oriented Type (TCP/IP)**: TCP/IP protocol suite supports a connection- oriented data transfer between two processes on two computers on the Internet. It calls for a logical connection to be established between the two processes before data is exchanged. However, TCP/IP depends on IP for efficient delivery of packets. IP neither guarantees delivery, nor it delivers packets in any particular sequence. It is responsibility of TCP to arrange the packets in the proper sequence, detecting whether or not errors have occurred, and requesting retransmission of packets in the case of an error e.g., Python Sock_stream – for reliable delivery in connection oriented TCP protocol.
   - **Connectionless Type (UDP)**: A connectionless data-exchange protocol does not require the sender and receiver to explicitly establish a connection. It is something like shouting for a friend in a crowd even without knowing for sure if he is there. UDP provides connectionless service for exchange of packets better known as datagrams. UDP neither guarantees that datagrams would reach the intended destination ever, nor it ensures that datagrams are delivered in the order they have been sent. UDP is used for exchange small amounts of data at a time or by applications that do not need the reliability and sequencing of data delivery as in Simple network management protocol (SNMP) e. g., Python Sock gram – for unreliable delivery in connectionless UDP protocol.
4. **MAC Address**: Major issues associated with MAC address unique identifier, for NIC are:
   - **MAC Address utility**: MAC address is important because it provides a secure way to find senders or receivers in the network and unwanted network access can be prevented. Similarly for wireless network it can be used as filter to prevent strangers from accessing a network. MAC address being unique, it is possible to track the device. MAC helps in detecting stolen device in network. Another use of MAC address could be data recovery. DHCP servers first identify device by the MAC address, then provide fixed IP addresses.
   - **MAC Address Format**: MAC address can be in hexadecimal format. Generally, it consists of a 48 bit (or 64-bit address) in 6 octets as shown in Fig. III 13.1b. As shown first three octets represent Organizationally unique identifier (OUI) by which the manufacturer, organization, or vendor is identified. IEEE registration authority committee assigns these MAC prefixes to its registered vendors. Last three octets are the number specific for the device and the number is assigned by manufacturer. First octet of the address, as shown contain 8 bits such as (say) b0, b1, b2, b3, b4, b5, b6, and b7. 0 in b0 indicates unicast with unique address of destination

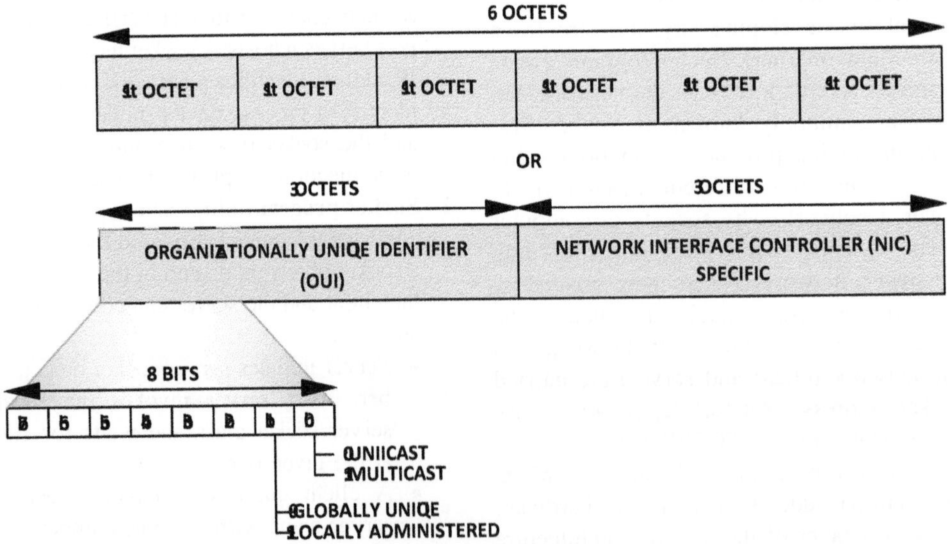


**FIGURE III 13.1B** MAC address.

server. One in b0 bit means multicast. Similarly, there could be 0 or one in b1. 0 indicates globally unique (OUI ENFORCED) to identify manufacturers and vendors worldwide or globally. One means LOCALLY ADMINISTERED, which is similar to LAN (Local Area Network) IP address.

5. **MAC with IP Address and Router**: Two issues here are:
   - **IP addresses and MAC addresses**: As discussed, in TCP/IP protocol, IP address is used to transport data from one network to another network. On the other hand, MAC address is used to deliver the data to the right device on a network. IP is a protocol that is used on a layer above Ethernet. MAC address helps computer to sense if the packet can be sent directly if the destination is in the same IP network as the computer itself, or it has to be sent through configured router. An important thing to note that routing cannot be done based on MAC address alone. MAC address at lower Ethernet, is possible only within one network. So, these are two addresses one is the original IP packet's target address and other is ID of the device to which we should send the packet. This is necessary as Ethernet uses MAC addresses; the sender needs to get the MAC address. Here role of ARP is very important. ARP is a procedure for mapping a dynamic IP address to a permanent physical machine address in a LAN. When a device wants to know the MAC address for a given IP address, it broadcasts MAC address asking response. All devices listen to the broadcast only the one with the address will respond. The asking device receives the answer and now knows that the source MAC address is the right MAC address to use. Of course, the result will be cached, so the device does not need to resolve the MAC address every time.
   - **MAC and Router [16]**: The router will receive packets for its own MAC address but for a different IP address. He will then check if he can directly reach the target IP address. If so, he sends the packet to the target. Otherwise, the router itself also has an upstream router configured and will send the packet to that route, when multiple routers are configured. Unlike home router, in the internet backbone the big routers have big routing tables so they know the best ways for all packets.

## 13.2 Network transmission IEEE 802 and vulnerability

There are various modes of various signal transmission methods. These could be unicast, multicast broadcast, polling, token passing, broadcasting etc.to name a few. In unicast there will be single sender and receiver. Instead of sending copies of packets individually to each intended recipient (unicast) or transmitting packets indiscriminately to all nodes on the local network segment (broadcast), multicast sends traffic only to registered end-points that comprise a specific multicast group. Star, Mesh, and Hybrid topology. On the other hand, Broadcast transfer (one-to-all) techniques i.e., here sender broadcast the information, for all other nodes to listen to. It is easy in Implementation but it creates network congestion and inefficient. It is used in Star and Bus topology. Channel allocation is another important issue. The Channel allocation problem arises in communication networks when multiple devices need to share a limited number of communication channels. The goal is to efficiently allocate channels to devices while avoiding interference, reducing congestion, and optimizing network performance. In this connection another ALOHA, which was invented at the University of Hawaii by Norman Abramson in the 1970's, have some contribution. The idea is applicable to systems in which uncoordinated users are competing for a single channel (shared resource). There are two types of ALOHA - Pure ALOHA (no global time synchronization) and Slotted ALOHA (global time synch. required). A subsection has been dedicated to discuss all these issues related to network transmission methods. IEEE 802 plays a great role here. IEEE 802 is a set of standards to define how network devices communicate in local and metropolitan area networks. These standards cover a range of protocols, including Ethernet and WiFi, ensuring devices can interoperate seamlessly. IEEE 802 sets the groundwork for modern networking by standardizing protocols for communication ensuring compatibility across different devices and networks to create a stable framework for innovations such as faster internet speeds and enhanced connectivity. The scope of IEEE 802 standards is limited to MAC and Physical layers (PHY). In the application domain, additional standardization of higher layer functionality is often required to ensure interoperability. Industry alliances build upon IEEE 802 standards and integrate multiple standards at multiple layers. IEEE 802 standards exist to address higher data rate applications such as network gateways, traffic aggregation points, data concentration (collectors), and internetwork routers. In order to complete the discussions on network, threat and vulnerability issues have been touched upon. Here, discussions are mainly restricted to vulnerability as threat issues are discussed at length in various chapters in Volume IV. For networking in modern days people are looking for open systems for system integrations so that there shall be smooth and seamless data exchange between OT involving plant automation and IT related to enterprise network for moving toward Industrie 4.0 and digital transformation. However, for such integration, there always exits scope for network vulnerability and security threats. Every business is under constant threat from multitude sources. Naturally this calls for suitable measure to prevent the same. In this section short discussions (to supplement the detailed discussions in volume III) will be presented to look into these security threats and their remedial measures. In this section, various security threats, firewall and Demilitarized Zone (DMZ) and cyber security issues including security issue for IoT will be covered with special reference to control systems and Supervisory control and data acquisition (SCADA). Let the section discussions start with methods of data transmission in network.

### 13.2.0 Methods of transmission in computer network

As already discussed, there are several ways and means for network transmission methods. Unicasting is simplest form, here data is sent from one source to one specific destination. Multicast networking refers to a method of sending data from one source device to specific multiple destination devices simultaneously. Multicast enables the network to handle replication of packets (to different specified nodes) through routes optimized for efficiency. "Multicast" is a network traffic distribution mode that enables one-to-many and many-to-many communication from data sources to multiple destinations simultaneously. Broadcast transfer (one-to-all) on the hand are techniques of sending information to every members of the network. Broadcasting can be classified into two types: Limited broadcasting and Direct broadcasting. In broadcasting mode, transmission happens from one host to all the other hosts connected on the LAN. In networking channel allocation is important for proper communication to take place. In network data communication channels are defined as the media or the frequency band utilized in transmitting data. The allocation of these channels can be managed in two primary ways. Channel allocation is a process in which a single channel is divided and allotted to multiple users to carry user specific tasks. The user's quantity may vary every time the process takes place. Channel allocation problem can be solved by two schemes: Static channel allocation in LANs and MANs, and Dynamic channel allocation. In computer networks, Controlled access protocol (CAP) controls how data packets are sent over a common communication medium. These protocols ensure that data is transmitted efficiently, without collisions, and with little interference from other data transmissions. Reservation, Polling and Token passing fall under CAP. Since there is competition for a single resource this kind of system is known as a contention system. As discussed, there are two types of ALOHA - Pure ALOHA (no global time synchronization) and Slotted ALOHA (global time synch. required). ALOHA permits users to transmit any time they feel like. Collisions will occur and therefore colliding frames will be destroyed. However, if feedback is available on the destruction, then users will be made aware of their frames have not been transmitted (received) successfully. So, from the above discussions it transpires that a number of methods are used in network transmission methods. This subsection has been dedicated to discuss these issues.

1. **Transmission types**: Various transmission types are:
   - **Unicasting**: Unicast is a single LAN interface where information transfer takes place between single sender and receiver. The Unicast address will have the value of the MAC address of the destination device e.g., basic mode supported by RS-232, since there are only two devices connected to the link. In a network, node A with an IP address transfers data to another node B with specific IP address in point-to-point communication. In this mode at any instant of time, there will be transmission between a pair of nodes (machines). Naturally, a packet in this network may have to travel intermediate nodes. So, routing algorithm plays a vital Role. Token passing ring or polling is also kind of point to point in LAN.
   - **Broadcasting**: Also known as one-to-all technique, where a node is able to transmit a message that can be received by all nodes connected to the network but only intended ones would receive it. Broadcasting method of data transfer are of two types viz. Limited broadcasting, Direct broadcasting.
   - **Multicasting**: Some networks allow group communication where data transmission is addressed to a group of receivers to receive only the messages that are important. This is known as multicast transmission.
   - **Broadcast and multicast elaborated**: Short message called Packet is sent by a machine (node) over the communication channel shared/listened by all other machines (nodes) in the network. Packet contains address, so, intended machine/node (s) check the address field (if it is for it!) and process the same, while others ignore the message. With the use of specific code, the transmitting machine (node) can send the message to all destinations at a time and each machine (node) shall check to process the call. In Multicasting a subnet of machines (nodes) can be addressed. Any machine (node) in the network can become member of the subnet to receive the group message viz. Group SMS/email [1,13]. So, multicast is a special case of broadcast mode only.

2. **Static channel allocation**: For more than one user for the channel, there has to be some means or algorithm for channel allocations among the users. In this classical approach of static channel allocation scheme, a fixed portion of the frequency channel is allotted to each user Naturally for N competing users, the bandwidth is divided into N channels, hence it is called is Frequency division multiplexing (FDM). This is good noninterfering for limited numbers of users but suitable, when there is large number of users with variable bandwidth requirements. If there are N channels then each capacity = C/N where C is the total capacity in bps. If arrival rate is λ frame/sec, then delay T will be given by

$$T_{FDM} = \frac{N}{\mu C - \lambda} = NT \qquad \text{(III 13.2a)}$$

where each μ represents mean length bits/frame.
3. **Dynamic channel allocation**: Here, frequency bands are allotted to users dynamically from central pool as needed, to optimize bandwidth usage and get fast transmissions. Here, assumption are:
   - **Station Model**: Assumes that each of N stations/terminals, independently produce frames. The probability of producing a frame in the interval Δt is μ. Δt where μ is the constant –arrival rate of new frames.
   - **Single Channel assumption**: A single is available for communication where all stations can send/receive. Also, all stations are equivalent and software and protocol can assign properties to them [1].
   - **Collision assumption**: When two frames transmitted simultaneously and they overlap in timewise, then collision happens. All stations can detect collision. Any collision is an error, and both frames must be retransmitted. Collisions are only possible error.
   - **Time**: Time can be divided into continuous or slotted. In continuous time there is no master clock for determining allocated interval, so frame transmission can begin at any instant. In slotted time, there will be discrete slot and frame transmission must start at the beginning of slotted time.
   - **Stations [1]**: There are two types of stations viz. Carrier sense and no carrier sense. In Carrier sense type, station can tell if the channel is in use before trying to use it. In no carrier sense type, the station cannot tell if the channel is in use before trying to use it. With this background, let us move on to discuss transmission methods.
4. **Polling**: Polling is a multipoint link consisting of on primary device and a number of secondary devices, each with address for identification, using single transmission medium. Here all data exchanges must be made through the controller. When primary device asks secondary device for data it is called polling, whereas when secondary sends something to primary it is called selecting. For any station, if there is no data, usually a "poll reject" (NAK) message is sent back to primary. Two types of polling are:
   - **Roll-Call Polling**: Each secondary device is polled in turn by the primary device (controller), by message (poll) containing the address of the device. Although all devices get the message, yet the addressed one sends data, (if any) or sends NAK. At the end of its backlog of messages it notifies the controller with a suffix to its last packet to enable controller to send a poll to the next device. Fig. III 13.2a shows the polling sequence.
     Fig. III 13.2a1 shows how one after one secondary device are polled. It also shows difference between polling and selection. All communications (even between two secondary devices) take through primary device. Various packet formats of polling has been depicted in Fig. III 13.2a2.
   - **Hub Polling**: Various packet formats of polling has been depicted in Fig I 2.0q2. Here, case go-ahead/suffix packet contains the next device address and essentially go-ahead is transmitted directly from one station to another. However, a monitoring channel is provided to indicate to the appropriate station that it should start transmitting. No wastage of slot, maximization of efficiency and bandwidth utilization are a few advantages. However, it has a few disadvantages are: Time wastage for polling lowers efficiency, starvation of a few stations in sending data, link sharing may not fair as each device has equal probability.
5. **ALHOA**: ALHOA is not an acronym. ALHOA is a multiple access protocol at the data link layer for multiple nodes' access. There are two types of ALHOA: Pure ALHOA and Slotted ALHOA (requiring global time synchronization). Fig. III 13.2b shows ALHOA transmission. ALHOA principle is explained in Fig. III 13.2b1.
   - **Pure ALHOA**: Users can transmit whenever they wish with the assumption is that when there will be collision and colliding frames may be damaged sender can determine if the frame is damaged on account of feedback. After transmission, waits for some time for acknowledgment from the receiving node. If no acknowledgment received, within specified time, then the transmission is unsuccessful. Sender deploys back off strategy to wait for random time period (to avoid more collision), after which it transmits the data packet again. Process goes on till back off time limit is reached. This has been explained in Fig. III 13.2b2. For G nodes willing to transmit data

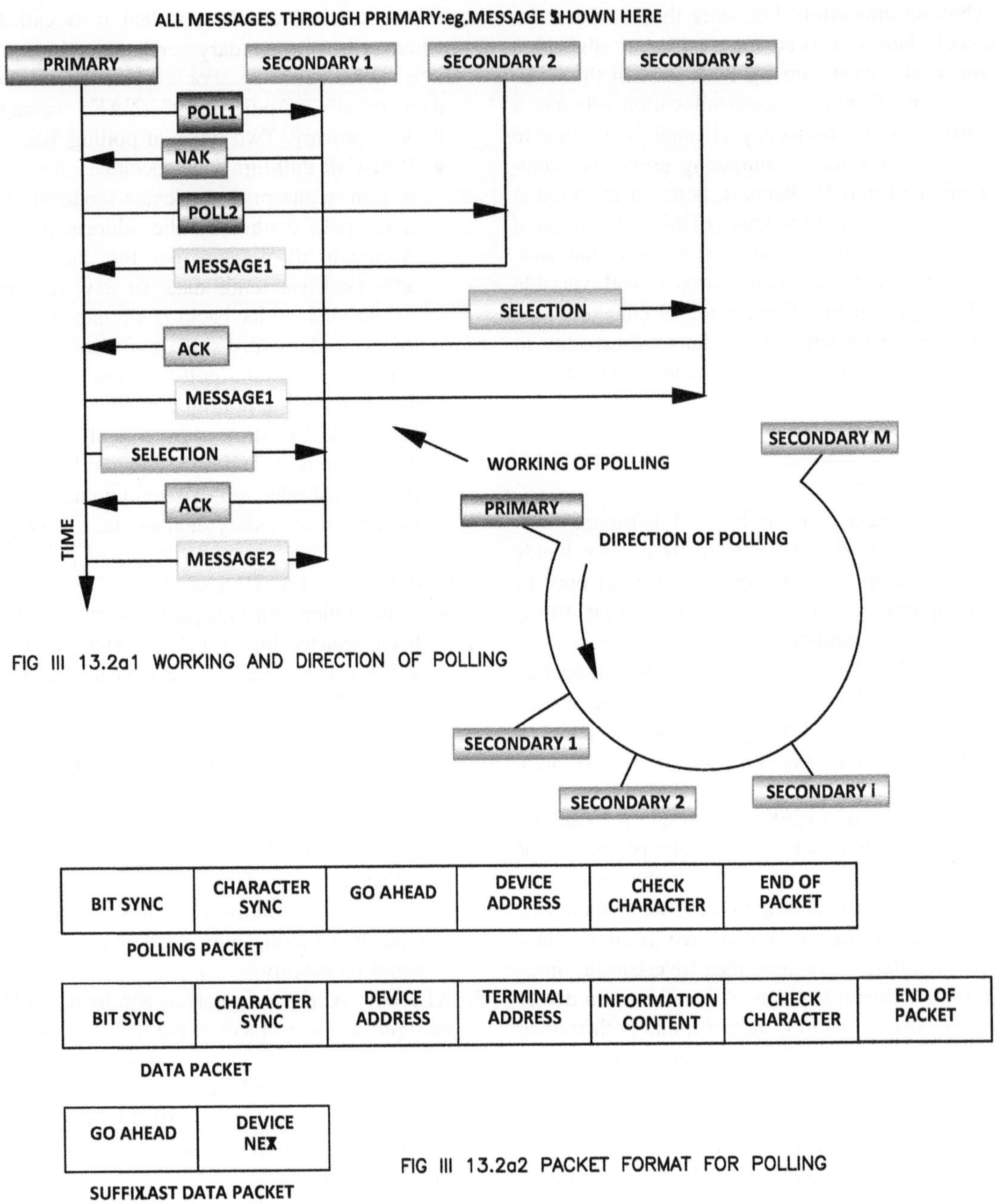

FIG III 13.2a1 WORKING AND DIRECTION OF POLLING

FIG III 13.2a2 PACKET FORMAT FOR POLLING

**FIGURE III 13.2A** Polling network.

Efficiency (η) of pure ALHOA: $\eta = G*_e{}^{-2G}$ (Eq.III 13.2b)

Maximum occurs at G = 1/2; and its value will be η = 18.4%.

- **Slotted ALHOA**: Slotted Aloha divides the time of shared channel into discrete intervals referred to as time slot, (here, synchronization is important). Here, any node can transmit its data in any time slot with a condition that the node has to start transmission from the beginning of the time slot. If missed, then node has to wait for the next time slot. Here a collision may occur if more than one node tries to transmit data at the beginning of the same time slot as explained in Fig. III 13.2b3. Here for G nodes willing to transmit data.

$$\text{Efficiency } \eta = G*_e{}^{-G} \quad \text{(III 13.2c)}$$

Maximum occurs at G = 1; and its value will be η = 36.8%.; so, it reduces collision to half and make efficiency double of pure ALHOA, which is simpler to implement.

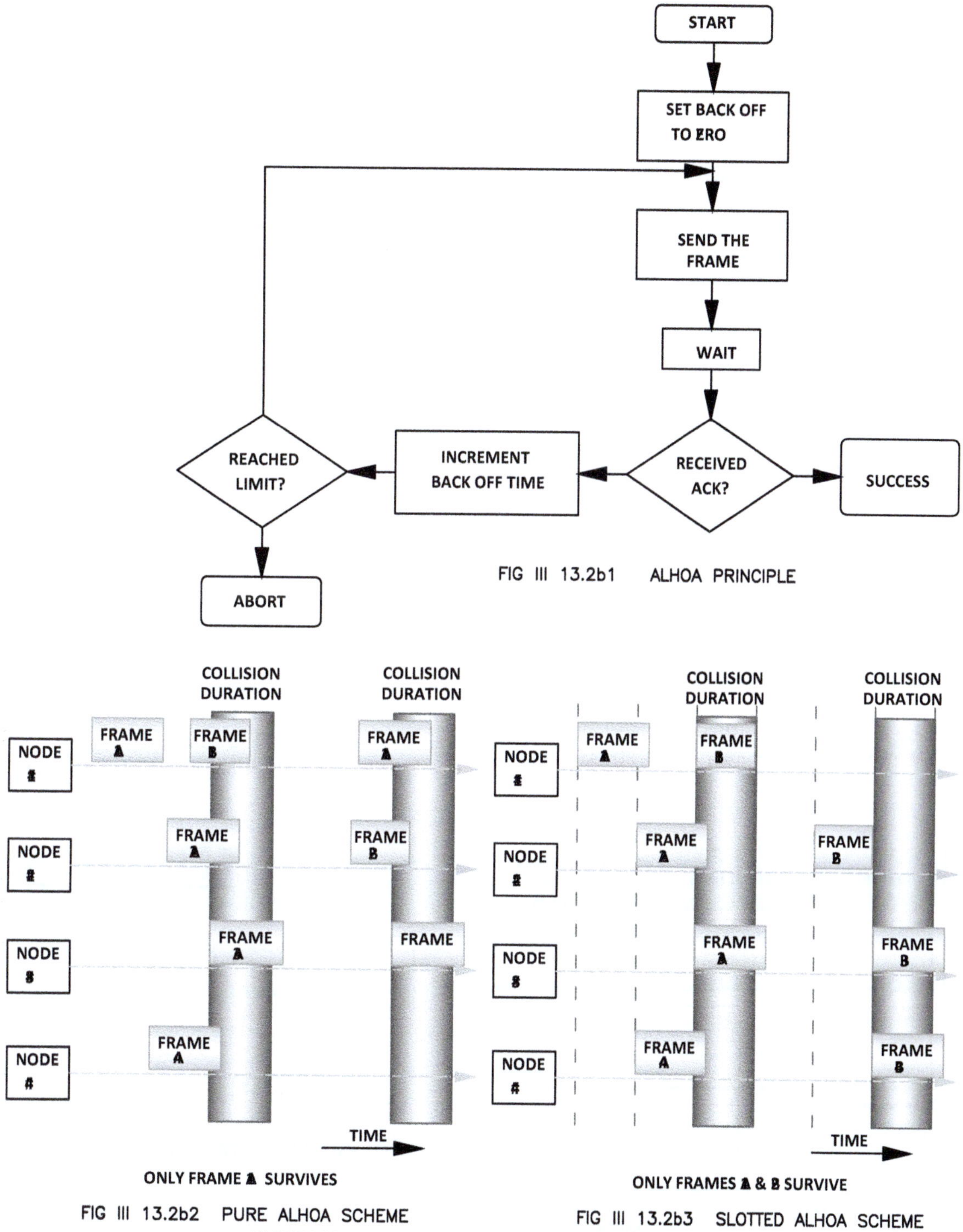


**FIGURE III 13.2B** ALHOA principle and collision of frames.

### 13.2.1 Discussions on IEEE standard 802

IEEE 802 is not a standard but a collection of networking standards to cover the physical and data-link layer specifications for different technologies such as Ethernet and wireless and applicable to Local area networks (LAN) and Metropolitan area networks (MAN) under LM standard committee (LMSC). With the standard, interoperability for vendors is possible. IEEE 802 helps in standardizing various protocols to ensure that multiple types of devices can be connected to different network types. It also helps in managing network. It is a part and parcel of internet services and technologies. The standard provides a set of recommended practices so network devices can all work together smoothly. Since IEEE 802 specifications are applicable to both hardware and software products, these specifications also define the connectivity infrastructure for networks. It aims for network integration enabling IoT and smart cities etc. In this subsection brief discussions on IEEE 802 has been covered starting from the basics of the standard. Functionally, IEEE 802.1 is responsible for: Enables Virtual LAN (VLAN) configuration, Ensures effective network management, Provides protocols for QoS etc. WiFi operates under the IEEE 802.11 standard, part of the IEEE 802 family. This standard specifies the protocols that wireless devices use to connect and share data over networks. IEEE 802 includes the Ethernet standard IEEE 802.3, which governs wired LANs. Ethernet defines how data packets move over cables. IEEE 802.3 covers Ethernet standards, It supports wired LAN communication, It defines protocols for consistent performance etc. Smart devices use WiFi (IEEE 802.11) or Ethernet (IEEE 802.3) for seamless data exchange within networks. By following the IEEE 802 standards, these devices guarantee reliable communication, which is essential for IoT environments and connected ecosystems. In the following subsection brief discussions will be presented on IEEE 802 standard and its influences in network transmission.

1. **Basics of IEEE 802 standard**: Major issue related to the basics of IEEE 802 (LLC and MAC) has been elaborated and explained in Fig. III 13.2c. In data link layer there are two parts: a common (to all 802 standards) upper interface to Logical link control (LLC) and Media access control (MAC) as shown Fig. III 13.2c1 (LLC and MAC positions).
   Lower MAC protocols are specified for use with different physical media. All 802 protocols use the same address format, with 48-bit, globally-unique addresses so that any interface can be plugged into any LAN worldwide without collision. First 24 bits are Organizationally unique identifier (OUI), available in web site, and last 24-bit MAC address are set by the organizer. Refer Table III 13.2a List of IEEE 802 standards.
2. **IEEE 802 Standard in Network**: Let us discuss basic application of IEEE 802 in network and IEEE 802.2 in LLC in Datalink layer
   - **IEEE 802 Application**: The IEEE 802 Standard comprises a family of networking standards that cover various the physical layer specifications of technologies from Ethernet to wireless. With 21 parts, IEEE 802 standards cover the physical and data-link aspects of networking i.e., it defines standards and protocols for wired LAN, MAN and wireless networks. However out of these only a few are active now. Now let us see various important network standards. Let it start with 802.2
   - **IEEE 802.2 LLC**: The definition for 802.2 is "the standard for the upper Data Link Layer sub layer also known as the Logical Link Control layer …." It is used with 802 standards such as the 802.3, 802.4, and 802.5 standards. LLC has been detailed

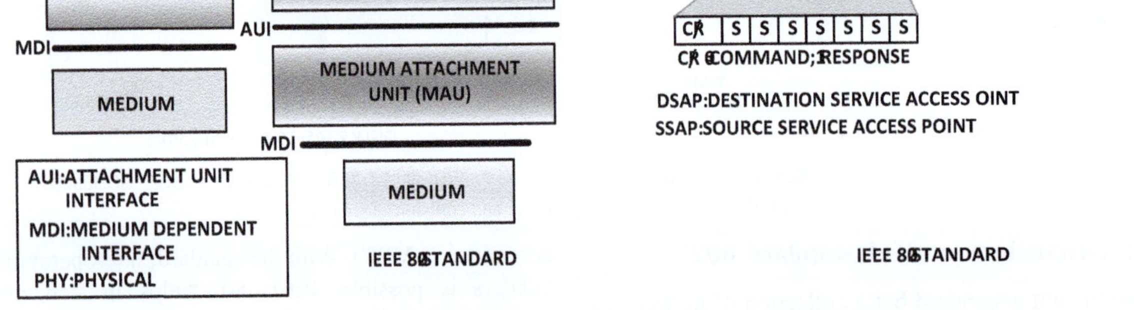


FIG III 13.2c1 LLC & MAC INTERFACE (POSITION)

FIG III 13.2c2 LLC FRAME FORMAT

**FIGURE III 13.2C** Explanation of basics of IEEE 802 (LLC and MAC).

**TABLE III 13.2A** List of IEEE 802 standards.

| No. | Standard name | Remarks |
|---|---|---|
| 802.1 | Interconnection (bridging) | |
| 802.2 | Logical link control | Inactive |
| 802.3 | Ethernet (CSMA/CD) LAN | |
| 802.4 | Token bus LAN | |
| 802.5 | Token ring LAN | Inactive |
| 802.11 | 802.11 wireless LAN [802.5 through 802.10 are dis-banned] | |
| 802.12 | Demand priority | Inactive |
| 802.15 | Wireless Personal area network (PAN) [802.14 dis-banded] | |
| 802.16 | Broadband wireless | |
| 802.17 | Resilient Packet | |
| 802.18 | Radio Regulatory WG | |
| 802.19 | Coexistence TA | |
| 802.20 | Mobile Broadband wireless | |
| 802.21 | Media independent handoff | |
| 802.22 | Wireless Regional area networks | |

out in Fig. III 13.2c1,2. 802.2 ("translator" for the Data Link Layer) specifies the general interface between the network layer and the data link layer with major responsibilities: Management of traffic over the physical network, flow and error control. LLC provides single format and interface to network layer (of OSI model) by hiding any differences among 802 networks [1]. When the network layer of the sending node passes a packet to LLC, it then adds LLC header to it, before it is transmitted. Naturally at receiver reverse process takes place. There are three fields in LLC header viz. Destination access point, Source access point and Control field. Access points tell the frame source and delivery details respectively. Control field accounts for sequence and acknowledge number. Various service types of are:

- Type 1: Unacknowledged and Connectionless (uni/multi/broadcasting mode)
- Type 2: Acknowledged and Connection oriented
- Type 3: Acknowledged and Connectionless

Let us now look into at MAX layer along with physical layers for Ethernet, Token passing wireless etc.

3. **Carrier sensed multiple access (CSMA) IEEE 802.3 (Ethernet)**: Let the discussions start with the function of MAC layer.
   - **Functions of MAC layer**: MAC layer, which lies between LLC and physical layer mainly performs the following functions.
     - It provides the idea about physical layer to the LLC and upper layers.
     - It resolves the addressing of source station as well as the destination station, or groups of destination stations.
     - It is responsible for encapsulating frames before transmission and frame parsing (analyzing), error detection after transmission.
     - It initiates frame recovery on transmission failure. It also performs collision resolution and initiating retransmission in case of collisions.
     - It performs multiple access resolutions for more than one data frame transmission and determines the channel access methods for transmission.
     - It generates the frame check sequences and thus contributes to protection against transmission errors.
   - **Physical Layer**: Typical physical layer for Ethernet is detailed out in Table III 13.2b
   - **Encoding and voltage**: it used +0.85V and encoding is Manchester/differential Manchester types shown.
   - **Frame details**: Typical IEEE802.3 CSMA frame details have been shown in Fig. III 13.2d. This should be read in conjunction with the discussions in III 16.2.0.1.

   Following explanation of terms shall be read in conjunction with Fig. III 13.2d.

**TABLE III 13.2B** Some cable details for ethernet types.

| Name | Cable | Max. Segment | Nodes /Segment | Remarks |
|---|---|---|---|---|
| 10Base5 | Thick Coax* | 500m | 100 | Good |
| 10Base2 | Thin Coax** | 200m | 30 | Cheap |
| 10Base-T | Twisted pair | 100m | 1024 | Easy set up |
| 10Base-F | Fiber optics | 2000m | 1024 | Long distance. |
| *Yellow coax vampire tap connection | | ** thin cable TV type: through T junction | | |

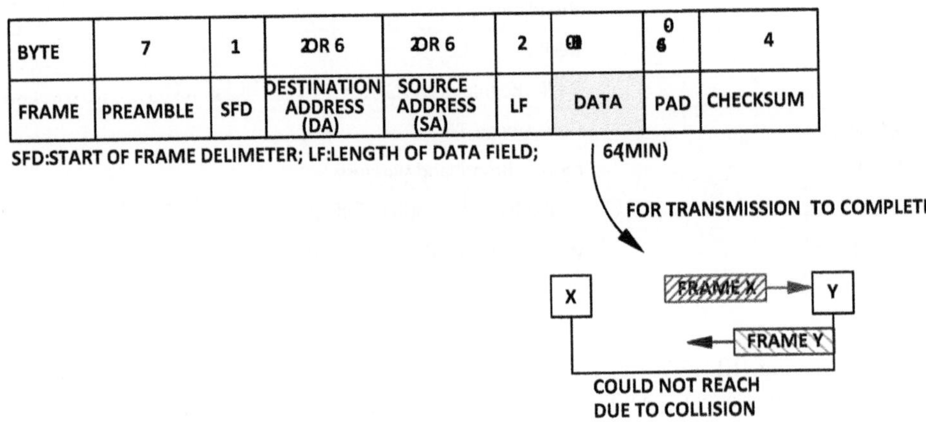


**FIGURE III 13.2D** IEEE 802.3 CSMA frame.

- Preamble: It is 7 byte (8 byte for DIX Ethernet) alternate pattern of 0 and 1 (101,010.) to tell receiving station a frame is coming and provides a means to synchronize frame for receiving physical layer. It is in a 10 MHz square wave for 5.6 μsec
- Start of frame delimiter (SFD): 1Byte: With pattern of 10,101,011 with two consecutive 1s at the end to indicate data is next.
- Source and destination addresses–MAC address: 6 byte long Standard Ethernet. Here first higher bit indicates 0 for Ordinary Address) and one for Group Addressing in Multicast mode for multiple machine addressing. If all the bits are 1s it is BROADCAST mode all stations to receive. Bit no. 46 (adjacent to higher order bit—— 6 byte meaning 6X8 = 48bit −2 = 46) distinguishes between local Address and Global Address. If it is 0 it is Local address decided by the LAN administrator and no body outside the LAN is interested about it. If 46 bit is one it stands for Global address assigned by IEEE to have unique address for each of the station
- Length Type: 2 bytes; It indicates of MAC data bytes in data field or frame type ID. There is also limit in the minimum length of frame (size fixed at 64 byte from destination station to checksum including both). Why minimum byte length? It is clear from Fig. III 13.2d (RHS part). If the frame size is too small then by the time first bit reaches distant receiver "y", frame transmission from x may be complete and at that time seeing no transmission "y" may like to transmit resulting.
- Padding Data (PAD): It is the field that is filled when data is less than 46 so it is padded to make total frame 64 (hence may have 0–46 bytes}.
- Checksum: 4 byte, Checksum – Cyclic redundancy check (CRC) of the data is added to the end. CRC provides error detection and protection of the MAC frame while being transmitted over physical medium from line errors (or transmission collisions in Ethernet)

- **Medium access: IEEE 802.3 CSMA/Collision Detection (CD)**: IEEE 802.3 CSMA/CD is operated in the medium access control layer. It senses if the shared channel is busy for broadcasting and interrupt the broadcast until the channel is free i.e., before transmission each node verifies that there is no communication in the line. If two nodes transmit simultaneously, there could be a collision. In such attempt, both the nodes send jam signal to all, preventing all from communication. Collision detector could be releasing own signal (24 mA), or error in CRC. Upon collision detection in CSMA/CD, nodes wait for a random time. After lapse of that period one may start checking the channel freeness and can communicate. Also, for one collision, there is random time, if there is another collision, so node shall wait for two waiting time and so on! CSMA/CD is applicable for Shared channel and for half duplex mode of communication. In this connection III 16.2.0.2 (shared media) may be referenced also. However, IEEE 802.3 is also applicable for switched Ethernet with full duplex transmission but treatment is different as detailed in III 16.1.1.5.

- **CSMA/Collision avoidance (CA)**: CSMA/CA is also operated in the medium access control layer, and is effective before a collision. IEEE802.11 DCF uses this technique. Some other Ethernet under IEEE 802.3 are 802.3u for fast Ethernet and 802.3x for Gigabit Ethernet discussed later.

4. **Token Passing system**: Token passing system is deterministic and priority (up to 6) can be set. It offers higher through put at higher network load (no. of nodes) compared with CSMA/CD. However, token passing systems have disadvantages such as: complex protocol and software, high cost and bandwidth. When a node does not have any data to transmit it passes the token to the neighbor. So, Token circulates amongst the nodes till one wants to communicate. When a node holds token for a specified time, only the node can transmit data and during that time following actions take place:
Sending node sets the token busy, adds Information, and a trailer packet.

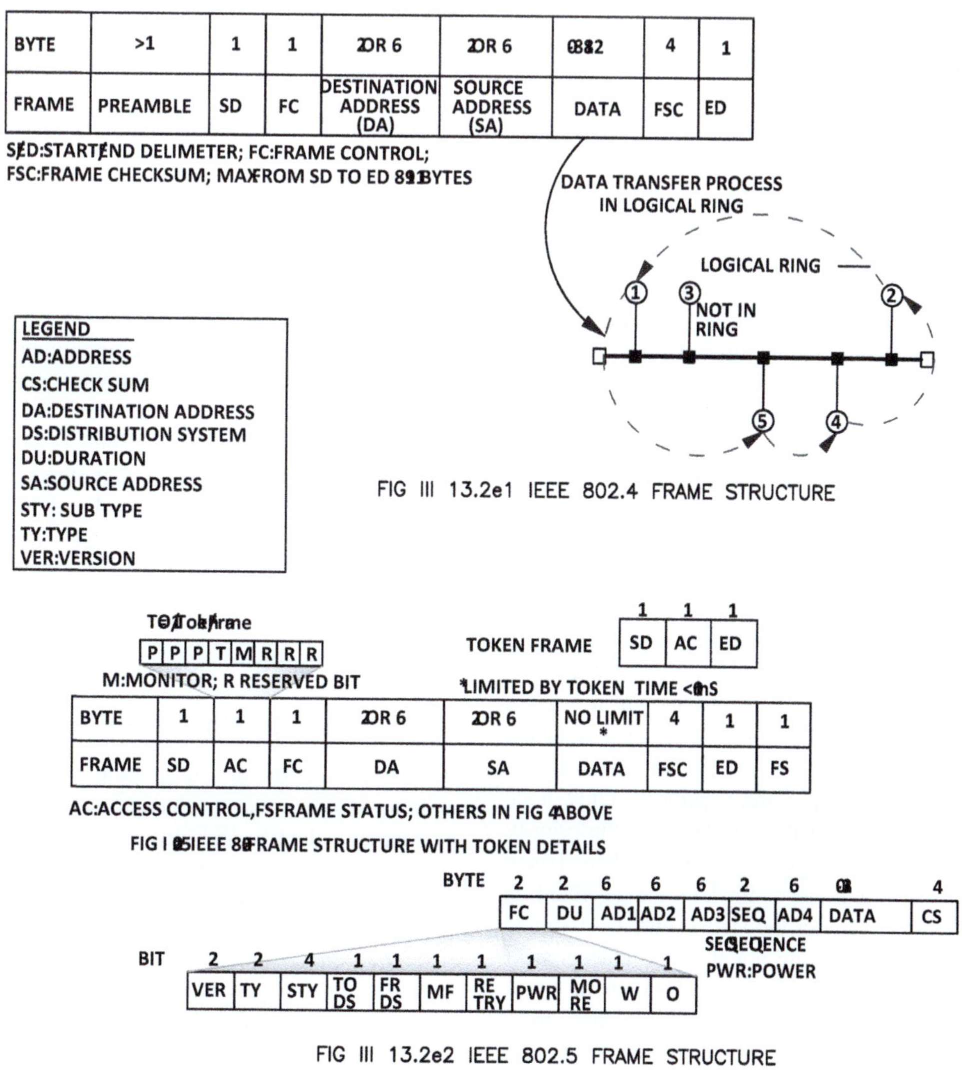


**FIGURE III 13.2E** IEEE 802.4 and 802.5 token passing frame.

Token passing frame for IEEE 802.4 and IEEE 802.5 have been depicted in Fig. III 13.2e. Entire message is sent over the channel for communication. Every node examines the header packet or frame, to check if belongs to it or to ignore in case not meant for it. Intended node copies the message set bit in the trailer field to acknowledge and sent back. The sender then checks that message received back to check it properly received. Then it frees the token in case it does not require.

- **Token Bus (IEEE 802.4)**: Fiber Distributed Data Interface (**FDDI**) is 100MBPS LAN based on Token passing bus (IEEE 802.4). Token bus is a protocol suit for Physical and Data link layers (MAC). Though it is physically it is Bus network but logically it has to be Ring network as shown in Fig. III 13.2e1. Stations are organized in Doubly linked list. It uses MAC address for distributed polling. *Token* is passed from highest to lowest node, each time a station receives the token, it may Broadcast (or may go for Point to point also) for a duration of time (token hold time). Self-explanatory data frame has been depicted in Fig. III 13.2e1. It has logical ring maintenance:
    - Initialization of ring
    - Periodic Token holder call to others to join, if outside ring.
    - Addition and deletion to and from ring is possible. Once the ring is formed, nodes need to keep address of predecessor and successor
- **Token Ring (IEEE 802.5)**: In token ring, when a station/node possessing Token wants to transmit information, seizes the token, alters 1 bit of the token (to change token into a start of frame), appends information and sends the same to the next node. Next node sees and sends it to next station. The information in this fashion circulates in the ring till it reaches the destination. Destination station copies the information and again circulates the token in the network till it reaches the sender. It then checks whether the information is copied! The token is then finally removed from the network. Token ring uses sophisticated means for priority management through Access control (AC). As shown in Fig. III 13.2e2, "*T*" bit in "AC" decides whether it is Token or Data frame. Access control (AC) has two fields, *priority* and *reserve* to control priority. Only node with priority value equal or *higher* than the priority value contained in *Token* can seize the token. Token ring deploys several means for fault management. A central station or any of the station may be assigned to do the timing and ring maintenance management function. When a station fails, its token may be circulating in the ring, which prevents other from continuing communication in the network. Such fault management can be done by station assigned for network maintenance. IEEE802.5 (inactive) frame format of Token Ring. and token ring has been shown in Fig. III 13.2e2. As shown, 1 byte Frame Status (FS) is for field terminating a command/data. It can go on sending data till it holds token, which can be held for a maximum time of 10 msec.

5. **802.11 Wireless Network:** 802.11 is active collection of standards for wireless networking. As per IEEE 802.11–1997, there have been three transmission techniques. These were: Infrared method (0.85–0.95 microns transmission speed of 1–2 Mbps) such as remote of TV. Frequency hopping spread spectrum (FHSS) (79 ch each 1 MHz wide) and Direct sequence spread spectrum (DSSS –CSMA/CA 1Mbps) are two other types covered in wireless communication. Two basic types in 802.11 are: *Infrastructure Basic service set* (Infrastructure BSS) with wireless Access Point (AP – used for all communications through BSS) and Independent BSS (ad-hoc network, MANET). Stations can communicate directly with one another. One important thing to be remembered that MAC of 802.11 is quite different from Ethernet MAC on account of complexities such as noise burst, *hidden* station and *exposed* station issues. There are different MACs viz. Point coordination function (PCF) and Distributed coordination function (DCF).
    - **PCF:** PCF is used for infrastructure BSS and access is coordinated by AP. It uses polling method. Once added, a station can send one frame per poll.
    - **DCF:** DCF Uses CSMA/CA and no central control. It uses physical and virtual channel sensing. Virtual sensing is done using *MACAW.*
    - One important thing to be remembered that all vendors offer DCF but PCF may not be available (even in chipset it may be inactive). Typical data frame and frame control for IEEE802.11 has been depicted in Fig I 2.0s6. Major popular standards are 802.11a (5Ghz radio frequency band/54Mbps speed), 802.11b (2.4Ghz/up to 11 Mbps), 802.11g (2.4Ghz/54Mbps) and 802.11n (similar as "g" but with receiver/transmitter quantity higher speed is possible).

## 13.2.2 Network vulnerability and threat

Any system, including automation system, handles data needs to deploy some mechanisms to monitor their cyber environment identify and close security holes as quick as possible. There are two different issues viz. Cyber threats and Vulnerability. Though two terms sound to bear similar meaning, yet they have some difference in them and it is crucial to understand the difference between them. There is another term called "cyber risk," which could be may be developed as a result of *combination of cyber threat and vulnerability*. It is better now to discuss a few terms prior o proceeding further.

- **Cyber Threat:** Cyber threats are security incidents or circumstances, which have negative impact on network or other data management systems. So, cyber threat represents potential danger to data or systems. A virus; a hacker, a disgruntled employee; a competitor, and other menaces could be a cyber agent.
- **Vulnerability:** Vulnerability is a gap or a weakness in a system through the hardware or software or security plan or organizational processes, that leaves a system or network open to a threat, which can result in a security breach. Attackers exploit this for unauthorized access or damage or destruction of data. Some vulnerability in network shall include but are not limited to SQL injections, server misconfigurations, nonencrypted data transmission. Network security vulnerabilities can be *physical* or nonphysical. Physical ones can be protected through data center security and preventing intruders' unauthorized access as well as keeping under lock and key or 24-hour surveillance. However, nonphysical vulnerabilities may arise out of data storage, data processing, software etc.
  There are two other terms used in connection with cyber security, which are very close but are not same. These are.
- **Attack:** Attack is an attempt to bypass security measures and/or unauthorized access to computer resources in a network.
- **Breach:** It represents successfully defeating security measures. Breaches are done with the intention to gain unauthorized access to data/resources for making them available to unauthorized persons, or to destroy computer/network data.

Having established, basic definition of all these terms it will be better to compare these terms, in tabular form (Table III 13.2c.

There can be physical and nonphysical network vulnerabilities. Latter one typically involves software/data e.g., Operating system (OS) could be vulnerable when it is not updated with security patches. Data servers especially when these are involved in revenue generation (e.g., e-commerce) and/or for trade secrets (Batch process formulation in production line) are provided with very strong physical security controls. Physical network vulnerabilities involve the physical protection of an asset such as locking a server in a rack closet or securing an entry point with a turnstile [17]. There are four major categories of network vulnerabilities. Apart from them there are several other security threats also. Some of these threats are: Malware (*virus, Ransomware, Trojans, Spyware)*, Social engineering attacks, Wire equivalent privacy (WEP), Eavesdropping, spam, man-in the middle attack, Denial of service (DOS) etc. to name a few. These are discussed in details in Vol IV of the book. Here, emphasis has been put mainly on vulnerability issues for network design.

1. **Security Vulnerabilities Types:** Network security vulnerability refers to broad category of flaws, potential exploits, and weaknesses in system hardware, software, design, administration, and associated policies/procedure. This is extremely crucial in assessment and prevention of a virus or malware i.e., threats that may enter/penetrate the system to infect the entire network and damage it. Three main categories of security vulnerabilities are:
   - **Hardware-Based:** Any improperly managed device connected to a network can pose a security vulnerability and risk of hackers gaining unauthorized access to install malicious software code etc. in the system viz. *Physical attacks, Firmware vulnerabilities.* By downloading the software to the device or using a USB thumb drive are very common ways malwares etc. can enter/infect the hardware. Major causes of vulnerabilities include: *Old version of systems/devices, Unprotected storage, Unencrypted devices* etc. Other than unauthorized devices, firewalls,

**TABLE III 13.2C Comparison among threat, vulnerability, and risk.**

| Feature | Threat | Vulnerability | Risk |
|---|---|---|---|
| Issue | Goal to steal and damage data with advantage of vulnerability advantage | Weakness in design hardware, software and | Loss or<br>Destruction of data is<br>Caused by cyber threats. |
| Intentional | Intentional and unintentional as well | Normally unintentional | Intentional |
| Detection | Antivirus/threat detection software | Penetration testing of hardware and scanner | Slower operation, unexpected results |
| Control | Cannot be controlled | Can be controlled | Can be controlled |
| Management | Manage through vulnerability control | Problem identification, categorization and prioritization for resolution | Reduction in data transfer rate. Use reliable sources only |

Wi-Fi routers, IoT devices, and employees' can create hardware vulnerability.

- **Software-Based:** Software vulnerabilities include operating systems and applications. Application plug-ins, downloadable apps, and addons for content management systems are especially vulnerable. Flaws or bugs in software can be used by hackers to compromise the system, often due to coding mistakes or outdated software viz. *Nonupdated software/Unpatched software, Insecure code, Buffer overflow* etc. Probable causes shall include: *Lack of input validation, Unverified uploads, Unencrypted data, Cross-site scripting* etc.
- **Design-Based:** As the name implies it refers to the weakness/flaw in the design, implementation, or configuration of a computer network that can be exploited by attackers to gain unauthorized access, steal data, or disrupt the system. These vulnerabilities can exist in hardware, software and network configurations. *Unsecured wireless networks*, Open port or Man -in the Middle are examples of examples. Unprotected communication, Malware/malicious software Worms, Social engineering attacks, Misconfigured firewalls etc. could be the probable causes.

2. **Security Vulnerabilities: Prevention:** Security vulnerabilities could be prevented by adapting suitable measures in respective types as discussed below:
   - **Hardware-Based:** Some **h**ardware-based vulnerability prevention methods are: Keeping devices secured physically, *Encrypting sensitive data***,** Keeping hardware upto date, Use of strong authentication etc.
   - **Software-Based:** *Regular patching and updates* (IT mainly), *Deployment of Antivirus software, Secured coding practice, Proper input validation* etc. are common preventive measure for this.
   - **Design-Based:** Strong encryption, *Secured wireless network, Limited open port, Use of strong Firewalls* with Intrusion detection systems (*IDS*) or Intrusion prevention systems (*IPS*) are major preventive measures in design vulnerability.
3. **Procedural Vulnerabilities:** This refers to *weaknesses* in the processes or rules to follow, e.g., using *Default passwords/Configuration, Improper access control, Inadequate monitoring.* Procedural vulnerabilities are handled in a systematic vulnerability management process comprising *Identification, Prioritization, Remedies, Monitoring* and *Verification. Changing of default configuration, Implementation of strong access controls—Rule-based access control, Regularization of logging and monitoring, Network segmentation, Conducting regular security audits, Use principle of least privilege,* User training and keeping updated on evolving security threats and best practices etc. all are part of key strategies.
4. **Human Vulnerabilities:** Human vulnerability refers to the possibilities for exploitation that can cause harm to an individual or company. The most vulnerable link in a system, program or team can bring down an entire structure. The weakest link concept is critical in information security and cybersecurity. Security risks caused by human behavior, such as falling for phishing attacks, using weak passwords, or not being aware of security threats, making it easier for hackers to exploit the system viz. *Social engineering, Theft and pilferage, Lack of security awareness, Negligence* etc. Major preventive measures include: *Security awareness training, Security audit at regular interval, Using strong password with regular updating, Multiple authentication strategy, Cautious on suspicious activities.*
5. **Vulnerability: Detection and Management:** It is important to detect/identify network security vulnerabilities, so that some proactive it could be mitigated/prevented. Similarly there should be certain procedure for vulnerability management as well. So, main issues here are:
   - **Vulnerability detection:** This refers to a systematic process of identifying security weaknesses before they can be exploited by attackers, using tools such as automated vulnerability scanners and penetration testing. This procedure involves defining the scope and follow the procedure in regular manner. Proactive efforts to identify security risks include: *Security audits at regular interval, Automated vulnerability scanning, Monitoring of network traffic*, Tracking trends using threat intelligence data, *Testing of penetration, Authorized permission management, Using extensive firewalls* etc.
   - **Vulnerability management:** In order to reduce risk exposure, protect critical asset and improved security arrangement, vulnerability management is important. Vulnerability management refers to the continuous, systematic process of identifying, assessing, prioritizing, and suggesting remedies security vulnerabilities. Major trend of activities are: *Risk and environmental controls, Vulnerability prioritization, Asset protection of critical paths, Defining scope of vulnerability and configuration management***,** *Identification of vulnerability management requires change.*

Let us conclude the discussions here to explore *Networking software.*

## List of abbreviations used

**AC** Access control
**AP** Access point
**APDU** Application protocol data unit
**API** Application programming interface
**ARP** Address resolution protocol
**ARPANET** Advanced research projects agency network
**BSS** Basic service set
**CA** Collision avoidance
**CAN** Campus area network
**CAP** Controlled access protocol
**CD** Collision detect
**CRC** Cyclic redundancy check
**CSLIP** Compressed SLIP
**CSMA** Carrier sensed multiple access
**DCF** Distributed coordination function
**DCS** Distributed control system
**DHCP** Dynamic host configuration protocol
**DMZ** Demilitarized Zone
**DS** Distribution system (IEEE 802.11)
**DSSS** Direct sequence spread spectrum
**EPN** Enterprise Private Network
**ESS** Extended service set
**FCE** Final Control element
**FH** Frequency hopping
**FHSS** FH spread spectrum
**ICMP** Internet control message protocol
**IDS** Intrusion detection systems
**IGMP** Internet group management protocol
**IP** Internet protocol
**IPS** Intrusion prevention systems
**ISP** Internet service provider
**IT** Information technology
**LAN** Local area network
**LLC** Logical link control
**MAC** Media access controller/Model adaptive controller
**MAN** Metropolitan area network
**MPI** Message passing interface
**NCS** Network control system
**NIC** Network interface card
**OS** Operating system
**OSI** Open systems interconnect
**OT** Operational technology
**OUI** Organizationally unique identifier
**P2P** Peer to Peer
**PAN** Personal area network
**PCF** Point coordination function
**PLC** Programmable logic controller
**POLAN** Passive optical local area network
**PPDU** Presentation protocol data unit
**PPP** Point to point protocol
**PSTN** Public switch telephone network
**QoP** Quality of performance
**QoS** Quality of service
**RARP** Reverse ARP
**RT** Real time
**SAN** Storage/System area network
**SCADA** Supervisory control and data acquisition
**SFD** Start of frame delimiter
**SLIP** Serial line interface protocol
**SMTP** Simple mail transfer protocol
**SOAP** Simple object access protocol
**SPDU** Session protocol data unit
**STA** Station
**TCP** Transmission control protocol
**TPDU** Transport protocol data unit
**UDP** User datagram protocol
**VPN** Virtual private network
**WAN** Wide area network

## References


[1] Computer Network; A.S. Tanenbaum & D.J. Weatherall; 5th Edition; 2011; Prentice Hall.

[2] Handbook of Networked and embedded Control systems (Network Protocols for Networked Control Systems –Ch4); F. -L. Lian, J. R. Moyne and D. M. Tilbury; Springer link;

[3] Network control systems Fundamentals; M.S. Mahmoud and Y.Xia ; Networked control systems 12019 ScienceDirect; ELsevier.

[4] Packet-Based Control;D. Georgiev and D. M. Tilbury; Proceeding of the 2004 American Control Conference Boston, Massachusetts; June - July' 2004; Research Gate document; https://www.researchgate.net/publication/4118497_Packet-based_control.

[5] Modeling of Random Delays in Networked Control Systems; Y. Ge, Q.Chen, M.Jiang and Y. Huang; Hindawi Publishing Corporation Journal of Control Science and Engineering Volume 2013, Article ID 383415; http://downloads.hindawi.com/journals/jcse/2013/383415.pdf.

[6] Networked Control Systems: The Communication Basics and Control Methodologies; Y.B.Zhao, X.M.Sun, J.Zhang and P.Shi; Hindawi Publishing Corporation, Mathematical Problems in Engineering,Volume 2015, Article ID 639793; Feb'2015.

[7] G. Koller, T. Sauter, T. Rauscher, Effect of Network Delay Quantization in Distributed Control Systems, ELevier IFAC publication, 2003.

[8] Variable sampling rate in Networked control systems; D. Xu, C-C. Chen, D Obradovic and B. Lang; Elsevier; IFAC Proceedings Volumes; Volume 40, Issue 13, 2007,

[9] Time delay modeling and sample time selection for Networked control Systems; F-L.Lian, J. Moyni and D. Tilbury; IMECE2001—DSC:6-5; Proceedings of ASME-DSC, Vol. XX; 2001 International Mechanical Engineering Congress and Exposition; November' 2001, New York.

[10] Control and Optimization of Network in Networked Control System; W. Zhiwen and S. Hongtao; Hindawi Publishing Corporation; Mathematical Problems in Engineering; Volume 2014, Article ID 237372; 2014.

[11] Wireless LAN and IEEE802.11; Tutorial point; internet document: https://www.tutorialspoint.com/Wireless-LAN-and-IEEE-802-11.

[12] How a VPN (Virtual Private Network) Works; J. Tyson, C. Pollette & S. Crawford; Howstuff works; https://computer.howstuffworks.com/vpn.htm.

[13] Power plant Instrumentation and control handbook; Rev 2; S. Basu & A.K. Debnath; Academy press –Elsevier B.V.; ISBN: 978-0-12-819504-8; 2019.

[14] 7 Layers of The OSI Model (A Complete Guide); What Is OSI Model?;Software testing Help; Sruthy; April'2025 Internet document; https://www.softwaretestinghelp.com/osi-model-layers/.

[15] Java – Networking; Tutorials point; Internet document; https://www.tutorialspoint.com/java/java_networking.htm.

[16] How to Geeek; What Exactly Is a MAC Address Used For? ; J. Fitzpatrick; September'2013; https://www.howtogeek.com/169540/what-exactly-is-a-mac-address-used-for/.

[17] J. Firch; What Are the Most Common Types of Network Vulnerabilities?; Network Vulnerabilities; Purplesec.

Chapter 14

# Networking software and Software-defined networking (SDN)

Network software encompasses entire range of programs be it Operating system (OS), application program, Application Program Interface (API) and protocols allowing communication, sharing data, and use network resources. Core Components mainly include but not limited to OS, Protocol suit, APIs, Application programs Network configuration/architecture, etc. Networking software refers to programs including various protocol suits, API, etc., used to operate, manage, monitor, or secure a network. Network software closely work with various network hardware and devices such as routers, switches, access points, firewalls, and other devices to give computer teams visibility and control. Without network software it is almost impossible to manage or secure at scale. For integrated operation, both Information technology (IT) and Operation technology (OT) work together. In modern days major enterprises opt for integrated approach. For integrated approach, two important networking issues are Internet and Intranet. *Internet* is global system of interconnected computer networks (with associated security dependent on of the user of *device connected* to network). Intranet is a private network specific to an organization (normally well secured through suitable firewall). A vital function of contemporary computer networks is the Network time protocol (NTP). It is NTP that makes computer network operate *efficiently*, *punctually* and *synchronously* (as necessary). In Real time (RT) operation, accurate timing and timestamps are essential in first out alarms, Sequence of event (SOE) recording for compliance, investigation, and troubleshooting. Refence clock, synchronizing protocols are essential items for this.

As the name suggests, authentication program is mainly concerned with the prevention of unauthorized access, to ensure that only authenticated users or systems can access network resources and data. So, network authentication is a process of verifying and confirming the identity of a user, device, or system trying to access a network. Network authentication helps in maintaining data integrity and confidentiality and also protect against cyber threats. In 1990s, Rockwell developed object oriented industrial network Common industrial protocol (CIP) for organizing and sharing data in industrial devices. CIP being an object-based technology, data exposed over a CIP network is presented as a *collection of attribute values* grouped in common categories called *objects*. Common objects with *common attributes and services* can be defined as a *classes of objects* with occurrences of that Class called *instances*. CIP also defines "*Connection and Messaging*" for connection management and messaging, which all CIPs would use. protocols use. CIP defines *specific objects* to manage connections and the connection types that specify how data moves over those connections. Like CIP, Time sensitive network (TSN) is an important type of network in industrial automation. The "Industrie 4.0" is underway, is marked by the pervasive deployment of Internet and Industrial Internet of things (IoT and IIoT) devices and services deployed in a self-organizing manner, relying on control and communication systems to manage their operation and interaction. Critical traffic flows generated by industrial automation applications require bounded low latency and low jitter to improve production efficiency and reduce communication costs. Typically, these critical traffic flows need to share the communication medium (e.g., Ethernet) with noncritical flows (with less severe timing constraints) originating from the same applications. The key to providing determinism is a shared concept of time. While IEEE 802.1AS provides standard for time and synchronization IEEE 802.1Qbv provides details on forwarding and queuing i.e., Qbv defines a means to transmit certain TSN Ethernet frames on a schedule, while allowing non-TSN Ethernet frames to be transmitted on a best effort basis around the TSN frames. Another important issue in automation is Manufacturing message specification (MMS), which is an application layer standard pertinent to Open system interconnect (OSI) of ISO. to define a messaging protocol for communication between programmable devices in manufacturing process run in a Computer integrated manufacturing (CIM) environment. It provides conceptual details. In 1990 it has been standardized as ISO/IEC 9506 (industrial communication to define MMS). The MMS protocol is structured so that subsets of protocol can be defined. MMS remains the only standardized protocol specification capable of supporting the IEC 61,850 (electrical substation) requirements for service, complex named

**Plant Intelligent Automation and Digital Transformation. https://doi.org/10.1016/B978-0-323-89892-8.00018-1**

data, and performance. It supports various objects such as: Virtual manufacturing device (VMD—a server that contains objects), Variable (self-describing variables), Named variable list (NVL—a collection of variables). Make/break connection, Associate, Conclude or Abort are some of the services offered. Though internet in present time has made everything at fingertip, yet traditional IP networks approach are complex hard to manage specially, when it is an integrated approach with vertical integration, i.e., control and data planes are bundled together. The Software-Defined Networking (SDN) has emerged as new paradigm with promises to get rid of this complex situation. SDN can separate the network's control logic from the underlying routers and switches, promoting (logical) centralization of network control, and introducing the ability to program the network. The control plane and the data plane are separated,3 can be realized by means of a well-defined programming interface between the switches and the SDN controller. Open southbound API is used between open flow switches and controller, while open northbound API interfaces controller and network application. Controller provides the working rules. *Simplified management, Enhanced flexibility, Improved scalability, Overall cost saving, Enhanced automation,* etc., have tremendous impact on industries. Major opportunities could be *Improved cost efficiencies, Innovative new network services, Higher security arrangement, Support for new technologies,* etc. However, it poses certain technical challenges also viz. increase in vulnerability in central control plane, limited scalability when the network grows big, latency in network due to separation of control and data plane. Also, there could be organizational challenges also. As a result, there will be certain advantages as well as risks, which could be circumvented through well planned strategic approach. Some industrial trends such as: *Emergence of 5G technology, Deployment of cloud applications, Remote operation* (work from home), *Revolution in IoT/IIoT,* make industries to adopt SDN.

In this chapter various software issues named and SDN shall be covered briefly.

## 14.0 Discussion on software issues in networking

Network software is an advanced, robust, and secure alternative to traditional networking, making the network easier to administer in terms of management, modifications, configuration, supply resources, and troubleshooting. The use of network software makes it possible to administer from one centralized user interface while completely eliminating the need to acquire additional hardware. It offers administrators the option to customize with greater flexibility to change and define the network speed, expand network capacity, and look after security.

Apart from software in network, there exists SDN. To understand how SDN works, let's look at the components that frame network software.

- **Application layer:** Application layer refers to the applications and services running on the network. This program conveys network information, the status of the network, and the network requirements for particular resource availability and application.
- **Controller layer:** Lying at the center of the architecture this layer is very important and acts as the brain in controlling the infrastructure layer/the data plane devices. is also done through the controller. The control layer is the intermediary that facilitates communication between the top and bottom layers through APIs interfaces.
- **Infrastructure layer:** The infrastructure layer, also known as the data plane, consists of the actual network devices residing in this layer and are primarily responsible for moving or forwarding the data packets after receiving due instructions from the control layer.

All these three components shown in Fig III 14.2b are tied together by API, which facilitates communication between these three layers. The *northbound APIs* are responsible for the communication between the application and the control layers, whereas *the southbound API* enables communication between the infrastructure and the control layers. Through north bound API Applications communicate to the controller to present the status of the network infrastructure and request resources based on availability, etc., and he control layer decides how the applications are allotted the resources available in the network. Through south bound API the control layer communicates with the infrastructure layer (routers and switches). The network infrastructure is informed about the route the application data based on the configurations made by the controller. Networks aren't static anymore, they are always changing with new users with newer applications, new devices, new threats also. Software plays a central role in network management, helping teams adapt without falling behind. Remote monitoring and control. Internet operation, cloud apps, Industrie 4.0 approach and global teams depend on reliable, secure networks, which have been made possible through tremendous advancements of software, which regulate networking operation. The *Internet, Intranet,* and *Extranet* are the most popular platforms, especially in e-commerce business. In our discussions on plant automation, extranet does not have any role. However, for integrated operation, both internet, intranet may be in use. Let us distinguish internet and intranet to understand their roles.

- **Internet:** Any individual with dial-up access/Wi-Fi (with access) or Local area network (LAN) can be the user with unlimited public access for general information. Usually, Internet service provider (ISP) provides connectivity.

- **Intranet:** Only authorized staff can access in a private and restricted for specific, enterprise and proprietary. The intranet links various servers, clients, databases, and application programs viz. Enterprise resource planning (ERP).

Network Time Protocol (NTP) synchronizes computer clocks across a network. It guarantees that the precise time of every device on the network is the same. The NTP protocol modifies the client's clock by exchanging time signals with the server. Network operations often need to synchronize tasks, protect data, and maintain the correct order of events. The four modes of operation for NTP are *Client/Server*, *Symmetric*(Peer), *Broadcast* and *Multicast*. Byzantine fault tolerance (BFT) described in volume II, enables the system to keep operating properly even if some of its parts malfunction or behave maliciously. Byzantine faults and BFT in NTP are important issues here. As discussed, network authentication helps in maintaining data integrity and confidentiality by allowing only authorized users and devices into the network, network to ensure that the data passing through the network remains secure and confidentiality is not lost. It offers benefits like.

*Accountability through traceable network activities*, *Well control over network resources*, *Fulfilling compliance requirements* and definite protection against threats such as phishing and malware. User name password, Multifactor authentication (MFA), etc., are some of the authentication methods commonly used as discussed in the dedicated subsystem. Now, let the discussions on this section begins with network software issues.

## 14.0.0 Basic network software (SW) issues

Network software refers to an overall term to describe a wide range of software functions, which streamline the *design*, the *operations*, *monitoring*, and *implementation* of computer networks. Network software is a fundamental element, like brain in a human being, the networking system for its administrations, reduction of network complexities, better control and management of data traffic as well as network securities including attack and vulnerabilities. This provides enabling seamless communication and interaction between various network devices. This also monitors various aspects of network such as time control, authenticity to improve network performance. Network software plays a crucial role in managing a network infrastructure and IT operations by facilitating communication, security, content, and data sharing. It has become an important tool in facilitating round-the-clock communication and allowing an uninterrupted exchange of information. Network software bears direct correlation with productivity and real time operation. The centralized nature of network software help to increases not only the productivity but in maintaining proper coordination among various components of the whole system. This helps in quick troubleshooting and reduction in end-user interface support problems. Therefore, the necessity of network software stems from its ability to manage complex networks efficiently, ensuring data is transmitted swiftly and securely. Network software streamlines various complex processes and also adapts to the evolving demands of technology, ensuring that organizations can maintain a competitive edge in a world where connectivity is key. While network hardware includes various devices such as router, switches for forwarding and physical connectivity, software in a network manages the data/information flows. It applies policies, controls bandwidth, detects anomalies, and automates device settings. It can run on servers or in the cloud and interacts directly with hardware to guide behavior. Network software also ensures reliable and secured operation of network and prevent from external attacks and reduce vulnerability of the network. Network software is mostly responsible for major challenges in computer network such as: Performance degradation, Host identification, Security issue, Configuration conflict, Slow connectivity, Capacity concern, Monitoring. In this subsection let us take up basic software issues.

1. **Software benefits:** Networking software brings about a number of advantages to enhance performance and operational efficiency coupled with strengthening network security.
   - **Network management:** Networking software simplifies this complexity of network and streamline network management and takes centralized approach also. In SDN approach there will be centralized control and automation of routine tasks.
   - **Mitigating Security risks:** Networking software addresses security issue in mitigating risks in diverse environments by implementing robust security protocols and real-time threat detection and to protect against cyber threats.
   - **Process Automation:** Through network software, various repeated process can be automated.
   - **Enhancing connectivity:** It is networking software, which helps in maintaining high uptime by enhancing connectivity to proactively identify and resolve potential issues. This also reduces down time.
   - **Compliance of regulatory requirements:** The challenges of maintaining compliance with various standards
   - **Central storage:** Network software makes it possible for centralized data storage and sharing,

which reduces data loss and simplifies data recovery in case of hardware issues.

- **Network Optimization:** It is networking software, which can optimize network optimization through suitable required resource allocation. Network software also provides analytics and insights for data-driven decision-making, enabling optimal resource utilization.

2. **Software functions:** Network software is multifaceted, to provide a number of valuable functions that improve management, security, and performance. Network software in traditional network is embedded directly into the hardware to manage all three planes viz. *Control*, *Data* and *Management* planes. A few relevant issues are:
   - **Hardware management:** Routing and switching managements by network software are major issues to efficiently manage the flow of data across the network by selecting the most effective paths in sending data packets.
   - **Setup and Virtualization:** Software also help in Installing the network. Virtualization of network is also the function of network software.
   - **User and File management:** This involves addition and removal of user in the network. File management is used for decision on the location of data storage and control user access to that data.
   - **Access control:** This is to regulate the access of the user and extent of access in the network to the user.
   - **Security management:** Network security enhancement and management is one of the most vital roles of network software to strengthen cybersecurity of network. Under this it safeguarding sensitive data against cyber threats this very important in critical control/automation.
   - **Monitoring issues:** Network performance monitoring is a vital function in Network software. Its function is to identify potential issues so, that IT people can proactively work on them so as to address the problems they escalate. This function is essential for maintaining high productivity and ensuring minimizing the network-related disruptions.
   - **Compliance to Regulation:** Adhering to regulatory standards is a nonnegotiable aspect of network management.
   - **Various Software issues:** As discussed, there will be software such as Control plane software: Routing software, Address resolve protocol (ARP). Data Plane software include: packet forwarding, Packet filtering. Management software are: Configuration management, Network OS (NOS).
3. **Software types:** Several network software types are:
   - **NOS:** As the backbone of network management, NOS allows control over resources and facilitating essential services. This also streamlines administrative tasks and enhancing efficiency.
   - **Patch management:** Patch management software aids in the smoother management of updates across numerous devices on the network through the installation of patches.
   - **Network configuration and change management (NCCM):** NCCM tools are essential for managing network configurations and tracking changes.
   - **Driver software:** This software provides an easy-to-use interface enabling the undertaking and maintenance of multiple tasks including the software, to set easily IP printing across networks or even deploy updates.
   - **Security Software:** This encompasses firewalls, antimalware tools, and intrusion detection systems, vital for protecting networks against cyber threats. With advanced feature it can include AI-based threat detection and analysis capability.
   - **Data Storage and Archiving:** Network storage software allows standard interface that manages countless databases between users or clients. Archiving software plays a crucial role in efficiently and securely data storing, securing, managing data. over extended periods. These tools ensure the data accessibility, future reference, compliance, or recovery needs. Often features such as data compression, deduplication, and encryption are part of the tool.
   - **Asset Management:** This software is assigned with the task of tracking and managing the network assets so that it is possible to optimize and maintain assets, reducing costs, and ensuring compliance with relevant regulations and standards.
   - **Network Monitoring Tools:** They provide insights into traffic flow, network utilization, and potential issues so that network can be managed proactively for better network performance and maintain optimal network health.
   - **Software Migration:** The use of migration software aids organizations in making processes such as upgrading systems hassle-free. The software provides an interface that enables easy monitoring of any deployment or data movement between the hardware and databases within the network.
   - **Network Virtualization:** This is used for the creation of virtual network resources to enhance resource utilization and network management efficiency. This also supports rapid network configuration and deployment, reducing the time and effort.

- **Cloud-based Software:** They are well suited for agile and cost-effective network management options. Cloud-based networking software can be integrated easily with other cloud services, to enhance collaboration and data-sharing capabilities.

4. **Software selection:** For network management, selecting the right networking software is crucial to ensure efficient and secure network management. Some key issues are:
   - **Assess requirements:** It is important to thoroughly understand the network specific needs; Major issues here are: *Network size, User count, Types of devices* connected, *Specific industry requirements.*
   - **Easy use and support:** The user-friendliness of the software and the quality of customer support are crucial.
   - **Scalability and flexibility:** It is better to opt for networking software that can grow and adapt with the needs. Scalability ensures that the software can handle an increasing number of users and devices, while flexibility allows for adjustments in response to changing organizational needs.
   - **Security Feature:** It is necessary to look for capabilities such as intrusion detection, zero-trust architecture, and encrypted. It's essential for the software to not only protect against current threats but also be adaptable to future security challenges.
   - **Cost analysis:** It is important to consider the total cost of ownership, including initial purchase price, ongoing maintenance and support costs. A cost-effective solution provides the necessary features and performance.
5. **Management practices:** Effective management of network software is key to implement, maintaining a robust, secure, and efficient network within an organization. Some pertinent issues here are:
   - **Ascertain requirement:** Some of the issues to be investigated are: *Benefit from network software solutions, The challenges that come along, Application details including impact on each dept, migration, Cost issue including maintenance and update cost* as well as *Return on investment (ROI).*
   - **Security policy:** Lack of a detailed security strategy is very dangerous, So detailed security strategy must be in place. Once a security strategy in place, then to develop and implement an ongoing security management policy.
   - **Pilot project:** To quickly gain the security benefits of enterprise network management, migrating an entire organization's network software in one go wouldn't be a smart move. Not only would that be risky but also quite challenging. Hence, organizations are advised to start small, plan and document processes, and take a step-by-step approach to mitigate risks [1].
   - **Train and educate staff:** to educate/train employees, especially key IT personnel, administrators, and operation engineers about the network's architecture, design, and interfaces, to be well prepared for a successful migration. Also, regularly train on the latest policies and procedures for utilizing your network software.
   - **Monitor network performance:** Implement continuous monitoring to quickly identify and address performance issues or security threats.
   - **Enforce strong access controls:** Implement strict access control policies to ensure only authorized personnel have access to network management tools.
   - **Review and audit:** Conduct regular audits of your network software to ensure compliance with internal policies and external regulations.
   - **Update and patch:** Keep your network software up to date with the latest patches and updates.
   - **Disaster recovery:** Have a robust disaster recovery and business continuity plan in place.

### 14.0.1 Discussions on internet-intranet software basics

From the discussions on OSI model and TCP/IP model, one common thing is noted that purpose of each layer is to provide service to the upper layers and shield them from implementation details. *Protocol* is the special set of rules that end points in a connection use when they communicate. Protocols specify interactions between the communicating entities. So, here a *protocol* stands to represent the agreement between two parties each with n layers. Hence it is n layer protocol. The entities of corresponding layers are referred to as *peer.* From Fig III 13.1a1 It is clear that in OSI model, the data flow from top to bottom and/or bottom to top in layers in two machine. So *peers communicate using protocol.* In software model between two layers there exists interface, which defines the *primitive service* lower layer to offer to the upper layer. This vertical communication is *real communication* whereas horizontal communication at a layer between hosts using protocols represents *virtual communication.* Protocols are *horizontal* lines while *services* are in vertical lines as shown in Fig III 14.0a. A set of layers and protocols together form the *network architecture* and list of protocols used by a system is known as *protocol stack.*

Except physical layer, each layer adds header to the message for identification. It is necessary at each layer these must be properly identified, through proper

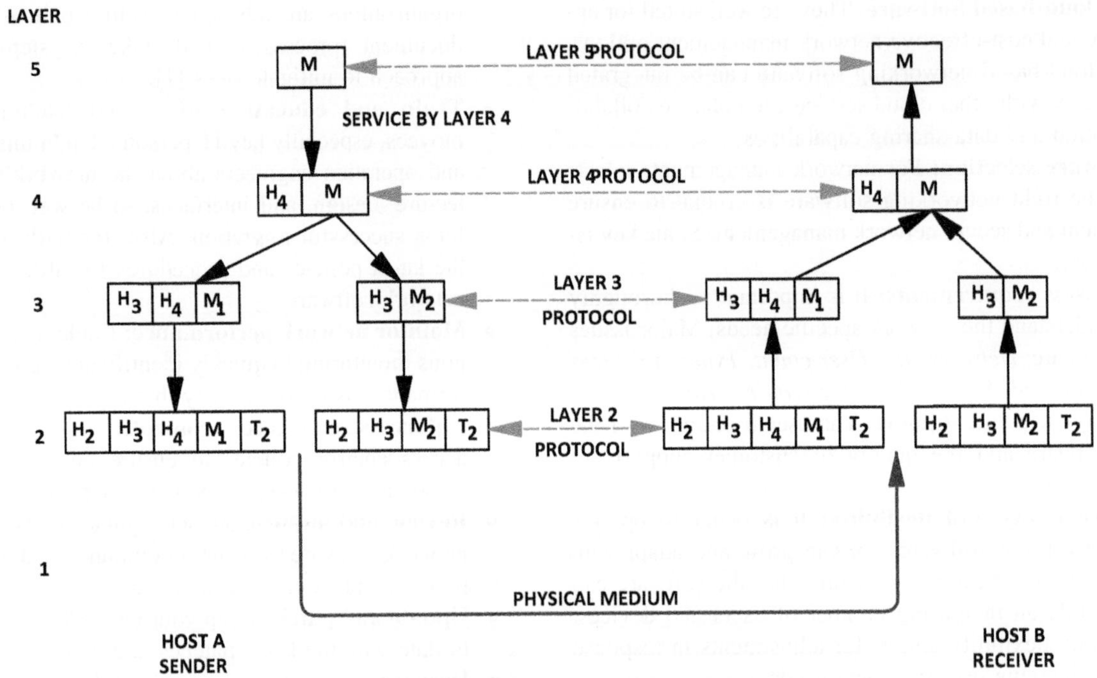


FIGURE III 14.0A Information flow in layers.

addressing. Only agreed upon mechanism for error detection and corrections, by both parties, can be adapted for *error control*. There is possibility that sender may be faster than the receiver. Therefore, suitable *flow control* is very important. Major design issues related to flow control are *congestion*, *starvation* and *QoS*. When there are several routes between sender and receiver, then best possible route is selected by *routing*. From previous discussions it has been noted that connection-oriented services and connectionless are two kinds of services. In this subsection very short discussions have been presented on software related to basics of internet and intranet. Let the discussions start with internet and intranet.

1. **Internet and Intranet:** While intranet refers to a private network and internet is a public network. These are detailed as follows:
   - **Internet:** The internet is TCP/IP protocol suite-based a global system to link various types of computers/computerized devices. The Internet network is built by the interconnection of a large number of computer networks not owned/admin by any entity. It also utilizes other protocols such as FTP, HTTP, and SMTP
   - **Intranet:** An intranet is a private network in a company/group to organize, manage, and share information/calendars and to-do list. It works in client server mode in LAN where computers are connected with MAC address. It also utilizes TCP/IP protocol suite to browse like internet.
2. **Point to point Protocol (PPP):** RFC1661 standard-based PPP allows data communication between two network nodes. Data link layer-based PPP is a full-duplex protocol that can be used on various physical media such as twisted pair copper wire, fiber optic lines, or satellite links. As per *RFC 1661*, three main components of PPP are: a) A method for encapsulating multiprotocol datagrams; b) A Link control protocol (LCP) for establishing, configuring, and testing the data-link connection; and c) A family of Network control protocols (NCPs) for establishing and configuring different network-layer protocols. PPP is very popular among Internet service providers (ISPs) for dial up connections and the Internet. PPP facilitates the transmission of data packets between point-to-point links. PPP utilizes LCP to establish a session. When PPP tries to communicate, it sends out LCP packets beforehand to the establishment of connections over the PP link. When PPP tries to communicate, it sends out LCP packets before the establishment of connections over the point—to—point link. The LCP packets check the communication line to ascertain whether it can sustain the data volume at the required speed. Accordingly, it agrees upon the size of the data frame.
3. **PPP over Ethernet (PPPoE):** PPPOE is similar to PPP as is based on RFC 2516 standard. Since TCP/IP do not support point-to-point connections, the use of PPP can enable them over Ethernet and other physical media. In PPPoE a virtual point to point connection over

multipoint Ethernet network architecture is established. PPPoE is widely used by ISPs to provision Digital subscriber line (DSL) high speed Internet services as seen in Asymmetric DSL.

4. **RFC1661 Standard:** Major terms of Standard RFC has been summarized as follows:
   - **Encapsulation:** PPP encapsulation provides for multiplexing of different network-layer protocols simultaneously over the same link. The default header and information fields fall on 32-bit boundaries, and the trailer may be padded to an arbitrary boundary.
   - **Link Control Protocol:** The LCP is used to automatically agree upon the encapsulation format options, handle varying limits on sizes of packets, detect a looped-back link and other common misconfiguration errors, and terminate the link.
   - **Network Control Protocols:** P–P links tend to exacerbate many problems with the current family of network protocols. These problems are handled by a family of NCPs, to manage the specific needs required by their respective network-layer protocols.
   - **Configuration:** It is intended that PPP links be easy to configure. The implementing personnel can specify improvements to the default configuration, which are automatically communicated to the peer without operator intervention. Finally, the operator may explicitly configure options for the link, which enable the link to operate in environments where it would otherwise be impossible. This self-configuration is implemented through an extensible option negotiation mechanism, wherein each end of the link describes to the other its capabilities and requirements.
   - **PPP Encapsulation:** The PPP encapsulation is used to disambiguate multiprotocol datagrams. This encapsulation requires framing to indicate the beginning and end of the encapsulation. These are 8/16-bit protocol followed by information and then padding.
   - **Protocol Field:** The Protocol field is one or two octets, and its value identifies the datagram encapsulated in the Information field of the packet. The field is transmitted and received most significant octet first.
5. **RFC2516 Standard:** Protocol overview and packet details have been elaborated below
   a. **Protocol Overview:** PPPoE has two distinct stages. There is a Discovery stage and a PPP Session stage. For a PPPoE session initiation, it first performs discovery to identify the Ethernet MAC address then to establish a PPPoE SESSION_ID.
   b. **Packet format and Payload:** Packet format and pay load has been detailed out in Fig III 14.0b, which is self-explanatory. For further detail RFC 2516 standard may be referenced.

### 14.0.2 Discussions on timing and authentication protocol

In any computer system, timing and control are crucial for ensuring orderly and efficient operations. Time

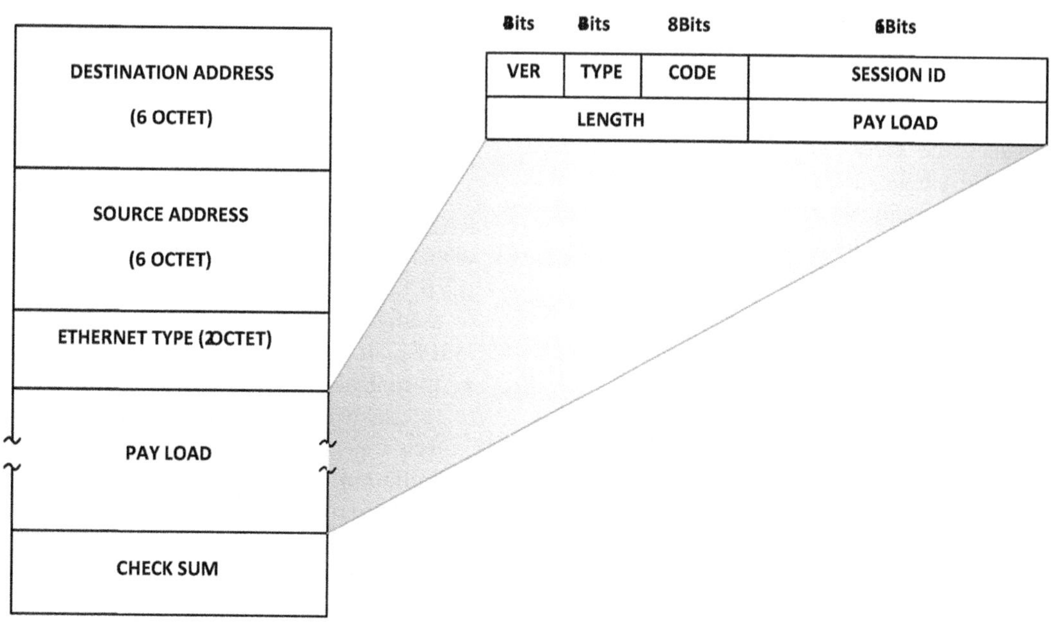


**FIGURE III 14.0B** PPPoE protocol structure RFC 2516.

synchronization serves a fundamental role in any network especially for control network, which in any case is time synchronized with plant master clock. Network time protocol (NTP), has been widely adopted as a means of network time keeping. There is another protocol Precision time protocol (PTP) is another network-based for time synchronization using hardware time stamping, with precision to achieve nanosecond- or even picoseconds-level synchronization. Highly fault-tolerant and scalable NTP (RFC 868) protocol is used to synchronize computer clock times in a public network utilizing TCP/Internet protocol (IP) suite. Another important issue in networking is authentication protocol, which authenticates an identity claim over the network. Good security design assumes that a network eavesdropper may sniff all packets sent between the client and authentication server, so the protocol should remain secure. [2]. The discussions shall also include authentication protocol, Virtual local area network (VLAN) and Media access control (MAC) authentication and other miscellaneous protocol issues. Let the discussions start with working of timing protocol.

1. **Working:** The NTP client initiates a time-request to the NTP server so as to calculate the link delay and its local offset, and adjust its local clock to match the clock at the server's computer. As a rule, *six exchanges* over a period of about five to 10 min are required to set the clock. Once synchronized, the client updates the clock about *once every 10 min*, with usually a single message. In addition to client server synchronization, this transaction occurs through the User Datagram Protocol on port 123 and NTP also supports broadcast mode for synchronization. The NTP client initiates a time-request to the NTP server so as to calculate the link delay and its local offset, and adjust its local clock to match the clock at the server's computer. As a rule, *six exchanges* over a period of about five to 10 min are required to set the clock. Once synchronized, the client updates the clock about *once every 10 min*, with usually a single message. In addition to client server synchronization, this transaction occurs through the User Datagram Protocol on port 123 and NTP also supports broadcast mode for synchronization.
2. **Time protocol: importance and Features:** Let us analyze importance and features:
   - **Importance:** Accurate time across a network is important for many reasons, because any discrepancies of even fractions of a second can cause problems precisely when people are opting for digitalization. Simplest example could be File-system updates carried out by a number of computers also depend on synchronized clock times only. So, air traffic control is one of the prime function of NTP. Therefore, Timing control protocols are important because they only ensure coordinated and synchronized operations of many systems. Apart from synchronization, this is necessary to prevent from data corruption and maintaining system stability.
   - **Features:** There are thousands around the world, have access to highly precise atomic clocks and GPS clocks of NTP servers. Specialized receivers are required to directly communicate with the NTP servers for these services. Therefore, *primary time servers* have been developed *as receivers.* They use protocols such as NTP to synchronize the clock times of networked computers i.e., all clocks are set toward that *time.* NTP uses Coordinated universal time (UTC) to synchronize computer clock times with extreme precision, offering greater accuracy on smaller networks—down to a single millisecond in a local area network. Highly scalable NTP is fault tolerant. It selects the best of several available time sources to synchronize to. NTP does not account for time zones, but associated hosts as required do that function. NTP can use measurements from the past to estimate current time and error even when network connection is temporarily unavailable. Simple network time protocol (SNTP) is another is basically NTP, for nonessential internal algorithms to make it simpler. So, it is a simplified version of NTP to avoid complexity.
3. **Authentication Protocol**: **IEEE 802.1X** It is part of the IEEE 802.1 group meant for Port-based network access control (PNAC) and provides an authentication mechanism for devices desires to get attached LAN or WLAN. The standard authentication protocol used on encrypted networks is referred to as Extensible authentication protocol (EAP). When an organization authenticates a user's identity and authorizes to access the network, the user's identity is determined based on their credentials or certificate. This credential is confirmed by Remote authentication dial-In user service (RADIUS server) representing a client/server protocol. The RADIUS server is able to do this by communicating with the organization's directory. Thus, we find that IEEE 802.1X gives access to the protected side based on username/password, certificates, OTP, etc. Four steps in IEEE 802.1X authentication process are:
   - **Initialization:** Detection of new device attempting to establish a connection by the authenticator. detects a new device and attempts to.
   - **Initiation:** The authenticator transmits EAP requests to the new device, for sending EAP responses back to the authenticator for identifying the new device. The authenticator received the EAP response and relays it to the authentication server in a RADIUS access request packet.

- **Negotiation:** On receipt of the request packet, it will respond with a RADIUS access challenge packet containing the *approved EAP authentication* method for the device for passing on the challenge packet to the device waiting authentication.
- **Authentication:** Once the EAP method is configured on the device, the authentication server will begin sending configuration profiles so the device will be authenticated. After the process is complete, the port will be set to "authorized" and the device is configured to the 802.1X network.
- **Accounting:** 802.1X RADIUS accounting involves recording the information of devices that are authenticated to the 802.1X network and the session duration. The device information, usually the MAC address and port number.

4. **VLAN explained and MAC authentication:** VLAN explanation and MAC authentication have been elaborated below:
   - **VLAN:** VLAN, refers to a method of configuring network to emulate a LAN with all of the management and security benefits. VLAN is a logical segmentation of a Data link layer network enabling devices to be grouped together regardless of their physical location. In place of physical topology in LAN, VLAN are implemented in switches using IEEE 802.1Q VLAN tagging. Also, VLAN can logically segment a physical LAN into multiple independent broadcast domains/groups based on function, department, or application regardless of their physical location to improve network performance. VLANs basically, segment the network to organize the security rules found on a network. In this way, it helps to make sure that devices and network resources that are on one VLAN aren't affected due to untoward incident.
   - **MAC authentication:** MAC authentication is used to authenticate devices on the basis of physical MAC addresses. MAC authentication grants access to a secure network by authenticating devices. When a device connects to the switch (direct link/through network), the switch forwards the device MAC address to the RADIUS server for authentication. The RADIUS server uses the device MAC address as the user name and password to grant or deny network access. MAC authentication, is a simple security measure in which you create a list of approved MAC addresses that are allowed network access. Unfortunately, it's not difficult to spoof MAC addresses, so MAC authentication is typically only used to supplement 802.1x authentication is enterprise environments. MAC authentication is typically combined with other forms of authentication.

5. **Miscellaneous protocols:** A few other authentication protocols are: Secure quick, reliable log in (SQRL) Open authorization (OAUTHO), to name a few. The most popular other protocols discussed below are: Password authentication protocol (PAP), Challenge-handshake authentication protocol (CHAP) and Extensible authentication protocol (EAP –used in wireless LAN also).
   - **PAP:** PAP sends the username and password in clear text. An attacker capable of sniffing the authentication process, easily log in. So, PAP is very insecure. It is noninteractive and supports both two way and one way authentication.
   - **CHAP:** This is a more secure authentication protocol and is less susceptible to reply attack, as it does not expose the clear text password. CHAP relies on a shared secret. At the time of account creation and enrollment the password is securely created and stored on the CHAP server. Since both the user and the CHAP server share a secret (the plaintext password), they can use that secret to securely communicate. To authenticate, the client first creates an initial (unauthenticated) connection through LCP. The server then begins CHAP authentication process: Server sends a challenge, with is a small random string (nonce), the user takes the challenge string and the password, uses a hash cipher to create a hash value, and sends that value back to the CHAP server as the response. The CHAP server also hashes the password and challenge, creating the expected response. The authenticator (CHAP server) checks the response against its own calculation of the expected hash value If the responses i.e., values are identical, the user must have entered the appropriate password, and is authenticated. If they are different, the user entered the wrong password, and access is denied. The CHAP server may reauthenticate by sending a new (and different) challenge. The challenges must be different each time; otherwise an attacker could authenticate by replaying an older encrypted response. A drawback of CHAP is that the server stores plaintext passwords of each client. An attacker who compromises a CHAP server may be able to steal all the passwords stored on it.
   - **EAP**: EAP is an authentication framework, not a specific authentication mechanism. Most enterprises desire to get for more security so a new authentication protocol, called the Extensible Authentication Protocol (EAP), was designed. EAP sits inside of PPP's authentication protocol to offer a generalized framework for several different authentication methods. 802.1x is the standard that is used for passing EAP over wired and Wireless LAN (WLAN). It

provides an encrypted EAP tunnel that prevents outside users from intercepting information. According to RFC 3748, EAP may be used on dedicated links, switched circuits. EAP has been implemented with hosts and routers that connects through switched circuits/dial up lines using PPP (RFC 1661). One of the advantages of EAP architecture is its flexibility. EAP is used to select a specific authentication mechanism typically after authentication requests more information in order to determine the specific authentication method to be used. In communications using EAP, a user requests connection to a wireless network through an access point. The access point requests identification (ID) data from the user and transmits that data to an authentication server. The authentication server asks the access point for proof of the validity of the ID. After the access point obtains that verification from the user and sends it back to the authentication server, the user is connected to the network as requested [3]. EAP can support multiple authentication mechanisms such as token cards, smart cards, certificates, one-time passwords, and public key encryption authentication.

## 14.1 Discussion on common networking types

As stated at the beginning that there will be defined CIP connections and message types. Explicit and Implicit are two connection types available in all CIP protocols:. *Explicit message connections* are *message/response* oriented and are used to *asynchronously* access data in a device. Implicit message connections (Input/Output, i.e., *I/O messages*) are used for *control*. Therefore, it transpires that *explicit messages*, are used to transfer noncontrol data, using a Transmission control protocol (TCP) connection. Implicit messages on the other hand are used to transfer inputs and outputs, using a User datagram protocol (UDP) connection. Inputs flow cyclically from a CIP target to a CIP message originator. Outputs flow cyclically from the CIP message originator to a CIP target device. Cyber Security is a critical topic that every manufacturer should consider addressing. CIP *security* is a new method of securely transmitting data at the protocol level rather than relying solely on additional hardware or applications to provide protection. Important issues here mainly include: *Authentication, Message integrity, Confidentiality,* etc. CIP security offers additional layer of protection at the protocol level also. CIP Safety extends the industry standard CIP network base services by adding CIP Safety services to transport data for CIP-based networks such as DeviceNet and EtherNet/IP with high integrity. CIP Safety is the TÜV certified extension to the standard CIP protocol. It extends the model by adding CIP Safety application layer functionality. CIP Safety is developed on the CIP that allows network independent routing of standard data. These base services were extended to allow high integrity safety services, by the addition of the CIP Safety protocol. For many critical industrial automation applications timing and IEEE guided TSN are important. There are wide varieties of systems included in industrial automation. Obviously, the functional requirements for different automation systems may vary across domains, but there would be similarities in terms of physical and logical organization complexity. Also, they may share common requirements for determinism, reliability, interoperability, and traffic convergence. Timing and determinism are some of the most important issues for industrial automation running on real-time applications with stringent functional dependence on temporal behavior and accuracy for responding to different events pertinent to the system be it internal or external. Naturally, performance of industrial automation highly depends on network functioning. Apart from throughput, the other the commonly used performance metrics, packet transmission latency, and its time variations (jitter) are critical concerns for many industrial control systems. Timing interactions can complicate different procedures and this is more so for many mission-critical industrial applications demanding high accuracy in timing. IEEE 802.1 TSN refers to TSN network applications where system must respond to external stimuli within a fixed, and often small, period of time i.e., synchronizing timing accuracy less than even 1 $\mu$S. While in resource sharing TSN can ensure zero congestion loss, bounded end-to-end latency to data streams, and bounded interference to non-TSN data streams. In this connection reference to IEEE standard 1588 and 802.1AS may be made. Worst case end to end latency is an important issue for zero congestion loss. For bounded latency, *Traffic shaping in output queues*, *Synchronized multibuffer swapping*, *Time-scheduled output queues*, congestion feedback, etc., may be adapted. MMS provides a *standardized, nonproprietary protocol* for *interdevice communication* in industrial automation and CIM. So, MMS could serve various purposes such as: *Standardized communication*, *RT data exchange*, *Interoperability*, *Device modeling*, *Remote monitoring and control*, *Integration of various devices*. ISO/IEC 9506 has several parts. Part 1 and 2 what is referred to as core of MMS. It is worth noting that there has been some modification MMS stack for incorporating TCP/IP from the original one (ISO/IEC 9506–1 and –2003). MMS provides a set of services for peer-to-peer real-time communications over a network. MMS is in use as a communication protocol for many common industrial automation devices viz. Programmable logic controller (PLC), Computerized numerical control (CNC) machines, and robots. Also, MMS finds its applications in the utility industry/electrical systems such as Remote terminal units (RTU), Intelligent electronic devices (IED), to name a few. MMS, as an application layer protocol, provides *application* services to the business functions. It is important to understand recognize the value provided by the network application functions in facilitating interoperability, independence, and access to data, so that its development cost could be justified. Following objects and devices are defined in MMS;

- A set of standard objects, which must exist in every device, on which operations such as *read, write, event signaling* can be executed.
- As discussed, VMD is the main object and all other objects such as *variables, domains, journals, files* comes under VMD.
- A set of standard messages exchanged between a client and a server for the purpose of monitoring and/or controlling these objects.
- A set of encoding rules for mapping these messages to bits and bytes when transmitted.

IEC 61,850 standard have been developed for communication between Intelligent electronic devices (IED)s and Supervisory control and data acquisition (SCADA) using logical devices and logical nodes. Previous subsection has been dedicated for MMS discussions. Now Let our discussions in this section begin with CIP.

### 14.1.0 Common industrial protocol (CIP)

Common Industrial Protocol (CIP) was developed by Rockwell and is now managed by the industry group, ODVA. CIP stands to represent a comprehensive message suites and services for a wide spectrum of industrial automation applications including information and network management for control, safety, energy, synchronization, and motion. One of the most useful features of CIP, which makes it very popular is its capability to integrate industrial automation applications with enterprise-level Ethernet networks and the Internet. For factory and process automation applications, CIP allows users many areas, which shall include but not limited to control, safety, synchronization, and motion, etc. CIP is truly media-independent and CIP provides users with unified communication architecture throughout the industrial enterprise. So, form too many vendors around the globe support CIP because, CIPs provide users the benefits of open networking facility without bringing about much change in existing automation systems. Also, CIP is extensible and upgradable communication architecture for the future [4].

CIP is media independent but object-oriented communication protocol. Each CIP object has attributes (data), services (commands), connections, and behaviors (relationship between attribute values and services). These are detailed in Fig III 14.1a. CIP includes object library to support communication and services. In addition to general purpose services such as FTP SNP, library also supports automation functions such as I/O, motion control, transducer as shown in Fig III 14.1b. In this subsection, basics of CIP have been covered as preamble to Chapter 18 where CIP networks are discussed in detail.

1. **Object and device profile:** The object library discussed above offer support to many common automation devices and functions. Therefore, the same object implemented in two or more devices will behave the same way. A grouping of objects in a device is referred to as the device's "object model." [5]. CIP defines device type and profile, which specifies objects implementation, configuration options and I/O data formats so that devices of a given type, all will have a common application interface. All devices that implement standard device profiles, will have the same network behavior as well as will respond to all the same commands other devices with same profile. Objects that define routing mechanisms also allow seamless message passing within different CIP networks such as EtherNet/IP and DeviceNet as shown in Fig III 14.1b.
2. **Object details:** CIP defines three types of objects. These are:
   - **Required objects:** This comprises: *the identity* object (–containing related identity data values referred to as attributes such as the vendor *ID, date of manufacturer, device serial number), message router* object (–meant to *route explicit request messages* from

*Object:* Very frequently in Networking we use the word "OBEC T" *what does it mean?*

**An object is charete rized b "attributes" and "behavior". Attribute refers to that ifi ormation which differentites the sid object from others d geera lly refers to da a of the object. Therefore, an "OBE CT" can b defined as a entity which contains da a d behaior ie . comb nation of da a d b ha ior in one pak age constitute a ob ect.**

**FIGURE III 14.1A** Understanding of object in networking.

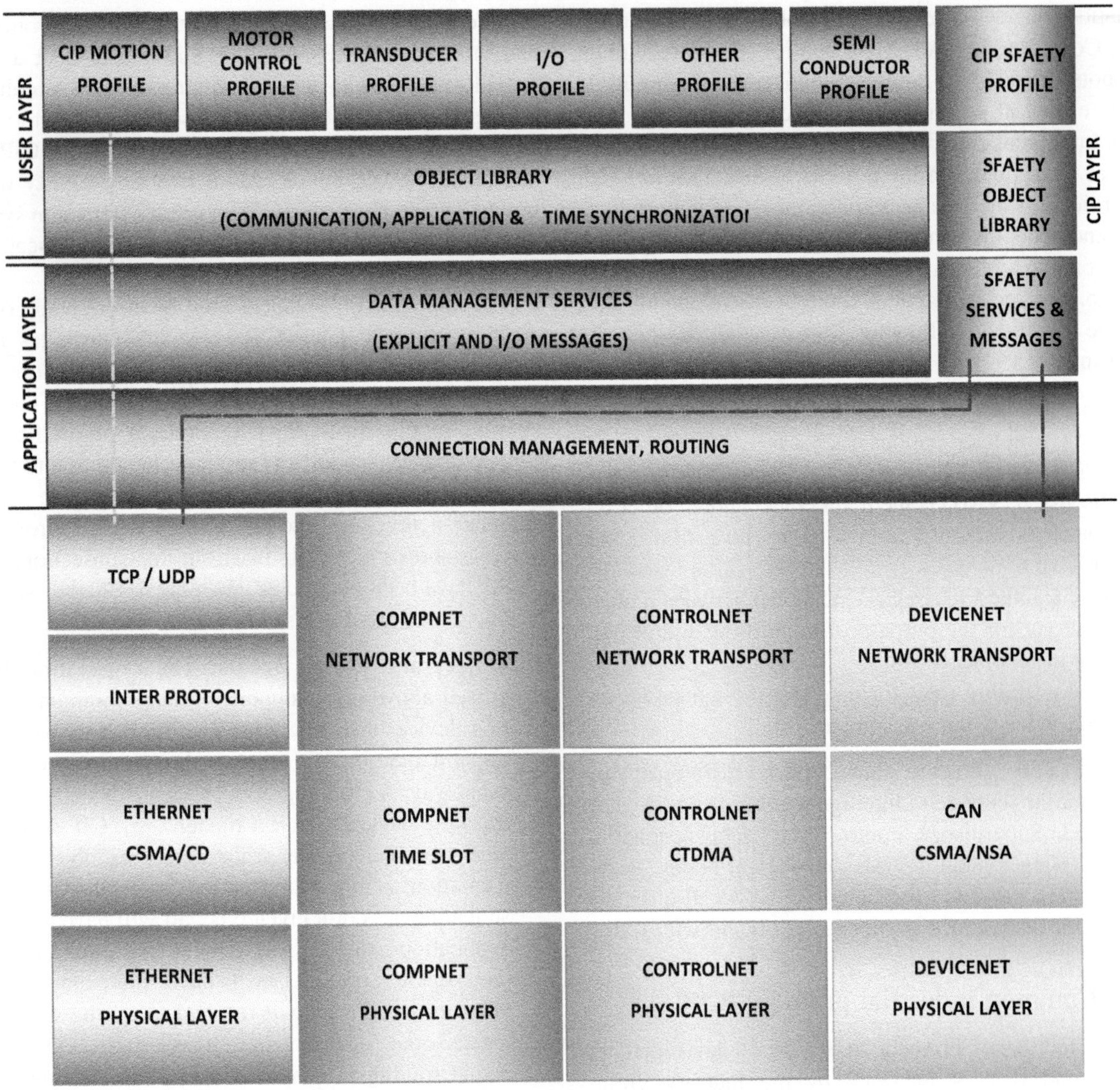


ABBREVIATIONS DETAILED AT THE END OF THE CHAPTER

FIGURE III 14.1B CIP layer with different protocols.

object to object in a device) and *network* object (–containing the *physical connection data* for the object).

- **Application objects:** These objects define the data encapsulated by the device i.e., the device type and function. These are predefined for a large number of common device types. All CIP devices with the same device type must contain an identical series of application objects. A device vendor can also group application layer objects into assembly objects– a convenient package for transporting data between devices.
- ***Vendor specific object*:** Objects not found in the profile for a device class are termed vendor specific to include additional features of the device. The CIP protocol provides access to these vendor extension objects also.

3. **CIP integration and Messaging:** The CIP defines device types, with each device type having a device profile discussed below. The device profiles specify the CIP objects to be implemented, and associated possible configurations and the formats of I/O data. CIP also helps integration and messaging. Object details already discussed above, here let us discuss integration and messaging.
   - **CIP Integration:** The Object Model in CIP as discussed above is based on multicast producer-consumer communication model. This model is more efficient use of network resources than other type i.e., a source-destination model because this allows the exchange of application information among one sending device (the producer) and many receiving devices (the consumers) avoiding multiple times transmission by single source to multiple

destinations. With CIP it is possible to integrate various applications with enterprise-level Ethernet networks and the Internet and is duly supported by many vendors around the world. At higher level, CIP support media independent object-oriented protocol with a producer-consumer communication model. CIP objects have attribute, service (command) connection and relational behavior between attribute and service. To support communication, it has a library also. So, CIP helps simple integration of Modbus, HART, and IO-Link devices through the use of native translation services.

- **Messaging:** CIP uses two messages types viz. explicit and implicit. Explicit messages sent through TCP/IP are used for data (information). Implicit messages sent through UDP are used for control data (inputs and outputs) with high speed. In producer-consumer model, a message is sent by *one* producer, *one* time, to many consumers as multicasting. A consumer takes decision on consumption of the data based on an identifier included in the data packet for more efficient use of network bandwidth and higher overall speed than the source-destination model.

4. **Pros and cons:** CIP offers following pros and cons:
   - Device profiles provide a common application interface.
   - Extremely stable and reliable ControlNet has flexibilities in topology selection choices
   - ControlNet has high speed deterministic data transfer capability
   - Multimedia option and designed safe cabling for process control
   - CIP is a comprehensive suite of messages and services for industrial automation (independent of media) providing the functionality for control, safety, energy, synchronization and motion, information and network management.
   - Multicast producer-consumer architecture allows efficient use of network bandwidth.
   - Better engineering with deterministic performance and product availability
   - Seamless bridging and routing allow flexible topology schemes for network architectures without need for program/configure intermediate devices.
   - Online node removal and less maintenance requirements are good features.
   - Separate cables are necessary for power
   - It is not only costly but complex too making troubleshooting very difficult.
5. **CIP Safety and Application:** Various networks under CIP umbrella have been discussed in details individually later in this volume. Let us conclude the discussions on CIP by touching upon safety and application issues.
   - **CIP Safety::** In addition to the family of networks CIP also includes CIP Safety for fail-safe communication in functional safety implementations. CIP Safety is the TÜV certified extension to the standard CIP protocol. CIP contains safety object profile, object library and messages and services. It extends the model by adding CIP Safety application layer functionality. These are shown in Fig III 14.1b. As the safety application layer extensions do not rely on the integrity of the underlying standard CIP services and data link layers, single channel (nonredundant) hardware can be used for the network would cause safety devices to go to a known state [6], [50]. Safety routing and Network routing in CIP network has been depicted in Fig III 14.0c

     Partitional functionality with standard routing for safety data sharing has been detailed Fig III 14.0c1. This routing capability allows quick reaction time. Using safety PLCs and two CIP network types have been detailed in Fig III 14.0c2.
   - **CIP Network Applications:** There are four industrial networks, where CIP have been incorporated at the session, presentation, and application layers. As shown in Fig III 14.0b, Ethernet I/P implements CIP over standard Ethernet. DeviceNet and ControlNet use CIP over a CAN network and a Concurrent time division multiple access (CTDMA) data link layer respectively. CompoNet implements CIP on a Time division multiple access (TDMA) data link layer.

### 14.1.1 Fundamentals of time sensitive networking (TSN)

TSN can be conceived as IEEE 802.1Q defined standard technology to provide deterministic messaging on standard Ethernet. TSN is not a protocol but it is an Ethernet standard that works at Layer 2 of OSI model. So, other protocols such as OPC UA may be implemented on top of TSN. Centrally managed TSN technology is able to offer guaranteed of delivery with minimum networking jitter by using scheduling for real-time applications requiring determinism implying predictability. Being predictable enables increased efficiency. This means that all networking devices implementing TSN has to have and share a common sense of time [7]. In line with IEEE 802.1AS, PTP is used for this purpose. It has been developed with the aim to enable deterministic communication on standard Ethernet, to meet the challenges of various industries e.g., Utilities, Aerospace, Automotive, manufacturing to name a few. One of the key a component

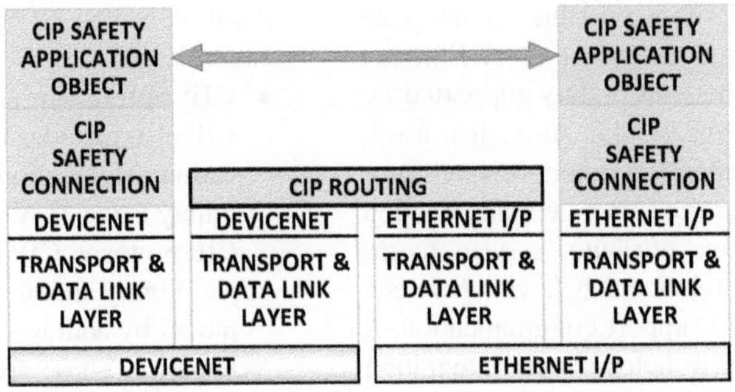


FIG III 14.1c1 CIP: SAFETY DATA SHARING

CIP: COMMON INDUSTRIAL PROTOCOL; I: INPUT; O: OUTPUT; PLC: PROGRAMMABLE LOGIC CONTROLLER

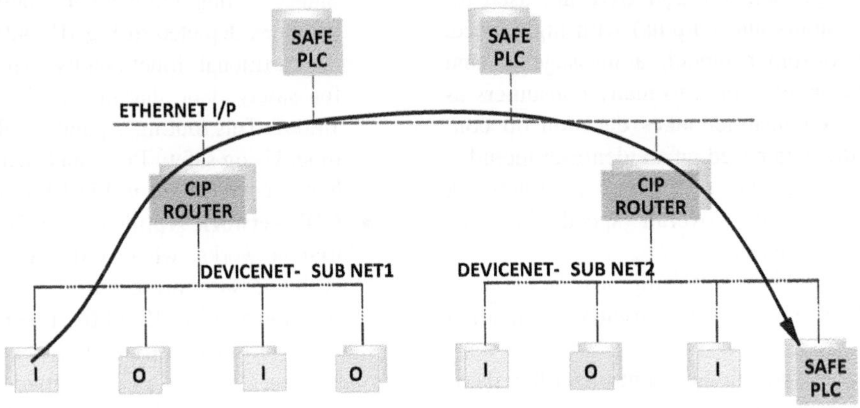


FIG III 14.1c2 CIP: NETWORK ROUTING

**FIGURE III 14.1C** CIP: Safety data and network routing.

in TSN is bridge. TSN bridges use the Ethernet header contents. Also, in IEEE 802.3 it could be seen that the payloads of the Ethernet frames can be anything. So, TSN can be used in any environment with the payload of any industrial application.

There are variants of industrial Ethernet protocols used in industrial devices by different vendors. So, users either have to choose devices from same vendor or have to face the considerable challenge of integrating protocol conversion. With the advent of the IIoT and Industrie 4.0, demands for full automation and more interoperable, flexible, and seamless in nature have increased. TSN Ethernet can therefore be the common communication protocol that connects industrial equipment from different vendors, simultaneously fulfilling the challenging requirements of industrial applications. TSN provides IEEE communication technology enabling interoperability between standard-conformant industrial devices from any vendor [8].

Another important aspect associated with TSN is that it lays the foundation for the convergence of IT and Industrial OT an attempt toward Industrie 4.0 concept. Traditionally in automation, Ethernet networks are based on the hierarchical automation pyramid separating IT from operational technology OT. Actually IT dependent on bandwidth and OT focused on high availability. Therefore data traffic in IT is labeled as noncritical, in contrast to real time OT data traffic, which is (time) critical. At IT level TCP/IP is mostly used in contrast to fieldbus in OT level for guaranteed latency times [9]. For digitalization, Industrie 4.0 and the IIoT now it has become a necessity to have connections and communication between IT and OT. For implementing TSN, FPGAs provides a significant advantage over other silicon (e.g., ASIC) due to their inherent feature of reprogrammability [8]. Let us start with feature and benefits.

1. **Features and Benefits:** TSN features mainly include: *Time synchronization* (IEEE 802.1AS), *Traffic scheduling* (IEEE 802.1Qbv), *Frame preemption* (IEEE

802.1Qbu), *Frame Replication and Elimination* (IEEE 802.1CB), Per-Stream Filtering and Policing, *Automatic Gate Schedule and Configuration,* etc. Major benefits of TSN network are:

- Guaranteed latency times of real-time critical data communication in the network as well as noncritical data traffic can be transmitted over the network.
- Open standard through IEEE 802.1, assures vendor's neutrality
- At higher level protocol layers, common network infrastructure can be used for exchanges of data between IT and OT to support IIoT and Industrie 4.0
- Reduced overall all implementation cost as well as reduced development lifecycle.
- Enhanced security on account of additional security mechanism coverage of IT through integration. It also supports safety applications.

2. **TSN Components:** Five main components are necessary for TSN operation.
   - **Centralized user configuration (CUC):** The CUC is vendor specific application to communicate with the CNC and the end devices. The CUC represents the control applications and the end devices. The CUC makes requests to the CNC for deterministic communication with specific details for those flows. Typically, the vendor of the TSN end devices will supply a CUC for those end devices [8].
   - **Computer numerical control/Central network controller (CNC):** For TSN, the CNC acts as a proxy for the Network to control applications requiring deterministic communication. It is responsibility of CNC to define and specify the schedule for all TSN frames transmissions. The CNC application is provided by the vendor of the TSN bridges.
   - **Bridges:** These are Ethernet switches with special capability by which they can transmit and receive the Ethernet frames of a TSN flow according to schedule. These Ethernet bridges also allow for use of the links to unscheduled traffic when there is no scheduled traffic [8].
   - **End devices:** The source and destination for time critical communication in TSN is referenced as end devices. End devices run the application, which calls for deterministic communication.
   - **TSN flow:** TSN flow stands to represent the time critical communication between two end devices in TSN. Each flow, which is uniquely identified in the network, has strict time requirement duly followed in TSN.
3. **Real time applications and shared time concept:** Guarantee on cycle times and fluctuations in cycle times are prerequisite for automation applications. In some cases of automation applications, data transfer times are significantly less than 1 ms, but in process control stringency may be less. TSN is a set of IEEE 802 standards aimed at addressing determinism and quality of service without compromising the strengths of Ethernet e.g., interoperability. TSN is an Local area network (LAN) but It can work non-TSN but timeliness is ensured in TSN-LAN. However, guaranteed latencies are must in all such automation applications i.e., deterministic communication is essential. The key to deterministic communication is a shared concept of time. The 802.1AS PTP profile allows all TSN network elements to share the same concept of time and adherence to that by all network elements is a necessity. IEEE 802.1Qbv defines a means to transmit certain TSN Ethernet frames on a schedule, while allowing non-TSN Ethernet frames to be transmitted on a best effort basis around the TSN frames. So, on account of sharing of same time by all network elements it is to deliver critical communication without jitter in delivery.
4. **Applications:** As per IEEE 802.1 major application heading for TSN shall include following major categories: Automotive RADAR, Machine tools, Mobile front haul networks, Industrial automation, automotive network and Utility. Some of the major beneficiaries are media networking, Grid control, Hardware in loop. Technologies such as IoT, IIoT Industrie 4.0 are also beneficiaries of TSN. Control systems with different in terms of scale, cycle time and accuracy, can use the same communication interface to transfer data deterministically. Sensors and actuators deployed in the control system can connect directly to real-time Ethernet or connect to a concentrator in the real-time Ethernet network using serial point-to-point connections. TSN with its deterministic networking performance is a good fit here, in the "field" level of a manufacturing floor. Table III 14.1a shows some network parameters in some application examples.
5. **IEEE Standards and TSN:** On account of IEEE standard, it is possible establish communication between multiple communication protocols through TSN, which guarantees compatibility and scales with Ethernet. Also it supports coexistence of different traffic classes e.g., mission-critical TSN-based systems and noncritical Ethernet-based systems. IEEE standard-based TSN allows convergence of network (IT - OT for seamless operation). Some IEEE standards for TSN are enlisted in Table III 14.1b.

### 14.1.2 Manufacturing message specification: discussions

Working toward MMS actually started way back in 1980, and finally came as standard in 2003. MMS, a protocol used in industrial networks. MMS is declared an

**TABLE III 14.1A** Network Parameters in some applications [10].

| Network feature | Machine tool | Automative radar |
|---|---|---|
| **Scale** | 64 axes | 4 sensors |
| **Bandwidth** | 100 MB | 1 GB |
| **Jitter** | 100 nano sec | 20 nano sec |
| **Latency** | 0.1 mS | 1 mS |
| **Cycle time** | <1 mS | 10 mS |
| **Time synchronized** | YES | YES |
| **Topology** | Line, ring | Star |

[Courtesy: Texas Instrument].

international standard, named ISO 9506 (Part1: services and Part 2: Protocol). The international standard MMS is an OSI application layer messaging protocol originally designed for the remote control and monitoring of devices such as Remote Terminal Units (RTU), Programmable Logic Controllers (PLC). The MMS is an application layer standard designed to support messaging communications to and from *programmable devices* in a Computer integrated manufacturing (CIM) environment, which is also referred to as the manufacturing environment. MMS supports transfer of simple and structured variables, program download and alarm handling. MMS is also used to define and make data available as access variables. In addition to protocol (Set of rules), the MMS standard also provides definitions for

- **Objects.** MMS defines a set of common objects (e.g., variables) and defines the network visible attributes of those objects (e.g., name, value, type).
- **Services.** MMS defines a set of communications services (e.g., read, write) for accessing and managing these objects in a networked environment.
- **Message:** A set of standard messages exchanged between a manager and an agent station for the purpose of controlling these objects
- ***Behavior.*** MMS defines the network visible behavior that a device should exhibit when processing these services.

Major conceptual issues in MMS include Client -server model, VMD, Object modeling and application of IEC 61,850. In Client -server, MMS describes the behavior of two communicating devices viz. the client as an operating and monitoring system, a control center or another intelligent device. The server represents one or several real devices or whole systems. VMD is a kind of server, hiding real devices in the VMD, which represents the behavior of a real device as far as it is visible. MMS uses an object-oriented to model approach for device resources and behavior. There are several object classes such as: Variable, Domain, File. Though initially it was developed based on OSI, but later on account of complexity, TCP/Internet protocol (TCP/IP). As discussed, within IEC 61,850, MMS has been used for client-server communication between IEDs SCADA, etc., utilizing MMS's object modeling and services to enable key substation functions also. Discussions on IEC 61,850 have already been

**TABLE III 14.1B** Brief list of IEEE standards for TSN.

| Standard | Area of definition | Standard title |
|---|---|---|
| **IEEE 802.1AS (rev), IEEE 1588** | Timing and synchronization | Enhancements and performance improvements |
| **IEEE 802.1Qbu and IEEE 802.3br** | Forwarding and queuing | Frame preemption |
| **IEEE 802.1Qbv** | Forwarding and queuing | Enhancements for scheduled traffic |
| **IEEE 802.1Qca** | Path control and reservation | Path control and reservation |
| **IEEE 802.1Qcc** | Central configuration method | Enhancements and performance improvements |
| **802.1Qch** | Cyclic queuing and forwarding | Bridges and bridged network (LAN and MAN) |
| **IEEE 802.1Qci** | Time-based ingress policing | Per-stream filtering and policing |
| **802.1Qcr** | Asynchronous traffic shaping | Bridges and bridged network (LAN/MAN–asynchronous traffic shaping) amendment |
| **IEEE 802.1CB** | Seamless redundancy | Frame replication and elimination for reliability |

covered in III 11.2.2. In this subsection brief discussions MMS has been presented.

1. **MMS basics:** MMS— an application layer protocol is designed to provide a generic messaging system for communication between heterogeneous industrial devices. MMS specifies services for exchange of RT data and supervisory control information between *networked devices* and/or *computer applications.* MMS composes a model from the definition of objects, services and behavior named the VMD model, which uses an object-oriented approach to represent different physical industrial (real) devices in a generic manner. So, major issues pertinent to MMS are:
   - **Objects:** Each conformant device must have a set of standard objects, on which operations can be executed.
   - **Messages:** For controlling aforesaid objects, a set of standard messages are exchanged between a client and a server station.
   - **Encoding Rules:** There will be a set of encoding rules for controlling these messages regarding mapping norms for *values and parameters* into bits and bytes for transmission.
   - **Protocol:** A set of rules for exchanging messages
   - **Applications:** MMS communication with access variables if we want exchange data between the two applications. MMS communication handles communication between applications within a controller or between applications allocated in different controllers.
   - **Application and access variable:** The *variable in the application,* which is mapped, can be any variable existing anywhere in the whole application such as: *Variable in a program or control module, Variable (global or local) in the application.* Again, an access variable is a variable that is shared on the network. For communication between two applications, access variables are defined. By sharing a local (or global) variable as an access variable, it is accessible by any system on the network.
   - **Naming Rule:** MMS variables can only be accessed by their name, which is a string. The naming rules are defined in MMS.
     - An MMS access variable name can be up to *32 characters long and contain letters, digits and the characters dollar and underscore.*
     - The access variable name cannot begin with a digit or the dollar ($) character. *No spaces* are permitted in the name of an MMS variable MMS variable names are case sensitive.
     - *The access variable* may be mapped to any type of simple variable or to any type of structured variable.
     - A structured variable can have mixed data type components
2. **Addressing scheme and Architecture:** After naming rule discussed above, let us take up addressing scheme. MMS has several modes of addressing. As clear from the above, MMS allows an address to have different syntax, based on the implementer's choice of what is most appropriate for that device. The *specification* separates between named and unnamed variables. The unnamed variables are identified by a fixed physical address in the VMD expressed as *Numeric (*A numeric address is an unsigned integer number), OR *Symbolic:* (A symbolic address is a character string) OR *Unconstrained* (An unconstrained address is an untyped string of bytes). As discussed, named variables are identified by an object name, VMD specific, domain specific or Association-specific [11]. Typical VMD communication model has been detailed in in Fig III 14.1d.
   The MMS architecture is based on a common client-server model. Real devices used in industrial networks often contain an MMS server allowing the device to be monitored and managed from an MMS client. An MMS client is typically part of a "control builder" application, Human -machine interface (HMI) or an MMS to OLE for Process Contro (OPC— Ref: III 11.0.2) Gateway (MMS/OPC GW).

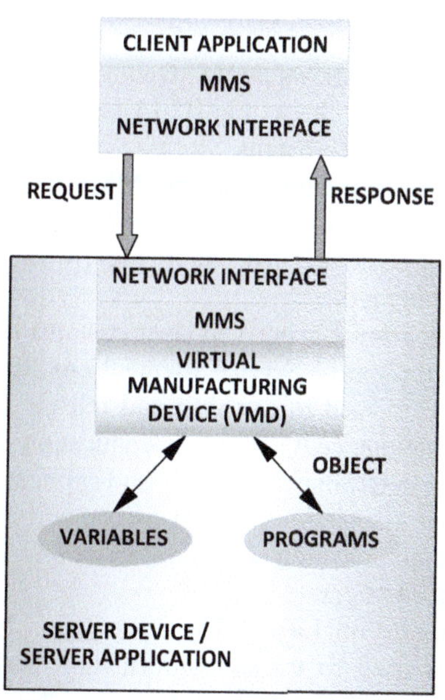


**FIGURE III 14.1D** VMD communication model [11].

3. **Objects and services:** Here let us discuss objects and services separately.
   - **Object detailing**: For each object there are corresponding MMS service primitives allowing client applications to access and manipulate those objects. The top-level object in the MMS is the VMD with at least one network-visible address. Real device is represented by a real object associated with vendor specific features. It is the VMD model, which maps the real object and devices onto virtual objects and devices, described in a generic manner in conformance to the VMD model. An MMS *variable is* part of a *virtual object* representing a mechanism for the MMS client to access the real variable. The MMS server containing the virtual MMS object can be understood as a communication driver, which hides the specifics of a real device from the client. All MMS objects contain an access method variable containing the information a device needs to identify the real variable. It contains values, which are necessary to find the memory location of the real variable with the contents that lie outside MMS.
   - **Network Services:** MMS is not a communication protocol but only defines messages to be transported by an unspecified network. MMS supports the use of both confirmed and unconfirmed services. MMS supports the use of both confirmed and unconfirmed services, but we will in this memo focus on the confirmed services. The MMS defines the following Protocol data units (PDU)s for a confirmed service exchange: *Confirmed-Request*, *Confirmed-Response*, *Confirmed-Error*, *Cancel-Request*, *Cancel-Response*, *Cancel-Error* and *Reject.*

   Fig III 14.1e shows MMS Confirmed services and Communication Stack. Fig III 14.1e1 illustrates MMS confirmed service details.
4. **MMS Communication layers:** MMS communication stack follows ISO OSI Model. MMS communication stack has been detailed out in Fig III 14.1e2. Major layer details are:
   - **Application Layer:** Top layer contains ISO's Association Control Service Element (ACSE) protocol. ACSE is used to establish and release Application associations (AA) between Application entity (AE) and to determine the identity and application context of that association [11]. ACSE supports two modes of communication service: *connection-oriented* and *connectionless.*
   - **Presentation Layer:** The presentation layer protocol is used to transmit information between open systems. Two functions here are: *Negotiation of transfer syntax* and *Transformation of syntax.*
   - **Transport Layer:** The OSI Transport layer protocol manages sequencing, error control and error recovery on an end-to-end basis to ensure complete and correct data transfer.
5. **Virtual Factory Communication:** For interconnections in a factory automation system comprising various sensors, controllers, and heterogeneous machines *MMS* standard have been developed. As seen, the MMS standard is composed of the common layered standards and the *MMS Companion standard* (CS), A virtual factory environment can be used to examine the correctness of an implementation before installing the factory automation system in a real environment. VMD models a real device's capabilities, allowing for testing and control in a virtual environment before implementation. Virtual MMS acts as a local or cloud-based instance of the MMS standard, enabling system testing, employee training, and software validation *without* affecting live production. A virtual factory is a digital twin of the physical facility. MMS is used as the communication protocol for various device interconnections and communication to connect different devices (represented by VMDs) within the virtual factory. A "*Virtual MMS*" instance is used to test and validate the communication and control logic between these VMDs without affecting the actual factory floor. By this testing new software, optimizing layouts, or training operators in a safe and controlled environment can be carried out. VMD service for the *Status acquisition (Identify*, and *Unsolicited Status)* of a networked machine, domain service for file transfer, Program invocation (PI) service for the remote execution of a program, and miscellaneous services for reading and writing services of device status and operations. PI services include: *creation*, *initialization*, *termination*, and *deletion* of a remotely executed program.

## 14.2 Discussions on software-defined network (SDN)

The computer networks is fast evolving and becoming more and more complex. So, managing them is posing serious challenge on the network operators. With addition of more devices in the network infrastructure facilitates the occurrence of various events in the network. SDN is a promising solution. SDN refers to a network model that has been developed to virtualize the network, having functionalities by separating the control plane from the forwarding data plane in a network device. As discussed, SDN controller makes use of *Southbound* and *Northbound* APIs, which are sets of standardized requests. The explosion of network devices, server virtualization, use of *cloud services* are among the trends driving the networking industry are facing especially in the areas such as: *Changing traffic patterns*, *Big data calls for more bandwidth*, *Rise in*

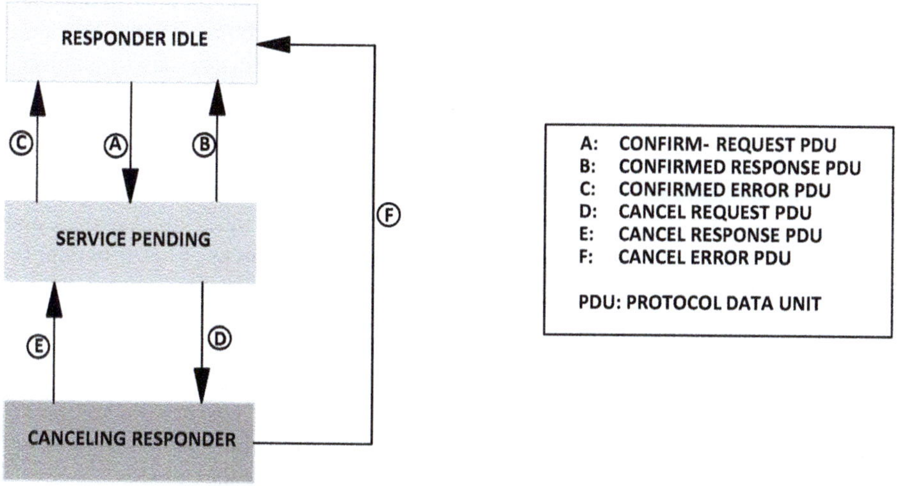


FIG III 14.1e1 MMS CONFIRMED SERVICES

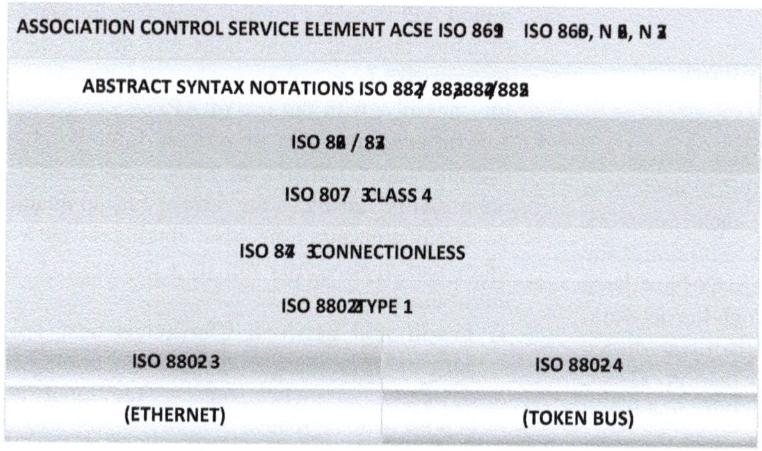


FIG III 14.1e2 MMS COMMUNICATION STYACK

DEVELOPED BASED ON IDEA FROM [1]

**FIGURE III 14.1E** MMS: Confirmed services and communication stack.

*cloud computing services*, *Consumerization of IT*, etc., the need to reexamine traditional network architectures was necessitated. In the SDN architecture, the control and data planes are decoupled, network intelligence and state are logically centralized, and the underlying network infrastructure is abstracted from the applications. As a result, enterprises and carriers gain unprecedented programmability, automation, and network control, etc. The control and data planes are tightly coupled in traditional IP networks in the same networking devices structurally highly decentralized. Naturally, this provides very complex and relatively static architecture. As a result, there could be network misconfigurations and related errors are common in complex networks. SDN with four following principles can be a solution:

- The control and data planes are *decoupled*.
- Forwarding decisions are *flow-based*, instead of destination-based. Flow programming enables unprecedented flexibility.
- Control logic is an external entity, as so called SDN *controller* or Network operating system (NOS) similar in nature as a traditional Operating system (OS).
- The network is *programmable* through software applications running on top of the NOS that interacts with the underlying *data plane* devices.

So, SDN can be defined by three fundamental abstractions: *Forwarding*, *Distribution*, and *Specification*. Traditional network architectures are built on key components working together for communication within and between networks. These key components are: Switches, Router, Load balancer, Firewall, Access point, etc. Traditional network architectures in the present context of operations at times viewed as limited in several ways such as: Poor Flexibility and Scalability, Complex Management. Limited Automation and Programming as well as Security. As discussed in initially, SDN is built on certain principles enlisted: *Separation of Control and Data planes*, *Central network control*, *Open standard protocol*, *Programmability*. So, SDN architecture comprises: Top application layer, Central control and Bottom data layer with Northbound API interfacing Application with control and Southbound API interfacing data plane with control. These give rise to innovative network functions with enhanced security functions, Better cost efficiency to support new technologies in the network. However, there could be challenges such as Technological challenges, Knowledge and skill gap, Fragmented market, Interoperability. Therefore, there will be challenges such as integrating with existing system, organizational and operational issues and some risks in SDN adoption also [12].

## 14.2.0 Basics of software driven network (SDN)

With growth of virtualization in data center, virtualized networks need to host many virtual and physical devices in more complex configuration. Maintaining such a network is expensive and time-consuming. In order to minimize the impact of above issues approaches using SDN have been used to control how data packets flow through the network. As discussed, traditional networks were hardware-based with software embedded. In contrast to that in SDN , software is separated from the hardware thus making it *more adaptable to the ever-changing nature of the computer network*, as well as making it easier to innovate and adapt the network to quickly meet changing network demands. The separation of functions from hardware, viz. firewalls or load balancing, is called Network functions virtualization (NFV). so, NFV is the process of decoupling network functions from hardware and running them on a software platform. As discussed, core principles based on which SDN is built, differentiate it from traditional network working and associated network architectures. These core principles are:

- **Separate Control and Data Planes:** SDN decouples the network's *control plane* (making decisions about the traffic destinations) from the data plane (responsible for moving packets from place to place). This separation allows centralized control and making the network more flexible for programming and network management.
- **Centralized Control:** In SDN, network intelligence is a centralized in software-based controllers, with a global view of the entire network. This centralization simplifies network management and allows for more efficient use of resources. This is because centralized control can gain global network view by collecting network-wide information, enabling dynamic resource allocation. Centralized intelligence also simplifies configuration. This in conjunction with programmability allows dynamic load balancing.
- **Programmability:** In SDN network administrators gains freedom for programmability, configuration, management including policy-based management and optimization of network resources. This programmability allows for more dynamic and automated control over the network, making it easier to adapt to changing needs.
- **Open Standards:** SDN is inherently designed to be open and interoperable, relying on standardized protocols such as OpenFlow to enable communication between the control plane and the data plane.
- **Interaction through API:** Northbound API and Southbound API in SDN also play great role. In SDN, instead of using physical connections and management interfaces applications connect over the network infrastructure flexibly, with the use of APIs.
- **Interoperability and openness:** Interoperability and openness are hallmarks of SDN architectures. Use of open APIs enables an array of applications, e.g., cloud orchestration. Also, intelligent network services and applications run in a common IT software environment.

1. **Traditional network Characteristics:** Traditional networking is based on fixed-functioning network devices e.g., switch/router. Traditional networks exhibit characteristics:
    - Dedicated devices such as switches/routers and application delivery controllers actually keep running the network whose performance also depends on suitable functioning of the devices.
    - Traditional hardware-centric networking is its limitations and software is embedded in hardware.
2. **Benefits of SDN:** SDN is capable of developing a framework, which bolsters data-intensive applications, such as big data and virtualization. It handles traffic better, offers more flexibility to meet the challenges of modern network demands by providing network automation through policy driven supervision. Other advantages shall include but not limited to:
    - **Centralized control:** In SDN with separate control plane the data plane, has centralized control, configure and management of network for simplified network management with dynamic and efficient traffic routing.
    - **Flexibility and scalability:** As SDN allows for the dynamic allocation and reallocation of network resources hence offers more flexibility and agility to meet the changing network

demands. SDN makes it easier to scale network infrastructure as addition and modification of new devices or services can be done quickly without major disruptions.

- **Enhanced Security**: SDN provides better unified security arrangements.
- **Improved performance:** Because of traffic optimization and load balancing, SDN offers better network performance and reduced congestion, and minimized downtime.
- **Cost savings:** SDN can reduce network operational costs through automation of many tasks and also reduce cost of commodity hardware, hence capital expenses.
- **Programmability and QoS:** APIs in SDN, enable network programmability, allowing developers to create custom applications and services. Also, SDN provides consistent QoS for users
- **Traffic engineering:** In SDN intelligent traffic engineering and path selection, optimizing application-based routes are allowed to ensure that critical applications receive the necessary bandwidth and low-latency connections.
- **Analytics:** SDN solutions often include built-in analytics and monitoring capabilities.
- **Compliance:** SDN allows for the automation of compliance policies and regulations.

3. **SDN definition and functionality:** As per IBM, SDN is defined by "the decoupling of control and packet forwarding planes in the network". Fig III 14.2a shows Network: Layered Functionality and SDN Concept. Fig III 14.2a1 shows Network: Layered concept. Network: Traditional and SDN architecture has been detailed out in Fig III 14.2a2. Thus, networks directly get connected to applications through application programming interfaces (APIs), bolstering application performance and security, and creating a flexible, dynamic network architecture that can be changed as needed [38]. On account of open APIs offered by SDN, maintaining of network control is easier. SDN decouples the network configuration and traffic engineering, separating them from their fundamental hardware infrastructure.

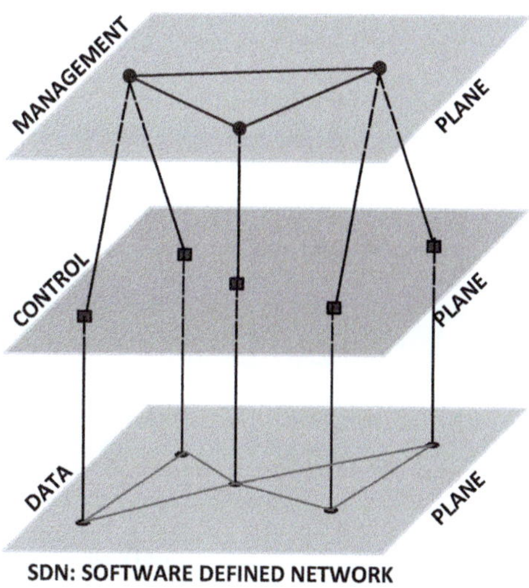


FIG III 14.2a1 NETWORK: LAYERED FUNCTIONALITY

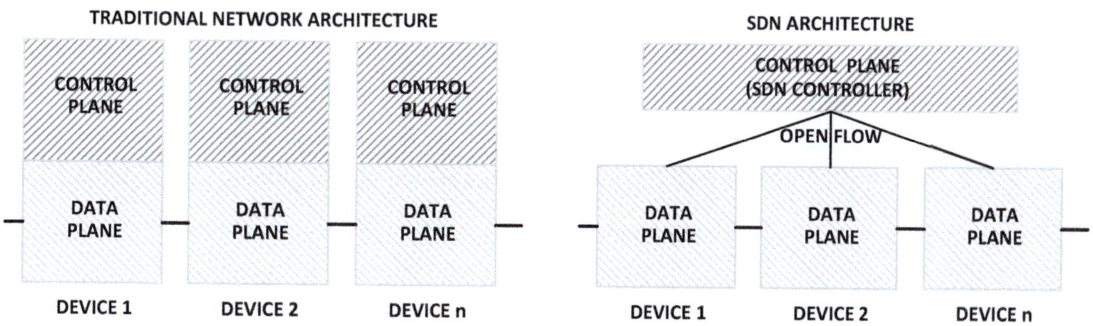


FIG III 14.2a2 NETWORK: SDN ARCHITECTURE CONCEPT

**FIGURE III 14.2A** Network: Layered functionality and SDN concept.

Major issue related to SDN is the virtualization of hardware. SDN consists of two separate planes viz. control and data plane:

- **Control plane:** The control plane looks after network performance and fault management of NetFlow. Like protocol It also looks after management of configurations of device, which are remotely connected to SDN. When a network administrator works with SDN to manage the network, the control plane dictates the path flow for the traffic (using flow protocol).
- **Data Plane:** The data plane forwards traffic to its desired destination [38].
- **Open flow control:** As a part of SDN, open flow protocol is used to communicate between an external OpenFlow controller and OpenFlow enabled switches

4. **SDN Frame work:** Typical frame work of SDN has been depicted in Fig III 14.2b.
   - **Application Layer:** Application Layer is the layer where SDN applications reside. It informs, control layer next to it, about its requirements of resources and services.
   - **Control Layer:** Control layer houses Network Control Software embedded in Network Operating System. It provides an abstract view of the underlying network infrastructure. As stated above, also It receives the requirements of the SDN applications to pass on to the network components residing in the next layer.
   - **Infrastructure Layer:** it is basically data plane. Here actual network components are housed. Network component capabilities are passed on through the Control to data-Plane Interface.

   Thus, as shown, business application interface and requirements are catered to through API. Services are handled in control plane through network components in data plane with the help of control to data plane interface. In this connection III 14.2.1.2 may be referred to also.

5. **SDN and Traditional network:** Here working principles of SDN and its difference with traditional network

- **SDN Working Principle:** SDN can be conceived anything virtualized. As discussed, the software is decoupled from the hardware. SDN moves the *control plane* that determines where to send traffic to software, and leaves the data plane that actually

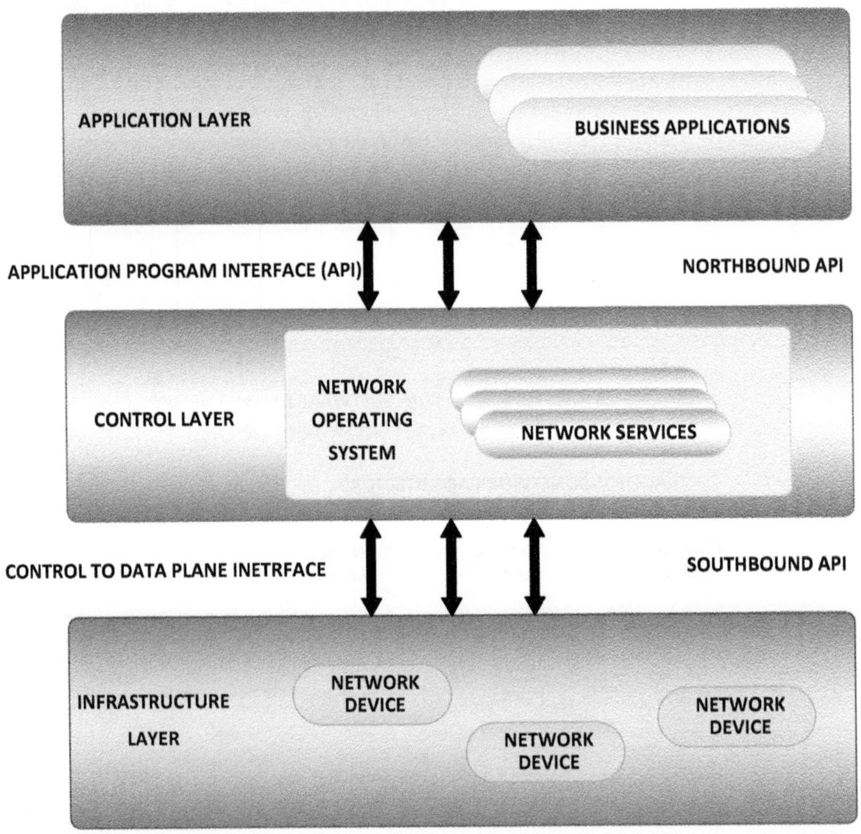


**FIGURE III 14.2B** Software-defined network (SDN) frame work.

*forwards* the traffic in the hardware allowing the programming and control the entire network through a single pane of glass in place of device basis. There are three parts to a typical SDN architecture, which may be located in different physical locations: Applications, which communicate resource requests or information about the network as a whole. Controllers, use the information from applications to decide how to route a data packet. Networking devices, which receive information from the controller about where to move the data Physical or virtual networking devices actually move the data through the network. In some cases, virtual switches, which may be embedded in either the software or the hardware, take over the responsibilities of physical switches and consolidate their functions into a single, intelligent switch. The switch checks the integrity of both the data packets and their virtual machine destinations and moves the packets along.

- **Comparisons: SDN and Traditional Network:** There are some similarities between SDN and traditional network such as:
  - Both types aim to establish networks connectivity between devices.
  - Standard networking protocols such as Ethernet or TCP/IP models are used.
  - They can provide Quality of service (QoS) and security in the network

  However, in spite of these similarities there are differences between them. The differences between SDN and traditional network has been tables as Table III 14.2a.

### 14.2.1 Detailed discussions on SDN

With SDN technology, the hardware layer including network routing and switching is

decoupled and Network function visualization (NFV) is combined with SDN. Network function visualization: NFV is the replacement of network appliance hardware with virtual machines. There is a requirement for a hypervisor to operate networking software and procedures such as load balancing and routing by virtual computers. NFV allows for the separation of communication services from dedicated hardware, viz. routers and firewalls; allowing new network operations and services dynamically and without installing new hardware.

There have been a number of innovations in the virtualization domain, which has benefited compute and storage. However, those benefits are constrained due to limitations in the traditional network model. It is becoming increasingly difficult task to manage infrastructure, which needs rapid network provisioning and also can be a time consuming. In addition, there is a risk associated with slow network provisioning as it may cause a performance issue. SDN refers to a piece of software to talk to the hardware to control network and physical devices. In modern time, business requirements are changing very quickly, SDN provides an ability to respond quickly as it uses software to control network router and switches and can provide centralized management of network. SDN Is extremely beneficial to IT architectures such as data centers, cloud. An organization does not need to purchase very expensive networking devices. SDN networks help in improving network efficiency and agility by deploying software-based functionality and a software-based network management layer on top of a physical network backbone. The SDN market is growing at a faster pace than anyone had anticipated as aninnovative architectural model, which is capable of enabling automated provisioning, network virtualization, and network programmability for data centers at cloud-providers and enterprise networks. Dynamic and flexible SDN architecture protects existing investments, with future proofing network in nature as it is capable of responding rapidly to changing business, end-user, and market needs. Some of the major features and benefits of SDN are:

- Open, centralized management and control of networking devices from multiple vendors
- Improved automation and management by using common APIs.

**TABLE III 14.2A** Difference between SDN and traditional network.

| Issue | SDN | Traditional network |
|---|---|---|
| Approach | Virtual approach | Traditional physical approach |
| Plane | Decoupled control and data plane | Control and data in same plane |
| Interface | Open interface | Close interface |
| Control | Centralized control | Distributed control |
| Program | Programmable | Nonprogrammable |
| Configuration | Automatic | Fixed manual |
| Prioritize | Prioritization possible | No prioritization possible |

- Quick innovation, the ability to deliver new network capabilities and services without reconfiguration and independent of vendor update requirements.
- Programmability by operators, enterprises, independent software vendors, and users
- More granular network control including policy-based control at the session, user, device, and application levels.
- Better end-user experience as applications exploit centralized network state information to seamlessly adapt network behavior to user needs.

In this subsection, discussions in previous sub section will be supplemented with detailed information and discussions on SDN.

1. **Importance of SDN:** SDN can potentially opens up vertically integrated distributed control networking gear and allows standard switch hardware to be controlled centrally through OpenFlow through an SDN controller. Open flow in conjunction with SDN creates an open-source cloud operating system with the mission to enable any organization to create and offer cloud computing services running on standard hardware. So, SDN has the ability to integrate seamlessly with cloud services, IoT devices, and remote access solutions, optimizing network efficiency and reducing operational costs. SDN centralized management and automation capabilities, enabling organizations to streamline network operations and improve resource utilization, thereby making it an ideal solution for network modernization. SDN also provides enhanced visibility and flexibility through the decoupling of software from hardware allowing them to dynamically adjust network policies and configurations with enhanced control over the network. SDN networks in brownfield environments and its implementation is much easier when compared with existing traditional systems. There is no need to use Virtual local area networks (VLANs) for segmentation so, it simplifies network configuration and management. Also, for digital transformation to improvize innovation and competitiveness, demands for networks to be scalable, secure, and able to support a variety of workloads and applications across distributed environments. So, SDN is increasingly important. So, one can infer that major reasons for importance of SDN lies with *customization of network, enhanced control with greater flexibility* and *robust security with better management.*
2. **Issues of SDN:** With reference to III 14.2.0.4 and Fig III 14.2b, we get that SDN architecture is made up of three layers that communicate using northbound APIs–enabling lower-level components to communicate with higher level components and southbound APIs—facilitating communication in the opposite direction. As shown, three layers are:

- **Application layer:** The application layer includes network applications and programs. The application layer communicates with the control layer through its northbound interface, informing the control layer of application resource needs.
- **Control layer:** The control layer the brain, or network operating system manages network infrastructure, the movement of traffic and data. The control layer plays a key role in *resource allocation* throughout the network and makes communication between the application layer and the infrastructure layer possible.
- **Infrastructure layer:** This layer consists of physical switches and routers, etc., for moving data packets and network traffic through the network. So, the data plane is responsible for forwarding network traffic.
- In addition to these layers, many other components constitute SDN such as:
- **Applications:** Applications work with information about the network or requests for specific resource availability or allocation.
- **Controller:** SDN controllers handle communication with the apps to determine the destination of data packets. The controllers are the *load balancers* within SDN.

  **Networking devices:** Networking devices receive instructions from the controllers for routing the packets.
- **Open-source technologies:** As shown in Fig III 14.2a2, networking protocols, such as OpenFlow, direct traffic among network devices in an SDN network. The Open networking foundation (ONF) standardizes the OpenFlow protocol and other open source SDN technologies. SDN network is an OpenFlow programmable networking. By combining these components, organizations get a simpler, centralized way to manage networks. SDN dissociates the routing and packet forwarding functions, known as the control plane, from the data plane/underlying infrastructure. SDN implements controllers and layers them above the network hardware in the *cloud* or *on-premises.* This is policy-based management—a kind of automation— to manage network control directly. SDN controllers tell switches where to send packets. Sometimes, virtual switches embedded in software or hardware replace the physical switches. This consolidates their functions into a single, intelligent switch that can check data packets and their virtual machine destinations to ensure there are no issues before moving packets along.

*Virtualization issue is another important issue and has been discussed separately below.*

3. **SDN Platform:** Function of both North and south bound APIs are extremely important and have been preliminary discussed above. Here, under SDN same will be discussed in details again. Different issues will be covered here. Fig III 14.2c shows typical SDN architecture with centralized control system. Various issues are discussed below:
   - **SDN Controller:** As discussed, in SDN there will be a centralized control (Ref: Fig III 14.2c), which is a brain and most strategic point. With a collection of modules, it can perform different network tasks including *network topology*, *network statistics,* etc. Different network applications including network policies are installed on the controllers for data communication.
   - **API discussions:** APIs play very crucial role in SDN by providing communication amongst the data, control and management planes as already detailed in Fig III 14.2a1 and III 14.2b. The well-known *Southbound API (SBI)*s and *Northbound API (NBI)*s, are applicable for Centralized control.

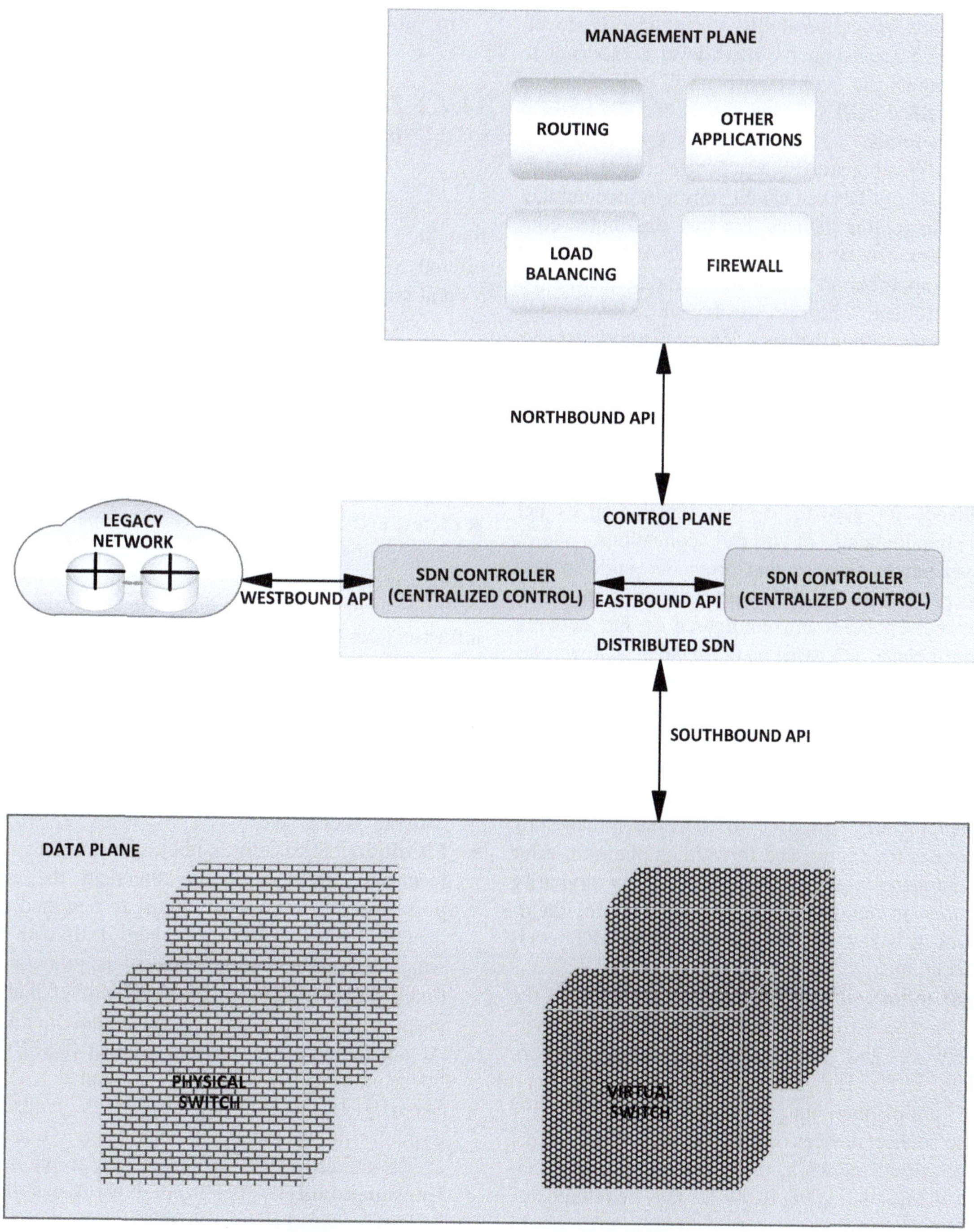


**FIGURE III 14.2C** SDN system architecture explained.

For distributed controllers, east/westbound APIs are used. These APIs are components of SDN, and they are used to configure forwarding devices or network applications, etc., shown in Fig III 14.2c.

- *SBI*: As discussed, SBI provides a communication protocol between the control plane and data plane to push *configuration information* and *install flow rules* at the data plane. It provides an abstraction of the network device's functionality to the control plane and is critical with respect to its availability and secure communication.
- *NBI*: Control and management planes use NBI to provide programmability to application developers i.e., it supports lower level component to communicate with higher level component as described. NBI supports variety of SDN applications forms.
- East/Westbound API [13]: Single central network control of SDN can handle only a limited number of forwarding devices. For this, distributed controllers can be used. Eastbound APIs are used to import/export information among distributed controllers whereas westbound APIs enable communication between legacy network devices and the controllers.

- **Configuration and Flow Rules:** Computerized networks are mainly configured based on Access control list (ACL) policies and routing protocols. The ACL policies refer to the set of rules and guidance for network devices functioning as per the requirements of users, applications, and/or organizations. The routing protocols guide to find best path between source and destination. In SDN, the ACL policies are configured at the network control plane, and based on those policies, flow rules are generated and installed at forwarding devices [13].
- **NFV:** NFV should not be confused with Virtualized network functions (VNF), which is meant to the specific function virtualization on the network. NFV is complementary approach to SDN in While SDN separates the control and forwarding planes to offer a centralized view of the network. NFV primarily focuses on optimizing the network services themselves. It is key part of cloud computing, 5G development, etc.

4. **Virtualization and SDN:** Network virtualization is the key to the success of *cloud computing*. The term "virtual network" and "SDN." are distinct, but can work together well. The term "virtual network" and "SDN." are distinct, but can work together well. NFV can also connect devices on different networks to create a single virtual network, often including virtual machines. In this context Fig III 14.2a1 may be referenced. NFV can also connect devices on different networks to create a single virtual network, often including virtual machines. The key reasons for virtualization are:
   - **Sharing:** for resource is too big to be divided it into multiple virtual pieces. Each processor can run multiple virtual machines each to be used by a different user.
   - **Isolation:** Multiple users sharing a resource may not trust each other hence to provide isolation among users. Hence SDN works well with NFV to assists NFV by refining the process of controlling data packet routing through a centralized server, for better visibility and control.
5. **Types of SDN:** Different SDN types have been tabled in Table III 14.2b.

### 14.2.2 SDN: industrial impact, challenges and trends

Adopting SDN to enable on-demand provisioning of networks and network services. By virtualizing the network through a programmable interface, better support can be offered. SDN can help us increase business value from the Virtual machines (VMs) in data centers and can offer benefits such as *Reduced network-provisioning time*, *Simple network creation process in a self-service environment*, *Reduction in service costs*, and Improved network management efficiency. Apart from these there will direct impact on industries due to several performance and scalability improvements in SDN components. On account of the programmability features SDN is gaining popularity in both academia and industry. However, this emerging paradigm has been facing diverse kinds of challenges during the SDN implementation process and with respect to adoption of existing technologies especially in enlarging network infrastructure. Major challenges include but not limited to:

- **Scalability**: This mainly comprises *Controller scalability* and, *Network-node scalability*. To increase scalability, the logically-centralized controller must be physically distributed. Here, the challenges will be interactions sharing among network devices hence network node issue.
- **Flexibility**: Flexibility is not alone but goes with performance also. Flexibility refers to the capability possessed by networks to adjust to new and unprecedented features. Performance here deals with speed at which network nodes at the data plane process information from control plane. It has been proven that General purpose processors provide the highest flexibility for Application specific standard product (ASSP) provide better performance. Hence the hindrance.
- **Security**: In the SDN architecture, the controller is a target for threats especially when open to unauthorized access and can cause serious damage to the network.
- **Interoperability:** Apart from above issues, there will be issues related to interoperability also.

**TABLE III 14.2B** Types of SDN.

| SDN Type | Description |
|---|---|
| **API SDN** | Often referred to as SBI. Application programming interfaces control the data moves through the network to and from each device. It allows orchestration platforms, cloud management tools and network management systems to integrate with SDN infrastructure. |
| **Hybrid SDN** | By combining SDN and traditional networking, the hybrid model supports different functions on a network and assigns the optimal protocol for each type of traffic. Hybrid SDN is often used as an incremental approach to SDN, enabling enterprises to integrate SDN into legacy environments i.e., traditional networking protocols continue to direct some traffic, while SDN takes on responsibility for other traffic. |
| **Open SDN** | Open protocols are used to control the virtual and physical devices responsible for routing the packets in data plane. Open SDN enables various teams of network operators, developers and vendors to work together on optimization. |
| **Overlay SDN** | Virtual networks run above existing hardware, dynamic tunnels to different on-premise and remote data centers. This model then allocates bandwidth and assigns devices to each channel. |

The evolving networking industry comes with new applications and innovations affecting networks. The network architectures must adapt to keep pace with modern digital trends inspiring innovations. SDN network operations aim to abstract control, automate workflows and optimize network resources to quickly acclimatize networks to changing business needs, applications and traffic patterns. We understand that SDN architectures provide improved system by separating the network's control plane and the data plane i.e., the control plane is a programmable controller infrastructure that is distinct from physical devices. Controllers may be based in the cloud or on premises. So, modern enterprises are quickly moving away from the days of device-by-device management and into a future where security protections. SDN solutions can also offer connectivity across clusters, stacks and runtimes other than containers and satisfy the latency obsessed. In short, software-defined networking is going to continue to evolve, but will always play an important role in modern networking [14]. So, in concluding subsection we discuss the industry impact, challenges and future trends for SDN.

1. **Role of SDN:** SDN play a crucial role in industrial networking in providing *enhanced flexibility*, *scalability*, *security* and efficiency. Major issues here are:
   - **Flexibility and agility:** This is achieved through *Dynamic network configuration*, *Customized networking*, *Dynamic resource allocation*, *Supportive of new technologies* (such as IIoT, 5G), *Device integration*, *Data management*, *Remote management*, etc.
   - **Scalability:** Improved scalability through *Resource optimization*, *Easy expansion* and *Handling of large-scale network* through virtual divisions.
   - **Reliability:** Reliability and redundancy are achieved through *Predictive maintenance* practice and *Back up facility.*
   - **Security:** *Centralized control*, Consistent *security policy adoption*, *Real-time monitoring*, *Supportive of advanced security measures,* etc., enhance security.
   - **Cost saving:** *Optimized* resource *allocation* and *Lower operating costs* are reasons behind cost saving.
   - **Performance:** Better performance is achieved through *Low latency*, *Bandwidth management*, *Faster recovery from failure*, Hand-in-hand *wired and wireless connections.*
   - **Computation:** *Supportive* of both cloud and edge computation
2. **Industry trends:** Five major Industrial trends for SDN are:
   - **Cloud-native Application:** With rising trends in adjacent technologies such as big data analytics, Artificial intelligence (AI) and Machine learning (ML), security, and IoT/edge computing, there will be rising trends in the cloud-native applications. SDN infrastructure with flexible and agile nature is well suited for cloud-native architecture. Cloud-native is much more than simply registering with a cloud provider and running cloud-based applications on their platforms. Cloud native is a software approach of building, deploying, and managing modern applications in cloud computing environments for highly scalable, flexible, and resilient applications, which can be updated quickly based on customer requirement. High availability, increased

efficiency with reduced cost are major reason for selecting cloud native.

- **Hybrid and Multi cloud adoption:** In hybrid cloud different cloud types are blended. In multi cloud different clouds of same types are blended. All size business types get benefitted from a hybrid and multicloud approach in terms of increased resilience, flexibility, freedom of choice, and cost reduction. The key challenge however is to interconnect these different cloud and on-premises environments efficiently on a regional and/or worldwide scale. SDN offers the ultimate connectivity concept for hybrid and multicloud solutions.
- **5G application:** 5G can bring lightning fast and utmost reliable connectivity options while being able to connect huge numbers of appliances in a low latency manner. So, 5G is now gradually replacing 4G networks paving the way for intelligently deploying industry applications supported by private 5G networks and reaping the benefits of the Industrie 4.0 concept.
- **Remote work:** Working from home/remote trend would persist. Organizations find it difficult to accommodate the massive inflow of remote worker Internet traffic streaming in over company firewalls to Virtual private network (VPN)s. They're up against the task of keeping business as normal in an unfamiliar environment, all while fending off fresh waves of phishing and DDoS attacks. SDN enables flexible alignment with evolving business needs with highly fluctuating levels of remote working. So, with very low latency and the ability to seamlessly scale the bandwidth capacity up or down, even in a segmented way, SDN allows for the infrastructural agility pretty much every CIO and CTO could wish for [ [15].
- **IoT application:** With rise of IoT applications, the demands on the network become more complex. SDN can provide for optimal connectivity and data transfer. With increase in applications of IIoT/IoT, there will be major challenges for traditional network especially in flexibility, latency, and security. SDN can uniquely solve these interconnectivity challenges. Because of its elastic nature, SDN makes it quite easy for an organization to manage the bandwidth that a network requires.

3. **SDN: Challenges:** When SDN are widely adopted and new solutions are being proposed, then there would be a few challenges from:
   - **Scalability:** Decoupling, in SDN has its own pitfalls. API between the two planes should be able to handle the needs of various architectures, and should be able to facilitate the independent evolution of both planes. SDN scalability can be classified into:
     - *Data plane*: Data plane scalability mainly depends on the processing power, memory/buffer capacity, and software implementation of the data plane devices. Another scalability challenge arises when the control traffic shares paths (and competes for resources) with the data plane traffic, and for a reliable management of the data plane.
     - *Control plane:* The most prominent SDN control plane performance metrics are throughput (single controller can handle <6 million flows per second.) and flow setup latency. Therefore, *distributed, hierarchical* and *hybrid* architectures are adopted. This challenge faced by the controller would be shared among network devices, both for distributed or peer-to-peer controller infrastructure. Scalability issues arise from the separation of the control and data planes, as now, network devices are managed by a *remote* controlling mechanism.
     - *Application plane:* SDN applications themselves may call for scaling to support fine-grained and optimized resource utilization in wide-area networks. Load balancing is an old applications envisioned for SDN.
     - *Cross-Cutting Scalability*: This may arise in the context of wireless software-defined. The wireless transmission medium introduces additional scopes for control and optimization.

     Onix is a distributed platform approach and Onix and HyperFlow are considered to be effective approaches toward achieving scalability. Onix is a distributed platform approach and HyperFlow is meant for OpenFlow.
   - **Flexibility and Performance**: Flexibility refers to the capability to adjust to new and unprecedented features. Performance here deals with speed at which network nodes at the data plane process information from control plane. It has been proven that general-purpose processors provide the highest flexibility whereas ASSPs lay the foundation for high-performance networks. The attributes of Application specific integrated circuit (ASIC)s providing the best requirement for implementing the SDN data plane and this is currently being incorporated into SDN products also.
   - **Interoperability**: Integration with legacy system and Integrating with existing network hardware and protocols can be difficult and challenging. So, ensuring interoperability and compatibility across diverse vendors and systems is essential for seamless operation.
   - **Security**: Central controller in the SDN architecture, is a potential target for threats. Attacks on the controller can cause serious damage to the network,

as it is responsible for controlling the entire network. Authorization and authentication are very complex as a group of controllers interacts with a single node or vice versa. So, SDN architecture must support a high-level security system. Various threat mitigation strategies would eventually arise even as the potential for unauthorized access increases.

4. **Applications:** SDN can be implemented in various application areas such as:
   - **Data Center**: The SDN architecture is targeted at Infrastructure-based networks and data centers fall under this category. Data center is a very large scale network operating based on policy enforcement and traffic management at a critical level.
   - **Enterprise Network:** An enterprise network with the connection of several temporary devices to the network.
   - **Optical Network:** Extending SDN to handle data traffic as flows enhances the integration of multiple network technologies. Consequently, communication between packet-switched and circuit-switched networks will be enhanced.
   - **Cloud computing:** SDN finds potential application in cloud computing for efficient management of virtualized resources, seamless multitenant environments, and efficient hybrid cloud integrations.

   Apart from these, WAN, network virtualization are some application areas for SDN.
5. **Future issues:** As network traffic grows, the mapping of flow rules between SDN controller and switches becomes overburdened, while a few others become unburdened. Therefore, for poor throughput and long reaction times, such an imbalance would impair the performance of the SDN network and it is difficult to manage the load across several controllers dynamically. SDN security addresses several SDN-based defensive primitives to cope up with contemporary cyber dangers. So, both novel primitives and modular protection systems that employ various primitives are required. Major bottlenecks in SDN include: controller scalability, separation of control and data planes, quantity of events/requests handled by a controller, and controller–switch communication delay. Hence due attention to these are desired. A few other issues are: the energy-efficient SDN networks, wireless networks, network virtualization techniques, cloud computing platforms, and SDN migration mechanisms [13]. Also, there will be need to make up the gap in new skill set requirement, operational complexities. AI/ML integration, Edge computation requirements must be met. These are in addition to requirements for interoperability, legacy system integration, etc., a few issues to name

   Let us now conclude the discussions on Networking software and SDN to explore Networking: OPC integration and migration issues in next chapter.

## List of abbreviations used

**AA** Application associations
**ACL** Access control list
**ACSE** Association Control Service Element
**ADSL** Asymmetric DSL
**AE** Application entity
**AI** Artificial intelligence
**API** Application programming interface
**ARP** Address resolution protocol
**ASSP** Application specific system processor
**BFT** Byzantine fault tolerance
**CHAP** Challenge-handshake authentication protocol
**CIM** Computer integrated manufacturing
**CIP** Common industrial protocol
**CNC** Computer numerical control
**CS** Companion standard
**CTDMA** Concurrent TDMA
**DSL** Digital subscriber line
**EAP** Extensible authentication protocol
**ERP** Enterprise resource planning
**GW** Gateway
**I/O** Input/Output
**IED** Intelligent electronic device
**IIoT** Industrial IoT
**IoT** Internet of things
**IP** Internet protocol
**ISP** Internet service provider
**IT** Information technology
**LAN** Local area network
**LCP** Link control protocol
**MAC** Media access controller
**MFA** Multi factor authentication
**ML** Machine learning
**MMS** Manufacturing message specifications
**NBI** North bound API
**NCCM** Network configuration and change management
**NCP** Network control protocol
**NFV** Network functions virtualization
**NOS** Network operating system
**NTP** Network time protocol
**NVL** Named variable list
**OAUTHO** Open authorization
**OLE** Object linking and embedding
**ONF** Open networking foundation
**OPC** OLE for Process control
**OS** Operating system
**OSI** Open systems interconnect
**OT** Operational technology
**PAP** Password authentication protocol
**PI** Program invocation
**PLC** Programmable logic controller
**PNAC** Port-based network access control
**PPP** Point to point protocol
**PPPOE** PPP over Ethernet
**PTP** Precision time protocol
**QoS** Quality of service
**RADIUS** Remote authentication dial-In user service
**RT** Real time
**RTU** Remote transmission unit
**SBI** South bound API

**SCADA** Supervisory control and data acquisition
**SDN** Software-defined network(ing)
**SOE(R)** Sequence of event (recording)
**SQRL** Secure quick, reliable log in
**TCP** Transmission control protocol
**TDMA** Time division multiple access
**TSN** Time sensitive network
**UDP** User datagram protocol
**VLAN** Virtual local area network
**VM** Virtual machines
**VMD** Virtual manufacturing device
**VNF** Virtualized network functions
**VPN** Virtual private network

## Tables (with clause reference)



## List of drawings (with clause reference)



## References


[1] R. Mohanakrishnan; What is Network Software? Definition, Types, Components, and Best Practices; Spicework; 2022; Internet document; https://www.spiceworks.com/tech/networking/articles/what-is-networksoftware/#:~:text=Network%20software%20plays%20a%20crucial,offers%20useful%20benefits%20to%20organizations.

[2] E. Conrad, J. Feldman, Authentication Protocol; Domain 4; Eleventh Hour CISSP®, third ed., Science Direct, 2017. https://www.sciencedirect.com/topics/computer-science/authentication-protocol.

[3] W. Burdick; Extensible Authentication Protocol (EAP); published by M. Rouse (what is com); https://searchsecurity.techtarget.com/definition/Extensible-Authentication-Protocol-EAP.

[4] Common industrial protocol (CIP™); ODVA; Internet Document; https://www.odva.org/technology-standards/key-technologies/common-industrial-protocol-cip/.

[5] D. Collins; What is the common industrial protocol (CIP)?; 2020; https://www.motioncontroltips.com/what-is-the-common-industrial-protocol-cip/.

[6] D. A. Vasko; CIP Safety: Safety Networking for Today and Beyond; Rockwell Automation; CIP safety; https://www.odva.org/Portals/0/Library/Publications_Numbered/PUB00110R1_CIP_Safety_White_Paper.pdf.

[7] Time-Sensitive Networking: A Technical Introduction; White Paper; Cisco Public; Internet document; https://www.cisco.com/c/dam/en/us/solutions/collateral/industry-solutions/white-paper-c11-738950.pdf.

[8] S. Brooks & E. Uludag; Time-Sensitive Networking: From Theory to Implementation in Industrial Automation ; TTTech;Intel; white paper; https://www.intel.com/content/dam/www/programmable/us/en/pdfs/literature/wp/wp-01279-time-sensitive-networking-from-theory-to-implementation-in-industrial-automation.pdf.

[9] D.A. Weder Time Sensitive Networking: An Introduction to TSN; Fraunhofer IPMS; Internet document; https://www.design-reuse.com/articles/46536/an-introduction-to-time-sensitive-networking.html.

[10] Time sensitive networking for industrial automation; Texas instrument; internet document; https://www.ti.com/lit/wp/spry316c/spry316c.pdf?ts=1724853139793&ref_url=https%253A%252F%252Fwww.google.com%252F.

[11] A description of the manufacturing message specification (MMS); SINTEF; stig ole johnsen, SINTEF T&S kai hansen, ABB'Aug'2007; https://sislab.no/MMS_Notat.pdf.

[12] M. Blessing, J. Olusegun, The Impact of Software-Defined Networking (SDN) on Traditional Network Architectures: Opportunities and Challenges, Research Gate, 2024. https://www.researchgate.net/publication/383467175_The_Impact_of_Software-Defined_Networking_SDN_on_Traditional_Network_Architectures_Opportunities_and_Challenges.

[13] M. Hussain, N. Shah, R. Amin, S. S. Alshamrani, A. Alotaibi and S. M. Raza; Software-Defined Networking: Categories, Analysis, and Future Directions; Sensor; MDPI; Internet document; https://www.mdpi.com/1424-8220/22/15/5551.

[14] What is the future of SDN? Ask juniper; 2019; Internet document;; https://blogs.juniper.net/en-us/ask-juniper/what-is-the-future-of-sdn.

[15] These 5 industry trends are driving SDN adoption; R. van der Zwan; WORLDSTREAM blog;; Internet document; https://www.worldstream.com/en/news/these-5-industry-trends-are-driving-sdn-adoption.

Chapter 15

# Networking: OPC, integration, and migration issues

For Industrie 4.0, aim is to integrate of Information technology (IT) with Operation Technology (OT) related to production, systems, solutions, and services. Once implemented, it opens the doors to unprecedented new business opportunities and customer value. This is possible if meaningful information can be shared openly and securely at every level. Open platform communication (OPC) standard makes this possible. Thus, initiatives like Industrie 4.0 Internet of Things (IoT) and Industrial IoT (IIoT) have created huge demand for data from manufacturing processes and the need for systems and devices to be connected. The network is obviously a key element of connectivity. The unique combination of two technologies viz. Publish-subscribe model of OPC foundation and Time-sensitive network (TSN) is paving the way for the emergence of a universal solution for Ethernet in industrial applications and may replace current industrial Ethernet solutions, that is, OPC Unified architecture (OPC UA) and TSN are catalysts for change in automation architectures. Present OPC particularly, OPC UA (unified the functionality of OPC specifications) for a single, comprehensive framework which can be characterized by *Platform independence*, *Enhanced security*, *Comprehensive information and* multiple communication *modeling*, *Transport flexibility*, *Scalability for Industrie 4.0*. At the heart of the IIoT/IoT, OPC addresses the need for standardized data connectivity and interoperability for both horizontal and vertical data communications and integration. OPC covers the communication and information layers (Ref: II 16.2a) So, OPC is at the heart of IIoT, Industrie 4.0 and Machine to machine (M2M) information exchange. OPC can provide solutions for almost all the needs of Industrie 4.0 and Digital transformation (DT) such as: *Independence of the communication technology*, *unplanned* ad hoc *communication*, *Secure transfer and authentication at application levels*, *Scalability for integrated networking*, *Service oriented architecture*, *Standard transportation*, *Mapping of information with all degrees of complexity*, *Secured transfer after authentication*, etc.

Unifying disparate data sources found in industrial automation environments can be a complex and challenging process as it involves establishing sometimes diverse *physical* and *digital* connections to process control equipment and machinery, and integrating multiple sets of data. Digital is more complex, and people take advantages of common platforms like OPC acting as a middleware and bridge that digital connectivity gap. Distinct OT requirements call for: Real-time (RT), reliable data connections, network supports control applications with security and distinct OT and IT networks with demilitarized zone (DMZ) Moving toward a fully connected facility is critical for being ready to integrate new networking tools. Industrie 4.0/Digital transformation is defined and established through connectivity. Therefore, with network integration, it is possible to have all devices on one singular flat network, machine learning data analytics can take over the tasks that were previously accomplished with more complex and inflexible network structures. Network integration helps manufacturers accommodate evolving networking requirements, such as decentralization of control, integrated diagnostics, and simplified maintenance, network protocols integrate with industrial equipment and control systems to communicate crucial status updates and production data. It not only maximizing production control, enterprise connectivity to improve product quality, customer satisfaction, and company profitability but also, enhances networking capabilities provides advantages ranging from complete process efficiency and remote access to system controls, to faster response times, to factory floor events and minimized downtime, to name a few. Major aspects of network integration shall include the following

- The convergence of IT and OT for industrial networks that support both deterministic control and IT-level data exchange.
- Open communication protocols and true system interoperability for machinery integrating with legacy industrial networks, that is, careful consideration of hardware, I/O, and existing factory standards to avoid future complications.
- Prioritizing hardware components and communication protocols aligning with the control requirements as

Plant Intelligent Automation and Digital Transformation. https://doi.org/10.1016/B978-0-323-89892-8.00016-8

well as the existing control platform for reduction in costs and learning curves.

Control networks, migration is the strategic transitioning from an existing control system to a new, upgraded system, that is, the process of moving a network's data, applications, and services from one environment to another with the goal is to improve performance, security, and scalability to meet evolving business needs. This applies to Industrial control systems (ICS) and may refer to replacement of hardware, updating software, or adopting new network architectures including adopting Software-defined networking (SDN). A major upgrade should not only focus process automation but also on workforce development. Therefore, a major migration project offers the best opportunity to train the next generation of automation engineers and plant operators in a way that can yield benefits for many years to come. This is considered an opportunity for mentoring, that is, to transfer personal and tribal knowledge that is difficult to capture and convey in any other way. A control system migration is an ideal time to learn about plant operations, automation, security, IT functions, and project management. Control system migration calls for complete understanding of all related issues like: *Instrumentation*, *Field wiring and device-level networks*, *Safety instrumented systems (SIS)*, *Control valve*, *Wireless network*, *Alarm management*, *Operator interface—Human machine interface (HMI)*, *Programmable logic controller (PLC)/Distributed control system (DCS)*, *Remote terminal unit (RTU)*, *Asset management*, Historical system, *Simulation*, *IT integration systems*, etc. Therefore, it calls for involvement from start to finish. So, control system migration is very exciting and transformational time a process plant will ever see. The new generation needs to take a challenge combining highly technical tasks with management of a major project [*courtesy ISA Blog*: https://blog.isa.org/control-system-migration-is-a-major-learning-opportunity]. This chapter has been dedicated to address these important issues on network.

## 15.0 Treatise on OPC in networking

Component object model (COM) makes the communication of objects on a computer possible. Objects which are located on different computers can also use Distributed COM (DCOM) to communicate with each other. So, DCOM is an object-oriented Remote procedure call (a protocol meant for distributed systems, allowing a client/program to execute functions procedure/subroutine on a remote server) system which is based on the DCE standard defined by Microsoft. DCOM is an expansion of COM and enables COM calls between distribution computers within one network. The acronym "OPC" comes from " Object linking and embedding (OLE) for Process Control" OPC is open connectivity in industrial automation for interoperability supported by creation and maintenance of open standards and specification. In OPC both COM and DCOM technologies are used to achieve network capability. DCOM Is available in all Windows platforms. From previous discussions we understand that because of COM, OPC use client server technology. Both OPC client and server can be in same of different computers. There are multiple standards under envelop of OPC to deal with multiple software-to-software and software-to-hardware problems. Each standard addresses a different problem type. Remember the original problem OPC set out to solve with OPC Data Access (OPC DA), The OPC DA standard defines function calls to read, write, and subscribe to data points. However, for network integration, OPC Unified architecture (OPC UA) is used and is very popular. OPC Coverage and OPC UA profile has been detailed out in Fig. III 15.0a. OPC UA can use a unified code-based and implement whatever profiles they need. So the classic OPC standards became OPC UA Profiles and an OPC UA server or client describes itself based on profiles it supports like OPC UA-DA profile, OPC Alarm and event (OPC A&E) profile, OPC Historical DA (HDA) profile, OPC Extendable mark-up language (OPC XML), etc. Fig. III 15.0a one shows OPC total coverage, while OPC UA profile has been detailed in. Fig. III 15.0a2. Preliminary knowledge on OPC has been acquired from III 11.0.2. Here, little deeper discussion on OPC with its various types shall be covered. This section covers detailed discussions on OPC.

### 15.0.0 Fundamentals of OPC

OPC solution represents an open solution for industry, in Client-Server architecture (COM) and computer–computer connection (DCOM). The control of OPC standard is now with The OPC Foundation run by members to offer scalable solution with flexible means of data access in secured manner as per Microsoft standard. OPC is an open standard for the uniform access of different Windows-based software including industrial applications: HMI, table calculation, archiving software, etc., on the various data sources at the automation level viz. PLC, scales, scanner, etc.). A uniform solution was required to the access of the software applications from various data sources. OPC is a vendor-independent interface standard for automation to simplify data communication significantly. The basis of OLE is Com with characteristics like

- Client/server architecture
- Definition of valid way to access software services,
- COM objects implemented within the server, offer services through interfaces
- Client applications use the services of the object through an interface.

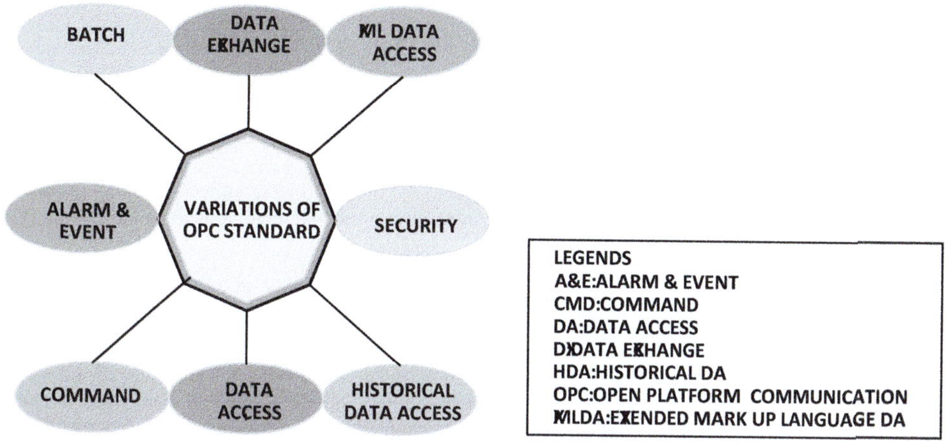


FIG III 15.0a1 OPC COVERAGE

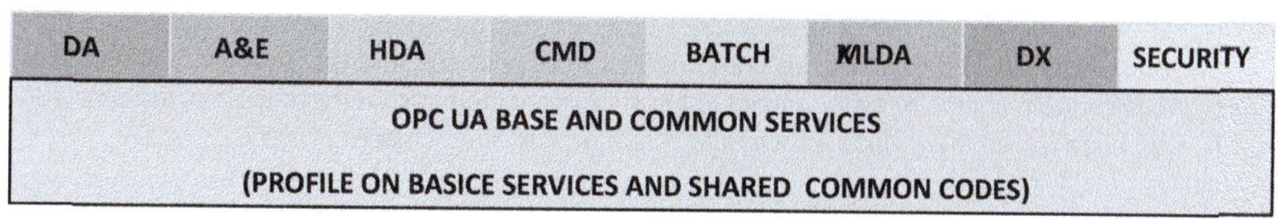


FIG III 15.0a2 OPC UA PROFILE

**FIGURE III 15.0A** OPC coverage and OPC UA profile.

Manufacturers of the devices providing process/service data offer an OPC server interface with the module for linking to the *data source* through OPC server responsible for it. In contrast to that, manufacturers of applications involved in recording, visualizing, etc., provide an OPC client interface with the application. OPC Servers are connectors and at as translators between the OPC world and a data source's interface/native communication protocol. Being bi-directional OPC Servers can read-from and write-to a *data source*. One OPC server will only transfer data to/from a data source if an OPC client commands it to. OPC Servers can communicate with *virtually* any *data source* whose output can be read or written to via electronic means. Common *data sources* are as follows: devices, PLCs, DCSs, RTUs, electronic scales, databases, historians, web-pages, etc.

The OPC Foundation has defined a series of specifications, whereby each specification is provided for a certain area of use. The OPC DA specification -the first specification of the OPC Foundation provides for process data communication by defining an interface between client programs and server programs for the data exchange. According to the OPC specification, three hierarchical classes in the class model are as follows

- **OPC Server:** An object of this class represents a manufacturer-specific OPC server.
- **OPC Group:** An object of this class *structures* the variables used by the OPC server.
- **OPC Item** An object of this class represents the actual process variable.

In this subsection fundamental issues of various OPC systems shall be discussed at length to supplement to the discussions in III 11.0.2

1. **Connectivity challenges** and **solutions** There are a number of factors which were major challenges for data sharing. These are detailed in Table III 15.0a which has been developed based on discussions in Ref. [1]. Table III 15.0a and in other aspects related to these challenges with solutions are discussed below:
   - **Architecture:** Fig. III 15.0b shows typical OPC architecture and communication method. OPC is a standard *software*-to-*software* interface as shown.

     As shown Fig. III 15.0b1 OPC client in HMI decouples but decoupled from communication infrastructure. Similarly, OPC server interfaces specific hardware. The network enjoys improved performances flexibility as well as Microsoft security. As shown in Fig. III 15.0b2, Server shield controller specific addresses and exposes PLC memory addresses as "Tags" (a friendly name, that maps to a memory register in the PLC). In this way, it decouples the OPC Client from the underlying device protocol and memory register knowledge. This architecture supports both Synchronous I/O (Blocking call/round-robin polling applications) and Asynchronous I/Os.

**TABLE III 15.0A** Major challenges of data sharing with OPC solution [1].

| Issue | Description |
|---|---|
| Proprietary Protocol | This allows communication among devices from particular product line, requiring particular driver to communicate with other. OPC does not require communication mode of *data source*. |
| Driver loading | Each driver establishes its own connection to the device to communicate with, hence driver loading can create additional traffic. OPC being a single device-specific, communicating many applications, traffic, and hence device loading is highly reduced. |
| Legacy system | With change in a version, device like HMI may require its own set of device drivers. OPC extends the useful life of legacy systems and allows the newest applications to continue communicating with the oldest systems. |
| Integration | For integration in traditional system is quite complex. Each node, for example, HMI needs a driver making so, version maintenance is a nightmare. OPC greatly simplifies integration as once an OPC for a particular *data source* is configured, all OPC-enabled applications can sharing data |
| Custom drivers | A custom driver is required for communication, historizing. OPC eliminates the need for custom drivers between each new application and *data source*. |
| Enterprise connectivity | With difficulties in establishing communication between IT and OT systems, establishing enterprise connectivity is very complex. OPC makes enterprise connectivity for data sharing possible by allowing approved applications to share data without the need for installing custom drivers |

- **Difference Server-client:** OPC Client is a software application requiring the data from the process control systems, or wanting to write something. OPC client can speak the OPC specified "language" to be able to get the data from an OPC Server. OPC servers talk to a device/devices using the specialized protocols of the devices, and then provide access to that data using the standardized formats defined by the OPC specifications by acting as protocol converter. Thus, the OPC server is the "*driver*" software that talks to particular *specific* hardware and reads/writes data and "serves it up" to the OPC client application.
- **Discussion on OPC server:** Here, let us concentrate on difference of OPC server from other systems developed by Microsoft and a few OPC server feature. As discussed in III 11.0, DDE was limited to one tag at a time and 50 concurrent connections over a period. OPC does not suffer from the performance and scalability issues. Similarly, ActiveX controls are typically user interface components and are also typically intended to be embedded inside of another application. In contrast to that OPC Servers are specialized applications, designed to get data from the plant floor and serve it to other applications. As per OPC Standard specification, OPC servers do not need server class hardware or Operating systems (OS). Hence, fairly lightweight OPC server software can easily co-exist with other software applications. OPC client-server can talk to each other also, one OPC server can talk to another OPC servers also. OPC *Bridging software* applications are OPC client applications that can connect to many different OPC servers and then allow to map data movement between the two *OPC servers*

2. **OPC communication** and **working:** OPC can be represented as an "*abstraction*" layer sitting between the *data source* and the *data sink*, allowing them to exchange data without knowing *anything about each other*. This device abstraction is possible by OPC client and OPC server discussed. It is important to note that neither respective native protocols are eliminated nor they are replaced by OPC as clearly shown in Fig. III 15.0b2, that is, the native protocols and/or interfaces communicate with one of the two OPC components. Data can travel from the application to the device without having one talk directly to the other. To understand OPC communication, it is better to refer to Fig. III 15.0c where these have been elaborated and all related issues have been discussed below in threadbare.
3. **Data exchange concept:** In control automation, OPC drivers help data sink (client say HMI) guide how the Data Source communicates or organizes its data so that it can pull data. As shown in Fig. III 15.0c1, in old system, each and every end-to-end connection required a custom driver for establishing communications. OPC abstracts only the implementation details for data exchange and eliminates unnecessary details. Therefore, based on Fig. III 15.0c2, one can conceive that OPC can be represented as an "abstraction" layer that sits between the *data source* and the *data sink*,

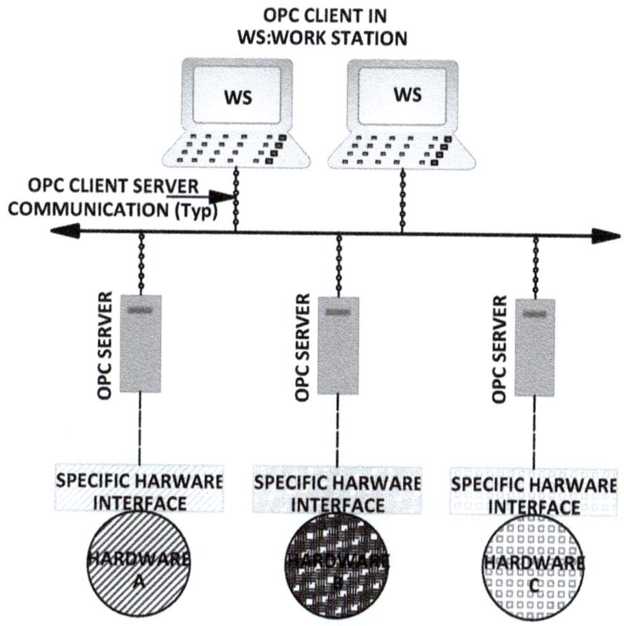


FIG III 15.0b1 OPC ARCHITECTURE

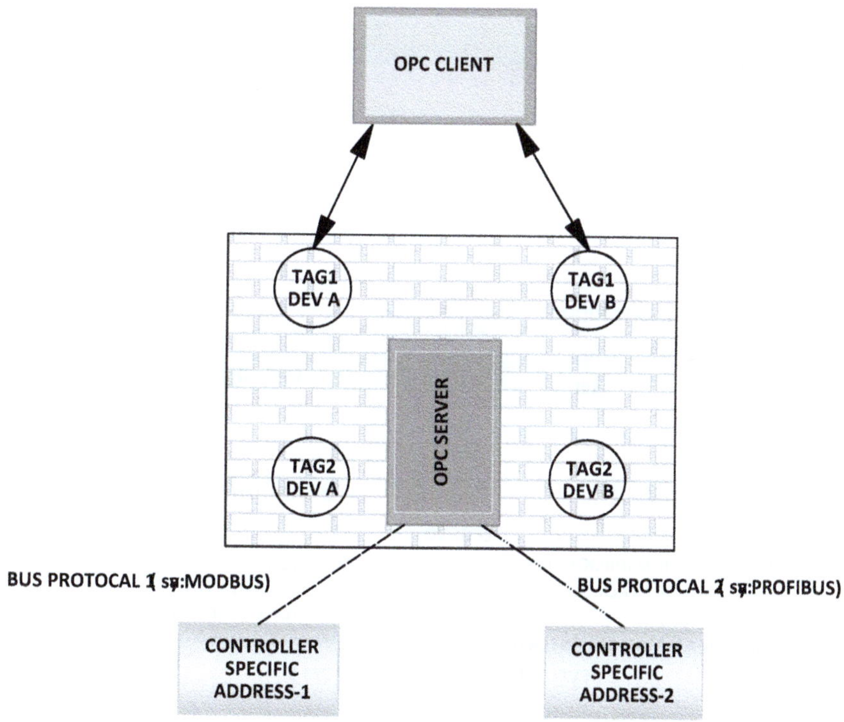


FIG III 15.0b2 OPC COMMUNICATION METHOD

**FIGURE III 15.0B** OPC architecture and Communication method.

allowing them to exchange data without knowing anything about each other [1]. Concept of OPC client–server communication has been elaborated in Fig. III 15.0c3.

4. **Server-Client Functioning:** Here, details on function concept of client and server are discussed.

- **Server functioning concept:** An OPC Server is a driver software application, developed to comply with one or more OPC specifications for different application classes. They act like convertors between the OPC world and interfaces of data Source. Being bi-directional OPC can read/write on Data

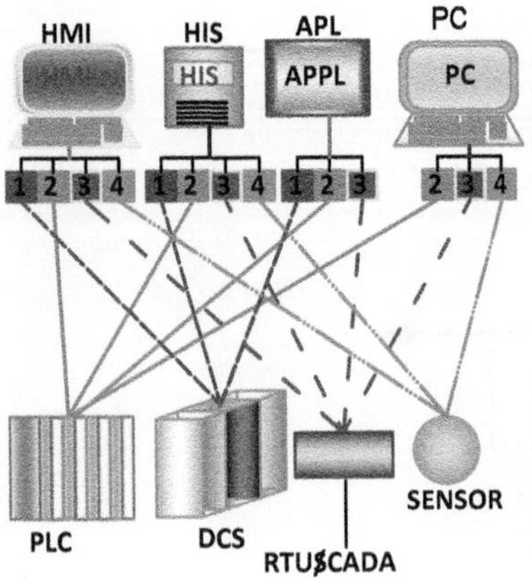


FIG III 15.0c1 DRIVER PROBLEMS

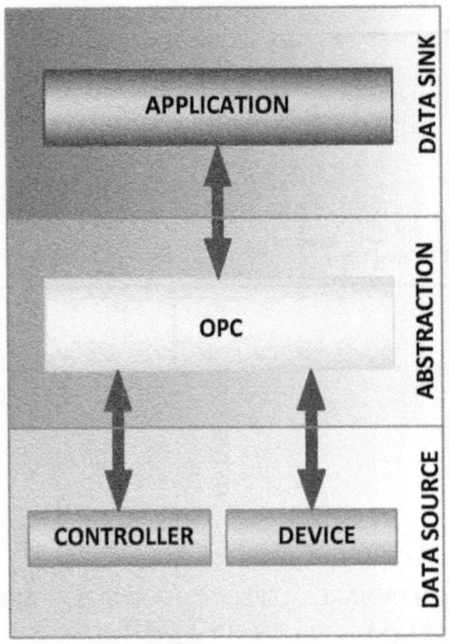


FIG III 15.0c2 OPC ABSTRACTION

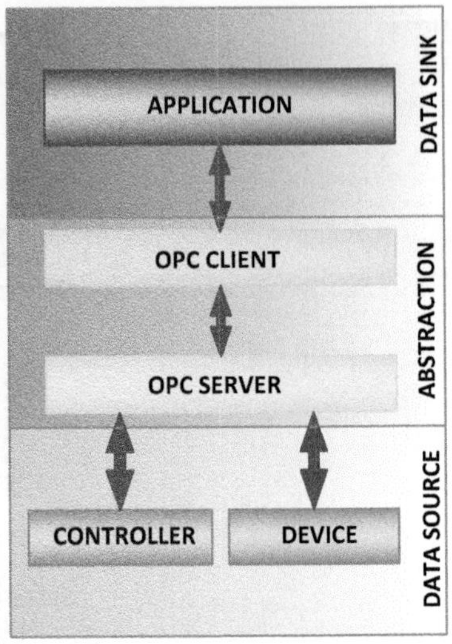


FIG III 15.0c3 OPC CLIENT SERVER

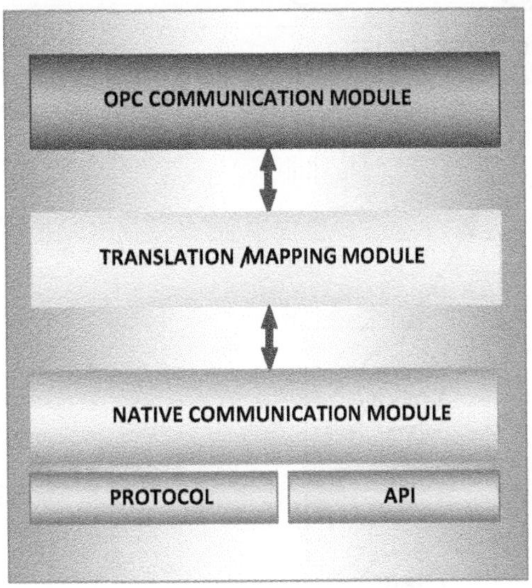


FIG III 4OPC SERVER CONCEPTUAL DETAILS

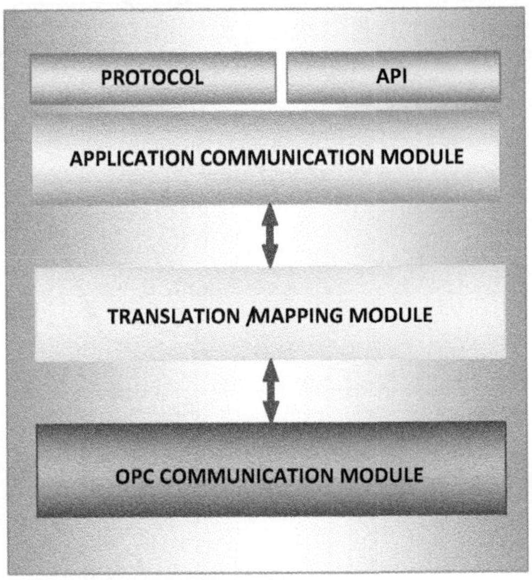


FIG III 5OPC CLIENT CONCEPTUAL DETAILS

LEGEND

APL:APPLICATION (PROGRAM)
DCS:DIGITAL CONTROL SYSTEM
HIS:HISTORIAN
HMI; HUMAN MACHINE INTERFACE
PLC:PROGAMMABLE LOGIC CONTROL
RTU:REMOTE TERMINAL UNIT
SCADA:SUPERVISORY CONTROL & DATA ACQISITION SYSTEM

DEVELOPED BASED ON IDEA FROM 1

**FIGURE III 15.0C** OPC Interface basics [1].

Source. OPC servers act as slave to their clients to act based on commands from clients. As shown in Fig. III 15.0c4, OPC server has three distinct parts viz. OPC communication module, translation and mapping module, and native communication module.

- *Communications module*: This is responsible for communication with OPC Client and is compliant with the OPC standard specification (s) for proper communication.
- *Translation/mapping module*: Main task of this module is to interpret properly the OPC requests from the Client (s). Another important task of this module is to translate/turn the client's requests into proper native requests that get sent to the *data source* for data and vice versa.

   - *Native communications module*: The OPC server should employ the most efficient method of communicating with the data source. In some cases, this means connecting to the data source directly via its native protocol, while in other cases, this means communicating with the data source via its custom driver via an Application programming interface (API) [1].
- **Client functioning concept:** An OPC client is software application developed to communicate with OPC connectors hence uses messaging compliant with OPC specification. OPC Client represents a data-sink to initiate and control communications with OPC server depending on the embedded application requests. Naturally, they need to translate application's communication requests into an OPC equivalent request and send it to the appropriate OPC server for processing. Again, when OPC data returns from the OPC server, the OPC Client translates it back into the application's native format. OPC Clients are embedded in applications such as HMIs, etc., for making them OPC-enabled. It is worth noting that OPC client *cannot* talk to another client as client–server communication is defined in OPC standard. However, OPC client can talk to a number of OPC servers simultaneously. Fig. III 15.0c5 shows the components of OPC client.
   - *OPC Communication Module*: It is not functional as the same in OPC server, and is responsible for OPC Client to behave correctly when it is connected to an OPC server, for data exchanges and disconnects without destabilizing the OPC Server [1].
   - *Translation/mapping module*: Bi-directional module translates information of embedded application requests data to be read-/written-on to data-source via server.
   - *Application communications module*: As the OPC Client is typically written to work within a specific application, so it relies on API to allow data to be passed from the application down to the OPC Server/Data-Source. However, it is also possible for some generic OPC Client to communicate with an application via a protocol [1].

5. **OPC Command:** In OPC, "commands" refers to actions to change the state of a server or of a device and execution time may be more time than that associated with reading and writing data. The OPC Commands specification defines the means to *discover, understand* and *control* commands a server provides. OPC Commands provides method of call capability. OPC Commands allow *control mechanisms* to be defined that go beyond *Read* and *Write* access. Using OPC Commands it is possible to provide a clear understanding of:
   - Preconditions before a command issuance
   - Result happening when a command is issued
   - Factors affecting the execution of the command
   - Progress of the execution of the command

   Commands can be exposed on different levels
   - Global level: Commands applying to the server itself/to the infrastructure represented by the server
   - Sublevel: representing only a part of the infrastructure behind a server

   Standards bodies and server vendors will define sets of commands for different types of application. These sets of commands will define different "application domains" like:. interfaces for programmable controls and complex devices viz. drives and valves. Also, a server vendor may choose to define a specific solution for a specific problem.

## 15.0.1 Various OPC types

The OPC Data access (DA), specification is the first of a group of specifications known as the OPC classic specifications. Also, The OPC DA Client/Server architecture was the first architecture defined by the OPC Foundation. OPC DA is an older version of OPC specification, and was built without security and firewalls in mind and has problems with authentication and security. OPC is based on the client/server architecture. OPC servers are the interface to the control devices that supply information. OPC client interfaces with the servers to read data from and write data to the control devices. As discussed, OPC clients can reside on the same device with the OPC server (local) or they can reside on a separate device (remote). The OPC Alarms event (AE) specification is intended for use with process automation systems to generate alarms and events information, that is, OPC AE server is written to expose for the same. The OPC AE server reports the alarms and events previously defined in the system only (not create) using a common communication interface. Once defined, the alarms and events in the system are automatically generated, based on the operating conditions and actions performed in the process plant. Once linked, OPC client can establish one or more subscriptions to that server. The client requests the server to monitor and report the alarms and events by call back method. It is possible to set filter to limit alarm flooding. OPC HDA refers to the specification that describes the OPC COM Objects and their interfaces implemented by OPC historical servers. Different vendors may provide OPC historian servers with vendor-supplied code to determine the access to the data by each server has access. An OPC Client can connect to OPC Historical servers provided by one or more vendors. Transport control protocol (TCP)/Internet protocol (IP) based OPC Unified architecture (UA) was developed in the mid-2000s as protocol with the ability to share data at the control system level in real-time with high reliability and efficiency. There are certain limitation with OPC classic like: its dependence on Windows platform only and has security vulnerability as it depends on DCOM for security. These are overcome in OPC UA which is well suited for other operating platform like Linux, etc., and has built in security, that is, platform-independent secured communication with robust connectivity. Also, OPC UA supports publisher-subscriber communication model in addition to classic Server-client communication model. Also, OPC UA supports object-oriented information data model. OPC UA has now become an open industry standard with easy migration from legacy system. After being familiar with basics of OPC, in this subsection, detailed discussions on OPC types with variations shall be discussed.

1. **Alarm event (AE) details:** As the name suggests here, alarm and event management through OPC shall be discussed briefly. Major issues here are as follows:
   - **AE explained:** In industries, alarms and events refer to describe meaningful occurrences in a plant. The OPC A&E specification identifies; an alarm is an abnormal condition requiring special attention. Alarm conditions are as follows:
     - *Enable/disable*: Indicates whether condition monitoring is active.
     - *Active/Inactive:* Indicates whether the condition is normal.
     - *Acknowledgment:* Alarm acknowledgment by operator
     - *Sub condition:* Specific subcondition defined for a condition

     An event may or may not be associated with a condition (but meaningful). Three event types are as follows: *OPC Simple event notification* (without condition), *OPC Condition event notification* (change condition), and *OPC Tracking event notification* (signaling for change). Various events have been elaborated in Table III 15.0b.
   - **AE Filtering:** Filtering enables an OPC A&E client to determine the events to receive and to adjust the number of delivered messages to its requirements. So, the OPC A&E client application can set filter conditions for each event subscription. Events may be selected using criteria like: *Type of event, Event categories, Process area* (list of areas), *Event source* (list of source), *Lowest/Highest Severity,* etc.
2. **OPC HDA:** OPC HAD is related with historical data access which lay an important role in DCS network.

**TABLE III 15.0B** OPC event description.

| Event issue | Description details |
|---|---|
| Simple event: This refers to informational messages requiring no action viz. Startup/shutdown, operator logon or logoff; | |
| Source | Reference of object generated this |
| Time | Time of occurrence |
| Category | This defines groupings of events by an OPC event server viz. System message; Device error |
| Severity | Urgency: Assigned value (e.g., 0-1000. 0–200 low; 801–1000 high). |
| Message | Text message describing the event |
| Condition event: This refers to the detection of some condition in the supplier system requiring some response/acknowledgment. | |
| Condition name | Name associated with OPC condition/subcondition. |
| Acknowledge | Indicated requirement (or not) for an acknowledgment |
| Mask change | Indicates properties of the condition have changed to the client |
| Bew state | New state of the condition |
| Active time | The transition time of the condition/subcondition with event notification |
| Quality issue | The quality of the underlying data items for the condition |
| Category | Alarm, system error, Deviation from set point |
| Actor ID | The OPC Client identification (ID) acknowledging the condition, |
| Tracking Event: Tracking events indicate that some specific action was performed on the source by some actor. So in addition to information of simple event it also contains an *'actor ID'* (OPC Client) to indicated who or what performed this action. Categories of message may include: *System configuration, Change in default value.* | |

According to OPC foundation OPC HDA specification describes the behavior of an OPC Server which stores data in some form of database, and how Clients can retrieve it using OPC, covering: *Overview and purpose of* Historical data technology,
*General architecture and methodology overview, Detailed descriptions of interfaces, methods, parameters, expected behaviors, aggregate functions, data types,* and *structures* as well as *sample code (OPC Server) containing interface definitions and error codes*. Fig. III 15.0d shows General HDA Architecture and Data Sources. Major issues are as follows

- **Architecture** and **component:** An OPC client application communicates to an OPC HDA server through the specified OPC custom and automation interfaces. OPC HDA servers must implement the custom interface, and optionally may implement the automation interface as shown in Fig. III 15.0d1. In this connection, Fig. III 15.0b1 may be referenced for general philosophy of OPC interface. The OPC Specification specifies COM interfaces (interfaces are), not the implementation of those interfaces. It specifies the behavior that the interfaces are expected to provide to the client applications that use them.
- **Sever overview:** OPC HDA server provides a read-only interface to the historical data located and collected in the *historian*. OPC HDA clients may use OPC HDA server to view all historical data, or request historical data for a given period of time. With HDA server and filtering criteria set, it is easier for HDA Client to view any or all historical data. HDA server use standardized ways to access and analyze historical data. Possible HDA servers have been shown in Fig. III 15.0d2.
- **Data sources:** The OPC HDA server provides a way to access and communicate to historical data sources whose types are functions of the server implemented as shown in Fig. III 15.0d2. The server may be implemented as a standalone OPC HDA server that collects data from data source or it may also be a *set of interfaces* layered on top of an existing proprietary historical data server. The clients referencing the OPC HDA server may be different kinds like simple trending packages or complex reporting system.
- **Object** and **interfaces:** The OPC HDA server objects provide the ability to read data from a historical server and write data to a historical server. All COM objects are accessed through Interfaces. The client

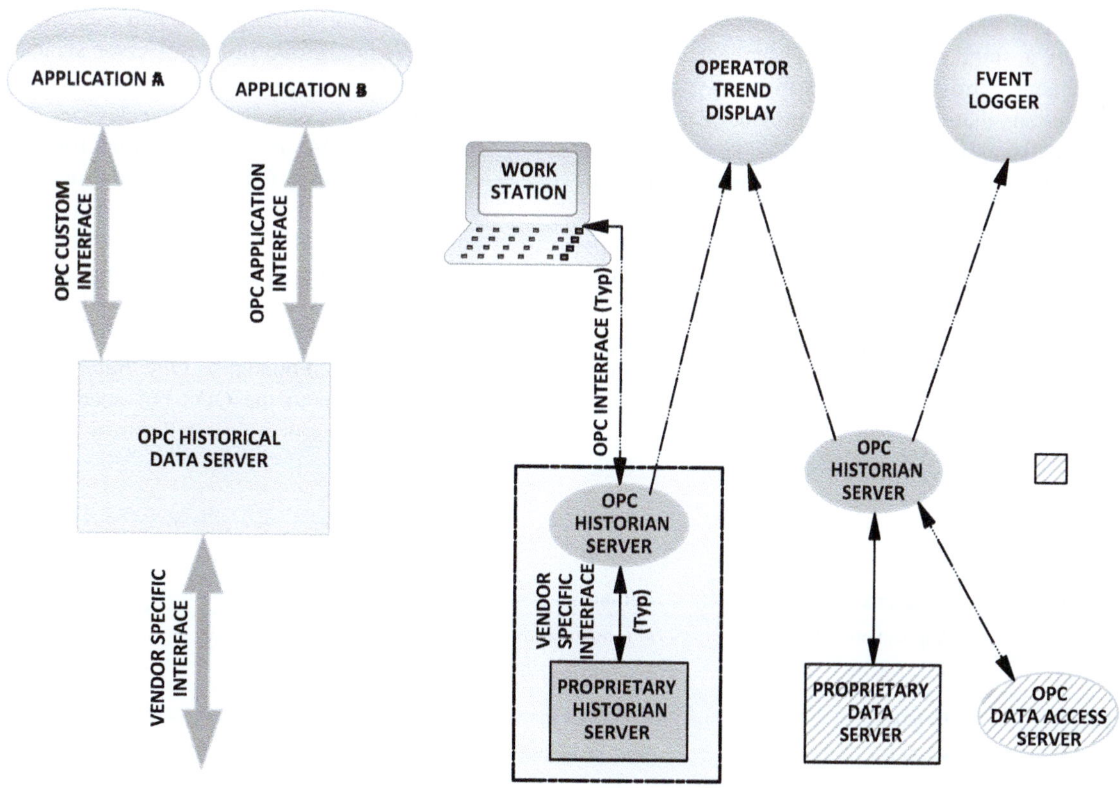


FIG III 12.0d1 GENERAL HDA ARCHITECTURE

FIG III 12.0d2 HDA DATA SOURCE DETAILS

**FIGURE III 15.0D** General HDA architecture and data sources.

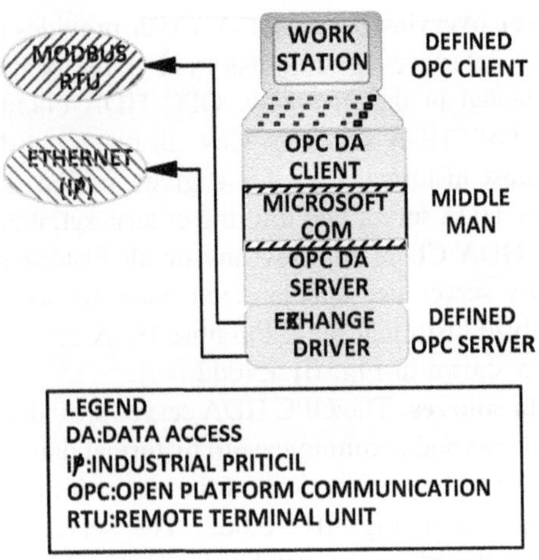


FIG III 15.0e1 OPC DA CLIENT SERVER

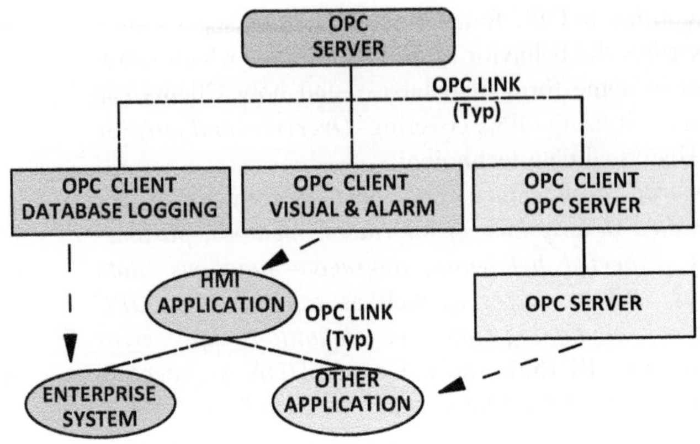


FIG III 15.0e2 OPC DA ARCHITECTURE

**FIGURE III 15.0E** OPC DA client-server and architecture.

sees only the interfaces. Thus, the objects here are "logical" representations having nothing to do with the actual internal implementation of the server. The browser interface provides a method for the client to review the address space of the historian which may be hierarchical or flat. The browser interface is essential in most large historical data servers, it allows a client to review the address space in a simple graphical manner. An OPC HDA client application must implement a callback interface to support a shutdown request. The client may also implement interfaces for the various asynchronous connections that a server may provide [2].

3. **OPC DA discussions:** OPC DA was developed first. Fig. III 15.0e Details out OPC DA Client-Server and Architecture.
   HMI represents OPC DA client and Ethernet types supplying data are OPC DA server and Microsoft COM is the common middleman as shown in Fig. III 15.0e1. In OPC DA architecture A server can support multiple Clients, concurrently and in similar manner, many OPC DA server may exist in network. This facilitates data exchange in the network as shown in Fig. III 15.0e2
   - **OPC DA and batch:** OPC-DA retrieves data (value, quality and timestamp) from a device. OPC-DA is not OS agnostic (independent of platform) but a Windows product, that is, OPC classic. With the help of converter, it is possible to convert OPC-DA messages to Message queuing telemetry transport (MQTT) which is preferred by developers due to its ease of use, speed, and user-friendliness. OPC DA is a group of client-server standards that provides specifications for communicating real-time data from data acquisition devices viz. PLCs to HMI/SCADA and may be used for Enterprise resource planning (ERP)/Manufacturing execution system (MES) also. OPC DA made it possible to connect to any RT *data source* without a custom connector written specifically for the *data source/ data sink* pair. Use of OPC DA has been tabled in Table III 15.0c OPC DA Usages.

     As per OPC Foundation OPC Batch specification is an extension to the OPC DA specification with ability to manage recipe definitions to send them

**TABLE III 15.0C** OPC DA Usages.

| Data source | Data sink | Remarks |
|---|---|---|
| Controller | Application (HMI) | No separate custom driver (Fig. III 15.0c1) |
| Controller A | Controller B | Controller to controller00 both vendors |
| Controller | Database | Structured query language (SQL) over the open database connectivity (ODBC) |
| Controller | Historian | OPC HDA and OPC DA Server |
| Redundancy | Controller/Application | Use of OPC Client for ODBC |

to a device, acquire information from a device, and report to a Client, covering: General overview and concepts, Detailed descriptions of interfaces, methods, parameters, behaviors, and return codes Detailed descriptions of data-types and structures and XML schema and Interface definition language (IDL)

- **OPC DX:** OPC DX has been designed for exchanging plant floor horizontal data between OPC DA servers. OPC DX does not specify a new method for these data transfers but it relies on OPC DA data transfer capabilities. OPC DX enables data interoperability between OPC based systems (including DCOM and XML-based systems running over Ethernet) including PLCs, HMI/SCADA, Devices, PCs. This specification is intended for developers of OPC compliant DX Clients and Servers. OPC DX is designed to provide for the direct transfer of data from one or more OPC DA and DX servers to an OPC DX server, without intermediate clients or servers to access the data from one server and forward it to another.
- **OPC XML:** The OPC XML-DA specification builds on the OPC DA specifications to communicate data in XML which is explained in I 11.0.1. OPC XML DA (OPC XML-DA) is the OPC Foundations' adoptions of the XML set of technologies to facilitate the exchange of plant data across the Internet, and upward into the enterprise domain. In this connection, III 9.0.1.5 may be referenced also. OPC XML encapsulates process control data making it available across all operating systems using XML—having a very few implementations.

4. **OPC UA Explained:** *Service-oriented architecture (SOA),* OPC UA can be conceived as: *Single, abstract set of objects with backward compatibility, Embeddable, within a device, Security* and *encryption built-in* and *Designed to take plant-floor data to the Enterp.* OPC UA finds main use in online data exchange between devices and HMI/SCADA systems for OPC DA functionality, that is, an OPC server data provides for consumption by an OPC client. As shown in Fig. III 15.0a2, OPC UA incorporates all features of classic OPC standards but defined in *platform-independent* communication mechanisms and generic, extensible, and object-oriented modeling capabilities for the information. Various kinds of systems and devices can communicate by sending *Messages* between *Clients* and *Servers* over various types of networks. The OPC UA network communication part defines different mechanisms optimized for different use cases. OPC UA is well-suited software in application areas to exchange information, use command and control for industrial processes. OPC UA defines a common infrastructure model to facilitate this information exchange by specifying: *The information model: structure, behavior* and *semantics, The message model: interaction between applications, The communication model: the data transfer between end-points* and *the conformance model: interoperability between systems.* It supports robust, secure communication that assures the identity of *Clients* and *Servers* and resists attacks. OPC UA defines sets of *Services* that *Servers* may provide, and individual *Servers* specify to *Clients* what *Service* sets they support. Information is conveyed using OPC UA-defined and vendor-defined data types, and *Servers* define object models that *Clients* can dynamically discover.
OPC UA provides, integrated *Address Space* and service model allowing a single OPC UA *Server* to integrate data, *Alarms* and *Events*, and history into its *Address Space*, and to provide access to them using an *Integrated set of Services* including an integrated security model. OPC UA also allows *Servers* to provide *Clients* with type definitions for the *Objects* accessed from the *Address Space*. OPC UA adds support for many relationships between *Nodes (The fundamental component of an Address* Space). An OPC UA Server can present data in a variety of hierarchies tailored in line with linking of Clients. This flexibility, combined with support for type definitions, makes OPC UA applicable to a wide array of problem domains. Fig. III 15.0f shows UA application area and Publisher-subscriber concept. This is illustrated in Fig. III 15.0f1 UA Application area. OPC-UA—also published as IEC 62,541—enables exchange of information models of any complexity—both instances and types (metadata). The following are Data exchange model, functionalities, characteristics, and architectural details specified for OPC-UA. OPC UA Transport and Meta model have been illustrated in Fig. III 15.0g. Let the discussions begin with brief discussions on OPC Client-Server and Publisher-Subscriber.
   - **Data Exchange**: Data exchange in OPC UA by two methods:
     - *Server-Client:* The Server and Client interface is request responses data exchange where the Client makes an explicit request for some specific data from the sever. Major issues in Client-Server communication are as follows: *Request-Response paradigm, Information is shared only on request* and *exchanges are principally bi-directional.* Fig. III 15.0f2 shows OPC UA Client -server concept
     - *Publisher-Subscriber:* With *Publisher-Subscribe*, OPC UA *Applications* do not directly exchange requests and responses but the *Publishers* send messages to a *Message Oriented*

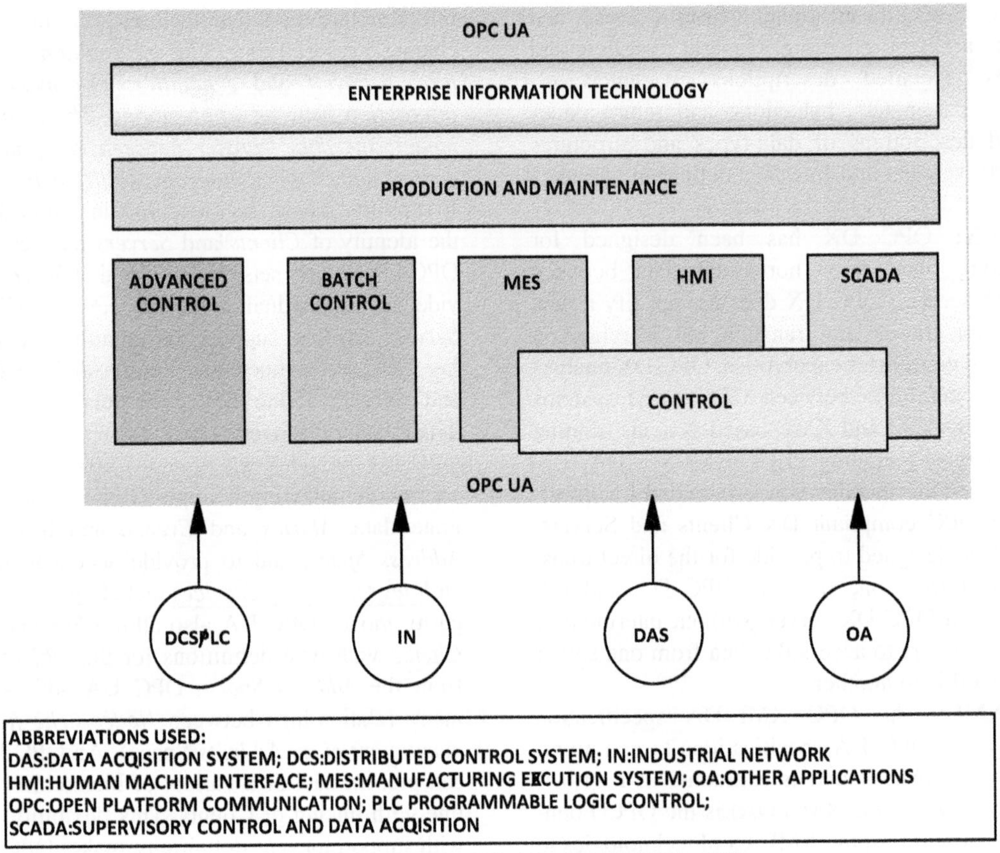


FIG III 15.0f1 OPC UA: APPLICATION AREA

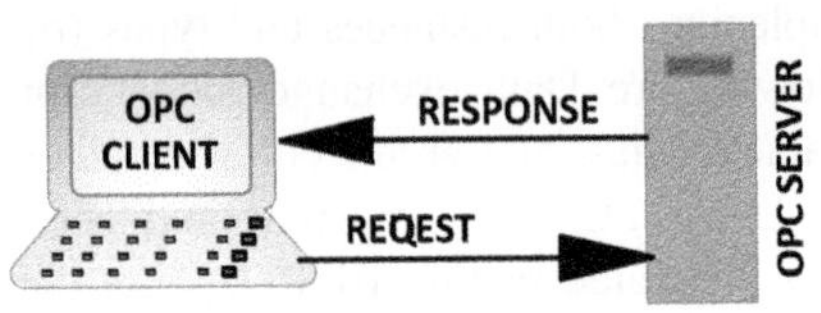


FIG III 15.0f2 OPC UA: CLIENT–SERVER

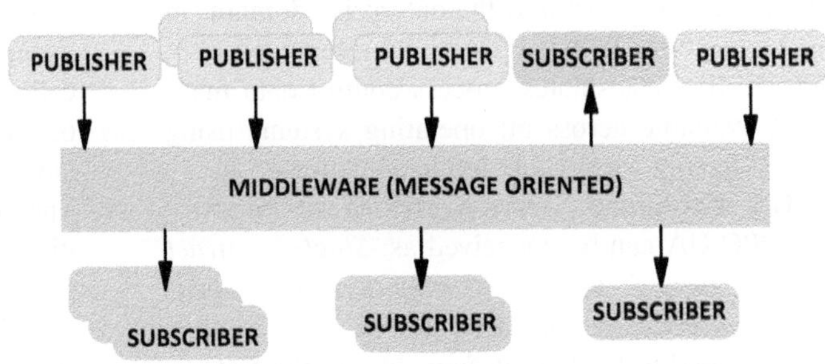


FIG III 15.0f3 OPC UA: SUBSCIBER– PUBLISHER

**FIGURE III 15.0F** OPC UA: Application area and data exchange concept.

*Middleware*, without knowledge of what, if any, *Subscribers* there may be. In similar way, *Subscribers* express interest in specific types of data, and process messages that contain this data, without knowledge of what *Publishers* there are. Based on OPC foundation details, Fig. III 15.0f3 shows Publisher-Subscription concept.

- **Transport**: For the data exchange mechanisms between OPC-UA applications—as discussed. In this connection, Fig. III 15.0g1 may be referenced.
- **Meta model**: Specifies the rules and basic components for publishing an information model via OPC-UA. OPC-UA services can be mapped to different technologies like UA-TCP and HTTPS. Fig. III 15.0g2
- **Services**: These constitute the interface between a server as information provider and clients as users of this information. Integrated services like: Secured Channel service, View service, Subscription service, etc., to name a few.
- **OPC UA Characteristics:** Through OPC UA, various kinds of data can be exchanged in line with the specification released in 2006. Some special features and characteristics of OPC UA are as follows:

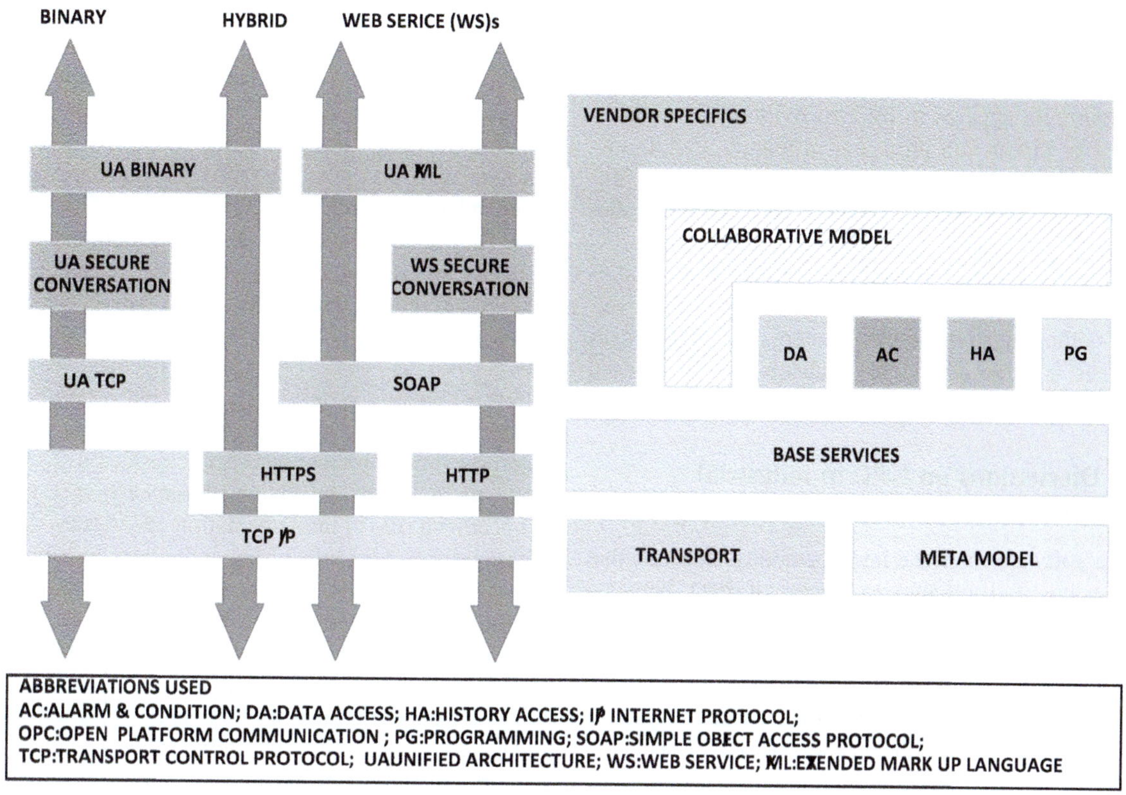


DEVELOPED BASED ON OPC FOUNDATION DOCUMRNT [COURTESY:OPC FOUNDATION]

FIG III 15.0g1 OPC UA: TRANSPORT PROFILE    FIG III 15.0g2 OPC UA: META MODEL

**FIGURE III 15.0G** OPC UA: Transport and meta model. *[Courtesy: OPC Foundation].*

- OPC UA—a platform independent, service-oriented architecture built on top of TCP/IP protocol, allows communication among industrial software, controllers, and field devices through standard TCP/IP network which is the backbone of Ethernet, and the internet.
- OPC UA is a client-server (or publisher subscriber) based application. The OPC UA server provides access to its data by exposing its data objects to an OPC UA client/subscriber viz. HMI application.
- Each configured OPC UA subscription has properties requesting data like: *Display name*, (data object or tag), *Tag Value*, *Quality or Status*, *Timestamp of the OPC UA Server and Client*, *Data Type*, and *Subscription ID*.
- The OPC UA TCP protocol uses essentially the same link structure as in an internet link, hence capable of directly discover, address, and subscribe to data in the OPC. UA server over any Ethernet network—hence independent of the OS and platform. All these make it well suitable for all industrial applications.
- OPC UA has built-in security measures like encryption, authentication, and authorization.

• **OPC UA Architecture:** To use OPC UA in control system, OPC UA Server software needs connection to the data sources. Also, an interaction of OPC UA Client software to the OPC UA server and to set up subscriptions for periodic data access. OPC UA can run on any device from desktops, to the internet cloud. There could be embedded OPC UA servers viz. Siemens S7-1500 PLC. Operation also includes application authentication, other security features, and data methods to access devices such as PLCs. The OPC UA Client software can be loaded on the same system as the OPC UA Server, on a different machine on the same network, or anywhere on the internet or in the cloud. Because of the use of TCP and Internet browsing protocols by the OPC UA Client, to connect to the OPC UA Server, the OPC UA Client can literally be located anywhere.

5. **OPC UA and TSN:** OPC UA TSN is the result of the combination of OPC UA, in the pub/sub communication model, as shown in Fig. III 15.0f3 and the performance capabilities of TSN with deterministic

transmission of RT data and offers the flexibility and openness inherent to OPC UA, while adhering to recognized, proven, and familiar IEEE standards. While this configuration appears to offer everything the industry needs for production processes in today's IIoT/Industrie 4.0. With an Ethernet-based network, multiple protocols can co-exist, so devices can support other protocols in parallel with OPC communication. OPC UA TSN is designed as a real-time Ethernet communication standard that provides open, deterministic, real-time communication between automation systems, as well as to the cloud for Industrie 4.0/IIoT-based production.

### 15.0.2 Discussions on OPC in industrial network

OPC is a software interface interoperability standard that allows secure and reliable exchange of data between various programs and industrial hardware devices. It is platform-independent and ensures the continuous flow of information across multiple vendor devices. The widespread adoption of OPC standards for interfacing systems on both the plant floor and the business network is a classic example of both the benefits and risks of adopting IT technologies. As discussed, OPC is used increasingly to interconnect HMI, Historians and other servers on the control network with enterprise databases and other business-oriented software. To assess the security risks in deploying OPC, it is important to define the set of inherent vulnerabilities, the threats that may exploit them, and the possible impact of compromised software components. The basic aim in the new Industrie 4.0 or Digital transformation approach is to marry IT and OT, that is, the integration of IT with products, systems, solutions, and services across the complete value chain spanning entire product and service life cycles. Once implemented, digitalization opens the doors to unprecedented new business opportunities and customer value. OPC UA is a standard that makes this possible through standard platform-independent interface. Also, at the heart of the IoT/IIoT, OPC UA addresses the need for standardized data connectivity and interoperability for both horizontal and vertical data communications. OPC is a very useful interoperability standard in industrial automation, with OPC UA. As OPC (–platform independent software interface) enables *seamless*, *secure*, and *reliable data exchange* among diverse industrial devices, systems, and enterprise-level applications, it is considered as a cornerstone for realizing the vision of *Industrie 4.0*. OPC UA plays crucial role in implementation of Industrie 4.0 which envisages smart manufacturing with all components—from sensors to enterprise systems are interconnected and communicating seamlessly. Also, the role of OPC in M2M cannot be overestimated. From these, it is seen that common factor in all these three pillar is OPC interface. Major issues for data interfaces by OPC are as follows: *Platform Independence* and *interoperability*, *Comprehensive information modeling*, *Greater security*, *Scalability with flexibility*, *Integration: Horizontal and Vertical,* etc. In this subsection, brief discussions will be presented on OPC in Industrial network with special reference to security and interoperability issues. Let the discussion start with OPC applications.

1. **OPC Application:** Broad application areas of OPC have been elaborated in Table III 15.0d
2. **OPC benefits:** Benefits of using OPC standard are as follows:

**TABLE III 15.0D** Application areas of OPC.

| Item | Description |
|---|---|
| OPC DA | Provides access to real-time data enables, that is, real-time communication between control devices and client applications |
| OPC A E | OPC AE servers are used to accept and exchange process alarms and events– handling and notification of alarms, system errors, and other significant events |
| OPC HDA | Used to retrieve RT and archived process data for quality control, regulatory compliance, analysis and performance optimization. |
| OPC DX XML | OPC server to server data exchange. XML represents schema and data representation formats based upon XML standard. |
| MES integration | MES integration to bridge the gap between the production equipment and higher-level enterprise systems for smooth flow of data across the enterprise. |
| IIoT-Industrie4.0 | OPC (UA) is back bone for smart manufacturing of Industrie 4.0 with IIoT, enabling secure data exchange between operation technology (OT) and cloud-based platforms. |

- Standardized approach, independent of platform. Also, it offers interoperability
- Because of OPC server, client applications need not know anything about hardware protocol details.
- Reduced load on the hardware device.
- Increased scalability of the system.
- OPC allows point addition or deletion while the server is up and running
- OPC offers ease of Integration. OPC standards open the door to an open exchange of information between applications throughout the plant floor so that the user can then focus on his business rather than system integration issues.
- Customer-developed applications can exchange data with any connected OPC server. Even with little to no knowledge of the customer's industrial network, that is, successful applications can be written and deployed successfully because of ease of connectivity/interoperability. In OPC application, users are free to choose through more and more plug-and-play options available to tailor make for their hardware and software to meet their needs from different vendors without fear of incompatibility.

3. **OPC Security issues:** IEC 62,541-2:2020 describes the OPC UA security model including the security threats for hardware and software environments in which OPC UA is expected to run. It describes how OPC UA relies upon other standards for security. It provides definition of common security terms used in OPC UA specification. It gives an overview of the security features, references services, mappings, and Profiles that are specified normatively in other parts of the OPC UA Specification. It provides suggestions or best practice guidelines on implementing security. General security model mainly covers authentication of *Clients* and *Servers*, which is of great concern for OPC UA security. Also, the authentication of users, the integrity and confidentiality of their communications, and the verifiability of claims of functionality are extremely important. Security was a central requirement in the development of OPC UA. it is addressed in various areas like:
   - **Authentication and authorization of users:** This is for establishing a connection, the user identifies him/herself
   - **Authentication and authorization of application:** OPC UA applications identify themselves (like user) via so-called software and application instance certificates.
   - **Integrity:** The signing of messages prevents a third party from changing the contents of a message.
   - **Confidentiality:** The confidentiality of the exchanged information is secured by the encryption of the exchanged messages. Modern cryptographic algorithms are used for this
   - **Types of vulnerability:** There are several kinds of vulnerabilities encountered in case OPC. Some of these are [to limit the volume only names]:*Unnecessary system services* (OPC servers directly monitors and controls process data, hence the computer containing the OPC server should not be used for any other purposes), *System enumeration and profiling [*here common tools: Traffic sniffing, Domain name system (DNS) query, Active Directory, TCP/UDP scanning, NetBIOS Enumeration*], Anonymous log[*for Account information, Policy setting, File Share], *Network management information, Password Vulnerability* (use suitable authentication), *Weak authentication method, Inadequate* log *in, Patching updates, Remote registry browsing, Use of Historically insecure transport, Lack of authentication in OPC Server Browser,* Lack in *Authorization policy, Protocol support* and *privilege in OPC Server Browser.* Also, any lack in *OPC communication, OPC Traffic, Fine-grained access control* are issues related to OPC vulnerabilities.
   - **Security specification:** Some common security specification issues shall include the following:
     - *Security: OPC DA:* OPC-DA could provide attacker the ability to gain unauthorized read access to process data, to modify data points on devices. Thus, attacker can provide erroneous values, message injections into the client application, and play with time stamps and quality metrics.
     - *Security: OPC AE:* OPC-AE provides direct access to another weak link, namely the human operator.
     - *Security: OPC HDA:* The vulnerability of this may not have the immediate consequences of impacts on other OPC applications.
     - *Security: OPC DX and XML:* As OPC-DX is used for OPC server-to-server data exchange, it provides multivendor control system interoperability. OPC XML does not use DCOM it has a different set of security concerns.
   - **Threat analysis:** Main threat analysis issues include the following:
   - *Attacker objective*: *Reconnaissance* (OPC server identity), *Confidentiality (*Gain unauthorized read access), *Denial of service (DoS)* for OPC server, *Alter OPC server data* to create rogue OPC server
   - *Attacker tools and techniques:* OS reconnaissance, Custom OPC attack tool, etc.
   - **OPC aware firewall:** Creating zone-based defenses with OPC-aware Firewalls: dividing a control system into security zones prevents issues in one area

from migrating to another area. Between the zones are "conduits" (typically firewalls) that carefully manage all the interzone network traffic.

4. **OPC UA: M2M, IIoT and Industrie 4.0:** IoT defines a series of technologies that have traditionally not been connected and will now be connected to an IP-based network. OPC Foundation with OPC-UA has been engaged in these standardization efforts. As discussed, M2M, Industrie 4.0, and IoT (IIoT) can be connected together through OPC UA interface. Major issues are as follows:
   - **M2M:** M2M typically defines the communication between two machines/the data exchange between a more or less intelligent device and a central computer. OPC provides standardized, secure, and platform-independent method for data exchange by breaking down communication silos by translating data into a common language, enabling seamless interoperability across different manufacturers and software.
   - **IoT/IIoT:** For IoT/IIoT realization, can be realized, when the of the central communication is based on a global communication standard. Also, publisher/subscriber model for the low-resource, one-to-many communication paradigm and client/server communication model of OPC UA is well-suited. It solves interoperability challenges by acting as a universal interface for different manufacturers' equipment, facilitating secure data integration from the shop floor to the cloud.
   - **Industrie 4.0:** Industrie 4.0 mainly demands flexibility, adaptability, transparency to be fulfilled by Industrie 4.0 components or systems to achieve horizontal and vertical interoperability as well as interoperability over the lifecycle. OPC UA provides secure, reliable, and interoperable data exchange between different machines, devices, and applications. Also, it offers unified communication standard to bridge the gap between OT and IT and to facilitate concepts of IIoT as discussed.
5. **Industrie 4.0 requirements:** With reference to Fig. II 16.2a, it can be said that OPC UA is the communication technology in RAMI 4.0 approach in Industrie 4.0. Based on OPC Foundation data, some of the various Industrie 4.0 requirements with OPC UA solution have been presented below [3]:
   - **Technological independence in communication:** Widely used OPC-UA is technologically sector-neutral, and runs on all OS—there are even chip layer implementations without an OS. OPC-UA can be implemented in various languages—currently, stacks in ANSI C/C++, .NET, and Java are available.
   - **Scalability for integrated networking:** OPC-UA scales small footprint to single- and multicore hardware having wide range of CPU architectures. OPC-UA is used in embedded field devices to virtually all controllers and SCADA/HMI products as well as MES/ERP systems as well as in Cloud applications.
   - **SOA and standardized communication:** OPC-UA is independent of the transport method. Current two protocol bindings are available: optimized TCP-based binary protocol and HTTP/HTTPS web service with binary or XML-coded messages. In addition, Publish/Subscribe communication model can be integrated. Also, historical data and their mathematical aggregation are standardized in OPC-UA.
   - **Authentication:** OPC-UA uses certificates, user/password for authentication of the application. Signed and encrypted transfer, as well as a rights concept at data point level with audit functionality is available in the stack.
   - **Verification:** OPC-UA is already an IEC 62,541standard with tools and test laboratories for testing and certifying conformity are available.
   - **Interoperability:** Platform independence, Companion specification, Standardization, and Rich information model are major issues related for interoperability requirements.

## 15.1 Integration issues in industrial networking

Advances in technologies, like IoT/IIoT and big data analytics, are systematically enabling the *digital information* world to understand and influence the *physical operational* world. When implemented properly, IT/OT convergence can merge business processes, insights, and controls into a single uniform environment. OT has the prime focus on the management and control of physical devices existing and operating in the physical world, that is, focused on behaviors and outcomes. On the contrary, IT focused on data and communication. There exists traditional physical world composed of machines, production systems, and other industrial equipment as well as the digital world using servers, storage, networking, and other devices used to run applications, process data, and even make critical decisions without human intervention. In recent times, two worlds have largely occupied separate domains, shared little. The benefits include greater connectivity and integration across plants, easier data sharing across the enterprise, and better visibility into real-time operations. The IoT—smart IP–enabled devices have created an opportunity to deploy a more flexible architecture with mobile access to data and connectivity within the production environment. Therefore, integration of both the world opened a new era including critical decision making and meeting customer needs. Because of the gray area between IT and OT roles and responsibilities, bringing together IT and OT networks is complex affair. IT and OT professionals need to have a common understanding of a host of techniques and technologies to overcome aforesaid complexities to supporting a converged infrastructure which is secure and manageable by all critical stakeholders. ISA/IEC-62,443 Zones and Conduits Model developed by the ISA, provides path for the same. The three design areas are as follows: *Cell or area zone*, *Production site operations,* and *Enterprise zone integration.* OPC Foundation standards are globally recognized and adopted by a wide range of organizations to enhance horizontal and vertical data interoperability including Industrie 4.0, Namur, IIoT, PLCopen, etc., a few to name. These are widely in use in various industries such as Oil and Gas, Machine Tools, Building automation, Business enterprise, etc. The Open process automation forum (OPAF), a relatively new initiative with a vision that calls for multivendor interoperability as one of its goals. OPC Foundation has joined this initiative to lend it many years of experience. OPC acts as a critical element in bridging IT and OT in an integrated network enabling secure data exchange, and providing a semantic layer for data understanding across disparate systems. It allows a standardized and secure way to connect OT with enterprise IT, facilitating modern initiatives viz. DT. OPC plays critical role in such integration through *Standardization, Rich semantic information modeling, Secured communication, Scalability* and *flexibility* and *Data integration.* Standardized, interoperable communication protocol for industrial automation systems, etc., have been made possible through OPC interface which plays great role in industrial network integration by acting as a bridge between different devices and software from various vendors, allowing them to exchange data seamlessly. *Elimination of proprietary drivers, Standardized communication for seamless data exchange, Interoperability, Support for SOA, Enhanced security, Scalability* and *Future readiness* are major contributions of OPC in network integration. OPC UA with Advanced physical layer (APL), TSN, etc., are major contributor for standardized Field level communications (FLC). OPC UA can then be used as a uniform and consistent communication solution for vertical and horizontal integration. This includes field, edge, and cloud. Critical network integration process involves combining various network systems, devices, and applications to function seamlessly as a unified infrastructure and computerized architecture. Common issue involved are as follows: *Compatibility issues, Performance issue, Troubleshooting problems, Cost factor, Vendor issues, Security challenges, Scalability limitations*, *Synchronization of data, Availability of expertise, Resistance for change in management.* In spite of all these challenges, network integration services which function by coordinating various network components into a unified system, are solutions enabling the seamless connection of various IT and OT systems and networks. Network integration and migration are complex processes used to upgrade, merge, or redesign an organization's network infrastructure. Having explained integration, it is essential to explain migration issue as well. Network migration is the physical and logical relocation of a network, or parts of it, from one infrastructure to another involving a one-time, bulk transfer of configurations, applications, and data viz. moving from an on-premise data center to a cloud environment. This calls for several phases for implementation discussed at length later in this chapter. Let us start discussions with integration in general.

## 15.1.0 Industrial network: OT–IT and general integration issues

The actual power of integration in automation lies with final achievement with the information available, in context, to all of the devices, systems, and individuals responsible for controlling, maintaining, and managing production. Realizing proper integration call for *seamless, transparent access* to data and information from all hierarchical levels of plant information from field level to devices enterprise level. IT references to use of *computers*, storage, *networking and other physical* devices, *infrastructure/processes* to create, process, store, secure, and exchange information in all forms of electronic data. Generally, OT refers to production/manufacturing and industrial environments, including industrial automation systems. Integrated operations in the industrial sector, particularly within OT frameworks, signify the seamless integration of systems, processes, people, and digital technologies to enhance operational efficiency, safety, and decision-making. Utilizing different technologies to integrate and interoperate efficiently as a single comprehensive connected system, can improve efficiency, enhance workflows coupled with reduced errors and costs to ensure gain in measurable competitive advantages. In modern days, equipment is fitted with sophisticated IoT/IIoT devices, and systems have advanced to realize M2M communications. All these made IT/OT convergence possible. The integration of automation, communications, and networking in industrial environments is an integral part of IoT and is often referred to as Industrie 4.0. The benefits of converging IT and OT have been discussed at the beginning of section. A truly converged IT and OT network architecture achieves seamless integration both horizontally, across multiple sites, and vertically, from the plant level to enterprise level. Also, cloud computing also helps integrate the enterprise. Moving data centers to the cloud simplifies *implementation, maintenance*, cost reductions, and improvements in organizational agility. A growing number of process manufacturers now have choices of automation architectures that can dramatically streamline integration tasks, enabling new levels of business benefits in the form of increased levels of collaboration and productivity (steps toward Industrie 4.0), improved operator effectiveness, as well as streamlined system engineering effort and lower lifecycle costs. Furthermore, because the proper infrastructure architecture abstracts process data and information itself from the underlying communications, the benefits of integration can more readily be preserved in the face of inevitable changes in technology, control strategy, and business processes. In this subsection, brief discussions have presented on general integration issues with reference to IT-OT integration.

1. **OT-IT convergence: types and benefit:** Let us look into the issues separately:
   - **IoT/IIoT on OT-IT Convergence:** IoT/IIoT paired with edge computing, enables IT/OT convergence. IIoT/IoT devices, are networked computing devices with the ability to collect, transfer, and analyze data. Newer smart sensors would be able to collect the data from the source, then transfer that information to an analytics application or an enterprise resource planning software platform to be integrated into an enterprise's unified system of business operations. In integrated network, an OT device can as an IIo/IoT device. The edge computing capabilities of IIoT devices enable real-time data processing. Typical IIoT infrastructure of IIoT has been presented as Fig. III 15.1a Developed based on idea from [5 courtesy: TechTarget].
   - **OT-IT Convergence Types:** OT-IT convergence can impact in varieties of ways depending on the needs and goals of the enterprise. These could be based on functional category or based on industrial applications. These are enumerated below
     - *Process convergence*: This refers to the convergence of workflows to accommodate each other and make sure important projects and information are communicated.
     - *Software and data convergence*: This deals with the software and data from the front office to plant to address OT needs. This technical convergence involves the network architecture of the business.
     - *Physical convergence*: This includes physical devices being converged or retrofitted with newer hardware to accommodate the addition of IT to OT.

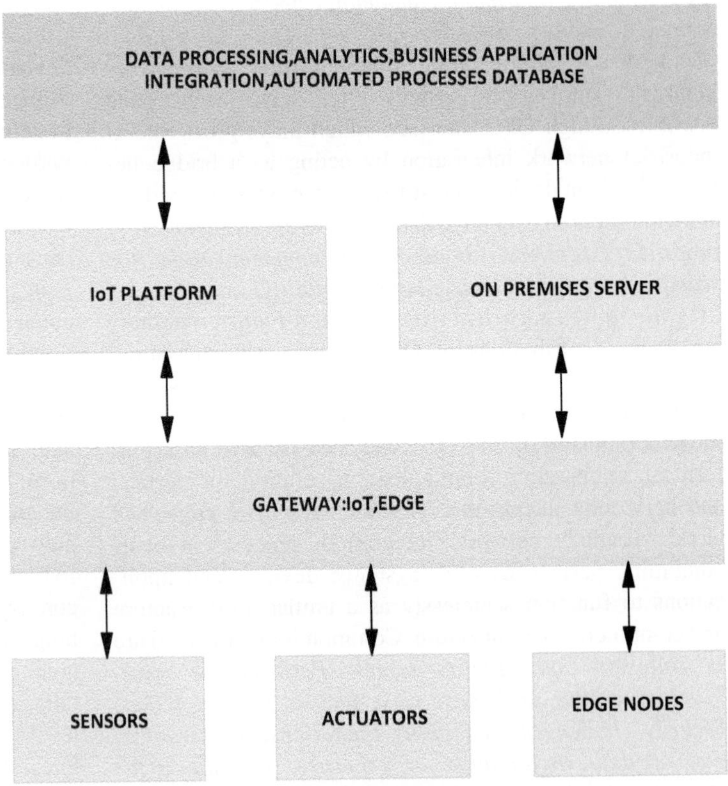


**FIGURE III 15.1A** IIoT infrastructure and edge computing [4].

  - *Manufacturing:* Through IT/OT convergence, the customer needs, sales, and inventory data can be utilized to drive manufacturing operations making enterprise to be more cost and resource-efficient
  - *Utility and energy companies*: Modern IT enables OT teams to access operational data remotely, helping industries utility and energy companies to optimize equipment inspections, make damage assessments, and handle inventory monitoring and distributions.
  - *Pharmaceuticals*: IT/OT integration enables improved medicine manufacturing and medical science as well
  - *Communications*: Regional and global communication providers can employ IT/OT convergence to oversee the *performance*, *operation of equipment*, *quality of service*, *faster troubleshooting*, etc., to achieve better user satisfaction.
  - *Transportation*: Integration of IT and OT helps transportation organizations to gain better visibility into the coordination and carry out asset management.
  - *Military*: IT/OT convergence can aid in the coordination and rapid deployment of resources.
  - *Retail*: The use of IoT devices such as start product tags, along with other OT devices such as cameras and point-of-sale terminals, can deliver more data to IT for analysis, leading to inventory and sales floor optimizations for cost savings,
- **OT-IT Convergence Benefits:** IT/OT convergence enables not only more direct control and complete monitoring, it also ensures easier analysis of data from these complex systems for decision making from any part of the world.
  - Less siloed IT and OT departments, to share their respective areas of expertise to manage converged technology. Integration aligns IT and OT goals with the broader organizational objectives. Also, easier and faster time to market for convergence.
  - Because of RT data analysis and stream-lined decision making, all decisions are well-informed, timely and aligned with organizational goals, ultimately helping organizations operate more efficiently and gain competitive advantage
  - Improved cybersecurity is established through protected a single unified environment
  - *Reduced* operational and support costs, with less unplanned downtime due to *predictive* maintenance enabled by IoT devices.
  - Improved compliance with regulatory standards, as the addition of IT to OT enables better visibility, management, and auditing.
  - Improved automation and visibility into distributed OT, enabling OT to transmit real-time maintenance data.
  - More efficient energy and resource usage in line with actual product needs. Data-driven insights help organizations allocate resources efficiently and minimize waste, leading to higher productivity across the organization.
  - More efficient asset management because of utilization of common methodology to manage all IT and OT systems.

2. **OT-IT convergence: challenges and practices:** Major challenging issues include but not limited to:
   - **Security:** The most challenging issue is the maintenance of security (especially related to OT) after OT-IT convergence. Inherently the security systems of OT and IT are different. OT system needs to incorporate standard communication/remote accessibility, and as a result, the risks of connectivity would be there. OT systems are often distributed in nature, and are often relied upon to uphold critical infrastructure. The vulnerabilities of OT systems can leave enterprises and critical infrastructure at risk of industrial espionage and sabotage.
   - **Process challenges:** Main issue is to handle previously siloed IT and OT departments to manage and operate newly converged technology.
   - **IoT Security:** Lethal combination of limited knowledge of security by OT, and limited knowledge of the projects currently underway by IT can create a dangerous security gap with IoT deployments.
   - **Training**: Proper training and trained people in converged OT-IT system may pose a big challenge.
   - **Integration with existing system:** The business/IT side of an organization might be tempted to replace technologies instead of modernizing existing OT with IT technology. Naturally, this would pose problem of compatibility.
3. **OT-IT convergence: security:** OT-IT convergence can pose challenges in IoT security and needs detailed attention on security issues in convergence, as security in IT and OT are not same. Major security threats for OT-IT convergence are as follows:
   - **Differing security priorities:** IT and OT have different security priorities—IT focuses on data protection and confidentiality, while OT emphasizes safety and reliability. Bridging these priorities in a unified security strategy is challenging.
   - **Collaboration Lacking:** To ensure security, it is necessary to prioritize, collaborate and communicate in ways that might not have been necessary or even possible in standalone systems in the past.
   - **Mission-critical demands.** OT production systems are often called upon to function in RT for 24/7.

Enterprises should never ignore potential security vulnerabilities simply because they can't afford the cost of downtime needed to remediate risks. Also, It is necessary to ensure that the integration process does not inadvertently disrupt operational processes while strengthening security. It is a delicate balance.

- **Complexity and insight.** The OT-IT convergence results in increased complexity, making it harder to monitor, manage, and secure all interconnected systems effectively. Processing power and memory required for modern security solutions, making it challenging to implement robust cybersecurity measures without impacting performance.IT routinely relies on asset discovery and configuration for the environment being managed. OT systems must be able to share this environment, offer discoverability, and remote configuration and management through necessary zone conduit protections. Gaps in visibility can leave blind spots that attackers can exploit.
- **Legacy and Update: OT systems.** Many OT systems rely on legacy technology with limited or no built-in security features, so due to convergence, these can get exposed to vulnerabilities that may be difficult to address. Also, when lifecycle of IT is a few years only, when OT systems like manufacturing machines, can have lifecycles over a few decades with less possibilities for upgrade. Every converged system must be evaluated for security. Systems that can't support security requirements may demand *updated OT equipment.* However, applying patches and updates to OT systems can be challenging for system stability and reliability. Delaying patches can leave vulnerabilities unaddressed. It is quite complex issue to handle.
- **Effect of disruption:** OT- IT convergence demands a careful assessment of resilience and fail-safe factors.
- **Supply Chain:** OT-IT converging environments can expose organizations to supply chain vulnerabilities. Malicious actors may target third-party suppliers or components that are part of the integrated system.
- **Regulatory compliance:** Many industries have specific regulatory requirements. Ensuring compliance across integrated systems can be *complex and time-consuming.*

4. **ICT and information management:** Acronym ICT stands for Information communication technology. Let us discuss them with little details:
   - **ICT Discussion:** As the name implies, ICT is the backbone of integrated operations with the necessary digital infrastructure for data collection, transmission, processing, and visualization, enabling RT monitoring and control to facilitate decision-making based on accurate and timely data, and enhancing collaboration across different operational domains and geographical locations. Relevant issues are as follows:
     - *Integration*: ICT ensures uniform and harmonious functions and communications among various systems and processes.
     - *Data Analytics*: Robust data management and analytics (including collecting, storing, processing, and analyzing large volumes of data from various sources) are critical components of ICT in for integrated operations.
     - *Cybersecurity*: Cybersecurity becomes paramount with the increasing interconnectivity and digitalization in integrated operations. ICT provides the tools and frameworks necessary to secure operational systems and data
     - *Decision-making*: ICT ensures to enhanced collaboration among stakeholders, with unified communication platforms and collaborative tools for decision making [5].
     - *Scalability and Flexibility*: ICT solutions in Integrated Operations must be scalable and flexible for adapting changes operational needs and technological advancements.
     - *Standard compliance*: ICT must adhere to relevant regulatory requirements and international standards.
     - *Training*: Implementing ICT for integrated operation, skilled workforce capable of managing and operating advanced technological solutions are essential.
   - **Information management:** Information management within OT environments is foundational to achieving operational excellence and security. This must adhere to well-defined information management principles. A few issues here are as follows:
     - *Accuracy and integrity*: Data collection from trusted sources, robust data validation and error-checking mechanisms, Integrity protection and secure storage.
     - *Relevant issues*: Prioritization, Regular update, Role-based access control, user friendliness, Training and support.
     - *Security and confidentiality*: **I**ntrusion detection, Industry standards and regulatory requirements, Regular security audits and risk assessments
     - *Standardization and Interoperability*: relevant standards compliant, Industry-standard protocols and data formats,
     - *Scalability and Flexibility*: Modular architectures, Cloud-based solutions and services, Future growth
     - *Continuity and resilience*: Robust disaster recovery and business continuity plans.

- ○ *Quality*: Metrics/benchmarks for the quality and performance, Feedback, Training.

5. **Integration factor and Power:** Let us now explore integration factors and power:
   - **Integration factors:** Several factors have to be taken into account when discussing *industrial control systems* integration. Major factors are as follows: *Software infrastructure*, *Process databases*, *Operator interface*, *Alarm management*, and *Network security*. Each of them has impact on integration implementations already discussed.
   - **Integration power:** In industrial automation, network integration significantly enhances efficiency, agility, scalability, flexibility in meeting user requirements as well as safety through RT information exchange, remote monitoring, with seamless communication both horizontally and vertically, that is, M2M and enterprise-level applications. Also, this paves the way to Industrie 4.0. Key powerful factors in these connections are as follows: *Enhanced operational efficiency*, *Greater scalability*, *agility* and *flexibility*, *Cost reduction*, *Improved safety with required security*, *RT monitoring and control*—including remote operation, *Smart manufacturing*, *Big data analytics* for data-driven insight and *decision making.*

   Let us now move on to see the role of OPC in network integration.

### 15.1.1 Role of OPC in network integration

An OPC Client can be connected to many OPC Servers provided by different vendors. There are two types of OPC Servers: in-process and out-process. In-process OPC servers are written in form of a Dynamic link library (DLL) loaded in the OPC Client's process space and only one client can be connected. Out-process OPC servers are separate applications. Multiple clients can be connected to one server. They offer both local and remote connections. Many OPC clients can be connected to the OPC Server. As discussed, the FLC initiative based on the OPC UA framework (IEC 62,541), enables a secure and reliable, manufacturer- and platform-independent information exchange in gaining the utilities of integrations. Connection-oriented client/server communication model and the publish/subscribe extensions, are indispensable for communication at the field level due to the corresponding requirements for flexibility, efficiency, and determinism. The security mechanisms specified in OPC UA are also used, which, among other things, support authentication, signing, and encryption of the data to be transported. So, OPC UA is much more than a protocol. Ethernet TSN facilitates deterministic data transmission via OPC UA, very much indispensable in automation. Ethernet APL is enabling technology for the use of OPC UA. OPC UA can be used as a uniform and consistent communication solution for vertical and horizontal integration from field level up to Edge cloud computing level. OPC UA is rapidly becoming the preferred standard for automation integrating controls, information, enterprise systems, and the supply chain. OPC UA standard serves are enabler for manufacturers, in achieving higher levels of quality, efficiency, and productivity to compete in a globalized economy. According to OPC foundation, the transition of the DCS to an IIoT-centric system is driven by the access and analysis of more data and that promote seamless interoperability of multivendor components. OPC UA is a key component of these standardizations and aid for integration in automation. Major issues need attention are as follows

- Enables access to leading-edge capability promoting innovation and value creation
- Keeps owner's application software—hence much lowers cost of replacement
- An adaptive intrinsic security model
- Applicable to brownfield/greenfield projects
- Allows integration of best-in-class components

So, with these basics in mind let us explore in details *role of OPC in system integration.*

1. **OPC Design philosophy:** The OPC UA architecture comprises data modeling and transport mechanisms with a comprehensive set of secure connectivity requirements. OPC UA meets the interoperability problems in Industrie 4.0 and proves to be a solution through its semantic communication standard. Choice of right Software development kit (SDK) vendor (if not in house) is an important factor as SDK should be *Scalable*, *Easy to use*, *CPU utility* of OPC UA SDK, *Good memory*, *Application compatibility*, etc. Also, SDK should offer support for *third-party library* and proper room to accommodate *future enhancement.* This ensures it can meet future challenges. Based on OPC foundation guideline, a number of other design issues are as follows:
   - **Dynamic data modeling**: OPC UA is an object-oriented, extensible information model with a complete set of associated services to act on the information models.
   - **Companion specifications**: OPC UA can assists to develop of companion specifications so as to unite the industrial automation landscape of multiple standards with the standardization of how *each is expressed* in OPC UA.
   - **Layered technology**: OPC UA is based on a set of layered and interchangeable technology mappings to meet the challenges of obsolescence due to continuous technical evolution. New transports, new

security, even new communication protocols can be easily added/enhanced without changes to information models or applications.

- **Profiles**: OPC UA functionality is broken down into defined groupings of functionality to allow implementations to be tested for specific behaviors during certification.

2. **OPC functional support:** Open process automation forum (OPAF) offers O-PAS standard for *an open, interoperable, and secure architecture for industrial process automation systems.* OPC UA functionality supports OPAF philosophy. A few functional issues are as follows:
   - **Leading edge capability:** The separation of the data model from its underlying technology is a core strength of OPC UA allowing it to adapt new technologies including those defined for security and communication protocols.
   - **Lower cost and integration:** Standardized information models lead to lower costs with ability for integration of best-in-class components. This information model defined in IEC 61,131-3 is a product of collaboration of the OPC Foundation and PLCopen.
   - **Intrinsic security model:** Security is a key concern of OPC UA which addresses security at all levels of the architecture. It was designed by security experts to ensure that it makes use of the latest technology and that it is designed with evolution in mind. Because of OPC UA, It can be used as plug-and-play interoperability system allowing new things to be *added* with *complete guarantee* for *reliable information integration* and *interoperability* between new and legacy applications and devices.
   - **Innovation and value creatio**n: Technological innovation and change will continue as end users and suppliers look for better ways to increase revenue and lower expenses.
   - **Open source:** OPC UA applications can utilize stacks and SDK to simplify implementation. Stacks provide the communication/security aspects of OPC UA which include the mapping to transports like TCP, HTTPS, etc. SDKs add the standard interactions like reads, writes, subscriptions provided by the services in line with the OPC UA specification.
3. **Integration: opportunity:** Information integration demands enterprises to work together for interoperability with harmonious opportunities to address *convergence* without overlapping complex information model architectures. As already discussed, OPC UA was developed with focus on developing strategies for platform independence and solutions allowing the convergence of OT and IT worlds to communicate with seamless interoperability for agreed upon syntactical and semantic data exchange formats [6]. OPC UA framework provides way for all standards organizations to seamlessly connect their data between the IT and OT worlds and simplifies the end user's task of digitalization. A few issues are as follows:
   - **Vertical Integration:** As discussed, IT and OT convergence is very important not only to the suppliers but to users also; because they want a strategy and a vertical integration from the plant level to the enterprise level. Integration is bidirectional between sensors and controllers or the enterprise and cloud, communicating all types of information, including control parameters, set points, operating parameters, real-time sensor data, asset information, real-time tracking, and device configurations.
   - **Horizontal Integration:** OPC UA companion specifications complete with templates for the standardized format of the data to be understood and consumed generically [6].
   - **Digitalization:** This is step toward DT, which is a burning issue now, calling for big data, machine learning, artificial intelligence, IIoT/IoT, cloud computing, edge computing, and the fog.
4. **Brownfield operation:** OPC UA is an important standard for diverse network and device integration in brownfield projects in existing plants with legacy equipment. Brownfield projects pose significant challenges on account of device heterogeneity, outdated systems, and operational constraints. OPC UA can act a kind of middleware to bridge these technological gaps. OPC UA can address issues like: *Bridge between legacy and modern systems*, *Enables a Unified namespace* (UNS—a centralized data layer), *Semantic information* and *Secured reliable data exchange*, OPC UA Publication-subscription is a framework for integrating Industrie 4.0 in brownfield facilities with connectivity and communication across multiple production sites with different technologies. With the integration of OPC UA and TSN, it is possible to collect and send the data securely and with guaranteed latency across production facilities with greenfield-level technologies and legacy brownfield equipment/technology. Major challenges in Brownfield are as follows: Limitation in hardware and software, Technological gap, high operational risk and cost, least documentation available. These challenges are met with: *Leverage of publisher-subscriber model* of OPC UA, Deployment of *middleware/bridge solution*, Edge computing facility, **etc.**
5. **Integration in Industrie 4.0:** The advent of Industry 4.0 has further accelerated the use of OPC UA in production systems. Because of the efficiency and security, OPC UA is an ideal choice for Industrie 4.0 initiatives through integrating into IIoT. OPC UA is a helpful tool

in integration, users of industrial automation community are using for implementing Industrie 4.0. Its open and secure communication protocol has revolutionized the way devices and systems interact, simplifying connectivity, improving efficiency, and enhancing data security. OPC provides unparalleled open connectivity across multiple platforms and applications. OPC servers decisively eliminate chances of compatibility issues with its universally understood language for seamless communication. This effectively streamlines the exchange of data across the entire automation infrastructure. OPC server plays a pivotal role in efficiently connecting and harmonizing various protocols and devices, effortlessly translating diverse data from numerous sources into an easily comprehensible standardized format. So there will be seamless communications and data exchanges between different applications, without compatibility issue. On account of incredibly robust functionality and unparalleled versatility, OPC servers have become indispensable tool for integration to achieve Industrie 4.0. Also, the flexibility of OPC technology empowers developers to effectively leverage their preferred tools, while harnessing the unparalleled power and capabilities of OPC. The OPC server application effortlessly provides the necessary data, while the client application confidently and assertively utilizes and harnesses that valuable data to effectively meet and exceed its objectives with absolute confidence. So, role of OPC in network integration to assist in implementing Industrie 4.0 cannot be overestimated.

Let us now move on to discuss *other pertinent issues* for network integration.

### 15.1.2 Network integration: miscellaneous issues and integrated operation

Modernizing information management environment enables enterprise to become more agile, flexible, scalable, and cost saving (in short and long term). System integration (SI) is a process to connects the various information systems and applications in an enterprise so that they work cohesively in a coordinated and unified manner. In present market, process manufacturers are under relentless pressure for improved operations. The challenges being experienced include globalization. When executing an order with customers' satisfaction, many things need to be well coordinated. The control and automation systems must be in the correct mode of operations at appropriate operating parameters and conditions, in defined automated sequences. Operator rounds may be required to collect process information. When production/manufacturing is completed, there may be final sequence like packaging. The entire process ends with material accounting needs to be performed so that the ERP inventory is updated to reflect raw material consumption, and the creation of finished goods, etc. All these shows how information flow is important in vertical direction. Five benefits of system integrations are as follows: *Simplicity*, *Cost saving*, *Security*, *RT visibility*, *Efficiency*, etc.

The benefits of control systems integration start at the plant floor and multiply all the way up the manufacturing value chain to the boardroom. The goal is to eliminate islands of automation by establishing channels for purposeful data exchange between technologies, networks, OEM-provided equipment, and other data sources. Network integration services facilitate the integration of *hardware*, *software*, and communication *protocols* (the rule) across different platforms, allowing different systems to work together effectively and efficiently. These services determine integration needs after assessment of the existing infrastructure so that systems can communicate effectively with each other. Normally, they cover configuring network devices, implementing security protocols, and optimizing performance. Major components of these services include the following: network design, implementation, and management. In network design, a blueprint that meets specific business needs is created. Implementation refers to the actual deployment of the network infrastructure. Management covers ongoing monitoring, maintenance, and optimization of the network. Network integration services enhance operational efficiency by streamlining communication and processes across various systems. Also, they facilitate the seamless integration of disparate technologies, allowing for improved data flow with minimum errors. All these facilitate enterprise to respond faster to the market changes after collaboration among teams leads to better decision-making. In network integration services, enterprises face many challenges viz. Compatibility issues including mismatch between different systems (legacy and modern systems). Solutions for optimizing these services include implementing automation tools to streamline processes, enhancing network visibility with minimum error, and adopting cloud-based solutions.

1. **Integration: important issues:** Control systems integration is facility design, or a component of it. Control systems integration team should work alongside process and facility engineers to align early design decisions with the plant's overall control philosophy. Following are some system issues and other major issues involved:
   - **System upgrade:** A newer hardware not touched for long may call for an upgrade.
   - **System Retrofitting:** Retrofitting outdated control systems increases both machine and process reliability while reducing unplanned downtime.

- **System migration:** Modernization strategy through a migration of your control system.
- **Efficiency:** This refers to eliminate wasted time and effort. Some ways and means include the following: Alignment of design with operational outcomes, Accelerate speed to market with automation, quick knowledge transfer.
- **Resilience:** This is important for reduced downtime under various operational requirements. Early warning for failure, Acceleration of Mean time to repair (MTTR), and use of automation in place of manual labor are a few issues could ensure this.
- **Flexibility:** This refers to prediction of future to adapt for it. Some related issues are as follows: Finding pathways future integration and rapid scaling, Setting up production target based on consumer demand and proactively regulate inventory through automation.
- **Profitability:** For better profitability new ways to be found: *All* critical business functions must be connected to share insights across the value chain. All bottlenecks between MES and ERP must be eliminated and if necessary, workflow may have to be reconfigured to maximize cost-effectiveness

2. **Integration and migration:** System integration is a process of combining various systems, subsystems, technologies, and software to form an integrated whole. It involves bringing together different components or modules from multiple sources into one comprehensive system that provides value-added functionality. Control System Migration is the upgrade or replacement of an existing control system with newer technology. This includes replacing outdated hardware and software, upgrading communication protocols, and integrating new features for increased efficiency and productivity.
   - **Benefits of integration and migration:** Some common benefits are as follows:
     - *Enhanced efficiency*: By integrating existing systems and migrating toward more advanced ones, businesses can streamline operation to reduce errors.
     - *Security*: Enhanced security in upgraded and migrated system.
     - *Better functionality*: With better integration and updated technology, improved capabilities, and expanded business is expected of the enterprise to offer better services and product features.
     - *Future-proven technology*: System integration and control system migration can offer the advancement of technology with a reliable IT infrastructure [7].
     - *Cost savings*: Replacing outdated systems helps prevent expensive downtime while
     - simultaneously improving operational efficiencies [7].
   - **Migration: Automation:** To reduce time and efforts, automation can be used to extract data from legacy systems and other data sources. Automation can be used to transform data into the format required by the target system to curtail the time and effort for manual data transform, Obviously, automation can also help to improve the speed and accuracy of data loading. Automation can be used to load data into the target system and reduce the time and effort required to manually load data [7].
   - **Migration and Integration:** Some common business migration integration issues are as follows:
     - *Upgrading/replacement*: This refers to business upgradation/replace an existing system, with data migration to transfer data from the old system to the new one.
     - *New software*: When an enterprise decides to implement a new software, data migration is often required for new system.
     - *Company acquisition*: Acquires a company, data integration is often required to combine the data from both companies and may call for migration to new system.
     - *Big data management*: For big data handling, data migration and integration become increasingly important for effective management.
     - *Data consolidation*: At time it is necessary to collect data from multiple sources and consolidate the same. Here, data migration is necessary.
     - *Governance and compliance*: Data migration and integration may be necessary to ensure compliance to strict regulations around data management.
     - *Cloud and data analytics*: Data migration is often required to transfer data to a cloud-based platform as well as deployment of data analytics.
     - *RT data integration*: Businesses relying on RT data, will call for required data integration.
   - **Cloud-based Data integration and Migration:** Pros and cons of cloud-based data integration and migration solutions have been enumerated ad-seriatim below:
     - *Scalability advantage*: Cloud-based solutions can be easily scaled up or down
     - *Flexibility advantage:* Cloud-based solutions can be accessed from anywhere with an internet connection.
     - *Automation*: Cloud-based systems provide automation optional advantage.
     - *Cost*: Due to less investment in expensive hardware/software for on premise solution cloud-based systems offer cost benefit.

- *Security*: Security of sensitive data needs in cloud solution needs special attention.
- *Internet dependence*: Cloud-based system is internet dependent
- *Limited control*: Enterprises have limited control over the infrastructure and environment.
- *Proprietary:* Cloud-based systems can be proprietary also.

3. **Data integration and operation:** ISA95 defines a systems architecture standard to provide guidance for the implementation of a digital manufacturing by implementing integrated software systems to eliminate paper-based processes. ISA-95 is the international standard for the integration of enterprise and control systems in a layered model as modeled in ISA 95 standard. Various levels are elaborated below:
   - **Level 0:** The actual physical processes.
   - **Level 1:** The activities for sensing and manipulating device (Intelligent device).
   - **Level 2:** The activities of monitoring and controlling the physical processes (DCS/PLC)
   - **Level 3**: The activities of workflow to produce the desired end products (MES).
   - **Level 4:** The business-related activities to manage a manufacturing operation (ERP).

   The ISA-95 standard primarily deals with the interface between levels 3 and 4. On the top, there will be corporate management of the company. Smart operations management software functions manage process manufacturing operations and functions complementary to the process control layers (ISA95 levels 1 and 2), providing solutions for material tracking, order management, manufacturing procedure workflow, systems integration, and data visualization, and document management. For detailed discussions in chapter 8 & 13 in volume I and Fig. II 16.0d may be referenced.
4. **Integration challenges and solution:** Integrating multiple data sources into a unified system can be a challenging task and major data-related challenges are as follows:
   - **Format and structure**: Different data sources may have data in different formats and
     - Structures. This is really challenging to combine the data into a single view.
   - **Quality and consistency**: It is really challenging to ensure that the data from different sources is accurate, complete, and consistent.
   - **Security and privacy**: Protection of sensitive data as well as keeping privacy, are big challenges especially, when integrating data from multiple sources.

   To overcome above issues some solutions adapted are as follows
   - **Extract, Transform, Load tool:** One solution is to use Extract, Transform, Load tool for extracting data from different sources, transforming into a consistent format and load it into the target system.
   - **Data virtualization:** Data virtualization tool allows to access and integrate data from multiple sources as if it were a *single, unified data source*, without the need to physically move or duplicate the data.
   - **Data quality and Security:** Enterprises can implement data governance policies with defined procedures to ensure data quality, consistency, and security.
5. **Plant floor integration:** Enterprises are putting effort to integrate the plant floor with their ERP systems for efficient coordination of the entire process viz. to match manufacturing capacity with the production demand. Such integrated operation of plant with the ERP would result in agile production responses with quickly and efficiently changing the production to meet the needs of the business. Also, the integrated process plant provides a better way to track material usage. The ERP creates a production order and allocated material to the order or process facilities raw materials may be allocated by the ERP. ERP to Plant floor integration enables a variety of capability as tabled in Table III 15.1a.

## 15.2 Control network: migration issues

System migration refers to the process of *transferring* data, applications, and other critical elements from one (operating) environment to another (relatively, modern). Such migration could be due to the need for improved performance, enhanced functionality, or cost reduction. In an industrial network, migration refers to the *transitioning* hardware, software, and data to a newer environment for improvement in terms of efficiency, enhancing security, ensuring long-term operational viability, etc. *Safety*, Minimum production loss, and compatibility issues are prioritized over other issues. with legacy systems. Key Considerations are as follows: *Safety*, *Compliance*, *Security robustness*, *Minimum downtime*, *Bridge with legacy system*, Stakeholder alignment, etc. Generally, networks migration takes place to take advantage of new technologies, boost performance, enhance security, save costs, and adapt to business growth and changing industry rules. With technological advancement, coupled with evolvement of business needs, enterprises especially mission-critical companies may come across some challenge which compel them to migrate to more advanced and efficient network. Complex processes of network migration consist of migrating the entire network to a new infrastructure, redesigning the entire network, replacing legacy equipment, upgrading hardware, software, or adopting new

**TABLE III 15.1A ERP integration: Common capability and business value.**

| Capability | Business value |
|---|---|
| Plant floor integration | Labor accounting to be based on actual labor |
| Automatic driving production order | Faster response to demands, better regulation of finished goods inventory |
| Automatic changes to equipment | Better capacity utilization, faster change least off spec production |
| Planning and scheduling | Improved planning, production, and capacity utilization |
| RT automatic ERP update | Accurate inventory management with<br>Better use of working capital |
| ERP: Automatic data entry from plant | Improves productivity and ERP with post no data entry errors |

protocols. Such transition could be huge involving millions of devices to meet the evolving needs of an organization. Operating an old DCS increases the risk of system failures, parts availability, and reduced support, among other operating inefficiencies. All of these issues can lead to plant downtime, and as a result, lost revenue and safety incidents. For control systems major migration issues are as follows: *Enhance system reliability*, *Reduction in risk*, *Enhanced early warning capabilities*, *Expanded access to DCS information from third-party systems*, *Better information management*, Improved communications, *Increased security*, etc. Migration involves the methodical planning and execution of moving existing system components to a new platform or infrastructure. Various steps for migration mainly cover: *Define objectives*, *Assessment—including risk assessment*, *Planning*, Design and engineering, *Testing* and *validation*, *Implementation*, *Execution (strategic execution in a phased manner as applicable) and verification*, *Post migration support*: including training and review of migration results. Major challenged faced in migration of modern days mainly include but not limited to: *Restrictions in legacy system*, *infrastructure complexity*, *Speed of deployment* Pace *of innovative adoption*, *Concerns for data center investment including connectivity, resilience, access to cloud*. In this section, short discussions on migration issue with special reference to control network shall be dealt with.

## 15.2.0 Basics of system migration

Globally, industries are facing an increasingly competitive business environment. Naturally, manufacturers need to find ways to produce *quality products meeting changing customer needs*, *efficiently* and *cost-effectively*. Also, there are competitions to *quickly market* the same. It is needless to say that control system performance can have a huge impact on Return on investment (ROI) throughout the lifecycle of the plant. In the modern days automation systems are no longer an isolated system, Now, they are connected with high level of integration to diverse systems. So, in case of mismatch or incompatibility between the systems, there may be necessity to go for migration of control network. Mostly, there is a specific rationale for migration. Some of them are as follows:

End of service life, Loading issues, Cost reduction, Security/Vulnerability, Amalgamation of operating consoles, Units or upgrades/modernization, Co-existence with multiple vendors and applications, Cross-unit closed-loop control, Value-added feature, Future-proofing system components, etc. Considering DCS as the "brain" of the facility, obviously migration carries a degree of risk analogous to brain surgery. In spite of the risk, people opt for migration because, they wish to avoid the enormous cost and risk associated with losing control of a process as well as to leverage innovative technologies that improve operational efficiency and performance. Even knowing they need to migrate to stay competitive, uncertainty and doubt can persist moving forward, especially when key personnel bandwidth is limited. Migration is a lot of work. Risks run high without the right team of experts to plan and implement. To this end, consider the risks inherent to a migration and how modernization can help overcome these hurdles. Obviously, enterprises need immediately do proper assessment of the situation, planning, and budget allocation for control system upgrades to avoid resource scarcity including human resource. All leading global automation suppliers indicating and taking actions to obsolete/phased out hundreds of thousands of its legacy DCS nodes and thousands of software licenses. The upgrade possibilities for DCS mainly include [8]:

Technology refresh [the replacement of legacy electronics with modern ones] for providing improved efficiency, low power consumption and environmental issue.

- Technology upgrades—replacement with newer generation technology for improved performance, reduced maintenance and new operational and functional capabilities.

- Intellectual property upgrades transitioning current control strategies, applications, and HMIs to more advanced technology

In this subsection brief discussions on basics of system migration shall be dealt.

1. **System migration:** Each of the different types of migration under system migration serves a specific purpose and addresses unique challenges in the process. These four common migration types under system migrations are as follows:
   - **Application migration:** Application migration is often driven by the need for improved performance, better scalability, or to take advantage of modern systems like microservices. This is the process of moving an application from its current operating environment to a new system. This can include moving an application from an on-premises server to a cloud environment or between cloud platforms also.
   - **Cloud migration:** As the name implies, this refers to the process of moving all business elements including, data, applications, etc., from an on-premises data center to a cloud-based infrastructure to take advantage of the scalability, flexibility, and cost-effectiveness of cloud computing. The migration process requires careful planning to ensure minimal downtime. Data security is a prime issue here.
   - **Data migration:** Data migration is crucial, when enterprises need to upgrade/consolidate their data storage infrastructure. This involves the movement data from one storage system to another. Extraction of the data from its original location, cleansing it to ensure data quality, and loading it into the new storage solution are the major steps necessary for the data transfer process.
   - **OS migration:** OS migration involves upgrading or changing the OS of a computer or network of computers. This could be moving from an older version of an OS to a newer target system or switching from one system to another (Fig. III 15.2a)
2. **Migration risks***:* While system migration offers significant benefits, it also presents a number of challenges and risks that must be carefully managed. The most important challenge is to ensure minimal disruption to ongoing business operations and safety for industrial process. Compatibility issues may arise if new operating systems do not seamlessly integrate with existing infrastructure, leading to operational inefficiencies. Risks from the risk of cost overruns and delays are quite common mostly, due to unexpected issues/changes in scope during the migration. Common system migration risks include but not limited to: *Depreciated data values, Security, Rollout strategy, Application dependencies, User training, Resistance to change.* In case of DT, data migration carries high risks which can entire DT initiatives and impact business continuity. Major challenging issues here are as follows: *Complexities of modern data ecosystems, Scale and performance, Legacy system constraints, Cloud and hybrid architectural structure, Business pressure/compressed timeline, Data loss/corruption, Poor data quality, Inadequate planning, Security and compliance risk, Downtime, business Interruption, Integration failure.*
3. **Migration reason and Modernization impact:** Major reasons and impacts are as follows:
   - **Reasoning:** Following are major reasons behind migration:
     - *Threat* of unscheduled downtime/incident.
     - *Not cost-effective*: old system/system dead-ended/phased out, no spare parts.
     - *Legacy system* without support for new IT.
     - *Emerging business opportunity* with new system.
     - *Inflexible old/legacy system* unable to meet customer demand.
     - *Lesser predictability/visibility* of old system to warn/prevent abnormal situations.
     - *Old system*: fail in capacity/inability to expand cost-effectively.

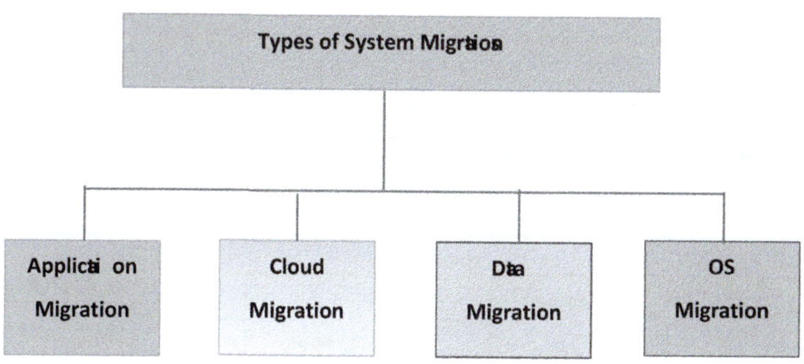


**FIGURE III 15.2A** Types of system migration.

- **Modernization Impact:** Legacy systems' reliance on outdated *user interfaces* and *unintuitive programming* styles can create a situation that increases *implementation time, costs,* and an obstructive barrier for new *personnel* to work. Modern technologies provide opportunity to improve asset utilization and fully realize the benefits of "Big Data" increased processing power, with powerful connectivity for information exchange. Many legacy control systems rely on proprietary data structures lacking in interoperability which is now essential in multivendor environment through network integration. Increasingly, manufacturers are utilizing data for a variety of functions, including data analytics, environmental calculation and monitoring, enterprise resource planning integration, asset management, predictive maintenance, digital twins, and various other IIoT applications.

4. **Migration Procedure:** Migration issues start with identification of current operational pains viz. *Slow performance/Frequent downtime/*Scalability limitations/Security vulnerabilities/High operating costs/Inadequate support. Various steps followed are as follows: *Goals to achieve, Current architecture review,* Migration strategy for components, Design future architecture (Scalability and flexibility, Modular design, Security), *Technology to achieve goals, Plan the migration, Metric framework.* These are depicted in Fig. III 15.2b.
Migration must be a well-documented system. A complex control/automation system migration requires a thoughtful well-planned, approach. A few steps are as follows
   - **Domain knowledge:** Domain and business expertise are very important especially for facility Input/output (I/O)s, equipment, operating conditions, safety risks, etc.
   - **Planning phase:** Emphasis can be placed on sound planning and budgeting through Front-end loading (FEL) to reduce operational project costs, minimize scheduling issues, decrease scope changes after authorization, and increase likelihood that a project meets or exceeds financial goals.
   - **Commercial issues:** FEL planning should accurately determine the effort and time required for a migration for quick buy in the idea.
   - **System and vendor:** Choosing an appropriate replacement system is important, and this is based on selection of technology and vendor. Sticking to the same or switch over each has merits and demerits to be well assessed. Issues like Advanced process control (APC) support, scalability, life-support track record, high-performance graphics, migration path, industry and region install base, IIoT, and control strategy portability are very important to be considered.

   According to ARC Advisory Group white paper [9]. Important issue for new system are as follows
   - Offer compelling value proposition and path to operational excellence
   - Minimize wiring and minimum or zero downtime
   - Minimize training costs for new system
   - Preserve hardware investment, graphics and assist with or perform graphics conversion and control strategies and assist with or perform control strategy conversion

5. **Miscellaneous Issues:** Strategic plan development for operational technology is basically a continuous process. Each time it is done, there will be some new opportunities uncovered. This is because of the fact that the exercise will include the observations and insights of a new mix of internal stakeholders, plus the observations and insights of a new mix of OT suppliers and the new technology capabilities they bring to the table [10].
   - **Total** versus **Phased Migration:** The hardware infrastructure, can become embedded in the plant, making it very difficult and extremely costly affair for complete replacement. In such cases of the OT strategy, a phased migration should be the correct direction. In a phased migration, only those components that need to be replaced are upgraded.

   There are cases, where the user may want to consider a single total replacement approach. Normally, opting for such approach includes *availability of a time window* during the regularly scheduled turnaround, the type of manufacturing process

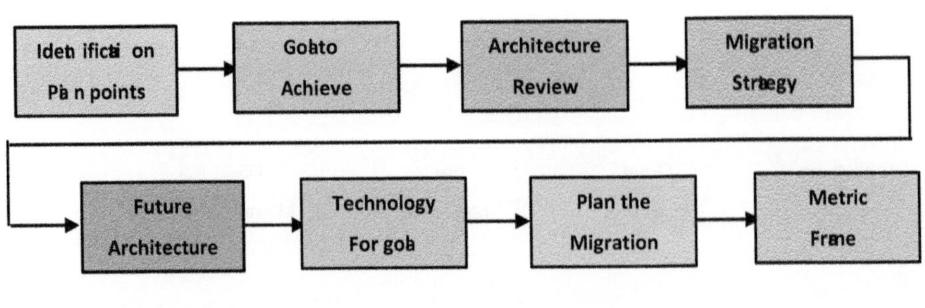


**FIGURE III 15.2B** Steps for system migration.

involved, and the experience of the supplier involved in the project. Another potential benefit of single total replacement is that it ensures a single, current generation of system and the associated reduced Total cost of ownership (TCO) [10].

- **Hot Swap** versus **Downtime Migration:** "Hot Swap" refers to the changing system components while the process is running. This can minimize the downtime at the cost of high risk, especially, if not carefully planned/executed. This involves significant preparation, planning, and testing with automations in some parts like database and graphics conversion. The supplier should offer both tools and services that speed the task of conversion as much as possible. Downtime migration refers to the process when the process is not operating. Here, the entire control system hardware and software can be successfully migrated without any chance of an accidental shutdown. The control system must be replaced and fully functioning within the downtime window that meets the plant's overall plan.

*After concluding this discussion, let us move on to network migration.*

### 15.2.1 Discussions on network migration

Network migration is a complex and comprehensive process involving various changes in infrastructure, that is, moving data, applications, and services from one network environment to another, or upgrading the existing infrastructure. It can be migrating the *entire network* to a new infrastructure, redesigning the entire network, replacing legacy equipment, upgrading hardware, software or adopting new protocols. The goal of network migration is to improve network performance, security, scalability, and adaptability to meet the evolving needs of an organization. If we look little in details we find that these migrations could involve: *Hardware upgrades* (replacement of older devices), *Software updates* (updating the software running on network devices), *Protocol changes* (transitioning to newer or more efficient network protocols), *Network redesign* (redefining the entire network architecture), *Virtualization and SDN* (adopting virtualization technologies or SDN approaches discussed), *Cloud migration* (movement of data and applications premise to cloud platforms), *Data center migration* (relocating to a different data center), etc. *Situation assessment, Planning, Design* and *engineering, Data back-up and security, Implementation, Testing for validation, Monitoring, Optimization,* and Decommissioning are major steps in migration process. In migration/modernization, making the transition to a modern DCS is a stressful challenge for users, from the increasingly difficult task of justifying the automation purchase, selecting right technology vis-a- vis a supplier, implementing the solution with a future roadmap. Naturally, there will be a number of challenges in network migration process as well. *Risk of disruption/performance degradation* during migration is common. Phase migration, strategies for fault tolerance, and failover mechanisms are common solutions. *Security and compliance risks* could be overcome by Implementing robust security measures like encryption, access controls, etc., to safeguard data and network assets. *For compatibility issues* It is necessary to ensure that the new network components are compatible. Important issues of *Resource allocation* and *Cost overruns* are addressed by required budget, skilled personnel, and time allocations by analyzing costs beforehand, using cost management tools, and optimizing resource use. Because of data migration, archival, and potential data retention legalities, Decommissioning legacy systems and hardware is complex. Roll back plan, proper Documentation and Postmigration review are parts of best migration practice. All these will be discussed in this subsection briefly.

1. **Migration reason:** Reasons for migration could be requirement of centralized repository devices used could be End of life (EoL) [10], and availability of advanced techniques best suited for the application now/in future. *Industrie 4.0/DT approach, Flexibility and Scalability*, improvement of operation and financial efficiency including ROI could be driving forces behind network migration. Major important issues are as follows:
    - **Hardware:** This refers to the replacement of older devices with newer and more advanced models with improved performance, enhanced features, and security aligning with modern networking requirements.
    - **Software:** This involves updating/upgrading the software running on network devices, for example, firmware upgrades on switches and routers or upgrading of OS on servers. These would ensure the network components are running the latest features, security patches, and bug fixes.
    - **EoL:** Many legacy networks component hardware/software which are not supported by vendors even spare parts/update, etc., scarce.
    - **Protocols:** Proprietary/old mostly protocols often struggle to integrate with modern network infrastructure, prompting network migration, that is, transitioning to open, standard, more efficient network protocols.
    - *Network redesign*: This refers to redefining subnets, segmenting networks for improved security, or optimizing the network topology for better traffic flow.
    - **Virtualization and SDN:** Network migration may involve adopting virtualization technologies

(multiple virtual network instances to run on a single physical device, to optimize resource utilization and increasing network flexibility) or SDN approach (SDN separates the control plane from the data plane in a network).

- **Network performance:** All the above issues would contribute toward performance improvements. However, specifically, there may be reasons to opt for migration viz. proliferation of IIoT devices and Industrie 4.0 call for higher bandwidth –demanding migration. Also, latency and reliability could be a driving force behind network migration.

2. **Migration challenges:** Enterprises often face a range of challenges for network migration. Some major challenges include but not limited to:
   - **Data Migration.** Without careful planning, data loss may occur due to format incompatibilities, schema mismatches, or network interruptions. It is recommended to adopt backup strategy and conduct migration testing to ensure that all data are accurately transferred and remains accessible in the target system.
   - **Service disruptions.** The risk of service disruptions or performance degradation during
   - migration is common. So, as discussed, fault tolerance and failover mechanisms are essential. to mitigate this challenge.
   - **Compatibility.** New network components should be compatible with existing infrastructure and systems is crucial. This is more so for data migration from on-premises to cloud-based systems, with significant different configuration. A well-structured data mapping and premigration compatibility assessment can identify potential issues.
   - **Resource and cost:** Adequate budget, skilled personnel, and time are essential for the successful completion of the migration process. Also, analyzing costs beforehand for realistic budgets, timelines, and contingency plans, use cost management tools, and optimize resource use.
   - **Security issues.** Migrating networks may expose vulnerabilities, making them potential targets for cyber-attacks. As discussed, implementing robust security measures is essential to safeguard data and network assets. To protect data, robust encryption and adherence to regulatory compliance measures should be integrated into every migration stage.
   - **Integration issue:** Assessing integration compatibility, updating APIs, and adjusting configurations are crucial for smooth connectivity and maintenance of functionality in postmigration period also.
   - **Decommissioning.** Decommissioning outdated systems and hardware can be complex, involving data migration, archival, and potential data retention legalities
3. **Best practices:** Networks are constantly changing due to new technologies, mergers, acquisitions, and resizing. These factors, combined with the growing demand for bandwidth, are forcing many carriers to modify their existing network to meet customer needs, increase network performance, and create additional revenue-generating opportunities. Basic idea about migration process has already been depicted in Fig. III 15.2b. To ensure a successful network migration, enterprises should consider the following best practices:
   - **Assessment:** First step is to evaluate the network to identify any potential problems, helping us establish the most cost-effective and least intrusive method for migration. This will also include complete detailed risk assessment and definition of scope and goal for migration.
   - **Planning**: Develop a detailed migration plan that includes a clear timeline, resource allocation, cost analysis, available experts, training, and different contingency measures and fallback arrangements.
   - **Design issue**: Design and development of new network architecture, including details on configuring the environment, preparing backups, and plugins for old pains.
   - **Security:** Implement robust data backup procedures and security measures including robust encryption, access controls, work with compliance teams to protect critical information throughout the migration process.
   - **Rollback Plan:** Keep a well-defined rollback plan ready in case the migration encounters significant challenges. This plan will allow reverting to the previous configuration to carry on business.
   - **Implementation:** Actual migrating network services and traffic, in a manner planned
   - **Validation:** Testing the new network for performance and security.
   - **Optimization:** Continuously checking the new environment and making adjustments to optimize operation and extract highest benefit. Also, one-on-one mentoring during the migration period, provides in-depth information and hands-on experience to handle future migrations in-house. Network optimization ensures a quick ROI.
   - **Documentation:** Maintain detailed documentation to aid troubleshooting for future.
   - **Postmigration Review:** Conduct a postmigration review with all stakeholders to evaluate the success of the migration, identify *lessons learned*, and plan for future improvements.
4. **Network migration result:** Industrial network migration, typically results: *Enhanced performance*, *Higher*

*efficiency and productivity*, *Robust security*, coupled with *greater scalability and flexibility*. Apart from these, other positive impacts include the following: *Data-driven decision making*, *Cost curtailment with reduced TCO*, *Better reliability and availability*, *Improved ROI*.

5. **ROI and Network monitoring tool:** Effective monitoring is essential for optimization as well as to calculate ROI. In its simplest form ROI on network migration refers to the relative (to the total *investment made in the migration*) value gained by an enterprise through *Cost savings*, *Revenue growth*, and *savings* or *earning* due to *operational efficiencies*. ROI is crucial for successful industrial network migration. For this specific Key performance index is tracked to ascertain the gain by comparing with costs. Migration process needs to be monitored continuously to identify and address issues proactively. It is continuous monitoring throughout the migration process, that helps in identifying and addressing the issues proactively, to ensure network activity smoothly. *Premigration* data are used for baselines of the existing network (performance, downtime operational costs, etc.). *During migration*, parameters like System availability, Network performance, resource utilization, etc., are monitored to assess the progress. *Postmigration measurements* include performance monitoring and cost metrics to ensure the new network meets long-term objectives and to identify areas for optimization. Regularly assess user satisfaction and adoption rates. Strategic planning, accurate TCO, etc., are helpful. Network monitoring tool is a software which checks and track network metrics, relieve personnel from tedious troubleshooting, lack of visibility, and productivity losses owing to period. Network monitoring tool can identify and alert on issues before they cause outages. All these could be utilized to calculate ROI. So, it is a helpful tool for network migration.
*Let us now conclude this discussion to address control network migration next.*

### 15.2.2 Control system-DCS migration practices

Automation systems are the brain for plant, for process control and output, and ultimately for profitability. So there will be a lot at stake, hence stakeholders. Although a relatively small portion of the overall investment in processing plants, automation systems can have a large impact on the successful outcome of the project and return of the investment throughout the lifecycle of the facility. This justification compares the cost of continued operation with the DCS to the costs and benefits of migration to a modern control solution. Together, these factors comprise the TCO. In general, for automation systems like DCS upgrade should always focus on both process automation as well as on workforce development. This is because, a major DCS migration offers the best opportunity to train the workforce and would yield benefits over the years to come. Major control system migrations are not at all a regular phenomenon for a plant and could be once-in-a-generation opportunity. Some of the major reasons behind DCS migration shall include but not limited to: *Obsolescence*, *End of service life*, *Loading issues*, *System security*, *Cost reduction*, *Cross-unit closed-loop control*, *Amalgamation of operating consoles*, *Future-proofing system components*, *New units or upgrades*, *New value-added features*, *Co-existence with multiple vendors and applications*, *Process data at the desktop*. A well-planned and executed migration to a modern control system not only improves plant availability and reliability, but can also provide a more flexible production platform. Flexibility is the key battleground on which companies seeking to make the most of business opportunities will thrive. For DCS/control planning it is always recommended to examine ways and means to manage DCS migration preparation with front-end loading design goals. It is needless to say that the benefit with new generation DCS well supported by user-friendly advanced high-performance HMI in mitigating abnormal situations and day to day plant operation. Globally, many DCS/Supervisory control and data acquisition (SCADA) systems have almost reached EoL. Also, older generation DCS/SCADA may not be equipped to support Industrie 4.0/DT operation through big data analytics/cloud connections. Therefore, control systems need to be replaced with more secure, reliable, and safe control system. In this subsection, brief discussions on control systems/DCS migration have been discussed.

1. **Migration concept:** A control system migration teaches nicely, about plant operations, automation, security, IT functions, and project management as a training to go a long way. DCS migration not only minimizes the risks associated with having to maintain older systems but also presents the opportunity to take advantage of new technologies. *Risk assessment and best ways to address them*, *Existing assets optimization*, *Quality maintenance*, *Minimum production loss*, and *Long-term support* these are the issues are mainly addressed to ensure a safe, cost-effective, and value-added migration process.
   Conceptually, automation level, risk assessment, planning issues, Value-added issues, etc., shall be discussed.
   - **Automation level:** Creating an automation system has to begin with a very deep knowledge of the process, instrumentation, and control-related issues. Some of them have enumerated below;

- *Sensors*: Here, the check points include the following: enough sensors to cover all and touching all application in suitable way. A device-by-device analysis. of field instrumentation can help new engineers understand how the process was designed initially and how it has changed over the years.
- *Field control element*: to look for feature, control capabilities as well as Accuracy in sizing and capability of controlling to offer improved performance.
- *Field wiring and device-level and wireless network*: This analysis is done to ensure well thought of maintenance friendly, reliable field/device I/O connections in adequate numbers for connection to control system. This also includes analysis for wireless network for proper applications to get the insights into the mechanics of network design.
- *Controller*: A study controllers to get good knowledge on the best approach and increase understanding of the entire automation system.
- I/O: The communication capabilities of field devices and I/O/smart devices for the effectiveness.
- *HMI*: Understanding operators functioning for effective and improved operational interfaces. This would also include alarm management strategy including warning handling capabilities.
- *SIS*: Hazard landscape and safety capabilities in new system without excessive false alarms and alarm management.
- *History and asset management*: This includes study on data history and asset management including studying maintenance history with diagnostic information for identifying and eliminate the bad actors causing the most outages.
- *Communication*: Everything needs to communicate, and each level needs the best method. Understanding and analysis communication capabilities.

- **Risk assessment:** Highly experienced migration experts need to migration risks by analyzing existing system to identify and prioritize potential issues, then with available database, risk mitigation solution should be devised.
- **Optimal strategy:** For develop an optimal migration strategy, issues like: *Migration plan and schedule*, *Benefit and ROI identification*, *Ways for safety improvement and flexibility*.
- **Value addition:** For migration—over a long period of time (phased migration), the blueprint to address specific needs is developed. It is important to ensure that the system is reliably operable during the entire process. Gap between targeted and actual results must be identified to devise the countermeasures. Also, it is necessary to make sure the implementation of these countermeasures.
- **Connectivity:** Connections to higher-level systems and subsystems are an important consideration In DCS migration, connectivity is extremely important for functioning of higher-level systems and subsystems.
- **Training:** It is also important to train the concerned persons
- **Coverage:** To complete the coverage, it is important to check back through *Process and instrumentation diagram*, *Loop diagrams*, *Complete database of existing system*, (*Field*) *Wiring drawings*, *Junction box detailed diagrams*, *Cable schedule with interconnections*, etc.
- **Reliable maintainability:** Reliability is directly connected to plant availability and must treated seriously. This is more so, with SIS and devices/systems certification by internationally recognized competent authority only should be entertained only.
- **Asset management:** Asset management can be improved using intelligent Highway addressable remote transducer (HART) and Fieldbus system by reducing wirings. Also, field device diagnoses would help in both maintenance and plant operation. Also, Remote system maintenance can be applied for identifying a problem quickly at a site, minimizing system downtime.
- **Involvement:** This is from start to finish of the entire migration process.

2. **Front-end loading (FEL):** In connection with DCS migration FEL refers to early planning phase that involving thorough upfront work to define project scope, risk identification, and take well thought of decision prior to significant resources commitment. Key issues include but not limited to Scope definition, Risk assessment, detailed planning with cost and time schedule, Lifecycle perspective, etc. The projects start with FEL with three primary goals: *Budgeting justification for capital expenditure with tollgate* (check point), *Reduction in the implementation risk* for *cost, schedule, and system performance* and *Reliable preliminary engineering and execution planning*. Three main levels are as follows:
   - **Explanation: of FEL phases:** With the above ideas in mind detailed explanation of FEL phases shall be elaborated. Also, FEL planning with capital funding has been elaborated in self-explanatory Fig. III 15.2c which will be helpful in understanding detailed explanation of various FEL phases. Fig. III 15.2c1 shows the capital funding issues.
     - *FEL0/1*: This is Justify and weigh the risk, that is, stakeholders evaluation of the risk of doing

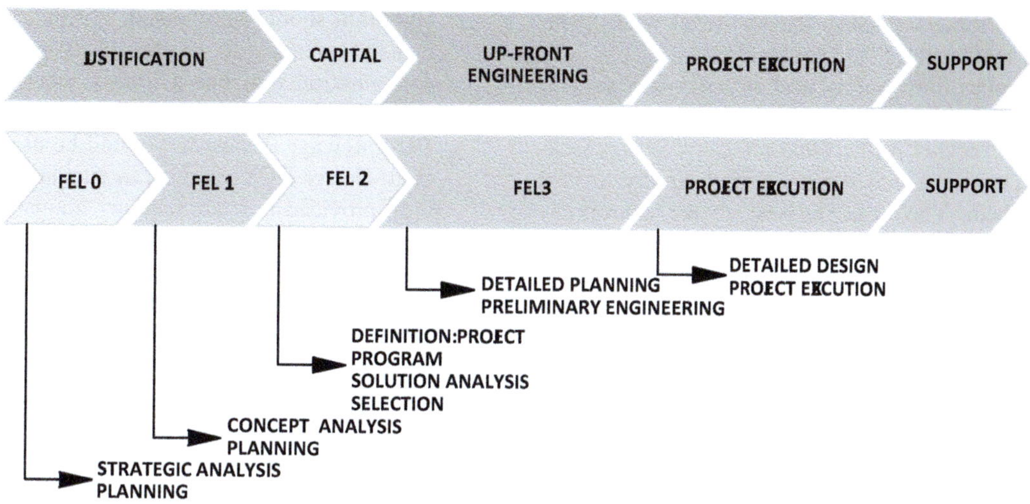


FIG IIII 15.2c1 FEL : CAPITAL FUNDING

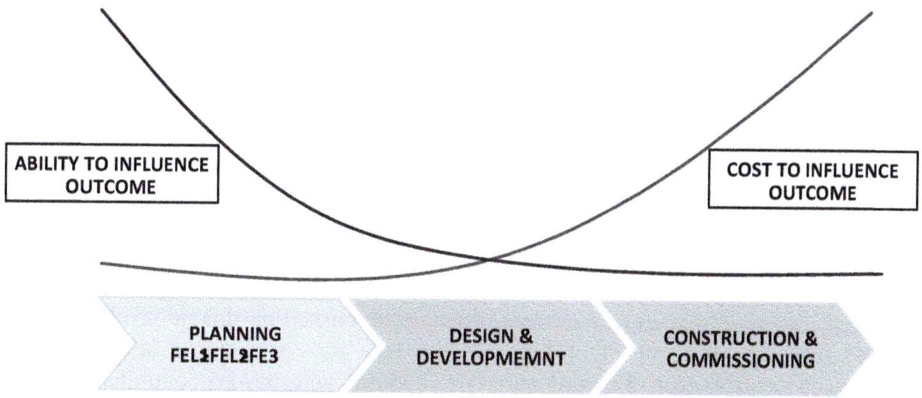


FIG IIII 15.2c2 FEL : OUTCOME INFLUENCE

**FIGURE III 15.2C** FEL: Capital funding and outcome influence.

the project versus doing nothing. Here, the decision on the rough total cost outweighs the cost of doing nothing. *Accuracy level*+50%

- *FEL2*: This is a major decisions phase, with the options and leveraging data. It focuses on various issues like: *Control platform selection*, *Breaking up of the project phases*, Enhancement of control strategies, and *Control room improvements.* Both the biggest risk and the biggest reward for the project can come from upgrading or improving control strategies. The task in the FEL2 is to weigh the rewards of the improvements against the risk of making the improvements. *Accuracy level*+30%
- *FEL3*: This deals with scope for project execution. Here, the focus is on defining the scope, costs, and schedule to ensure successful project implementation with objective to minimize risk. This estimates accuracy of a ±10% of TIC. Some major issues in this phase includes but not limited to are as follows: *Execution plan with definitions of responsibilities*, *DCS equipment architecture*, *Bill of materials and specifications*, *Final TIC and TCO* estimates, *Some detailed designs. Accuracy level*+10% (of estimated TIC)

- **Time-line impact:** This is obvious that as we progress from FEL1 to FEL three or in design stage and finally, construction commissioning stage, ability to influence outcome will fall asymptotically toward zero. On the contrary, cost increases exponentially with time as elaborated in Fig. III 15.2c2
- **Estimating, planning:** This starts with scope definition in line with facility requirements. Based on requirement, scope may need to be parsed out into multiple projects in FEL3. At early-stage risk analysis with mitigation plan is framed followed by critical activity inputs from the stakeholders for a

brainstorming session at the site level with operations, maintenance, and engineering.

- **DCS migration: Safety and security safety:** This deals with critical parts related to *Resource availability*, *Network traffic*, *Data integrity*, *Safety*, *Security*, *Downtime* and *the greatest flexibility* to deal with them. The FEL2 phases are the appropriate time to evaluate and select the best control system platform to be migrated to and evaluate project options.
- **DCS migration: Functionality:** It is necessary to ensure that no functionality is left behind as well as how the new system will be cut over. Total installed cost (TIC) estimate is critical for the development of the migration schedule. A proper migration schedule should be all inclusive. The outcome of a DCS migration, can be altered throughout the effort (even toward the end). Two important issues are as follows: *DCS migration schedule* (the current and subsequent FELs, design, engineering, testing, installation, startup, and commissioning) and *Successful migration plan* (an early and strong effort spent on good effective planning)

3. **Planning engineering:** There may be requirements for replacement of I/Os, I/O interface without touching field terminal assemblies. These are rather complex issues. However, some important aspects at planning engineering could be:
   - **Control Strategy:** This has big impact. Taking advantage of advanced controls, can Support and strengthen the bottom line while new control methods to improvise turnaround time and automate responses to critical situations.
   - **Operator experience:** New system(s) will enhance the operator experience. However, implementing high-performance HMI graphics can bring more benefits for controlling the station or alarm and abnormal situation management standard.
   - **Intellectual investment:** The goal is to save intellectual investments while migrating to new system through use of special software. For special software, it is better to discuss and decide with selected vendor, third-party items when to be included reconfiguration issue must be in mind.
   - **Security:** Security should be Strengthened to justify the migration project.
   - **System Code:** Migration of existing code from one system to another is a big risk. It is better to extract the old code into a readable format for evaluation by operating personnel for improvements and coded back into the new system.
   - **Engineering library:** Prefer to select the vendor supplying the engineering tool for the new controller a library of IEC61131 function blocks. Old algorithm should be retained. For this, it is recommended to ask the vendor if its new controller configuration tool has a library that emulates the old controller's code.
   - **I/O gateway:** Easiest way could be use of fieldbus card. Newer DCS vendor can supply the I/O modules, provided they are fieldbus compatible.
4. **Components and systems:** Brief discussions on modifications associated with various DCS/control system and components. These mainly include the following:
   - **HMI system:** There may be may types of HMIs, some difficult to get replica/part replacement may be very costly. It is always recommended to replace multiple, proprietary HMIs with open system can significantly cut and need to learn multiple interfaces by operator. Depending upon system selected it may be difficult to move the graphical data from one HMI to another. Here also, it is better to have the graphic package with desired features than to replace. The new system may also use an object-oriented approach to the entire HMI. HMI high-performance graphics, and the second is alarm rationalization are important features. Therefore, it is better to look at those innovations during a DCS migration.
   - **Controller system:** New controllers support IEC 61,131 which might not have been supported by old ones. For this to avoid abundance, better to look for conversion available with vendor. Mix of older and newer controller may cause complex peer-to-peer communication. In case newer controller supplier cannot provide translating software, better to look out for third-party software. OPC UA may be a way to handle the situation.
   - **Safety and security:** It is recommended to Identify new technologies to be incorporated based on anticipated benefits under given situation. Establishing integration across multiple platforms, including MES, has become crucial to being competitive.
5. **Time and procedure** [11]: Impending crisis drives the need for a control system migration at times it could be a massive replacement project including replacement of OT system. For OT replacement, strategic planning aligning with enterprise's business strategy is very important. ARC advisory recommends that the multidisciplinary team of stakeholders needs to discuss the issue to develop a strategic plan for OT, taking into account the "current situation," followed by a determination of the "desired situation." This is not an one-time issue, but a continuous process as, each time it is done, there would be a new one. The exercise will include the observations and insights of a new mix of internal stakeholders, *plus* the observations and insights of a new mix of OT suppliers and the new technology

capabilities they bring to the table [11]. Because of embedded hardware infrastructure, like wiring, I/O, it is not possible to do a wholesale replacement of the system. If this is the conclusion of the OT Strategy phase, then a *phased migration* could be the direction selected. In a phased migration, only those components that need to be replaced are upgraded. There are cases, however, where the user may want to consider a single total replacement approach. Primary factors driving this approach include availability of a time window during the regularly scheduled turnaround, the type of manufacturing process involved, and the experience of the supplier and/or the system integrator partner involved in the project.

*Let us now conclude the discussions on OPC integration and migration to move on to important issue of Ethernet.*

## List of abbreviations used

**AE** Alarm event (OPC)
**APC** Advanced process control
**API** Application programming interface
**APL** Advanced physical layer –Ethernet APL
**COM** Component object model
**DA** Data Access (OPC)
**DCOM** Distributed COM
**DCS** Distributed control system
**DLL** Dynamic-link library
**DMZ** Demilitarized Zone
**DNS** Domain name system
**DoS** Denial-of-service
**DT** Digital transformation
**DX** Data exchange (OPC)
**EoL** End of life
**ERP** Enterprise resource planning
**FEL** Front-end loading
**FLC** Field level communication
**HART** Highway addressable remote transducer
**HDA** Historical DA (OPC)
**HMI** Human machine interface
**I/O** Input/Output
**ICS** Industrial control system
**ICT** Information and communication technologies
**IDL** Interface definition language
**IIoT** Industrial IoT
**IoT** Internet of Things
**IT** Information technology
**M2M** Machine to Machine
**MES** Manufacturing execution systems
**MQTT** Message queuing telemetry transport
**MTTR** Mean time to repair
**O PAS** Open Process automation standard
**ODBC** Open database connectivity
**OLE** Object linking and embedding
**OPAF** Open process automation forum
**OPC** Open Platform communication
**OT** Operation technology
**PLC** Programmable logic controller
**ROI** Return on investment
**RT** Real time
**RTU** Remote transmission unit
**SCADA** Supervisory control and data acquisition
**SDN** Software defined network(ing)
**SI** System integration
**SIS** Safety instrumentation system
**SOA** Service-oriented architecture
**SQL** Structured query language
**TCO** Total cost of ownership
**TIC** Total installed cost
**TSN** Time-sensitive network
**UA** Unified architecture (OPC)
**UNS** Unified namespace
**XML** Extended markup language

## References

[1] OPC, The Ins and Outs to what It's About–"The Every Man's Guide to OPC", MatroKon OPC; D. Kominek, Canada, 2009. https://www.automation.com/pdf_articles/Guide_to_OPC.pdf.

[2] OPC Historical Data Access Specification; Version 1.20; OPC Foundation; Dec'2003; https://advosol.com/OpcSpecs/OPC%20HDA%201.20%20Specification.pdf.

[3] OPC Unified Architecture Interoperability for Industrie 4.0 and the Internet of Things; T.J. Burke; OPC Foundation; 2016; https://opcfoundation.org/wp-content/uploads/2016/05/OPC-UA-Interoperability-For-Industrie4-and-IoT-EN-v5.pdf.

[4] What is IT/OT convergence? Everything you need to know; Tech-Target; April 2024; https://www.techtarget.com/searchitoperations/definition/IT-OT-convergence.

[5] Five critical factors in industrial control system integration; A. A. Amin; Control Engineering; August' 2016; www.controleng.com/articles/five-critical-factors-in-industrial-control-systemintegration/ https://www.intechww.com/five-critical-factors-in-industrial-control-system-integration/.

[6] OPC: Interoperability Standard for Industrial Automation; T. J. Burke; InTech; ISA; https://www.isa.org/intech-home/2018/november-december/features/opc-interoperability-for-industrial-automation.

[7] PLC Programming, System Integration and Control System Migration; Global Automation Group; Internet document; https://globalautomationgroup.com/system-integration.php.

[8] Control System Migration: Surviving the "Perfect Storm" to Improve Performance; Honeywell; White paper; Internet document; https://prod-edam.honeywell.com/content/dam/honeywell-edam/pmt/hps/products/pmc/modular-systems/experion-lcn/pmt-hps-control-system-migration-surviving-the-perfect-storm-wp-16-05-eng.pdf?download=false.

[9] Emerson's Flexible Approach to Control System Migration; ARC Advisory Group; White Paper; January'2008; https://www.emerson.com/documents/automation/emerson-s-flexible-approach-to-control-system-migration-en-55954.pdf.

[10] Network Migration Strategies; V. Nagpure; International Journal of Engineering Research & Technology (IJERT); Vol. 3 Issue 11,

November' 2014; https://www.ijert.org/research/network-migration-strategies-IJERTV3IS110835.pdf.

[11] Control System Migration Best Practices; ARC Advisory group; Dec' 2017.

# Further reading

OPC Commands; T. Hadlich (OPC Commands Working Group); OPC Connect.com; Internet document; www.opcconnect.com/commands intro.php#:~:text=In%20OPC%20terms%2C%20%22commands%22, with%20reading%20and%20 writing%20data.

Chapter 16

# Discussions on Ethernet networking

High speed, consistent performance, reliability, security, determinism in Time-Sensitive Networking (TSN), standards compliance, and low cost with simplicity are major areas of focus in Ethernet technology, which continues to evolve to meet new industrial application requirements. Ethernet is a family of networking technologies used in Local Area Networks (LANs) and is defined under the IEEE 802.2 and IEEE 802.3 standards. Some Ethernet standards use copper cabling, whereas others use fiber-optic media (IEEE 802.3). The Ethernet family of protocols operates at the physical and data-link layers of the Open Systems Interconnection (OSI) model and is defined by the IEEE 802.2 and IEEE 802.3 standards. While the Logical Link Control (LLC) sublayer is implemented in software and is hardware independent, the Media Access Control (MAC) sublayer is implemented in hardware. IEEE 802.11 is the standard for wireless networking. Ethernet is the most widely used LAN technology and enables computers, switches, and other devices to transmit and receive data reliably at high speeds. Transmission Control Protocol/Internet Protocol (TCP/IP) is a suite of communication protocols used to interconnect network devices over the Internet and other networks. TCP/IP and Ethernet are complementary technologies for computer communication. TCP/IP provides the framework for transmitting data across networks, including the Internet, whereas Ethernet provides the communication protocol for LANs. Ethernet primarily defines the lower two layers of the OSI model: the physical layer and the data-link layer (Fig. III 16.1b). IP is the third Network layer and TCP is the fourth Transport layer. EtherNet/IP, PROFINET, and similar technologies are examples of Industrial Ethernet. Although PROFINET uses Ethernet as its underlying network technology and TCP/IP for noncritical functions such as parameterization, configuration, and diagnostics, it does not rely on TCP/IP for real-time process-data transmission. Similarly, EtherCAT does not use TCP/IP by default. Ethernet itself has evolved into many variants based on factors such as speed, transmission medium, topology, and specialized features such as Power over Ethernet (PoE). Speed-based variants include Fast Ethernet (100 Mbps), Gigabit Ethernet (1 Gbps), and 10-Gigabit Ethernet (10 Gbps), with even higher speeds available for large data centers and cloud infrastructures. Fast Ethernet, operating at 100 Mbps, is well suited for small office and home networks and typically uses twisted-pair or fiber-optic cabling, although its bandwidth may be insufficient for some applications. Gigabit Ethernet, also based primarily on twisted-pair and fiber-optic cabling, provides data-transfer rates of 1 Gbps and is increasingly popular. Cat 5e cable is commonly used to support Gigabit Ethernet. Ten-Gigabit Ethernet is widely used in enterprise networks, cloud-computing environments, and data centers, utilizing Cat 6a, Cat 7, or fiber-optic cabling over distances up to 10,000 m. Forty-Gigabit and 100-Gigabit Ethernet are also available for large-scale data centers and cloud infrastructures and typically employ OM3 or OM4 fiber-optic cabling.

Ethernet implementations also vary according to network architecture and connection media. Switch-based Ethernet configurations employ switches or hubs to interconnect devices. Ethernet ports, Ethernet network adapters, and Ethernet cables are common Ethernet connection elements. A network switch primarily transfers data between devices on the same network. Wired Ethernet, which uses physical cabling, remains the most common form of LAN and WAN communication. Wireless Ethernet is another form of Ethernet-based communication that relies on high-frequency radio signals rather than physical cables. Suitable modems may be used with both wired and wireless Ethernet systems. Coaxial cables, capable of carrying high-frequency signals with low loss, are used in Ethernet standards such as 10BASE2 and 10BASE5; examples include RG-6 and triaxial cables. Twisted-pair cables, including Cat 5, Cat 5e, Cat 6, Cat 6a, and Cat 7, are commonly used for Ethernet applications ranging from 10 Mbps to 40 Gbps. Single-mode and multimode fiber-optic cables support much greater transmission distances and are used in Ethernet variants such as 10BASE-F, 100BASE-FX, 100BASE-BX, 100BASE-SX, 1000BASE-FX, 1000BASE-SX, and 1000BASE-BX.

Ethernet networks can be configured in different topologies, including bus, star, ring, and mesh. There are also functional variations of Ethernet, such as Power over Ethernet (PoE), which enables both data and electrical power to be transmitted through a single Ethernet cable. Another emerging technology is Single-Pair Ethernet (SPE), which is gaining importance in industrial Machine-

Plant Intelligent Automation and Digital Transformation. https://doi.org/10.1016/B978-0-323-89892-8.00008-9

to-Machine (M2M), Internet of Things (IoT), and Industrial Internet of Things (IIoT) applications. Industrial Ethernet technologies based on serial communication interfaces include PROFINET, POWERLINK, and Modbus TCP. Successful implementation of Industrial Ethernet requires careful planning, proper network design, and the use of high-quality components. Planning involves assessing the number of users and connected devices and collecting appropriate application requirements. Ethernet speeds and standards should be selected based on current needs while allowing for future expansion to ensure a sufficient infrastructure life cycle. Selection of suitable transmission media and high-quality industrial-grade cables is essential for durability and performance. Optimization of cable infrastructure is also important, with emphasis placed on application-specific cabling, reliable redundancy to ensure high availability and eliminate single points of failure, network security through proper switchport configuration that restricts access to authorized devices, and effective ongoing network management. Common network issues that must be addressed during design include connectivity, performance, configuration, and protocol-related problems. Detailed discussions of Industrial Ethernet and Ethernet applications in the field are presented in this section. Let the discussion begin with Ethernet technology.

## 16.0 Discussions on Ethernet technology

Ethernet covers a specification of *Physical* and *Data link* layers in the OSI model. It is not a complete network protocol but a subnet on which other protocols such as the *TCP/IP* suite can work. Functionally, Ethernet is responsible for operations at both the physical and data-link layers of the OSI model. At the physical layer, Ethernet supports the exchange of serial bit streams over the transmission medium and provides collision detection (CD) or, in wireless systems, collision avoidance (CA). Physical-layer functions include bit-by-bit transmission, encoding and decoding, signal transmission, modulation and demodulation, transmission-mode management, and data control. At the data-link layer, Ethernet incorporates both the Media Access Control (MAC) sublayer and the Logical Link Control (LLC) sublayer. The MAC sublayer provides access-control mechanisms to the network, including Carrier Sense Multiple Access with Collision Detection (CSMA/CD), and is responsible for frame transmission. The LLC sublayer provides data reliability and makes data available to higher protocol layers. Major data-link-layer functions include framing, error detection using parity checks, checksums, and Cyclic Redundancy Check (CRC), error correction, flow control, and addressing. As discussed earlier, Ethernet has evolved into several variants based on transmission speed and transmission media. These variations, together with their basic characteristics, are described later in the main text.

- Thick Ethernet (10BASE5)
- Thick Ethernet (10BASE2)
- Broadband Ethernet (10Broad36)
- Ethernet over twisted pair (10BASE-T)
- Ethernet over Fiber (10BASE-F)
- Fast Ethernet (100BASE-T/100BASE-F)
- Gigabit Ethernet (1000BASE-T)
- Wireless Ethernet

Though Ethernet was originally designed around the concept of bus topology, there can be variations in topology. The three main types of Ethernet transmission media in LANs are: *coaxial* cables (coax is that only half duplex communication can be applied); *twisted*-pair cables; and *fiber-optic* cables. Apart from these, IEEE802.11 describes wireless Ethernet connections. Triaxial and RG-6 are examples of coaxial cables that use various connector types, including BNC and TNC connectors. Twisted-pair cables are generally classified into two categories: Shielded Twisted Pair (STP) cables, which are protected by a copper braid or foil shield with a drain wire, and Unshielded Twisted Pair (UTP) cables, which do not include shielding. UTP cables, such as Category 5 (Cat 5) cables, support data-transmission rates of up to 100 Mbps. The cable consists of eight wires, arranged in four pairs. Cat 5e and Cat 6 cables (up to 100 m) are commonly used for Gigabit Ethernet applications. Similarly, Cat 6a and Cat 7 cables can be used for 10-Gbps Ethernet, whereas Cat 8 cables support 25-Gbps or 40-Gbps Ethernet over distances of up to 30 m. In IEEE 802.11 wireless LANs, radio communication takes place in the 2.4-GHz Industrial, Scientific, and Medical (ISM) band and/or the 5-GHz band. Several standards have been defined within the IEEE 802.11 family. The frequency bands and corresponding data rates for some common standards are as follows: IEEE 802.11a (5 GHz, 54 Mbps), IEEE 802.11b (2.4 GHz, 11 Mbps), IEEE 802.11g (2.4 GHz, 54 Mbps), IEEE 802.11h (5 GHz, 54 Mbps), and IEEE 802.11n (2.4 GHz and/or 5 GHz, up to 600 Mbps). This section presents a detailed discussion of Ethernet technology.

### 16.0.0 Ethernet technology at a glance

In 1970 the Xerox company developed Ethernet technology. In Ethernet packet switching is used to send messages. Packet switching is mostly applied for computer-to-computer communication. On a shared transmission medium in a LAN, every station must have a unique address. Accordingly, each device connected to the network is assigned a unique Ethernet, or physical, address known as a Media Access Control (MAC) address. In IEEE

802.3–based Ethernet networks, the Carrier Sense Multiple Access with Collision Detection (CSMA/CD) protocol is used for access control in wired transmission systems. With CSMA/CD, two or more stations can use a common transmission medium. The CSMA/CD technology cannot be applied to wireless Ethernet. As mentioned before, in wireless Ethernet CSMA/CA is used. The Distributed Coordination Function (DCF) is a MAC protocol used in the IEEE 802.11 standard for Wireless Local Area Networks (WLANs). It is based on a contention-access mechanism in which stations compete for access to the transmission medium. The Point Coordination Function (PCF), on the other hand, is a polling-based access scheme that operates without contention. In PCF, the access point controls medium access by polling stations and divides channel-access time into periodic beacon intervals. This approach provides more deterministic access to the wireless medium and can support time-sensitive communication requirements. These are detailed in Fig. III 16.0a.

DCF is fundamental in Wi-Fi. Here, the DCF Interframe Space (DIFS), PCF Interframe Space (PIFS), and Short Interframe Space (SIFS) are described in detail. As discussed earlier, Ethernet has evolved into a number of categories and variants based on speed, transmission medium, and application requirements. Common Ethernet categories are summarized in Table III 16.0a.

The Protocol Data Unit (PDU) at the data-link layer is the frame, and medium access control is achieved through CSMA/CD, which manages data collisions on shared media. Manchester encoding is commonly used in Ethernet communication. In addition to high-speed operation, Ethernet offers advantages such as reliability, cost-effectiveness, interoperability, security, scalability, and broad compatibility with protocols such as TCP/IP and FTP. The Ethernet protocol, Ethernet ports (supporting RJ-45 connectors), Ethernet adapters, and Ethernet cables are the major components required for Ethernet network operation. This subsection discusses the fundamentals of Ethernet technology.

1. **Data link-MAC:** In the network system, the data are divided into small blocks/packets for sending over network separately. There are two main reasons for using packets in data communication. First, in the event of transmission errors, it is easier to determine upon arrival which packets have been received correctly and which have been corrupted. Second, in a shared network environment, packet-based communication ensures that all computers have fair access to the communication medium. Thus, no single computer can occupy a shared resource for longer than the time required to transmit one packet. The Data-Link Layer (DLL) is primarily responsible for node-to-node delivery in an error-free manner. It also performs functions such as encoding and decoding of incoming and outgoing data. According to the IEEE 802 standards, the DLL is divided into two sublayers: the Logical Link Control (LLC) sublayer and the Media Access Control (MAC) sublayer. One of the major functions of the DLL is MAC addressing,

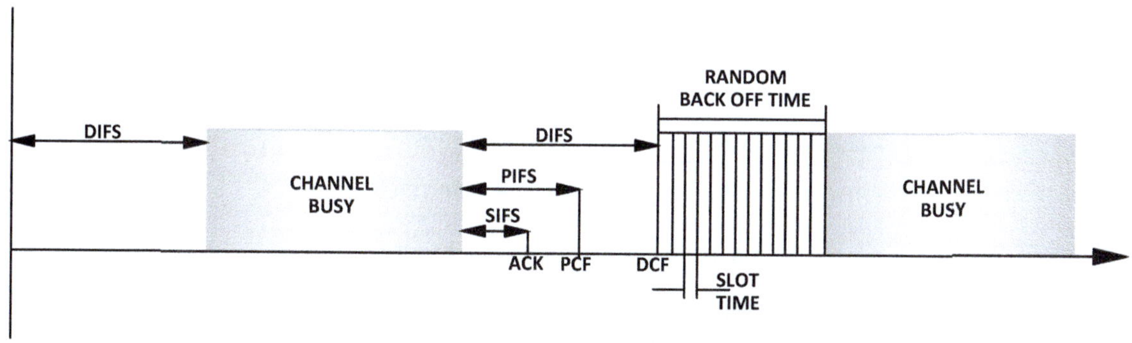


**FIGURE III 16.0A** Ethernet CSMA CA access details.

**TABLE III 16.0A** Common Ethernet categories.

| Ethernets | Speed | Media | Other features |
|---|---|---|---|
| Fast | 100 Mbps | Cat 5. Fiber optics (FO) | Variants: 100BASE-TX, 100BASE- FX, 100BASE-T4 |
| Gigabit | 1000 Mbps | Cat 5e, Cat 6, FO | Mid office/home application |
| 10Gigabit | 10,000 Mbps | Cat 6a, Cat 7, FO | Supports long distances (<10 km FO)<br>Application: Data centers and enterprise backbones |

which provides a unique physical address for each network device and operates at the data-link layer. A MAC address uniquely identifies network interfaces. MAC addresses are unique 48-bit hardware numbers of a computer that are embedded into a network interface card (NIC). Fig. III 16.0b shows MAC address formats and types.

A MAC address is a 12-digit hexadecimal number (48-bit binary number) that is typically represented using colon-separated hexadecimal notation. The first six hexadecimal digits of the MAC address identify the manufacturer and are known as the Organizationally Unique Identifier (OUI). The remaining six hexadecimal digits uniquely identify the specific network interface assigned by the manufacturer.

This is detailed in Fig. III 16.0b1. The three types of MAC addresses (ref: III 13.2.0.1) shown in Fig. III 16.0b2 are:

- **Unicast:** A unicast-addressed frame is sent out to the interface leading to a specific NIC and the least significant bit (LSB) of the first octet of an address is set to zero.
- **Multicast:** The multicast address allows the source to send a frame to *a group of devices*. In multicast addresses, the LSB of the first octet of an address is set to one. With Group destination addresses.
- **Broadcast:** Broadcast is also possible on DLL. Ethernet frames with ones in all bits of the destination address (FF-FF-FF-FF-FF-FF) are referred to as the broadcast addresses.

There are both MAC and IP addresses for specific purposes like:

- **MAC address function:** The primary function of MAC addresses is to manage how data is transported from one network node to another on a direct, physical basis.

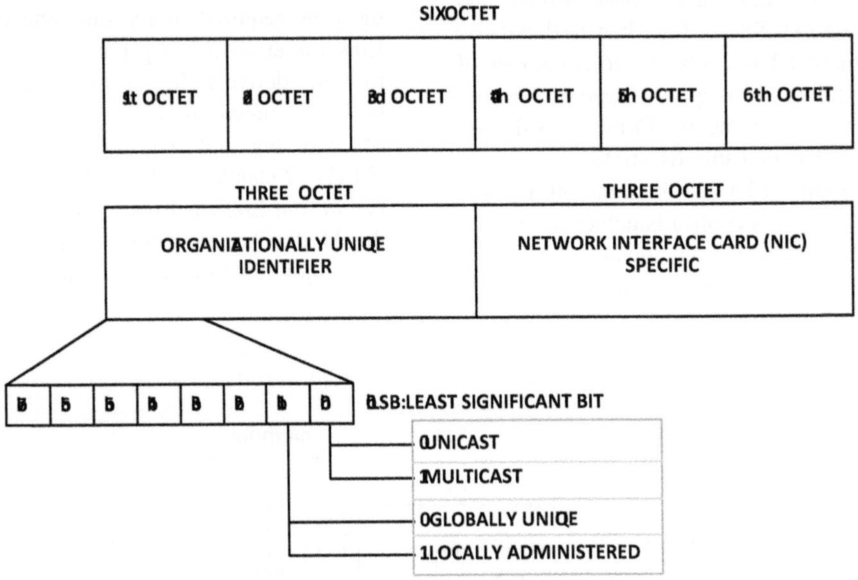


FIG III 16.0b1 MAC ADDRESS FORMAT

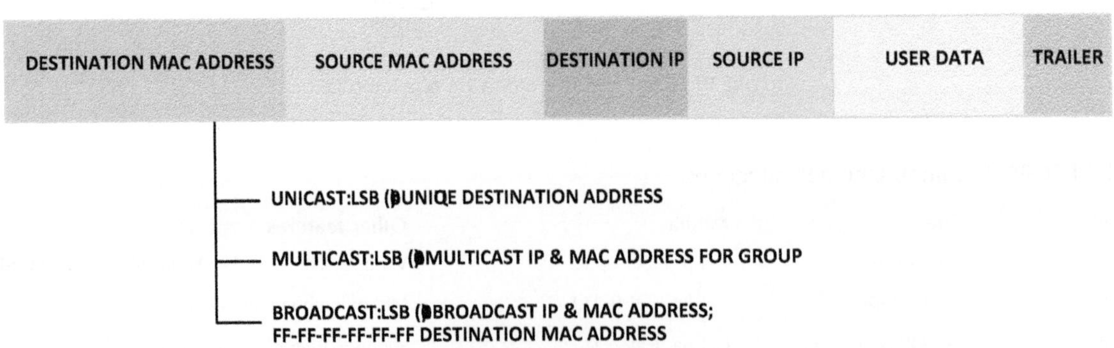


FIG III 16.0b2 MAC ADDRESS TYPE

**FIGURE III 16.0B** MAC address format and types.

- **IP address functions**: IP addresses are used to identify devices on a network for routing traffic between networks. IP addresses ensure that data transmitted from its original source reaches its intended final destination, thereby providing end-to-end delivery of data across interconnected networks.

2. **Ethernet Frame**: The basic frame format is required for all MAC implementation and is defined in the IEEE 802.3 standard. The Ethernet frame consists of several fields, including the preamble and Start Frame Delimiter (SFD), which operate at the physical layer. These are followed by the Ethernet header containing the destination and source MAC addresses. The header is followed by the payload field, which carries the actual data being transmitted. Additional fields, such as the Frame Check Sequence (FCS), are used for error detection.

   There are two frame types: data frame and tagged frame. The IEEE 802.3 standard data frame has payload >46 bytes but <1500 bytes. *Jumbo frames* are frames with a payload >1500 bytes. Jumbo frames can increase network throughput by reducing the overhead associated with transmitting a large number of small frames. Fig. III 13.2d shows the Ethernet data frame. Ethernet tagged-frame structure and the CSMA/CD flow process (discussed below) are illustrated in Fig. III 16.0c based on idea from Ref. [1].

   - **Data frame:** The Ethernet data frame, including the PAD field used to add extra bytes when necessary to ensure a minimum payload of 46 data bytes, together with the fixed protocol overhead, is illustrated in Fig. III 13.2d. The minimum payload size is required to satisfy the Ethernet slot-time requirement used in collision detection. The various fields of the Ethernet data frame are described in Section III 13.2.1.3 (Frame Details).
   - **Tagged frame:** Fig. III 16.0c1 shows the tagged frame with four extra bytes. A tagged frame is an Ethernet frame that has a virtual local area network (VLAN) added to identify the specific VLAN it belongs to. With tagging allows the segmentation for a single physical network into multiple logical networks is possible for better traffic management, security, and scalability. The IEEE 802.1Q standard defines four additional bytes in the Ethernet frame, organized into two extra fields, to support advanced networking applications such as Virtual Local Area Networks (VLANs) and traffic prioritization. The details of these fields are as follows:
     - *TYPE(TAG)*: These two bytes (value 8100 h) TYPE identifies as a tagged frame contains an extra information field
     - *VLAN TPID*: 2 bytes: VLAN Tag Protocol Identifier with three bits user priority (detailed in IEEE802.1p), Canonical format indicator (CFI) in line with IEEE802.1Q and VLAN ID: Identification of the VLAN, (4094 possibilities).

3. **Collision:** Further to the discussions in III 13.2.1, here also CSMA shall be detailed out also.
   - **CSMA CD**: Before sending a data frame, a station must wait for an idle period, that is, a period during which no other participant is transmitting data. If two or more stations transmit simultaneously, a collision occurs and is detected by the participating stations. The station that first detects the collision transmits an error (jam) signal, as discussed in Section III 13.2.1. The CSMA CD flow is shown in Fig. III 16.0c2. Normal collisions occur during the first 512-bit of frame transmission. If a collision occurs *after* 512-bit times, then it is considered an error and called a *late collision.* A collision is a serious transmission error that indicates a problem in network access and causes the frame being transmitted to be discarded. According to the CSMA/CD protocol, a collision domain is a network segment or multisegment configuration in which a collision occurs when two or more participants transmit data frames simultaneously. Before transmitting, a participant first senses the carrier to determine whether another station is already sending data. If an active carrier is detected, transmission is deferred. If no active carrier is detected for a period equal to or greater than the interframe gap, the station may begin transmitting its message. During transmission, the participant continues to monitor the medium for collisions. Therefore, a network interface must be capable of transmitting data and monitoring the medium simultaneously. A few important considerations are as follows:
     - *Interframe gap*: Ethernet participants must observe a minimum period of inactivity, known as the idle period or interframe gap, between the transmission of two consecutive frames. The minimum interframe gap is 9.6 μs for 10-Mbps Ethernet, 960 ns for 100-Mbps Ethernet, and 96 ns for Gigabit Ethernet. This interval allows network devices sufficient time to process received frames and prepare for subsequent transmissions.
     - *Slot time*: This parameter is defined as 512-bit times for the 10 Mbps and the 100 Mbps versions, and 4096-bit times for gigabit Ethernet. The slot time is an important Gigabit Ethernet parameter because it determines the minimum length of a data frame and the maximum length of a collision domain, thereby helping to avoid late collisions and ensure reliable collision detection.

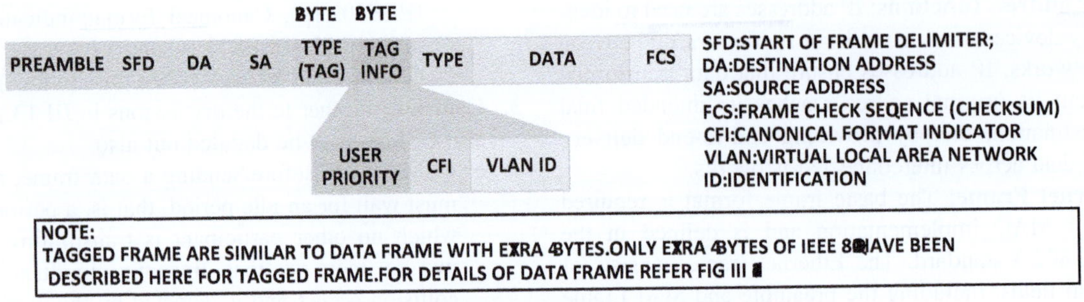


FIG III 16.0c1 ETHERNET:TAGGED FRAME DETAILS

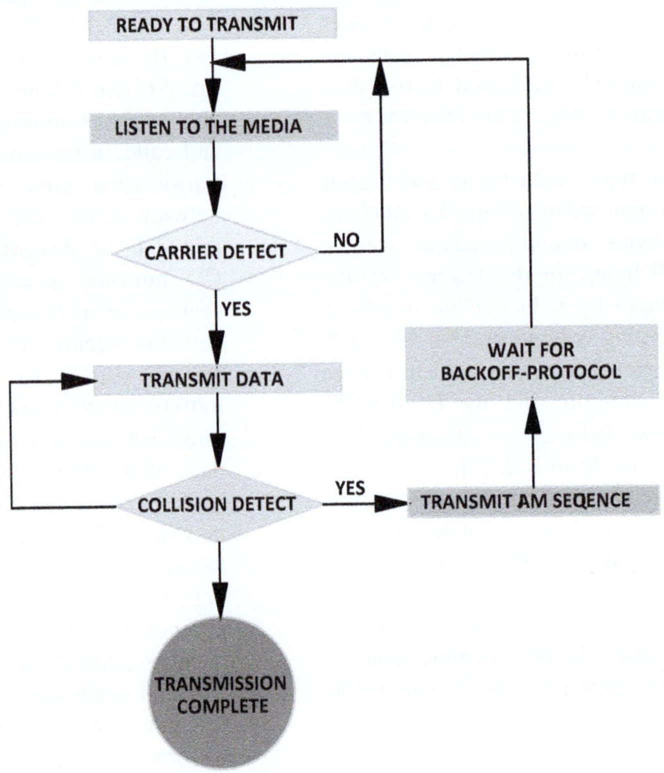


FIG III 16.0c2 ETHERNET: CSMA/CD FLOW

DEVELOPED BASED ON IDEA FROM [COURTESY:PHOENIXCONTACT]

**FIGURE III 16.0C** Ethernet: tagged frame and CSMA/CD flow.

- **CSMA CA**: This has already been discussed in Section III 13.2.1.3. Fig. III 16.0a shows some important parameters about waiting times for access to the medium. Slot time-related parameters are:
  - *SIFS*: This is the shortest waiting time for medium access and therefore has the highest priority. The access point uses this waiting time to send ACK messages.
  - *PIFS*: Medium-priority waiting time for medium access. This interval is used for the polling actions of an access point.
  - *DIFS*: Lowest-priority waiting time for medium access. It is used by normal participants on the wireless segment.

4. **Structural elements:** Ethernet repeaters, hubs, switches, etc., are major structural elements and are shown in Fig. III 16.0d.

   In a generalized way these have been discussed in Section III 12.1.1. Here, these will be discussed with special reference to Ethernet.

   - **Repeaters** and **Hubs:** The maximum segment length of a LAN is determined by the transmission

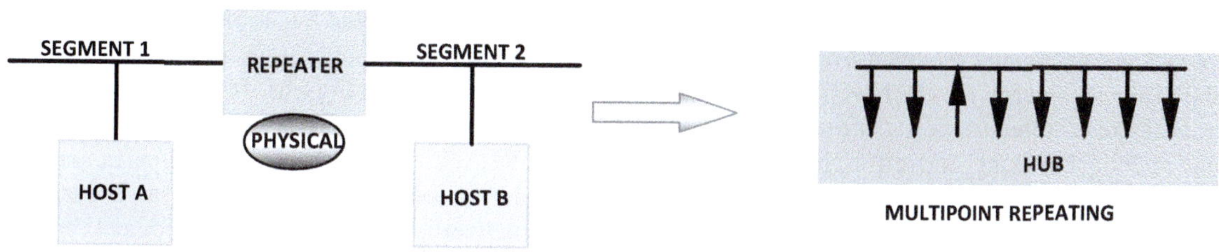


FIG III 16.0d1 REPEATER & HUB

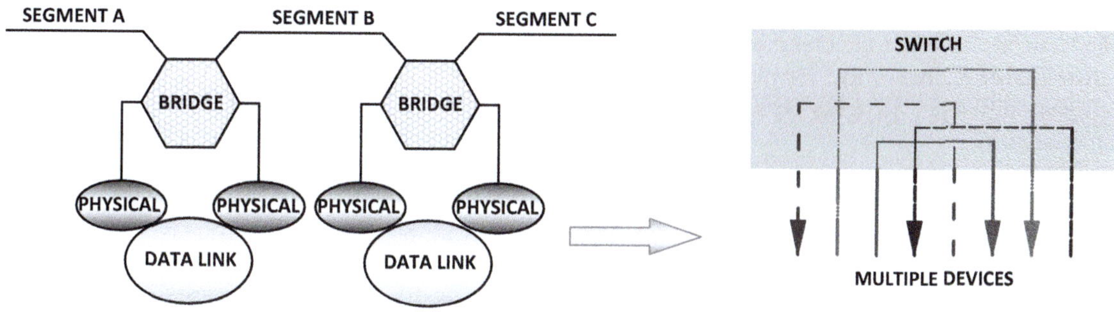


FIG III 16.0d2 BRIDGE & SWITCH

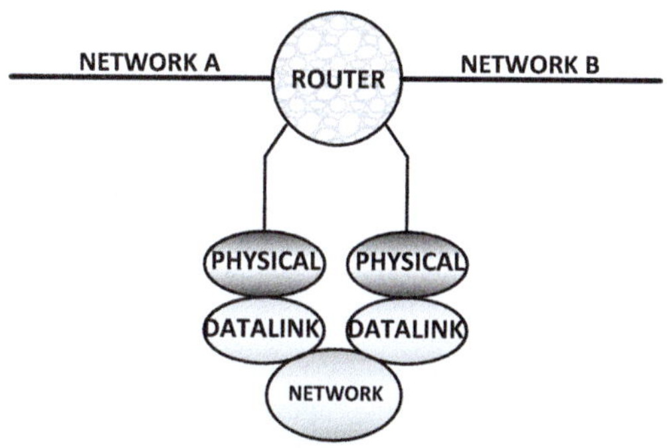


FIG III 16.0d3 ROUTER

**FIGURE III 16.0D** Ethernet structural elements connection details.

medium and the access mechanism used. As discussed in Section III 12.1.1.4, a repeater is a signal-regeneration device that transparently retransmits packets independent of their content. Repeaters are used to connect two or more Ethernet segments. The connected segments may use different transmission media, and the repeater forwards not only data signals but also collision and error signals. A multiport repeater, commonly known as a hub, regenerates incoming signals and forwards them to all other ports. All segments interconnected through a hub belong to the same collision domain. An important feature of modern hubs is support for network management, including the ability to disable individual ports and detect network failures. Refer also to Section III 12.1.1.5.

- **Bridges** and **Switches**: A bridge is more than a simple signal-regeneration device such as a repeater. A bridge divides a single network into multiple network segments; this process is known as network bridging. When a data packet is to be transmitted from one segment to another, the bridge examines the destination MAC address and determines whether the packet should be forwarded to the other segment. Both bridges and switches are intelligent networking devices, and a switch can be considered a multiport bridge. In a switch, a MAC address table is maintained separately for each port. This table is populated by monitoring network traffic on the connected segments and recording the MAC addresses observed on each segment. Each address is retained for a limited period and is removed when a predefined hold time expires. As a result, inactive stations are eventually removed from the table. Unlike bridges, which typically have only two or four ports, switches often support a large number of ports and include buffering and error-checking capabilities. Bridges generally do not provide extensive buffering or error-checking functions.

- **Routers**: A router is a device that connects two or more distinct computer networks with each other. As shown in Fig. III 16.0d, a router acts as a packet-switching device that operates at Layer 3 (the Network Layer) of the OSI model. It examines network-layer addresses and forwards data packets between different networks based on routing information.

  Various router types can be distinguished by their physical form, the router connections they support, and the additional functions built into the router (see Section III 12.1.1.4 and Table III 12).
- **New Ethernet switch:** The introduction of Ethernet switching technology, particularly in high-speed Ethernet networks, eliminated the use of shared media by employing MAC-address-based forwarding tables for each port. A MAC address is a unique identifier permanently assigned to every network-enabled device. As the switch learns the locations of connected devices, it builds source and destination address tables that map each device to a specific port. This creates a point-to-point communication environment. Switching technology greatly improves network efficiency by reducing congestion and minimizing packet loss caused by collisions. It also enables full-duplex communication in Ethernet networks.

5. **Redundancy and VLAN:** Let us conclude the discussions on Ethernet with:
   - **Ethernet redundancy:** Network redundancy refers to the integration of hardware and software to ensure optimum network availability in the event of any single failure. The communication system is the core of every project. To handle network faults, various protocols can be incorporated into the network architecture. Common approaches include Spanning Tree Protocol (STP) and Rapid Spanning Tree Protocol (RSTP) (IEEE 802.1D) for mesh topologies, Media Redundancy Protocol (MRP) for ring topologies, Parallel Redundancy Protocol (PRP), Bridge Protocol Data Units (BPDUs), and Multiple Spanning Tree Protocol (MSTP) (IEEE 802.1s). Fault-Tolerant Ethernet (FTE) is an example of such an approach. FTE connects a group of nodes through multiple communication paths, enabling the network to tolerate all single faults and many multiple faults. FTE can rapidly detect faults and, in the event of a communication failure, achieve switchover times of approximately 1 s. This is detailed in Fig. III 16.0e, which is based on a document from Honeywell.
   - **VLAN:** A VLAN is a group of participants within a large network that forms a separate logical network, allowing multiple logical groups, each with its own broadcast domain, to be created on a single physical network. VLANs may be organized based on department, hierarchy, or usage. VLANs can be either static or dynamic. Static VLANs are easy to configure because they use fixed switch ports, whereas dynamic VLANs allow connection through any port but are more complex and costly to implement. Trunking is a method used to transmit data from multiple VLANs between switches, for example, using Inter-Switch Link (ISL) or IEEE 802.1Q. Major advantages of VLANs include simplified movement of devices within the network, enhanced security, and restriction of network traffic.

     Next, we discuss Ethernet TCP-IP.

## 16.0.1 Networking with TCP-IP

TCP is a hardware-independent transport layer protocol and IP is a network layer protocol meant mainly responsible for addressing and routing data packets in connectionless packet delivery service. Ethernet provides the physical and data link layers of the OSI model to form the foundation for the *TCP/IP protocol suite* for data communication over a network and the internet. An internet is not limited by size; there are internets with only a few networks, and there are also internets comprising hundreds of networks. TCP is one of the basic standards that defines the rules of the Internet and is one of the most commonly used protocols in digital network communications, ensuring end-to-end data delivery. TCP accomplishes difficult tasks by utilizing the unreliable packet service of IP to provide reliable data delivery to different application programs. The Internet protocol suite also includes a connectionless transport protocol, namely the User Datagram Protocol (UDP), which sends IP packets without establishing a connection. An internet consists of different networks connected to each other by routers, which are structural elements used to connect networks. Every router has a processor, a certain amount of memory, and a separate interface for each network to which it is connected. From a software perspective, a universal communication service must be active on every host. Although many software protocols are adapted for internetworking, only one suite is widely used. This is the TCP/IP suite [1]. The separate TCP and IP protocols work together to ensure that data are delivered to their intended destination within a

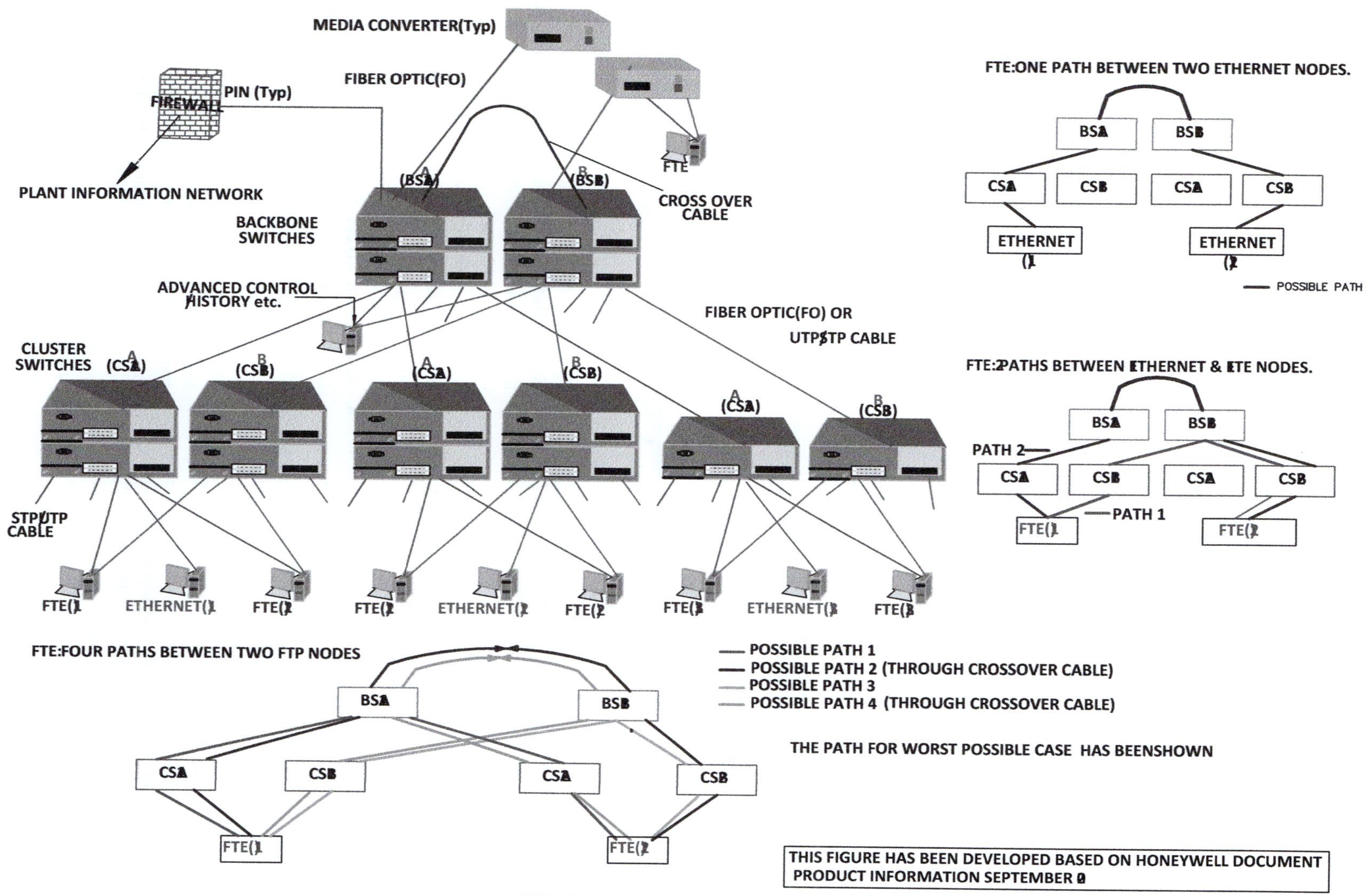


FIGURE III 16.0E Fault tolerant Ethernet.

network. IP obtains and defines the address (IP address) of the application or device to which the data must be sent. Then it is the responsibility of the TCP to transport and route data through the network ensuring it gets delivered to the destination application/device defined by IP in the most efficient way possible. TCP/IP divides communication tasks into layers, making the communication process standardized and independent of hardware and software vendors. Data packets must pass through four layers before being received by the destination device. At the destination, the reverse order is followed to reconstruct the message in its original format. TCP/IP uses a three-way handshake to establish a connection between a device and a server, ensuring that multiple TCP socket connections can be transferred concurrently in both directions. In contrast to connection-oriented TCP, UDP is connectionless so, it is unreliable compared to TCP which guarantees delivery. UDP often is referred to as "fire & forget." However, UDP is much faster than TCP. In time-sensitive applications, a connectionless alternative to TCP, namely UDP, is used when speed is prioritized over guaranteed delivery of every packet. This is common in applications such as streaming and real-time I/O data transmission in industrial automation. Every IP address is unique. IP addresses are divided into different classes (A, B, C, D, and E) for Host ID and Network ID identification. A distinction is drawn between public networks and private networks (corporate networks). A few other features of the IP protocol are as follows:

- This connectionless protocol can follow different routes to the same target host when transmitting different IP packets.
- A universal data packet is constructed, consisting of a header (including the addresses of the sender and destination) and a data field. IP also does not verify correct delivery.
- The IP header is at least 20 bytes long. When the options field is used, the header can be a maximum of 60 bytes long.

In this subsection discussions will be presented on TCP/IP details and this is further to the discussions in Section III 13.1.1.

1. **TCP services** [1]**:** The TCP model has four layers (OSI model): Application, Transport, Network/Internet, and Network Access. The TCP protocol ensures the reliable transmission of information across one or more networks. This is a *connection-oriented* protocol, meaning that a logical connection is established and used and then stopped again. TCP is an end-to-end protocol. HTTP, HTTPS, FTP, SMTP, DNS, and Telnet are common web services. TCP uses various techniques to ensure service reliability. Fig. III 16.0f shows TCP services and three-way handshaking. Fig. III 16.0f1 details the TCP end-to-end service.
2. **TCP reliability:** TCP/IP reliability refers to the assurance that data are delivered completely, without errors, and in the correct order. Of the several methods used to achieve this, the following are important.
   - **Datagram:** After receiving the data, an acknowledgment is sent back. Upon sending the data, a timer is started. If an acknowledgment is not received within a stipulated time, the data are retransmitted.
   - **Windows:** After a connection is established, both ends reserve storage (buffers, referred to as windows) for incoming and outgoing messages and share the buffer size with the other end. The receiver sends a window advertisement as an acknowledgment. When the receiving application can read the data as soon as they arrive, a positive window advertisement is sent for every message. If the sending end is faster, the incoming data will eventually fill the receiver's buffer. In that case, the receiver sends a zero-window advertisement until buffer space becomes available.
   - **Handshaking:** As detailed in Fig. III 16.0f2, TCP uses the term Synchronization Segment (SYN segment) for messages in the three-way handshake used to establish a connection, and the term FIN segment for messages in the three-way handshake used to close a connection.
3. **Protocol: Stack** and **encapsulation:** For Ethernet role, refer to Fig. III 16.0g.
   The layered approach in IP stack is detailed in Fig. III 16.0g1. Important issues are:
   - **Encapsulation:** The packet encapsulation describes how Ethernet works in the protocol stack. Each layer of the protocol stack is responsible for a particular level of functionality, with each higher layer utilizing the underlying layers independently. This layered approach is implemented through the use of encapsulation, as detailed in Fig. III 16.0g2.
   - **Application layer:** As discussed, the Application layer sits on top of the Transport layer in TCP/IP and contains applications like HTTP, FTP SMTP, etc.
   - **Transport layer:** The Transport layer shields the upper layers from network-dependent details, including transport-address-to-network-address translation, sequencing, error detection and recovery, and related functions. It may use either the connection-oriented TCP protocol or the connectionless UDP protocol.
   - **Network layer:** This layer determines message routing, including Quality of Service (QoS) functions. Common protocols include IP, Address

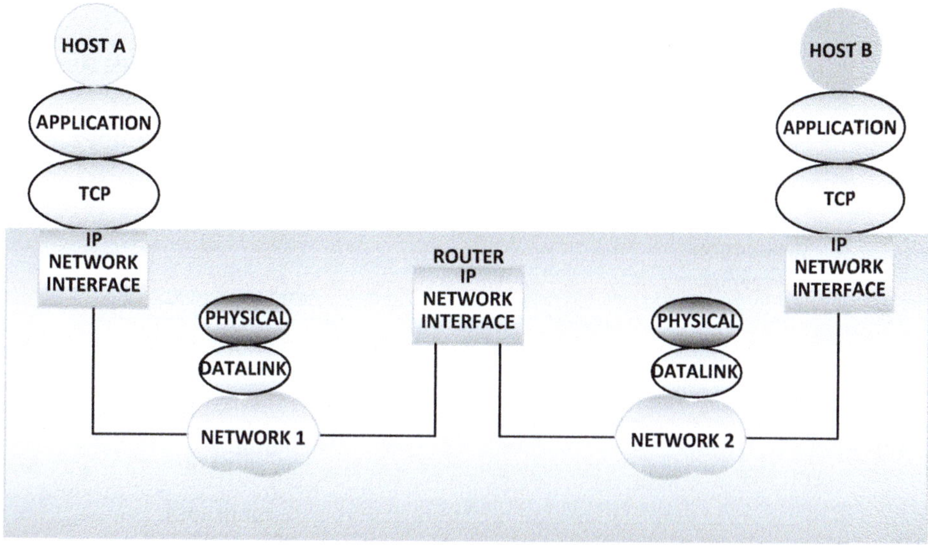


FIG III 16.0f1 TCP (TRANSPORT) SERVICE

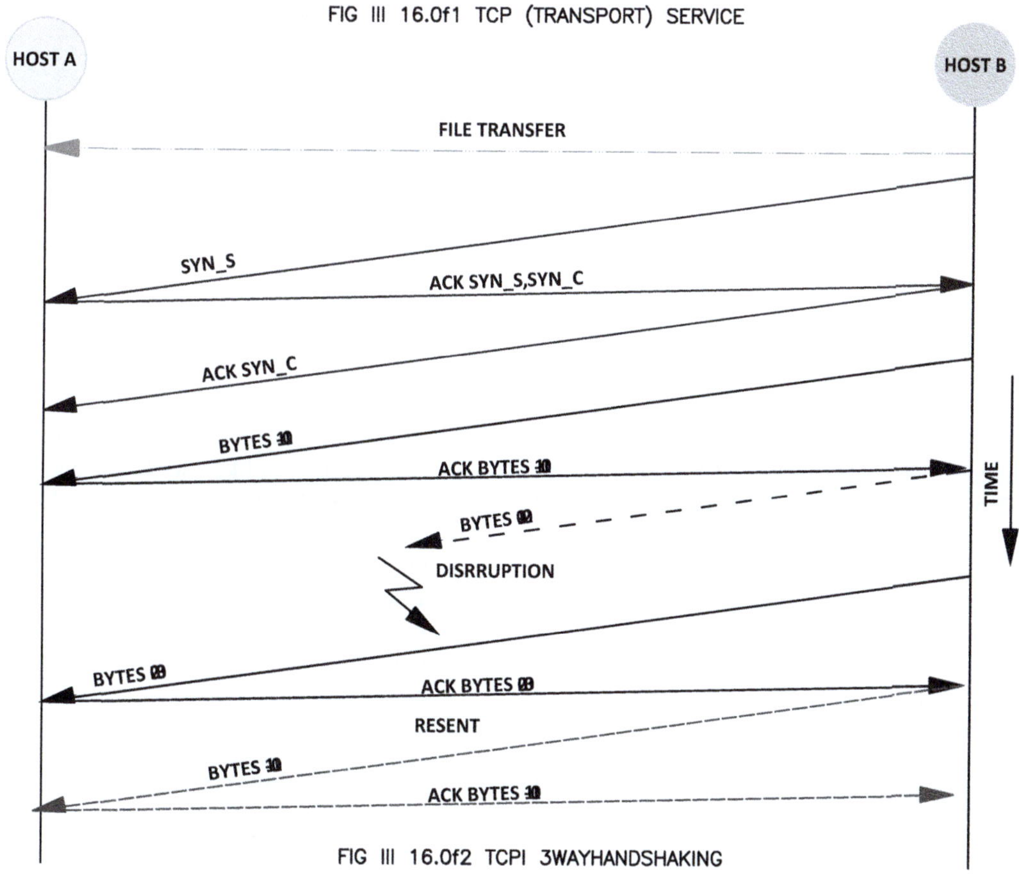


FIG III 16.0f2 TCPI 3WAYHANDSHAKING

**FIGURE III 16.0F** TCP services and three-way handshaking.

Resolution Protocol (ARP), Reverse Address Resolution Protocol (RARP), and others.

4. **Discussions on TCP and UDP:** Here, we offer brief details on TCP and UDP. Both TCP and UDP segments are detailed in Fig. III 116.0h
   - **TCP segment:** The data to be sent are transmitted by the Application layer to the Transport layer. The Transport layer packs the information in the data field and then adds a TCP header. The whole is then transmitted to the Internet layer for further processing. the header contains 20 bytes comprising the Source Port and Destination Port (for different upper-layer applications), Sequence Number (identifying the first data byte in a TCP segment), Header

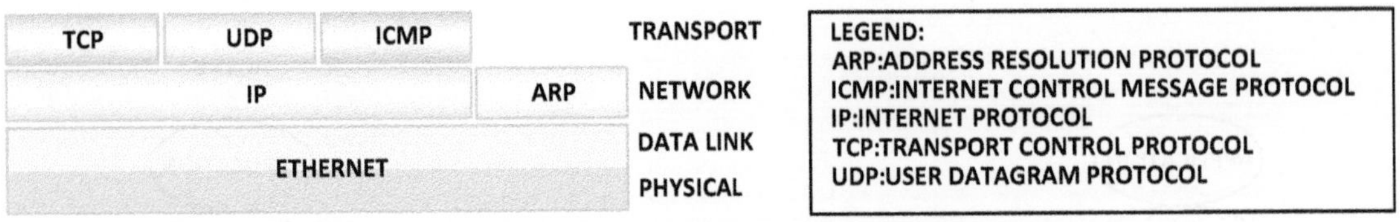


FIG III 16.0g1 LAYER APPROACH IP STACK

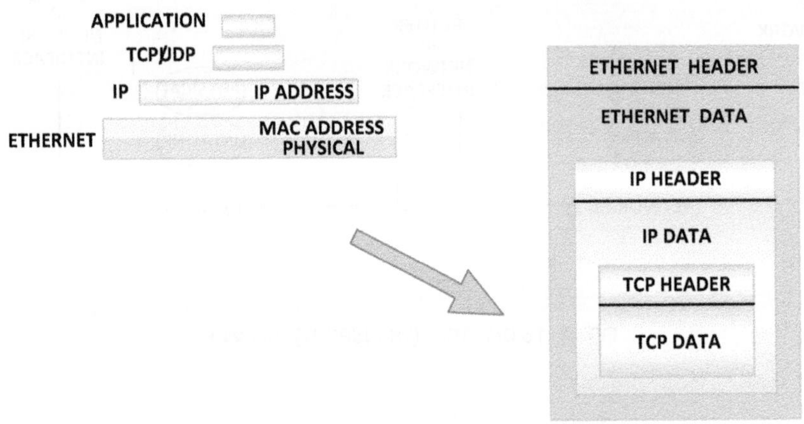


FIG III 16.0g2 DATA ENCAPSULATION

**FIGURE III 16.0G** IP stack and data encapsulation.

Length, Code Bits (for initialization and control), Window Size (maximum number of bytes), Checksum, and Urgent Pointer (for urgent information in TCP).

- **UDP segment:** Simpler than TCP, the UDP (connectionless) header is more or less a minimal protocol. The header comprises the Source Port, Destination Port, Length, and Checksum fields, each 2 bytes long, for a total header length of 8 bytes. Fig. III 16.0h2 shows the UDP segment.

5. **Miscellaneous issues:** Various topics, including IP packets, IPv6, and communication formats, are discussed below.
   - **IP packet:** The Internet layer packs the information in the data field and then adds an IP header. On the basis of IP packets, the whole is then transmitted to the Network layer for further processing. When a router receives an IP packet that is too large for a subnet, IPv4 divides the packet into smaller packets that fit within the data frames of the relevant subnet. At the final destination, IPv4 reassembles these packets into their original order. IP packets are detailed in Fig. III 16.0i.

     The following issues are involved in packet fragmentation:
     - Each packet has an IP header.
     - All divided messages have the original identification field. The "More Fragments" flag indicates that additional fragments follow and is therefore not set in the last fragment.
     - The *fragment offset field* specifies the position of this fragment in the total message.

       Fig. III 16.0i1 shows the IP packet with header which has been explained in a self-explanatory Fig. III 16.0i2.
   - **IPv6:** In the TCP/IP model, IPv4 and IPv6 are network-layer protocols that handle addressing and routing. Unlike the 32-bit addressing system used in IPv4, IPv6 uses a 128-bit addressing system to provide a much larger address space and improved support for security. IPv4 suffered from a shortage of available IP addresses. Therefore, IPv6 was introduced to provide a larger address space and to allow new functionalities to be integrated more easily with improved performance. A dual-stack approach was adopted for the introduction of the new system, enabling nodes to process both IPv4 and IPv6 datagrams. Currently, the automation industry has shown limited interest in the integration of IPv6 [1].
   - **Communication forms:** Communication over TCP (UDP)/IP uses various forms, including the client-server model, endpoints and Internet sockets, dynamic servers, unambiguous communication, status on a socket, connection-oriented communication, and connectionless communication.

   In the following sections we discuss more Ethernet networking technology.

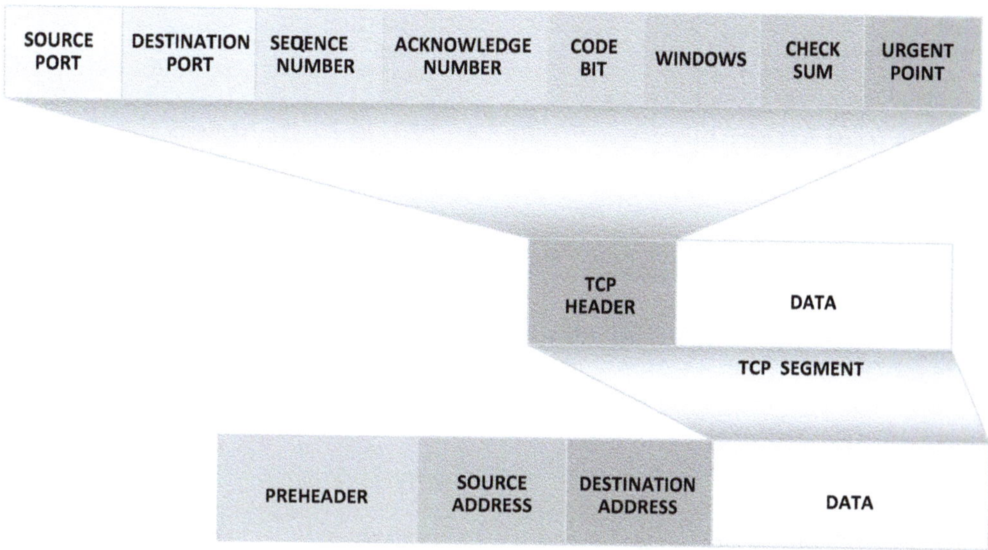


FIG III 16.0g1TCP SEGMENT

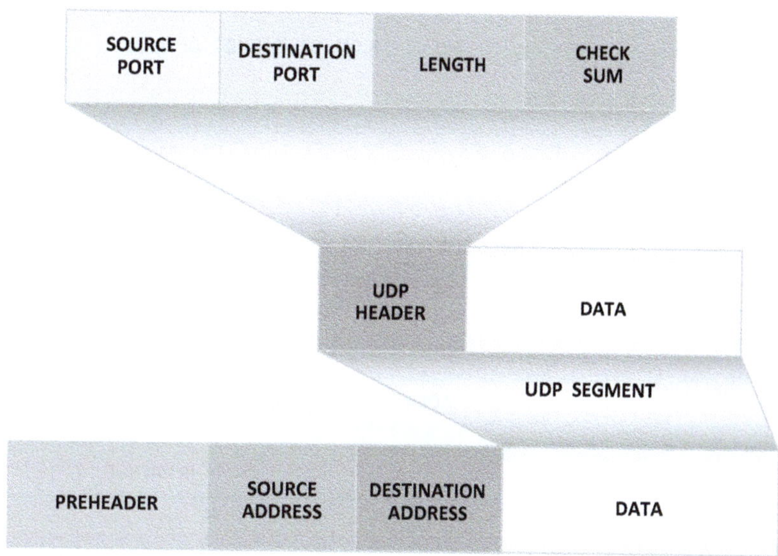


FIG III 16.0g2 UDP SEGMENT

**FIGURE III 16.0H** Segment details of TCP and UDP.

## 16.0.2 Miscellaneous Ethernet networking issues

From the discussion in Section III 13.1.1, we understand that the IP address is a virtual address processed by software. To transport an IP packet, the data must be encapsulated in a frame that can be delivered by the local hardware to the correct participant. Several network communication protocols operate at different layers of the OSI model and the practical TCP/IP model, making network communication easier. Operating between the Network Layer and the Data-Link Layer, ARP is one of the crucial protocols for smooth network communication. When the IP protocol sends a message over Ethernet, the MAC address of the destination must also be known in addition to the destination IP address. ARP maps a known IP address to its corresponding physical MAC address within a network. Data packets can be lost in IP communication, their delivery can be significantly delayed, or they may arrive out of order. The supporting protocol ICMP, which is part of the IP suite, is used by network devices to send error messages and operational information and primarily handles network diagnostics, error reporting, and control functions. Besides IP addresses, there are user-friendly host names that are easier to remember. However, host names do not contain information about the

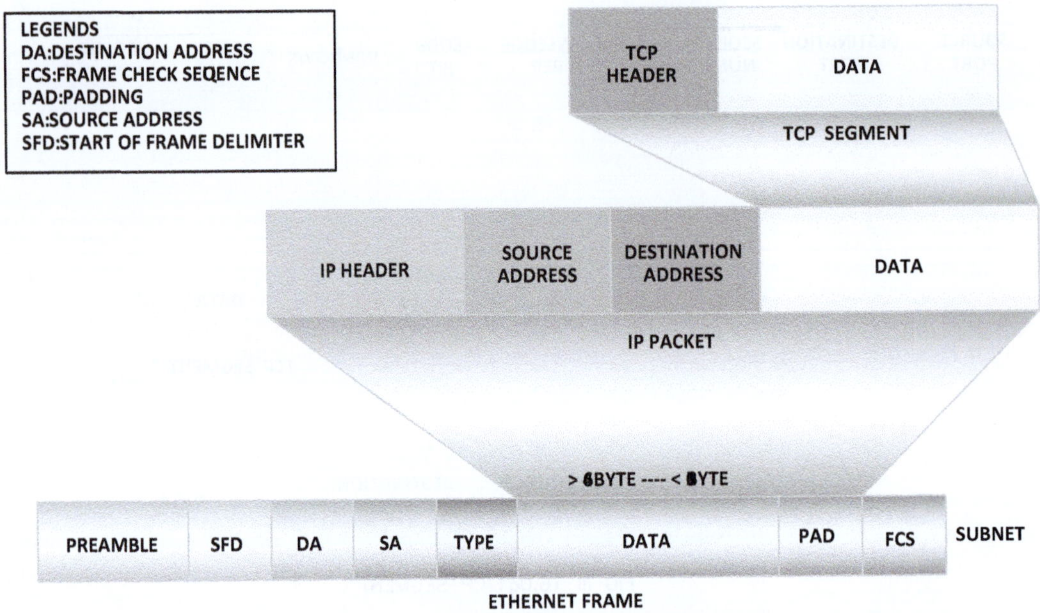


FIG III 16.0i1 IP PACKET DETAILS

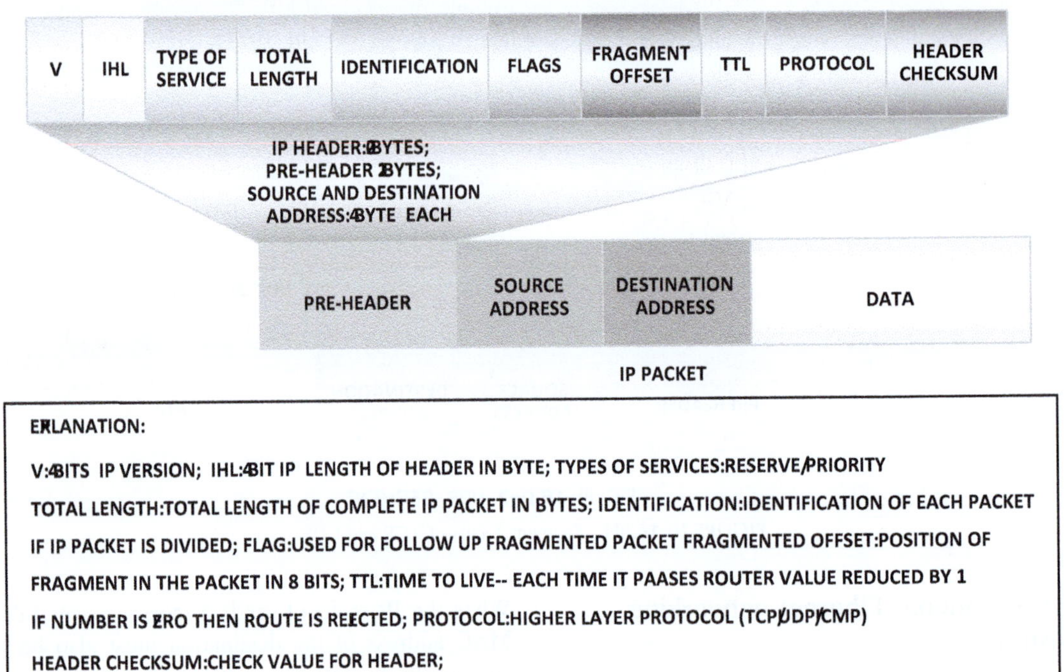


FIG III 16.0i2 IP PACKET HEADER DETAILS

**FIGURE III 16.01** IP packet and packet header.

location of a host within the Internet. Therefore, a link must be established between host names and IP addresses because TCP/IP protocols are based on IP addressing. The application-layer Bootstrap Protocol (BOOTP) is primarily used by devices to obtain their IP address and other network information during startup. BOOTP allows a device with only a MAC address to discover its logical (IP) address from a centralized server that maintains a static mapping between MAC and IP addresses. The Dynamic Host Configuration Protocol (DHCP) is an extension of BOOTP that automates the configuration of devices on IP networks. DHCP can automatically assign IP addresses, subnet masks, default gateways, and DNS server information to devices as they connect to a network. A Virtual Private Network (VPN) is a private communication channel that functions as an overlay on a public infrastructure, most commonly the Internet. Data transmitted through a VPN are secured to ensure integrity, authorization, and authentication. IPsec (IP Security) is a group of protocols used to secure communications between devices and to protect data transmitted over public networks. It is commonly used to establish VPNs and works by encrypting IP packets and authenticating their source. A firewall is a security barrier in an Ethernet network that protects a trusted internal network from threats originating on an untrusted external network, such as the Internet, based on predefined security rules. It monitors and regulates data packets before permitting transmission, thereby preventing unauthorized access and malicious activity. Firewalls may be hardware-based, software-based, or cloud-based and can provide advanced features such as VPN support. This subsection addresses various issues related to Ethernet and concludes the discussion of Ethernet technology.

1. **Common network protocols:** A few of the protocols listed below are important for network communication:
   - **ARP:** When communicating with a device on the same LAN, a host first checks its ARP cache. If the MAC address is missing, it broadcasts an ARP request asking for the MAC address corresponding to the target IP address. The device with the matching IP sends an *ARP response* with the corresponding MAC address which it is registered for
   - **Bootstrap (BOOTP):** This application-layer protocol can assign an IP address based on a MAC address. In the TCP/IP suite, it can combine a number of dynamic configuration steps into a single step. Upon recognizing a broadcast message, a Bootstrap server replies with all the required configuration information requested. It uses a client-server model and communicates through UDP. BOOTP has now been largely replaced by DHCP.
   - **DHCP:** As noted above, DHCP is used to automate configuration. DHCP assists a host on a new network by automatically and dynamically assigning IP addresses, subnet masks, gateways, DNS information, and other network parameters without manual intervention.
   - **ICMP:** TCP/IP includes ICMP, which is primarily intended for network diagnostics, error reporting, and control functions associated with the connectionless IP protocol. ICMP reports when a particular network feature is unavailable or when a host or router cannot be reached. Network diagnostics and error reporting (e.g., Source Quench, Time Exceeded, Redirect, and Unreachable messages) are among its primary applications.
2. **Domain name system (DNS):** The DNS architecture was established in 1983 and described in RFC 882 and RFC 883.When an application on the host of a user need to convert a host name this application will call the client component of the DNS with the host name, which then takes over and sends a request message to the network. All DNS request and reply messages are sent in UDP segments to port 53. After a short delay, the DNS client component receives a DNS reply message containing the requested reference on the user host. This reference is then transmitted to the application. DNS refers to
   - A distributed database in DNS servers as per hierarchy.
   - An Application layer protocol for hosts and DNS servers can communication during the translation session (from an IP address to a host name and vice versa).
3. **IP security (IPSec): VPN:** IPsec is the most widely used VPN protocol. A Virtual Private Network (VPN) is an encrypted connection between two or more computers over public networks, allowing data to be exchanged privately through encryption and enabling the transfer of confidential information. IPsec encrypts data exchanged between two communication partners to protect primarily against man-in-the-middle and spoofing attacks. It provides services such as integrity, authentication, and acknowledgment. Through password-based VPN logins to a VPN application, users can access an IPsec VPN, provided that the application is installed on the device. Only if the password is compromised and an attacker gains access to the VPN can encrypted data be intercepted or stolen. IPsec provides security at the Network Layer. It is optional in IPv4 but mandatory in IPv6. IPsec is built on protocols such as the Authentication Header (AH), Encapsulating Security Payload (ESP), IP Payload Compression Protocol (IPComp), and Internet Key Exchange (IKE). Two modes of IPsec operation are Tunnel Mode and Transport Mode. Three VPN implementation methods are security gateway-to-security gateway, host-to-security gateway, and host-to-host.
   The following steps are involved in the functioning of a VPN:
   - **Key exchange:** A string of random character-based keys is necessary for encryption.
   - **Packets:** All data are broken down into smaller pieces called packets, which contain both a payload and a header. IPsec adds several headers to the data packets containing authentication and encryption information.

- **Authentication:** IPsec provides authentication for each packet to ensure that packets are from a trusted source and not an attacker.
- **Encryption:** IPsec encrypts the payloads within each packet with its IP header.
- **Transmission:** Encrypted IPsec packets travel across one or more networks to their destination using a transport protocol.
- **Decryption:** At the other end of the communication, the packets are decrypted, and applications can now use the delivered data.

4. **Automation cell:** Generally, an automation cell refers to a set of computers, data servers, controllers, I/O devices, sensors, and actuators required to execute the different functionalities of an automation concept, that is, a self-contained automated workspace. An automation project may comprise production lines, Programmable Logic Controller (PLC) systems, Distributed Control Systems (DCSs), Emergency Shutdown (ESD) and safety controller systems, Supervisory Control and Data Acquisition (SCADA) systems, and related components. Switches are structural elements used to build a complete automation cell and form a self-contained unit [1].
5. **Firewall in Ethernet:** A firewall application regulates access to the data on the network. A firewall is designed to allow only traffic that is explicitly permitted to pass. This is not limited to Internet connections. Hardware and software firewalls are the two main types of firewalls. A hardware firewall is a separate device, such as a router with an integrated firewall. A software firewall is installed as a program on a computer. Two important issues are as follows:
   - **Packet filter:** Based on a set of rules, the firewall decides whether to allow or disallow an IP packet to pass. Such rules are defined using IP addresses, domain names, protocols, and port numbers. Packet-filtering firewalls use two main components of each data packet to determine its legitimacy: the header and the payload. Packet-filtering firewalls can be configured to manage both inbound and outbound traffic.
   - **Stateful inspection:** This is the latest type of firewall. Stateful Packet Inspection (SPI) means that each packet is examined in the context of the session established after sign-on and the subsequent handshaking between the communicating hosts. This method is used by firewalls to monitor and track the characteristics of network connections, making it possible to detect patterns indicative of potential threats. The process involves building and maintaining a state table that records every incoming and outgoing packet.

## 16.1 Discussions on Ethernet varieties

From Section II 18.0.2, readers are familiar with the concept of PoE. PoE was developed primarily for WLAN and Bluetooth access points, RFID units, touch panels, and similar devices. PoE technology describes a system for safely transferring electrical power, along with data, to remote devices over standard Ethernet data cables (Cat 3, Cat 5, Cat 5e, and Cat 6) 2. In accordance with IEEE 802.3af, PoE can deliver 15.4 W at the source, with approximately 12.95 W available at the powered device (PD), within a voltage range of 44–57 V and a current range of 10–350 mA. In PoE, the device supplying power is referred to as Power Sourcing Equipment (PSE). A PoE solution does not require an additional AC power cord at the product location. Therefore, it minimizes the amount of cabling required and eliminates the difficulties and costs associated with installing additional power outlets.

Wireless Ethernet, known as Wi-Fi, is popular because of its mobility, speed, and performance. Major features of wireless Ethernet include wireless connectivity, speed and performance, mobility and flexibility, scalability, security, ease of setup and configuration, and wide compatibility. Advances in modern technology and increasing network speeds, together with the need for more flexible networking to support IIoT applications, are driving innovations in industrial wireless communication. Industrial wireless technology offers enabling solutions that have the potential to be game changers for modern manufacturing [3].

Fast Ethernet (IEEE 802.3u), Gigabit Ethernet (IEEE 802.3z/ab-1999), and 10 Gigabit Ethernet (IEEE 802.3ae-2002) represent the evolution of wired networking speeds under different IEEE standards, as indicated in parentheses. Gigabit Ethernet was originally designed as a switched technology and used fiber-optic media for uplinks and interbuilding connections. The Gigabit Ethernet specification defines three transmission schemes for different transmission media; therefore, some infrastructure changes are also required. Fast Ethernet uses Cat 5 or higher twisted-pair copper cabling for distances up to 100 m. Similarly, Gigabit Ethernet uses Cat 5e/Cat 6 copper cabling for distances up to 100 m, whereas 10 Gigabit Ethernet uses Cat 6a/Cat 7 copper cabling for distances up to 100 m. In all cases, however, fiber-optic (FO) cabling may also be used.

Compared with Fast Ethernet, Gigabit Ethernet is more complex to configure but offers lower latency and greater coverage. Compared with traditional Ethernet, 10 GbE provides a more efficient and less expensive approach to moving data over backbone connections between networks. Ten-Gigabit Ethernet operates using full-duplex point-to-point links. Industrial Ethernet (IE) is an environmentally enhanced version of the IEEE 802.3 Ethernet standard. It is also deterministic in nature and is designed to withstand harsh industrial environments. IE can meet the demanding, time-critical requirements of industrial automation and control systems. It incorporates specialized protocols and ruggedized hardware to ensure reliability, real-time deterministic performance, and seamless integration across different operational levels. Whereas conventional Ethernet is primarily based on TCP/UDP-IP, IE is designed around specialized protocols such as PROFINET, EtherNet/IP, EtherCAT, and Modbus TCP. IE is used to describe Ethernet-based industrial communication protocols. IIoT, Industry 4.0, and smart manufacturing are closely related concepts that combine industrial production and operations with real-time digital data, machine learning, and artificial intelligence.

IE enables a new level of interconnectivity and communication between people and machines and is one of the key drivers of Industry 4.0 and the IIoT. IE provides access to real-time information that can improve control and visibility across the supply chain, support automated and streamlined maintenance, and enhance collaboration and productivity, all of which are essential for smart manufacturing. IE also facilitates improved information sharing between the factory floor and the corporate office, thereby integrating Operational Technology (OT) with Information Technology (IT). Support is also simplified through the availability of Ethernet-trained IT personnel and tools 4.

For field PoE implementation details, see Section II 18.0.2. The discussion of wireless technologies in this section should also be read in conjunction with Chapter 10 of this volume. From the preceding discussion, it is evident that Ethernet has evolved into a number of variants based on infrastructure requirements and application needs. These Ethernet variants are discussed in detail in this section.

## 16.1.0 Discussions on power over Ethernet (poE) and wireless Ethernet

PoE technology can be considered an extension of Industrial Ethernet (IE). Initially, PoE was developed soon after offices began adopting Voice over IP (VoIP) phone systems; naturally, the use of this technology in other devices followed. PoE was quickly adopted for remote wireless access points, IP-based cameras, and other devices commonly found in Industrial Ethernet network deployments [6]. In PoE technology, power is supplied remotely from a central location over distances of up to 100 m. Without PoE, these devices must be powered individually through AC outlets. Therefore, the obvious benefits of PoE include simplified installation and cost savings, centralized power backup, centralized power management, and safe power delivery.

Because of their flexibility, mobility, ease of setup, and reliability, wireless solutions are increasingly being used in industrial automation. Wireless Ethernet can easily integrate mobile or moving systems into a control network. Movement often also introduces safety concerns. Therefore, the ability to establish functionally safe wireless data transmission with moving subsystems is often an essential prerequisite for placing these systems in a safe state during an emergency. Manufacturers also require that functional-safety requirements be met through wireless Ethernet. Wireless PROFIsafe is a technology that allows the wireless transmission of safety-related data. This is achieved by embedding the safety data from the input module into the standard communication telegrams using an independent safety mechanism. In this connection, Fig. II 5.1b may be referenced. The data are then unpacked in the corresponding safe output module. The safety communication stacks verify whether an error or transmission overrun has occurred. If an error is detected, the system is immediately returned to a safe state 5.

Certain wireless Ethernet applications require appropriate hardware, such as routers, access points, NICs, and modems. Wi-Fi also requires proper equipment placement, taking into account factors such as location and the use of range extenders or mesh networks. Interference management is an important consideration, and factors such as channel selection and router configuration must be addressed. For security, password protection, VPNs, firewalls, and encryption are commonly used. Reliable power supplies are essential for proper operation. Other important considerations include device compatibility, regulatory compliance (frequency and legal requirements), adequate bandwidth, and bandwidth management. Maintenance and updates (firmware updates) are also important requirements. This subsection provides a brief discussion of PoE and wireless Ethernet. The discussion begins with the basics of the PoE scheme.

1. **PoE scheme:** PoE technology enables electrical power to pass over Ethernet cabling at distances up to 100 m. Power supply schemes in PoE are shown in Fig. III 16.1a. A basic PoE scheme is depicted in Fig. III 16.1a1

   Generally, a PoE system comprises of a minimum of four pieces of equipment:

   - **PSE**: A device that draws power from a power source and distributes it to the rest of the PoE system.
   - **PD:** A PD receives power from the PSE and requires no separate power source of its own.
   - **Ethernet cable:** The Ethernet cable is the power and data transmission medium of a PoE system. It provides the link between two devices for bidirectional communication and unidirectional power supply.

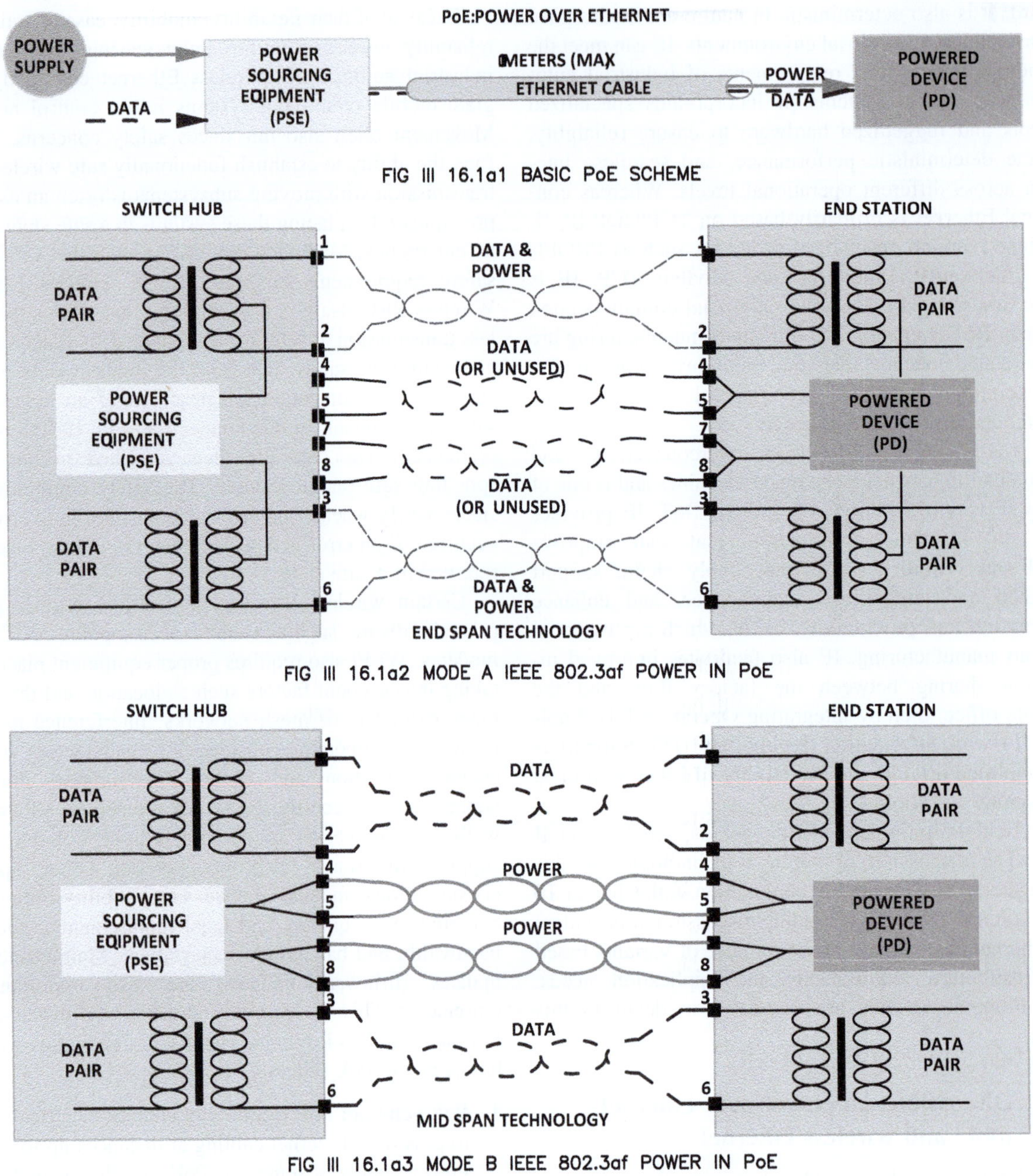


**FIGURE III 16.1A** Basic PoE scheme with standard alternatives.

- **Power supply:** This refers to the power source from which the PSE draws power and is typically a nominal 48VDC power source.

There are two ways to supply power:

- **End-span technology:** Power originates from a powered port on an Ethernet switch and is superimposed on data-transmission wire pairs 1, 2, 3, and 6. In the IEEE 802.3af PoE standard, Mode A uses this approach to combine power and data on the same wire pairs through a technique known as phantom powering (see Fig. III 16.1a2).
- **Mid-span technology:** Power is supplied from an external device to the non-data wire pairs 4, 5, 7, and 8 through a patch-panel-style hub. This approach is used in IEEE 802.3af PoE Mode B. In this mode, the Ethernet spare pairs 4/5 and 7/8 carry power, whereas pairs 1/2 and 3/6 are used exclusively for data transmission (see Fig. III 16.1a3).

The PSE applies a voltage of 20.5 V to the load at different power levels. There are various PoE classes corresponding to different power ratings: Class 0, 0.44–12.95 W; Class 1, 0.44–3.84 W; Class 2, 3.84–6.49 W; and Class 3, 6.49–12.95 W.

2. **IEEE PoE standard:** The major standards of IEEE on PoE include:
   - **802.3af:** IEEE 802.3af-2003 is the standard for PoE and facilitates the supply of up to 15.4 W of DC power. IEEE 802.3af, approved in 2003, provides 15.4 W per port for PDs up to 12.95 W (350 mA) at 48 V DC. General requirements call for a handshake before power is applied, and the PSE transmits power over two pairs (four of eight wires) of various categories of Ethernet cable. It also provides a means of detecting devices that require power and determining the power needed, when to turn power on, and when to remove power. An enhancement to IEEE 802.3af enables higher-power PDs, such as wireless access points; higher-power PSEs (at least 30.0 W at the PSE port); full backward compatibility and interoperability with existing IEEE 802.3af equipment; support for 1000BASE-T connections through mid-span PSEs; restricted cost increases for PSE ports and PD equipment; improved power-management granularity and power-budgeting capability; and resolution of specification-clarity issues inherent in the IEEE 802.3af specification.
   - **802.3 at:** IEEE 802.3at-2009, commonly known as PoE+, facilitates the supply of higher power levels of up to 34.2 W of DC power. This standard was developed to ensure at least 24 W of power to the PD, provide backward compatibility with IEEE 802.3af-2003, and support Gigabit mid-span implementations. The standard specifies Cat 5e cabling, a voltage range of 44–57 V, and a current level of 600 mA, with specified polarity and power delivery over either two pairs or four pairs. The mid-span power interface can be Type A (4,5,7,8) or Type B (1,2,3,6).
   - **Comparison** [6]**:** Here, the comparison of PoE & PoE+ i.e., comparison between IEEE 802.3af and IEEE 802.3 at as Table III 16.1a
   - **Classification** and **type 802.3bt:** In order to meet the demand for higher power, in 2018 IEEE released a new PoE standard IEEE 802.3bt to allow power up to 60–100 W of power at the source by providing power over *four* pairs of wires. This allows the use of PoE for a greater range of devices and applications. The new standard also improves energy efficiency by minimizing power loss in the cable (reducing power loss by half). The standard supports up to 10GBASE-T. During classification, the PSE and PD perform an analog handshake in which the PD requests a power class and the PSE responds with the class granted to the PD. Type defines the kind of analog handshake between the PSE and the PD. Class refers to the minimum power the PSE guarantees the PD will receive and the maximum power the PD is allowed to draw from the cable.
3. **PoE components:** The important PoE components include:
   - **Ethernet switch:** Apart from usual function, PoE Ethernet switches can act as a PSE to inject power into the Ethernet cable to power the attached PoE-enabled devices (e.g., as wireless access points).
   - **Splitter:** The splitter receives a PoE signal containing both data and power and separates them onto two different lines to support non-PoE devices.
   - **Injectors:** Injectors are used to feed power into a non-PoE cable run, combine that power with an Ethernet data input, and create a PoE output. Thus, PoE injectors carry data and power simultaneously over a single Ethernet cable.
   - **Repeater** and **extender:** Enable extension of PoE connections beyond the 100-meter limit.
   - **Media converter:** Media converters are used to transmit Ethernet data over long distances by converting the Ethernet transmission medium from electrical signals over copper wire to optical signals over fiber-optic cable. They are an ideal way to extend

**TABLE III 16.1A** Comparison of PoE and PoE+.

| Item | IEEE 802.3af | IEEE 802.3 at |
|---|---|---|
| Max power at PD | 12.95 W | 25.50 W |
| Max power PSE delivery | 15.4 W | 30 W |
| PSE voltage range | 44.0–57.0 V | 50.0–57.0 V |
| PD voltage range | 37.0–57.0 V | 42.5–57.0 V |
| Max. current | 350 mA | 600 mA |
| Cable support | Cat 3, Cat 5 | Cat 5 above |
| Supported node | Alternate A and B | Alternate A and B |
| Power class | 3 power class levels | 4 power class levels |

networks over fiber or copper media and can also provide power to remote PoE devices over long-distance data links.

4. **Wireless Ethernet applications:** There are a number of applications of wireless technology in industrial control systems. The major ones are listed below:
   - **Remote operation:** Remote monitoring and control of industrial process and equipment with enhanced efficiency without physical presence.
   - **IIoT application:** Wireless communication is an important enabler of IIoT, providing network connectivity and facilitating real-time (RT) data collection and analysis, predictive maintenance, and the optimization of industrial processes.
   - **Industry 4.0 application:** Industry 4.0 has created opportunities for expanded industrial wireless solutions and capabilities. IoT device sensors can provide early warning signs of potential malfunctions, thereby creating a safer workplace.
   - **Flexible production:** Flexible production allows quick reconfiguration of production systems, reduces downtime during production-line changes, and lowers reconfiguration costs.
   - **Industrial Ethernet:** Industrial Ethernet (IE) technologies, such as EtherNet/IP, support wireless Ethernet and offer the benefits of Industrial Ethernet.
   - **Wireless sensor networks (WSNs):** Wireless Ethernet enables WSNs, providing a cost-effective solution for monitoring hard-to-reach or hazardous areas without the need for extensive cabling.
   - **Safety** and **security:** Wireless Ethernet can be used to connect safety devices, such as gas detectors, located at remote sites to central control systems to enhance safety and security.
   - **Mobile and handheld device:** Wireless Ethernet supports the connectivity of mobile and handheld devices, enabling remote operating personnel to access real-time data, log maintenance activities, and update inventory levels on the go, thereby increasing productivity and reducing errors.
   - **Augmented reality (AR) and virtual reality (VR):** Wireless Ethernet provides the connectivity for AR/VR devices to access real-time data and interact with control systems and can enhance on-the-job training.
   - **Miscellaneous RT applications:** Asset tracking for inventory control, remote theft detection, data acquisition, and historical data storage for future analysis are all facilitated through wireless Ethernet.
5. **Wired and wireless Ethernet:** Table III 16.1b compares wired and wireless Ethernet.

   *Let us now move on to discuss variation of Ethernet based on speed.*

## 16.1.1 Fast, Giga, and 10 GiagaBIT Ethernet approaches

In the 1980s, 10 Mb/s seemed very fast. Introduced in 1995, Fast Ethernet has a transmission speed of 100 Mbps. Covered under the IEEE 802.3u standard, a few Fast Ethernet types include 100BASE-FX, 100BASE-T4, and 100BASE-TX. During this time, network designers had to determine which customers needed the extra bandwidth provided by the transition from the 10BASE-T to the 100BASE-T standard. However, in the present context, fast

**TABLE III 16.1B** Comparison of wire and wireless Ethernet.

| Issue | Wired Ethernet | Wireless Ethernet |
|---|---|---|
| Medium | Physical cable/FO | Guideless medium RF wave |
| Speed | 1 or 10 Gbit Ethernet | Wi-Fi is much lower in speed |
| Reliability and stability | Reliable and stable connection | Susceptible to interference |
| Mobility and flexibility, | Limited mobility | High mobility and flexibility, |
| Security | Generally, more secured | Robust security warranted |
| Installation | Complex and costlier | Simple and easy |
| Maintenance | Complex and costlier | Comparatively easier |
| Latency | Lower latency suitable for RT application | Higher latency due to signal processing, interference. |
| Power | No additional power | Power required for access point, Wi-Fi maintenance, etc. |
| Compatible | All Ethernet enabled device | May require network adapter |
| Application criteria | Hi speed, stability, and security | Mobility and convenience |
| Accessibility | Physical | Proximity |

Ethernet with limited bandwidth and scalability is inexpensive and simpler to configure. Introduced in 1999, Gigabit Ethernet (as per standard IEEE 802.3z/ab) has a transmission speed of 1 *Gigabit per second*. It is today one of the widely used Ethernet types and supports full-duplex and half-duplex modes. It offers a good balance of performance and cost to replace Fast Ethernet as the standard for most LAN, capable of handling large file. Transfers. Gigabit Ethernet uses the same IEEE 802.3 frame format, full duplex, and flow control methods. Gigabit Ethernet takes advantage of jumbo frames to reduce the frame rate to the end host. Standard Ethernet frame sizes are between 64 and 1518 bytes. Gigabit Ethernet can be transmitted over CAT 5 cable and optical fiber:

- 1000BASE-CX—Short distance transport (copper)
- 1000BASESX—850 nm wavelength (fiber optics)
- 1000BASE-LX—1300 nm wavelength (fiber optics)

Released in 2002, 10 Gigabit Ethernet (10 GbE), as defined by IEEE 802.3ae/an, provides a very large bandwidth available on a single segment. 10 GbE supports both copper and fiber-optic cabling. This high-performance solution for demanding applications can be used in large data centers to avoid data bottlenecks that would otherwise occur. It commonly runs over fiber-optic cables for longer distances but can also use Cat 6a or higher copper cables for shorter runs. This section presents a brief discussion of various Ethernet types based on speed.

1. **PHY sublayer** and **SONET:** The PHY sublayer refers to the Physical Layer sublayer, whereas SONET is a digital communication standard used in North America. Both are important aspects of Ethernet technology.
   - **Physical sublayer:** As shown in Fig. III 16.1b the PHY is a circuit block at the Physical Layer that includes the following sublayers: Physical Coding Sublayer (PCS), Physical Medium Attachment (PMA), and Physical Medium Dependent (PMD).
     - PCS: The PCS encodes and decodes the data stream between the MAC and PHY layers. Major functions include encoding/decoding, synchronization, MAC interfacing, link management, carrier sensing, and collision detection. Various PCS categories include 10GBASE-R—serially encoded, operating at a 10.3 Gb/s rate and not compatible with SONET (described below); 10GBASE-X—serially encoded and used for Wavelength Division Multiplexing (WDM) transmission; and 10GBASE-W—serially encoded and compatible with SONET standards for a 10 Gb/s WAN (WAN PHY).
     - PMA: Acting as an intermediary between the PCS and PMD, in Ethernet physical layer, the PMA sublayer handles the conversion of data symbols for transmission and reception over the physical medium, performing functions such as bit-to-symbol conversion, carrier detection, and link status monitoring. The PMA transmits and receives high-speed serial data. The PMA transmitter serializes parallel data, and the PMA receiver deserializes serial data. It also does carrier detection, clock recovery, link status monitoring, etc.
     - PMD: The PMD defines the specifics of the physical medium. Its main functions include medium specifications, physical-medium interfacing, and bit timing. It also maps TX and RX symbol streams to signal values appropriate for the medium being used. Common PMD interfaces include the 10 Gigabit Ethernet Attachment Unit Interface and the 10 Gigabit Ethernet 16-bit Interface. In addition, special 10 Gigabit Ethernet optical modules and system backplanes are available.
   - **SONET:** SONET, an acronym for Synchronous Optical Network, is a North American standardized digital communication protocol for synchronous data transmission over optical fibers. SONET standards are defined in ANSI T1.105 and T1.117. SONET contains network elements such as Synchronous Transport Signal (STS) multiplexers and demultiplexers, regenerators, and add-drop multiplexers. It also consists of four layers: Path, Line, Section, and Photonic. SONET is structured to provide a flexible synchronous optical hierarchy capable of carrying signals at various rates. Because of the proliferation of LANs that require intercommunication with other LANs across the country and around the world, SONET, developed by Bellcore, is used to transmit large amounts of data over long distances through fiber-optic (FO) networks.
2. **Repeater classes:** As a network device working at the physical layer of the OSI model, the repeater regenerates weakened/corrupted signals to *restore* them to original form by cleaning and amplifying signal before retransmission to ensure reliable data delivery over longer distances. Based on various factors, repeaters can be classified as:
   - **Signal type:** Analog repeaters, digital repeaters
   - **Connection type:** Wired, optical, and wireless (microwave, radio wave)
   - **Location based:** Local repeater, remote repeater
   - **Ethernet Class I repeater:** The Class I repeater translates line signals received on the incoming port into digital signals, allowing translation between different Fast Ethernet types, such as 100BASE-TX and 100BASE-FX. A Class I repeater

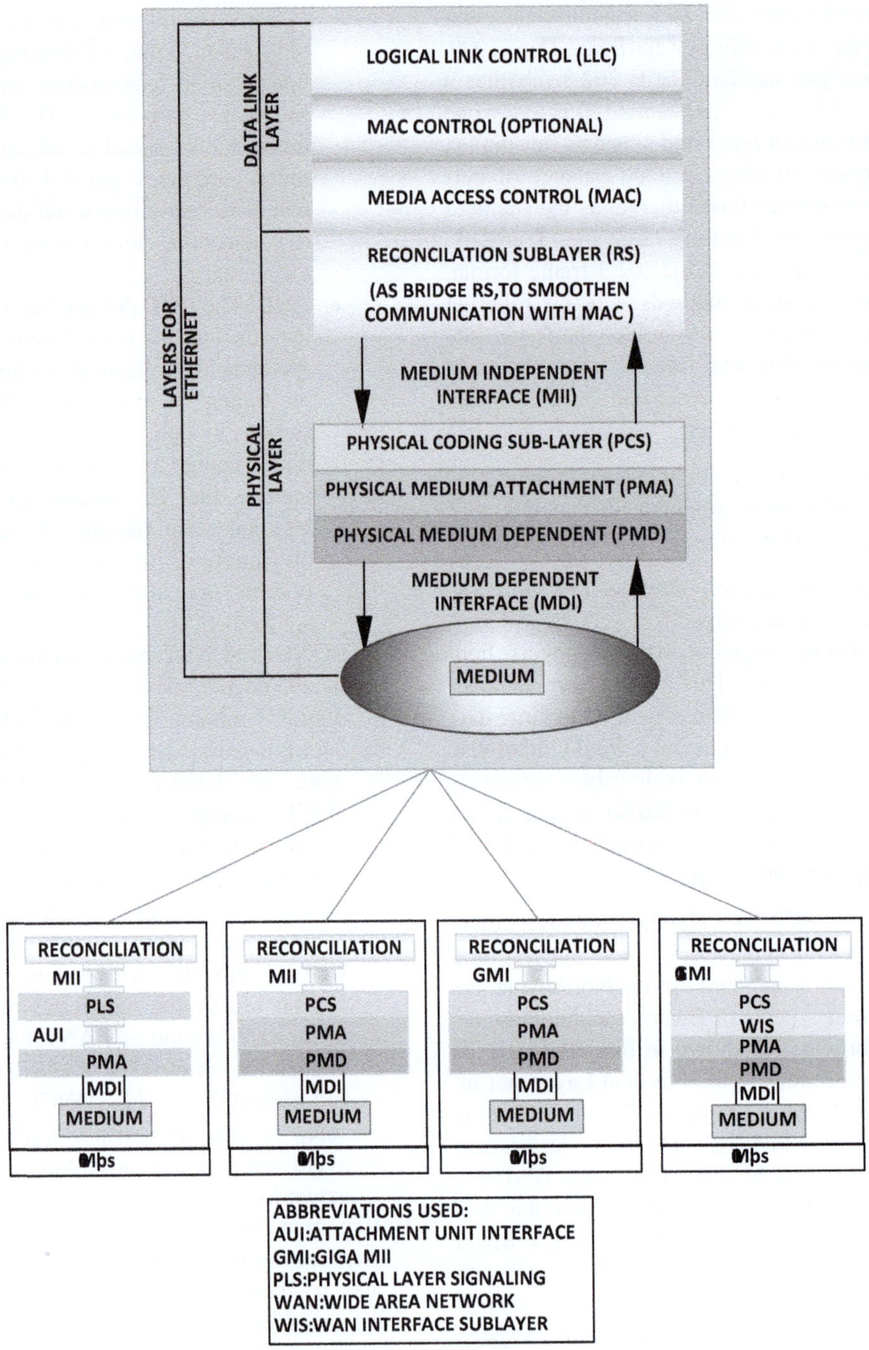


**FIGURE III 16.1B** Physical sublayers in Ethernet.

introduces delays during this conversion process. Class I repeaters have a maximum delay of 140 bit-times. Therefore, only one repeater can be used in a single Fast Ethernet LAN segment.

- **Ethernet Class II repeater:** The Class II repeater repeats the signal on an incoming port to all the ports on the repeater. This movement of data across the repeater introduces very little delay. Class II repeaters have a maximum delay of 92 bit-times. Thus, two Class II repeaters are allowed per fast Ethernet segment.

3. **Variations in types:** Since its inception, Ethernet has evolved through four generations based on speed: Standard Ethernet (10 Mbps), Fast Ethernet (100 Mbps), Gigabit Ethernet (1000 Mbps), and 10 Gigabit Ethernet (10,000 Mbps). Standard Ethernet defines several Physical layer implementations. Four of the most common are listed in Table III 16.1c

**TABLE III 16.1C** Common classes of standard Ethernet.

| Characteristics | 10BASE5 | 10BASE2 | 10BASET | 10BASEF |
|---|---|---|---|---|
| Topology | Bus | Bus | Star | Star |
| Media cable/FO | Thick coaxial | Thin coaxial | 2UTP | 2 FO |
| Max. length | 500 m | 200 (185) m | 100 m | 2000 m |
| Line coding | Manchester | Manchester | Manchester | Manchester |

Bridge Ethernet, switch Ethernet, and full-duplex Ethernet are a few common types of standard Ethernet. Next-generation Fast Ethernet, developed by the IEEE under the IEEE 802.3u standard, was designed to be backward compatible with Standard Ethernet. Gigabit Ethernet was developed subsequently. Let us now concentrate on these technologies.

Both Fast Ethernet and Gigabit Ethernet have variations, as listed below:

- **Fast Ethernet:** Fast Ethernet, utilizing a 100 Mbps data transmission rate (IEEE 802.3u), is well suited for many networking implementations that require an affordable solution. It has been commonly implemented in LANs, SMEs, schools, and similar organizations. Major features include increased speed, backward compatibility, media flexibility, cost-effectiveness, and collision detection. As mentioned previously, it is typically implemented in a star topology through an Ethernet switch. It is important to note that variations in fiber-optic (FO) length limitations depend on whether half-duplex or full-duplex mode is used. Full-duplex mode provides a longer reach because it avoids the collision-detection delays associated with half-duplex operation. 100BASE-TX operates on pins 1, 2, 3, and 6. 100BASE-FX also supports full-duplex communication. Speed limitations, scalability constraints, limited future-proofing, cable-distance limitations, fiber-optic cost, power consumption, and hardware-support requirements are some of the limitations of Fast Ethernet that led to the evolution of Gigabit Ethernet. Common issues with fast Ethernet are listed in Table III 16.1d.
- **Gigabit Ethernet:** Gigabit Ethernet has both half-duplex (in shared media) and full-duplex (in switched connections) modes of operation, with the majority operating in full-duplex mode. It is designed to connect two or more stations in a star or hierarchical-star topology through Ethernet hubs or switches. Higher speed, standardization (IEEE 802.3z for FO and STP copper and IEEE 802.3ab for UTP copper), backward compatibility, media support (UTP and STP copper and FO), full-duplex communication, auto-negotiation, flow control, and collision management are major features of Gigabit Ethernet. Apart from providing seamless migration, Gigabit Ethernet is, at present, a reliable and cost-effective way to improve connectivity between switches and between switches and servers. Gigabit Ethernet is implemented in different physical-layer standards: 1000BASE-CX (balanced twin-axial/STP cabling up to 25 m), 1000BASE-SX (uses FO up to 220–550 m), 1000BASE-LX (uses FO up to 5 km), 1000BASE-T (UTP Cat 5, Cat 5e, Cat 6, and Cat 7 copper cables up to 100 m), 1000BASE-T1 (STP copper cables up to 15 m), 1000BASE-TX (UTP copper cables up to 100 m), and 1000BASE-KX (UTP-type cables up to 1 m). Common issues related to Gigabit Ethernet implementation are listed in Table III 16.1e.

4. **Fast** versus **Giga Ethernet:** Let us now compare Fast and Giga Ethernet types.

**TABLE III 16.1D** Fast Ethernet implementation.

| Characteristics | 100BASE-Tx | 100BASE-T4 | 100BASE-Fx |
|---|---|---|---|
| Media | Cat 5 UTP or STP | Cat 4 UTP | FO |
| No. of wire/FO | 2 | 4 | 2/multi-mode FO |
| Line coding | MLT-3 | 8B/6T | NRZ-I |
| Max distance | 100m | 100m | 2000 duplex/412(half duplex) |

**TABLE III 16.1E** Giga Ethernet implementation.

| Characteristics | 1000BASE-Sx | 1000BASE-Lx | 1000BASE-Cx | 1000BASEß-T |
|---|---|---|---|---|
| Media | FO SW[a] | FO LW[b] | STP | Cat 5 UTP |
| No. of Wire/FO | 2 | 2 | 2 | 4 |
| Max length | 550m | 5000m | 25m | 10 |
| Line coding | NRZ | NRZ | NRZ | 4D-PAM-5 |

[a]*Short wave.*
[b]*Long wave.*

- **Pros and cons:** The advantages and disadvantages of both fast Ethernet and Gigabit Ethernet are given in Table III 16.1f.

**5.** 10 Gigabit Ethernet: Used in data centers and enterprise networks, 10 Gigabit Ethernet may use copper or FO for short-distance connections, whereas longer connections typically use FO. Ten-Gigabit Ethernet supports only full-duplex operation. It focuses on physical-layer details that can operate at very high speeds. CSMA/CD is no longer part of 10 Gigabit Ethernet. Compatibility still matters; therefore, 10 Gigabit Ethernet interfaces auto-negotiate and fall back to the highest speed supported by both ends of the link. Here, discussions are presented on (Table III 16.1g):

- **Full-duplex Ethernet:** In full-duplex Ethernet, each node is cabled to a switch with a pair of circuits: one for transmissions to the switch and the other for transmissions from the switch. In full-duplex Ethernet, simultaneous transmission is the default between the Ethernet switch and each node operation between a pair of stations. The link between the stations must use a point-to-point media segment, such as twisted-pair or fiber-optic media, to provide independent transmit and receive data paths, allowing stations to transmit and receive simultaneously and thereby doubling link capacity. An advantage of full-duplex operation is that the maximum segment length is no longer limited by the timing requirements of shared-channel half-duplex Ethernet. In full-duplex mode, the only limits are those imposed by the signal-carrying capabilities of the media segment. This is especially useful for fiber-optic segments. Ethernet operating at speeds $\geq 1$ Gbps uses full-duplex links

**TABLE III 16.1F** Pros and cons of Fast and Giga Ethernet.

| Ethernet type | Advantages | Disadvantages |
|---|---|---|
| Fast Ethernet | **1.** High data transfer speed<br>**2.** Easy simple expansion<br>**3.** Cost effective | **(a)** Data transfer limitation<br>**(b)** Scalability limited<br>**(c)** No future proofing |
| Giga Ethernet | **1.** Higher speed and duplex mode<br>**2.** Improved performance<br>**3.** More cost effective<br>**4.** High-volume data handling | **(a)** Upgraded infrastructure<br>**(b)** Higher power consumption<br>**(c)** Limited backward compatibility |

**TABLE III 16.1G** Comparison between Fast Ethernet and Giga Ethernet.

| Features | Fast Ethernet | Giga Ethernet |
|---|---|---|
| Speed | 100 Mbps | 1000 Mbps |
| Delay | More | Less |
| Round trip delay (time between browser sending signal and receiving acknowledgment) | 100–500-bit time | 4000-bit time |
| Transmission in half and full duplex | Yes | Yes |
| Layer standards | Discussed individually above | |

(transmitting and receiving simultaneously on separate communication channels) and offers features such as no shared medium, elimination of collision domains, increased efficiency, no timing or speed constraints, and no propagation-delay limitations. Therefore, Gigabit Ethernet was designed for full-duplex operation, ensuring optimal speed and performance, and modern switches eliminate collisions, making half-duplex operation largely obsolete.

- **10GBASE cabling** and **interfaces:** Table III 16.1h shows details of 10GBASE interface cabling. It is worth noting that 10GBASE-SR, 10GBASE-LR, and 10GBASE-ER are intended for LAN applications. When used in WAN applications, they are referred to as 10GBASE-SW, 10GBASE-LW, and 10GBASE-EW, respectively.

Next, we address miscellaneous issues related to industrial Ethernet.

### 16.1.2 Industrial Ethernet with Modbus TCP/IP

Industrial Ethernet (IE) is deployed in production lines and carries out important, time-sensitive, and sometimes dangerous tasks. Obviously, a small amount of error or incorrect movement can damage equipment, injure personnel, or adversely affect product quality or the production process. For example, if a few Ethernet packets are delayed by <1 s, the machine or production line may have to be halted to avoid a potential safety issue. This, in turn, may require hours to restore the system to a safe state and restart operation. By introducing IE into an enterprise and building IE networks to industrial cabling standards, interoperability and cross-vendor support can be ensured, including support for future applications and the use of proven installation, verification, and maintenance methods that ultimately ensure quality and reliability throughout the lifetime of the network. From Section III 9.0.2, we understand that a few factors influence Industrial Ethernet, including communication needs, process concerns, environmental challenges, and hardware requirements. Industrial Ethernet (IE) has been developed to meet the demands of industrial automation by delivering enhanced reliability, determinism, and data visibility across the plant floor. The higher speeds, greater bandwidth, and seamless integration of Operational Technology (OT) and Information Technology (IT) provided by IE have made it one of the driving forces behind Industrie 4.0. IE addresses various constraints of traditional Ethernet, such as probabilistic data delivery, through modifications to the Physical Layer and Data-Link Layer protocols to ensure deterministic and reliable communication. Naturally, IE is an attractive option for industrial applications because of the inherent ubiquity, cost-effectiveness, and high bandwidth of Ethernet. Key characteristics of IE include environmental adaptability, reliable determinism, support for industrial communication protocols, and network topologies with redundancy. Ethernet profiles covered in IEC 61784-5 include EtherCAT, PROFINET, Modbus TCP/IP, and EtherNet/IP. TSN and Ethernet Advanced Physical Layer (APL) represent future developments aimed at further improving standardization, determinism, reliability, and ease of use in the most demanding process-automation environments. Modbus, originally developed for serial communication, was later adapted for IE infrastructure. Modbus TCP/IP operates on a client-server model supporting multiple clients and a single server. License-free Modbus TCP/IP is an open, vendor-neutral standard that operates over standard Ethernet infrastructure and provides interoperability, data reliability, and routing capabilities. Modbus TCP/IP encapsulates Modbus messages within standard TCP/IP frames. It is widely used in industrial automation because of its simplicity in connecting various devices. However, it does not provide built-in security and transmits data in plain text. In this subsection, brief discussions on IE are presented. As part of the IE discussion, Modbus TCP/IP is included to supplement the material presented in Section III 8.1.0.

1. **IE influencing factors:** Apart from environmental adoptability and other factors mentioned at the beginning of this subsection, there are a few other

**TABLE III 16.1H** 10Gbase Ethernet Interface media.

| Interface | Media | Max segment | Remarks |
|---|---|---|---|
| 10GBASE-SR | FO | <300 m | Multimode fiber (0.85 μm) |
| 10GBASE-LR | FO | 10 Km | Single-mode fiber (1.3 μm) |
| 10GBASE-ER | FO | 40 Km | Single-mode fiber (1.5 μm) |
| 10GBASE-ZR | FO | 80 Km | Single-mode fiber (1.5 μm) |
| 10GBASE-CX4 | 4P* Twinax | 15m | P*-Pair Twinaxial copper |
| 10GBASE-T | 4 pairs of UTP | 100m | Cat 6a UTP |

influencing factors. Some of them are purely technical and a few others are market driven. Also, integration of OT and IT is a driving factor behind increased use of IE. All these are discussed in the following.

- **Environmental conditions:** Important environmental conditions influencing IE are:
  - *Temperature*: Temperature has a major effect on a network. Under near-freezing conditions, cables may break, protective jackets may be damaged, and signal attenuation may increase. Similarly, excessive heat may melt cable materials, leading to shorts and increased vulnerability.
  - *Chemical:* Industrial environments are also harsh because of corrosive chemicals. Therefore, it is necessary to provide physical protection for cables and connectors.
  - *Noise and vibration*: Electromagnetic noise generated in the environment also needs to be handled for IE. Vibration can cause degradation of the jacket and disconnection if poor connectors are used.
- **Components grade:** Heavy-duty and light-duty components can influence IE availability. In harsh industrial environments, heavy-duty components may be a better choice. Heavy-duty cables, together with cable jackets, metallic components, and heavy-duty connector types, are important and, in certain cases, essential. The need for specialized, durable hardware capable of continuous operation under harsh conditions is a critical influencing factor, even though heavy-duty components are more costly.
- **Topology:** IE has many different topology options to meet the needs of diverse industrial applications.
- **Operational issues:** In a production setting, downtime can be very costly. A continuous process can be rendered nonfunctional if part of the automation system fails, for example, in a batch process involving repeated time-bound heating and cooling cycles. Critical processes could be ruined, leading to the loss of material and money. Therefore, when designing an IE network, provisions must be made to ensure that the system remains available and reliable in operation, although this may increase cost.
- **Security:** IE must be developed to guard against unauthorized use as well as tampering with security measures. Suitable measures must also be built in so that, in the event of accidental damage by an employee, bringing the device or network to a complete halt can be avoided as far as possible.
- **Enhanced demand for automation:** Globally, industries are adopting industrial automation, including robotics, to enhance operational productivity and efficiency.
- **Open standards:** IEs like PROFINET, EtherNet/IP, EtherCAT, and Modbus TCP/IP are based on standard Ethernet and thus allow seamless communication between devices from different vendors, promoting flexibility and innovation.
- **OT/IT convergence:** For better data flow and improved decision-making across the enterprise, that is, for the efficient operation of the enterprise, integration of OT and IT is essential to ensure unified and interoperable communication through Industrial Ethernet (IE).
- **Data processing:** Real-time (RT) data processing and low-latency communication are now pressing demands for modern industrial applications. With IE, not only are these requirements met, but process optimization, predictive maintenance, and timely decision-making are also possible.
- **Industry 4.0 approach:** With the introduction of IIoT/IoT, there has been a steep rise in the deployment of smart factories and the Industrie 4.0 approach. This calls for seamless connectivity and data exchange across various industrial applications, remote-monitoring systems, and the integration of OT and IT through IE networks.
- **Advancements:** Advancements such as TSN, 5G integration, edge computing, and cloud-based solutions are continuously enhancing the capabilities and applications of Industrial Ethernet. Therefore, these are also influencing factors.

2. **Modbus TCP/IP:** Modbus TCP/IP (Modbus TCP) is essentially the Modbus RTU protocol (see Section III 8.1.0) with a TCP interface that allows it to run over Ethernet. The Modbus messaging structure is the application protocol that defines the rules for organizing and interpreting data independently of the data-transmission medium. The TCP/IP protocol suite does not define the meaning of the data or how it is to be interpreted, because this is the role of the Modbus application protocol. Therefore, Modbus TCP combines a physical Ethernet network with a standard method of representing data encapsulated within the TCP/IP standard, as detailed in Fig. III 16.1c.

   Drawbacks such as low speed and the nondeterministic nature inherent in TCP/IP Ethernet are largely resolved in Modbus TCP through the use of Fast Ethernet switches to interconnect devices. These switches minimize network traffic by facilitating a direct connection between the sender and receiver so that only the intended receiver obtains the data, rather than the entire network. The switch learns and stores the MAC address of each connected device along with its associated port number [7]. The Ethernet collision domain is

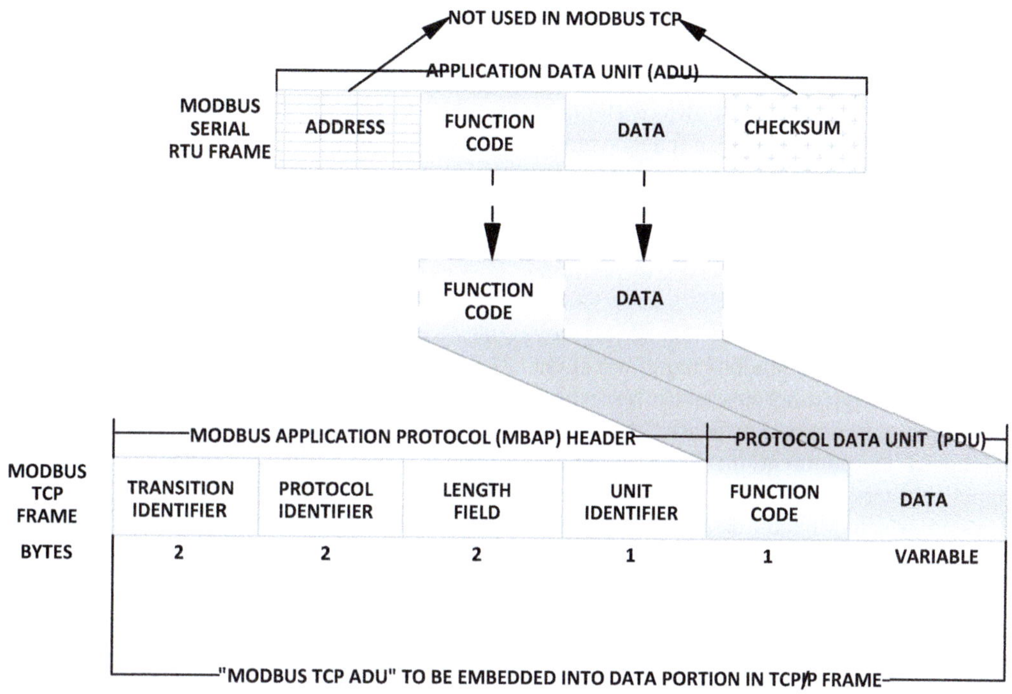


**FIGURE III 16.1C** Modbus data frame.

terminated at the switch port, and the switch effectively divides the network into separate data links, or collision domains, one at each switch port. Modbus/TCP devices can easily connect and communicate over existing Ethernet and FO networks. Modbus/TCP also allows many more addresses than RS485 and supports multiple masters and speeds in the gigabit range. The RTU limitation of 247 nodes per network is overcome, and Modbus/TCP networks can support as many slaves as the physical layer can handle. The rapid adoption of Ethernet in the process-control industry has allowed Modbus/TCP to become the most widely used, fastest-growing, and best-supported industrial protocol over Ethernet [8].

3. **Modbus TCP data frame:** By this point, the reader is well aware of the Modbus TCP protocol from Figs. III 8.1b1 and III 8.1b3. The Modbus TCP/IP architecture represents a layered approach with two main components: the TCP/IP stack and the Modbus Application Protocol (MBAP), as shown in Fig. III 16.1c. The MBAP, positioned above the TCP/IP stack, is responsible for encapsulating the Modbus message inside a TCP/IP. This architecture allows for seamless communication of Modbus data over standard network infrastructure. Modbus TCP messages contain an MBAP header, with a function code for the message operation, and the Data field, which varies depending on the function code. As shown in Fig. III 16.1c, the MBAP header is a 7-byte structure prefixed to the standard Modbus message. This header includes:

- **Transaction Identifier (2 bytes)**: unique ID number
- **Protocol Identifier (2 bytes)**:0 set for Modbus services
- **Length (2 bytes)**: indicates the number of remaining bytes in the message
- **Unit Identifier (1-byte)**: identifies the target device/unit within a Modbus network

  The MBAP encapsulation enables the Modbus TCP to utilize the robust transport mechanism of TCP/IP while maintaining the simplicity and straightforwardness of the Modbus messaging structure. Fig. III 16.1c shows both Modbus serial bus (RTU) frames and Modbus TCP frame. In the Modbus TCP frame, "ADDRESS" and "CHECK SUM" of the RTU frame are missing because:
- **Address:** In RTU, the slave address identifies the specific device on a serial network. In Modbus TCP, addressing is performed through IP addresses at the network level. In addition, a Unit Identifier field is included in the MBAP for Modbus TCP when acting through a gateway (refer to Fig. III 8.1b3).
- **Checksum**: Modbus TCP relies on the underlying TCP/IP protocol for error checking; hence, there is no need for the checksum required in RTU to ensure data integrity.

4. **Modbus TCP/IP:** From discussions in Section III 8.1.0.5, we understand that Modbus TCP/IP is a protocol specifically designed for industrial Ethernet for use in plant automation and control systems. Modbus TCP/

IP was developed with a paramount emphasis on reliability, efficiency, and simplicity in industrial automation applications. The Modbus TCP/IP framework is based on a client-server model. As shown, each client request contains an MBAP header, a function code, and data fields. The MBAP header is unique to Modbus TCP/IP and includes details like the transaction identifier and protocol identifier. The function code denotes the operation the client wants the server to perform. Major pillars of Modbus TCP/IP are:

- **Client/Server architecture:** Command/response messaging structure of the Modbus requires a client (also TCP client) requesting data and a server that processes requests to return a response.
- **Register-based communication:** Discrete inputs, coils, input registers, and holding registers are the four ways in which Modbus data are stored and are essential for the organization and interpretation of data in a Modbus system. The first two types represent binary information, whereas the other two are used for numerical data.
- **Function code:** The function code in the Modbus data frame represents the data access in Modbus. Function codes specify the type of action, ranging from reading and writing to data manipulation functions to be performed on a data entity.
- **Addressing:** Simple addressing typically uses numerical values to specify the starting register address and the number of registers to read or write.
- **Transaction orientation:** Each request made by a Modbus client is independent of all others. This transactional property allows Modbus to be employed in a broad spectrum of applications and use cases [9].

5. **Major IE consortium:** Durability and environmental adaptability, rugged hardware, fault tolerance through redundancy, PoE, integration of OT and IT, overall cost-effectiveness, scalability, flexible topology, ease of installation, future-proofing, and protocol support have made IE very popular. Many vendors are collaborating to develop several IE networks based on open standards. Therefore, they form consortia. The major IE consortia are listed in Table III 16.1i.

With this the discussions on Ethernet varieties come to end to look forward discussions on Miscellaneous issues on Ethernet including implementation issues.

## 16.2 Ethernet implementation and networking issues

Ethernet communications have become pervasive in today's technological environment. Gigabit Ethernet serves as a versatile, standards-based interconnect in many systems. There is now a new option that provides all the benefits of Ethernet while offering higher performance, namely Energy-Efficient Ethernet (EEE) and Backplane Ethernet. The evolution of Ethernet has led to higher speeds of 10 Gbit/s (IEEE Std 802.3ae) and 40 Gbit/s and 100 Gbit/s Ethernet (IEEE 802.3ba), Green Energy-Efficient Ethernet (IEEE 802.3az), Power over Ethernet (IEEE 802.3at), Carrier Ethernet technologies (implemented by carriers and service providers), and others. PCS, PMA, and PMD in the PHY, along with the backplane and copper-cabling PHYs, also include an auto-negotiation sublayer and a Forward Error Correction (FEC) sublayer. New PDs can take advantage of higher power levels.

There has been a growing focus on reducing IT energy consumption. Although the majority of Ethernet links remain idle most of the time, they consume power at a nearly constant level. Energy-Efficient Ethernet (EEE) provides a mechanism and a standard for reducing this energy consumption without compromising essential functionality. Carrier Ethernet has contributed to the development of MANs and WANs through improved service management, QoS, standardized services, and scalability. Ethernet transmits structured data in the form of frames, which are encapsulated within packets at the network layer for processing at different layers.

Backplane Ethernet, a specialized application of Ethernet technology, is defined in specific IEEE 802.3 standards (e.g., IEEE 802.3bj) tailored to a unique physical medium that differs from standard twisted-pair or fiber-

**TABLE III 16.1I** Major IE consortia.

| Item | Profinet | Ethernet/IP | EtherCAT | Modbus TCP |
|---|---|---|---|---|
| Major supplier | Siemen | Rockwell | Beckhoff | Schneider |
| Consortium | Profibus international | ODVA | ETG | Modbus IDA |
| Technology | Profinet IO/Td | CIP[a] | Shared frame | RTPS[b] |
| PHY MAC | Yes ASIC | Yes | Shared frame | None |

[a]*CIP: Common industrial protocol.*
[b]*RTPS: RT publish subscribe.*

optic cabling because of its electrical characteristics. Backplane and copper PHYs often include auto-negotiation to determine the highest common data rate and supported operational parameters, as well as FEC to maintain signal integrity and ensure reliable transmission.

The success of industrial automation depends on a reliable and efficient communication network that seamlessly connects all components of the production system. Ethernet technology dominates the industry because it is cost-effective and ubiquitous, offering common physical links with increased speed. From the discussion in Section III 9.2.0, we understand that IE differs from standard Ethernet and is designed to operate reliably in harsh industrial environments while providing deterministic, real-time communication for automation and control systems. It forms the foundation of networking technology for IIoT and Industrie 4.0 initiatives. IE can operate at high speed in full-duplex mode, providing high bandwidth for critical real-time control and high-volume data exchange between connected devices. IE also provides the deterministic networking required for real-time operational control and automation.

Robustness, interoperability, standardization, scalability, and flexibility are important contributions of IE. It also integrates OT and IT, which is extremely helpful for enterprise management and represents a major step toward Industrie 4.0. In the future, IE may contribute to the development of TSN for enhanced real-time communication and integration with IIoT, AI, and cloud platforms for advanced data analytics and progress toward Digital Twins (DTs). New physical layers, such as Single Pair Ethernet (SPE), a new Ethernet communication standard for factory, process, and building automation, will enable more rapid deployment of integrated strategies such as IIoT.

Ethernet at the field level in industrial automation primarily refers to Ethernet-APL, which provides a unified, high-speed network infrastructure from the sensor to the cloud. The reader is already familiar with this from the discussions in Volume II, Chapter 5 (see Fig. II 5.0e and Table II 5.0b), where details of the scheme and specifications have been elaborated. Many requirements, such as higher bandwidth, real-time data analysis, connectivity for IIoT/IoT devices, and integration of OT and IT networks within Industrie 4.0, are addressed by Ethernet-APL in industrial automation. Therefore, Ethernet-APL can be considered a facilitator of Industrie 4.0. Unified network architecture, high speed, interoperability, higher bandwidth, simplified SPE cabling, enhanced diagnostics, and Ethernet standards are among its major features.

Ethernet-APL is a managed switched-network architecture (see Fig. II 5.0e). With data traffic of 10 Mbit/s and a power budget of $\leq$60 W, it provides excellent scalability and flexibility with regard to the number of Ethernet-APL switches and instruments that can be connected. Additional features include automatic neighborhood detection at the switch, multiple communication paths that can operate in parallel, and ring-redundancy or resiliency concepts at the network layer that reroute communications if an Ethernet-APL trunk segment fails, thereby improving plant availability. In addition, Ethernet-APL transports and enables functions from higher levels of the ISO-OSI model [10]. In this concluding section on Ethernet, the above topics have been discussed in detail. Let us begin with an additional issue pertinent to Ethernet technology.

## 16.2.0 More on Ethernet technology

In Ethernet technology, the MAC protocol encapsulates payload data by adding a 14-byte header, called Protocol Control Information (PCI), before the data and appending an integrity checksum. The fundamental idea of EEE is that a communication link should consume power only when real data are being transmitted. Earlier wireline communication systems consumed power continuously. EEE is designed to work with backplanes to reduce power consumption during periods of low data activity. New link standards have been developed to meet unique needs ranging from lightweight cabling for automotive applications to 100 Gigabit optical links for data centers. Rack-based systems of various shapes, sizes, and descriptions have been the norm in DCS and SIS environments. A backplane refers to a circuit board that contains sockets or slots for connecting other printed circuit boards (PCBs). It acts as a backbone that facilitates communication between different components of a system. In Ethernet and networking contexts, a backplane is used to connect multiple network devices within a single chassis, allowing high-speed data transfer and efficient management of network traffic. In this subsection, the discussion starts with a brief recapitulation of frame-size calculations to supplement the discussion in Section III 16.0 and then concentrates on backplane issues in Ethernet communication.

1. **Packet data with computation:** In Section III 13.2.1.3, we discussed IEEE 802.3 data frames at length. Here, let us recapitulate them to gain insight into traditional Ethernet frames. The idle time before transmission provides a short interval for the receiver electronics in each node to settle after completion of the previous frame. A node then starts transmission by sending an 8-byte (64-bit) preamble sequence to allow time for the receiver in each node to achieve lock of the receiver (Digital Phase-Locked Loop) and synchronize the receive data clock with the transmit data clock. At the point when the first bit of the preamble is received, each receiver may be in an arbitrary state. From Fig. III 13.2d, we understand that the minimum frame payload is 46 bytes (dictated by the slot

time of the Ethernet LAN architecture). Let us now calculate the maximum frame rate of a node on an Ethernet LAN. The maximum frame rate is achieved by a single transmitting node without any collisions. This implies a frame consisting of 72 bytes; the inter-frame gap is 12 bytes (9.6 μs). Referring to Fig. III 13.2d, we obtain: Preamble + SFD = 8 bytes; Source and Destination Address = 12 bytes (6 bytes each); Type = 2 bytes; Minimum Payload = 46 bytes; Checksum = 4 bytes. Therefore, the total number of bytes = $(12 + 8 + 12 + 2 + 46 + 4) = 84$ bytes. The maximum number of frames per second is: Ethernet Data Rate (bits per second) / Total Physical Frame Size (bits) = $10{,}000{,}000 / (84 \times 8) = 14{,}880$ frames per second.

2. **Ethernet shared medium:** The protocol ALOHA (Refer III 13.2.0.5) decides which NICs are allowed to transmit at any given time. ALOHA does not guarantee that the frame of data will actually reach the remote recipient without corruption. It therefore relies on Automatic repeat request (ARQ—refer III 6.0.2) protocols to retransmit any data which is corrupted. Also, each NIC must add a checksum/CRC for maintaining data integrity. When a NIC has data to transmit, the NIC *first* listens to the cable (using a transceiver) to see if a carrier (signal) is being transmitted by another node, sensed by monitoring current flowing in the cable (each bit ~ 18–20 mA). The individual bits are sent by encoding them with a 10/100 MHz clock using *Manchester encoding*. Data is only sent when *no carrier* is observed (no current) i.e., the physical medium is therefore idle. Any NIC which does not need to transmit, listens to see if other NICs have started to transmit information to it. All the NICs receive the frame and each examines it to check its length and checksum. The header destination MAC address is next examined. MAC destination address ensures that only the specified destination actually should be accepted, received and forwarded to the network-layer software in the computer. Now, if two NICs *simultaneously* try to transmit, then both may see an idle physical medium so, both will then decide to transmit and a *collision* will occur. The collision will result in the corruption of the frame being sent, and will subsequently be discarded by the receiver. When there is data waiting to be sent, each transmitting NIC also monitors its own transmission. If it observes a collision (excess current), i.e., collision detection, it stops transmission immediately and instead transmits a 32-bit jam sequence to cause receivers to discard the frame. To ensure that all NICs start to receive a frame before the transmitting NIC has finished sending it, Ethernet defines a minimum frame size (see Fig. III 13.2d), which is related to the distance over which the network spans, the type of media being used, and the number of repeaters through which the signal may have to pass to reach the furthest part of the LAN. The collision sequence and collision-detection process are illustrated in Fig. III 16.2a
3. **Backplane Ethernet:** As discussed, a backplane is a circuit board that contains interconnections and facilitates communication between different components of a system. In Ethernet networking, it acts as a backbone to connect multiple network devices. Essential details of a backplane are:
   - **Functional details:** A few functional details of backplane include:
     - *Interconnection:* Primary function is to provide interconnection.
     - *Transmission and distribution*: Backplanes support high-speed data transmission; they use parallel or serial communication protocols to transfer

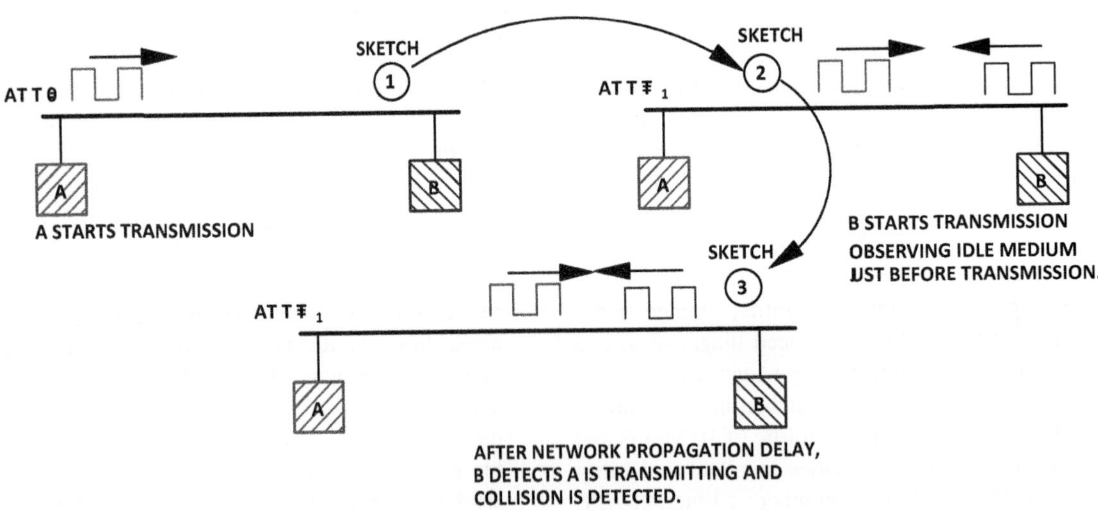


**FIGURE III 16.2A** Ethernet collision detection.

data. They also play active role in power distribution, supplying power to the connected modules.
  - *Signal integrity*: Backplanes minimize signal degradation and interference to maintain signal integrity, which is crucial in high-speed networks.
  - *Scalability*: Backplanes facilitate scalability by providing a standardized interface for connecting additional modules.
- **Types of backplanes:** Backplanes can be of two types: passive (having no active components and meant only for interconnection) and active (with active elements for storage, driver functions, etc.).
- **Features:** New backplane features include but are not limited to the following:
  - *Fewer parts*: Many of the latest processors feature integrated backplane Ethernet transceivers, meaning they can be connected without a separate controller or PHY device, the bulky transformers
  - *Higher performance*: Newer devices feature 10GBASE-KR, which provides 10 Gbps over one link pair.
  - *Fewer w*ires: Backplane Ethernet uses serializer/deserializer technology with one pair in each direction (Rx/Tx), compared to four pairs for 1000BASE-T Gigabit Ethernet.
  - *Ethernet types*: Gigabit, 10 Gbps and 40 Gbps backplane Ethernet standards.
  - *Support and management*: Fully managed multilayer switching services including multicast, QoS, and security. With a backplane-based architecture, network management becomes more centralized.
  - *Modular design*: Backplanes supports a modular design approach in network.
  - *Reliability and redundancy*: Backplane is a very reliable approach that can provide redundant paths as well.
- **Application:** Low-latency 10 Gbps control-plane switches and high-throughput 40 Gbps data-plane fabrics are some of the common applications of Ethernet.

4. **Backplane Ethernet benefits:** Some common benefits available from backplane include but are not limited to:
   - **Scalability:** Ethernet backbone architecture in backplane design allows complete scalability from a single I/O card to hundreds.
   - **Redundancy:** A redundant Ethernet switch fabric between the controller and I/O levels allows different levels of redundancy to coexist within the same controller, enabling controllers and I/O devices to be distributed freely to minimize overall cost and maximize system value.
   - **Low-cost cabling:** Redundant Ethernet communications on every I/O card in the Ethernet-based system can replace dozens or hundreds of individual conductors.
   - **Digital bus integration:** Digital buses can be integrated with traditional I/O.
   - **DCS-Safety instrumentation system (SIS) integration:** With proper application layer security, SIS and DCS I/O can be shared on the same Ethernet network. This can allow the DCS to access that portion of SIS I/O information that does not interfere with the safety function of the SIS [11].
   - **SIS requirements:** With this approach, based on individual requirement, redundancy can be designed in any safety instrumentation function (SIF) on an individual need basis for SIS.
5. **Ethernet backplanes:** Backplane can be conceived as a circuit board serving as a central communication highway for multiple modules in a system, e.g., network switch. As discussed, a backplane provides centralized connections, a shared bus (a common physical connection and communication pathway for all modules), and a modular design (a standard interface for different components). In Ethernet, it provides a high-speed shared electrical bus for internal communication between modules, offering modularity, support for specialized functions, scalability, and high-speed data transfer without individual cables. There are a few challenges as well. Common challenges include maintenance of signal integrity, management of heat dissipation, management of overall cost, and compatibility of modules with the backplane. With the evolution of network technology, backplanes will also advance. These trends may include optical backplanes for higher data-transfer rates and the integration of advanced technologies.

### 16.2.1 Industrial Ethernet: Benefits, applications, and future

The use of shielded cabling, proper grounding, and mitigation of electromagnetic and radio-frequency interference are common practices for IE deployment in industrial environments. For network reliability and desired performance, industrial-grade components with the required IP ratings are deployed. Ethernet/IP supports IP65 and higher-rated connectors, shielded cabling, and devices. In addition, cabincts can provide good protection for electronics and may facilitate locating devices closer to equipment, with higher IP ratings offering a suitable alternative. OPC UA and IE can work together to provide a semantic information model over an IE physical infrastructure. OPC UA can provide resilience against cyber threats and enable cross-vendor interoperability, both of which are essential for network reliability and performance in industrial settings. IE technologies such as EtherCAT offer short cycle times, reduced bandwidth requirements, and superior diagnostics, ensuring consistent performance even in large networks. Although star topology is the most popular, bus and ring topologies are also used. In a star topology, a switch connects devices to a central access point. Both unmanaged and managed Ethernet switches are used, depending on the application. Unmanaged switches are plug-and-play devices with fixed settings and are suitable for small networks. Managed switches are fully configurable and provide a range of performance data, making them well suited for larger networks and networks supporting critical activities. Managed and unmanaged switches differ in capabilities, security, and cost. Managed switches can also prevent network shutdown if someone accidentally creates a switch loop. Remote-access speeds are increasing, reducing the need for on-site monitoring, adjustment, and troubleshooting activities. A real-time deterministic response can be achieved with industrial Ethernet technologies such as TSN. TSN adds real-time capabilities to standard IEEE Ethernet. IEEE TSN defines Layer 2 functionality and LAN-level switching, including the concept of time. However, the software interface required to configure these hardware features is not specified, and an appropriate Application Programming Interface (API) must be used. In addition, issues such as jitter and latency associated with TSN are not specified. With these features, IE is expected to dominate the automation-connectivity market for the foreseeable future because of its high speed, robustness, and extensibility. A growing focus on sustainability will further drive the need to make data-driven business decisions more quickly and effectively than ever before. Owing to its highly reliable deterministic networking, high speed, and capability for seamless integration, IE has become a pillar of modern industrial automation. In this subsection, brief discussions are presented on the benefits, applications, and future of IE.

1. **Controls: Realtime** and **determinism:** In industrial automation, "real-time" and "determinism" are two distinct concepts. Real time refers to the requirement that a system or event complete its tasks within a specific time-sensitive deadline and is therefore characterized by timeliness. The required real-time response may vary from one application to another. For example, in motion control or electrical tripping, it may be measured in milliseconds, whereas in temperature control it may be measured in seconds or even minutes. Determinism in IE refers to the predictability of system behavior, rather than speed. Therefore, a deterministic response is defined by user requirements for response time. Both concepts are application dependent. Higher bandwidth allows more data to be transferred simultaneously, reducing delay and latency and thereby improving response time. Deterministic communication enables real-time performance in Ethernet networks. In real-time Industrial Ethernet (IE), protocols modify standard Ethernet to provide deterministic communication, guaranteeing that data arrive within a specified cycle time. Precise time synchronization across devices is a cornerstone of real-time Ethernet. For this purpose, protocols such as IEEE 1588 Precision Time Protocol (PTP) are used. Real-time Ethernet (RTE) may be classified as either hard RTE or soft RTE.
2. **Networked safety:** Networked safety refers to the integration of functional safety into Ethernet to ensure reliable and predictable communication essential for time-critical, real-time industrial control systems. From the above discussion, we find that the need for implementing a real-time deterministic network depends on the specific application. As seen, one benefit of a real-time deterministic network is the maintenance of functional safety. Another potential benefit is improved operational efficiency. If communication data are time-sensitive and mission-critical, migration to a real-time deterministic network can help products and services remain competitive and future-proof.
3. **Determinism and TSN:** Determinism, latency, and TSN approaches in Ethernet are as follows:
   - **IE determinism:** IE determinism refers to the guaranteed delivery of critical data packets within a specific, predictable timeframe to ensure the precise and synchronized operation of industrial automation systems. Some transmission delays are acceptable for general IT tasks but are not acceptable in industrial automation because they may cause production-line shutdowns, product defects, or safety risks. From the previous discussion, it is clear that determinism is a significant driving force behind IE communication networks intended for automation. Modern factory automation demands increasingly powerful communication systems that support real-time, time-critical transmission of control data. A set of IEEE 802.1 standards operating at Layer 2 of the Data-Link Layer has been developed, referred to as TSN, to help deliver deterministic communications and ensure reliable data delivery between endpoints. Apart from TSN, IE can achieve determinism through a combination of specialized protocols, dedicated hardware, and network-management techniques such as industrial protocols (used even at the field level in place of fieldbus systems), prioritization, master-slave communication, managed switches, traffic scheduling, synchronization, and bandwidth reservation.
   - **Latency-TSN approach:** Data-transfer delay is referred to as latency, which creates unpredictability

in standard Ethernet, making it unsuitable for industrial automation. With the inclusion of smart devices, the amount of data communicated over the network increases exponentially. Ethernet physical-layer device latency occurs when a frame is being transmitted or received. Higher latency results in lower production throughput, ultimately affecting revenue. TSN addresses this issue by using a set of IEEE standards to provide deterministic, low-latency communication over standard Ethernet through features such as precise time synchronization, traffic scheduling, and frame preemption. The TSN approach helps improve redundancy and traffic control and provides time stamps for messages in an Ethernet network. It allows flexibility, scalability, and support for both small devices and big data in time-critical communication. Deterministic communication, time synchronization, scheduled traffic (IEEE 802.1Qbv), frame preemption (IEEE 802.1Qbu), and integration with other technologies are some of the features that help reduce latency in Ethernet through the TSN approach.

4. **Future IE:** With SPE, TSN, and IIoT/IoT, the future of IE is very strong and will have a direct impact on industrial automation. Major issues here are:
   - **Challenges:** General control philosophy demands continuous operation of a system, however complex it may be, on a very demanding schedule. Downtime is rarely acceptable, regardless of the cost of lost production per hour at a given plant. IE provides additional diagnostics to help troubleshoot issues more quickly, which can be used in conjunction with sensor data to detect deviations from normal operation at the edge or in the cloud and determine whether premature device failure may be imminent [12]. In addition, asset management using the open, vendor-neutral Link Layer Discovery Protocol (LLDP) allows automatic topology and device identification. Major challenges of IE are:
     - Fast, reliable communication with bounded latencies
     - Moving from fragmented, diverse and siloed communication systems to a unified, ubiquitous, packet-based system built on different (flexible) physical layers
     - Handling of convergent traffic from RT data and non-RT data
     - Seamless connectivity to the internet without complicated to configure gateways
     - Increased complexities and cost for maintaining performance with huge increased data
     - Achieving this migration in a cost effective and sustainable way
     - Integrating new technologies like SPE, 5G communication, etc.
     - Management of cyber security in complex system requiring advanced security measures
     - Knowledge gap to cover up differences in OT with IT
   - **SPE** and **TSN:** SPE, a new system discussed below, is being increasingly adopted in process-control systems. Its long reach (>1 km), ability to support field devices, including intrinsically safe applications, and the cost benefits of providing data and power over a single pair of wires are highly attractive. TSN is also gaining traction and is becoming the enabling connectivity technology for fully realizing the Intelligent Edge [12]. 10BASE-T1L is an example of SPE, operating at 10 Mbps over 1000 m and providing cost savings through the use of a single-pair cable. Connectors are simplified and easier to wire, reducing material and installation costs. TSN is possible in Gigabit Ethernet, and the increased bandwidth is crucial for industrial automation and facilitates the implementation of Industrie 4.0. IEEE 802.1AS provides time synchronization in participating LAN-attached stations to an accuracy of within 1 μs and establishes a common time base across a network, enabling synchronization of all devices. This means that the time required for traffic to flow across the network is precisely defined and understood, providing the basis for determinism. The second standard is IEEE 802.1Qbv. This works with IEEE 802.1AS to define queuing mechanisms that determine the priorities of different traffic types and hence their access to the network. This ensures that different traffic types do not conflict and further enhances determinism [12].
   - **IIoT/IoT issues:** Because of the high bandwidth and shared IT architectures of IE, IIoT/IoT can offer all the associated benefits. IE is the foundational infrastructure for the realization of IIoT/IoT. IIoT and IoT are driving a transformation in manufacturing by connecting machines, sensors, actuators, monitoring devices, and other systems to data streams from business-specific solutions to coordinate and optimize operations. Connecting all of these solutions to a common Ethernet ecosystem creates a wide range of new opportunities and challenges. Data connectivity, deterministic communication, high-bandwidth communication, connectivity from sensors to the cloud, reliability and performance, and IT/OT convergence are major capabilities provided by IE that contribute to the successful implementation of IIoT/IoT and support the advancement of Industrie 4.0 and DT.

5. **SPE:** SPE is an Ethernet network that uses a unique physical-layer transceiver over a single pair of wires instead of two or four pairs of wires. SPE reduces system cost, weight, and wiring complexity compared with standard Ethernet multi-pair cabling. OSI model Layers 2–7 remain the same for all physical-layer types. Also, no software changes are required for SPE. SPE is represented by "T1" and can support multiple bandwidths and distances. In Ethernet over a single-pair cable, communication takes place in duplex mode (10BASE-T1S half-duplex is also available), and the distance limit may vary depending on the use of UTP or STP. SPE enables communication from the field level to the cloud and is therefore an ideal infrastructure for IIoT applications. Variations in SPE are listed in Table III 16.2a. SPE in multipoint and point-to-point modes are shown in Fig. III 16.2b
   - **Features:** Cost-effectiveness, space-saving design, power and data integration, lower power consumption, seamless integration (easy integration with existing systems), communication across IE protocols such as EtherNet/IP, PROFINET, and EtherCAT, and versatility for many applications are major features.
   - **Benefits:** Some common benefits are:
     - Reduction in cable size and weight by 25% and 35%, respectively, in comparison to standard Ethernet cabling
     - SPE cabling can communicate across standard IE protocols
     - SPE implementation is simple; existing control devices can interface with SPE networks through field switches
     - SPE is most suitable for supporting IIoT applications
     - Compared to standard Ethernet SPE has reduced power consumption
     - No patent license fees
     - Versatile solution with applications ranging from industrial settings to smart buildings
   - **Applications**: Building automation, factory automation, elevator control, infotainment, legacy network retrofit, etc.
   - **PoDL–PoE**: SPE can also deliver power to end devices through Power over Data Line (PoDL). PoDL is similar to PoE; however, in PoDL, power is delivered over the same wire pair as the Ethernet signals, unlike PoE, where power is delivered over separate

**TABLE III 16.2A** Single-pair Ethernet types.

| Standard | Bandwidth and protocol | Description |
|---|---|---|
| IEEE 802.3cg | 10BASE-T1S | 10MBP; point to point and multipoint: <25 m |
| IEEE 802.3cg | 10BASE-T1L | 10MBPS; mainly point to point; <1000 m |
| IEEE 802.3bw | 100BASE-T1 | 100MBPS; point to point; <15 m |
| IEEE 802.3bp | 1000Base-T1 | 100MBPS; point and multipoint; <15 m |
| IEEE 802.3bp | NGBBase-T1 | 2/5/10 Gbps; point and multipoint; <15 m |
| IEEE 802.3cy | 25Base-T1 | 25 Gbps multipoint; <15 m |

Let us now look forward for discussions on another Ethernet type in field level.

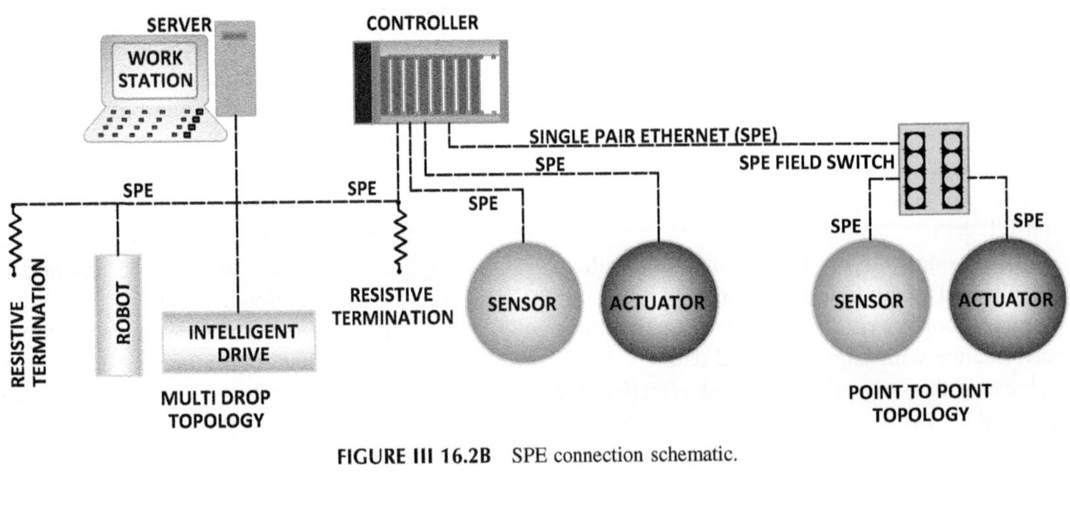


**FIGURE III 16.2B** SPE connection schematic.

pairs. PoE (IEEE standards 802.3af, 802.3at, and 802.3bt), which requires special network switches and Cat 5 cabling, has physical network limitations that restrict its usefulness for integrating plant-level sensors and field devices. Based on IEEE 802.3bu, PoDL for SPE can deliver 0.5–52 W across multiple power classes. Various cable configurations and power capacities are available, enabling delivery of up to 16 A to connected devices.

- **Hazardous areas:** SPE, together with advances in intrinsically safe barrier technology for communication networks, provides an opportunity to extend Ethernet to these plant-level areas.
- **Cable** and **connector:** Special cables and connectors allow both power and data to be transmitted over a single cable. Cables and connectors defined by IEC 63171-6 and IEC 63171-7 have been developed for SPE to ensure interoperability between SPE devices. SPE technology uses a standardized connector interface. Power and data are delivered to devices over a single cable using 4-pin M8 connectors or 6-pin M12 connectors.
- **SPE types:** Various SPE types have been tabled in Table III 16.2a

## 16.2.2 Ethernet at field level

Ethernet technology can also open up the field of process plants in standard applications and in areas requiring functional safety, both in plants with hazardous-area requirements and in those without such requirements. Ethernet APL (referred to as APL in this book) brings two-wire Ethernet, including power supply, into the field for all plant areas and applications. Ethernet at the field level primarily refers to the adoption of a new physical-layer standard known as Ethernet APL, which was developed mainly for the demanding environments of process automation, particularly hazardous areas. Uniform communication, real-time deterministic information flow at all levels, enhanced data availability, easier IIoT implementation, and enablement of Industrie 4.0 are major reasons for its development and adoption. SPE, discussed in the previous subsection, is directly related to Ethernet APL. In fact, Ethernet APL is a specialized version of SPE and is built on 10BASE-T1L (IEEE 802.3cg) (Table III 16.2b) standard to add specialized features for hazardous environments, i.e., intrinsic safety and power delivery over the same two-wire cable for longer distance. While SPE is intended for a broader range of applications, including factory automation and IIoT, with coverage distances of up to 1000 m, APL is primarily intended for process control, including hazardous areas requiring intrinsic safety. Ethernet APL meets the basic objectives required for digital transformation, including a standardized solution based on IEEE 802.3, support for process-control requirements such as trunk-spur topology, two-wire data and power transmission, use in hazardous areas, and intrinsic-safety support. Major features of APL include fast real-time deterministic data transmission through stable Ethernet communication, easy and flexible installation, and simple and fast commissioning. APL faces strong competition from traditional fieldbus systems, which currently occupy around 58% of the market. Although developed later, field-level Ethernet is not far behind, occupying about 40% of the market, with APL predominating in process industries. The major differentiation has focused on deterministic hard real-time performance for manufacturing in discrete automation, where safety and security are very important. In this subsection, brief discussions on Ethernet APL are presented. It is important to note that Ethernet APL, especially intrinsic-safety and powering schemes, has already been covered in Sections III 5.0.2, II 8.1, and II 18.2; hence, there may be some overlap in the discussion presented here.

1. **Ethernet APL technology:** Ethernet is standard for data transmission, at the upper levels of the automation pyramid. As only a physical-layer technology, Ethernet

**TABLE III 16.2B** Difference between OPC UA and IE.

| Issue | OPC UA | IE |
|---|---|---|
| Systematization | Information model | Protocol based data flow |
| Communication | Both client = server and publisher-subscriber model for information exchange | Client-server for data exchange |
| Technology and purpose | Platform independent information exchange model from field to cloud | Real-time, deterministic protocol for data exchange from field level to cloud |
| Data model | Sophisticated hierarchical information model | Flat address space and object model with focus on data |
| OSI model | Lower two layers | Higher layer 7 |

APL is compatible with IE protocols such as PROFINET, EtherNet/IP, and OPC UA. NAMUR NE 168 describes the "Requirements for an Ethernet communication system at the field level in process plants" and outlines specific requirements and objectives:

- The standardized solution should be based on the IEEE 802.3 family.
- Process automation specific requirements must be met; these are: The key requirements include support for trunk (1000 m) and spur (200 m) topologies in process automation, coverage of up to 1000 m, two-wire cabling for data and power supply, use in potentially explosive areas of process plants, including intrinsic safety, verification of intrinsic safety according to FISCO, and sufficient power supply for up to 60 field devices on a single segment. For further details, see Section II 5.0.2.2.
- Uniform procedure for the certification of the Ethernet-based solution.
- Supportive of documentation for engineering and installation.

  APL is a specialized version of SPE and is defined by various standards and documents, including 10BASE-T1L (IEEE 802.3cg-2019) (Table III 16.2a). APL can use intrinsic safety through 2-Wire Intrinsically Safe Ethernet (2-WISE; IEC TS 60079-47); Section II 5.0.2.4 may be referenced for details. Simplicity and robustness, high data rates, interoperability, long-distance real-time communication, low energy consumption, application in hazardous areas, and integrated power over two wires are common benefits; for details, see Section II 5.0.2.3. The APL port-profile specification for both hazardous and nonhazardous areas describes installation rules, including cable usage, connection technology, shielding, and grounding. APL port profiles define power classes. For PoDL in APL, power classes 10–12 are defined. As APL is a type of SPE, M8 and M12 connectors are defined in the specification. Engineering guidelines and conformance-test specifications are important aspects of APL technology. The reader is advised to read this section in conjunction with Section II 5.0.2 and Fig. II 5.0e for a proper understanding.

2. **APL IE** and **IIoT:** Let us next discuss two prominent applications of Ethernet APL:
   - **APL in IE:** Of the different IEs, PROFINET is the most popular platform. The appeal of PROFINET has further expanded with the arrival of SPE-based Ethernet-APL devices. This is implemented PROFINET with PA-Profile 4.02. The PA-Profile 4.02 offers greater freedom for users to achieve improved engineering efficiency through true interoperability between devices from different vendors. This is an IE communication standard for process automation as a special profile to standardize field devices, primarily through the use of Ethernet APL allowing intelligent devices at the field level to communicate with all levels of automation pyramid via a unified Ethernet network.
   - **APL in IIoT implementation**: From the discussion on SPE, we understand that SPE is a key enabler for IIoT. Naturally, Ethernet APL, being a specialized version of SPE, is also a key enabler for IIoT by providing a standardized physical layer that bridges the field level with higher-level systems and the cloud, while offering ruggedized, high-speed, long-distance real-time communication with lower energy consumption and support for intrinsic safety in hazardous areas. Major features that support IIoT implementation include seamless end-to-end connectivity, high-speed data access, simple two-wire transmission of both power and data, suitability for harsh environments, operation in hazardous areas, interoperability, flexibility, and enhanced diagnostics and predictive maintenance. With APL in place, rapid data transmission and seamless communication from field devices used in process control to the cloud are achieved through Ethernet. This capability extends to hazardous areas up to Zone 0/Class 1, Div. 1.
3. **Fieldbus** and **IE:** Ethernet is generally faster than fieldbus because of its higher communication speed, despite the additional protocol overhead. Ethernet does, however, face distance limitations from the hub. Process skids are particularly well suited to fieldbus because all wiring can be completed at the factory. IE technologies such as PROFINET and EtherNet/IP are faster, offer larger address spaces, and provide greater bandwidth than serial fieldbuses. They can also serve as bridges between different automation levels while offering enhanced diagnostics and other benefits. The tools available for IE can, in some cases, make installation simpler than fieldbus setup. There are many operational reasons why the total cost of ownership (TCO) of an Ethernet infrastructure can be lower than that of fieldbus systems [13]. Fieldbus systems are often considered more reliable than nondeterministic Ethernet. Ethernet offers several advantages over fieldbus, including support for vertical integration, future-proofing and investment protection, reduced overall cost, and access to higher bandwidth through full-duplex communication. OPC UA can work with IE as well as with almost all fieldbus systems. Therefore, it provides a common solution, which is discussed next.
4. **OPC UA and IE:** Four of the most important data models in OPC UA are Data Access (DA), Alarms

and Conditions (AC), Historical Access (HA), and Programs (Prog). OPC UA provides secure, vendor-independent data exchange with semantic meaning, whereas TSN provides the deterministic, high-performance network infrastructure required for guaranteed on-time data delivery. Thus, the Pub-Sub version, which supports one-to-many communications, can be implemented with TSN for hard real-time communication. OPC UA and PROFINET can work together in production environments, with overlapping capabilities but different areas of focus; hence, it is unlikely that one will replace the other in the near future. The strength of OPC UA, enabled by TSN, is that it allows different networks to communicate, especially at the factory and enterprise levels. Consequently, Industrial Ethernet, Industrie 4.0, and IIoT implementations become easier [14].

- **Similarities:** Similarities between OPC UA and IE include the use of standard Ethernet and TCP/IP, a client-server model (with the client making a connection to the server), robust security based on Public Key Infrastructure (PKI), data represented as attributes grouped into objects, and certification by recognized certification centers.
- **Differences:** Difference between OPC UA and IE are listed in Table III 16.2b.

5. **Miscellaneous issues:** A few common issues associated with APL include:
   - **Transparent communication:** APL supports simple, cost-effective plant modernization through spur-and-trunk technology. It enables simple field installation through screw-type terminals. By allowing transmission of process values and device parameters, APL supports efficient asset management, predictive maintenance, configuration, and parameterization. Therefore, a common physical layer allows seamless, transparent communication across all hierarchical levels, involving devices from hazardous areas to the cloud. Interoperability is ensured through certification of both the protocol and the physical layer, meaning that devices can work together.
   - **Open infrastructure:** For an open infrastructure, interoperability across devices from different vendors is important. Interoperability within the system is ensured through IEEE 802.3cg, which defines data communication as 10BASE-T1L; the APL Port Profile, which defines power supply, connectors, cables, and accessories; and intrinsic-safety specifications such as FISCO (see IEC TS 60079-47, 2-WISE).

## List of abbreviations

**AH** Authentication header
**APL** Advanced physical layer –Ethernet APL
**AR** Augmented reality
**BOOTP** BOOT protocol
**BPDU** Bridge protocol data units
**CA** Collision avoidance
**CD** Collision detect
**CFI** Canonical format indicator
**CIP** Common industrial protocol
**DCF** Distributed coordination function
**DCS** Distributed control system
**DHCP** Dynamic host configuration protocol
**DIFS** DCF interframe spacing
**DLL** Data link layer
**EEE** Energy efficient Ethernet
**ESD** Emergency shutdown
**ESP** Encapsuling security protocol
**FEC** Forward error correction
**FO** Fiber optics
**FTE** Fault tolerant Ethernet
**IE** Industrial Ethernet
**IIoT** Industrial IoT
**IKE** Internal key exchange
**IoT** Internet of things
**IP** Internet protocol
**IT** Information technology
**LAN** Local area network
**LLC** Logical link control
**LSB** Least significant bit
**M2M** Machine to Machine
**MAC** Media access controller
**MBAP** Modbus application protocol
**MSTP** Multiple spanning tree protocol
**NIC** Network interface card
**OSI** Open systems interconnect
**OT** Operational technology
**PCF** Point coordination function
**PCI** Protocol control information
**PCS** Physical coding sublayer
**PD** Powered device (PD)
**PDU** Protocol data unit
**PISF** PCF interspacing space
**PKI** Public key infrastructure
**PLC** Programmable logic controller
**PMA** Physical medium attachment
**PMD** Physical media dependent
**PoDL** Power over data line
**PoE** Power over Ethernet
**PSE** Power source equipment (PoE)
**PTP** Precision time protocol
**QoS** Quality of service
**RSTP** Rapid STP
**RT** Real time
**RTE** Real time Ethernet
**SCADA** Supervisory control and data acquisition
**SIF** Safety instrumentation function
**SIFS** Short interframe spacing
**SIS** Safety instrumentation system
**SONET** Synchronous optical network
**SPE** Single Pair Ethernet

**SPI** Stateful packet Inspection
**STP** Shielded twisted pair/Spanning tree protocol
**STS** Synchronous transport signal
**TCO** Total cost of ownership
**TCP** Transmission control protocol
**TSN** Time sensitive network
**UDP** User datagram protocol
**UTP** Unshielded twisted pair
**VLAN** Virtual local area network
**VPN** Virtual private network
**VR** Virtual reality
**WAN** Wide area network
**WSN** Wireless sensor network

## References

[1] Ethernet Basics Rev2; Phoenix Contact; E-Book; Internet document. https://www.mouser.com/pdfdocs/Ethernet_Basics_rev2_en.pdf.

[2] Introduction to PoE and the IEEE802.3af and 802.3at Standards; M. Eisen; Marcum Technology; IEEE; Oct 2009. https://www.ieee.li/pdf/viewgraphs/introduction_to_poe_802.3af_802.3at.pdf.

[3] Industrial Wireless Technology Special Report; Industrial Ethernet Book; Nov 2022. https://iebmedia.com/technology/industrial-5g-wireless/industrial-wireless-technology-2022- special-report/.

[4] Industrial Ethernet Standards: Paving the Way for the Next Industrial Revolution' Fluke network; White Paper. https://www.flukenetworks.com/edocs/white-paper-industrial-ethernet-standards-paving-way-next-industrial-revolution.

[5] Functional Safety via Wireless Ethernet; J. Weczerek; Phoenix Contact; White paper. https://www.phoenixcontact.com/assets/downloads_ed/global/web_dwl_promotion/EN_HQ_Whitepaper_Functional_Safety_LoRes.pdf.

[6] Power over Ethernet technology for industrial Ethernet networks; Amplicon.com; White Paper. https://www.saelig.com/supplier/amplicon/PoE-white-paper.pdf.

[7] BusWorks® 900EN Series 10/100M Industrial Ethernet I/O Modules w/Modbus; Technical Reference—Modbus TCP/IP; Acromag; White paper. https://www.amplicon.com/docs/White-Papers/Intro_modbusTCP_765a-white-paper.pdf.

[8] Process Control and Automation using Modbus Protocol; Amplicon; White paper. https://www.amplicon.com/docs/white-papers/MODBUS-in-Process-control.pdf.

[9] Understanding Modbus TCP-IP: An Indepth Exploration; S. Sharma; Wevolver; June 2023. https://www.wevolver.com/article/modbus-tcp-ip.

[10] Ethernet To The Field; Ethernet APL; POC foundation White paper; Internet document. https://opcfoundation.org/wp-content/uploads/2020/06/Ethernet-APL_Ethernet-To-The-Field_EN.pdf.

[11] Ethernet backbone improves design, implementation, and lifecycle management of safety and basic process control systems; GE Intelligent platforms; Internet Document. https://scadahacker.com/library/Documents/ICS_Protocols/GE-IP%20-%20Ethernet%20Backbones%20in%20BPCS-SIS.pdf.

[12] Looking at the future of Industrial Ethernet technology; Industrial Ethernet book; Industrial Networking & IIoT; May 2023. https://iebmedia.com/technology/industrial-ethernet/looking-at-the-future-of-industrial-ethernet-technology/.

[13] Ethernet vs. fieldbus: the right network for the right application; M. Bacidore; Control Design; Oct 2016. https://www.controldesign.com/articles/2016/ethernet-vs-fieldbus-the-right-network-for-the-right-application/?show=all.

[14] What is OPC UA and how does it compare with Industrial Ethernet?; D. Collins; Motion Control Tips; May 2020. https://www.motioncontroltips.com/what-is-opc-ua-and-how-does-it-compare-with-industrial-ethernet/.

Chapter 17

# Industrial Ethernet: EtherCAT, Ethernet I/P, and PROFINET

After having generalized knowledge on Ethernet and Industrial Ethernet (IE), Specific discussions on selected IE have been covered here. EtherCAT stands for "Ethernet for Control Automation Technology." This protocol, standardized under IEC61158, is popular for "determinism and time accuracy." Thus, EtherCAT is an open, high-performance real-time (RT) IE *fieldbus* system for automation applications. EtherCAT operates on the famous "*Processing on the fly*" technology. It operates on a single-master, many-slaves architecture with the master initiating the frames for slaves to forward them. While passing through, each slave reads the address, processes the data frame, and inserts its input data into it to minimize hardware delay (a few nanoseconds). EtherCAT uses standard IEEE 802.3 frames utilizing inexpensive, off-the-shelf Ethernet cables and hardware in a flexible topology. It has built in robust diagnostic feature. High precision synchronization (single reference clock) with jitter in microsecond and Integrated safety with Safety over EtherCAT" (FSoE) are special features of the protocol to meet high safety integrity levels (SIL 3) according to IEC 61,508 and IEC 61,784–3. CAN application protocol over EtherCAT (CoE), Ethernet over EtherCAT (EoE), and File Access over EtherCAT (FoE) are its specialties. Motion controls, Robotics, Assembly lines, etc., are major areas EtherCAT finds its applications. Ethernet/IP is an open standard, common industrial protocol (CIP) offering high-speed, large package size data exchange for industrial automation. Ethernet/IP is an application layer protocol (not physical) that runs *on* standard Ethernet allowing devices to communicate in a unified network ready for Internet of Things (IoT)/Industrial IoT. This offers flexible topology including Device-level ring (DLR). Devices on an Ethernet/IP network present their data as attributes grouped into objects as per Ethernet/IP specification [identity, Transport Control Protocol (TCP), router] with application objects for specific device. Two message types in Ethernet/IP are Asynchronous initiated on-demand *Explicit* type mainly on TCP protocol and initiated continuously *Implicit* type mainly based on User Datagram Protocol (UDP) protocol. Ethernet/IP is a certifiable standard. Ethernet/IP has well-defined (by CIP) object structure and message structure. On account of data-driven efficiency seamless integration Ethernet/IP not only very popular in manufacturing industries but in process industries like Oil & gas, Food Beverage Waste treatment, etc. Leading, open international standard for IE communication, PROFINET is designed to meet the various demands of industrial automation. Ethernet-based PROFINET built on standard IEEE 802.3, includes features: *RT communication types, viz. RT, Isochronous* RT (*IRT*)**,** *Time sensitive network (TSN)*, *Flexible topologies*, *Diagnostics & redundancy*, *Safety* and *Management*. For time-critical applications, it deploys real-time channel to deliver data in a fast and deterministic fashion. For increased performance, PROFINET IRT eliminates variable data delays by enhancing the rules used to switch Ethernet traffic and creating special rules for PROFINET traffic. PROFINET defines cyclic and acyclic communication between components, including diagnostics, functional safety, alarms, and additional information. Ethernet cables connect PROFINET components within a network, allowing other Ethernet protocols to coexist within the same infrastructure. Open Platform Communication (OPC) Universal Architecture (UA) is a classic example. PROFINET can employ TCP or UDP communication for nontime-critical tasks, such as configuration and parameterization. Apart from Factory & process automation, PROFINET is well suited in applications like: *Robotics & motion control*, *Automotive assembly lines*, Power and other process plants like Waste water treatment, etc. For high popularity of PROFINET, it could be considered as foundational IE technology for Industrie 4.0 and allowing IIoT implementation. In this chapter, we deal with these IE types in details.

## 17.0 Industrial Ethernets: EtherCat

EtherCAT far very powerful Ethernet approach, and the one that is best tailored to automation requirements! Apart from different transmission speeds of 10, 100, or 1000 MBaud, EtherCAT can operate in half and full duplex communication depending on network configuration & hardware. Compared with full duplex, half duplex is less efficient in data transfer since the system has to pause after

**Plant Intelligent Automation and Digital Transformation. https://doi.org/10.1016/B978-0-323-89892-8.00019-3**

each transfer in one direction to compensate for propagation and response times. Utilizing the standard IEEE 802.3 physical layer, EtherCAT technology overcomes many limitations of other Ethernet solutions. The newly developed Fieldbus memory management unit (FMMU) in each I/O terminal reads the data addressed to it, while the telegram continues through the device. Similarly, input data are inserted while the telegram passes through. The telegrams are only delayed by a few nanoseconds. EtherCAT's key functional principle lies in how its nodes process Ethernet frames: each node reads the data addressed to it and writes its data back to the frame all while the frame is moving downstream. This leads to improved bandwidth utilization (one frame per cycle is often sufficient for communication) while also eliminating the need for separate switches. With Ethernet frames reaching the data of many devices in both directions, the data rate increases to ~ 80%, that is, the full-duplex features of 100BaseTx are fully utilize to make effective data rates of up to 80% of 2 x 100 Mbit/s. Instead of external switches EtherCAT device embeds a switch. It conforms to IEEE 802.3 right up to the individual terminals but the transfer physics is converted in the coupler to E-bus, for electronic terminal box. The Window CAT or TWINCAT EtherCAT master — the core software suite to turn Personal computer (PC) into a real-time control system. TwinCAT (Y driver) is integrated into the system as an Operating system (OS) compatible network driver, and as a TwinCAT fieldbus card. At the receiving end, Ethernet frames received are examined by the TwinCAT I/O system, to filter out those with real-time relevance and other frames are passed on to the OS. Windows device driver Network driver interface specification (NDIS), enables a single Network Interface Card (NIC) to support multiple network protocols (TCP/IP, IPX ISDN). Apart from the decentralized I/Os, operating devices, complex systems such as fieldbus masters, fast serial interfaces, gateways, and other communication interfaces even further Ethernet devices (other protocol variant through decentralized port terminal) can be addressed. The EtherCAT is fully Ethernet-compatible, and open design to tolerate other Ethernet-based services and protocols with minimum loss of performance. Devices with fieldbus interface are integrated via EtherCAT fieldbus master terminals. There is no restriction on the type of Ethernet device that can be connected within the EtherCAT strand via a switch port.

## 17.0.0 Fundamentals of EtherCAT

Open, high-performance, real-time Ethernet fieldbus (RT down to I/O level) EtherCAT optimized for automation applications requiring short data update times with very low jitter. Standard Ethernet frame, High precision synchronization with distributed clocks, flexible topology, Diagnostics, etc., are major features. EtherCAT is also an IE, and any Ethernet device can be connected to a switch port It can interface to any Ethernet device, and it supports vertical integration for IT-OT integration. The use of one frame per network node per cycle leads to the use of a small bandwidth, and consequently, a weak overall performance of the network. EtherCAT approaches the matter differently: namely by "processing on the fly" (each node receives, interprets, and copies a packet before sending it) to achieve deterministic high efficient data exchange (12,000 digital I/O in 350 μS. Also, direct memory access saves time. Media access control (MAC) is Peripheral Component Interconnect (PCI) Bus master, data are provided by Direct memory access (DMA) directly to PC Random access memory (RAM). The unique way EtherCAT processes frames makes it the fastest Industrial Ethernet (IE).

Technology and has the highest utilization of bandwidth hence the corresponding performance and efficiency. EtherCAT has the maximum deterministic properties. Also, EtherCAT has less device granularity to accommodate complex nodes/2 bit Input/Output (I/O)s should be able to be used economically as EtherCAT slaves. The development of EtherCAT mainly focused on various issues like:

- **Broad applicability**: All commercially available Ethernet controller should be able to be used as an EtherCAT master. Also, any process, PC, or Industrial PC (IPC) could be turned into EtherCAT through TWinCAT as discussed above.
- **Conformity:** EtherCAT is able to coexist with other Ethernet devices/protocols on the same bus. Standard structural components such as Ethernet switches are usable for EtherCAT.
- **Short cycle times:** Typical cycle times of significantly < 100 μs?
  In this subsection, brief discussions have been presented on these fundamental issues related to EtherCAT.

1. **Functional principle**: The EtherCAT applications read the data addressed to them while the telegram runs through the application. Input data also are inserted while the telegram is running through. The frames are only delayed by a fraction of 1 microsecond at each network node, and many nodes can be addressed with a single frame. Effective data rate is increased by over 90%, hence for duplex over 2X90%. Since the master uses a standard MAC, a master can be implemented on any hardware platform with Ethernet port. EtherCAT slave controller (ESC-EtherCAT communication chip) processes frames on the fly in hardware. Here, important issues are as follows:

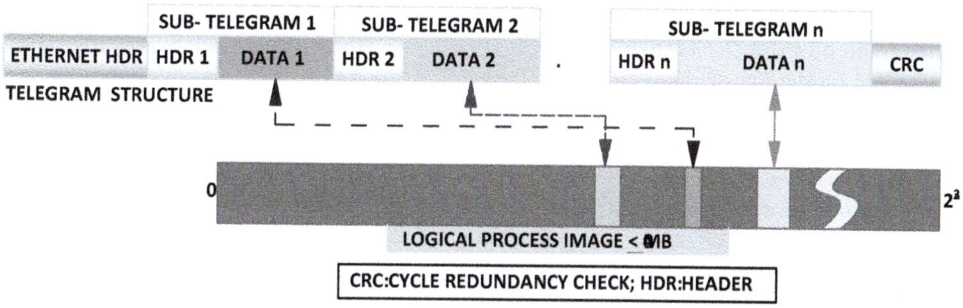


FIG III 17.0a1 TYPICAL ETHERCAT PROTOCOL STRUCTURE

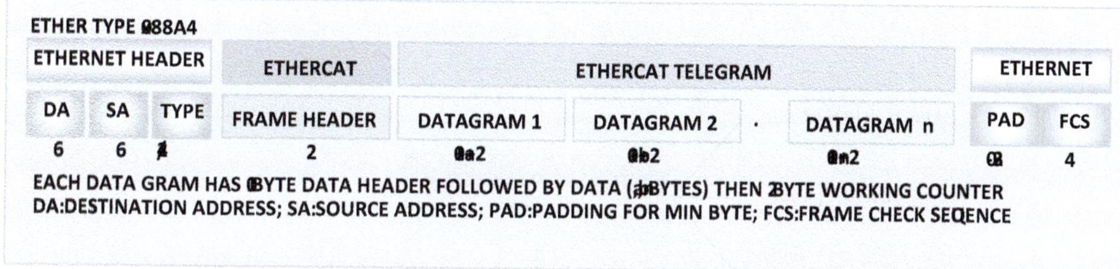


FIG III 17.0a2 ETHERCAT PROTOCOL IN ETHERNET FRAME

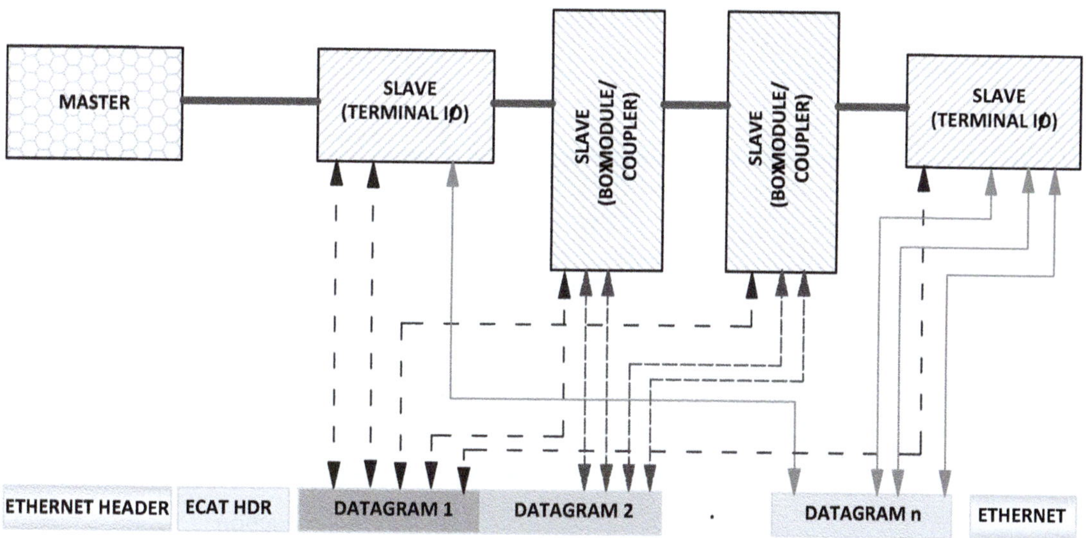


FIG III 17.0a3 ETHERCAT DATAGRAM DETAILS

**FIGURE III 17.0A** EtherCAT protocol structure & ethernet frame.

- **Protocol:** Fig III 17.0a shows EtherCAT protocol structure and Ethernet frame IEEE 802.3.

  When the components are in same subnet, the EtherCAT protocol is transported directly within the Ethernet frame, comprising several subtelegrams individually serving a particular memory area of the logical process images (up to 4 gigabytes in size). However, EtherCAT applications are not limited to a subnet; communication across routers into other subnets is possible. Broadcast, Multicast and communication between slaves are possible with addressing in any order independent of the physical order of the Ethernet terminals. Protocol structure has been detailed in Fig III 17.0a1. In freely configurable the process image, data are copied directly in the I/O terminal to the desired location without any additional mapping. EtherCAT embeds its payload in a standard Ethernet frame as shown in Fig III

17.0a2. EtherCAT protocol is optimized for cyclic process data, the protocol stacks, like TCP.UDP-IP can be tunneled on the top of the stack. The EtherCAT frame contains several datagrams with datagram header indicating the type of access of the master device viz. *Read*, *Write*, *Read-write*, or *Access to a specific slave device* (direct addressing, or access to multiple slave devices through logical addressing). Fig III 17.0a3 shows each *datagram* addresses a specific part of the process image in 4 GBytes address space available. The master device can decide when and which data to access. This also provides much relieves to the master device which only needs to fill via DMA to MAC controller. This configuration is well-suited for slave-to-slave communication (multicast).

- **Components and topology:** Fig III 17.0b shows EtherCAT Components and Topologies.

EtherCAT has one master node/network. The master can be implemented on a MAC in a full-duplex communication mode. As shown in Fig III 17.0b1, various components are as follows: Lowest physical layer uses 100BASE-TX (copper)/-FX (optical fiber), or E-bus (power & communication)

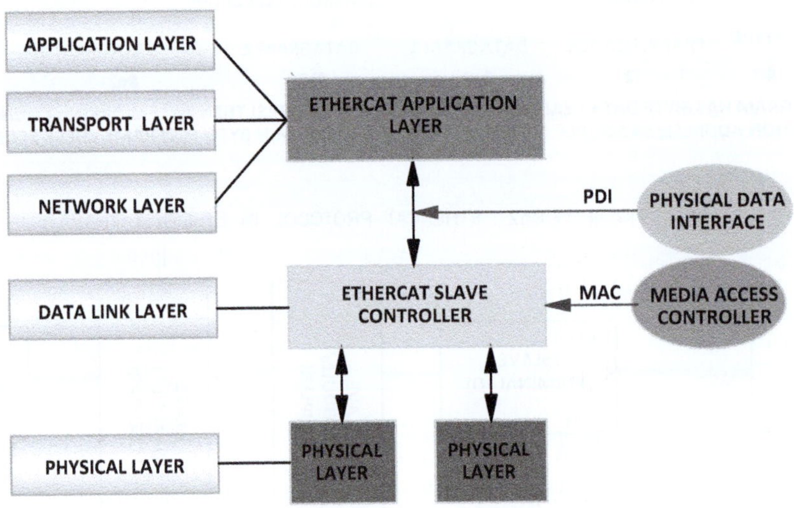


FIG III 17.0b1 ETHERCAT LAYERED COMPONENTS

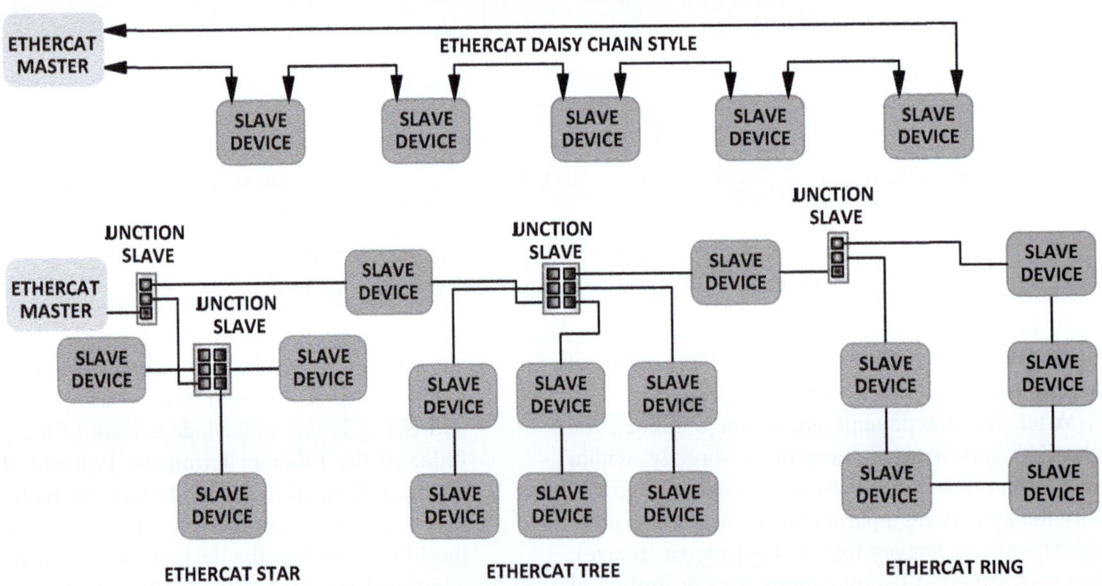


FIG III 17.0b2 ETHERCAT TOPOLOGY VARIATIONS

**FIGURE III 17.0B** EtherCAT layered components and topologies.

based on Low voltage differential signal (LVDS–IEEE P1596.3–1995) signaling as a transmit medium then the MAC layer is implemented as per specification IEC6115 and finally, the application layer comprising different Process data interfaces (PDIs) are available, depending on ESC, may be parallel IO interfaces or Serial peripheral interface (SPI).

EtherCAT supports almost any topology most popular ones are Line (typical fieldbus with junction box), tree or star (switch connections) even ring or daisy-chain, etc., are common. EtherCAT offers a lot of flexibility regarding cable types, so each segment can use the exact type of cable that best meets its needs. Also, use of different optical fibers and copper cables in combination with switches or media converters is worth noting. Typical EtherCAT topologies have been depicted in Fig III 17.0b2

- **Distributed clock (DC)-synchronization:** Accurate synchronization is important for spatially distributed processes requiring simultaneous actions. DC as per IEEE 1588 standard is applied. EtherCAT, the data exchange is based on a pure hardware machine and the communication utilizes a logical ring structure. DC mechanism is initiating by the master, by sending a message to all slaves at a specific address. On receipt, all slaves change the value of their internal clock *twice*, once when the message is received and once when it returns. The master can read all of the changed values and can calculate the delay for each slave. This process can be repeated to prevent zero values. Finally, the time from the first DC slave device is cyclically distributed to all other devices in the system for the slave device clocks for precise adjustment to this reference clock. The distributed clocks are adjusted based on this value,—network-wide time base with a jitter of significantly less than 1 µs is available.
- **Performance:** Short cycle times are possible because the host microprocessors in the slave applications are not involved in the data processing, and the entire process takes place in the slave controller hardware. Purely hardware-based protocol processing through FMMU chip & DMA access to the master, is independent of protocol stack run-times, CPU performance, and software implementation. The transfer of huge data quantity (12,000 digital I/O only takes 300 µs? Distributed clocks enable the axes to be synchronized with a deviation of significantly less than 1 µs. These are a few examples only.

2. **Cable Issue:** Here, let us cover cable redundancy and single cable issue. Cabling issues have been illustrated in Fig III 17.0c
   - **Cable redundancy:** Cable redundancy is designed to take care of the failure of a communication cable in the EtherCAT system by using a logical ring operated in both directions. Both branches can nevertheless still be reached if the ring is interrupted at some point. A second network port is used for ring closure at the EtherCAT Master control. Both cyclic and acyclic frames are sent simultaneously through both ports, for transporting. In the absence of any fault, all the EtherCAT slaves are reached in the forward direction from the primary port meaning, they are processed, and the ESC is only passed through in the forward sense [1]. When there is no fault, all the EtherCAT slaves are reached from the secondary port in the reverse direction—the data in the "*redundancy*" frame is therefore not changed [1]. In case of a cable break, *both frames are processed*—each one on the respective side of the failure, that is, both frames contain a part of the input data. The master has to combine the data of both frames, and gets one frame with all the input data. The EtherCAT master has to consider that a frame on one side is lost and the other frame returns. This is depicted in Fig III 17.0c1, which is also shown in Fig II 8.1b1. Fig II 8.1b2 describes in details how signal/slave failure is taken care of. The working counters from both frames are added to check for its validity.
   - **Single cable: Two** electrically isolated, individually switchable 24VDC supplies power the new EtherCAT P devices as detailed out in Fig III 17.0c2. Both voltages, are directly coupled into the 100 Mbps EtherCAT communication line. In this system, several EtherCAT devices can be used in cascade using single cable to reduce cabling and making system compact, cost-effective wiring & system costs with smaller footprint for devices.
3. **Hot connect:** EtherCAT "Hot Connect" function allows *preconfigured sections* to be added/removed in the data traffic before start/during operation of the system. This can be made effective by modification of connection of "*the communication line,*" "*switching device,*" etc. After connection establishment, to a "Hot Connect" group, some commissioning is carried out resulting a boot up time of several seconds, viz. Establishment of an Ethernet link, Device parameterization, EtherCAT boot up, DC synchronization (as necessary). "Hot Connect" is an extension to the generally applicable rule. No separate license is required for setup, just the EtherCAT devices (couplers and branches) conceived for the purpose.

   **Communication & Direct data exchange:** Communication is always initiated by the EtherCAT master (say IPC). A telegram sent by the master passes through all EtherCAT slaves. The last slave of the communication

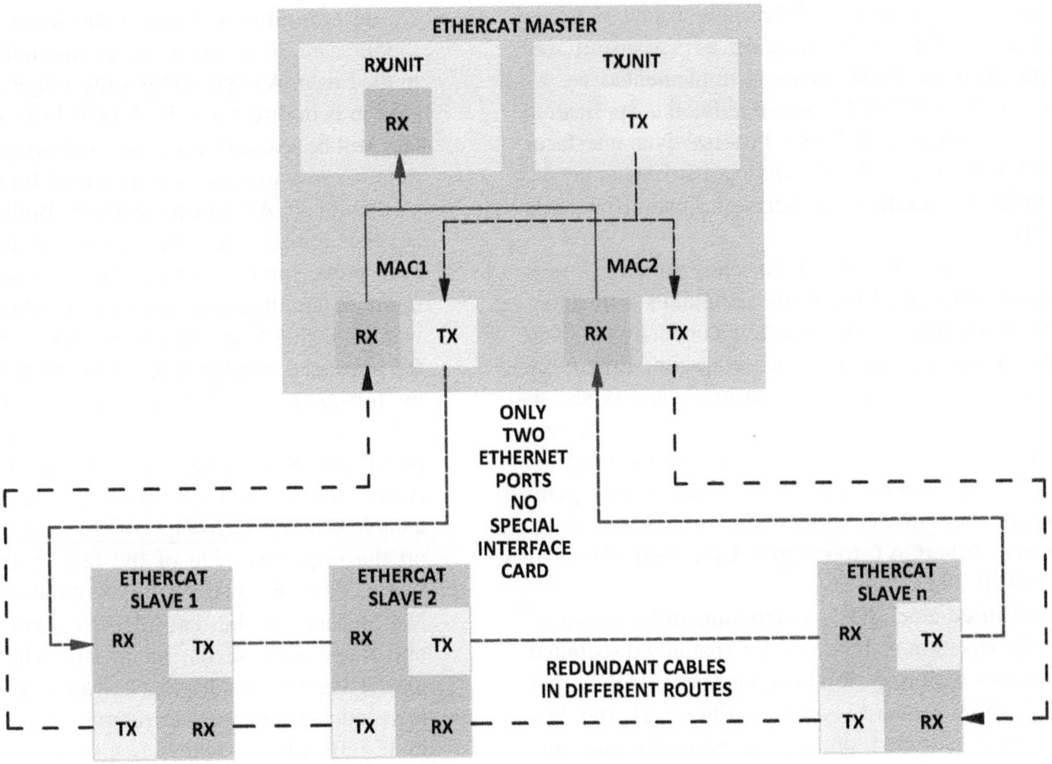


FIG III 17.0c1 ETHERCAT: CABLE REDUNDANCY

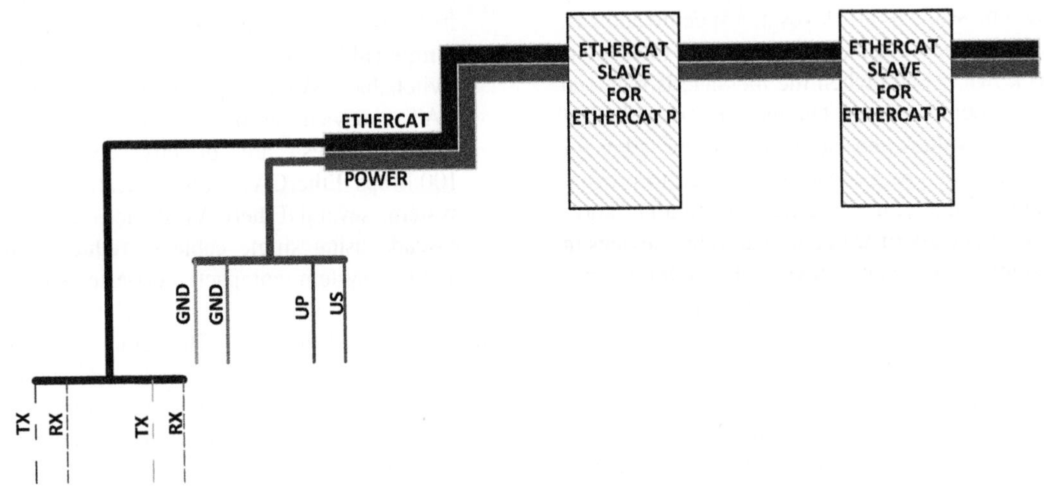


FIG III 17.0c2 ETHERCAT: SINGLE CABLE

**FIGURE III 17.0C** EtherCAT: Cable redundancy and single cable.

chain sends the telegram back to the EtherCAT master. EtherCAT utilizes primarily, *Cyclic* and *Acyclic communication* types mainly for Process data object (PDO) & Service data object (SDO), respectively. PDO *cyclic* communication is the foundation for high-speed and deterministic performance to handle the real-time, time-critical data exchange in cycle times and master mapping the logical process data space to each slave. Acyclic communications related to SDO mainly cover: Nontime-critical data, various mailbox protocols, device configuration, diagnostic data, etc. Apart from these, there are data exchange schemes in Twin EtherCAT. A range of components is available for implementing direct synchronous data exchange between two EtherCAT systems. Depending on the application requirements, the appropriate method can be selected based on the various criteria like: *Synchronous data exchange* (predefined process data in configuration) including synchronization of the DC in the two systems, *Asynchronous* data exchange and support for Automation device specification (ADS) over EtherCAT (AoE), that is, communication within TwinCAT control running on standard Ethernet to exchange data and control devices.

4. **Possible applications**: Because of lowers delay in communication, EtherCAT gives terrific performance for fast machine control. Also, EtherCAT not only suitable for the position control loop, but also excellent for the velocity control loop and even the current control loop for drive controls as cost-effective control solutions. As protocol processing takes place in hardware, it only places low demands on the master connection: a commercially available Ethernet Controller is sufficient. Apart from applications in robotics, CNC machines, packaging, and semiconductor manufacturing, EtherCAT also finds applications in all kinds of power plants, process & food industries.
   *With fundamental knowledge on EtherCAT let us now explore structural and architectural details of EtherCAT.*

### 17.0.1 Structural and architectural issues on EtherCAT

Structurally, EtherCAT comprises Master-slave connections, Ethernet frame, No requirement of Ethernet switch, Use of ESC, and vast choice of physics. "On the fly" processing, flexible topology including logical ring topology are major architectural aspects of EtherCAT. EtherCAT is suitable for both centralized and decentralized system architectures supporting master-to-master, master to-slave, and slave-to-slave communications. An EtherCAT network does not have any topological restrictions in supporting over 65,000 devices. As Fast Ethernet, it can support device up to 100 m (330 ft) apart–greater distance with Fiber optics (FO). Apart from topology flexibility EtherCAT supports Hot Connect/Swap for devices, cable redundancy, single cable options discussed in previous subsection. For parameter data transferred acyclically in large quantities and the timing requirements are relatively noncritical except for diagnostic & event-driven data for which timing requirements are more demanding. However, process data for which the timing requirements are very stringent, are transferred cyclically with different cycle times. It is important to ensure that there are no "dropped cycles" (some slaves fail to communicate during EtherCAT frame). EtherCAT has different addressing options for different types of communication, optimized for the particular requirements. There are different basic approaches for transferring acyclic Ethernet telegrams in cyclic fieldbus mode. In one approach, an appropriate time slice is allocated, in which the acyclic Ethernet telegrams can be embedded. This time slice has to be chosen sufficiently large to be able to accommodate complete Ethernet telegrams. The common Maximum transmission unit (MTU) is 1514 bytes, that is, 125 μs @ 100 MBaud. The minimum resulting cycle time in this variant is about 200–250 μs. EtherCAT takes a different route and combines the advantages of fieldbus technology with the otherwise indisputable advantages of the Ethernet world. The available bandwidth is almost fully utilized at reduced cost [2]. Let us discuss little bit details these issues in this subsection.

1. **Hardware and Motion control:** Major hardware & motion control issues are as follows:
   - **Hardware Issue:** Hardware **design** issues in EtherCAT network are mainly centered on the implementation of specialized slave controllers with respect to physical layer and the use of standard. Physical layer issues are as follows: Major issues for EtherCAT master include the following: Cabling/FO network, flexibility of topology, EtherCAT P. ESC design issues mainly cover: Dedicated ESC, Phy interface, FMMU, and synchronization. Master design considerations are as follows: MAC and DMA capability, etc. Apart from design issues pertinent to *master controller* and *ESC* others issues include the following: Ethernet physics, power management, and host processor interfaces. Power management. The power requirements for each active component must be computed to calculate power requirement. To save footprint, highly integrated solutions mounted in multiple rails, it is a real challenge for power supply system to maintain design operating temperature range. The host processor interface in an EtherCAT refers to the physical and software connections between the main host CPU & the EtherCAT communication hardware, viz. ESC. As shown in Fig III 17.0b1, the host processor interfaces with the ESC via a specific Process data interface (PDI). SPI, on chip bus are common PDI.
   - **Motion control:** EtherCAT is very much suited for motion control applications on account of "Deterministic" performance, "Flexible" topology, and other diverse set features. It can optimize machine performance with low cycle times. Also, *High-speed, RT communication, DC, Master-slave architecture* of EtherCAT mainly make EtherCAT well suited for motion control. The on-the-fly processing of EtherCAT can offer consistent predictable performance necessary for motion control applications. EtherCAT goes beyond the strict physical limitations of centralized PCI-based communications. Distributed processor architecture of EtherCAT offers the bandwidth, synchronization, and physical

flexibility matching with the power of centralized control. High performance, scalability, IIoT/IoT, and machine integration, reduced wiring are some of the advantages of EtherCAT in motion control. Multiaxis control, Advanced interpolations, High-speed tasks are some of the features it can offer. Motion profiling and IO control are handled in the EtherCAT Master [3]. Robotics, Computer numeral control (CNC) machines, Filling & Packaging machines are common examples where motion control done through EtherCAT network.

2. **Communication profile:** EtherCAT **profile** refers to specific functions for a device on EtherCAT enabling interoperability & supporting a wide variety of devices and applications as listed below: Fig III 17.0d shows communication profile and fieldbus integration in EtherCAT. Communication profile has been depicted in Fig III 17.0d1.
   - **CAN over EtherCAT (CoE): With** the CoE protocol, EtherCAT provides the same communication mechanisms as in CANopen-Standard EN 50,325–4: object dictionary, mapping of PDO and SDO [4].
   - **Servo-drive (SoE):** As discussed, SERCOS is a real-time communication protocol especially suited for motion control applications. The SERCOS profile for servo drives is included in the international standard IEC 61,800–7. SoE is responsible for mapping SERCOS profile to EtherCAT.
   - **Ethernet over EtherCAT (EoE):** EtherCAT uses physical layers and the frame of Ethernet. Ethernet refers mainly IT data transfer based on a TCP/IP connection. Using EoE protocol, any Ethernet data traffic can be transported within an EtherCAT segment. Switch port can be used to connect Ethernet devices, and Ethernet frame can be tunneled through EtherCAT (as in Internet).
   - **File over EtherCAT (FoE):** It is similar to Trivial FTP (TFTP) that enables file access in a device. FoE supports a uniform firmware upload to devices across a network. It also includes uniform firmware upload to devices across the network.
   - **ADS over EtherCAT (AoE):** Automation device specification (ADS) is a protocol developed and disclosed by Beckhoff for data exchange between hardware- or software-based devices. ADS over EtherCAT (AoE) is meant to support for different mailbox sizes on the primary/secondary side. AoE complements communication profile such as Co with routable and parallel services.
3. **EtherCAT Addressing:** The EtherCAT system uses two types of addressing for *the slaves,* viz. Auto-increment & Fixed-address addressing. Former one is used during initialization stage, then when *Preoperational* state has been reached, fixed address can be used [ Ref: III 17.0.2.1]. Core of EtherCAT addressing is the extended memory space where slave device data are mapped as logical process image (4 GB space). *ESC*, *FMMU,* and *Working counter* are major components. Several addressing types are as follows:
   - **Physical addressing:** Physical addressing is used particularly for transferring parameter data. A telegram invariably addresses precisely one slave. for Slave addressing can be either fixed-address or auto-increment type. During startup auto-increment types are mainly used whereas fixed one is used once topology id mapped.
   - **Logical addressing:** With logical addressing, a section of a 4 GB logical address space is addressed that may include any number of slaves. The assignment of physical addresses to logical one is configured in the master and transferred to the FMMU of the slaves at the startup phase. Each FMMU configuration contains: *Bit-oriented logical address*, *Physical start address*, *Bit length*, and specification for the direction of the mapping (I/O). Upon receipt of a telegram, the slave checks for matching logical addressing to insert or extract data at the associated position of the data field (for I/O, respectively). Telegrams can therefore be assembled flexibly and optimized to the requirements of the control. This is primarily used for high-performance cyclic data exchange.
   - **Multiple addressing:** With multiple addressing, physical address areas of several slaves can be read. The first slave is addressed via the station address. Through multiple read flag the subsequent slaves in the ring can recognize that they are being addressed. Each slave that has the required physical address area enters its station address and the associated data into the data field, as long as there is space available.
   - **Broadcast:** Broadcast write can be used to write physical address areas of all slaves; state transitions. With a broadcast read, physical address areas of all slaves are read.
4. **Interface and Configuration:** Here, brief discussions have been presented EtherCAT interface and configuration. So, there are as follows:
   - **Interface & integration:** EtherCAT interface involves standard Ethernet hardware like PHYs,& switches with processing frames on the fly at the MAC layer. Integration refers to connecting diverse devices via couplers or direct integration using ESCs. For enough bandwidth of EtherCAT, it is possible to embed conventional fieldbus networks as an underlying system via an EtherCAT Gateway,

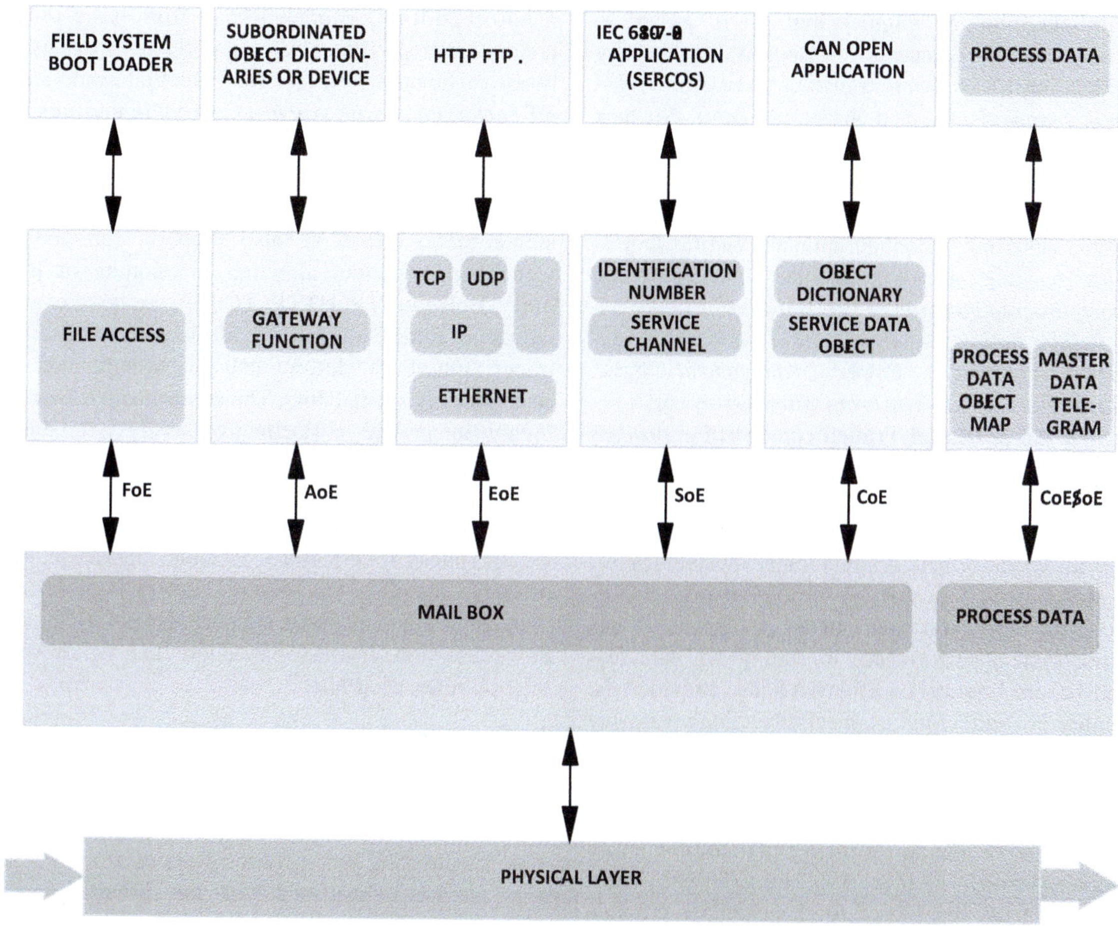


FIG III 17.0d1ETHERCAT: COMMUNICATION PROFILE

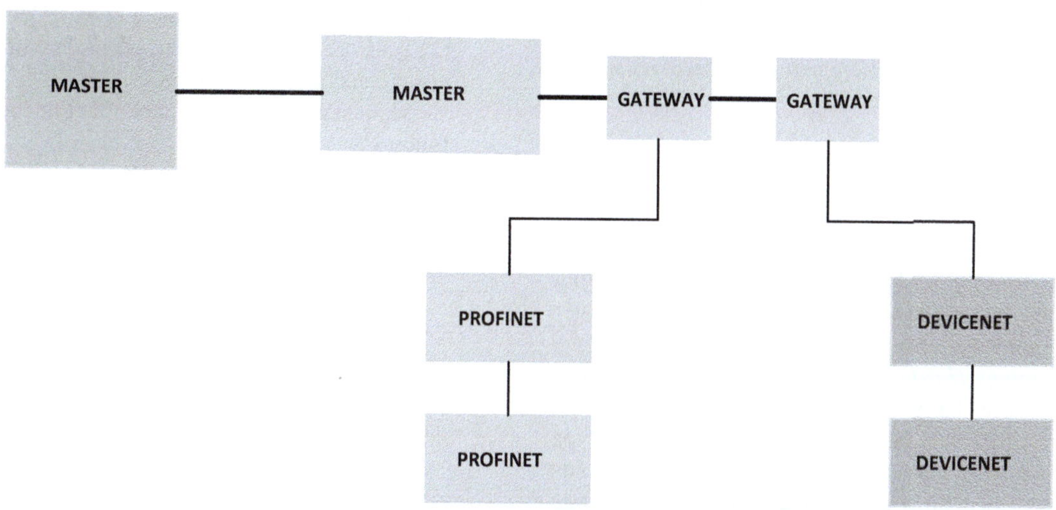


FIG III 17.0d2 ETHERCAT: FIELDBUS INTEGRATION

**FIGURE III 17.0D** EtherCAT: Communication profile and integration [4].

which is particularly helpful when migrating from a conventional fieldbus to EtherCAT. The ability to integrate decentralized gateways also reduces the physical size of the industrial PC, sometimes even to an embedded industrial PC. Fig III 17.0d2 shows integration in EtherCAT.

- **Configuration & engineering:** It is easier to configure and engineer for various issues like: for addressing (*No manual address*, *Addresses assignment automatically,* and *Addresses kept*), No new addressing for new added node. Topology (with target/actual comparison), Diagnostics with exact location, Proper network planning, Slave implementation. Other major configuration and engineering issues needing attention include the following: *Complex configuration of PDOs*, *EtherCAT subdivision information file mismatches*, *Dynamic configuration challenge*, *Power management issue*, *Hardware conflict*, *synchronization error,* etc.

5. **Industrie 4.0 & Digital Transformation:** For *process optimizations*, *predictive maintenance*, *IoT/IIoT implementation*, *Industrie 4.0,* etc., there exists need for *seamless*, *continuous,* and *standardized communication across* all levels. EtherCAT inherently meets these requirements of the Industrie 4.0 through its high performance, flexibility, and open interfaces: EtherCAT has superior system performance in adding big data features to control networks. EtherCAT also provides the flexibility to add cloud connectivity. Edge gateway can access any data within any EtherCAT field device via the mailbox gateway feature of the EtherCAT master. Also, EtherCAT supports open IT interfaces like OPC UA, MQTT, etc., to the master or to the connected devices— providing direct link to IoT/IIoT. Thus, connection from the sensor to the cloud is possible through EtherCAT. TSN is another new area support for EtherCAT for improved RT communication between controller to controller.

   *Let us now explore design and functional issues for EtherCAT in last subsection.*

## 17.0.2 Design and functional issues on EtherCAT

We are aware that the core design issues of EtherCAT enabling it to achieve high speed and efficiency using standard Ethernet principles, but with determinism to cater to RT communication are as follows: *Processing on the fly*, *Flexible topology Hardware implementation* (Slave nodes are dedicated ESC hardware chip), *Scalability* and standard compliance (IEEE 802 Ethernet frames— EtherCAT type identifier 88A4). As far as its design is concerned it avoids stack delays, switching delays, etc. As against average stack delay of EtherCAT of about *0.11 mS,* same for PROFINET and Ethernet/IP will be around 0.58 mS and 1.89 mS, respectively. Similarly, jitter (determinism) in case of EtherCAT will be very low –less than 1 µs (µs) whereas same for Ethernet/IP or PROFINET will be quite high. Instead of relying on time synchronization techniques, it uses DC for synchronization with nanosecond-level accuracy. Apart from basic I/O or data communication, EtherCAT supports & provides a wide range of functions critical for industrial automation like: *RT performance with synchronization*, *Integration of safety* (FSoE), Protocol integration viz. CoE, Ethernet integration EoE, Hot-swap & Hot-connect, Extensive diagnostics, Redundancy including cable redundancy as well as use of single cable. EtherCAT also supports common internet technologies without affecting functioning of RT capability. EoE for FTP, HTTP, TCP/IP, are good example for the same. EtherCAT delivers all benefits of IE at fieldbus prices. EtherCAT doesn't call for specific active infrastructure, any special NIC. The connected hardware used is highly integrated, cost-effective easily available chips. Also, for commissioning and maintenance, there is no need IT experts. Functional safety is built directly into the bus with options for both centralized and decentralized safety logic. Thanks to the Black Channel approach, it is also available for other bus systems. Let us explore in details the design and functional support issues for EtherCAT in this subsection. In this connection II, 8.1.2 of volume II may be referenced also.

1. **Operational issues:** Various areas in basic operational principles are as follows:
   - **State machine:** The EtherCAT state machine (ESM) refers to the standardized sequence of states controlling the operational phase of each EtherCAT device. Thus, state of the EtherCAT slave is controlled via the ESM. Based on the state and specific commands sent by the EtherCAT master, various functions are accessed & executed at EtherCAT slave as detailed in Fig III 17.0e with different arrow head and designation.

     Main states with associated identifications are as follows:
     - *Init (I)*: After switch-on the EtherCAT slave in the *Init* state, no mailbox or process data communication is possible.
     - *Preoperational (P)*: Between I & P slave check proper initiation. In *Pre-Op* state mailbox communication is possible, but not process data communication (initialize process data– sync manager PDO assignment FMMU channel).
     - *Safe-operational (S)*: Between P & S, slave checks the readiness and correctness of Sync. Manager, DC, etc. In *Safe-Op* state mailbox, and process data communication is possible.
     - *Operational (O):* The master switches the slave from *Safe-Op* to *Op* after transfer of valid output data. In this state, the slave copies the output data of the masters to its outputs and mailbox and process data communications are possible.

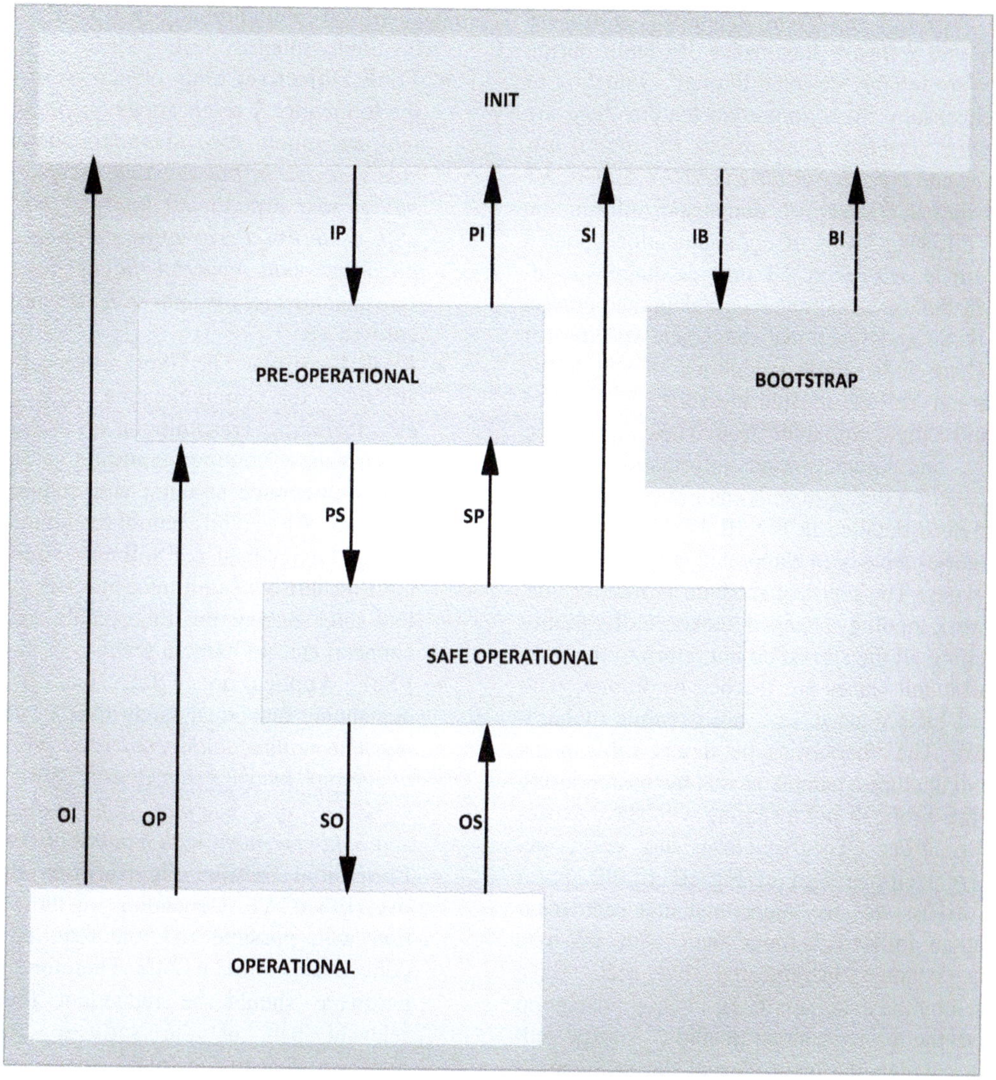


**FIGURE III 17.0E** EtherCAT state machine states.

  - *Boot (B)*: In the *Boot* state the slave firmware can be updated and can be reached only via the *Init* state.
- **COE interface**: Different devices like controllers, shaft encoders, servo drives, etc., used in automation, can be used separately or together in a group on a bus system. For creating order and exchangeability between devices, CAN automation, defined so-called profiles. Some characteristic issues are as follows: *Index number*, *Official name*, *Specification of changeability* and a value (text/number). The CoE interface is used for parameter management of EtherCAT devices. EtherCAT slaves or the EtherCAT master manage fixed (read only) or variable parameters which they require for operation, diagnostics, or commissioning.

2. **Diagnostic discussions:** The EtherCAT slave in each node checks the moving frame for errors with a checksum to supply the slave application only if the frame is received correctly. The master device is able to detect where the fault originally occurred by analyzing the nodes' error counters. EtherCAT can detect and localize occasional disturbances before the issue impacts the machine's operation. In EtherCAT's single frames, the likelihood of bit errors induced by disturbances within a frame is substantially lower for the same cycle time. For fieldbus systems, the availability and commissioning times depend highly on the diagnostic capability. If faults are detected quickly & located precisely with high accuracy, then fault can be corrected quickly. This was philosophy during the development of EtherCAT. There are various broad

diagnostic categories viz. *Cyclic* diagnostic, *Hardware* diagnostics, and *Software* diagnostics. Bit faults during the transfer are reliably detected through evaluation of the CRC checksum. Some important features here are *Breaking point detection, Localization, Physical transfer behavior,* and *topology of the EtherCAT.* EtherCAT Slave provides a variety of diagnostic information relating to differing levels of communication updated at various times. All EtherCAT provide diagnostic information in the exchange of process data, and correct operating mode and functional diagnostic specific for channel. There exist three categories of diagnostic alerts. Red and Yellow updated in every cycle related to input and output variables from slave to master, respectively, and Green master information acyclic updated. EtherCAT diagnostic categories and error types have been detailed in Fig III 17.0f. Slave diagnostic categories have been elaborated in Fig III 17.0f1.

- **Error types:** The physical medium is interruption, the network topology changed unexpectedly, frames not reaching all the slaves, do not return to the master at all or all slaves are reached by frames, with corrupted bit sequence, etc., are example of hardware error. The other error type, that is, software error mainly includes: parameter sent by master during startup are wrong or not matching with slave expectation or earlier error-free slaves any error, etc. These are well depicted in Fig III 17.0f2. From earlier discussions, we understand that each datagram in an EtherCAT frame ends with a 2-byte working counter incremented by each slave addressed by the datagram itself. In case a datagram returns to the master with an invalid (=unexpected) working counter, the input data of the datagram is discarded by master as shown in Fig III 17.0f3.
- **Hardware Diagnostics:** The diagnostic information at hardware level consists of error counters provided by slave devices at standard memory addresses which could be accessed by the master device to provide to the application. Master considers a frame is lost when it does not return or it is corrupted.
- **Software Diagnostics:** As shown in Fig III 17.0e slave operation is controlled by ESM. Each slave reports its current state, and the flag of an error condition in ESM which provides the basic diagnostic information at software level.

3. **EtherCAT safety:** EtherCAT offers the possibility of safety-related data transmission in parallel with standard data on the same network with the help of the FSoE protocol. FSoE is based on "Black Channel" principle but is not considered to be part of the data safeguarding but allows to tunnel safe data via a nonsafe communication channel. FSoE is a TÜV-certified technology developed according to IEC 61,508 and standardized internationally in IEC 61,784–3. Also, it meets Safety integrity level –SIL3.

- **FSoE Objective:** Main objectives of FSoE include the following: *A single communication for both control and safety data, Seamless integration of the safety design in the machine design, The ability to modify and expand the safety system architecture with flexibility, Precertified solutions for safety applications with powerful diagnostics of EtherCAT, Use of same tool for both standard and safety applications,* etc.
- **FSoE Features:** The FSoE specification has no restrictions for communication layer and interface, etc., including Transmission speed, Length of safe process data, Routing via unsafe gateways. It offers totally integrated solution with reduced fieldbus & interface. Also, it has central configuration, diagnostics, and maintenance. FSoE makes full use of EtherCAT in terms of unlimited number of nodes, very low jitter & reaction time, cable redundancy, hot connect, etc., to name a few.
- **FSoE Application:** FSoE finds its applications demanding functional safety like: *Protection against machine malfunctioning, Operator protection against dangerous machine movements, Monitoring of the workspace of a machine, Safe feeding of material,* and *Safe movement with manual intervention* [5].
- **Functional Details:** Fig III 17.0g shows Failsafe over EtherCAT. Depending on the target application with possible risk the required SIL can be defined as per IEC 61,508. Therefore, typical safety hardware should be redundant for the safety-relevant part of the software and hardware. However, due to the "black channel principle" the communication interface is single channel, hence and therefore a standard EtherCAT Sub-Device Controller can be used in line with Fig III 17.0g1, that is, the safety containers are transported as part of the communication's process data. Safe data transfer does not depend on the underlying communication technology, and isn't restricted to EtherCAT only and can make use of copper cables, fiber optics, and even wireless connections. FSoE architecture with frame has been detailed in Fig III 17.0g1
- **FSoE frame:** As shown in Fig III 17.0g2, FSoE Frame is embedded as a container in the process data of the device. Each device detects a new FSoE Frame if *at least 1 bit* in the frame has changed. The minimum FSoE frame length is 6 bytes. The FSoE frame includes a command (CMD), Safe Data, CRC, and connection ID.
- **Decentralized FSoE:** Self-explanatory Fig III 17.0g3 shows multiple FSoE devices for

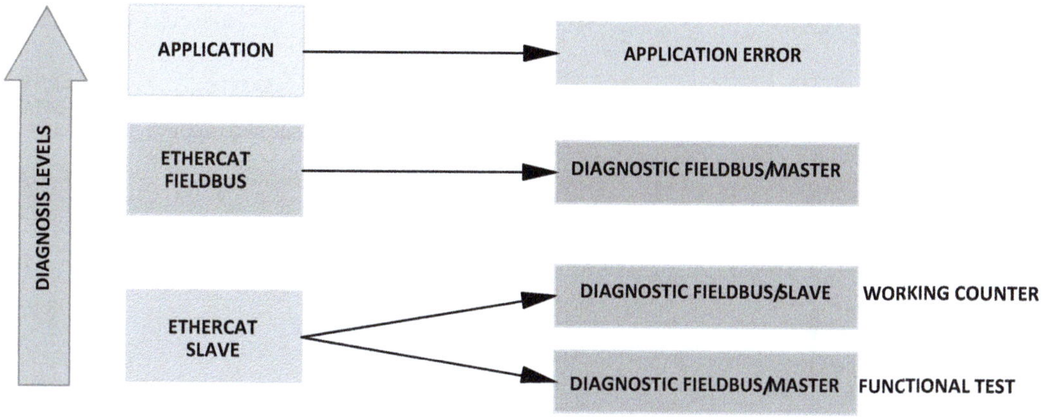


FIG III 17.0f1 SLAVE DIAGNOSTIC CATEGORIES

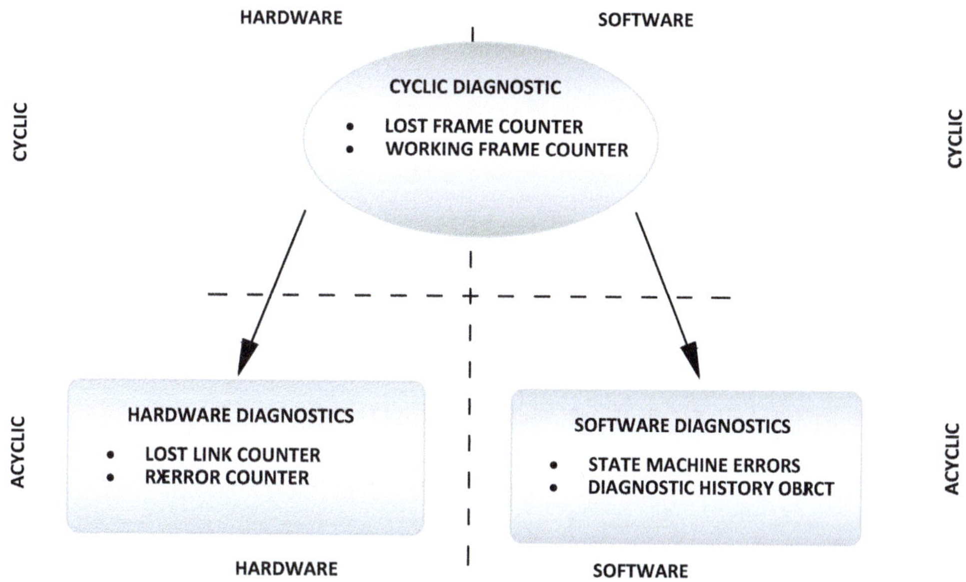


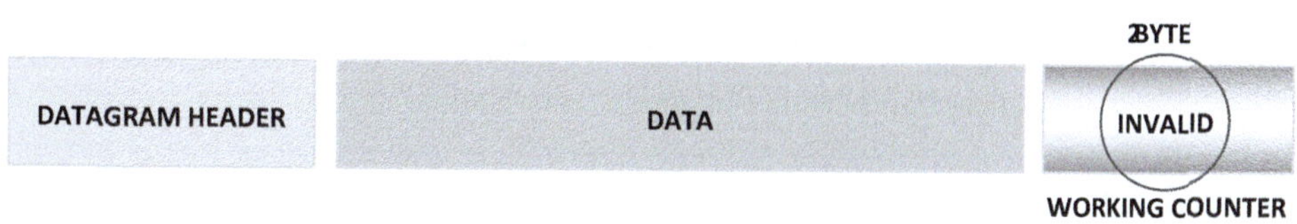


FIG III 17.0f3 WORKING COUNTER ERROR

**FIGURE III 17.0F** Slave diagnostic categories and error types.

decentralized operation. The great advantage here that main controller can be implemented in nonsafe device.

4. **TwinCAT I/O features:** TwinCAT I/O is a *real-time driver* for Windows programs. *Program variables*, the *I/O devices* present in the system, and the *connected I/O channels* are brought into the proper relationship. *Hardware independence with flexibility*, *Seamless EtherCAT integration*, *Automatic configuration*, *Process image mapping* are features worth noting. A few other features of TwinCAT are as follows:
   - It provides variable-oriented linkage of I/O devices & tasks as well as variable-oriented linkage of tasks to other tasks.
   - With 1 bit as smallest unit, it supports both synchronous and asynchronous relationships.
   - Data regions and process images are exchanged consistently.

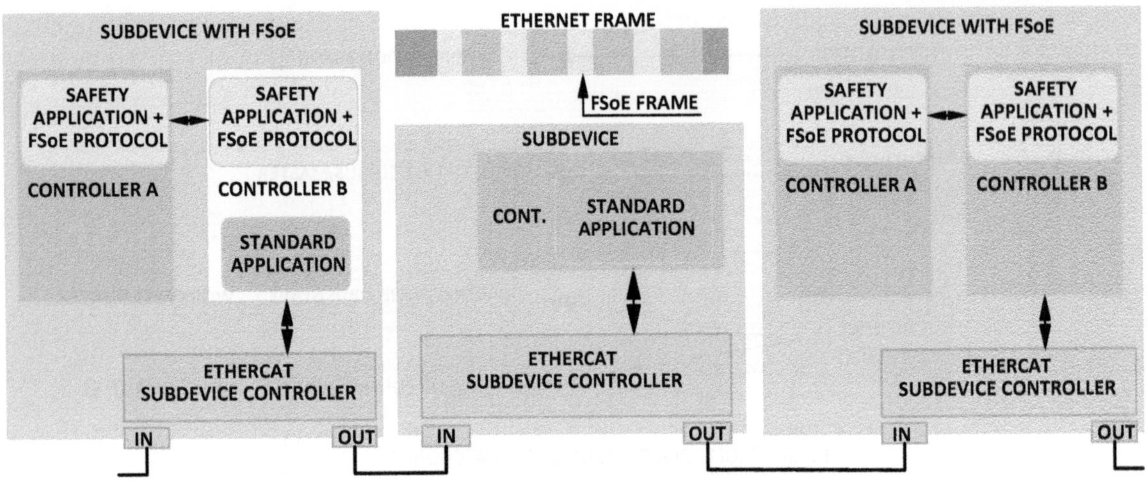


FIG III 17.0g1 FSoE ARCHITECTURE WITH FRAME

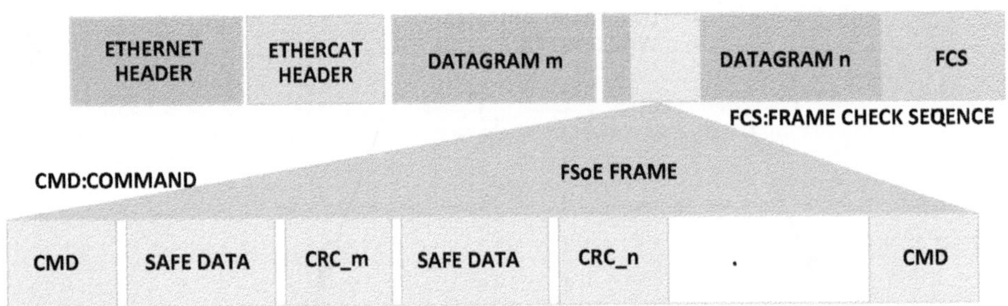


FIG III 17.0g2 FSoE FRAME DETAILS

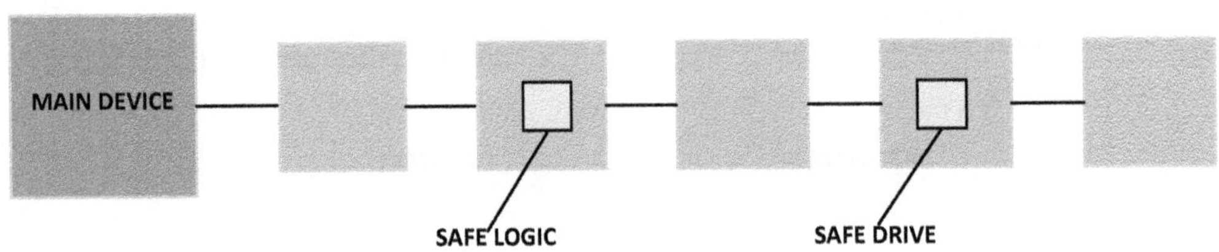


FIG III 17.0g3 DECENTRALIZED FSoE

**FIGURE III 17.0G** FailSafe over EtherCAT.

- TwinCAT is open and supports all fieldbuses and other interfaces like: EtherCAT, Ethernet, Lightbus, PROFIBUS DP, PROFIBUS MC (Motion control), Interbus, SERCOS, CAN Open, DevicNet, USB, Printer port, System management bus, etc.
- TwinCAT I/O provides the connected I/O channels as a structure to organize the mapping of the data transfer.
- TwinCAT I/O permits server process images to be linked bit-wise to I/O channels and provides tools which can, for instance, link a lot of channels continuously with a single command. The connections can be moved by drag and drop.
- With the aid of the TwinCAT I/O system, the I/O of a task can be cyclically exchanged with the outputs and inputs of another task—the data consistency remains intact. The TwinCAT I/O system is variable-oriented, which means that the smallest unit which can be addressed and linked is a variable, and the smallest size of a variable is 1 bit.

5. **TwinCAT Mapping:** TwinCAT I/O generates address relationships, that is, mappings between tasks and I/O devices to provide powerful online tools for:
   - Online display in a directory tree, watch window,
   - Force and write to commission and test task variables and I/O devices.

- Consistent diagnostic data are displayed for all I/O devices.
- By managing variables, TwinCAT I/O makes import and export functions available.

*Now, let us conclude the discussions on EtherCAT and move on to discuss Ethernet/IP.*

## 17.1 Industrial Ethernets: Ethernet/IP

Automation architectures are expected to provide primary services viz. *Control, Configure,* and *Collection*. Control services involving the exchange of time-critical data (with some level of priority and interruption handling capability) between controlling devices such as Programmable logic controllers (PLCs), Distributed control system (DCS), and I/O devices including drive controls. Also, networks must provide users configuration capabilities to set up and maintain their automation systems. Finally, services with capability to collect data for monitoring, alarms, analysis, troubleshooting, and maintenance. Introduced in 2001, EtherNet/IP is a CIP, implemented over standard Ethernet. EtherNet/IP and CIP are managed by ODVA, publish EtherNet/IP Specification, and help ensure compliance through conformance testing. Since this is a CIP network, application and user layers of CIP are on the top of the Ethernet comprising physical, data link, and transport layer. So, as a core concept, three main parts of EtherNet/IP are *Standard Ethernet, CIP,* and seamless integration/convergence of plant Operational technology (OT) with Information technology (IT) networks. ODVA-based EtherNet/IP is compliant to IEEE 60,802 for TSN, and Namur 107 for diagnostics. Also, it can offer high speed up 1 Gbps as Ethernet. EtherNet/IP utilizes standard TCP for *explicit* messaging and UDP for implicit I/O messaging. *Flexible topologies, Scalability, Improved performance,* and *Integrated services* like *CIP– Safety*, Security, Synchronization, and Motion control are major advantages of EtherNet/IP. All devices on an EtherNET/IP present their data as attributes grouped into objects. There are EtherNet/IP required objects every device must have as defined by EtherNet/IP. Also, there will be application objects supporting specific device. As a part of CIP, it has the defined object structure and detailed message transfer. Open flexible industrial network protocol EtherNet/IP enables devices from different manufacturers to easily communicate with each other viz. PLCs, DCSs, etc. Application layer protocol. Ethernet/IP captures about 30% of the industrial Ethernet market. EtherNet/IP can integrate applications with one another and the enterprise creating an active infrastructure using standard Ethernet and Internet protocols. The EtherNet/IP security approach aligns with the size and connectivity of the network. The media independence of EtherNet/IP also contributes to support for IEEE 10BASE-T1L/T1S SPE, which is growing in importance. As a media independent network, EtherNet/IP can use Single pair Ethernet (SPE) as another physical layer offering benefits of SPE including cost & size reductions in Ethernet PHYs, cable, and connectors that enable cost-effective connectivity for smaller devices as well as more efficient long cable runs. This also reduces overall labor for panel installations. The possibility to connect simple devices to Ethernet is opened up. SPE also reduces the and provides information for concepts such as prognostics. Major functional issues influence EtherNet/IP mainly include determinism and RT communication which is well managed through use of managed switches, TSN, and well-planned network to get desired Quality of service (QoS). With underlying Ethernet technology can support high speeds. With added CIP security features and diagnostic support EtherNet/IP functionally is highly secured and well suited for diagnostic feature developments. CIP protocol ensures basic interoperability, vendors can define their own application-specific objects and functions as a part of implementation issue. Proper network planning configuration, etc., are essential. For proper implementation selection of the correct industrial-grade switches, routers are very important. In this section, we discuss EtherNet/IP at length to cover all the above issues.

### 17.1.0 Overview of industrial Ethernet: Ethernet/IP

Using the Ethernet infrastructure, EtherNet/IP manages the connection between various automation devices, for example, Robots, PLCs, Sensors, CNCs, etc. Robust, but little complex with steep learning curve, EtherNet/IP is based on CIP and is managed by the ODVA. EtherNet/IP offers "Producer-consumer" services enabling users to control, configure, and collect data from over a single network. It can be used as a backbone for multiple distributed networks. EtherNet/IP uses existing standard Ethernet physical and data link layers with TCP/IP suite—the Ethernet standard. In case if we recall discussions in III 14.1.0 we find that the CIP is a media-independent, connection-based, object-oriented protocol designed for automation applications. It encompasses a comprehensive set of communication services for automation applications: control, safety, synchronization, motion, configuration, and information. As discussed, it integrates plant-level OT with enterprise-level IT networks & the Internet. Supported by hundreds of vendors around the world CIP provides users with a unified communication architecture throughout the manufacturing enterprise. *Coherent integration of I/O control, Device configuration data collection, Seamless flow of information across multiple networks, Ability to implement multilayer networks,* etc., are some of the services offered by CIP. This standard organizes networked devices as a collection of objects. It defines the access, object behavior, and extensions that allow widely disparate devices to be accessed using a common mechanism. The core of EtherNet/IP's operation and strength lies in its object-oriented approach to device communication and data representation. Objects orientation helps in standardization, interoperability, and modularity. Object orientation in EtherNet/IP mainly encompasses Objects, Classes, Instances, Attributes, Predefined object, Application-specific object, etc. In this connection III, 14.1.0 may be referenced also. Building on standard PC technologies, EtherNet/IP works transparently with all the standard off the-shelf Ethernet devices found in today's market place and EtherNet/IP can be easily supported on standard PCs. Ethernet/IP is a certifiable standard. With this in mind let us briefly discuss overview of EtherNet/IP covering fundamental issues.

1. **Working principle:** Fig III 17.1a shows EtherNet/IP in models for simplified Object, OSI, and CIP. The CIP application layer refers to a set of application objects & device profiles that define common interfaces and behaviors. EtherNet/IP maps the CIP communication services to Ethernet, enabling multivendor interoperability between devices on Ethernet as well as with the other CIP networks. As discussed, in CIP application layer, devices are represented by an object model, defining device data representation & accessed. There would be application-specific objects for parameters like IP address configuration and other and specific functions [Ref: III 14.1.0 also]. Communication objects and services provide the means to establish communication associations and access device data and services over the network as in Fig III 17.1a1.
   EtherNet/IP uses a client-server architecture to ensure efficient & reliable communication between devices, for exchanging data over an Ethernet network. EtherNet/IP follows OSI model, used for network communication. For the upper layers of the OSI model (levels 5, 6 & 7), as shown in Fig III 17.1a2, EtherNet/IP uses the CIP object-oriented protocol, providing a consistent framework for data exchange between devices. CIP defines a set of rules and conventions for devices to follow to communicate effectively. It ensures that devices can understand each other's messages and interpret the data being exchanged. EtherNet/IP combines CIP with standard Ethernet (IEEE 802.3), that is, mainly for the protocols' transport, network, data link, and physical layers as shown in Fig III 17.1a3.
2. **Features & characteristics:** Features and characteristics of Ethernet IP (EIP) are as follows:
   - **Features:** EtherNet/IP, is specifically tailored for industrial applications with features like: *RT performance*, *Multicast support*, *Integration with other protocols.* All These make EtherNet/IP more reliable and efficient choice for industries with features like:
     - In EtherNet/IP, data can be transmitted at extremely high speeds, for ensuring efficient communication.
     - Real-time performance of EtherNet/IP is excellent allowing immediate exchange of data, for RT monitoring & control of industrial processes. This is extremely crucial for time critical applications process industries/manufacturing or robotics.
     - EtherNet/IP supports multicast communication, meaning that a single message can be sent to multiple devices simultaneously, reducing network traffic to improve efficiency. This is particularly useful, if data needs to be distributed to multiple recipients, as in a factory setting.
     - EtherNet/IP supports a wide range of data types like integers and floats, as well as more complex data structures, allowing devices to exchange different types of data, and work together seamlessly.
     - EtherNet/IP also provides mechanisms for device discovery, configuration, and diagnostics, making it easier for devices to join the network, configure their settings, and troubleshoot any issues that may rise.
   - **Characteristics issues:** Major EtherNet/IP characteristics are as follows:
     - EtherNet/IP has two types of network connections: explicit and implicit.
     - Basically, EtherNet/IP is an application layer protocol on a TCP/IP packet.
     - Devices on EtherNet/IP network present their data as attributes grouped into objects.
     - EtherNet/IP specification defines required objects (every device must have) are as follows: Identity, TCP, Router.
     - EtherNet/IP supports application-specific objects for specific device. EtherNet/IP devices supporting specific devices (viz. motor) have the same set of application objects.
     - EtherNet/IP is part of the CIP that defines the object structure and specifies the message transfer.
3. **Device and communication types:** Various device and communication types are as follows:
   - **Device types and related issues:** Several device classifications, based on their general behavior and types of EtherNet/IP communications they support, as detailed in Fig III 17.1b. Apart from the core EtherNet/IP devices, there are additional device-related issues also. All of them have been enumerated below:

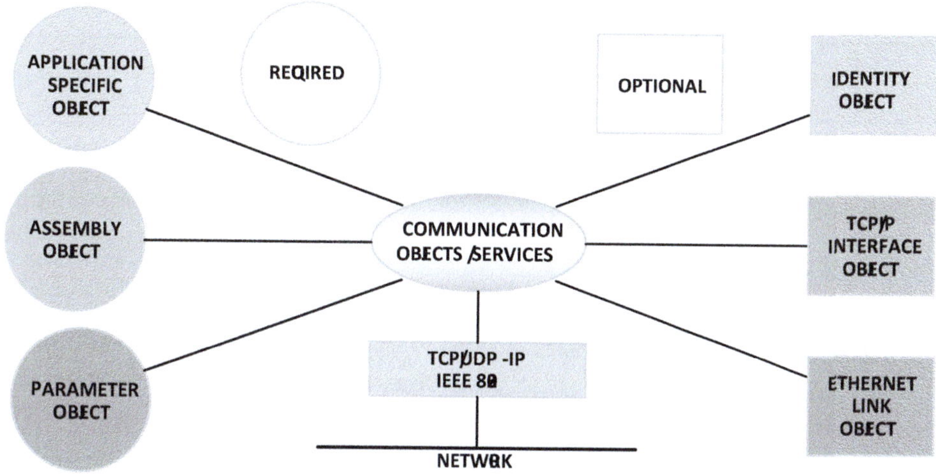


FIG III 17.1a1 ETHERNET/IP MODEL: OBJECT

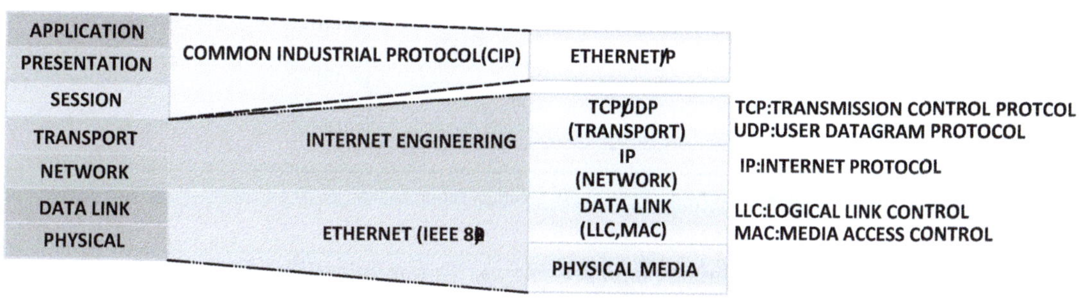


FIG III 17.1a2 ETHERNET/IP MODEL: ISO

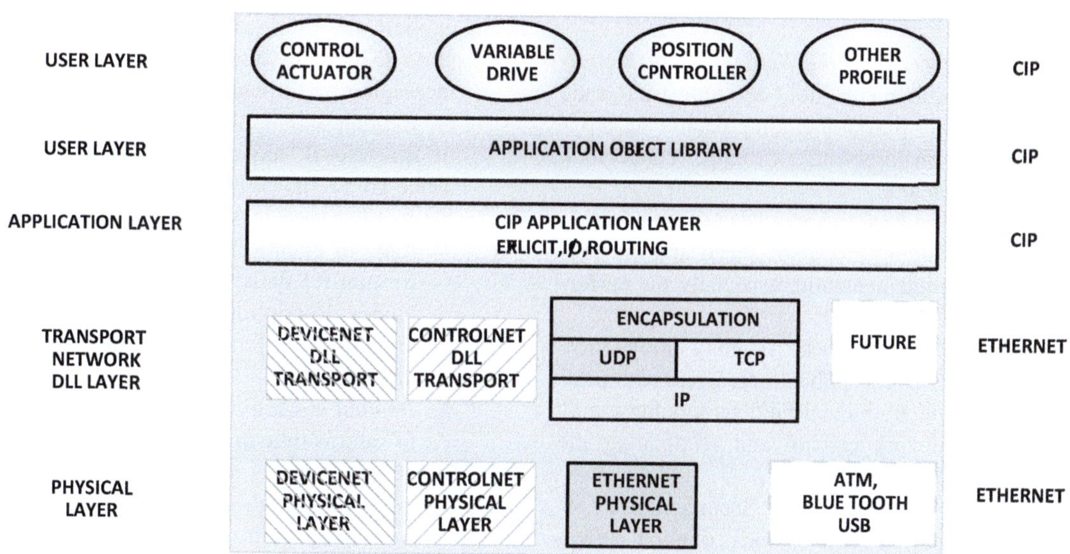


FIG III 17.1a3 ETHERNET/IP MODEL: CIP

**FIGURE III 17.1A** EtherNet/IP model: Object OSI and CIP.

- *Explicit message server:* This server responds to request/response-oriented communications initiated by explicit message clients.
- *Explicit message client:* This client initiates request/response-oriented communications with other devices. Here, message rates and latency are typically not too demanding.
- *I/O Adapter*: An I/O adapter receives *implicit* communication connection requests from an I/O scanner then produces its I/O data at the requested rate.

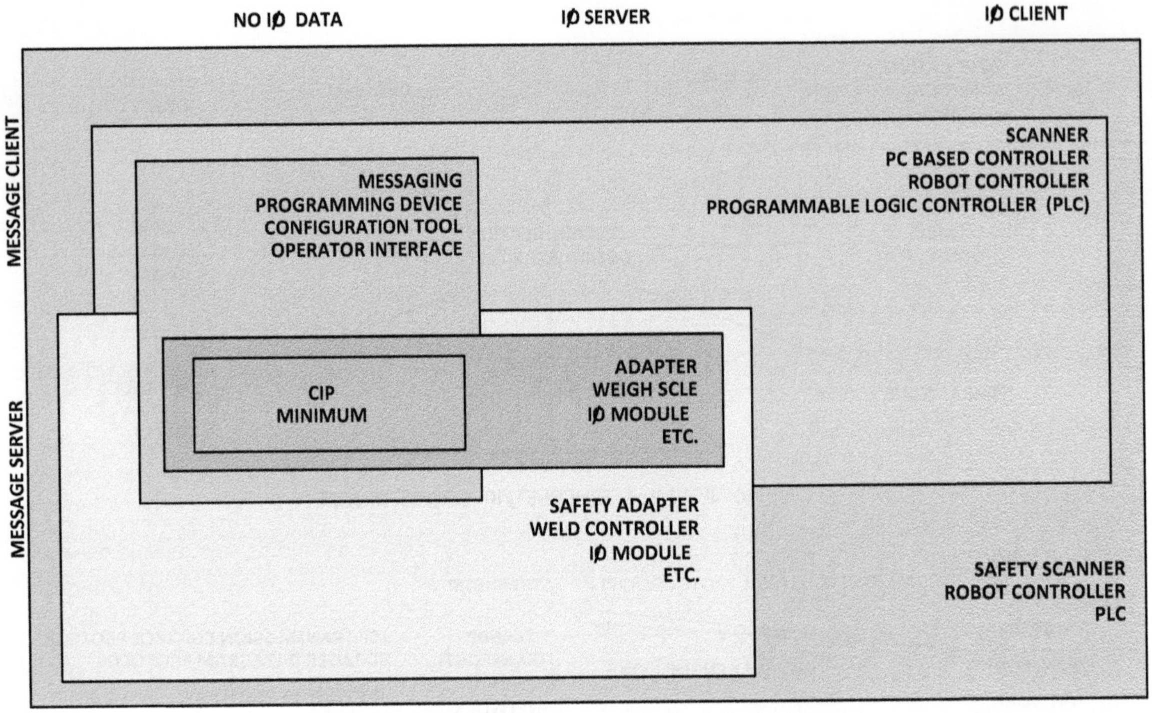


**FIGURE III 17.1B** EtherNet/IP communication and device types.

- *I/O Scanner*: An I/O scanner initiates *implicit* communications with I/O adapter devices. It deals with issues like configuration, connections, etc.
- *Physical layer*: *The EtherNet/IP specification* defines requirements for an industrial-grade physical layer design including details on. environmental requirements, cabling, connectors, and other factors.
- *Performance:* Performance depends on several factors, including processor speed and the efficiency of implementation, especially the performance of TCP/IP stack.
- *Linear topology:* A linear topology enjoys the benefits for certain applications. Users with preference for the traditional fieldbus topology, may simplify network cabling and eliminating the cost of switches.
- *Indicators:* The EtherNet/IP Specification (Volume 2, Chapter 9) also defines common indicators and behaviors for EtherNet/IP devices.
- *Security consideration:* For security fundamental issues include the following: Risk analysis, Risk Reduction, and cost trade off. For device robustness in complex network minimum security issues are as follows: Port Scanning. Resource Starvation Known Flaw Checks, Multicast/broadcast overload, etc., are minimum security checks.

- **Communication types:** EtherNet/IP uses TCP/IP and can manage client-server types of transactions between nodes (any IP address)—this is known as explicit messaging. Equally, EtherNet/IP also supports TCP/UDP (implicit messaging), enabling RT messaging by through multicast.

  Explicit and Implicit are two primary messages of EtherNet/IP having characteristics described in Table III 17.1a.
  - *Explicit messages*: Generally, this has a request/reply or client/server nature approach, and used for non-RT data, normally for information. It includes description of their meaning (expressed explicitly), offering high flexibility with less efficiency.
  - *Implicit messaging*: This is also referred to as "I/O" and is time-critical in nature. This communication type is used for RT data exchange demanding high speed and low latency. Implicit messages, with little information about their meaning, are more efficient but less flexible than Explicit message. With *implicit messaging,* it is possible to establish an association between two devices (CIP) to produce the *implicit message* according to a predetermined trigger mechanism.

4. **DLR operation and nodes:** DLR (an EtherNet/IP) protocol defined by ODVA, provides a means to detect,

**TABLE III 17.1A** Explicit and implicit messaging.

| Issue | Explicit messaging | Implicit messaging |
|---|---|---|
| **Communication type** | Request/response | I/O data transfer |
| **CIP communication** | Connected/Unconnected | Connected |
| **Transport protocol** | TCP/IP | UDP/IP |
| **Usage** | Noncritical information | RT I/O data |
| **Typical use** | Read/Write configuration | RT I/O control data |

manage, and recover from single fault in ring-based network. In DLR technology the network can continue to operate with a single node fault, that is, a single fault will not destroy the entire network. Device on the ring can interrogate & monitor the ring status to act when disruptions occur to the ring. During normal network operation, an active *ring supervisor* uses beacon and other DLR protocol frames to monitor the health of the network and manage the traffic. There could be backup ring supervisor also. The backup ring supervisor and other ring participants monitor the beacon frames to track transitions between normal and faulted connections in the ring [6]. These are explained in details in Fig III 17.1c.

There will be optional redundant gateways which are multiple switches used to connect a DLR network and also connect together through the rest of the network. DLR demands that at least on node to perform as ring supervisor. The ring supervisor needs to perform functions like management of traffic on DLR, collection of diagnostic information. Ring participants also perform functions like: provide these functions: Process data transmission over the network, Passing on data to next node, reporting fault location to ring supervisor.

5. **Object model:** EtherNet/IP's object model uses the CIP to represent device components as objects. Objects within a device are groups of related data & behavior associated with this data. Object structure mainly comprises *Class* (type of component), Instance (unique occurrence), and attribute (Specific data). In its core, CIP objects have three things viz. *Identity object* (device type, vendor ID, serial no.), *Message router* (explicit message between objects in a device), and *Network object* (Network interface, for example, IP Address, Gateway, etc.). Therefore, all these form the Required object (CIP). Two other CIP objects are as follows: Application object describing how data are encapsulated in the device. Vendor-specific object is the third CIP object and is concerned with vendor specifics. Refer Fig III 17.1a1 also.

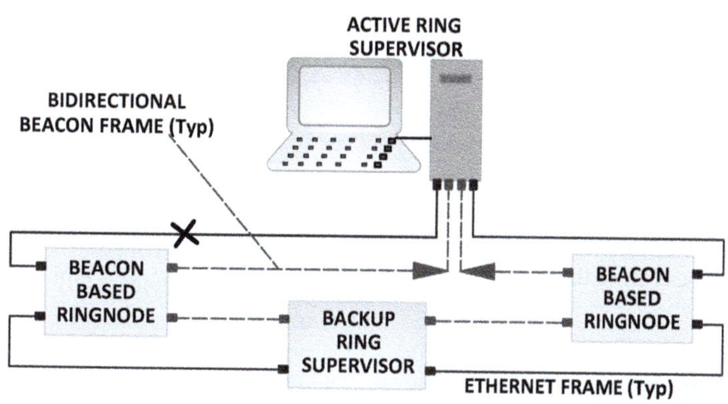


NOTE
BEACON FRAMES ARE SENT VIA BOTH PORTS VIA PRIMARY & SECONDARY; DATA NORMALLY ARE SENT THROUGH PRIMARY ONLY; WHEN THERE IS BREAK (AT X'POINT RING SUPERVISOR WILL UNBLOCK INTERFACE DURING FAILURE), ETHERNET PACKETS ARE SENT THROUGH BOTH PORTS SO THAT SINGLE FAILURE DOES NOT AFFECT THE COMMUNICATION. DURING THIS TIME,IT DOES NOT OPERATE IN RING MODE BUT IN LINE MODE.AFTER IT IS REPAIRED ieWHEN LINK STATUS MESSAGE IS RECEIVED VIA SECONDARY PORT; NORMALCY RETURNED; DATA ARE SENT VIA PRIMARY PORT ALONE.

**FIGURE III 17.1C** DLR operation.

### 17.1.1 Ethernet/IP infrastructure, applications, and practices

EtherNet/IP provides a seamless network communication infrastructure from shop floor to enterprise-level with the help of CIP application layer with user layer as shown in Fig III 17.1a3. EtherNet/IP is a communication network designed for use in industrial environments and time-critical applications. Using the CIP, it has many unique features to offer some advantages of it over other Ethernet-based industrial networks. Some of them are as follows: Provides robust physical layer options for industrial environments and use sealed RJ-45 and M12-4 "D" Coded connectors. Provides flexible installation options leveraging commercially available industrial infrastructure products, including copper, fiber, fiber ring, and wireless solutions. Industrial-grade switches/routers, to manage data traffic efficiently, are backbone component of EtherNet/IP. DLR, QoS Virtual local area network (VLAN), etc., are supported as a part of infrastructure. As shown III 17.1a3 in TCP/UDP- IP are well encapsulated in EtherNet/IP with CIP application layer describing network devices as set of objects with predefined behaviors, facilitating interoperability. EtherNet/IP is used across diverse industries which normally include RT control requiring high-speed time sensitive, synchronized motion applications. Application areas also include critical process industries, Supervisory control and data acquisition (SACDA) including Human machine interface (HMI) systems. Also, on account of CIP safety security coverage EtherNet/IP is well suited for IIoT and essential element in Industrie 4.0/ Digital transformation (DT). EtherNet/IP best practices involve utilization of object-oriented CIP for seamless device communication, utilizing both time-critical UDP-based implicit messaging (UDP) for I/O and TCP-based explicit message for configuration, etc. EtherNet/IP also uses DLR network design. EtherNet/IP uses Electronic data sheets (EDS) for device configuration, defining data types and formats for clients. Some of the common practices include the following: Network planning & assessment, Managed switch, Proper cabling & grounding, suitable Planning for Scalability and integrated security. In this subsection, brief discussions have been placed on infrastructural, application issues in EtherNet/IP with common practices for this.

1. **Infrastructure details:** Here, discussions will be put forward on EtherNet/IP application infrastructure and variations in application areas.
   - **Applicational infrastructure:** Being an IE, is not just an ordinary fieldbus with passive infrastructure, EtherNet/IP networks have an active infrastructure. There exists an interconnection of Ethernet data link layer and network layer switches, Layer two switches allow all protocols to pass, but Layer three switches behave like routers, allowing only designated protocols to pass through to certain locations [7]. For control applications, it is necessary to satisfy end-to-end response time requirements for each transaction. Processing delays in the network infrastructure are usually negligible in comparison with delays in end-devices (nodes. But there is no need to create two physical networks to separate devices with time-critical data and for nontime-critical traffic. This can be achieved by applying switches that support virtual LAN (VLAN) functionality, thus making it possible to create separate virtual networks on the same wire. There are several issues in applicational infrastructure. Some prominent issues include the following: Multicast traffic management within control system, Isolation of enterprise & control traffic, Control device sensitivity to extraneous traffic (unwanted) concerning traffic overload, variation of sensitivity with device type, etc.
   - **Application cases:** We know that EtherNet/IP (also referred to as EIP in the book) supports both the time-critical implicit as well as nontime-critical explicit message transfer services of CIP. The exchange of RT or I/O control (time-critical messages) is based on the producer-consumer model. In this model, a transmitting device produces data on the network, and many receiving devices consume this data simultaneously. This functionality is provided through multicast mechanisms. There is no standard approach to infrastructure implementation for the right balance of performance, data connectivity, and installed cost for all applications. These applications are handled by dividing the situations into following types [7]:
     - *Isolated EtherNet/IP network with a single controller.*
     - *Isolated EtherNet/IP network with multiple controllers.*
     - *EtherNet/IP network connected to the enterprise network.*
     - *EtherNet/IP network integrated with the enterprise network.*
2. **Safety EIP device:** CIP Safety is a high-integrity communication protocol built on top of standard CIP services. CIP safety is an extension of CIP duly certified by bodies like TUV. CIP Safety device using EtherNet/IP must meet the requirements of IEC 61,508 up to and including SIL3, IEC 61,784–3 specifies Functional safety communication profiles (FSCPs). The important requirements for development of EtherNet/IP safety are as follows:
   - **Safety Validator Object:** The safety validator object offers core functions like Error detection,

Authentication, Data integrity check, Time management, etc., using black channel, safety CRC. It executes the safety protocol behavior in the end devices and encapsulates and unencapsulates the safety protocol when it is being transmitted using standard CIP services. In this connection, Fig III 17.1d may be referenced.

- **Safety Supervisor Object:** *Configuration management*, *Coordination*, *Integrity assurance*, *Status* and *Diagnostics* are major functions of supervisor object. So, the safety supervisor object provides the high-integrity configuration behavior for safety devices and assumes all mode control for the device.
- **High integrity:** High-integrity safety systems are typically systems with redundant hardware and well defined and controlled design processes [8].
- **Certification:** Typically, safety designs are certified to ensure they meet the applicable standards by an independent authorized body. CIP Safety devices and typically certified to meet IEC 61,508 and EN 954–1 by TÜV.

3. **Motion in EIP:** CIP Motion is also an extension of standard CIP meant to realize high-performance motion control. For development of CIP Motion on EtherNet/IP, CIP Motion drive device profile is defined and to implement a CIP Motion-compliant drive device. CIP Motion, integrates high-performance *Servo*/Stepper control, I/O, and HMIs onto a single standard Ethernet network. CIP motion refers to the standard layer in EtherNet/IP which defines motion commands & feedback exchange, ensuring interoperability determinism and integration. CIP Motion defines [8]:

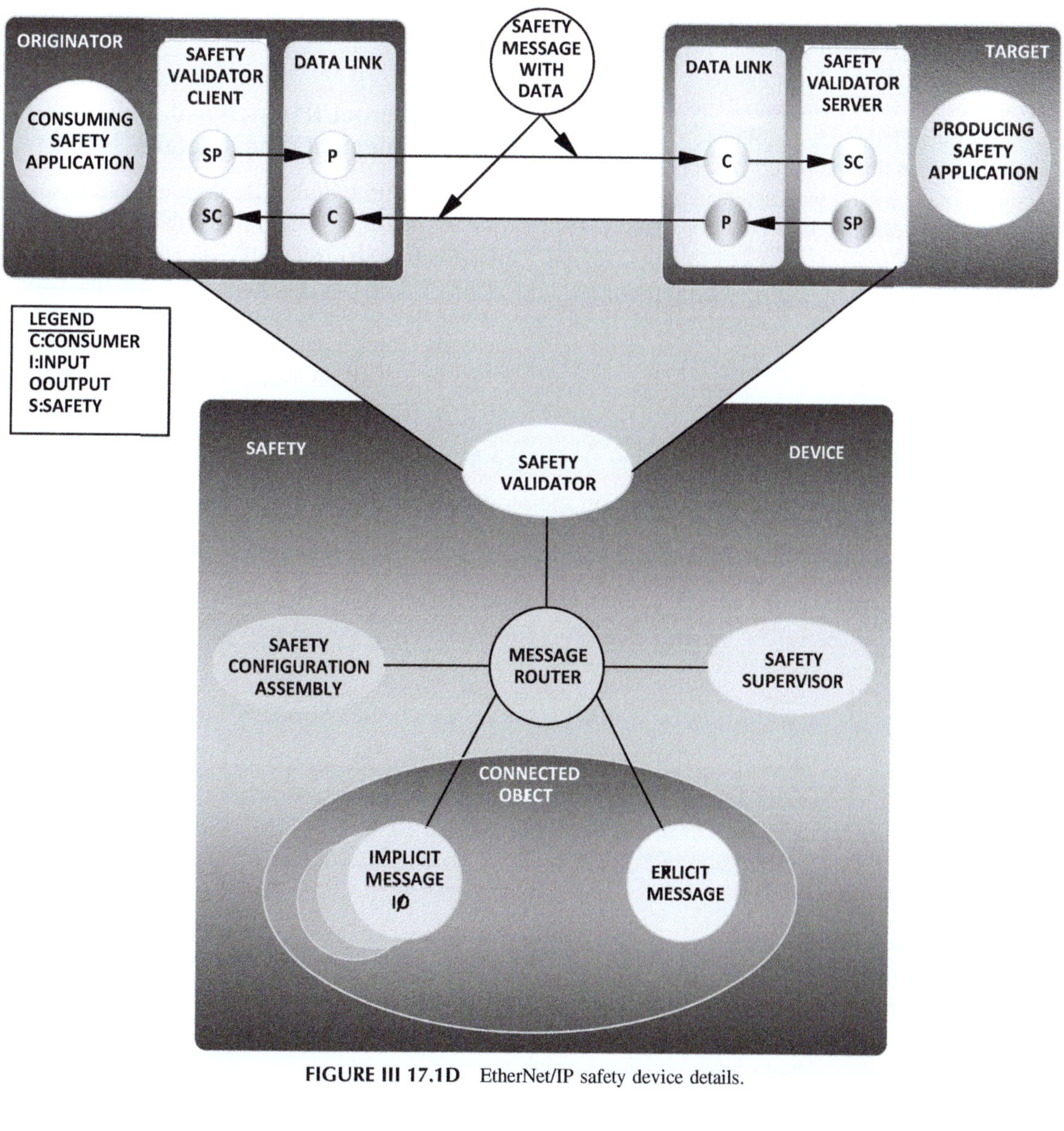


FIGURE III 17.1D EtherNet/IP safety device details.

- Torque, velocity, or position control of servo and variable speed drives
- Drive configuration, status, and diagnostic services
- Control-to-drive communications
- Centralized & distributed motion support
- Multicast peer-to-peer communications for position & velocity synchronization for
- multiple distributed controllers
- Configuration, status, and diagnostic services and common application instruction support for variable speed and servo drives

4. **Communication specification:** Based on Ethernet (IEEE 802.3) and the CIP, EtherNet/IP is an industrial protocol for high-level control, with fast, reliable device communication for industrial automation. It also defines specific communication access methods at the application layer CIP. EtherNet/IP uses CIP services like *CIP* Security, *CIP Safety*, CIP Motion, and CIP Synchronization. This also supports SPE. All these are specified In a generalized manner these are specified in Table III 17.1b.
5. **EIP practices:** Common industrial network design practices for EtherNet/IP are as follows:
   - Proper understanding proposed networked device's functional requirements with associated application.
   - Use of CIP objects for better interoperability and future readiness.
   - Use redundancy such as DLR is a good practice.
   - Prioritization of controls, management of IP addresses, use QoS with performance monitoring are part of good practices.
   - To reduce jitter, prioritization of traffic mentioned above, use CAT6 cable for better resistance to interference and use of buffer to smoothen out delivery.
   - Segment the logical topology into modular building blocks (based on zones as applicable).
   - Determine application and network security requirements, that is, integrated security. Also, division into zones, isolation, and using firewalls, encrypt data are good practices.
   - Design and implement a robust physical layer ensuring availability and resiliency requirements.
   - Robust network infrastructure, Reduction of unnecessary data, Regular monitoring & maintenance, etc., would result better availability of more information.
   - Use managed industrial switches.

*With this, let us conclude the discussions on infrastructure and application pertinent to EtherNet/IP and move to discuss EtherNet/IP functional and implementation details.*

### 17.1.2 Ethernet/IP functional and implementational details

In EtherNet/IP and all CIP Networks, *exchange of time-critical data* is based on the *producer-consumer* messaging model with efficient use of bandwidth. The transmitting device produces data on the network, for consumption of this data by one or more receiving simultaneously. Implementation of this model in CIP is supported by the IP, is supported by the Ethernet multicast

**TABLE III 17.1B** EtherNet/IP network communication specification.

| Technical Issue | Specification | Remarks |
|---|---|---|
| Standard | Ethernet IEEE802.3 TCP.UDP-IP & CIP | CIP: 3 top layers |
| OSI model | All layers as detailed above | |
| Physical media | Copper (Cat 5/6/7 UTP) and fiber optics | @100/1000 Mbps |
| Media access | Carrier sense multiple access (CSMAs) with collision detection (CD) | |
| Addressing | MAC & IP | |
| Communication types | Implicit (I/O) & explicit (client server*), configuration, diagnostics, and large data transfers. | *TCP/IP |
| Implicit messaging characteristic | Use UDP, producer/consumer model, cyclic/periodic transfer, | Network communication |
| Other data access | Polled, change in state (COS), cyclic | |
| Object model | Standard objects with attributes, object types: *Identity, Connection, Network specific link, Assembly* | |
| Network | Star, line & ring (DLR) | |
| CIP services | Security, safety, motion, synchronization | |

service. With the source-destination model, nodes receive only the packets that contain their destination node number. CIP security defines the security requirements, and capabilities of CIP devices with special reference to EtherNet/IP with beneficial features like *Data integrity*, *Authentication* and *Authorization*. It provides standard cryptographic protocol Transport layer security (TLS) and Datagram TLS (DTLS) to secure internet communications & online traffic with. Secure TLS and DTLS traffic is only possible if two entities trust one another. CIP Security for EtherNet/IP supports two mechanisms for entities to trust another: Pre-Shared Key and X.509 Certificates. CIP security in industrial automation using EtherNet/IP, leverages TLS for *explicit message* using TCP and DTLS for *implicit message* using UDP, to ensure secured communications at the device level. *Device identity & authentication*, *Data integrity & confidentiality (Encryption)*, *Authorization* are major features of CIP security. It also includes a number of security profiles viz. *EtherNet/IP confidentiality*, *CIP user authentication*, *Resource constraint*, etc., profiles. Implementation & practices of EtherNet/IP mainly focus on creation of a robust, secure, and interoperable industrial network with Commercial off-the-shelf (COTS) components. This calls for proper planning for physical and logical layers to ensure performance better QoS, reliability, and security. *Requirement assessment*, *Topology selection* with redundancy criteria, *Infrastructural assessment*, *Hardware & software set up* including *Physical installation*, *Use of configuration tool*, *IP addressing*, *Network segmentation*, etc., in this subsection all these issues as well as synchronization issue have been covered.

1. **Control network security:** When Ethernet is deployed for manufacturing applications, security is very important. On account of the coexistence of multiple protocols and IP addressing, additional steps are essential to ensure secured operation of manufacturing applications. Also, appropriate action must be initiated so that inappropriate changes in controller configurations and programs are prevented. Proper installation of managed (configurable) switches, routers, and firewalls is important to achieve a more secure network installation. EtherNet/IP, & CIP Security, includes constrained device (with restricted processing power, memory, etc.) and option to use Link layer discovery protocol (LLDP) enabling easy adoption of TSN into the EtherNet/IP Specification.

   With reference to Fig III 17.1e showing connections for isolated single and multiple controller, let us now look into the details with various probable cases as listed:

   - **Isolated single controller:** Whether a small, single-operator station, or a large number of isolated work cells, when there is a single controller as shown in Fig III 17.1e1, they may have a large number of adapters and I/O points requiring multiple switches. An attacker would have to be in direct contact with the network to affect its operation, that is, the threat has to be from the connected infected computer resource(s). There could be threat also from the damage/manipulation of the controller code. Likewise, common Operating system (OS) can introduce malware. Antivirus software, which usually protects against a range of known malware attacks, and regular patching are all common mechanisms to reduce the potential of an attack, and end devices should have security protection.
   - **Isolated multiple controller:** This type of network can have a multilayered architecture using managed switches and VLANs to segment the larger number of devices including local servers. Controllers can be at different VLANs, as shown in Fig III 17.1e2, to improve system performance and availability by separating traffic between devices. This type of architecture is complex in nature, & it is possible that an incident at one end of a facility could affect a device on the other end of the facility but VLANs help to limit that impact.
   - **Integrated Network:** The industrial network may exist throughout an entire facility in an integrated network and may have many susceptible network ports. Enterprise networks are composed of Local area networks (LANs) that in turn connect to Wide area networks (WANs) and the cloud. So, enterprise networking refers to the architecture, technologies, and process that organizations use to connect their data centers, cloud services, remote office, etc. An enterprise-integrated industrial network refers to the industrial network need the enterprise network to function properly, for exchanging RT data. The industrial network(s) may pass low-frequency data like inventory statistics to an ordering system or high-frequency data like real-time I/O data between work cells on a plant floor, both are considered enterprise connected as detailed out in Fig III 17.1f. The risk of attack to the industrial network increases greatly when it is integrated enterprise IT system. This risk can be reduced if both networks are designed with security in mind and security zoning is applied.
   - **Recommended practice:** Zero trust model, Defense-in-depth are fundamental principle upon which security systems are arranged. Some of the best practice precautions include the following:
     - *Network segmentation*: Using network infrastructure to highly segment the enterprise and industrial networks, for example with firewalls.

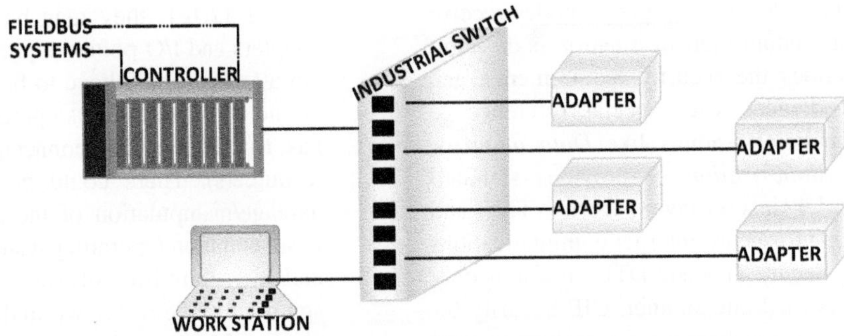


FIG III 17.1e1 ISOLATED NETWORK FOR SINGLE CONTROLLER

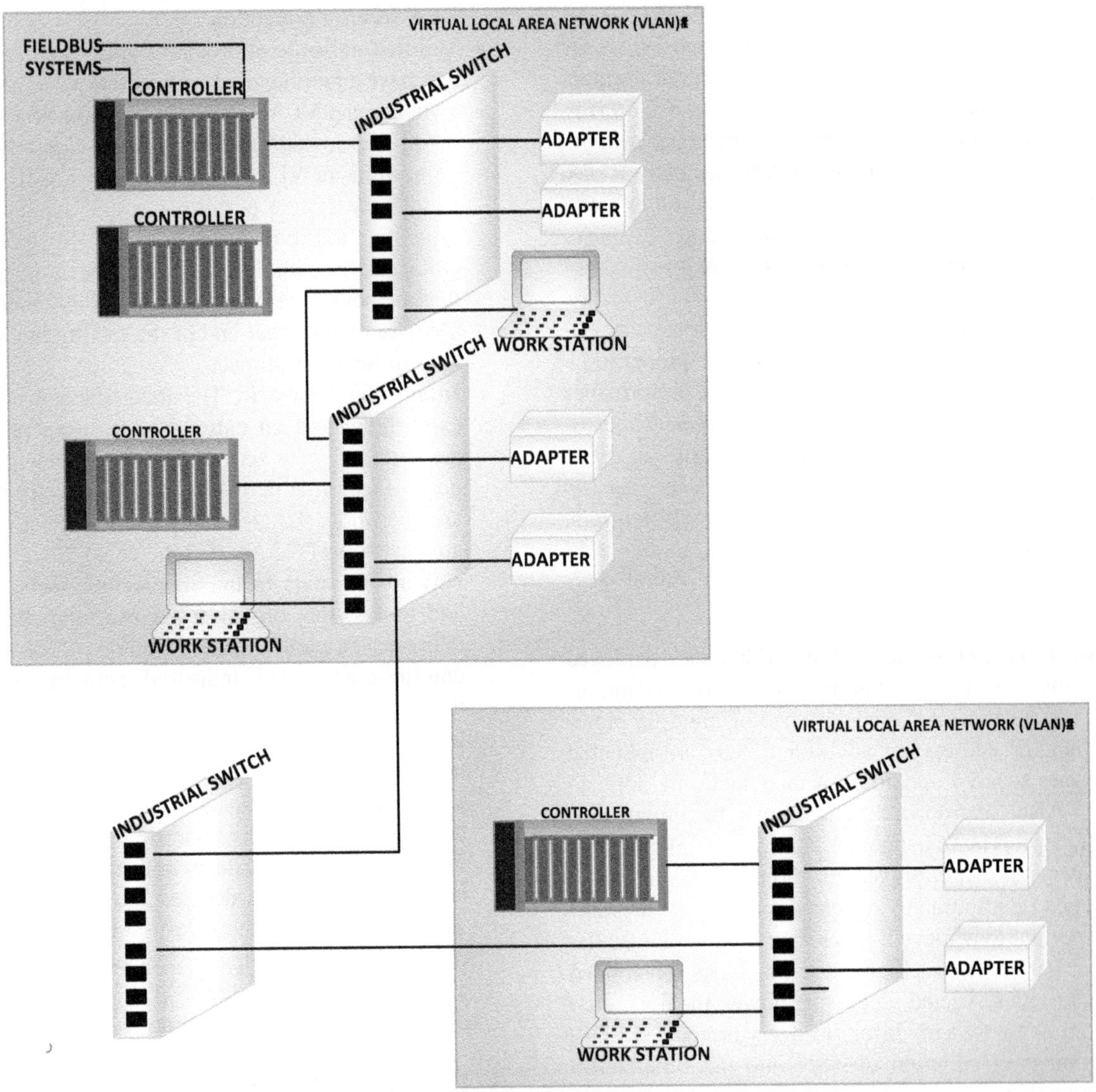


FIG III 17.1e1 ISOLATED NETWORK FOR MULTIPLE CONTROLLERS

**FIGURE III 17.1E** Isolated network controllers.

- *De-Militarized Zone* (DMZ): Development of the system a where data and services can be shared, but with separation.
- *Firewalls and Intrusion prevention/detection systems (IDPS):* Development and deployment of these in network
- *Techniques and remote access*: Development techniques like encryption, patch management, and services and provide Remote Access.

- Replication: Replicate or place critical network-based services into the industrial networks (e.g., active directory servers, DHCP servers, print servers, etc.).

2. **Additional wireless security:** Industrial wireless communications are becoming increasingly popular. IEEE 802.11 Wireless LAN (WLAN) is especially straightforward for installations using EtherNet/IP due to TCP/IP use by EtherNet/IP, that is, WLAN is common. So, some additional security measures shall include the following:
   - **Rogue access point & device detection:** Two-step process of "Rogue access point and device detection comprises discovering the presence of an *access Point/wireless device* in the network and then identifying whether it is a rogue or not.
   - **Disable Service set identifier (SSID):** As SSIDs are not encrypted, it becomes easy to grab one by snooping the WLAN. Knowing the SSID brings hackers one step closer to a successful intrusion.
3. **EIP synchronization:** CIP Sync (synchronization) following IEEE 1588 standard provides precise time synchronization for EtherNet/IP for the increased control coordination mechanism needed for industrial automation where absolute time synchronization is extremely important and vital issue to achieve real-time synchronization between distributed intelligent devices and systems. CIP Sync allows synchronization accuracy between two devices of fewer than 100 ns. RT synchronization can be achieved over conventional Ethernet systems with a switch-based architecture. Key features include the following: High accuracy, RT capability, Use with standard Ethernet, CIP Motion integration. Time-stamped data. CIP Sync consists of a time Sync object and associated services that allow users to synchronize devices, including motion axes, using "time." On account of flexibility and scalability for network architecture, this synchronization method is ideal for large-scale motion systems and is not restricted by many of the constraints imposed by event-based or time-sliced implementations.
4. **Network implementation:** Major points for network implementation are as follows:
   - **Implementation steps:** Generalized issues related to EtherNet/IP implementation are as follows:
     - Understanding & definition of the application with requirements
     - Determination of the level of EtherNet/IP capability including device type & communication service types (devices or application *specifics*, communication service *specifics* including *frequency* & *quantity* of data exchanges envisioned, allowable *latencies*, *diagnostic* capabilities, etc.).
     - Development of an EDS to describe the device capabilities
     - Selection of an implementation approach including hardware or software
     - Development the EtherNet/IP implementation method/approach
     - Testing of implementation.
   - **Implementation types:** There are several hardware and software ways EtherNet/IP can be implemented. With reference to the project background checking may be carried out to take care of Hardware development, Ethernet communications, Embedded device development, CIP & EtherNet/IP knowledge. Important types of implementations are as follows:
     - *Hardware basis*: *Microprocessor with embedded EtherNet/IP, Addition of interface daughter card, Complete custom hardware, Use external protocol gateway device.*
     - *Software basis*: *Writing EtherNet/IP stack, Integrate an EtherNet/IP stack into an implementation* and *Complete outsourced custom development*
5. **Industrie 4.0 issue:** ODVA ensures through its recommendations that *Object-oriented, Interoperable, Highly standardized,* and *Proven* industrial automation network EtherNet/IP meets all requirements of Industry 4.0 and IIoT. EtherNet/IP also leverages TCP/IP and COTS hardware with capabilities to get connected to the broader infrastructure. The object-oriented foundation of EtherNet/IP has been a key factor in its current global industrial success and will play an important role in continuing its growth with accelerating IT and OT convergence [9]. Also, continuous enhancement efforts are on to meet future control & communication across industries including the discrete and process industries. Unparallel device level security along with best-in-class safety, motion/synchronization, and energy performance within industrial control with CIP support combined with inherent TCP/IP capability for higher-level communication, make EtherNet/IP an ideal solution for tomorrow also, that is, CIP support connectivity and standardized hardware. Factory floor to cloud connectivity is a key component of IIoT. ODVA is developing an OPC UA Companion Specification for CIP as one way to help address this need. Learn more about OPC UA [9]. *Device connection, Data aggregation,*

*Cloud integration* and *Actionable insights* with object-oriented CIP support are working force behind Industrie 4.0 implementation.

*The discussions on EtherNet/IP come to an end to look forward to discussing PROFINET.*

## 17.2 Industrial Ethernet: PROFINET

The PROFINET had been developed (Siemens GmbH) as an extension to the popular PROFIBUS fieldbus technology. PROFINET is an RT Ethernet standard for the high-speed, deterministic communications catering to requirements in industrial and building automation. This Adapting Ethernet to support industrial applications creates a technology bridge between IT and plant OT system. However, it falls short on several counts for factory automation. PROFINET minimizes the costs for installation, engineering, and commissioning. The mandatory certification for PROFINET devices also ensures a high-quality standard. This is based on IEEE 802 standard in collaboration with IEC 61,158 and IEC 61,784 as detailed in Fig III 17.2a which is developed based on Idea from Ref. [10].

While PROFINET IO uses Ethernet hardware and software for the task of exchanging, data, alarms, and diagnostics with automation controllers. PROFINET Component-based automation (CBA) focuses on distributed automation systems providing a Distributed component object model-based system for networks of peer devices to automatically exchange data between the interfaces of the automation components [11]. So, PROFINET is a comprehensive automation concept offering modular, reusable machines and plants with distributed intelligence. It offers vendor-independent plantwide engineering model with consistent communications from field level to enterprise level having openness and integration capability. PROFINET's efficiency ensures the optimal use of available resources to increase system availability significantly. PROFINET provides consistent end-to-end communication with unified standard, and an open system.

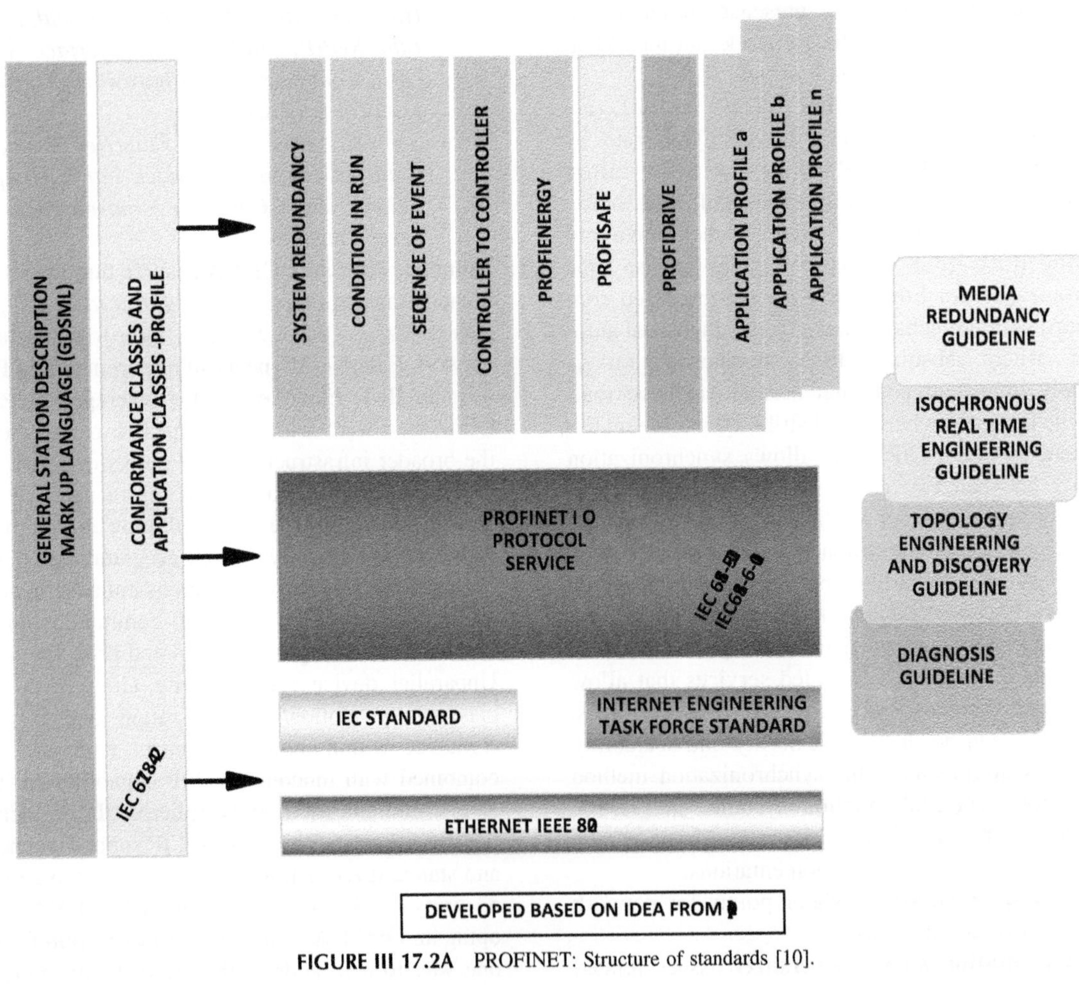


FIGURE III 17.2A PROFINET: Structure of standards [10].

PROFINET offers simple integration of various existing systems. *Deterministic responses*, *RT capability*, *Full conformity with IEEE standard*, *Integrated diagnostic capability*, and *Harmonious working* with cutting-edge standards including as OPC UA could make PROFINET as the ideal foundations for Industrie 4.0 communication. PROFINET's short cycle times increase the productivity of machines and plants. PROFIsafe (same family) offers safety-related communication, and PROFINET allows the combination of safety and industrial WLAN. In this section, brief discussions have been presented on PROFINET covering overview, system description, engineering details structural details.

### 17.2.0 Overview of industrial Ethernet: PROFINET

PROFINET Stands for Process Field Network. PROFINET, offering the speed, performance, reliability required for industrial applications, is one of the very popular IE taking 27% market share followed by Ethernet/IP and EtherCAT. *Robust* PORFINET communication includes features like: *Fast real-time automation*, *Uniformly distributed automation*, *Communication in RT, IRT & TSN*, *Utilization of Wireless local area network* (WLAN), *Flexible* topology, *Seamless integration*, *Comprehensive diagnostics* and *Capability to handle large volume of data*. PROFINET is meant for Operational technology but can integrate Information technology (IT). PROFINET specification was also copied over from PROFIBUS DPV1 and contains basic identification and maintenance information about the device viz. Device type, Hardware/Software version, Media access control (MAC) address, etc. Simple device replacement is possible in PROFINET. After PROFINET network set up, device xan be replaced through the use of standard Ethernet protocols. With the IT protocol called *Link layer discovery protocol* (LLDP) each device and switch can identify their neighbors. PROFINET network diagnostics uses standard IT protocols, specifically the Simple network management protocol (SNMP) to provide a vendor-neutral reporting for network failures/status. PROFINET has many functions viz. cyclic data, simple device replacement, and IRT, etc. However, it is not mandatory to be used for PROFINET implementation. Some added features for PROFINET in process automation include the following: *Configuration in run*, *Time-stamped data*, *Redundancy method*, *Easy integration to existing fieldbus*, etc. Though there are many commonality like common organization, application profile, Engineering GSD between PROFIBUS and PROFINET, yet they have many differences between them as speed of RS 485-based PROFIBUS is limited to 12 Mbps, whereas Ethernet-based PROFINET can have speed of 1000 Mbps. In comparison with 126 address space of PROFIBUS, PROFINET has unlimited address space. PROFINET can carry out vertical integration, Machine-to-machine (M2M) integration but for PROFIBUS these are not possible. In this subsection brief discussion on overview of PROFINET shall be covered.

1. **PROFINET concept:** PROFINET is an open standard based on industrial Ethernet, enabling application from control to communication at the field level uniformly in vertical network structure where IT systems can also be integrated. PROFINET fundamentals with device types & conformance class have been detailed in Fig III 17.2b. PROFINET It can be viewed as an extension of the PROFIBUS. According to PI international, PROFINET consists of Ethernet standard (offering high transmission rate, Wireless communication possibilities, and Flexible topologies) and PROFIBUS as fieldbus standard (offering High-speed IO communication, Safety & Diagnostics). Fig III 17.2a may be referenced.
   - **Physical layer:** Following are the features of PROFINET Physical layer [12]:
     - *Standard:* IEC 8802/IEEE 802.3
     - *Power and communication:* No
     - *Speed:* 1Gbps/100 Mbps
     - *Distance:* 100m
     - *Number of devices:* unlimited
   - **Data link layer:** Following are the features of PROFINET data link layer [12]:
     - *Standard:* IEC 8802/IEEE 802.3
     - *Link type:* Token Passing
     - *Deterministic:* Yes
     - *Communication technology:* Provider-consumer
   - **Application layer:** Following are the features of PROFINET application layer [12]:
     - *Standard:* IEC 61,158
     - *Transfer:* PROFIBUS DP type
     - *Alarm trend:* yes in field device also
     - *Field control:* Not direct
   - **Comparison with EtherNet/IP:** Let us now compare two common IEs PROFINET and EtherNet/IP in Table III 17.2a.
2. **PROFINET benefits:** A few advantages of PROFINET, are as follows:
   - **Openness & vendor neutrality:** Consistent openness and end-to-end communication via a unified standard. PROFINET is vendor neutral.
   - **Connectivity**: Connectivity and interoperability with components from various manufacturers.
   - **Availability:** PROFINET integrated redundancy solutions for media, controllers, devices.
   - **Scalability:** It is highly scalable so that communication takes place over the same cable in all

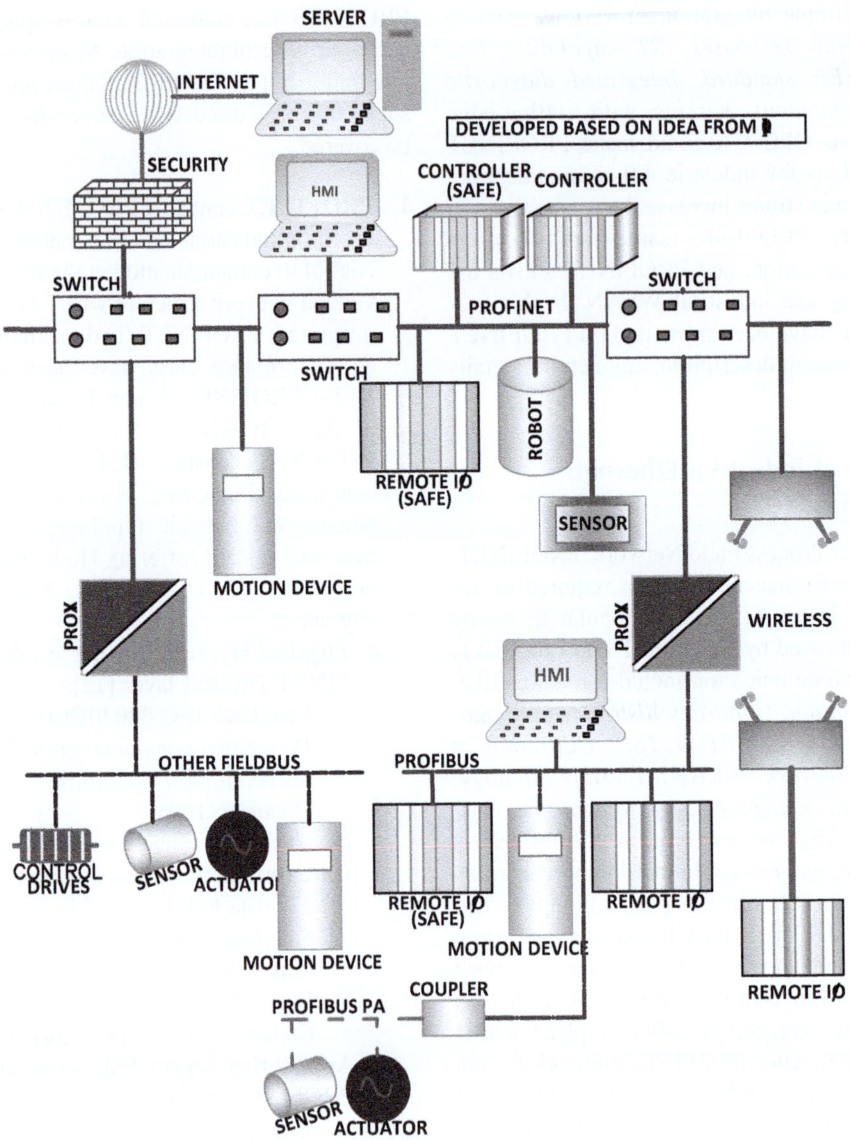


FIG III 17.2b1 PROFINET INTEGRATION

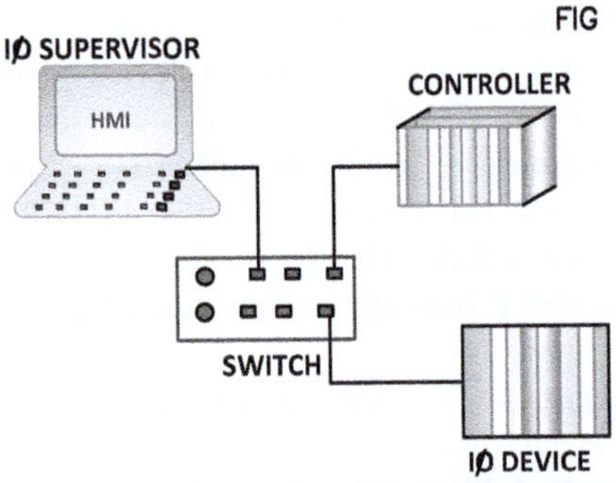


FIG III 17.2b2 SOME DEVICE TYPES

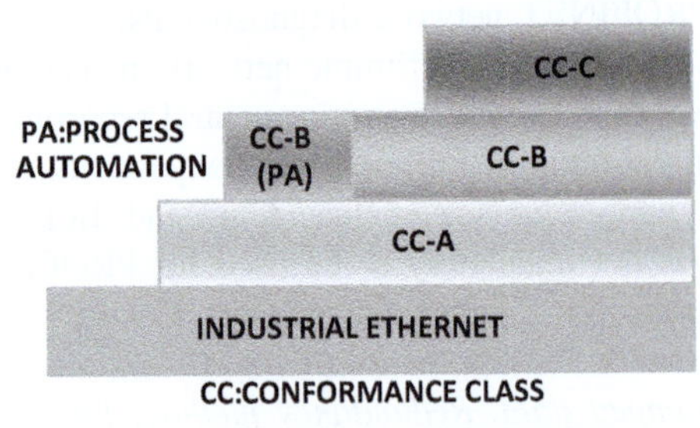


FIG III 17.2b3 STRUCTURE OF CONFORMANCE CLASS

**FIGURE III 17.2B** PROFINET fundamentals.

applications, from simple control tasks to highly demanding motion control.

- **One cable integrated communication:** On account of highly integrated Ethernet-based communication, in one cable it can perform wide range of tasks.
- **Cycle Time:** Cycle time of PROFINET is *Min. 31,25μs* (for PROFIBUS, it is *Min. 300μs.)*
- **Redundancy:** Redundancy is possible directly on the Products, no OLM, no FO!!

**TABLE III 17.2A** Comparison PROFINET and EtherNET/IP.

| Technical Issue | PROFINET | EtherNet/IP |
|---|---|---|
| Authority | PROFINET international (PI) | ODVA |
| Controller I O | Device-side devices | EtherNet/IP slave |
| Cyclic I/O messaging | Both use cyclic messaging between I/Os and controller | |
| Acyclic message | Both use for unscheduled/on demand data | |
| Control files | EDS | GSDML |
| Standard services | Both support; with some device specific services | |
| Communication | TCP/IP -for use in ethernet; producer consumer model | |
| Device class | Controller & device side | Scanner & adapter |
| Device identity | Vendor details device type | Vendor details device type |
| Protocol standard | Open; IEC 61158 | Open; IEC 61158 |
| I O jitter | Very low < 1 μS | Comparatively higher |
| Transmission | Unicast; multicast optional | Multicast; unicast optional |
| Device name | Mandatary | Optional |
| COTS switch | Yes | Not recommended |
| Full duplex switch | >100 Mbps | >1000 Mbps |
| IP address | Automatic discovery and basic configuration protocol (DCP) & DHCP | Dynamic host configuration protocol (DHCP) & managed |
| Redundancy | Possible: Bumpless, controller, device & media | Possible: Media |
| Fieldbus integration | PROFIBUS, ASI, DeviceNet | CompoNet, DeviceNet, ControlNet, MODBUS TCP |
| IEEE 1588 & timing | Yes & timing calculation | IEEE 1588 |
| Determinism | Yes | No |
| Synchronization | Yes | Yes |

- **Wireless LAN:** Industrial Wireless LAN with precise, reliable, and deterministic data transmission coupled with media redundancy makes it possible to venture new areas of application, for example, IIoT.
- **Integration of Existing system:** Simple integration of existing systems: IO-Modules with a PROFINET Interface are exactly the same costs as on PROFIBUS and only the head module needs to be changed, the I/Os can be kept. Also, there is no difference between PROFIBUS and PROFINET IO-Access!
- **Web service and implementation:** PROFINET offers standard web services with and Maximum flexibility for quick implementation of modular design concepts with no waiting periods.
- **Energy optimization:** With the PROFIenergy profile integrated in PROFINET devices, the energy use in an automation system can be measured and controlled by selectively switching functions.
- **Efficiency:** High performance of the network boosts productivity and efficiency for economic advantage. Single cable for Equipment and plant communication.
- **Investment Protection:** Because of easy integration, conservation of existing devices without additional special tool PROFINET protects investment in terms of cost and time.
- **Diagnostics and safe service:** PROFINET has easy, fast, and reliable Diagnostics and for safe plant operation it has Profisafe [Ref: II 5.1.1].
- **Equipment replacement:** PROFINET builds industrial communication networks quickly and reliably to optimize processes during equipment replacement. A typical PROFINET integration has been depicted in Fig III 17.2b1. Proxy is basically a part of PROFIBUS and detailed in next subsection. For safety applications, shown in Fig III 17.2b1, sub-section II 5.1.1 may be referenced.

3. **Device type and class:** PROFINET device types and device class conformance have been described separately below:
   - **Device Type:** Systems can be assembled in various configurations in combination of any number of IO-Controllers with any number of IO-Devices. The network representation of a device is the view of the device from the network. PROFINET I O consists of a device with slots, subslots, and channels to accommodate subcomponents of a slot. Each subslot is assigned a number of I/O points or channels. A channel in PROFINET IO refers to one physical discrete/analog IO/. A device can have almost any number of slots, subslots, and channels. Based on this, various device types have been detailed below. Some of these have been illustrated in Fig III 17.2b2.
     - *IO-Controller:* Various controllers to control process could be PLC/DCS or a Programmable automation controller (PAC) equivalent to Class 1 master in PROFIBUS. Different services offered are as follows: *Cyclic data exchange* between I/O-controllers and I/O-devices, *Acyclic data exchange* viz. configuration/diagnostic data, *Alarms data exchange* from an I/O-device to an I/O controller and *Context management.*
     - *IO-Supervisors* These could be an engineering station (like Class 2 master in PROFIBUS).
     - Used for commissioning & diagnostic data collection, IO-Supervisors can read and write internal diagnostic data associated with the PROFINET IO/diagnostic data provided by the application program of a device. IO-Supervisors use acyclic data exchange services.
     - *IO-Device*: These refer to physical I/O device-(like slave device in PROFIBUS). These are input data to the controller and consumer of output from controller.
     - *Switches*: These are meant to connect and manage these traffics.
     - *Configuring devices*: PROFINET IO devices are configured using a configuration tool that acts as the IO-Supervisor which uses General station description (GSD) file. PROFINET IO GSD files (GSDML) are Extensible markup language (XML) based and carry much more information than the PROFIBUS GSD. Also, PROFINET extends this to include some information on acyclic information as well.
   - **Conformance classes:** There are three conformance classes that build upon one another as shown in Fig III 17.2b3. Various Conformane classes (CC) are as follows:
     - *CC-A:* This provides basic functions for PROFINET IO with RT communication. All IT services can be used without restriction and can be implemented with standard Ethernet hardware. VLAN Tag is used in RT frames with a high priority according to IEEE 802.1Q to optimize their processing within a network component viz. Standard Ethernet component, cyclic data exchange, Acyclic parameter, etc.
     - *CC-B*: As shown CC-B includes all of the functionalities of CC A and additionally includes functionality like network diagnostics and topology detection utilizing SNMP which is mandatory to conform to CC-B. Media redundancy protocol (MRP) is included in an extended version of CC-B. CC-B in process automation is named CC-B(PA), for example, Topology detection, SNMP.
     - *CC-C*: As shown, it includes all of the functionalities of CC B and adds the critical feature of synchronization between IO-Devices. This is the basis for isochronous applications. CC-C describes the basic functions for devices with hardware-supported bandwidth reservation and synchronization (IRT communication) and is thus the basis for isochronous applications like synchronization, Bandwidth reservation.
4. **Message types:** As seen in PROFIBUS, two different message types of PROFINET are as follows:
   - **Cyclic message:** Cyclic messages (PROFINET with RT communication) are a group of commands that pass time-sensitive input and output information between IO-Device and an IO-Controller. The cycle time can be specified individually for connections to the individual devices and is thus adapted to the requirements of the application. This can be realized with standard Ethernet controllers and takes place without any TCP/IP information directly. Cyclic messages are sent electronically in two ways. For RT I/O uses standard Ethernet switches and high-speed IRT applications sent via PROFINET IRT switches. For CC-A infrastructure, switches meet 100 Mbps Full duplex as per IEEE 802.1D, as well as conforms to IEEE 802.1Q and neighborhood detection in line with LLDP as per IEEE 802.1AB. For cyclic addressing (like PROFIBUS DPV1), PROFINET also uses a slot and index format for information access easy.
   - **Acyclic Message:** Functionally, it is used for read and write record data—Parameter assignment or I/O configuration. It finds its use in diagnostic information, identification, and maintenance information. Apart from nontime critical information listed above, this group of commands is used for time-

critical information like Alarm messages. Addressing is similar to that discussed for cyclic message.

5. **Device diagnostics concept:** PROFINET's device diagnostic refers to the detailed, manufacturer-independent fault information from IO devices to the controller for quick troubleshooting. This is related to Integrated diagnostics, Detailed information and Vendor independence are core issues for this. PROFINET uses the real-time channel to send diagnostic & process alarm based on different priority levels. When there is either a device-related or process-related *alarm* or *diagnostic*, the IO-Device can immediately send a message to the IO-Controller notifying it of the problem—no waiting involved. Instant and detailed information is provided. In PROFINET IO, alarms & diagnostic data are related. Diagnostic data are data associated with a channel or I/O point. Diagnostic status (not alarm) always acyclic messaging using non-RT record data communications. An I/O-supervisor needs to request diagnostic data from the recorded data object. Alarm data including diagnostic alert data exists on a specific channel is very different. Diagnostic alarms monitor the hardware or physical wiring, such as wire-break or failed module electronics. There are watchdog timers in place to ensure that all the communications occur within a set period of time. The IO-Supervisor configures and diagnoses IO-Devices and communicates to other devices on the network using acyclic messages which are defined in the EDD file or DTM. A status-based maintenance approach is currently gaining relevance for operation and maintenance. It is based on the capability of devices to determine their states and to communicate them over agreed mechanisms.

### 17.2.1 Structure and system description

PROFINET has been designed & developed for RT communication in automation, integrating control systems with field devices as well as with enterprise IT protocols & technology. Apart from IO controller, supervisor, devices, proxy there are I-device– an *intelligent field device* that can function controller to its own local devices and IO device to a higher-level controller. As discussed, XML-based GSDML files in PROFINET provide some extended functions when compared to GSD file in PROFIBUS. PROFINET belongs to same family with PROFIBUS. Obviously, there will be many similarities with PROFIBUS already discussed in III 7.2. PROFINET supervisor has the similar role of PROFIBUS DP Class 2 Master. In they can run concurrently with a controller and connect to devices to read or write parameters/data. Also in PROFINET, they can operate without an active controller to work with devices that aren't communicating yet. Supervisors use a special type of connection, the Supervisor application relationship (containing information one of the submodules plugged into the device) to communicate with devices. PROFINET IO is very similar to PROFIBUS on Ethernet. Like PROFIBUS, PROFINET IO uses cyclic data transfer to exchange data with programmable controllers over Ethernet. A programmable controller and a device must both have a prior understanding of the data structure and meaning. Data are organized as slots containing modules with the total number of I/O points for a system the sum of the I/O points for the individual modules.

Industrie 4.0, demand seamless integration of device and production data to business systems to make more use of data generated on the plant floor, and to harness the power of modern analytics. PROFINET with IRT has opened up new technological frontiers by allowing hundreds of machine axes to be *synchronized electronically* rather than mechanically. PROFINET takes the advantages of OPC for communications at operator level. It also supports safety applications through Profisafe which goes hand to hand with PROFINET. With the PROFIenergy into PROFINET devices, the energy use in an automation system can be measured & controlled by selectively switching functions on and off without additional hardware. In this subsection brief description and structural details of PROFINET shall be discussed.

1. **Proxy and Gateway:** For moving data from the specialized systems up to the PROFINET network, an interpreter is necessary. This could be done by either a Proxy or a Gateway. The difference between the two is nuanced but with big impact on automation functions.
   - **Gateway:** In gateways, I/O data are moved between protocols, but may not be with "special features" from one protocol to another viz. gateway may map data from its slaves up to slots and subslots in the PROFINET but without diagnostic data. Each gateway's "data mapping" must be configured for each individual gateway in a vendor-specific format. Also, there will be loss of data, for example, some protocols don't pass all the alarm data. Gateways are relatively easier to develop but with costlier implementation.
   - **Proxy:** Proxy is a representative of Fieldbus (Major) in PROFINET. Proxy is used to produce invisible transition between PROFINET and the foreign protocol (including *alarms*, *diagnostics*—even network topology and health). Proxy, maps I/O data, alarms, diagnostics, network topology, and network health. It implements the full data map defined by PI. For Proxy, PROFINET can utilize fieldbus features such as high dynamic response, pinpoint diagnostics, and automatic system configuration without

settings on devices. Also, commissioning and operation are made easier through the comprehensive diagnostics properties of the fieldbus system. The devices and software tools are also supported in the usual manner and integrated into the handling of the PROFINET system [10]. Proxies are already defined for major fieldbuses such as *PROFIBUS DP/PA*, *Interbus*, *DeviceNet*, *Foundation Filed bus*, *CANopen*, *Modbus*, *AS-I*, *IO link* and protocol like HART.

2. **Redundancy in PROFINET:** Fig III 17.2c, depicts redundancies in PROFINET (Fig III 17.2c1 shows redundancies for controller, device, and media, and Fig III 17.2c2 shows network redundancy. Various redundancies in PROFINET are as follows:

- **Media Redundancy** (Fig III 17.2c1)**:** Media redundancy helps ensure network availability. PROFINET implemented the MRP to handle media failures gracefully without requiring the intervention of any external application. IEC 62,439 describes PROFINET redundancy with a typical reconfiguration time < 200 ms for communication paths with TCP/IP and RT frames after a fault. IEC 61,185 describes the redundancy concept of Media redundancy planned duplication.
- **Device Redundancy:** As shown in Fig III 17.2c1, the controller establishes a connection to both devices simultaneously. and sends the same output data to both devices. However, data sent to secondary device is referred to as *inactive,* and the data sent to the primary as *active.* For a failure change over,

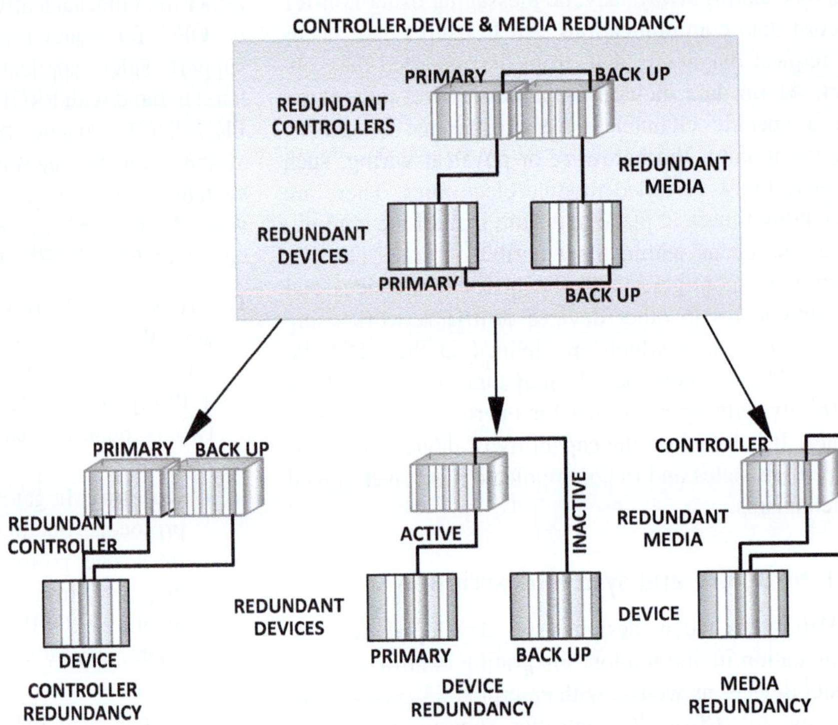


FIG III 17.2c1 PROFINET REDUNDANCY: CONTROLLER, DEVICE & MEDIA

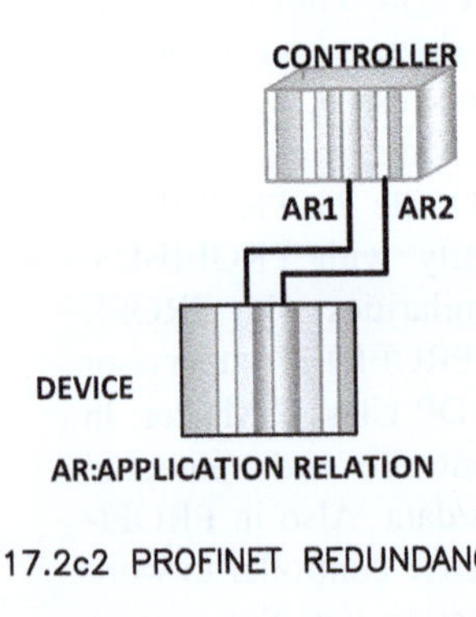


FIG III 17.2c2 PROFINET REDUNDANCY: NETWORK RDUNDANCY

**FIGURE III 17.2C** Redundancy in Profinet.

controller has to mark the secondary output data active and secondary device can take over from the primary one seamlessly.

- **Controller Redundancy:** In controller redundancy, the first one establishes a primary connection, the second controller establishes a special, "backup" connection, as shown in Fig III 17.2c1. If the primary controller fails, the device waits for the backup controller to request a transition to primary status.
- **Network Redundancy:** As shown in Fig III 17.2c2, each network module has an independent Application relation (AR) with the PROFINET controller. With network redundancy, a network continue running after a subnet failure by swapping communication to a second subnet.

3. **Runtime structure:** Runtime structure of PROFINET utilizes standard Ethernet with specialized layers for fast, deterministic communication (digressing from TCP/IP), for *RT* communication necessary for industrial automation (EtherType 0x8892) and *IRT* for ultra-fast motion control, featuring time synchronization, scheduling, and dedicated bandwidth for microsecond cycles. User application, PROFINET IO Kermel, RT Operating system (RTOS) and interface modules are major components for this. The PROFINET Runtime Software refers to the software structure prepared by PROFINET International. This software is an essential component to implement a PROFINET device.

As shown in Fig III 17.2d, various components are as follows:

- **User application:** The user application plugs the modules into the PROFINET slots to start the PROFINET tasks and generate diagnostic data, issues & releases alarms and the I/O data mapping between device I/O and PROFINET IO.
- **PROFINET IO:** As shown, this software consists of: *IO control program* for IO-Device exchanges I/O data with the application, *Context manager* creates & manages the application & communication, relationships with programmable controllers and IO-Supervisors. *Remote procedure call* (RPC) to manage acyclic communication over UDP (read/write), *Discovery and configuration protocol (DCP)* and *Alarm control protocol (ACP)* are meant to identify PROFINET devices and manage alarms, respectively. *Sockets* supports the sending/receiving of UDP messages, *Ethernet device driver (EDD)* provides services to send and receive cyclic, acyclic and non-RT messages and checker when provided manages frame and routing services.
- **RTOS:** The RTOS performs task management function.

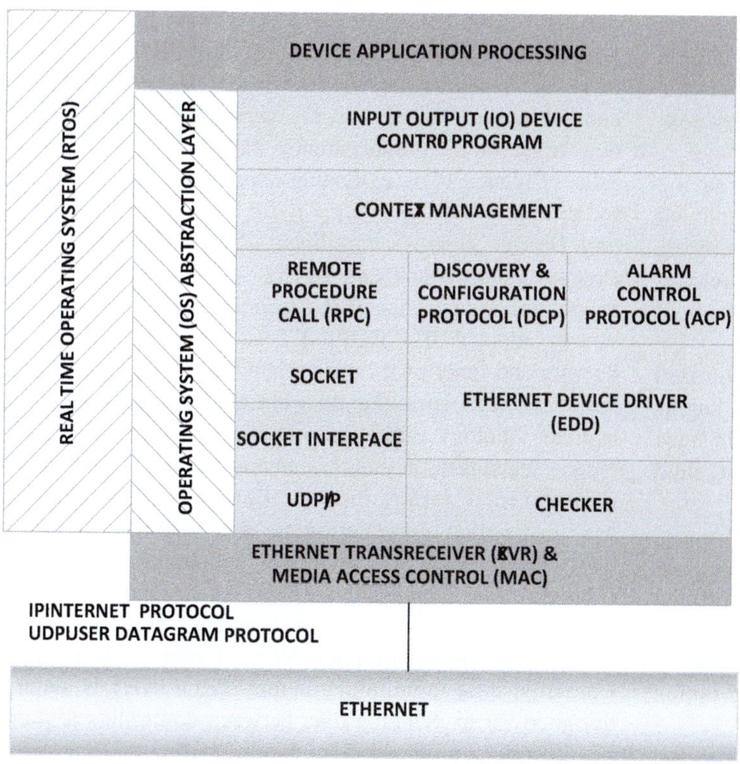


**FIGURE III 17.2D** PROFINET: Runtime structure.

- **Interface modules:** They are placed between the PROFINET IO kernel and the RTOS to support PROFINET IO operation.

4. **Process automation:** Process automation has a few special characteristics in defining automation to a large extent. Plants can have a long service life hence there can be coexistence of older and newer technologies. Reliability is another pertinent issue here. For the optimal use of PROFINET in process automation, PI has created requirement catalogs with an aim to replace PROFIBUS DP with PROFINET. Major functional requirements include the following: *Cyclic & acyclic data exchange*, *Integration of fieldbuses* (PROFIBUS PA, HART, and FF), *Integration* & parameterization of devices including configuration in run, Redundancy, Time stamping, Diagnostics, and Maintenance.
5. **Miscellaneous issues:** PROFINET offers a variety of optional functions (not by default). Some of these are *Shared device/input, Configuration in run, Time synchronization,* etc. Use of additional functions should be checked on case to case basis (data sheet, manuals, GSDML file). Some detailings are as follows:
   - **Device identification:** Information for standardized and simplified identification and maintenance is defined in additional I&M data records. Vendor ID, MAC address, Device ID and Device name are major types of unique IDs.
   - **Device access:** Device access refers to a feature (optional) by which a field device could be accessed by configuring tool or even by controllers. Multiple access is the starting point for shared devices is the parallel and independent access of two different controllers to the same device.
   - **Parameter back up:** This is the process of saving configuration settings (parameters) of intelligent field devices to a central controller or engineering station which is available for backing up and reloading of other optional individual parameters of a field device.
   - **Fast Startup:** Fast startup refers to an optimized system power-up in which data exchange begins much faster starting with the second power-up since many parameters are already stored in the field devices.
   - **Configuration in Run and time stamping:** Configuration in Run (CiR) refers to crucial feature allowing changing, adding, removing, or replacing devices/modules live network without any interruption and without adversely affecting network communication. *Time stamping* is assigning timing in alarm/sequence of event, etc.
   - **Fieldbus integration:** PROFINET specifies model for integrating existing various fieldbus systems. Major issues here are as follows: *Possible to integrate existing installations into a newly installed PROFINET system easily, Easy use of their proven and familiar devices* and *ability to integrate existing field devices into PROFINET systems without costly modification.*
   - **PROFINET certification:** PI has established a certification process for PROFINET devices to ensure that devices are compliant and interoperable with other devices in machines and plants. According to PROFIBUS & PROFINET International (PI), PROFINET certification test of the PROFINET devices is an essential part of the PI quality assurance. PROFINET certification is important for customers and manufacturers of PROFINET devices, to ensure that devices are compliant and interoperable with other devices in the system.

*Let us conclude the discussions here and move on to discuss engineering and functional issues of PROFINET.*

### 17.2.2 Engineering and functional issues of PROFINET

With its high data rate and large bandwidth, Ethernet is a stark contrast to established process fieldbuses. IE solution like PROFINET has been adapted for its high performance in the discrete industries, as well as in process applications. IE may not likely replace process fieldbuses directly. PROFIBUS PA segments can be connected directly to PROFINET, meaning that PROFINET can serve as a plant-wide backbone network. *PROFINET's connectivity to legacy process networks* will open up a new level of data transparency and data availability in process plants. PI is extending *support for process* users by adding PROFINET support for the profiles *Weighing and Dosage Devices*, *Intelligent Pumps*, *Remote IO for Process Control,* and *Lab Devices.* These are in addition to existing support for *PROFIsafe*, *PROFIdrive*. Robust, widely adopted IE PROFINET addresses specific engineering and functional issues related to network design;—including key issues like: *Network design and planning* to address topology limitation, network loading and performance scalability, *physical implementation* to take care of environmental factor, distance limitation, and adopting best installation commissioning and maintenance practices. PROFINET also addresses functional issues like: *Configuration complexity*, *Diagnostics with troubleshooting*, *Device replacement Security*, *Integration,* and other issues like *Interoperability*, *Cyber security,* and *Coexistence* on same infrastructure. PROFINET is mainly designed for exchanging data between controllers as well as devices within an automation setting. PROFINET simply defines cyclic & acyclic communication between different components like diagnostics, alarms, functional safety, and additional information. PROFINET uses an RT channel for delivering data in a very quick and deterministic way to use in time-critical-based applications. PROFINET RT fulfills the necessities of the majority of time-critical applications. Overall performance mainly depends on the design of the network, however, cycle times from 250 µs to 512 ms are well expected. PROFINET IRT is used to increase performance so that it reduces changeable data delays by increasing the rules utilized to switch Ethernet traffic & also creating special rules for the traffic of PROFINET. Fig III 17.2e shows PROFINET Functional Details. In this subsection, briefly these engineering and functional issues on PROFINET shall be discussed.

1. **Technological detailing:** Like PROFIBUS, PROFINET is based on master/slave. IO-Controller (master) can exchange frames with all devices. Process data are exchanged between controller and slaves cyclically as real-time data. IO-Device as slave can exchange frames with IO-Controller but traffic to other IO-Devices is only possible *noncyclically*. In PROFINET, for RT communication (1-10 mS RT performance), the EtherType 0x8892 (PROFINET RT over Ethernet) is used at layer 2 [ref: Fig III 17.2e1]. It uses a special tag to bypasses slow TCP/IP layers and send the packet directly to the application (Layer 7) through Ethernet. Device classes defined for PROFINET standard are as follows: IO-Controller (say PLC-like Class 1 master in PROFIBUS), IO-Device (distributed field devices),

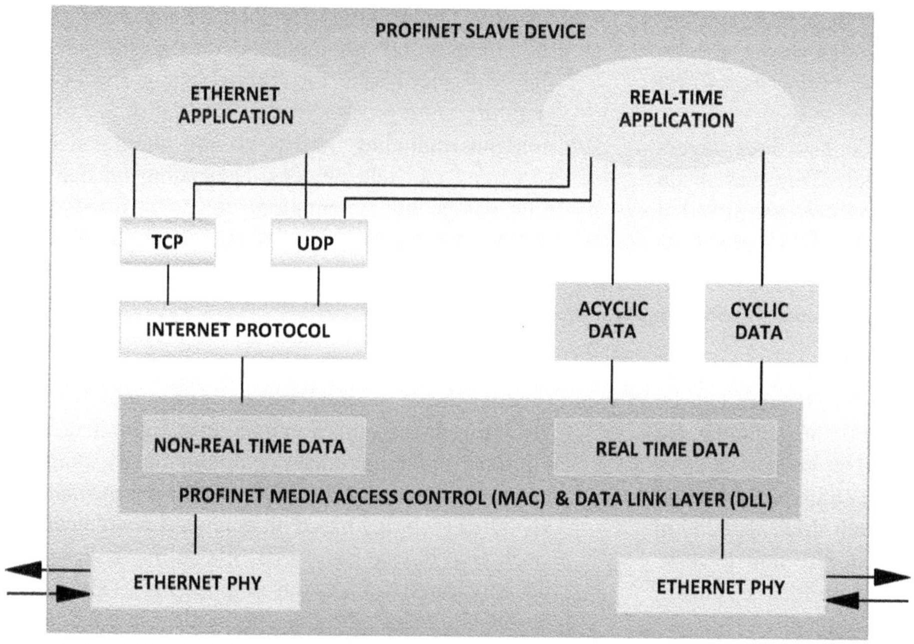


TCP:TRANSPORT CONTROL PROTOCOL; UDP USER DATAGRAM PROTOCOL

FIGURE III 17.2E PROFINET functional details.

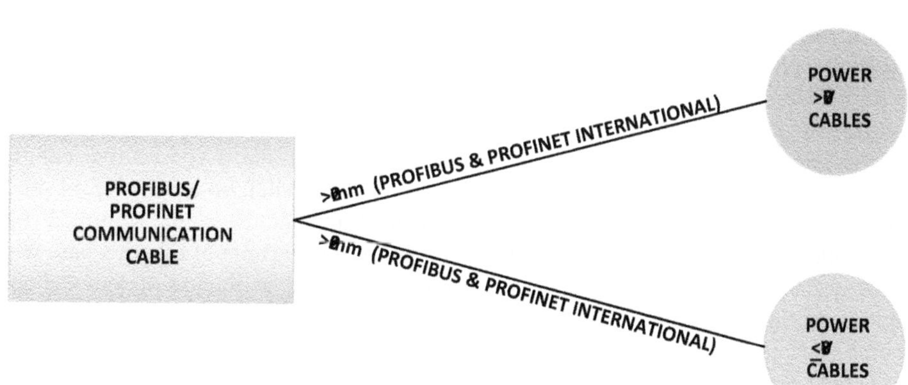


FIG III 17.2e2 PROFINET CABLE DISTANCES

and supervisor (Programming device) as shown in Fig III 17.2b2.

- **Device detail:** The *Slot* holds an I/O module when inserted in a modular IO-Device. In a configured modules can be one or more subslots used for data exchange and are addressed according to the different slots. *Subslots* implement the actual interface to the process with bit-wise/byte-wise/word-wise data granularity is determined by the manufacturer. Data content of a subslot is always accompanied by *status information* for the validity of the data can be determined. The *Index* specifies the data within a slot/subslot that can be read or written acyclically via read/write services.
- **Communication Basis:** IO-controller sets up communication paths between the IO-controller and an IO-device during system startup based on the configuration data in the engineering system. There will be cyclic and acyclic data exchange between IO devices and controllers. However, acyclic data exchange takes place between supervisor to IO device/controller.
- **Node details:** PROFINET node has three layers conforming to the OSI model. This has been depicted in Fig III 17.2e1 which shows are included in each PROFINET node. The physical layer is simply the medium, however as PROFINET is Ethernet compatible, so the physical layer can be twisted pair copper or fiber optic cabling to support 100 Mbps data rates. MAC layer can be implemented by an ASIC/FPGA can be used. Here, also there can be two options PROFINET switch (manage/unmanage types) with ASIC and external processor or integrated processor (say ARM) with PROFINET switch. The restriction on the industrial application is that it supports a standard TCP/IP and UDP/IP stack and Ethernet-based device profiles as shown. Also, application profile can contain PROFIsafe/PROFIenergy, etc.

2. **Engineering & modeling:** Major issues related to engineering and modeling are as follows:
   - **Design rule:** PROFINET uses low-voltage communication grade cables as in PROFIBUS. Each region has specific electrical codes that state how far apart different classes of cables need to be from one another. Fig III 17.2e2 shows cable separation distances as per PI. Basic design rules for PROFINET on copper mainly includes the following:
     - There are several topologies available for PROFINET and these are as follows: star, line, ring, tree, or any combination thereof. Selection of topology with subnetworks (if any) based on application.
     - Selection of appropriate switch based on the application requirements.
     - Distance between the switch and device should be $< 100$ m. So, PROFINET copper cables between switch to switch/device should be $< 100$ m. For more distance, a repeater or fiber optics could be used.
     - For selection of switches issues like: Managed/unmanaged type, Number of ports, Redundancy support, Transmission rate, Voltage level, Approval, PROFINET IRT, etc., must be taken into account.
   - **Engineering issues:** Here, engineering issues, are mainly developed around IO. IO controllers are provided with an engineering tool for configuration. During configuration, the IO controller(s) and the IO devices suitably defined and relations established including defining the desired properties of the cyclic data exchange. Also, the assignment of modules & submodules for the slots & subslots of the IO device must be specified for each IO device based on the modules and submodules defined in the *GSDML* file. Ethernet-based PROFINET supports TCP/IP enabling among other things, the use of web technologies including access to an integrated web server on the field devices. Depending on the specific device implementation, diagnostics and other information can be easily called up using a standard web browser, even across network boundaries [6].
   - **Addressing:** Each field device in a PROFINET IO receives a symbolic name to uniquely identify the field device within this IO system and also for relating the IP address to the MAC address using DCP during commissioning (device initialization). Optionally, the name can be automatically assigned to the IO device by the IO controller by means of a specified topology based on neighborhood detection. Each field device has (GSD) file so the devices can be configured for system engineering. This *XML*-based GSD describes the properties and functions and contains all data relevant for engineering as well as for data exchange with the field device. The field device manufacturer must supply the GSDML specification. Globally unique MAC address of each PROFINET device consists of a company code (bits 24 … 47) as Organizationally unique identifier (OUI) and a consecutive number (bits 0 … 23). With an OUI, up to 16,777,214 products of a single manufacturer can be identified [6].
   - **Web Integration:** On account of Ethernet base with support for TCP/IP enables the use of web technologies, that is, access to an integrated web server on the field devices. Depending on the specific device implementation, diagnostics and other information can be easily called up using a standard web browser, even across network boundaries.

- **Engineering model:** Engineering model provides defined ways to configure, connect automation components like controllers, I/Os, etc., for communicate are structured utilizing standard Ethernet so that it is modular, manufacturer neutral, deterministic with high performance. PROFINET engineering model it includes different communication channels like TCP/IP for configuration/diagnostics, EtherType 0x8892 for RT/IRT data. So, an engineering object model enables the development of configuration tools & the specification of manufacturer/customized functional expansions. This should distinguish between the control logic programming of individual module and the configuration of the overall plant for an application. The software created during programming is encapsulated in the form of a PROFINET component using the component editor interface that is also to be integrated in the tool. PROFINET specification defines the XML file in which the component editor interface generates the component description. The PROFINET engineering tool helps to configure the plant by interconnecting the PROFINET components to an application. PROFINET components are transferred to the connection editor by importing their XML files and the relationship is established over graphical lines [13].
- **Migration model:** As shown in Fig III 17.2b1 PROFIBUS segments are integrated into PROFINET using proxies. These assume a proxy function explained in III 17.2.1.1 for all the devices connected to PROFIBUS. With this, it is ensured that when rebuilding or expanding plants, the entire spectrum of PROFIBUS devices, including products of PROFIdrive and PROFIsafe can be implemented unchanged.

3. **Communication & Configuration issues:** Communication & configuration details are as follows:
   - **Communication model:** The PROFINET communications model defines a vendor-independent standard for communication on Ethernet with conventional IT mechanisms including TCP/IP & Component object model (COM)/Distributed COM (DCOM), common standards in PC world. It provides direct access between the office world to the automation level, that is, vertical integration. With PROFINET, the DCOM wire protocol, together with the afore-mentioned standards, defines the data exchange between the components of different manufacturers over Ethernet [13]. In this connection Fig III 17.2f1 may be referenced. COM/DCOM rely on functioning of TCP/IP. The runtime ensures that COM/DCOM traffic (slow, configuration) doesn't interfere with high-speed I/O data (RT/IRT). Built on standard DCOM and RPC (Remote Procedure Call) technology, PROFINET CBA provides an integrated system from the operational network to the controller and ERP systems.
   - **Communication Channel:** PROFINET communication channels have been detailed out in Fig III 17.2f showing IRT, RT, and non-RT channels and AR detailed.

     Two main problems associated with PROFINET data using the entire OSI stack are as follows: Each layer of the stack calls for extra work for packing unpacking PROFINET data at the source and destination and Using network layer introduce additional delay. Basic communication model has been described above, from there we find that there are two main types of data exchanges as shown in Fig III 17.2f1. Also, communication between IO controller and device are established through AR as shown in Fig III 17.2f2.
     - *RT data*: As shown for RT data in PROFINET uses OSI layers one and two to transmit real-time data. Only the destination & source MAC addresses are packed in the frame, with the Ethertype. The RT channel skips the encapsulation steps in the network, transport & session layers so as to have low latency and jitter. However, as there is no IP address, RT frames can't be routed between LANs. IRT divides bus cycle into open interval and reserve interval.
     - *Non-RT data*: PROFINET uses the non-RT channel for configuration and diagnostic data from an external network. Routing restrictions can be a real problem on large networks, where diagnostic tools need to have access to PROFINET devices.
     - *AR*: The communication between IO controller and device is established during system startup based on the configuration data received from the engineering system. This specifies the data exchange explicitly. Every data exchange is embedded into an AR. Within the AR, Communication relations (CRs) specify the data explicitly.
   - **Network configuration:** PROFINET supports various topologies for Ethernet communication like: *Line topology*, connecting terminals with integrated switches, *Star topology*, incorporates a central switch located preferably in the control cabinet and *Ring topology*, that is, a line is closed to form a ring to achieve media redundancy. It also supports Tree topology.

4. **Diagnostics & Security:** Important issues like diagnostics and security are as follows:
   - **Network Diagnostic:** Various diagnostic requirements especially for data access from anywhere are as follows: *Quick faults localization*, *Plane text*

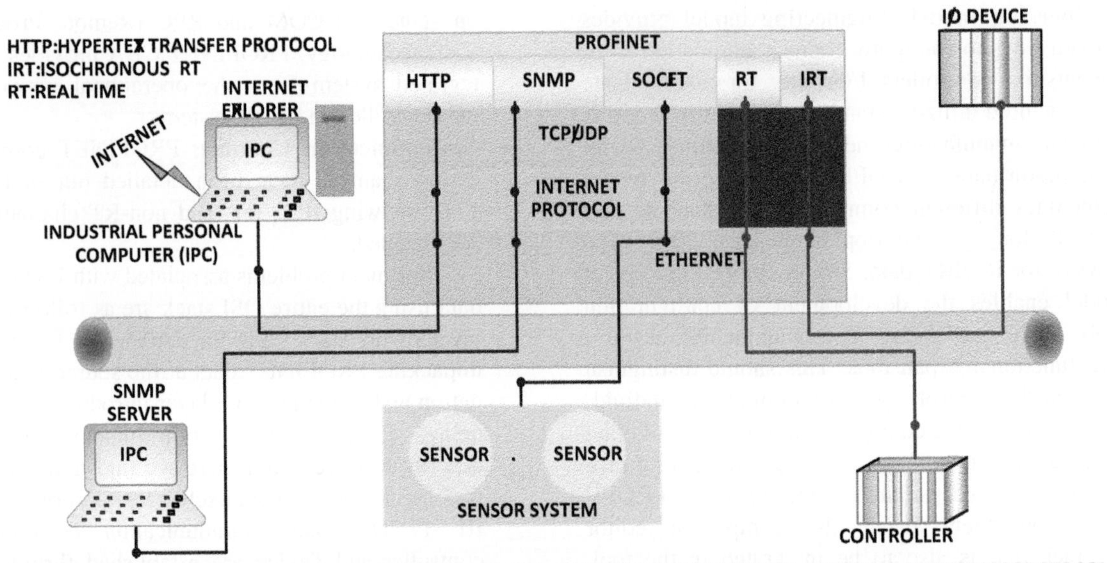


FIG III 17.2f1 PROFINET IEEE 802.3, RT AND IRT DTATA EXCHANGES

FIG III 17.2f2 PROFINET: APPLICATION RELATION WITH COMMUNICATION RELATION

**FIGURE III 17.2F** PROFINET communication details.

*fault messages*, *Comprehensive diagnostics*, *Use of common IT standards*, *Preventative diagnostics & maintenance,* etc. Diagnostic information is structured in hierarchically. PROFINET provides comprehensive diagnostics for application and network. Network management is done through DHCP, DCP, SNMP & LLDP. The network diagnostics of all PROFNET devices Is compiled in the Management information base (MIB) and the extensions to the LLDP-EXT MIB. MIB contains information like Device name, location, IP address, status, etc., and via vendor-independent LLDP, connected devices specify their identity and properties. Let us now throw some light on important issue like:

- o *Network management protocol*: SNMP is the de facto standard for maintenance and monitoring of network components and their functions. For diagnostic purposes, SNMP can read access network components, read out statistical data pertaining to the network, port-specific data, and information for neighborhood detection. For devices of conformance Class B and C implementation of SNMP is mandatory.
- o *Neighborhood Detection*: PROFINET field devices use the LLDP according to IEEE 802.1AB to exchange the available addressing information via each port. This allows the respective port neighbor to be explicitly identified. With this neighborhood detection, changes of the topology during operation can be recognized immediately.
- o *Device Replacement*: If a field device fails in a known topology, it is possible to check whether the replacement device has been reconnected in the proper position.
- o *I/O Diagnostics*: For integration of the network diagnostics into the IO system diagnostics, a switch must also be used as a PROFINET IO device, so that the switch can signal identified network errors of a lower-level Ethernet line and specific operating states to its IO controller.

- **Security issue:** To enhance security in the system, PROFINET offers protection from: *Unauthorized access, Address errors and Improper operation including address error, Manipulation & espionage,* etc. Also, proven standard security offers unique features like scalability, requirement of preliminary

IT security knowledge only and use in all levels of automation.

5. **Application profile:** Application profiles refer to specifications concerning certain properties, performance characteristics, behavior of devices & systems that are developed by manufacturers and users. *General application profiles* for different applications including PROFIsafe & PROFIenergy. *Specific application profiles* are developed for individual type such as *PROFIdrive* or *devices for process automation.* Since PROFIsafe has been discussed in II 5.1.1 hence not repeated here.
   - **Profi-energy:** The high energy cost & legal obligations are compelling industry to engage in energy conservation. PROFIenergy enables an active and effective energy management. In PROFINET energy saving is possible through production pause saving and organizing data. PROFIenergy is used for "Switch off during *Short/Long break*, during *unplanned break*" reading of measured data, etc. These can be used for shutting off power when demand exceeded or reducing energy use.
   - **PROFIdrive:** The PROFIdrive refers to the standardized drive interface specification, developed by PI, for PROFIBUS and PROFINET. This vendor-neutral application profile has been standardized in IEC 61,800–7. Here, the primary focus is on the realm of drives, encoders, motors, and their versatile applications, spanning from simple to highly demanding motion control tasks. PROFIdrive serves as the cornerstone for defining device behavior and techniques for accessing internal device data for the connected electric drives. This covers from basic frequency converters to high-performance servo controllers having key strengths of interoperability. PROFIdrive uses standardized drive IO data sets, messages, and parameter sets, while also allowing manufacturers to incorporate proprietary features, fostering healthy competition and innovation.
   - **Process automation:** For the optimal use of PROFINET in process automation, PI has developed one requirements catalog in collaboration with users to ensure that plants having an existing future-proof system based on PROFIBUS can change to PROFINET at any time, that is, the use of PROFIBUS DP can be replaced with PROFINET [10]. The requirements mainly include the functions for cyclic and acyclic data exchange, integration of fieldbuses integration and parameter assignment of devices including *Configuration in run, Diagnostics, Redundancy, Time stamping,* and *Maintenance.*

*With this, the discussions on PROFINET is not only concluded, but discussions on main IE have been concluded as well to move forward for other network types.*

## List of abbreviations used

**ACP** Alarm control protocol
**ADS** Automation device specification
**AoE** ADS over EtherCAT
**APL** Advanced physical layer –Ethernet APL
**AR** Application relation
**CBA** Component-based automation (PROFINET)
**CC** Conformance class
**CD** Collision detect
**CIP** Common industrial protocol
**CiR Configuration in Run PROFINET**
**CNC** Computer numerical control
**CoE** CAN application protocol over EtherCAT
**COM** Component object model
**COS** Change of state
**COTS** Commercial off-the-shelf
**CSMA** Carrier sensed multiple access
**DC** Distributed clock
**DCOM** Distributed COM
**DCP** Discovery and basic configuration protocol (PROFINET)
**DCP** Discovery and configuration protocol
**DCS** Distributed control system
**DHCP** Dynamic host configuration protocol
**DLR** Device-level ring
**DMA** Direct memory access
**DTLS** Datagram TLS
**EDD** Ethernet device driver
**EDS** Electronic Data Sheet
**EIP** EtherNet/IP (in the book)
**EoE** Ethernet over EtherCAT
**ERP** Enterprise resource planning
**ESC** EtherCAT slave controller
**ESM** EtherCAT state machine
**FoE** File Access over EtherCAT
**FSCP** Functional safety communication profile
**FSoE** **Fail**safe/Safety over EtherCAT
**FTP** File transfer protocol
**GSD** General station descriptions
**GSDML** General station descriptions markup language
**HMI** Human machine interface
**I/O** Input/Output
**IDPS** Intrusion prevention detection systems
**IE** Industrial Ethernet
**IIoT** Industrial IoT
**IoT** Internet of things
**IRT** Isochronous RT
**LAN** Local area network
**LLDP** Link layer discovery protocol
**LVDS** Low voltage differential signal
**MAC** Media access control
**MIB** Management information base
**MRP** Media redundancy protocol
**NDIS** Network driver interface specification
**NIC** Network interface card
**OPC** Open platform communication
**OS** Operating system
**OT** Operational technology
**OUI** Organizationally unique identifier

**PAC** Process automation control
**PC** Personal computer
**PCI** Peripheral component interconnect
**PDI** Process data interface
**PDO** Process data object
**PHY** (Ethernet) Physical
**PI** PROFIBUS/PROFINET International
**PLC** Programmable logic controller
**QoS** Quality of service
**RAM** Random Access memory
**RPC** Remote procedure call
**RT** Real time
**RTOS** RT operating system
**SCADA** Supervisory control and data acquisition
**SDO** Service data object
**SIL** Safety integrity level
**SNMP** Simple network management protocol
**SPI** Serial peripheral interface
**TCP** Transmission control protocol
**TFTP** Trivial FTP
**TLS** Transport layer security
**UDP** User datagram protocol
**VLAN** Virtual local area network
**WAN** Wide area network
**WLAN** Wireless local area network
**XML** Extended markup language

## References


[1] EtherCAT System Description; Beckhoff New Automation Technology; Version 5.4; April'2020; Internet document. https://download.beckhoff.com/download/document/io/ethercat-terminals/ethercatsystem_en.pdf.

[2] EtherCAT –the Ethernet fieldbus; The technical response to the question: "Why EtherCAT?"; Internet Document; EtherCAT org; https://www.ethercat.org/pdf/ethercat_e.pdf.

[3] Ethernet or EtherCAT for Motion Control: Choosing the Right Network for Your Application; GALIL; White Paper; https://www.galil.com/news/white-paper-ethernet-or-ethercat-motion-control-choosing-right-network-your-application.

[4] EtherCAT – The Ethernet Fieldbus; EtherCAT Technology group; Internet document; https://www.ethercat.org/download/documents/ETG_Brochure_EN.pdf.

[5] Safety with FSoE; acontis blogs; Acontis; Internet document; https://www.acontis.com/en/safety-with-fsoe-failsafe-over-ethercat.html.

[6] EtherNet/IP Device Level Ring; Application Technique; Allen Bradley; Rockwell automation Catalog; https://literature.rockwellautomation.com/idc/groups/literature/documents/at/enet-at007_-en-p.pdf.

[7] Network infrastructure for EtherNet/IP: Introduction and Considerations: EtherNet/IP; ODVA; Internet document; https://www.odva.org/Portals/0/Library/Publications_Numbered/PUB00035R0_Infrastructure_Guide.pdf.

[8] EtherNet/IP Quick Start for Vendors Handbook; A Guide for EtherNet/IP Developers; ODVA; PUB00213R0; 2008; https://www.odva.org/wp-content/uploads/2020/05/PUB00213R0_EtherNetIP_Developers_Guide.pdf.

[9] IIoT/Industry 4.0' ODVA publication; Internet document; https://www.odva.org/technology-standards/market-drivers/process-automation/iiot-industry-4-0/.

[10] PROFINET System Description – Technology and Application; Profinet; US. Profinet: https://us.profinet.com/wp-content/uploads/2012/11/PROFINET_SystemDescription_ENG_2014_web.pdf.

[11] Introduction to PROFINET I/O; Real Time Automation; Internet document; https://www.rtautomation.com/technologies/profinet%20-io/.

[12] S. Basu, A.K. Debnath, Power plant Instrumentation and control handbook, 2nd ed., Elsevier; Academic Press, 2019. https://www.elsevier.com/books/power-plant-instrumentation-and-control-handbook/basu/978-0-08-102804-9.

[13] PROFIBUS Technology and Application - System Description; PI; Internet document; https://www.profibus.com/download/profibus-technology-and-application-system-description.

Chapter 18

# CANbus and CIP networks: ControlNet & DeviceNet

The message-based Controller Area Network (CAN) bus protocol enables the electronic control units (ECUs) in automobiles and other devices. Messages/"frames" are received by all devices in the network without a host supporting International Standards Organization (ISO) 11898. Flexible data rate–based CAN is referred to as CAN FD. The standard length of each message has been increased by 800% to 64 bytes, and the maximum data rate has similarly been increased from 1 to 8 Mbps. This *decentralized*, *standardized*, *priority-driven* protocol is highly reliable and robust. CANopen is a CAN-based communication protocol, that is, it is a "higher-layer protocol" based on CANbus that adds several core concepts like communication models such as master-slave, client-server, and producer-consumer; communication protocols such as service data object (SDO) and process data object (PDO); device profile standards such as vendor independence; and device states, which support various states viz. a master node changing the state of a slave. Each device has one object directory (OD), and electronic data sheet (EDS) is used as the file format for the OD.

The "token passing"–based Common Industrial Protocol (CIP) network ControlNet is managed by ODVA. ControlNet talks to devices like programmable logic controllers (PLCs) and human-machine interface (HMI) drives. ControlNet uses various connectors and taps to support various topologies. This high-speed CIP network offers features like *determinism*, *scheduled communication* with a network update time (NUT), support for *producer-consumer* models, various device types, and *redundancy*. With an integrated architecture model, it finds applications in wide industrial areas. ControlNet uses a communication model in which, other than the user's application objects, all other blocks reside within either a ControlNet application-specific integrated circuit or ControlNet class-specific object. ControlNet can have 99 nodes per link and supports a data rate of 5 Mbps with an undetected error rate of less than 1 error in 20 years. For long distances it supports repeaters, whose quantity depends on the cable type used.

The object-oriented, CIP-based networks CompoNet and DeviceNet are both managed by ODVA and share the same application layer framework. Being members of the CIP family, both also support *object modeling*, *messaging*, *producer-consumer models up to 64 nodes*, and others. Naturally, these networks can be easily integrated as well as integrated with other members of CIP. However, they are divergent in many issues, especially in their physical and data link layers, with different industrial automation application areas. CompoNet is characterized by its high speeds of up to 1000 Mbps and simplicity in cabling (it uses RS 485 as a physical medium). This is very much suitable for machine to machine–level networking because of the high-speed data transfer. Because of its robustness, flexibility, and power on one cable, DeviceNet finds its major use in drive-level networks. DeviceNet uses CAN as a physical layer. The DeviceNet object model is a template for organizing and implementing the *attributes* (data), *services* (procedures), and *behaviors* of the components of a DeviceNet product. The model provides an addressing scheme consisting of the *node address* (media access control [MAC] identification [ID]), *object class identifier*, *instance number*, and *attribute number*. These four-level addresses are used in conjunction with an *explicit messaging connection* to move data from one place to another on a DeviceNet network.

In this chapter, all these advanced issues on CAN, ControlNet, CompoNet, and DeviceNet will be discussed in detail, with some information overlapping with III 8.0.2 and III 9.0.2.

## 18.0 Advanced discussion on Controller Area Network bus and Controller Area Network open

The Bosch CAN specification became an ISO standard (ISO 11898) in 1993 (CAN 2.0A) and was extended in 1995. Typically, CAN interconnects a network nodes using a two-wire, twisted pair cable. Classical CAN has a maximum data rate of 1 Mbps with an 8-byte data payload. As a result, advanced CANs like CAN FD (5–8 Mbps with a payload of 64 bytes) and CAN with extended data-field length (CAN-XL, offering speeds of 10 Mbps or higher

**Plant Intelligent Automation and Digital Transformation. https://doi.org/10.1016/B978-0-323-89892-8.00011-9**

with a high payload of 2048 bytes) were introduced. CAN provides the basis of communication, so a set of *standardized protocols* exist to specify data exchange and communication between CAN nodes. The most common are:

- **SAE J1939:** J1939 is the standard for in-vehicle networks pertinent to heavy-duty vehicles. Here, parameters are identified by a suspect parameter number and grouped in a parameter group (PG) identified by a PG number.
- **OBD2:** On-board diagnostics (OBD), as per ISO 15765, is a self-diagnostic and reporting capability used to identify car issues. The OBD2 protocol is a standardized system that was defined along with the connector specification in 1994 by the California Air Resources Board. The OBD 2 protocol has become an essential tool for vehicle diagnostics. OBD2 specifies Diagnostic trouble codes (DTCs) and Real-time (RT) data. OBD2 is a vehicle's built-in self-diagnostic system with the OBD2 16-pin connector placed near the steering wheel.
- **CANopen:** CANopen uses the Open Systems Interconnection (OSI) model as a conceptual model to standardize communication function across diverse communication technologies. CANbus represents the two lowest layers. Based on CAN, CANopen is used widely in embedded control applications, which is key in machine diagnostics and optimizing production.
- **CAN FD:** This is an extension of the classical CAN data link layer that increases the payload from 8 to 64 bytes and allows for a higher data bit rate, depending on the CAN transceiver. It is used in electrical vehicles (EVs).

There also exist universal diagnostic service (UDS) on CANbus and diagnostics over CAN. With the introduction of ISOBUS, it is possible to standardize communication between modern vehicles like tractors. In this section, CANbus advanced systems will be discussed.

## 18.0.0 Further discussions on Controller Area Network bus

From III 8.0.2, it is clear that CAN is a two-wired, half-duplex, high-speed serial network technology used in communication in a low-radius region, viz. automobile. A CAN protocol is a Carrier Sense Multiple Access with Collision Detection (CSMA/CD) method, which was discussed in arbitrations on message priority protocol. Collision detection ensures that the collision is avoided by selecting the messages based on their prescribed priority. A signaling rate from 125 kbps to 1 Mbp is possible. There are about 2048 different message identifiers. It is an ISO-11898 standard. From discussions in III 8.0.2, we know that there are four types of CAN messages/frames. These are the *data*, *remote*, *error*, and *overload frames*. Various frame details have already been detailed in Fig. III 8.0g. The data frame refers to the standard CAN message that broadcasts data from the transmitter to the other nodes. A remote frame is broadcasted by a transmitter to request data from a specific node. An error frame may be transmitted by the node that detects a bus error. Overload frames are used to introduce additional delay between data or remote frames. Standard and extended frames may exist on the same bus. Based on previous information, we find that the CAN data frame is composed of seven fields: start of frame, arbitration, control, data, cyclic redundancy check (CRC), acknowledgment (ACK), and end of frame (EOF). The arbitration field of the CAN message consists of an 11- or 29-bit identifier and a remote transmission request (RTR) bit. The CAN arbitration scheme is CSMA/CD to assure the highest priority message is broadcast. The dominance of the RTR bit distinguishes the data frame (logical 0) from the remote frame. The control field consists of 6 bits (only the lower 4 are used) that indicate the amount of data in the message. Let us now recapitulate error handling. CAN implements five levels of error detection, some of which are elaborated here. At the message level, it performs CRCs, frame checks, and ACK checks. Bit-level checks are monitoring and stuffing. Cyclical redundancy errors are detected using a 15-bit CRC (by the transmitter). Each receiver accepting the message recalculates the CRC to compare against the transmitted value. A discrepancy between the two *flags* an error by a receiver for the detection of *an invalid bit in the CRC delimiter*, *ACK delimiter*, *EOF*, or 3-bit interframe space. Finally, each receiving node writes a dominant bit into the ACK slot of the message frame that is read by the transmitting node [1]. In this subsection, details on CANbus will be provided that build on discussions in III 8.0.2, with some overlap.

1. **Message types, physical layer, and transceiver:** The three issues addressed here are:
   - **Message types:** The five message types are:
     - *Data frame*: The various fields are the arbitrary, data, CRC, and ACK fields.
     - *Remote frame*: This is a request for transmission of data from another node with a recessive RTR bit.
     - *Error frame*: This is sent when an error is detected.
     - *Overload frame*: This is used when the node is too busy to show the delay between messages.
     - *Valid frame*: A message is valid when the EOF field is recessive; otherwise, the message is transmitted again.

- **Physical layer:** As shown in Fig. III 8.0e2, there is a two-wire serial link—CAN_H and CAN_L—and their voltage levels relative to each other (for differential signaling) determine whether a 1 or 0 is transmitted. The current flow in each signal line is equal and opposite in direction, resulting in low noise emissions due to a field-canceling effect. This balanced differential signaling with low noise coupling allows a high rate of transmission over the wires. Generally, shielded or unshielded twisted-pair cables (with a characteristic impedance 120 ohms) are used for a bus length of 40 m, accommodating a maximum of 30 nodes.
- **Transceiver:** The two wires—CANH and CANL—are normally at 2.5 V, which is determined by the two transistors and the 2.5 V voltage source. Basically, the difference between the two wires should always be 0. The driver control determines the voltage applied to the CAN_H and CAN_L wires. When both the transistors are conducting, the voltage drop across the first transistor and the diode is 1.5 V, making the CAN_H wire pull up to 3.5 V. This is elaborated in Fig. III 8.0e2.

2. **CAN frames:** The following important CAN frames are discussed in further detail.
   - **Error frame details:** When the transmitting/receiving node detects an error, the transmission is immediately aborted. Also, an error frame comprising an error flag (made up of six dominant bits) and an error flag delimiter (made up of eight recessive bits) are broadcasted. As bit string violates the *bit stuffing rule*, all other nodes respond by transmitting error flags as well. When a sufficient number of errors are detected, a node will eventually turn itself off. Robustness, especially in automotive environments, requires that the network *determines* whether errors are *transient* (due to voltage spikes, noise or, some other condition) or a *permanent failure* of the node due to defective hardware. Consequently, nodes store and track the number of errors detected. If the count in buffer of a node is >0 but <128, the node is "*error active*"; an error count between 128 and 255 puts the node in "*error passive*" mode. Error counts above 255 will cause the node to enter "bus off" mode and take itself offline.
   - **Remote frame details:** The intended purpose of the remote frame is to solicit the transmission of the corresponding data frame. It is like the data frame, but with some differences, for example, it is explicitly marked as an RTR bit in the arbitration field and has no data field. It can be used to implement a request-response type of bus traffic management. In the remote frame the data length code (DLC) must be set to the length of the expected response message. The remote frame may be accepted by any number of nodes in the network but can only be sent by one node associated with the requested message. If a data frame and remote frame using the same message ID try to access the bus simultaneously, the data frame will gain bus access over the remote frame since data frame uses a dominant (logical 0) RTR bit. Remote frames can only be transmitted with a DLC identical to the DLC of the corresponding data frame [2].
   - **Overload frame details:** If a CAN node receives messages faster than it can process, then an overload frame will be generated to provide extra time between successive data or remote frames. The overload frame has two fields viz. an overload flag (six dominant bits) and an overload delimiter (eight recessive bits). Unlike error frames, error counters are not incremented.
3. **Bus loading and time protocol:** Bus loading and time protocols have been discussed.
   - **Bus loading:** In the case of CANbus, the message latency is *not deterministic* on account of the existence of error frames, overload frames, and retransmissions. Latency increases with the amount of traffic on the bus. Generally, bus utilization should be within 30% of the bus capacity so that low-priority messages do not experience unacceptable delay.
   - **Time protocol:** For RT control, it is always better to implement a communication protocol that guarantees timing deadlines for messages irrespective of the load on the bus. "*Time-triggered CAN,*" (TTCAN) is one of that kind (ISO 11898-4). A TTCAN message frame incorporates two types of "time windows": exclusive time windows and arbitrating time windows. Exclusive time windows are assigned to specific messages that are transmitted periodically [1].
4. **Automotive CANbus types:** CANbus is defined by the ISO 11898 standard, which details issues like how communication happens, how wiring is configured, and messages transactions. For automotive networks, there exist a number of different network types like:
   - **High-speed CAN:** ISO 11898 mainly focuses on high-speed CANbus. This supports bit rates from 40 kbps to 1 Mbps (classical CAN) and provides simple cabling for automotive applications. OBD2, J1939, National Marine Electronics Association (NMEA) 2000, CANopen, and others are developed from this. The next generation of this is referred to as CAN FD.
   - **Low-speed CANbus:** Here, speed is in the range of bit rates from 40 to 125 kbps. CANbus can communicate even if there is a fault on one of the two wires. Therefore, it is called "fault-tolerant CAN."

- **Local Interconnect Network (LIN) bus:** LIN bus is a lower-cost supplement to CANbus networks, with less harness and cheaper nodes. LIN bus clusters typically consist of an LIN master acting as gateway and up to 16 slave nodes [3]. This single-wire network is used for simple tasks like window motors, seat adjustments, and mirrors.
- **Automotive Ethernet:** To meet the high bandwidth requirements of advanced driver assistance systems, this was developed to support cameras and other applications. However, it lacks some of the safety/performance features of CANbus and CAN FD.
- **FlexRay**: This is a deterministic, fault-tolerant, and high-speed bus system. FlexRay delivers the error tolerance and time determinism performance requirements for x-by-wire applications (i.e., drive-by-wire, steer-by-wire, brake-by-wire). It offers a speed of 10 Mbps in 2 or 4 wires for high-performance powertrain, safety, and other features.

5. **Comparison between CANbus and CANopen:** Table III 18.0a provides a comparison between CANbus and CANopen.

### 18.0.1 Discussion on Controller Area Network open and Controller Area Network Flexible Data Rate

Message-oriented, multimaster protocol CAN is very useful for quick serial data exchange. It finds its application in numerous areas of industrial automation, having a high degree of robustness and security, *low cost*, and *wide range of suppliers of components, associated software*, and *tools*. Its application areas mainly include automotive, automation, aerospace, and medical engineering. CANopen is a device- and manufacturer-independent protocol for communication on CANbus and covers the application layer (layer 7) of the OSI reference model. CANopen was initially designed for motion-oriented machine control systems, but now CANopen is extensively used not only in motor control but also in a wide range of other applications. The CANopen standard is very useful for interoperability between devices (nodes), especially in industrial machinery. Also, it provides standard methods for configuring devices. Fig. III 18.0a depicts details on CANopen and CAN FD.

The standardization of CANopen by "CAN in Automation (CiA)" in EN 50325-4 defines the following key points: *uniform basis for the exchange of commands and data between CANbus nodes through communication objects*, *mechanism for exchanging process data in RT (transferring large data volumes or sending alarm frames)*, and *definition of defined interfaces for addressing certain parameters of a device through profiles*. The CANopen and OSI models are detailed in Fig. III 18.0a1. The recommendations and specifications for CAN FD networks are detailed in Fig. III 18.0a2. The new CAN FD frame format that is currently being integrated into ISO 11898-1 introduces a second bit time configuration into CAN networks. Although the nominal CAN bit time is still used at the beginning and end of the frames, for arbitration and ACK, called the arbitration phase, the new data bit time may be used for the main part of the frames, called the data phase. Error detection is a crucial functionality in communication protocols and works well in CAN FD. A receiving node must be able to judge with a high probability whether a frame was received with or without bit errors. In this subsection, CANopen and CAN FD were briefly discussed.

**TABLE III 18.0A** Comparison between Controller Area Network bus and Controller Area Network open.

| Issue | CANbus | CANopen |
|---|---|---|
| Purpose | Robust RT comm. bus | Seamless device integration |
| Operating layers | Physical and data link (1 and 2) | Application (7) |
| Communication type | Low-level message protocol | High-level device communication |
| Message format | Data and control bits | Object with data, attributes |
| Device profile | No specific profile customer to configure | Standardized profile to simplify device integration |
| Interoperability | Device to follow CAN in layers 1 and 2 | Device to follow specific CANopen device profile |
| Error handling | Basic: CRC/ACK error | Advanced error detection |
| Standards | ISO 11898-1, 11898-2,<br>ISO 16845, ISO 16840 | CiA 303-1, 303-2, 303-3 |
| Message structures | Data and remote frame | Object dictionary PDO/SDO |
| Network management | NIL for diagnostics, etc. | Provides these services |
| Application areas | Automotive, aerospace, industrial automation | Industrial automation (device integration) |

*ACK*, Acknowledgement; *CAN*, Controller Area Network; *comm.*, communication; *CRC*, cyclic redundancy check; *ISO*, International Standards Organization; *Mngt*, management; *PDO*, process data object; *RT*, real time, *SDO*, service data object.

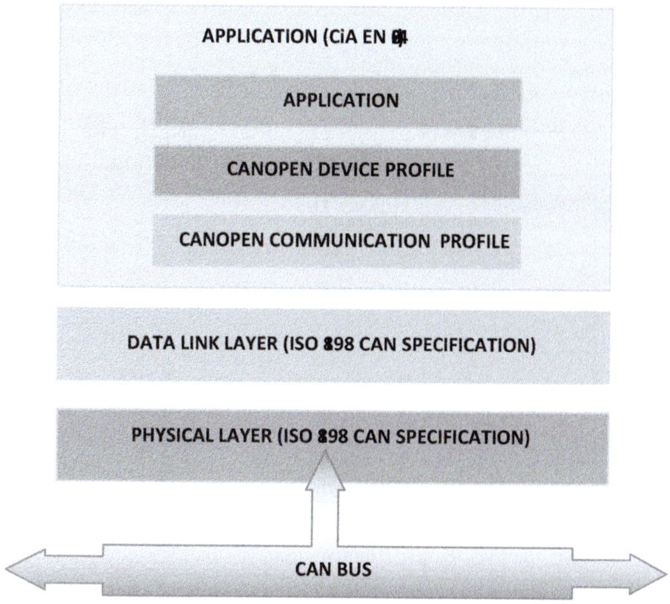


FIG III 18.0a1 CANOPEN OSI MODEL

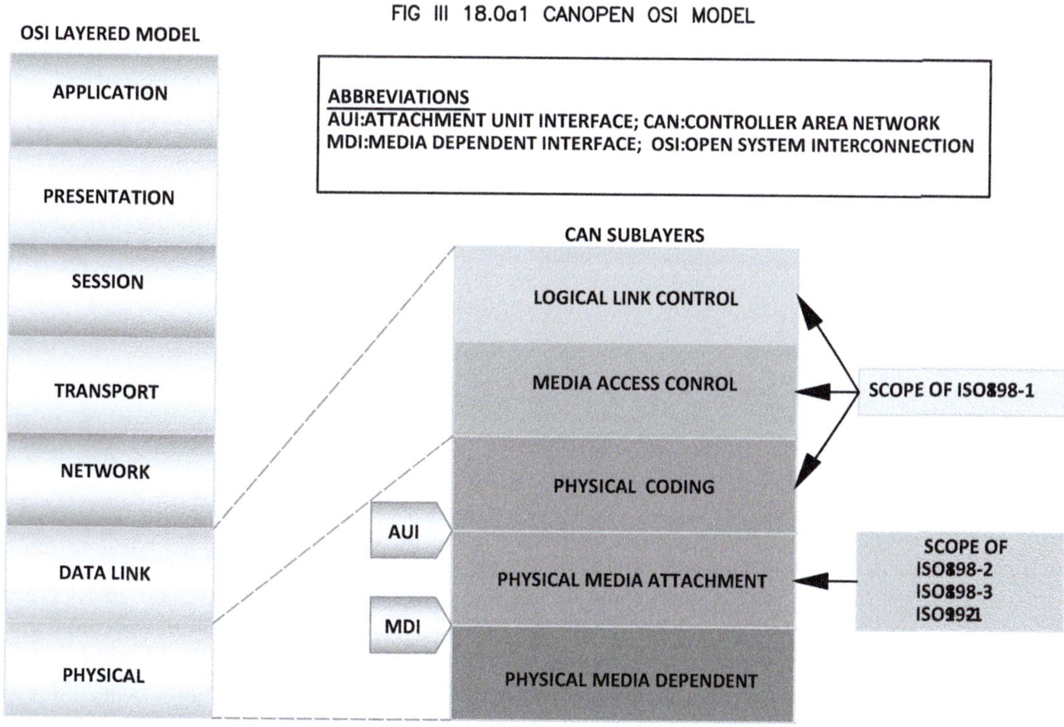


FIG III 18.0a2 CAN FD OSI MODEL

**FIGURE III 18.0A** Controller Area Network (CAN) open and CAN Flexible Data Rate (FD) details.

1. **CANopen explained:** At the very beginning of the chapter, CANopen was briefly explained. The standard open architecture configuration of CANopen allows access to all devices and synchronization with cyclic, event-driven, and process-driven data. It is characterized by interoperability and interchangeability, and thus is supported by a large number of international suppliers. The CANopen configuration and its disposition in the OSI model are detailed in Fig. III 18.0b. CANopen defines a system configuration that is supported by standard personal computer (PC) EDS file formats, nonvolatile central data storage, and distributed data storage, as depicted in Fig. III 18.0b1. CANopen mainly comprises:

**FIGURE III 18.0B** Controller Area Network open configuration and disposition in the Open Systems Interconnection (OSI) model.

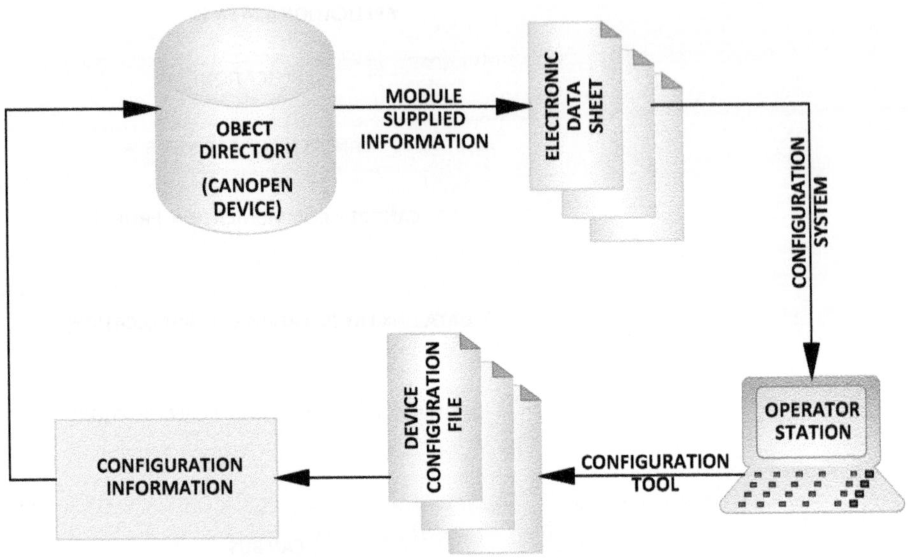


FIG III 18.0b1 CANOPEN CONFIGURATION (Typical)

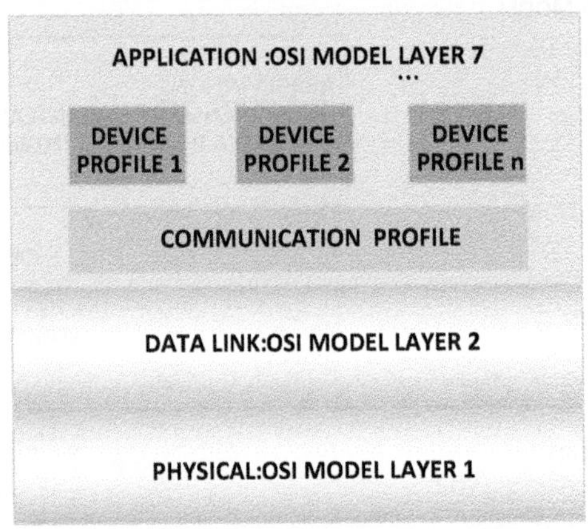


FIG III 18.0b2 CANOPEN IN OSI MODEL

- **Structural standard:** As shown in Table III 18.0a and Fig. III 18.0a1, CANopen is a "layer 7" communication profile developed on CANbus (ISO 11898). Layer 7 is supplemented by a set of device and communication profiles, as shown in Fig. III 18.0b2. CiA manages the documents viz. "CiA DS 301—CANopen Application Layer And Communication Profile" (EN 50325-4). This describes all basic principles, data types, encoding rules, and services. As discussed, the physical medium for CANopen devices is a differentially driven two-wire bus line with a common return (ISO 11898). The CANopen standard does not mandate a specific physical layer, allowing the use of CANopen in specific environments. CANopen recommends the specifications for industrial use. CiA DR 303-1 defines pin assignments for many connectors.
- **Object profile and OD:** Processes in CANopen are performed using objects. The object profiles define the structure and content of specific regions within a CANopen device's OD. *Profiles dictate content, ODs implement the profile.* Here, these are elaborated separately:
  - Objects: Objects can perform various tasks like identifying device information, displaying error conditions for a node (error objects), transporting data between nodes (communication objects), assigning node configuration data (network configuration objects), and monitoring and control (network management [NMT] objects). PDOs are used for RT process data and are located in a specific place in the OD. These are flexible in data length, without overhead in a preagreed data frame. However, they have

high-priority identifiers. Table III 18.0b provides the characteristics of PDO and SDO. This may be read in conjunction with III 8.0.2.4 and Fig. III 8.0g1.

- *Object profile*: There can be two types of profiles viz. *standardized profiles and manufacturer-specific profile*. Standardized profiles comprise objects to be used on different devices without customization. These are communication profiles and different device profiles, which are elaborated below.
- *OD*: CANopen is built around the concept of an OD—the interface between the application and communication within *each* device—so each function, variable, and data type seen via the network must be described in the OD with characteristics like *standardized structure*, *collection of communication and application-relevant parameters*, *reference to data types*, and *access to structured arrays* and *records* to facilitate the OD's definition of a standard addressing scheme.

- **Communication profile:** There exist distinctions between RT data and other data. Basically, a distinction is made between the communication objects: *SDOs for transferring data from and to the OD*, *PDOs for exchanging RT data*, *objects for NMT*, and *objects for controlling the CAN messages*. Three relationships between the network nodes in CANopen are depicted in Fig. III 18.0c. These relationships are:
  - *Master-slave*: A master controls the message traffic and the slaves only respond to master requests. As shown in Fig. III 18.0c1, messages are sent on an unacknowledged or acknowledged basis.
  - *Client-server*: A client-server relationship is always established and initiated by the client between two nodes and is bidirectional, as shown in Fig. III 18.0c2
  - *Producer-consumer*: As shown in Fig. III 18.0c3, a producer-consumer relationship is used where there is quick data exchange. The producer sends a frame that can be received by one or more nodes.
- **Device profile:** CiA defines device profiles containing objects for the basic functions and parameters of the respective standard device for important modules used in industrial automation. These profiles are unique and used on a manufacturer-independent basis.

2. **CANopen communication:** CANopen is an open, high-level communication protocol based on CAN for embedded systems. It uses an OD to define device parameters and is accessed via SDOs for configuration and PDOs. Fig. III 18.0d shows PDO, SDO, and emergency (EMCY) frames.
   - **PDO frame:** PDOs are used for RT data transmissions. It supports both synchronous and asynchronous transmissions. The PDO frame is detailed in Fig. III 18.0d1
   - **SDO frame:** SDOs and the OD entries of another network node can be accessed and the values of the objects can be read. The SDO frame is detailed in Fig. III 18.0d2
   - **Error detection and EMCY frame:** EMCY messages are sent when a critical error situation has occurred/been resolved in the device or if other devices must be provided with important information. The EMCY frames are sent according to the consumer/provider relationship, as detailed in Fig. III 18.0d3.
3. **NMT:** As shown in Fig. III 18.0e and detailed below, NMT has two parts.
   - **State-machine management:** NMT state machine are device control services for *initializing* the network and network nodes and *controlling* the operating states of the nodes. A CANopen network requires an NMT master and NMT slave devices to implement a state machine. The NMT master controls the status of all the NMT slaves on the network. Each slave device may be configured and parameterized via SDO (not PDO) communication. The NMT master may switch one or all nodes to their "*operational*" state. In the "*operational*" state, all communication objects are active.

**TABLE III 18.0B** Characteristics of process data objects and service data objects [4].

| Process data object | Service data object |
|---|---|
| High priority RT data exchange by broadcast | OD access |
| No overhead, no acknowledgment | Acknowledged, low priority |
| Synchronous/asynchronous, cyclic/acyclic, event driven | Peer-to-peer communication<br>Transfer of domain |

*OD*, Object direction; *RT*, real time.

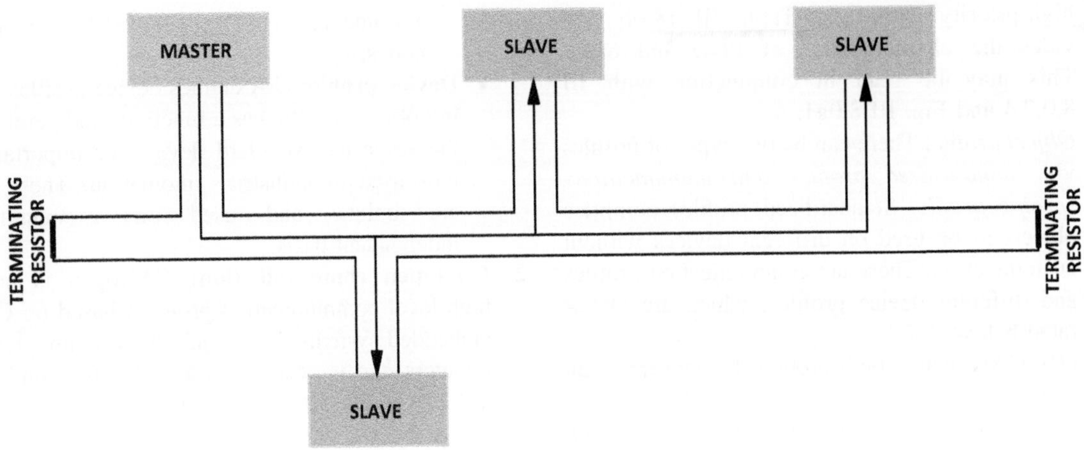


FIG III 18.0c1 MASTER SLAVE CONNECTION IN NODES

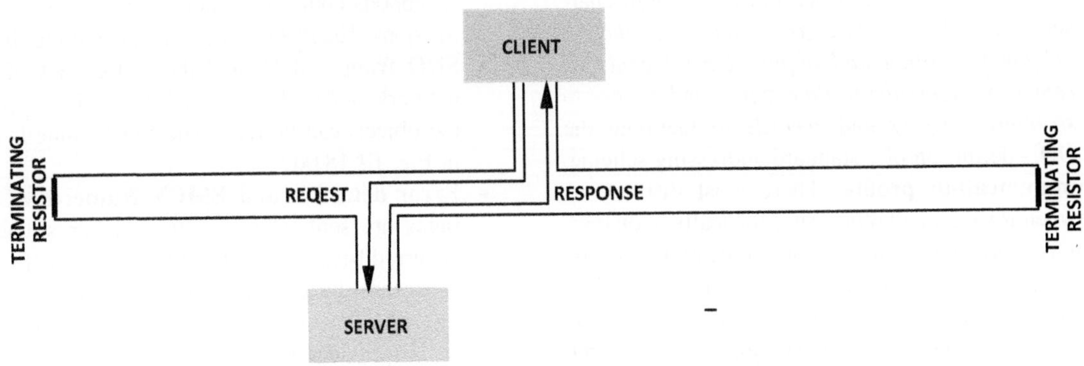


FIG III 18.0c2 CLIENT SERVER CONNECTION IN NODES

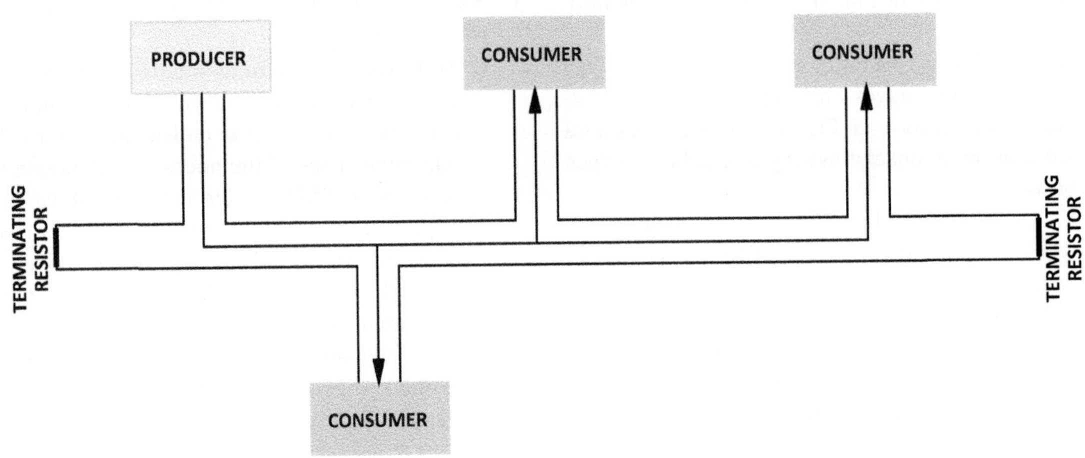


FIG III 18.0c3 PRODUCER CONSUMER CONNECTION IN NODES

**FIGURE III 18.0C** Controller Area Network open: three node connection types.

Switching a node into a "*stopped*" state terminates all PDO and SDO communication, as shown in Fig. III 18.0e1. Various functions are described in Table III 18.0c.

- **Monitoring NMT:** Connection monitoring services are used to monitor the nodes during network operation viz. *guarding function* and *heartbeat function.* Guarding functions can be node guarding (master) and life guarding (slave). Modern system designs use the heartbeat protocol, as detailed in Fig. III 18.0e2. CANopen devices have signal error conditions when an event occurs instead of having to be polled by a master. The EMCY message contains standard and profile-specific error codes.

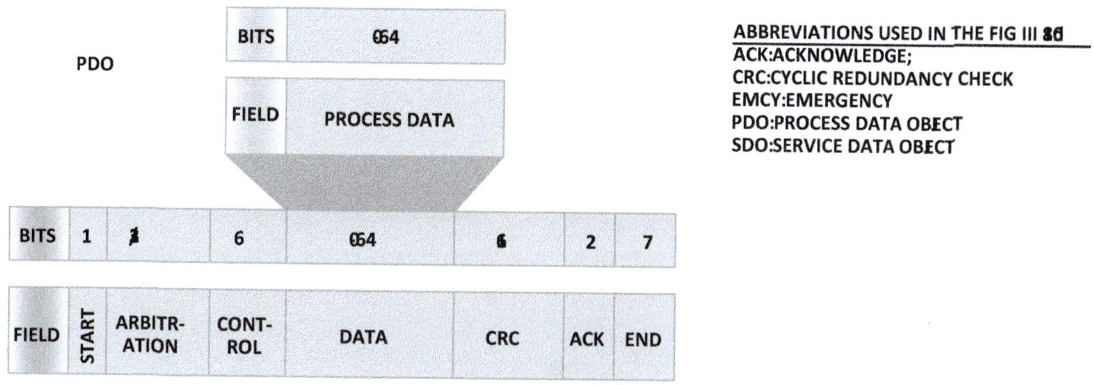


FIG III 18.0d1 CANOPEN PDO FRAME

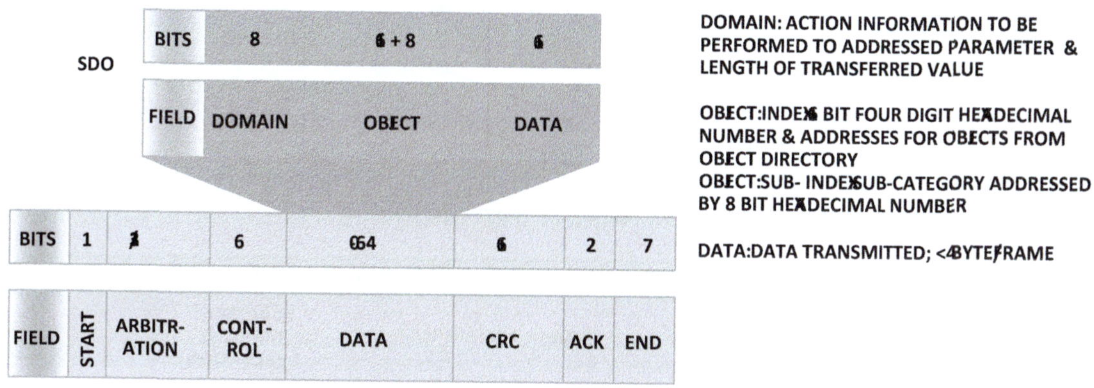


FIG III 18.0d2 CANOPEN SDO FRAME

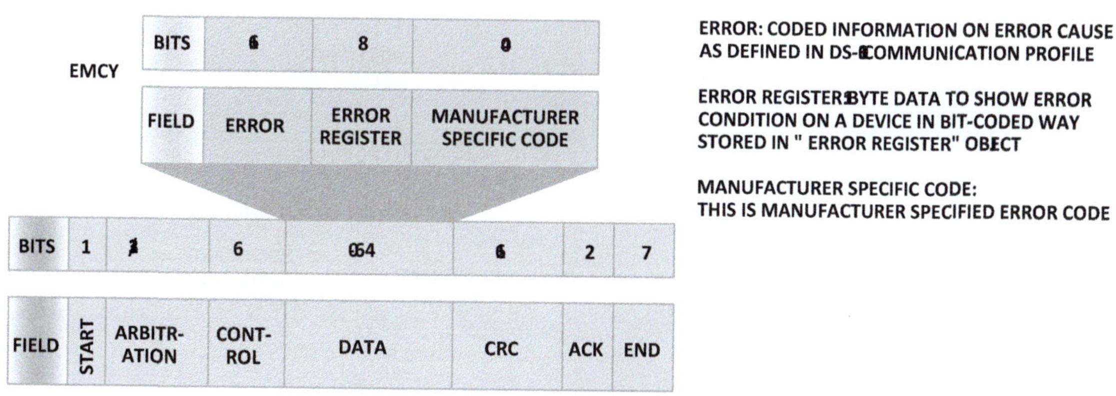


FIG III 18.0d3 CANOPEN EMCY FRAME

**FIGURE III 18.0D** Controller Area Network open: process data object (PDO), service data object (SDO), and emergency (EMCY) frames.

4. **Advanced discussion: CAN FD and LIN bus:** Here, both CANbus types have been explained at length, starting with FD.
   - **CAN FD recap:** As discussed, CAN FD was standardized in ISO 11898, offers 64 bytes, and supports a dual bit rate of 1 and 5 Mbps. FD not only supports improved CRC but also offers a "protected stuff bit counter" to reduce undetected errors and increase system reliability. ECUs of CAN FD and classical CAN may be mixed under certain conditions, which allows a gradual introduction of CAN FD and greatly reduces cost and complexity. "Avoidance of message delay" and "maintenance of CAN wire length" are two challenges for CAN FD.
   - **FD frame format:** ISO 11898-1 has been expanded to include the CAN FD frame format. The existing

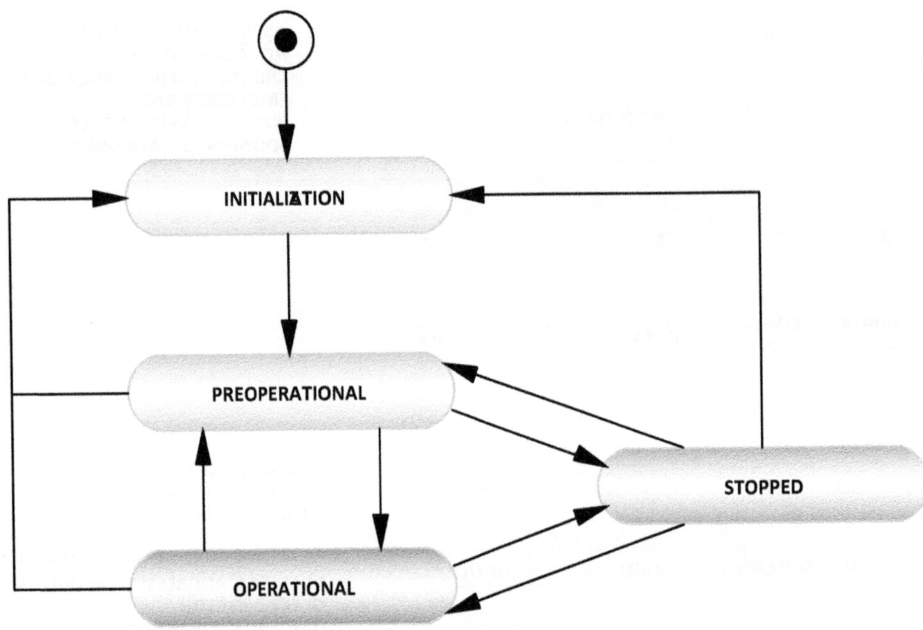


FIG III 18.0e1 NETWORK MANAGEMENT: STATE MACHINE

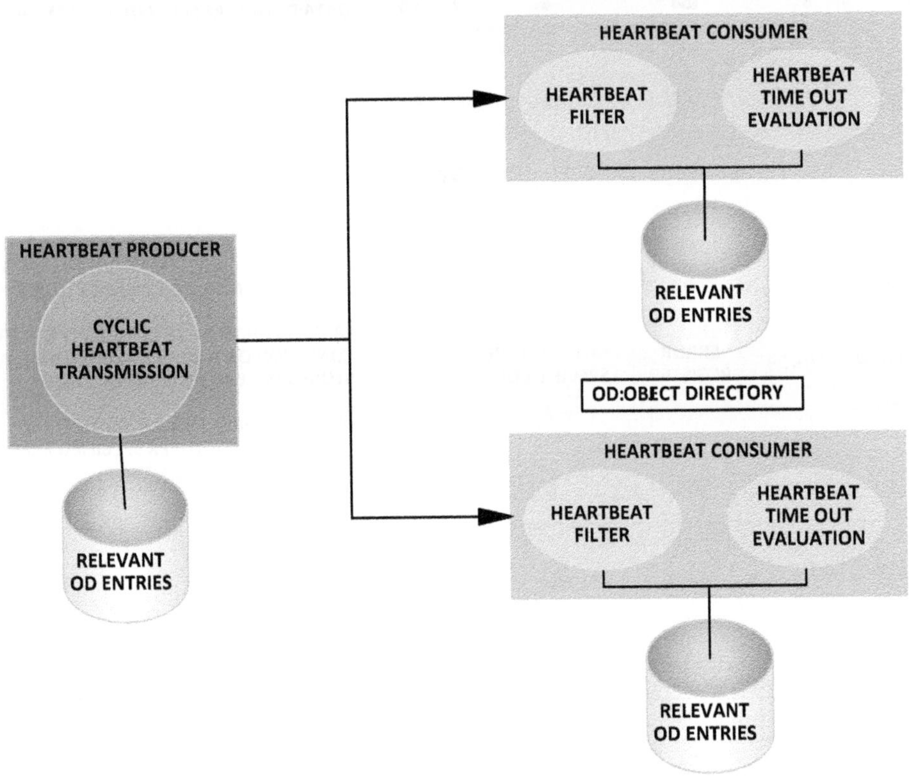


FIG III 18.0e2 NETWORK MANAGEMENT: MONITORING (HEARTBEAT)

**FIGURE III 18.0E** Controller Area Network: open network management.

frame formats underwent some editorial changes when the FD frame was included. The extended format shown in III 8.0g4 is now named FDF (FD format) and the two dominant reserved bits r1 and r0 are now named remote request substitutes. Let us now take a close look at the issue. The three main differences between CAN FD and classical CAN are *bit rate switching* (BRS), *maximum data*

**TABLE III 18.0C** Network management: state machine.

| State | Description | Used object |
|---|---|---|
| Initialization | Switched on, devices ready for operation | |
| Preoperational | Boot up to NMT master for taking control; configuration via SDO; PDO mapping; synchronization; connection monitoring | SDO, NMT, EMCY |
| Operational | Normal bus operation, devices functional (SYNC= synchronization) | SDO, NMT, EMCY; PDO, SYNC |
| Stopped | Communication stopped but connection monitoring for configuration may be on | |

*EMCY*, Emergency; *NMT*, network management; *PDO*, process data object; *SDO*, service data object.

*payload size*, and coverage of the CRC. The first two of these are relevant and will be explained, because they are directly visible and controllable. The difference pertinent to the size of the CRC is largely invisible and not controllable. Fig. III 18.0f shows classical CAN and CAN FD frames and explains BRS functioning.

- In a classical CAN frame, all the data are sent at 1 bit rate. In CAN FD, as the name implies, two different bit rates can be utilized for two different parts of the frame. This feature is BRS enabled by a new control bit called BRS (not shown), added to the existing control bits between the CAN ID and DLC. It is important to note that a CAN FD network does not have to enable BRS. In a system without bandwidth limitation, it would be better to upgrade to CAN FD for the longer data payloads. During the arbitration phase, any nodes on the network can be concurrently transmitting, which can be done at a nominal bitrate. The arbitration phase is transmitted at the nominal bit rate, and with the BRS control bit enabled (set to recessive), the data phase is transmitted at a higher bit rate: the data bit rate.
- **FD bit timing:** The FDF was introduced to increase net throughput. Two ways this is achieved are by extending the range of the data field (allowing up to 64 data bytes in one frame) and improving the frame's header/payload ratio and reducing the need for transport protocol frames. The second is the option to transmit part of the frame (the data phase) at a higher bit rate than the nominal bit rate of CAN communication.
- **LIN bus details:** In LIN communication, a master node loops through each of the slave nodes sending a request for information and each slave responds with data when polled. The data bytes contain LIN bus signals (in raw form). However, with each specification update, new features have been added to the LIN specification, making it more complex. LIN features and applications are shown in Table III 18.0d.

5. **Fault, error, and detection:** Fault and error refer to two different issues and error detection is very crucial in CAN, so these are treated separately.
   - **Fault and error:** Fault refers to an *erroneous frame modification* seen by a CAN node, whereas error is a *fault detected* by the protocol's *error detection*

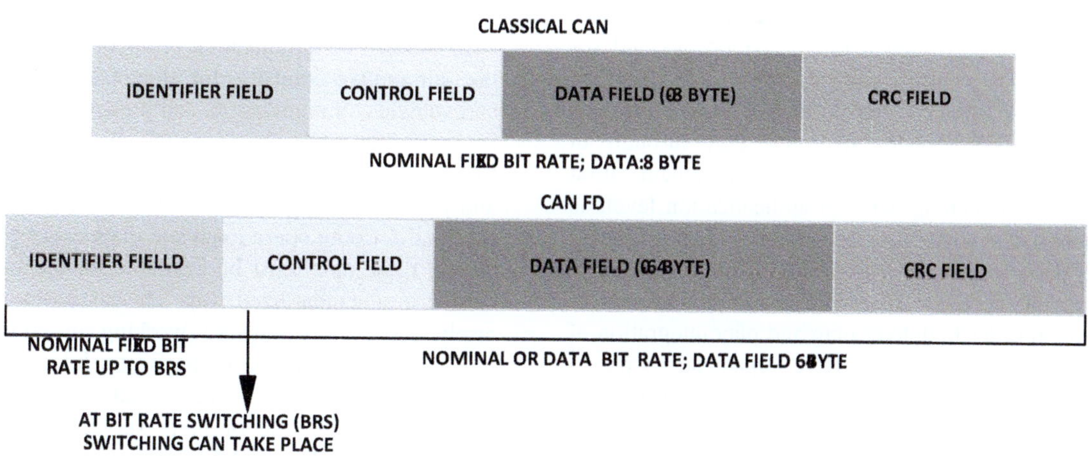


**FIGURE III 18.0F** Classical Controller Area Network (CAN) and CAN Flexible Data Rate frame format.

**TABLE III 18.0D** Some Local Interconnect Network features and applications.

| LIN features | LIN application |
|---|---|
| Low-cost option used in vehicles where speed and fault tolerance is noncritical | Steering wheel: cruise control, wiper, climate control |
| Time-tagged scheduling (deterministic) and variable bit length | Comfort: sensors for temperature, sunroof, light, humidity |
| One master with 16 slaves. Operate at 12 V | Engine: small motors, cooling fan |
| ISO 9741; sleep mode, wake up | Doors, seat for mirror, position |

*mechanisms.* Fault types include fault type A: *bit flip* (1 bit inversed/flipped) and fault type B: *bit drop/ insertion* (in received bit sequence). Error classes included *class 1: normal bit errors, class 2: encoding-related errors*, and *class 3: message length modifying errors.*

- **Error detection issue:** In CAN communication, all nodes check the validity of each frame, including the transmitter of the current frame. The checks are based on a combination of several protocol mechanisms for error detection. If at least one node detects an error, it signals this error to the other nodes by sending an error frame. Various error detection capabilities include *bit monitoring, frame format checks, CRC, ACK, stuff rule checks, stuff count checks*, and *interaction between error detection and error signaling.*

*To limit the volume of this book, let us conclude the discussions here and discuss various protocols related to CAN in the next subsection.*

### 18.0.2 Various Controller Area Network–related protocols

J1939 offers a standardized method for communication across ECUs, with a common language across manufacturers. The UDS offers standardized diagnostics and security via session control (authentication). As UDS has extensive control over the vehicle ECUs, for security reasons, critical UDS services are made only through an authentication process. There are several authentication levels, and the switching between authentication levels is done through diagnostic session control, which is one of the UDS services available. Both NMEA 2000 and ISOBUS state the need for strict protocol conformance requirements in order to deliver plug and play integration of products across manufacturers. Here, organizations play a key role in facilitating a standardized process for product certification and in providing a database for identifying conforming manufacturers/products. The NMEA standards are NMEA 0183 v4.00 2008 and NMEA 2000 v3.000-2022. In the automotive industry, an effective and standardized diagnostic protocol is crucial. For the advanced automotive industry, effective diagnostics are essential for ensuring optimal performance and diagnosing issues accurately. The complexity of modern vehicles, with numerous ECUs and advanced systems, requires a robust communication protocol that can handle the diverse diagnostic requirements, and UDS is the key protocol. The Automotive UDS Protocol fulfills this need by providing a standardized approach to diagnostics, ensuring compatibility across different vehicle models and manufacturers. There are two types of interfaces: type A (regular vehicles at 12 V) and type B (heavy-duty vehicles at 24 V). The OBD-II protocol supports multiple signal protocols, which determine how data are transferred between the vehicle and the diagnostic tool. Increasingly, modern marine electronic equipment requires data from multiple sources to enable the host of features and functions that are available to the mariner. NMEA 0183 (IEC 61162-1) provides serial data distribution from a single transmitter to multiple receivers. The other associated standard is NMEA 2000 (IEC 61162-3). Here, we will take a closer look at the basics of different (other than those discussed in previous subsections) protocols used.

1. **UDS explanation:** Fig. III 18.0g depicts UDS. The ISO 14229–based communication protocol UDS is used in the automotive industry for diagnostics, debugging, and vehicle communication, and it provides a standardized way for ECUs to communicate with diagnostic tools and software applications. Each ECU consumes and generates a large amount of information, which is essential for both operational and diagnostics purposes. So, all these data must be available for all those concerned in a standardized way. The UDS protocol precisely serves this purpose working in client-server mode. The client requests diagnostic information and can be communicating with a standalone server/a server as a gateway for one or more servers, as shown in Fig. III 18.0g1. The ISO 14229–based UDS works on

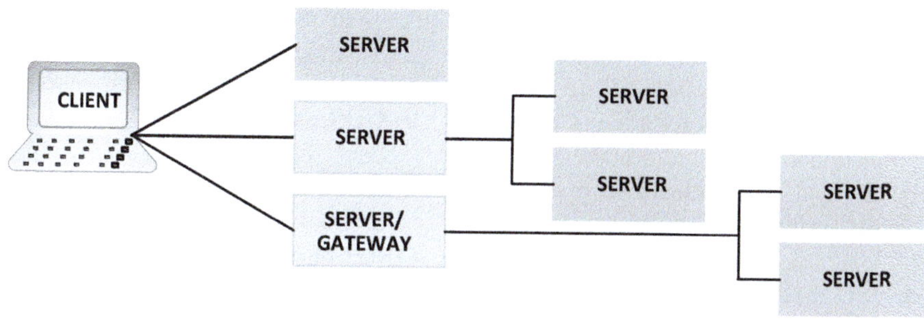


FIG III 18.0g1 UDS: MULTIPLE SERVER CONFIGURATION

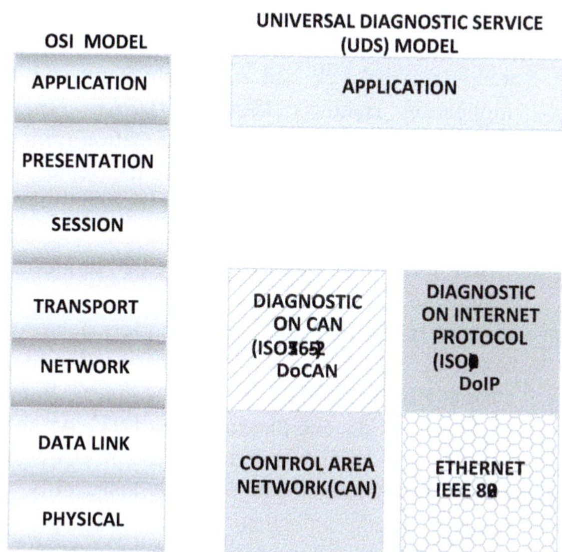


FIG III 18.0g2 UDS: OSI MODEL

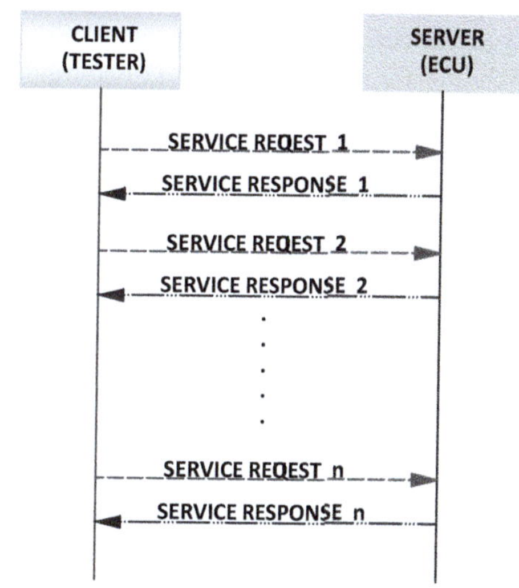


FIG III 18.0g3 UDS:CLIENT–SEVER COMMUNICATION

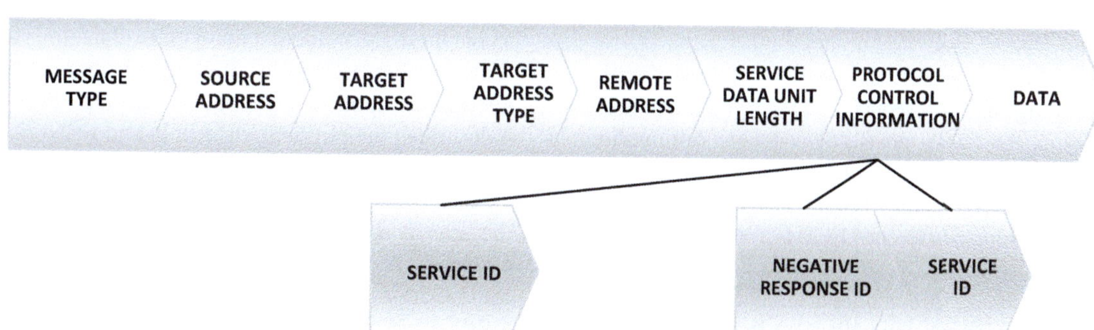


FIG III 18.0g4 UDS: FRAME FORMAT

**FIGURE III 18.0G** Universal diagnostic service (UDS) explained.

the application layer in the OSI model, as depicted in Fig. III 18.0g2. The UDS can run on any of the underlying physical layers such as CAN, Ethernet, LIN, or FlexRay. The core UDS communication protocol specification, ISO 14229-1, is independent of the transport, with few extensions defined viz. ISO 14229-3 for CAN and ISO 14229-7 for LIN. The entity originating the communication is called the *client* or *tester* and the responding ECU is the server, as detailed in Fig. III 18.0g3. UDS (ISO 14229) is implemented in ECUs of top manufacturers across all tiers to enable diagnostics, firmware updates, routine testing, and others.

- **Diagnostic format:** The UDS frame format is detailed in Fig. III 18.0g4. As shown, message types that could be diagnostic (same network), remote diagnostic (remote request response through gateway), secure diagnostic, or remote secure diagnostic are very important. The target address could be physical or functional. The remote address signifies the encoded client-server identifier, protocol control information service ID, negative response ID, and others.

- **ISO standard:** ISO 14229-1 specifies data link–independent requirements of diagnostic services, allowing a diagnostic tester (client) to control diagnostic functions. ISO 14229-2 specifies common session layer services and requirements to provide independence for all transport protocols and network layer services. ISO 14229-2 specifies USD on CAN. ISO 15031-5 and 6 are for data reporting requirements and DTC service-related issues.
- **Security issue:** Security accesses provide a means to access data/diagnostic services, with restricted access for security and safety reasons. Improper routines/data downloaded into a server could potentially damage the electronics or other vehicle components or risk the vehicle's compliance with emission, safety, security standards [5].

2. **UDS features and usages:** Briefly, UDS features and applications include:
   - **UDS features:** Key features pertinent to UDS include:
     - *Security and authentication*: The protocol incorporates security mechanisms to secure communication. It implements challenge-response authentication to prevent unauthorized access to sensitive ECU functions. Authentication and encryption techniques are used to prevent unauthorized access.
     - *DTC management*: The protocol allows for the reading and clearing of DTCs (alphanumeric codes for fault), providing valuable information about vehicle faults and malfunctions.
     - RT data monitoring: Technicians can retrieve RT data from different ECUs for performance monitoring.
     - *Bidirectional communication*: The protocol supports two-way communication for tests, configuration of parameters, and activation of specific components.
     - *Programming and reprogramming*: With the UDS, ECUs can be programmed/reprogrammed, allowing software updates and enhancements.
     - *Diagnostic services*: A range of diagnostic services are available, enabling specific diagnostic operations like reading and writing data, executing tests, and retrieving information.
     - *Request-response model:* It follows the client (test)/server request-response model for both positive and negative responses.
   - **UDS usages:** *Troubleshooting, data parameter access, security access, security and session management, configuration and programming*, and others are major uses of UDS.
3. **OBD2 discussion:** The OBD2 signal on the CAN protocol is the most advanced and widely adopted, allowing fast and reliable communication between various ECUs and enabling the exchange of data related to engine performance, emissions, and transmission easily. The lowest two layers (physical and data link) of OBD2 are as per CAN (ISO 11898). ISO 15765-4 stands for network and transport layers and the top application layer is ISO 15031/SAE J1979. The OBD2 system gathers crucial information from various sensors. The car's ECU, acting as the vehicle's brain, processes this information to pinpoint issues like engine problems, exhaust emissions, and fuel efficiency. In OBD2, when an issue is detected it produces DTC, which can be read using an OBD2 scanner, which helps in understanding and fixing performance glitches. Some services offered by OBD2 include requesting current diagnostic data, requesting freeze frame data, requesting DTC, and clearing reset DTC. All these services have a defined packet format. The OBD2 also defines a set of parameters with IDs that can be used to retrieve specific information. OBD2 can provide RT diagnostic data like engine RPM, vehicle speed, and coolant temperature. These data can be utilized for overall health monitoring and identifying potential issues. OBD2 frame and DTC details are shown in Fig. III 18.0h.
   Typical OBD2 frame details with explanations are presented in Fig. III 18.0h1. ISO 15765 refers to a set of restrictions applied to the CAN standard. ISO 15765-4 describes the physical, data link, and network layers, seeking to standardize the CANbus interface for external test equipment. ISO 15765-2 and 3 refer to the transport and session (UDS) layer. Five OBD2-compliant protocol standards are *ISO 15765* (CANbus), *ISO14230-4* (KWP2000), *ISO9141-2*, *SAE J1850 (VPW)*, and *SAE J1850 (PWM)*.
4. **OBD II DTC and future:** Here, important issues include:
   - **OBD II DTC:** OBD-II DTCs are alphanumeric codes that indicate specific issues within a vehicle's systems. When a problem is detected, the vehicle's onboard computer generates a DTC and stores it in its memory. By utilizing an OBD2 scanner or tool, users can retrieve these codes and decipher their meaning. Each DTC corresponds to a specific fault

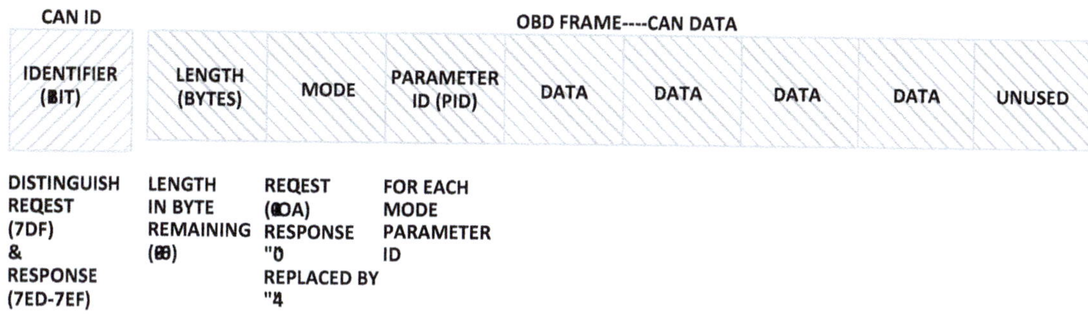


FIG III 18.0h1 OBD2 FRAME DETAILS

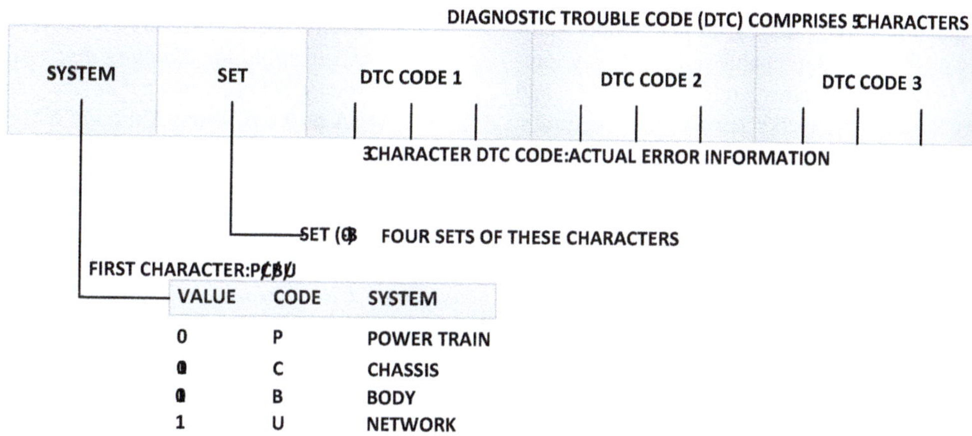


FIG III 18.0h2 DTC CHARACTERS

**FIGURE III 18.0H** OBD2 frame and diagnostic trouble code details.

or malfunction, providing valuable insights into the root cause of the problem. The DTCs are made up of 5 characters, as detailed in Fig. III 18.0h2. The OBD2 protocol can provide RT diagnostic data, and by connecting an OBD2 scanner/tool to the vehicle's OBD2 port, users can access valuable information.

- **Future OBD II:** The future OBD 2 protocol looks bright with issues like:
  - *Wireless connectivity*: Wireless communication through Bluetooth and Wi-Fi are integrated into OBD 2 to support the current trend toward connected vehicles, allowing seamless communication with smartphones, tablets, and other devices, enhancing the accessibility and convenience of OBD 2 diagnostics.
  - *Data analysis:* OBD2 data can be analyzed with big data analytics and machine learning to enable *predictive maintenance*, *anomaly detection*, and *proactive diagnostics* for improved reliability and reduced downtime.
  - *Cybersecurity:* To ensure the security of OBD2 systems, manufacturers are actively working on implementing robust cybersecurity measures to protect against potential threats and unauthorized access.
  - *Integration*: OBD 2 is being integrated more closely with other vehicle systems.

5. **NMEA and ISOBUS explained:** Here, NMEA and ISOBUS will be briefly discussed. Let the discussion start with NMEA.
   - **NMEA explained:** CAN-based NMEA 2000 is a communication standard (IEC 61162-3) used in the maritime industry for connecting, for example, engines, instruments, and sensors on boats to exchange data between devices over a single network "backbone" cable. NMEA 2000 connects GPS, autopilots, engine monitors, wind instruments, and others and shares vital information, creating a unified view of the boat's status. It has *interoperability*, *plug and play*, *a backbone network*, *multi talk-listen*, and other features. It offers a few important features

like *faster data transmission, improved device interoperability with backward compatibility, a simple low-cost network*, large network size, and robust communication network. In 1994, for heavy-duty vehicles, the Society of Automotive Engineers (SAE) published the J1939 protocol, which is well connected with NMEA2000. NMEA 2000 (2001) has references from both SAE J1939 and ISO 11783 meant for agriculture. All these are based on CAN.

- **ISOBUS explained:** The CAN-based ISOBUS (ISO 11783), derived from SAE J1939, is a standardized communication protocol used in agriculture and forestry machinery (i.e., mostly for heavy-duty vehicles). It was developed to enable plug and play integration between vehicles (e.g., tractors) and implements (e.g., sprayers) across manufacturers. Certified ISOBUS end users avoid cluttering cabins with implement-specific terminals, and instead control any implement from a single universal terminal.
- **NMEA characteristics:** Important characteristics of the NMEA 2000 network are [6]:
  - *Network issues*: NMEA uses four twisted-pair wires as a system bus in a linear network with end terminations and multiple drop cables. It supports 50 connections (254 addresses) in a 200-meter maximum cable length. Based on CAN, the network can be accessed by CSMA. This multimaster network supports self-configuration and diagnostics through special tools.
  - *Physical layer*: In the physical layer, it utilizes CAN (ISO11898) for media access and bit transmission. Because of differential signaling, there is high noise immunity, but time delays and network loading limit bit rate and network length.
  - *Data link layer*: NMEA 2000 references the data link layer of the ISO 11783-3 (identical to the SAE J1939 data link layer). However, the NMEA 2000 data link layer provides some additional features and functionality.
  - *Network layer*: NMEA 2000 is conceptualized as a future adaption to define data routing through a network from source to destination when more than one network segment is included in a system.
  - *Application layer*: The application layer refers to the approved messages, both NMT and data messages, for transmission on the NMEA 2000 network. Also, proprietary messages may be transmitted as provided for by the data link layer protocol.
  - *NMT*: NMT in NMEA 2000 is in line with ISO 11783-5 (SAE J1939-81). NMT regulates *network initialization, the claiming and assignment of addresses, ID of connected devices*, and others.
- **NMEA and CANbus:** Although NMEA is based on CANbus, the two are not identical. Table III 18.0e shows a comparison between CAN and NMEA.

*With this, let us conclude the discussion on CANbus, CANopen, and others and proceed to an advanced discussion on CIP networks.*

## 18.1 Advanced discussion on Common Industrial Protocol network—ControlNet

The CIP network was introduced in III 14.1.0. Here, an advanced discussion on the CIP Network, including *CIP safety*, shall be provided. ControlNet technology was first developed in 1995 by Rockwell Automation because it strives to be ahead of recent technology trends. However, the management and support of ControlNet was transferred to ODVA. Both ControlNet and DeviceNet belong to CIP, with a unified application layer and object-oriented approach in the industrial automation network. ControlNet is a network used for industrial automation when

**TABLE III 18.0E Comparison between Controller Area Network bus and National Marine Electronics Association.**

| Issue | CANbus | NMEA 2000 |
|---|---|---|
| Physical layer | Two-wire bus CAN system | CAN in backbone network |
| Data link | CAN protocol (ISO 11898) | ISO 11783-3 |
| Message | CAN messaging system | On CAN message frame |
| Data | Message SDO, PDO | Binary marine data on CAN |
| Device profile | CiA for CAN | Through unique identifier |
| Application | Communication protocol used in automotive, industrial automation, medical | NMEA 2000 network, multiple marine devices |

*CAN*, Controller Area Network; *CiA*, CAN in Automation; *ISO*, International Standards Organization; *PDO*, process data object; *SDO*, service data object.

the required volume of data is very high and critical in timing, including applications like robotics, motion sensing, and motion control. It is popular in advanced automation since it uses artificial intelligence (AI), the internet of things (IoT), and the industrial IoT (IIoT) in sophisticated applications. DeviceNet is used to deliver an open, low-cost, effective communication network for devices at the field level. DeviceNet allows up to 64 devices on a single node; hence, it is easier to troubleshoot and maintain. DeviceNet is ideal for factories requiring simple data control and for the initial IIoT journey. ControlNet allows up to 99 devices per node. ControlNet is not restricted to sensors and on/off commands, but also connects PLCs and HMIs and could be used for business management. Although ControlNet utilizes Concurrent Time Domain Multiple Access (CTDMA), it is specifically designed to ensure a *deterministic, repeatable performance* for time-critical applications. DeviceNet adapts CIP to low-cost CAN technology. ControlNet is highly deterministic with an associated NUT to ensure all critical input/output (I/O) data are dealt with at precise, scheduled intervals, making it well suited for motion and robotics control applications. In that sense, DeviceNet uses CSMA with a nondestructive bit arbitration method, and hence can be used on field-level devices for less critical applications. Naturally, the speed and bandwidth of the two networks are different; the speeds in ControlNet and DeviceNet are 5 Mbps and 500 Kbps, respectively. DeviceNet can use a single wire for both power and signal. Its high speed, determinism, and scheduled data handling as well as its cable redundancy for availability are major advantages of ControlNet. However, its high hardware costs, signal-only (no power) cabling, and difficulties in troubleshooting in complex networks are major drawback of ControlNet with respect to DeviceNet. Similarly, its low cost, wide acceptability, efficient use of bandwidth, reliability, and power-on-signal cable are major pros for DeviceNet, but it has a limited bandwidth, message size, and cable distance. The above discussion is based on a comparison of the two network systems and form part of the advanced discussions on CIP. In this section, the discussion has been restricted to ControlNet.

## 18.1.0 Common Industrial Protocol network: Advanced features and safety

CIP is a protocol with an *object-oriented* approach, so that different industrial automation devices can communicate with each other effectively and efficiently. CIP enables the transfer of control-oriented data: the configuration and diagnostics information. With peer-to-peer communication, it is possible to seamlessly monitor and manage field devices by the controllers. CIP *safety* is the *Technischer Überwachungsverein* (TÜV)-certified extension to the standard CIP protocol, which adds *safety application layer* functionality to the existing protocol. No special hardware is used for CIP *safety*. It is important to regulate latency of packets between the safety controller and its safety modules to maintain the required response time to changes in the memory of the device. CIP *safety* has been certified by TÜV as a "black channel" protocol, meaning the safety integrity is independent of the physical media. As a black channel protocol, CIP *safety* can be communicated over different wired Ethernet platforms, fiber optics, and wireless systems. Both wired and wireless communication networks have complexities, so any change to any part of the network could require revalidation from a functional safety point of view. Fig. III 18.1a shows CIP communication layer and safety. Since the safety application layer extensions do not rely on the integrity of the underlying standard CIP services and data link layers, single-channel (nonredundant) hardware can be used for the data link communication interface. This is shown in Fig. III 18.1a1. In addition to safety, in this subsection the advanced features of CIP will be discussed. Let the discussion begin with a description of CIP *safety*.

1. **CIP safety description:** When configuring automation lines, it is important to protect employees from equipment in motion by controlling energy for the employee to interact with the machine. Also, these devices must have multiple layers of redundancy, including I/O. For very complex safety situations, like robotic safety zones/multiple human presence safety devices, software solutions can be adapted. CIP safety can travel through the same network media with an additional security level for *safety* data for seamless integration and increased flexibility. So, CIP *safety* can coexist with application layer standards like CIP *motion* and CIP *security*. Safety data verifications like timestamps/CRC are used to validate the integrity of the received data. The *safety PLC* must claim ownership of the safety device to exchange safety data back and forth. CIP *safety* provides fail-safe communication between nodes like safety I/O blocks, safety interlock switches, and safety controllers. Both machine and process automation safety applications may reach up to *Safety Integrity Level (SIL) 3* according to IEC 61508 standards. CIP *safety* comprises high-integrity safety services and diagnostics in the application layer without need for special communications hardware. *CIP safety* cannot prevent errors but it ensures transmission integrity by detecting errors and devices so appropriate actions can be taken, like:
   - The CIP *safety* protocol is present only in safety devices.

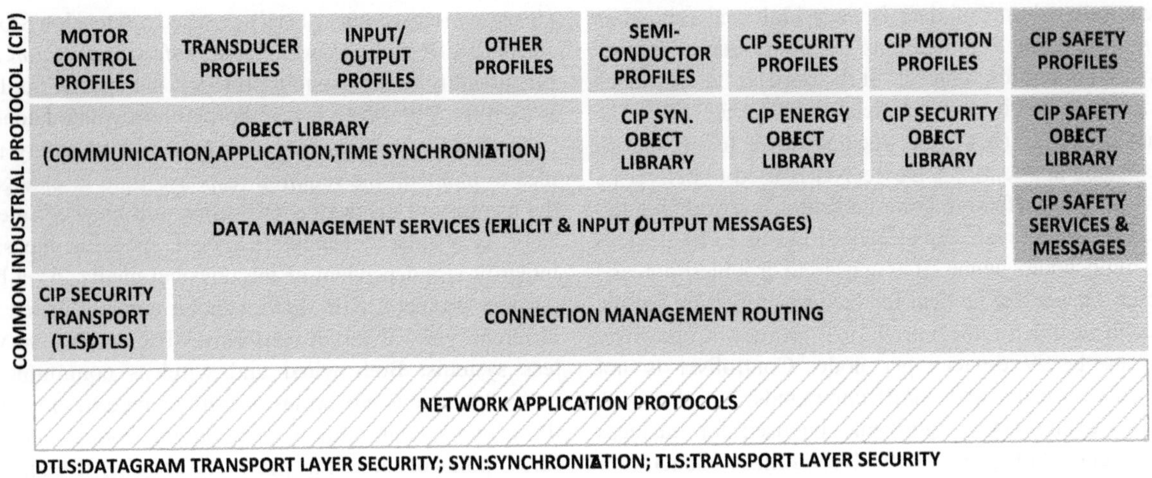


FIG III 18.1a1 CIP COMMUNICATION LAYER

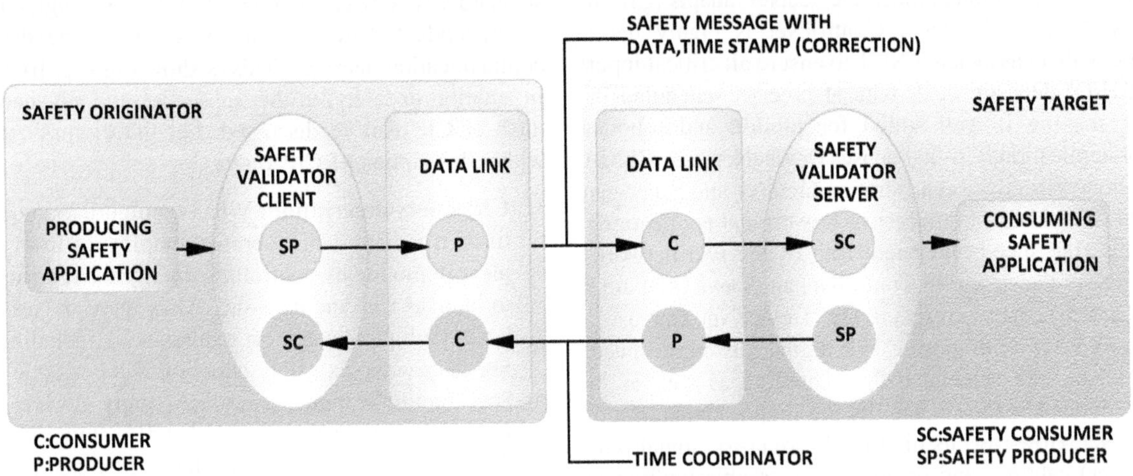


FIG III 18.1a2 CIP SAFETY VALIDATOR

DEVELOPED IDEA FROM [7]

**FIGURE III 18.1A** Common Industrial Protocol communication layers and safety validators [7].

- A production identifier (PID) is encoded in all data produced to ensure that the message is received by the correct consumer.
- CIP *safety* data is produced with a timestamp, allowing safety consumers to ascertain the produced data age.
- All safety transfers on CIP *safety* use *safety CRCs*/ checksums to maintain and ensure the integrity of the transferred information. Redundancy in data and CRC/checksum with cross checking provides additional protection against possible corruption of data by detection.

For a CIP *safety* device over the network, there are measures to ensure integrity of the configuration viz. a *safety network number* (identifying network path) for a unique ID.

2. **Safety implementation:** Safety devices can be used in a safety system only after the devices are configured (i. e., placing configuration data from a configuration tool into a safety device) and connections are established. There are two possible sequences for configuration: The configuration tool may be connected *directly* to the device or via an *intermediate* device. The CIP *safety* application layer is specified using a safety validator object, which is responsible for managing the CIP *safety* connections and serves as the interface between the safety *application objects* and the *link layer connections*, as detailed in Fig. III 18.1a2. The Safety

Validator also ensures the integrity of the safety data transfers. From Fig. III 18.1a2 we see that:

- The producing safety application uses a client validator to *produce* safety data and time coordination.
- The client utilizes a *link data producer* to *transmit* the data and a *link data consumer* to *receive* the time coordination messages.
- As shown, the consumer uses a server *validator* to receive and check data.
- The server uses a link consumer to receive data and a link producer to transmit *time coordination messages.*

It is important to note that the link producers and consumers are unaware of *knowledge* of the safety packet and fulfill *no safety function*. The safety validator is responsible for detecting these communication errors. Error detection and actions are shown in Table III 18.1a

**3.** Specific implementation: Various implementation (Table III 18.1a) issues are:

- **Time stamp:** Independent of the network technologies, CIP *safety* devices can create a logical connection to each other. IEC 61784-3-2 deals with various error mitigation techniques. *Time stamps* are used with expected times to detect packets that are lost, delayed, repeated, or transmitted out of order. CIP safety data are produced with a time stamp to allow safety consumers to determine the age of the produced data. This measure is superior to conventional reception and watchdog timers since they do not convey any information about the actual age of the data. This can also be used to detect media access/arbitration, queuing, retry, and routing delays. CIP *safety* provides tolerance to minor disturbances by allowing retransmissions. Unique device identifiers are used to authenticate the communication between two safety devices. Additional diagnostics and checks are included to validate the messages. These features are separate from standard communication methods. These mitigations are put together as CIP *safety*, a single wired/wireless connection between two devices, and can be used for certification up to SIL 3 per IEC 61508 and category 4/PLe per ISO 13849-1.
- **PID:** The PID is derived from an *electronic key*, the *device serial number*, and the *CIP connection serial number*. PID refers to an encoding in each safety packet produced that ensures that each received message arrives at the correct consumer. When any device inadvertently receives a message with the *incorrect PID*/any device does not receive a message within the expected time interval with the correct PID, then the device will go to a safe state.
- **Redundancy checks** [7]**:** Data and CRC redundancy with cross checking provides an additional measure of protection by detecting possible corruption of transmitted data, effectively increasing the hamming distance for improved error detection.
- **Safety CRC** [7]**:** All safety transfers on CIP *safety* use *safety* CRCs to confirm the integrity of the transfer of information, and they serve as the primary reassurance to detect possible corruption of the transmitted data. Safety CRCs are generated in the safety producers and checked in the safety consumers.
- **Safety connection:** Fig. III 18.1b depicts CIP *safety* connections, which are of two types: *unicast* and *multicast.*

**TABLE III 18.1A** Error detection and actions (developed based on [7]).

| Communication error | Action 1 | Action 2 | Action 3 | Action 4 | Action 5 |
|---|---|---|---|---|---|
| Message repetition | X | | X | | |
| Message loss | X | | X | | |
| Message insertion | X | X | X | | |
| Incorrect sequence | X | | X | | |
| Message corruption | | | X | X | |
| Message delay | X | | | | |
| Coupling of safety and safety data | | X | | | |
| Coupling of safety and standard data | X | X | X | X | X |
| Increased age of data in bridge | X | | | | |

Action 1: time expectation via time stamp; action 2: send and receive ID; action 3: safety CRC; action 4: redundancy with cross checking; action 5: diverse measure.

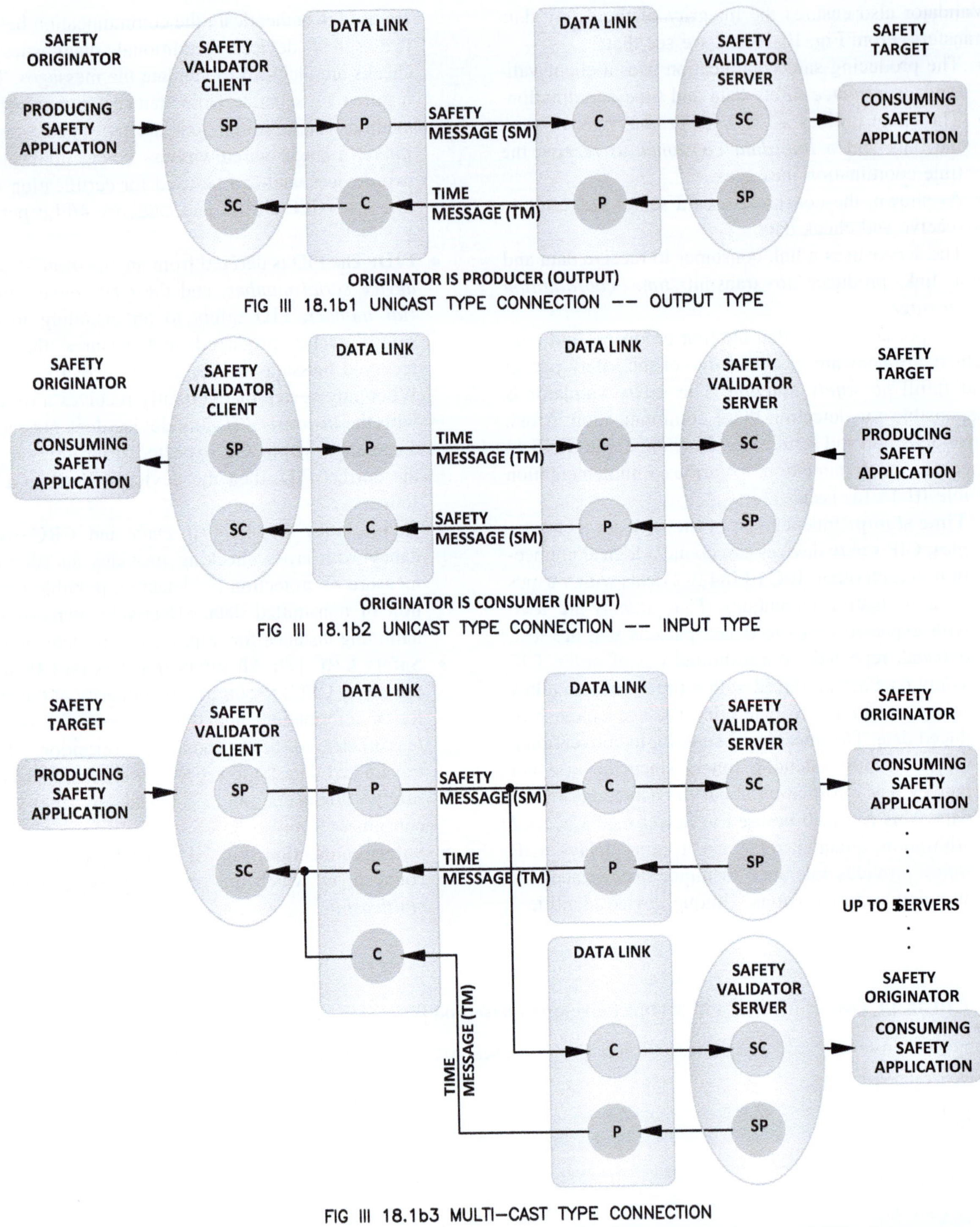


**FIGURE III 18.1B** Common Industrial Protocol safety connection types.

As shown, Fig. III 18.1b1 and 2 show a unicast connection, which allows a safety validator client to be connected to a safety validator server. As shown, there are two variations of this type viz. the originator could be the producer for output and the target could be the producer for input, as shown respectively in Fig. III 18.1b1 and 2. A multicast connection is detailed in Fig. III 18.1b3, which allows up to 15 safety validator servers to consume safety data from a safety validator client.

4. **Message packet:** Key components/parts of CIP *safety* are *data*, a *time stamp, time correction, time coordination, safety CRC, and PID.* CIP safety supports two different-sized ranges for the data section viz. the short data size and the long data size. The short data size provides high-integrity transmissions for up to 2 bytes of safety data. It also includes an instance of the safety data, the time stamp, and a 24-bit *safety* CRC for the entire message (the 3 bytes of the safety CRC are not contiguous). The long data provides high integrity transmission for up to 250 bytes of safety data. In the long data size, the original safety data are sent along with a 16- bit *safety* CRC, an inverted copy of the safety data, the time stamp, and a 24-bit safety CRC to cover the complemented data and time stamp. Like the short data section, the 3 bytes of the 24-bit safety CRC are not contiguous [7].
5. **Wireless CIP safety:** In wireless networks, CIP *safety* utilizes the "black channel" principle (Ref: II 5.1.2), wherein the protocol itself ensures data integrity over the underlying network, including wireless (e.g., Wi-Fi, 5G), by embedding strong diagnostics (CRC, timestamps, and unique IDs). Thus, the two safety devices must have enough intelligence in them, with enough diagnostics in their communications, so that the entire communication network has zero impact on the ability of the device to detect communication errors. Also, if the communication fails, the device can revert back to the safe state.

*With this in mind, let us further explore ControlNet in the next subsection.*

### 18.1.1 ControlNet communication network: overview

ControlNet is a high-speed, deterministic, open industrial networking protocol designed specifically for time-critical, high-throughput applications in manufacturing and process automation. This is managed by CIP ODVA. This protocol was mainly designed to provide consistent, high-speed control and I/O data transfer with programming that sets the logic to a particular time above the network. ControlNet is an open industrial network protocol that uses the CIP for the higher layers of the OSI model. The ControlNet designers adapted the CIP transport, network, data link, and physical layers to fit the ControlNet network requirements. Different devices that communicate with this network mainly include PLCs, HMIs, I/O chassis, drives, robots, and PCs. It is normally utilized for applications that work excellently with scheduled communications. *Determinism, high speed, media redundancy, hot swap ability,* and *support for a large number of nodes* are some of the features of this network. CIP adapts CTDMA technology in communication. It utilizes the producer-consumer model and uses NUT for scheduled/unscheduled and guardband phases. ControlNet utilizes a "token-passing" bus control network, which implements a logical "token ring" through a coaxial cable. In ControlNet, the nodes are assigned an MAC ID address from 1 to 99, and each node also knows the address of its predecessor and successor. Each node has a scheduled time determined by a "token". While the node has possession of the token, it transmits data frames until it reaches the *end of its transfer* or the token reaches its *time limit.* Then, a new token is created and passed on to the next logical successor. There may be nodes that do not have any data to transfer and, in this case, they just pass the token on to the next logical successor. While it operates on a "token bus" structure for managing network access, it does not use a traditional "token passing" method in which a physical token is passed from node to node; rather, access is determined by time-based scheduling. In ControlNet, there are communication stacks/libraries, which refer to the software protocols and toolkits enabling ControlNet. For deterministic, high-speed device communication, CIP and CTDMA management enable unified messaging across vendors for robust automation. Some of the popular ControlNet communication stacks/libraries include vendor stacks, vendor-specific implementation drivers/actual packet internals (APIs), any bus ControlNet stack, the Softing ControlNet stack, and vendor support. In this subsection, a brief overview of ControlNet will be provided.

1. **Working of ControlNet:** In ControlNet, one node at a time can send, which prevents data collisions from occurring. Once the node stops sending and does not get ahead of the token, then ControlNet has written conditions into it to generate the token again and transmit it to the next logical successor. The node is assigned a specific, unique orthogonal code sequence for its transmission. So, with a unique code, multiple nodes can *technically* send data at the same time without interference, as the receiver can separate them. ControlNet is an open control network with RT high-throughput application demands. This protocol supports interlocking from one controller to another controller and RT I/O control, valves, and drives.
   - **NUT and timing issue:** In ControlNet, *NUT* is a crucial, fixed time cycle (2–100 ms) that controls data flow to ensure deterministic delivery for RT control by dividing bandwidth into *guaranteed scheduled, flexible unscheduled* (configuration), and guardband for synchronization. For synchronization, each node has a timer synchronized to the NUT, ensuring all devices agree on the timing, preventing

collisions and guaranteeing timely data access. The NUT is a repetitive time interval. In ControlNet, the nodes include timers that are synchronized by the NUT, which separates the entrée time among nodes consistently. This is shown in Fig. III 18.1c1.

- **Messaging issue** [8]**:** Connected messaging in this protocol mainly uses resources in every node, which are used for particular purposes like transfer of RT I/O data and transactions of frequent explicit messages. The resources of a connection are set aside and arranged with available communications services using the unconnected message manager (UCMM). Unconnected messaging is utilized in the establishment procedure of the connection. This kind of messaging always uses unscheduled bandwidth.

2. **ControlNet architecture:** ControlNet network CTDMA, architecture, and MAC frame details are depicted in Fig. III 18.1c.
   As depicted in Fig. III 18.1c2, ControlNet uses the CIP for facilitating an interconnection between different OSI layers. This protocol is designed to provide consistent data transfer and high-speed control through programming. It sets the logic to a particular timing above the network and helps ensure quick data transfer. As shown, this network includes different devices such as PLCs, I/O chassis, PCs/industrial PCs, and drivers. The nodes in ControlNet are simply allocated through an MAC ID address. Furthermore, every node also identifies the successor and predecessor addresses. Every node that is available within the interface of ControlNet includes a scheduled time (token). Thus, just one node can transmit at a time to avoid collisions of data. ControlNet supports different topologies, which are elaborated in III 18.1.1.4. ControlNet uses RG-6 coaxial cables through BNC connectors for easy connection or disconnection. These cables can also be used in any environment. Other physical layer hardware include taps, segments/dropline terminating resistors, bridges, and repeaters.
3. **Packet frame:** In ControlNet, when a node sends data over the network, it packages that data into an MAC frame. Each MAC frame can transmit multiple together. This is shown in Fig. III 18.1c3
   - **MAC frame**: The following items are inserted into MAC frames before transmission: *preamble*, *start delimiter*, *source network address*, and then after L packets the *CRC* and *end delimiter*.
   - **Link packet/L packet:** Link packet refers to data that has been packaged and labeled by a node in preparation for transmission, that is, it is the fundamental unit of data sent.
4. **ControlNet topology:** ControlNet supports various topologies like trunkline-dropline, tree, and star.
   - **Trunkline-dropline topology:** This topology utilizes only one thin/thick cable. The main distance between network limits varies with the length of the cable and data speed. In this topology, the nodes can be directly connected to the main line through short branches. Trunkline refers to the main coaxial cable carrying he network signal, and dropline refers to a short, 1-meter cable that branches off the main trunk line to connect individual devices (nodes) through tap (T/Y-shaped connector).
   - **Star topology:** In star topology, each device is connected individually to a central node, called the hub/switch. This topology looks like a star. This topology helps in reducing the impact of transmission line failure by connecting every host independently.
   - **Tree topology**: Like a tree, this topology includes one central node, and each node is simply connected to this node throughout a single lane. Since this is a combination of star and bus topologies, error detection, durability, device support, network growth, and others are easier and advantageous.
5. **Controller configuration:** Here, let us discuss setting up a microcontroller as a ControlNet node. The following are the major steps:
   - **Microcontroller selection:** The chosen microcontroller needs to support ControlNet protocol, CIP, or similar hardware like CAN with a compatible transceiver.
   - **ControlNet communication stack:** Acquire the ControlNet communication stack/library discussed for the microcontroller.
   - **Hardware setup:** Connect the microcontroller to a ControlNet network using the CAN interface. Arrange the power supply and other required connections as per its microcontroller specifications.
   - **Firmware development:** Write the microcontroller firmware using the ControlNet communication stack/library and implementation of functions.
   - **Device configuration:** As discussed, each node on the ControlNet network is assigned a unique node address, so it is necessary to ensure that the microcontroller's node address is correctly configured for identification in the network.
   - **Network integration:** In the final lap, the microcontroller is integrated with the rest of the control system.

*With this, the discussions is concluded, to explore NMT for ControlNet.*

## 18.1.2 Network management for ControlNet

The ControlNet network provides reliable, high-speed transport of two basic types of application information: *control and I/O data and non–time-critical messaging data related to the controlled system*, with control and I/O

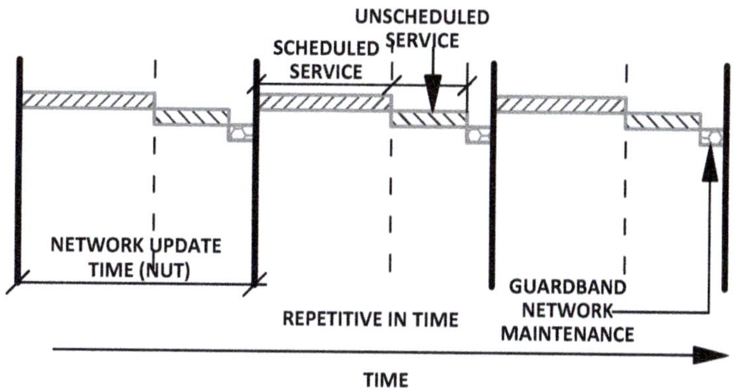


FIG III 18.1c1 CONTROLNET: CTDMA DETAILING

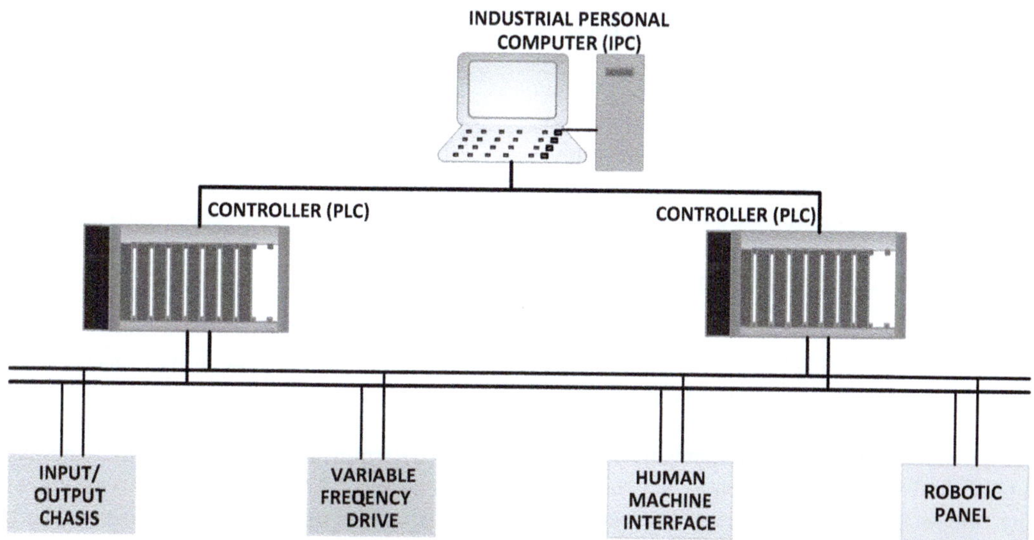


FIG III 18.1c2 CONTROLNET ARCHITECTURE

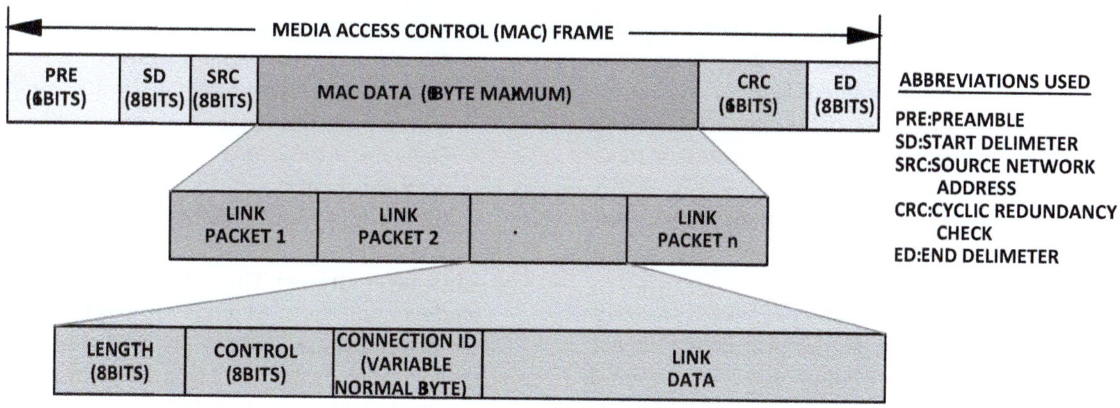


FIG III 18.1c3 CONTROLNET: MAC FRAME

**FIGURE III 18.1C** ControlNET: Concurrent Time Domain Multiple Access (CTDMA), architecture, and media access control (MAC) frame.

data given the highest priority. Other information, like programming/parameter upload and download operations, does not interfere with the transport of control and I/O data. In addition to the CIP features discussed in III 14.1.0, like deterministic data transfer, it also offers *a high-speed control and I/O network, transparent media redundancy, easy configuration, maintenance, trouble-shooting*, and other features. It is clear from previous discussions that the core principles of the NMT of ControlNet have been developed around three basic issues: the producer-consumer model, CTDMA with repetitive NUT, and support for a large number of nodes (99 maximum). NUT, NUT structure (Fig. III 18.1c1), and media redundancy are major NMT components of ControlNet for providing NMT services such as *configuration* (planning and scheduling), *topology management, fault management*, and *maintenance*. ControlNet is well characterized as having high *reliability* (due to determinism), *efficiency* (for producer-consumer model), and transparency (to bridge between CIP network types). An important issue in ControlNet is the requested packet interval (RPI). The RPI refers to the update rate specified for a particular piece of data on the network. A faster RPI consumes more network bandwidth, and hence is set as necessary to avoid wasting bandwidth. Another term, API, is also important. API refers to the actual update rate for a particular piece of data on the network. It is possible for ControlNet to set this rate equal to or faster than the RPI, based on the binary multiple of the NUT. When planning a ControlNet, various factors like *topology, number of nodes, distances*, and *connections* are important. This is demonstrated in Fig. III 18.1d Communication Model and Producer Consumer Connections. In this subsection, the discussions on ControlNet will be concluded with a detailed discussion on NMT of ControlNet. Let the discussions start with the basics of NMT.

1. **NMT:** NMT in ControlNet is based around "core functions" (e.g., high-speed deterministic communication in controller, I/O), "management tasks" (e.g., CTDMA, node configuration), and "key mechanisms" (mainly adapting CIP functionalities). In ControlNet configuration, any *single node* cannot disrupt the network's normal operations since all the nodes agree to a common set of operating rules.
   - All nodes on a link must agree on a *common set of network configuration parameters* prior to transmitting scheduled or unscheduled data.
   - The connection originator (CO) must have specific scheduling information prior to requesting or transmitting scheduled data.
   - The connection target (CT) can transmit unscheduled data immediately after powerup and a request from a connected originator.

   In ControlNet there exists a "*keeper node*"—a designated device that is responsible for distributing the network's configuration data to other devices to ensure a consistent network setup, hot swapping of devices, acting master coordinator, and others. There will be one *active keeper* (normally with the lowest node address). The ControlNet configuration keeper node is responsible for maintaining and distributing the network and scheduling information.
2. **Communication model:** Fig. III 18.1d1 depicts the communication model for ControlNet. Based on this, let us discuss the following issues.

- **Physical layer interface:** The physical layer interface consists of the hardware components to get on and off the network's physical media and is responsible for:
  - Providing transceiver components for the redundant coaxial media ports and
  - Providing transceiver components for the network access port

  A tap connects the node to the trunk cable. There exists a direct relationship between trunk cable length and the number of taps allowable within the specified distance (i.e., indicating that beyond that *repeaters* are necessary). The relationship between the maximum number of nodes, data rate, and number of repeaters was discussed at the beginning of the chapter (chapter write up). The allowed cable length can be calculated empirically from Eq. III 18.1a.

$$\begin{aligned}\text{Max segment length} &= 1000 - 16.3 * (\text{no. of taps} - 2) \\ &\quad \text{in meters } \mathbf{OR} \\ &= 3280 - 53.4 * (\text{no. of taps} - 2) \text{ in Feet}\end{aligned} \quad \text{(III 18.1a)}$$

- **Access services:** Media access in ControlNet is strictly scheduled, which prevents collisions by managing how frames are sent and received. In the communication model, it can be characterized by determinism, node flexibility, and NMT, as discussed. The media access service allows an application to transmit on the network's media. It is responsible for:
  - *Providing transmit and* receive *data and flow control on and off the media*
  - *Checking received data for* errors
  - *Adding the proper header* and *trailer information on transmit frames*
  - *Deleting header and trailer information from* received *frames*
  - *Controlling the media redundancy switch-over algorithm*
- **Link layer driver:** This refers to the software and hardware interface, which is responsible for managing data transmission over physical media. Overall, the key roles of the link layer driver in a communication model include *CTDMA, handling of NUT, error checking and reliability, redundancy, and token management.* The link layer driver manages data in the correct format in ControlNet frames. It is responsible for *setting up, controlling data flow, holding received data for checking*, and *moving data to and from* the media access services.
- **Message manager:** The full name of the message manager is *UCMM.* The UCMM allows a message

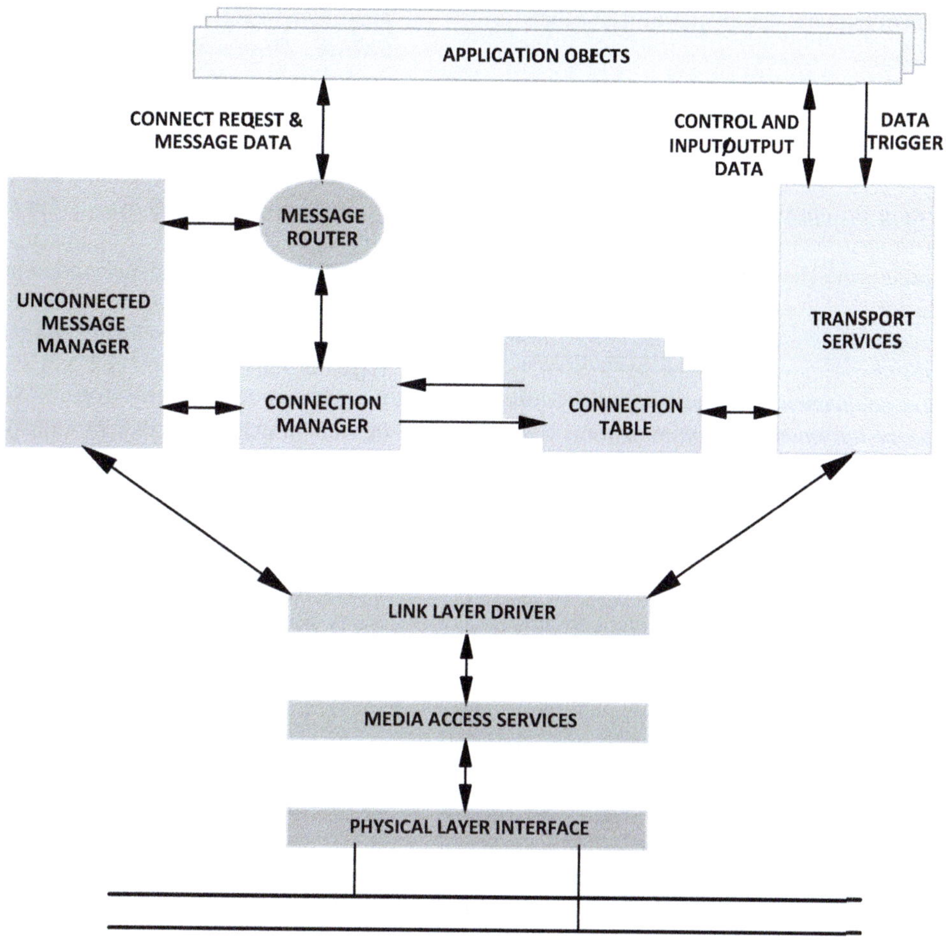


FIG III 18.1d1 CONTROLNET: COMMUNICATION MODEL

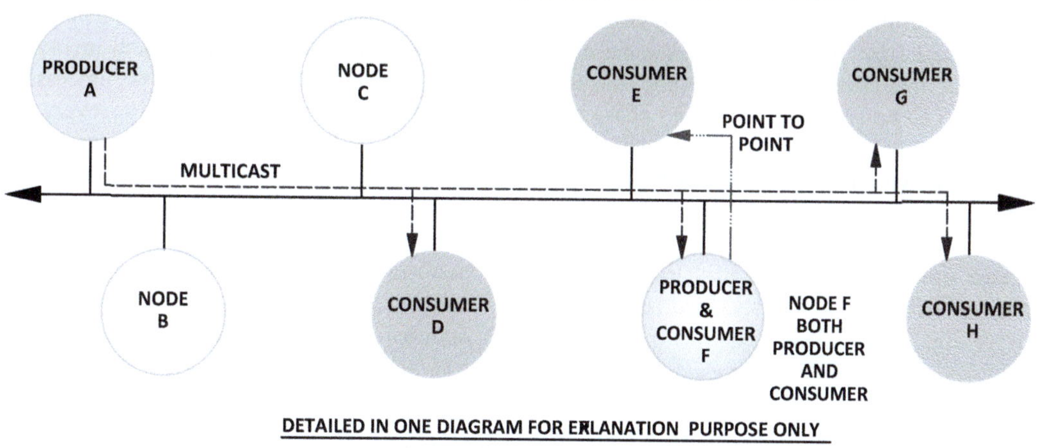


FIG III 18.1d2 CONTROLNET: NETWORK MODEL

**FIGURE III 18.1D** Communication and network models of ControlNet.

to be sent *without* an established connection, that is, the sending and receiving of unconnected messages—UCMM objects on other nodes and messages from a local router. It provides duplicate detection, automatic retries, and message time out.

- **Message router:** This allows an application to open connections to objects within the same node to interpret the part of the message with a local destination and rout it to the appropriate object. It allows multiple connections for multiple objects within the same node.

- **Connection manager:** The connection manager is responsible for:
  - *Establishing a connection to a target node*
  - *Setting up a connection within its node*
  - *Processing all connection requests on the local service*
  - *Forwarding an open connection service via the UCMM to next the node*
- **Application** object: The application object is mainly responsible for:
  - *Data generation as requested by other applications*
  - *Making a connection request to establish a path to exchange information across objects*
  - *Selection of connection type to define the number of application objects that use the data*
  - *Selection of connection priority to define time criticality of the nature of the data*
  - *Selection of transport services*

3. **Network model:** The producer-consumer model is utilized by ControlNet to exchange application information. So, producer-consumer is the network model for all ControlNet transactions, and the basic building blocks are *object messages*, *connection IDs*, *producer-consumer connection types*, *transport services*, and *transport connection types.* There can be different producer-consumer model types. The easiest type is point to point, that is, one producer and another consumer between two nodes. The concept of more than one consumer per message, or multicast, is shown in Fig. III 18.1d2. This figure shows that node A produces a message on the network. Nodes D, G, F, and H consume the message. Nodes B and C see the message but are not interested in it and do not consume it. Also, a node can be both a producer and consumer. Fig. III 18.1d2 shows point-to-point communication with different line types between producer F (both producer and consumer), which produces a message for consumer E.
4. **Media access method:** Each node may transmit only during its assigned turn, within a specified time frame. The software algorithm CTDMA controls the chances to communicate. This opportunity *repeats* itself at precise intervals. The NUT is the *fixed* and *known rate* at which the network update interval (NUI) is repeated. This is shown in Fig. III 18.1c1, in which the repeating NUI is divided into three sections: *scheduled*, *unscheduled*, and *guardband* (*network* maintenance)
   - **Scheduled services:** Scheduled services refer to scheduled traffic, which are time-critical data meant to be sent at a fixed repetitive rate. As shown in Fig. III 18.1c1, the first part of the NUI is reserved for scheduled services and message delivery; this portion is deterministic and repetitive. The maximum node of the schedule is referred to as SMAX. This is the highest address that can be used to reserve the bandwidth in the scheduled services of the NUI. So, any address below is assured one chance per NUT. Digital/analog and interlock data can be sent as scheduled data, and each node can transmit up to 510 bytes.
   - **Unscheduled services:** The unscheduled portion of the NUI begins after all scheduled nodes have had a chance to transmit. As the name implies, Unscheduled data are not associated with time criticality. Here, an important term is UMAX, which refers to the highest address (by default: SMAX +8) that can communicate in this portion of the NUI. So, the time remaining before the beginning of the guardband is available on a sequentially rotating basis to all nodes with a network address between 0 and UMAX. Programming data, peer-to-peer messaging data, and connection establishment data are typically sent in this portion of the NUI.
   - **Guardband:** The guardband is the final part of the NUI and is reserved for network maintenance. An important term here is moderator, which represents the node with the lowest address. During this portion of the NUI, a moderator frame comprising NUT, SMAX, UMAX, and others is sent.
5. **Connection establishment:** A connection provides a path between two endpoints. Once the connection manager (III 18.1.1.2) determines the virtual circuit, the route between endpoints is fixed. The connection establishment process is handled at the data link and network/transport layers with the help of the CIP and CTMA. Connection establishment involves both physical layer and data link layer processes using the UCMM. To establish a connection, every node containing a *UCMM*, *message router*, *connection manager*, *ControlNet object*, and *identity object* is utilized. In addition, a CO node contains a connect scheduler object. The required issues here are a *brief functional description on each connection object listed above* and a *graphic and written explanation of the interaction between components while a connection is initiated and established.*

*With this, let us conclude the discussion on ControlNet and proceed to discuss other CIP-type networks.*

## 18.2 Advanced discussion on CompoNet and DeviceNet

CompoNet, a member of the CIP family, is a field network mainly for applications with distributed field devices, each with few I/O points. Its high-speed communication capability and flexible topology reduces wiring costs and decreases troubleshooting and maintenance time. CompoNet, which has a comprehensive suite of messages and services for a variety of manufacturing automation applications, implements CIP at its upper layers. DeviceNet is built around open standard; DeviceNet devices can be developed without paying a royalty and enjoy the associated benefits of competitive pricing and a wide availability of components.

The low-cost, open communication network DeviceNet links intelligent control devices over a byte (0–8 bytes) level system that can be optimized for machine control, balancing network speed and data throughput. DeviceNet is a simple networking solution that reduces cost and the time spent on wiring and installation and provides interchangeability of "like" components among various vendors. DeviceNet can be seen in two different ways:

- **Layered details**: DeviceNet *Physical Layer and Media* (-Physical layer), *CAN and its use in* DeviceNet (Data link layer) and *Communication protocol & Application* - Application layer
- **Device profile**: for interoperability and interchange ability

The DeviceNet protocol dramatically reduces costs by integrating all I/O devices on a four-wire trunk network with data and power conductors in the same cable. This direct connectivity reduces costly and time-consuming wiring. As a CIP member, DeviceNet is a truly media-independent protocol that encompasses a variety of manufacturing automation applications, including control, safety, security, energy, synchronization, motion, configuration, diagnostics, and information. DeviceNet, adapts CIP to CAN technology that is mostly used in automotive vehicles. By leveraging the economies of scale in this proven commercial technology, DeviceNet provides users with the ability to distribute and manage simple devices throughout their architecture in a cost-effective manner. The simple trunk and dropline networking solution not only reduces wiring costs, installation time, and downtime, but it also enhances the ease of interchanging conventional and complex devices. In this section, CompoNet and DeviceNet are discussed in detail. Let the discussion begin with CompoNet.

## 18.2.0 Discussion on Common Industrial Protocol network: CompoNet

A simple network cable and connector, scheme, and others not only reduces the overall system cost and time, but CompoNet can also maximize network throughput for applications that need to transmit small packets of data quickly between controllers, sensors, and actuators. CompoNet utilizes CIP for its upper layers. CIP networks follow the OSI model to define a complete suite of network functionality from the physical implementation through the application or user interface layer. As a truly media-independent protocol, it is supported by hundreds of vendors around the world. CIP provides users with a unified communication architecture throughout the manufacturing enterprise. CompoNet adapts CIP to Time Domain Multiple Access (TDMA) on a simple two-wire physical layer. TDMA assures each device a fixed time slot in the communication cycle, resulting in low overhead and strict cycle times. This efficiency allows large numbers of nodes to share small amounts of data, without requiring extremely high baud rates. As a consequence, CompoNet can reliably run on inexpensive, unshielded two-wire cable [9]. In this subsection, CompoNet is briefly discussed.

1. **CompoNet with OSI model:** CompoNet, like other CIP networks, follows the OSI model, which defines a framework for implementing network protocols in seven layers. Fig. III 18.2a shows the CompoNet-OSI model and frame block. Networks that follow this model define a complete suite of network functionality from the physical implementation through the application or user interface layer. CompoNet implements CIP at the session layer and above while adopting CompoNet specific definitions at the transport layer and below. For the CIP the communication layer, please see Fig. III 18.1a1. For the CompoNet in OSI model and frame block, please see Fig. III 18.2a.
   Based on the OSI model, the CompoNet layers are detailed in Fig. III 18.2a1, which will be referenced in later discussions to describe different layers.
2. **Advantages and application area:** CompoNet offers some special advantages in plant automation. The main advantages and application profiles are:
   - **Advantages:** Several unique advantages for automation applications offered by CompoNet are:
     - It supports fast cyclic exchange of bit and word-sized data for high-performance networks for sensors and actuators.
     - CompoNet supports up to 384 distributed nodes for cyclic I/O data exchange.
     - With CIP messaging, CompoNet allows for remote configuration and maintenance of intelligent field devices.
     - CompoNet supports a wide range of data rates (93.75 kbps to 4 Mbps), with flexibility for trading off between speed and distance.
     - It uses four-wire flat cables/unshielded two-wire cables for easy installation and cost efficiency.
     - CompoNet supports up to 64 repeaters, enabling a large number of nodes and coverage.
   - **Application area:** CompoNet has specific advantages for application in:

- *Manufacturing automation*: It provides wire saving in modular machines coupled with sub-millisecond cycle times at higher baud rates. A single, flat cable can link machine modules without loss of performance, making CompoNet well suited for manufacturing automation and enhancing efficiency on factory floors.
- *Machine maintenance*: The simple wiring and cable approach make it convenient for troubleshooting and machine maintenance.
- *Conveyor and warehouse applications*: With I/O points few and far apart, the free network topology at the low baud rate will cut down wiring cost. Repeaters can extend the covered area [9].
- *System integration*: CompoNet belongs to CIP networks; hence, it can seamlessly integrate devices for smooth communication between various automation devices from a variety of vendors.
- *Device connectivity*: It can access and exchange data with smart sensors and devices in RT.

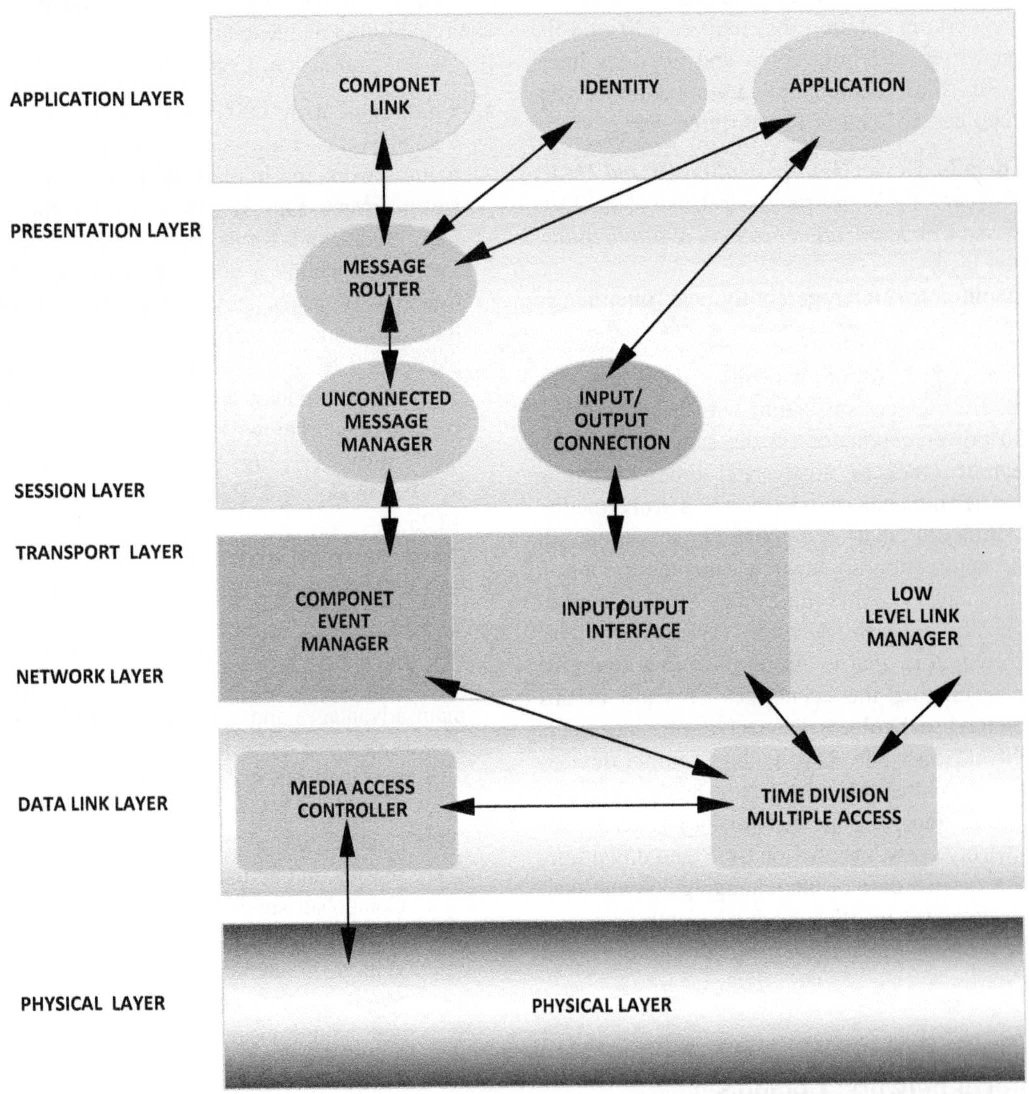


FIG III 18.1a1 COMPONET–OSI LAYER

CRC:CYCLIC REDUNDANCY CHECK

| PREAMBLE | COMMAND CODE | COMMAND DEPENDENT BLOCK | CRC |
|---|---|---|---|

FIG III 18.1a2 COMPONET–FRAME BLOCK

DEVELOPED BASED ON [9]COURTESY:ODVA

**FIGURE III 18.2A** CompoNet—Open Systems Interconnection model and frame block [9].

3. **Physical layer:** The CompoNet physical layer utilizes RS485 specifications along with Manchester signal coding technology and a pulse transformer to improve the reliability of communication, even with the use of cheap unshielded cables. The pulse transformer in each device creates an isolation in reducing signal disturbances. A special "*receive mask*" kind of *digital signal filtering* used by CompoNet allows the use of cables of different electrical characteristics, allowing transmission at various data rates. The "*receive mask*" allows reduced voltage, so skewed signals can be received readily in a reliable manner with a high rejection rate of noise spikes and reflected waves. With dedicated repeaters, CompoNet can extend the maximum trunk line length to three times the standard length. Also, with the use of repeaters (maximum 64), the installation area can be expanded to accommodate a large number of node connections. TDMA minimizes the transmission delay due to repeaters and decreases its influence on the communication cycle [9].
4. **Data link layer:** MAC and TDMA are two major parts of the data link layer, as shown in Fig. III 18.2a1. MAC generates and analyzes the CompoNet communication frames, as detailed in Fig. III 18.2a2. To satisfy CompoNet's transmission quality and high speed, a few functions performed at the MAC layer include *Manchester coding*, *CRC generation/check*, and *addition of a special preamble to accomplish zero-interframe delay* between frames, meaning no idle time on the wire. CompoNet frame transmission and reception is regulated through TDMA. In CompoNet, the master controls frames at optimal states for better network efficiency. Seven types of frames supported by CompoNet are *OUT*, *TRG*, *BEACON*, *CN*, *IN*, *A_EVENT*, and *B_EVENT* frames. The master directly controls *OUT*, *TRG*, *BEACON*, *A_EVENT*, and *B_EVENT* frames but not *CN* and *IN* frames, which are sent by slaves. One of the main characteristics of TDMA is the minimization of transmission delay due to repeaters. To improve *maintainability*, some functions of CompoNet include *gathering slave information by CN frame* and *getting slave information by status read by B_EVENT frame*, which also indicates node connection details to repeaters. Baud rate synchronization is done by BEACON frame.
5. **Higher layers:** As shown, the following are major higher layers:
   - **Network and transport layer:** As shown in Fig. III 18.2a1, CompoNet uses UCMM to send or receive the explicit message protocol; the master must be aware of the existence of the destination in advance. Furthermore, the master must establish the transmission timing of IN and CN frames to the slaves/destination prior to sending the frames. The master detects the existence of slaves through CN frames. On receiving a CN frame from a nonparticipating slave, it automatically processes the participation state of the slave. After the participation process, an explicit message then implements allocation of the predefined connection set. As shown, the I/O interface passes on IN/OUT data between TDMA and the upper layers. Explicit message fragmentation is regulated by an event manager. TDMA notifies the CompoNet event manager of the explicit message response to be monitored.
   - **Application layers:** CompoNet uses CIP, a strictly object-oriented protocol, at the upper layers, as illustrated in III 18.1a1. As discussed, a CIP object has attributes (data), services (commands), and behaviors (reactions to events), and seamless bridging and routing is the most significant advantage for CIP networks users. A message originating from one CIP network can be smoothly passed on to another CIP network.

## 18.2.1 DeviceNet communication network: overview

DeviceNet is an application-level protocol used in the automation environment as a communication tool. The DeviceNet communication tool allows logical talks between the controller (PLC) and other control devices like motors, conveyors, flowmeters, and level sensors. Thus, in place of direct exchange between a controller and discrete I/O modules, DeviceNet talks via a piece of hardware called a *DeviceNet scanner*. DeviceNet follows the seven-layered OSI model. The three upper layers of DeviceNet are based on CIP. The bottom four layers as per the OSI model have been adapted to the DeviceNet application. DeviceNet uses CAN for the physical and data link layers. DeviceNet uses the trunkline-dropline topology, that is, a single, main "trunk" cable carries both signal and 24 V DC power, with "drop" cables connecting individual devices to the main line. DeviceNet can be characterized by various factors like nodes and addresses (supports up to 64 nodes), selectable speeds that vary with distance, hot swapping (addition and deletion of nodes online), and support for power cable. DeviceNet supports both implicit and explicit messages and both peer-to-peer and master-slave access modes. DeviceNet communication is covered briefly in III 9.0.2.4 and III 9.0.2.5 as well to supplement the discussions here. Although DeviceNet was developed by Rockwell, it is now managed and maintained by ODVA. In this subsection, an overview of DeviceNet communication is provided. Let the discussions start with the main technology.

1. **Technology discussion:** The CIP network DeviceNet follows the OSI model. DeviceNet implements CIP at the session layer and above and adapts CIP to the specific DeviceNet technology at the transport layer and below. This network architecture is shown in Fig. III 18.2b. Please also see Fig. III 18.1a.

Major technological issues are:

- The DeviceNet network infrastructure is passive, with functionality independent of physical location and the network itself, and it has tolerance to lost node connections.

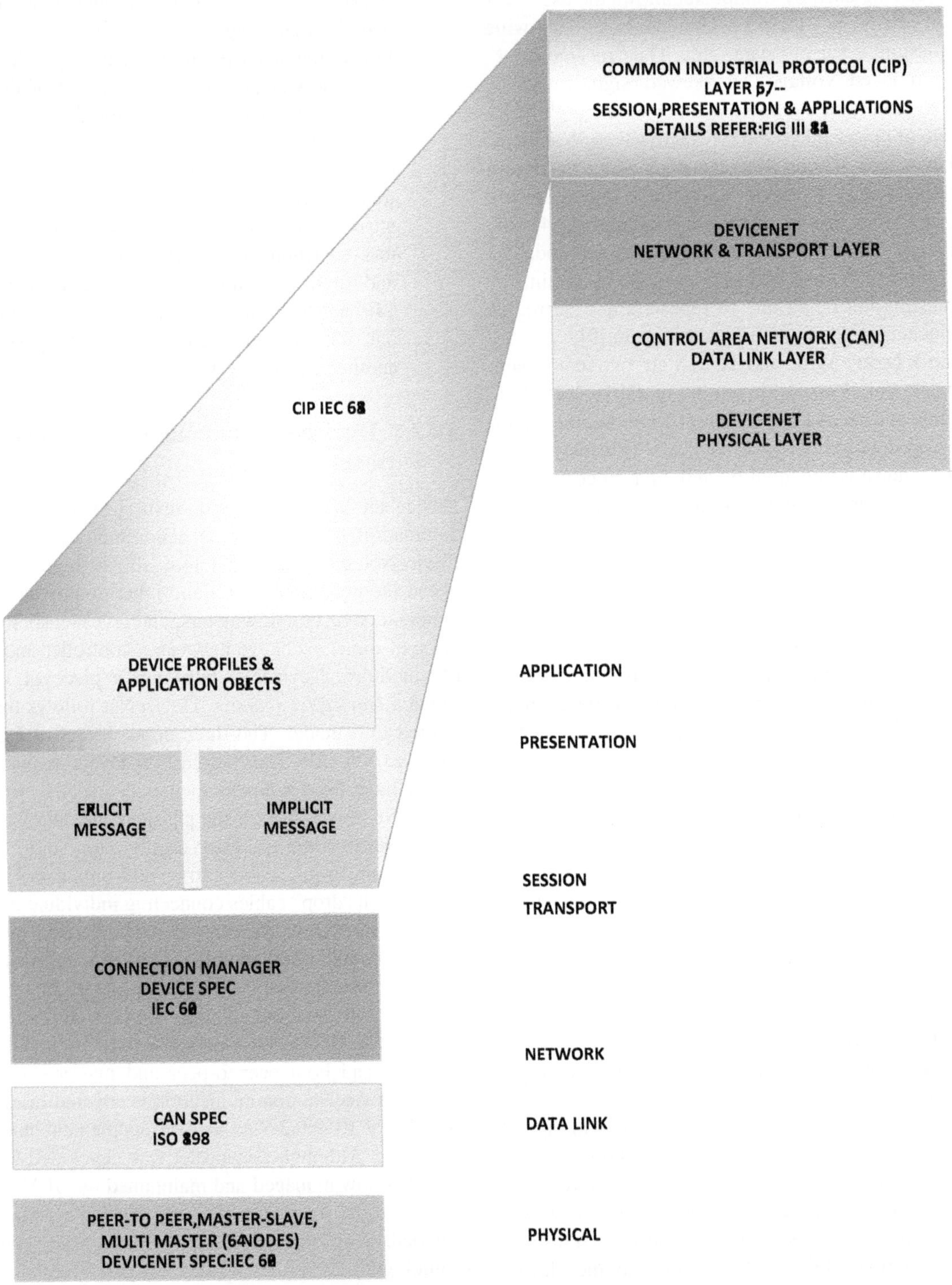


**FIGURE III 18.2B** DeviceNet layered model.

- Even with passive network infrastructure, it is possible to convey power over the same cable with communication messages. This simplifies the number of system components and connections.
- DeviceNet supports both implicit (I/O—time-critical control data) and explicit messages (carrying protocol information and specific service requests). Thus, a single point of connection is required for both configuration and control in DeviceNet.
- DeviceNet is a producer-consumer network that supports multiple communication hierarchies and message prioritization.
- DeviceNet uses bandwidth more efficiently than a device network based on a source destination model [10].
- DeviceNet systems can be configured to operate either in a controller/device or distributed control architecture using peer-to-peer communication [10].

2. **Functional details:** Fig. III 18.2c shows DeviceNet topology and frame format details. DeviceNet topology is detailed in Fig. III 18.2c1. The DeviceNet I/O bus network is basically meant to communicate information and power supply in a single network cable to the field devices connected to the bus. The controller viz. PLC can drive the field devices directly using a network scanner instead of I/O modules to carry out various functions like *reading/writing I/O from or to a device, monitoring the operational status of a device, configuring data downloading,* and *communicating with the controller to exchange information like device I/O data, status information, and configuration data.* The unique feature of DeviceNet is having power on the network to limit the power requirement and simplifying wiring through a reduction of wiring in terms of both connection points and physical size. Since DeviceNet uses a CIP network, it is object oriented and has two

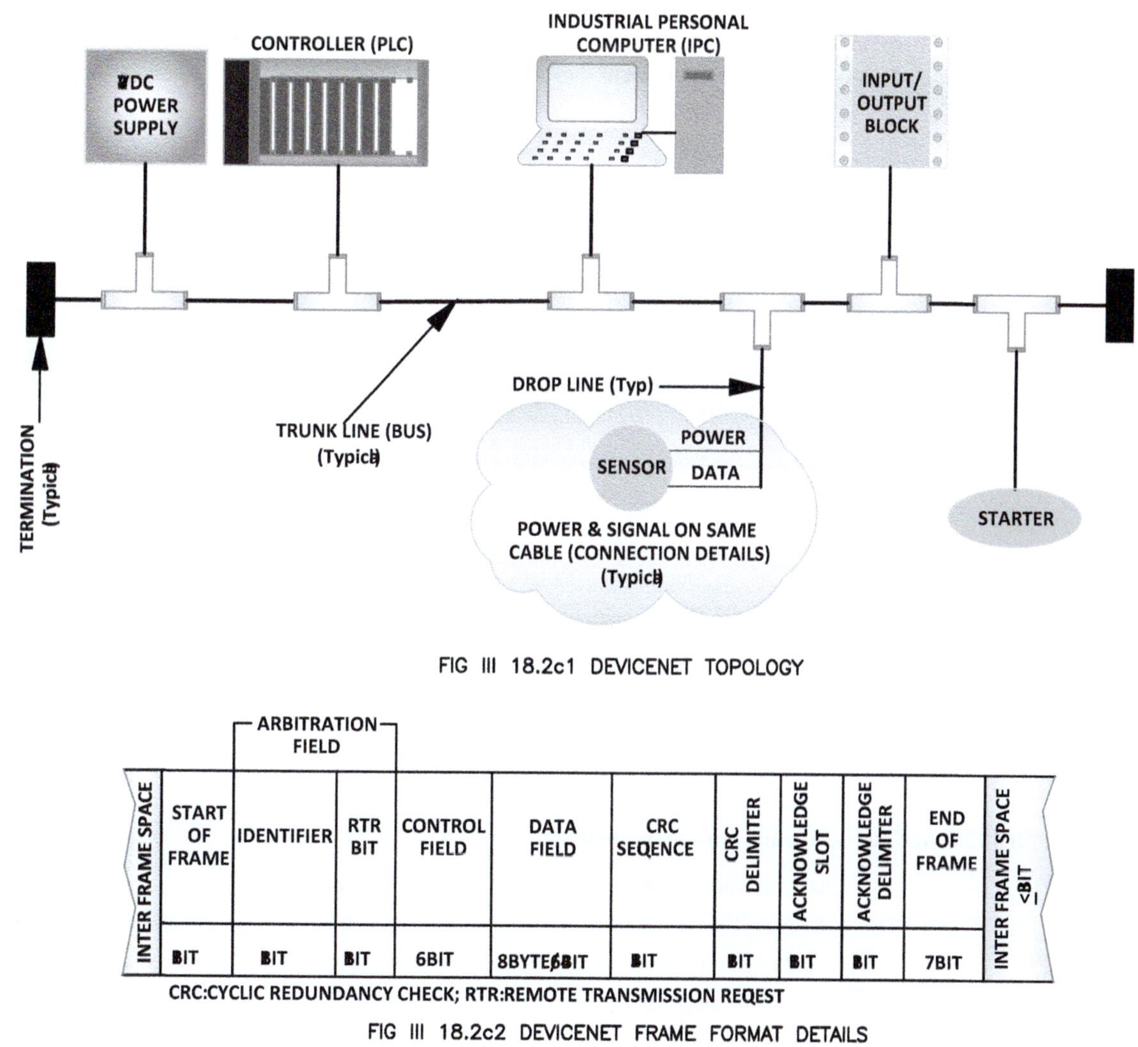


**FIGURE III 18.2C** DeviceNet topology and frame format.

different types of objects viz. communication objects and application-specific objects, as defined by CIP (Ref: III 14.1.0). The field devices connected to the network contain intelligence devices and can communicate not only the device status but also diagnostic information about their operating. DeviceNet features and functionality are elaborated in Table III 18.2a.

Also, we understand that DeviceNet is a (0–8)* byte-–level network. It is better to understand what different device network architectures mean. This is shown in Table III 18.2b.

3. **Layer distribution:** DeviceNet follows the seven layers of ISO. It is based on a CIP protocol at the upper level and CAN at the bottom layer. Hence, this description may contain some repetition from previous sections. Major layer details are:
   - **Physical layer:** DeviceNet uses trunkline-dropline topology, allowing the use of *separate* twisted-pair buses for both signal and power distribution. Generally, the end-to-end network length varies based on the cable type and the data rate used. The maximum lengths are indicated in Table III 18.2a. DeviceNet supports *both* isolated and nonisolated physical layers. However, since the DeviceNet physical layer *must be optically isolated* from the rest of the device, externally powered devices can share the same bus cable. DeviceNet also provides a choice of screw terminal or pluggable connector types.
   - **Data link layer:** DeviceNet uses CAN for its data link layer for relatively minimal overhead, making it efficient in messaging. Also, DeviceNet uses bandwidth efficiently to package and transmit CIP messages. The format for the CAN data frame is shown in Fig. III 18.2c2. In line with the CAN specification discussed, any transmitter can drive the bus to a dominant state, and the bus can be in the recessive state only when no transmitter is in the dominant state. Also, it uses a "carrier sense" network with peer-to-peer connection capabilities. Transmission in line with the CAN specification has been discussed previously. When a device transmits, it also monitors to validate the data transmitted and enable immediate detection of simultaneous transmissions. If a node transmitting a recessive bit receives a dominant bit while sending the arbitration field, it stops transmitting. The winner of arbitration

**TABLE III 18.2A** Features and functions of DeviceNet.

| Network feature | Functional support |
|---|---|
| Number of nodes | <64 nodes |
| Baud rate 125 kbps | 500 m/1640 ft max; variation with cable type |
| Baud rate 250 kbps | 250 m/820 ft max; variation with cable type |
| Baud rate 500 kbps | 100 m/328 ft max; variation with cable type |
| Topology | Bus: trunk (as above) and drop line (6 m/20 ft) |
| Power and signal | Power and signal on same network cable (Fig. III 18.2c1) |
| Power details | 24 V DC power dropline length <3 m/10 ft |
| Bus addressing | Peer to peer and multicast |
| Master slave | Multimaster—master-slave |
| Hot swapping | Hot swapping supported—online addition and deletion |
| Other features | Supports: mix of open-screw terminal and sealed connectors, multiple power connection and interoperability |

**TABLE III 18.2B** Various device network architectures.

| Type | Characteristics/usage | Description |
|---|---|---|
| Bit | High speed/low diagnostics | Bit-sized information processed |
| Byte | /Suited for machine control | Byte-size* information passed |
| Process | /Plant and building control | Complex system with a lot of information |

between all nodes transmitting simultaneously is the one with the lowest-numbered 11-bit identifier. The CAN data frame is followed by the CRC field to detect frame errors as well as several frame-formatting delimiters.

- **Network and transport layer:** DeviceNet is a connection-based network, that is, a connection with a device is established first to exchange data with the device. Connections are established via UCMM or unconnected port. DeviceNet supports two types of messages: *explicit* and *implicit* (I/O messages). Explicit messages are used for request/response-oriented transactions for configuration or data collection, that is, they are not time critical. Implicit messages are used to communicate RT I/O data. Devices may be *clients*, *servers*, or *both* and may be *producers*, *consumers*, or both, as previously discussed (e.g., similar to Fig. III 18.1d2). DeviceNet uses a device address inside the CAN identifier field that automatically incorporates a mechanism for detecting nodes with duplicate addresses. The uniqueness of each connection ID is strictly controlled to take advantage of DeviceNet's producer-consumer capabilities. DeviceNet uses an 11-bit CAN identifier for the connection ID and divides the CAN identifier into four groups [10].
- **Upper layers:** DeviceNet uses object-oriented CIP at the upper layer with CIP objects like *attributes/data*, *services*, and *behaviors* (*reactions* to events). Also, CIP's producer-consumer communication model provides efficient use of network resources by allowing the exchange of application information between a *producer* and many *consumer* devices. This model provides advantages like [10]:
  - If a node wants to receive data, it only needs to ask for it once, and it is made available each time it is produced.
  - If any node wants the same data, all it needs to know is the connection ID to receive the same data simultaneously with all other nodes.

4. DeviceNet benefits: The following are a few major benefits of DeviceNet:
   - Producer-consumer services allow data to be controlled, configured, and collected from intelligent devices over a single network.
   - It has a robust physical layer with high noise immunity for challenging environments and robust signal transmission.
   - Supporting up to 64 nodes, DeviceNet is less expensive with fast installation times.
   - DeviceNet offers a flexible network architecture with data rates of 125, 250, and 500 kbps and maximum trunk-line distances up to 500 m (125 kbps with thick cable).
   - Power (24 V DC) and signal are on the same wire with hot swapping facilities.
   - DeviceNet technology can be used with either controller, like PLC/PC-based systems.
   - Devices can provide diagnostic information for troubleshooting to minimize downtime.
   - Rugged installation options include round cable for flexible topologies (including daisy chain and trunk-line) with a range of connector options available (including screw-terminal and hard-wired [IP20] and mini and microstyle plugs [IP67]) or flat cable with flat trunk connectors (IP67).
5. Bus comparison: Two CIP-based protocols, DeviceNet and ControlNet, are compared in Table III 18.2c

**TABLE III 18.2C** Comparison between DeviceNet and ControlNet.

| Issue | DeviceNet | ConrolNet |
|---|---|---|
| Topology | Trunk and dropline | Trunk, star, tree, ring |
| Cost | Low cost | High-cost hardware |
| Speed—distance | Max 500 kbps; limited bandwidth and distance | Max 5 Mbps—max 1000 m |
| Cable type | Four wire for power and signal | Coaxial fiber optic—signal only |
| Nodes | Up to 64 | Up to 99 |
| Determinism | No priority based | Yes |
| Utilization | Highly reliable with efficient bandwidth utilization | Scheduled data and redundant cable |
| Market | Clear market spot for device-level communication | Well accepted for AI and IoT integration |

*AI*, Artificial intelligence; *IoT*, internet of things.

Now let us explore the details of DeviceNet communication in the next subsection.

## 18.2.2 DeviceNet: Communication networking details

With CAN technology, DeviceNet provides users with the ability to distribute and manage simple devices throughout their architecture in a cost-effective manner. The DeviceNet network protocol used within automation is used for interconnecting industrial controllers/PLCs, sensors, actuators, and automation systems from different vendors. The protocol utilizes CIP over a CAN media layer to cover various device profiles. The main applications of the DeviceNet protocol include safety devices, exchange of data, and large I/O control networks. Carrying both the signal and power on a similar cable and hot swapping are two important features. Thick (trunk line <8 A) and thin (dropline <3 A) are two standard cable types used in DeviceNet. The cables are color coded (0/24 V DC are black and red; CAN high and low are blue and white), as is standard. As discussed, this network can include a single master and up to 63 slaves to support master-slave and peer-to-peer communication by using I/O and explicit messaging for monitoring, controlling, and configuring. The master device first connects to slave devices with the connection of explicit message to describe the connection object. Apart from explicit messages, there exist four I/O connections: bit strobe connection, poll connection, change of status, and cyclic connection. Since DeviceNet utilizes object-oriented CIP, the specification of CIP is included in every device. These objects include an *identity object* and *message router object*. There will also be *DeviceNet objects* and *application objects*. Devices are basically treated as objects with attributes (data) and services (methods). It also covers high diagnostic capabilities. EDSs are the files from vendors that define the configuration and integration and are important aspects of DeviceNet *implicit explicit*. By this time, the reader is well aware of all these *aspects of* DeviceNet. These will be discussed at length in this subsection to conclude the discussion on DeviceNet and the network system. The last subsection of the volume will begin with a discussion on device types.

1. **Device types:** Any device can support any one or all of the device types simultaneously. Three basic DeviceNet device types are:
   - **Master: scanner:** Master devices are also known as scanners and occasionally as client devices. A master can own many/all the slave devices but *a slave* can only be owned by one master at a time in DeviceNet. A master device must first "*allocate*" a DeviceNet slave device for ownership by a set of handshaking messages. In handshaking messages (*Predefined Master Slave Connection Set*), the master requests control of the slave and configures the slave to transfer a particular set of data at a data rate *specified* by the master. In the allocation process, an "*open explicit request message*" is issued on the *unconnected message port* requesting ownership of the slave device. If the slave accepts *ownership*, it replies affirmative to create the connections requested by the master device in the request message. Thus, we understand that the master requests both *an explicit* and an *I/O connection.*
   - **Slave device:** Slave devices, also called servers, are devices that receive and transmit application-specific data to and from a master device. Slave devices, by definition, implement the *Predefined Master Slave Connection Set* described. Slave devices support at least one I/O message connection type such as *polling* (slaves respond to polling message by producing the data), *cyclic* (slave devices may not be configured for immediate response; slave devices produce their I/O data at a configured frequency), and *change of state* message (slaves produce I/O data *only* when the application layer indicates data has changed). This is shown in Table III 18.2d.
   - **Peer device:** In DeviceNet, a peer device is a DeviceNet product that can directly exchange messages with other DeviceNet products over a nonmaster-slave connection. Peer-to-peer communication is one of the features of DeviceNet, providing flexibility for distributed control architectures. Key issues include *direct communication*, *UCMM capability*, applications such as direct reading of data by an intelligent HMI from a PLC driver, and versatility.
2. **Object model:** We understand the object model for DeviceNet. Fig. III 18.2d shows the DeviceNet object model, comprising:
   - **Identity object:** As the name suggests, this typically comprises attributes like:
     vendor ID, device type, product code, revision, status, serial number, product name, and state.
   - **Connection object:** A DeviceNet product typically has a minimum of two connection objects, with each connection object representing one endpoint of a virtual connection between two nodes on a DeviceNet network. The two types of connections are:
     - *Explicit messaging*: The attributes are address, values, and a service code for the desired action.
     - *Implicit (I/O) messaging*: All the information about the action with the data is contained in the *connection object* with the I/O message.

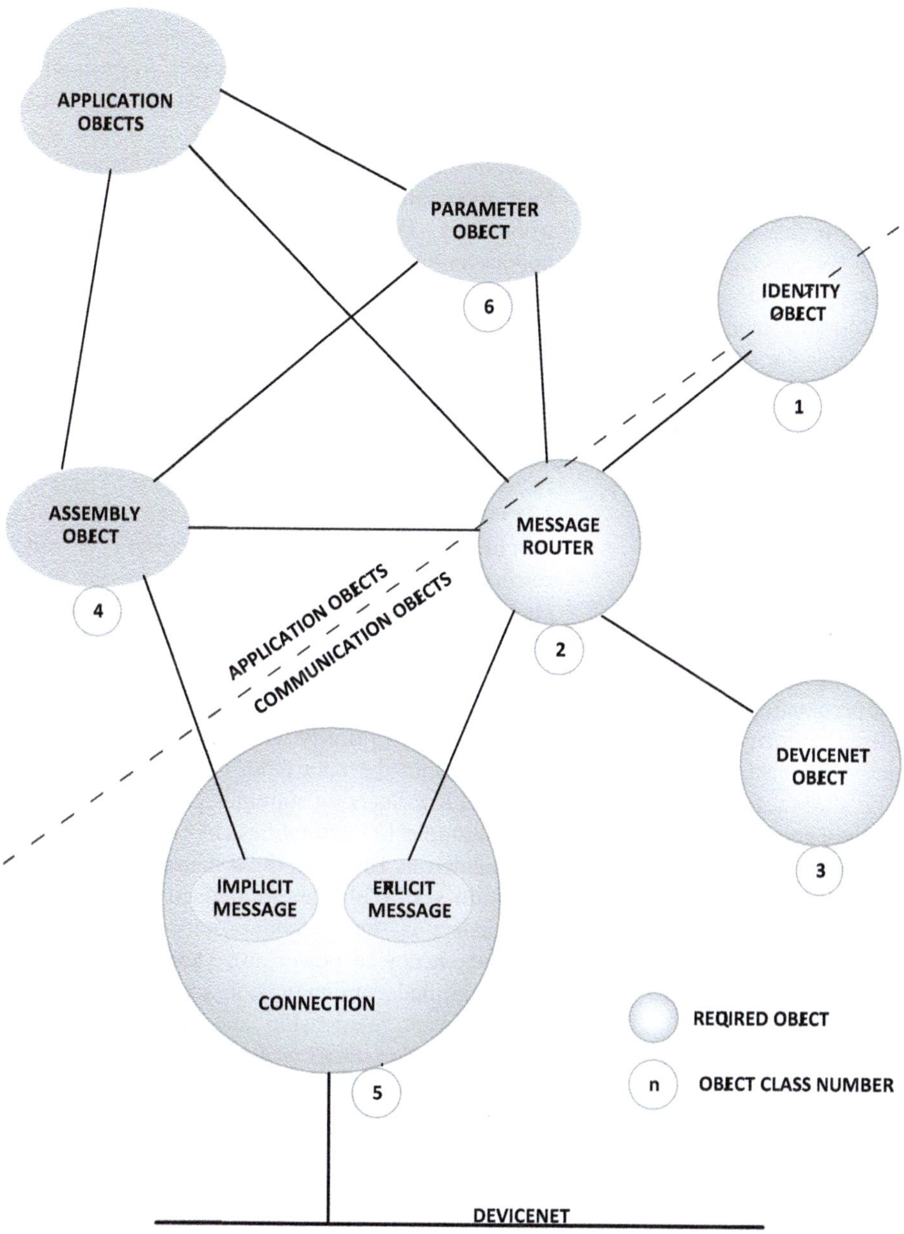


**FIGURE III 18.2D** DeviceNet object model.

- **Message router:** This object represents the component of a product that passes *explicit messages* to other objects.
- **DeviceNet object:** A DeviceNet product will have a *node address/MAC ID*, *baud rate*, *bus-off action*, *bus-off counter*, *allocation choice*, and *master's MAC ID* as attributes.
- **Assembly object:** A DeviceNet product will typically have one or more optional *assembly object(s)* with the primary purpose *of grouping* different attributes (data) from different application objects as a single attribute that can be moved with a single message.
- **Parameter object:** Optional parameter objects in DeviceNet products are used with configurable parameters.
- **Application object:** Usually at least one application object besides *assembly* or *parameter* will be present in a device.

3. **Messaging issue:** As shown in Fig. III 18.0f, the identifier is an 11-bit field in a CAN frame header that plays a critical role in messaging. The identifier provides the means for multiple priority levels (used in arbitration),

more efficient transfer of I/O data, and multiple consumers. The DeviceNet application layer *defines the assignment of identifiers for controlling priorities, uses the CAN data field to specify services, moves data*, and others. DeviceNet uses an efficient *producer-consumer* model, requiring packets with identifier fields for the data. The device with data produces the data on the network with the proper identifier. All consumers requiring data listen for messages and consume the data when devices recognize the appropriate identifier. In the producer-consumer model, the message is no longer specific to a particular source or destination.

4. **Message types:** DeviceNet primarily uses two main message types: implicit (*I/O messaging*) and *explicit* (for configuration and diagnostics) messaging, which are built on the CAN protocol. Also, there exists unformatted messages between two nodes, which are referred to as *peer* messaging. So, the three message types are:
    - **Explicit messaging:** Explicit messages are request/response-type messages from a DeviceNet master to a slave that "get"/"set" an attribute in a DeviceNet object to configure, diagnose, or monitor devices. Characteristically, these are lower priority with a need for specific, explicit request-response communication to read or write specific attributes in an object. The *service code* field in the explicit message identifies the requested service, with the remainder of the message containing any data required to execute the service. The GET_ATTRIBUTE Service Code (typical request) returns the value of an attribute to the device issuing the explicit message. Explicit messages are used by all devices, including configuration tools. *Object-class, instance-number, attribute- ID*, and *value* are some of the service code data associated with these messages. Explicit messages can be used to read I/O data. An explicit message connection may support a *fragmentation* mechanism for messages >8 bytes (fragmented messages contain only 6 bytes of data), and *fragmentation* in messages must be acknowledged.
    - **Peer messaging:** These are messages sent directly from one DeviceNet node to another (i.e., nonmaster-slave connection). When two devices support unconnected messaging and one of the devices can issue an unconnected message, a peer communication channel can be created between them. There is no well-defined rule or organization for the data exchanged on a peer communication channel. These are mainly between devices manufactured by the same vendor.
    - **I/O messaging:** Implicit (I/O) messaging mainly refers to *time-critical, specialized messages* for transferring RT, predefined data between a master and a slave. So, input data are produced by an I/O device for transferring to a scanner and output data are consumed by an I/O device as transferred from a scanner over the DeviceNet network. These messages have high priority without any protocol information within the data field, resulting in efficient, fast, and low-latency communication. A device can have multiple I/O assemblies. I/O messaging types are detailed in Table III 18.2d below.
5. **Connection set:** With the "Predefined Controller/Device Connection Set," (connection objects mostly configured on device power-up), DeviceNet can provide a simplified communication scheme based on a controller/device relationship and simplify both the *packaging* and *movement of data* contained in the I/O messages. Because of "power" on a network, many

**TABLE III 18.2D** DeviceNet: predefined message types.

| Message type | Description |
|---|---|
| Polled | Polled messages refer to request/reply messages issued to the polled connection. A polled input/output (I/O)-configured device will receive "output" from the controller sequentially in an order as per the controller's scan list. Also, it will transmit its "input" data in response to the controller's poll at the polling rate determined by the number of nodes in the scan list. |
| Cyclic | This refers to scheduled messaging. A device configured to produce a cyclic I/O message will produce its data at a precisely defined interval. This type of I/O messaging allows each user to configure the system to produce data at a rate appropriate for the application. Depending on the application, this can use bandwidth efficiently to reduce the amount of traffic. |
| Change of state (COS) | COS messages are produced on the I/O connection on an event. A device is configured to produce COS messages whenever it changes, or at a base "heartbeat" rate mentioned above. |

DeviceNet-compliant sensors and actuators can perform a predetermined function on power-up. Therefore, the type and amount of data a compliant device will produce/consume is also known. Additionally, the DeviceNet slave device issues a heartbeat message to let the scanner know that the slave device is still alive. Devices can produce the data types based on the implicit messaging methods described in Table III 18.2d. Also, there exists another type of broadcast messaging, Bitstrobe (8 bytes), for 64 nodes.

*With that, the discussion on DeviceNet comes to an end. This also marks the end of this chapter and Volume III of the book. Now let us look forward to detailed discussions on network and cybersecurity, IoT, IIoT, and plant applications of intelligent automations in a variety of industrial plants covering continuous/discrete and batch processing. I sincerely feel that these will be helpful for them and keenly look forward to honest feedback on the volume.*

## List of abbreviations

**ACK** Acknowledgment
**ADAS** Advanced driver assistance systems
**AI** Artificial intelligence
**API** Actual packet internal
**ASIC** Application-specific integrated circuit
**BRS** Bit rate switching
**CAN** Controller Area Network
**CD** Collision detect
**CIP** Common Industrial Protocol
**CO** Connection originator
**COB** Communication objective
**COS** Change of state
**CRC** Cyclic redundancy check
**CSMA** Carrier Sense Multiple Access
**CT** Connection target
**CTDMA** Concurrent Time Domain Multiple Access
**DLC** Data length code
**DTC** Diagnostic trouble code
**ECU** Electronic control unit
**EDS** Electronic data sheet
**EOF** End of frame
**HMI** Human-machine interface
**I/O** Input/output
**ID** Identification
**IIoT** Industrial IoT
**IoT** Internet of things
**IPC** Industrial personal computer
**MAC** Media access control
**NAP** Network access port
**NBA** Nondestructive bit arbitration
**NMT (protocol)** Network management (protocol)
**NUI** Network update interval
**NUT** Network update time
**OBD** Onboard diagnostics
**OD** Object directory
**PC** Personal computer
**PCI** Protocol control information
**PDO** Process data object
**PG** Parameter group
**PGN** PG number
**PID** Production identifier
**PLC** Programmable logic controller
**RPI** Request packet interval
**RRS** Remote request substitute
**RT** Real time
**RTR** Remote transmission request
**SDO** Service data object
**SOF** Start of frame
**SPN** Suspect parameter number
**TDMA** Time Division Multiple Access
**UCMM** Unconnected message manager
**UDS** Universal diagnostic service

## References


[1] Controller Area Network (CAN); J. A. Cook and J. S. Freudenberg; EECS 461 Fall 2008; University of Michigan. https://www.eecs.umich.edu/courses/eecs461/doc/CAN_notes.pdf.

[2] Controller Area Network (CAN Bus) Tutorial—Remote Frame; Posted by Wilfried Voss Jun 4th 2021; Copper Hill Technologies. https://copperhilltech.com/blog/controller-area-network-can-bus-tutorial-remote-frame/.

[3] A simple intro to CAN Open; CSS electronics; June 2023. https://www.csselectronics.com/pages/canopen-tutorial-simple-intro#:~:text=The%20following%20is%20important%20to,CANopen%20messages%20(like%20containers).

[4] Introduction To The CANopen Protocol; Application Note AN-ION-1-1100; Vector; Version 1.1; Internet Document. https://cdn.vector.com/cms/content/know-how/_application-notes/canopen/AN-ION-1-1100_Introduction_to_the_CANopen_Protocol.pdf.

[5] Overview of Unified Diagnostic Services Protocol; automotive; Dec 2021; Internet Document. https://nvdungx.github.io/unified-diagnostic-protocol-overview/.

[6] NMEA 2000 Explained—The Latest Word; F.Cassidy; Kvaser; Internet document. https://kvaser.com/wp-content/uploads/2014/08/nmea2000-explained-cassidy.pdf; https://kvaser.com/about-can/higher-layer-protocols/nmea-2000/.

[7] CIP Safety: Safety Networking for Today and Beyond; CIP Safety; Technology Overview Series; ODVA; Internet document. https://www.odva.org/wp-content/uploads/2020/11/PUB00110R4_CIP_Safety.pdf.

[8] ControlNet: Architecture, Working, Differences & Its Applications; Electronics Projects Focus; El Pro Cus; Internet document. https://www.elprocus.com/controlnet/.

[9] CompoNet—CIP on TDMA technology; Technology overview series; ODVA; Internet document; Pub66161R1; March 2016. https://www.odva.org/wp-content/uploads/2020/05/PUB00161R1-Tech-Series-CompoNet.pdf.

[10] DeviceNet—ODVA; CIP on CAN technology; Technology overview series; ODVA; April 2021. Internet document. https://www.odva.org/wp-content/uploads/2021/05/PUB00026R5_Tech-Series-DeviceNet.pdf.

# Index

'*Note:* Page numbers followed by "f" indicate figures and "t" indicate tables.'

## I



## J



## K



## L

## X



## Z

Printed and bound by CPI Group (UK) Ltd, Croydon, CR0 4YY
04/08/2026
02185506-0002